1.8 Absolute Value

$$|x| = \begin{cases} x & \text{when } x \geq 0 \\ -x & \text{when } x < 0 \end{cases}$$

$$|a| = \sqrt{a^2}$$

If $a > 0$, then

$|x| = a$ is equivalent to $x = a$ or $x = -a$.

$|x| < a$ is equivalent to $-a < x < a$.

$|x| > a$ is equivalent to $x > a$ or $x < -a$.

$$|ab| = |a||b| \qquad \left|\frac{a}{b}\right| = \frac{|a|}{|b|} \qquad |a + b| \leq |a| + |b|$$

2.1 The Rectangular Coordinate System

Distance formula: $d(PQ) = \sqrt{(x_2 - x_1)^2 + (y_2 - y_1)^2}$

Midpoint formula: $\left(\dfrac{x_1 + x_2}{2}, \dfrac{y_1 + y_2}{2}\right)$

2.2 The Slope of a Nonvertical Line

Formula for slope: $m = \dfrac{y_2 - y_1}{x_2 - x_1} \quad (x_2 \neq x_1)$

Lines with equal slopes are parallel.

Lines with slopes that are negative reciprocals are perpendicular.

2.3 Writing Equations of Lines

Point–slope form: $y - y_1 = m(x - x_1)$

Slope–intercept form: $y = mx + b$

General form: $Ax + By = C$

2.4 Graphs of Equations

Tests for symmetry:

If $(-x, y)$ lies on a graph whenever (x, y) does, the graph is symmetric about the y-axis.

If $(x, -y)$ lies on a graph whenever (x, y) does, the graph is symmetric about the x-axis.

If $(-x, -y)$ lies on a graph whenever (x, y) does, the graph is symmetric about the origin.

Equations of a circle with radius r:

$(x - h)^2 + (y - k)^2 = r^2$; center at (h, k)

$x^2 + y^2 = r^2$; center at origin

2.5 Proportion and Variation

If k is a constant,

$y = kx \quad$ y varies directly with x

$y = \dfrac{k}{x} \quad$ y varies inversely with x

$y = kxz \quad$ y varies jointly with x and z

3.2 Quadratic Functions

The graph of

$y = a(x - h)^2 + k \quad (a \neq 0)$

is a parabola with vertex at (h, k).

The graph of

$y = ax^2 + bx + c \quad (a \neq 0)$

is a parabola with vertex at

$$\left(-\frac{b}{2a}, c - \frac{b^2}{4a}\right).$$

3.4 Translating and Stretching Graphs

If $k > 0$, the graph of $\begin{cases} y = f(x) + k \\ y = f(x) - k \end{cases}$ is identical to the graph of $y = f(x)$, except it is translated k units $\begin{cases} \text{up} \\ \text{down} \end{cases}$.

If $k > 0$, the graph of $\begin{cases} y = f(x - k) \\ y = f(x + k) \end{cases}$ is identical to the graph of $y = f(x)$, except it is translated k units to the $\begin{cases} \text{right} \\ \text{left} \end{cases}$.

3.6 Operations on Functions

If the ranges of functions f and g are subsets of the real numbers, then

$(f + g)(x) = f(x) + g(x)$

$(f - g)(x) = f(x) - g(x)$

$(f \cdot g)(x) = f(x) \cdot g(x)$

$(f/g)(x) = \dfrac{f(x)}{g(x)} \quad (g(x) \neq 0)$

$(f \circ g)(x) = f(g(x))$

To our wives, Carol and Martha;
and our children:
Kristy and Steven,
Sarah, Heidi, and David

Books in the Gustafson/Frisk Series

Beginning Algebra, Seventh Edition
Beginning and Intermediate Algebra: An Integrated Approach, Fourth Edition
Intermediate Algebra, Seventh Edition
Algebra for College Students, Seventh Edition
College Algebra, Ninth Edition

College Algebra

9TH EDITION

R. David Gustafson
Rock Valley College

Peter D. Frisk
Rock Valley College

THOMSON
BROOKS/COLE

Australia • Brazil • Canada • Mexico • Singapore • Spain
United Kingdom • United States

THOMSON
BROOKS/COLE

College Algebra, **Ninth Edition**
Gustafson and Frisk

Acquisition Editor: *John-Paul Ramin*
Assistant Editor: *Katherine Brayton/Katherine Cook*
Editorial Assistant: *Leata Holloway/Dianna Muhammad*
Associate Developmental Editor: *Shona Burke*
Senior Marketing Manager: *Karin Sandberg*
Marketing Assistant: *Jennifer Velasquez*
Marketing Communications Manager: *Darlene Amidon-Brent*
Technology Project Manager: *Fiona Chong*
Senior Project Manager, Editorial Production: *Janet Hill*
Senior Art Director: *Vernon Boes*
Print Buyer: *Rebecca Cross*
Production Service: *Chapter Two, Ellen Brownstein*
Text Designer: *Diane Beasley*
Cover Designer: *Denise Davidson*
Cover Image: *Vincenzo Lombardo/Getty Images, Galleria Vittorio Emanuelle II, Milan, Italy*
Cover Printer: *Coral Graphic Services*
Composition: *Graphic World, Inc.*
Interior Printer: *R.R. Donnelley/Crawfordsville*

Printed in the United States of America
1 2 3 4 5 6 7 09 08 07 06 05

Library of Congress Control Number: 2005928321

Student Edition: ISBN 0-495-01266-1

Thomson Higher Education
10 Davis Drive
Belmont, CA 94002-3098
USA

Contents

Preface

To the Instructor

College Algebra, Ninth Edition maintains the same philosophy as the highly successful previous editions. It is designed to prepare students for business mathematics, finite mathematics, trigonometry, or calculus. *College Algebra* will also prepare students for employment in the 21st century. Since it contains all of the topics associated with a college algebra or a precalculus course, the book allows you to pick and choose those topics that are relevant to your students' needs.

As before, our goal is to write a text that

- is relevant and easy to understand,
- stresses the concept of function,
- uses real-life applications to motivate problem solving,
- develops critical-thinking skills in all students, and
- develops the skills students need to continue their study of mathematics.

We believe that we have accomplished this goal through a successful blending of content and pedagogy. We present a straight-forward, comprehensive, in-depth, precise coverage of the topics of college algebra, incorporated into a framework of tested teaching strategies with carefully selected pedagogical features. In keeping with the spirit of the NCTM standards and the recommendations of the American Mathematical Association of Two-Year Colleges, this book emphasizes conceptual understanding, problem solving, and the use of technology. The coverage of skill topics remains comprehensive and thorough.

■ Changes in the Ninth Edition

The overall effects of the changes made to the Ninth Edition are as follows:

1. *To continue the emphasis on solving problems through realistic applications* by increasing the wide variety of application problems. All application problems have titles.

APPLICATIONS *Use a calculator to solve each problem.*

7. **Gain of an amplifier** An amplifier produces an output of 17 volts when the input signal is 0.03 volt. Find the decibel voltage gain. 55 db
8. **Transmission lines** A 4.9-volt input to a long transmission line decreases to 4.7 volts at the other end. Find the decibel voltage loss. 0.36 db
9. **Gain of an amplifier** Find the db gain of an amplifier whose input voltage is 0.71 volt and whose output voltage is 20 volts. 29 db
10. **Gain of an amplifier** Find the db gain of an amplifier whose output voltage is 2.8 volts and whose input voltage is 0.05 volt. 35 db

18. **Battery charge** If $k = 0.201$, how long will it take a battery to reach a 40% charge? Assume that the battery was fully discharged when it began charging. about 2.5 min
19. **Population growth** A town's population grows at the rate of 12% per year. If this growth rate remains constant, how long will it take the population to double? about 5.8 yr
20. **Fish population growth** One thousand bass were stocked in Catfish Lake in Eagle River, Wisconsin, a lake with no bass population. If the population of bass is expected to grow at a rate of 25% per year, how long will it take the population to double?

2. *To increase the relevance of the material to students' lives* by introducing a new feature called Everyday Connections. The Everyday Connections boxes relate the mathematics in the chapter to students' experiences in practical ways.

Everyday Connections **Carbon-14 Dating**

"With these things [cleanliness, care, seriousness, and practice], it is possible to obtain radiocarbon dates which are consistent and which may indeed help roll back the pages of history and reveal to mankind something more about his ancestors, and in this way perhaps about his future." Willard F. Libby, Physical Chemist, pioneered radiocarbon (Carbon-14) dating method, awarded Nobel Prize for Chemistry in 1960.

Turin Shroud Older Than Thought

The Shroud of Turin, the piece of linen long believed to have been wrapped around Jesus's body after the crucifixion, is much older than the date suggested by radiocarbon tests, according to new microchemical research.

Published in the current issue of *Thermochimica Acta,* a chemistry peer-reviewed scientific journal, the study dismisses the results of the 1988 Carbon-14 dating.

At that time, three reputable laboratories in Oxford, Zurich, and Tucson, Ariz., concluded that the cloth on which the smudged outline of the body of a man is indelibly impressed, was a medieval fake dating from 1260 to 1390, and not the burial cloth wrapped around the body of Christ.

Barrie M. Schwortz/AP/Wide World Photos

3. *To fine-tune the presentation of topics* for better flow of ideas and for clarity.
4. *To increase visual interest with a new design* that is student friendly. We include color not just as a design feature, but to highlight terms that instructors would point to in a classroom discussion.

EXAMPLE 1 Graph: $x + 2y = 5$.

Solution We pick values for x or y, substitute them into the equation, and solve for the other variable. If $y = -1$, we find x as follows.

$$x + 2y = 5$$
$$x + 2(-1) = 5 \quad \text{Substitute } -1 \text{ for } y.$$
$$x - 2 = 5 \quad \text{Simplify.}$$
$$x = 7 \quad \text{Add 2 to both sides.}$$

The pair $(7, -1)$ satisfies the equation. To find another, we let $x = 0$ and find y.

$$x + 2y = 5$$
$$0 + 2y = 5 \quad \text{Substitute 0 for } x.$$
$$2y = 5 \quad \text{Simplify.}$$
$$y = \frac{5}{2} \quad \text{Divide both sides by 2.}$$

We have made a number of changes to make the book more useful to students. We have:

- Integrated *College AlgebraNow,* an on-line assessment-centered study tool, with the text. For each chapter, *College AlgebraNow* includes diagnostic questions that create a *Personalized Learning Plan* for students. The *Personalized Learning Plan* allows students to focus their study time and improve their understanding of key content for each chapter. A *Post-Test* will check their knowledge of the chapter material and help prepare for exams. As students read the text, they will see references like this:

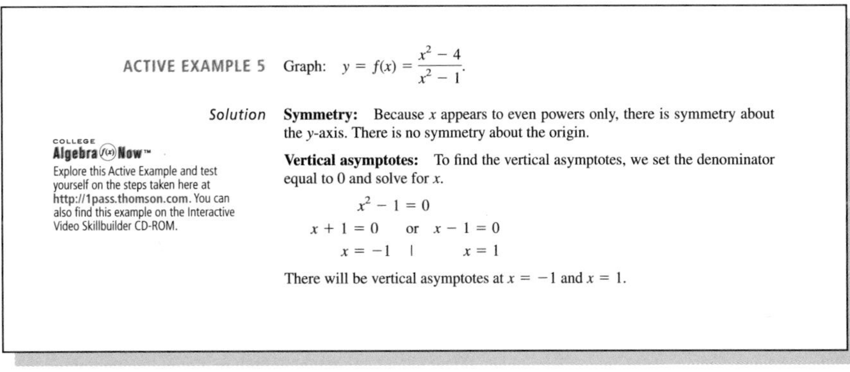

ACTIVE EXAMPLE 5 Graph: $y = f(x) = \frac{x^2 - 4}{x^2 - 1}$.

Solution **Symmetry:** Because x appears to even powers only, there is symmetry about the y-axis. There is no symmetry about the origin.

Vertical asymptotes: To find the vertical asymptotes, we set the denominator equal to 0 and solve for x.

$$x^2 - 1 = 0$$
$$x + 1 = 0 \quad \text{or} \quad x - 1 = 0$$
$$x = -1 \quad | \quad x = 1$$

There will be vertical asymptotes at $x = -1$ and $x = 1$.

COLLEGE Algebra Now™
Explore this Active Example and test yourself on the steps taken here at http://1pass.thomson.com. You can also find this example on the Interactive Video Skillbuilder CD-ROM.

These references direct students to corresponding Active Examples (also found on the Interactive Video Skillbuilder CD-ROM) and Active Figures on *College AlgebraNow.*

- Up-dated information in the Careers in Mathematics feature. Each chapter opens with a job description and the job outlook for various careers that require the study of mathematics. This feature utilizes information from the *Occupational Outlook Handbook,* published by the Bureau of Labor Statistics, an agency of the United States Department of Labor.

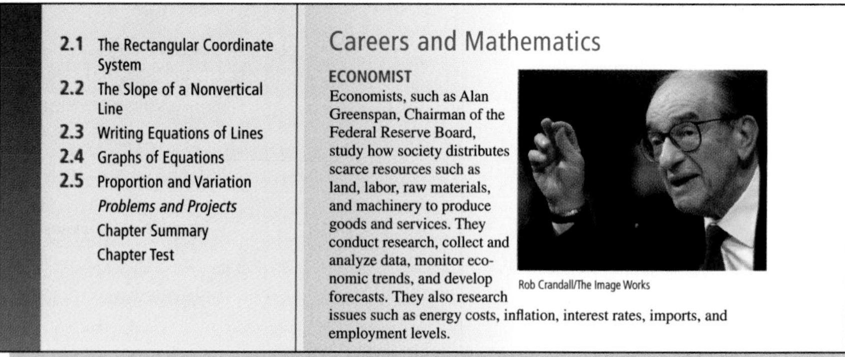

2.1 The Rectangular Coordinate System
2.2 The Slope of a Nonvertical Line
2.3 Writing Equations of Lines
2.4 Graphs of Equations
2.5 Proportion and Variation
Problems and Projects
Chapter Summary
Chapter Test

Careers and Mathematics

ECONOMIST

Economists, such as Alan Greenspan, Chairman of the Federal Reserve Board, study how society distributes scarce resources such as land, labor, raw materials, and machinery to produce goods and services. They conduct research, collect and analyze data, monitor economic trends, and develop forecasts. They also research issues such as energy costs, inflation, interest rates, imports, and employment levels.

Rob Crandall/The Image Works

- Revised several sections to make the presentation easier to read.
- Added more review and application problems throughout the text. An index of applications is included.

Some changes made to specific chapters include:

Chapter 0 The first chapter is called Chapter 0 to indicate that all of the material in the chapter is a review of basic algebra and can be omitted.

Chapter 1 Section 1.2 introduces the idea of mathematical modeling. Some new examples and many exercises have been added to the chapter.

Chapter 2 Chapter 2 discusses graphing lines, slope, writing equations of lines, and ratio and proportion. Additional review problems have been added throughout the chapter. The x-coordinates of graphs are better related to solutions of equations.

Chapter 3 Chapter 3 continues to emphasize defining functions by tables, graphs, and equations. The definitions of polynomial functions and their degrees have been rewritten, and a table of basic functions is now included. Many new exercises have been added.

Chapter 4 Chapter 4 covers exponential and logarithmic functions. Example 3 in Section 4.2 has been updated. Many problems were revised to provide more than just routine drill.

Chapter 5 Chapter 5 discusses the solution of polynomial equations. We now connect roots of an equation to zeros of a polynomial, and the discussion on integer bounds has been improved. For better continuity, Example 1 in Section 5.4 comes from Section 5.3. This chapter contains many new exercises.

Chapter 6 Chapter 6 discusses systems of linear equations. In Section 6.3, the excessive algebraic notation has been toned down. This chapter contains many new exercises.

Chapter 7 Chapter 7 discusses conic sections and quadratic systems. In Section 7.1, Example 3 has been reworked to make it clearer. Many more application and drill exercises have been added. There is a new appendix (Appendix II) that provides an alternate approach to circles and parabolas. In this treatment, circles and parabolas are presented in separate sections and the treatment of parabolas does not use the idea of directrix.

Chapter 8 Chapter 8 covers the binomial theorem, sequences, mathematical induction, permutations and combinations, and probability. The statement of the binomial theorem now includes the term involving b^r and this method is now used to write a particular term of a binomial expansion. Some easier examples are included using summation notation, the notation for the last term of a sequence has been changed from l to a_n, and the Tower of Hanoi problem has been added to Section 8.5. This chapter contains many new exercises.

Chapter 9 Chapter 9 covers interest, effective rates of interest, present value, annuities and future value, and amortization. The topic of simple interest has been added in Section 9.1. The rest of the chapter contains only minor revisions.

Appendices New Appendix II offers an alternate approach to circles and parabolas.

FEATURES

We have kept the pedagogical features that made the previous editions of the book so successful:

Solid Mathematics The treatment of college algebra remains direct and straightforward. Although the mathematics is sound, it is not so rigorous that it will confuse students. All exercise sets have Vocabulary and Concept problems, Practice problems, Discovery and Writing problems, and Review problems. The book contains more than 4,000 exercises.

Emphasis on Applications To show that mathematics is useful, we include a large number of word problems and applications throughout the book. An index of applications is included.

Accessibility to Students The book is written for students to read and understand. The numerous problems within each exercise set are carefully keyed to more than 400 worked examples in which author's notes explain many of the steps used in the problem-solving process. Most examples are followed by a Self Check problem, whose answers appear at the end of each section. Answers to the odd-numbered exercises appear in an appendix in the Student Edition.

Use of Calculators Throughout the text, there are an abundant number of exercises requiring calculators. These are marked with a calculator icon. Accent on Technology boxes present graphing calculator material by providing clear and concise instructions and examples. Although graphing calculators are incorporated into the book, their use is not required. All of the graphing topics are fully discussed in traditional ways. Of course, we recommend that instructors use the graphing calculator material.

Accent on Technology USING CALCULATORS TO GRAPH LOGARITHMIC FUNCTIONS

Figure 4-20

Graphing calculators can draw graphs of logarithmic functions when the base of the logarithmic function is 10 or e. To use a calculator to graph $f(x) = -2 + \log_{10}\left(\frac{1}{2}x\right)$, we enter the right-hand side of the equation after the symbol $Y_1 =$. The display will show the equation

$$Y_1 = -2 + \log(1/2{*}x)$$

If we use window settings of $[-1, 5]$ for x and $[-5, 1]$ for y and press GRAPH, we will obtain the graph shown in Figure 4-20.

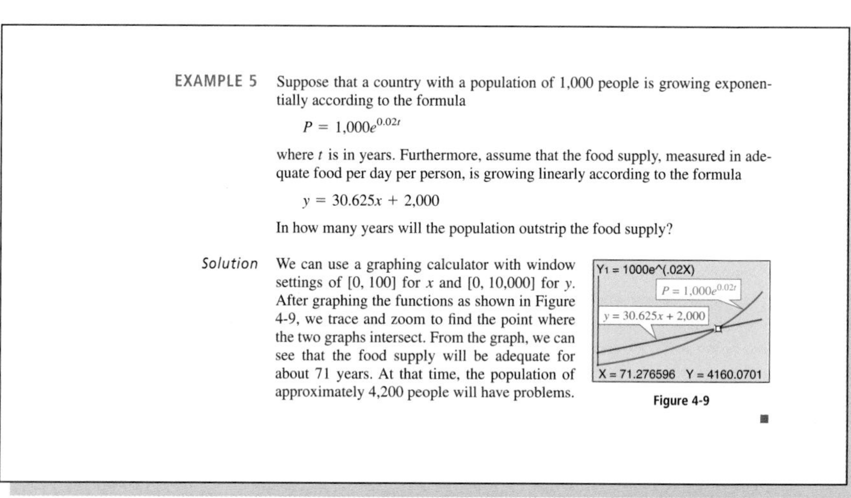

EXAMPLE 5 Suppose that a country with a population of 1,000 people is growing exponentially according to the formula

$$P = 1{,}000e^{0.02t}$$

where t is in years. Furthermore, assume that the food supply, measured in adequate food per day per person, is growing linearly according to the formula

$$y = 30.625x + 2{,}000$$

In how many years will the population outstrip the food supply?

Solution We can use a graphing calculator with window settings of [0, 100] for x and [0, 10,000] for y. After graphing the functions as shown in Figure 4-9, we trace and zoom to find the point where the two graphs intersect. From the graph, we can see that the food supply will be adequate for about 71 years. At that time, the population of approximately 4,200 people will have problems.

Figure 4-9

■

Problems and Projects For group work or extended assignments, each chapter includes a Problems and Projects feature.

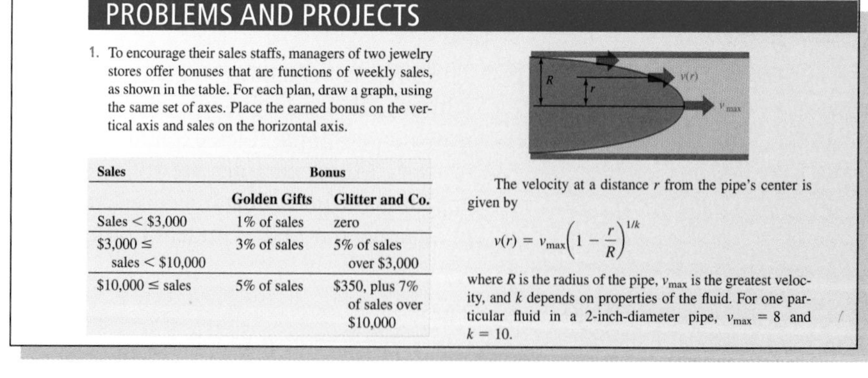

PROBLEMS AND PROJECTS

1. To encourage their sales staffs, managers of two jewelry stores offer bonuses that are functions of weekly sales, as shown in the table. For each plan, draw a graph, using the same set of axes. Place the earned bonus on the vertical axis and sales on the horizontal axis.

Sales	Bonus	
	Golden Gifts	Glitter and Co.
Sales < \$3,000	1% of sales	zero
\$3,000 ≤ sales < \$10,000	3% of sales	5% of sales over \$3,000
\$10,000 ≤ sales	5% of sales	\$350, plus 7% of sales over \$10,000

The velocity at a distance r from the pipe's center is given by

$$v(r) = v_{max}\left(1 - \frac{r}{R}\right)^{1/k}$$

where R is the radius of the pipe, v_{max} is the greatest velocity, and k depends on properties of the fluid. For one particular fluid in a 2-inch-diameter pipe, $v_{max} = 8$ and $k = 10$.

Ample Review Material Review is incorporated into the book in many ways: There is a Chapter Summary with review exercises and a Chapter Test at the end of each chapter. Cumulative Review Exercises appear after every two chapters. Endpapers list the important formulas developed in the book.

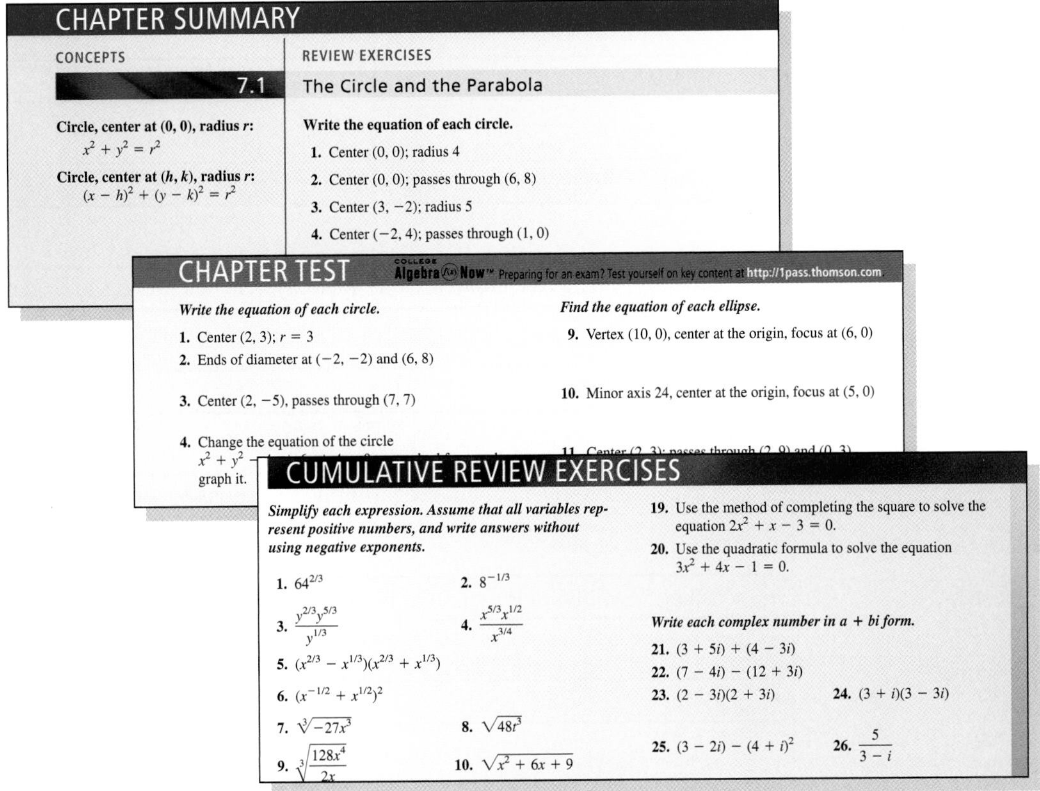

CHAPTER SUMMARY

CONCEPTS	REVIEW EXERCISES
7.1	The Circle and the Parabola
Circle, center at (0, 0), radius r: $x^2 + y^2 = r^2$	**Write the equation of each circle.** 1. Center (0, 0); radius 4
Circle, center at (h, k), radius r: $(x - h)^2 + (y - k)^2 = r^2$	2. Center (0, 0); passes through (6, 8)
	3. Center (3, −2); radius 5
	4. Center (−2, 4); passes through (1, 0)

CHAPTER TEST

College Algebra Now™ Preparing for an exam? Test yourself on key content at http://1pass.thomson.com

Write the equation of each circle.

1. Center (2, 3); $r = 3$
2. Ends of diameter at (−2, −2) and (6, 8)
3. Center (2, −5), passes through (7, 7)
4. Change the equation of the circle $x^2 + y^2$ … graph it.

Find the equation of each ellipse.

9. Vertex (10, 0), center at the origin, focus at (6, 0)
10. Minor axis 24, center at the origin, focus at (5, 0)

CUMULATIVE REVIEW EXERCISES

Simplify each expression. Assume that all variables represent positive numbers, and write answers without using negative exponents.

1. $64^{2/3}$
2. $8^{-1/3}$
3. $\dfrac{y^{2/3}y^{5/3}}{y^{1/3}}$
4. $\dfrac{x^{5/3}x^{1/2}}{x^{3/4}}$
5. $(x^{2/3} - x^{1/3})(x^{2/3} + x^{1/3})$
6. $(x^{-1/2} + x^{1/2})^2$
7. $\sqrt[3]{-27x^3}$
8. $\sqrt{48t^3}$
9. $\sqrt[3]{\dfrac{128x^4}{2x}}$
10. $\sqrt{x^2 + 6x + 9}$

19. Use the method of completing the square to solve the equation $2x^2 + x - 3 = 0$.
20. Use the quadratic formula to solve the equation $3x^2 + 4x - 1 = 0$.

Write each complex number in a + bi form.

21. $(3 + 5i) + (4 - 3i)$
22. $(7 - 4i) - (12 + 3i)$
23. $(2 - 3i)(2 + 3i)$
24. $(3 + i)(3 - 3i)$
25. $(3 - 2i) - (4 + i)^2$
26. $\dfrac{5}{3 - i}$

Historical Notes Interspersed throughout the text are pictures and captions honoring people who have made contributions to mathematics.

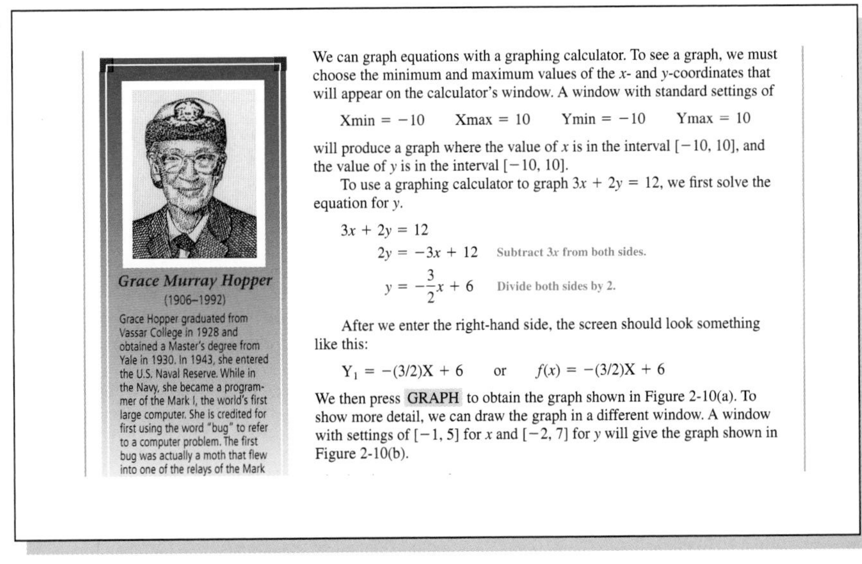

Grace Murray Hopper (1906–1992)

Grace Hopper graduated from Vassar College in 1928 and obtained a Master's degree from Yale in 1930. In 1943, she entered the U.S. Naval Reserve. While in the Navy, she became a programmer of the Mark I, the world's first large computer. She is credited for first using the word "bug" to refer to a computer problem. The first bug was actually a moth that flew into one of the relays of the Mark

We can graph equations with a graphing calculator. To see a graph, we must choose the minimum and maximum values of the x- and y-coordinates that will appear on the calculator's window. A window with standard settings of

$$\text{Xmin} = -10 \quad \text{Xmax} = 10 \quad \text{Ymin} = -10 \quad \text{Ymax} = 10$$

will produce a graph where the value of x is in the interval $[-10, 10]$, and the value of y is in the interval $[-10, 10]$.

To use a graphing calculator to graph $3x + 2y = 12$, we first solve the equation for y.

$$3x + 2y = 12$$
$$2y = -3x + 12 \quad \text{Subtract } 3x \text{ from both sides.}$$
$$y = -\frac{3}{2}x + 6 \quad \text{Divide both sides by 2.}$$

After we enter the right-hand side, the screen should look something like this:

$$Y_1 = -(3/2)X + 6 \quad \text{or} \quad f(x) = -(3/2)X + 6$$

We then press GRAPH to obtain the graph shown in Figure 2-10(a). To show more detail, we can draw the graph in a different window. A window with settings of $[-1, 5]$ for x and $[-2, 7]$ for y will give the graph shown in Figure 2-10(b).

Organization and Coverage

This text can be used in a variety of ways. To maintain optimum flexibility, many chapters are sufficiently independent to allow you to pick and choose topics that are relevant to your students' needs. The following diagram shows how the chapters are related.

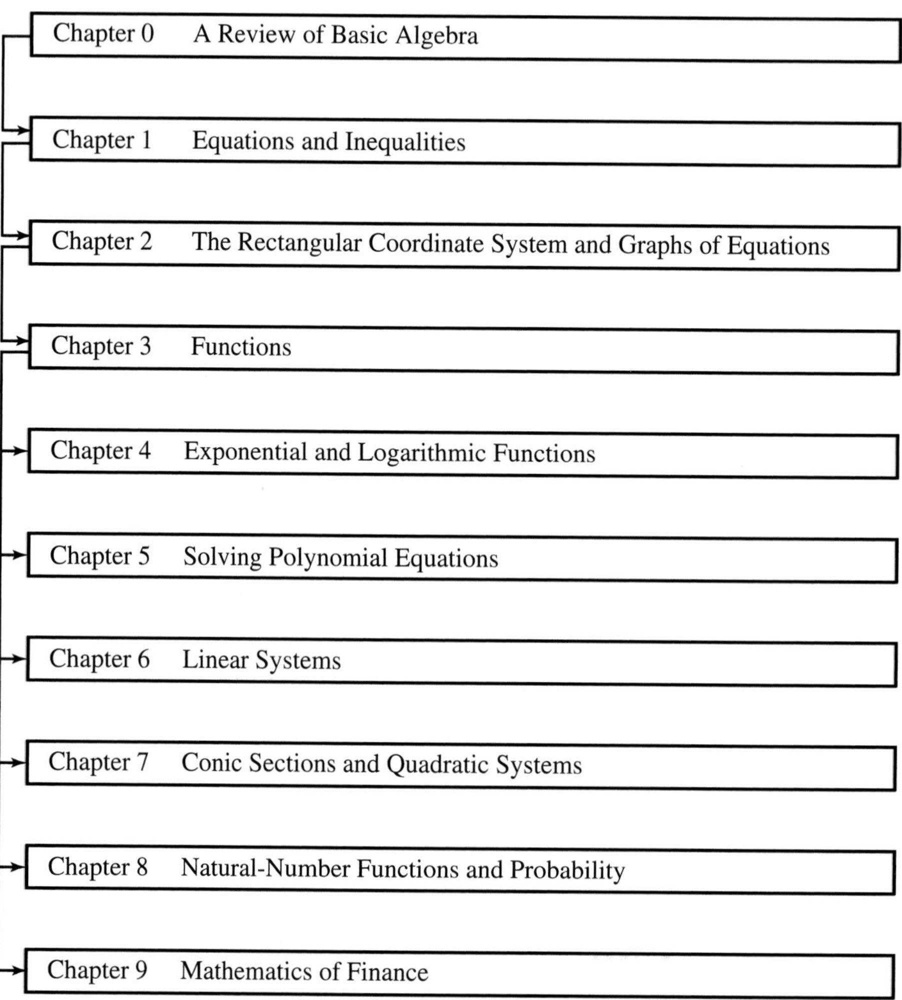

Ancillaries for the Instructor

Annotated Instructor's Edition (0-495-01836-8) This instructor's version of the complete student text includes the answers to each exercise printed in blue next to the respective exercise.

Instructor's Resource CD-ROM (0-495-11056-6) The CD provides the instructor with dynamic media tools for teaching college algebra. PowerPoint lecture slides and art slides of figures from the text, together with electronic files for the test bank and solutions manual, are available.

Test Bank (0-495-01797-3) The *Test Bank* includes 6 tests per chapter as well as 3 final exams. The tests are made up of a combination of multiple-choice, free-response, true/false, and fill-in-the-blank questions.

Complete Solutions Manual (0-495-01796-5) This manual contains solutions to all exercises from the text, including Chapter Review Exercises, Chapter Tests, and Cumulative Review Exercises.

iLrn Providing instructors and students with unsurpassed control, variety, and all-in-one utility, *iLrn* is a powerful and fully integrated teaching and learning system. *iLrn* ties together five fundamental learning activities: diagnostics, tutorials, homework, quizzing, and testing. Easy to use, *iLrn* offers instructors complete control when creating assessments in which they can draw from the wealth of exercises provided or create their own questions. *iLrn* features the greatest variety of problem types—allowing instructors to assess the way they teach! A real time-saver for instructors, *iLrn* offers automatic grading of homework, quizzes, and tests with results flowing directly into the gradebook. The auto-enrollment feature also saves time with course setup as students self-enroll into the course gradebook. *iLrn* provides seamless integration with Blackboard and WebCT. Contact your local Thomson sales representative for instructor access to *iLrn*.

Text-Specific Videotapes (0-495-01799-X) These text-specific videotape sets, available at no charge to qualified adopters of the text, feature 10- to 20-minute problem-solving lessons that cover each section of every chapter.

JoinIn on Turning Point (0-495-10637-2) Thomson Brooks/Cole is pleased to offer you book-specific *JoinIn* content for electronic response systems tailored to *College Algebra,* Ninth Edition. You can transform your classroom and assess your students' progress with instant in-class quizzes and polls. *Turning Point* software lets you pose book-specific questions and display students' answers seamlessly within Microsoft PowerPoint slides of your own lecture, in conjunction with the "clicker" hardware of your choice. Enhance the way your students interact with you, your lecture, and each other.

Ancillaries for the Student

COLLEGE Algebra $f(x)$ Now™

College AlgebraNow is a powerful on-line learning companion that helps students gauge their unique study needs and provides them with a *Personalized Learning Plan.* The individualized resources in *College AlgebraNow* give your students all the learning tools they need to master course concepts. Completely integrated with the textbook, *College AlgebraNow* and this new edition of Gustafson and Frisk's *College Algebra* enhance each other, providing students with a seamless, integrated learning system. Icons in the text direct students to Active Examples and Active Figures on *College AlgebraNow.*

College AlgebraNow consists of three powerful, easy-to-use assessment components:

- **A diagnostic: Gateway and Challenge Questions** for each chapter give students an initial assessment of their knowledge.
- **A Personalized Learning Plan,** based on the student's answers to the *Gateway and Challenge Questions,* outlines key elements for review.
- **A Post-Test** for each text chapter assesses students' mastery of core chapter concepts. The results can be e-mailed to the instructor.

College AlgebraNow is accessed via a *1pass™* access card included with every new copy of Gustafson/Frisk's text. If your students do not buy a new text but would like access to *College AlgebraNow,* lead them to http://1pass.thomson.com to purchase electronic access.

Student Solutions Manual (0-495-01795-7) This manual features worked out solutions to all odd-numbered exercises in the text, as well as solutions to the Chapter Review Exercises, Chapter Tests, and Cumulative Review Exercises. It also offers hints and additional problems for practice, similar to those in the text.

iLrn Tutorial (0-495-01803-1) Featuring a variety of approaches that connect with all types of learners, *iLrn Tutorial* offers text-specific tutorials that require no set-up by instructors. Students can begin exploring examples from the text by using the *1pass* access code packaged with each new book. *iLrn Tutorial* supports students with explanations from the text, examples, step-by-step problem-solving help, unlimited practice, and chapter-by-chapter video lessons. With this self-paced system, students can even check their comprehension along the way by taking quizzes and receiving feedback.

vMentor™ When students get stuck on a particular problem or concept, they need only log on to *vMentor.* Accessed through *1pass, vMentor* allows students to talk (using their own computer microphones) to tutors who will skillfully guide them through the problem using an interactive whiteboard for illustration. Students have access to up to 40 hours of live tutoring a week!

Interactive Video Skillbuilder CD-ROM (0-495-01800-7) Think of it as portable office hours! The *Interactive Video Skillbuilder* CD-ROM contains more than eight hours of video instruction. The problems worked during each video lesson are shown next to the viewing screen so that students can try working them before watching the solution. To help students evaluate their progress, each section contains a 10-question Web quiz, the results of which can be e-mailed to the instructor, and each chapter contains a chapter test, with the answer to each problem on each test. This interactive learning resource also includes *MathCue* tutorial and quizzing software.

Web Site http://mathematics.brookscole.com/gfca9e Access book- and course-specific resources, such as tutorial quizzes for each chapter, a glossary of important terms, and links to mathematics sites on the Web.

To the Student

Congratulations. You now own a state-of-art textbook that has been written especially for you. We have tried to write a book that you can read and understand. The book includes carefully written narrative and an extensive number of worked examples with Self Checks. The answers to the Self Check problems appear at the end of each section.

To get the most out of this course, you must read and study the textbook properly. We recommend that you work the examples on paper first, and then work the Self Checks. Only after you thoroughly understand the concepts taught in the examples should you attempt to work the exercises.

You may purchase a *Student Solutions Manual,* which contains detailed solutions to all the odd-numbered exercises in the text, as well as solutions to the Chapter Review Exercises, Chapter Tests, and Cumulative Review Exercises.

Since the material presented in *College Algebra,* Ninth Edition will be of value to you in later years, we suggest that you keep this book. It will be a good source of reference and will keep at your fingertips the material that you have learned here.

We wish you well.

Acknowledgments

We are grateful to the following people who reviewed the manuscript in its various stages. All of them had valuable suggestions that have been incorporated into this book.

Ebrahim Ahmadizadeh, Northampton Community College
Ricardo Alfaro, University of Michigan–Flint
Richard Andrews, University of Wisconsin
James Arnold, University of Wisconsin
Ronald Atkinson, Tennessee State University
Wilson Banks, Illinois State University
Chad Bemis, Riverside Community College
Anjan Biswas, Tennessee State College
Jerry Bloomberg, Essex Community College
Elaine Bouldin, Middle Tennessee State University
Dale Boye, Schoolcraft College
Lee R. Clancy, Golden West College
Krista Blevins Cohlmia, Odessa College
Jan Collins, Embry Riddle College
Cecilia Cooper, William & Harper College
John S. Cross, University of Northern Iowa
M. Hilary Davies, University of Alaska–Anchorage
Elias Deeba, University of Houston–Downtown
Grace DeVelbiss, Sinclair Community College
Lena Dexter, Faulkner State Junior College
Emily Dickinson, University of Arkansas
Mickey P. Dunlap, University of Tennessee, Martin
Eric Ellis, Essex Community College
Eunice F. Everett, Seminole Community College
Dale Ewen, Parkland College
Harold Farmer, Wallace Community College–Hanceville
Ronald J. Fischer, Evergreen Valley College
Mary Jane Gates, University of Arkansas at Little Rock
Marvin Goodman, Monmouth College
Edna Greenwood, Tarrant County College
Jerry Gustafson, Beloit College
Jerome Hahn, Bradley University
Douglas Hall, Michigan State University
Robert Hall, University of Wisconsin
David Hansen, Monterey Peninsula College
Kevin Hastings, University of Delaware
William Hinrichs, Rock Valley College
Arthur M. Hobbs, Texas A & M University
Jack E. Hofer, California Polytechnic State University
Ingrid Holzner, University of Wisconsin
Warren Jaech, Tacoma Community College
Nancy Johnson, Broward Community College
Patricia H. Jones, Methodist College
William B. Jones, University of Colorado
Barbara Juister, Elgin Community College
David Kinsey, University of Southern Indiana
Helen Kriegsman, Pittsburg State University
Marjorie O. Labhart, University of Southern Indiana
Jaclyn LeFebvre, Illinois Central College
Susan Loveland, University of Alaska–Anchorage
James Mark, Eastern Arizona College
Marcel Maupin, Oklahoma State University, Oklahoma City
Judy McKinney, California Polytechnic Institute at Pomona
Sandra McLaurin, University of North Carolina
Robert O. McCoy, University of Alaska–Anchorage
Marcus McWaters, University of Southern Florida
Donna Menard, University of Massachusetts, Dartmouth
James W. Mettler, Pennsylvania State University
Eldon L. Miller, University of Mississippi
Stuart E. Mills, Louisiana State University, Shreveport
Mila Mogilevskaya, Wichita State University
Gilbert W. Nelson, North Dakota State
Marie Neuberth, Catonsville City College
Anthony Peressini, University of Illinois
David L. Phillips, University of Southern Colorado
William H. Price, Middle Tennessee State University
Ronald Putthoff, University of Southern Mississippi
Janet P. Ray, Seattle Central Community College
Barbara Riggs, Tennessee Technological University
Renee Roames, Purdue University
Paul Schaefer, SUNY, Geneseo
Vincent P. Schielack, Jr., Texas A & M University
Robert Sharpton, Miami Dade Community College
L. Thomas Shiflett, Southwest Missouri State University
Richard Slinkman, Bemidji State University
Merreline Smith, California Polytechnic Institute at Pomona
John Snyder, Sinclair Community College
Warren Strickland, Del Mar College
Paul K. Swets, Angelo State College
Ray Tebbetts, San Antonio College
Faye Thames, Lamar State University
Douglas Tharp, University of Houston–Downtown
Carolyn A. Wailes, University of Alabama, Birmingham
Carol M. Walker, Hinds Community College
William Waller, University of Houston–Downtown
Carroll G. Wells, Western Kentucky University
William H. White, University of South Carolina at Spartanburg
Charles R. Williams, Midwestern State University
Harry Wolff, University of Wisconsin
Clifton Whyburn, University of Houston
Roger Zarnowski, Angelo State University
Albert Zechmann, University of Nebraska

We wish to thank the staff at Brooks/Cole, especially John-Paul Ramin, Shona Burke, and Janet Hill, for their support in the production process. We also thank Lori Heckleman for her fine artwork, Mike Welden for his problem checking, and Diane Koenig for her proof reading. Special thanks go to Ellen Brownstein for her superb patience and copyediting skills and to Graphic World for their excellent typesetting skills. Finally, we wish to thank Mark McCombs, at University of North Carolina, Chapel Hill, for contributing the Everyday Connections problems, Robert McCoy, at University of Alaska, Anchorage, for his contribution of the Tower of Hanoi problem, and Mehrdad Simkani, at University of Michigan, Flint, for his contribution of the Excel problems.

R. David Gustafson
Peter D. Frisk

Index of Applications

Examples that are applications are shown with boldface numbers.
Exercises that are applications are shown with lightface numbers.

0 A Review of Basic Algebra

COLLEGE **Algebra $f(x)$ Now™**

Throughout the chapter, this icon introduces resources on the College AlgebraNow Web Site, accessed through **http://1pass.thomson.com**, that will:

- Help you test your knowledge of the material in this chapter prior to reading it,
- Allow you to take an exam-prep quiz, and
- Provide a Personalized Learning Plan targeting areas you should study.

Careers and Mathematics

MECHANICAL ENGINEER

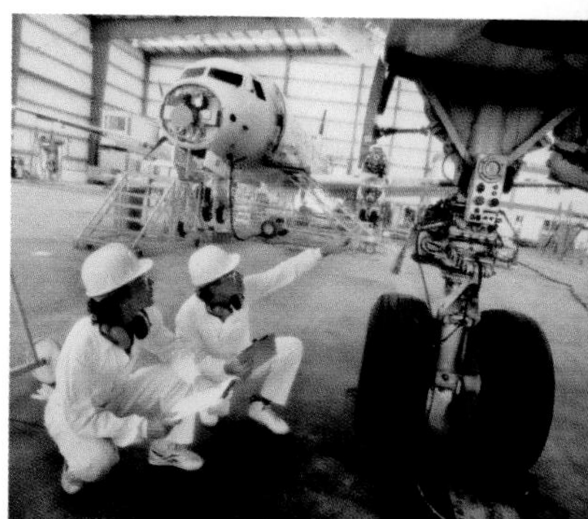

age fotostock/SuperStock

Mechanical engineers research, develop, design, manufacture, and test tools, engines, machines, and other mechanical devices. They also work on power-producing machines such as electrical generators, internal combustion engines, and steam and gas turbines. In addition, they develop power-using machines such as refrigeration and air-conditioning equipment, machine tools, material-handling systems, elevators and escalators, industrial production equipment, and robots used in manufacturing. Mechanical engineers held about 215,000 jobs in 2002. More than 1 out of 2 jobs were in manufacturing—mostly in machinery, transportation equipment, electrical equipment, instruments, and fabricated metal products industries.

JOB OUTLOOK

Employment of mechanical engineers is projected to grow more slowly than the average for all occupations through 2012. Median annual salaries of mechanical engineers were $62,880 in 2002.

For a sample application, see Problem 97 in Section 0.6.

In this preliminary chapter, we review many concepts and skills learned in previous algebra courses. Be sure to master this material now, because it is the basis for the rest of this course.

0.1 Sets of Real Numbers

In this section, you will learn about

- Sets of Real Numbers
- Properties of Real Numbers
- Graphing Subsets of Real Numbers
- Inequality Symbols
- Intervals
- Absolute Value
- Distance on a Number Line

A **set** is a collection of objects, such as a set of dishes or a set of golf clubs. The set of vowels of the English language can be denoted as $\{a, e, i, o, u\}$, where the braces { } are read as "the set of."

Sets of Real Numbers

There are several sets of numbers that we use in everyday life.

Basic Sets of Numbers

Natural numbers
The numbers that we use for counting: $\{1, 2, 3, 4, 5, 6, \ldots\}$

Whole numbers
The set of natural numbers including 0: $\{0, 1, 2, 3, 4, 5, 6, \ldots\}$

Integers
The set of whole numbers and their negatives:

$$\{\ldots, -5, -4, -3, -2, -1, 0, 1, 2, 3, 4, 5, \ldots\}$$

Each group of three dots, called an **ellipsis,** indicates that the numbers continue forever in the indicated direction.

When every number in one set is included in a second set, we say that the first set is a **subset** of the second set. Two important subsets of the natural numbers are the *prime* and the *composite* numbers. A **prime number** is a natural number greater than 1 that is divisible only by itself and 1. A **composite number** is a natural number greater than 1 that is not prime.

The set of prime numbers: $\{2, 3, 5, 7, 11, 13, 17, 19, 23, 29, 31, \ldots\}$

The set of composite numbers: $\{4, 6, 8, 9, 10, 12, 14, 15, 16, 18, 20, 21, \ldots\}$

Two important subsets of the set of integers are the *even* and the *odd* integers. The **even integers** are the integers that are exactly divisible by 2. The **odd integers** are the integers that are not exactly divisible by 2.

The set of even integers: $\{\ldots, -10, -8, -6, -4, -2, 0, 2, 4, 6, 8, 10, \ldots\}$

The set of odd integers: $\{\ldots, -9, -7, -5, -3, -1, 1, 3, 5, 7, 9, \ldots\}$

So far, we have listed numbers inside braces to specify sets. This method is called the **roster method.** When we give a rule to tell which numbers are in a set, we are using **set-builder notation.** To use set-builder notation to denote the set of prime numbers, we write

$$\{x \mid x \text{ is a prime number}\}$$

variable ↑ such that ↑ rule ↑

Read as "the set of all numbers x such that x is a prime number." Recall that when a letter stands for a number, it is called a variable.

The fractions of arithmetic are called *rational numbers.*

Rational Numbers

Rational numbers are fractions that have an integer numerator and a nonzero integer denominator. Using set-builder notation, the rational numbers are

$$\left\{\frac{a}{b} \,\middle|\, a \text{ is an integer and } b \text{ is a nonzero integer}\right\}$$

Comment Remember that the denominator of a fraction can never be 0.

Rational numbers can be written as fractions or decimals. Some examples of rational numbers are

$$5 = \frac{5}{1}, \quad \frac{3}{4} = 0.75, \quad \frac{1}{3} = 0.333. \ . \ . , \quad \frac{5}{11} = 0.454545. \ . \ .$$

The = sign indicates that two quantities are equal.

These examples suggest that the decimal forms of all rational numbers are either *terminating decimals* or *repeating decimals.*

EXAMPLE 1 Determine whether the decimal form of each fraction terminates or repeats: **a.** $\frac{7}{16}$ and **b.** $\frac{65}{99}$.

Solution **a.** To change $\frac{7}{16}$ to a decimal, we do a long division to get $\frac{7}{16} = 0.4375$. Thus, we can write $\frac{7}{16}$ as a terminating decimal.

b. To change $\frac{65}{99}$ to a decimal, we do a long division to get $\frac{65}{99} = 0.656565. \ . \ . \ .$ Thus, we can write $\frac{65}{99}$ as a repeating decimal. We can write repeating decimals in compact form by using an overbar. For example, $0.656565. \ . \ . = 0.\overline{65}$.

Self Check Determine whether the decimal form of each fraction terminates or repeats: **a.** $\frac{38}{99}$ and **b.** $\frac{7}{8}$. ■

Some numbers have decimal forms that neither terminate nor repeat. These nonterminating, nonrepeating decimals are called **irrational numbers.** Two examples of irrational numbers are

$$\sqrt{2} = 1.414213562. \ . \ . \quad \text{and} \quad \pi = 3.141592654. \ . \ .$$

The set of rational numbers (the terminating and repeating decimals) and the set of irrational numbers (the nonterminating, nonrepeating decimals) combine to form the set of *real numbers* (the set of all decimals).

Real Numbers

A **real number** is any number that is rational or irrational. Using set-builder notation, the set of real numbers is

$$\{x \mid x \text{ is a rational or an irrational number}\}$$

Figure 0-1 shows how many of the previous sets of numbers are related.

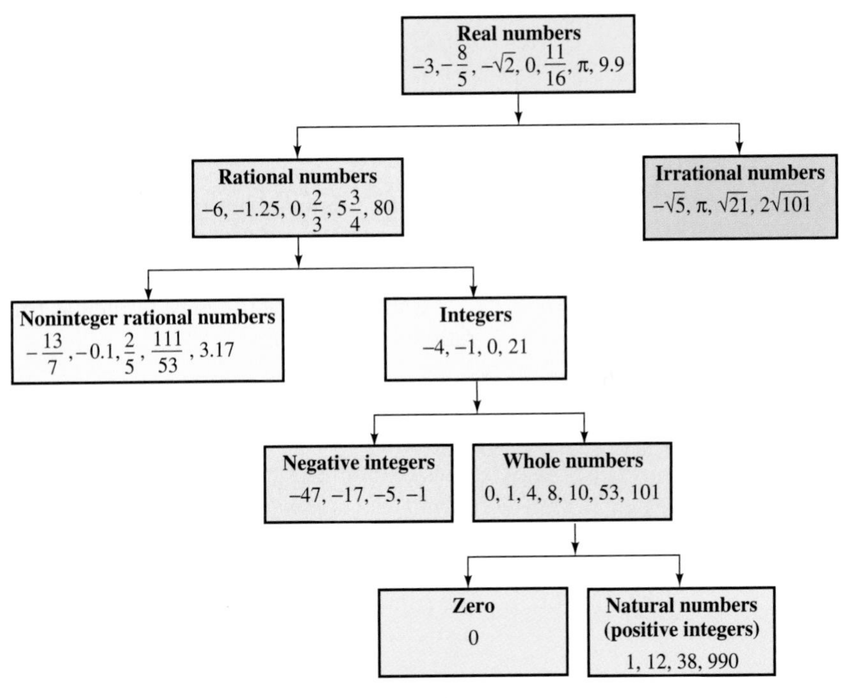

Figure 0-1

EXAMPLE 2 In the set $\left\{-3, -2, 0, \frac{1}{2}, 1, \sqrt{5}, 2, 4, 5, 6\right\}$, list all **a.** even integers, **b.** prime numbers, and **c.** rational numbers.

Solution **a.** $-2, 0, 2, 4, 6$ **b.** $2, 5$ **c.** $-3, -2, 0, \frac{1}{2}, 1, 2, 4, 5, 6$

Self Check In the set in Example 2, list all **a.** odd integers, **b.** composite numbers, and **c.** irrational numbers. ■

■ Properties of Real Numbers

When we work with real numbers, we will use the following properties.

Properties of Real Numbers

If a, b, and c are real numbers,

The Associative Properties for Addition and Multiplication

$$(a + b) + c = a + (b + c) \qquad (ab)c = a(bc)$$

The Commutative Properties for Addition and Multiplication

$$a + b = b + a \qquad ab = ba$$

The Distributive Property of Multiplication over Addition

$$a(b + c) = ab + ac \quad \text{or} \quad a(b - c) = ab - ac$$

The Double Negative Rule

$$-(-a) = a$$

The distributive property also applies when more than two terms are within parentheses.

EXAMPLE 3 Determine which property of real numbers justifies each statement.

a. $(9 + 2) + 3 = 9 + (2 + 3)$ **b.** $3(x + y + 2) = 3x + 3y + 3 \cdot 2$

Solution **a.** Associative property of addition **b.** Distributive property

Self Check Determine which property of real numbers justifies each statement.
a. $mn = nm$ **b.** $(xy)z = x(yz)$ **c.** $p + q = q + p$ ■

■ Graphing Subsets of Real Numbers

We can graph subsets of real numbers on the **number line.** The number line shown in Figure 0-2 continues forever in both directions. The **positive numbers** are represented by the points to the right of 0, and the **negative numbers** are represented by the points to the left of 0.

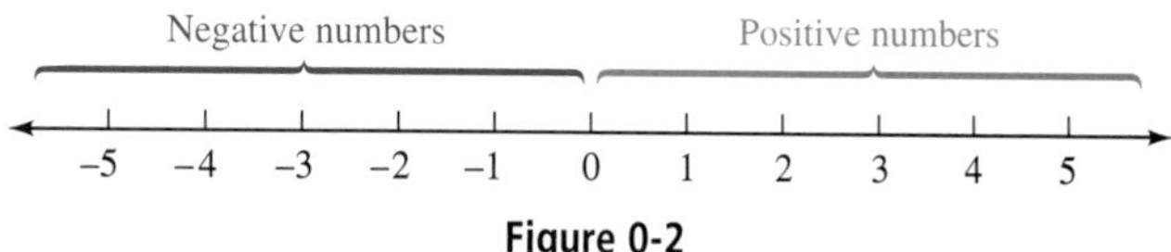

Figure 0-2

Comment Zero is neither positive nor negative.

Figure 0-3(a) shows the graph of the natural numbers from 1 to 5. The point associated with each number is called the **graph** of the number, and the number is called the **coordinate** of its point. Figure 0-3(b) shows the graph of the prime numbers that are less than 10. Figure 0-3(c) shows the graph of the integers from -4 to 3. Figure 0-3(d) shows the graph of the real numbers $-\frac{3}{4}$, $-\frac{7}{3}$, $0.\overline{3}$, and $\sqrt{2}$.

Comment $\sqrt{2}$ can be shown as the diagonal of a square with sides of length 1:

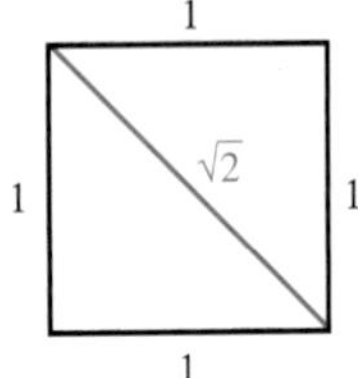

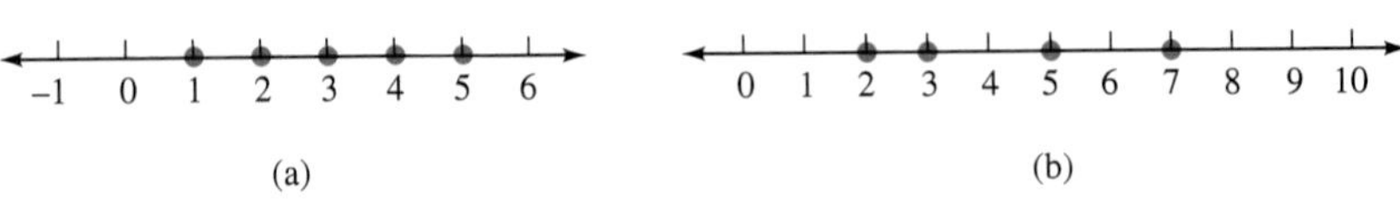

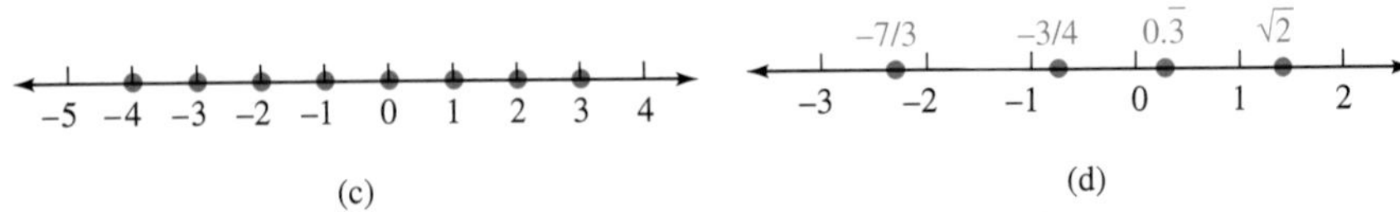

Figure 0-3

The graphs in Figure 0-3 suggest that there is a **one-to-one correspondence** between the set of real numbers and the points on a number line. This means that to each real number there corresponds exactly one point on the number line, and to each point on the number line there corresponds exactly one real-number coordinate.

EXAMPLE 4 Graph the set $\left\{-3, -\frac{4}{3}, 0, \sqrt{5}\right\}$.

Solution Note that to the nearest tenth, $\sqrt{5} = 2.2$.

−4/3 √5
−3 −2 −1 0 1 2 3

Self Check Graph the set $\left\{-2, \frac{3}{4}, \sqrt{3}\right\}$, (*Hint:* To the nearest tenth, $\sqrt{3} = 1.7$.) ■

■ Inequality Symbols

To show that two quantities are not equal, we use an **inequality symbol.**

Symbol	Read as	Examples		
$\neq$	"is not equal to"	$5 \neq 8$	and	$0.25 \neq \frac{1}{3}$
$<$	"is less than"	$12 < 20$	and	$0.17 < 1.1$
$>$	"is greater than"	$15 > 9$	and	$\frac{1}{2} > 0.2$
$\leq$	"is less than or equal to"	$25 \leq 25$	and	$1.7 \leq 2.3$
$\geq$	"is greater than or equal to"	$19 \geq 19$	and	$15.2 \geq 13.7$
$\approx$	"is approximately equal to"	$\sqrt{2} \approx 1.414$	and	$\sqrt{3} \approx 1.732$

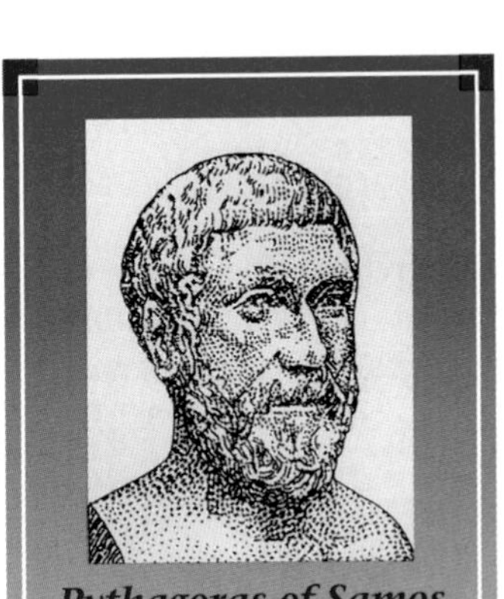

Pythagoras of Samos

(569?–475?BC)

Pythagoras is thought to be the world's first pure mathematician. Although he is famous for the theorem that bears his name, he is often called "the father of music," because a society he led discovered some of the fundamentals of musical harmony. This secret society had numerology as its religion. The society is also credited with the discovery of irrational numbers.

It is possible to write an inequality with the inequality symbol pointing in the opposite direction. For example,

$12 < 20$ is equivalent to $20 > 12$

$2.3 \geq -1.7$ is equivalent to $-1.7 \leq 2.3$

In Figure 0-2, the coordinates of points get larger as we move from left to right on a number line. Thus, if a and b are the coordinates of two points, the one to the right is the greater. This suggests the following facts:

If $a > b$, point a lies to the right of point b on a number line.

If $a < b$, point a lies to the left of point b on a number line.

Intervals

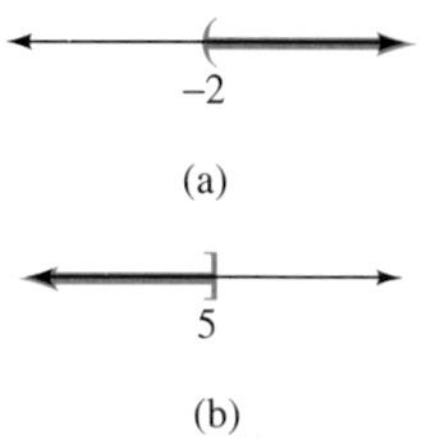

Figure 0-4

Figure 0-4(a) shows the graph of the **inequality** $x > -2$ (or $-2 < x$). This graph includes all real numbers x that are greater than -2. The parenthesis at -2 indicates that -2 is not included in the graph. Figure 0-4(b) shows the graph of $x \le 5$ (or $5 \ge x$). The bracket at 5 indicates that 5 is included in the graph.

Sometimes two inequalities can be written as a single expression called a **compound inequality.** For example, the compound inequality

$$5 < x < 12$$

is a combination of the inequalities $5 < x$ and $x < 12$. It is read as "5 is less than x, and x is less than 12," and it means that x is between 5 and 12. Its graph is shown in Figure 0-5.

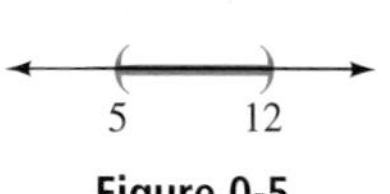

Figure 0-5

The graphs shown in Figures 0-4 and 0-5 are portions of a number line called **intervals.** The interval shown in Figure 0-6(a) is denoted by the inequality $-2 < x < 4$, or in *interval notation* as $(-2, 4)$. The parentheses indicate that the endpoints are not included. The interval shown in Figure 0-6(b) is denoted by the inequality $x > 1$, or as $(1, \infty)$ in interval notation.

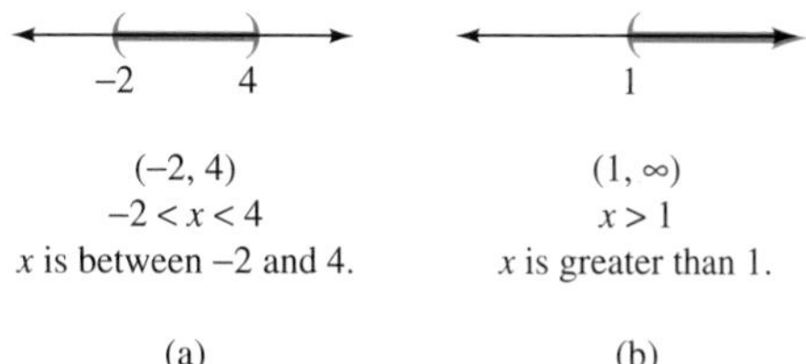

Figure 0-6

The symbol ∞ (infinity) is not a real number. It is used to indicate that the graph in Figure 0-6(b) extends infinitely far to the right.

Comment A compound inequality such as $-2 < x < 4$ can be written as two separate inequalities:

$$x > -2 \qquad \text{and} \qquad x < 4$$

This expression represents the **intersection** of two intervals. In interval notation, this expression can be written as

$$(-2, \infty) \cap (-\infty, 4) \qquad \text{Read the symbol } \cap \text{ as "intersection."}$$

Since the graph of $-2 < x < 4$ will include all points whose coordinates satisfy both $x > -2$ and $x < 4$ at the same time, its graph will include all points that are larger than -2 but less than 4. This is the interval $(-2, 4)$, whose graph is shown in Figure 0-6(a).

COLLEGE **Algebra** $f(x)$ **Now**™
Explore this Active Example and test yourself on the steps followed here at **http://1pass.thomson.com**. You can also find this example on the Interactive Video Skillbuilder CD-ROM.

ACTIVE EXAMPLE 5 Write the inequality $-3 < x < 5$ in interval notation and graph it.

Solution This is the interval $(-3, 5)$. Its graph includes all real numbers between -3 and 5, as shown in Figure 0-7.

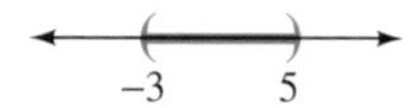

Figure 0-7

Self Check Write the inequality $x < 5$ in interval notation and graph it. ■

If an interval extends forever in one direction, it is called an **unbounded interval.**

Unbounded Intervals

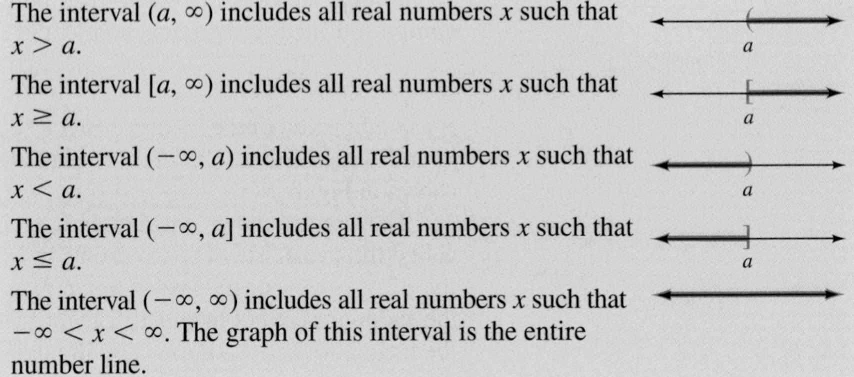

The interval (a, ∞) includes all real numbers x such that $x > a$.

The interval $[a, \infty)$ includes all real numbers x such that $x \geq a$.

The interval $(-\infty, a)$ includes all real numbers x such that $x < a$.

The interval $(-\infty, a]$ includes all real numbers x such that $x \leq a$.

The interval $(-\infty, \infty)$ includes all real numbers x such that $-\infty < x < \infty$. The graph of this interval is the entire number line.

A bounded interval with no endpoints is called an **open interval.** Figure 0-8(a) shows the open interval between -3 and 2. A bounded interval with one endpoint is called a **half-open interval.** Figure 0-8(b) shows the half-open interval between -2 and 3, including -2.

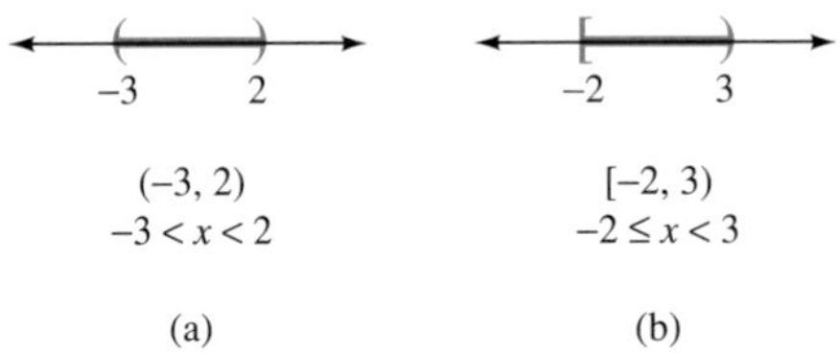

Figure 0-8

Intervals that include two endpoints are called **closed intervals.** Figure 0-9 shows the graph of a closed interval from -2 to 4.

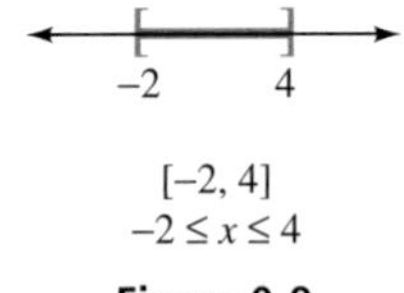

Figure 0-9

Open Intervals The interval (a, b) includes all real numbers x such that $a < x < b$.

Half-Open Intervals The interval $[a, b)$ includes all real numbers x such that $a \leq x < b$.

The interval $(a, b]$ includes all real numbers x such that $a < x \leq b$.

Closed Intervals The interval $[a, b]$ includes all real numbers x such that $a \leq x \leq b$.

EXAMPLE 6 Write the inequality $3 \leq x$ in interval notation and graph it.

Solution The inequality $3 \leq x$ can be written in the form $x \geq 3$. This is the interval $[3, \infty)$. Its graph includes all real numbers greater than or equal to 3, as shown in Figure 0-10.

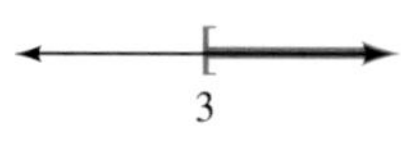

Figure 0-10

Self Check Write the inequality $-2 < x \leq 5$ in interval notation and graph it. ■

ACTIVE EXAMPLE 7 Write the inequality $5 \geq x \geq -1$ in interval notation and graph it.

Solution The inequality $5 \geq x \geq -1$ can be written in the form

$$-1 \leq x \leq 5$$

COLLEGE **Algebra** $f(x)$ **Now**™
Go to **http://1pass.thomson.com** or your CD to practice this example.

This is the interval $[-1, 5]$. Its graph includes all real numbers from -1 to 5. The graph is shown in Figure 0-11.

Figure 0-11

Self Check Write the inequality $0 \leq x \leq 3$ in interval notation and graph it. ■

The expression

$x < -2$ or $x \geq 3$ Read as "x is less than -2 or x is greater than or equal to 3."

represents the **union** of two intervals. In interval notation, it is written as

$(-\infty, -2) \cup [3, \infty)$ Read the symbol $\cup$ as "union."

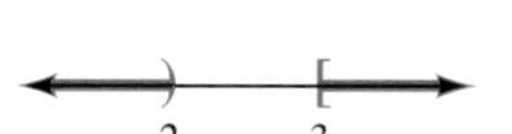

Figure 0-12

Its graph is shown in Figure 0-12.

■ Absolute Value

The **absolute value** of a real number x (denoted as $|x|$) is the distance on a number line between 0 and the point with a coordinate of x. For example, points with coordinates of 4 and -4 both lie four units from 0, as shown in Figure 0-13. Therefore, it follows that

$$|-4| = |4| = 4$$

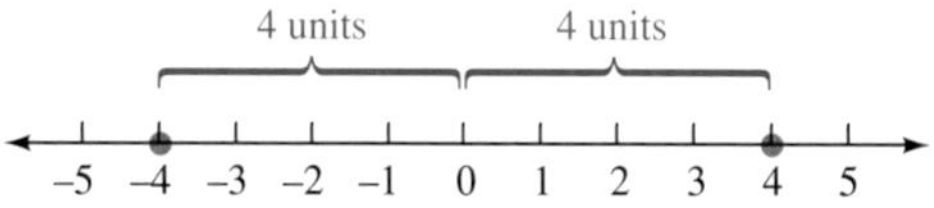

4 units 4 units

–5 –4 –3 –2 –1 0 1 2 3 4 5

Figure 0-13

In general, for any real number x,

$$|-x| = |x|$$

We can define absolute value algebraically as follows.

Absolute Value

If x is a real number, then

$$|x| = x \qquad \text{when } x \geq 0$$
$$|x| = -x \qquad \text{when } x < 0$$

This definition indicates that when x is positive or 0, then x is its own absolute value. However, when x is negative, then $-x$ (which is positive) is its absolute value. Thus, $|x|$ is always nonnegative.

$|x| \geq 0$ **for all real numbers x**

Comment Remember that x is not always positive and $-x$ is not always negative.

EXAMPLE 8 Write each number without using absolute value symbols: **a.** $|3|$, **b.** $|-4|$, **c.** $|0|$, **d.** $-|-8|$.

Solution **a.** $|3| = 3$ **b.** $|-4| = 4$ **c.** $|0| = 0$ **d.** $-|-8| = -(8) = -8$

Self Check Write each number without using absolute value symbols: **a.** $|-10|$, **b.** $|12|$, **c.** $-|6|$. ■

ACTIVE EXAMPLE 9 Write each number without using absolute value symbols: **a.** $|\pi - 1|$, **b.** $|2 - \pi|$, **c.** $|2 - x|$ if $x \geq 5$.

Solution **a.** Since $\pi \approx 3.1416$, $\pi - 1$ is positive, and $\pi - 1$ is its own absolute value.

$$|\pi - 1| = \pi - 1$$

b. Since $2 - \pi$ is negative, its absolute value is $-(2 - \pi)$.

$$|2 - \pi| = -(2 - \pi) = -2 - (-\pi) = -2 + \pi = \pi - 2$$

c. Since $x \geq 5$, the expression $2 - x$ is negative, and its absolute value is $-(2 - x)$.

$$|2 - x| = -(2 - x) = -2 + x = x - 2 \qquad \text{provided } x \geq 5$$

COLLEGE **Algebra** $f(x)$ **Now**™
Go to **http://1pass.thomson.com** or your CD to practice this example.

Self Check Write each number without using absolute value symbols. (*Hint*: $\sqrt{5} \approx 2.236$.) **a.** $|2 - \sqrt{5}|$ **b.** $|2 - x|$ if $x \leq 1$ ■

■ Distance on a Number Line

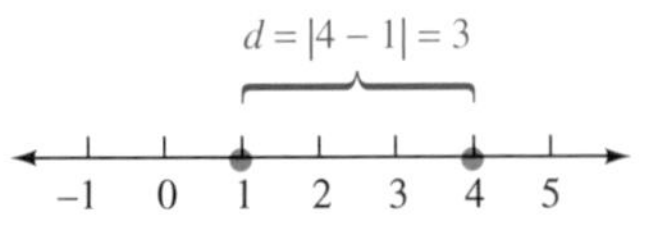

Figure 0-14

On the number line shown in Figure 0-14, the distance between the points with coordinates of 1 and 4 is $4 - 1$, or 3 units. However, if the subtraction were done in the other order, the result would be $1 - 4$, or -3 units. To guarantee that the distance between two points is always positive, we can use absolute value symbols. Thus, the distance d between two points with coordinates of 1 and 4 is

$$d = |4 - 1| = |1 - 4| = 3$$

In general, we have the following definition for the distance between two points on the number line.

Distance between Two Points

If a and b are the coordinates of two points on the number line, the distance between the points is given by the formula

$$d = |b - a|$$

EXAMPLE 10 Find the distance on a number line between points with coordinates of **a.** 3 and 5, **b.** -2 and 3, and **c.** -5 and -1.

Solution

a. $d = |5 - 3| = |2| = 2$

b. $d = |3 - (-2)| = |3 + 2| = |5| = 5$

c. $d = |-1 - (-5)| = |-1 + 5| = |4| = 4$

Self Check Find the distance on a number line between points with coordinates of **a.** 4 and 10 and **b.** -2 and -7. ■

Self Check Answers

1. a. repeats **b.** terminates **2. a.** $-3, 1, 5$ **b.** 4, 6 **c.** $\sqrt{5}$ **3. a.** commutative property of multiplication **b.** associative property of multiplication **c.** commutative property of addition

4. **5.** $(-\infty, 5)$, **6.** $(-2, 5]$,

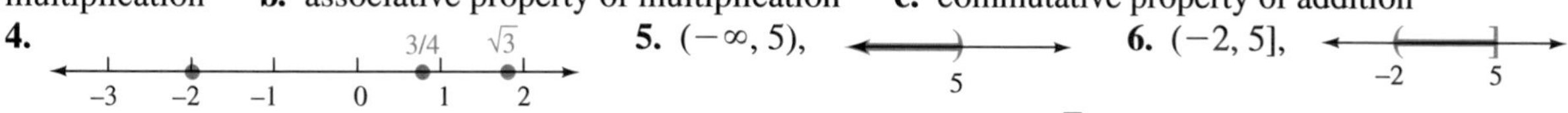

7. $[0, 3]$, (0, 3) **8. a.** 10 **b.** 12 **c.** -6 **9. a.** $\sqrt{5} - 2$ **b.** $2 - x$ **10. a.** 6 **b.** 5

0.1 Exercises

VOCABULARY AND CONCEPTS ***Fill in the blanks.***

1. A real number is any number that can be expressed as a decimal.

2. A variable is a letter that is used to represent a number.

3. The smallest prime number is 2.

4. All integers that are exactly divisible by 2 are called even integers.

5. Natural numbers greater than 1 that are not prime are called composite numbers.

6. Fractions such as $\frac{2}{3}$, $\frac{8}{2}$, and $-\frac{7}{9}$ are called rational numbers.

7. Irrational numbers are decimals that don't terminate and don't repeat.

8. The symbol $\leq$ is read as "is less than or equal to."

9. On a number line, the negative numbers are to the left of 0.

10. The only integer that is neither positive nor negative is 0.

11. The associative property of addition states that $(x + y) + z = x + (y + z)$.

12. The commutative property of multiplication states that $xy = yx$.

13. Use the distributive property to complete the statement: $5(m + 2) = 5m + 5 \cdot 2$.

14. The statement $(m + n)p = p(m + n)$ illustrates the commutative property of multiplication.

15. The graph of an interval is a portion of a number line.

16. The graph of an open interval has no endpoints.

17. The graph of a closed interval has two endpoints.

18. The graph of a half-open interval has one endpoint.

19. Except for 0, the absolute value of every number is positive.

20. The distance between two distinct points on a number line is always positive.

Let

N = the set of natural numbers
W = the set of whole numbers
Z = the set of integers
Q = the set of rational numbers
R = the set of real numbers

Determine whether each statement is true or false. Read the symbol ⊆ as "is a subset of."

21. $\mathbf{N} \subseteq \mathbf{W}$ true

22. $\mathbf{Q} \subseteq \mathbf{R}$ true

23. $\mathbf{Q} \subseteq \mathbf{N}$ false

24. $\mathbf{Z} \subseteq \mathbf{Q}$ true

25. $\mathbf{W} \subseteq \mathbf{Z}$ true

26. $\mathbf{R} \subseteq \mathbf{Z}$ false

PRACTICE ***Consider the following set:*** $\{-5, -4, -\frac{2}{3}, 0, 1, \sqrt{2}, 2, 2.75, 6, 7\}$.

27. Which numbers are natural numbers? 1, 2, 6, 7

28. Which numbers are whole numbers? 0, 1, 2, 6, 7

29. Which numbers are integers? −5, −4, 0, 1, 2, 6, 7

30. Which numbers are rational numbers? $-5, -4, -\frac{2}{3}, 0, 1, 2, 2.75, 6, 7$

31. Which numbers are irrational numbers? $\sqrt{2}$

32. Which numbers are prime numbers? 2, 7

33. Which numbers are composite numbers? 6

34. Which numbers are even integers? −4, 0, 2, 6

35. Which numbers are odd integers? −5, 1, 7

36. Which numbers are negative numbers? $-5, -4, -\frac{2}{3}$

Graph each subset of the real numbers on a number line.

37. The natural numbers between 1 and 5

38. The composite numbers less than 10

39. The prime numbers between 10 and 20

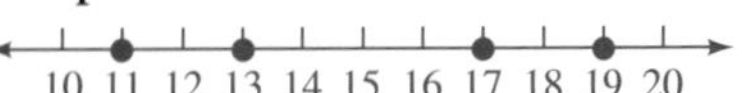

40. The integers from −2 to 4

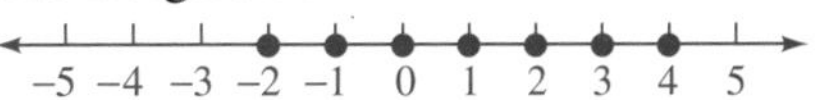

41. The integers between −5 and 0

42. The even integers between −9 and −1

43. The odd integers between −6 and 4

44. -0.7, 1.75, and $3\frac{7}{8}$

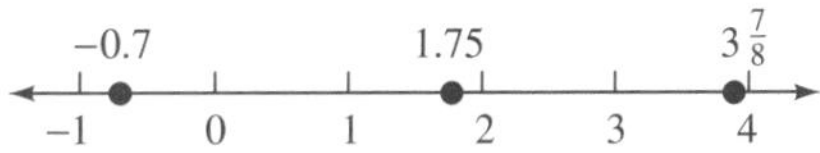

Write each inequality in interval notation and graph the interval.

45. $x > 2$ $(2, \infty)$

46. $x < 4$ $(-\infty, 4)$

47. $0 < x < 5$ $(0, 5)$

48. $-2 < x < 3$ $(-2, 3)$

49. $x > -4$ $(-4, \infty)$

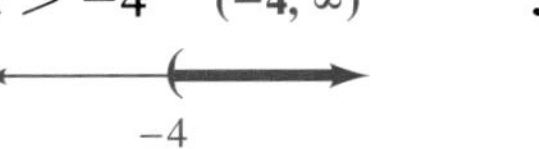

50. $x < 3$ $(-\infty, 3)$

51. $-2 \le x < 2$ $[-2, 2)$

52. $-4 < x \le 1$ $(-4, 1]$

53. $x \le 5$ $(-\infty, 5]$

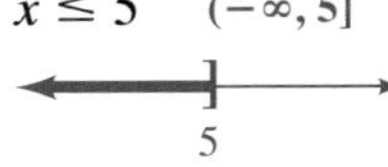

54. $x \ge -1$ $[-1, \infty)$

55. $-5 < x \le 0$ $(-5, 0]$

56. $-3 \le x < 4$ $[-3, 4)$

57. $-2 \le x \le 3$ $[-2, 3]$

58. $-4 \le x \le 4$ $[-4, 4]$

59. $6 \ge x \ge 2$ $[2, 6]$

60. $3 \ge x \ge -2$ $[-2, 3]$

Write each pair of inequalities as the intersection of two intervals and graph the result.

61. $x > -5$ and $x < 4$ $(-5, \infty) \cap (-\infty, 4)$

62. $x \geq -3$ and $x < 6$ $[-3, \infty) \cap (-\infty, 6)$

63. $x \geq -8$ and $x \leq -3$ $[-8, \infty) \cap (-\infty, -3]$

64. $x > 1$ and $x \leq 7$ $(1, \infty) \cap (-\infty, 7]$

Write each inequality as the union of two intervals and graph the result.

65. $x < -2$ or $x > 2$ $(-\infty, -2) \cup (2, \infty)$

66. $x \leq -5$ or $x > 0$ $(-\infty, -5] \cup (0, \infty)$

67. $x \leq -1$ or $x \geq 3$ $(-\infty, -1] \cup [3, \infty)$

68. $x < -3$ or $x \geq 2$ $(-\infty, -3) \cup [2, \infty)$

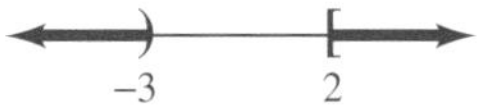

Write each expression without using absolute value symbols.

69. $|13|$ 13

70. $|-17|$ 17

71. $|0|$ 0

72. $-|63|$ -63

73. $-|-8|$ -8

74. $|-25|$ 25

75. $-|32|$ -32

76. $-|-6|$ -6

77. $|\pi - 5|$ $5 - \pi$

78. $|8 - \pi|$ $8 - \pi$

79. $|\pi - \pi|$ 0

80. $|2\pi|$ 2π

81. $|x + 1|$ and $x \geq 2$ $x + 1$

82. $|x + 1|$ and $x \leq -2$ $-(x + 1)$

83. $|x - 4|$ and $x < 0$ $-(x - 4)$

84. $|x - 7|$ and $x > 10$ $x - 7$

Find the distance between each pair of points on the number line.

85. 3 and 8 5

86. -5 and 12 17

87. -8 and -3 5

88. 6 and -20 26

APPLICATIONS

89. What subset of the real numbers would you use to describe the populations of several cities?
natural numbers

90. What subset of the real numbers would you use to describe the subdivisions of an inch on a ruler?
rational numbers

91. What subset of the real numbers would you use to report temperatures in several cities?
integers

92. What subset of the real numbers would you use to describe the financial condition of a business?
rational numbers

DISCOVERY AND WRITING

93. Explain why $-x$ could be positive.

94. Explain why every integer is a rational number.

95. Is the statement $|ab| = |a| \cdot |b|$ always true? Explain.

96. Is the statement $\left|\frac{a}{b}\right| = \frac{|a|}{|b|}$ $(b \neq 0)$ always true? Explain.

97. Is the statement $|a + b| = |a| + |b|$ always true? Explain.

98. Under what conditions will the statement given in Exercise 97 be true?

99. Explain why it is incorrect to write $a < b > c$ if $a < b$ and $b > c$.

100. Explain why $|b - a| = |a - b|$.

0.2 Integer Exponents and Scientific Notation

In this section, you will learn about

- Natural-Number Exponents
- Rules of Exponents
- Order of Operations
- Evaluating Expressions
- Scientific Notation
- Using Scientific Notation to Simplify Computations

Natural-Number Exponents

When two or more quantities are multiplied together, each quantity is called a **factor** of the product. The exponential expression x^4 indicates that x is to be used as a factor four times.

$$x^4 = x \cdot x \cdot x \cdot x$$

In general, the following is true.

Natural-Number Exponents

For any natural number n,

$$x^n = \underbrace{x \cdot x \cdot x \cdot \cdots \cdot x}_{n \text{ factors of } x}$$

In the **exponential expression** x^n, x is called the **base,** and n is called the **exponent** or the **power** to which the base is raised. The expression x^n is called a **power of *x***. From the definition, we see that a natural-number exponent tells how many times the base of an exponential expression is to be used as a factor in a product. If an exponent is 1, the 1 is usually not written:

$$x^1 = x$$

EXAMPLE 1 Write each expression without using exponents: **a.** 4^2, **b.** $(-4)^2$, **c.** 5^3, **d.** $(-5)^3$, **e.** $3x^4$, and **f.** $(3x)^4$.

Solution

a. $4^2 = 4 \cdot 4 = 16$ — Read 4^2 as "four squared."

b. $(-4)^2 = (-4)(-4) = 16$ — Read $(-4)^2$ as "negative four squared."

c. $5^3 = 5 \cdot 5 \cdot 5 = 125$ — Read 5^3 as "five cubed."

d. $(-5)^3 = (-5)(-5)(-5) = -125$ — Read $(-5)^3$ as "negative five cubed."

e. $3x^4 = 3 \cdot x \cdot x \cdot x \cdot x$ — Read $3x^4$ as "3 times x to the fourth power."

f. $(3x)^4 = (3x)(3x)(3x)(3x) = 81 \cdot x \cdot x \cdot x \cdot x$ — Read $(3x)^4$ as "$3x$ to the fourth power."

Self Check Write each expression without using exponents: **a.** 7^3, **b.** $(-3)^2$, **c.** $5a^3$, and **d.** $(5a)^3$. ■

Comment Note the distinction between ax^n and $(ax)^n$:

$$ax^n = a \cdot \overbrace{x \cdot x \cdot x \cdot \cdots \cdot x}^{n \text{ factors of } x} \qquad (ax)^n = \overbrace{(ax)(ax)(ax) \cdot \cdots \cdot (ax)}^{n \text{ factors of } ax}$$

Also note the distinction between $-x^n$ and $(-x)^n$:

$$-x^n = -\overbrace{(x \cdot x \cdot x \cdot \cdots \cdot x)}^{n \text{ factors of } x} \qquad (-x)^n = \overbrace{(-x)(-x)(-x) \cdot \cdots \cdot (-x)}^{n \text{ factors of } -x}$$

Accent on Technology — USING A CALCULATOR TO FIND POWERS

We can use calculators to find powers of numbers. For example, to find 2.35^3 with a scientific calculator, we enter these numbers and press these keys:

2.35 y^x 3 =

The display will read 12.977875 .

To find 2.35 with a graphing calculator, we enter these numbers and press these keys:

2.35 ^ 3 ENTER

The display will read

2.35^3
12.977875

In either case, $2.35^3 = 12.977875$.

Rules of Exponents

We begin to review the rules of exponents by considering the product $x^m x^n$. Since x^m indicates that x is to be used as a factor m times, and since x^n indicates that x is to be used as a factor n times, there are $m + n$ factors of x in the product $x^m x^n$.

$$x^m x^n = \overbrace{\overbrace{x \cdot x \cdot x \cdot \cdots \cdot x}^{m \text{ factors of } x} \cdot \overbrace{x \cdot x \cdot x \cdot \cdots \cdot x}^{n \text{ factors of } x}}^{m + n \text{ factors of } x}$$

This suggests that to multiply exponential expressions with the same base, we *keep the base and add the exponents.*

Product Rule for Exponents

If m and n are natural numbers, then

$$x^m x^n = x^{m+n}$$

Comment The product rule applies to exponential expressions with the same base. A product of two powers with different bases, such as x^4y^3, cannot be simplified.

To find another property of exponents, we consider the exponential expression $(x^m)^n$. The exponent n indicates that x^m is to be used as a factor n times. This implies that x is to be used as a factor mn times.

Emmy Amalie Noether
(1882–1935)

Emmy Noether is best known for her work in abstract algebra. Her work in the study of invariants led to concepts that Albert Einstein used in his theory of relativity. Einstein described her as the most creative female genius since the beginning of higher education for women. In 1933, she lost her position at the University of Göttingen because of Nazi pressure. She came to the United States and taught at Bryn Mawr College and lectured at the Institute for Advanced Study at Princeton.

$$(x^m)^n = \overbrace{\overbrace{(x^m)(x^m)(x^m)\cdot\cdots\cdot(x^m)}^{n \text{ factors of } x^m}}^{mn \text{ factors of } x} = x^{mn}$$

This suggests that to raise an exponential expression to a power, we *keep the base and multiply the exponents.*

To raise a product to a power, we raise each factor to that power.

$$(xy)^n = \overbrace{(xy)(xy)(xy)\cdot\cdots\cdot(xy)}^{n \text{ factors of } xy} = \overbrace{(x\cdot x\cdot x\cdot\cdots\cdot x)}^{n \text{ factors of } x}\overbrace{(y\cdot y\cdot y\cdot\cdots\cdot y)}^{n \text{ factors of } y} = x^n y^n$$

To raise a fraction to a power, we raise both the numerator and the denominator to that power. If $y \neq 0$, then

$$\left(\frac{x}{y}\right)^n = \overbrace{\left(\frac{x}{y}\right)\left(\frac{x}{y}\right)\left(\frac{x}{y}\right)\cdot\cdots\cdot\left(\frac{x}{y}\right)}^{n \text{ factors of } \frac{x}{y}}$$
$$= \frac{\overbrace{xxx\cdot\cdots\cdot x}^{n \text{ factors of } x}}{\underbrace{yyy\cdot\cdots\cdot y}_{n \text{ factors of } y}}$$
$$= \frac{x^n}{y^n}$$

The previous three results are called the **power rules of exponents.**

Power Rules of Exponents

If m and n are natural numbers, then

$$(x^m)^n = x^{mn} \qquad (xy)^n = x^n y^n \qquad \left(\frac{x}{y}\right)^n = \frac{x^n}{y^n} \quad (y \neq 0)$$

ACTIVE EXAMPLE 2 Simplify: **a.** x^5x^7, **b.** $x^2y^3x^5y$, **c.** $(x^4)^9$, **d.** $(x^2x^5)^3$, **e.** $\left(\frac{x}{y^2}\right)^5$, and **f.** $\left(\frac{5x^2y}{z^3}\right)^2$.

Solution

a. $x^5x^7 = x^{5+7} = x^{12}$

b. $x^2y^3x^5y = x^{2+5}y^{3+1} = x^7y^4$

c. $(x^4)^9 = x^{4\cdot 9} = x^{36}$

d. $(x^2x^5)^3 = (x^7)^3 = x^{21}$

e. $\left(\frac{x}{y^2}\right)^5 = \frac{x^5}{(y^2)^5} = \frac{x^5}{y^{10}} \quad (y \neq 0)$

f. $\left(\frac{5x^2y}{z^3}\right)^2 = \frac{5^2(x^2)^2y^2}{(z^3)^2} = \frac{25x^4y^2}{z^6} \quad (z \neq 0)$

COLLEGE **Algebra *f(x)* Now™**
Explore this Active Example and test yourself on the steps followed here at **http://1pass.thomson.com.** You can also find this example on the Interactive Video Skillbuilder CD-ROM.

Self Check Simplify: **a.** $(y^3)^2$, **b.** $(a^2a^4)^3$, **c.** $(x^2)^3(x^3)^2$, and **d.** $\left(\frac{3a^3b^2}{c^3}\right)^3$ $(c \neq 0)$. ■

If we assume that the rules for natural-number exponents hold for exponents of 0, we can write

$$x^0x^n = x^{0+n} = x^n = 1x^n$$

Since $x^0x^n = 1x^n$, it follows that if $x \neq 0$, then $x^0 = 1$.

Zero Exponent

$$x^0 = 1 \qquad (x \neq 0)$$

If we assume that the rules for natural-number exponents hold for exponents that are negative integers, we can write

$$x^{-n}x^n = x^{-n+n} = x^0 = 1 \qquad (x \neq 0)$$

However, we know that

$$\frac{1}{x^n} \cdot x^n = 1 \qquad (x \neq 0)$$

$\frac{1}{x^n} \cdot x^n = \frac{x^n}{x^n}$, and any nonzero number divided by itself is 1.

Since $x^{-n}x^n = \frac{1}{x^n} \cdot x^n$, it follows that $x^{-n} = \frac{1}{x^n}$ $(x \neq 0)$.

Negative Exponents

If n is an integer and $x \neq 0$, then

$$x^{-n} = \frac{1}{x^n} \quad \text{and} \quad \frac{1}{x^{-n}} = x^n$$

Because of the previous definitions, all of the rules for natural-number exponents hold for integer exponents.

EXAMPLE 3 Simplify: **a.** $(3x)^0$, **b.** $3(x^0)$, **c.** x^{-4}, **d.** $\frac{1}{x^{-6}}$, **e.** $x^{-3}x$, and **f.** $(x^{-4}x^8)^{-5}$. Write all answers without using negative exponents.

Solution **a.** $(3x)^0 = 1$ **b.** $3(x^0) = 3(1) = 3$ **c.** $x^{-4} = \frac{1}{x^4}$

d. $\frac{1}{x^{-6}} = x^6$

e. $x^{-3}x = x^{-3+1}$
$= x^{-2}$
$= \frac{1}{x^2}$

f. $(x^{-4}x^8)^{-5} = (x^4)^{-5}$
$= x^{-20}$
$= \frac{1}{x^{20}}$

Self Check Simplify: **a.** $7a^0$, **b.** $3a^{-2}$, **c.** $a^{-4}a^2$, and **d.** $(a^3a^{-7})^3$. Write all answers without using negative exponents. ■

To develop the quotient rule for exponents, we proceed as follows:

$$\frac{x^m}{x^n} = x^m\left(\frac{1}{x^n}\right) = x^mx^{-n} = x^{m+(-n)} = x^{m-n} \qquad (x \neq 0)$$

This suggests that to divide two exponential expressions with the same nonzero base, we *keep the base and subtract the exponent in the denominator from the exponent in the numerator.*

Quotient Rule for Exponents

If m and n are integers, then

$$\frac{x^m}{x^n} = x^{m-n} \qquad (x \neq 0)$$

EXAMPLE 4 Simplify: **a.** $\frac{x^8}{x^5}$ and **b.** $\frac{x^2x^4}{x^{-5}}$. Write all answers without using negative exponents.

Solution **a.** $\frac{x^8}{x^5} = x^{8-5}$
$= x^3$

b. $\frac{x^2x^4}{x^{-5}} = \frac{x^6}{x^{-5}}$
$= x^{6-(-5)}$
$= x^{11}$

Self Check Simplify: **a.** $\frac{x^{-6}}{x^2}$ and **b.** $\frac{x^4x^{-3}}{x^2}$. Write all answers without using negative exponents. ■

EXAMPLE 5 Simplify: **a.** $\left(\frac{x^3y^{-2}}{x^{-2}y^3}\right)^{-2}$ and **b.** $\left(\frac{x}{y}\right)^{-n}$. Write all answers without using negative exponents.

Solution **a.** $\left(\frac{x^3y^{-2}}{x^{-2}y^3}\right)^{-2} = (x^{3-(-2)}y^{-2-3})^{-2}$
$= (x^5y^{-5})^{-2}$
$= x^{-10}y^{10}$
$= \frac{y^{10}}{x^{10}}$

b. $\left(\frac{x}{y}\right)^{-n} = \frac{x^{-n}}{y^{-n}}$

$= \frac{x^{-n}x^ny^n}{y^{-n}x^ny^n}$ Multiply numerator and denominator by 1 in the form $\frac{x^ny^n}{x^ny^n}$.

$= \frac{x^0y^n}{y^0x^n}$ $x^{-n}x^n = x^0$ and $y^{-n}y^n = y^0$.

$= \frac{y^n}{x^n}$ $x^0 = 1$ and $y^0 = 1$.

$= \left(\frac{y}{x}\right)^n$

Self Check Simplify: **a.** $\left(\frac{x^4y^{-3}}{x^{-3}y^2}\right)^2$ and **b.** $\left(\frac{2a}{3b}\right)^{-3}$. Write all answers without using negative exponents. ■

Part b of Example 5 establishes the following rule.

A Fraction to a Negative Power

If n is a natural number, then

$$\left(\frac{x}{y}\right)^{-n} = \left(\frac{y}{x}\right)^{n} \qquad (x \neq 0 \quad \text{and} \quad y \neq 0)$$

Order of Operations

When several operations occur in an expression, they should be done in the following order to get the correct result.

Order of Operations

If an expression does not contain grouping symbols such as parentheses or brackets, follow these steps:

1. Find the values of any exponential expressions.
2. Perform all multiplications and/or divisions, working from left to right.
3. Perform all additions and/or subtractions, working from left to right.

If an expression contains grouping symbols, use the rules above to perform the calculations within each pair of grouping symbols, working from the innermost pair to the outermost pair.

In a fraction, simplify the numerator and the denominator separately. Then simplify the fraction, if possible.

Comment Some students remember the order of operations with the acronym PEMDAS:

parentheses
exponents
multiplication
division
addition
subtraction

For example, to simplify $\frac{3[4-(6+10)]}{2^2-(6+7)}$, we proceed as follows:

$$\frac{3[4-(6+10)]}{2^2-(6+7)} = \frac{3(4-16)}{2^2-(6+7)}$$ Simplify within the inner parentheses: $6 + 10 = 16$.

$$= \frac{3(-12)}{2^2-13}$$ Simplify within the parentheses: $4 - 16 = -12$, and $6 + 7 = 13$.

$$= \frac{3(-12)}{4-13}$$ Evaluate the power: $2^2 = 4$.

$$= \frac{-36}{-9}$$ $3(-12) = -36$. $4 - 13 = -9$.

$$= 4$$

Evaluating Expressions

ACTIVE EXAMPLE 6 If $x = -2$, $y = 3$, and $z = -4$, evaluate **a.** $-x^2 + y^2z$ and **b.** $\frac{2z^3 - 3y^2}{5x^2}$.

Solution **a.** $-x^2 + y^2z = -(-2)^2 + 3^2(-4)$

$= -(4) + 9(-4)$ Evaluate the powers.

$= -4 + (-36)$ Do the multiplication.

$= -40$ Do the addition.

COLLEGE Algebra $f(x)$ Now™
Go to http://1pass.thomson.com or your CD to practice this example.

b. $\dfrac{2z^3 - 3y^2}{5x^2} = \dfrac{2(-4)^3 - 3(3)^2}{5(-2)^2}$

$= \dfrac{2(-64) - 3(9)}{5(4)}$ Evaluate the powers.

$= \dfrac{-128 - 27}{20}$ Do the multiplications.

$= \dfrac{-155}{20}$ Do the subtraction.

$= -\dfrac{31}{4}$ Simplify the fraction.

Self Check If $x = 3$ and $y = -2$, evaluate $\dfrac{2x^2 - 3y^2}{x - y}$. ■

■ Scientific Notation

Scientists often work with numbers that are very large or very small. These numbers can be written compactly by expressing them in *scientific notation.*

Scientific Notation

> A number is written in **scientific notation** when it is written in the form
>
> $N \times 10^n$
>
> where $1 \leq |N| < 10$ and n is an integer.

Light travels 29,980,000,000 centimeters per second. To express this number in scientific notation, we must write it as the product of a number between 1 and 10 and some integer power of 10. The number 2.998 lies between 1 and 10. To get 29,980,000,000, the decimal point in 2.998 must be moved ten places to the right. This is accomplished by multiplying 2.998 by 10^{10}.

Standard notation ⟶ $29{,}980{,}000{,}000 = 2.998 \times 10^{10}$ ⟵ Scientific notation

One meter is approximately 0.0006214 mile. To express this number in scientific notation, we must write it as the product of a number between 1 and 10 and some integer power of 10. The number 6.214 lies between 1 and 10. To get 0.0006214, the decimal point in 6.214 must be moved four places to the left. This is accomplished by multiplying 6.214 by $\frac{1}{10^4}$ or by multiplying 6.214 by 10^{-4}.

Standard notation ⟶ $0.0006214 = 6.214 \times 10^{-4}$ ⟵ Scientific notation

To write each of the following numbers in scientific notation, we start to the right of the first nonzero digit and count to the decimal point. The exponent gives the number of places the decimal point moves, and the sign of the exponent indicates the direction in which it moves.

a. $372000 = 3.72 \times 10^5$ 5 places to the right.

b. $0.000537 = 5.37 \times 10^{-4}$ 4 places to the left.

c. $7.36 = 7.36 \times 10^0$ No movement of the decimal point.

Everyday Connections

Ocean CO_2 May 'Harm Marine Life'

"The problems that exist in the world today cannot be solved by the level of thinking that created them." Albert Einstein

Nearly 50% of the carbon dioxide that humans have pumped into the atmosphere over the last 200 years has been absorbed by the sea, scientists say. Consequently, atmospheric levels of the potent greenhouse gas are not nearly as high as they might have been.

But the heavy concentration of carbon dioxide in the oceans has changed their chemistry, making it hard for some marine animals to form shells.

Source: **BBC NEWS:** http://news.bbc.co.uk/1/hi/sci/tech/3896425.stm Published: 2004/07/15

	Carbon Dioxide (CO_2) Emissions				
	Total		Per capita		Cumulative 1800-2000
	(million metric tons) 1999	(percent change since 1990)	(metric tons per person) 1999	(percent change since 1990)	(million metric tons)
WORLD	23,172.2	8.9	3.9	(4.2)	1,017,359
ASIA (EXCL. MIDDLE EAST)	6,901.7	38.0	2.1	19.3	
Armenia	3.0	. .	0.8	. .	290
Azerbaijan	33.2	. .	4.2	. .	2,300
Bangladesh	26.3	83.4	0.2	46.2	442
Bhutan	. .	. .	. .	. .	4
Cambodia	. .	. .	. .	. .	16
China	3,051.1	25.6	2.5	16.6	72,615
Georgia	5.3	. .	1.0	. .	380
India	903.8	52.9	0.9	31.9	20,275
Indonesia	244.9	76.9	1.2	56.0	4,872
Japan	1,158.5	10.5	9.1	7.7	36,577
Kazakhstan	114.5	. .	7.0	. .	8,264
Korea, Dem People's Rep	214.3	(1.2)	9.7	(10.8)	6,114
Korea, Rep	410.4	75.5	8.8	62.2	7,120
Kyrgyzstan	4.7	. .	1.0	. .	440
Lao People's Dem Rep	. .	. .	. .	. .	11
Malaysia	101.3	90.4	4.6	55.7	1,832
Mongolia	. .	. .	. .	. .	237
Myanmar	9.0	122.2	0.2	90.0	257
Nepal	3.0	234.4	0.1	225.0	32
Pakistan	92.2	48.9	0.7	17.9	1,952
Philippines	66.3	69.0	0.9	39.1	1,555
Singapore	53.2	53.1	13.6	17.9	1,690
Sri Lanka	9.6	141.5	0.5	121.7	220
Tajikistan	5.7	. .	0.9	. .	270
Thailand	155.8	95.5	2.5	73.1	2,535
Turkmenistan	33.9	. .	7.3	. .	910
Uzbekistan	117.5	. .	4.8	. .	5,020
Viet Nam	36.6	103.7	0.5	74.1	1,061

Source: EarthTrends Data Tables: Climate and Atmosphere: http://earthtrends.wri.org

Continued

The table shows data for carbon dioxide (CO_2) emissions for Asia. Use the table to express each of the following using scientific notation:

1. Total CO_2 emissions in 1999 for India. **9.038×10^8 metric tons**
2. Per capita CO_2 emissions in 1999 for Bangladesh. **2.0×10^{-1} metric tons**
3. Cumulative CO_2 emissions for the world. **1.017359×10^{12} metric tons**

EXAMPLE 7 Write **a.** 62,000 and **b.** -0.0027 in scientific notation.

Solution **a.** We must express 62,000 as a product of a number between 1 and 10 and some integer power of 10. This is accomplished by multiplying 6.2 by 10^4.

$$62{,}000 = 6.2 \times 10^4$$

b. We must express -0.0027 as a product of a number whose absolute value is between 1 and 10 and some integer power of 10. This is accomplished by multiplying -2.7 by 10^{-3}.

$$-0.0027 = -2.7 \times 10^{-3}$$

Self Check Write **a.** $-93{,}000{,}000$ and **b.** 0.0000087 in scientific notation. ■

EXAMPLE 8 Write **a.** 7.35×10^2 and **b.** 3.27×10^{-5} in standard notation.

Solution **a.** The factor of 10^2 indicates that 7.35 must be multiplied by 2 factors of 10. Because each multiplication by 10 moves the decimal point one place to the right, we have

$$7.35 \times 10^2 = 735$$

b. The factor of 10^{-5} indicates that 3.27 must be divided by 5 factors of 10. Because each division by 10 moves the decimal point one place to the left, we have

$$3.27 \times 10^{-5} = 0.0000327$$

Self Check Write **a.** 6.3×10^3 and **b.** 9.1×10^{-4} in standard notation. ■

■ Using Scientific Notation to Simplify Computations

Another advantage of scientific notation becomes evident when we multiply and divide combinations of very large and very small numbers.

COLLEGE **Algebra** $f(x)$ **Now**™
Go to **http://1pass.thomson.com** or your CD to practice this example.

ACTIVE EXAMPLE 9 Use scientific notation to calculate $\dfrac{(3{,}400{,}000)(0.00002)}{170{,}000{,}000}$.

Solution After changing each number to scientific notation, we can do the arithmetic on the numbers and the exponential expressions separately.

$$\begin{aligned}\frac{(3{,}400{,}000)(0.00002)}{170{,}000{,}000} &= \frac{(3.4 \times 10^6)(2.0 \times 10^{-5})}{1.7 \times 10^8} \\ &= \frac{6.8}{1.7} \times 10^{6+(-5)-8} \\ &= 4.0 \times 10^{-7} \\ &= 0.0000004\end{aligned}$$

Self Check Use scientific notation to simplify $\frac{(192{,}000)(0.0015)}{(0.0032)(4{,}500)}$. ■

Accent on Technology SCIENTIFIC NOTATION

Calculators often give answers in scientific notation. For example, if we use a calculator to find 21^8, the display will read

On a scientific calculator	***On a graphing calculator***
3.782285936 10	21^ 8 3.782285936E10

In either case, the result is given in scientific notation, and it means $3.782285936 \times 10^{10}$.

We can enter numbers into a calculator in scientific form. For example, to enter 0.000000000061 (which is 6.1×10^{-11}), we enter these numbers and press these keys:

On a scientific calculator	***On a graphing calculator***
6.1 EXP 11 +/−	6.1 EE (−) 11

To use a scientific calculator to simplify $\frac{21^8}{0.000000000061}$, we must enter the denominator in scientific notation, because there are too many digits to enter it directly. To do the calculation with a scientific calculator, we enter these numbers and press these keys:

21 y^x 8 = ÷ 6.1 EXP 11 +/− =

The display will read 6.200468748 20. In standard notation, the answer is approximately 620,046,874,800,000,000,000.

The steps are similar using a graphing calculator.

Self Check Answers

1. a. $7 \cdot 7 \cdot 7 = 343$ **b.** $(-3)(-3) = 9$ **c.** $5 \cdot a \cdot a \cdot a$ **d.** $(5a)(5a)(5a) = 125 \cdot a \cdot a \cdot a$ **2. a.** y^6 **b.** a^{18} **c.** x^{12} **d.** $\frac{27a^9b^6}{c^9}$ **3. a.** 7 **b.** $\frac{3}{a^2}$ **c.** $\frac{1}{a^2}$ **d.** $\frac{1}{a^{12}}$ **4. a.** $\frac{1}{x^8}$ **b.** $\frac{1}{x}$ **5. a.** $\frac{x^{14}}{y^{10}}$ **b.** $\frac{27b^3}{8a^3}$ **6.** $\frac{6}{5}$ **7. a.** -9.3×10^7 **b.** 8.7×10^{-6} **8. a.** 6,300 **b.** 0.00091 **9.** 20

0.2 Exercises

VOCABULARY AND CONCEPTS *Fill in the blanks.*

1. Each quantity in a product is called a factor of the product.
2. A natural number exponent tells how many times a base is used as a factor.
3. In the expression $(2x)^3$, 3 is the exponent and $2x$ is the base.
4. The expression x^n is called an exponential expression.
5. A number is in scientific notation when it is written in the form $N \times 10^n$, where $1 \leq |N| < 10$ and n is an integer.
6. Unless parentheses indicate otherwise, multiplications are done before additions.

Complete each formula.

7. $x^m x^n = x^{m+n}$
8. $(x^m)^n = x^{mn}$
9. $(xy)^n = x^n y^n$
10. $\dfrac{x^m}{x^n} = x^{m-n}$
11. $x^0 = 1$
12. $x^{-n} = \dfrac{1}{x^n}$

PRACTICE *Write each number or expression without using exponents.*

13. 13^2 169
14. 10^3 1,000
15. -5^2 -25
16. $(-5)^2$ 25
17. $4x^3$ $4 \cdot x \cdot x \cdot x$
18. $(4x)^3$ $(4x)(4x)(4x)$
19. $(-5x)^4$ $(-5x)(-5x)(-5x)(-5x)$
20. $-6x^2$ $-6 \cdot x \cdot x$

Write each expression using exponents.

21. $7xxx$ $7x^3$
22. $-8yyyy$ $-8y^4$
23. $(-x)(-x)$ x^2
24. $(2a)(2a)(2a)$ $8a^3$
25. $(3t)(3t)(-3t)$ $-27t^3$
26. $-(2b)(2b)(2b)(2b)$ $-16b^4$
27. $xxxyy$ x^3y^2
28. $aaabbbb$ a^3b^4

Use a calculator to simplify each expression.

29. 2.2^3 10.648
30. 7.1^4 2,541.1681
31. -0.5^4 -0.0625
32. $(-0.2)^4$ 0.0016

Simplify each expression. Write all answers without using negative exponents. Assume that all variables are restricted to those numbers for which the expression is defined.

33. x^2x^3 x^5
34. y^3y^4 y^7
35. $(z^2)^3$ z^6
36. $(t^6)^7$ t^{42}
37. $(y^5y^2)^3$ y^{21}
38. $(a^3a^6)a^4$ a^{13}
39. $(z^2)^3(z^4)^5$ z^{26}
40. $(t^3)^4(t^5)^2$ t^{22}
41. $(a^2)^3(a^4)^2$ a^{14}
42. $(a^2)^4(a^3)^3$ a^{17}
43. $(3x)^3$ $27x^3$
44. $(-2y)^4$ $16y^4$
45. $(x^2y)^3$ x^6y^3
46. $(x^3z^4)^6$ $x^{18}z^{24}$
47. $\left(\dfrac{a^2}{b}\right)^3$ $\dfrac{a^6}{b^3}$
48. $\left(\dfrac{x}{y^3}\right)^4$ $\dfrac{x^4}{y^{12}}$
49. $(-x)^0$ 1
50. $4x^0$ 4
51. $(4x)^0$ 1
52. $-2x^0$ -2
53. z^{-4} $\dfrac{1}{z^4}$
54. $\dfrac{1}{t^{-2}}$ t^2
55. $y^{-2}y^{-3}$ $\dfrac{1}{y^5}$
56. $-m^{-2}m^3$ $-m$
57. $(x^3x^{-4})^{-2}$ x^2
58. $(y^{-2}y^3)^{-4}$ $\dfrac{1}{y^4}$
59. $\dfrac{x^7}{x^3}$ x^4
60. $\dfrac{r^5}{r^2}$ r^3
61. $\dfrac{a^{21}}{a^{17}}$ a^4
62. $\dfrac{t^{13}}{t^4}$ t^9
63. $\dfrac{(x^2)^2}{x^2x}$ x
64. $\dfrac{s^9s^3}{(s^2)^2}$ s^8
65. $\left(\dfrac{m^3}{n^2}\right)^3$ $\dfrac{m^9}{n^6}$
66. $\left(\dfrac{t^4}{t^3}\right)^3$ t^3
67. $\dfrac{(a^3)^{-2}}{aa^2}$ $\dfrac{1}{a^9}$
68. $\dfrac{r^9r^{-3}}{(r^{-2})^3}$ r^{12}
69. $\left(\dfrac{a^{-3}}{b^{-1}}\right)^{-4}$ $\dfrac{a^{12}}{b^4}$
70. $\left(\dfrac{t^{-4}}{t^{-3}}\right)^{-2}$ t^2
71. $\left(\dfrac{r^4r^{-6}}{r^3r^{-3}}\right)^2$ $\dfrac{1}{r^4}$
72. $\dfrac{(x^{-3}x^2)^2}{(x^2x^{-5})^{-3}}$ $\dfrac{1}{x^{11}}$
73. $\left(\dfrac{x^5y^{-2}}{x^{-3}y^2}\right)^4$ $\dfrac{x^{32}}{y^{16}}$
74. $\left(\dfrac{x^{-7}y^5}{x^7y^{-4}}\right)^3$ $\dfrac{y^{27}}{x^{42}}$

75. $\left(\dfrac{5x^{-3}y^{-2}}{3x^2y^{-3}}\right)^{-2}$ $\dfrac{9x^{10}}{25y^2}$

76. $\left(\dfrac{3x^2y^{-5}}{2x^{-2}y^{-6}}\right)^{-3}$ $\dfrac{8}{27x^{12}y^3}$

77. $\left(\dfrac{3x^5y^{-3}}{6x^{-5}y^3}\right)^{-2}$ $\dfrac{4y^{12}}{x^{20}}$

78. $\left(\dfrac{12x^{-4}y^3z^{-5}}{4x^4y^{-3}z^5}\right)^3$ $\dfrac{27y^{18}}{x^{24}z^{30}}$

79. $\dfrac{(8^{-2}z^{-3}y)^{-1}}{(5y^2z^{-2})^3(5yz^{-2})^{-1}}$ $\dfrac{64z^7}{25y^6}$

80. $\dfrac{(m^{-2}n^3p^4)^{-2}(mn^{-2}p^3)^4}{(mn^{-2}p^3)^{-4}(mn^2p)^{-1}}$ $\dfrac{m^{13}p^{17}}{n^{20}}$

Simplify each expression.

81. $-\dfrac{5[6^2 + (9 - 5)]}{4(2 - 3)}$ 50

82. $\dfrac{6[3 - (4 - 7)^2]}{-5(2 - 4^2)}$ $-\dfrac{18}{35}$

Let $x = -2, y = 0,$ and $z = 3$ and evaluate each expression.

83. x^2 4

84. $-x^2$ −4

85. x^3 −8

86. $-x^3$ 8

87. $(-xz)^3$ 216

88. $-xz^3$ 54

89. $\dfrac{-(x^2z^3)}{z^2 - y^2}$ −12

90. $\dfrac{z^2(x^2 - y^2)}{x^3z}$ $-\dfrac{3}{2}$

91. $5x^2 - 3y^3z$ 20

92. $3(x - z)^2 + 2(y - z)^3$ 21

93. $\dfrac{-3x^{-3}z^{-2}}{6x^2z^{-3}}$ $\dfrac{3}{64}$

94. $\dfrac{(-5x^2z^{-3})^2}{5xz^{-2}}$ $-\dfrac{40}{81}$

Express each number in scientific notation.

95. 372,000 3.72×10^5

96. 89,500 8.95×10^4

97. −177,000,000 -1.77×10^8

98. −23,470,000,000 -2.347×10^{10}

99. 0.007 7×10^{-3}

100. 0.00052 5.2×10^{-4}

101. −0.000000693 -6.93×10^{-7}

102. −0.000000089 -8.9×10^{-8}

103. one trillion 1×10^{12}

104. one millionth 1×10^{-6}

Express each number in standard notation.

105. 9.37×10^5 937,000

106. 4.26×10^9 4,260,000,000

107. 2.21×10^{-5} 0.0000221

108. 2.774×10^{-2} 0.02774

109. 0.00032×10^4 3.2

110. $9{,}300.0 \times 10^{-4}$ 0.93

111. -3.2×10^{-3} −0.0032

112. -7.25×10^3 −7,250

Use the method of Example 9 to do each calculation. Write all answers in scientific notation.

113. $\dfrac{(65{,}000)(45{,}000)}{250{,}000}$ 1.17×10^4

114. $\dfrac{(0.000000045)(0.00000012)}{45{,}000{,}000}$ 1.2×10^{-22}

115. $\dfrac{(0.00000035)(170{,}000)}{0.00000085}$ 7×10^4

116. $\dfrac{(0.0000000144)(12{,}000)}{600{,}000}$ 2.88×10^{-10}

117. $\dfrac{(45{,}000{,}000{,}000)(212{,}000)}{0.00018}$ 5.3×10^{19}

118. $\dfrac{(0.00000000275)(4{,}750)}{500{,}000{,}000{,}000}$ 2.6125×10^{-17}

APPLICATIONS ***Use scientific notation to compute each answer. Write all answers in scientific notation.***

119. The speed of sound in air is 3.31×10^4 centimeters per second. Compute the speed of sound in meters per minute. 1.986×10^4 meters per min

120. Calculate the volume of a box that has dimensions of 6,000 by 9,700 by 4,700 millimeters. 2.7354×10^{11} mm^3

121. The mass of one proton is 0.00000000000000000000000167248 gram. Find the mass of one billion protons. 1.67248×10^{-15} g

122. The speed of light in a vacuum is approximately 30,000,000,000 centimeters per second. Find the speed of light in miles per hour. (160,934.4 cm = 1 mile.) 6.711×10^8 mph

123. **Astronomy** (See the illustration on the next page.) The distance d, in miles, of the nth planet from the sun is given by the formula

$$d = 9{,}275{,}200[3(2^{n-2}) + 4]$$

To the nearest million miles, find the distance of Earth and the distance of Mars from the sun. Give each answer in scientific notation.
9.3×10^7 mi, 1.48×10^8 mi

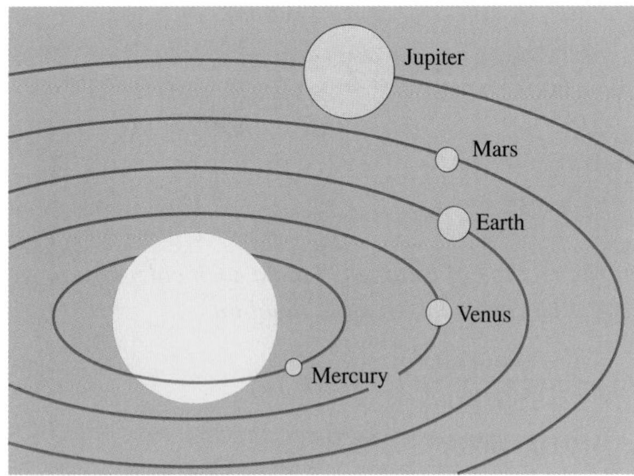

124. License Plates The number of different license plates of the form three digits followed by three letters, as in the illustration, is $10 \cdot 10 \cdot 10 \cdot 26 \cdot 26 \cdot 26$. Write this expression using exponents. Then evaluate it and express the result in scientific notation. $10^3 \cdot 26^3$; 1.7576×10^7

DISCOVERY AND WRITING ***Write each expression with a single base.***

125. $x^n x^2$ x^{n+2}

126. $\dfrac{x^m}{x^3}$ x^{m-3}

127. $\dfrac{x^m x^2}{x^3}$ x^{m-1}

128. $\dfrac{x^{3m+5}}{x^2}$ x^{3m+3}

129. $x^{m+1}x^3$ x^{m+4}

130. $a^{n-3}a^3$ a^n

131. Explain why $-x^4$ and $(-x)^4$ represent different numbers.

132. Explain why 32×10^2 is not in scientific notation.

REVIEW

133. Graph the interval $(-2, 4)$.

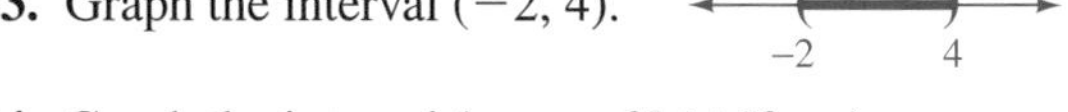

134. Graph the interval $(-\infty, -3] \cup [3, \infty)$.

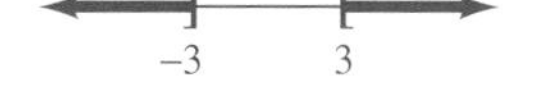

135. Evaluate $|\pi - 5|$. $5 - \pi$

136. Find the distance between -7 and -5 on the number line. 2

0.3 Rational Exponents and Radicals

In this section, you will learn about

- Rational Exponents Whose Numerators Are 1
- Rational Exponents Whose Numerators Are Not 1
- Radical Expressions ■ Simplifying and Combining Radicals
- Rationalizing Denominators and Numerators

Rational Exponents Whose Numerators Are 1

If we apply the rule $(x^m)^n = x^{mn}$ to $(25^{1/2})^2$, we obtain

$$\begin{aligned}(25^{1/2})^2 &= 25^{(1/2)2} && \text{Keep the base and multiply the exponents.}\\ &= 25^1 && \tfrac{1}{2} \cdot 2 = 1.\\ &= 25\end{aligned}$$

Thus, $25^{1/2}$ is a real number whose square is 25. Although $5^2 = 25$ and $(-5)^2 = 25$, we define $25^{1/2}$ to be the positive real number whose square is 25:

$$25^{1/2} = 5 \quad \text{Read } 25^{1/2} \text{ as "the square root of 25."}$$

In general, we have the following definition.

Rational Exponents

If $a \geq 0$ and n is a natural number, then $a^{1/n}$ (read as "the nth root of a") is the nonnegative real number b such that

$$b^n = a$$

Comment In the expression $a^{1/n}$, there is no real-number nth root of a when n is even and $a < 0$. For example, $(-64)^{1/2}$ is not a real number, because the square of no real number is -64.

EXAMPLE 1

a. $16^{1/2} = 4$ — Because $4^2 = 16$. Read $16^{1/2}$ as "the square root of 16."

b. $27^{1/3} = 3$ — Because $3^3 = 27$. Read $27^{1/3}$ as "the cube root of 27."

c. $\left(\frac{1}{81}\right)^{1/4} = \frac{1}{3}$ — Because $\left(\frac{1}{3}\right)^4 = \frac{1}{81}$. Read $\left(\frac{1}{81}\right)^{1/4}$ as "the fourth root of $\frac{1}{81}$."

d. $-32^{1/5} = -\left(32^{1/5}\right)$ — Read $32^{1/5}$ as "the fifth root of 32."

$= -(2)$ — Because $2^5 = 32$.

$= -2$

Self Check Simplify: **a.** $100^{1/2}$ and **b.** $243^{1/5}$. ■

If n is even in the expression $a^{1/n}$ and the base contains variables, we often use absolute value symbols to guarantee that an even root is nonnegative.

$(49x^2)^{1/2} = 7|x|$ — Because $(7|x|)^2 = 49x^2$. Since x could be negative, absolute value symbols are necessary to guarantee that the square root is nonnegative.

$(16x^4)^{1/4} = 2|x|$ — Because $(2|x|)^4 = 16x^4$. Since x could be negative, absolute value symbols are necessary to guarantee that the fourth root is nonnegative.

$(729x^{12})^{1/6} = 3x^2$ — Because $(3x^2)^6 = 729x^{12}$. Since x^2 is always nonnegative, no absolute value symbols are necessary. Read $(729x^{12})^{1/6}$ as "the sixth root of $729x^{12}$."

If n is an odd number in the expression $a^{1/n}$, the base a can be negative.

EXAMPLE 2

a. $(-8)^{1/3} = -2$ — Because $(-2)^3 = -8$.

b. $(-3{,}125)^{1/5} = -5$ — Because $(-5)^5 = -3{,}125$.

c. $\left(-\frac{1}{1{,}000}\right)^{1/3} = -\frac{1}{10}$ — Because $\left(-\frac{1}{10}\right)^3 = -\frac{1}{1{,}000}$.

Self Check Simplify: **a.** $(-125)^{1/3}$ and **b.** $(-100{,}000)^{1/5}$. ■

If n is odd in the expression $a^{1/n}$, we don't need to use absolute value symbols, because odd roots can be negative.

$(-27x^3)^{1/3} = -3x$ — Because $(-3x)^3 = -27x^3$.

$(-128a^7)^{1/7} = -2a$ — Because $(-2a)^7 = -128a^7$.

We summarize the definitions concerning $a^{1/n}$ as follows.

If n is a natural number and a is a real number in the expression $a^{1/n}$, then

If $a \geq 0$, then $a^{1/n}$ is the nonnegative real number b such that $b^n = a$.

$$\text{If } a < 0 \begin{cases} \text{and } n \text{ is odd, then } a^{1/n} \text{ is the real number } b \text{ such that } b^n = a. \\ \text{and } n \text{ is even, then } a^{1/n} \text{ is not a real number.} \end{cases}$$

Rational Exponents Whose Numerators Are Not 1

The definition of $a^{1/n}$ can be extended to include rational exponents whose numerators are not 1. For example, $4^{3/2}$ can be written as either

$$(4^{1/2})^3 \quad \text{or} \quad (4^3)^{1/2} \qquad \text{Because of the power rule, } (x^m)^n = x^{mn}.$$

This suggests the following rule.

Rule for Rational Exponents

If m and n are positive integers, the fraction $\frac{m}{n}$ is in lowest terms, and $a^{1/n}$ is a real number, then

$$a^{m/n} = (a^{1/n})^m = (a^m)^{1/n}$$

In the previous rule, we can view the expression $a^{m/n}$ in two ways:

1. $(a^{1/n})^m$: the mth power of the nth root of a
2. $(a^m)^{1/n}$: the nth root of the mth power of a

For example, $(-27)^{2/3}$ can be simplified in two ways:

$$\begin{aligned}(-27)^{2/3} &= [(-27)^{1/3}]^2 \\ &= (-3)^2 \\ &= 9\end{aligned} \qquad \text{or} \qquad \begin{aligned}(-27)^{2/3} &= [(-27)^2]^{1/3} \\ &= (729)^{1/3} \\ &= 9\end{aligned}$$

As this example suggests, it is usually easier to take the root of the base first to avoid large numbers.

Negative Rational Exponents

If m and n are positive integers, the fraction $\frac{m}{n}$ is in lowest terms, and $a^{1/n}$ is a real number, then

$$a^{-m/n} = \frac{1}{a^{m/n}} \quad \text{and} \quad \frac{1}{a^{-m/n}} = a^{m/n} \quad (a \neq 0)$$

EXAMPLE 3

a. $$\begin{aligned}25^{3/2} &= (25^{1/2})^3 \\ &= 5^3 \\ &= 125\end{aligned}$$

b. $$\begin{aligned}\left(-\frac{x^6}{1{,}000}\right)^{2/3} &= \left[\left(-\frac{x^6}{1{,}000}\right)^{1/3}\right]^2 \\ &= \left(-\frac{x^2}{10}\right)^2 \\ &= \frac{x^4}{100}\end{aligned}$$

c. $32^{-2/5} = \dfrac{1}{32^{2/5}} = \dfrac{1}{(32^{1/5})^2} = \dfrac{1}{2^2} = \dfrac{1}{4}$

d. $\dfrac{1}{81^{-3/4}} = 81^{3/4} = (81^{1/4})^3 = 3^3 = 27$

Self Check Simplify: **a.** $49^{3/2}$, **b.** $16^{-3/4}$, and **c.** $\dfrac{1}{(27x^3)^{-2/3}}$. ■

Because of the definition, rational exponents follow the same rules as integer exponents.

ACTIVE EXAMPLE 4 Assume that all variables represent positive numbers and write all answers without using negative exponents.

COLLEGE **Algebra** $f(x)$ **Now**™

Explore this Active Example and test yourself on the steps followed here at **http://1pass.thomson.com**. You can also find this example on the Interactive Video Skillbuilder CD-ROM.

a. $(36x)^{1/2} = 36^{1/2}x^{1/2} = 6x^{1/2}$

b. $\dfrac{(a^{1/3}b^{2/3})^6}{(y^3)^2} = \dfrac{a^{6/3}b^{12/3}}{y^6} = \dfrac{a^2b^4}{y^6}$

c. $\dfrac{a^{x/2}a^{x/4}}{a^{x/6}} = a^{x/2+x/4-x/6} = a^{6x/12+3x/12-2x/12} = a^{7x/12}$

d. $\left[\dfrac{-c^{-2/5}}{c^{4/5}}\right]^{5/3} = (-c^{-2/5-4/5})^{5/3} = [(-1)(c^{-6/5})]^{5/3} = (-1)^{5/3}(c^{-6/5})^{5/3} = -1c^{-30/15} = -c^{-2} = -\dfrac{1}{c^2}$

Self Check Use the directions for Example 4.

a. $\left(\dfrac{y^2}{49}\right)^{1/2}$, **b.** $\dfrac{b^{3/7}b^{2/7}}{b^{4/7}}$, and **c.** $\dfrac{(9r^2s)^{1/2}}{rs^{-3/2}}$ ■

■ Radical Expressions

Radical signs can also be used to express roots of numbers.

Definition of $\sqrt[n]{a}$

If n is a natural number greater than 1 and if $a^{1/n}$ is a real number, then

$$\sqrt[n]{a} = a^{1/n}$$

In the **radical expression** $\sqrt[n]{a}$, the symbol $\sqrt{}$ is the **radical sign,** a is the **radicand,** and n is the **index** (or the **order**) of the radical expression. If the order is 2, the expression is a **square root,** and we do not write the index.

$$\sqrt{a} = \sqrt[2]{a}$$

If the index of a radical is 3, the radical is called a **cube root.**

nth Root of a Nonnegative Number

If n is a natural number greater than 1 and $a \geq 0$, then $\sqrt[n]{a}$ is the nonnegative number whose nth power is a.

$$\left(\sqrt[n]{a}\right)^n = a$$

Comment In the expression $\sqrt[n]{a}$, there is no real-number nth root of a when n is even and $a < 0$. For example, $\sqrt{-64}$ is not a real number, because the square of no real number is -64.

If 2 is substituted for n in the equation $\left(\sqrt[n]{a}\right)^n = a$, we have

$$\left(\sqrt[2]{a}\right)^2 = \left(\sqrt{a}\right)^2 = \sqrt{a}\sqrt{a} = a$$

This shows that if a number a can be factored into two equal factors, either of those factors is a square root of a. Furthermore, if a can be factored into n equal factors, any one of those factors is an nth root of a.

If n is an odd number greater than 1 in the expression $\sqrt[n]{a}$, the radicand can be negative.

EXAMPLE 5

a. $\sqrt[3]{-27} = -3$ Because $(-3)^3 = -27$.

b. $\sqrt[3]{-125} = -5$ Because $(-5)^3 = -125$.

c. $\sqrt[3]{-\dfrac{27}{1{,}000}} = -\dfrac{3}{10}$ Because $\left(-\frac{3}{10}\right)^3 = -\frac{27}{1{,}000}$.

d. $-\sqrt[5]{-243} = -\left(\sqrt[5]{-243}\right)$
$= -(-3)$
$= 3$

Self Check Find each root: **a.** $\sqrt[3]{216}$ and **b.** $\sqrt[5]{-\dfrac{1}{32}}$. ■

We summarize the definitions concerning $\sqrt[n]{a}$ as follows.

If n is a natural number greater than 1 and a is a real number, then

If $a \geq 0$, then $\sqrt[n]{a}$ is the nonnegative real number such that $\left(\sqrt[n]{a}\right)^n = a$.

If $a < 0$ $\begin{cases} \text{and } n \text{ is odd, then } \sqrt[n]{a} \text{ is the real number such that } \left(\sqrt[n]{a}\right)^n = a. \\ \text{and } n \text{ is even, then } \sqrt[n]{a} \text{ is not a real number.} \end{cases}$

We have seen that $a^{m/n} = (a^{1/n})^m = (a^m)^{1/n}$. This same fact can be stated in radical notation.

$$a^{m/n} = \left(\sqrt[n]{a}\right)^m = \sqrt[n]{a^m}$$

Thus, *the mth power of the nth root of a* is the same as *the nth root of the mth power of a*. For example, to find $\sqrt[3]{27^2}$, we can proceed in either of two ways:

$$\sqrt[3]{27^2} = \left(\sqrt[3]{27}\right)^2 = 3^2 = 9 \qquad \text{or} \qquad \sqrt[3]{27^2} = \sqrt[3]{729} = 9$$

By definition, $\sqrt{a^2}$ represents a nonnegative number. If a could be negative, we must use absolute value symbols to guarantee that $\sqrt{a^2}$ will be nonnegative. Thus, if a is unrestricted,

$$\sqrt{a^2} = |a|$$

A similar argument holds when the index is any even natural number. The symbol $\sqrt[4]{a^4}$, for example, means the *positive* fourth root of a^4. Thus, if a is unrestricted,

$$\sqrt[4]{a^4} = |a|$$

EXAMPLE 6 If x is unrestricted, simplify **a.** $\sqrt[6]{64x^6}$, **b.** $\sqrt[3]{x^3}$, and **c.** $\sqrt{9x^8}$.

Solution

a. $\sqrt[6]{64x^6} = 2|x|$ — Use absolute value symbols to guarantee that the result will be nonnegative.

b. $\sqrt[3]{x^3} = x$ — Because the index is odd, no absolute value symbols are needed.

c. $\sqrt{9x^8} = 3x^4$ — Because $3x^4$ is always nonnegative, no absolute value symbols are needed.

Self Check Use the directions for Example 6: **a.** $\sqrt[4]{16x^4}$, **b.** $\sqrt[3]{27y^3}$, and **c.** $\sqrt[4]{x^8}$. ■

■ Simplifying and Combining Radicals

Many properties of exponents have counterparts in radical notation. For example, since $a^{1/n}b^{1/n} = (ab)^{1/n}$ and $\dfrac{a^{1/n}}{b^{1/n}} = \left(\dfrac{a}{b}\right)^{1/n}$ $(b \neq 0)$, we have the following.

Multiplication and Division Properties of Radicals

If all expressions represent real numbers,

$$\sqrt[n]{a}\sqrt[n]{b} = \sqrt[n]{ab} \qquad \frac{\sqrt[n]{a}}{\sqrt[n]{b}} = \sqrt[n]{\frac{a}{b}} \quad (b \neq 0)$$

In words, we say

The product of two nth roots is equal to the nth root of their product.

The quotient of two nth roots is equal to the nth root of their quotient.

Comment These properties involve the nth root of the product of two numbers or the nth root of the quotient of two numbers. There is no such property for sums or differences. For example, $\sqrt{9+4} \neq \sqrt{9} + \sqrt{4}$, because

$$\sqrt{9+4} = \sqrt{13} \qquad \text{but} \qquad \sqrt{9} + \sqrt{4} = 3 + 2 = 5$$

and $\sqrt{13} \neq 5$. In general,

$$\sqrt{a+b} \neq \sqrt{a} + \sqrt{b} \qquad \text{and} \qquad \sqrt{a-b} \neq \sqrt{a} - \sqrt{b}$$

Numbers that are squares of positive integers, such as 1, 4, 9, 16, 25, and 36, are called **perfect squares.** Expressions such as x^2 and $\frac{1}{9}x^6$ are also perfect squares, because each one is the square of another expression with integer exponents and rational coefficients.

Numbers that are cubes of positive integers, such as 1, 8, 27, 64, 125, and 216, are called **perfect cubes.** Expressions such as x^3 and $\frac{1}{27}x^9$ are also perfect cubes, because each one is the cube of another expression with integer exponents and rational coefficients. There are also perfect fourth powers, perfect fifth powers, and so on.

We can use perfect powers and the multiplication property of radicals to simplify many radical expressions. For example, to simplify $\sqrt{12x^5}$, we factor $12x^5$ so that one factor is the largest perfect square that divides $12x^5$. In this case, it is $4x^4$. We then rewrite $12x^5$ as $4x^4 \cdot 3x$ and simplify.

$$\begin{aligned} \sqrt{12x^5} &= \sqrt{4x^4 \cdot 3x} && \text{Factor } 12x^5 \text{ as } 4x^4 \cdot 3x. \\ &= \sqrt{4x^4}\sqrt{3x} && \text{Use the multiplication property of radicals: } \sqrt{ab} = \sqrt{a}\sqrt{b}. \\ &= 2x^2\sqrt{3x} && \sqrt{4x^4} = 2x^2. \end{aligned}$$

To simplify $\sqrt[3]{432x^9y}$, we find the largest perfect cube factor of $432x^9y$ (which is $216x^9$) and proceed as follows:

$$\begin{aligned} \sqrt[3]{432x^9y} &= \sqrt[3]{216x^9 \cdot 2y} && \text{Factor } 432x^9y \text{ as } 216x^9 \cdot 2y. \\ &= \sqrt[3]{216x^9}\sqrt[3]{2y} && \text{Use the multiplication property of radicals: } \sqrt[3]{ab} = \sqrt[3]{a}\sqrt[3]{b}. \\ &= 6x^3\sqrt[3]{2y} && \sqrt[3]{216x^9} = 6x^3. \end{aligned}$$

Radical expressions with the same index and the same radicand are called **like** or **similar radicals.** We can combine the like radicals in $3\sqrt{2} + 2\sqrt{2}$ by using the distributive property.

$$\begin{aligned} 3\sqrt{2} + 2\sqrt{2} &= (3 + 2)\sqrt{2} \\ &= 5\sqrt{2} \end{aligned}$$

This example suggests that to combine like radicals, we *add their numerical coefficients and keep the same radical.*

When radicals have the same index but different radicands, we can often change them to equivalent forms having the same radicand. We can then combine them. For example, to simplify $\sqrt{27} - \sqrt{12}$, we simplify both radicals and combine like radicals.

$$\begin{aligned} \sqrt{27} - \sqrt{12} &= \sqrt{9 \cdot 3} - \sqrt{4 \cdot 3} \\ &= \sqrt{9}\sqrt{3} - \sqrt{4}\sqrt{3} \\ &= 3\sqrt{3} - 2\sqrt{3} && \sqrt{9} = 3 \text{ and } \sqrt{4} = 2. \\ &= \sqrt{3} \end{aligned}$$

ACTIVE EXAMPLE 7 Simplify: **a.** $\sqrt{50} + \sqrt{200}$ and **b.** $3z\sqrt[5]{64z} - 2\sqrt[5]{2z^6}$.

Solution **a.**
$$\begin{aligned} \sqrt{50} + \sqrt{200} &= \sqrt{25 \cdot 2} + \sqrt{100 \cdot 2} \\ &= \sqrt{25}\sqrt{2} + \sqrt{100}\sqrt{2} \\ &= 5\sqrt{2} + 10\sqrt{2} \\ &= 15\sqrt{2} \end{aligned}$$

b. $$\begin{aligned} 3z\sqrt[5]{64z} - 2\sqrt[5]{2z^6} &= 3z\sqrt[5]{32 \cdot 2z} - 2\sqrt[5]{z^5 \cdot 2z} \\ &= 3z\sqrt[5]{32}\sqrt[5]{2z} - 2\sqrt[5]{z^5}\sqrt[5]{2z} \\ &= 3z(2)\sqrt[5]{2z} - 2z\sqrt[5]{2z} \\ &= 6z\sqrt[5]{2z} - 2z\sqrt[5]{2z} \\ &= 4z\sqrt[5]{2z} \end{aligned}$$

Self Check Simplify: **a.** $\sqrt{18} - \sqrt{8}$ and **b.** $2\sqrt[3]{81a^4} + a\sqrt[3]{24a}$. ■

■ Rationalizing Denominators and Numerators

By **rationalizing the denominator,** we can write a fraction such as

$$\frac{\sqrt{5}}{\sqrt{3}}$$

as a fraction with a rational number in the denominator. All that we must do is multiply both the numerator and the denominator by $\sqrt{3}$. (Note that $\sqrt{3}\sqrt{3}$ is the rational number 3.)

$$\frac{\sqrt{5}}{\sqrt{3}} = \frac{\sqrt{5}\sqrt{3}}{\sqrt{3}\sqrt{3}} = \frac{\sqrt{15}}{3}$$

To rationalize the numerator, we multiply both the numerator and the denominator by $\sqrt{5}$. (Note that $\sqrt{5}\sqrt{5}$ is the rational number 5.)

$$\frac{\sqrt{5}}{\sqrt{3}} = \frac{\sqrt{5}\sqrt{5}}{\sqrt{3}\sqrt{5}} = \frac{5}{\sqrt{15}}$$

ACTIVE EXAMPLE 8 Rationalize each denominator and simplify: **a.** $\frac{1}{\sqrt{7}}$, **b.** $\sqrt[3]{\frac{3}{4}}$, **c.** $\sqrt{\frac{3}{x}}$, and **d.** $\sqrt{\frac{3a^3}{5x^5}}$. Assume that all variables represent positive numbers.

Solution

a. $$\frac{1}{\sqrt{7}} = \frac{1\sqrt{7}}{\sqrt{7}\sqrt{7}} = \frac{\sqrt{7}}{7}$$

b. $$\sqrt[3]{\frac{3}{4}} = \frac{\sqrt[3]{3}}{\sqrt[3]{4}}$$

$$= \frac{\sqrt[3]{3}\sqrt[3]{2}}{\sqrt[3]{4}\sqrt[3]{2}}$$ Multiply numerator and denominator by $\sqrt[3]{2}$, because $\sqrt[3]{4}\sqrt[3]{2} = \sqrt[3]{8} = 2$.

$$= \frac{\sqrt[3]{6}}{\sqrt[3]{8}}$$

$$= \frac{\sqrt[3]{6}}{2}$$

COLLEGE **Algebra** $f(x)$ **Now**™
Go to **http://1pass.thomson.com** or your CD to practice this example.

c. $\sqrt{\frac{3}{x}} = \frac{\sqrt{3}}{\sqrt{x}}$
$= \frac{\sqrt{3}\sqrt{x}}{\sqrt{x}\sqrt{x}}$
$= \frac{\sqrt{3x}}{x}$

d. $\sqrt{\frac{3a^3}{5x^5}} = \frac{\sqrt{3a^3}}{\sqrt{5x^5}}$
$= \frac{\sqrt{3a^3}\sqrt{5x}}{\sqrt{5x^5}\sqrt{5x}}$ **Multiply numerator and denominator by $\sqrt{5x}$, because $\sqrt{5x^5}\sqrt{5x} = \sqrt{25x^6} = 5x^3$.**
$= \frac{\sqrt{15a^3x}}{\sqrt{25x^6}}$
$= \frac{\sqrt{a^2}\sqrt{15ax}}{5x^3}$
$= \frac{a\sqrt{15ax}}{5x^3}$

Self Check Use the directions for Example 8: **a.** $\frac{6}{\sqrt{6}}$ and **b.** $\sqrt[3]{\frac{2}{5x}}$. ■

EXAMPLE 9 Rationalize each numerator and simplify (assume that all variables represent positive numbers): **a.** $\frac{\sqrt{x}}{7}$ and **b.** $\frac{2\sqrt[3]{9x}}{3}$.

Solution **a.** $\frac{\sqrt{x}}{7} = \frac{\sqrt{x}\cdot\sqrt{x}}{7\sqrt{x}}$
$= \frac{x}{7\sqrt{x}}$

b. $\frac{2\sqrt[3]{9x}}{3} = \frac{2\sqrt[3]{9x}\cdot\sqrt[3]{3x^2}}{3\sqrt[3]{3x^2}}$
$= \frac{2\sqrt[3]{27x^3}}{3\sqrt[3]{3x^2}}$
$= \frac{2(3x)}{3\sqrt[3]{3x^2}}$
$= \frac{2x}{\sqrt[3]{3x^2}}$ **Divide out the 3's.**

Self Check Use the directions for Example 9: **a.** $\frac{\sqrt{2x}}{5}$ and **b.** $\frac{3\sqrt[3]{2y^2}}{6}$. ■

After rationalizing denominators, we often can simplify an expression.

EXAMPLE 10 Simplify: $\sqrt{\frac{1}{2}} + \sqrt{\frac{1}{8}}$.

Solution

$$\sqrt{\frac{1}{2}} + \sqrt{\frac{1}{8}} = \frac{1}{\sqrt{2}} + \frac{1}{\sqrt{8}}$$ $\sqrt{\frac{1}{2}} = \frac{\sqrt{1}}{\sqrt{2}} = \frac{1}{\sqrt{2}}; \sqrt{\frac{1}{8}} = \frac{\sqrt{1}}{\sqrt{8}} = \frac{1}{\sqrt{8}}$.

$$= \frac{1\sqrt{2}}{\sqrt{2}\sqrt{2}} + \frac{1\sqrt{2}}{\sqrt{8}\sqrt{2}}$$

$$= \frac{\sqrt{2}}{2} + \frac{\sqrt{2}}{\sqrt{16}}$$

$$= \frac{\sqrt{2}}{2} + \frac{\sqrt{2}}{4}$$

$$= \frac{3\sqrt{2}}{4}$$

Self Check Simplify: $\sqrt[3]{\frac{x}{2}} - \sqrt[3]{\frac{x}{16}}$. ■

Another property of radicals can be derived from the properties of exponents. If all of the expressions represent real numbers,

$$\sqrt[n]{\sqrt[m]{x}} = \sqrt[n]{x^{1/m}} = (x^{1/m})^{1/n} = x^{1/(mn)} = \sqrt[mn]{x}$$
$$\sqrt[m]{\sqrt[n]{x}} = \sqrt[m]{x^{1/n}} = (x^{1/n})^{1/m} = x^{1/(nm)} = \sqrt[mn]{x}$$

These results are summarized in the following *theorem* (a fact that can be proved).

Theorem If all of the expressions involved represent real numbers, then

$$\sqrt[m]{\sqrt[n]{x}} = \sqrt[n]{\sqrt[m]{x}} = \sqrt[mn]{x}$$

We can use the previous theorem to simplify many radicals. For example,

$$\sqrt[3]{\sqrt{8}} = \sqrt{\sqrt[3]{8}} = \sqrt{2}$$

Rational exponents can be used to simplify many radical expressions, as shown in the following example.

EXAMPLE 11 Simplify: **a.** $\sqrt[6]{4}$, **b.** $\sqrt[12]{x^3}$, and **c.** $\sqrt[9]{8y^3}$. (Assume that x and y are positive numbers.)

Solution **a.** $\sqrt[6]{4} = 4^{1/6} = (2^2)^{1/6} = 2^{2/6} = 2^{1/3} = \sqrt[3]{2}$

b. $\sqrt[12]{x^3} = x^{3/12} = x^{1/4} = \sqrt[4]{x}$

c. $\sqrt[9]{8y^3} = (2^3y^3)^{1/9} = (2y)^{3/9} = (2y)^{1/3} = \sqrt[3]{2y}$

Self Check Simplify: **a.** $\sqrt[4]{4}$ and **b.** $\sqrt[9]{27x^3}$. ■

Self Check Answers

1. a. 10 **b.** 3 **2. a.** -5 **b.** -10 **3. a.** 343 **b.** $\frac{1}{8}$ **c.** $9x^2$ **4. a.** $\frac{y}{7}$ **b.** $b^{1/7}$ **c.** $3s^2$
5. a. 6 **b.** $-\frac{1}{2}$ **6. a.** $2|x|$ **b.** $3y$ **c.** x^2 **7. a.** $\sqrt{2}$ **b.** $8a\sqrt[3]{3a}$ **8. a.** $\sqrt{6}$ **b.** $\frac{\sqrt[3]{50x^2}}{5x}$
9. a. $\frac{2x}{5\sqrt{2x}}$ **b.** $\frac{y}{\sqrt[3]{4y}}$ **10.** $\frac{\sqrt[3]{4x}}{4}$ **11. a.** $\sqrt{2}$ **b.** $\sqrt[3]{3x}$

0.3 Exercises

VOCABULARY AND CONCEPTS *Fill in the blanks.*

1. If $a = 0$ and n is a natural number, then $a^{1/n} = \underline{0}$.
2. If $a > 0$ and n is a natural number, then $a^{1/n}$ is a $\underline{\text{positive}}$ number.
3. If $a < 0$ and n is an even number, then $a^{1/n}$ is $\underline{\text{not}}$ a real number.
4. $6^{2/3}$ can be written as $\underline{(6^2)^{1/3}}$ or $\underline{(6^{1/3})^2}$.
5. $\sqrt[n]{a} = \underline{a^{1/n}}$
6. $\sqrt{a^2} = \underline{|a|}$
7. $\sqrt[n]{a}\sqrt[n]{b} = \underline{\sqrt[n]{ab}}$
8. $\sqrt[n]{\dfrac{a}{b}} = \underline{\dfrac{\sqrt[n]{a}}{\sqrt[n]{b}}}$
9. $\sqrt{x+y} \underline{\neq} \sqrt{x} + \sqrt{y}$
10. $\sqrt[m]{\sqrt[n]{x}}$ or $\sqrt[n]{\sqrt[m]{x}}$ can be written as $\underline{\sqrt[mn]{x}}$.

PRACTICE *Simplify each expression.*

11. $9^{1/2}$ 3
12. $8^{1/3}$ 2
13. $\left(\dfrac{1}{25}\right)^{1/2}$ $\dfrac{1}{5}$
14. $\left(\dfrac{16}{625}\right)^{1/4}$ $\dfrac{2}{5}$
15. $-81^{1/4}$ -3
16. $-\left(\dfrac{8}{27}\right)^{1/3}$ $-\dfrac{2}{3}$
17. $(10{,}000)^{1/4}$ 10
18. $1{,}024^{1/5}$ 4
19. $-64^{1/3}$ -4
20. $\left(-\dfrac{27}{8}\right)^{1/3}$ $-\dfrac{3}{2}$
21. $64^{1/3}$ 4
22. $(-125)^{1/3}$ -5

Simplify each expression. Use absolute value symbols when necessary.

23. $(16a^2)^{1/2}$ $4|a|$
24. $(25a^4)^{1/2}$ $5a^2$
25. $(16a^4)^{1/4}$ $2|a|$
26. $(-64a^3)^{1/3}$ $-4a$
27. $(-32a^5)^{1/5}$ $-2a$
28. $(64a^6)^{1/6}$ $2|a|$
29. $(-216b^6)^{1/3}$ $-6b^2$
30. $(256t^8)^{1/4}$ $4t^2$
31. $\left(\dfrac{16a^4}{25b^2}\right)^{1/2}$ $\dfrac{4a^2}{5|b|}$
32. $\left(-\dfrac{a^5}{32b^{10}}\right)^{1/5}$ $-\dfrac{a}{2b^2}$
33. $\left(-\dfrac{1{,}000x^6}{27y^3}\right)^{1/3}$ $-\dfrac{10x^2}{3y}$
34. $\left(\dfrac{49t^2}{100z^4}\right)^{1/2}$ $\dfrac{7|t|}{10z^2}$

Simplify each expression. Write all answers without using negative exponents.

35. $4^{3/2}$ 8
36. $8^{2/3}$ 4
37. $-16^{3/2}$ -64
38. $(-8)^{2/3}$ 4
39. $-1{,}000^{2/3}$ -100
40. $100^{3/2}$ 1,000
41. $64^{-1/2}$ $\dfrac{1}{8}$
42. $25^{-1/2}$ $\dfrac{1}{5}$
43. $64^{-3/2}$ $\dfrac{1}{512}$
44. $49^{-3/2}$ $\dfrac{1}{343}$
45. $-9^{-3/2}$ $-\dfrac{1}{27}$
46. $(-27)^{-2/3}$ $\dfrac{1}{9}$
47. $\left(\dfrac{4}{9}\right)^{5/2}$ $\dfrac{32}{243}$
48. $\left(\dfrac{25}{81}\right)^{3/2}$ $\dfrac{125}{729}$
49. $\left(-\dfrac{27}{64}\right)^{-2/3}$ $\dfrac{16}{9}$
50. $\left(\dfrac{125}{8}\right)^{-4/3}$ $\dfrac{16}{625}$

Simplify each expression. Write all answers without using negative exponents. Assume that all variables represent positive numbers.

51. $(100s^4)^{1/2}$ $10s^2$
52. $(64u^6v^3)^{1/3}$ $4u^2v$
53. $(32y^{10}z^5)^{-1/5}$ $\dfrac{1}{2y^2z}$
54. $(625a^4b^8)^{-1/4}$ $\dfrac{1}{5ab^2}$
55. $(x^{10}y^5)^{3/5}$ x^6y^3
56. $(64a^6b^{12})^{5/6}$ $32a^5b^{10}$
57. $(r^8s^{16})^{-3/4}$ $\dfrac{1}{r^6s^{12}}$
58. $(-8x^9y^{12})^{-2/3}$ $\dfrac{1}{4x^6y^8}$
59. $\left(-\dfrac{8a^6}{125b^9}\right)^{2/3}$ $\dfrac{4a^4}{25b^6}$
60. $\left(\dfrac{16x^4}{625y^8}\right)^{3/4}$ $\dfrac{8x^3}{125y^6}$
61. $\left(\dfrac{27r^6}{1{,}000s^{12}}\right)^{-2/3}$ $\dfrac{100s^8}{9r^4}$
62. $\left(-\dfrac{32m^{10}}{243n^{15}}\right)^{-2/5}$ $\dfrac{9n^6}{4m^4}$
63. $\dfrac{a^{2/5}a^{4/5}}{a^{1/5}}$ a
64. $\dfrac{x^{6/7}x^{3/7}}{x^{2/7}x^{5/7}}$ $x^{2/7}$

Simplify each radical expression.

65. $\sqrt{49}$ 7
66. $\sqrt{81}$ 9
67. $\sqrt[3]{125}$ 5
68. $\sqrt[3]{-64}$ -4
69. $-\sqrt[4]{81}$ -3
70. $\sqrt[5]{-243}$ -3
71. $\sqrt[5]{-\dfrac{32}{100{,}000}}$ $-\dfrac{1}{5}$
72. $\sqrt[4]{\dfrac{256}{625}}$ $\dfrac{4}{5}$

Simplify each expression, using absolute value symbols when necessary. Write answers without using negative exponents.

73. $\sqrt{36x^2}$ $6|x|$
74. $-\sqrt{25y^2}$ $-5|y|$

75. $\sqrt{9y^4}$ $3y^2$

76. $\sqrt{a^4b^8}$ a^2b^4

77. $\sqrt[3]{8y^3}$ $2y$

78. $\sqrt[3]{-27z^9}$ $-3z^3$

79. $\sqrt[4]{\dfrac{x^4y^8}{z^{12}}}$ $\dfrac{|x|y^2}{|z^3|}$

80. $\sqrt[5]{\dfrac{a^{10}b^5}{c^{15}}}$ $\dfrac{a^2b}{c^3}$

Simplify each expression. Assume that all variables represent positive numbers, so that no absolute value symbols are needed.

81. $\sqrt{8} - \sqrt{2}$ $\sqrt{2}$

82. $\sqrt{75} - 2\sqrt{27}$ $-\sqrt{3}$

83. $\sqrt{200x^2} + \sqrt{98x^2}$ $17x\sqrt{2}$

84. $\sqrt{128a^3} - a\sqrt{162a}$ $-a\sqrt{2a}$

85. $2\sqrt{48y^5} - 3y\sqrt{12y^3}$ $2y^2\sqrt{3y}$

86. $y\sqrt{112y} + 4\sqrt{175y^3}$ $24y\sqrt{7y}$

87. $2\sqrt[3]{81} + 3\sqrt[3]{24}$ $12\sqrt[3]{3}$

88. $3\sqrt[4]{32} - 2\sqrt[4]{162}$ 0

89. $\sqrt[4]{768z^5} + \sqrt[4]{48z^5}$ $6z\sqrt[4]{3z}$

90. $-2\sqrt[5]{64y^2} + 3\sqrt[5]{486y^2}$ $5\sqrt[5]{2y^2}$

91. $\sqrt{8x^2y} - x\sqrt{2y} + \sqrt{50x^2y}$ $6x\sqrt{2y}$

92. $3x\sqrt{18x} + 2\sqrt{2x^3} - \sqrt{72x^3}$ $5x\sqrt{2x}$

93. $\sqrt[3]{16xy^4} + y\sqrt[3]{2xy} - \sqrt[3]{54xy^4}$ 0

94. $\sqrt[4]{512x^5} - \sqrt[4]{32x^5} + \sqrt[4]{1{,}250x^5}$ $7x\sqrt[4]{2x}$

Rationalize each denominator and simplify. Assume that all variables represent positive numbers.

95. $\dfrac{3}{\sqrt{3}}$ $\sqrt{3}$

96. $\dfrac{2}{\sqrt{x}}$ $\dfrac{2\sqrt{x}}{x}$

97. $\dfrac{2}{\sqrt[3]{2}}$ $\sqrt[3]{4}$

98. $\dfrac{5a}{\sqrt[3]{25a}}$ $\sqrt[3]{5a^2}$

99. $\dfrac{2b}{\sqrt[4]{3a^2}}$ $\dfrac{2b\sqrt[4]{27a^2}}{3a}$

100. $\sqrt{\dfrac{x}{2y}}$ $\dfrac{\sqrt{2xy}}{2y}$

101. $\sqrt[3]{\dfrac{2u^4}{9v}}$ $\dfrac{u\sqrt[3]{6uv^2}}{3v}$

102. $\sqrt[3]{-\dfrac{3s^5}{4r^2}}$ $-\dfrac{s\sqrt[3]{6rs^2}}{2r}$

Rationalize each numerator and simplify. Assume that all variables are positive numbers.

103. $\dfrac{\sqrt{5}}{10}$ $\dfrac{1}{2\sqrt{5}}$

104. $\dfrac{\sqrt{y}}{3}$ $\dfrac{y}{3\sqrt{y}}$

105. $\dfrac{\sqrt[3]{9}}{3}$ $\dfrac{1}{\sqrt[3]{3}}$

106. $\dfrac{\sqrt[3]{16b^2}}{16}$ $\dfrac{b}{4\sqrt[3]{4b}}$

107. $\dfrac{\sqrt[5]{16b^3}}{64a}$ $\dfrac{b}{32a\sqrt[5]{2b^2}}$

108. $\sqrt{\dfrac{3x}{57}}$ $\dfrac{x}{\sqrt{19x}}$

Rationalize each denominator and simplify.

109. $\sqrt{\dfrac{1}{3}} - \sqrt{\dfrac{1}{27}}$ $\dfrac{2\sqrt{3}}{9}$

110. $\sqrt[3]{\dfrac{1}{2}} + \sqrt[3]{\dfrac{1}{16}}$ $\dfrac{3\sqrt[3]{4}}{4}$

111. $\sqrt{\dfrac{x}{8}} - \sqrt{\dfrac{x}{2}} + \sqrt{\dfrac{x}{32}}$ $-\dfrac{\sqrt{2x}}{8}$

112. $\sqrt[3]{\dfrac{y}{4}} + \sqrt[3]{\dfrac{y}{32}} - \sqrt[3]{\dfrac{y}{500}}$ $\dfrac{13\sqrt[3]{2y}}{20}$

Simplify each radical expression.

113. $\sqrt[4]{9}$ $\sqrt{3}$

114. $\sqrt[6]{27}$ $\sqrt{3}$

115. $\sqrt[10]{16x^6}$ $\sqrt[5]{4x^3}$

116. $\sqrt[6]{27x^9}$ $x\sqrt{3x}$

DISCOVERY AND WRITING ***We often can multiply and divide radicals with different indices. For example, to multiply $\sqrt{3}$ by $\sqrt[3]{5}$, we first write each radical as a sixth root***

$$\sqrt{3} = 3^{1/2} = 3^{3/6} = \sqrt[6]{3^3} = \sqrt[6]{27}$$

$$\sqrt[3]{5} = 5^{1/3} = 5^{2/6} = \sqrt[6]{5^2} = \sqrt[6]{25}$$

and then multiply the sixth roots.

$$\sqrt{3}\sqrt[3]{5} = \sqrt[6]{27}\sqrt[6]{25} = \sqrt[6]{(27)(25)} = \sqrt[6]{675}$$

Division is similar. Use this idea to write each of the following expressions as a single radical.

117. $\sqrt{2}\sqrt[3]{2}$ $\sqrt[6]{32}$

118. $\sqrt{3}\sqrt[3]{5}$ $\sqrt[6]{675}$

119. $\dfrac{\sqrt[4]{3}}{\sqrt{2}}$ $\dfrac{\sqrt[4]{12}}{2}$

120. $\dfrac{\sqrt[3]{2}}{\sqrt{5}}$ $\dfrac{\sqrt[6]{500}}{5}$

121. For what values of x does $\sqrt[4]{x^4} = x$? Explain.

122. If all of the radicals involved represent real numbers and $y \neq 0$, explain why

$$\sqrt[n]{\frac{x}{y}} = \frac{\sqrt[n]{x}}{\sqrt[n]{y}}$$

123. If all of the radicals involved represent real numbers and there is no division by 0, explain why

$$\left(\frac{x}{y}\right)^{-m/n} = \sqrt[n]{\frac{y^m}{x^m}}$$

124. The definition of $x^{m/n}$ requires that $\sqrt[n]{x}$ be a real number. Explain why this is important. (*Hint:* Consider what happens when n is even, m is odd, and x is negative.)

REVIEW

125. Write $-2 < x \leq 5$ using interval notation. $(-2, 5]$

126. Write the expression $|3 - x|$ without using absolute value symbols. Assume that $x > 4$. $x - 3$

Evaluate each expression when $x = -2$ and $y = 3$.

127. $x^2 - y^2$ -5

128. $\dfrac{xy + 4y}{x}$ -3

129. Write 617,000,000 in scientific notation. 6.17×10^8

130. Write 0.00235×10^4 in standard notation. 23.5

0.4 Polynomials

In this section, you will learn about

- Polynomials
- Adding and Subtracting Polynomials
- Multiplying Polynomials
- Conjugate Binomials
- Dividing Polynomials

A **monomial** is a number or the product of a number and one or more variables with whole-number exponents. The number is called the **coefficient** of the variables. Some examples of monomials are

$$3x, \quad 7ab^2, \quad -5ab^2c^4, \quad x^3, \quad \text{and} \quad -12$$

with coefficients of 3, 7, -5, 1, and -12, respectively.

The **degree** of a monomial is the sum of the exponents of its variables. All nonzero constants (except 0) have a degree of 0.

The degree of $3x$ is 1.
The degree of $7ab^2$ is 3.
The degree of $-5ab^2c^4$ is 7.
The degree of x^3 is 3.
The degree of -12 is 0 (since $-12 = -12x^0$).
0 has no defined degree.

Polynomials

A monomial or a sum of monomials is called a **polynomial.** Each monomial in that sum is called a **term** of the polynomial. A polynomial with two terms is called a **binomial,** and a polynomial with three terms is called a **trinomial.**

Monomials	Binomials	Trinomials
$3x^2$	$2a + 3b$	$x^2 + 7x - 4$
$-25xy$	$4x^3 - 3x^2$	$4y^4 - 2y + 12$
a^2b^3c	$-2x^3 - 4y^2$	$12x^3y^2 - 8xy - 24$

The **degree of a polynomial** is the degree of the term in the polynomial with highest degree. The only polynomial with no defined degree is 0, which is called the **zero polynomial.** Here are some examples.

- $3x^2y^3 + 5xy^2 + 7$ is a trinomial of 5th degree, because its term with highest degree (the first term) is 5.
- $3ab + 5a^2b$ is a binomial of degree 3.
- $5x + 3y^2 + \sqrt[3]{3}z^4 - \sqrt{7}$ is a polynomial, because its variables have whole-number exponents. It is of degree 4.
- $-7y^{1/2} + 3y^2 + \sqrt[5]{3}z$ is not a polynomial, because some of its variables do not have whole-number exponents.

If two terms of a polynomial have the same variables with the same exponents, they are **like** or **similar** terms. To combine the like terms in $3x^2y + 5x^2y$, we use the distributive property:

$$\begin{aligned} 3x^2y + 5x^2y &= (3 + 5)x^2y \\ &= 8x^2y \end{aligned}$$

This illustrates that *to combine like terms, we add their coefficients and keep the same variables and the same exponents.*

Adding and Subtracting Polynomials

Recall that we can use the distributive property to remove parentheses enclosing the terms of a polynomial. When the sign preceding the parentheses is +, we simply drop the parentheses:

$$\begin{aligned} +(a + b - c) &= +1(a + b - c) \\ &= 1a + 1b - 1c \\ &= a + b - c \end{aligned}$$

When the sign preceding the parentheses is −, we drop the parentheses and the − sign and change the sign of each term within the parentheses.

$$\begin{aligned} -(a + b - c) &= -1(a + b - c) \\ &= -1a + (-1)b - (-1)c \\ &= -a - b + c \end{aligned}$$

We can use these facts to add and subtract polynomials. *To add (or subtract) polynomials, we remove parentheses (if necessary) and combine like terms.*

EXAMPLE 1 Add: $(3x^3y + 5x^2 - 2y) + (2x^3y - 5x^2 + 3x)$.

Solution To add the polynomials, we remove parentheses and combine like terms.

$$\begin{aligned} &(3x^3y + 5x^2 - 2y) + (2x^3y - 5x^2 + 3x) \\ &\quad = 3x^3y + 5x^2 - 2y + 2x^3y - 5x^2 + 3x \\ &\quad = 3x^3y + 2x^3y + 5x^2 - 5x^2 - 2y + 3x && \text{Use the commutative property to rearrange terms.} \\ &\quad = 5x^3y - 2y + 3x && \text{Combine like terms.} \end{aligned}$$

Self Check Add: $(4x^2 + 3x - 5) + (3x^2 - 5x + 7)$. ■

EXAMPLE 2 Subtract: $(2x^2 + 3y^2) - (x^2 - 2y^2 + 7)$.

Solution To subtract the polynomials, we remove parentheses and combine like terms.

$$
\begin{aligned}
&(2x^2 + 3y^2) - (x^2 - 2y^2 + 7) \\
&\quad = 2x^2 + 3y^2 - x^2 + 2y^2 - 7 \\
&\quad = 2x^2 - x^2 + 3y^2 + 2y^2 - 7 && \text{Use the commutative property to rearrange terms.} \\
&\quad = x^2 + 5y^2 - 7 && \text{Combine like terms.}
\end{aligned}
$$

Self Check Subtract: $(4x^2 + 3x - 5) - (3x^2 - 5x + 7)$. ■

We also use the distributive property to remove parentheses enclosing several terms that are multiplied by a constant. For example,

$$
\begin{aligned}
4(3x^2 - 2x + 6) &= 4(3x^2) - 4(2x) + 4(6) \\
&= 12x^2 - 8x + 24
\end{aligned}
$$

This example suggests that *to add multiples of one polynomial to another, or to subtract multiples of one polynomial from another, we remove parentheses and combine like terms.*

EXAMPLE 3 Simplify: $7x(2y^2 + 13x^2) - 5(xy^2 - 13x^3)$.

Solution

$$
\begin{aligned}
&7x(2y^2 + 13x^2) - 5(xy^2 - 13x^3) \\
&\quad = 14xy^2 + 91x^3 - 5xy^2 + 65x^3 && \text{Use the distributive property to remove parentheses.} \\
&\quad = 14xy^2 - 5xy^2 + 91x^3 + 65x^3 && \text{Use the commutative property to rearrange terms.} \\
&\quad = 9xy^2 + 156x^3 && \text{Combine like terms.}
\end{aligned}
$$

Self Check Simplify: $3(2b^2 - 3a^2b) + 2b(b + a^2)$. ■

■ Multiplying Polynomials

To find the product of $3x^2y^3z$ and $5xyz^2$, we proceed as follows:

$$
\begin{aligned}
(3x^2y^3z)(5xyz^2) &= 3 \cdot x^2 \cdot y^3 \cdot z \cdot 5 \cdot x \cdot y \cdot z^2 \\
&= 3 \cdot 5 \cdot x^2 \cdot x \cdot y^3 \cdot y \cdot z \cdot z^2 && \text{Use the commutative property to rearrange terms.} \\
&= 15x^3y^4z^3
\end{aligned}
$$

This illustrates that *to multiply two monomials, we multiply the coefficients and then multiply the variables.*

To find the product of a monomial and a polynomial, we use the distributive property.

$$
\begin{aligned}
3xy^2(2xy + x^2 - 7yz) &= 3xy^2(2xy) + (3xy^2)(x^2) - (3xy^2)(7yz) \\
&= 6x^2y^3 + 3x^3y^2 - 21xy^3z
\end{aligned}
$$

This illustrates that *to multiply a polynomial by a monomial, we multiply each term of the polynomial by the monomial.*

To multiply one binomial by another, we use the distributive property twice.

EXAMPLE 4 Multiply: **a.** $(x + y)(x + y)$, **b.** $(x - y)(x - y)$, and **c.** $(x + y)(x - y)$.

Solution

a.
$$\begin{aligned}(x + y)(x + y) &= (x + y)x + (x + y)y\\ &= x^2 + xy + xy + y^2\\ &= x^2 + 2xy + y^2\end{aligned}$$

b.
$$\begin{aligned}(x - y)(x - y) &= (x - y)x - (x - y)y\\ &= x^2 - xy - xy + y^2\\ &= x^2 - 2xy + y^2\end{aligned}$$

c.
$$\begin{aligned}(x + y)(x - y) &= (x + y)x - (x + y)y\\ &= x^2 + xy - xy - y^2\\ &= x^2 - y^2\end{aligned}$$

Self Check Multiply: **a.** $(x + 2)(x + 2)$, **b.** $(x - 3)(x - 3)$, and **c.** $(x + 4)(x - 4)$. ■

The products in Example 4 are called **special products.** Because they occur so often, it is worthwhile to learn their forms.

Special Product Formulas

$$\begin{aligned}(x + y)^2 &= (x + y)(x + y) = x^2 + 2xy + y^2\\ (x - y)^2 &= (x - y)(x - y) = x^2 - 2xy + y^2\\ (x + y)(x - y) &= x^2 - y^2\end{aligned}$$

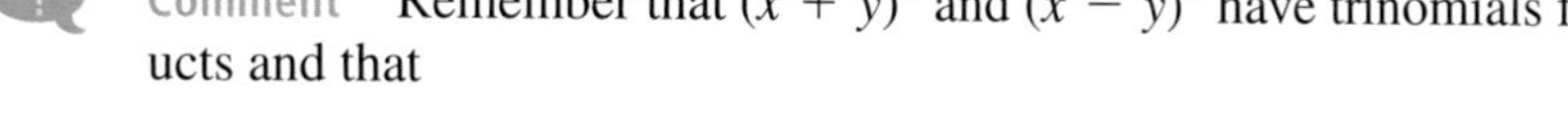

Comment Remember that $(x + y)^2$ and $(x - y)^2$ have trinomials for their products and that

$$(x + y)^2 \neq x^2 + y^2 \qquad \text{and} \qquad (x - y)^2 \neq x^2 - y^2$$

We can use the **FOIL method** to multiply one binomial by another. The word FOIL is an acronym for **F**irst terms, **O**uter terms, **I**nner terms, and **L**ast terms. To use this method to multiply $3x - 4$ by $2x + 5$, we write

First terms Last terms

$$\begin{aligned}(3x - 4)(2x + 5) &= 3x(2x) + 3x(5) - 4(2x) - 4(5)\\ &= 6x^2 + 15x - 8x - 20\\ &= 6x^2 + 7x - 20\end{aligned}$$

Inner terms

Outer terms

In this example,

- the product of the first terms is $6x^2$,
- the product of the outer terms is $15x$,
- the product of the inner terms is $-8x$, and
- the product of the last terms is -20.

The resulting like terms of the product are then combined.

ACTIVE EXAMPLE 5 Use the FOIL method to multiply: $(\sqrt{3} + x)(2 - \sqrt{3}x)$.

COLLEGE **Algebra f(x) Now™**
Explore this Active Example and test yourself on the steps taken here at **http://1pass.thomson.com**. You can also find this example on the Interactive Video Skillbuilder CD-ROM.

Solution

$$\begin{aligned}(\sqrt{3} + x)(2 - \sqrt{3}x) &= 2\sqrt{3} - \sqrt{3}\sqrt{3}x + 2x - x\sqrt{3}x \\ &= 2\sqrt{3} - 3x + 2x - \sqrt{3}x^2 \\ &= 2\sqrt{3} - x - \sqrt{3}x^2\end{aligned}$$

Self Check Multiply: $(2x + \sqrt{3})(x - \sqrt{3})$. ■

To multiply a polynomial with more than two terms by another polynomial, we multiply each term of one polynomial by each term of the other polynomial and combine like terms whenever possible.

EXAMPLE 6 Multiply: **a.** $(x + y)(x^2 - xy + y^2)$ and **b.** $(x + 3)^3$.

Solution **a.**

$$\begin{aligned}(x + y)(x^2 - xy + y^2) &= x^3 - x^2y + xy^2 + yx^2 - xy^2 + y^3 \\ &= x^3 + y^3\end{aligned}$$

b.

$$\begin{aligned}(x + 3)^3 &= (x + 3)(x + 3)^2 \\ &= (x + 3)(x^2 + 6x + 9) \\ &= x^3 + 6x^2 + 9x + 3x^2 + 18x + 27 \\ &= x^3 + 9x^2 + 27x + 27\end{aligned}$$

Self Check Multiply: $(x + 2)(2x^2 + 3x - 1)$. ■

If n is a whole number, the expressions $a^n + 1$ and $2a^n - 3$ are polynomials and we can multiply them as follows:

$$\begin{aligned}(a^n + 1)(2a^n - 3) &= 2a^{2n} - 3a^n + 2a^n - 3 \\ &= 2a^{2n} - a^n - 3 \qquad \text{Combine like terms.}\end{aligned}$$

We can also use the methods previously discussed to multiply expressions that are not polynomials, such as $x^{-2} + y$ and $x^2 - y^{-1}$.

$$\begin{aligned}(x^{-2} + y)(x^2 - y^{-1}) &= x^{-2+2} - x^{-2}y^{-1} + x^2y - y^{1-1} \\ &= x^0 - \frac{1}{x^2y} + x^2y - y^0 \\ &= 1 - \frac{1}{x^2y} + x^2y - 1 \qquad x^0 = 1 \text{ and } y^0 = 1. \\ &= x^2y - \frac{1}{x^2y}\end{aligned}$$

Conjugate Binomials

If the denominator of a fraction is a binomial containing square roots, we can use the product formula $(x + y)(x - y)$ to rationalize the denominator. For example, to rationalize the denominator of

$$\frac{6}{\sqrt{7} + 2}$$

we multiply the numerator and the denominator by $\sqrt{7} - 2$ and simplify.

$$\begin{aligned}\frac{6}{\sqrt{7} + 2} &= \frac{6(\sqrt{7} - 2)}{(\sqrt{7} + 2)(\sqrt{7} - 2)} \\ &= \frac{6(\sqrt{7} - 2)}{7 - 4} \\ &= \frac{6(\sqrt{7} - 2)}{3} \\ &= 2(\sqrt{7} - 2)\end{aligned}$$

In this example, we multiplied both the numerator and the denominator of the given fraction by $\sqrt{7} - 2$. This binomial is the same as the denominator of the given fraction $\sqrt{7} + 2$, except for the sign between the terms. Such binomials are called *conjugate binomials* or *radical conjugates.*

Conjugate Binomials

Conjugate binomials are binomials that are the same except for the sign between their terms. The conjugate of $a + b$ is $a - b$, and the conjugate of $a - b$ is $a + b$.

ACTIVE EXAMPLE 7 Rationalize the denominator: $\dfrac{\sqrt{3x} - \sqrt{2}}{\sqrt{3x} + \sqrt{2}}$ $(x > 0)$.

Solution We multiply the numerator and the denominator by $\sqrt{3x} - \sqrt{2}$ (the conjugate of $\sqrt{3x} + \sqrt{2}$) and simplify.

$$\begin{aligned}\frac{\sqrt{3x} - \sqrt{2}}{\sqrt{3x} + \sqrt{2}} &= \frac{(\sqrt{3x} - \sqrt{2})(\sqrt{3x} - \sqrt{2})}{(\sqrt{3x} + \sqrt{2})(\sqrt{3x} - \sqrt{2})} \\ &= \frac{\sqrt{3x}\sqrt{3x} - \sqrt{3x}\sqrt{2} - \sqrt{2}\sqrt{3x} + \sqrt{2}\sqrt{2}}{(\sqrt{3x})^2 - (\sqrt{2})^2} \\ &= \frac{3x - \sqrt{6x} - \sqrt{6x} + 2}{3x - 2} \\ &= \frac{3x - 2\sqrt{6x} + 2}{3x - 2}\end{aligned}$$

COLLEGE **Algebra** f(x) **Now**™
Go to **http://1pass.thomson.com** or your CD to practice this example.

Self Check Rationalize the denominator: $\dfrac{\sqrt{x} + 2}{\sqrt{x} - 2}$. ■

In calculus, we often rationalize a numerator.

EXAMPLE 8 Rationalize the numerator: $\dfrac{\sqrt{x+h}-\sqrt{x}}{h}$.

Solution Multiply the numerator and the denominator by the conjugate of the numerator and simplify.

$$\begin{aligned}\frac{\sqrt{x+h}-\sqrt{x}}{h} &= \frac{\left(\sqrt{x+h}-\sqrt{x}\right)\left(\sqrt{x+h}+\sqrt{x}\right)}{h\left(\sqrt{x+h}+\sqrt{x}\right)}\\ &= \frac{x+h-x}{h\left(\sqrt{x+h}+\sqrt{x}\right)}\\ &= \frac{h}{h\left(\sqrt{x+h}+\sqrt{x}\right)}\\ &= \frac{1}{\sqrt{x+h}+\sqrt{x}} \quad \text{Divide out the common factor of } h.\end{aligned}$$

Self Check Rationalize the numerator: $\dfrac{\sqrt{4+h}-2}{h}$. ■

Gottfried Wilhelm Leibniz
(1646–1716)

Leibniz, a German philosopher and logician, is best known as one of the inventors of calculus, along with Isaac Newton. He also developed the binary numeration system that is basic to modern computers. He even invented a calculating machine that would add, subtract, multiply, divide, and find roots of numbers.

■ Dividing Polynomials

To divide monomials, we write the quotient as a fraction and simplify by using the rules of exponents. For example,

$$\begin{aligned}\frac{6x^2y^3}{-2x^3y} &= -3x^{2-3}y^{3-1}\\ &= -3x^{-1}y^2\\ &= -\frac{3y^2}{x}\end{aligned}$$

To divide a polynomial by a monomial, we write the quotient as a fraction, write the fraction as a sum of separate fractions, and simplify each one. For example, to divide $8x^5y^4 + 12x^2y^5 - 16x^2y^3$ by $4x^3y^4$, we proceed as follows:

$$\begin{aligned}\frac{8x^5y^4 + 12x^2y^5 - 16x^2y^3}{4x^3y^4} &= \frac{8x^5y^4}{4x^3y^4} + \frac{12x^2y^5}{4x^3y^4} + \frac{-16x^2y^3}{4x^3y^4}\\ &= 2x^2 + \frac{3y}{x} - \frac{4}{xy}\end{aligned}$$

To divide two polynomials, we can use long division. To illustrate, we consider the division

$$\frac{2x^2 + 11x - 30}{x + 7}$$

which can be written in long division form as

$$x + 7\overline{)2x^2 + 11x - 30}$$

The binomial $x + 7$ is called the **divisor,** and the trinomial $2x^2 + 11x - 30$ is called the **dividend.** The final answer, called the **quotient,** will appear above the long division symbol.

We begin the division by asking "What monomial, when multiplied by x, gives $2x^2$?" Because $x \cdot 2x = 2x^2$, the answer is $2x$. We place $2x$ in the quotient, multiply each term of the divisor by $2x$, subtract, and bring down the -30.

$$\begin{array}{r} 2x\phantom{{}+11x-30} \\ x + 7\overline{)2x^2 + 11x - 30} \\ \underline{2x^2 + 14x}\phantom{{}-30} \\ -3x - 30 \end{array}$$

We continue the division by asking "What monomial, when multiplied by x, gives $-3x$?" We place the answer, -3, in the quotient, multiply each term of the divisor by -3, and subtract. This time, there is no number to bring down.

$$\begin{array}{r} 2x - 3\phantom{{}-30} \\ x + 7\overline{)2x^2 + 11x - 30} \\ \underline{2x^2 + 14x}\phantom{{}-30} \\ -3x - 30 \\ \underline{-3x - 21} \\ -9 \end{array}$$

Because the degree of the remainder, -9, is less than the degree of the divisor, the division process stops, and we can express the result in the form

$$\text{quotient} + \frac{\text{remainder}}{\text{divisor}}$$

Thus,

$$\frac{2x^2 + 11x - 30}{x + 7} = 2x - 3 + \frac{-9}{x + 7}$$

EXAMPLE 9 Divide $6x^3 - 11$ by $2x + 2$.

Solution We set up the division, leaving spaces for the missing powers of x in the dividend.

$$2x + 2\overline{)6x^3 \qquad\qquad\qquad - 11}$$

Comment In Example 9, we could write the missing powers of x using a coefficient of 0:

$$2x + 2\overline{)6x^3 + 0x^2 + 0x - 11}$$

The division process continues as usual, with the following results:

$$\begin{array}{r} 3x^2 - 3x + 3\phantom{{}-11} \\ 2x + 2\overline{)6x^3 \qquad\qquad\quad - 11} \\ \underline{6x^3 + 6x^2}\phantom{{}-6x-11} \\ -6x^2\phantom{{}-6x-11} \\ \underline{-6x^2 - 6x}\phantom{{}-11} \\ +6x - 11 \\ \underline{+6x + 6} \\ -17 \end{array}$$

Thus, $\dfrac{6x^3 - 11}{2x + 2} = 3x^2 - 3x + 3 + \dfrac{-17}{2x + 2}$.

Self Check Divide: $3x + 1\overline{)9x^2 - 1}$. ■

EXAMPLE 10 Divide $-3x^3 - 3 + x^5 + 4x^2 - x^4$ by $x^2 - 3$.

Solution The division process works best when the terms in the divisor and dividend are written with their exponents in descending order.

$$\begin{array}{r} x^3 - x^2 + 1 \\ x^2 - 3\overline{)x^5 - x^4 - 3x^3 + 4x^2 - 3} \\ \underline{x^5 - 3x^3} \\ - x^4 + 4x^2 \\ \underline{- x^4 + 3x^2} \\ x^2 - 3 \\ \underline{x^2 - 3} \\ 0 \end{array}$$

Thus, $\dfrac{-3x^3 - 3 + x^5 + 4x^2 - x^4}{x^2 - 3} = x^3 - x^2 + 1$.

Self Check Divide: $x^2 + 1\overline{)3x^2 - x + 1 - 2x^3 + 3x^4}$. ■

Self Check Answers

1. $7x^2 - 2x + 2$ **2.** $x^2 + 8x - 12$ **3.** $8b^2 - 7a^2b$ **4. a.** $x^2 + 4x + 4$ **b.** $x^2 - 6x + 9$ **c.** $x^2 - 16$ **5.** $2x^2 - x\sqrt{3} - 3$ **6.** $2x^3 + 7x^2 + 5x - 2$ **7.** $\dfrac{x + 4\sqrt{x} + 4}{x - 4}$ **8.** $\dfrac{1}{\sqrt{4 + h} + 2}$ **9.** $3x - 1$ **10.** $3x^2 - 2x + \dfrac{x + 1}{x^2 + 1}$

0.4 Exercises

VOCABULARY AND CONCEPTS ***Fill in the blanks.***

1. A monomial is a number or the product of a number and one or more variables.
2. The degree of a monomial is the sum of the exponents of its variables.
3. A trinomial is a polynomial with three terms.
4. A binomial is a polynomial with two terms.
5. A monomial is a polynomial with one term.
6. The constant 0 is called the zero polynomial.
7. Terms with the same variables with the same exponents are called like terms.
8. The degree of a polynomial is the same as the degree of its term of highest degree.
9. To combine like terms, we add their coefficients and keep the same variables and the same exponents.
10. The conjugate of $3\sqrt{x} + 2$ is $3\sqrt{x} - 2$.

Determine whether the given expression is a polynomial. If so, tell whether it is a monomial, a binomial, or a trinomial, and give its degree.

11. $x^2 + 3x + 4$ yes, trinomial, 2nd degree

12. $5xy - x^3$ yes, binomial, 3rd degree

13. $x^3 + y^{1/2}$ no

14. $x^{-3} - 5y^{-2}$ no

15. $4x^2 - \sqrt{5}x^3$ yes, binomial, 3rd degree

16. x^2y^3 yes, monomial, 5th degree

17. $\sqrt{15}$ yes, monomial, 0th degree

18. $\frac{5}{x} + \frac{x}{5} + 5$ no

19. 0 yes, monomial, no defined degree

20. $3y^3 - 4y^2 + 2y + 2$ yes, 3rd degree

PRACTICE ***Perform the operations and simplify.***

21. $(x^3 - 3x^2) + (5x^3 - 8x)$ $6x^3 - 3x^2 - 8x$

22. $(2x^4 - 5x^3) + (7x^3 - x^4 + 2x)$ $x^4 + 2x^3 + 2x$

23. $(y^5 + 2y^3 + 7) - (y^5 - 2y^3 - 7)$ $4y^3 + 14$

24. $(3t^7 - 7t^3 + 3) - (7t^7 - 3t^3 + 7)$ $-4t^7 - 4t^3 - 4$

25. $2(x^2 + 3x - 1) - 3(x^2 + 2x - 4) + 4$ $-x^2 + 14$

26. $5(x^3 - 8x + 3) + 2(3x^2 + 5x) - 7$
$5x^3 + 6x^2 - 30x + 8$

27. $8(t^2 - 2t + 5) + 4(t^2 - 3t + 2) - 6(2t^2 - 8)$
$-28t + 96$

28. $-3(x^3 - x) + 2(x^2 + x) + 3(x^3 - 2x)$ $2x^2 - x$

29. $y(y^2 - 1) - y^2(y + 2) - y(2y - 2)$ $-4y^2 + y$

30. $-4a^2(a + 1) + 3a(a^2 - 4) - a^2(a + 2)$
$-2a^3 - 6a^2 - 12a$

31. $xy(x - 4y) - y(x^2 + 3xy) + xy(2x + 3y)$
$2x^2y - 4xy^2$

32. $3mn(m + 2n) - 6m(3mn + 1) - 2n(4mn - 1)$
$-15m^2n - 2mn^2 - 6m + 2n$

33. $2x^2y^3(4xy^4)$ $8x^3y^7$

34. $-15a^3b(-2a^2b^3)$ $30a^5b^4$

35. $-3m^2n(2mn^2)\left(-\frac{mn}{12}\right)$ $\frac{m^4n^4}{2}$

36. $-\frac{3r^2s^3}{5}\left(\frac{2r^2s}{3}\right)\left(\frac{15rs^2}{2}\right)$ $-3r^5s^6$

37. $-4rs(r^2 + s^2)$ $-4r^3s - 4rs^3$

38. $6u^2v(2uv^3 - y)$ $12u^3v^4 - 6u^2vy$

39. $6ab^2c(2ac + 3bc^2 - 4ab^2c)$
$12a^2b^2c^2 + 18ab^3c^3 - 24a^2b^4c^2$

40. $-\frac{mn^2}{2}(4mn - 6m^2 - 8)$ $-2m^2n^3 + 3m^3n^2 + 4mn^2$

41. $(a + 2)(a + 2)$
$a^2 + 4a + 4$

42. $(y - 5)(y - 5)$
$y^2 - 10y + 25$

43. $(a - 6)^2$
$a^2 - 12a + 36$

44. $(t + 9)^2$
$t^2 + 18t + 81$

45. $(x + 4)(x - 4)$
$x^2 - 16$

46. $(z + 7)(z - 7)$
$z^2 - 49$

47. $(x - 3)(x + 5)$
$x^2 + 2x - 15$

48. $(z + 4)(z - 6)$
$z^2 - 2z - 24$

49. $(u + 2)(3u - 2)$
$3u^2 + 4u - 4$

50. $(4x + 1)(2x - 3)$
$8x^2 - 10x - 3$

51. $(5x - 1)(2x + 3)$
$10x^2 + 13x - 3$

52. $(4x - 1)(2x - 7)$
$8x^2 - 30x + 7$

53. $(3a - 2b)^2$
$9a^2 - 12ab + 4b^2$

54. $(4a + 5b)(4a - 5b)$
$16a^2 - 25b^2$

55. $(3m + 4n)(3m - 4n)$
$9m^2 - 16n^2$

56. $(4r + 3s)^2$
$16r^2 + 24rs + 9s^2$

57. $(2y - 4x)(3y - 2x)$
$6y^2 - 16xy + 8x^2$

58. $(-2x + 3y)(3x + y)$
$-6x^2 + 7xy + 3y^2$

59. $(9x - y)(x^2 - 3y)$
$9x^3 - 27xy - yx^2 + 3y^2$

60. $(8a^2 + b)(a + 2b)$
$8a^3 + 16a^2b + ab + 2b^2$

61. $(5z + 2t)(z^2 - t)$
$5z^3 - 5tz + 2tz^2 - 2t^2$

62. $(y - 2x^2)(x^2 + 3y)$
$-2x^4 - 5x^2y + 3y^2$

63. $(3x - 1)^3$
$27x^3 - 27x^2 + 9x - 1$

64. $(2x - 3)^3$
$8x^3 - 36x^2 + 54x - 27$

65. $(3x + 1)(2x^2 + 4x - 3)$ $6x^3 + 14x^2 - 5x - 3$

66. $(2x - 5)(x^2 - 3x + 2)$ $2x^3 - 11x^2 + 19x - 10$

67. $(3x + 2y)(2x^2 - 3xy + 4y^2)$
$6x^3 - 5x^2y + 6xy^2 + 8y^3$

68. $(4r - 3s)(2r^2 + 4rs - 2s^2)$
$8r^3 + 10r^2s - 20rs^2 + 6s^3$

Multiply the expressions as you would multiply polynomials.

69. $2y^n(3y^n + y^{-n})$ $6y^{2n} + 2$

70. $3a^{-n}(2a^n + 3a^{n-1})$ $6 + \frac{9}{a}$

71. $-5x^{2n}y^n(2x^{2n}y^{-n} + 3x^{-2n}y^n)$ $-10x^{4n} - 15y^{2n}$

72. $-2a^{3n}b^{2n}(5a^{-3n}b - ab^{-2n})$ $-10b^{2n+1} + 2a^{3n+1}$

73. $(x^n + 3)(x^n - 4)$ $x^{2n} - x^n - 12$

74. $(a^n - 5)(a^n - 3)$ $a^{2n} - 8a^n + 15$

75. $(2r^n - 7)(3r^n - 2)$ $6r^{2n} - 25r^n + 14$

76. $(4z^n + 3)(3z^n + 1)$ $12z^{2n} + 13z^n + 3$

77. $x^{1/2}(x^{1/2}y + xy^{1/2})$ $xy + x^{3/2}y^{1/2}$

78. $ab^{1/2}(a^{1/2}b^{1/2} + b^{1/2})$ $a^{3/2}b + ab$

79. $(a^{1/2} + b^{1/2})(a^{1/2} - b^{1/2})$ $a - b$

80. $(x^{3/2} + y^{1/2})^2$ $x^3 + 2x^{3/2}y^{1/2} + y$

Rationalize each denominator.

81. $\dfrac{2}{\sqrt{3} - 1}$ $\sqrt{3} + 1$

82. $\dfrac{1}{\sqrt{5} + 2}$ $\sqrt{5} - 2$

83. $\dfrac{3x}{\sqrt{7} + 2}$ $x(\sqrt{7} - 2)$

84. $\dfrac{14y}{\sqrt{2} - 3}$ $-2y(\sqrt{2} + 3)$

85. $\dfrac{x}{x - \sqrt{3}}$ $\dfrac{x(x + \sqrt{3})}{x^2 - 3}$

86. $\dfrac{y}{2y + \sqrt{7}}$ $\dfrac{y(2y - \sqrt{7})}{4y^2 - 7}$

87. $\dfrac{y + \sqrt{2}}{y - \sqrt{2}}$ $\dfrac{(y + \sqrt{2})^2}{y^2 - 2}$

88. $\dfrac{x - \sqrt{3}}{x + \sqrt{3}}$ $\dfrac{x^2 - 2x\sqrt{3} + 3}{x^2 - 3}$

89. $\dfrac{\sqrt{2} - \sqrt{3}}{1 - \sqrt{3}}$ $\dfrac{\sqrt{3} + 3 - \sqrt{2} - \sqrt{6}}{2}$

90. $\dfrac{\sqrt{3} - \sqrt{2}}{1 + \sqrt{2}}$ $\sqrt{6} + \sqrt{2} - \sqrt{3} - 2$

91. $\dfrac{\sqrt{x} - \sqrt{y}}{\sqrt{x} + \sqrt{y}}$ $\dfrac{x - 2\sqrt{xy} + y}{x - y}$

92. $\dfrac{\sqrt{2x} + y}{\sqrt{2x} - y}$ $\dfrac{2x + 2y\sqrt{2x} + y^2}{2x - y^2}$

Rationalize each numerator.

93. $\dfrac{\sqrt{2} + 1}{2}$ $\dfrac{1}{2(\sqrt{2} - 1)}$

94. $\dfrac{\sqrt{x} - 3}{3}$ $\dfrac{x - 9}{3(\sqrt{x} + 3)}$

95. $\dfrac{y - \sqrt{3}}{y + \sqrt{3}}$ $\dfrac{y^2 - 3}{y^2 + 2y\sqrt{3} + 3}$

96. $\dfrac{\sqrt{a} - \sqrt{b}}{\sqrt{a} + \sqrt{b}}$ $\dfrac{a - b}{a + 2\sqrt{ab} + b}$

97. $\dfrac{\sqrt{x + 3} - \sqrt{x}}{3}$ $\dfrac{1}{\sqrt{x + 3} + \sqrt{x}}$

98. $\dfrac{\sqrt{2 + h} - \sqrt{2}}{h}$ $\dfrac{1}{\sqrt{2 + h} + \sqrt{2}}$

Perform each division and write all answers without using negative exponents.

99. $\dfrac{36a^2b^3}{18ab^6}$ $\dfrac{2a}{b^3}$

100. $\dfrac{-45r^2s^5t^3}{27r^6s^2t^8}$ $\dfrac{-5s^3}{3r^4t^5}$

101. $\dfrac{16x^6y^4z^9}{-24x^9y^6z^0}$ $-\dfrac{2z^9}{3x^3y^2}$

102. $\dfrac{32m^6n^4p^2}{26m^6n^7p^2}$ $\dfrac{16}{13n^3}$

103. $\dfrac{5x^3y^2 + 15x^3y^4}{10x^2y^3}$ $\dfrac{x}{2y} + \dfrac{3xy}{2}$

104. $\dfrac{9m^4n^9 - 6m^3n^4}{12m^3n^3}$ $\dfrac{3mn^6}{4} - \dfrac{n}{2}$

105. $\dfrac{24x^5y^7 - 36x^2y^5 + 12xy}{60x^5y^4}$ $\dfrac{2y^3}{5} - \dfrac{3y}{5x^3} + \dfrac{1}{5x^4y^3}$

106. $\dfrac{9a^3b^4 + 27a^2b^4 - 18a^2b^3}{18a^2b^7}$ $\dfrac{a}{2b^3} + \dfrac{3}{2b^3} - \dfrac{1}{b^4}$

Perform each division. If there is a nonzero remainder, write the answer in quotient + $\frac{\text{remainder}}{\text{divisor}}$ form.

107. $x + 3\overline{)3x^2 + 11x + 6}$ $3x + 2$

108. $3x + 2\overline{)3x^2 + 11x + 6}$ $x + 3$

109. $2x - 5\overline{)2x^2 - 19x + 37}$ $x - 7 + \dfrac{2}{2x - 5}$

110. $x - 7\overline{)2x^2 - 19x + 35}$ $2x - 5$

111. $x^2 + x - 1\overline{)x^3 - 2x^2 - 4x + 3}$ $x - 3$

112. $x^2 - 3\overline{)x^3 - 2x^2 - 4x + 5}$ $x - 2 + \dfrac{-x - 1}{x^2 - 3}$

113. $\dfrac{x^5 - 2x^3 - 3x^2 + 9}{x^3 - 2}$ $x^2 - 2 + \dfrac{-x^2 + 5}{x^3 - 2}$

114. $\dfrac{x^5 - 2x^3 - 3x^2 + 9}{x^3 - 3}$ $x^2 - 2 + \dfrac{3}{x^3 - 3}$

115. $\dfrac{x^5 - 32}{x - 2}$ $x^4 + 2x^3 + 4x^2 + 8x + 16$

116. $\dfrac{x^4 - 1}{x + 1}$ $x^3 - x^2 + x - 1$

117. $11x - 10 + 6x^2\overline{)36x^4 - 121x^2 + 120 + 72x^3 - 142x}$ $6x^2 + x - 12$

118. $x + 6x^2 - 12\overline{)-121x^2 + 72x^3 - 142x + 120 + 36x^4}$ $6x^2 + 11x - 10$

APPLICATIONS

119. Geometry Find an expression that represents the area of the brick wall. **$(x^2 + 3x - 10)$ ft^2**

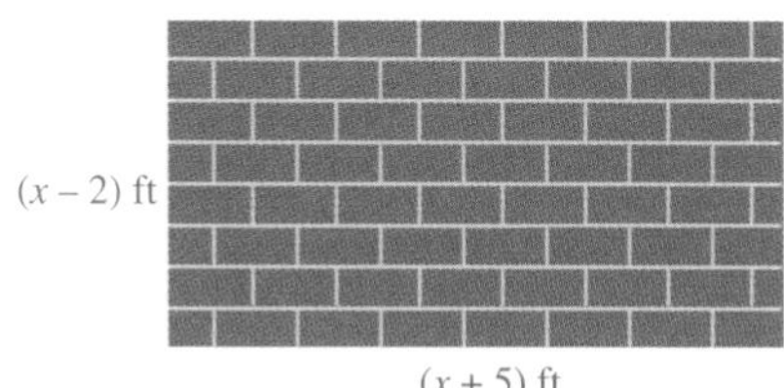

120. Geometry The area of the triangle shown in the illustration is represented as $(x^2 + 3x - 40)$ square feet. Find an expression that represents its height. **$(2x - 10)$ ft**

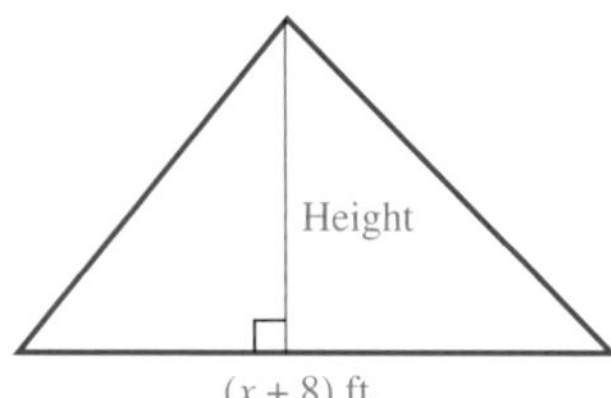

121. Gift Boxes The corners of a 12 in.-by-12 in. piece of cardboard are folded inward and glued to make a box. Write a polynomial that represents the volume of the resulting box shown in the next column. **$(144x - 48x^2 + 4x^3)$ in.^{3}**

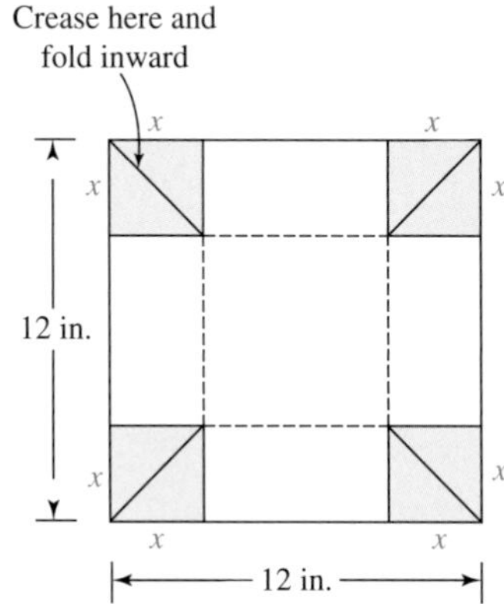

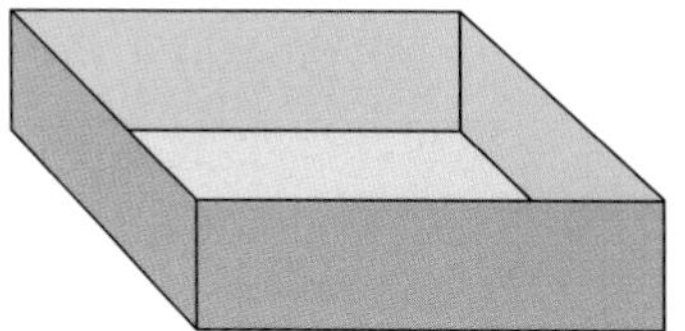

122. Travel Complete the following table, which shows the rate (mph), time traveled (hr), and distance traveled (mi) by a family on vacation.

r	$\cdot$	t	$=$	d
$3x + 4$		$x + 5$		$3x^2 + 19x + 20$

DISCOVERY AND WRITING

123. Show that a trinomial can be squared by using the formula $(a + b + c)^2 = a^2 + b^2 + c^2 + 2ab + 2bc + 2ac$.

124. Show that $(a + b + c + d)^2 = a^2 + b^2 + c^2 + d^2 + 2ab + 2ac + 2ad + 2bc + 2bd + 2cd$.

125. Explain the FOIL method.

126. Explain how to rationalize the numerator of $\frac{\sqrt{x} + 2}{x}$.

127. Explain why $(a + b)^2 \neq a^2 + b^2$.

128. Explain why $\sqrt{a^2 + b^2} \neq \sqrt{a^2} + \sqrt{b^2}$.

REVIEW ***Simplify each expression. Assume that all variables represent positive numbers.***

129. $9^{3/2}$ **27**

130. $\left(\frac{8}{125}\right)^{-2/3}$ **$\frac{25}{4}$**

131. $\left(\frac{625x^4}{16y^8}\right)^{3/4}$ **$\frac{125x^3}{8y^6}$**

132. $\sqrt{80x^4}$ **$4x^2\sqrt{5}$**

133. $\sqrt[3]{16ab^4} - b\sqrt[3]{54ab}$ **$-b\sqrt[3]{2ab}$**

134. $x\sqrt[4]{1,280x} + \sqrt[4]{80x^5}$ **$6x\sqrt[4]{5x}$**

0.5 Factoring Polynomials

In this section, you will learn about

- Factoring Out a Common Monomial
- Factoring by Grouping
- Factoring the Difference of Two Squares
- Factoring Trinomials
- Factoring Trinomials by Grouping
- Factoring the Sum and Difference of Two Cubes
- Miscellaneous Factoring

When two or more polynomials are multiplied together, each polynomial is called a **factor** of the resulting product. For example, the factors of $7(x + 2)(x + 3)$ are

$$7, \qquad x + 2, \qquad \text{and} \qquad x + 3$$

The process of writing a polynomial as the product of several factors is called **factoring.**

In this section, we will discuss factoring where the coefficients of the polynomial and the polynomial factors are integers. If a polynomial cannot be factored by using integers only, we call it a **prime polynomial.**

Factoring Out a Common Monomial

EXAMPLE 1 Factor: $3xy^2 + 6x$.

Solution We note that each term contains a factor of $3x$:

$$3xy^2 + 6x = 3x(y^2) + 3x(2)$$

We can then use the distributive property to factor out the common factor of $3x$:

$$3xy^2 + 6x = 3x(y^2 + 2)$$

Self Check Factor: $4a^2 - 8ab$. ■

EXAMPLE 2 Factor: $x^2y^2z^2 - xyz$.

Solution We factor out the common factor of xyz:

$$\begin{aligned} x^2y^2z^2 - xyz &= xyz(xyz) - xyz(1) \\ &= xyz(xyz - 1) \end{aligned}$$

The last term in the expression $x^2y^2z^2 - xyz$ has an understood coefficient of 1. When the xyz is factored out, the 1 must be written.

Self Check Factor: $a^2b^2c^2 + a^3b^3c^3$. ■

Factoring by Grouping

EXAMPLE 3 Factor: $ax + bx + a + b$.

Solution Although there is no factor common to all four terms, we can factor x out of the first two terms and write the expression as

$$ax + bx + a + b = x(a + b) + (a + b)$$

We can now factor out the common factor of $a + b$.

$$\begin{aligned} ax + bx + a + b &= x(a + b) + (a + b) \\ &= x(a + b) + 1(a + b) \\ &= (a + b)(x + 1) \end{aligned}$$

Self Check Factor: $x^2 + xy + 2x + 2y$. ■

Factoring the Difference of Two Squares

EXAMPLE 4 Factor: $49x^2 - 4$.

Solution We observe that each term is a perfect square:

$$49x^2 - 4 = (7x)^2 - 2^2$$

The difference of the squares of two quantities is the product of two factors. One is the sum of the quantities, and the other is the difference of the quantities. Thus, $49x^2 - 4$ factors as

$$\begin{aligned} 49x^2 - 4 &= (7x)^2 - 2^2 \\ &= (7x + 2)(7x - 2) \end{aligned}$$

Self Check Factor: $9a^2 - 16b^2$. ■

Example 4 suggests a formula for factoring the difference of two squares.

Factoring the Difference of Two Squares

$$x^2 - y^2 = (x + y)(x - y)$$

Comment If you are limited to integer coefficients, the sum of two squares cannot be factored. For example, $x^2 + y^2$ is a prime polynomial.

EXAMPLE 5 Factor: $16m^4 - n^4$.

Solution The binomial $16m^4 - n^4$ can be factored as the difference of two squares:

$$\begin{aligned} 16m^4 - n^4 &= (4m^2)^2 - (n^2)^2 \\ &= (4m^2 + n^2)(4m^2 - n^2) \end{aligned}$$

The first factor is the sum of two squares and is prime. The second factor is a difference of two squares and can be factored:

$$\begin{aligned} 16m^4 - n^4 &= (4m^2 + n^2)[(2m)^2 - n^2] \\ &= (4m^2 + n^2)(2m + n)(2m - n) \end{aligned}$$

Self Check Factor: $a^4 - 81b^4$. ■

EXAMPLE 6 Factor: $18t^2 - 32$.

Solution We begin by factoring out the common monomial factor of 2.

$$18t^2 - 32 = 2(9t^2 - 16)$$

Since $9t^2 - 16$ is the difference of two squares, it can be factored.

$$\begin{aligned} 18t^2 - 32 &= 2(9t^2 - 16) \\ &= 2(3t + 4)(3t - 4) \end{aligned}$$

Self Check Factor: $-3x^2 + 12$. ■

■ Factoring Trinomials

Trinomials that are squares of binomials can be factored by using the following formulas.

Factoring Trinomial Squares

(1) $x^2 + 2xy + y^2 = (x + y)(x + y) = (x + y)^2$
(2) $x^2 - 2xy + y^2 = (x - y)(x - y) = (x - y)^2$

For example, to factor $a^2 - 6a + 9$, we note that it can be written in the form

$$a^2 - 2(3a) + 3^2 \quad x = a \text{ and } y = 3.$$

which matches the left-hand side of Equation 2 above. Thus,

$$\begin{aligned} a^2 - 6a + 9 &= a^2 - 2(3a) + 3^2 \\ &= (a - 3)(a - 3) \\ &= (a - 3)^2 \end{aligned}$$

Factoring trinomials that are not squares of binomials often requires some guesswork. If a trinomial with no common factors is factorable, it will factor into the product of two binomials.

EXAMPLE 7 Factor: $x^2 + 3x - 10$.

Solution To factor $x^2 + 3x - 10$, we must find two binomials $x + a$ and $x + b$ such that

$$x^2 + 3x - 10 = (x + a)(x + b)$$

where the product of a and b is -10 and the sum of a and b is 3.

$$ab = -10 \qquad \text{and} \qquad a + b = 3$$

To find such numbers, we list the possible factorizations of -10:

$$10(-1) \qquad 5(-2) \qquad -10(1) \qquad -5(2)$$

Only in the factorization $5(-2)$ do the factors have a sum of 3. Thus, $a = 5$ and $b = -2$, and

$$x^2 + 3x - 10 = (x + a)(x + b)$$

$$x^2 + 3x - 10 = (x + 5)(x - 2) \tag{3}$$

Because of the commutative property of multiplication, the order of the factors in Equation 3 is not important. Equation 3 can also be written as

$$x^2 + 3x - 10 = (x - 2)(x + 5)$$

Self Check Factor: $p^2 - 5p - 6$. ■

EXAMPLE 8 Factor: $2x^2 - x - 6$.

Solution Since the first term is $2x^2$, the first terms of the binomial factors must be $2x$ and x:

$$2x^2 - x - 6 = (2x + ?)(x + ?)$$

The product of the last terms must be -6, and the sum of the products of the outer terms and the inner terms must be $-x$. The only factorization of -6 that will cause this to happen is $3(-2)$.

$$2x^2 - x - 6 = (2x + 3)(x - 2)$$

Self Check Factor: $6x^2 - x - 2$. ■

It is not easy to give specific rules for factoring trinomials, because some guesswork is often necessary. However, the following hints are helpful.

Strategy for Factoring a General Trinomial with Integer Coefficients

1. Write the trinomial in descending powers of one variable.
2. Factor out any greatest common factor, including -1 if that is necessary to make the coefficient of the first term positive.
3. When the sign of the first term of a trinomial is + and the sign of the third term is +, the sign between the terms of each binomial factor is the same as the sign of the middle term of the trinomial.

   When the sign of the first term is + and the sign of the third term is −, one of the signs between the terms of the binomial factors is + and the other is −.
4. Try various combinations of first terms and last terms until you find one that works. If no possibilities work, the trinomial is prime.
5. Check the factorization by multiplication.

ACTIVE EXAMPLE 9 Factor: $10xy + 24y^2 - 6x^2$.

Solution We write the trinomial in descending powers of x and then factor out the common factor of -2.

$$\begin{aligned} 10xy + 24y^2 - 6x^2 &= -6x^2 + 10xy + 24y^2 \\ &= -2(3x^2 - 5xy - 12y^2) \end{aligned}$$

COLLEGE **Algebra** $f(x)$ **Now**™

Explore this Active Example and test yourself on the steps followed here at **http://1pass.thomson.com**. You can also find this example on the Interactive Video Skillbuilder CD-ROM.

Since the sign of the third term of $3x^2 - 5xy - 12y^2$ is $-$, the signs between the binomial factors will be opposite. Since the first term is $3x^2$, the first terms of the binomial factors must be $3x$ and x:

$$-2(3x^2 - 5xy - 12y^2) = -2(3x \qquad ?)(x \qquad ?)$$

The product of the last terms must be $-12y^2$, and the sum of the outer terms and the inner terms must be $-5xy$. Of the many factorizations of $-12y^2$, only $4y(-3y)$ leads to a middle term of $-5xy$.

$$\begin{aligned} 10xy + 24y^2 - 6x^2 &= -6x^2 + 10xy + 24y^2 \\ &= -2(3x^2 - 5xy - 12y^2) \\ &= -2(3x + 4y)(x - 3y) \end{aligned}$$

Self Check Factor: $-6x^2 - 15xy - 6y^2$. ■

Factoring Trinomials by Grouping

Another method of factoring trinomials involves factoring by grouping. This method can be used to factor trinomials of the form $ax^2 + bx + c$. For example, to factor $6x^2 + 5x - 6$, we proceed as follows:

1. Find the product ac: $6(-6) = -36$. This number is called the **key number.**
2. Find two factors of the key number (-36) whose sum is $b = 5$. Two such numbers are 9 and -4.

   $$9(-4) = -36 \qquad \text{and} \qquad 9 + (-4) = 5$$

3. Use the factors 9 and -4 as coefficients of two terms to be placed between $6x^2$ and -6.

   $$6x^2 + 5x - 6 = 6x^2 + 9x - 4x - 6$$

4. Factor by grouping:

   $$\begin{aligned} 6x^2 + 9x - 4x - 6 &= 3x(2x + 3) - 2(2x + 3) \\ &= (2x + 3)(3x - 2) \qquad \text{Factor out } 2x + 3. \end{aligned}$$

EXAMPLE 10 Factor: $15x^2 + x - 2$.

Solution Since $a = 15$ and $c = -2$ in the trinomial, $ac = -30$. We now find factors of -30 whose sum is $b = 1$. Such factors are 6 and -5. We use these factors as coefficients of two terms to be placed between $15x^2$ and -2.

$$15x^2 + 6x - 5x - 2$$

Finally, we factor by grouping.

$$3x(5x + 2) - (5x + 2) = (5x + 2)(3x - 1)$$

Self Check Factor: $15a^2 + 17a - 4$. ■

We can often factor polynomials with variable exponents. For example, if n is a natural number,

$$a^{2n} - 5a^n - 6 = (a^n + 1)(a^n - 6)$$

because

$$(a^n + 1)(a^n - 6) = a^{2n} - 6a^n + a^n - 6$$
$$= a^{2n} - 5a^n - 6 \qquad \text{Combine like terms.}$$

Factoring the Sum and Difference of Two Cubes

Two other types of factoring involve binomials that are the sum or the difference of two cubes. Like the difference of two squares, they can be factored by using a formula.

Factoring the Sum and Difference of Two Cubes

$$x^3 + y^3 = (x + y)(x^2 - xy + y^2)$$
$$x^3 - y^3 = (x - y)(x^2 + xy + y^2)$$

EXAMPLE 11 Factor: $x^3 - 8$.

Solution This binomial can be written as $x^3 - 2^3$, which is the difference of two cubes. Substituting into the formula for the difference of two cubes gives

$$x^3 - 2^3 = (x - 2)(x^2 + 2x + 2^2)$$
$$= (x - 2)(x^2 + 2x + 4)$$

Self Check Factor: $64 - p^3$. ■

EXAMPLE 12 Factor: $27x^6 + 64y^3$.

Solution We can write this expression as the sum of two cubes and factor it as follows:

$$27x^6 + 64y^3 = (3x^2)^3 + (4y)^3$$
$$= (3x^2 + 4y)[(3x^2)^2 - (3x^2)(4y) + (4y)^2]$$
$$= (3x^2 + 4y)(9x^4 - 12x^2y + 16y^2)$$

Self Check Factor: $8a^3 + 1{,}000b^6$. ■

Miscellaneous Factoring

ACTIVE EXAMPLE 13 Factor: $x^2 - y^2 + 6x + 9$.

Solution Here we will factor a trinomial and a difference of two squares.

$$x^2 - y^2 + 6x + 9 = x^2 + 6x + 9 - y^2 \qquad \text{Use the commutative property to rearrange terms.}$$
$$= (x + 3)^2 - y^2 \qquad \text{Factor } x^2 + 6x + 9.$$
$$= (x + 3 + y)(x + 3 - y) \qquad \text{Factor the difference of two squares.}$$

COLLEGE Algebra f(x) Now™
Go to http://1pass.thomson.com or your CD to practice this example.

We could try to factor this expression in another way.

$$x^2 - y^2 + 6x + 9 = (x + y)(x - y) + 3(2x + 3) \qquad \text{Factor } x^2 - y^2 \text{ and } 6x + 9.$$

However, we are unable to finish the factorization. If grouping in one way doesn't work, try various other ways.

Self Check Factor: $a^2 + 8a - b^2 + 16$. ■

ACTIVE EXAMPLE 14 Factor: $z^4 - 3z^2 + 1$.

COLLEGE **Algebra** $f(x)$ **Now**™
Go to **http://1pass.thomson.com** or your CD to practice this example.

Solution This trinomial cannot be factored as the product of two binomials, because no combination will give a middle term of $-3z^2$. However, if the middle term were $-2z^2$, the trinomial would be a perfect square, and the factorization would be easy:

$$\begin{aligned} z^4 - 2z^2 + 1 &= (z^2 - 1)(z^2 - 1) \\ &= (z^2 - 1)^2 \end{aligned}$$

We can change the middle term in $z^4 - 3z^2 + 1$ to $-2z^2$ by adding z^2 to it. To ensure that adding z^2 does not change the value of the trinomial, however, we must also subtract z^2. We can then proceed as follows.

$$\begin{aligned} z^4 - 3z^2 + 1 &= z^4 - 3z^2 + z^2 + 1 - z^2 && \text{Add and subtract } z^2. \\ &= z^4 - 2z^2 + 1 - z^2 && \text{Combine } -3z^2 \text{ and } z^2. \\ &= (z^2 - 1)^2 - z^2 && \text{Factor } z^4 - 2z^2 + 1. \\ &= (z^2 - 1 + z)(z^2 - 1 - z) && \text{Factor the difference of two squares.} \end{aligned}$$

In this type of problem, we will always try to add and subtract a perfect square in the hope of making a perfect square trinomial that will lead to factoring a difference of two squares.

Self Check Factor: $x^4 + 3x^2 + 4$. ■

It is helpful to identify the problem type when we must factor polynomials that are given in random order.

Identifying Factoring Problem Types

1. Factor out all common monomial factors.
2. If an expression has two terms, check whether the problem type is
   a. **The difference of two squares:** $x^2 - y^2 = (x + y)(x - y)$
   b. **The sum of two cubes:** $x^3 + y^3 = (x + y)(x^2 - xy + y^2)$
   c. **The difference of two cubes:** $x^3 - y^3 = (x - y)(x^2 + xy + y^2)$
3. If an expression has three terms, attempt to factor it as a **trinomial.**
4. If an expression has four or more terms, try factoring by **grouping.**
5. Continue until each individual factor is prime.
6. Check the results by multiplying.

Self Check Answers

1. $4a(a - 2b)$ **2.** $a^2b^2c^2(1 + abc)$ **3.** $(x + 2)(x + y)$ **4.** $(3a + 4b)(3a - 4b)$ **5.** $(a^2 + 9b^2)(a + 3b)(a - 3b)$ **6.** $-3(x + 2)(x - 2)$ **7.** $(p - 6)(p + 1)$ **8.** $(3x - 2)(2x + 1)$ **9.** $-3(x + 2y)(2x + y)$ **10.** $(3a + 4)(5a - 1)$ **11.** $(4 - p)(16 + 4p + p^2)$ **12.** $8(a + 5b^2)(a^2 - 5ab^2 + 25b^4)$ **13.** $(a + 4 + b)(a + 4 - b)$ **14.** $(x^2 + 2 + x)(x^2 + 2 - x)$

0.5 Exercises

VOCABULARY AND CONCEPTS ***Fill in the blanks.***

1. When polynomials are multiplied together, each polynomial is a factor of the product.
2. If a polynomial cannot be factored using integer coefficients, it is called a prime polynomial.

Complete each factoring formula.

3. $ax + bx = x(a + b)$.
4. $x^2 - y^2 = (x + y)(x - y)$.
5. $x^2 + 2xy + y^2 = (x + y)(x + y)$.
6. $x^2 - 2xy + y^2 = (x - y)(x - y)$.
7. $x^3 + y^3 = (x + y)(x^2 - xy + y^2)$.
8. $x^3 - y^3 = (x - y)(x^2 + xy + y^2)$.

PRACTICE ***Factor each expression completely. If an expression is prime, so indicate.***

9. $3x - 6$ $3(x - 2)$
10. $5y - 15$ $5(y - 3)$
11. $8x^2 + 4x^3$ $4x^2(2 + x)$
12. $9y^3 + 6y^2$ $3y^2(3y + 2)$
13. $7x^2y^2 + 14x^3y^2$ $7x^2y^2(1 + 2x)$
14. $25y^2z - 15yz^2$ $5yz(5y - 3z)$
15. $3a^2bc + 6ab^2c + 9abc^2$ $3abc(a + 2b + 3c)$
16. $5x^3y^3z^3 + 25x^2y^2z^2 - 125xyz$ $5xyz(x^2y^2z^2 + 5xyz - 25)$
17. $a(x + y) + b(x + y)$ $(x + y)(a + b)$
18. $b(x - y) + a(x - y)$ $(x - y)(b + a)$
19. $4a + b - 12a^2 - 3ab$ $(4a + b)(1 - 3a)$
20. $x^2 + 4x + xy + 4y$ $(x + 4)(x + y)$
21. $3x^3 + 3x^2 - x - 1$ $(x + 1)(3x^2 - 1)$
22. $4x + 6xy - 9y - 6$ $(3y + 2)(2x - 3)$
23. $2txy + 2ctx - 3ty - 3ct$ $t(y + c)(2x - 3)$
24. $2ax + 4ay - bx - 2by$ $(x + 2y)(2a - b)$
25. $ax + bx + ay + by + az + bz$ $(a + b)(x + y + z)$
26. $6x^2y^3 + 18xy + 3x^2y^2 + 9x$ $3x(xy^2 + 3)(2y + 1)$
27. $4x^2 - 9$ $(2x + 3)(2x - 3)$
28. $36z^2 - 49$ $(6z + 7)(6z - 7)$
29. $4 - 9r^2$ $(2 + 3r)(2 - 3r)$
30. $16 - 49x^2$ $(4 + 7x)(4 - 7x)$
31. $(x + z)^2 - 25$ $(x + z + 5)(x + z - 5)$
32. $(x - y)^2 - 9$ $(x - y + 3)(x - y - 3)$
33. $25x^4 + 1$ prime
34. $36t^4 + 121$ prime
35. $x^2 - (y - z)^2$ $(x + y - z)(x - y + z)$
36. $z^2 - (y + 3)^2$ $(z + y + 3)(z - y - 3)$
37. $(x - y)^2 - (x + y)^2$ $-4xy$
38. $(2a + 3)^2 - (2a - 3)^2$ $24a$
39. $x^4 - y^4$ $(x^2 + y^2)(x + y)(x - y)$
40. $z^4 - 81$ $(z^2 + 9)(z + 3)(z - 3)$
41. $3x^2 - 12$ $3(x + 2)(x - 2)$
42. $3x^3y - 3xy$ $3xy(x + 1)(x - 1)$
43. $18xy^2 - 8x$ $2x(3y + 2)(3y - 2)$
44. $27x^2 - 12$ $3(3x + 2)(3x - 2)$
45. $x^2 + 8x + 16$ $(x + 4)^2$
46. $a^2 - 12a + 36$ $(a - 6)^2$
47. $b^2 - 10b + 25$ $(b - 5)^2$
48. $y^2 + 14y + 49$ $(y + 7)^2$
49. $m^2 + 4mn + 4n^2$ $(m + 2n)^2$
50. $r^2 - 8rs + 16s^2$ $(r - 4s)^2$
51. $x^2 + 10x + 21$ $(x + 7)(x + 3)$
52. $x^2 + 7x + 10$ $(x + 5)(x + 2)$
53. $x^2 - 4x - 12$ $(x - 6)(x + 2)$
54. $x^2 - 2x - 63$ $(x - 9)(x + 7)$
55. $x^2 - 2x + 15$ prime
56. $x^2 + x + 2$ prime
57. $12x^2 - xy - 6y^2$ $(4x - 3y)(3x + 2y)$
58. $8x^2 - 10xy - 3y^2$ $(4x + y)(2x - 3y)$
59. $-15 + 2a + 24a^2$ $(6a + 5)(4a - 3)$
60. $-32 - 68x + 9x^2$ $(9x + 4)(x - 8)$
61. $6x^2 + 29xy + 35y^2$ $(3x + 7y)(2x + 5y)$
62. $10x^2 - 17xy + 6y^2$ $(5x - 6y)(2x - y)$
63. $12p^2 - 58pq - 70q^2$ $2(6p - 35q)(p + q)$
64. $3x^2 - 6xy - 9y^2$ $3(x - 3y)(x + y)$
65. $-6m^2 + 47mn - 35n^2$ $-(6m - 5n)(m - 7n)$
66. $-14r^2 - 11rs + 15s^2$ $-(7r - 5s)(2r + 3s)$
67. $-6x^3 + 23x^2 + 35x$ $-x(6x + 7)(x - 5)$
68. $-y^3 - y^2 + 90y$ $-y(y + 10)(y - 9)$
69. $6x^4 - 11x^3 - 35x^2$ $x^2(2x - 7)(3x + 5)$
70. $12x + 17x^2 - 7x^3$ $-x(x - 3)(7x + 4)$
71. $x^4 + 2x^2 - 15$ $(x^2 + 5)(x^2 - 3)$
72. $x^4 - x^2 - 6$ $(x^2 - 3)(x^2 + 2)$
73. $a^{2n} - 2a^n - 3$ $(a^n - 3)(a^n + 1)$
74. $a^{2n} + 6a^n + 8$ $(a^n + 4)(a^n + 2)$
75. $6x^{2n} - 7x^n + 2$ $(3x^n - 2)(2x^n - 1)$
76. $9x^{2n} + 9x^n + 2$ $(3x^n + 2)(3x^n + 1)$
77. $4x^{2n} - 9y^{2n}$ $(2x^n + 3y^n)(2x^n - 3y^n)$
78. $8x^{2n} - 2x^n - 3$ $(4x^n - 3)(2x^n + 1)$
79. $10y^{2n} - 11y^n - 6$ $(5y^n + 2)(2y^n - 3)$
80. $16y^{4n} - 25y^{2n}$ $y^{2n}(4y^n + 5)(4y^n - 5)$
81. $8z^3 - 27$ $(2z - 3)(4z^2 + 6z + 9)$
82. $125a^3 - 64$ $(5a - 4)(25a^2 + 20a + 16)$

83. $2x^3 + 2{,}000$
$2(x + 10)(x^2 - 10x + 100)$

84. $3y^3 + 648$
$3(y + 6)(y^2 - 6y + 36)$

85. $(x + y)^3 - 64$
$(x + y - 4)(x^2 + 2xy + y^2 + 4x + 4y + 16)$

86. $(x - y)^3 + 27$
$(x - y + 3)(x^2 - 2xy + y^2 - 3x + 3y + 9)$

87. $64a^6 - y^6$
$(2a + y)(2a - y)(4a^2 - 2ay + y^2)(4a^2 + 2ay + y^2)$

88. $a^6 + b^6$ $(a^2 + b^2)(a^4 - a^2b^2 + b^4)$

89. $a^3 - b^3 + a - b$ $(a - b)(a^2 + ab + b^2 + 1)$

90. $(a^2 - y^2) - 5(a + y)$ $(a + y)(a - y - 5)$

91. $64x^6 + y^6$ $(4x^2 + y^2)(16x^4 - 4x^2y^2 + y^4)$

92. $z^2 + 6z + 9 - 225y^2$ $(z + 3 + 15y)(z + 3 - 15y)$

93. $x^2 - 6x + 9 - 144y^2$ $(x - 3 + 12y)(x - 3 - 12y)$

94. $x^2 + 2x - 9y^2 + 1$ $(x + 1 + 3y)(x + 1 - 3y)$

95. $(a + b)^2 - 3(a + b) - 10$ $(a + b - 5)(a + b + 2)$

96. $2(a + b)^2 - 5(a + b) - 3$
$(2a + 2b + 1)(a + b - 3)$

97. $x^6 + 7x^3 - 8$
$(x + 2)(x^2 - 2x + 4)(x - 1)(x^2 + x + 1)$

98. $x^6 - 13x^4 + 36x^2$ $x^2(x + 3)(x - 3)(x + 2)(x - 2)$

99. $x^4 + 3x^2 + 4$ $(x^2 + x + 2)(x^2 - x + 2)$

100. $x^4 + x^2 + 1$ $(x^2 + 1 + x)(x^2 + 1 - x)$

101. $x^4 + 7x^2 + 16$ $(x^2 + x + 4)(x^2 - x + 4)$

102. $y^4 + 2y^2 + 9$ $(y^2 + 3 + 2y)(y^2 + 3 - 2y)$

103. $4a^4 + 1 + 3a^2$ $(2a^2 + a + 1)(2a^2 - a + 1)$

104. $x^4 + 25 + 6x^2$ $(x^2 + 5 + 2x)(x^2 + 5 - 2x)$

APPLICATIONS

105. Candy To find the amount of chocolate used in the outer coating of the malted-milk ball shown, we can find the volume V of the chocolate shell

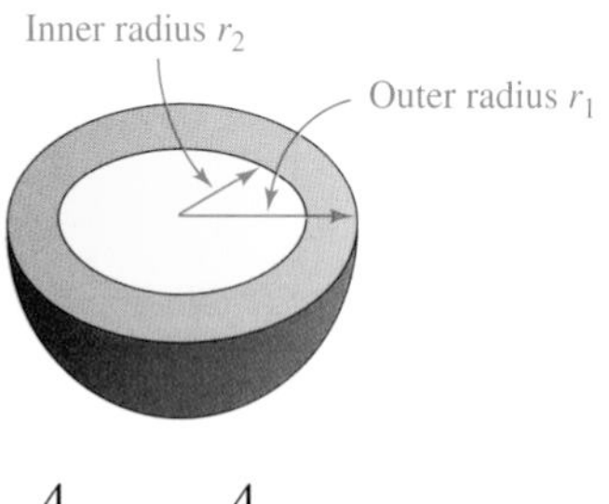

using the formula $V = \frac{4}{3}\pi r_1^{\,3} - \frac{4}{3}\pi r_2^{\,3}$. Factor the expression on the right-hand side of the formula.
$\frac{4}{3}\pi(r_1 - r_2)(r_1^2 + r_1r_2 + r_2^2)$

106. Movie Stunts The formula that gives the distance a stuntwoman is above the ground t seconds after she falls over the side of a 144-foot tall building is $f = 144 - 16t^2$. Factor the right-hand side.
$16(3 - t)(3 + t)$

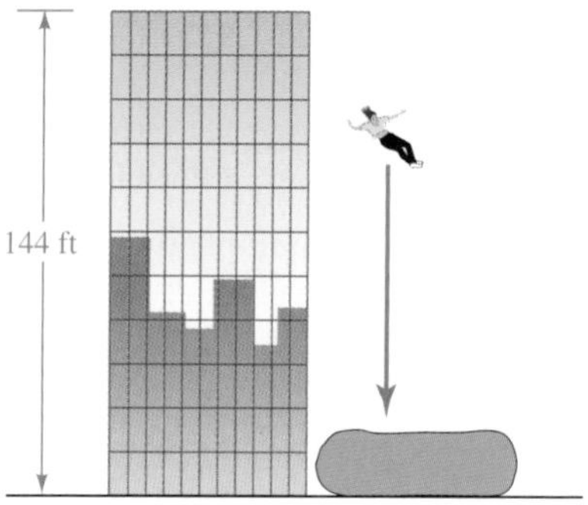

DISCOVERY AND WRITING

107. Explain how to factor the difference of two squares.

108. Explain how to factor the difference of two cubes.

109. Explain how to factor $a^2 - b^2 + a + b$.

110. Explain how to factor $x^2 + 2x + 1$.

Factor the indicated monomial from the given expression.

111. $3x + 2; 2$ $2\left(\frac{3}{2}x + 1\right)$

112. $5x - 3; 5$ $5\left(x - \frac{3}{5}\right)$

113. $x^2 + 2x + 4; 2$ $2\left(\frac{1}{2}x^2 + x + 2\right)$

114. $3x^2 - 2x - 5; 3$ $3\left(x^2 - \frac{2}{3}x - \frac{5}{3}\right)$

115. $a + b; a$ $a\left(1 + \frac{b}{a}\right)$

116. $a - b; b$ $b\left(\frac{a}{b} - 1\right)$

117. $x + x^{1/2}; x^{1/2}$ $x^{1/2}(x^{1/2} + 1)$

118. $x^{3/2} - x^{1/2}; x^{1/2}$ $x^{1/2}(x - 1)$

119. $2x + \sqrt{2}y; \sqrt{2}$ $\sqrt{2}(\sqrt{2}x + y)$

120. $\sqrt{3}a - 3b; \sqrt{3}$ $\sqrt{3}(a - \sqrt{3}b)$

121. $ab^{3/2} - a^{3/2}b; ab$ $ab(b^{1/2} - a^{1/2})$

122. $ab^2 + b; b^{-1}$ $b^{-1}(ab^3 + b^2)$

Factor each expression by grouping three terms and two terms.

123. $x^2 + x - 6 + xy - 2y$ $(x - 2)(x + 3 + y)$

124. $2x^2 + 5x + 2 - xy - 2y$ $(x + 2)(2x + 1 - y)$

125. $a^4 + 2a^3 + a^2 + a + 1$ $(a + 1)(a^3 + a^2 + 1)$

126. $a^4 + a^3 - 2a^2 + a - 1$ $(a - 1)(a^3 + 2a^2 + 1)$

REVIEW

127. Which natural number is neither prime nor composite? 1

128. Graph the interval $[-2, 3)$.

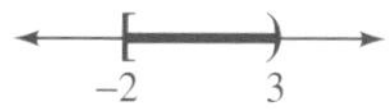

129. Simplify: $(x^3x^2)^4$ x^{20}

130. Simplify: $\dfrac{(a^3)^3(a^2)^4}{(a^2a^3)^3}$. a^2

131. $\left(\dfrac{3x^4x^3}{6x^{-2}x^4}\right)^0$ 1

132. Simplify: $\sqrt{20x^5}$. $2x^2\sqrt{5x}$

133. Simplify: $\sqrt{20x} - \sqrt{125x}$. $-3\sqrt{5x}$

134. Rationalize the denominator: $\dfrac{3}{\sqrt[3]{3}}$ $\sqrt[3]{9}$

0.6 Algebraic Fractions

In this section, you will learn about

- Algebraic Fractions
- Simplifying Fractions
- Multiplying and Dividing Fractions
- Adding and Subtracting Fractions
- Complex Fractions

If x and y are real numbers, the quotient $\frac{x}{y}$ $(y \neq 0)$ is called a **fraction.** The number x is called the **numerator,** and y is called the **denominator.**

Algebraic Fractions

Algebraic fractions are quotients of algebraic expressions. If the expressions are polynomials, the fraction is called a **rational expression.** The first two of the following algebraic fractions are rational expressions. The third is not, because the numerator and denominator are not polynomials.

$$\frac{5y^2 + 2y}{y^2 - 3y - 7} \qquad \frac{8ab^2 - 16c^3}{2x + 3} \qquad \frac{x^{1/2} + 4x}{x^{3/2} - x^{1/2}}$$

Comment Remember that the denominator of a fraction cannot be zero.

We summarize some of the properties of fractions as follows.

If a, b, c, and d are real numbers and no denominators are 0, then

Equality of Fractions

$$\frac{a}{b} = \frac{c}{d} \quad \text{if and only if} \quad ad = bc$$

Fundamental Property of Fractions

$$\frac{ax}{bx} = \frac{a}{b}$$

Multiplication and Division of Fractions

$$\frac{a}{b} \cdot \frac{c}{d} = \frac{ac}{bd} \quad \text{and} \quad \frac{a}{b} \div \frac{c}{d} = \frac{a}{b} \cdot \frac{d}{c} = \frac{ad}{bc}$$

Addition and Subtraction of Fractions

$$\frac{a}{b} + \frac{c}{b} = \frac{a + c}{b} \quad \text{and} \quad \frac{a}{b} - \frac{c}{b} = \frac{a - c}{b}$$

The first two examples illustrate each of the previous properties of fractions.

EXAMPLE 1 Assume that no denominators are 0.

a. $\frac{2a}{3} = \frac{4a}{6}$ Because $2a(6) = 3(4a)$.

b. $\frac{6xy}{10xy} = \frac{3(\cancel{2xy})}{5(\cancel{2xy})}$ Factor the numerator and denominator and divide out the common factors.

$= \frac{3}{5}$

Self Check **a.** Is $\frac{3y}{5} = \frac{15z}{25}$? **b.** Simplify: $\frac{15a^2b}{25ab^2}$. ■

EXAMPLE 2 Assume that no denominators are 0.

a. $\frac{2r}{7s} \cdot \frac{3r}{5s} = \frac{2r \cdot 3r}{7s \cdot 5s}$

$= \frac{6r^2}{35s^2}$

b. $\frac{3mn}{4pq} \div \frac{2pq}{7mn} = \frac{3mn}{4pq} \cdot \frac{7mn}{2pq}$

$= \frac{21m^2n^2}{8p^2q^2}$

c. $\frac{2ab}{5xy} + \frac{ab}{5xy} = \frac{2ab + ab}{5xy}$

$= \frac{3ab}{5xy}$

d. $\frac{6uv^2}{7w^2} - \frac{3uv^2}{7w^2} = \frac{6uv^2 - 3uv^2}{7w^2}$

$= \frac{3uv^2}{7w^2}$

Self Check Perform each operation: **a.** $\frac{3a}{5b} \cdot \frac{2a}{7b}$, **b.** $\frac{2ab}{3rs} \div \frac{2rs}{4ab}$, **c.** $\frac{5pq}{3t} + \frac{3pq}{3t}$, and **d.** $\frac{5mn^2}{3w} - \frac{mn^2}{3w}$. ■

To add or subtract fractions with unlike denominators, we write each fraction as an equivalent fraction with a common denominator. We can then add or subtract the fractions. For example,

$$\begin{aligned} \frac{3x}{5} + \frac{2x}{7} &= \frac{3x(7)}{5(7)} + \frac{2x(5)}{7(5)} \\ &= \frac{21x}{35} + \frac{10x}{35} \\ &= \frac{21x + 10x}{35} \\ &= \frac{31x}{35} \end{aligned}$$

$$\begin{aligned} \frac{4a^2}{15} - \frac{3a^2}{10} &= \frac{4a^2(2)}{15(2)} - \frac{3a^2(3)}{10(3)} \\ &= \frac{8a^2}{30} - \frac{9a^2}{30} \\ &= \frac{8a^2 - 9a^2}{30} \\ &= \frac{-a^2}{30} \\ &= -\frac{a^2}{30} \end{aligned}$$

A fraction is **in lowest terms** if all factors common to the numerator and the denominator have been removed. To **simplify a fraction** means to write it in lowest terms.

Simplifying Fractions

To simplify a fraction, we use the fundamental property of fractions. This enables us to divide out all factors that are common to the numerator and the denominator.

EXAMPLE 3 Simplify: $\dfrac{x^2 - 9}{x^2 - 3x}$ $(x \neq 0, 3)$.

Solution We factor the difference of two squares in the numerator, factor out x in the denominator, and divide out the common factor of $x - 3$.

$$\frac{x^2 - 9}{x^2 - 3x} = \frac{(x + 3)\cancel{(x - 3)}}{x\cancel{(x - 3)}}$$
$$= \frac{x + 3}{x}$$

Self Check Simplify: $\dfrac{a^2 - 4a}{a^2 - a - 12}$ $(a \neq 4, -3)$. ■

We will encounter the following properties of fractions in the next examples.

If a and b represent real numbers and there are no divisions by 0, then

$$\frac{a}{1} = a \qquad \frac{a}{a} = 1$$

$$\frac{a}{b} = \frac{-a}{-b} = -\frac{a}{-b} = -\frac{-a}{b} \qquad -\frac{a}{b} = \frac{a}{-b} = \frac{-a}{b} = -\frac{-a}{-b}$$

EXAMPLE 4 Simplify: $\dfrac{x^2 - 2xy + y^2}{y - x}$ $(x \neq y)$.

Solution We factor the trinomial in the numerator, factor -1 from the denominator, and divide out the common factor of $x - y$.

$$\frac{x^2 - 2xy + y^2}{y - x} = \frac{(x - y)\cancel{(x - y)}}{-\cancel{(x - y)}}$$
$$= \frac{x - y}{-1}$$
$$= -\frac{x - y}{1}$$
$$= -(x - y)$$

Self Check Simplify: $\dfrac{a^2 - ab - 2b^2}{2b - a}$ $(2b - a \neq 0)$. ■

EXAMPLE 5 Simplify: $\dfrac{x^2 - 3x + 2}{x^2 - x - 2}$ $(x \neq 2, -1)$.

Solution We factor the numerator and denominator and divide out the common factor of $x - 2$.

$$\frac{x^2 - 3x + 2}{x^2 - x - 2} = \frac{(x-1)\cancel{(x-2)}}{(x+1)\cancel{(x-2)}}$$
$$= \frac{x-1}{x+1}$$

Self Check Simplify: $\dfrac{a^2 + 3a - 4}{a^2 + 2a - 3}$ $(a \neq 1, -3)$. ■

■ Multiplying and Dividing Fractions

EXAMPLE 6 Multiply: $\dfrac{x^2 - x - 2}{x^2 - 1} \cdot \dfrac{x^2 + 2x - 3}{x - 2}$ $(x \neq 1, -1, 2)$.

Solution To multiply the fractions, we multiply the numerators, multiply the denominators, and divide out the common factors.

$$\frac{x^2 - x - 2}{x^2 - 1} \cdot \frac{x^2 + 2x - 3}{x - 2} = \frac{(x^2 - x - 2)(x^2 + 2x - 3)}{(x^2 - 1)(x - 2)}$$
$$= \frac{\cancel{(x-2)}\cancel{(x+1)}\cancel{(x-1)}(x+3)}{\cancel{(x+1)}\cancel{(x-1)}\cancel{(x-2)}}$$
$$= x + 3$$

Self Check Simplify: $\dfrac{x^2 - 9}{x^2 - x} \cdot \dfrac{x - 1}{x^2 - 3x}$ $(x \neq 0, 1, 3)$. ■

EXAMPLE 7 Divide: $\dfrac{x^2 - 2x - 3}{x^2 - 4} \div \dfrac{x^2 + 2x - 15}{x^2 + 3x - 10}$ $(x \neq 2, -2, 3, -5)$.

Solution To divide the fractions, we multiply by the reciprocal of the divisor. We then factor the numerator and denominator and divide out the common factors.

$$\frac{x^2 - 2x - 3}{x^2 - 4} \div \frac{x^2 + 2x - 15}{x^2 + 3x - 10} = \frac{x^2 - 2x - 3}{x^2 - 4} \cdot \frac{x^2 + 3x - 10}{x^2 + 2x - 15}$$
$$= \frac{(x^2 - 2x - 3)(x^2 + 3x - 10)}{(x^2 - 4)(x^2 + 2x - 15)}$$
$$= \frac{\cancel{(x-3)}(x+1)\cancel{(x-2)}\cancel{(x+5)}}{(x+2)\cancel{(x-2)}\cancel{(x+5)}\cancel{(x-3)}}$$
$$= \frac{x+1}{x+2}$$

Self Check Simplify: $\dfrac{a^2 - a}{a + 2} \div \dfrac{a^2 - 2a}{a^2 - 4}$ $(a \neq 0, 2, -2)$. ■

ACTIVE EXAMPLE 8 Simplify: $\dfrac{2x^2 - 5x - 3}{3x - 1} \cdot \dfrac{3x^2 + 2x - 1}{x^2 - 2x - 3} \div \dfrac{2x^2 + x}{3x} \quad \left(x \neq \dfrac{1}{3}, -1, 3, 0, -\dfrac{1}{2}\right).$

Solution We can change the division to a multiplication, factor, and simplify.

$$\begin{aligned} \frac{2x^2 - 5x - 3}{3x - 1} \cdot \frac{3x^2 + 2x - 1}{x^2 - 2x - 3} \div \frac{2x^2 + x}{3x} &= \frac{2x^2 - 5x - 3}{3x - 1} \cdot \frac{3x^2 + 2x - 1}{x^2 - 2x - 3} \cdot \frac{3x}{2x^2 + x} \\ &= \frac{(2x^2 - 5x - 3)(3x^2 + 2x - 1)(3x)}{(3x - 1)(x^2 - 2x - 3)(2x^2 + x)} \\ &= \frac{\cancel{(x - 3)}\cancel{(2x + 1)}\cancel{(3x - 1)}\cancel{(x + 1)}3\cancel{x}}{\cancel{(3x - 1)}\cancel{(x + 1)}\cancel{(x - 3)}\cancel{x}\cancel{(2x + 1)}} \\ &= 3 \end{aligned}$$

COLLEGE **Algebra** f(x) **Now**™

Explore this Active Example and test yourself on the steps followed here at **http://1pass.thomson.com**. You can also find this example on the Interactive Video Skillbuilder CD-ROM.

Self Check Simplify: $\dfrac{x^2 - 25}{x - 2} \div \dfrac{x^2 - 5x}{x^2 - 2x} \cdot \dfrac{x^2 + 2x}{x^2 + 5x} \quad (x \neq 0, 2, 5, -5).$ ■

■ Adding and Subtracting Fractions

To add fractions with like denominators, we add the numerators and keep the common denominator.

EXAMPLE 9 Add: $\dfrac{2x + 5}{x + 5} + \dfrac{3x + 20}{x + 5} \quad (x \neq -5).$

Solution

$$\begin{aligned} \frac{2x + 5}{x + 5} + \frac{3x + 20}{x + 5} &= \frac{5x + 25}{x + 5} \\ &= \frac{5\cancel{(x + 5)}}{1\cancel{(x + 5)}} \qquad \text{Factor out 5 and divide out the common factor of } x + 5. \\ &= 5 \end{aligned}$$

Self Check Add: $\dfrac{3x - 2}{x - 2} + \dfrac{x - 6}{x - 2} \quad (x \neq 2).$ ■

To add or subtract fractions with unlike denominators, we must find a common denominator, called the **least** (or lowest) **common denominator (LCD).** Suppose the unlike denominators of three fractions are 12, 20, and 35. To find the LCD, we first find the prime factorization of each number.

$$\begin{aligned} 12 &= 4 \cdot 3 & 20 &= 4 \cdot 5 & 35 &= 5 \cdot 7 \\ &= 2^2 \cdot 3 & &= 2^2 \cdot 5 & & \end{aligned}$$

Because the LCD is the smallest number that can be divided by 12, 20, and 35, it must contain factors of 2^2, 3, 5, and 7. Thus,

$$\text{LCD} = 2^2 \cdot 3 \cdot 5 \cdot 7 = 420$$

That is, 420 is the smallest number that can be divided without remainder by 12, 20, and 35.

When finding an LCD, we always factor each denominator and then create the LCD by using each factor the greatest number of times that it appears in any one denominator. The product of these factors is the LCD.

EXAMPLE 10 Add: $\dfrac{1}{x^2 - 4} + \dfrac{2}{x^2 - 4x + 4}$ $(x \neq 2, -2)$.

Solution We factor each denominator and find the LCD.

$$x^2 - 4 = (x + 2)(x - 2)$$
$$x^2 - 4x + 4 = (x - 2)(x - 2) = (x - 2)^2$$

The LCD is $(x + 2)(x - 2)^2$. We then write each fraction with its denominator in factored form, convert each fraction into an equivalent fraction with a denominator of $(x + 2)(x - 2)^2$, add the fractions, and simplify.

$$\begin{aligned}
\frac{1}{x^2 - 4} + \frac{2}{x^2 - 4x + 4} &= \frac{1}{(x + 2)(x - 2)} + \frac{2}{(x - 2)(x - 2)} \\
&= \frac{1(x - 2)}{(x + 2)(x - 2)(x - 2)} + \frac{2(x + 2)}{(x - 2)(x - 2)(x + 2)} \\
&= \frac{1(x - 2) + 2(x + 2)}{(x + 2)(x - 2)(x - 2)} \\
&= \frac{x - 2 + 2x + 4}{(x + 2)(x - 2)(x - 2)} \\
&= \frac{3x + 2}{(x + 2)(x - 2)^2}
\end{aligned}$$

Comment Always attempt to simplify the final result. In this case, the final fraction is already in lowest terms.

Self Check Add: $\dfrac{3}{x^2 - 6x + 9} + \dfrac{1}{x^2 - 9}$ $(x \neq 3, -3)$. ■

ACTIVE EXAMPLE 11 Simplify: $\dfrac{x - 2}{x^2 - 1} - \dfrac{x + 3}{x^2 + 3x + 2} + \dfrac{3}{x^2 + x - 2}$ $(x \neq 1, -1, -2)$.

Solution We factor the denominators to find the LCD.

COLLEGE **Algebra f(x) Now™**
Go to **http://1pass.thomson.com** or your CD to practice this example.

$$x^2 - 1 = (x + 1)(x - 1)$$
$$x^2 + 3x + 2 = (x + 2)(x + 1)$$
$$x^2 + x - 2 = (x + 2)(x - 1)$$

The LCD is $(x + 1)(x + 2)(x - 1)$. We now write each fraction as an equivalent fraction with this LCD, and proceed as follows:

$$\begin{aligned}
\frac{x - 2}{x^2 - 1} - \frac{x + 3}{x^2 + 3x + 2} + \frac{3}{x^2 + x - 2} &= \frac{x - 2}{(x + 1)(x - 1)} - \frac{x + 3}{(x + 1)(x + 2)} + \frac{3}{(x - 1)(x + 2)} \\
&= \frac{(x - 2)(x + 2)}{(x + 1)(x - 1)(x + 2)} - \frac{(x + 3)(x - 1)}{(x + 1)(x + 2)(x - 1)} + \frac{3(x + 1)}{(x - 1)(x + 2)(x + 1)}
\end{aligned}$$

$$= \frac{(x^2 - 4) - (x^2 + 2x - 3) + (3x + 3)}{(x + 1)(x + 2)(x - 1)}$$

$$= \frac{x^2 - 4 - x^2 - 2x + 3 + 3x + 3}{(x + 1)(x + 2)(x - 1)}$$

$$= \frac{x + 2}{(x + 1)(x + 2)(x - 1)}$$

$$= \frac{1}{(x + 1)(x - 1)} \quad \text{Divide out the common factor of } x + 2.$$

Self Check Simplify: $\dfrac{4y}{y^2 - 1} - \dfrac{2}{y + 1} + 2 \quad (y \neq 1, -1)$. ■

■ Complex Fractions

A **complex fraction** is a fraction that has a fractional numerator or a fractional denominator.

ACTIVE EXAMPLE 12 Simplify: $\dfrac{\frac{1}{x} + \frac{1}{y}}{\frac{x}{y}} \quad (x, y \neq 0)$.

COLLEGE **Algebra** $f(x)$ **Now**™
Go to **http://1pass.thomson.com** or your CD to practice this example.

Method 1 We note that the LCD of the three fractions in the complex fraction is xy. So we multiply the numerator and denominator of the complex fraction by xy and simplify:

$$\frac{\frac{1}{x} + \frac{1}{y}}{\frac{x}{y}} = \frac{xy\left(\frac{1}{x} + \frac{1}{y}\right)}{xy\left(\frac{x}{y}\right)} = \frac{\frac{xy}{x} + \frac{xy}{y}}{\frac{xxy}{y}} = \frac{y + x}{x^2}$$

Method 2 We combine the fractions in the numerator of the complex fraction to obtain a single fraction over a single fraction.

$$\frac{\frac{1}{x} + \frac{1}{y}}{\frac{x}{y}} = \frac{\frac{1(y)}{x(y)} + \frac{1(x)}{y(x)}}{\frac{x}{y}} = \frac{\frac{y + x}{xy}}{\frac{x}{y}}$$

Then we use the fact that any fraction indicates a division:

$$\frac{\frac{y + x}{xy}}{\frac{x}{y}} = \frac{y + x}{xy} \div \frac{x}{y} = \frac{y + x}{xy} \cdot \frac{y}{x} = \frac{(y + x)y}{xyx} = \frac{y + x}{x^2}$$

Self Check Simplify: $\dfrac{\frac{1}{x} - \frac{1}{y}}{\frac{1}{x} + \frac{1}{y}} \quad (x, y \neq 0)$. ■

Self Check Answers

1. a. no **b.** $\frac{3a}{5b}$ **2. a.** $\frac{6a^2}{35b^2}$ **b.** $\frac{4a^2b^2}{3r^2s^2}$ **c.** $\frac{8pq}{3t}$ **d.** $\frac{4mn^2}{3w}$ **3.** $\frac{a}{a+3}$ **4.** $-(a+b)$ **5.** $\frac{a+4}{a+3}$ **6.** $\frac{x+3}{x^2}$ **7.** $a-1$ **8.** $x+2$ **9.** 4 **10.** $\frac{4x+6}{(x+3)(x-3)^2}$ **11.** $\frac{2y}{y-1}$ **12.** $\frac{y-x}{y+x}$

0.6 Exercises

VOCABULARY AND CONCEPTS ***Fill in the blanks.***

1. In the fraction $\frac{a}{b}$, a is called the numerator.

2. In the fraction $\frac{a}{b}$, b is called the denominator.

3. $\frac{a}{b} = \frac{c}{d}$ if and only if $ad = bc$.

4. The denominator of a fraction can never be zero.

Complete each formula.

5. $\frac{a}{b} \cdot \frac{c}{d} = \frac{ac}{bd}$

6. $\frac{a}{b} \div \frac{c}{d} = \frac{ad}{bc}$

7. $\frac{a}{b} + \frac{c}{b} = \frac{a+c}{b}$

8. $\frac{a}{b} - \frac{c}{b} = \frac{a-c}{b}$

Determine whether the fractions are equal. Assume that no denominators are 0.

9. $\frac{8x}{3y}, \frac{16x}{6y}$ equal

10. $\frac{3x^2}{4y^2}, \frac{12y^2}{16x^2}$ not equal

11. $\frac{25xyz}{12ab^2c}, \frac{50a^2bc}{24xyz}$ not equal

12. $\frac{15rs^2}{4rs^2}, \frac{37.5a^3}{10a^3}$ equal

PRACTICE ***Simplify each fraction. Assume that no denominators are 0.***

13. $\frac{7a^2b}{21ab^2}$ $\frac{a}{3b}$

14. $\frac{35p^3q^2}{49p^4q}$ $\frac{5q}{7p}$

Perform the operations and simplify, whenever possible. Assume that no denominators are 0.

15. $\frac{4x}{7} \cdot \frac{2}{5a}$ $\frac{8x}{35a}$

16. $\frac{-5y}{2z} \cdot \frac{4}{y^2}$ $\frac{-10}{yz}$

17. $\frac{8m}{5n} \div \frac{3m}{10n}$ $\frac{16}{3}$

18. $\frac{15p}{8q} \div \frac{-5p}{16q^2}$ $-6q$

19. $\frac{3z}{5c} + \frac{2z}{5c}$ $\frac{z}{c}$

20. $\frac{7a}{4b} - \frac{3a}{4b}$ $\frac{a}{b}$

21. $\frac{15x^2y}{7a^2b^3} - \frac{x^2y}{7a^2b^3}$ $\frac{2x^2y}{a^2b^3}$

22. $\frac{8rst^2}{15m^4t^2} + \frac{7rst^2}{15m^4t^2}$ $\frac{rs}{m^4}$

Simplify each fraction. Assume that no denominators are 0.

23. $\frac{2x-4}{x^2-4}$ $\frac{2}{x+2}$

24. $\frac{x^2-16}{x^2-8x+16}$ $\frac{x+4}{x-4}$

25. $\frac{25-x^2}{x^2+10x+25}$ $-\frac{x-5}{x+5}$

26. $\frac{4-x^2}{x^2-5x+6}$ $\frac{x+2}{3-x}$

27. $\frac{6x^3+x^2-12x}{4x^3+4x^2-3x}$ $\frac{3x-4}{2x-1}$

28. $\frac{6x^4-5x^3-6x^2}{2x^3-7x^2-15x}$ $\frac{x(3x+2)(2x-3)}{(2x+3)(x-5)}$

29. $\frac{x^3-8}{x^2+ax-2x-2a}$ $\frac{x^2+2x+4}{x+a}$

30. $\frac{xy+2x+3y+6}{x^3+27}$ $\frac{y+2}{x^2-3x+9}$

Perform the operations and simplify, whenever possible. Assume that no denominators are 0.

31. $\frac{x^2-1}{x} \cdot \frac{x^2}{x^2+2x+1}$ $\frac{x(x-1)}{x+1}$

32. $\frac{y^2-2y+1}{y} \cdot \frac{y+2}{y^2+y-2}$ $\frac{y-1}{y}$

33. $\frac{3x^2+7x+2}{x^2+2x} \cdot \frac{x^2-x}{3x^2+x}$ $\frac{x-1}{x}$

34. $\frac{x^2+x}{2x^2+3x} \cdot \frac{2x^2+x-3}{x^2-1}$ 1

35. $\frac{x^2+x}{x-1} \cdot \frac{x^2-1}{x+2}$ $\frac{x(x+1)^2}{x+2}$

36. $\frac{x^2+5x+6}{x^2+6x+9} \cdot \frac{x+2}{x^2-4}$ $\frac{x+2}{(x+3)(x-2)}$

37. $\frac{2x^2+32}{8} \div \frac{x^2+16}{2}$ $\frac{1}{2}$

38. $\dfrac{x^2+x-6}{x^2-6x+9} \div \dfrac{x^2-4}{x^2-9}$ $\quad \dfrac{(x+3)^2}{(x-3)(x+2)}$

39. $\dfrac{z^2+z-20}{z^2-4} \div \dfrac{z^2-25}{z-5}$ $\quad \dfrac{z-4}{(z+2)(z-2)}$

40. $\dfrac{ax+bx+a+b}{a^2+2ab+b^2} \div \dfrac{x^2-1}{x^2-2x+1}$ $\quad \dfrac{x-1}{a+b}$

41. $\dfrac{3x^2+5x-2}{x^3+2x^2} \div \dfrac{6x^2+13x-5}{2x^3+5x^2}$ $\quad 1$

42. $\dfrac{x^2+13x+12}{8x^2-6x-5} \div \dfrac{2x^2-x-3}{8x^2-14x+5}$ $\quad \dfrac{(x+12)(2x-1)}{(2x+1)(2x-3)}$

43. $\dfrac{x^2+7x+12}{x^3-x^2-6x} \cdot \dfrac{x^2-3x-10}{x^2+2x-3} \cdot \dfrac{x^3-4x^2+3x}{x^2-x-20}$ $\quad 1$

44. $\dfrac{x^2-2x-3}{21x^2-50x-16} \cdot \dfrac{3x-8}{x-3} \div \dfrac{x^2+6x+5}{7x^2-33x-10}$

$\dfrac{x-5}{x+5}$

45. $\dfrac{x^3+27}{x^2-4} \div \left(\dfrac{x^2+4x+3}{x^2+2x} \div \dfrac{x^2+x-6}{x^2-3x+9}\right)$

$\dfrac{x(x+3)}{x+1}$

46. $\dfrac{x(x-2)-3}{x(x+7)-3(x-1)} \cdot \dfrac{x(x+1)-2}{x(x-7)+3(x+1)}$ $\quad \dfrac{x+2}{x+3}$

47. $\dfrac{3}{x+3} + \dfrac{x+2}{x+3}$ $\quad \dfrac{x+5}{x+3}$

48. $\dfrac{3}{x+1} + \dfrac{x+2}{x+1}$ $\quad \dfrac{x+5}{x+1}$

49. $\dfrac{4x}{x-1} - \dfrac{4}{x-1}$ $\quad 4$

50. $\dfrac{6x}{x-2} - \dfrac{3}{x-2}$ $\quad \dfrac{3(2x-1)}{x-2}$

51. $\dfrac{2}{5-x} + \dfrac{1}{x-5}$ $\quad \dfrac{-1}{x-5}$

52. $\dfrac{3}{x-6} - \dfrac{2}{6-x}$ $\quad \dfrac{5}{x-6}$

53. $\dfrac{3}{x+1} + \dfrac{2}{x-1}$ $\quad \dfrac{5x-1}{(x+1)(x-1)}$

54. $\dfrac{3}{x+4} + \dfrac{x}{x-4}$ $\quad \dfrac{x^2+7x-12}{(x+4)(x-4)}$

55. $\dfrac{a+3}{a^2+7a+12} + \dfrac{a}{a^2-16}$ $\quad \dfrac{2(a-2)}{(a+4)(a-4)}$

56. $\dfrac{a}{a^2+a-2} + \dfrac{2}{a^2-5a+4}$ $\quad \dfrac{a^2-2a+4}{(a+2)(a-1)(a-4)}$

57. $\dfrac{x}{x^2-4} - \dfrac{1}{x+2}$ $\quad \dfrac{2}{(x+2)(x-2)}$

58. $\dfrac{b^2}{b^2-4} - \dfrac{4}{b^2+2b}$ $\quad \dfrac{b^3-4b+8}{b(b+2)(b-2)}$

59. $\dfrac{3x-2}{x^2+2x+1} - \dfrac{x}{x^2-1}$ $\quad \dfrac{2(x^2-3x+1)}{(x+1)^2(x-1)}$

60. $\dfrac{2t}{t^2-25} - \dfrac{t+1}{t^2+5t}$ $\quad \dfrac{t^2+4t+5}{t(t+5)(t-5)}$

61. $\dfrac{2}{y^2-1} + 3 + \dfrac{1}{y+1}$ $\quad \dfrac{3y-2}{y-1}$

62. $2 + \dfrac{4}{t^2-4} - \dfrac{1}{t-2}$ $\quad \dfrac{2t+3}{t+2}$

63. $\dfrac{1}{x-2} + \dfrac{3}{x+2} - \dfrac{3x-2}{x^2-4}$ $\quad \dfrac{1}{x+2}$

64. $\dfrac{x}{x-3} - \dfrac{5}{x+3} + \dfrac{3(3x-1)}{x^2-9}$ $\quad \dfrac{x+4}{x-3}$

65. $\left(\dfrac{1}{x-2} + \dfrac{1}{x-3}\right) \cdot \dfrac{x-3}{2x}$ $\quad \dfrac{2x-5}{2x(x-2)}$

66. $\left(\dfrac{1}{x+1} - \dfrac{1}{x-2}\right) \div \dfrac{1}{x-2}$ $\quad \dfrac{-3}{x+1}$

67. $\dfrac{3x}{x-4} - \dfrac{x}{x+4} - \dfrac{3x+1}{16-x^2}$ $\quad \dfrac{2x^2+19x+1}{(x+4)(x-4)}$

68. $\dfrac{7x}{x-5} + \dfrac{3x}{5-x} + \dfrac{3x-1}{x^2-25}$ $\quad \dfrac{4x^2+23x-1}{(x+5)(x-5)}$

69. $\dfrac{1}{x^2+3x+2} - \dfrac{2}{x^2+4x+3} + \dfrac{1}{x^2+5x+6}$ $\quad 0$

70. $\dfrac{-2}{x-y} + \dfrac{2}{x-z} - \dfrac{2z-2y}{(y-x)(z-x)}$ $\quad 0$

71. $\dfrac{3x-2}{x^2+x-20} - \dfrac{4x^2+2}{x^2-25} + \dfrac{3x^2-25}{x^2-16}$

$\dfrac{-x^4+3x^3-43x^2-58x+697}{(x+5)(x-5)(x+4)(x-4)}$

72. $\dfrac{3x+2}{8x^2-10x-3} + \dfrac{x+4}{6x^2-11x+3} - \dfrac{1}{4x+1}$

$\dfrac{7x^2+31x-1}{(4x+1)(2x-3)(3x-1)}$

Simplify each complex fraction. Assume that no denominators are 0.

73. $\dfrac{\frac{3a}{b}}{\frac{6ac}{b^2}}$ $\quad \dfrac{b}{2c}$

74. $\dfrac{\frac{3t^2}{9x}}{\frac{t}{18x}}$ $\quad 6t$

75. $\dfrac{3a^2b}{\dfrac{ab}{27}}$ $81a$

76. $\dfrac{\dfrac{3u^2v}{4t}}{3uv}$ $\dfrac{u}{4t}$

77. $\dfrac{\dfrac{x-y}{ab}}{\dfrac{y-x}{ab}}$ -1

78. $\dfrac{\dfrac{x^2-5x+6}{2x^2y}}{\dfrac{x^2-9}{2x^2y}}$ $\dfrac{x-2}{x+3}$

79. $\dfrac{\dfrac{1}{x}+\dfrac{1}{y}}{xy}$ $\dfrac{y+x}{x^2y^2}$

80. $\dfrac{xy}{\dfrac{11}{x}+\dfrac{11}{y}}$ $\dfrac{x^2y^2}{11(y+x)}$

81. $\dfrac{\dfrac{1}{x}+\dfrac{1}{y}}{\dfrac{1}{x}-\dfrac{1}{y}}$ $\dfrac{y+x}{y-x}$

82. $\dfrac{\dfrac{1}{x}-\dfrac{1}{y}}{\dfrac{1}{x}+\dfrac{1}{y}}$ $\dfrac{y-x}{y+x}$

83. $\dfrac{\dfrac{3a}{b}-\dfrac{4a^2}{x}}{\dfrac{1}{b}+\dfrac{1}{ax}}$ $\dfrac{a^2(3x-4ab)}{ax+b}$

84. $\dfrac{1-\dfrac{x}{y}}{\dfrac{x^2}{y^2}-1}$ $\dfrac{-y}{x+y}$

85. $\dfrac{x+1-\dfrac{6}{x}}{x+5+\dfrac{6}{x}}$ $\dfrac{x-2}{x+2}$

86. $\dfrac{2z}{1-\dfrac{3}{z}}$ $\dfrac{2z^2}{z-3}$

87. $\dfrac{3xy}{1-\dfrac{1}{xy}}$ $\dfrac{3x^2y^2}{xy-1}$

88. $\dfrac{x-3+\dfrac{1}{x}}{-\dfrac{1}{x}-x+3}$ -1

89. $\dfrac{3x}{x+\dfrac{1}{x}}$ $\dfrac{3x^2}{x^2+1}$

90. $\dfrac{2x^2+4}{2+\dfrac{4x}{5}}$ $\dfrac{5x^2+10}{2x+5}$

91. $\dfrac{\dfrac{x}{x+2}-\dfrac{2}{x-1}}{\dfrac{3}{x+2}+\dfrac{x}{x-1}}$ $\dfrac{x^2-3x-4}{x^2+5x-3}$

92. $\dfrac{\dfrac{2x}{x-3}+\dfrac{1}{x-2}}{\dfrac{3}{x-3}-\dfrac{x}{x-2}}$ $-\dfrac{2x^2-3x-3}{x^2-6x+6}$

Write each expression without using negative exponents, and simplify the resulting complex fraction. Assume that no denominators are 0.

93. $\dfrac{1}{1+x^{-1}}$ $\dfrac{x}{x+1}$

94. $\dfrac{y^{-1}}{x^{-1}+y^{-1}}$ $\dfrac{x}{y+x}$

95. $\dfrac{3(x+2)^{-1}+2(x-1)^{-1}}{(x+2)^{-1}}$ $\dfrac{5x+1}{x-1}$

96. $\dfrac{2x(x-3)^{-1}-3(x+2)^{-1}}{(x-3)^{-1}(x+2)^{-1}}$ $2x^2+x+9$

APPLICATIONS

97. **Engineering** The stiffness k of the shaft shown in the illustration is given by the formula

$$k=\frac{1}{\dfrac{1}{k_1}+\dfrac{1}{k_2}}$$

Section 1 Section 2

where k_1 and k_2 are the individual stiffnesses of each section. Simplify the complex fraction on the right-hand side of the formula. $\dfrac{k_1k_2}{k_2+k_1}$

98. **Electronics** The combined resistance R of three resistors with resistances of R_1, R_2, and R_3 is given by the formula

$$R=\frac{1}{\dfrac{1}{R_1}+\dfrac{1}{R_2}+\dfrac{1}{R_3}}$$

Simplify the complex fraction on the right-hand side of the formula. $\dfrac{R_1R_2R_3}{R_2R_3+R_1R_3+R_1R_2}$

DISCOVERY AND WRITING

Simplify each complex fraction. Assume that no denominators are 0.

99. $\dfrac{x}{1+\dfrac{1}{3x^{-1}}}$ $\dfrac{3x}{3+x}$

100. $\dfrac{ab}{2+\dfrac{3}{2a^{-1}}}$ $\dfrac{2ab}{4+3a}$

101. $\dfrac{1}{1+\dfrac{1}{1+\dfrac{1}{x}}}$ $\dfrac{x+1}{2x+1}$

102. $\dfrac{y}{2+\dfrac{2}{2+\dfrac{2}{y}}}$ $\dfrac{y(y+1)}{3y+2}$

103. Explain why the formula $\frac{a}{b} + \frac{c}{d} = \frac{ad + bc}{bd}$ is valid.

104. Explain why the formula $\frac{a}{b} \div \frac{c}{d} = \frac{a}{b} \cdot \frac{d}{c}$ is valid.

105. Explain the commutative property of addition and tell why it is useful.

106. Explain the distributive property and tell why it is useful.

REVIEW ***Write each expression without using absolute value symbols.***

107. $|-6|$ **6**

108. $|5 - x|$, given that $x < 0$ **$5 - x$**

Simplify each expression.

109. $\left(\frac{x^3y^{-2}}{x^{-1}y}\right)^{-3}$ **$\frac{y^9}{x^{12}}$**

110. $(27x^6)^{2/3}$ **$9x^4$**

111. $\sqrt{20} - \sqrt{45}$ **$-\sqrt{5}$**

112. $2(x^2 + 4) - 3(2x^2 + 5)$ **$-4x^2 - 7$**

PROBLEMS AND PROJECTS

1. Find the highest power of 2 that can be evaluated with a scientific or graphing calculator. **332**

2. Find the highest power of 7 that can be evaluated with a scientific or graphing calculator. **118**

3. While in Switzerland, you see a Rolex watch that a close friend wants. You send him the following e-mail message:

   E-Mail

   ROLEX WATCH $5,800. SHOULD I BUY IT FOR YOU?

   Your friend responds as follows:

   E-Mail

   NO PRICE TOO HIGH! REPEAT... NO! PRICE TOO HIGH.

   Would you buy the watch? Why or why not? What is wrong with your friend's message? What mathematical principle does this example illustrate?

4. Note that $\frac{2}{3} = \frac{4}{6} = \frac{6}{9} = \frac{2+4+6}{3+6+9}$, which is $\frac{12}{18}$. Is this principle always true? That is, is

$$\frac{a}{b} = \frac{c}{d} = \frac{e}{f} = \frac{a+c+e}{b+d+f}$$

   Can you prove it?

   Is $\frac{a}{b} = \frac{c}{d} = \frac{e}{f} = \frac{g}{h} = \frac{a+c+e+g}{b+d+f+h}$?

   For how many fractions is this principle true?

Project 1

Complete the table and work the following problems.

Figure	Name	Perimeter/ circumference
	Square	$P = 4s$
	Rectangle	$P = 2l + 2w$
	Triangle	$P = a + b + c$
	Trapezoid	$P = a + b + c + d$
	Circle	$C = \pi D$

(π is approximately 3.1416)

1. Find the perimeter of a garden with dimensions of 12 meters by 18 meters. **60 m**

2. Find the circumference of a circle with a radius of 3.8 feet. Give the answer to the nearest hundredth. **23.88 ft**

Project 2

Complete the table and work the problems that follow.

Figure	Name	Area
	Square	$A = s^2$
	Rectangle	$A = lw$
	Circle	$A = \pi r^2$
	Triangle	$A = \frac{1}{2}bh$
	Trapezoid	$A = \frac{1}{2}h(b_1 + b_2)$

Figure	Name	Area
	Cube	$V = s^3$
	Rectangular solid	$V = lwh$
	Sphere	$V = \frac{4}{3}\pi r^3$

Figure	Name	Area
	Cylinder	$V = Bh$*
	Cone	$V = \frac{1}{3}Bh$*
	Pyramid	$V = \frac{1}{3}Bh$*

*B represents the area of the base.

Give each answer to the nearest hundredth.

1. Find the area of a circle with a diameter of 21 feet. **346.36 ft²**
2. Find the area of a triangle with a base of 21.3 centimeters and a height of 7.5 centimeters. **79.88 cm²**
3. Find the area of a trapezoid with a height of 9.3 inches and bases of 7.2 inches and 10.1 inches. **80.45 in.²**
4. Find the volume of a rectangular solid with dimensions of 8.5 meters, 10.3 meters, and 12.7 meters. **1,111.89 m³**
5. Find the volume of a sphere with a radius of 20.5 feet. **36,086.95 ft³**
6. Find the volume of a 10-foot-long cylinder whose base is a circle with a radius of 1.6 feet. **80.42 ft³**
7. Find the volume of a cone with a circular base 15 centimeters in diameter and a height of 12.5 centimeters. **736.31 cm³**
8. Find the volume of a pyramid whose rectangular base has dimensions of 8.7 meters by 9.3 meters and whose height is 15.8 meters. **426.13 m³**

Project 3

A village council is debating whether or not to build an emergency water reservoir for use in drought conditions. The tentative plan calls for a conical reservoir with a diameter of 173 feet and height of 87.2 feet.

During drought conditions, the reservoir will lose water to evaporation as well as supply the village with water. The company that designed the reservoir provided the following information.

If D equals the number of consecutive days the reservoir is used to provide water, the total amount E (in cubic feet) of water lost to evaporation will be

$$E = 0.1V\left(\frac{D - 0.7}{D}\right)^2$$

The water left in the reservoir after it has been used for D days will be

$$\text{Water left} = \text{original volume} - D \cdot \text{(water used per day)} - E$$

Under emergency conditions, the village estimates that it will use about 61,000 cubic feet of water per day. The majority of the council believe that building the reservoir is a good idea if it will supply the city with water for at least 10 days. Otherwise, they will vote against the plan.

1. Find the volume of water that the reservoir will hold. Express the result in scientific notation.
   6.832463461×10^5 ft^3
2. Will the reservoir supply the village for 10 days? **yes**
3. Would you vote for the plan? Explain.
4. How much water could the village use per day if the supply must last for two weeks?
   about 44,399 ft^3 per day

CHAPTER SUMMARY

0.1 Sets of Real Numbers

CONCEPTS

Natural numbers:
$\{1, 2, 3, 4, \ldots\}$

Whole numbers:
$\{0, 1, 2, 3, 4, \ldots\}$

Integers:
$\{\ldots, -3, -2, -1, 0, 1, 2, 3, \ldots\}$

Rational numbers: $\{x \mid x$ can be written in the form $\frac{a}{b}$ $(b \neq 0)$, where a and b are integers.$\}$ All decimals that either terminate or repeat.

Irrational numbers: nonrational real numbers. All decimals that neither terminate nor repeat.

Real numbers: any number that can be expressed as a decimal.

REVIEW EXERCISES

Consider the set $\{-6, -3, 0, \frac{1}{2}, 3, \pi, \sqrt{5}, 6, 8\}$. List the numbers in this set that are

1. natural numbers. **3, 6, 8**
2. whole numbers. **0, 3, 6, 8**
3. integers. **$-6, -3, 0, 3, 6, 8$**
4. rational numbers. **$-6, -3, 0, \frac{1}{2}, 3, 6, 8$**
5. irrational numbers. **$\pi, \sqrt{5}$**
6. real numbers. **$-6, -3, 0, \frac{1}{2}, 3, \pi, \sqrt{5}, 6, 8$**

Prime numbers:
$\{2, 3, 5, 7, 11, 13, \ldots\}$

Composite numbers:
$\{4, 6, 8, 9, 10, \ldots\}$

Even integers:
$\{\ldots, -4, -2, 0, 2, 4, \ldots\}$

Odd integers:
$\{\ldots, -3, -1, 1, 3, 5, \ldots\}$

Consider the set $\{-6, -3, 0, \frac{1}{2}, 3, \pi, \sqrt{5}, 6, 8\}$. **List the numbers in this set that are**

7. prime numbers. **3**
8. composite numbers. **6, 8**
9. even integers. **−6, 0, 6, 8**
10. odd integers. **−3, 3**

Associative properties:
$(a + b) + c = a + (b + c)$
of addition
$(ab)c = a(bc)$
of multiplication

Commutative properties:
$a + b = b + a$
of addition
$ab = ba$
of multiplication

Distributive property:
$a(b + c) = ab + ac$

Double negative rule:
$-(-a) = a$

Determine which property of real numbers justifies each statement.

11. $(a + b) + 2 = a + (b + 2)$ **assoc. prop. of add.**
12. $a + 7 = 7 + a$ **comm. prop. of add.**
13. $4(2x) = (4 \cdot 2)x$ **assoc. prop. of mult.**
14. $3(a + b) = 3a + 3b$ **distrib. prop.**
15. $(5a)7 = 7(5a)$ **comm. prop. of mult.**
16. $(2x + y) + z = (y + 2x) + z$ **comm. prop. of add.**
17. $-(-6) = 6$ **double negative rule**

Graph each subset of the real numbers:

18. the prime numbers between 10 and 20

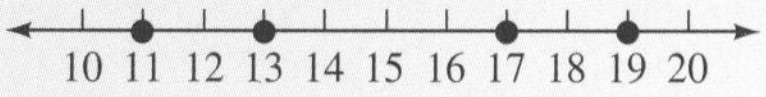

10 11 12 13 14 15 16 17 18 19 20

19. the even integers from 6 to 14

6 7 8 9 10 11 12 13 14

Open intervals have no endpoints.

Closed intervals have two endpoints.

Half-open intervals have one endpoint.

Graph each interval on the number line.

20. $-3 < x \leq 5$ (−3, 5)
21. $x \geq 0$ or $x < -1$ (−1, 0)
22. $(-2, 4]$ (−2, 4) 
23. $(-\infty, 2) \cap (-5, \infty)$ (−5, 2)
24. $(-\infty, -4) \cup [6, \infty)$ (−4, 6)

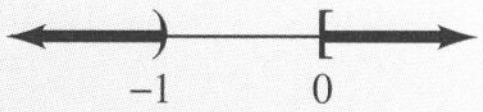

If $x \geq 0$, then $|x| = x$.
If $x < 0$, then $|x| = -x$.
$|x| \geq 0$

The **distance between points** a and b on a number line is $d = |b - a|$.

Write each expression without absolute value symbols.

25. $|6|$ **6**
26. $|-25|$ **25**
27. $|1 - \sqrt{2}|$ $\sqrt{2} - 1$
28. $|\sqrt{3} - 1|$ $\sqrt{3} - 1$

29. On a number line, find the distance between points with coordinates of −5 and 7. **12**

0.2 Integer Exponents and Scientific Notation

Natural-number exponents:

$$x^n = \overbrace{x \cdot x \cdot x \cdot \cdots \cdot x}^{n \text{ factors of } x}$$

Rules of exponents: If there are no divisions by 0,

$$x^m x^n = x^{m+n} \qquad (x^m)^n = x^{mn}$$

$$(xy)^n = x^n y^n \qquad \left(\frac{x}{y}\right)^n = \frac{x^n}{y^n}$$

$$x^0 = 1 \qquad x^{-n} = \frac{1}{x^n}$$

$$\frac{x^m}{x^n} = x^{m-n} \qquad \left(\frac{x}{y}\right)^{-n} = \left(\frac{y}{x}\right)^n$$

A number is written in **scientific notation** when it is written in the form $N \times 10^n$, where $1 \le |N| < 10$.

Write each expression without using exponents.

30. $-5a^3$ $\quad -5aaa$

31. $(-5a)^2$ $\quad (-5a)(-5a)$

Write each expression using exponents.

32. $3ttt$ $\quad 3t^3$

33. $(-2b)(3b)$ $\quad -6b^2$

Simplify each expression.

34. n^2n^4 $\quad n^6$

35. $(p^3)^2$ $\quad p^6$

36. $(x^3y^2)^4$ $\quad x^{12}y^8$

37. $\left(\frac{a^4}{b^2}\right)^3$ $\quad \frac{a^{12}}{b^6}$

38. $(m^{-3}n^0)^2$ $\quad \frac{1}{m^6}$

39. $\left(\frac{p^{-2}q^2}{2}\right)^3$ $\quad \frac{q^6}{8p^6}$

40. $\frac{a^5}{a^8}$ $\quad \frac{1}{a^3}$

41. $\left(\frac{a^2}{b^3}\right)^{-2}$ $\quad \frac{b^6}{a^4}$

42. $\left(\frac{3x^2y^{-2}}{x^2y^2}\right)^{-2}$ $\quad \frac{y^8}{9}$

43. $\left(\frac{a^{-3}b^2}{ab^{-3}}\right)^{-2}$ $\quad \frac{a^8}{b^{10}}$

44. $\left(\frac{-3x^3y}{xy^3}\right)^{-2}$ $\quad \frac{y^4}{9x^4}$

45. $\left(-\frac{2m^{-2}n^0}{4m^2n^{-1}}\right)^{-3}$ $\quad -\frac{8m^{12}}{n^3}$

46. If $x = -3$ and $y = 3$, evaluate $-x^2 - xy^2$ $\quad 18$

Write each number in scientific notation.

47. 6,750 $\quad 6.75 \times 10^3$

48. 0.00023 $\quad 2.3 \times 10^{-4}$

Write each number in standard notation.

49. 4.8×10^2 $\quad 480$

50. 0.25×10^{-3} $\quad 0.00025$

51. Use scientific notation to simplify $\frac{(45{,}000)(350{,}000)}{0.000105}$. $\quad 1.5 \times 10^{14}$

0.3 Rational Exponents and Radicals

If $a \ge 0$, then $a^{1/n}$ is the nonnegative number b such that $b^n = a$.

If $a < 0$ and n is odd, then $a^{1/n}$ is the real number b such that $b^n = a$.

If $a < 0$ and n is even, then $a^{1/n}$ is not a real number.

Simplify each expression, if possible.

52. $121^{1/2}$ $\quad 11$

53. $\left(\frac{27}{125}\right)^{1/3}$ $\quad \frac{3}{5}$

54. $(32x^5)^{1/5}$ $\quad 2x$

55. $(81a^4)^{1/4}$ $\quad 3|a|$

56. $(-1{,}000x^6)^{1/3}$ $\quad -10x^2$

57. $(-25x^2)^{1/2}$ $\quad$ not a real number

58. $(x^{12}y^2)^{1/2}$ $\quad x^6|y|$

59. $\left(\frac{x^{12}}{y^4}\right)^{-1/2}$ $\quad \frac{y^2}{x^6}$

60. $\left(\frac{-c^{2/3}c^{5/3}}{c^{-2/3}}\right)^{1/3}$ $\quad -c$

61. $\left(\frac{a^{-1/4}a^{3/4}}{a^{9/2}}\right)^{-1/2}$ $\quad a^2$

If m and n are positive integers, $\frac{m}{n}$ is in lowest terms, and $a^{1/n}$ is a real number, then

$$a^{m/n} = (a^{1/n})^m = (a^m)^{1/n}$$

Simplify each expression.

62. $64^{2/3}$ 16

63. $32^{-3/5}$ $\frac{1}{8}$

64. $\left(\frac{16}{81}\right)^{3/4}$ $\frac{8}{27}$

65. $\left(\frac{32}{243}\right)^{2/5}$ $\frac{4}{9}$

66. $\left(\frac{8}{27}\right)^{-2/3}$ $\frac{9}{4}$

67. $\left(\frac{16}{625}\right)^{-3/4}$ $\frac{125}{8}$

68. $(-216x^3)^{2/3}$ $36x^2$

69. $\frac{p^{a/2}p^{a/3}}{p^{a/6}}$ $p^{2a/3}$

$$\sqrt[n]{a} = a^{1/n}$$

Simplify each expression.

70. $\sqrt{36}$ 6

71. $-\sqrt{49}$ -7

72. $\sqrt{\frac{9}{25}}$ $\frac{3}{5}$

73. $\sqrt[3]{\frac{27}{125}}$ $\frac{3}{5}$

74. $\sqrt{x^2y^4}$ $|x|y^2$

75. $\sqrt[3]{x^3}$ x

76. $\sqrt[4]{\frac{m^8n^4}{p^{16}}}$ $\frac{m^2|n|}{p^4}$

77. $\sqrt[5]{\frac{a^{15}b^{10}}{c^5}}$ $\frac{a^3b^2}{c}$

If all radicals are real numbers and there are no divisions by 0, then

$$\sqrt[n]{ab} = \sqrt[n]{a}\sqrt[n]{b}$$

$$\sqrt[n]{\frac{a}{b}} = \frac{\sqrt[n]{a}}{\sqrt[n]{b}}$$

$$\sqrt[m]{\sqrt[n]{a}} = \sqrt[n]{\sqrt[m]{a}} = \sqrt[mn]{a}$$

Simplify and combine terms.

78. $\sqrt{50} + \sqrt{8}$ $7\sqrt{2}$

79. $\sqrt{12} + \sqrt{3} - \sqrt{27}$ 0

80. $\sqrt[3]{24x^4} - \sqrt[3]{3x^4}$ $x\sqrt[3]{3x}$

Rationalize each denominator.

81. $\frac{\sqrt{7}}{\sqrt{5}}$ $\frac{\sqrt{35}}{5}$

82. $\frac{8}{\sqrt{8}}$ $2\sqrt{2}$

83. $\frac{1}{\sqrt[3]{2}}$ $\frac{\sqrt[3]{4}}{2}$

84. $\frac{2}{\sqrt[3]{25}}$ $\frac{2\sqrt[3]{5}}{5}$

Rationalize each numerator.

85. $\frac{\sqrt{2}}{5}$ $\frac{2}{5\sqrt{2}}$

86. $\frac{\sqrt{5}}{5}$ $\frac{1}{\sqrt{5}}$

87. $\frac{\sqrt{2x}}{3}$ $\frac{2x}{3\sqrt{2x}}$

88. $\frac{3\sqrt[3]{7x}}{2}$ $\frac{21x}{2\sqrt[3]{49x^2}}$

0.4 Polynomials

Monomial: A polynomial with one term.

Binomial: A polynomial with two terms.

Trinomial: A polynomial with three terms.

Give the degree of each polynomial and tell whether the polynomial is a monomial, a binomial, or a trinomial.

89. $x^3 - 8$ **3rd degree, binomial**

90. $8x - 8x^2 - 8$ **2nd degree, trinomial**

91. $\sqrt{3}x^2$ **2nd degree, monomial**

92. $4x^4 - 12x^2 + 1$ **4th degree, trinomial**

The **degree of a monomial** is the sum of the exponents on its variables.

The **degree of a polynomial** is the degree of the term in the polynomial with highest degree.

$a(b + c + d + \cdots) = ab + ac + ad + \cdots$

$(x + y)^2 = x^2 + 2xy + y^2$

$(x - y)^2 = x^2 - 2xy + y^2$

$(x + y)(x - y) = x^2 - y^2$

Perform the operations and simplify.

93. $2(x + 3) + 3(x - 4)$ $5x - 6$

94. $3x^2(x - 1) - 2x(x + 3) - x^2(x + 2)$ $2x^3 - 7x^2 - 6x$

95. $(3x + 2)(3x + 2)$ $9x^2 + 12x + 4$

96. $(3x + y)(2x - 3y)$ $6x^2 - 7xy - 3y^2$

97. $(4a + 2b)(2a - 3b)$ $8a^2 - 8ab - 6b^2$

98. $(z + 3)(3z^2 + z - 1)$ $3z^3 + 10z^2 + 2z - 3$

99. $(a^n + 2)(a^n - 1)$ $a^{2n} + a^n - 2$

100. $\left(\sqrt{2} + x\right)^2$ $2 + 2x\sqrt{2} + x^2$

101. $\left(\sqrt{2} + 1\right)\left(\sqrt{3} + 1\right)$ $\sqrt{6} + \sqrt{2} + \sqrt{3} + 1$

102. $\left(\sqrt[3]{3} - 2\right)\left(\sqrt[3]{9} + 2\sqrt[3]{3} + 4\right)$ -5

The **conjugate** of $a + b$ is $a - b$.

Rationalize each denominator.

103. $\dfrac{2}{\sqrt{3} - 1}$ $\sqrt{3} + 1$

104. $\dfrac{-2}{\sqrt{3} - \sqrt{2}}$ $-2\left(\sqrt{3} + \sqrt{2}\right)$

105. $\dfrac{2x}{\sqrt{x} - 2}$ $\dfrac{2x\left(\sqrt{x} + 2\right)}{x - 4}$

106. $\dfrac{\sqrt{x} - \sqrt{y}}{\sqrt{x} + \sqrt{y}}$ $\dfrac{x - 2\sqrt{xy} + y}{x - y}$

Rationalize each numerator.

107. $\dfrac{\sqrt{x} + 2}{5}$ $\dfrac{x - 4}{5\left(\sqrt{x} - 2\right)}$

108. $\dfrac{1 - \sqrt{a}}{a}$ $\dfrac{1 - a}{a\left(1 + \sqrt{a}\right)}$

Perform each division.

109. $\dfrac{3x^2y^2}{6x^3y}$ $\dfrac{y}{2x}$

110. $\dfrac{4a^2b^3 + 6ab^4}{2b^2}$ $2a^2b + 3ab^2$

111. $2x + 3\overline{)2x^3 + 7x^2 + 8x + 3}$ $x^2 + 2x + 1$

112. $x^2 - 1\overline{)x^5 + x^3 - 2x - 3x^2 - 3}$ $x^3 + 2x - 3 - \dfrac{6}{x^2 - 1}$

0.5 Factoring Polynomials

$x^2 - y^2 = (x + y)(x - y)$

$x^2 + 2xy + y^2 = (x + y)^2$

$x^2 - 2xy + y^2 = (x - y)^2$

$x^3 + y^3 = (x + y)(x^2 - xy + y^2)$

$x^3 - y^3 = (x - y)(x^2 + xy + y^2)$

Factor each expression completely, if possible.

113. $3t^3 - 3t$ $3t(t + 1)(t - 1)$

114. $5r^3 - 5$ $5(r - 1)(r^2 + r + 1)$

115. $6x^2 + 7x - 24$ $(3x + 8)(2x - 3)$

116. $3a^2 + ax - 3a - x$ $(3a + x)(a - 1)$

117. $8x^3 - 125$
$(2x - 5)(4x^2 + 10x + 25)$

118. $6x^2 - 20x - 16$
$2(3x + 2)(x - 4)$

119. $x^2 + 6x + 9 - t^2$
$(x + 3 + t)(x + 3 - t)$

120. $3x^2 - 1 + 5x$
prime

121. $8z^3 + 343$
$(2z + 7)(4z^2 - 14z + 49)$

122. $1 + 14b + 49b^2$
$(7b + 1)^2$

123. $121z^2 + 4 - 44z$
$(11z - 2)^2$

124. $64y^3 - 1{,}000$
$8(2y - 5)(4y^2 + 10y + 25)$

125. $2xy - 4zx - wy + 2zw$
$(y - 2z)(2x - w)$

126. $x^8 + x^4 + 1$
$(x^2 + 1 + x)(x^2 + 1 - x)(x^4 + 1 - x^2)$

0.6 Algebraic Fractions

If there are no divisions by 0, then

$\frac{a}{b} = \frac{c}{d}$ if and only if $ab = bc$.

$\frac{a}{b} = \frac{ax}{bx}$

$\frac{a}{b} \cdot \frac{c}{d} = \frac{ac}{bd}$ $\qquad \frac{a}{b} \div \frac{c}{d} = \frac{ad}{bc}$

$\frac{a}{b} + \frac{c}{b} = \frac{a + c}{b}$

$\frac{a}{b} - \frac{c}{b} = \frac{a - c}{b}$

$a \cdot 1 = a \qquad \frac{a}{1} = a \qquad \frac{a}{a} = 1$

$\frac{a}{b} = \frac{-a}{-b} = -\frac{a}{-b} = -\frac{-a}{b}$

$-\frac{a}{b} = \frac{a}{-b} = \frac{-a}{b} = -\frac{-a}{-b}$

Perform each operation and simplify. Assume that no denominators are 0.

127. $\frac{x^2 - 4x + 4}{x + 2} \cdot \frac{x^2 + 5x + 6}{x - 2}$ $\quad (x - 2)(x + 3)$

128. $\frac{2y^2 - 11y + 15}{y^2 - 6y + 8} \cdot \frac{y^2 - 2y - 8}{y^2 - y - 6}$ $\quad \frac{2y - 5}{y - 2}$

129. $\frac{2t^2 + t - 3}{3t^2 - 7t + 4} \div \frac{10t + 15}{3t^2 - t - 4}$ $\quad \frac{t + 1}{5}$

130. $\frac{p^2 + 7p + 12}{p^3 + 8p^2 + 4p} \div \frac{p^2 - 9}{p^2}$ $\quad \frac{p(p + 4)}{(p^2 + 8p + 4)(p - 3)}$

131. $\frac{x^2 + x - 6}{x^2 - x - 6} \cdot \frac{x^2 - x - 6}{x^2 + x - 2} \div \frac{x^2 - 4}{x^2 - 5x + 6}$ $\quad \frac{(x - 2)(x + 3)(x - 3)}{(x - 1)(x + 2)^2}$

132. $\left(\frac{2x + 6}{x + 5} \div \frac{2x^2 - 2x - 4}{x^2 - 25}\right)\frac{x^2 - x - 2}{x^2 - 2x - 15}$ $\quad 1$

133. $\frac{2}{x - 4} + \frac{3x}{x + 5}$ $\quad \frac{3x^2 - 10x + 10}{(x - 4)(x + 5)}$

134. $\frac{5x}{x - 2} - \frac{3x + 7}{x + 2} + \frac{2x + 1}{x + 2}$ $\quad \frac{2(2x^2 + 3x + 6)}{(x + 2)(x - 2)}$

135. $\frac{x}{x - 1} + \frac{x}{x - 2} + \frac{x}{x - 3}$ $\quad \frac{3x^3 - 12x^2 + 11x}{(x - 1)(x - 2)(x - 3)}$

136. $\frac{x}{x + 1} - \frac{3x + 7}{x + 2} + \frac{2x + 1}{x + 2}$ $\quad \frac{-5x - 6}{(x + 1)(x + 2)}$

137. $\frac{3(x + 1)}{x} - \frac{5(x^2 + 3)}{x^2} + \frac{x}{x + 1}$ $\quad \frac{-x^3 + x^2 - 12x - 15}{x^2(x + 1)}$

138. $\frac{3x}{x + 1} + \frac{x^2 + 4x + 3}{x^2 + 3x + 2} - \frac{x^2 + x - 6}{x^2 - 4}$ $\quad \frac{3x}{x + 1}$

Simplify each complex fraction. Assume that no denominators are 0.

139. $\dfrac{\frac{5x}{2}}{\frac{3x^2}{8}}$ $\frac{20}{3x}$

140. $\dfrac{\frac{3x}{y}}{\frac{6x}{y^2}}$ $\frac{y}{2}$

141. $\dfrac{\frac{1}{x}+\frac{1}{y}}{x-y}$ $\frac{y+x}{xy(x-y)}$

142. $\dfrac{x^{-1}+y^{-1}}{y^{-1}-x^{-1}}$ $\frac{y+x}{x-y}$

CHAPTER TEST

Consider the set $\{-7, -\frac{2}{3}, 0, 1, 3, \sqrt{10}, 4\}$.

1. List the numbers in the set that are odd integers. $-7, 1, 3$

2. List the numbers in the set that are prime numbers. 3

Determine which property justifies each statement.

3. $(a + b) + c = (b + a) + c$ **comm. prop. of add.**

4. $a(b + c) = ab + ac$ **distrib. prop.**

Graph each interval on a number line.

5. $-4 < x \leq 2$

6. $(-\infty, -3) \cup [6, \infty)$

Write each expression without using absolute value symbols.

7. $|-17|$ 17

8. $|x - 7|$, when $x < 0$. $-(x - 7)$

Find the distance on a number line between points with the following coordinates.

9. -4 and 12 16

10. -20 and -12 8

Simplify each expression. Assume that all variables represent positive numbers, and write all answers without using negative exponents.

11. $x^4x^5x^2$ x^{11}

12. $\dfrac{r^2r^3s}{r^4s^2}$ $\frac{r}{s}$

13. $\dfrac{(a^{-1}a^2)^{-2}}{a^{-3}}$ a

14. $\left(\dfrac{x^0x^2}{x^{-2}}\right)^6$ x^{24}

Write each number in scientific notation.

15. 450,000 4.5×10^5

16. 0.000345 3.45×10^{-4}

Write each number in standard notation.

17. 3.7×10^3 3,700

18. 1.2×10^{-3} 0.0012

Simplify each expression. Assume that all variables represent positive numbers, and write all answers without using negative exponents.

19. $(25a^4)^{1/2}$ $5a^2$

20. $\left(\dfrac{36}{81}\right)^{3/2}$ $\frac{216}{729}$

21. $\left(\dfrac{8t^6}{27s^9}\right)^{-2/3}$ $\frac{9s^6}{4t^4}$

22. $\sqrt[3]{27a^6}$ $3a^2$

23. $\sqrt{12} + \sqrt{27}$ $5\sqrt{3}$

24. $2\sqrt[3]{3x^4} - 3x\sqrt[3]{24x}$ $-4x\sqrt[3]{3x}$

25. Rationalize the denominator: $\dfrac{x}{\sqrt{x} - 2}$. $\frac{x(\sqrt{x}+2)}{x-4}$

26. Rationalize the numerator: $\dfrac{\sqrt{x} - \sqrt{y}}{\sqrt{x} + \sqrt{y}}$.

$\frac{x-y}{x+2\sqrt{xy}+y}$

Perform each operation.

27. $(a^2 + 3) - (2a^2 - 4)$ $-a^2 + 7$

28. $(3a^3b^2)(-2a^3b^4)$ $-6a^6b^6$

29. $(3x - 4)(2x + 7)$ $6x^2 + 13x - 28$

30. $(a^n + 2)(a^n - 3)$ $a^{2n} - a^n - 6$

31. $(x^2 + 4)(x^2 - 4)$ $x^4 - 16$

32. $(x^2 - x + 2)(2x - 3)$ $2x^3 - 5x^2 + 7x - 6$

33. $x - 3\overline{)6x^2 + x - 23}$ $6x + 19 + \frac{34}{x-3}$

34. $2x - 1\overline{)2x^3 + 3x^2 - 1}$ $x^2 + 2x + 1$

Factor each polynomial.

35. $3x + 6y$ $3(x + 2y)$

36. $x^2 - 100$ $(x + 10)(x - 10)$

37. $10t^2 - 19tw + 6w^2$ $(5t - 2w)(2t - 3w)$

38. $3a^3 - 648$ $3(a - 6)(a^2 + 6a + 36)$

39. $x^4 - x^2 - 12$ $(x + 2)(x - 2)(x^2 + 3)$

40. $6x^4 + 11x^2 - 10$ $(3x^2 - 2)(2x^2 + 5)$

Perform each operation and simplify if possible. Assume that no denominators are 0.

41. $\dfrac{x}{x + 2} + \dfrac{2}{x + 2}$ 1

42. $\dfrac{x}{x + 1} - \dfrac{x}{x - 1}$ $\dfrac{-2x}{(x + 1)(x - 1)}$

43. $\dfrac{x^2 + x - 20}{x^2 - 16} \cdot \dfrac{x^2 - 25}{x - 5}$ $\dfrac{(x + 5)^2}{x + 4}$

44. $\dfrac{x + 2}{x^2 + 2x + 1} \div \dfrac{x^2 - 4}{x + 1}$ $\dfrac{1}{(x + 1)(x - 2)}$

Simplify each complex fraction. Assume that no denominators are 0.

45. $\dfrac{\frac{1}{a} + \frac{1}{b}}{\frac{1}{b}}$ $\dfrac{b + a}{a}$

46. $\dfrac{x^{-1}}{x^{-1} + y^{-1}}$ $\dfrac{y}{y + x}$

1 Equations and Inequalities

COLLEGE **Algebra** $f(x)$ **Now™**

Throughout the chapter, this icon introduces resources on the College AlgebraNow Web Site, accessed through **http://1pass.thomson.com** that will:

- Help you test your knowledge of the material in this chapter prior to reading it,
- Allow you to take an exam-prep quiz, and
- Provide a Personalized Learning Plan targeting areas you should study.

Careers and Mathematics

REGISTERED NURSE

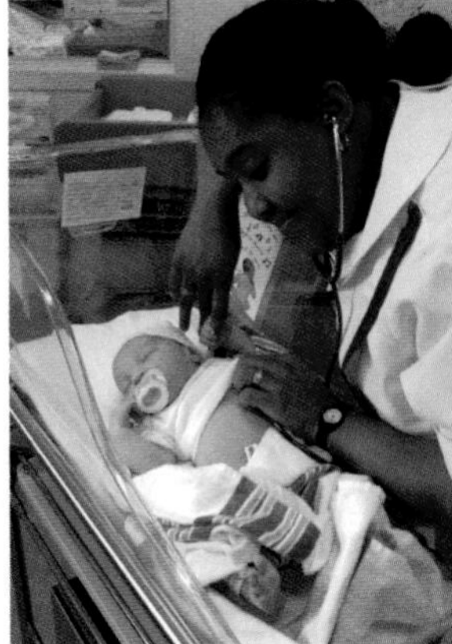

Sondra Dawes/The Image Works

Registered nurses (RNs) work to promote health, prevent disease, and help patients cope with illness. When providing direct patient care, they observe, assess, and record symptoms, reactions, and progress; assist physicians during treatments and examinations; administer medications; and assist in convalescence and rehabilitation.

In the largest health care occupation, registered nurses held about 2.2 million jobs in 2000. About 3 in 5 jobs were in hospitals. In all states and the District of Columbia, students must graduate from an approved nursing program and pass a national licensing examination to obtain a nursing license.

JOB OUTLOOK

Job opportunities for RNs are expected to be very good. Employment of registered nurses is expected to grow faster than the average for all occupations through 2012. Median annual earnings of registered nurses in 2002 were $48,090.

For a sample application, see Problem 41 in Section 1.2.

The topic of this chapter is equations—one of the most important concepts in algebra. Equations are used in almost every academic discipline and vocational area, especially in chemistry, physics, medicine, economics, and business.

1.1 Equations

In this section, you will learn about

- Properties of Equality
- Linear Equations
- Rational Equations
- Formulas

François Vieta (Viête)

(1540–1603)

By using letters in place of unknown numbers, Vieta simplified algebra and brought its notation closer to the notation that we use today. The one symbol he didn't use was the equal sign.

An **equation** is a statement indicating that two quantities are equal. An equation can be either true or false. For example, the equation $2 + 2 = 4$ is true, and the equation $2 + 3 = 6$ is false. An equation such as $3x - 2 = 10$ can be true or false depending on the value of x, which is called a **variable.** If $x = 4$, the equation is true, because 4 satisfies the equation.

$$
\begin{aligned}
3x - 2 &= 10 \\
3(4) - 2 &\stackrel{?}{=} 10 \quad \text{Substitute 4 for } x. \\
12 - 2 &\stackrel{?}{=} 10 \\
10 &= 10
\end{aligned}
$$

This equation is false for all other values of x.

Any number that satisfies an equation is called a **solution** or **root** of the equation. The set of all solutions of an equation is called its **solution set.** We have seen that the solution set of $3x - 2 = 10$ is $\{4\}$. To **solve** an equation means to find its solution set.

There can be restrictions on the values of a variable. For example, in the fraction

$$\frac{x^2 + 4}{x - 2}$$

we cannot replace x with 2, because that would make the denominator equal to 0.

EXAMPLE 1 Find the restrictions on the values of b in the equation: $\sqrt{b} = \dfrac{2}{b - 1}$.

Solution For $\sqrt{b}$ to be a real number, b must be nonnegative, and for $\dfrac{2}{b - 1}$ to be a real number, b cannot be 1. Thus, the values of b are restricted to the set of all nonnegative real numbers except 1.

Self Check Find the restrictions on a: $\sqrt{a} = \dfrac{3}{a - 2}$. ■

For some equations, called **identities,** every acceptable replacement for the variable is a solution. For example, the equation $x^2 - 9 = (x + 3)(x - 3)$ is an identity, because every real number x is a solution. For other equations, called **impossible equations** or **contradictions,** no real number is a solution. The equation $x = x + 1$ is an impossible equation. It has no solution, because no real number can be 1 greater than itself.

Equations whose solution sets contain some but not all numbers are called **conditional equations.** The equation $3x - 2 = 10$ is a conditional equation with one solution, the number 4.

If two equations have the same solution set, they are called **equivalent equations.**

Properties of Equality

There are certain properties of equality we can use to transform equations into equivalent but less complicated equations. If we use these properties, the resulting equations will be equivalent and will have the same solution set.

Properties of Equality

The Addition and Subtraction Properties
If a, b, and c are real numbers and $a = b$, then

$$a + c = b + c \quad \text{and} \quad a - c = b - c$$

The Multiplication and Division Properties
If a, b, and c are real numbers and $a = b$, then

$$ac = bc \quad \text{and} \quad \frac{a}{c} = \frac{b}{c} \quad (c \neq 0)$$

The Substitution Property
In an equation, a quantity may be substituted for its equal without changing the truth of the equation.

Linear Equations

The easiest equations to solve are the **first-degree** or **linear equations.** Since these equations involve a first-degree polynomial, they are also called **first-degree polynomial equations.**

Linear Equations

A **linear equation in one variable** (say, x) is any equation that can be written in the form

$$ax + b = 0 \quad (a \text{ and } b \text{ are real numbers and } a \neq 0)$$

To solve the linear equation $2x + 3 = 0$, we subtract 3 from both sides of the equation and then divide both sides by 2.

$$2x + 3 = 0$$

$$2x + 3 - 3 = 0 - 3 \quad \text{To undo the addition of 3, subtract 3 from both sides.}$$

$$2x = -3$$

$$\frac{2x}{2} = -\frac{3}{2} \qquad \text{To undo the multiplication by 2, divide both sides by 2.}$$

$$x = -\frac{3}{2}$$

To show that $-\frac{3}{2}$ satisfies the equation, we substitute $-\frac{3}{2}$ for x and simplify:

$$2x + 3 = 0$$

$$2\left(-\frac{3}{2}\right) + 3 \stackrel{?}{=} 0 \qquad \text{Substitute } -\tfrac{3}{2} \text{ for } x.$$

$$-3 + 3 \stackrel{?}{=} 0 \qquad 2\left(-\tfrac{3}{2}\right) = -3.$$

$$0 = 0$$

Because both sides of the equation are equal, the solution checks.

In Exercise 74, you will be asked to solve the general linear equation $ax + b = 0$ for x, thereby showing that every conditional linear equation has exactly one solution.

EXAMPLE 2 Find the solution set: $3(x + 2) = 5x + 2$.

Solution We proceed as follows:

$$3(x + 2) = 5x + 2$$

$$3x + 6 = 5x + 2 \qquad \text{Use the distributive property and remove parentheses.}$$

$$3x + 6 - 3x = 5x - 3x + 2 \qquad \text{Subtract } 3x \text{ from both sides.}$$

$$6 = 2x + 2 \qquad \text{Combine like terms.}$$

$$6 - 2 = 2x + 2 - 2 \qquad \text{Subtract 2 from both sides.}$$

$$4 = 2x \qquad \text{Simplify.}$$

$$\frac{4}{2} = \frac{2x}{2} \qquad \text{Divide both sides by 2.}$$

$$2 = x \qquad \text{Simplify.}$$

Because all of the above equations are equivalent, the solution set of the original equation is $\{2\}$. Verify that 2 satisfies the equation.

Self Check Find the solution set: $4(x - 3) = 7x - 3$. ■

Everyday Connections — Using the Internet

"I am sometimes something of a lazy person, so when I end up spending a lot of time using something myself, as I did with Google in the earliest of days, I knew it was a big deal."–Sergey Brin, Cofounder of Google

Continued

Have you ever gone online?

	College students (1,092 surveyed)	General population (2,501 surveyed)
All respondents	86%	59%
Men	87%	62%
Women	85%	56%
Whites	90%	61%
Blacks	74%	45%
Hispanics	82%	60%

Source: Pew Internet & American Life Project Survey June 26–July 26, 2002 from *The Internet Goes to College,* 09/15/2002 http://www.pewinternet.org/pdfs/PIP_College_Report.pdf

According to the Pew Internet & American Life Project: "College students are early adopters and heavy users of the Internet." We can use *variables* and equations to analyze the data from this survey. We can let the variable x represent the desired quantity. That is, let $x =$ the number of college students surveyed who have ever gone online.

1. Given the data in the table, determine how many college students surveyed have ever gone online. **939 students**
2. Given the data in the table, determine how many non-students surveyed have ever gone online. **1,476 non-students**

ACTIVE EXAMPLE 3 Find the solution set: $\frac{3}{2}y - \frac{2}{3} = \frac{1}{5}y$.

Solution To clear the equation of fractions, we multiply both sides by the LCD of the three fractions and proceed as follows:

COLLEGE **Algebra** $f(x)$ **Now**™

Explore this Active Example and test yourself on the steps taken here at **http://1pass.thomson.com**. You can also find this example on the Interactive Video Skillbuilder CD-ROM.

$$\frac{3}{2}y - \frac{2}{3} = \frac{1}{5}y$$

$$30\left(\frac{3}{2}y - \frac{2}{3}\right) = 30\left(\frac{1}{5}y\right)$$ **Multiply both sides by 30, the LCD of $\frac{3}{2}$, $\frac{2}{3}$, and $\frac{1}{5}$.**

$$45y - 20 = 6y$$ **Remove parentheses and simplify.**

$$45y - 20 + 20 = 6y + 20$$ **Add 20 to both sides.**

$$45y = 6y + 20$$ **Simplify.**

$$45y - 6y = 6y - 6y + 20$$ **Subtract 6y from both sides.**

$$39y = 20$$ **Combine like terms.**

$$\frac{39y}{39} = \frac{20}{39}$$ **Divide both sides by 39.**

$$y = \frac{20}{39}$$ **Simplify.**

The solution set is $\left\{\frac{20}{39}\right\}$. Verify that $\frac{20}{39}$ satisfies the equation.

Self Check Find the solution set: $\frac{2}{3}p - 3 = \frac{p}{6}$. ■

EXAMPLE 4 Solve: **a.** $3(x + 5) = 3(1 + x)$ and **b.** $5 + 5(x + 2) - 2x = 3x + 15$.

Solution **a.**

$$\begin{aligned} 3(x + 5) &= 3(1 + x) \\ 3x + 15 &= 3 + 3x && \text{Remove parentheses.} \\ 3x - 3x + 15 &= 3 + 3x - 3x && \text{Subtract } 3x \text{ from both sides.} \\ 15 &= 3 && \text{Combine like terms.} \end{aligned}$$

Since $15 = 3$ is false, the equation has no roots. Its solution set is the empty set, which is denoted as $\varnothing$.

b.

$$\begin{aligned} 5 + 5(x + 2) - 2x &= 3x + 15 \\ 5 + 5x + 10 - 2x &= 3x + 15 && \text{Remove parentheses.} \\ 3x + 15 &= 3x + 15 && \text{Simplify.} \end{aligned}$$

Because both sides of the final equation are identical, every value of x will make the equation true. The solution set is the set of all real numbers. This equation is an identity.

Self Check Solve: **a.** $2(x + 1) + 4 = 2(x + 3)$ and **b.** $-2(x - 4) + 6x = 4(x + 1)$. ■

■ Rational Equations

Rational equations are equations that contain rational expressions. Some examples of rational equations are

$$\frac{2}{x - 3} = 7, \qquad \frac{x + 1}{x - 2} = \frac{3}{x - 2}, \qquad \text{and} \qquad \frac{x + 2}{x + 3} + \frac{1}{x^2 + 2x - 3} = 1$$

Comment Be sure to exclude from the solution set of an equation any value that makes the denominator of a fraction equal to 0.

When solving these equations, we will multiply both sides by a quantity containing a variable. When we do this, we could inadvertently multiply both sides of an equation by 0 and obtain a solution that makes the denominator of a fraction 0. In this case, we have found a false solution, called an **extraneous solution.** These solutions do not satisfy the equation and must be discarded.

The following equation has an extraneous solution.

$$\begin{aligned} \frac{x + 1}{x - 2} &= \frac{3}{x - 2} \\ (x - 2)\left(\frac{x + 1}{x - 2}\right) &= (x - 2)\left(\frac{3}{x - 2}\right) && \text{Multiply both sides by } x - 2. \\ x + 1 &= 3 && \frac{x - 2}{x - 2} = 1. \\ x &= 2 && \text{Subtract 1 from both sides.} \end{aligned}$$

If we check by substituting 2 for x, we obtain 0's in the denominator. Thus, 2 is not a root. The solution set is $\varnothing$.

ACTIVE EXAMPLE 5 Solve: $\dfrac{x+2}{x+3} + \dfrac{1}{x^2+2x-3} = 1.$

Solution Note that x cannot be -3, because that would cause the denominator of the first fraction to be 0. To find other restrictions, we factor the trinomial in the denominator of the second fraction.

COLLEGE **Algebra** $f(x)$ **Now**™
Go to **http://1pass.thomson.com** or your CD to practice this example.

$$x^2 + 2x - 3 = (x+3)(x-1)$$

Since the denominator will be 0 when $x = -3$ or $x = 1$, x cannot be -3 or 1.

$$\frac{x+2}{x+3} + \frac{1}{x^2+2x-3} = 1$$

$$\frac{x+2}{x+3} + \frac{1}{(x+3)(x-1)} = 1 \qquad \text{Factor } x^2 + 2x - 3.$$

$$(x+3)(x-1)\left[\frac{x+2}{x+3} + \frac{1}{(x+3)(x-1)}\right] = (x+3)(x-1)1 \qquad \text{Multiply both sides by } (x+3)(x-1).$$

$$(x+3)(x-1)\left(\frac{x+2}{x+3}\right) + (x+3)(x-1)\frac{1}{(x+3)(x-1)} = (x+3)(x-1)1 \qquad \text{Remove brackets.}$$

$$(x-1)(x+2) + 1 = (x+3)(x-1) \qquad \text{Simplify.}$$

$$x^2 + x - 2 + 1 = x^2 + 2x - 3 \qquad \text{Multiply the binomials.}$$

$$x - 1 = 2x - 3 \qquad \text{Subtract } x^2 \text{ from both sides and combine like terms.}$$

$$2 = x \qquad \text{Add 3 and subtract } x \text{ from both sides.}$$

Because 2 is a meaningful replacement for x, it is a root. However, it is a good idea to check it.

$$\frac{x+2}{x+3} + \frac{1}{x^2+2x-3} = 1$$

$$\frac{2+2}{2+3} + \frac{1}{2^2+2(2)-3} \stackrel{?}{=} 1 \qquad \text{Substitute 2 for } x.$$

$$\frac{4}{5} + \frac{1}{5} \stackrel{?}{=} 1$$

$$1 = 1$$

Since 2 satisfies the equation, it is a root.

Self Check Solve: $\dfrac{3}{5} + \dfrac{7}{x+2} = 2.$ ■

■ Formulas

Many equations, called **formulas,** contain several variables. For example, the formula that converts degrees Celsius to degrees Fahrenheit is $F = \frac{9}{5}C + 32$. If we want to change a large number of Fahrenheit readings to degrees Celsius, it would be tedious to substitute each value of F into the formula and then repeatedly solve it for C. It is better to solve the formula for C, substitute the values for F, and evaluate C directly.

EXAMPLE 6 Solve $F = \frac{9}{5}C + 32$ for C.

Solution We use the same methods as for solving linear equations.

$$F = \frac{9}{5}C + 32$$

$$F - 32 = \frac{9}{5}C \qquad \text{Subtract 32 from both sides.}$$

$$\frac{5}{9}(F - 32) = \frac{5}{9}\left(\frac{9}{5}C\right) \qquad \text{Multiply both sides by } \tfrac{5}{9}.$$

$$\frac{5}{9}(F - 32) = C \qquad \text{Simplify.}$$

This result can be written in the alternate form $C = \frac{5F - 160}{9}$.

Self Check Solve $C = \frac{5}{9}(F - 32)$ for F. ■

ACTIVE EXAMPLE 7 The formula $A = p + prt$ is used to find the amount of money in a savings account at the end of a specified time. A represents the amount, p represents the principal (the original deposit), r represents the rate of simple interest per unit of time, and t represents the number of units of time. Solve this formula for p.

Solution We factor p from both terms on the right-hand side of the equation and proceed as follows:

COLLEGE **Algebra** $f(x)$ **Now**™
Go to **http://1pass.thomson.com** or your CD to practice this example.

$$A = p + prt$$

$$A = p(1 + rt) \qquad \text{Factor out } p.$$

$$\frac{A}{1 + rt} = p \qquad \text{Divide both sides by } 1 + rt.$$

$$p = \frac{A}{1 + rt}$$

Self Check Solve $pq = fq + fp$ for f. ■

Self Check Answers

1. all nonnegative real numbers but 2 **2.** $\{-3\}$ **3.** $\{6\}$ **4. a.** all real numbers **b.** no solution **5.** 3 **6.** $F = \frac{9}{5}C + 32$ **7.** $f = \frac{pq}{q + p}$

1.1 Exercises

VOCABULARY AND CONCEPTS ***Fill in the blanks.***

1. If a number satisfies an equation, it is called a root or a solution of the equation.
2. If an equation is true for all values of its variable, it is called an identity.
3. A contradiction is an equation that is true for no values of its variable.
4. A conditional equation is true for some values of its variable and is not true for others.
5. An equation of the form $ax + b = 0$ is called a linear equation.
6. If an equation contains rational expressions, it is called a rational equation.
7. A conditional linear equation has one root.
8. The denominator of a fraction can never be 0.

PRACTICE ***Each quantity represents a real number. Find any restrictions on x.***

9. $x + 3 = 1$ no restrictions
10. $\frac{1}{2}x - 7 = 14$ no restrictions
11. $\frac{1}{x} = 12$ $x \neq 0$
12. $\frac{3}{x-2}$ $x \neq 2$
13. $\sqrt{x} = 4$ $x \geq 0$
14. $\sqrt[3]{x} = 64$ no restrictions
15. $\frac{1}{x-3} = \frac{5}{x+2}$ $x \neq 3$ and $x \neq -2$
16. $\frac{24}{\sqrt{x-3}}$ $x > 3$

Solve each equation, if possible. Classify each one as an identity, a conditional equation, or an equation with no solutions.

17. $2x + 5 = 15$ 5; conditional equation
18. $3x + 2 = x + 8$ 3; conditional equation
19. $2(n + 2) - 5 = 2n$ no solution
20. $3(m + 2) = 2(m + 3) + m$ identity
21. $\frac{x+7}{2} = 7$ 7; conditional equation
22. $\frac{x}{2} - 7 = 14$ 42; conditional equation
23. $2(a + 1) = 3(a - 2) - a$ no solution
24. $x^2 = (x + 4)(x - 4) + 16$ identity
25. $3(x - 3) = \frac{6x - 18}{2}$ identity
26. $x(x + 2) = (x + 1)^2$ no solution
27. $\frac{3}{b-3} = 1$ 6; conditional equation
28. $x^2 - 8x + 15 = (x - 3)(x + 5)$ 3; conditional equation
29. $2x^2 + 5x - 3 = (2x - 1)(x + 3)$ identity
30. $2x^2 + 5x - 3 = 2x\left(x + \frac{19}{2}\right)$ $-\frac{3}{14}$; conditional equation

Solve each equation. If an equation has no solution, so indicate.

31. $2x + 7 = 10 - x$ 1
32. $9a - 3 = 15 + 3a$ 3
33. $\frac{5}{3}z - 8 = 7$ 9
34. $\frac{4}{3}y + 12 = -4$ −12
35. $\frac{z}{5} + 2 = 4$ 10
36. $\frac{3p}{7} - p = -4$ 7
37. $\frac{3x-2}{3} = 2x + \frac{7}{3}$ −3
38. $\frac{7}{2}x + 5 = x + \frac{15}{2}$ 1
39. $5(x - 2) = 2x + 8$ 6
40. $5(r - 4) = -5(r - 4)$ 4
41. $2(2x + 1) - \frac{3x}{2} = \frac{-3(4 + x)}{2}$ −2
42. $(x - 2)(x - 3) = (x + 3)(x + 4)$ $-\frac{1}{2}$
43. $7(2x + 5) - 6(x + 8) = 7$ $\frac{5}{2}$
44. $(t + 1)(t - 1) = (t + 2)(t - 3) + 4$ −1
45. $(x - 2)(x + 5) = (x - 3)(x + 2)$ 1
46. $\frac{3x+1}{20} = \frac{1}{2}$ 3
47. $\frac{3}{2}(3x - 2) - 10x - 4 = 0$ $-\frac{14}{11}$
48. $a(a - 3) + 5 = (a - 1)^2$ 4
49. $x(x + 2) = (x + 1)^2 - 1$ identity
50. $\frac{3+x}{3} + \frac{x+7}{2} = 4x + 1$ $\frac{21}{19}$
51. $\frac{(y+2)^2}{3} = y + 2 + \frac{y^2}{3}$ 2

52. $2x - \frac{7}{6} + \frac{x}{6} = \frac{4x + 3}{6}$ $\frac{10}{9}$

53. $2(s + 2) + (s + 3)^2 = s(s + 5) + 2\left(\frac{17}{2} + s\right)$ 4

54. $\frac{3}{x} + \frac{1}{2} = \frac{4}{x}$ 2

55. $\frac{2}{x + 1} + \frac{1}{3} = \frac{1}{x + 1}$ -4

56. $\frac{3}{x - 2} + \frac{1}{x} = \frac{3}{x - 2}$ no solution

57. $\frac{9t + 6}{t(t + 3)} = \frac{7}{t + 3}$ no solution

58. $x + \frac{2(-2x + 1)}{3x + 5} = \frac{3x^2}{3x + 5}$ -2

59. $\frac{2}{(a - 7)(a + 2)} = \frac{4}{(a + 3)(a + 2)}$ 17

60. $\frac{2}{n - 2} + \frac{1}{n + 1} = \frac{1}{n^2 - n - 2}$ $\frac{1}{3}$

61. $\frac{2x + 3}{x^2 + 5x + 6} + \frac{3x - 2}{x^2 + x - 6} = \frac{5x - 2}{x^2 - 4}$ $-\frac{2}{5}$

62. $\frac{3x}{x^2 + x} - \frac{2x}{x^2 + 5x} = \frac{x + 2}{x^2 + 6x + 5}$ no solution

63. $\frac{3x + 5}{x^3 + 8} + \frac{3}{x^2 - 4} = \frac{2(3x - 2)}{(x - 2)(x^2 - 2x + 4)}$ $\frac{2}{3}$

64. $\frac{1}{n + 8} - \frac{3n - 4}{5n^2 + 42n + 16} = \frac{1}{5n + 2}$ 2

65. $\frac{1}{11 - n} - \frac{2(3n - 1)}{-7n^2 + 74n + 33} = \frac{1}{7n + 3}$ 3

66. $\frac{4}{a^2 - 13a - 48} - \frac{2}{a^2 - 18a + 32} = \frac{1}{a^2 + a - 6}$ -2

67. $\frac{5}{y + 4} + \frac{2}{y + 2} = \frac{6}{y + 2} - \frac{1}{y^2 + 6y + 8}$ 5

68. $\frac{6}{2a - 6} - \frac{3}{3 - 3a} = \frac{1}{a^2 - 4a + 3}$ $\frac{7}{4}$

69. $\frac{3y}{6 - 3y} + \frac{2y}{2y + 4} = \frac{8}{4 - y^2}$ no solution

70. $\frac{3 + 2a}{a^2 + 6 + 5a} - \frac{2 - 3a}{a^2 - 6 + a} = \frac{5a - 2}{a^2 - 4}$ $-\frac{2}{5}$

71. $\frac{a}{a + 2} - 1 = -\frac{3a + 2}{a^2 + 4a + 4}$ 2

72. $\frac{x - 1}{x + 3} + \frac{x - 2}{x - 3} = \frac{1 - 2x}{3 - x}$ 0

Solve each formula for the specified variable.

73. $k = 2.2p; p$
$p = \frac{k}{2.2}$

74. $ax + b = 0; x$
$x = -\frac{b}{a}$

75. $p = 2l + 2w; w$
$w = \frac{p - 2l}{2}$

76. $V = \frac{1}{3}\pi r^2 h; h$
$h = \frac{3V}{\pi r^2}$

77. $V = \frac{1}{3}\pi r^2 h; r^2$
$r^2 = \frac{3V}{\pi h}$

78. $z = \frac{x - \mu}{\sigma}; \mu$
$\mu = x - z\sigma$

79. $P_n = L + \frac{si}{f}; s$
$s = \frac{f(P_n - L)}{i}$

80. $P_n = L + \frac{si}{f}; f$
$f = \frac{si}{P_n - L}$

81. $F = \frac{mMg}{r^2}; m$
$m = \frac{r^2F}{Mg}$

82. $\frac{1}{f} = \frac{1}{p} + \frac{1}{q}; f$
$f = \frac{pq}{q + p}$

83. $\frac{x}{a} + \frac{y}{b} = 1; y$
$y = b\left(1 - \frac{x}{a}\right)$

84. $\frac{x}{a} - \frac{y}{b} = 1; a$
$a = \frac{bx}{b + y}$

85. $\frac{1}{r} = \frac{1}{r_1} + \frac{1}{r_2}; r$
$r = \frac{r_1 r_2}{r_1 + r_2}$

86. $\frac{1}{r} = \frac{1}{r_1} + \frac{1}{r_2}$ for r_1
$r_1 = \frac{rr_2}{r_2 - r}$

87. $l = a + (n - 1)d; n$
$n = \frac{l - a + d}{d}$

88. $l = a + (n - 1)d; d$
$d = \frac{l - a}{n - 1} = \frac{a - l}{1 - n}$

89. $a = (n - 2)\frac{180}{n}; n$
$n = \frac{360}{180 - a}$

90. $S = \frac{a - lr}{1 - r}; a$
$a = S - Sr + lr$

91. $R = \frac{1}{\frac{1}{r_1} + \frac{1}{r_2} + \frac{1}{r_3}}; r_1$ $r_1 = \frac{-Rr_2r_3}{Rr_3 + Rr_2 - r_2r_3}$

92. $R = \frac{1}{\frac{1}{r_1} + \frac{1}{r_2} + \frac{1}{r_3}}; r_3$ $r_3 = \frac{Rr_1r_2}{r_1r_2 - r_2R - r_1R}$

DISCOVERY AND WRITING

93. Explain why a conditional linear equation always has exactly one root.

94. Define an extraneous solution and explain how such a solution occurs.

REVIEW ***Simplify each expression. Use absolute value symbols when necessary.***

95. $(25x^2)^{1/2}$ $5|x|$

96. $\left(\frac{25p^2}{16q^4}\right)^{1/2}$ $\frac{5|p|}{4q^2}$

97. $\left(\frac{125x^3}{8y^6}\right)^{-2/3}$ $\frac{4y^4}{25x^2}$

98. $\left(-\frac{27y^3}{1{,}000x^6}\right)^{1/3}$ $-\frac{3y}{10x^2}$

99. $\sqrt{25y^2}$ $5|y|$

100. $\sqrt[3]{-125y^9}$ $-5y^3$

101. $\sqrt[4]{\frac{a^4b^{12}}{z^8}}$ $\frac{|ab^3|}{z^2}$

102. $\sqrt[5]{\frac{x^{10}y^5}{z^{15}}}$ $\frac{x^2y}{z^3}$

1.2 Applications of Linear Equations

In this section, you will learn about

- Number Problems
- Geometric Problems
- Investment Problems
- Break-Point Analysis
- Shared Work Problems
- Mixture Problems
- Uniform Motion Problems

In this section, we will apply the equation-solving techniques discussed in the previous section to solve applied problems (often called word problems). To solve these problems, we must translate the verbal description of the problem into an equation. The process of finding the equation that describes the words of the problem is called **mathematical modeling.** The equation itself is often called a **mathematical model** of the situation described in the word problem.

The following list of steps provides a strategy to follow when we try to find the equation that models an applied problem.

Strategy for Modeling with Equations

1. Analyze the problem to see what information is given and what you are to find. Often, drawing a diagram or making a table will help you visualize the facts.
2. Pick a variable to represent the quantity that is to be found, and write a sentence telling what that variable represents. Express all other quantities mentioned in the problem as expressions involving this single variable.
3. Find a way to express a quantity in two different ways. This might involve a formula from geometry, finance, or physics.
4. Form an equation indicating that the two quantities found in Step 3 are equal.
5. Solve the equation.
6. Answer the questions asked in the problem.
7. Check the answers in the words of the problem.

This list does not apply to all situations, but it can be used for a wide range of problems with only slight modifications.

Number Problems

EXAMPLE 1 A student has scores of 74%, 78%, and 70% on three exams. What score is needed on a fourth exam for the student to earn an average grade of 80%?

Solution To find an equation that models the problem, we can let x represent the required grade on the fourth exam. The average grade will be one-fourth of the sum of the four grades. We know this average is to be 80.

The average of the four grades	equals	the required average grade.
$\dfrac{74 + 78 + 70 + x}{4}$	$=$	80

We can solve this equation for x.

$$\frac{222 + x}{4} = 80 \qquad 74 + 78 + 70 = 222.$$

$$222 + x = 320 \qquad \text{Multiply both sides by 4.}$$

$$x = 98 \qquad \text{Subtract 222 from both sides.}$$

To earn an average of 80%, the student must score 98% on the fourth exam. ■

Geometric Problems

EXAMPLE 2 A city ordinance requires a man to install a fence around the swimming pool shown in Figure 1-1. He wants the border around the pool to be of uniform width. If he has 154 feet of fencing, find the width of the border.

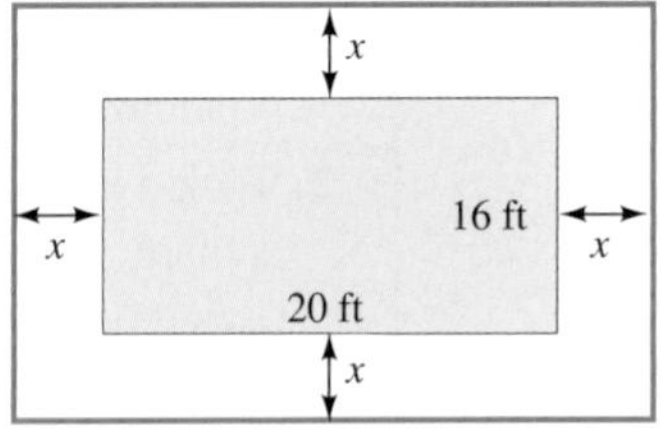

Figure 1-1

Solution We can let x represent the width of the border. The distance around the large rectangle, called its **perimeter,** is given by the formula $P = 2l + 2w$, where l is the length, $20 + 2x$, and w is the width, $16 + 2x$. Since the man has 154 feet of fencing, the perimeter will be 154 feet. To find an equation that models the problem, we substitute these values into the formula for perimeter.

$P = 2l + 2w$ — The formula for the perimeter of a rectangle.

$154 = 2(\mathbf{20 + 2x}) + 2(\mathbf{16 + 2x})$ — Substitute 154 for P, $20 + 2x$ for l, and $16 + 2x$ for w.

$154 = 40 + 4x + 32 + 4x$ — Use the distributive property to remove parentheses.

$154 = 72 + 8x$ — Combine like terms.

$82 = 8x$ — Subtract 72 from both sides.

$10\frac{1}{4} = x$ — Divide both sides by 8.

This border will be $10\frac{1}{4}$ feet wide. ■

Investment Problems

ACTIVE EXAMPLE 3 A woman invested \$10,000, part at 9% and the rest at 14%. If the annual income from these investments is \$1,275, how much was invested at each rate?

Solution We can let x represent the amount invested at 9%. Then $10{,}000 - x$ represents the amount invested at 14%. Since the annual income from any investment is the product of the interest rate and the amount invested, we have the following information.

COLLEGE **Algebra f(x) Now™**

Explore this Active Example and test yourself on the steps taken here at **http://1pass.thomson.com**. You can also find this example on the Interactive Video Skillbuilder CD-ROM.

Type of investment	Rate	Amount invested	Interest earned
9% investment	0.09	x	$0.09x$
14% investment	0.14	$10{,}000 - x$	$0.14(10{,}000 - x)$

The total income from these two investments can be expressed in two ways: as \$1,275 and as the sum of the incomes of the two investments.

The income from the 9% investment	plus	the income from the 14% investment	equals	the total income.
$0.09x$	+	$0.14(10{,}000 - x)$	=	1,275

We can solve this equation for x.

$0.09x + 0.14(10{,}000 - x) = 1{,}275$

$9x + 14(10{,}000 - x) = 127{,}500$ — Multiply both sides by 100 to eliminate the decimal points.

$9x + 140{,}000 - 14x = 127{,}500$ — Use the distributive property to remove parentheses.

$-5x + 140{,}000 = 127{,}500$ — Combine like terms.

$-5x = -12{,}500$ — Subtract 140,000 from both sides.

$x = 2{,}500$ — Divide both sides by −5.

The amount invested at 9% was \$2,500, and the amount invested at 14% was \$7,500 (\$10,000 − \$2,500). These amounts are correct, because 9% of \$2,500 is \$225, and 14% of \$7,500 is \$1,050, and the sum of these amounts is \$1,275. ■

Break-Point Analysis

Running a machine involves two costs—**setup costs** and **unit costs.** Setup costs include the cost of installing a machine and preparing it to do a job. Unit cost is the cost to manufacture one item, which includes the costs of material and labor.

EXAMPLE 4 Suppose that one machine has a setup cost of \$400 and a unit cost of \$1.50, and a second machine has a setup cost of \$500 and a unit cost of \$1.25. Find the **break point** (the number of units manufactured at which the cost on each machine is the same).

Solution We can let x represent the number of items to be manufactured. The cost c_1 of using machine 1 is

$$c_1 = 400 + 1.5x$$

and the cost c_2 of using machine 2 is

$$c_2 = 500 + 1.25x$$

The break point occurs when these two costs are equal.

The cost of using machine 1	equals	the cost of using machine 2.
$400 + 1.5x$	$=$	$500 + 1.25x$

We can solve this equation for x.

$$400 + 1.5x = 500 + 1.25x$$
$$1.5x = 100 + 1.25x \quad \text{Subtract 400 from both sides.}$$
$$0.25x = 100 \quad \text{Subtract } 1.25x \text{ from both sides.}$$
$$x = 400 \quad \text{Divide both sides by 0.25.}$$

The break point is 400 units. This result is correct, because it will cost the same amount to manufacture 400 units with either machine.

$$c_1 = \$400 + \$1.5(400) = \$1{,}000 \quad \text{and} \quad c_2 = \$500 + \$1.25(400) = \$1{,}000$$ ■

George Polya
(1888–1985)
Polya, a Hungarian, became a professor of mathematics at Stanford University. His approach to problem solving made him very popular with faculty and students. Polya's book *How to Solve It* became a best-seller. His problem-solving approach involves four steps:

1. Understand the problem.
2. Devise a plan.
3. Carry out the plan.
4. Check back.

Shared Work Problems

ACTIVE EXAMPLE 5 The Tollway Authority needs to pave 100 miles of interstate highway before freezing temperatures come in about 60 days. Sjostrom and Sons has estimated that it can do the job in 110 days. Scandroli and Sons has estimated that it can do the job in 140 days. If the authority hires both contractors, will the job get done in time?

Solution We can let n represent the number of days it will take to pave the highway if both contractors are hired. In one day, the contractors working together can do $\frac{1}{n}$ of the job. In one day, Sjostrom can do $\frac{1}{110}$ of the job. In one day, Scandroli can do $\frac{1}{140}$ of the job. The work that they can do together in one day is the sum of what each can do in one day.

COLLEGE Algebra $f(x)$ Now™
Go to **http://1pass.thomson.com** or your CD to practice this example.

The part Sjostrom can pave in one day	plus	the part Scandroli can pave in one day	equals	the part they can pave together in one day.
$\frac{1}{110}$	$+$	$\frac{1}{140}$	$=$	$\frac{1}{n}$

We can solve this equation for n.

$$\frac{1}{110} + \frac{1}{140} = \frac{1}{n}$$

$$(110)(140)n\left(\frac{1}{110} + \frac{1}{140}\right) = (110)(140)n\left(\frac{1}{n}\right)$$ **Multiply both sides by $(110)(140)n$ to eliminate the fractions.**

$$\frac{(110)(140)n}{110} + \frac{(110)(140)n}{140} = \frac{(110)(140)n}{n}$$ **Use the distributive property to remove parentheses.**

$$140n + 110n = 15{,}400$$ **$\frac{110}{110} = 1$, $\frac{140}{140} = 1$ and $\frac{n}{n} = 1$.**

$$250n = 15{,}400$$ **Combine like terms.**

$$n = 61.6$$ **Divide both sides by 250.**

It will take the contractors about 62 days to pave the highway. If the Tollway Authority is lucky, the job will be done in time. ■

■ Mixture Problems

ACTIVE EXAMPLE 6 A container is partially filled with 20 liters of whole milk containing 4% butterfat. How much 1% milk must be added to obtain a mixture that is 2% butterfat?

Solution Since the first container shown in Figure 1-2(a) contains 20 liters of 4% milk, it contains 0.04(20) liters of butterfat. To this amount, we will add the contents of the second container, which holds $0.01l$ of butterfat.

COLLEGE **Algebra** f(x) **Now**™
Go to **http://1pass.thomson.com** or your CD to practice this example.

The sum of these two amounts will equal the amount of butterfat in the third container, which is $0.02(20 + l)$ liters of butterfat. This information is presented in table form in Figure 1-2(b).

The butterfat in the 4% milk	plus	the butterfat in the 1% milk	equals	the butterfat in the 2% milk.
4% of 20 liters	$+$	1% of l liters	$=$	2% of $(20 + l)$ liters

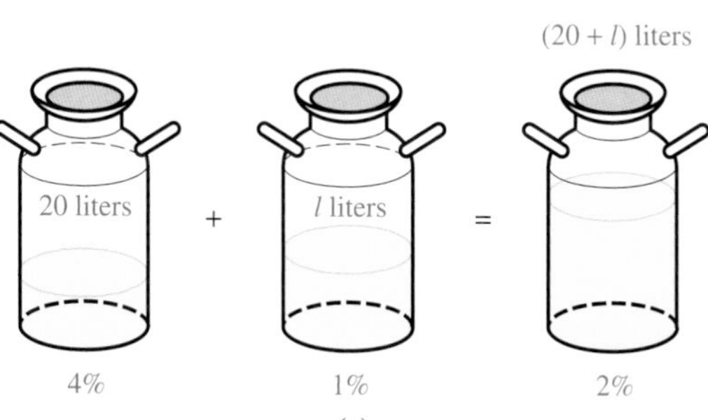

(a)

	Percentage of butterfat	·	Amount of milk	=	Amount of butterfat
4% milk	0.04		20		0.04(20)
1% milk	0.01		l		$0.01(l)$
2% milk	0.02		$20 + l$		$0.02(20 + l)$

(b)

Figure 1-2

We can solve this equation for l.

$$0.04(20) + 0.01(l) = 0.02(20 + l)$$
$$4(20) + l = 2(20 + l) \quad \text{Multiply both sides by 100.}$$
$$80 + l = 40 + 2l \quad \text{Remove parentheses.}$$
$$40 = l \quad \text{Subtract 40 and } l \text{ from both sides.}$$

To dilute the 20 liters of 4% milk to a 2% mixture, 40 liters of 1% milk must be added. To check, we note that the final mixture contains $0.02(60) = 1.2$ liters of pure butterfat, and that this is equal to the amount of pure butterfat in the 4% milk and the 1% milk;$0.04(20) + 0.01(40) = 1.2$ liters. ■

■ Uniform Motion Problems

EXAMPLE 7 A man leaves home driving at the rate of 50 mph. When his daughter discovers that he has forgotten his wallet, she drives after him at the rate of 65 mph. How long will it take her to catch her dad if he has a 15-minute head start?

Solution Uniform motion problems are based on the formula $d = rt$, where d is the distance, r is the rate, and t is the time. We can organize the information given in the problem in a chart like the one shown in Figure 1-3. In the chart, t represents the number of hours the daughter must drive to overtake her father. Because the father has a 15-minute, or $\frac{1}{4}$ hour, head start, he has been on the road for $\left(t + \frac{1}{4}\right)$ hours.

	d	=	r	·	t
Man	$50\left(t + \frac{1}{4}\right)$		50		$t + \frac{1}{4}$
Daughter	$65t$		65		t

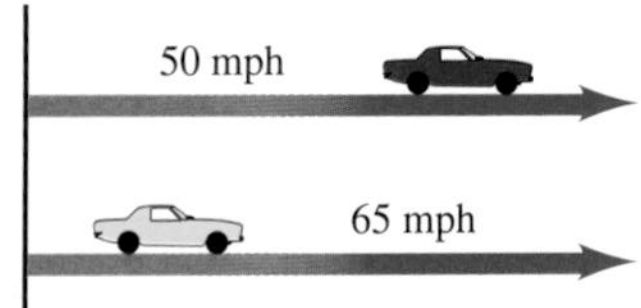

Figure 1-3

We can set up the following equation and solve it for t.

The distance the man drives	equals	the distance the daughter drives.
$50\left(t + \frac{1}{4}\right)$	=	$65t$

We can solve this equation for t.

$$50\left(t + \frac{1}{4}\right) = 65t$$
$$50t + \frac{25}{2} = 65t$$
$$\frac{25}{2} = 15t$$
$$\frac{5}{6} = t$$

It will take the daughter $\frac{5}{6}$ hours, or 50 minutes, to overtake her father. ■

1.2 Exercises

VOCABULARY AND CONCEPTS ***Fill in the blanks.***

1. To average n scores, add the scores and divide by n.
2. The formula for the perimeter of a rectangle is $P = 2l + 2w$.
3. The simple interest earned on an investment is the product of the interest rate and the amount invested.
4. The number of units manufactured at which the cost on two machines is equal is called the break point.
5. Distance traveled is the product of the rate and the time.
6. 5% of 30 liters is 1.5 liters.

PRACTICE ***Solve each problem.***

7. **Test scores** Juan scored 5 points higher on his midterm and 13 points higher on his final than he did on his first exam. If his mean (average) score was 90, what was his score on the first exam? **84**
8. **Test scores** Sally took four tests in science class. On each successive test, her score improved by 3 points. If her mean score was 69.5%, what did she score on the first test? **65%**
9. **Teacher certification** On the Illinois certification test for teachers specializing in learning disabilities, a teacher earned the scores shown in the accompanying table. What was the teacher's score in program development? **94**

Human development with special needs	82
Assessment	90
Program development and instruction	?
Professional knowledge and legal issues	78
AVERAGE SCORE	86

10. **Golfing** Par on a golf course is 72. If a professional golfer shot rounds of 76, 68, and 70 in a tournament, what will she need to shoot on the final round to average par? **74**
11. **Replacing locks** A locksmith charges \$40 plus \$28 for each lock installed. How many locks can be replaced for \$236? **7**
12. **Delivering ads** A college student earns \$20 per day delivering advertising brochures door-to-door, plus 75¢ for each person he interviews. How many people did he interview on a day when he earned \$56? **48**
13. **Width of a picture frame** The picture frame with the dimensions shown in the illustration was built with 14 feet of framing material. Find its width. **$2\frac{1}{2}$ ft**

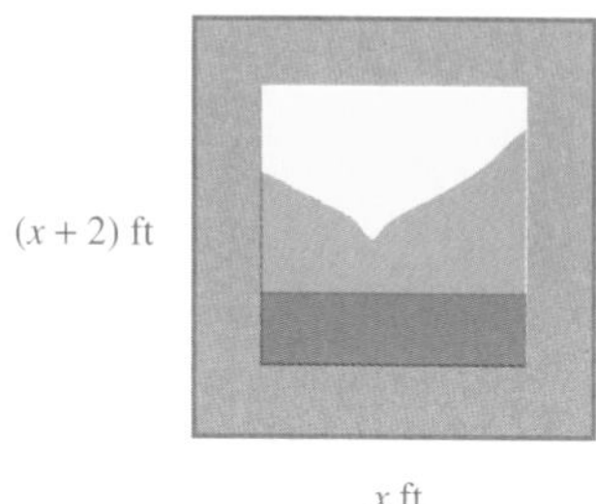

14. **Fencing a garden** If a gardener fences in the total rectangular area shown in the illustration instead of just the square area, he will need twice as much fencing to enclose the garden. How much fencing will he need? **96 ft**

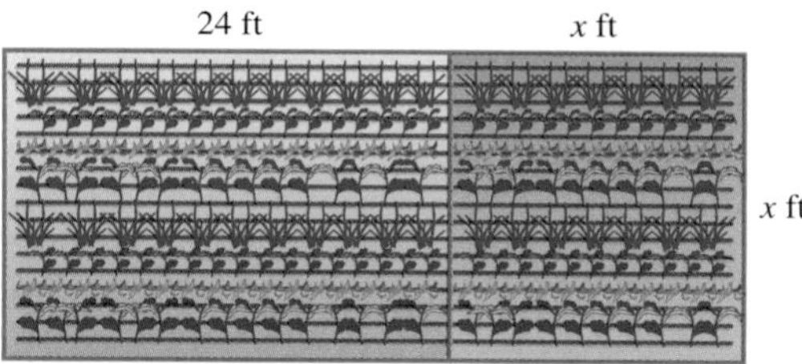

15. **Wading pool dimensions** The area of the triangular swimming pool shown in the illustration is doubled by adding a rectangular wading pool. Find the dimensions of the pool. (*Hint:* The area of a triangle $= \frac{1}{2}bh$, and the area of a rectangle $= lw$.) **20 ft by 8 ft**

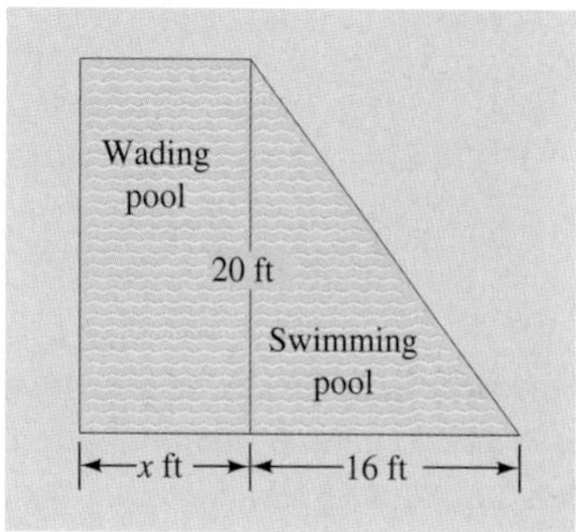

16. House construction A builder wants to install a triangular window with the angles shown in the illustration. What angles will he have to cut to make the window fit? (*Hint:* The sum of the angles in a triangle equals 180°.) **70°, 70°, 40°**

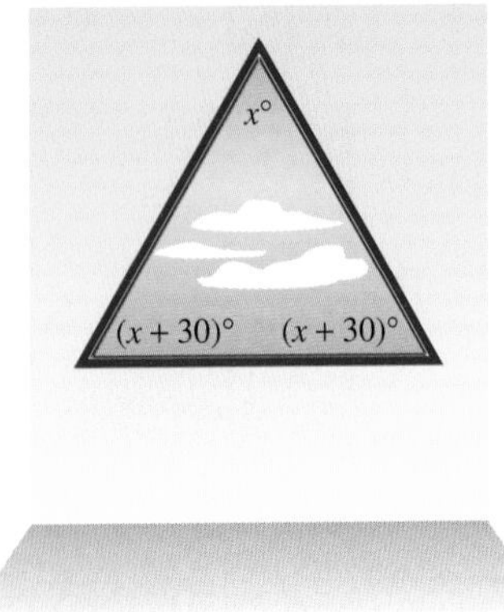

17. Length of a living room If a carpenter adds a porch with dimensions shown in the illustration to the living room, the living area will be increased by 50%. Find the length of the living room. **20 ft**

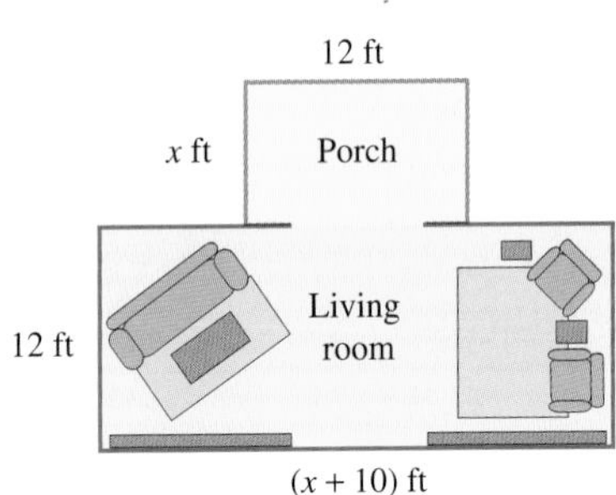

18. Depth of water in a trough The trough in the illustration has a cross-sectional area of 54 square inches. Find the depth, d, of the trough. (*Hint:* Area of a trapezoid $= \frac{1}{2}h(b_1 + b_2)$.) **$d = 5.4$ in.**

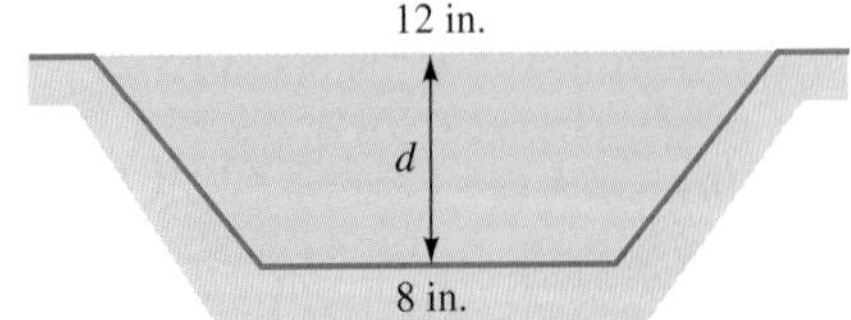

19. Investment problem An executive invests $22,000, some at 7% and some at 6% annual interest. If he receives an annual return of $1,420, how much is invested at each rate? **$10,000 at 7%, $12,000 at 6%**

20. Financial planning After inheriting some money, a woman wants to invest enough to have an annual income of $5,000. If she can invest $20,000 at 9% annual interest, how much more will she have to invest at 7% to achieve her goal? (See the table.) **$45,714.29**

Type	Rate	Amount	Income
9% investment	0.09	20,000	.09(20,000)
7% investment	0.07	x	$.07x$

21. Ticket sales A full-price ticket for a college basketball game costs $2.50, and a student ticket costs $1.75. If 585 tickets were sold, and the total receipts were $1,217.25, how many tickets were student tickets? **327**

22. Ticket sales Of the 800 tickets sold to a movie, 480 were full-price tickets costing $3 each. If the gate receipts were $2,080, what did a student ticket cost? **$2**

23. Investment problem A woman invests $37,000, part at 8% and the rest at $9\frac{1}{2}$% annual interest. If the $9\frac{1}{2}$% investment provides $452.50 more income than the 8% investment, how much is invested at each rate? **$17,500 at 8%, $19,500 at $9\frac{1}{2}$%**

24. Investment problem Equal amounts are invested at 6%, 7%, and 8% annual interest. If the three investments yield a total of $2,037 annual interest, find the total investment. **$29,100**

25. Discounts After being discounted 20%, a radio sells for $63.96. Find the original price. **$79.95**

26. Markups A merchant increases the wholesale cost of a washing machine by 30% to determine the selling price. If the washer sells for $588.90, find the wholesale cost. **$453**

27. Break-point analysis A machine to mill a brass plate has a setup cost of $600 and a unit cost of $3 for each plate manufactured. A bigger machine has a setup cost of $800 but a unit cost of only $2 for each plate manufactured. Find the break point. **200 units**

28. Break-point analysis A machine to manufacture fasteners has a setup cost of $1,200 and a unit cost of $0.005 for each fastener manufactured. A newer machine has a setup cost of $1,500 but a unit cost of only $0.0015 for each fastener manufactured. Find the break point. **about 85,714 units**

29. Computer sales A computer store has fixed costs of $8,925 per month and a unit cost of $850 for every computer it sells. If the store can sell all the computers it can get for $1,275 each, how many must be sold for the store to break even? (*Hint:* The break-even point occurs when costs equal income.) **21**

30. Restaurant management A restaurant has fixed costs of $137.50 per day and an average unit cost of $4.75 for each meal served. If a typical meal costs $6, how many customers must eat at the restaurant each day for the owner to make a profit? **110**

31. Roofing houses A man estimates that it will take him 7 days to roof his house. A professional roofer estimates that it will take him 4 days to roof the same house. How long will it take if they work together? **$2\frac{6}{11}$ days**

32. Sealing asphalt One crew can seal a parking lot in 8 hours and another in 10 hours. How long will it take to seal the parking lot if the two crews work together? **$4\frac{4}{9}$ hr**

33. Mowing lawns If a woman can mow a lawn with a lawn tractor in 2 hours, and her husband can mow the same lawn with a push mower in 4 hours, how long will it take to mow the lawn if they work together? **$\frac{4}{3}$ hr**

34. Filling swimming pools A garden hose can fill a swimming pool in 3 days, and a larger hose can fill the pool in 2 days. How long will it take to fill the pool if both hoses are used? **$1\frac{1}{5}$ days**

35. Filling swimming pools An empty swimming pool can be filled in 10 hours. When full, the pool can be drained in 19 hours. How long will it take to fill the empty pool if the drain is left open? **$21\frac{1}{9}$ hr**

36. Preparing seafood Sam stuffs shrimp in his job as a seafood chef. He can stuff 1,000 shrimp in 6 hours. When his sister helps him, they can stuff 1,000 shrimp in 4 hours. If Sam gets sick, how long will it take his sister to stuff 500 shrimp? **6 hr**

37. Diluting solutions How much water should be added to 20 ounces of a 15% solution of alcohol to dilute it to a 10% solution? **10 oz**

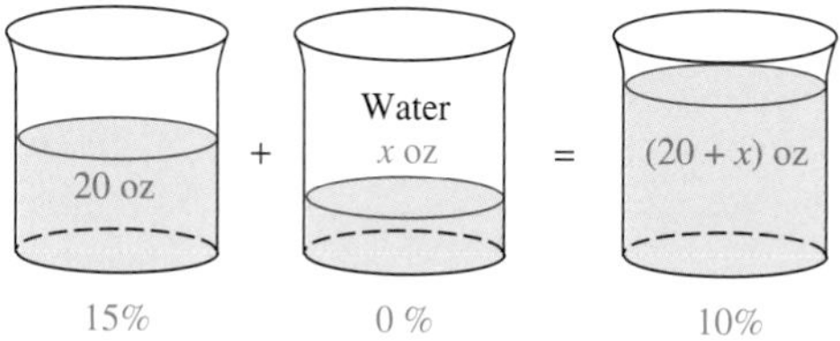

38. Increasing concentrations The beaker shown below contains a 2% saltwater solution.

a. How much water must be boiled away to increase the concentration of the salt solution from 2% to 3%? **100 ml**

b. Where on the beaker would the new water level be? **200-ml mark**

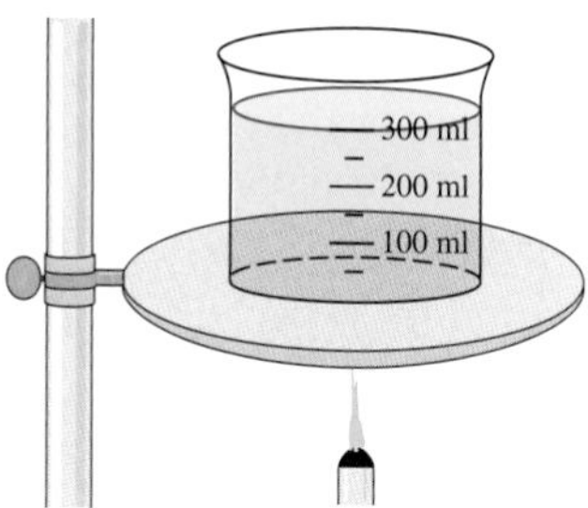

39. Winterizing cars A car radiator has a 6-liter capacity. If the liquid in the radiator is 40% antifreeze, how much liquid must be replaced with pure antifreeze to bring the mixture up to a 50% solution? **1 liter**

40. Mixing milk If a bottle holding 3 liters of milk contains $3\frac{1}{2}$% butterfat, how much skimmed milk must be added to dilute the milk to 2% butterfat? **$2\frac{1}{4}$ liters**

41. Preparing solutions A nurse has 1 liter of a solution that is 20% alcohol. How much pure alcohol must she add to bring the solution up to a 25% concentration? **$\frac{1}{15}$ liter**

42. Diluting solutions If there are 400 cubic centimeters of a chemical in 1 liter of solution, how many cubic centimeters of water must be added to dilute it to a 25% solution? (*Hint:* 1,000 cc = 1 liter.) **600 cm^3**

43. Cleaning swimming pools A swimming pool contains 15,000 gallons of water. How many gallons of chlorine must be added to "shock the pool" and bring the water to a $\frac{3}{100}$% solution? **about 4.5 gal**

44. Mixing fuels A new automobile engine can run on a mixture of gasoline and a substitute fuel. If gas costs $1.50 per gallon and the substitute fuel costs 40¢ per gallon, what percent of a mixture must be substitute fuel to bring the cost down to $1 per gallon? **about $45\frac{1}{2}$%**

45. Evaporation How many liters of water must evaporate to turn 12 liters of a 24% salt solution into a 36% solution? **4 liters**

46. Increasing concentrations A beaker contains 320 ml of a 5% saltwater solution. How much water should be boiled away to increase the concentration to 6%? **$53\frac{1}{3}$ ml**

47. Lowering fat How many pounds of extra-lean hamburger that is 7% fat must be mixed with 30 pounds of hamburger that is 15% fat to obtain a mixture that is 10% fat? **50 lb**

48. Dairy foods How many gallons of cream that is 22% butterfat must be mixed with milk that is 2% butterfat to get 20 gallons of milk containing 4% butterfat? **2 gal**

49. Mixing solutions How many gallons of a 5% alcohol solution must be mixed with 90 gallons of 1% solution to obtain a 2% solution? **30 gal**

50. Preparing medicines A doctor prescribes an ointment that is 2% hydrocortisone. A pharmacist has 1% and 5% concentrations in stock. How much of each should the pharmacist use to make a 1-ounce tube? **$\frac{3}{4}$ oz of 1%; $\frac{1}{4}$ oz of 5%**

51. Driving rates John drove to a distant city in 5 hours. When he returned, there was less traffic, and the trip took only 3 hours. If John averaged 26 mph faster on the return trip, how fast did he drive each way? **39 mph going; 65 mph returning**

52. Distance problem Suzi drove home at 60 mph, but her brother Jim, who left at the same time, could drive at only 48 mph. When Suzi arrived, Jim still had 60 miles to go. How far did Suzi drive? **300 mi**

53. Distance problem Two cars leave Pima Community College traveling in opposite directions. One car travels at 60 mph and the other at 64 mph. In how many hours will they be 310 miles apart? **$2\frac{1}{2}$ hr**

54. Bank robberies Some bank robbers leave town, speeding at 70 mph. Ten minutes later, the police give chase, traveling at 78 mph. How long will it take the police to overtake the robbers? **1.625 hr**

55. Jogging problem Two cross-country runners are 440 yards apart and are running toward each other, one at 8 mph and the other at 10 mph. In how many seconds will they meet? **50 sec**

56. Driving rates One morning, John drove 5 hours before stopping to eat. After lunch, he increased his speed by 10 mph. If he completed a 430-mile trip in 8 hours of driving time, how fast did he drive in the morning? **50 mph**

57. Boating problem A motorboat goes 5 miles upstream in the same time it requires to go 7 miles downstream. If the river flows at 2 mph, find the speed of the boat in still water. **12 mph**

58. Wind velocity A plane can fly 340 mph in still air. If it can fly 200 miles downwind in the same amount of time it can fly 140 miles upwind, find the velocity of the wind. **60 mph**

59. Feeding cattle A farmer wants to mix 2,400 pounds of cattle feed that is to be 14% protein. Barley (11.7% protein) will make up 25% of the mixture. The remaining 75% will be made up of oats (11.8% protein) and soybean meal (44.5% protein). How many pounds of each will he use?
600 lb barley; 1,637 lb oats; 163 lb soybean meal

60. Feeding cattle If the farmer in Exercise 59 wants only 20% of the mixture to be barley, how many pounds of each should he use?
480 lb barley; 1,757 lb oats; 163 lb soybean meal

Use a calculator to help solve each problem.

61. Machine tool design 712.51 cubic millimeters of material was removed by drilling the blind hole as shown in the illustration. Find the depth of the hole. (*Hint:* The volume of a cylinder is given by $V = \pi r^2 h$.) **about 11.2 mm**

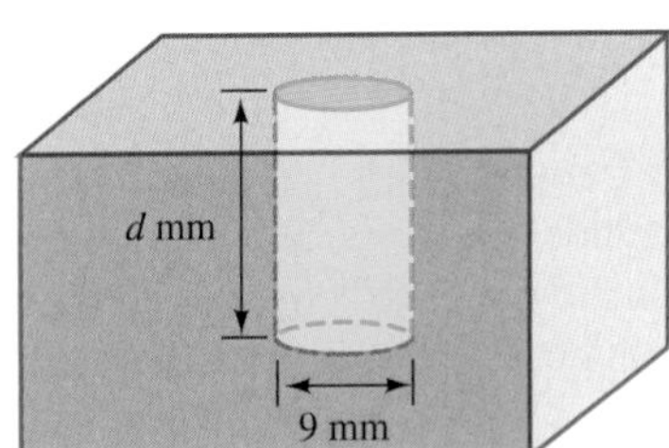

62. Architecture The Norman window with dimensions as shown is a rectangle topped by a semicircle. If the area of the window is 68.2 square feet, find its height h. **about 12 ft**

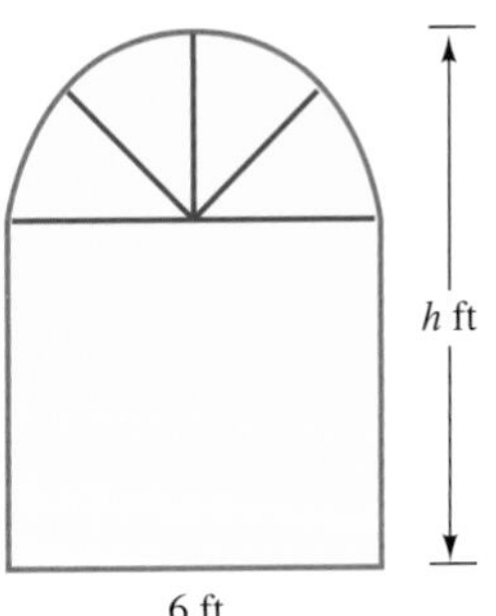

DISCOVERY AND WRITING

63. Which type of problem was easiest for you to solve? Why?

64. Which type of problem was hardest for you to solve? Why?

REVIEW *Factor each expression.*

65. $x^2 - 2x - 63$ $(x + 7)(x - 9)$

66. $2x^2 + 11x - 21$ $(2x - 3)(x + 7)$

67. $9x^2 - 12x - 5$ $(3x - 5)(3x + 1)$

68. $9x^2 - 2x - 7$ $(9x + 7)(x - 1)$

69. $x^2 + 6x + 9$ $(x + 3)^2$

70. $x^2 - 10x + 25$ $(x - 5)^2$

71. $x^3 + 8$ $(x + 2)(x^2 - 2x + 4)$

72. $27a^3 - 64$ $(3a - 4)(9a^2 + 12a + 16)$

1.3 Quadratic Equations

In this section, you will learn about

- Quadratic Equations
- Completing the Square
- The Quadratic Formula
- Formulas
- The Discriminant
- Writing Equations in Quadratic Form

We have previously solved linear equations in one variable, such as $2x + 3 = 0$. In this section, we will solve equations that contain second-degree polynomials, such as $2x^2 - 11x - 21 = 0$.

Quadratic Equations

Polynomial equations such as $2x^2 - 11x - 21 = 0$ and $3x^2 - x - 2 = 0$ are called *quadratic* or *second-degree* equations.

Quadratic Equations

A **quadratic equation** is an equation that can be written in the form $ax^2 + bx + c = 0$, where a, b, and c are real numbers and $a \neq 0$.

To solve quadratic equations by factoring, we can use the following theorem.

Zero-Factor Theorem

If a and b are real numbers, and if $ab = 0$, then

$$a = 0 \quad \text{or} \quad b = 0$$

Proof Suppose that $ab = 0$. If $a = 0$, we are finished, because at least one of a or b is 0.

If $a \neq 0$, then a has a reciprocal $\frac{1}{a}$, and we can multiply both sides of the equation $ab = 0$ by $\frac{1}{a}$ to obtain

$$ab = 0$$

$$\frac{1}{a}(ab) = \frac{1}{a}(0) \qquad \text{Multiply both sides by } \tfrac{1}{a}.$$

$$\left(\frac{1}{a} \cdot a\right)b = 0 \qquad \text{Use the associative property to group } \tfrac{1}{a} \text{ and } a \text{ together.}$$

$$1b = 0 \qquad \tfrac{1}{a} \cdot a = 1.$$

$$b = 0$$

Thus, if $a \neq 0$, then b must be 0, and the theorem is proved. ■

EXAMPLE 1 Solve: $2x^2 - 9x - 35 = 0$.

Solution The left-hand side can be factored and written as

$$(2x + 5)(x - 7) = 0$$

If either factor is 0, the product will be 0. So we can use the zero-factor theorem and set each factor equal to 0. Then we can solve for x.

$$\begin{array}{rcl} 2x + 5 = 0 & \text{or} & x - 7 = 0 \\ 2x = -5 & & x = 7 \\ x = -\dfrac{5}{2} & & \end{array}$$

Because $(2x + 5)(x - 7) = 0$ only if one of its factors is zero, $-\frac{5}{2}$ and 7 are the only solutions of the equation.

Verify that each one satisfies the equation.

Self Check Solve: $6x^2 + 7x - 3 = 0$. ■

In many quadratic equations, the quadratic expression does not factor over the set of integers. For example, the left-hand side of $x^2 - 5x + 3 = 0$ is a prime polynomial and cannot be factored over the set of integers.

To develop a method to solve these equations, we consider the equation $x^2 = c$. If c is positive, its two real roots can be found by adding $-c$ to both sides, factoring $x^2 - c$ over the set of real numbers, setting each factor equal to 0, and solving for x.

$$x^2 = c$$

$$x^2 - c = 0 \quad \text{Subtract } c \text{ from both sides.}$$

$$x^2 - (\sqrt{c})^2 = 0 \quad (\sqrt{c})^2 = c.$$

$$(x - \sqrt{c})(x + \sqrt{c}) = 0 \quad \text{Factor the difference of two squares.}$$

$$x - \sqrt{c} = 0 \quad \text{or} \quad x + \sqrt{c} = 0 \quad \text{Set each factor equal to 0.}$$

$$x = \sqrt{c} \quad \Big| \quad x = -\sqrt{c}$$

The roots of $x^2 = c$ are $x = \sqrt{c}$ and $x = -\sqrt{c}$. This fact is called the **square root property.**

Square Root Property

If $c > 0$, the equation $x^2 = c$ has two real roots:

$$x = \sqrt{c} \quad \text{or} \quad x = -\sqrt{c}$$

EXAMPLE 2 Solve: $x^2 - 8 = 0$.

Solution We solve for x^2 and apply the square root property.

$$x^2 - 8 = 0$$
$$x^2 = 8$$
$$x = \sqrt{8} \quad \text{or} \quad x = -\sqrt{8}$$
$$x = 2\sqrt{2} \quad | \quad x = -2\sqrt{2} \qquad \sqrt{8} = \sqrt{4}\sqrt{2} = 2\sqrt{2}.$$

Verify that each root satisfies the equation.

Self Check Solve: $x^2 - 12 = 0$. ■

EXAMPLE 3 Solve: $(x + 4)^2 = 1$.

Solution

$$(x + 4)^2 = 1$$
$$x + 4 = \sqrt{1} \quad \text{or} \quad x + 4 = -\sqrt{1}$$
$$x + 4 = 1 \quad | \quad x + 4 = -1$$
$$x = -3 \quad | \quad x = -5$$

Verify that each root satisfies the equation.

Self Check Solve: $(x + 5)^2 = 4$. ■

■ Completing the Square

Another method used to solve quadratic equations is called **completing the square.** This method is based on the following products:

$$x^2 + 2ax + a^2 = (x + a)^2 \qquad \text{and} \qquad x^2 - 2ax + a^2 = (x - a)^2$$

The trinomials $x^2 + 2ax + a^2$ and $x^2 - 2ax + a^2$ are perfect square trinomials, because each one factors as the square of a binomial. In each case, the coefficient of the first term is 1. If we take one-half of the coefficient of x in the middle term and square it, we obtain the third term.

$$\left[\frac{1}{2}(2a)\right]^2 = a^2 \qquad \text{and} \qquad \left[\frac{1}{2}(-2a)\right]^2 = (-a)^2 = a^2$$

This suggests that to make $x^2 + bx$ a perfect square trinomial, we find one-half of b, square it, and add the result to the binomial. For example, to make $x^2 + 10x$ a perfect square trinomial, we find one-half of 10 to get 5, square 5 to get 25, and add 25 to $x^2 + 10x$.

$$x^2 + \mathbf{10}x + \left[\frac{1}{2}(\mathbf{10})\right]^2 = x^2 + 10x + (5)^2$$
$$= x^2 + 10x + 25 \qquad \text{Note that } x^2 + 10x + 25 = (x + 5)^2.$$

To make $x^2 - 11x$ a perfect square trinomial, we find one-half of -11 to get $-\frac{11}{2}$, square $-\frac{11}{2}$ to get $\frac{121}{4}$, and add $\frac{121}{4}$ to $x^2 - 11x$.

$$x^2 - \mathbf{11}x + \left[\frac{1}{2}(\mathbf{-11})\right]^2 = x^2 - 11x + \left(-\frac{11}{2}\right)^2$$
$$= x^2 - 11x + \frac{121}{4} \qquad \text{Note that } x^2 - 11x + \tfrac{121}{4} = \left(x - \tfrac{11}{2}\right)^2$$

To solve a quadratic equation in x by completing the square, we follow these steps.

Completing the Square

1. If the coefficient of x^2 is not 1, change it to a 1 by dividing both sides of the equation by the coefficient of x^2.
2. If necessary, add a number to both sides of the equation to get the constant on the right-hand side of the equation.
3. Complete the square on x:
   a. Identify the coefficient of x, take one-half of it, and square the result.
   b. Add the number found in part a to both sides of the equation.
4. Factor the perfect square trinomial and combine like terms.
5. Solve the resulting quadratic equation by using the square root property.

To use completing the square to solve $x^2 - 10x + 24 = 0$, we note that the coefficient of x^2 is 1. We move on to Step 2 and subtract 24 from both sides to get the constant term on the right-hand side of the equal sign.

$$x^2 - 10x = -24$$

We then complete the square by adding $\left[\frac{1}{2}(-10)\right]^2 = 25$ to both sides.

$$x^2 - 10x + 25 = -24 + 25$$

$$x^2 - 10x + 25 = 1 \quad \text{Simplify on the right-hand side.}$$

We then factor the perfect square trinomial on the left-hand side.

$$(x - 5)^2 = 1$$

Finally, we use the square root property to solve this equation.

$$x - 5 = 1 \quad \text{or} \quad x - 5 = -1$$

$$x = 6 \quad | \quad x = 4$$

ACTIVE EXAMPLE 4 Use completing the square to solve $x^2 + 4x - 6 = 0$.

Solution Here the coefficient of x^2 is already 1. We move to Step 2 and add 6 to both sides to isolate the binomial $x^2 + 4x$.

$$x^2 + 4x = 6$$

We then find the number to add to both sides by completing the square: One-half of 4 (the coefficient of x) is 2, and $2^2 = 4$.

$$x^2 + 4x + 4 = 6 + 4 \quad \text{Add 4 to both sides.}$$

$$x^2 + 4x + 4 = 10$$

$$(x + 2)^2 = 10 \quad \text{Factor } x^2 + 4x + 4.$$

$$x + 2 = \sqrt{10} \quad \text{or} \quad x + 2 = -\sqrt{10} \quad \text{Use the square root property.}$$

$$x = -2 + \sqrt{10} \quad | \quad x = -2 - \sqrt{10}$$

Verify that each root satisfies the original equation.

COLLEGE Algebra f(x) Now™
Explore this Active Example and test yourself on the steps taken here at **http://1pass.thomson.com**. You can also find this example on the Interactive Video Skillbuilder CD-ROM.

Self Check Solve: $x^2 - 2x - 9 = 0$. ■

EXAMPLE 5 Solve by completing the square: $x(x + 3) = 2$.

Solution We remove parentheses to get

$$x^2 + 3x = 2$$

Since the coefficient of x^2 is 1 and the constant is on the right-hand side, we move to Step 3 and find the number to be added to both sides to complete the square: One-half of 3 (the coefficient of x) is $\frac{3}{2}$, and the square of $\frac{3}{2}$ is $\frac{9}{4}$.

$$x^2 + 3x + \frac{9}{4} = 2 + \frac{9}{4} \quad \text{Add } \tfrac{9}{4} \text{ to both sides.}$$

$$\left(x + \frac{3}{2}\right)^2 = \frac{17}{4} \quad \text{Factor } x^2 + 3x + \tfrac{9}{4}.$$

$$x + \frac{3}{2} = \frac{\sqrt{17}}{2} \quad \text{or} \quad x + \frac{3}{2} = -\frac{\sqrt{17}}{2} \quad \text{Use the square root property.}$$

$$x = \frac{-3 + \sqrt{17}}{2} \quad \Big| \quad x = \frac{-3 - \sqrt{17}}{2} \quad \text{Subtract } \tfrac{3}{2} \text{ from both sides.}$$

Verify that each root satisfies the original equation.

Self Check Solve: $x(x + 5) = 1$. ■

EXAMPLE 6 Solve by completing the square $6x^2 + 5x - 6 = 0$.

Solution We begin by dividing both sides of the equation by 6 to make the coefficient of x^2 equal to 1. Then we proceed as follows:

$$6x^2 + 5x - 6 = 0$$

$$x^2 + \frac{5}{6}x - 1 = 0 \quad \text{Divide both sides by 6.}$$

$$x^2 + \frac{5}{6}x = 1 \quad \text{Add 1 to both sides.}$$

$$x^2 + \frac{5}{6}x + \frac{25}{144} = 1 + \frac{25}{144} \quad \text{Add } \left(\tfrac{1}{2} \cdot \tfrac{5}{6}\right)^2\text{, or } \tfrac{25}{144}\text{, to both sides.}$$

$$\left(x + \frac{5}{12}\right)^2 = \frac{169}{144} \quad \text{Factor } x^2 + \tfrac{5}{6}x + \tfrac{25}{144}.$$

Now we apply the square root property.

$$x + \frac{5}{12} = \sqrt{\frac{169}{144}} \quad \text{or} \quad x + \frac{5}{12} = -\sqrt{\frac{169}{144}}$$

$$x + \frac{5}{12} = \frac{13}{12} \quad \Big| \quad x + \frac{5}{12} = -\frac{13}{12}$$

$$x = \frac{8}{12} \quad \Big| \quad x = -\frac{18}{12}$$

$$x = \frac{2}{3} \quad \Big| \quad x = -\frac{3}{2}$$

Verify that each root satisfies the original equation.

Self Check Solve: $2x^2 - 5x - 3 = 0.$ ■

■ The Quadratic Formula

We can solve the equation $ax^2 + bx + c = 0$ $(a \neq 0)$ by completing the square. The result will be a formula that we can use to solve quadratic equations.

$$ax^2 + bx + c = 0$$

$$x^2 + \frac{b}{a}x + \frac{c}{a} = \frac{0}{a} \quad \text{Divide both sides by } a.$$

$$x^2 + \frac{b}{a}x = -\frac{c}{a} \quad \text{Subtract } \tfrac{c}{a} \text{ from both sides.}$$

$$x^2 + \frac{b}{a}x + \frac{b^2}{4a^2} = \frac{b^2}{4a^2} - \frac{4ac}{4aa} \quad \text{Add } \tfrac{b^2}{4a^2} \text{ to both sides and multiply the numerator and denominator of } \tfrac{c}{a} \text{ by } 4a.$$

$$\left(x + \frac{b}{2a}\right)^2 = \frac{b^2 - 4ac}{4a^2} \quad \text{Factor the left-hand side and add the fractions on the right-hand side.}$$

We can now apply the square root property.

$$x + \frac{b}{2a} = \sqrt{\frac{b^2 - 4ac}{4a^2}} \quad \text{or} \quad x + \frac{b}{2a} = -\sqrt{\frac{b^2 - 4ac}{4a^2}}$$

$$x = -\frac{b}{2a} + \frac{\sqrt{b^2 - 4ac}}{2a} \quad\Big|\quad x = -\frac{b}{2a} - \frac{\sqrt{b^2 - 4ac}}{2a}$$

$$x = \frac{-b + \sqrt{b^2 - 4ac}}{2a} \quad\Big|\quad x = \frac{-b - \sqrt{b^2 - 4ac}}{2a}$$

These values of x are the two roots of the equation $ax^2 + bx + c = 0$. They are usually combined into a single expression, called the **quadratic formula.**

Quadratic Formula The solutions of the general quadratic equation, $ax^2 + bx + c = 0$, are

$$x = \frac{-b \pm \sqrt{b^2 - 4ac}}{2a} \qquad (a \neq 0)$$

The quadratic formula should be read twice, one using the + sign and once using the − sign. The quadratic formula implies that

$$x = \frac{-b + \sqrt{b^2 - 4ac}}{2a} \quad \text{or} \quad x = \frac{-b - \sqrt{b^2 - 4ac}}{2a}$$

Comment Be sure to write the quadratic formula correctly. Do not write it as

$$x = -b \pm \frac{\sqrt{b^2 - 4ac}}{2a}$$

EXAMPLE 7 Use the quadratic formula to solve $x^2 - 5x + 3 = 0$.

Solution In this equation $a = 1$, $b = -5$, and $c = 3$.

$$x = \frac{-b \pm \sqrt{b^2 - 4ac}}{2a}$$ The quadratic formula.

$$x = \frac{-(-5) \pm \sqrt{(-5)^2 - 4(1)(3)}}{2(1)}$$ Substitute 1 for a, -5 for b, and 3 for c.

$$x = \frac{5 \pm \sqrt{13}}{2}$$ $(-5)^2 - 4(1)(3) = 25 - 12 = 13$.

Both values satisfy the original equation.

Self Check Solve: $3x^2 - 5x + 1 = 0$. ■

ACTIVE EXAMPLE 8 Use the quadratic formula to solve $2x^2 + 8x - 7 = 0$.

Solution In this equation, $a = 2$, $b = 8$, and $c = -7$.

COLLEGE **Algebra** $f(x)$ **Now**™
Go to **http://1pass.thomson.com** or your CD to practice this example.

$$x = \frac{-b \pm \sqrt{b^2 - 4ac}}{2a}$$ The quadratic formula.

$$x = \frac{-8 \pm \sqrt{8^2 - 4(2)(-7)}}{2(2)}$$ Substitute 2 for a, 8 for b, and -7 for c.

$$x = \frac{-8 \pm \sqrt{120}}{4}$$ $8^2 - 4(2)(-7) = 64 + 56 = 120$.

$$x = \frac{-8 \pm 2\sqrt{30}}{4}$$ $\sqrt{120} = \sqrt{4 \cdot 30} = 2\sqrt{30}$.

$$x = -2 + \frac{\sqrt{30}}{2} \quad \text{or} \quad x = -2 - \frac{\sqrt{30}}{2}$$

Both values satisfy the original equation.

Self Check Solve: $4x^2 + 16x - 13 = 0$. ■

■ Formulas

Many formulas involve quadratic equations. For example, if an object is fired straight up into the air with an initial velocity of 88 feet per second, its height is given by the formula $h = 88t - 16t^2$, where h represents its height (in feet) and t represents the elapsed time (in seconds) since it was fired.

To solve this formula for t, we use the quadratic formula.

$$h = 88t - 16t^2$$

$$16t^2 - 88t + h = 0$$ Add $16t^2$ and $-88t$ to both sides.

$$t = \frac{-(-88) \pm \sqrt{(-88)^2 - 4(16)(h)}}{2(16)}$$ Substitute into the quadratic formula.

$$t = \frac{88 \pm \sqrt{7{,}744 - 64h}}{32}$$ Simplify.

The Discriminant

We can predict what type of numbers will be roots of a quadratic equation before we solve it. Suppose that the coefficients a, b, and c in the equation $ax^2 + bx + c = 0$ $(a \neq 0)$ are real numbers. Then the two roots of the equation are given by the quadratic formula

$$x = \frac{-b \pm \sqrt{b^2 - 4ac}}{2a} \qquad (a \neq 0)$$

The value of $b^2 - 4ac$, called the **discriminant,** determines the nature of the roots. The possibilities are summarized as follows.

Nature of the Roots of a Quadratic Equation

If a, b, and c are real numbers and $b^2 - 4ac$ is . . .	***then the solutions are***
positive	unequal real numbers
0	equal real numbers
negative	not real numbers
If a, b, and c are rational numbers and $b^2 - 4ac$ is . . .	***then the solutions are***
0	equal rational numbers
a nonzero perfect square	unequal rational numbers
a positive nonperfect square	unequal irrational numbers

EXAMPLE 9 Determine the nature of the roots of $3x^2 + 4x + 1 = 0$.

Solution We calculate the discriminant $b^2 - 4ac$.

$$\begin{aligned} b^2 - 4ac &= \mathbf{4}^2 - 4(\mathbf{3})(\mathbf{1}) \quad \text{Substitute 4 for } b\text{, 3 for } a\text{, and 1 for } c. \\ &= 16 - 12 \\ &= 4 \end{aligned}$$

Since a, b, and c are rational numbers and the discriminant is a non-zero perfect square, the two roots will be unequal rational numbers.

Self Check Determine the nature of the roots of $4x^2 - 3x - 2 = 0$. ■

EXAMPLE 10 If k is a constant, many quadratic equations are represented by the equation

$$(k - 2)x^2 + (k + 1)x + 4 = 0$$

Find the values of k that will give an equation with roots that are equal real numbers.

Solution We calculate the discriminant $b^2 - 4ac$ and set it equal to 0.

$$\begin{aligned}
b^2 - 4ac &= (k + 1)^2 - 4(k - 2)(4) \\
0 &= k^2 + 2k + 1 - 16k + 32 \\
0 &= k^2 - 14k + 33 \\
0 &= (k - 3)(k - 11)
\end{aligned}$$

$$k - 3 = 0 \quad \text{or} \quad k - 11 = 0$$
$$k = 3 \quad | \quad k = 11$$

When $k = 3$ or $k = 11$, the equation will have equal roots. As a check, we let $k = 3$ and note that the equation $(k - 2)x^2 + (k + 1)x + 4 = 0$ becomes

$$\begin{aligned}
(3 - 2)x^2 + (3 + 1)x + 4 &= 0 \\
x^2 + 4x + 4 &= 0
\end{aligned}$$

The roots of this equation are equal real numbers, as expected:

$$\begin{aligned}
x^2 + 4x + 4 &= 0 \\
(x + 2)(x + 2) &= 0
\end{aligned}$$

$$x + 2 = 0 \quad \text{or} \quad x + 2 = 0$$
$$x = -2 \quad | \quad x = -2$$

Similarly, $k = 11$ will give an equation with equal real roots.

Self Check Find k such that $(k - 2)x^2 - (k + 3)x + 9 = 0$ will have equal roots. ■

■ Writing Equations in Quadratic Form

If an equation can be written in quadratic form, it can be solved with the techniques used for solving quadratic equations.

ACTIVE EXAMPLE 11 Solve: $\dfrac{1}{x - 1} + \dfrac{3}{x + 1} = 2.$

Solution Since neither denominator can be zero, $x \neq 1$ and $x \neq -1$. If either number appears as a root, it must be discarded.

COLLEGE **Algebra** $f(x)$ **Now**™
Go to http://1pass.thomson.com or your CD to practice this example.

$$\frac{1}{x - 1} + \frac{3}{x + 1} = 2$$

$$(x - 1)(x + 1)\left[\frac{1}{x - 1} + \frac{3}{x + 1}\right] = (x - 1)(x + 1)2 \qquad \text{Multiply both sides by } (x - 1)(x + 1).$$

$$(x + 1) + 3(x - 1) = 2(x^2 - 1) \qquad \text{Remove brackets and simplify.}$$

$$4x - 2 = 2x^2 - 2 \qquad \text{Remove parentheses and simplify.}$$

$$0 = 2x^2 - 4x \qquad \text{Add } 2 - 4x \text{ to both sides.}$$

The resulting equation is a quadratic equation that we can solve by factoring.

$$\begin{aligned}
2x^2 - 4x &= 0 \\
2x(x - 2) &= 0 \qquad \text{Factor } 2x^2 - 4x.
\end{aligned}$$

$$2x = 0 \quad \text{or} \quad x - 2 = 0$$
$$x = 0 \quad | \quad x = 2$$

Verify these results by checking each root in the original equation.

Self Check Solve: $\dfrac{1}{x-1} + \dfrac{2}{x+1} = 1.$ ■

Self Check Answers

1. $\frac{1}{3}, -\frac{3}{2}$ **2.** $2\sqrt{3}, -2\sqrt{3}$ **3.** $-3, -7$ **4.** $1 + \sqrt{10}, 1 - \sqrt{10}$ **5.** $\dfrac{-5 \pm \sqrt{29}}{2}$ **6.** $3, -\dfrac{1}{2}$
7. $\dfrac{5 \pm \sqrt{13}}{6}$ **8.** $-2 \pm \dfrac{\sqrt{29}}{2}$ **9.** unequal and irrational **10.** 3, 27 **11.** 0, 3

1.3 Exercises

VOCABULARY AND CONCEPTS *Fill in the blanks.*

1. A quadratic equation is an equation that can be written in the form **$ax^2 + bx + c = 0$**, where $a \neq 0$.
2. If a and b are real numbers and **$ab = 0$**, then $a = 0$ or $b = 0$.
3. If $c > 0$, the equation $x^2 = c$ has two roots. They are $x =$ **$\sqrt{c}$** and $x =$ **$-\sqrt{c}$**.
4. The quadratic formula is $x =$ **$\dfrac{-b \pm \sqrt{b^2 - 4ac}}{2a}$** $(a \neq 0)$.
5. If a, b, and c are real numbers and if $b^2 - 4ac = 0$, the roots of the quadratic equation are **equal real numbers**.
6. If a, b, and c are real numbers and $b^2 - 4ac < 0$, the roots of the quadratic equation are **not real numbers**.

PRACTICE *Solve each equation by factoring. Check all answers.*

7. $x^2 - x - 6 = 0$ **3, −2**
8. $x^2 + 8x + 15 = 0$ **−5, −3**
9. $x^2 - 144 = 0$ **12, −12**
10. $x^2 + 4x = 0$ **0, −4**
11. $2x^2 + x - 10 = 0$ **$2, -\frac{5}{2}$**
12. $3x^2 + 4x - 4 = 0$ **$\frac{2}{3}, -2$**
13. $5x^2 - 13x + 6 = 0$ **$2, \frac{3}{5}$**
14. $2x^2 + 5x - 12 = 0$ **$-4, \frac{3}{2}$**
15. $15x^2 + 16x = 15$ **$\frac{3}{5}, -\frac{5}{3}$**
16. $6x^2 - 25x = -25$ **$\frac{5}{3}, \frac{5}{2}$**
17. $12x^2 + 9 = 24x$ **$\frac{3}{2}, \frac{1}{2}$**
18. $24x^2 + 6 = 24x$ **$\frac{1}{2}, \frac{1}{2}$**

Use the square root property to solve each equation. You may need to factor an expression.

19. $x^2 = 9$ **3, −3**
20. $x^2 = 20$ **$2\sqrt{5}, -2\sqrt{5}$**
21. $y^2 - 50 = 0$ **$5\sqrt{2}, -5\sqrt{2}$**
22. $x^2 - 75 = 0$ **$5\sqrt{3}, -5\sqrt{3}$**
23. $(x - 1)^2 = 4$ **3, −1**
24. $(y + 2)^2 - 49 = 0$ **5, −9**
25. $a^2 + 2a + 1 = 9$ **2, −4**
26. $x^2 - 6x + 9 = 25$ **8, −2**

Complete the square to make each binomial a perfect square trinomial.

27. $x^2 + 6x$ **$x^2 + 6x + 9$**
28. $x^2 + 8x$ **$x^2 + 8x + 16$**
29. $x^2 - 4x$ **$x^2 - 4x + 4$**
30. $x^2 - 12x$ **$x^2 - 12x + 36$**
31. $a^2 + 5a$ **$a^2 + 5a + \frac{25}{4}$**
32. $t^2 + 9t$ **$t^2 + 9t + \frac{81}{4}$**
33. $r^2 - 11r$ **$r^2 - 11r + \frac{121}{4}$**
34. $s^2 - 7s$ **$s^2 - 7s + \frac{49}{4}$**
35. $y^2 + \dfrac{3}{4}y$ **$y^2 + \frac{3}{4}y + \frac{9}{64}$**
36. $p^2 + \dfrac{3}{2}p$ **$p^2 + \frac{3}{2}p + \frac{9}{16}$**
37. $q^2 - \dfrac{1}{5}q$ **$q^2 - \frac{1}{5}q + \frac{1}{100}$**
38. $m^2 - \dfrac{2}{3}m$ **$m^2 - \frac{2}{3}m + \frac{1}{9}$**

Solve each equation by completing the square.

39. $x^2 - 8x + 15 = 0$ **5, 3**

40. $x^2 + 10x + 21 = 0$ **−3, −7**

41. $x^2 + x - 6 = 0$ **2, −3**

42. $x^2 - 9x + 20 = 0$ **5, 4**

43. $x^2 - 25x = 0$ **0, 25**

44. $x^2 + x = 0$ **0, −1**

45. $3x^2 + 4x = 4$ $\frac{2}{3}, -2$

46. $2x^2 + 5x = 12$ $-4, \frac{3}{2}$

47. $x^2 + 5 = -5x$ $\frac{-5 \pm \sqrt{5}}{2}$

48. $x^2 + 1 = -4x$ $-2 \pm \sqrt{3}$

49. $3x^2 = 1 - 4x$ $\frac{-2 \pm \sqrt{7}}{3}$

50. $2x^2 = 3x + 1$ $\frac{3 \pm \sqrt{17}}{4}$

Use the quadratic formula to solve each equation.

51. $x^2 - 12 = 0$ $\pm 2\sqrt{3}$

52. $x^2 - 20 = 0$ $\pm 2\sqrt{5}$

53. $2x^2 - x - 15 = 0$ $3, -\frac{5}{2}$

54. $6x^2 + x - 2 = 0$ $\frac{1}{2}, -\frac{2}{3}$

55. $5x^2 - 9x - 2 = 0$ $2, -\frac{1}{5}$

56. $4x^2 - 4x - 3 = 0$ $\frac{3}{2}, -\frac{1}{2}$

57. $2x^2 + 2x - 4 = 0$ **1, −2**

58. $3x^2 + 18x + 15 = 0$ **−1, −5**

59. $-3x^2 = 5x + 1$ $\frac{-5 \pm \sqrt{13}}{6}$

60. $2x(x + 3) = -1$ $\frac{-3 \pm \sqrt{7}}{2}$

61. $5x\left(x + \frac{1}{5}\right) = 3$ $\frac{-1 \pm \sqrt{61}}{10}$

62. $7x^2 = 2x + 2$ $\frac{1 \pm \sqrt{15}}{7}$

Solve each formula for the indicated variable.

63. $h = \frac{1}{2}gt^2; t$ $t = \pm\sqrt{\frac{2h}{g}}$

64. $x^2 + y^2 = r^2; x$ $x = \pm\sqrt{r^2 - y^2}$

65. $h = 64t - 16t^2; t$ $t = \frac{8 \pm \sqrt{64 - h}}{4}$

66. $y = 16x^2 - 4; x$ $x = \frac{\pm\sqrt{y + 4}}{4}$

67. $\frac{x^2}{a^2} + \frac{y^2}{b^2} = 1; y$ $y = \pm\frac{b\sqrt{a^2 - x^2}}{a}$

68. $\frac{x^2}{a^2} - \frac{y^2}{b^2} = 1; x$ $x = \pm\frac{a\sqrt{b^2 + y^2}}{b}$

69. $\frac{x^2}{a^2} - \frac{y^2}{b^2} = 1; a$ $a = \pm\frac{bx\sqrt{b^2 + y^2}}{b^2 + y^2}$

70. $\frac{x^2}{a^2} - \frac{y^2}{b^2} = 1; b$ $b = \pm\frac{ay\sqrt{x^2 - a^2}}{x^2 - a^2}$

71. $x^2 + xy - y^2 = 0; x$ $x = \frac{-y \pm y\sqrt{5}}{2}$

72. $x^2 - 3xy + y^2 = 0; y$ $y = \frac{3x \pm x\sqrt{5}}{2}$

Use the discriminant to determine the nature of the roots of each equation. **Do not solve the equation.**

73. $x^2 + 6x + 9 = 0$ **rational and equal**

74. $x^2 - 5x + 2 = 0$ **irrational and unequal**

75. $3x^2 - 2x + 5 = 0$ **not real numbers**

76. $9x^2 + 42x + 49 = 0$ **rational and equal**

77. $10x^2 + 29x = 21$ **rational and unequal**

78. $10x^2 + x = 21$ **rational and unequal**

79. $-3x^2 + 2x = 21$ **not real numbers**

80. $-8x^2 - 2x = 13$ **not real numbers**

81. Does $1{,}492x^2 + 1{,}984x - 1{,}776 = 0$ have any roots that are real numbers? **yes**

82. Does $2{,}004x^2 + 10x + 1{,}994 = 0$ have any roots that are real numbers? **no**

83. Find two values of k such that $x^2 + kx + 3k - 5 = 0$ will have two roots that are equal. **2, 10**

84. For what value(s) of b will the solutions of $x^2 - 2bx + b^2 = 0$ be equal? **all values of *b***

Change each equation to quadratic form and solve it by any method.

85. $x + 1 = \frac{12}{x}$ **3, −4**

86. $x - 2 = \frac{15}{x}$ **5, −3**

87. $8x - \frac{3}{x} = 10$ $\frac{3}{2}, -\frac{1}{4}$

88. $15x - \frac{4}{x} = 4$ $\frac{2}{3}, -\frac{2}{5}$

89. $\frac{5}{x} = \frac{4}{x^2} - 6$ $\frac{1}{2}, -\frac{4}{3}$

90. $\frac{6}{x^2} + \frac{1}{x} = 12$ $\frac{3}{4}, -\frac{2}{3}$

91. $x\left(30 - \frac{13}{x}\right) = \frac{10}{x}$ $\frac{5}{6}, -\frac{2}{5}$

92. $x\left(20 - \frac{17}{x}\right) = \frac{10}{x}$ $-\frac{2}{5}, \frac{5}{4}$

93. $(a - 2)(a + 4) = 2a(a - 3)$ $4 \pm 2\sqrt{2}$

94. $\frac{4 + a}{2a} = \frac{a - 2}{3}$ $\frac{7 \pm \sqrt{145}}{4}$

95. $\frac{1}{x} + \frac{3}{x + 2} = 2$ **1, −1**

96. $\frac{1}{x - 1} + \frac{1}{x - 4} = \frac{5}{4}$ $5, \frac{8}{5}$

97. $\frac{1}{x + 1} + \frac{5}{2x - 4} = 1$ $-\frac{1}{2}, 5$

98. $\frac{x(2x + 1)}{x - 2} = \frac{10}{x - 2}$ $-\frac{5}{2}$

99. $x + 1 + \frac{x+2}{x-1} = \frac{3}{x-1}$ -2

100. $\frac{1}{4-y} = \frac{1}{4} + \frac{1}{y+2}$ $2, -8$

101. $\frac{24}{a} - 11 = \frac{-12}{a+1}$ $3, -\frac{8}{11}$

102. $\frac{36}{b} - 17 = \frac{-24}{b+1}$ $\frac{43 \pm \sqrt{4{,}297}}{34}$

DISCOVERY AND WRITING

103. If r_1 and r_2 are the roots of $ax^2 + bx + c = 0$, show that $r_1 + r_2 = -\frac{b}{a}$.

104. If r_1 and r_2 are the roots of $ax^2 + bx + c = 0$, show that $r_1 r_2 = \frac{c}{a}$.

In Exercises 105 and 106, a stone is thrown upward, higher than the top of a tree. The stone is even with the top of the tree at times t_1 on the way up and t_2 on the way down. If the height of the tree is h feet, both t_1 and t_2 are solutions of $h = v_0 t - 16t^2$.

105. Show that the tree is $16t_1t_2$ feet tall.

106. Show that v_0 is $16(t_1 + t_2)$ feet per second.

107. Explain why the zero-factor theorem is true.

108. Explain how to complete the square on $x^2 - 17x$.

***Quadratic equations can be solved automatically by using a computer program, such as* Excel.**

109. Solve: $2x^2 - 3x - 4 = 0$.

a. Open an Excel spreadsheet. In cell B1, enter the left-hand side of the equation as =2*A1^2−3*A1−4. (Cell A1 is reserved for the value of x.)

b. After pressing the ENTER key, you will see the number −4 in cell B1. Since cell A1 is empty, its value is considered to be 0 and the value of the quadratic is −4 when $x = 0$.

c. To solve the equation, enter a guess for the solution in cell A1. To find a positive solution, enter 1 as a guess in cell A1.

d. Click inside cell B1. On the Menu bar, look under the Tools menu for SOLVER. (If there is no SOLVER, choose Add-Ins and then, inside the dialog box, select Solver Add-In and click OK.) After choosing SOLVER, a parameters window will appear. Inside the window, in front of Equal To, select Value Of, and be sure it is followed by the number 0. In the GUESS box, enter A1. Then click Solve. When the Solver Results dialog box opens, be sure that Keep Solver Solution is selected, and click OK. The solution of the equation appears in cell A1.

e. Find the negative solution of the equation.

110. Use Excel to solve $6x^2 + 13x - 5 = 0$.

REVIEW ***Simplify each expression.***

111. $5x(x-2) - x(3x-2)$ $2x^2 - 8x$

112. $(x+3)(x-9) - x(x-5)$ $-x - 27$

113. $(m+3)^2 - (m-3)^2$ $12m$

114. $[(y+z)(y-z)]^2$ $y^4 - 2y^2z^2 + z^4$

115. $\sqrt{50x^3} - x\sqrt{8x}$ $3x\sqrt{2x}$

116. $\frac{2x}{\sqrt{5} - 2}$ $2x(\sqrt{5} + 2)$

1.4 Applications of Quadratic Equations

In this section, you will learn about

- Geometric Problems
- Uniform Motion Problems
- Flying Object Problems
- Business Problems

The solutions of many problems involve quadratic equations.

Geometric Problems

EXAMPLE 1 The length of a rectangle exceeds its width by 3 feet. If its area is 40 square feet, find its dimensions.

Solution To find an equation that models the problem, we can let w represent the width of the rectangle. Then, $w + 3$ will represent its length (see Figure 1-4). Since the formula for the area of a rectangle is $A = lw$ (area = length × width), the area of the rectangle is $(w + 3)w$, which is equal to 40.

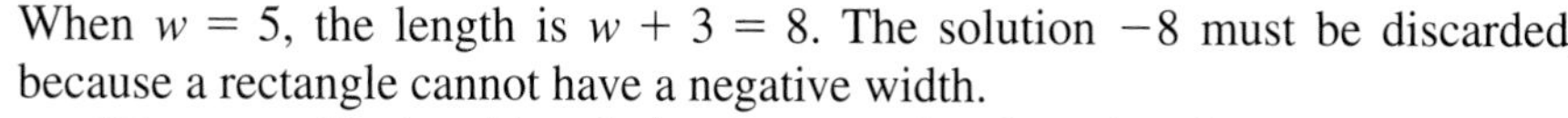

The length of the rectangle	times	the width of the rectangle	equals	the area of the rectangle.
$(w + 3)$	$\cdot$	w	$=$	40

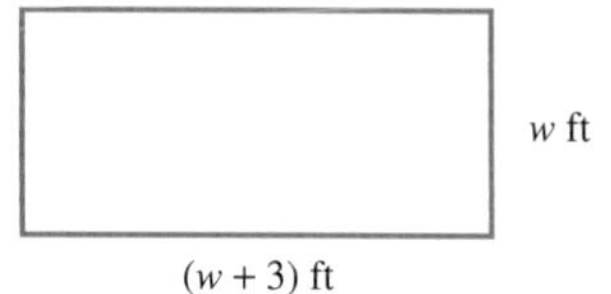

Figure 1-4

We can solve this equation for w.

$$(w + 3)w = 40$$
$$w^2 + 3w = 40$$
$$w^2 + 3w - 40 = 0 \quad \text{Subtract 40 from both sides.}$$
$$(w - 5)(w + 8) = 0 \quad \text{Factor.}$$
$$w - 5 = 0 \quad \text{or} \quad w + 8 = 0$$
$$w = 5 \quad | \quad w = -8$$

When $w = 5$, the length is $w + 3 = 8$. The solution -8 must be discarded, because a rectangle cannot have a negative width.

We can verify that this solution is correct by observing that a rectangle with dimensions of 5 feet by 8 feet has an area of 40 square feet. ■

ACTIVE EXAMPLE 2 On a college campus, a sidewalk 85 meters long (represented by the red line in Figure 1-5) joins a dormitory building D with the student center C. However, the students prefer to walk directly from D to C. If segment DC is 65 meters long, how long is each piece of the existing sidewalk?

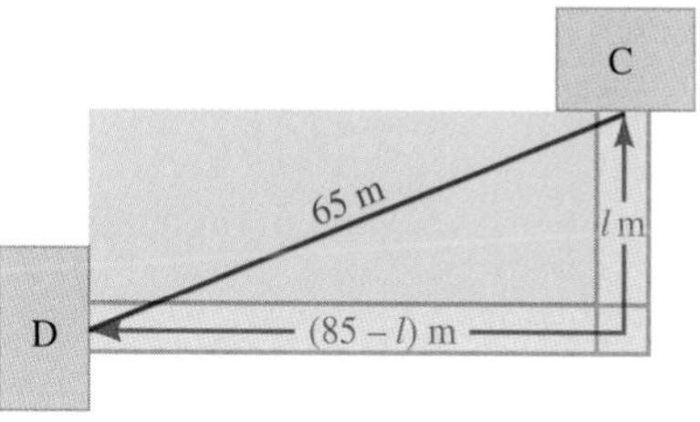

Figure 1-5

Solution We note that the triangle shown in the figure is a right triangle, with a **hypotenuse** that is 65 meters long. If we let the shorter leg of the triangle be l meters long, the length of the longer leg will be $(85 - l)$ meters. By the **Pythagorean theorem,** we know that the sum of the squares of the two legs of a right triangle is equal to the square of the hypotenuse. Thus, we can form the equation

$$l^2 + (85 - l)^2 = 65^2 \quad \text{In a right triangle, } a^2 + b^2 = c^2.$$

which we can solve as follows.

$$l^2 + 7{,}225 - 170l + l^2 = 4{,}225 \quad \text{Expand } (85 - l)^2.$$
$$2l^2 - 170l + 3{,}000 = 0 \quad \text{Combine like terms and subtract 4,225 from both sides.}$$
$$l^2 - 85l + 1{,}500 = 0 \quad \text{Divide both sides by 2.}$$

We can solve this equation using the quadratic formula.

COLLEGE **Algebra** $f(x)$ **Now**™

Explore this Active Example and test yourself on the steps taken here at **http://1pass.thomson.com**. You can also find this example on the Interactive Video Skillbuilder CD-ROM.

$$l = \frac{-b \pm \sqrt{b^2 - 4ac}}{2a}$$

$$l = \frac{-(-85) \pm \sqrt{(-85)^2 - 4(1)(1{,}500)}}{2(1)}$$

$$l = \frac{85 \pm \sqrt{1{,}225}}{2}$$

$$l = \frac{85 \pm 35}{2}$$

$$l = \frac{85 + 35}{2} \quad \text{or} \quad l = \frac{85 - 35}{2}$$

$$= 60 \qquad\qquad = 25$$

The length of the shorter leg is 25 meters. The length of the longer leg is (85 − 25) meters, or 60 meters. ■

■ Uniform Motion Problems

ACTIVE EXAMPLE 3 A man drives 600 miles to a convention. On the return trip, he is able to increase his speed by 10 mph and save 2 hours of driving time. How fast did he drive in each direction?

Solution We can let s represent the car's speed (in mph) driving to the convention. On the return trip, his speed was $s + 10$ mph. Recall that the distance traveled by an object moving at a constant rate for a certain time is given by the formula $d = rt$. If we divide both sides of this formula by r, we will have a formula for time.

COLLEGE **Algebra** $f(x)$ **Now**™
Go to **http://1pass.thomson.com** or your CD to practice this example.

$$t = \frac{d}{r}$$

We can organize the information given in this problem as in Figure 1-6.

	d	=	r	·	t
Outbound trip	600		s		$\frac{600}{s}$
Return trip	600		$s + 10$		$\frac{600}{s + 10}$

Figure 1-6

Although neither the outbound nor the return travel time is given, we know the difference of those times.

The longer time of the outbound trip	minus	the shorter time of the return trip	equals	the difference in travel times.
$\frac{600}{s}$	$-$	$\frac{600}{s + 10}$	$=$	2

Sofia Kovalevskaya
(1850–1891)

In addition to being a great mathematician, Kovalevskaya was also an early advocate for women's rights. She hoped to study mathematics at the University of Berlin, but strict rules prohibited women from attending lectures there. Undaunted, she studied with Karl Weierstrauss, who taught at the university. In 1874, she was granted a Ph.D. from the University of Göttingen. Her work produced the basis for future discoveries by other mathematicians.

We can solve this equation for s.

$$\frac{600}{s} - \frac{600}{s+10} = 2$$

$$s(s+10)\left(\frac{600}{s} - \frac{600}{s+10}\right) = s(s+10)2 \quad \text{Multiply both sides by } s(s+10) \text{ to clear the equation of fractions.}$$

$$600(s+10) - 600s = 2s(s+10) \quad \text{Simplify.}$$

$$600s + 6{,}000 - 600s = 2s^2 + 20s \quad \text{Remove parentheses.}$$

$$6{,}000 = 2s^2 + 20s \quad \text{Combine like terms.}$$

$$0 = 2s^2 + 20s - 6{,}000 \quad \text{Subtract 6,000 from both sides.}$$

$$0 = s^2 + 10s - 3{,}000 \quad \text{Divide both sides by 2.}$$

$$0 = (s-50)(s+60) \quad \text{Factor.}$$

$$s - 50 = 0 \quad \text{or} \quad s + 60 = 0 \quad \text{Set each factor equal to 0.}$$

$$s = 50 \quad | \quad s = -60$$

The solution $s = -60$ must be discarded. The man drove 50 mph to the convention and $50 + 10$, or 60 mph, on the return trip.

These answers are correct, because a 600-mile trip at 50 mph would take $\frac{600}{50}$, or 12 hours. At 60 mph, the same trip would take only 10 hours, which is 2 hours less time. ■

■ Flying Object Problems

EXAMPLE 4 If an object is thrown straight up into the air with an initial velocity of 144 feet per second, its height is given by the formula $h = 144t - 16t^2$, where h represents its height (in feet) and t represents the time (in seconds) since it was thrown. How long will it take for the object to return to the point from which it was thrown?

Solution When the object returns to its starting point, its height is again 0. Thus, we can set h equal to 0 and solve for t.

$$h = 144t - 16t^2$$

$$0 = 144t - 16t^2 \quad \text{Let } h = 0.$$

$$0 = 16t(9 - t) \quad \text{Factor.}$$

$$16t = 0 \quad \text{or} \quad 9 - t = 0 \quad \text{Set each factor equal to 0.}$$

$$t = 0 \quad | \quad t = 9$$

At $t = 0$, the object's height is 0, because it was just released. When $t = 9$, the height is again 0, and the object has returned to its starting point. ■

■ Business Problems

ACTIVE EXAMPLE 5 A bus company shuttles 1,120 passengers daily between Rockford, Illinois and O'Hare airport. The current one-way fare is \$10. For each 25¢ increase in the fare, the company predicts that it will lose 48 passengers. What increase in fare will produce daily revenue of \$10,208?

Solution Let q represent the number of quarters the fare will be increased. Then the new fare will be $\$(10 + 0.25q)$. Since the company will lose 48 passengers for each 25¢ increase, $48q$ passengers will be lost when the rate increases by q quarters. The passenger load will then be $(1{,}120 - 48q)$ passengers.

COLLEGE **Algebra f(x) Now™**
Go to http://1pass.thomson.com or your CD to practice this example.

Since the daily revenue of \$10,208 will be the product of the rate and the number of passengers, we have

$$(10 + 0.25q)(1{,}120 - 48q) = 10{,}208$$
$$11{,}200 - 480q + 280q - 12q^2 = 10{,}208 \quad \text{Remove parentheses.}$$
$$-12q^2 - 200q + 992 = 0 \quad \text{Combine like terms and subtract 10,208 from both sides.}$$
$$3q^2 + 50q - 248 = 0 \quad \text{Divide both sides by } -4.$$

We can solve this equation with the quadratic formula.

$$q = \frac{-b \pm \sqrt{b^2 - 4ac}}{2a}$$
$$q = \frac{-50 \pm \sqrt{50^2 - 4(3)(-248)}}{2(3)} \quad \text{Substitute 3 for } a\text{, 50 for } b\text{, and } -248 \text{ for } c.$$
$$q = \frac{-50 \pm \sqrt{2{,}500 + 2{,}976}}{6}$$
$$q = \frac{-50 \pm \sqrt{5{,}476}}{6}$$
$$q = \frac{-50 \pm 74}{6}$$

$$q = \frac{-50 + 74}{6} = \frac{24}{6} = 4 \quad \text{or} \quad q = \frac{-50 - 74}{6} = \frac{-124}{6} = -\frac{62}{3}$$

Since the number of riders cannot be negative, the result of $-\frac{62}{3}$ must be discarded. To generate \$10,208 in daily revenues, the company should raise the fare by 4 quarters, or \$1, to \$11. ■

1.4 Exercises

VOCABULARY AND CONCEPTS ***Fill in the blanks.***

1. The formula for the area of a rectangle is $\underline{A = lw}$.
2. The formula that relates distance, rate, and time is $\underline{d = rt}$.

PRACTICE ***Solve each problem.***

3. **Geometric problem** A rectangle is 4 feet longer than it is wide. If its area is 32 square feet, find its dimensions. 4 ft by 8 ft
4. **Geometric problem** A rectangle is 5 times as long as it is wide. If the area is 125 square feet, find its perimeter. 60 ft
5. **Geometric problem** The side of a square is 4 centimeters shorter than the side of a second square. If the sum of their areas is 106 square centimeters, find the length of one side of the larger square. 9 cm
6. **Geometric problem** The base of a triangle is one-third as long as its height. If the area of the triangle is 24 square meters, how long is its base? 4 m

7. Flags In 1912, an order by President Taft fixed the width and length of the U.S. flag in the ratio 1 to 1.9. If 100 square feet of cloth are to be used to make a U.S. flag, estimate its dimensions to the nearest $\frac{1}{4}$ foot. **width: $7\frac{1}{4}$ ft; length: $13\frac{3}{4}$ ft**

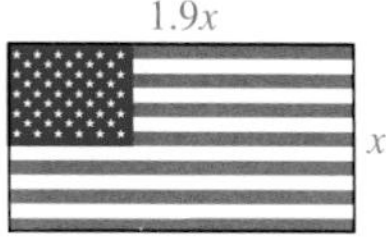

8. Imax screens A large permanent movie screen is in the Panasonic Imax theater at Darling Harbor, Sydney, Australia. The rectangular screen has an area of 11,349 square feet. Find the dimensions of the screen if it is 20 feet longer than it is wide. **97 ft by 117 ft**

9. Metal fabrication A piece of tin, 12 inches on a side, is to have four equal squares cut from its corners, as in the illustration. If the edges are then to be folded up to make a box with a floor area of 64 square inches, find the depth of the box. **2 in.**

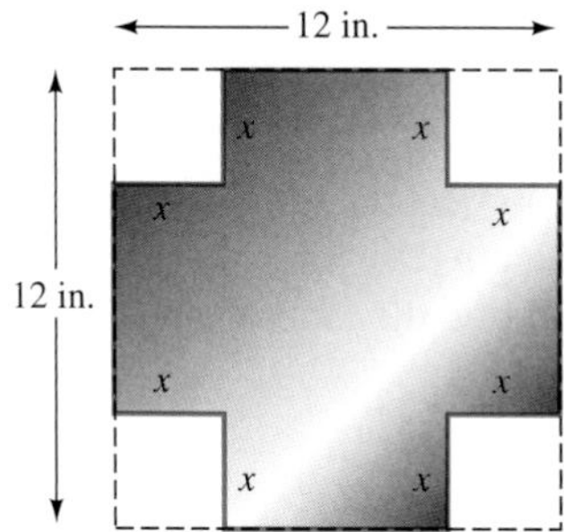

10. Making gutters A piece of sheet metal, 18 inches wide, is bent to form the gutter shown in the illustration. If the cross-sectional area is 36 square inches, find the depth of the gutter. **3 in. or 6 in.**

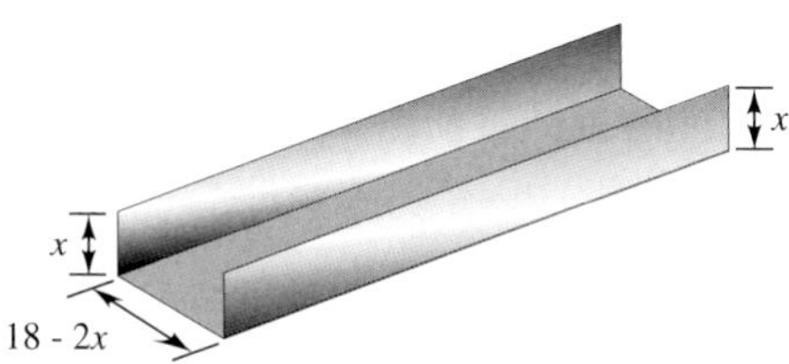

11. Manufacturing A manufacturer of television sets received an order for sets with a 46-inch screen (measured along the diagonal), as shown in the illustration). If the tubes are to be rectangular in shape and $17\frac{1}{2}$ inches wider than they are high, find the dimensions of the screen to the nearest tenth of an inch. **40.1 in. by 22.6 in.**

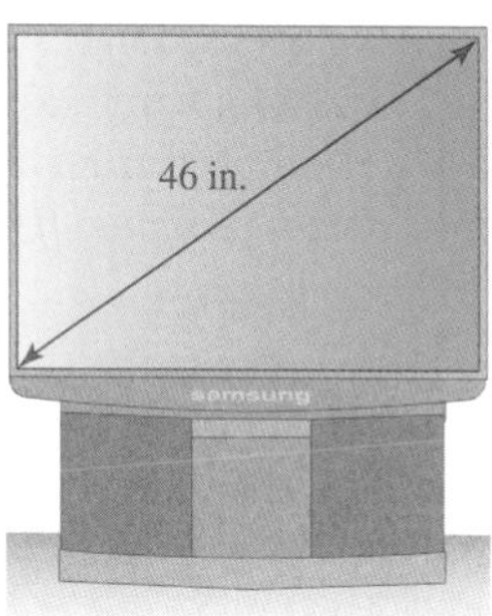

12. Finding dimensions The oriental rug shown is 2 feet longer than it is wide. To the nearest tenth of a foot, find its dimensions. **7.4 ft by 9.4 ft**

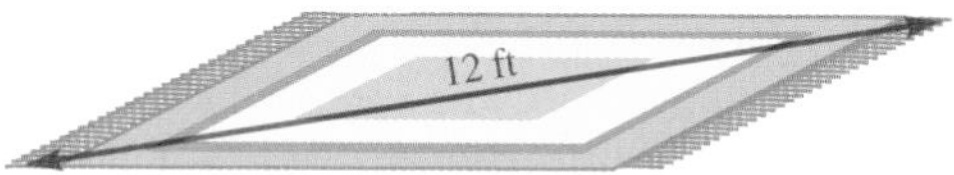

13. Cycling rates A cyclist rides from DeKalb to Rockford, a distance of 40 miles. His return trip takes 2 hours longer, because his speed decreases by 10 miles per hour. How fast does he ride each way? **20 mph going and 10 mph returning**

14. Travel times A farmer drives a tractor from one town to another, a distance of 120 kilometers. He drives 10 kilometers per hour faster on the return trip, cutting 1 hour off the time. How fast does he drive each way? **30 kph going and 40 kph returning**

15. Uniform motion problem If the speed were increased by 10 mph, a 420-mile trip would take 1 hour less time. How long will the trip take at the slower speed? **7 hr**

16. Uniform motion problem By increasing her usual speed by 25 kilometers per hour, a bus driver decreases the time on a 25-kilometer trip by 10 minutes. Find the usual speed. **50 kph**

17. Ballistics The height of a projectile fired upward with an initial velocity of 400 feet per second is given by the formula $h = -16t^2 + 400t$, where h is the height in feet and t is the time in seconds. Find the time required for the projectile to return to earth. **25 sec**

18. Ballistics The height of an object tossed upward with an initial velocity of 104 feet per second is given by the formula $h = -16t^2 + 104t$, where h is the height in feet and t is the time in seconds. Find the time required for the object to return to its point of departure. **6.5 sec**

19. **Falling coins** An object will fall s feet in t seconds, where $s = 16t^2$. How long will it take for a penny to hit the ground if it is dropped from the top of the Sears Tower in Chicago? (*Hint:* The tower is 1,454 feet tall.) **about 9.5 sec**

20. **Movie stunts** According to the *Guinness Book of World Records, 1998,* stuntman Dan Koko fell a distance of 312 feet into an airbag after jumping from the Vegas World Hotel and Casino. The distance d in feet traveled by a free-falling object in t seconds is given by the formula $d = 16t^2$. To the nearest tenth of a second, how long did the fall last? **4.4 sec**

21. **Accidents** The height h (in feet) of an object that is dropped from a height of s feet is given by the formula $h = s - 16t^2$, where t is the time the object has been falling. A 5-foot-tall woman on a sidewalk looks directly overhead and sees a window washer drop a bottle from 4 stories up. How long does she have to get out of the way? Round to the nearest tenth. (A story is 12 feet.) **1.6 sec**

22. **Ballistics** The height of an object thrown upward with an initial velocity of 32 feet per second is given by the formula $h = -16t^2 + 32t$, where t is the time in seconds. How long will it take the object to reach a height of 16 feet? **1 sec**

23. **Setting fares** A bus company has 3,000 passengers daily, paying a 25¢ fare. For each nickel increase in fare, the company projects that it will lose 80 passengers. What fare increase will produce \$994 in daily revenue? **10¢**

24. **Jazz concerts** A jazz group on tour has been drawing average crowds of 500 persons. It is projected that for every \$1 increase in the \$12 ticket price, the average attendance will decrease by 50. At what ticket price will nightly receipts be \$5,600? **\$14**

25. **Concert receipts** Tickets for the annual symphony orchestra pops concert cost \$15, and the average attendance at the concerts has been 1,200 persons. Management projects that for each 50¢ decrease in ticket price, 40 more patrons will attend. How many people attended the concert if the receipts were \$17,280? **1,440**

26. **Projecting demand** The *Vilas County News* earns a profit of \$20 per year for each of its 3,000 subscribers. Management projects that the profit per subscriber would increase by 1¢ for each additional subscriber over the current 3,000. How many subscribers are needed to bring a total profit of \$120,000? **4,000**

27. **Architecture** A **golden rectangle** is said to be one of the most visually appealing of all geometric forms. The front of the Parthenon, built in Athens in the 5th century B.C. and shown in the illustration, is a golden rectangle. In a golden rectangle, the length l and the height h of the rectangle must satisfy the equation

$$\frac{l}{h} = \frac{h}{l - h}$$

If a rectangular billboard is to have a height of 15 feet, how long should it be if it is to form a golden rectangle? Round to the nearest tenth of a foot. **24.3 ft**

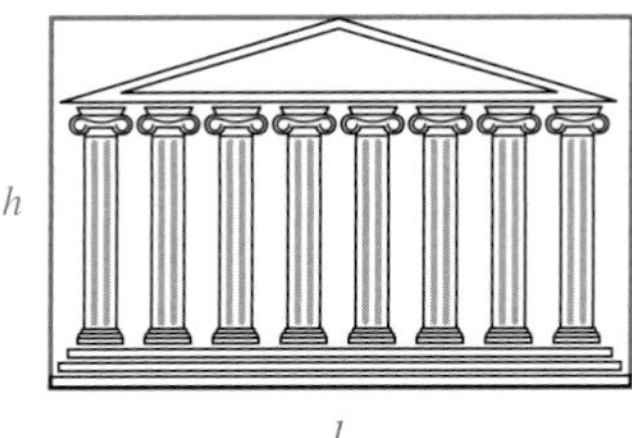

28. **Golden ratio** Rectangle $ABCD$, shown here, will be a **golden rectangle** if $\frac{AB}{AD} = \frac{BC}{BE}$ where $AE = AD$. Let $AE = 1$ and find the ratio of AB to AD. **about 1.618 to 1**

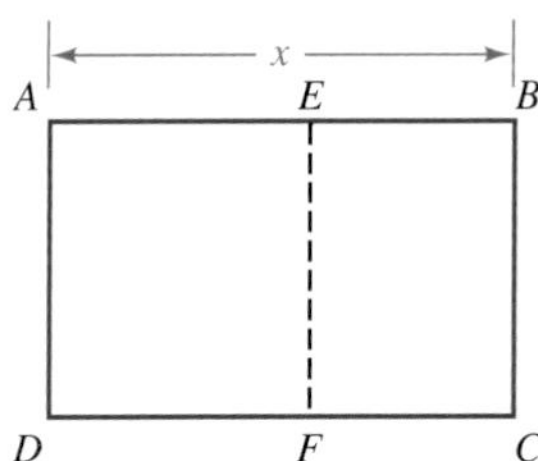

29. **Filling storage tanks** Two pipes are used to fill a water storage tank. The first pipe can fill the tank in 4 hours, and the two pipes together can fill the tank in 2 hours less time than the second pipe alone. How long would it take for the second pipe to fill the tank? **4 hr**

30. **Filling swimming pools** A hose can fill a swimming pool in 6 hours. Another hose needs 3 more hours to fill the pool than the two hoses combined. How long would it take the second hose to fill the pool? **6 hr**

31. **Mowing lawns** Kristy can mow a lawn in 1 hour less time than her brother Steven. Together they can finish the job in 5 hours. How long would it take Kristy if she worked alone? **about 9.5 hr**

32. Milking cows Working together, Sarah and Heidi can milk the cows in 2 hours. If they work alone, it takes Heidi 3 hours longer than it takes Sarah. How long would it take Heidi to milk the cows alone?
6 hr

33. Geometric problem Is it possible for a rectangle to have a width that is 3 units shorter than its diagonal and a length that is 4 units longer than its diagonal?
no

34. Geometric problem If two opposite sides of a square are increased by 10 meters and the other sides are decreased by 8 meters, the area of the rectangle that is formed is 63 square meters. Find the area of the original square. **121 m^2**

35. Investment problem Maude and Matilda each have a bank CD. Maude's is \$1,000 larger than Matilda's, but the interest rate is 1% less. Last year Maude received interest of \$280, and Matilda received \$240. Find the rate of interest for each CD.
Matilda at 8%; Maude at 7%

36. Investment problem Scott and Laura have both invested some money. Scott invested \$3,000 more than Laura and at a 2% higher interest rate. If Scott received \$800 annual interest and Laura received \$400, how much did Scott invest?
either \$8,000 or \$15,000

37. Buying microwave ovens Some mathematics professors would like to purchase a \$150 microwave oven for the department workroom. If four of the professors don't contribute, everyone's share will increase by \$10. How many professors are in the department? **10**

38. Planting windscreens A farmer intends to construct a windscreen by planting trees in a quarter-mile row. His daughter points out that 44 fewer trees will be needed if they are planted 1 foot farther apart. If her dad takes her advice, how many trees will be needed? A row starts and ends with a tree. (*Hint:* 1 mile = 5,280 feet.) **221**

39. Puzzle problem If a wagon wheel had 10 more spokes, the angle between spokes would decrease by 6°. How many spokes does the wheel have? **20**

40. Puzzle problem A merchant could sell one model of digital cameras at list price for \$180. If he had 3 more cameras, he could sell each one for \$10 less and still receive \$180. Find the list price of each camera. **\$30**

41. Geometric problem If one leg of a right triangle is 14 meters shorter than the other leg, and the hypotenuse is 26 meters, find the length of the two legs.
10 m and 24 m

42. Geometric problem Find the dimensions of a rectangle whose area is 180 cm^2 and whose perimeter is 54 cm. **12 cm by 15 cm**

43. Automobile engines As the piston shown moves upward, it pushes a cylinder of a gasoline/air mixture that is ignited by the spark plug. The formula that gives the volume of a cylinder is $V = \pi r^2 h$, where r is the radius and h the height. Find the radius of the piston (to the nearest hundredth of an inch) if it displaces 47.75 cubic inches of gasoline/air mixture as it moves from its lowest to its highest point. **1.70 in.**

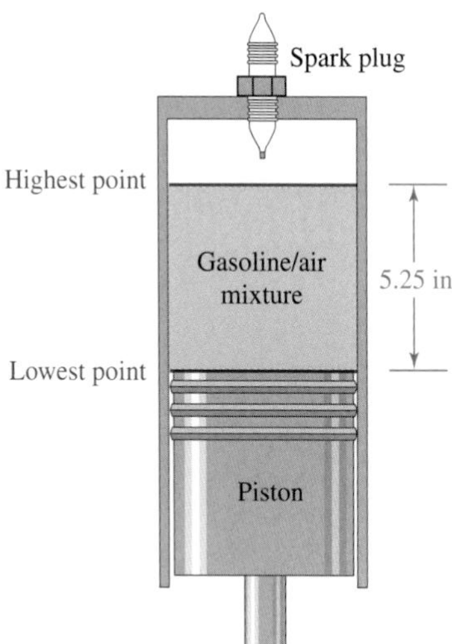

44. History One of the important cities of the ancient world was Babylon. Greek historians wrote that the city was square-shaped. Its area numerically exceeded its perimeter by about 124. Find its dimensions in miles. (Round to the nearest tenth.)
13.3 mi by 13.3 mi

DISCOVERY AND WRITING

45. Which of the preceding problems did you find the hardest? Why?

46. Which of the preceding problems did you find the easiest? Why?

REVIEW *Perform the operations and simplify.*

47. $\frac{2}{x} - \frac{1}{x-3}$ $\quad \frac{x-6}{x(x-3)}$

48. $\frac{1}{x} \cdot \frac{x^2-5x}{x-3}$ $\quad \frac{x-5}{x-3}$

49. $\frac{x+3}{x^2-x-6} \div \frac{x^2+3x}{x^2-9}$ $\quad \frac{x+3}{x(x+2)}$

50. $\frac{\frac{1}{x}-\frac{1}{2}}{x-2}$ $\quad -\frac{1}{2x}$

51. $\frac{\frac{1}{x}+\frac{1}{y}}{\frac{1}{x}-\frac{1}{y}}$ $\quad \frac{y+x}{y-x}$

52. $\frac{x}{x+1} + \frac{x+1}{x} \cdot \frac{x^2-x}{3}$ $\quad \frac{x^3+x^2+2x-1}{3(x+1)}$

1.5 Complex Numbers

In this section, you will learn about

- Imaginary Numbers
- Simplifying Imaginary Numbers
- Complex Numbers
- The Arithmetic of Complex Numbers
- Powers of *i*
- Absolute Value of a Complex Number
- Solving Quadratic Equations with Complex Roots
- Factoring the Sum of Two Squares

Solutions of quadratic equations are not always real numbers. For example, if we use the quadratic formula to solve $x^2 + x + 2 = 0$, we get solutions that are nonreal.

$$x = \frac{-b \pm \sqrt{b^2 - 4ac}}{2a}$$

$$x = \frac{-1 \pm \sqrt{1^2 - 4(1)(2)}}{2(1)} \qquad \text{Substitute 1 for } a, \text{ 1 for } b, \text{ and 2 for } c.$$

$$x = \frac{-1 \pm \sqrt{1-8}}{2}$$

$$x = \frac{-1 \pm \sqrt{-7}}{2}$$

Each solution involves $\sqrt{-7}$. This is not a real number, because the square of no real number is -7.

Imaginary Numbers

For years, mathematicians believed that numbers such as

$$\sqrt{-1}, \quad \sqrt{-4}, \quad \sqrt{-5}, \quad \text{and} \quad \sqrt{-7}$$

were nonsense. Even the great English mathematician Sir Isaac Newton (1642–1727) called them "impossible numbers." In the 17th century, these symbols were called **imaginary numbers** by René Descartes. Today, they have important uses, such as describing the behavior of alternating current in electronics.

The imaginary numbers are based on the **imaginary unit i,** where

$$i^2 = -1$$

Because i represents the square root of -1, we also write

$$i = \sqrt{-1}$$

Simplifying Imaginary Numbers

Because imaginary numbers follow the rules for exponents, we have

$$(3i)^2 = 3^2 i^2 = 9(-1) = -9 \qquad i^2 = -1.$$

Since $(3i)^2 = -9$, $3i$ is a square root of -9, and we can write

$$\sqrt{-9} = 3i$$

This result can also be obtained by using the multiplication property of radicals.

$$\begin{aligned} \sqrt{-9} &= \sqrt{9(-1)} & \\ &= \sqrt{9}\sqrt{-1} & \sqrt{ab} = \sqrt{a}\sqrt{b}. \\ &= 3i & \sqrt{9} = 3. \end{aligned}$$

We can use the multiplication property of radicals to simplify imaginary numbers.

$$\sqrt{-25} = \sqrt{25(-1)} = \sqrt{25}\sqrt{-1} = 5i$$

$$\sqrt{-7} = \sqrt{7(-1)} = \sqrt{7}\sqrt{-1} = \sqrt{7}i$$

$$\sqrt{\frac{-100}{49}} = \sqrt{\frac{100}{49}(-1)} = \sqrt{\frac{100}{49}}\sqrt{-1} = \frac{10}{7}i$$

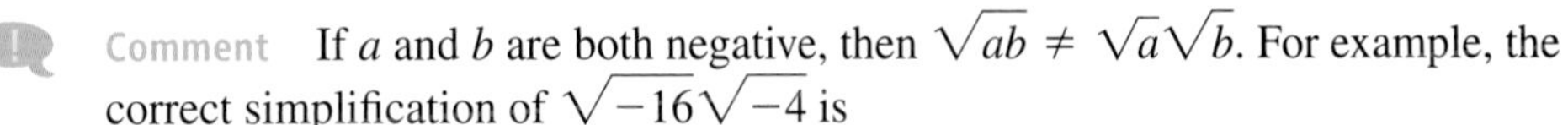

Comment If a and b are both negative, then $\sqrt{ab} \neq \sqrt{a}\sqrt{b}$. For example, the correct simplification of $\sqrt{-16}\sqrt{-4}$ is

$$\sqrt{-16}\sqrt{-4} = (4i)(2i) = 8i^2 = 8(-1) = -8$$

The following simplification is incorrect, because we get a different result.

$$\sqrt{-16}\sqrt{-4} = \sqrt{(-16)(-4)} = \sqrt{64} = 8$$

Here, a and b are both negative, and the multiplication property of radicals does not apply.

Complex Numbers

Numbers that are the sum or difference of a real number and an imaginary number, such as $3 + 4i$, $-5 + 7i$, and $-1 - 9i$, are called *complex numbers.*

Complex Numbers

A **complex number** is a number that can be written in the form $a + bi$, where a and b are real numbers and $i = \sqrt{-1}$.

The number a is called the **real part,** and b is called the **imaginary part.**

If $b = 0$, the complex number $a + bi$ is the real number a. If $a = 0$ and $b \neq 0$, the complex number $a + bi$ is the imaginary number bi. It follows that the set of real numbers and the set of imaginary numbers are subsets of the set of complex numbers.

Figure 1-7 illustrates how the various sets of numbers are related.

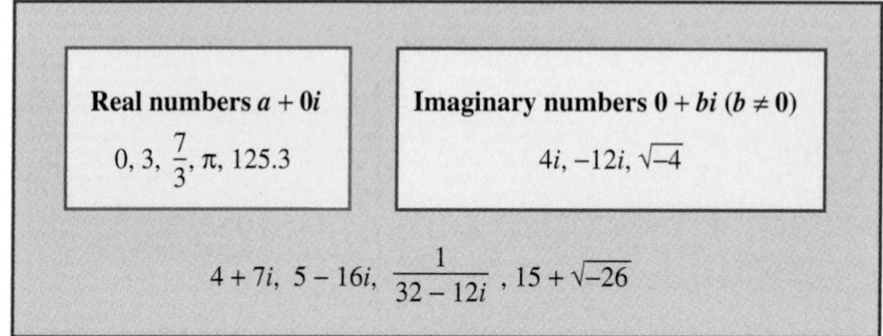

Figure 1-7

To decide whether two complex numbers are equal, we can use the following definition.

Equality of Complex Numbers

Two complex numbers are equal if their real parts are equal and their imaginary parts are equal. If $a + bi$ and $c + di$ are two complex numbers, then

$$a + bi = c + di \quad \text{if and only if} \quad a = c \quad \text{and} \quad b = d$$

EXAMPLE 1 For what numbers x and y is $3x + 4i = (2y + x) + xi$?

Solution Since the numbers are equal, their imaginary parts must be equal: $x = 4$. Since their real parts are equal, $3x = 2y + x$. We can solve the system

$$\begin{cases} x = 4 \\ 3x = 2y + x \end{cases}$$

by substituting 4 for x in the second equation and solving for y. We find that $y = 4$. The solution is $x = 4$ and $y = 4$.

Self Check Find x: $a + (x + 3)i = a - (2x - 1)i$. ■

■ The Arithmetic of Complex Numbers

Complex numbers can be added and subtracted as if they were binomials.

Addition and Subtraction of Complex Numbers

Two complex numbers such as $a + bi$ and $c + di$ are added and subtracted as if they were binomials:

$$(a + bi) + (c + di) = (a + c) + (b + d)i$$
$$(a + bi) - (c + di) = (a - c) + (b - d)i$$

Because of the preceding definition, the sum or difference of two complex numbers is another complex number.

EXAMPLE 2 Simplify: **a.** $(3 + 4i) + (2 + 7i)$ and **b.** $(-5 + 8i) - (2 - 12i)$.

Solution **a.**
$$\begin{aligned}(3 + 4i) + (2 + 7i) &= 3 + 4i + 2 + 7i\\ &= 3 + 2 + 4i + 7i\\ &= 5 + 11i\end{aligned}$$

b.
$$\begin{aligned}(-5 + 8i) - (2 - 12i) &= -5 + 8i - 2 + 12i\\ &= -5 - 2 + 8i + 12i\\ &= -7 + 20i\end{aligned}$$

Self Check Simplify: **a.** $(5 - 2i) + (-3 + 9i)$ and **b.** $(2 + 5i) - (6 + 7i)$. ■

Complex numbers can be multiplied as if they were binomials.

Multiplication of Complex Numbers

The numbers $a + bi$ and $c + di$ are multiplied as if they were binomials, with $i^2 = -1$:

$$(a + bi)(c + di) = (ac - bd) + (ad + bc)i$$

Because of this definition, the product of two complex numbers is another complex number.

EXAMPLE 3 Multiply: **a.** $(3 + 4i)(2 + 7i)$ and **b.** $(5 - 7i)(1 + 3i)$.

Solution **a.**
$$\begin{aligned}(3 + 4i)(2 + 7i) &= 6 + 21i + 8i + 28i^2\\ &= 6 + 21i + 8i + 28(-1)\\ &= 6 - 28 + 29i\\ &= -22 + 29i\end{aligned}$$

b.
$$\begin{aligned}(5 - 7i)(1 + 3i) &= 5 + 15i - 7i - 21i^2\\ &= 5 + 15i - 7i - 21(-1)\\ &= 5 + 21 + 8i\\ &= 26 + 8i\end{aligned}$$

Self Check Multiply: $(2 - 5i)(3 + 2i)$. ■

Comment To avoid errors in determining the sign of the result, always express numbers in $a + bi$ form before attempting any algebraic manipulations.

ACTIVE EXAMPLE 4 Multiply: $\left(-2 + \sqrt{-16}\right)\left(4 - \sqrt{-9}\right)$.

Solution We change each number to $a + bi$ form:

$$-2 + \sqrt{-16} = -2 + \sqrt{16}\sqrt{-1} = -2 + 4i$$
$$4 - \sqrt{-9} = 4 - \sqrt{9}\sqrt{-1} = 4 - 3i$$

COLLEGE **Algebra** $f(x)$ **Now**™
Explore this Active Example and test yourself on the steps taken here at **http://1pass.thomson.com**. You can also find this example on the Interactive Video Skillbuilder CD-ROM.

and then find the product.

$$\begin{aligned}(-2 + 4i)(4 - 3i) &= -8 + 6i + 16i - 12i^2 \\ &= -8 + 6i + 16i - 12(-1) \\ &= -8 + 12 + 22i \\ &= 4 + 22i\end{aligned}$$

Self Check Multiply: $(3 + \sqrt{-25})(2 - \sqrt{-9})$. ■

Before we discuss the division of complex numbers, we introduce the concept of a complex conjugate.

Complex Conjugates

The complex numbers $a + bi$ and $a - bi$ are called **complex conjugates** of each other.

For example,

$2 + 5i$ and $2 - 5i$ are complex conjugates.

$-\frac{1}{2} + 4i$ and $-\frac{1}{2} - 4i$ are complex conjugates.

What makes this concept important is the fact that the product of two complex conjugates is always a real number. For example,

$$\begin{aligned}(2 + 5i)(2 - 5i) &= 4 - 10i + 10i - 25i^2 \\ &= 4 - 25(-1) \\ &= 4 + 25 \\ &= 29\end{aligned}$$

In general, we have

$$\begin{aligned}(a + bi)(a - bi) &= a^2 - abi + abi - b^2i^2 \\ &= a^2 - b^2(-1) \\ &= a^2 + b^2\end{aligned}$$

To divide complex numbers, we use the concept of complex conjugates to rationalize the denominator.

EXAMPLE 5 Divide and write the result in $a + bi$ form: $\frac{3}{2 + i}$.

Solution To divide, we rationalize the denominator and simplify.

$$\frac{3}{2 + i} = \frac{3(2 - i)}{(2 + i)(2 - i)} \qquad \text{Multiply the numerator and denominator by the complex conjugate of } 2 + i \text{, which is } 2 - i.$$

$$= \frac{6 - 3i}{4 - 2i + 2i - i^2} \qquad \text{Multiply.}$$

$$= \frac{6 - 3i}{4 + 1} \qquad \text{Simplify the denominator.}$$

$$= \frac{6 - 3i}{5}$$

$$= \frac{6}{5} - \frac{3}{5}i$$

It is common to accept $\frac{6}{5} - \frac{3}{5}i$ as a substitute for $\frac{6}{5} + \left(-\frac{3}{5}\right)i$.

Self Check Divide and write the result in $a + bi$ form: $\frac{3}{3 - i}$. ■

ACTIVE EXAMPLE 6 Divide and write the result in $a + bi$ form: $\frac{2 - \sqrt{-16}}{3 + \sqrt{-1}}$.

Solution

$$\frac{2 - \sqrt{-16}}{3 + \sqrt{-1}} = \frac{2 - 4i}{3 + i}$$ Change each number to $a + bi$ form.

$$= \frac{(2 - 4i)(3 - i)}{(3 + i)(3 - i)}$$ Multiply the numerator and denominator by $3 - i$.

$$= \frac{6 - 2i - 12i + 4i^2}{9 - 3i + 3i - i^2}$$ Remove parentheses.

$$= \frac{2 - 14i}{9 + 1}$$ Combine like terms; $i^2 = -1$.

$$= \frac{2}{10} - \frac{14i}{10}$$

$$= \frac{1}{5} - \frac{7}{5}i$$

COLLEGE Algebra $f(x)$ Now™
Go to http://1pass.thomson.com or your CD to practice this example.

Self Check Divide and write the result in $a + bi$ form: $\frac{3 + \sqrt{-25}}{2 - \sqrt{-1}}$. ■

Examples 5 and 6 illustrate that the quotient of two complex numbers is another complex number.

■ Powers of *i*

The powers of i with natural number exponents produce an interesting pattern.

$i^1 = \sqrt{-1} = i$	$i^5 = i^4 i = 1i = i$
$i^2 = \left(\sqrt{-1}\right)^2 = -1$	$i^6 = i^4 i^2 = 1(-1) = -1$
$i^3 = i^2 i = -1i = -i$	$i^7 = i^4 i^3 = 1(-i) = -i$
$i^4 = i^2 i^2 = (-1)(-1) = 1$	$i^8 = i^4 i^4 = 1(1) = 1$

The pattern continues: $i, -1, -i, 1, \ldots$.

EXAMPLE 7 Simplify: i^{365}.

Solution Since $i^4 = 1$, each occurrence of i^4 is a factor of 1. To determine how many factors of i^4 are in i^{365}, we divide 365 by 4. The quotient is 91, and the remainder is 1.

$$\begin{aligned} i^{365} &= (i^4)^{91} \cdot i^1 \\ &= 1^{91} \cdot i && i^4 = 1. \\ &= i && 1^{91} = 1 \text{ and } 1 \cdot i = i. \end{aligned}$$

Self Check Simplify: $i^{1,999}$. ■

The result of Example 7 illustrates the following theorem.

Powers of i

If n is a natural number that has a remainder of r when divided by 4, then

$$i^n = i^r$$

When n is divisible by 4, the remainder r is 0 and $i^0 = 1$.

We can also simplify powers of i that involve negative integer exponents.

$$i^{-1} = \frac{1}{i} = \frac{1 \cdot i}{i \cdot i} = \frac{i}{-1} = -i \qquad i^{-2} = \frac{1}{i^2} = \frac{1}{-1} = -1$$

$$i^{-3} = \frac{1}{i^3} = \frac{1 \cdot i}{i^3 \cdot i} = \frac{i}{i^4} = \frac{i}{1} = i \qquad i^{-4} = \frac{1}{i^4} = \frac{1}{1} = 1$$

■ Absolute Value of a Complex Number

Absolute Value of a Complex Number

If $a + bi$ is a complex number, then

$$|a + bi| = \sqrt{a^2 + b^2}$$

Because of the previous definition, the absolute value of a complex number is a real number. For this reason, i does not appear in the result.

EXAMPLE 8 Write without absolute value symbols: **a.** $|3 + 4i|$ and **b.** $|4 - 6i|$.

Solution **a.**
$$\begin{aligned} |3 + 4i| &= \sqrt{3^2 + 4^2} \\ &= \sqrt{9 + 16} \\ &= \sqrt{25} \\ &= 5 \end{aligned}$$

b.
$$\begin{aligned} |4 - 6i| &= \sqrt{4^2 + (-6)^2} \\ &= \sqrt{16 + 36} \\ &= \sqrt{52} \\ &= \sqrt{4 \cdot 13} \\ &= \sqrt{4}\sqrt{13} \\ &= 2\sqrt{13} \end{aligned}$$

Self Check Write without absolute value symbols: $|2 - 5i|$. ■

ACTIVE EXAMPLE 9 Write without absolute value symbols: **a.** $\left|\dfrac{2i}{3+i}\right|$ and **b.** $|a + 0i|$.

Solution **a.** We first write $\dfrac{2i}{3+i}$ in $a + bi$ form:

COLLEGE Algebra f(x) Now™
Go to **http://1pass.thomson.com** or your CD to practice this example.

$$\frac{2i}{3+i} = \frac{2i(3-i)}{(3+i)(3-i)} = \frac{6i - 2i^2}{9 - i^2} = \frac{6i + 2}{10} = \frac{1}{5} + \frac{3}{5}i$$

and then find the absolute value of $\frac{1}{5} + \frac{3}{5}i$.

$$\left|\frac{2i}{3+i}\right| = \left|\frac{1}{5} + \frac{3}{5}i\right| = \sqrt{\left(\frac{1}{5}\right)^2 + \left(\frac{3}{5}\right)^2} = \sqrt{\frac{10}{25}} = \frac{\sqrt{10}}{5}$$

b. $|a + 0i| = \sqrt{a^2 + 0^2} = \sqrt{a^2} = |a|$

From part b, we see that $|a| = \sqrt{a^2}$.

Self Check Write without absolute value symbols: $\left|\dfrac{3i}{2-i}\right|$. ■

■ Solving Quadratic Equations with Complex Roots

The roots of many quadratic equations are complex numbers, as the following example shows.

EXAMPLE 10 Solve: $x^2 - 4x + 5 = 0$.

Solution In this equation, $a = 1$, $b = -4$, and $c = 5$.

$$x = \frac{-b \pm \sqrt{b^2 - 4ac}}{2a}$$

$$= \frac{-(-4) \pm \sqrt{(-4)^2 - 4(1)(5)}}{2(1)}$$ Substitute 1 for a, -4 for b, and 5 for c.

$$= \frac{4 \pm \sqrt{16 - 20}}{2}$$

$$= \frac{4 \pm \sqrt{-4}}{2}$$

$$= \frac{4 \pm 2i}{2}$$ $\sqrt{-4} = \sqrt{4}\sqrt{-1} = 2i$.

$$= 2 \pm i$$ $\dfrac{4 \pm 2i}{2} = \dfrac{2(2 \pm i)}{2} = 2 \pm i$.

The roots $x = 2 + i$ and $x = 2 - i$ both satisfy the equation. Note that the roots are complex conjugates.

Self Check Solve: $x^2 + 3x + 4 = 0$. ■

Factoring the Sum of Two Squares

We have seen that the sum of two squares cannot be factored over the set of integers. However, it is possible to factor the sum of two squares over the set of complex numbers. For example, to factor $9x^2 + 16y^2$, we proceed as follows:

$$\begin{aligned} 9x^2 + 16y^2 &= 9x^2 - (-1)16y^2 \\ &= 9x^2 - i^2(16y^2) && i^2 = -1. \\ &= 9x^2 - 16y^2i^2 \\ &= (3x + 4yi)(3x - 4yi) && \text{Factor the difference of two squares.} \end{aligned}$$

Self Check Answers

1. $-\frac{2}{3}$ **2. a.** $2 + 7i$ **b.** $-4 - 2i$ **3.** $16 - 11i$ **4.** $21 + i$ **5.** $\frac{9}{10} + \frac{3}{10}i$ **6.** $\frac{1}{5} + \frac{13}{5}i$ **7.** $-i$ **8.** $\sqrt{29}$ **9.** $\frac{3\sqrt{5}}{5}$ **10.** $-\frac{3}{2} \pm \frac{\sqrt{7}}{2}i$

1.5 Exercises

VOCABULARY AND CONCEPTS ***Fill in the blanks.***

1. $\sqrt{-3}$, $\sqrt{-9}$, and $\sqrt{-12}$ are examples of imaginary numbers.
2. In the complex number $a + bi$, a is the real part, and b is the imaginary part.
3. If $a = 0$ and $b \neq 0$ in the complex number $a + bi$, the number is an imaginary number.
4. If $b = 0$ in the complex number $a + bi$, the number is a real number.
5. The complex conjugate of $2 + 5i$ is $2 - 5i$.
6. By definition, $|a + bi| = \sqrt{a^2 + b^2}$.
7. The absolute value of a complex number is a real number.
8. The product of two complex conjugates is a real number.

PRACTICE ***Find the values of x and y.***

9. $x + (x + y)i = 3 + 8i$ $\quad x = 3; y = 5$
10. $x + 5i = y - yi$ $\quad x = -5; y = -5$
11. $3x - 2yi = 2 + (x + y)i$ $\quad x = \frac{2}{3}; y = -\frac{2}{9}$
12. $\begin{cases} 2 + (x + y)i = 2 - i \\ x + 3i = 2 + 3i \end{cases}$ $\quad x = 2; y = -3$

Perform all operations. Give all answers in a + bi form.

13. $(2 - 7i) + (3 + i)$ $\quad 5 - 6i$
14. $(-7 + 2i) + (2 - 8i)$ $\quad -5 - 6i$
15. $(5 - 6i) - (7 + 4i)$ $\quad -2 - 10i$
16. $(11 + 2i) - (13 - 5i)$ $\quad -2 + 7i$
17. $(14i + 2) + (2 - \sqrt{-16})$ $\quad 4 + 10i$
18. $(5 + \sqrt{-64}) - (23i - 32)$ $\quad 37 - 15i$
19. $(3 + \sqrt{-4}) - (2 + \sqrt{-9})$ $\quad 1 - i$
20. $(7 - \sqrt{-25}) + (-8 + \sqrt{-1})$ $\quad -1 - 4i$
21. $(2 + 3i)(3 + 5i)$ $\quad -9 + 19i$
22. $(5 - 7i)(2 + i)$ $\quad 17 - 9i$
23. $(2 + 3i)^2$ $\quad -5 + 12i$
24. $(3 - 4i)^2$ $\quad -7 - 24i$
25. $(11 + \sqrt{-25})(2 - \sqrt{-36})$ $\quad 52 - 56i$
26. $(6 + \sqrt{-49})(6 - \sqrt{-49})$ $\quad 85 + 0i$
27. $(\sqrt{-16} + 3)(2 + \sqrt{-9})$ $\quad -6 + 17i$
28. $(12 - \sqrt{-4})(-7 + \sqrt{-25})$ $\quad -74 + 74i$
29. $\frac{1}{i^3}$ $\quad 0 + i$
30. $\frac{3}{i^5}$ $\quad 0 - 3i$
31. $\frac{-4}{i^{10}}$ $\quad 4 + 0i$
32. $\frac{-10}{i^{24}}$ $\quad -10 + 0i$
33. $\frac{1}{2 + i}$ $\quad \frac{2}{5} - \frac{1}{5}i$
34. $\frac{-2}{3 - i}$ $\quad -\frac{3}{5} - \frac{1}{5}i$
35. $\frac{2i}{7 + i}$ $\quad \frac{1}{25} + \frac{7}{25}i$
36. $\frac{-3i}{2 + 5i}$ $\quad -\frac{15}{29} - \frac{6}{29}i$
37. $\frac{2 + i}{3 - i}$ $\quad \frac{1}{2} + \frac{1}{2}i$
38. $\frac{3 - i}{1 + i}$ $\quad 1 - 2i$

39. $\dfrac{4-5i}{2+3i}$ $-\frac{7}{13}-\frac{22}{13}i$

40. $\dfrac{34+2i}{2-4i}$ $3+7i$

41. $\dfrac{5-\sqrt{-16}}{-8+\sqrt{-4}}$ $-\frac{12}{17}+\frac{11}{34}i$

42. $\dfrac{3-\sqrt{-9}}{2-\sqrt{-1}}$ $\frac{9}{5}-\frac{3}{5}i$

43. $\dfrac{2+i\sqrt{3}}{3+i}$ $\dfrac{6+\sqrt{3}}{10}+\dfrac{3\sqrt{3}-2}{10}i$

44. $\dfrac{3+i}{4-i\sqrt{2}}$ $\dfrac{12-\sqrt{2}}{18}+\dfrac{4+3\sqrt{2}}{18}i$

Simplify each expression.

45. i^9 i

46. i^{27} $-i$

47. i^{38} -1

48. i^{99} $-i$

49. i^{-6} -1

50. i^0 1

51. i^{-10} -1

52. i^{-31} i

Write without absolute value symbols.

53. $|3+4i|$ 5

54. $|5+12i|$ 13

55. $|2+3i|$ $\sqrt{13}$

56. $|5-i|$ $\sqrt{26}$

57. $|-7+\sqrt{-49}|$ $7\sqrt{2}$

58. $|-2-\sqrt{-16}|$ $2\sqrt{5}$

59. $\left|\dfrac{1}{2}+\dfrac{1}{2}i\right|$ $\dfrac{\sqrt{2}}{2}$

60. $\left|\dfrac{1}{2}-\dfrac{1}{4}i\right|$ $\dfrac{\sqrt{5}}{4}$

61. $|-6i|$ 6

62. $|5i|$ 5

63. $\left|\dfrac{2}{1+i}\right|$ $\sqrt{2}$

64. $\left|\dfrac{3}{3+i}\right|$ $\dfrac{3\sqrt{10}}{10}$

65. $\left|\dfrac{-3i}{2+i}\right|$ $\dfrac{3\sqrt{5}}{5}$

66. $\left|\dfrac{5i}{i-2}\right|$ $\sqrt{5}$

67. $\left|\dfrac{i+2}{i-2}\right|$ 1

68. $\left|\dfrac{2+i}{2-i}\right|$ 1

Use the quadratic formula to solve each equation. Simplify all solutions and write them in a + bi form.

69. $x^2+2x+2=0$
$-1\pm i$

70. $a^2+4a+8=0$
$-2\pm 2i$

71. $y^2+4y+5=0$
$-2\pm i$

72. $x^2+2x+5=0$
$-1\pm 2i$

73. $x^2-2x=-5$
$1\pm 2i$

74. $z^2-3z=-8$
$\dfrac{3}{2}\pm\dfrac{\sqrt{23}}{2}i$

75. $x^2-\dfrac{2}{3}x=-\dfrac{2}{9}$
$\dfrac{1}{3}\pm\dfrac{1}{3}i$

76. $x^2+\dfrac{5}{4}=x$
$\dfrac{1}{2}\pm i$

Factor each expression over the set of complex numbers.

77. x^2+4
$(x+2i)(x-2i)$

78. $16a^2+9$
$(4a+3i)(4a-3i)$

79. $25p^2+36q^2$
$(5p+6qi)(5p-6qi)$

80. $100r^2+49s^2$
$(10r+7si)(10r-7si)$

81. $2y^2+8z^2$
$2(y+2zi)(y-2zi)$

82. $12b^2+75c^2$
$3(2b+5ci)(2b-5ci)$

83. $50m^2+2n^2$
$2(5m+ni)(5m-ni)$

84. $64a^4+4b^2$
$4(4a^2+bi)(4a^2-bi)$

APPLICATIONS *In electronics, the forumla $V = IR$ is called* **Ohm's law.** *It gives the relationship in a circuit between the voltage V (in volts), the current I (in amperes), and the resistance R (in ohms).*

85. **Electronics** Find V when $I=3-2i$ amperes and $R=3+6i$ ohms. $21+12i$

86. **Electronics** Find R when $I=2-3i$ amperes and $V=21+i$ volts. $3+5i$

87. **Electronics** The impedance Z in an AC (alternating current) circuit is a measure of how much the circuit impedes (hinders) the flow of current through it. The impedance is related to the voltage V and the current I by the formula

$$V = IZ$$

If a circuit has a current of $(0.5+2.0i)$ amps and an impedance of $(0.4-3.0i)$ ohms, find the voltage.
$6.2-0.7i$

88. **Fractals** Complex numbers are fundamental in the creation of the intricate geometric shape shown below, called a *fractal.* The process of creating this image is based on the following sequence of steps, which begins by picking any complex number, which we will call z.

1. Square z, and then add that result to z.
2. Square the result from step 1, and then add it to z.
3. Square the result from step 2, and then add it to z.

If we begin with the complex number i, what is the result after performing steps 1, 2, and 3? $-1+i$

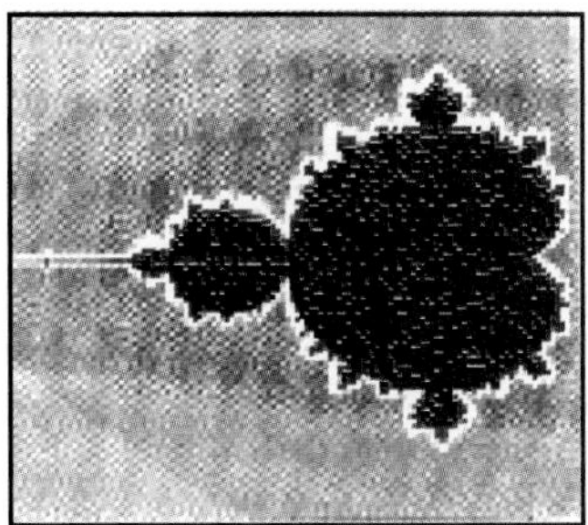

DISCOVERY AND WRITING

89. Show that the addition of two complex numbers is commutative by adding the complex numbers $a + bi$ and $c + di$ in both orders and observing that the sums are equal.

90. Show that the multiplication of two complex numbers is commutative by multiplying the complex numbers $a + bi$ and $c + di$ in both orders and observing that the products are equal.

91. Show that the addition of complex numbers is associative.

92. Find three examples of complex numbers that are reciprocals of their own conjugates.

93. Tell how to decide whether two complex numbers are equal.

94. Define the complex conjugate of a complex number.

REVIEW ***Simplify each expression. Assume that all variables represent nonnegative numbers.***

95. $\sqrt{8x^3}\sqrt{4x}$ $\quad 4x^2\sqrt{2}$

96. $(\sqrt{x} - 5)^2$ $\quad x - 10\sqrt{x} + 25$

97. $(\sqrt{x+1} - 2)^2$ $\quad x - 4\sqrt{x+1} + 5$

98. $(-3\sqrt{2x+1})^2$ $\quad 18x + 9$

99. $\dfrac{4}{\sqrt{5} - 1}$ $\quad \sqrt{5} + 1$

100. $\dfrac{x - 4}{\sqrt{x} + 2}$ $\quad \sqrt{x} - 2$

1.6 Polynomial and Radical Equations

In this section, you will learn about

- Solving Polynomial Equations by Factoring
- Other Equations That Can Be Solved by Factoring
- Radical Equations

Solving Polynomial Equations by Factoring

The equation $ax^2 + bx + c = 0$ is a polynomial equation of second degree, because its left-hand side contains a second-degree polynomial. Many polynomial equations of higher degree can be solved by factoring.

EXAMPLE 1 Solve: **a.** $6x^3 - x^2 - 2x = 0$ and **b.** $x^4 - 5x^2 + 4 = 0$.

Solution **a.**

$$6x^3 - x^2 - 2x = 0$$
$$x(6x^2 - x - 2) = 0 \quad \text{Factor out } x.$$
$$x(3x - 2)(2x + 1) = 0 \quad \text{Factor out } 6x^2 - x - 2.$$

We set each factor equal to 0.

$$x = 0 \quad \text{or} \quad 3x - 2 = 0 \quad \text{or} \quad 2x + 1 = 0$$
$$x = \frac{2}{3} \qquad\qquad x = -\frac{1}{2}$$

Verify that each solution satisfies the original equation.

b.

$$x^4 - 5x^2 + 4 = 0$$
$$(x^2 - 4)(x^2 - 1) = 0 \quad \text{Factor } x^4 - 5x^2 + 4.$$
$$(x + 2)(x - 2)(x + 1)(x - 1) = 0 \quad \text{Factor each difference of two squares.}$$

We set each factor equal to 0.

$$x + 2 = 0 \quad \text{or} \quad x - 2 = 0 \quad \text{or} \quad x + 1 = 0 \quad \text{or} \quad x - 1 = 0$$
$$x = -2 \quad | \quad x = 2 \quad | \quad x = -1 \quad | \quad x = 1$$

Verify that each solution satisfies the original equation.

Self Check Solve: $2x^3 + 3x^2 - 2x = 0$. ■

■ Other Equations That Can Be Solved by Factoring

To solve another type of equation by factoring, we use a property that states that equal powers of equal numbers are equal.

Power Property of Real Numbers

If a and b are numbers, n is an integer, and $a = b$, then

$$a^n = b^n$$

When we raise both sides of an equation to the same power, the resulting equation might not be equivalent to the original one. For example, if we raise both sides of

(1) $x = 4$ with a solution set of $\{4\}$

to the second power, we obtain

(2) $x^2 = 16$ with a solution set of $\{4, -4\}$

Equations 1 and 2 have different solution sets, and the solution -4 of Equation 2 does not satisfy Equation 1. Because raising both sides of an equation to the same power often introduces **extraneous solutions** (false solutions that don't satisfy the original equation), we must check all suspected roots to be certain that they satisfy the original equation.

The following equation has an extraneous solution.

$$x - x^{1/2} - 6 = 0$$
$$(x^{1/2} - 3)(x^{1/2} + 2) = 0 \qquad \text{Factor } x - x^{1/2} - 6.$$
$$x^{1/2} - 3 = 0 \quad \text{or} \quad x^{1/2} + 2 = 0 \qquad \text{Set each factor equal to 0.}$$
$$x^{1/2} = 3 \quad | \quad x^{1/2} = -2$$

Because equal powers of equal numbers are equal, we can square both sides of the previous equations to get

$$(x^{1/2})^2 = (3)^2 \quad \text{or} \quad (x^{1/2})^2 = (-2)^2$$
$$x = 9 \quad | \quad x = 4$$

The number 9 satisfies the equation $x - x^{1/2} - 6 = 0$, but 4 does not, as the following check shows:

***If x* = 9**	***If x* = 4**
$x - x^{1/2} - 6 = 0$	$x - x^{1/2} - 6 = 0$
$9 - 9^{1/2} - 6 \stackrel{?}{=} 0$	$4 - 4^{1/2} - 6 \stackrel{?}{=} 0$
$9 - 3 - 6 \stackrel{?}{=} 0$	$4 - 2 - 6 \stackrel{?}{=} 0$
$0 = 0$	$-4 \neq 0$

The number 9 is the only root.

ACTIVE EXAMPLE 2 Solve: $2x^{2/5} - 5x^{1/5} - 3 = 0.$

COLLEGE **Algebra** $f(x)$ **Now**™

Explore this Active Example and test yourself on the steps taken here at **http://1pass.thomson.com**. You can also find this example on the Interactive Video Skillbuilder CD-ROM.

Solution

$$2x^{2/5} - 5x^{1/5} - 3 = 0$$

$$(2x^{1/5} + 1)(x^{1/5} - 3) = 0 \qquad \text{Factor } 2x^{2/5} - 5x^{1/5} - 3.$$

$$2x^{1/5} + 1 = 0 \quad \text{or} \quad x^{1/5} - 3 = 0 \qquad \text{Set each factor equal to 0.}$$

$$2x^{1/5} = -1 \quad \Big| \quad x^{1/5} = 3$$

$$x^{1/5} = -\frac{1}{2}$$

We can raise both sides of each of the previous equations to the fifth power to get

$$(x^{1/5})^5 = \left(-\frac{1}{2}\right)^5 \quad \text{or} \quad (x^{1/5})^5 = (3)^5$$

$$x = -\frac{1}{32} \quad \Big| \quad x = 243$$

Verify that each solution satisfies the original equation.

Self Check Solve: $x^{2/5} - x^{1/5} - 2 = 0.$ ■

■ Radical Equations

Radical equations are equations containing radicals with variables in the radicand. To solve such equations, we use the power property of real numbers.

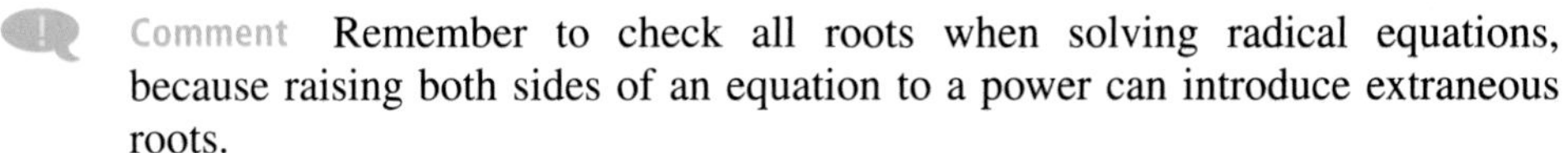

Comment Remember to check all roots when solving radical equations, because raising both sides of an equation to a power can introduce extraneous roots.

EXAMPLE 3 Solve: $\sqrt{x + 3} - 4 = 7.$

Solution

$$\sqrt{x + 3} - 4 = 7$$

$$\sqrt{x + 3} = 11 \qquad \text{Add 4 to both sides to isolate the radical.}$$

$$\left(\sqrt{x + 3}\right)^2 = (11)^2 \qquad \text{Square both sides.}$$

$$x + 3 = 121 \qquad \text{Simplify.}$$

$$x = 118 \qquad \text{Subtract 3 from both sides.}$$

Since squaring both sides might introduce extraneous roots, we must check the result of 118.

$$\sqrt{x + 3} - 4 = 7$$

$$\sqrt{118 + 3} - 4 \stackrel{?}{=} 7 \qquad \text{Substitute 118 for } x.$$

$$\sqrt{121} - 4 \stackrel{?}{=} 7$$

$$11 - 4 \stackrel{?}{=} 7$$

$$7 = 7$$

Because it checks, 118 is a root of the equation.

Self Check Solve: $\sqrt{x-3}+4=7$. ■

ACTIVE EXAMPLE 4 Solve: $\sqrt{x+3}=3x-1$.

Solution

COLLEGE **Algebra** $f(x)$ **Now**™
Go to **http://1pass.thomson.com** or your CD to practice this example.

$$\begin{aligned}
\sqrt{x+3} &= 3x-1 \\
\left(\sqrt{x+3}\right)^2 &= (3x-1)^2 && \text{Square both sides.} \\
x+3 &= 9x^2-6x+1 && \text{Remove parentheses.} \\
0 &= 9x^2-7x-2 && \text{Add } -x-3 \text{ to both sides.} \\
0 &= (9x+2)(x-1) && \text{Factor } 9x^2-7x-2.
\end{aligned}$$

$$9x+2=0 \quad \text{or} \quad x-1=0 \qquad \text{Set each factor equal to 0.}$$

$$x=-\frac{2}{9} \quad \Big| \quad x=1$$

Since squaring both sides can introduce extraneous roots, we must check each result.

$$\begin{array}{c|c}
\sqrt{x+3}=3x-1 & \sqrt{x+3}=3x-1 \\
\sqrt{-\frac{2}{9}+3} \stackrel{?}{=} 3\left(-\frac{2}{9}\right)-1 & \sqrt{1+3} \stackrel{?}{=} 3(1)-1 \\
\sqrt{\frac{25}{9}} \stackrel{?}{=} -\frac{2}{3}-1 & \sqrt{4} \stackrel{?}{=} 3-1 \\
\frac{5}{3} \neq -\frac{5}{3} & 2=2
\end{array}$$

(or)

Since $-\frac{2}{9}$ does not satisfy the equation, it is extraneous; 1 is the only solution.

Self Check Solve: $\sqrt{x-2}=2x-10$. ■

EXAMPLE 5 Solve: $\sqrt[3]{x^3+56}=x+2$.

Solution To eliminate the radical, we cube both sides of the equation.

$$\begin{aligned}
\sqrt[3]{x^3+56} &= x+2 \\
\left(\sqrt[3]{x^3+56}\right)^3 &= (x+2)^3 && \text{Cube both sides.} \\
x^3+56 &= x^3+6x^2+12x+8 && \text{Remove parentheses.} \\
0 &= 6x^2+12x-48 && \text{Simplify.} \\
0 &= x^2+2x-8 && \text{Divide both sides by 6.} \\
0 &= (x+4)(x-2) && \text{Factor } x^2+2x-8.
\end{aligned}$$

$$x+4=0 \quad \text{or} \quad x-2=0 \qquad \text{Set each factor equal to 0.}$$

$$x=-4 \quad | \quad x=2$$

We check each suspected solution to see whether either is extraneous.

For x = −4	***For x = 2***
$\sqrt[3]{x^3 + 56} = x + 2$	$\sqrt[3]{x^3 + 56} = x + 2$
$\sqrt[3]{(-4)^3 + 56} \stackrel{?}{=} -4 + 2$	$\sqrt[3]{2^3 + 56} \stackrel{?}{=} 2 + 2$
$\sqrt[3]{-64 + 56} \stackrel{?}{=} -2$	$\sqrt[3]{8 + 56} \stackrel{?}{=} 4$
$\sqrt[3]{-8} \stackrel{?}{=} -2$	$\sqrt[3]{64} \stackrel{?}{=} 4$
$-2 = -2$	$4 = 4$

Since both values satisfy the equation, −4 and 2 are roots.

Self Check Solve: $\sqrt[3]{x^3 + 7} = x + 1$. ■

ACTIVE EXAMPLE 6 Solve: $\sqrt{2x + 3} + \sqrt{x - 2} = 4$.

Solution We can write the equation in the form

COLLEGE **Algebra** $f(x)$ **Now**™
Go to **http://1pass.thomson.com** or your CD to practice this example.

$$\sqrt{2x + 3} = 4 - \sqrt{x - 2} \quad \text{Subtract } \sqrt{x - 2} \text{ from both sides.}$$

so that the left-hand side contains one radical. We then square both sides to get

$$\left(\sqrt{2x + 3}\right)^2 = \left(4 - \sqrt{x - 2}\right)^2$$
$$2x + 3 = 16 - 8\sqrt{x - 2} + x - 2$$
$$2x + 3 = 14 - 8\sqrt{x - 2} + x \quad \text{Combine like terms.}$$
$$x - 11 = -8\sqrt{x - 2} \quad \text{Subtract 14 and } x \text{ from both sides.}$$

We then square both sides again to eliminate the radical.

$$(x - 11)^2 = \left(-8\sqrt{x - 2}\right)^2$$
$$x^2 - 22x + 121 = 64(x - 2)$$
$$x^2 - 22x + 121 = 64x - 128$$
$$x^2 - 86x + 249 = 0$$
$$(x - 3)(x - 83) = 0$$
$$x - 3 = 0 \quad \text{or} \quad x - 83 = 0 \quad \text{Set each factor equal to 0.}$$
$$x = 3 \quad | \quad x = 83$$

Substituting these results into the equation will show that 83 doesn't check; it is extraneous. However, 3 does satisfy the equation and is a root.

Self Check Solve: $\sqrt{2x + 1} + \sqrt{x + 5} = 6$. ■

Accent on Technology HIGHWAY ENGINEERING

A highway curve banked at 8° will accommodate traffic traveling s mph if the radius of the curve is r feet, according to the formula $s = 1.45\sqrt{r}$. To find what radius is necessary to accommodate 70-mph traffic, highway engineers substitute 70 for s in the formula and solve for r. See Figure 1-8.

Continued

$$s = 1.45\sqrt{r}$$
$$70 = 1.45\sqrt{r}$$ Substitute 70 for s.
$$\frac{70}{1.45} = \sqrt{r}$$ Divide both sides by 1.45.
$$\left(\frac{70}{1.45}\right)^2 = r$$ Square both sides.

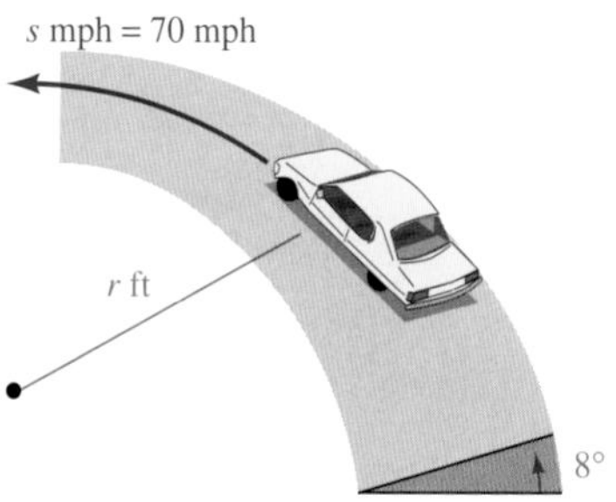

Figure 1-8

We can use a calculator to find r by entering these numbers and pressing these keys:

Scientific calculator	*Graphing calculator*
(70 ÷ 1.45) x^2	(70 ÷ 1.45) x^2 ENTER

The display will read 2330.558859. The radius of the curve should be 2,331 feet.

Self Check Answers

1. $0, \frac{1}{2}, -2$ **2.** $-1, 32$ **3.** 12 **4.** 6 **5.** $1, -2$ **6.** 4

1.6 Exercises

VOCABULARY AND CONCEPTS *Fill in the blanks.*

1. Equal powers of equal real numbers are __equal__.
2. If a and b are real numbers and $a = b$, then $a^2 =$ __b^2__.
3. False solutions that don't satisfy the equation are called __extraneous__ solutions.
4. __Radical__ equations contain radicals with variables in their __radicands__.

PRACTICE *Use factoring to solve each equation for real values of the variable.*

5. $x^3 + 9x^2 + 20x = 0$ **$0, -5, -4$**
6. $x^3 + 4x^2 - 21x = 0$ **$0, -7, 3$**
7. $6a^3 - 5a^2 - 4a = 0$ **$0, \frac{4}{3}, -\frac{1}{2}$**
8. $8b^3 - 10b^2 + 3b = 0$ **$0, \frac{1}{2}, \frac{3}{4}$**
9. $y^4 - 26y^2 + 25 = 0$ **$5, -5, 1, -1$**
10. $y^4 - 13y^2 + 36 = 0$ **$2, -2, 3, -3$**
11. $x^4 - 37x^2 + 36 = 0$ **$6, -6, 1, -1$**
12. $x^4 - 50x^2 + 49 = 0$ **$1, -1, 7, -7$**
13. $2y^4 - 46y^2 = -180$ **$3\sqrt{2}, -3\sqrt{2}, \sqrt{5}, -\sqrt{5}$**
14. $2x^4 - 102x^2 = -196$ **$-\sqrt{2}, \sqrt{2}, -7, 7$**
15. $z^{3/2} - z^{1/2} = 0$ **$0, 1$**
16. $r^{5/2} - r^{3/2} = 0$ **$0, 1$**
17. $2m^{2/3} + 3m^{1/3} - 2 = 0$ **$\frac{1}{8}, -8$**
18. $6t^{2/5} + 11t^{1/5} + 3 = 0$ **$-\frac{1}{243}, -\frac{243}{32}$**
19. $x - 13x^{1/2} + 12 = 0$ **$1, 144$**
20. $p + p^{1/2} - 20 = 0$ **16**
21. $2t^{1/3} + 3t^{1/6} - 2 = 0$ **$\frac{1}{64}$**
22. $z^3 - 7z^{3/2} - 8 = 0$ **4**
23. $6p + p^{1/2} - 1 = 0$ **$\frac{1}{9}$**
24. $3r - r^{1/2} - 2 = 0$ **1**

Find all real solutions of each equation.

25. $\sqrt{x - 2} = 5$ **27**
26. $\sqrt{a - 3} - 5 = 0$ **28**
27. $3\sqrt{x + 1} = \sqrt{6}$ **$-\frac{1}{3}$**
28. $\sqrt{x + 3} = 2\sqrt{x}$ **1**
29. $\sqrt{5a - 2} = \sqrt{a + 6}$ **2**
30. $\sqrt{16x + 4} = \sqrt{x + 4}$ **0**

31. $2\sqrt{x^2+3} = \sqrt{-16x-3}$ $-\frac{3}{2}, -\frac{5}{2}$

32. $\sqrt{x^2+1} = \dfrac{\sqrt{-7x+11}}{\sqrt{6}}$ $-\frac{5}{3}, \frac{1}{2}$

33. $\sqrt[3]{7x+1} = 4$ 9

34. $\sqrt[3]{11a-40} = 5$ 15

35. $\sqrt[4]{30t+25} = 5$ 20

36. $\sqrt[4]{3z+1} = 2$ 5

37. $\sqrt{x^2+21} = x+3$ 2

38. $\sqrt{5-x^2} = -(x+1)$ -2

39. $\sqrt{y+2} = 4-y$ 2

40. $\sqrt{3z+1} = z-1$ 5

41. $x - \sqrt{7x-12} = 0$ 3, 4

42. $x - \sqrt{4x-4} = 0$ 2

43. $x+4 = \sqrt{\dfrac{6x+6}{5}} + 3$ $\frac{1}{5}, -1$

44. $\sqrt{\dfrac{8x+43}{3}} - 1 = x$ 4

45. $\sqrt{\dfrac{x^2-1}{x-2}} = 2\sqrt{2}$ 3, 5

46. $\dfrac{\sqrt{x^2-1}}{\sqrt{3x-5}} = \sqrt{2}$ 3

47. $\sqrt[3]{x^3+7} = x+1$ $-2, 1$

48. $\sqrt[3]{x^3-7} + 1 = x$ $2, -1$

49. $\sqrt[3]{8x^3+61} = 2x+1$ $2, -\frac{5}{2}$

50. $\sqrt[3]{8x^3-37} = 2x-1$ $2, -\frac{3}{2}$

51. $\sqrt{2p+1} - 1 = \sqrt{p}$ 0, 4

52. $\sqrt{r} + \sqrt{r+2} = 2$ $\frac{1}{4}$

53. $\sqrt{x+3} = \sqrt{2x+8} - 1$ -2

54. $\sqrt{x+2} + 1 = \sqrt{2x+5}$ $2, -2$

55. $\sqrt{y+8} - \sqrt{y-4} = -2$ no solution

56. $\sqrt{z+5} - 2 = \sqrt{z-3}$ 4

57. $\sqrt{2b+3} - \sqrt{b+1} = \sqrt{b-2}$ 3

58. $\sqrt{a+1} + \sqrt{3a} = \sqrt{5a+1}$ 0

59. $\sqrt{\sqrt{b} + \sqrt{b+8}} = 2$ 1

60. $\sqrt{\sqrt{x+19} - \sqrt{x-2}} = \sqrt{3}$ 6

APPLICATIONS

61. Height of a bridge The distance d (in feet) that an object will fall in t seconds is given by the formula

$$t = \sqrt{\frac{d}{16}}$$

To find the height of a bridge above a river, a man drops a stone into the water (see the illustration). If it takes the stone 5 seconds to hit the water, how high is the bridge? 400 ft

62. Horizon distance The higher a lookout tower, the farther an observer can see. (See the illustration.) The distance d (called the **horizon distance,** measured in miles) is related to the height h of the observer (measured in feet) by the formula

$$d = 1.4\sqrt{h}$$

How tall must a tower be for the observer to see 30 miles? about 460 ft

63. Carpentry During construction, carpenters often brace walls, as shown in the illustration. The appropriate length of the brace is given by the formula

$$l = \sqrt{f^2 + h^2}$$

If a carpenter nails a 10-foot brace to the wall 6 feet above the floor, how far from the base of the wall should he nail the brace to the floor? 8 ft

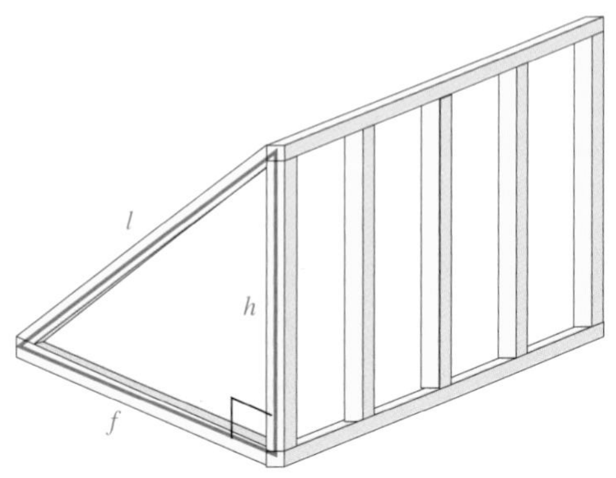

64. Windmills The power generated by a windmill is related to the velocity of the wind by the formula

$$v = \sqrt[3]{\frac{P}{0.02}}$$

where P is the power (in watts) and v is the velocity of the wind (in mph). To the nearest 10 miles, find the power generated when the velocity of the wind is 31 mph. **600 watts**

65. Diamonds The *effective rate of interest r* earned by an investment is given by the formula

$$r = \sqrt[n]{\frac{A}{P}} - 1$$

where P is the initial investment that grows to value A after n years. If a diamond buyer got $4,000 for a 1.03-carat diamond that he had purchased 4 years earlier, and earned an annual rate of return of 6.5% on the investment, what did he originally pay for the diamond? **about $3,109**

66. Theater productions The ropes, pulleys, and sandbags shown in the illustration are part of a mechanical system used to raise and lower scenery for a stage play. For the scenery to be in the proper position, the following formula must apply:

$$w_2 = \sqrt{w_1^2 + w_3^2}$$

If $w_2 = 12.5$ lb and $w_3 = 7.5$ lb, find w_1. **10 lb**

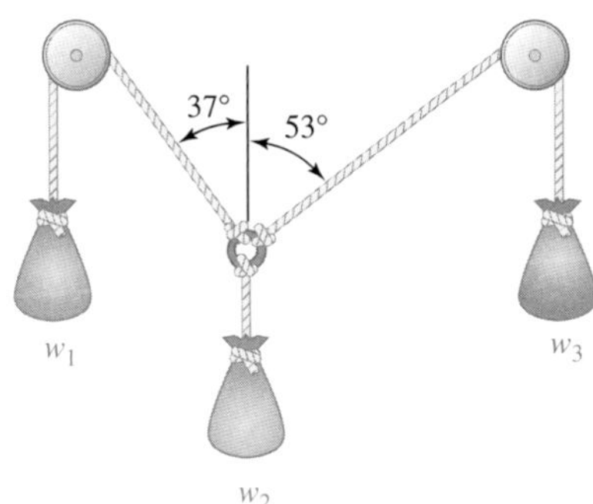

DISCOVERY AND WRITING

67. Explain why squaring both sides of an equation might introduce extraneous roots.

68. Can cubing both sides of an equation introduce extraneous roots? Explain.

REVIEW ***Graph each subset of the real numbers on the number line.***

69. The natural numbers between -4 and 4

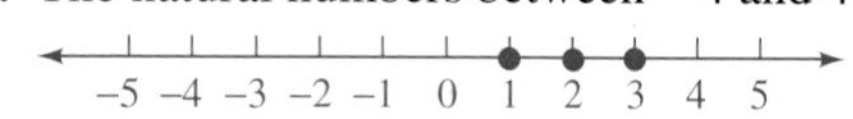

70. The integers between -4 and 4

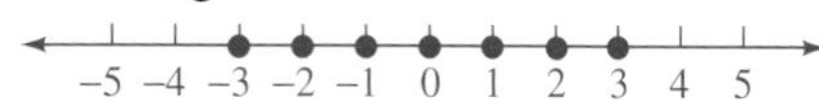

Write each inequality in interval notation and graph the interval.

71. $x \geq 3$

[3, ∞)

3

72. $x < -5$

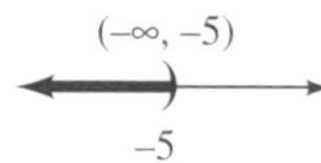

73. $-2 \leq x < 1$

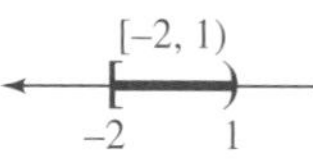

74. $-3 \leq x \leq 3$

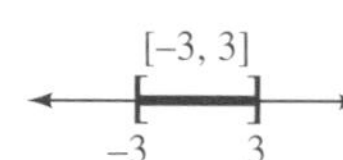

75. $x < 1$ or $x \geq 2$

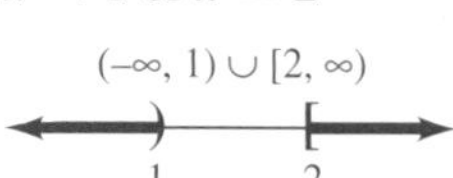

76. $x \leq 1$ or $x > 2$

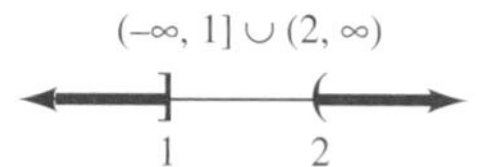

1.7 Inequalities

In this section, you will learn about

- Properties of Inequalities
- Solving Linear Inequalities
- Solving Compound Inequalities
- Solving Quadratic Inequalities
- Solving Rational Inequalities

We previously introduced the following symbols.

Symbol	Read as	Examples
$\neq$	"is not equal to"	$8 \neq 10$ and $25 \neq 12$
$<$	"is less than"	$8 < 10$ and $12 < 25$
$>$	"is greater than"	$30 > 10$ and $100 > -5$
$\leq$	"is less than or equal to"	$-6 \leq 12$ and $-8 \leq -8$
$\geq$	"is greater than or equal to"	$12 \geq -5$ and $9 \geq 9$
$\approx$	"is approximately equal to"	$7.49 \approx 7.5$ and $\frac{1}{3} \approx 0.33$

Since the coordinates of points get larger as we move from left to right on the number line,

$a > b$ if point a lies to the right of point b on a number line.

$a < b$ if point a lies to the left of point b on a number line.

EXAMPLE 1 **a.** $5 > -3$, because 5 lies to the right of -3 on the number line.

b. $-7 < -2$, because -7 lies to the left of -2 on the number line.

c. $3 \leq 3$, because $3 = 3$.

d. $2 \leq 3$, because $2 < 3$.

e. $x + 1 > x$, because $x + 1$ lies one unit to the right of x on the number line.

Self Check Write an inequality symbol to make a true statement: **a.** 25 ___ 12, **b.** -5 ___ -5, and **c.** -12 ___ -20. ■

■ Properties of Inequalities

The Trichotomy Property

For any real numbers a and b, one of the following statements is true.

$a < b$, $a = b$, or $a > b$

This property indicates that one of the following statements is true about two real numbers. Either the first is less than the second, or the first is equal to the second, or the first is greater than the second.

The Transitive Property

If a, b, and c are real numbers, then

if $a < b$ and $b < c$, then $a < c$.

if $a > b$ and $b > c$, then $a > c$.

The first part of the transitive property indicates that if a first number is less than a second and the second number is less than a third, then the first number is less than the third.

The second part is similar, with the words "is greater than" substituted for "is less than."

Addition and Subtraction Properties of Inequality

Let a, b, and c represent real numbers.

If $a < b$, then $a + c < b + c$.

If $a < b$, then $a - c < b - c$.

Similar properties exist for $>$, $\leq$, and $\geq$.

This property states that *any real number can be added to (or subtracted from) both sides of an inequality to obtain another inequality with the same order (direction).*

For example, if we add 4 to (or subtract 4 from) both sides of $8 < 12$, we get

$$\begin{aligned} 8 &< 12 \\ 8 + 4 &< 12 + 4 \\ 12 &< 16 \end{aligned} \qquad \begin{aligned} 8 &< 12 \\ 8 - 4 &< 12 - 4 \\ 4 &< 8 \end{aligned}$$

and the $<$ symbol is unchanged.

Multiplication and Division Properties of Inequality

Let a, b, and c represent real numbers.

Part 1: If $a < b$ and $c > 0$, then $ca < cb$.

If $a < b$ and $c > 0$, then $\frac{a}{c} < \frac{b}{c}$.

Part 2: If $a < b$ and $c < 0$, then $ca > cb$.

If $a < b$ and $c < 0$, then $\frac{a}{c} > \frac{b}{c}$.

Similar properties exist for $>$, $\leq$, and $\geq$.

This property has two parts. Part 1 states that *both sides of an inequality can be multiplied (or divided) by the same positive number to obtain another inequality with the same order.*

For example, if we multiply (or divide) both sides of $8 < 12$ by 4, we get

$$\begin{aligned} 8 &< 12 \\ 8(4) &< 12(4) \\ 32 &< 48 \end{aligned} \qquad \begin{aligned} 8 &< 12 \\ \frac{8}{4} &< \frac{12}{4} \\ 2 &< 3 \end{aligned}$$

and the $<$ symbol is unchanged.

Part 2 states that *both sides of an inequality can be multiplied (or divided) by the same negative number to obtain another inequality with the opposite order.*

For example, if we multiply (or divide) both sides of $8 < 12$ by -4, we get

$$\begin{aligned} 8 &< 12 \\ 8(-4) &> 12(-4) \\ -32 &> -48 \end{aligned} \qquad \begin{aligned} 8 &< 12 \\ \frac{8}{-4} &> \frac{12}{-4} \\ -2 &> -3 \end{aligned}$$

and the $<$ symbol is changed to a $>$ symbol.

Comment Note that unless we are multiplying or dividing by a negative number, the properties of inequality are the same as the properties of equality.

Solving Linear Inequalities

Linear inequalities are inequalities such as $ax + c < 0$ or $ax - c \geq 0$, where $a \neq 0$. Numbers that make an inequality true when substituted for the variable are solutions of the inequality. Inequalities with the same solution set are called **equivalent inequalities.** Because of the previous properties, we can solve inequalities as we do equations. However, we must always remember to change the order of an inequality when multiplying (or dividing) both sides by a negative number.

EXAMPLE 2 Solve: $3(x + 2) < 8$.

Solution We proceed as with equations.

$$3(x + 2) < 8$$

$$3x + 6 < 8 \quad \text{Remove parentheses.}$$

$$3x < 2 \quad \text{Subtract 6 from both sides.}$$

$$x < \frac{2}{3} \quad \text{Divide both sides by 3.}$$

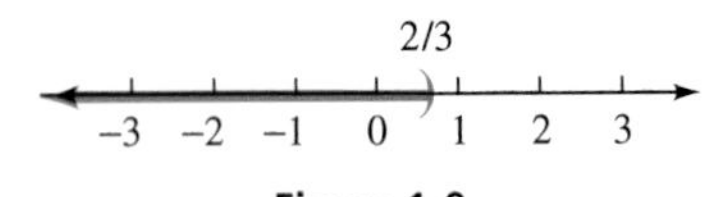

Figure 1-9

All numbers that are less than $\frac{2}{3}$ are solutions of the inequality. The solution set can be expressed in interval notation as $\left(-\infty, \frac{2}{3}\right)$ and be graphed as in Figure 1-9.

Self Check Solve: $5(p - 4) > 25$. ■

ACTIVE EXAMPLE 3 Solve: $-5(x - 2) \leq 20 + x$.

COLLEGE Algebra f(x) Now™
Explore this Active Example and test yourself on the steps taken here at **http://1pass.thomson.com**. You can also find this example on the Interactive Video Skillbuilder CD-ROM.

Solution

$$-5(x - 2) \leq 20 + x$$

$$-5x + 10 \leq 20 + x \quad \text{Remove parentheses.}$$

$$-6x + 10 \leq 20 \quad \text{Subtract } x \text{ from both sides.}$$

$$-6x \leq 10 \quad \text{Subtract 10 from both sides.}$$

We now divide both sides of the inequality by -6, which changes the order of the inequality.

$$x \geq \frac{10}{-6} \quad \text{Divide both sides by } -6.$$

$$x \geq -\frac{5}{3} \quad \text{Simplify the fraction.}$$

Figure 1-10

The graph of the solution set is shown in Figure 1-10. It is the interval $\left[-\frac{5}{3}, \infty\right)$.

Self Check Solve: $-4(x + 3) \geq 16$. ■

EXAMPLE 4 An empty truck with driver weighs 4,350 pounds. It is loaded with feed corn weighing 31 pounds per bushel. Between farm and market is a bridge with a 10,000-pound load limit. How many bushels can the truck legally carry?

Solution The empty truck with driver weighs 4,350 pounds, and corn weighs 31 pounds per bushel. If we let b represent the number of bushels in a legal load, the weight of the corn will be $31b$ pounds. Since the combined weight of the truck, driver, and cargo cannot exceed 10,000 pounds, we can form the following inequality.

The weight of the empty truck	plus	the weight of the corn	must be less than or equal to	10,000 pounds.
4,350	$+$	$31b$	$\leq$	10,000

We can solve the inequality as follows.

$$4{,}350 + 31b \leq 10{,}000$$

$$31b \leq 5{,}650 \quad \text{Subtract 4,350 from each side.}$$

$$b \leq 182.2580645 \quad \text{Divide both sides by 31.}$$

The truck can legally carry $182\frac{1}{4}$ bushels or less. ■

■ Solving Compound Inequalities

The statement that x is between 2 and 5 implies two inequalities,

$$x > 2 \quad \text{and} \quad x < 5$$

It is customary to write both inequalities as one **compound inequality:**

$$2 < x < 5 \quad \text{Read as "2 is less than } x \text{ and } x \text{ is less than 5."}$$

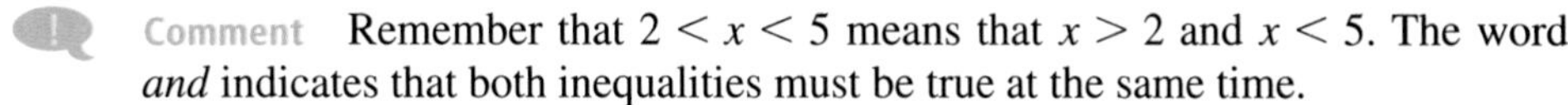

Comment Remember that $2 < x < 5$ means that $x > 2$ and $x < 5$. The word *and* indicates that both inequalities must be true at the same time.

To express that x is not between 2 and 5, we must convey the idea that either x is greater than or equal to 5, or that x is less than or equal to 2. This is equivalent to the statement

$$x \geq 5 \quad \text{or} \quad x \leq 2$$

This inequality is satisfied by all numbers x that satisfy one or both of its parts.

Comment It is incorrect to write $x \geq 5$ or $x \leq 2$ as $2 \geq x \geq 5$, because this would mean that $2 \geq 5$, which is false.

EXAMPLE 5 Solve: $5 < 3x - 7 \leq 8$.

Solution We can isolate x between the inequality symbols by adding 7 to each part of the inequality to get

$$5 + 7 < 3x - 7 + 7 \leq 8 + 7 \quad \text{Add 7 to each part.}$$

$$12 < 3x \leq 15 \quad \text{Do the additions.}$$

and dividing all parts by 3 to get

$$4 < x \leq 5$$

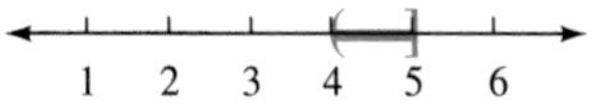

1 2 3 4 5 6

Figure 1-11

The solution set is the interval (4, 5], whose graph appears in Figure 1-11.

Self Check Solve: $-5 \leq 2x + 1 < 9$. ■

EXAMPLE 6 Solve: $3 + x \leq 3x + 1 < 7x - 2$.

Solution Because it is impossible to isolate x between the inequality symbols, we must solve each inequality separately.

$$\begin{array}{rl|rl} 3 + x \leq 3x + 1 & \text{and} & 3x + 1 < 7x - 2 & \\ 3 \leq 2x + 1 & & 1 < 4x - 2 & \\ 2 \leq 2x & & 3 < 4x & \\ 1 \leq x & & \frac{3}{4} < x & \\ x \geq 1 & & x > \frac{3}{4} & \end{array}$$

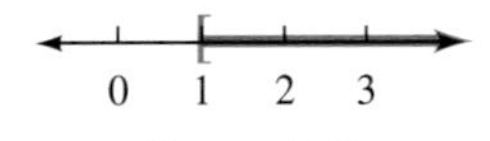

Figure 1-12

Since the connective in this inequality is "and," the solution set is the intersection (or overlap) of the intervals $[1, \infty)$ and $\left(\frac{3}{4}, \infty\right)$, which is $[1, \infty)$. The graph is shown in Figure 1-12.

Self Check Solve: $x + 1 < 2x - 3 \leq 3x - 5$. ■

It is possible for an inequality to be true for all values of its variable. It is also possible for an inequality to have no solutions.

$x < x + 1$ is true for all numbers x.

$x > x + 1$ is true for no numbers x.

■ Solving Quadratic Inequalities

If $a \neq 0$, inequalities like $ax^2 + bx + c < 0$ and $ax^2 + bx + c > 0$ are called **quadratic inequalities.** We will begin by discussing two methods for solving the quadratic inequality $x^2 - x - 6 > 0$.

EXAMPLE 7 Solve: $x^2 - x - 6 > 0$.

Method 1 First we solve the equation $x^2 - x - 6 = 0$.

$$\begin{aligned} x^2 - x - 6 &= 0 \\ (x + 2)(x - 3) &= 0 \end{aligned}$$

$$x + 2 = 0 \quad \text{or} \quad x - 3 = 0$$
$$x = -2 \quad | \quad x = 3$$

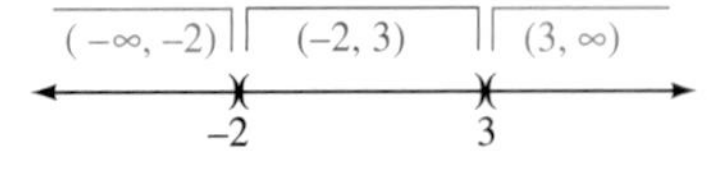

Figure 1-13

The graphs of these solutions establish the three intervals shown in Figure 1-13. To decide which intervals are solutions, we test a number in each one and see whether it satisfies the inequality. Read $-6 \in (-\infty, -2)$ as "-6 is in $(-\infty, -2)$."

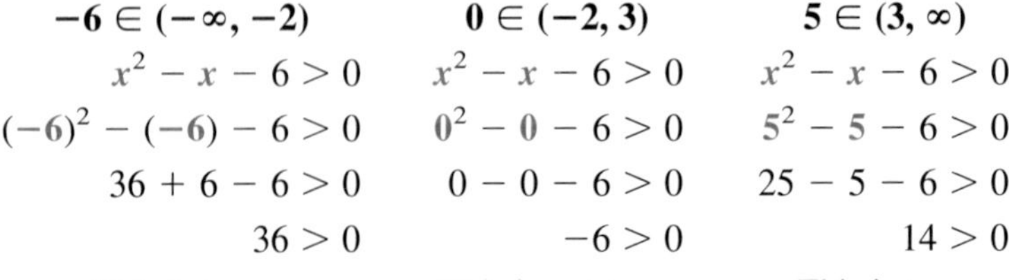

$\mathbf{-6 \in (-\infty, -2)}$	$\mathbf{0 \in (-2, 3)}$	$\mathbf{5 \in (3, \infty)}$
$x^2 - x - 6 > 0$	$x^2 - x - 6 > 0$	$x^2 - x - 6 > 0$
$(-6)^2 - (-6) - 6 > 0$	$0^2 - 0 - 6 > 0$	$5^2 - 5 - 6 > 0$
$36 + 6 - 6 > 0$	$0 - 0 - 6 > 0$	$25 - 5 - 6 > 0$
$36 > 0$	$-6 > 0$	$14 > 0$
This is true.	This is not true.	This is true.

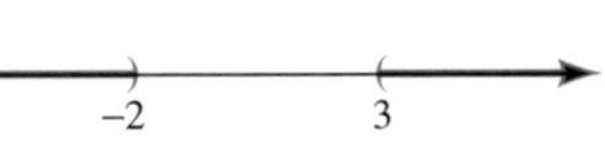

Figure 1-14

The solutions are in the intervals $(-\infty, -2)$ or $(3, \infty)$, as shown in Figure 1-14.

Method 2 Another method relies on the number line and a notation that keeps track of the signs of the factors of $x^2 - x - 6$, which are $(x - 3)(x + 2)$. First, we consider the factor $x - 3$.

If $x = 3$, then $x - 3 = 0$.

If $x < 3$, then $x - 3$ is negative.

If $x > 3$, then $x - 3$ is positive.

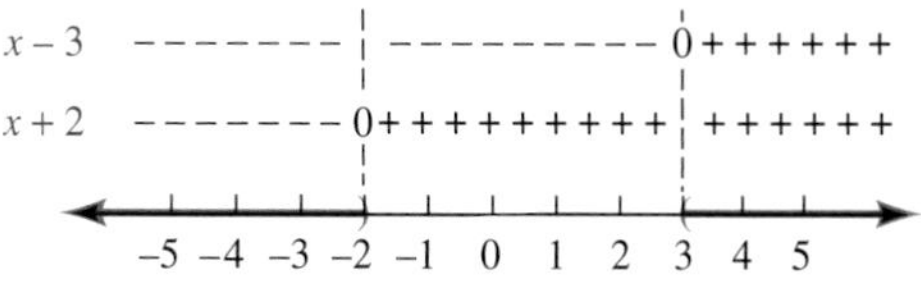

Figure 1-15

Then we consider the factor $x + 2$.

If $x = -2$, then $x + 2 = 0$.

If $x < -2$, then $x + 2$ is negative.

If $x > -2$, then $x + 2$ is positive.

We place this information on the **sign graph** shown in Figure 1-15 by using + and − signs. Only to the left of −2 and to the right of 3 do the signs of both factors agree. Only there is the product positive.

Self Check Solve: $x(x + 1) - 6 \geq 0$. ■

EXAMPLE 8 Solve: $x(x + 3) < -2$.

We remove parentheses and add 2 to both sides to make the right-hand side equal to 0. We then solve $x^2 + 3x + 2 < 0$.

Method 1 First we solve the equation $x^2 + 3x + 2 = 0$.

$$x^2 + 3x + 2 = 0$$
$$(x + 2)(x + 1) = 0$$
$$x + 2 = 0 \quad \text{or} \quad x + 1 = 0$$
$$x = -2 \quad | \quad x = -1$$

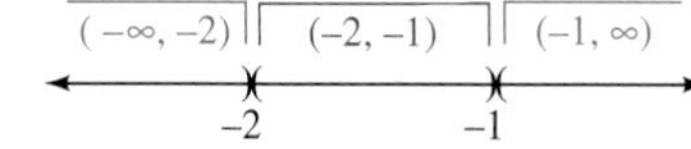

Figure 1-16

These solutions establish the intervals shown in Figure 1-16.

The solutions of $x^2 + 3x + 2 < 0$ will be the numbers in one or more of these intervals. To decide which ones are solutions, we test any number in each interval to see whether it satisfies the inequality.

$\mathbf{-7 \in (-\infty, -2)}$	$\mathbf{-\frac{3}{2} \in (-2, -1)}$	$\mathbf{0 \in (-1, \infty)}$
$x^2 + 3x + 2 < 0$	$x^2 + 3x + 2 < 0$	$x^2 + 3x + 2 < 0$
$(-7)^2 + 3(-7) + 2 < 0$	$\left(-\frac{3}{2}\right)^2 + 3\left(-\frac{3}{2}\right) + 2 < 0$	$0^2 + 3(0) + 2 < 0$
$49 - 21 + 2 < 0$	$\frac{9}{4} + \left(-\frac{9}{2}\right) + 2 < 0$	$0 + 0 + 2 < 0$
$30 < 0$	$-\frac{1}{4} < 0$	$2 < 0$
This is not true.	This is true.	This is not true.

The solution is the interval $(-2, -1)$, whose graph appears in Figure 1-17.

−2 −1

Figure 1-17

Method 2 We construct the sign graph shown in Figure 1-18. Only between -2 and -1 do the factors have opposite signs. Here, the product is negative.

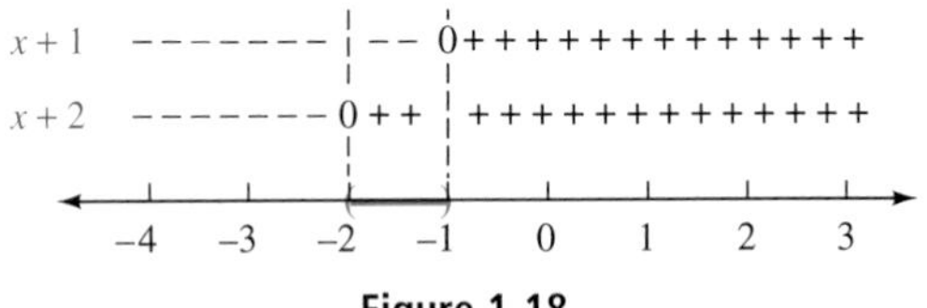

Figure 1-18

Self Check Solve: $x^2 - 5x - 6 < 0$.

■ Solving Rational Inequalities

Inequalities that contain fractions with polynomial numerators and denominators are called **rational inequalities.** To solve them, we can use the same techniques that we used to solve quadratic inequalities.

ACTIVE EXAMPLE 9 Solve the rational inequality: $\dfrac{x^2 - x - 2}{x^2 - 4x + 3} \leq 0$.

Method 1 The intervals are found by solving $x^2 - x - 2 = 0$ and $x^2 - 4x + 3 = 0$. The solutions of the first equation are -1 and 2, and the solutions of the second equation are 1 and 3. These solutions establish the five intervals shown in Figure 1-19.

COLLEGE **Algebra** $f(x)$ **Now**™
Go to **http://1pass.thomson.com** or your CD to practice this example.

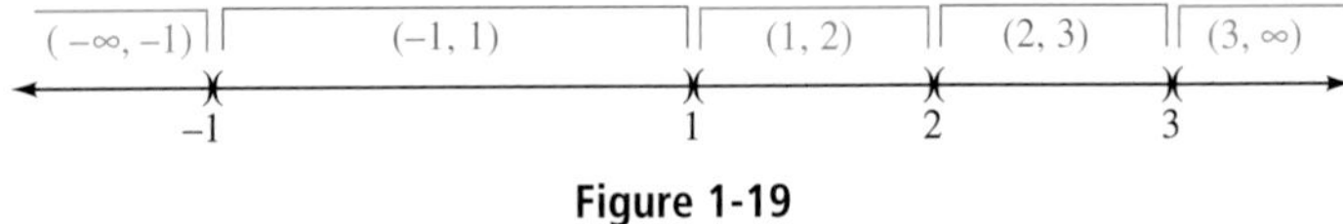

Figure 1-19

We test some numbers to see that numbers in the intervals $(-1, 1)$ and $(2, 3)$ satisfy the inequality, but numbers in the intervals $(-\infty, -1)$, $(1, 2)$, and $(3, \infty)$ do not. The graph of the solution set is shown in Figure 1-20.

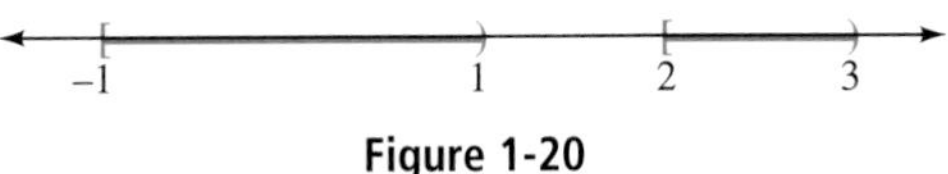

Figure 1-20

Because $x = -1$ and $x = 2$ make the numerator 0, they satisfy the inequality. Thus, their graphs are drawn with brackets to show that -1 and 2 are included. Because 1 and 3 give 0's in the denominator, the parentheses at $x = 1$ and $x = 3$ show that 1 and 3 are not in the solution set.

Method 2 We factor each trinomial and write the inequality in the form

$$\frac{(x - 2)(x + 1)}{(x - 3)(x - 1)} \leq 0$$

We then construct the sign graph shown in Figure 1-21. The value of the fraction will be 0 when $x = 2$ and $x = -1$. The value will be negative when there is an odd number of negative factors. This happens between -1 and 1 and between 2 and 3.

The graph of the solution set also appears in Figure 1-21. The brackets at -1 and 2 show that these numbers are in the solution set. The parentheses at 1 and 3 show that these numbers are not in the solution set.

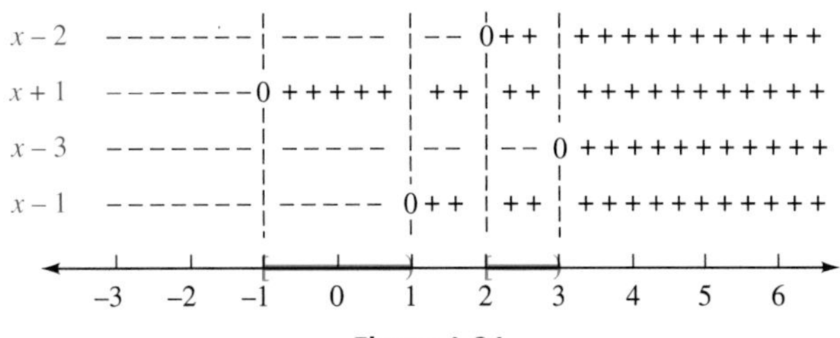

Figure 1-21

Self Check Solve: $\dfrac{x^2 + 2x - 3}{x^2 + 4x + 3} > 0.$ ■

ACTIVE EXAMPLE 10 Solve: $\dfrac{6}{x} > 2.$

Solution To get a 0 on the right-hand side, we add -2 to both sides. We then combine like terms on the left-hand side.

COLLEGE **Algebra** *f(x)* **Now**™
Go to **http://1pass.thomson.com** or your CD to practice this example.

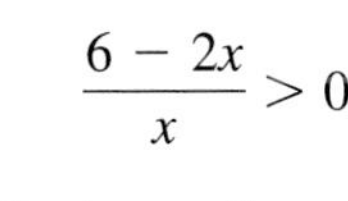

$$\frac{6}{x} > 2$$

$$\frac{6}{x} - 2 > 0$$

$$\frac{6}{x} - \frac{2x}{x} > 0$$

$$\frac{6 - 2x}{x} > 0$$

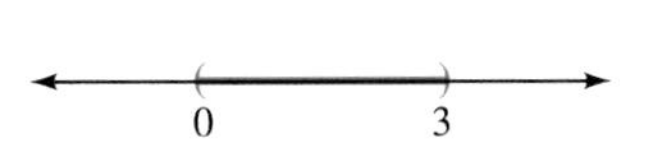

Figure 1-22

The inequality now has the form of a rational inequality. The intervals are found by solving $6 - 2x = 0$ and $x = 0$. The solution of the first equation is 3, and the solution of the second equation is 0. This determines the intervals $(-\infty, 0)$, $(0, 3)$, and $(3, \infty)$. Because only the numbers in the interval $(0, 3)$ satisfy the original inequality, the solution set is $(0, 3)$. The graph is shown in Figure 1-22.

We could construct a sign graph as in Figure 1-23 and obtain the same solution set.

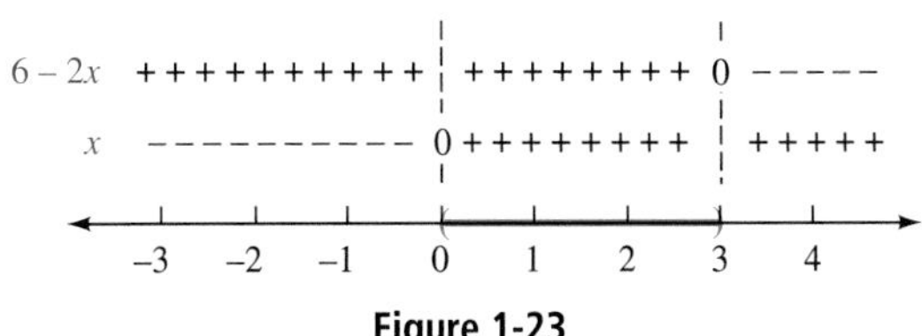

Figure 1-23

Self Check Solve: $\dfrac{2}{x} < 4.$ ■

Comment It is tempting to solve Example 10 by multiplying both sides by x and solving the inequality $6 > 2x$. However, multiplying both sides by x gives $6 > 2x$ only when x is positive. If x is negative, multiplying both sides by x will reverse the direction of the $>$ symbol, and the inequality $\frac{6}{x} > 2$ will be equivalent to $6 < 2x$. If you fail to consider both cases, you will get a wrong answer.

Self Check Answers

1. a. $>$ or $\geq$ **b.** $\leq$ or $\geq$ **c.** $>$ or $\geq$ **2.** 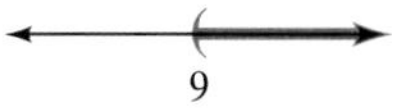 (open at 9, shaded right) **3.** 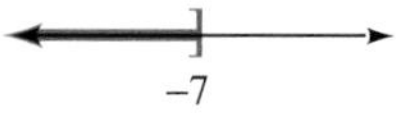 (shaded left, closed at −7)

5. [−3, 4) **6.** (open at 4, shaded right) **7.** (shaded left to −3 closed, closed at 2 shaded right) **8.** (−1, 6)

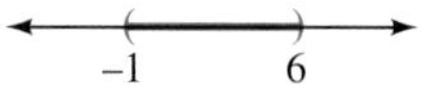

9.  (shaded to −3 open, −3 to −1 open, open at 1 shaded right) **10.**  (shaded left to 0 open, open at 1/2 shaded right)

1.7 Exercises

VOCABULARY AND CONCEPTS ***Fill in the blanks. Assume that all variables represent real numbers.***

1. If $x > y$, then x lies to the <u>right</u> of y on a number line.
2. $a < b$, <u>$a = b$</u>, or $a > b$.
3. If $a < b$ and $b < c$, then <u>$a < c$</u>.
4. If $a < b$, then $a + c <$ <u>$b + c$</u>.
5. If $a < b$, then $a - c <$ <u>$b - c$</u>.
6. If $a < b$ and $c > 0$, then ac <u>$<$</u> bc.
7. If $a < b$ and $c < 0$, then ac <u>$>$</u> bc.
8. If $a < b$ and $c < 0$, then $\frac{a}{c}$ <u>$>$</u> $\frac{b}{c}$.
9. $3x - 5 < 12$ and $ax + c > 0$ $(a \neq 0)$ are examples of <u>linear</u> inequalities.
10. $ax^2 + bx - c \geq 0$ and $3x^2 - 6x < 0$ are examples of <u>quadratic</u> inequalities.
11. If two inequalities have the same solution set, they are called <u>equivalent</u> inequalities.
12. An inequality that contains a fraction with a polynomial numerator and denominator is called a <u>rational</u> inequality.

PRACTICE ***Solve each inequality, graph the solution set, and write the answer in interval notation. Do not worry about drawing your graphs exactly to scale.***

13. $3x + 2 < 5$ — $(-\infty, 1)$ (graph at 1)
14. $-2x + 4 < 6$ — $(-1, \infty)$ (graph at −1)
15. $3x + 2 \geq 5$ — $[1, \infty)$ (graph at 1)
16. $-2x + 4 \geq 6$ — $(-\infty, -1]$ (graph at −1)
17. $-5x + 3 > -2$ — $(-\infty, 1)$ (graph at 1) 
18. $4x - 3 > -4$ — $(-1/4, \infty)$ (graph at −1/4)
19. $-5x + 3 \leq -2$ — $[1, \infty)$ (graph at 1)
20. $4x - 3 \leq -4$ — $(-\infty, -1/4]$ (graph at −1/4)
21. $2(x - 3) \leq -2(x - 3)$ — $(-\infty, 3]$ (graph at 3)
22. $3(x + 2) \leq 2(x + 5)$ — $(-\infty, 4]$ (graph at 4)
23. $\frac{3}{5}x + 4 > 2$ — $(-10/3, \infty)$ (graph at −10/3)
24. $\frac{1}{4}x - 3 > 5$ — $(32, \infty)$ (graph at 32) 
25. $\frac{x + 3}{4} < \frac{2x - 4}{3}$ — $(5, \infty)$ (graph at 5)
26. $\frac{x + 2}{5} > \frac{x - 1}{2}$ — $(-\infty, 3)$ (graph at 3) 
27. $\frac{6(x - 4)}{5} \geq \frac{3(x + 2)}{4}$ — $[14, \infty)$ (graph at 14)
28. $\frac{3(x + 3)}{2} < \frac{2(x + 7)}{3}$ — $(-\infty, 1/5)$ (graph at 1/5)

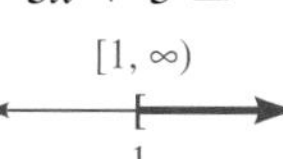

29. $\frac{5}{9}(a+3) - a \geq \frac{4}{3}(a-3) - 1$ $(-\infty, 15/4]$

30. $\frac{2}{3}y - y \leq -\frac{3}{2}(y-5)$ $(-\infty, 45/7]$

31. $\frac{2}{3}a - \frac{3}{4}a < \frac{3}{5}\left(a + \frac{2}{3}\right) + \frac{1}{3}$ $(-44/41, \infty)$

32. $\frac{1}{4}b + \frac{2}{3}b - \frac{1}{2} > \frac{1}{2}(b+1) + b$ $(-\infty, -12/7)$

33. $4 < 2x - 8 \leq 10$ $(6, 9]$

34. $3 \leq 2x + 2 < 6$ $[1/2, 2)$

35. $9 \geq \frac{x-4}{2} > 2$ $(8, 22]$

36. $5 < \frac{x-2}{6} < 6$ $(32, 38)$

37. $0 \leq \frac{4-x}{3} \leq 5$ $[-11, 4]$

38. $0 \geq \frac{5-x}{2} \geq -10$ $[5, 25]$

39. $-2 \geq \frac{1-x}{2} \geq -10$ $[5, 21]$

40. $-2 \leq \frac{1-x}{2} < 10$ $(-19, 5]$

41. $-3x > -2x > -x$ $(-\infty, 0)$

42. $-3x < -2x < -x$ $(0, \infty)$

43. $x < 2x < 3x$ $(0, \infty)$

44. $x > 2x > 3x$ $(-\infty, 0)$

45. $2x + 1 < 3x - 2 < 12$ $(3, 14/3)$

46. $2 - x < 3x + 5 < 18$ $(-3/4, 13/3)$

47. $2 + x < 3x - 2 < 5x + 2$ $(2, \infty)$

48. $x > 2x + 3 > 4x - 7$ $(-\infty, -3)$

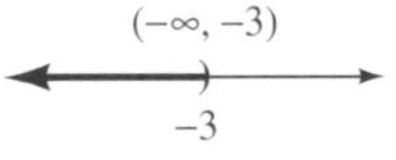

49. $3 + x > 7x - 2 > 5x - 10$ $(-4, 5/6)$

50. $2 - x < 3x + 1 < 10x$ $(1/4, \infty)$

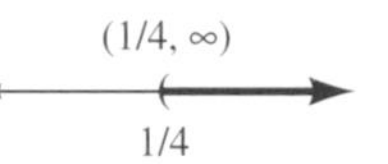

51. $x \leq x + 1 \leq 2x + 3$ $[-2, \infty)$

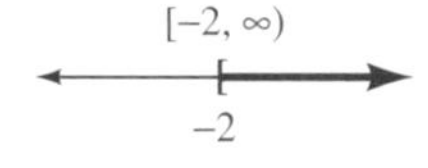

52. $-x \geq -2x + 1 \geq -3x + 1$ $[1, \infty)$

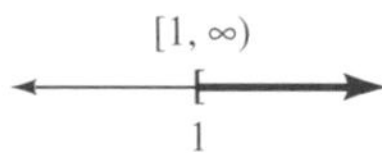

53. $x^2 + 7x + 12 < 0$ $(-4, -3)$

54. $x^2 - 13x + 12 \leq 0$ $[1, 12]$

55. $x^2 - 5x + 6 \geq 0$ $(-\infty, 2] \cup [3, \infty)$

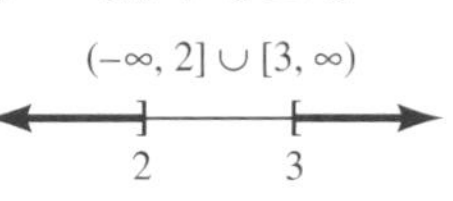

56. $6x^2 + 5x - 6 > 0$ $(-\infty, -3/2) \cup (2/3, \infty)$

57. $x^2 + 5x + 6 < 0$ $(-3, -2)$

58. $x^2 + 9x + 20 \geq 0$ $(-\infty, -5] \cup [-4, \infty)$

59. $6x^2 + 5x + 1 \geq 0$ $(-\infty, -1/2] \cup [-1/3, \infty)$

60. $x^2 + 9x + 20 < 0$ $(-5, -4)$

61. $6x^2 - 5x < -1$ $(1/3, 1/2)$

62. $9x^2 + 24x > -16$ $(-\infty, -4/3) \cup (-4/3, \infty)$

63. $2x^2 \geq 3 - x$ $(-\infty, -3/2] \cup [1, \infty)$

64. $9x^2 \leq 24x - 16$ $[4/3, 4/3]$

65. $\frac{x+3}{x-2} < 0$ $(-3, 2)$

66. $\frac{x+3}{x-2} > 0$ $(-\infty, -3) \cup (2, \infty)$

67. $\frac{x^2 + x}{x^2 - 1} > 0$ $(-\infty, -1) \cup (-1, 0) \cup (1, \infty)$

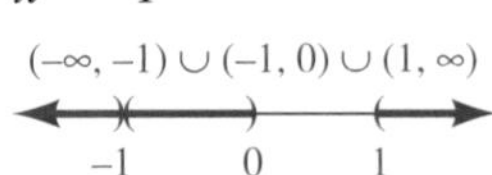

68. $\frac{x^2 - 4}{x^2 - 9} < 0$ $(-3, -2) \cup (2, 3)$

69. $\frac{x^2 + 5x + 6}{x^2 + x - 6} \geq 0$ $(-\infty, -3) \cup (-3, -2] \cup (2, \infty)$

70. $\frac{x^2 + 10x + 25}{x^2 - x - 12} \leq 0$ $[-5, -5] \cup (-3, 4)$

71. $\dfrac{6x^2 - x - 1}{x^2 + 4x + 4} > 0$ $(-\infty, -2) \cup (-2, -1/3) \cup (1/2, \infty)$

72. $\dfrac{6x^2 - 3x - 3}{x^2 - 2x - 8} < 0$ $(-2, -1/2) \cup (1, 4)$

73. $\dfrac{3}{x} > 2$ $(0, 3/2)$

74. $\dfrac{3}{x} < 2$ $(-\infty, 0) \cup (3/2, \infty)$

75. $\dfrac{6}{x} < 4$ $(-\infty, 0) \cup (3/2, \infty)$

76. $\dfrac{6}{x} > 4$ $(0, 3/2)$

77. $\dfrac{3}{x - 2} \leq 5$ $(-\infty, 2) \cup [13/5, \infty)$

78. $\dfrac{3}{x + 2} \leq 4$ $(-\infty, -2) \cup [-5/4, \infty)$

79. $\dfrac{6}{x^2 - 1} < 1$ $(-\infty, -\sqrt{7}) \cup (-1, 1) \cup (\sqrt{7}, \infty)$

80. $\dfrac{6}{x^2 - 1} > 1$ $(-\sqrt{7}, -1) \cup (1, \sqrt{7})$

APPLICATIONS *Solve each problem.*

81. Long distance A long-distance telephone call costs 36¢ for the first three minutes and 11¢ for each additional minute. How long can a person talk for less than $2? **17 min**

82. Buying a computer A student who can afford to spend up to $2,000 sees the ad shown in the illustration. If she buys a computer, how many games can she buy? **15**

83. Buying CDs Andy can spend up to $275 on a CD player and some CDs. If he can buy a disk player for $150 and disks for $9.75, what is the greatest number of disks that he can buy? **12**

84. Buying DVDs Mary wants to spend less than $600 for a DVD recorder and some DVDs. If the recorder of her choice costs $425 and DVDs cost $7.50 each, how many DVDs can she buy? **23**

85. Buying a refrigerator A woman who has $1,200 to spend wants to buy a refrigerator. Refer to the following table and write an inequality that shows how much she can pay for the refrigerator. $p \leq \$1{,}124.12$

State sales tax	6.5%
City sales tax	0.25%

86. Renting a rototiller The cost of renting a rototiller is $17.50 for the first hour and $8.95 for each additional hour. How long can a person have the rototiller if the cost must be less than $75? **7 hr**

87. Real estate taxes A city council has proposed the following two methods of taxing real estate:

Method 1	$2,200 + 4% of assessed value
Method 2	$1,200 + 6% of assessed value

For what range of assessments a would the first method benefit the taxpayer? $a > \$50{,}000$

88. Medical plans A college provides its employees with a choice of the two medical plans shown in the following table. For what size hospital bills is Plan 2 better for the employee than Plan 1? (*Hint:* The cost to the employee includes both the deductible payment and the employee's coinsurance payment.) **anything over $900**

Plan 1	Plan 2
Employee pays $100	Employee pays $200
Plan pays 70% of the rest	Plan pays 80% of the rest

89. Medical plans To save costs, the college in Exercise 88 raised the employee deductible, as shown in the following table. For what size hospital bills is Plan 2 better for the employee than Plan 1? (*Hint:* The cost to the employee includes both the deductible payment and the employee's coinsurance payment.)
anything over \$1,800

Plan 1	Plan 2
Employee pays \$200 Plan pays 70% of the rest	Employee pays \$400 Plan pays 80% of the rest

90. Geometry The perimeter of a rectangle is to be between 180 inches and 200 inches. Find the range of values for its length when its width is 40 inches.
50 in. $< l <$ 60 in.

91. Geometry The perimeter of an equilateral triangle is to be between 50 centimeters and 60 centimeters. Find the range of lengths of one side.
$16\frac{2}{3}$ cm $< s <$ 20 cm

92. Geometry The perimeter of a square is to be from 25 meters to 60 meters. Find the range of values for its area. $\frac{625}{16}\text{ m}^2 \le A \le 225\text{ m}^2$

DISCOVERY AND WRITING

93. Express the relationship $20 < l < 30$ in terms of P, where $P = 2l + 2w$. $40 + 2w < P < 60 + 2w$

94. Express the relationship $10 < C < 20$ in terms of F, where $F = \frac{9}{5}C + 32$. $50 < F < 68$

95. The techniques used for solving linear equations and linear inequalities are similar, yet different. Explain.

96. Explain why the relation $\geq$ is transitive.

REVIEW ***In Exercises 97–102,*** $A = \left\{-9, -\pi, -2, -\frac{1}{2}, 0, 1, 2, \sqrt{7}, \frac{21}{2}\right\}$.

97. Which numbers are even integers? $-2, 2$

98. Which numbers are natural numbers? 1, 2

99. Which numbers are prime numbers? 2

100. Which numbers are irrational numbers? $-\pi, \sqrt{7}$

101. Which numbers are real numbers? All are real.

102. Which numbers are rational numbers?
$-9, -2, -\frac{1}{2}, 0, 1, 2, \frac{21}{2}$

1.8 Absolute Value

In this section, you will learn about

- Absolute Value
- Equations of the Form $|x| = k$
- Equations with Two Absolute Values
- Inequalities of the Form $|x| < k$
- Inequalities of the Form $|x| > k$
- Inequalities with Two Absolute Values

In this section, we will review absolute value and examine its consequences in greater detail.

Absolute Value

Absolute Value

The **absolute value** of the real number x, denoted as $|x|$, is defined as follows:

If $x \geq 0$, then $|x| = x$.
If $x < 0$, then $|x| = -x$.

This definition provides a way to associate a nonnegative real number with any real number.

- If $x \geq 0$, then x (which is positive or 0) is its own absolute value.
- If $x < 0$, then $-x$ (which is positive) is the absolute value.

Either way, $|x|$ is positive or 0:

$$|x| \geq 0 \qquad \textbf{for all real numbers } x$$

EXAMPLE 1 Write without absolute value symbols: **a.** $|7|$, **b.** $|-3|$, **c.** $-|-7|$, and **d.** $|x - 2|$.

Solution **a.** Because 7 is positive, $|7| = 7$.

b. Because -3 is negative, $|-3| = -(-3) = 3$.

c. The expression $-|-7|$ means "the negative of the absolute value of -7." Thus, $-|-7| = -(7) = -7$.

d. To denote the absolute value of a variable quantity, we must give a conditional answer.

If $x - 2 \geq 0$, then $|x - 2| = x - 2$.
If $x - 2 < 0$, then $|x - 2| = -(x - 2) = 2 - x$.

Self Check Write without absolute value symbols: **a.** $|0|$, **b.** $|-17|$, and **c.** $|x + 5|$. ■

■ Equations of the Form $|x| = k$

In the equation $|x| = 8$, x can be either 8 or -8, because $|8| = 8$ and $|-8| = 8$. In general, the following is true.

Absolute Value Equations

If $k \geq 0$, then

$$|x| = k \quad \text{is equivalent to} \quad x = k \quad \text{or} \quad x = -k$$

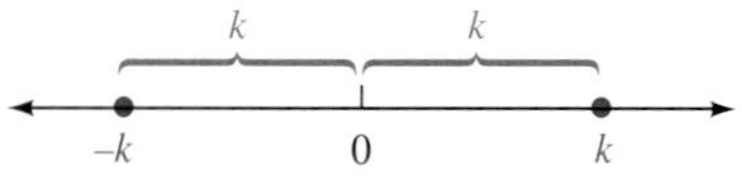

Figure 1-24

The absolute value of a number represents the distance on the number line from a point to the origin. The solutions of $|x| = k$ are the coordinates of the two points that lie exactly k units from the origin. (See Figure 1-24.)

The equation $|x - 3| = 7$ indicates that a point on the number line with a coordinate of $x - 3$ is 7 units from the origin. Thus, $|x - 3|$ can be 7 or -7.

$$x - 3 = 7 \quad \text{or} \quad x - 3 = -7$$
$$x = 10 \quad \Big| \quad x = -4$$

The solutions of 10 and -4 are shown in Figure 1-25. Both of these numbers satisfy the equation.

$\lvert x - 3\rvert = 7$ and	$\lvert x - 3\rvert = 7$
$\lvert 10 - 3\rvert = 7$	$\lvert -4 - 3\rvert = 7$
$\lvert 7\rvert = 7$	$\lvert -7\rvert = 7$
$7 = 7$	$7 = 7$

–4 0 10

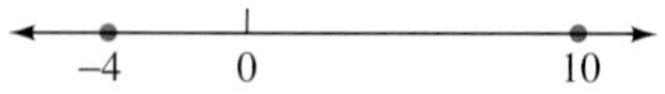

Figure 1-25

EXAMPLE 2 Solve: $|3x - 5| = 7$.

Solution The equation $|3x - 5| = 7$ is equivalent to two equations

$$3x - 5 = 7 \quad \text{or} \quad 3x - 5 = -7$$

which can be solved separately:

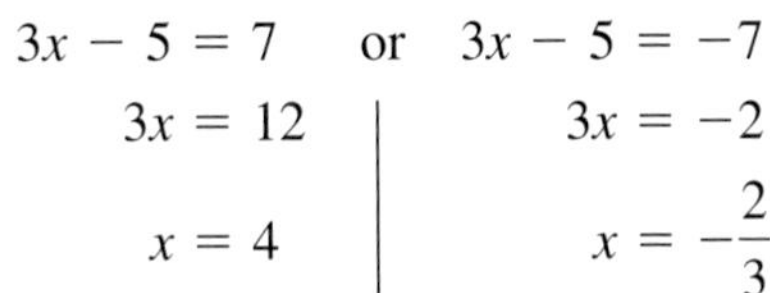

$$\begin{array}{r|r} 3x - 5 = 7 \quad \text{or} & 3x - 5 = -7 \\ 3x = 12 & 3x = -2 \\ x = 4 & x = -\frac{2}{3} \end{array}$$

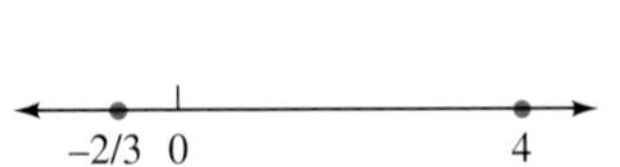

Figure 1-26

The solution set consists of the points shown in Figure 1-26.

Self Check Solve: $|2x + 3| = 7$.

■ Equations with Two Absolute Values

The equation $|a| = |b|$ is true when $a = b$ or when $a = -b$. For example,

$$\begin{array}{c|c} |3| = |3| \quad \text{or} & |3| = |-3| \\ 3 = 3 & 3 = 3 \end{array}$$

In general, the following is true.

Equations with Two Absolute Values

If a and b represent algebraic expressions, the equation $|a| = |b|$ is equivalent to

$$a = b \quad \text{or} \quad a = -b$$

ACTIVE EXAMPLE 3 Solve: $|2x| = |x - 3|$.

Solution The equation $|2x| = |x - 3|$ will be true when $2x$ and $x - 3$ are equal or when they are negatives. This gives two equations, which can be solved separately:

COLLEGE **Algebra** $f(x)$ **Now**™

Explore this Active Example and test yourself on the steps taken here at **http://1pass.thomson.com**. You can also find this example on the Interactive Video Skillbuilder CD-ROM.

$$\begin{array}{r|l} 2x = x - 3 \quad \text{or} & 2x = -(x - 3) \\ x = -3 & 2x = -x + 3 \\ & 3x = 3 \\ & x = 1 \end{array}$$

Verify that -3 and 1 satisfy the equation.

Self Check Solve: $|3x + 1| = |5x - 3|$.

■ Inequalities of the Form $|x| < k$

The inequality $|x| < 5$ indicates that a point with coordinate x is less than 5 units from the origin. (See Figure 1-27.) Thus, x is between -5 and 5, and

$$|x| < 5 \quad \text{is equivalent to} \quad -5 < x < 5$$

In general, the inequality $|x| < k$ $(k > 0)$ indicates that a point with coordinate x is less than k units from the origin. (See Figure 1-28.)

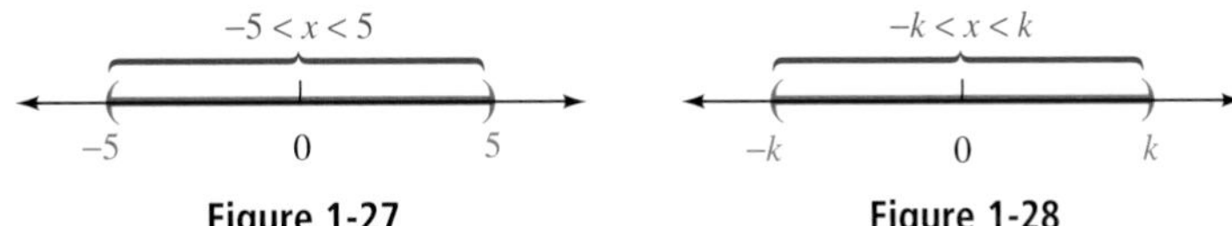

Figure 1-27 Figure 1-28

Inequalities of the Form $|x| < k$

If $k > 0$, then

$|x| < k$ is equivalent to $-k < x < k$

If $k \geq 0$,

$|x| \leq k$ is equivalent to $-k \leq x \leq k$

ACTIVE EXAMPLE 4 Solve: $|x - 2| < 7$.

Solution The inequality $|x - 2| < 7$ is equivalent to

$$-7 < x - 2 < 7$$

We can add 2 to each part of this inequality to get

$$-5 < x < 9$$

The solution set is the interval $(-5, 9)$, shown in Figure 1-29.

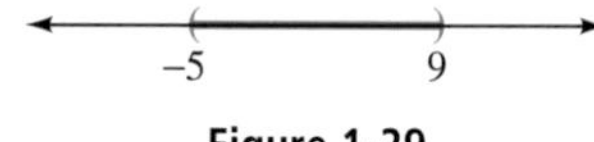

Figure 1-29

COLLEGE **Algebra** $f(x)$ **Now**™
Go to **http://1pass.thomson.com** or your CD to practice this example.

Self Check Solve: $|x + 3| < 9$. ■

■ Inequalities of the Form $|x| > k$

The inequality $|x| > 5$ indicates that a point with coordinate x is more than 5 units from the origin. (See Figure 1-30.) Thus, $x < -5$ or $x > 5$.

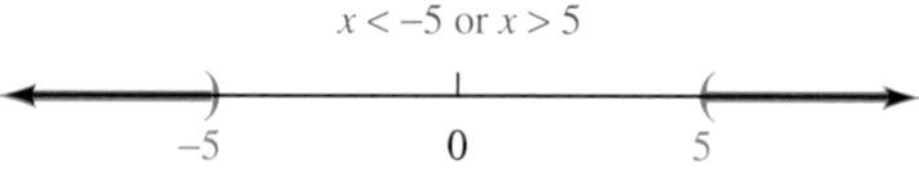

Figure 1-30

In general, the inequality $|x| > k$ $(k \geq 0)$ indicates that a point with coordinate x is more than k units from the origin. (See Figure 1-31.)

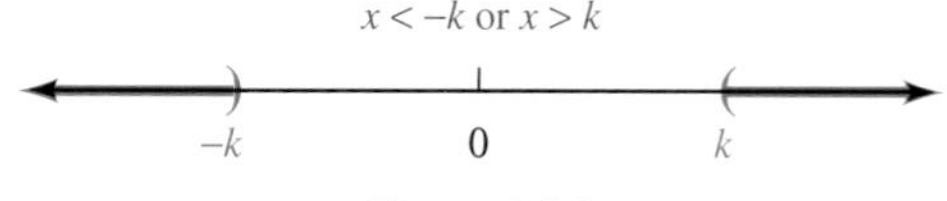

Figure 1-31

Inequalities of the Form $|x| > k$

If $k \geq 0$, then

$|x| > k$ is equivalent to $x < -k$ or $x > k$

$|x| \geq k$ is equivalent to $x \leq -k$ or $x \geq k$

ACTIVE EXAMPLE 5 Solve: $\left| \dfrac{2x + 3}{2} \right| + 7 \geq 12$.

Solution We begin by subtracting 7 from both sides of the inequality to isolate the absolute value on the left-hand side.

$$\left| \frac{2x + 3}{2} \right| \geq 5$$

COLLEGE **Algebra** $f(x)$ **Now**™
Go to **http://1pass.thomson.com** or your CD to practice this example.

This result is equivalent to two inequalities that can be solved separately.

$$\frac{2x+3}{2} \le -5 \quad \text{or} \quad \frac{2x+3}{2} \ge 5$$

$$\begin{array}{r|r} 2x+3 \le -10 & 2x+3 \ge 10 \\ 2x \le -13 & 2x \ge 7 \\ x \le -\dfrac{13}{2} & x \ge \dfrac{7}{2} \end{array}$$

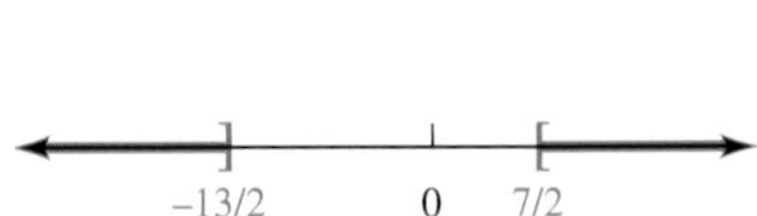

Figure 1-32

The solution set is the union of the intervals $\left(-\infty, -\frac{13}{2}\right]$ and $\left[\frac{7}{2}, \infty\right)$. Its graph appears in Figure 1-32.

Self Check Solve: $\left|\dfrac{3x-6}{3}\right| + 2 \ge 12.$ ■

EXAMPLE 6 Solve: $0 < |x - 5| \le 3.$

Solution The inequality $0 < |x - 5| \le 3$ consists of two inequalities that can be solved separately. The solution will be the intersection of the inequalities

$$0 < |x - 5| \quad \text{and} \quad |x - 5| \le 3$$

The inequality $0 < |x - 5|$ is true for all x except 5. The inequality $|x - 5| \le 3$ is equivalent to the inequality

$$-3 \le x - 5 \le 3$$

$$2 \le x \le 8 \qquad \text{Add 5 to each part.}$$

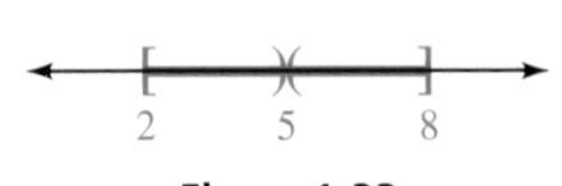

Figure 1-33

The solution set is the intersection of these two solutions, which is the interval [2, 8], except 5. This is the union of the intervals [2, 5) and (5, 8], as shown in Figure 1-33.

Self Check Solve: $0 < |x + 2| \le 5.$ ■

■ Inequalities with Two Absolute Values

In Example 9 of Section 1.5, we saw that $|a|$ could be defined as

$$|a| = \sqrt{a^2}$$

We will use this fact in the next example.

EXAMPLE 7 Solve $|x + 2| > |x + 1|$ and give the result in interval notation.

Solution

$$|x+2| > |x+1|$$

$$\sqrt{(x+2)^2} > \sqrt{(x+1)^2} \qquad \text{Use } |a| = \sqrt{a^2}.$$

$$(x+2)^2 > (x+1)^2 \qquad \text{Square both sides.}$$

$$x^2 + 4x + 4 > x^2 + 2x + 1 \qquad \text{Expand each binomial.}$$

$$4x > 2x - 3 \qquad \text{Subtract } x^2 \text{ and 4 from both sides.}$$

$$2x > -3 \qquad \text{Subtract } 2x \text{ from both sides.}$$

$$x > -\frac{3}{2} \qquad \text{Divide both sides by 2.}$$

The solution set is the interval $\left(-\frac{3}{2}, \infty\right)$. Check several solutions to verify that this interval does not contain any extraneous solutions.

Self Check Solve $|x - 3| \leq |x + 2|$ and give the result in interval notation. ■

Three other properties of absolute value are sometimes useful.

Properties of Absolute Value

If a and b are real numbers, then

1. $|ab| = |a||b|$

2. $\left|\dfrac{a}{b}\right| = \dfrac{|a|}{|b|} \quad (b \neq 0)$

3. $|a + b| \leq |a| + |b|$

Properties 1 and 2 state that the absolute value of a product (or a quotient) is the product (or the quotient) of the absolute values.

Property 3 states that the absolute value of a sum is either equal to or less than the sum of the absolute values.

Self Check Answers

1. a. 0 **b.** 17 **c.** if $x + 5 \geq 0$, $|x + 5| = x + 5$; if $x + 5 < 0$, $|x + 5| = -x - 5$

2. 

3. $2, \frac{1}{4}$ **4.** 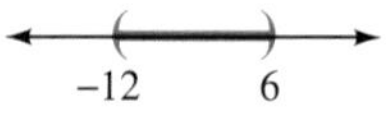

5. 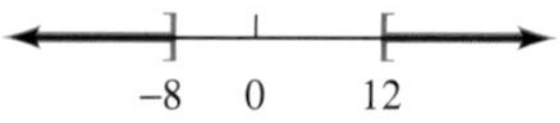

6. 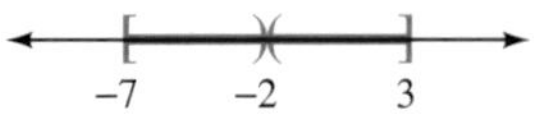

7. $\left[\frac{1}{2}, \infty\right)$

1.8 Exercises

VOCABULARY AND CONCEPTS ***Fill in the blanks.***

1. If $x \geq 0$, then $|x| =$ $\underline{x}$.

2. If $x < 0$, then $|x| =$ $\underline{-x}$.

3. $|x| = k$ is equivalent to $\underline{x = k \text{ or } x = -k}$.

4. $|a| = |b|$ is equivalent to $a = b$ or $\underline{a = -b}$.

5. $|x| < k$ is equivalent to $\underline{-k < x < k}$.

6. $|x| > k$ is equivalent to $\underline{x < -k \text{ or } x > k}$.

7. $|x| \geq k$ is equivalent to $\underline{x \leq -k \text{ or } x \geq k}$.

8. $\sqrt{a^2} =$ $\underline{|a|}$.

PRACTICE ***Write each expression without absolute value symbols.***

9. $|7|$ 7

10. $|-9|$ 9

11. $|0|$ 0

12. $|3 - 5|$ 2

13. $|5| - |-3|$ 2

14. $|-3| + |5|$ 8

15. $|\pi - 2|$ $\pi - 2$

16. $|\pi - 4|$ $4 - \pi$

17. $|x - 5|$ and $x > 5$
$x - 5$

18. $|x - 5|$ and $x < 5$
$5 - x$

19. $|x^3|$
x^3 if $x \geq 0$; $-x^3$ if $x < 0$

20. $|2x|$
$2x$ if $x \geq 0$; $-2x$ if $x < 0$

Solve for x.

21. $|x + 2| = 2$
$0, -4$

22. $|2x + 5| = 3$
$-1, -4$

23. $|3x - 1| = 5$
$2, -\frac{4}{3}$

24. $|7x - 5| = 3$
$\frac{8}{7}, \frac{2}{7}$

25. $\left|\dfrac{3x - 4}{2}\right| = 5$
$\frac{14}{3}, -2$

26. $\left|\dfrac{10x + 1}{2}\right| = \dfrac{9}{2}$
$\frac{4}{5}, -1$

27. $\left|\frac{2x-4}{5}\right| = 2$
7, −3

28. $\left|\frac{3x+11}{7}\right| = 1$
$-\frac{4}{3}, -6$

29. $\left|\frac{x-3}{4}\right| = -2$
no solution

30. $\left|\frac{x-5}{3}\right| = 0$
5

31. $\left|\frac{4x-2}{x}\right| = 3$
$\frac{2}{7}, 2$

32. $\left|\frac{2(x-3)}{3x}\right| = 6$
$-\frac{3}{8}, \frac{3}{10}$

33. $|x| = x$ $x \geq 0$

34. $|x| + x = 2$ 1

35. $|x+3| = |x|$
$-\frac{3}{2}$

36. $|x+5| = |5-x|$
0

37. $|x-3| = |2x+3|$
0, −6

38. $|x-2| = |3x+8|$
$-5, -\frac{3}{2}$

39. $|x+2| = |x-2|$
0

40. $|2x-3| = |3x-5|$
$2, \frac{8}{5}$

41. $\left|\frac{x+3}{2}\right| = |2x-3|$
$\frac{3}{5}, 3$

42. $\left|\frac{x-2}{3}\right| = |6-x|$
5, 8

43. $\left|\frac{3x-1}{2}\right| = \left|\frac{2x+3}{3}\right|$ $-\frac{3}{13}, \frac{9}{5}$

44. $\left|\frac{5x+2}{3}\right| = \left|\frac{x-1}{4}\right|$ $-\frac{11}{17}, -\frac{5}{23}$

Solve each inequality, express the solution set in interval notation, and graph it.

45. $|x-3| < 6$
(−3, 9)
−3 9

46. $|x-2| \geq 4$
(−∞, −2] ∪ [6, ∞)
−2 6

47. $|x+3| > 6$
(−∞, −9) ∪ (3, ∞)
−9 3

48. $|x+2| \leq 4$
[−6, 2]
−6 2

49. $|2x+4| \geq 10$
(−∞, −7] ∪ [3, ∞)
−7 3

50. $|5x-2| < 7$
(−1, 9/5)
−1 9/5

51. $|3x+5| + 1 \leq 9$
[−13/3, 1]
−13/3 1

52. $|2x-7| - 3 > 2$
(−∞, 1) ∪ (6, ∞)
1 6

53. $|x+3| > 0$
(−∞, −3) ∪ (−3, ∞)
−3

54. $|x-3| \leq 0$
[3, 3]
3

55. $\left|\frac{5x+2}{3}\right| < 1$
(−1, 1/5)
−1 1/5

56. $\left|\frac{3x+2}{4}\right| > 2$
(−∞, −10/3) ∪ (2, ∞)
−10/3 2

57. $3\left|\frac{3x-1}{2}\right| > 5$
(−∞, −7/9) ∪ (13/9, ∞)
−7/9 13/9

58. $2\left|\frac{8x+2}{5}\right| \leq 1$
[−9/16, 1/16]
−9/16 1/16

59. $\frac{|x-1|}{-2} > -3$
(−5, 7)
−5 7

60. $\frac{|2x-3|}{-3} < -1$
(−∞, 0) ∪ (3, ∞)
0 3

61. $0 < |2x+1| < 3$ (−2, −1/2) ∪ (−1/2, 1)
−2 −1/2 1

62. $0 < |2x-3| < 1$ (1, 3/2) ∪ (3/2, 2)
1 3/2 2

63. $8 > |3x-1| > 3$ (−7/3, −2/3) ∪ (4/3, 3)
−7/3 −2/3 4/3 3

64. $8 > |4x-1| > 5$ (−7/4, −1) ∪ (3/2, 9/4)
−7/4 −1 3/2 9/4

65. $2 < \left|\frac{x-5}{3}\right| < 4$ (−7, −1) ∪ (11, 17)
−7 −1 11 17

66. $3 < \left|\frac{x-3}{2}\right| < 5$ (−7, −3) ∪ (9, 13)
−7 −3 9 13

67. $10 > \left|\frac{x-2}{2}\right| > 4$ (−18, −6) ∪ (10, 22)
−18 −6 10 22

68. $5 \geq \left|\frac{x+2}{3}\right| > 1$ [−17, −5) ∪ (1, 13]
−17 −5 1 13

69. $2 \leq \left|\frac{x+1}{3}\right| < 3$ (−10, −7] ∪ [5, 8)
−10 −7 5 8

70. $8 > \left|\frac{3x+1}{2}\right| > 2$ (−17/3, −5/3) ∪ (1, 5)
−17/3 −5/3 1 5

Solve each inequality and express the solution using interval notation.

71. $|x + 1| \geq |x|$ $\left[-\frac{1}{2}, \infty\right)$

72. $|x + 1| < |x + 2|$ $\left(-\frac{3}{2}, \infty\right)$

73. $|2x + 1| < |2x - 1|$ $(-\infty, 0)$

74. $|3x - 2| \geq |3x + 1|$ $\left(-\infty, \frac{1}{6}\right]$

75. $|x + 1| < |x|$ $\left(-\infty, -\frac{1}{2}\right)$

76. $|x + 2| \leq |x + 1|$ $\left(-\infty, -\frac{3}{2}\right]$

77. $|2x + 1| \geq |2x - 1|$ $[0, \infty)$

78. $|3x - 2| < |3x + 1|$ $\left(\frac{1}{6}, \infty\right)$

APPLICATIONS

79. Finding temperature ranges The temperatures on a summer day satisfy the inequality $|t - 78°| \leq 8°$, where t is the temperature in degrees Fahrenheit. Express this range without using absolute value symbols. $70° \leq t \leq 86°$

80. Finding operating temperatures A car CD player has an operating temperature of $|t - 40°| < 80°$, where t is the temperature in degrees Fahrenheit. Express this range without using absolute value symbols. $-40° < t < 120°$

81. Range of camber angles The specifications for a certain car state that the camber angle c of its wheels should be $0.6° \pm 0.5°$. Express this range with an inequality containing an absolute value.
$|c - 0.6°| \leq 0.5°$

82. Tolerance of a sheet of steel A sheet of steel is to be 0.25 inch thick, with a tolerance of 0.015 inch. Express this specification with an inequality containing an absolute value. $|x - 0.25| \leq 0.015$

83. Humidity level A Steinway piano should be placed in an environment where the relative humidity h is between 38% and 72%. Express this range with an inequality containing an absolute value.
$|h - 55| < 17$

84. Life of a lightbulb A lightbulb is expected to last h hours, where $|h - 1{,}500| \leq 200$. Express this range without using absolute value symbols.
$1{,}300 \leq h \leq 1{,}700$

85. Error analysis In a lab, students measured the percent of copper p in a sample of copper sulfate. The students know that copper sulfate is actually 25.46% copper by mass. They are to compare their results to the actual value and find the amount of *experimental error.*

Lab 4 Section A
Title:
"Percent copper (CU) in copper sulfate ($CuSO_4 \cdot 5H_2O$)"

Results

	% Copper
Trial #1:	22.91%
Trial #2:	26.45%
Trial #3	26.49%
Trial #4:	24.76%

a. Which measurements shown in the illustration satisfy the absolute value inequality $|p - 25.46| \leq 1.00$? 26.45%, 24.76%

b. What can be said about the amount of error for each of the trials listed in part a?
It is less than or equal to 1%.

86. Error analysis See Exercise 85.

a. Which measurements satisfy the absolute value inequality $|p - 25.46| > 1.00$? 22.91%, 26.49%

b. What can be said about the amount of error for each of the trials listed in part a?
It is more than 1%.

DISCOVERY AND WRITING

87. Explain how to find the absolute value of a number.

88. Explain why the equation $|x| + 9 = 0$ has no solution.

89. Explain the use of parentheses and brackets when graphing inequalities.

90. If $k > 0$, explain the differences between the solution sets of $|x| < k$ and $|x| > k$.

REVIEW *Write each number in scientific notation.*

91. 37,250 3.725×10^4

92. 0.0003725 3.725×10^{-4}

Write each number in standard notation.

93. 5.23×10^5 523,000

94. 7.9×10^{-4} 0.00079

Simplify each expression.

95. $(x - y)^2 - (x + y)^2$ $-4xy$

96. $(p + q)^2 + (p - q)^2$ $2p^2 + 2q^2$

PROBLEMS AND PROJECTS

1. A total of 736 people paid \$7,772 to attend a concert. Senior citizen tickets cost \$7, and all others cost \$12. For the rights to perform the music, a 5% royalty on the receipts from senior citizen tickets and a 9% royalty on the remaining receipts must be paid. How much is owed in royalties? **\$640.12**

2. Find the distance x required to balance the lever shown in the illustration. Research beginning and intermediate algebra books to find the principles from physics that are necessary to solve the problem. **4 ft**

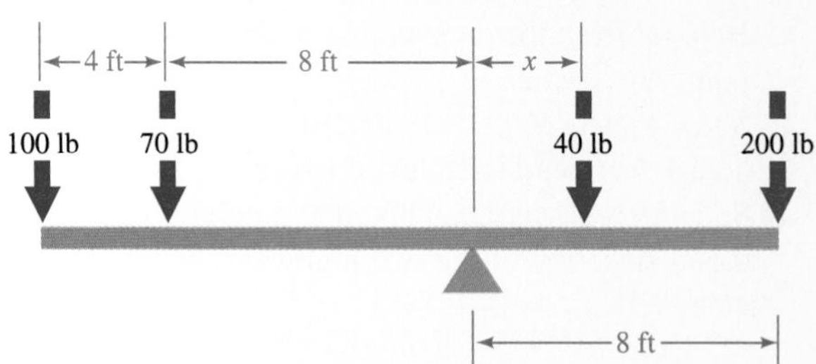

3. Present to a friend the proof of the quadratic formula given in Section 1.3. Then present the following proof.

$$\begin{aligned} ax^2 + bx + c &= 0 \\ ax^2 + bx &= -c \\ 4a(ax^2) + 4a(bx) &= 4a(-c) \\ 4a^2x^2 + 4abx &= -4ac \\ 4a^2x^2 + 4abx + b^2 &= b^2 - 4ac \\ (2ax + b)^2 &= b^2 - 4ac \\ 2ax + b &= \pm\sqrt{b^2 - 4ac} \\ 2ax &= -b \pm \sqrt{b^2 - 4ac} \\ x &= \frac{-b \pm \sqrt{b^2 - 4ac}}{2a} \end{aligned}$$

Which proof did your friend understand better? Why? Which proof did you prefer?

4. A researcher wants to estimate the mean real estate tax paid by homeowners living in Rockford, Illinois. To do so, he decides to select a random sample of homeowners and compute the mean tax paid by the homeowners in that sample. How large must the sample be for the researcher to be 95% certain that his computed sample mean will be within \$35 of the true population mean—that is, within \$35 of the mean tax paid by all homeowners in the city? Assume that the standard deviation, σ, of all tax bills in the city is \$120.

From statistics, the researcher has the formula

$$\frac{3.84\sigma^2}{N} < E^2$$

where E is the maximum acceptable error and N is the sample size. **46 homeowners**

Project 1

Use a geometry textbook to find the necessary facts to find x in each problem.

1. **40**

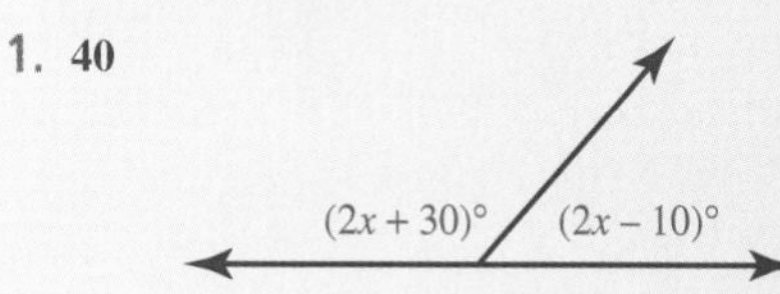

2. **25**

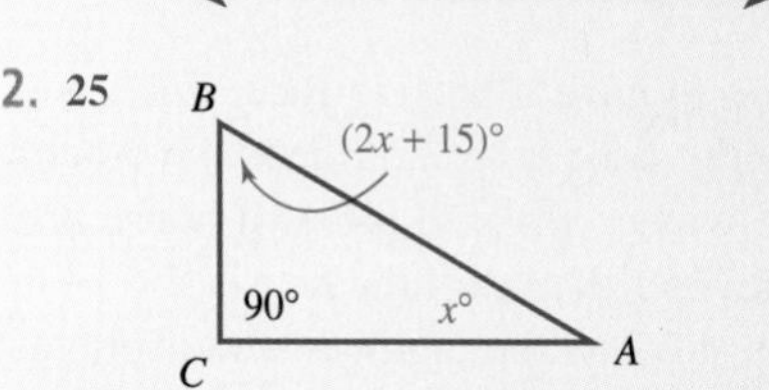

3. **50**

4. **10**

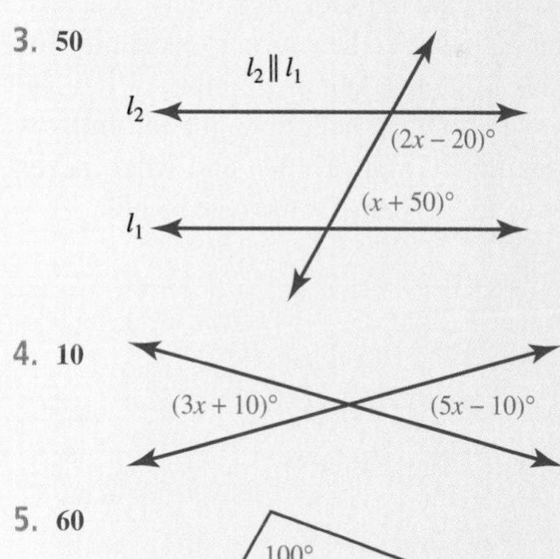

5. **60**

6. **50**

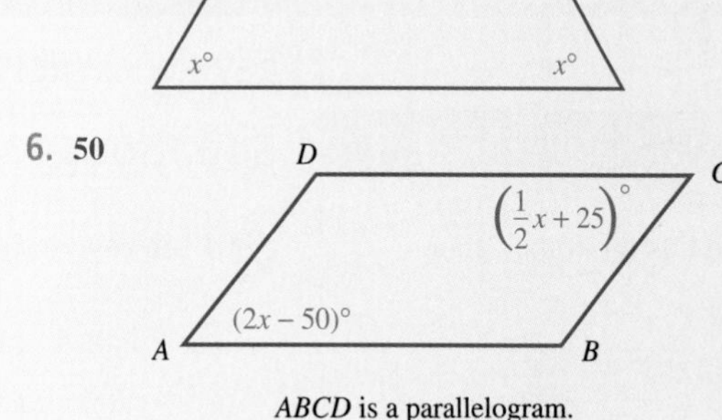

ABCD is a parallelogram.

Project 2

Find the lengths of sides a, b, c, and d shown in the illustration. What pattern do you notice? How many triangles will be necessary to produce a side of length 3?
$\sqrt{2}$, $\sqrt{3}$, 2, $\sqrt{5}$; 8

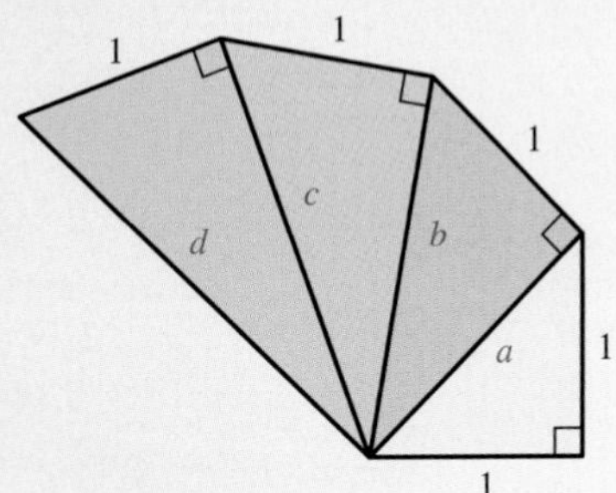

Project 3

Carlos and Mike agree to have a bicycle race. Each will leave his own house at the same time and ride to the other's house, whereupon the winner will call his own house and leave a message for the loser. A map of the race is shown in the illustration. Carlos stays on the highway and averages 21 mph. Mike knows that the two are evenly matched when biking on the highway, so he cuts across country for the first part of his trip, averaging 15 mph. When he reaches the highway at point A, he turns right and follows the highway, averaging 21 mph. Carlos and Mike never meet during the race, and amazingly, the race is a tie.

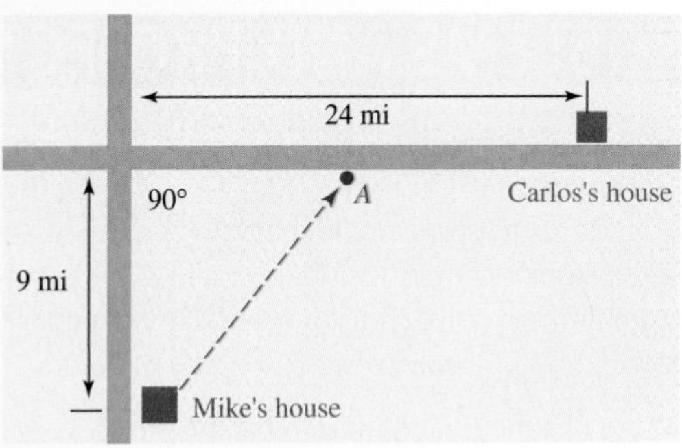

a. To the nearest second, how long did it take each person to complete the race?
$\frac{33}{21}$ hr, which is 1 hr 34 min 17 sec

b. How far from the intersection of the two highways is point A? **12 mi**

c. Show that if Mike had started straight for Carlos's house, he would have lost the race.

d. Show that if Mike had biked to a point 9 miles from the intersection of the two highways, he would have won.

Project 4

1. The number $\frac{a+b}{2}$ is called the **arithmetic mean** between a and b. If $a < b$, show that $a < \frac{a+b}{2} < b$.

2. The number $\sqrt{ab}$ is called a **geometric mean** between a and b. If $0 < a < b$, show that $a < \sqrt{ab} < b$.

3. The number $\frac{2ab}{a+b}$ is called a **harmonic mean** between a and b. If $0 < a < b$, show that $a < \frac{2ab}{a+b} < b$.

CHAPTER SUMMARY

CONCEPTS	REVIEW EXERCISES
1.1	**Equations**
If $a = b$ and c is a number, then $a + c = b + c$ $a - c = b - c$ $ac = bc$ and $\frac{a}{c} = \frac{b}{c}\ (c \neq 0)$	**Find the restrictions on x, if any.** **1.** $3x + 7 = 4$ **no restrictions** **2.** $x + \frac{1}{x} = 2$ **$x \neq 0$** **3.** $\sqrt{x} = 4$ **$x \geq 0$** **4.** $\frac{1}{x-2} = \frac{2}{x-3}$ **$x \neq 2, x \neq 3$**

A **linear equation** is an equation that can be written in the form

$$ax + b = 0$$

Solve each equation and classify it as an identity, a conditional equation, or an equation with no solution.

5. $3(9x + 4) = 28$
   $\frac{16}{27}$; conditional equation

6. $\frac{3}{2}a = 7(a + 11)$
   -14; conditional equation

7. $8(3x - 5) - 4(x + 3) = 12$
   $\frac{16}{5}$; conditional equation

8. $\frac{x + 3}{x + 4} + \frac{x + 3}{x + 2} = 2$
   no solution

9. $\frac{3}{x - 1} = \frac{1}{2}$
   7; conditional equation

10. $\frac{8x^2 + 72x}{9 + x} = 8x$
    identity

11. $\frac{3x}{x - 1} - \frac{5}{x + 3} = 3$
    7; conditional equation

12. $x + \frac{1}{2x - 3} = \frac{2x^2}{2x - 3}$
    $\frac{1}{3}$; conditional equation

Solve each formula for the indicated variable.

13. $C = \frac{5}{9}(F - 32)$; F
    $F = \frac{9}{5}C + 32$

14. $P_n = l + \frac{si}{f}$; f
    $f = \frac{is}{P_n - l}$

15. $\frac{1}{f} = \frac{1}{f_1} + \frac{1}{f_2}$; f_1
    $f_1 = \frac{ff_2}{f_2 - f}$

16. $S = \frac{a - lr}{1 - r}$; l
    $l = \frac{a - S + Sr}{r}$

1.2 Applications of Linear Equations

Use the following steps to find a mathematical model of a problem:

1. Read the problem.
2. Pick a variable to represent the quantity to be found.
3. Form an equation.
4. Solve the equation.
5. Check the solution in the words of the problem.

17. **Preparing a solution** A liter of fluid is 50% alcohol. How much water must be added to dilute it to a 20% solution? **1.5 liters**

18. **Washing windows** Scott can wash 37 windows in 3 hours, and Bill can wash 27 windows in 2 hours. How long will it take the two of them to wash 100 windows? **about 3.9 hr**

19. **Filling a tank** A tank can be filled in 9 hours by one pipe and in 12 hours by another. How long will it take both pipes to fill the empty tank? **$5\frac{1}{7}$ hr**

20. **Producing brass** How many ounces of pure zinc must be alloyed with 20 ounces of brass that is 30% zinc and 70% copper to produce brass that is 40% zinc? **$3\frac{1}{3}$ oz**

21. **Lending money** A bank lends \$10,000, part of it at 11% annual interest and the rest at 14%. If the annual income is \$1,265, how much was lent at each rate? **\$4,500 at 11%; \$5,500 at 14%**

22. **Producing oriental rugs** An oriental rug manufacturer can use one loom with a setup cost of \$750 that can weave a rug for \$115. Another loom, with a setup cost of \$950, can produce a rug for \$95. How many rugs are produced if the costs are the same on each loom? **10**

1.3 Quadratic Equations

Zero-factor theorem:
If $ab = 0$, then $a = 0$ or $b = 0$.

Square root property:
If $c > 0$, $x^2 = c$ has two real roots:

$$x = \sqrt{c} \quad \text{and} \quad x = -\sqrt{c}$$

Quadratic formula:

$$x = \frac{-b \pm \sqrt{b^2 - 4ac}}{2a} \quad (a \neq 0)$$

Discriminant:
If $b^2 - 4ac > 0$, the roots of $ax^2 + bx + c = 0$ are unequal real numbers.

If $b^2 - 4ac = 0$, the roots of $ax^2 + bx + c = 0$ are equal real numbers.

If $b^2 - 4ac < 0$, the roots of $ax^2 + bx + c = 0$ are nonreal numbers.

Solve each equation by factoring.

23. $2x^2 - x - 6 = 0$ **$2, -\frac{3}{2}$**

24. $12x^2 + 13x = 4$ **$\frac{1}{4}, -\frac{4}{3}$**

25. $5x^2 - 8x = 0$ **$0, \frac{8}{5}$**

26. $27x^2 = 30x - 8$ **$\frac{2}{3}, \frac{4}{9}$**

Solve each equation by completing the square.

27. $x^2 - 8x + 15 = 0$ **$3, 5$**

28. $3x^2 + 18x = -24$ **$-4, -2$**

29. $5x^2 - x - 1 = 0$ **$\frac{1 \pm \sqrt{21}}{10}$**

30. $5x^2 - x = 0$ **$0, \frac{1}{5}$**

Use the quadratic formula to solve each equation.

31. $x^2 + 5x - 14 = 0$ **$2, -7$**

32. $3x^2 - 25x = 18$ **$9, -\frac{2}{3}$**

33. $5x^2 = 1 - x$ **$\frac{-1 \pm \sqrt{21}}{10}$**

34. $-5 = a^2 + 2a$ **$-1 \pm 2i$**

35. Find the value of k that will make the roots of $kx^2 + 4x + 12 = 0$ equal. **$\frac{1}{3}$**

36. Find the values of k that will make the roots of $4y^2 + (k + 2)y = 1 - k$ equal. **10, 2**

37. Solve: $\dfrac{4}{a - 4} + \dfrac{4}{a - 1} = 5.$ **$\frac{8}{5}, 5$**

1.4 Applications of Quadratic Equations

38. Fencing a field A farmer wishes to enclose a rectangular garden with 300 yards of fencing. A river runs along one side of the garden, so no fencing is needed there. Find the dimensions of the rectangle if the area is 10,450 square yards. **either 95 by 110 yd or 55 by 190 yd**

39. Flying rates A jet plane, flying 120 mph faster than a propeller-driven plane, travels 3,520 miles in 3 hours less time than the propeller plane requires to fly the same distance. How fast does each plane fly? **320 mph for prop plane; 440 mph for jet plane**

40. Flight of a ball A ball thrown into the air reaches a height h (in feet) according to the formula $h = -16t^2 + 64t$, where t is the time elapsed since the ball was thrown. Find the shortest time it will take the ball to reach a height of 48 feet. **1 sec**

41. Width of a walk A man built a walk of uniform width around a rectangular pool. If the area of the walk is 117 square feet and the dimensions of the pool are 16 feet by 20 feet, how wide is the walk? **$1\frac{1}{2}$ ft**

1.5 Complex Numbers

$a + bi = c + di$
if and only if
$a = c$ and $b = d$

$(a + bi) + (c + di) =$
$(a + c) + (b + d)i$

$(a + bi) - (c + di) =$
$(a - c) + (b - d)i$

$(a + bi)(c + di) =$
$(ac - bd) + (ad + bc)i$

$|a + bi| = \sqrt{a^2 + b^2}$

The **complex conjugate** of $a + bi$ is $a - bi$.

Perform all operations and express all answers in $a = bi$ form.

42. $(2 - 3i) + (-4 + 2i)$ $-2 - i$

43. $(2 - 3i) - (4 + 2i)$ $-2 - 5i$

44. $(3 - \sqrt{-36}) + (\sqrt{-16} + 2)$ $5 - 2i$

45. $(3 + \sqrt{-9})(2 - \sqrt{-25})$ $21 - 9i$

46. $\frac{3}{i}$ $0 - 3i$

47. $-\frac{2}{i^3}$ $0 - 2i$

48. $\frac{3}{1 + i}$ $\frac{3}{2} - \frac{3}{2}i$

49. $\frac{2i}{2 - i}$ $-\frac{2}{5} + \frac{4}{5}i$

50. $\frac{3 + i}{3 - i}$ $\frac{4}{5} + \frac{3}{5}i$

51. $\frac{3 - 2i}{1 + i}$ $\frac{1}{2} - \frac{5}{2}i$

52. Simplify: i^{53}. $0 + i$

53. Simplify: i^{103}. $0 - i$

54. $|3 - i|$ $\sqrt{10}$

55. $\left|\frac{1 + i}{1 - i}\right|$ 1

1.6 Polynomial and Radical Equations

If $a = b$, then $a^2 = b^2$.

Solve each equation.

56. $\frac{3x}{2} - \frac{2x}{x - 1} = x - 3$ $2, -3$

57. $\frac{12}{x} - \frac{x}{2} = x - 3$ $4, -2$

58. $x^4 - 2x^2 + 1 = 0$ $1, 1 -1, -1$

59. $x^4 + 36 = 37x^2$ $1, -1, 6, -6$

60. $a - a^{1/2} - 6 = 0$ 9

61. $x^{2/3} + x^{1/3} - 6 = 0$ $8, -27$

62. $\sqrt{x - 1} + x = 7$ 5

63. $\sqrt{a + 9} - \sqrt{a} = 3$ 0

64. $\sqrt{5 - x} + \sqrt{5 + x} = 4$ $4, -4$

65. $\sqrt{y + 5} + \sqrt{y} = 1$ no solution

1.7 Inequalities

If a, b, and c are real numbers:
If $a < b$, $a + c < b + c$ and $a - c < b - c$.
If $a < b$ and $c > 0$, $ac < bc$ and $\frac{a}{c} < \frac{b}{c}$.

Solve each inequality.

66. $2x - 9 < 5$

$(-\infty, 7)$

7

67. $5x + 3 \geq 2$

$[-1/5, \infty)$

$-1/5$

If $a < b$ and $c < 0$, $ac > bc$ and $\frac{a}{c} > \frac{b}{c}$.

If $a < b$ and $b < c$, then $a < c$.

68. $\frac{5(x-1)}{2} < x$

$(-\infty, 5/3)$

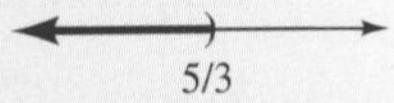

69. $0 \le \frac{3+x}{2} < 4$

$[-3, 5)$

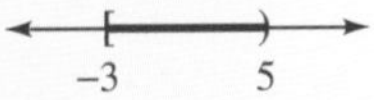

70. $(x+2)(x-4) > 0$

$(-\infty, -2) \cup (4, \infty)$

71. $(x-1)(x+4) < 0$

$(-4, 1)$

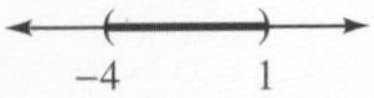

72. $x^2 - 2x - 3 < 0$

$(-1, 3)$

73. $2x^2 + x - 3 > 0$

$(-\infty, -3/2) \cup (1, \infty)$

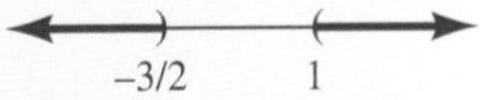

74. $\frac{x+2}{x-3} \ge 0$

$(-\infty, -2] \cup (3, \infty)$

75. $\frac{x-1}{x+4} \le 0$

$(-4, 1]$

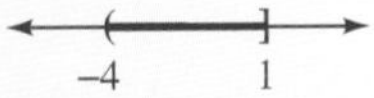

76. $\frac{x^2+x-2}{x-3} \ge 0$

$[-2, 1] \cup (3, \infty)$

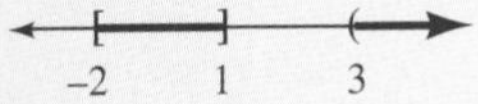

77. $\frac{5}{x} < 2$

$(-\infty, 0) \cup (5/2, \infty)$

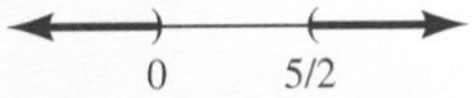

1.8 Absolute Value

$$|x| = \begin{cases} x \text{ when } x \ge 0 \\ -x \text{ when } x < 0 \end{cases}$$

If $a \ge 0$, then $|x| = a$ is equivalent to $x = a$ or $x = -a$.

If $a > 0$, then $|x| < a$ is equivalent to $-a < x < a$.

If $a > 0$, then $|x| > a$ is equivalent to $x > a$ or $x < -a$.

$|a| = \sqrt{a^2}$

Properties of absolute value:

1. $|ab| = |a||b|$

2. $\left|\frac{a}{b}\right| = \frac{|a|}{|b|}$ $(b \ne 0)$

3. $|a + b| \le |a| + |b|$

Solve each equation or inequality.

78. $|x+1| = 6$ 5, −7

79. $|2x-1| = |2x+1|$ **0**

80. $|x+3| < 3$

$(-6, 0)$

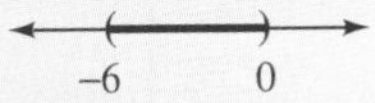

81. $|3x-7| \ge 1$

$(-\infty, 2] \cup [8/3, \infty)$

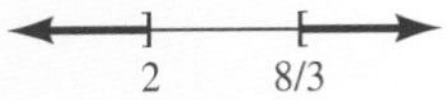

82. $\left|\frac{x+2}{3}\right| < 1$

$(-5, 1)$

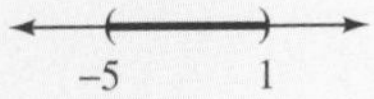

83. $\left|\frac{x-3}{4}\right| > 8$

$(-\infty, -29) \cup (35, \infty)$

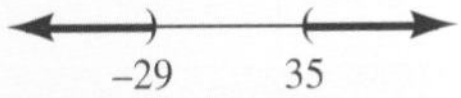

84. $1 < |2x+3| < 4$

$(-7/2, -2) \cup (-1, 1/2)$

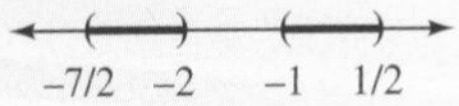

85. $0 < |3x-4| < 7$

$(-1, 4/3) \cup (4/3, 11/3)$

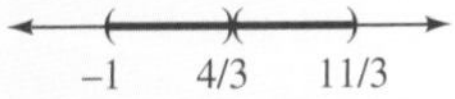

CHAPTER TEST

Find all restrictions on x.

1. $\dfrac{x}{x(x-1)}$ $x \neq 0, x \neq 1$

2. $\sqrt{x}$ $x \geq 0$

Solve each equation.

3. $7(2a+5) - 7 = 6(a+8)$ $\frac{5}{2}$

4. $\dfrac{3}{x^2 - 5x - 14} = \dfrac{4}{x^2 + 5x + 6}$ 37

5. Solve for x: $z = \dfrac{x - \mu}{\sigma}$. $x = \mu + z\sigma$

6. Solve for a: $\dfrac{1}{a} = \dfrac{1}{b} + \dfrac{1}{c}$. $a = \dfrac{bc}{c+b}$

7. A student's average on three tests is 75. If the final is to count as two one-hour tests, what grade must the student make to bring the average up to 80? 87.5

8. A woman invested part of \$20,000 at 6% interest and the rest at 7%. If her annual interest is \$1,260, how much did she invest at 6%? \$14,000

Solve each equation.

9. $4x^2 - 8x + 3 = 0$ $\frac{1}{2}, \frac{3}{2}$

10. $2b^2 - 12 = -5b$ $\frac{3}{2}, -4$

11. Write the quadratic formula. $x = \dfrac{-b \pm \sqrt{b^2 - 4ac}}{2a}$

12. Use the quadratic formula to solve $3x^2 - 5x - 9 = 0$. $\dfrac{5 \pm \sqrt{133}}{6}$

13. Find k such that $x^2 + (k+1)x + k + 4 = 0$ will have two equal roots. 5, −3

14. The height of a projectile shot up into the air is given by the formula $h = -16t^2 + 128t$. Find the time t required for the projectile to return to its starting point. 8 sec

Perform each operation and write all answers in a + bi form.

15. $(4 - 5i) - (-3 + 7i)$ $7 - 12i$

16. $(4 - 5i)(3 - 7i)$ $-23 - 43i$

17. $\dfrac{2}{2 - i}$ $\frac{4}{5} + \frac{2}{5}i$

18. $\dfrac{1 + i}{1 - i}$ $0 + i$

Simplify each expression.

19. i^{13} i

20. i^0 1

Find each absolute value.

21. $|5 - 12i|$ 13

22. $\left|\dfrac{1}{3 + i}\right|$ $\dfrac{\sqrt{10}}{10}$

Solve each equation.

23. $z^4 - 13z^2 + 36 = 0$ 2, −2, 3, −3

24. $2p^{2/5} - p^{1/5} - 1 = 0$ $1, -\frac{1}{32}$

25. $\sqrt{x + 5} = 12$ 139

26. $\sqrt{2z + 3} = 1 - \sqrt{z + 1}$ −1

Solve each inequality.

27. $5x - 3 \leq 7$ $(-\infty, 2]$

28. $\dfrac{x + 3}{4} > \dfrac{2x - 4}{3}$ $(-\infty, 5)$

29. $1 + x < 3x - 3 < 4x - 2$ $(2, \infty)$

30. $\dfrac{x + 2}{x - 1} \leq 0$ $[-2, 1)$

Solve each equation.

31. $\left|\dfrac{3x + 2}{2}\right| = 4$ $2, -\frac{10}{3}$

32. $|x + 3| = |x - 3|$ 0

Solve each inequality and graph the solution set.

33. $|2x - 5| > 2$ $(-\infty, 3/2) \cup (7/2, \infty)$

34. $\left|\dfrac{2x + 3}{3}\right| \leq 5$ $[-9, 6]$

CUMULATIVE REVIEW EXERCISES

Consider the set $\{-5, -3, -2, 0, 1, \sqrt{2}, 2, \frac{5}{2}, 6, 11\}$.

1. Which numbers are even integers? $-2, 0, 2, 6$

2. Which numbers are prime numbers? $2, 11$

Write each inequality as an interval and graph it.

3. $-4 \le x < 7$ $[-4, 7)$

−4 7

4. $x \ge 2$ or $x < 0$ $(-\infty, 0) \cup [2, \infty)$

0 2

Determine which property of the real numbers justifies each expression.

5. $(a + b) + c = c + (a + b)$
comm. prop. of addition

6. If $x < 3$ and $3 < y$, then $x < y$. transitive prop.

Simplify each expression. Assume that all variables represent positive numbers. Give all answers with positive exponents.

7. $(81a^4)^{1/2}$ $9a^2$

8. $81(a^4)^{1/2}$ $81a^2$

9. $(a^{-3}b^{-2})^{-2}$ a^6b^4

10. $\left(\dfrac{4x^4}{12x^2y}\right)^{-2}$ $\dfrac{9y^2}{x^4}$

11. $\left(\dfrac{4x^0y^2}{x^2y}\right)^{-2}$ $\dfrac{x^4}{16y^2}$

12. $\left(\dfrac{4x^{-5}y^2}{6x^{-2}y^{-3}}\right)^2$ $\dfrac{4y^{10}}{9x^6}$

13. $(a^{1/2}b)^2(ab^{1/2})^2$
a^3b^3

14. $(a^{1/2}b^{1/2}c)^2$
abc^2

Rationalize each denominator and simplify.

15. $\dfrac{3}{\sqrt{3}}$ $\sqrt{3}$

16. $\dfrac{2}{\sqrt[3]{4x}}$ $\dfrac{\sqrt[3]{2x^2}}{x}$

17. $\dfrac{3}{y - \sqrt{3}}$
$\dfrac{3(y + \sqrt{3})}{y^2 - 3}$

18. $\dfrac{3x}{\sqrt{x} - 1}$
$\dfrac{3x(\sqrt{x} + 1)}{x - 1}$

Simplify each expression and combine like terms.

19. $\sqrt{75} - 3\sqrt{5}$
$5\sqrt{3} - 3\sqrt{5}$

20. $\sqrt{18} + \sqrt{8} - 2\sqrt{2}$
$3\sqrt{2}$

21. $(\sqrt{2} - \sqrt{3})^2$
$5 - 2\sqrt{6}$

22. $(3 - \sqrt{5})(3 + \sqrt{5})$
4

Perform the operations and simplify when necessary.

23. $(3x^2 - 2x + 5) - 3(x^2 + 2x - 1)$ $-8x + 8$

24. $5x^2(2x^2 - x) + x(x^2 - x^3)$ $9x^4 - 4x^3$

25. $(3x - 5)(2x + 7)$ $6x^2 + 11x - 35$

26. $(z + 2)(z^2 - z + 2)$ $z^3 + z^2 + 4$

27. $3x + 2\overline{)6x^3 + x^2 + x + 2}$ $2x^2 - x + 1$

28. $x^2 + 2\overline{)3x^4 + 7x^2 - x + 2}$ $3x^2 + 1 - \dfrac{x}{x^2 + 2}$

Factor each polynomial.

29. $3t^2 - 6t$ $3t(t - 2)$

30. $3x^2 - 10x - 8$ $(3x + 2)(x - 4)$

31. $x^8 - 2x^4 + 1$ $(x + 1)^2(x - 1)^2(x^2 + 1)^2$

32. $x^6 - 1$ $(x + 1)(x^2 - x + 1)(x - 1)(x^2 + x + 1)$

Perform the operations and simplify.

33. $\dfrac{x^2 - 4}{x^2 + 5x + 6} \cdot \dfrac{x^2 - 2x - 15}{x^2 + 3x - 10}$ $\dfrac{x - 5}{x + 5}$

34. $\dfrac{6x^3 + x^2 - x}{x + 2} \div \dfrac{3x^2 - x}{x^2 + 4x + 4}$ $(2x + 1)(x + 2)$

35. $\dfrac{2}{x + 3} + \dfrac{5x}{x - 3}$ $\dfrac{5x^2 + 17x - 6}{(x + 3)(x - 3)}$

36. $\dfrac{x - 2}{x + 3}\left(\dfrac{x + 3}{x^2 - 4} - 1\right)$ $\dfrac{-x^2 + x + 7}{(x + 2)(x + 3)}$

37. $\dfrac{\frac{1}{a} + \frac{1}{b}}{\frac{1}{ab}}$ $b + a$

38. $\dfrac{x^{-1} - y^{-1}}{x - y}$ $-\dfrac{1}{xy}$

Solve each equation.

39. $\dfrac{3x}{x + 5} = \dfrac{x}{x - 5}$ $0, 10$

40. $8(2x - 3) - 3(5x + 2) = 4$ 34

Solve each formula for the indicated variable.

41. $\dfrac{1}{R} = \dfrac{1}{R_1} + \dfrac{1}{R_2}$; R $R = \dfrac{R_1R_2}{R_1 + R_2}$

42. $S = \dfrac{a - lr}{1 - r}$; r $r = \dfrac{a - S}{l - S}$ or $r = \dfrac{S - a}{S - l}$

43. Gardening A gardener wishes to enclose her rectangular raspberry patch with 40 feet of fencing. The raspberry bushes are planted along the garage, so no fencing is needed on that side. Find the dimensions if the total area is to be 192 square feet.
either 8 ft by 24 ft or 12 ft by 16 ft

44. Financial planning A college student invested part of a $25,000 inheritance at 7% interest and the rest at 6%. If his annual interest is $1,670, how much did he invest at 6%? **$8,000**

Perform the operations. If the result is not real, express the answer in a + bi form.

45. $\dfrac{2+i}{2-i}$ $\quad \dfrac{3}{5}+\dfrac{4}{5}i$

46. $\dfrac{i(3-i)}{(1+i)(1+i)}$ $\quad \dfrac{3}{2}-\dfrac{1}{2}i$

47. $|3+4i|$ **5**

48. $\dfrac{5}{i^7}+5i$ **$0+10i$**

Solve each equation.

49. $\dfrac{x+3}{x-1}-\dfrac{6}{x}=1$ **3**

50. $x^4+36=13x^2$ **2, −2, 3, −3**

51. $\sqrt{y+2}+\sqrt{11-y}=5$ **2, 7**

52. $z^{2/3}-13z^{1/3}+36=0$ **64, 729**

Graph the solution set of each inequality.

53. $5x-7\le 4$ 11/5

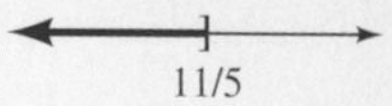

54. $x^2-8x+15>0$ 3 5

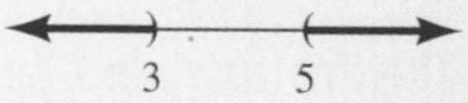

55. $\dfrac{x^2+4x+3}{x-2}\ge 0$ −3 −1 2

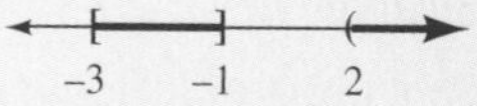

56. $\dfrac{9}{x}>x$ −3 0 3

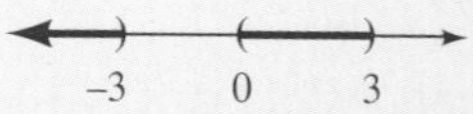

57. $|2x-3|\ge 5$ −1 4

58. $\left|\dfrac{3x-5}{2}\right|<2$ 1/3 3

2 The Rectangular Coordinate System and Graphs of Equations

COLLEGE **Algebra $f(x)$ Now™**

Throughout the chapter, this icon introduces resources on the College AlgebraNow Web Site, accessed through **http://1pass.thomson.com**, that will:

- Help you test your knowledge of the material in this chapter prior to reading it,
- Allow you to take an exam-prep quiz, and
- Provide a Personalized Learning Plan targeting areas you should study.

Careers and Mathematics

ECONOMIST

Rob Crandall/The Image Works

Economists, such as Alan Greenspan, Chairman of the Federal Reserve Board, study how society distributes scarce resources such as land, labor, raw materials, and machinery to produce goods and services. They conduct research, collect and analyze data, monitor economic trends, and develop forecasts. They also research issues such as energy costs, inflation, interest rates, imports, and employment levels.

Economists use mathematical models to help predict answers to questions such as the nature and length of business cycles, the effects of a specific rate of inflation on the economy, and the effects of tax legislation on unemployment levels.

JOB OUTLOOK

Employment of economists and market researchers is expected to grow about as fast as the average for all occupations through 2012. Economists held about 160,000 jobs in 2002.

Median annual earnings of economists were $68,550 in 2002.

For a sample application, see Example 10 in Section 2.3.

Mathematical expressions often indicate relationships between two variables. To visualize these relationships, we draw graphs of their equations.

2.1 The Rectangular Coordinate System

In this section, you will learn about

- The Rectangular Coordinate System
- Graphing Linear Equations
- Graphing Horizontal and Vertical Lines
- Applications
- The Distance Formula
- The Midpoint Formula

In this section, we will discuss equations containing two variables. For example, the equation $y = -\frac{1}{2}x + 4$ contains the variables x and y. The solutions of such equations are ordered pairs of real numbers (x, y) that satisfy the equation. To find some ordered pairs that satisfy the equation, we substitute **input values** of x into the equation and compute the corresponding **output values** of y. If we substitute 2 for x, we obtain

$$y = -\frac{1}{2}x + 4$$

$$y = -\frac{1}{2}(2) + 4 \quad \text{Substitute 2 for } x.$$

$$= -1 + 4$$

$$= 3$$

Since $y = 3$ when $x = 2$, the ordered pair (2, 3) is a solution of the equation. The first coordinate, 2, of the ordered pair is usually called the ***x*-coordinate**. The second coordinate, 3, is usually called the ***y*-coordinate**. The solution (2, 3) and several others are listed in the table of values shown in Figure 2-1.

$y = -\frac{1}{2}x + 4$

x	y	(x, y)
−4	6	(−4, 6)
−2	5	(−2, 5)
0	4	(0, 4)
2	3	(2, 3)
4	2	(4, 2)

Pick values for x. ↑ Compute each y-value. ↑ Write each solution as an ordered pair. ↑

Note that we choose x-values that are multiples of the denominator, 2. This makes the computations easier when multiplying the x-value by $\frac{1}{2}$ to find the corresponding y-value.

Figure 2-1

Accent on Technology — GENERATING TABLES WITH A GRAPHING CALCULATOR

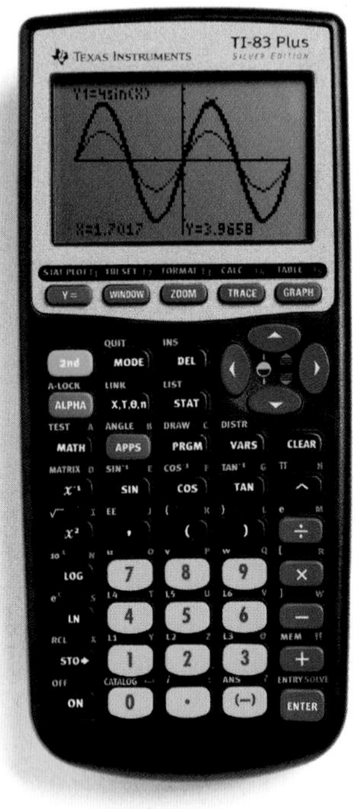

Courtesy of Texas Instruments Incorporated

If an equation in x and y is solved for y, we can use a graphing calculator to generate a table of values. The instructions in this discussion are for a TI-83 Plus graphing calculator. For details about other brands, please consult the owner's manual.

To construct a table of values for $x + 2y = 8$, we first solve the equation for y.

$$x + 2y = 8$$

$$2y = -x + 8 \quad \text{Subtract } x \text{ from both sides.}$$

$$y = -\frac{1}{2}x + 4 \quad \text{Divide both sides by 2 and simplify.}$$

Note that this is the equation shown in Figure 2-1.

To construct a table of values for $y = -\frac{1}{2}x + 4$, we press 2nd TBLSET and enter one value for x on the line labeled TblStart = . In Figure 2-2(a), -4 has been entered on this line. Other values for x that will appear in the table are determined by setting an **increment value** on the line labeled ΔTbl = . In Figure 2-2(a), an increment value of 2 has been entered. This means that each x-value in the table will be 2 units larger than the previous one.

To enter the equation, we press $y =$ and enter $-(1/2)x + 4$, as shown in Figure 2-2(b). Finally, we press 2nd TABLE to obtain the table of values shown in Figure 2-2(c). This table contains all of the solutions listed in Figure 2-1, plus the two additional solutions (6, 1) and (8, 0). To see other values, we simply scroll up and down the screen by pressing the up and down arrow keys.

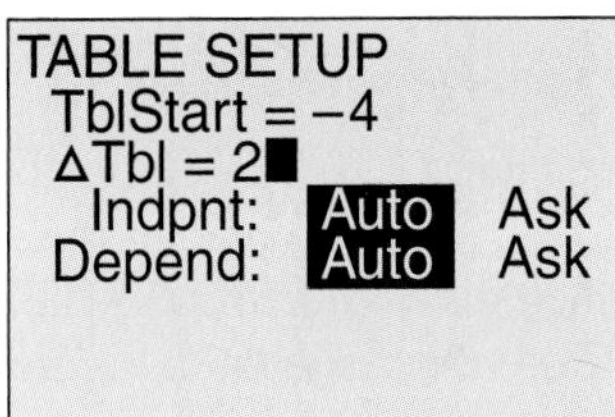

(a)

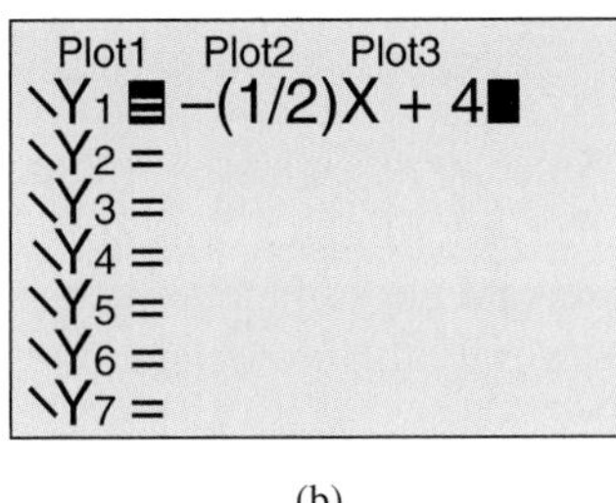

(b)

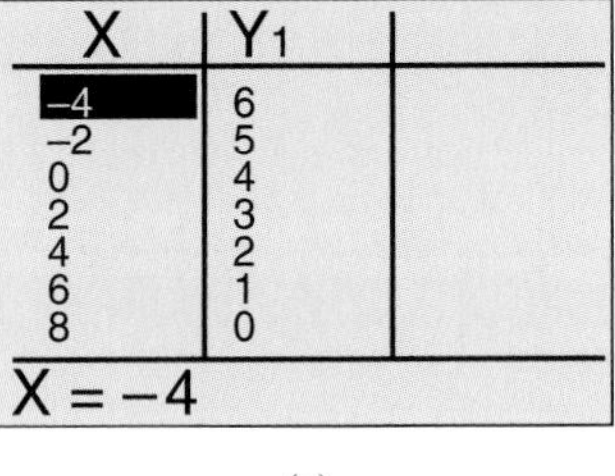

(c)

Figure 2-2

Before we can present the information shown in Figure 2-1 in graphical form, we need to discuss the rectangular coordinate system.

■ The Rectangular Coordinate System

The **rectangular coordinate system** consists of two perpendicular number lines that divide the plane into four **quadrants,** numbered as shown in Figure 2-3. The horizontal number line is called the ***x*-axis,** and the vertical number line is called the ***y*-axis.** These axes intersect at a point called the **origin,** which is the 0 point on each axis. The positive direction on the x-axis is to the right, the positive direction on the y-axis is upward, and the same unit distance is used on both axes, unless otherwise indicated.

To plot (or graph) the point associated with the pair $x = 2$ and $y = 3$, denoted as (2, 3), we start at the origin, count 2 units to the right, and then count 3 units up. (See Figure 2-4.) Point P (which lies in the first quadrant) is the graph of the pair (2, 3). The pair (2, 3) gives the **coordinates** of point P.

To plot point Q with coordinates $(-4, 6)$, we start at the origin, count 4 units to the left, and then count 6 units up. Point Q lies in the second quadrant. Point R with coordinates $(6, -4)$ lies in the fourth quadrant.

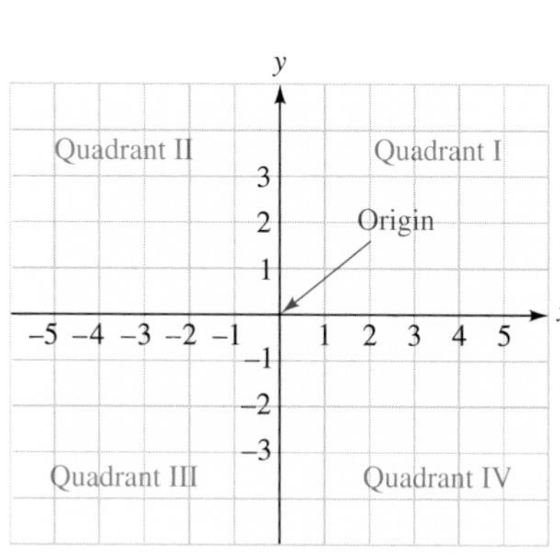

Figure 2-3

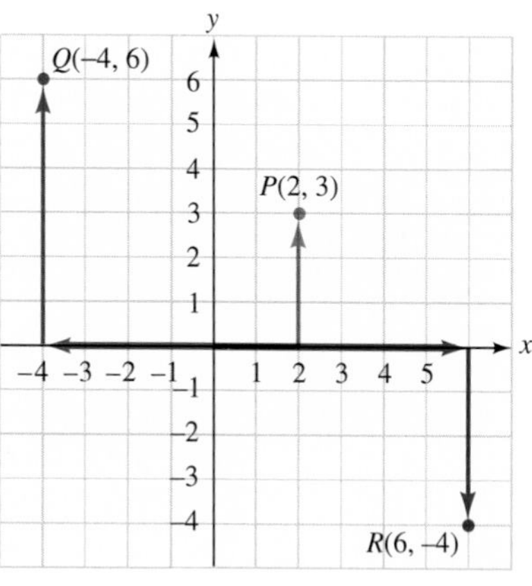

Figure 2-4

Comment The pairs $(-4, 6)$ and $(6, -4)$ represent different points. One is in the second quadrant, and one is in the fourth quadrant.

Graphing Linear Equations

The **graph of the equation** $y = -\frac{1}{2}x + 4$ is the graph of all points (x, y) on the rectangular coordinate system whose coordinates satisfy the equation. To graph $y = -\frac{1}{2}x + 4$, we plot the pairs listed in the table shown in Figure 2-5. These points lie on the line shown in the figure.

When we say that the graph of an equation is a line, we imply two things:

1. Every point with coordinates that satisfy the equation will lie on the line.
2. Any point on the line will have coordinates that satisfy the equation.

René Descartes

(1596–1650)

Descartes is famous for his work in philosophy as well as for his work in mathematics. His philosophy is expressed in the words "I think, therefore I am." He is best known in mathematics for his invention of a coordinate system and his work with conic sections.

$y = -\frac{1}{2}x + 4$

x	y	(x, y)
-4	6	$(-4, 6)$
-2	5	$(-2, 5)$
0	4	(0, 4)
2	3	(2, 3)
4	2	(4, 2)

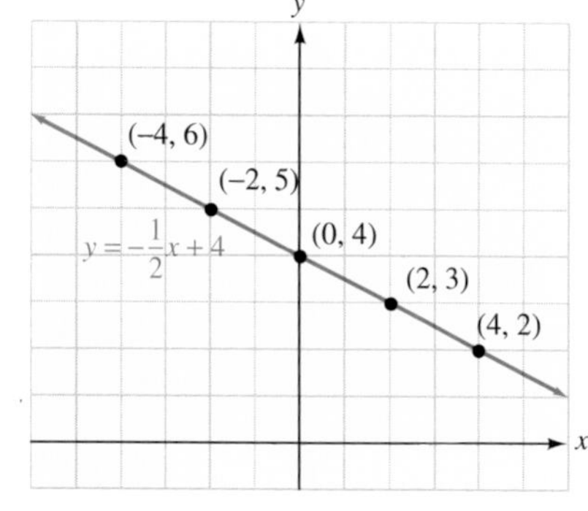

Figure 2-5

When the graph of an equation is a line, we call the equation a **linear equation.** These equations are often written in the form $Ax + By = C$, where A, B, and C stand for specific numbers (called **constants**) and x and y are variables.

EXAMPLE 1 Graph: $x + 2y = 5$.

Solution We pick values for x or y, substitute them into the equation, and solve for the other variable. If $y = -1$, we find x as follows.

$$x + 2y = 5$$
$$x + 2(-1) = 5 \quad \text{Substitute } -1 \text{ for } y.$$
$$x - 2 = 5 \quad \text{Simplify.}$$
$$x = 7 \quad \text{Add 2 to both sides.}$$

The pair $(7, -1)$ satisfies the equation. To find another, we let $x = 0$ and find y.

$$x + 2y = 5$$
$$0 + 2y = 5 \quad \text{Substitute 0 for } x.$$
$$2y = 5 \quad \text{Simplify.}$$
$$y = \frac{5}{2} \quad \text{Divide both sides by 2.}$$

The pair $\left(0, \frac{5}{2}\right)$ also satisfies the equation.

These pairs and others that satisfy the equation are shown in Figure 2-6. We plot the points and join them with a line to get the graph of the equation.

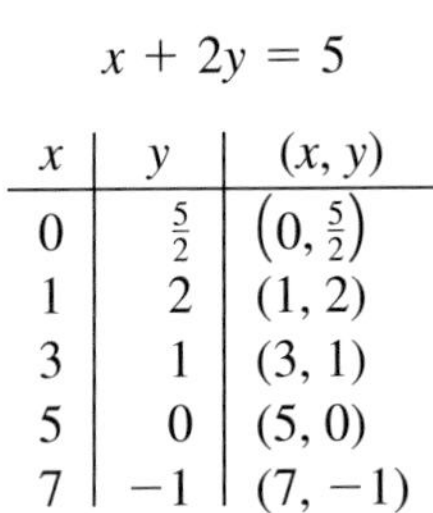

$x + 2y = 5$

x	y	(x, y)
0	$\frac{5}{2}$	$\left(0, \frac{5}{2}\right)$
1	2	$(1, 2)$
3	1	$(3, 1)$
5	0	$(5, 0)$
7	-1	$(7, -1)$

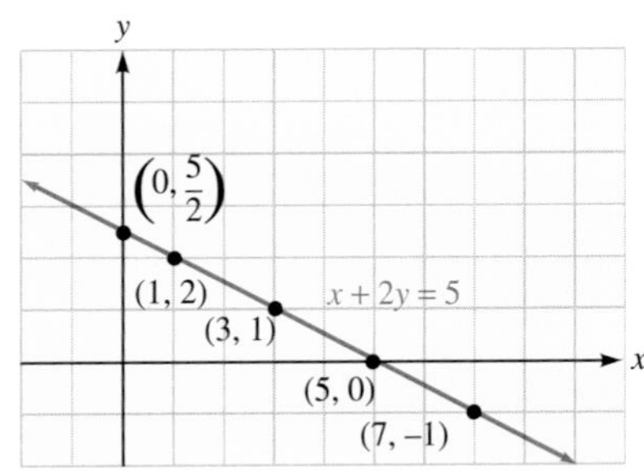

Figure 2-6

Self Check Graph: $3x - 2y = 6$. ■

In Example 1, the graph intersects the y-axis at the point $\left(0, \frac{5}{2}\right)$, called the ***y*-intercept.** It intersects the x-axis at the point $(5, 0)$, called the ***x*-intercept.**

Intercepts of a Line

The ***y*-intercept** of a line is the point $(0, b)$, where the line intersects the y-axis. To find b, substitute 0 for x in the equation of the line and solve for y.

The ***x*-intercept** of a line is the point $(a, 0)$, where the line intersects the x-axis. To find a, substitute 0 for y in the equation of the line and solve for x.

EXAMPLE 2 Use the x- and y-intercepts to graph the equation $3x + 2y = 12$.

Solution To find the y-intercept, we substitute 0 for x and solve for y.

$$3x + 2y = 12$$
$$3(0) + 2y = 12 \quad \text{Substitute 0 for } x.$$
$$2y = 12 \quad \text{Simplify.}$$
$$y = 6 \quad \text{Divide both sides by 2.}$$

The y-intercept is the point (0, 6). To find the x-intercept, we substitute 0 for y and solve for x.

$$3x + 2y = 12$$
$$3x + 2(0) = 12 \quad \text{Substitute 0 for } y.$$
$$3x = 12 \quad \text{Simplify.}$$
$$x = 4 \quad \text{Divide both sides by 3.}$$

The x-intercept is the point (4, 0). Although two points are enough to draw the line, it is a good idea to find and plot a third point as a check. If we let $x = 2$, we will find that $y = 3$.

We plot each pair (as in Figure 2-7) and join them with a line to get the graph of the equation.

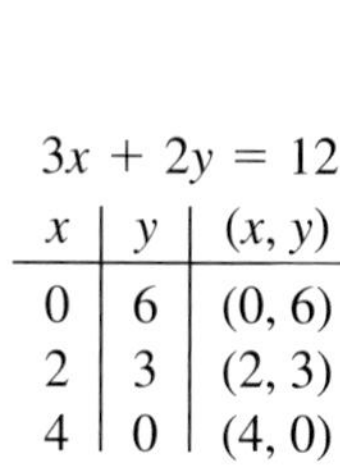

$3x + 2y = 12$

x	y	(x, y)
0	6	(0, 6)
2	3	(2, 3)
4	0	(4, 0)

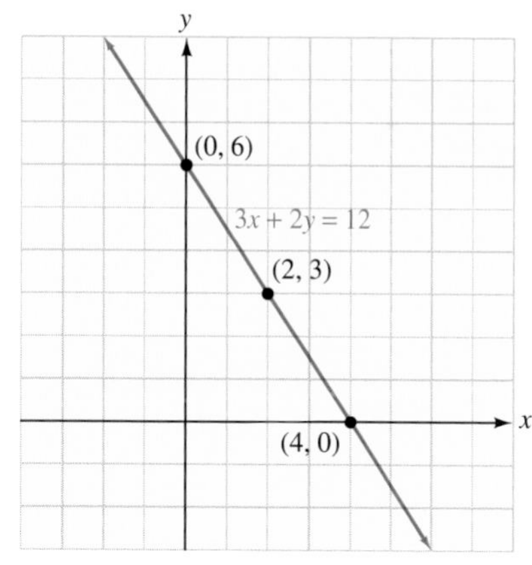

Figure 2-7

Self Check Graph: $2x - 3y = 12$. ■

ACTIVE EXAMPLE 3 Graph: $3(y + 2) = 2x - 3$.

Solution To make it easier to find pairs (x, y) that satisfy the equation, we solve it for y.

$$3(y + 2) = 2x - 3$$
$$3y + 6 = 2x - 3 \quad \text{Use the distributive property to remove parentheses.}$$
$$3y = 2x - 9 \quad \text{Subtract 6 from both sides.}$$
$$y = \frac{2}{3}x - 3 \quad \text{Divide both sides by 3.}$$

COLLEGE **Algebra** *f(x)* **Now**™

Explore this Active Example and test yourself on the steps taken here at **http://1pass.thomson.com**. You can also find this example on the Interactive Video Skillbuilder CD-ROM.

We now substitute numbers for x to find the corresponding values of y. If we let $x = 3$, we get

$$y = \frac{2}{3}x - 3$$

$$y = \frac{2}{3}(3) - 3 \quad \text{Substitute 3 for } x.$$

$$y = 2 - 3 \quad \text{Simplify.}$$

$$y = -1$$

The point $(3, -1)$ lies on the graph. To find the y-intercept of the graph, we let $x = 0$ and find y:

$$y = \frac{2}{3}x - 3$$

$$y = \frac{2}{3}(0) - 3 \quad \text{Substitute 0 for } x.$$

$$y = -3 \quad \text{Simplify.}$$

The point $(0, -3)$ lies on the graph. We plot these points and others, as in Figure 2-8, and draw the line.

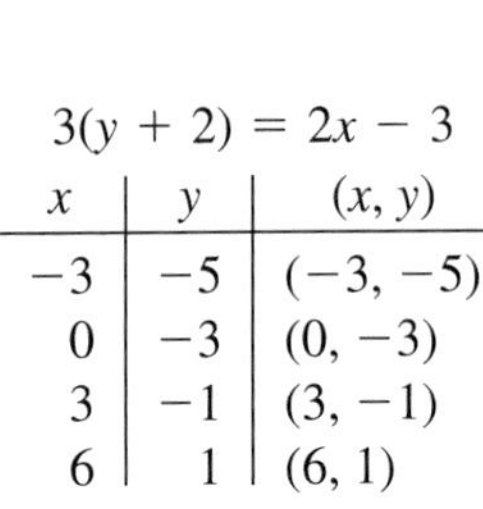

$3(y + 2) = 2x - 3$

x	y	(x, y)
-3	-5	$(-3, -5)$
0	-3	$(0, -3)$
3	-1	$(3, -1)$
6	1	$(6, 1)$

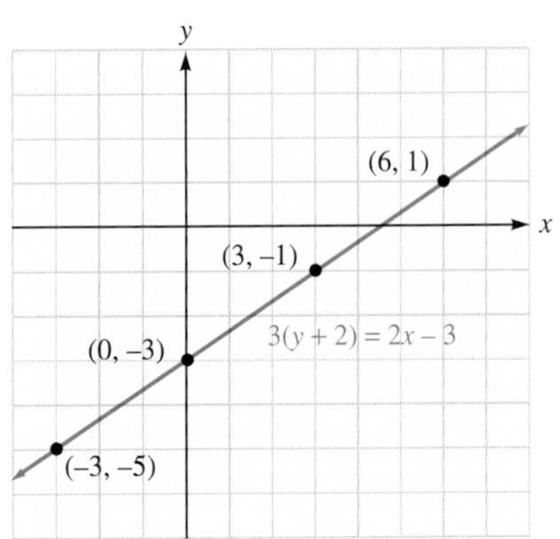

Figure 2-8

Self Check Graph: $2(x - 1) = 6 - 8y$. ■

■ Graphing Horizontal and Vertical Lines

EXAMPLE 4 Graph: **a.** $y = 2$ and **b.** $x = -3$.

Solution **a.** In the equation $y = 2$, the value of y is always 2. After plotting the pairs shown in Figure 2-9, we see that the graph is a horizontal line, parallel to the x-axis and having a y-intercept of $(0, 2)$. The line has no x-intercept.

b. In the equation $x = -3$, the value of x is always -3. After plotting the pairs shown in Figure 2-9, we see that the graph is a vertical line, parallel to the y-axis and having an x-intercept of $(-3, 0)$. The line has no y-intercept.

Self Check Graph: **a.** $x = 2$ and **b.** $y = -3$.

$y = 2$

x	y	(x, y)
-3	2	$(-3, 2)$
0	2	$(0, 2)$
2	2	$(2, 2)$
4	2	$(4, 2)$

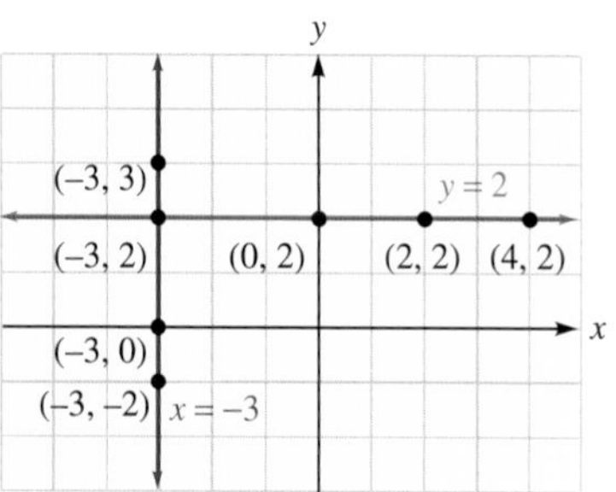

$x = -3$

x	y	(x, y)
-3	-2	$(-3, -2)$
-3	0	$(-3, 0)$
-3	2	$(-3, 2)$
-3	3	$(-3, 3)$

Figure 2-9

Example 4 suggests the following facts.

Equations of Vertical and Horizontal Lines

If a and b are real numbers, then

- The graph of the equation $x = a$ is a vertical line with x-intercept of $(a, 0)$. If $a = 0$, the line is the y-axis.
- The graph of the equation $y = b$ is a horizontal line with y-intercept of $(0, b)$. If $b = 0$, the line is the x-axis.

Accent on Technology — GRAPHING CALCULATORS

Grace Murray Hopper
(1906–1992)

Grace Hopper graduated from Vassar College in 1928 and obtained a Master's degree from Yale in 1930. In 1943, she entered the U.S. Naval Reserve. While in the Navy, she became a programmer of the Mark I, the world's first large computer. She is credited for first using the word "bug" to refer to a computer problem. The first bug was actually a moth that flew into one of the relays of the Mark II. From then on, locating computer problems was called "debugging" the system.

We can graph equations with a graphing calculator. To see a graph, we must choose the minimum and maximum values of the x- and y-coordinates that will appear on the calculator's window. A window with standard settings of

$$\text{Xmin} = -10 \qquad \text{Xmax} = 10 \qquad \text{Ymin} = -10 \qquad \text{Ymax} = 10$$

will produce a graph where the value of x is in the interval $[-10, 10]$, and the value of y is in the interval $[-10, 10]$.

To use a graphing calculator to graph $3x + 2y = 12$, we first solve the equation for y.

$$3x + 2y = 12$$

$$2y = -3x + 12 \qquad \text{Subtract } 3x \text{ from both sides.}$$

$$y = -\frac{3}{2}x + 6 \qquad \text{Divide both sides by 2.}$$

After we enter the right-hand side, the screen should look something like this:

$$Y_1 = -(3/2)X + 6 \qquad \text{or} \qquad f(x) = -(3/2)X + 6$$

We then press GRAPH to obtain the graph shown in Figure 2-10(a). To show more detail, we can draw the graph in a different window. A window with settings of $[-1, 5]$ for x and $[-2, 7]$ for y will give the graph shown in Figure 2-10(b).

Finding intercepts of a graph

We can trace to find the coordinates of any point on a graph. After TRACE is pressed, a flashing cursor will appear on the screen. The coordinates of the cursor will also appear at the bottom of the screen.

Continued

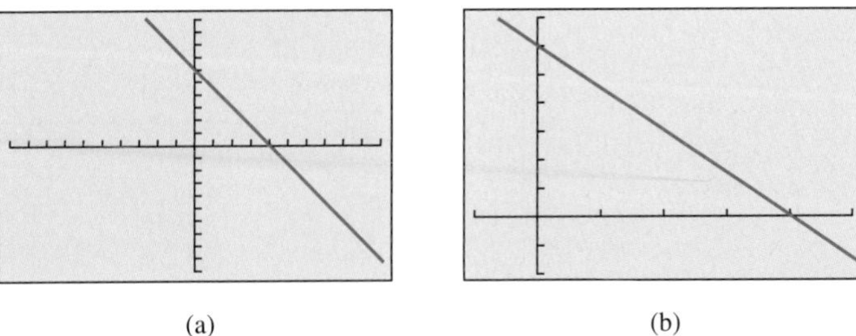

(a) (b)

Figure 2-10

To find the x-intercept of the graph of $2y = -5x - 7$ $\left(\text{or } y = -\frac{5}{2}x - \frac{7}{2}\right)$, we graph the equation, using $[-10, 10]$ for x and $[-10, 10]$ for y, and press TRACE to get Figure 2-11(a). We can then move the cursor along the line toward the x-intercept until we arrive at a point with the coordinates shown in Figure 2-11(b).

To get better results, we can zoom in to get a magnified picture, trace again, and move the cursor to the point with coordinates shown in Figure 2-11(c). Since the y-coordinate is almost 0, this point is nearly the x-intercept.

We can achieve better results with repeated zooms.

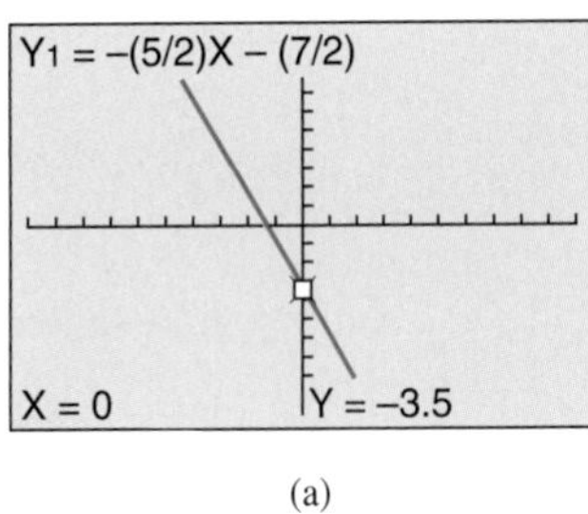

(a)

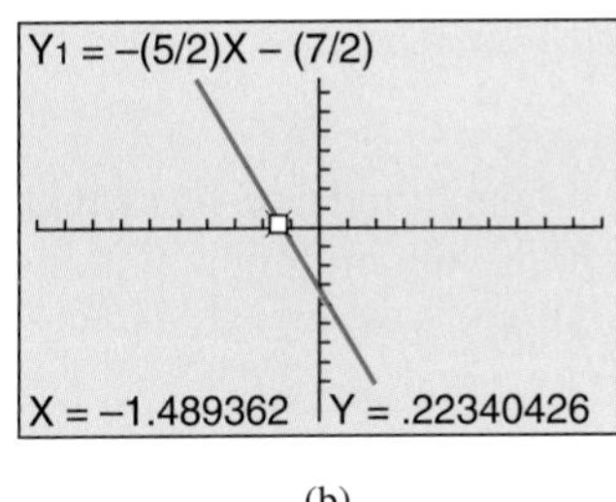

(b)

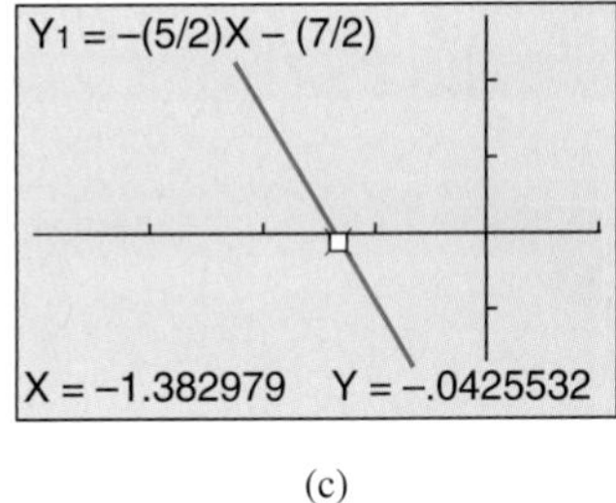

(c)

Figure 2-11

We can also find the coordinates of the x-intercept of the graph of $\text{Y} = -\left(\frac{5}{2}\right)\text{X} - \left(\frac{7}{2}\right)$ by using the ZERO command, found under the CALC menu. After we guess left and right bounds and press ENTER after the prompt GUESS, the cursor automatically moves to the x-intercept of the graph, and the coordinates of that point are displayed on the screen. See Figure 2-12(a).

To find the y-intercept of the graph, we can use the VALUE command, which is also found under the CALC menu. With this option, we enter an x-value of 0, as shown in Figure 2-12(b). After we press ENTER, the cursor highlights the y-intercept, and its coordinates are displayed. See Figure 2-12(c).

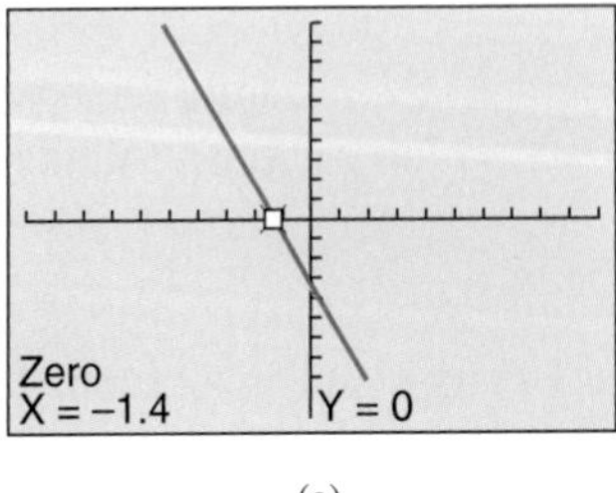

(a)

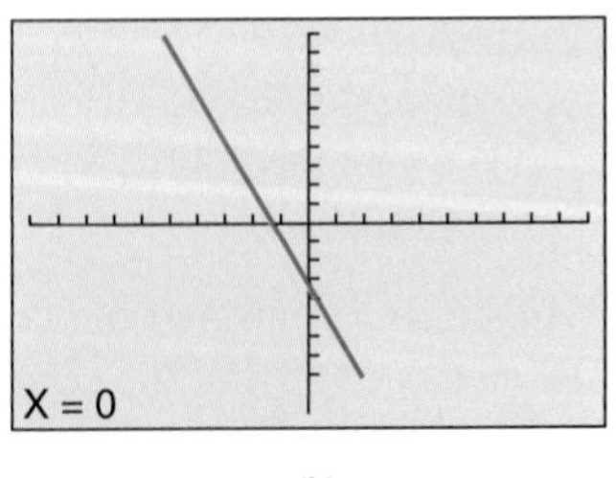

(b)

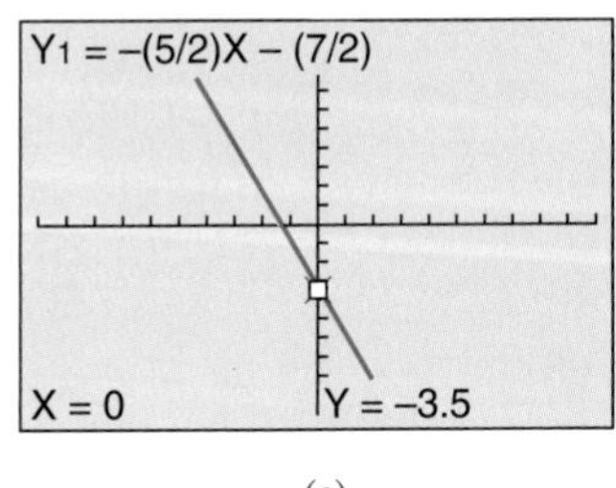

(c)

Figure 2-12

Applications

EXAMPLE 5 A computer purchased for \$2,750 is expected to depreciate according to the formula $y = -550x + \$2{,}750$, where y is the value of the computer after x years. When will the computer have no value?

Solution The computer will have no value when its value (y) is 0. To find x when $y = 0$, we substitute 0 for y and solve for x.

$$y = -550x + 2{,}750$$
$$0 = -550x + 2{,}750$$
$$-2{,}750 = -550x \quad \text{Subtract 2,750 from both sides.}$$
$$5 = x \quad \text{Divide both sides by } -550.$$

The computer will have no value in 5 years. ■

Accent on Technology — DEPRECIATION

To solve Example 5 with a graphing calculator, we graph $y = -550x + 2{,}750$ in the window X = [−10, 10] and Y = [−10, 3,000], as shown in Figure 2-13(a). We chose the maximum y-value of 3,000 because y is almost 3,000 when $x = 0$. We then trace to get Figure 2-13(b). We then zoom and trace again to get Figure 2-13(c), which shows that $y = 0$ when $x = 5$.

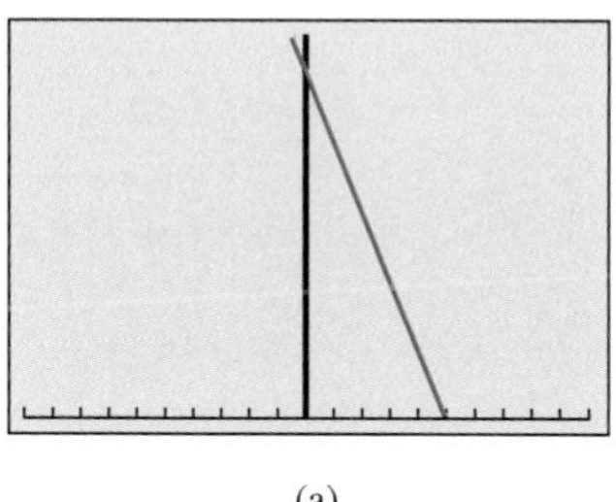

(a)

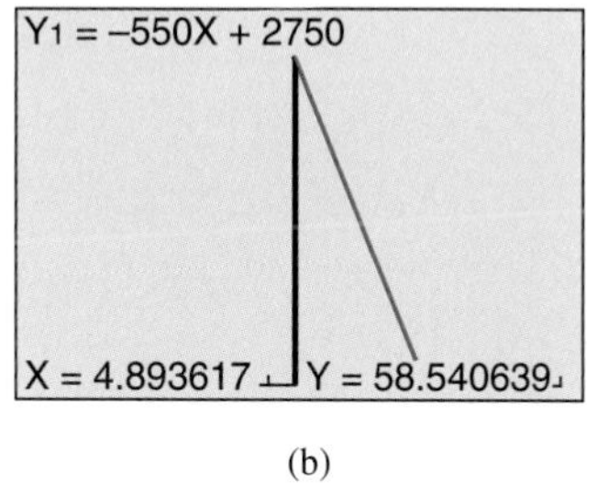

(b)

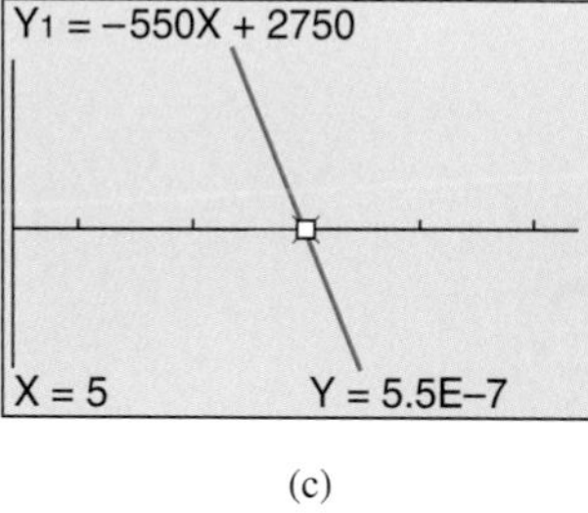

(c)

Figure 2-13

We can obtain the same result using the ZERO command.

The Distance Formula

To derive the formula used to find the distance between two points on a rectangular coordinate system, we use **subscript notation** and denote the points as

$P(x_1, y_1)$ Read as "point P with coordinates of x sub 1 and y sub 1."

$Q(x_2, y_2)$ Read as "point Q with coordinates of x sub 2 and y sub 2."

If $P(x_1, y_1)$ and $Q(x_2, y_2)$ are two points in Figure 2-14 and point R has coordinates (x_2, y_1), triangle PQR is a right triangle. By the Pythagorean theorem, the square of the hypotenuse of right triangle PQR is equal to the sum of the squares of the two legs. Because leg RQ is vertical, the square of its length is $(y_2 - y_1)^2$. Since leg PR is horizontal, the square of its length is $(x_2 - x_1)^2$. Thus, we have

$$d^2 = (x_2 - x_1)^2 + (y_2 - y_1)^2 \tag{1}$$

Because equal positive numbers have equal positive square roots, we can take the positive square root of both sides of Equation 1 to obtain the **distance formula.**

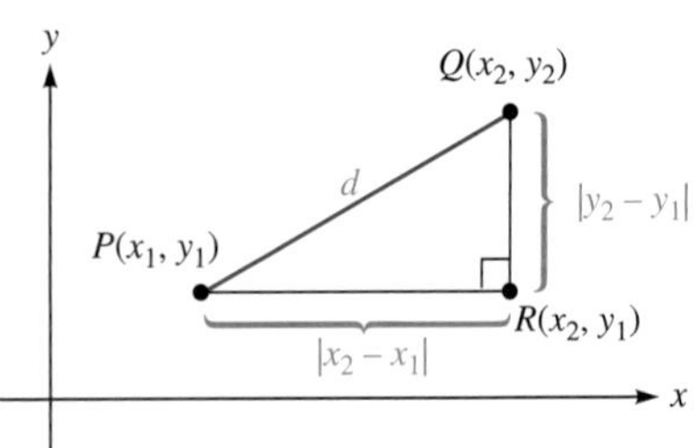

Figure 2-14

The Distance Formula

The distance d between points (x_1, y_1) and (x_2, y_2) is given by

$$d = \sqrt{(x_2 - x_1)^2 + (y_2 - y_1)^2}$$

ACTIVE EXAMPLE 6 Find the distance between $P(-1, -2)$ and $Q(-7, 8)$.

Solution If we let $P(-1, -2) = P(x_1, y_1)$ and $Q(-7, 8) = Q(x_2, y_2)$, we can substitute -1 for x_1, -2 for y_1, -7 for x_2, and 8 for y_2 into the formula and simplify.

COLLEGE **Algebra** f(x) **Now**™
Go to **http://1pass.thomson.com** or your CD to practice this example.

$$d(PQ) = \sqrt{(x_2 - x_1)^2 + (y_2 - y_1)^2}$$

Read $d(PQ)$ as "the length of segment PQ."

$$\begin{aligned} d(PQ) &= \sqrt{[-7 - (-1)]^2 + [8 - (-2)]^2} \\ &= \sqrt{(-6)^2 + (10)^2} \\ &= \sqrt{36 + 100} \\ &= \sqrt{136} \\ &= \sqrt{4 \cdot 34} \\ &= 2\sqrt{34} \end{aligned}$$

$\sqrt{4 \cdot 34} = \sqrt{4}\sqrt{34} = 2\sqrt{34}$.

Self Check Find the distance between $P(-2, -5)$ and $Q(3, 7)$. ■

■ The Midpoint Formula

If point M in Figure 2-15 lies midway between points $P(x_1, y_1)$ and $Q(x_2, y_2)$, point M is called the **midpoint** of segment PQ. To find the coordinates of M, we find the average of the x-coordinates and the average of the y-coordinates of P and Q.

The Midpoint Formula

The midpoint of the line segment with endpoints at $P(x_1, y_1)$ and $Q(x_2, y_2)$ is the point M with coordinates of

$$M\left(\frac{x_1 + x_2}{2}, \frac{y_1 + y_2}{2}\right)$$

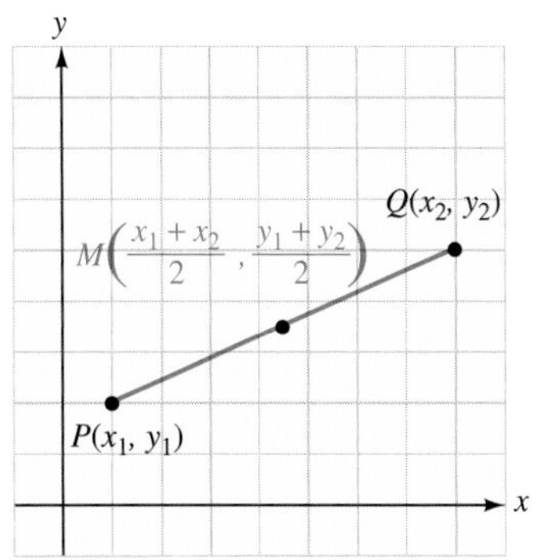

Figure 2-15

We can prove this formula by using the distance formula to show that $d(PM) + d(MQ) = d(PQ)$.

EXAMPLE 7 Find the midpoint of the segment joining $P(-7, 2)$ and $Q(1, -4)$.

Solution We substitute $P(-7, 2)$ for $P(x_1, y_1)$ and $Q(1, -4)$ for $Q(x_2, y_2)$ into the midpoint formula to get

$$x_M = \frac{x_1 + x_2}{2} \quad \text{and} \quad y_M = \frac{y_1 + y_2}{2}$$

$$= \frac{-7 + 1}{2} \qquad\qquad = \frac{2 + (-4)}{2}$$

$$= \frac{-6}{2} \qquad\qquad = \frac{-2}{2}$$

$$= -3 \qquad\qquad = -1$$

The midpoint is $M(-3, -1)$.

Self Check Find the midpoint of the segment joining $P(-7, -8)$ and $Q(-2, 10)$. ■

ACTIVE EXAMPLE 8 The midpoint of the segment joining $P(-3, 2)$ and $Q(x_2, y_2)$ is $M(1, 4)$. Find the coordinates of Q.

Solution We can let $P(x_1, y_1) = P(-3, 2)$ and $M(x_M, y_M) = M(1, 4)$, and then find the coordinates x_2 and y_2 of point $Q(x_2, y_2)$.

COLLEGE **Algebra** f(x) **Now**™
Go to **http://1pass.thomson.com** or your CD to practice this example.

$$x_M = \frac{x_1 + x_2}{2} \quad \text{and} \quad y_M = \frac{y_1 + y_2}{2}$$

$$1 = \frac{-3 + x_2}{2} \qquad\qquad 4 = \frac{2 + y_2}{2}$$

$$2 = -3 + x_2 \qquad\qquad 8 = 2 + y_2 \quad \text{Multiply both sides by 2.}$$

$$5 = x_2 \qquad\qquad 6 = y_2$$

The coordinates of point Q are $(5, 6)$.

Self Check If the midpoint of a segment PQ is $M(2, -5)$ and one endpoint is $Q(6, 9)$, find P. ■

Self Check Answers

1. 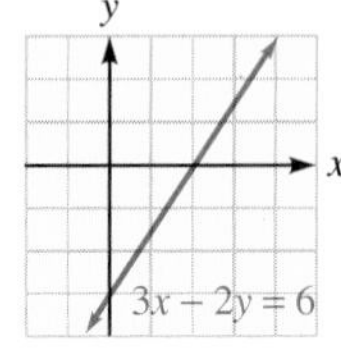

2. 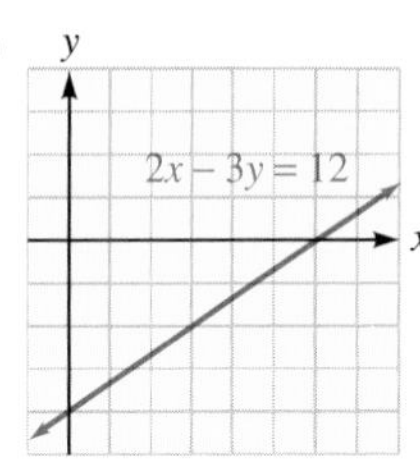

3. 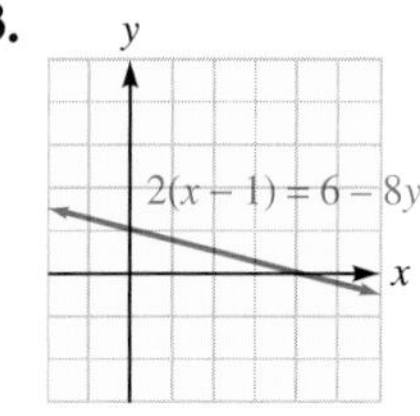

4. 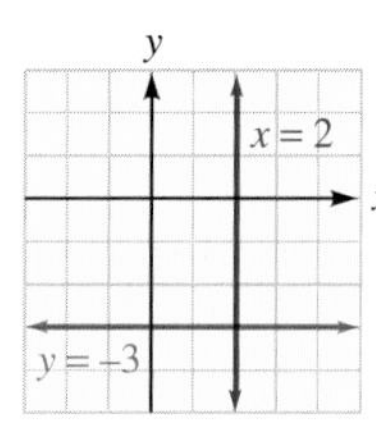

6. 13
7. $M\left(-\frac{9}{2}, 1\right)$
8. $(-2, -19)$

2.1 Exercises

VOCABULARY AND CONCEPTS ***Fill in the blanks.***

1. The coordinate axes divide the plane into four **quadrants**.
2. The coordinate axes intersect at the **origin**.
3. The positive direction on the x-axis is **to the right**.
4. The positive direction on the y-axis is **upward**.
5. The x-coordinate is the **first** coordinate in an ordered pair.
6. The y-coordinate is the **second** coordinate in an ordered pair.
7. A **linear** equation is an equation whose graph is a line.
8. The point where a line intersects the **y-axis** is called the y-intercept.
9. The point where a line intersects the x-axis is called the **x-intercept**.
10. The graph of the equation $x = a$ will be a **vertical** line.
11. The graph of the equation $y = b$ will be a **horizontal** line.
12. Complete the distance formula: $d = \sqrt{(x_2 - x_1)^2 + (y_2 - y_1)^2}$.
13. If a point divides a segment into two equal segments, the point is called the **midpoint** of the segment.
14. The midpoint of the segment joining $P(x_1, y_1)$ and $Q(x_2, y_2)$ is $M\left(\frac{x_1 + x_2}{2}, \frac{y_1 + y_2}{2}\right)$.

PRACTICE ***Refer to the illustration and determine the coordinates of each point.***

15. A **A(2, 3)**
16. B **B(−3, 5)**
17. C **C(−2, −3)**
18. D **D(4, −5)**
19. E **E(0, 0)**
20. F **F(−4, 0)**
21. G **G(−5, −5)**
22. H **H(2, −2)**

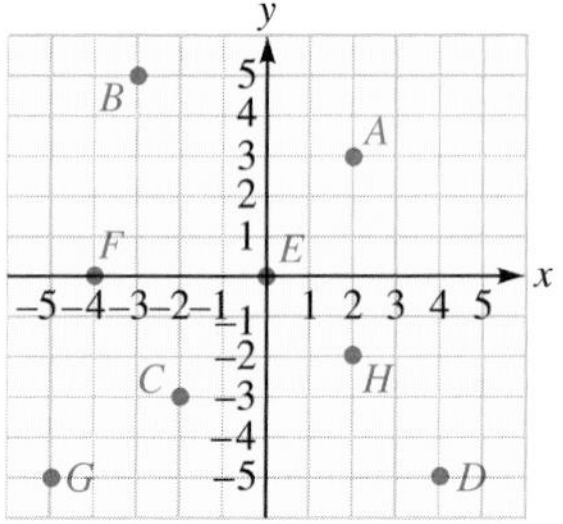

Graph each point. Indicate the quadrant in which the point lies, or the axis on which it lies.

23. (2, 5) **QI**
24. (−3, 4) **QII**
25. (−4, −5) **QIII**
26. (6, 2) **QI**
27. (5, 2) **QI**
28. (3, −4) **QIV**
29. (4, 0) **positive x-axis**
30. (0, 2) **positive y-axis**

Find the x- and y-intercepts and use them to graph each equation.

31. $x + y = 5$

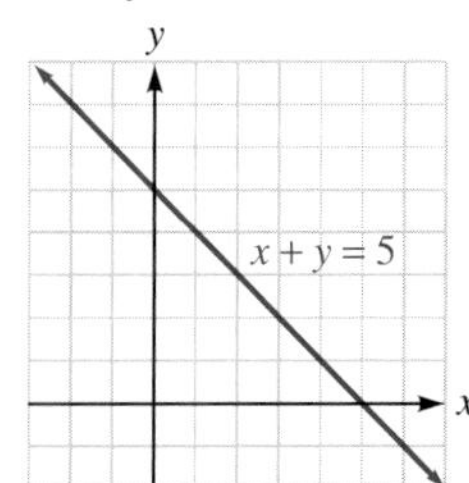

32. $x - y = 3$

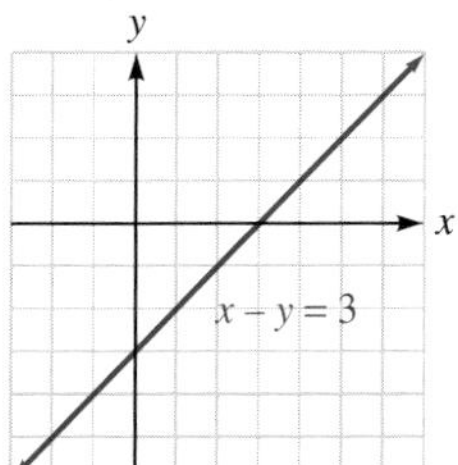

33. $2x - y = 4$

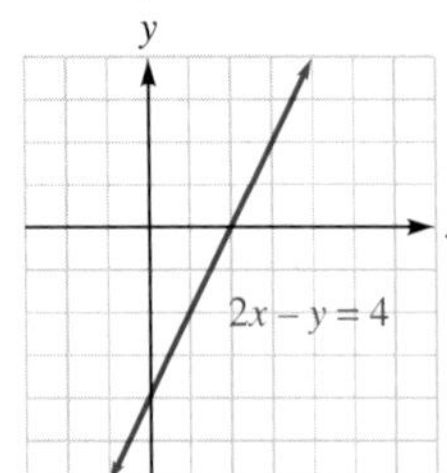

34. $3x + y = 9$

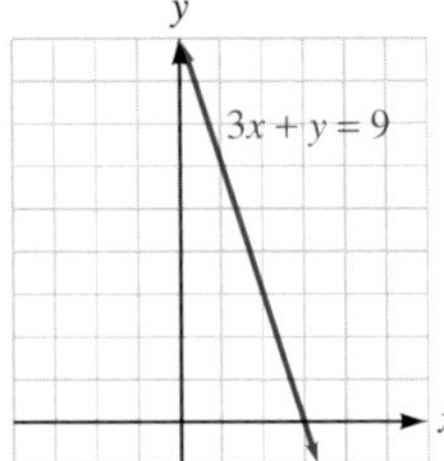

35. $3x + 2y = 6$

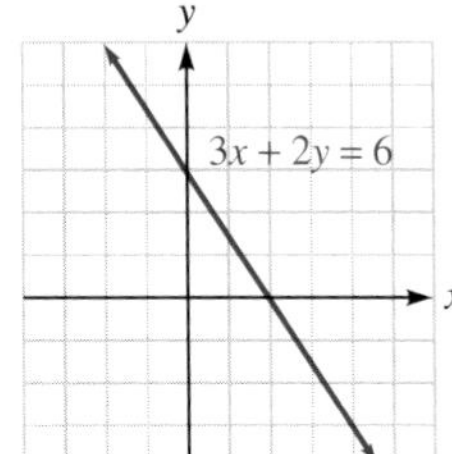

36. $2x - 3y = 6$

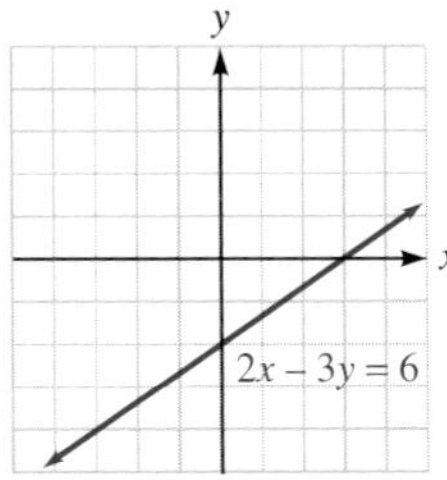

37. $4x - 5y = 20$

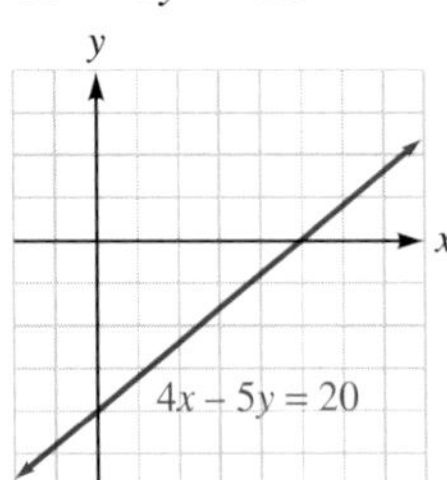

38. $3x - 5y = 15$

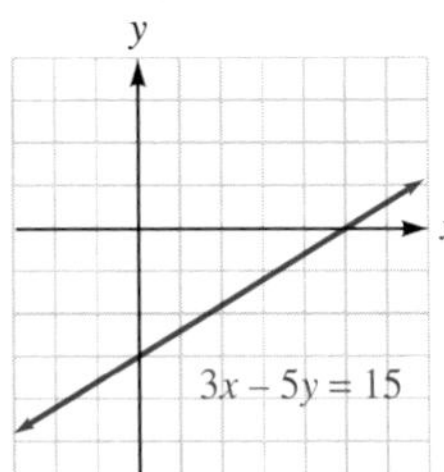

Solve each equation for y and graph the equation. Then check your graph with a graphing calculator.

39. $y - 2x = 7$

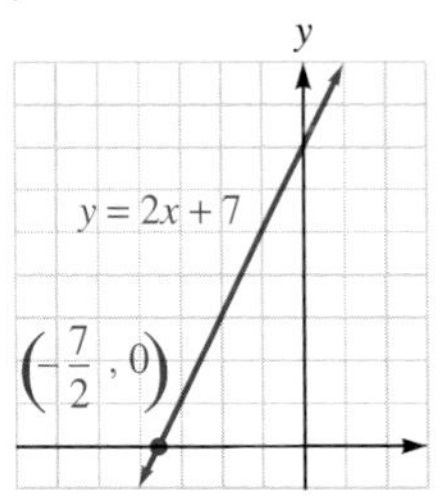

40. $y + 3 = -4x$

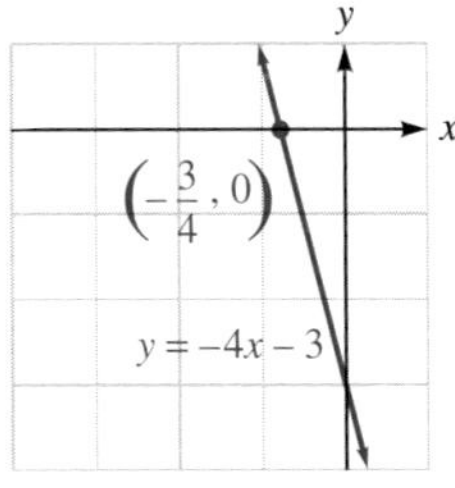

41. $y + 5x = 5$

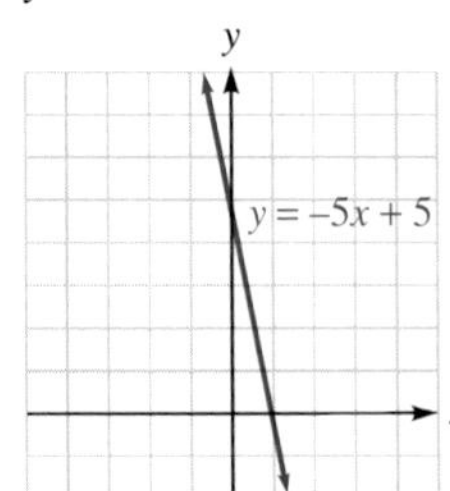

42. $y - 3x = 6$

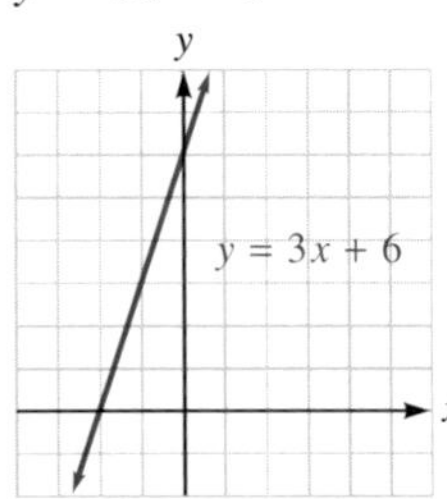

43. $6x - 3y = 10$

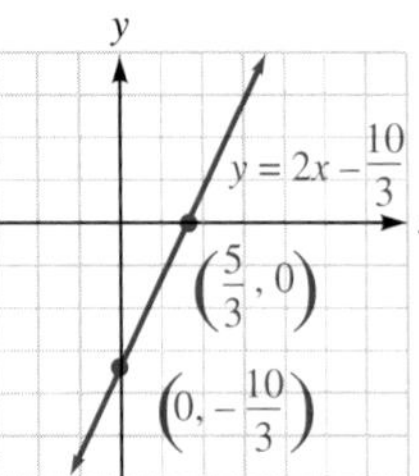

44. $4x + 8y - 1 = 0$

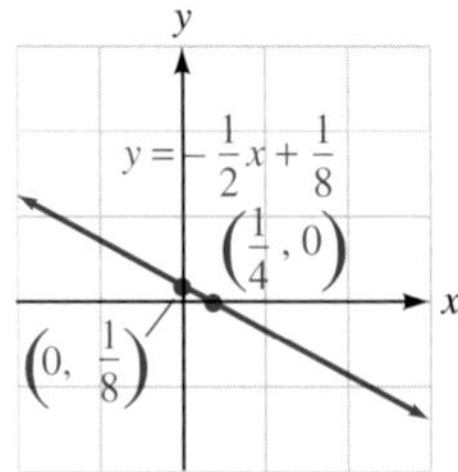

45. $3x = 6y - 1$

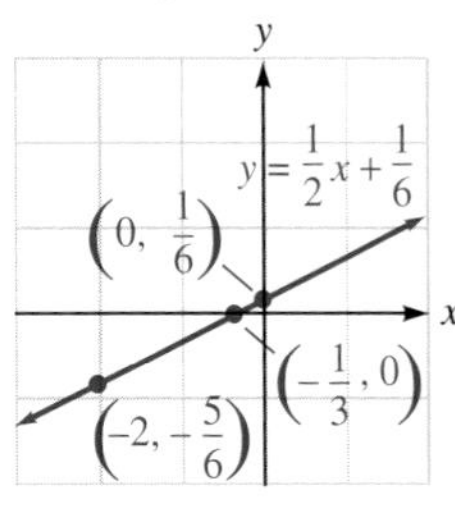

46. $2x + 1 = 4y$

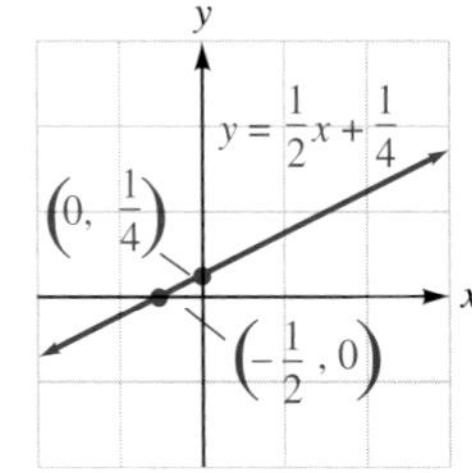

47. $2(x + y + 1) = x + 2$

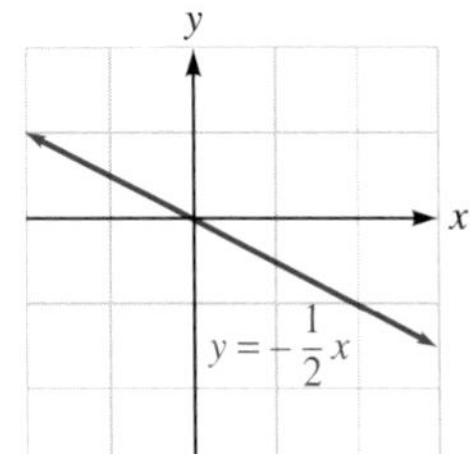

48. $5(x + 2) = 3y - x$

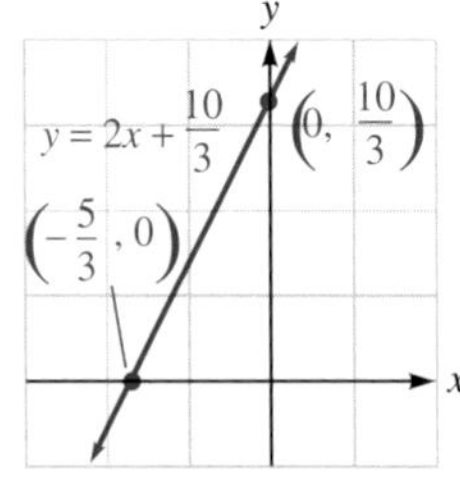

Graph each equation.

49. $y = 3$

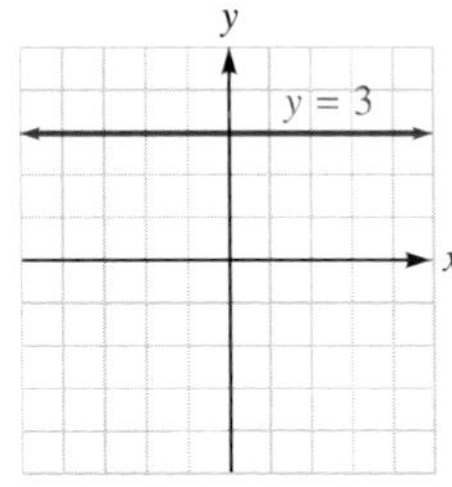

50. $x = -4$

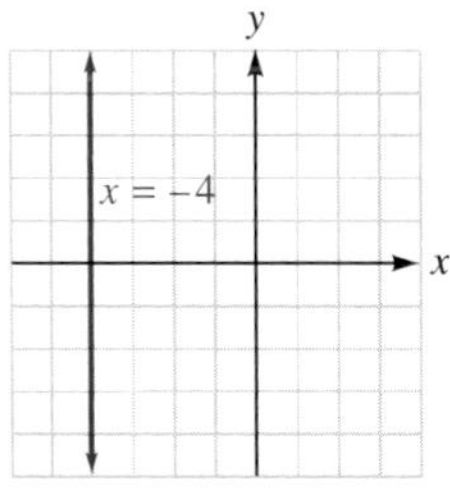

51. $3x + 5 = -1$

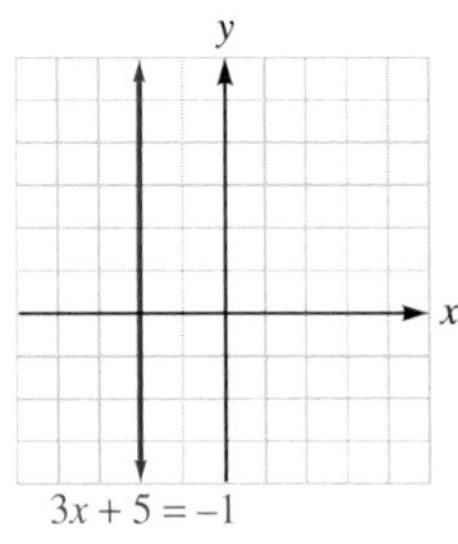

52. $7y - 1 = 6$

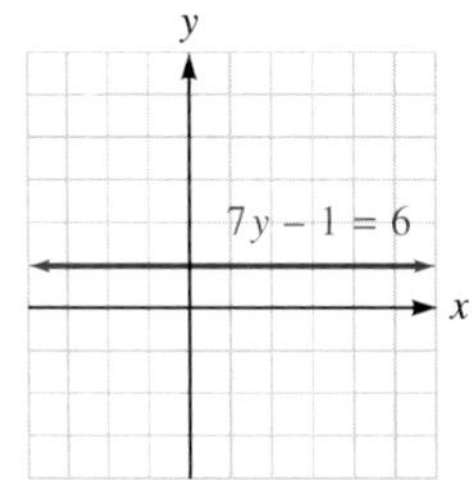

53. $3(y + 2) = y$

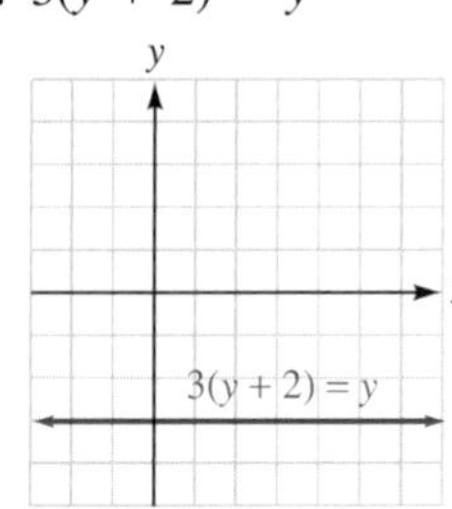

54. $4 + 3y = 3(x + y)$

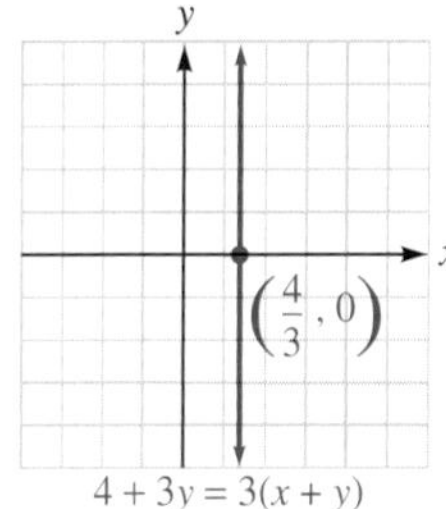

55. $3(y + 2x) = 6x + y$

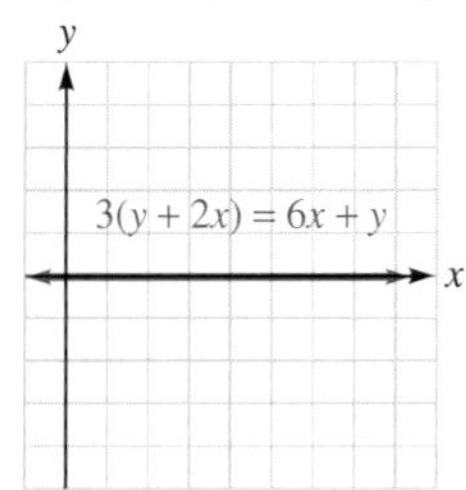

56. $5(y - x) = x + 5y$

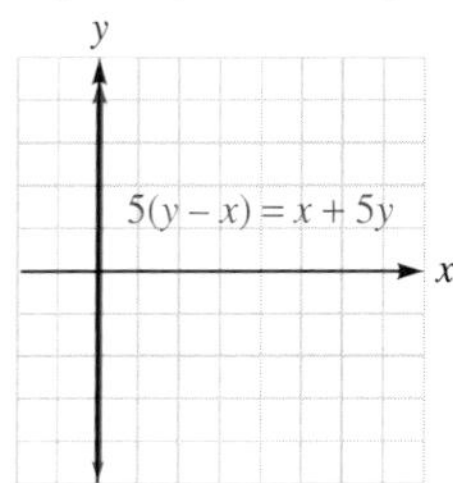

Use a graphing calculator to graph each equation and then find the x-coordinate of the x-intercept to the nearest hundredth.

57. $y = 3.7x - 4.5$ **1.22**

58. $y = \frac{3}{5}x + \frac{5}{4}$ **−2.08**

59. $1.5x - 3y = 7$ **4.67**

60. $0.3x + y = 7.5$ **25.00**

Find the distance between P and O(0, 0).

61. $P(4, -3)$ **5**

62. $P(-5, 12)$ **13**

63. $P(-3, 2)$ $\sqrt{13}$

64. $P(5, 0)$ **5**

65. $P(1, 1)$ $\sqrt{2}$

66. $P(6, -8)$ **10**

67. $P(\sqrt{3}, 1)$ **2**

68. $P(\sqrt{7}, \sqrt{2})$ **3**

Find the distance between P and Q.

69. $P(3, 7)$; $Q(6, 3)$ **5**

70. $P(4, 9)$; $Q(9, 21)$ **13**

71. $P(4, -6)$; $Q(-1, 6)$ **13**

72. $P(0, 5)$; $Q(6, -3)$ **10**

73. $P(-2, -15)$; $Q(-9, -39)$ **25**

74. $P(-7, 11)$; $Q(3, -13)$ **26**

75. $P(3, -3)$; $Q(-5, 5)$ $8\sqrt{2}$

76. $P(6, -3)$; $Q(-3, 2)$ $\sqrt{106}$

77. $P(\pi, -2)$; $Q(\pi, 5)$ **7**

78. $P(\sqrt{5}, 0)$; $Q(0, 2)$ **3**

Find the midpoint of the line segment PQ.

79. $P(2, 4)$; $Q(6, 8)$ **(4, 6)**

80. $P(3, -6)$; $Q(-1, -6)$ **(1, −6)**

81. $P(2, -5)$; $Q(-2, 7)$ **(0, 1)**

82. $P(0, 3)$; $Q(-10, -13)$ **(−5, −5)**

83. $P(-8, 5)$; $Q(8, -5)$ **(0, 0)**

84. $P(3, -2)$; $Q(2, -3)$ $\left(\frac{5}{2}, -\frac{5}{2}\right)$

85. $P(0, 0)$; $Q(\sqrt{5}, \sqrt{5})$ $\left(\frac{\sqrt{5}}{2}, \frac{\sqrt{5}}{2}\right)$

86. $P(\sqrt{3}, 0)$; $Q(0, -\sqrt{5})$ $\left(\frac{\sqrt{3}}{2}, -\frac{\sqrt{5}}{2}\right)$

One endpoint P and the midpoint M of line segment PQ are given. Find the coordinates of the other endpoint, Q.

87. $P(1, 4)$; $M(3, 5)$ **(5, 6)**

88. $P(2, -7)$; $M(-5, 6)$ **(−12, 19)**

89. $P(5, -5)$; $M(5, 5)$ **(5, 15)**

90. $P(-7, 3)$; $M(0, 0)$ **(7, −3)**

91. Show that a triangle with vertices at $(13, -2)$, $(9, -8)$, and $(5, -2)$ is isosceles.

92. Show that a triangle with vertices at $(-1, 2)$, $(3, 1)$, and $(4, 5)$ is isosceles.

93. In the illustration, points M and N are the midpoints of AC and BC, respectively. Find the length of MN. $\sqrt{2}$ **units**

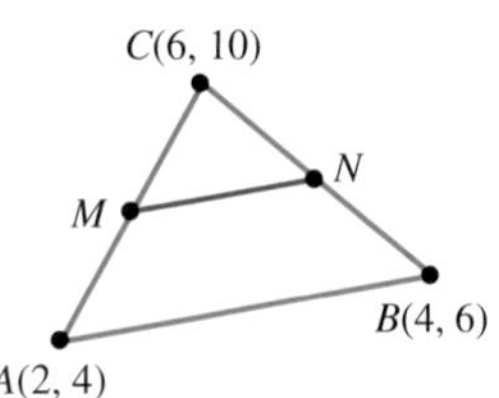

94. In the illustration, points M and N are the midpoints of AC and BC, respectively. Show that $d(MN) = \frac{1}{2}[d(AB)]$.

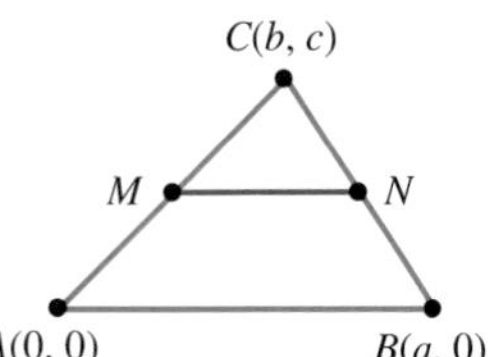

95. In the illustration, point M is the midpoint of the hypotenuse of right triangle AOB. Show that the area of rectangle $OLMN$ is one-half of the area of triangle AOB.

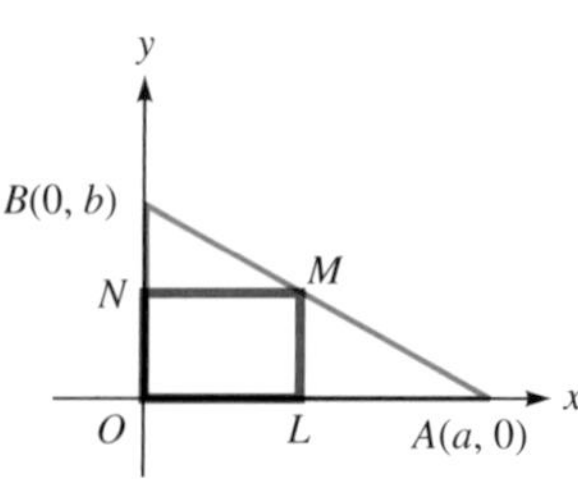

96. Rectangle $ABCD$ in the illustration is twice as long as it is wide, and its sides are parallel to the coordinate axes. If the perimeter is 42, find the coordinates of point C. **(11, 5)**

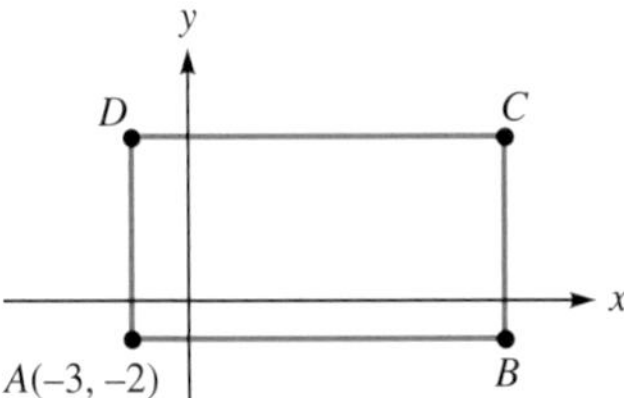

APPLICATIONS

97. House appreciation A house purchased for \$125,000 is expected to appreciate according to the formula $y = 7{,}500x + 125{,}000$, where y is the value of the house after x years. Find the value of the house 5 years later. **\$162,500**

98. Car depreciation A car purchased for \$17,000 is expected to depreciate according to the formula $y = -1{,}360x + 17{,}000$. When will the car be worthless? **12.5 yr**

99. Demand equations The number of television sets that consumers buy depends on price. The higher the price, the fewer TVs people will buy. The equation that relates price to the number of TVs sold at that price is called a **demand equation.** If the demand equation for a 25-inch TV is $p = -\frac{1}{10}q + 170$, where p is the price and q is the number of TVs sold at that price, how many TVs will be sold at a price of \$150? **200**

100. Supply equations The number of television sets that manufacturers produce depends on price. The higher the price, the more TVs manufacturers will produce. The equation that relates price to the number of TVs produced at that price is called a **supply equation.** If the supply equation for a 25-inch TV is $p = \frac{1}{10}q + 130$, where p is the price and q is the number of TVs produced for sale at that price, how many TVs will be produced if the price is \$150? **200**

101. Meshing gears The rotational speed V of a large gear (with N teeth) is related to the speed v of the smaller gear (with n teeth) by the equation $V = \frac{nv}{N}$. If the larger gear in the illustration is making 60 revolutions per minute, how fast is the smaller gear spinning? **100 rpm**

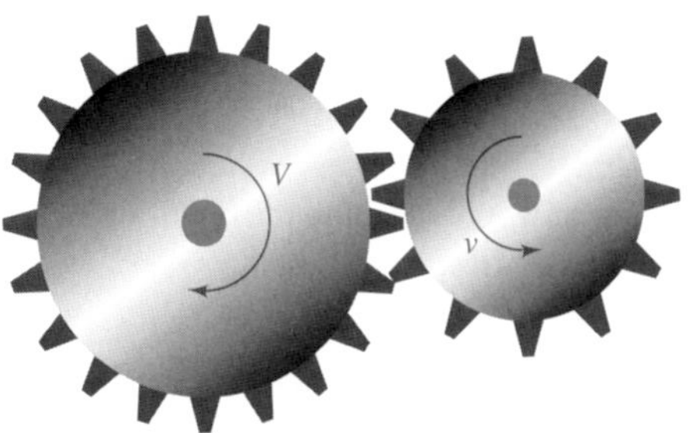

102. Crime prevention The number n of incidents of family violence requiring police response appears to be related to d, the money spent on crisis intervention, by the equation $n = 430 - 0.005d$. What expenditure would reduce the number of incidents to 350? **\$16,000**

103. Navigation See the illustration. An ocean liner is located 23 miles east and 72 miles north of Pigeon Cove Lighthouse, and its home port is 47 miles west and 84 miles south of the lighthouse. How far is the ship from port? **approx. 170 mi**

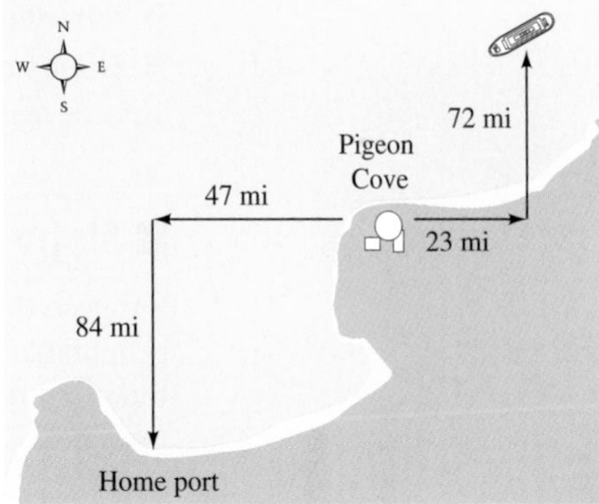

104. Engineering Two holes are to be drilled at locations specified by the engineering drawing shown in the illustration. Find the distance between the centers of the holes. **approx. 10.4 mm**

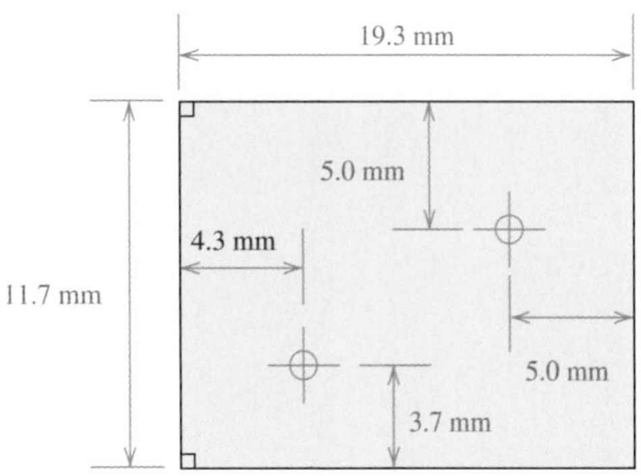

DISCOVERY AND WRITING

105. Explain how to graph a line using the intercept method.

106. Explain how to determine the quadrant in which the point $P(a, b)$ lies.

107. In Figure 2-15, show that $d(PM) + d(MQ) = d(PQ)$.

108. Use the result of Exercise 107 to explain why point M is the midpoint of segment PQ.

REVIEW ***Graph each interval on the number line.***

109. $[-3, 2) \cup (-2, 3]$

−3 3

110. $(-1, 4) \cap [-2, 2]$

−1 2

111. $[-3, -2) \cap (2, 3]$

no graph; the intersection is the empty set

112. $[-4, -3) \cup (2, 3]$

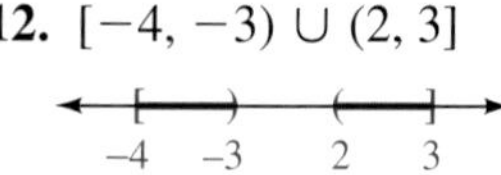

Solve each equation.

113. $\dfrac{3}{y+6} = \dfrac{4}{y+4}$ **−12**

114. $\dfrac{z+4}{z^2+z} - \dfrac{z+1}{z^2+2z} = \dfrac{8}{z^2+3z+2}$ $\mathbf{\frac{7}{4}}$

2.2 The Slope of a Nonvertical Line

In this section, you will learn about

- Slope of a Line
- Applications of Slope
- Horizontal and Vertical Lines
- Slopes of Parallel Lines
- Slopes of Perpendicular Lines

Slope of a Line

Suppose that a college student rents a room for \$300 per month, plus a \$200 nonrefundable deposit. The table shown in Figure 2-16(a) gives the cost (y) for different numbers of months (x). If we construct a graph from these data, we get the line shown in Figure 2-16(b).

Time in months (x)	Total cost (y)
0	200
1	500
2	800
3	1,100
4	1,400

(a)

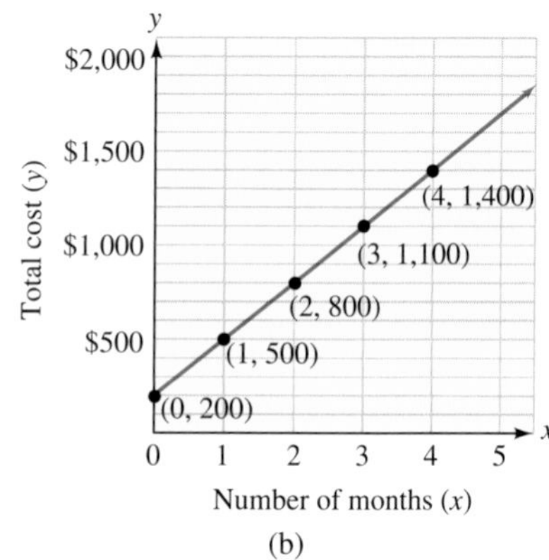

(b)

Figure 2-16

From the graph, we can see that if x changes from 0 to 1, y changes from 200 to 500. As x changes from 1 to 2, y changes from 500 to 800, and so on. The ratio of the change in y divided by the change in x is the constant 300.

$$\frac{\text{Change in } y}{\text{Change in } x} = \frac{500 - 200}{1 - 0} = \frac{800 - 500}{2 - 1} = \frac{1{,}100 - 800}{3 - 2} = \frac{1{,}400 - 1{,}100}{4 - 3} = \frac{300}{1} = 300$$

The ratio of the change in y divided by the change in x between any two points on any line is always a constant. This constant rate of change is called the **slope** of the line.

The Slope of a Nonvertical Line

The **slope of the nonvertical line** (see Figure 2-17) passing through points $P(x_1, y_1)$ and $Q(x_2, y_2)$ is

$$m = \frac{\text{change in } y}{\text{change in } x} = \frac{y_2 - y_1}{x_2 - x_1} \quad (x_2 \neq x_1)$$

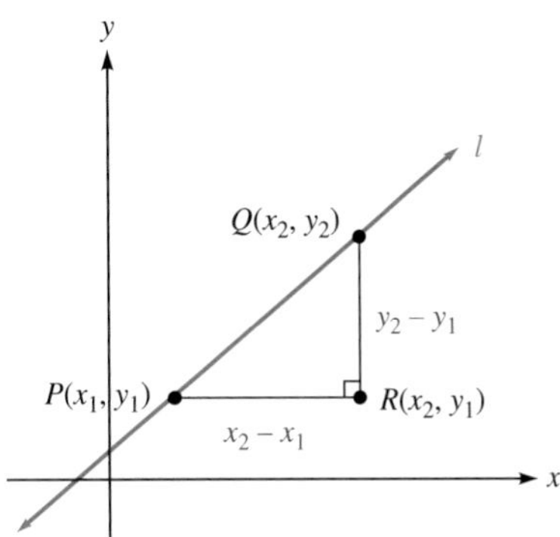

Figure 2-17

Comment Slope is often considered to be a measure of the steepness or tilt of a line. Note that you can use the coordinates of any two points on a line to compute the slope of the line.

ACTIVE EXAMPLE 1 Find the slope of the line passing through $P(-1, -2)$ and $Q(7, 8)$. (See Figure 2-18.)

Solution We can let $P(x_1, y_1) = P(-1, -2)$ and $Q(x_2, y_2) = Q(7, 8)$. Then we substitute -1 for x_1, -2 for y_1, 7 for x_2, and 8 for y_2 to get

$$m = \frac{\text{change in } y}{\text{change in } x}$$

$$m = \frac{y_2 - y_1}{x_2 - x_1}$$

$$= \frac{8 - (-2)}{7 - (-1)}$$

$$= \frac{10}{8}$$

$$= \frac{5}{4}$$

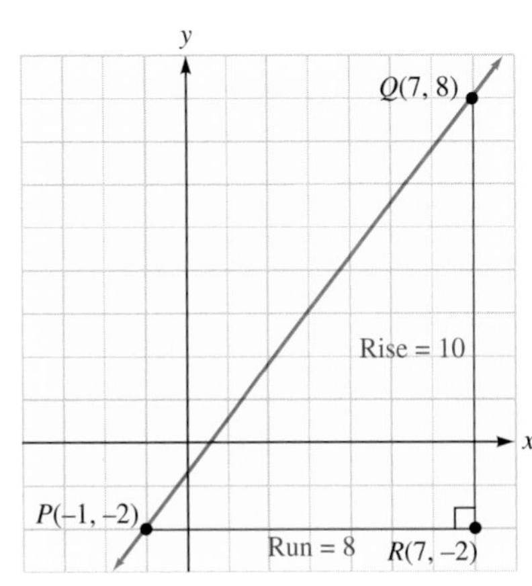

Figure 2-18

COLLEGE **Algebra** $f(x)$ **Now™**

Explore this Active Example and test yourself on the steps taken here at **http://1pass.thomson.com**. You can also find this example on the Interactive Video Skillbuilder CD-ROM.

The slope of the line is $\frac{5}{4}$. We would have obtained the same result if we had let $P(x_1, y_1) = P(7, 8)$ and $Q(x_2, y_2) = Q(-1, -2)$.

Self Check Find the slope of the line passing through $P(-3, -4)$ and $Q(5, 9)$. ■

Comment When calculating slope, always subtract the y-values and the x-values in the same order.

$$m = \frac{y_2 - y_1}{x_2 - x_1} \quad \text{or} \quad m = \frac{y_1 - y_2}{x_1 - x_2}$$

Otherwise, you will obtain an incorrect result.

The change in y (often denoted as Δy) is the **rise** of the line between points P and Q. The change in x (often denoted as Δx) is the **run.** Using this terminology, we can define slope to be the ratio of the rise to the run:

$$m = \frac{y_2 - y_1}{x_2 - x_1} = \frac{\Delta y}{\Delta x} = \frac{\text{rise}}{\text{run}} \quad (\Delta x \neq 0)$$

EXAMPLE 2 Find the slope of the line determined by $5x + 2y = 10$. (See Figure 2-19.)

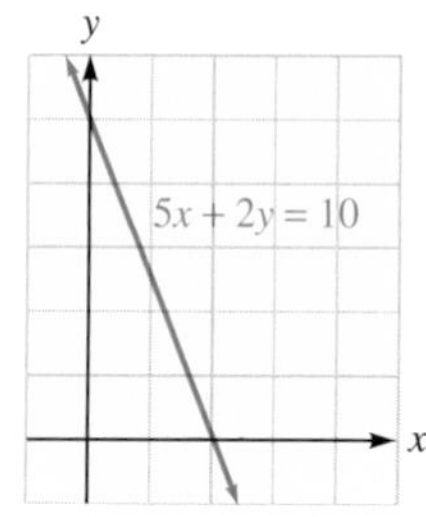

Figure 2-19

Solution We first find the coordinates of two points on the line. Two convenient points are the y- and x-intercepts.

- If $y = 0$, then $x = 2$, and the point $(2, 0)$ lies on the line.
- If $x = 0$, then $y = 5$, and the point $(0, 5)$ lies on the line.

We then find the slope of the line between $P(2, 0)$ and $Q(0, 5)$.

$$m = \frac{\text{change in } y}{\text{change in } x}$$

$$m = \frac{y_2 - y_1}{\mathbf{x_2} - \mathbf{x_1}}$$

$$= \frac{5 - 0}{\mathbf{0} - \mathbf{2}}$$

$$= -\frac{5}{2}$$

The slope is $-\frac{5}{2}$.

Self Check Find the slope of the line determined by $3x - 2y = 9$. ■

Applications of Slope

EXAMPLE 3 If carpet costs \$25 per square yard plus a delivery charge of \$30, the total cost c of n square yards is given by the formula

Total cost	equals	cost per square yard	times	the number of square yards purchased	plus	the delivery charge.
c	$=$	25	$\cdot$	n	$+$	30

Graph the equation $c = 25n + 30$ and interpret the slope of the line.

Solution We can graph the equation on a coordinate system with a vertical c-axis and a horizontal n-axis. Figure 2-20 shows a table of ordered pairs and the graph.

$c = 25n + 30$

x	y	(x, y)
10	280	(10, 280)
20	530	(20, 530)
30	780	(30, 780)
40	1,030	(40, 1,030)
50	1,280	(50, 1,280)

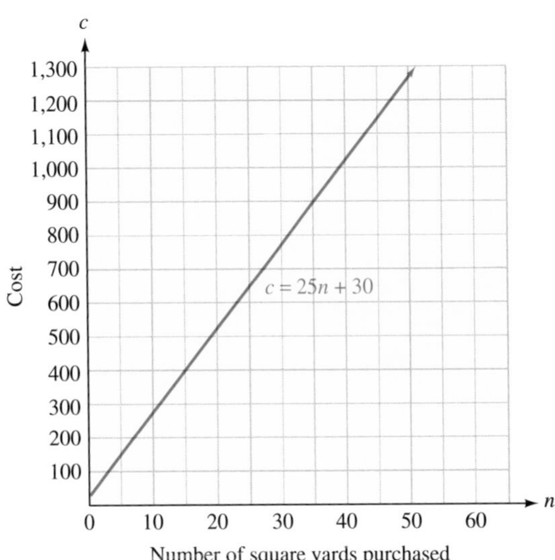

Figure 2-20

If we pick the points (30, 780) and (50, 1,280) to find the slope, we have

$$\begin{aligned} m &= \frac{\Delta c}{\Delta n} \\ &= \frac{c_2 - c_1}{n_2 - n_1} \\ &= \frac{1{,}280 - 780}{50 - 30} && \text{Substitute 1,280 for } c_2\text{, 780 for } c_1\text{, 50 for } n_2\text{, and 30 for } n_1. \\ &= \frac{500}{20} \\ &= 25 \end{aligned}$$

The slope of 25 (in dollars/square yard) is the cost per square yard of the carpet. ■

Everyday Connections — Sales of CDs and Cassettes

"It's important to note that a lot of people are beholden to the old media world and, by definition, the new world changes things and the old world is always going to resist that change. I think that there is skepticism from people and that's OK. I accept that as a challenge. Hopefully, that makes us stronger." Michael Robertson, founder of MP3.com

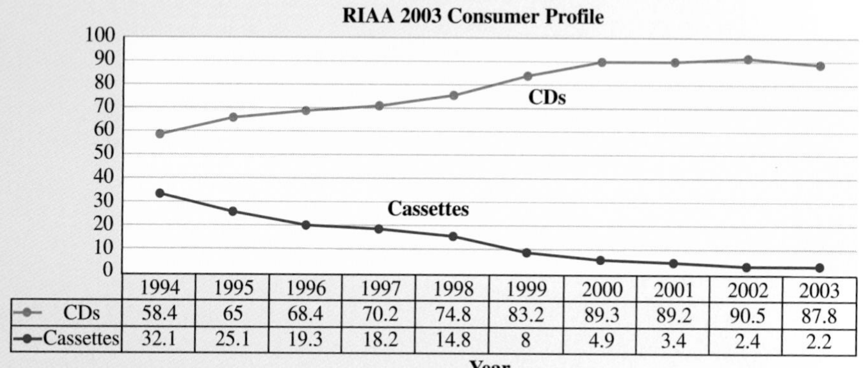

	1994	1995	1996	1997	1998	1999	2000	2001	2002	2003
CDs	58.4	65	68.4	70.2	74.8	83.2	89.3	89.2	90.5	87.8
Cassettes	32.1	25.1	19.3	18.2	14.8	8	4.9	3.4	2.4	2.2

Source: www.riaa.com

We can approximate the percent **rate of growth (or decrease)** of a quantity during a given time interval by calculating the slope of the line segment that connects the endpoints of the graph on the given interval.

Use the data from the graph to compute the percent rate of growth (or decrease) of the following:

1. CD sales from 1996–1997. **1.8% growth**
2. Cassette sales from 1994–1995. **7% decrease**
3. During which one-year interval was the rate of growth of CD sales the greatest? **1998–1999**

ACTIVE EXAMPLE 4 It takes a skier 25 minutes to complete the course shown in Figure 2-21. Find his average rate of descent in feet per minute.

COLLEGE **Algebra** $f(x)$ **Now**™
Go to **http://1pass.thomson.com** or your CD to practice this example.

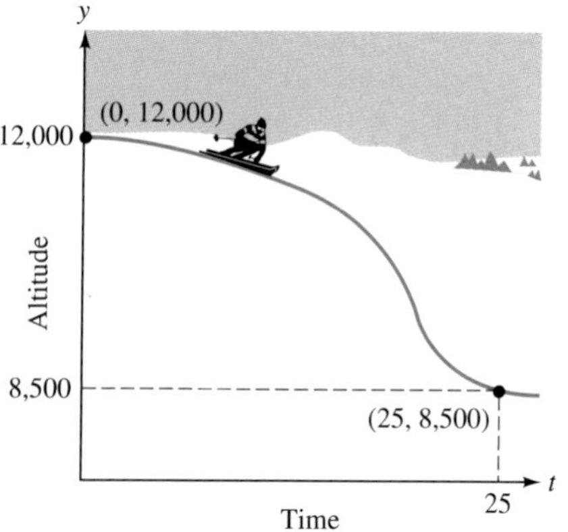

Figure 2-21

Solution To find the average rate of descent, we must find the ratio of the change in altitude to the change in time. To find this ratio, we calculate the slope of the line passing through the points (0, 12,000) and (25, 8,500).

$$\begin{aligned} \text{Average rate of descent} &= \frac{12{,}000 - 8{,}500}{0 - 25} \\ &= \frac{3{,}500}{-25} \\ &= -140 \end{aligned}$$

The average rate of descent is -140 ft/min. ■

■ Horizontal and Vertical Lines

If $P(x_1, y_1)$ and $Q(x_2, y_2)$ are points on the horizontal line shown in Figure 2-22(a), then $y_1 = y_2$, and the numerator of the fraction

$$\frac{y_2 - y_1}{x_2 - x_1} \quad \text{On a horizontal line, } x_2 \neq x_1.$$

is 0. Thus, the value of the fraction is 0, and the slope of the horizontal line is 0.

If $P(x_1, y_1)$ and $Q(x_2, y_2)$ are points on the vertical line shown in Figure 2-22(b), then $x_1 = x_2$, and the denominator of the fraction

$$\frac{y_2 - y_1}{x_2 - x_1} \quad \text{On a vertical line, } y_2 \neq y_1.$$

is 0. Since the denominator of a fraction cannot be 0, the slope of a vertical line is not defined.

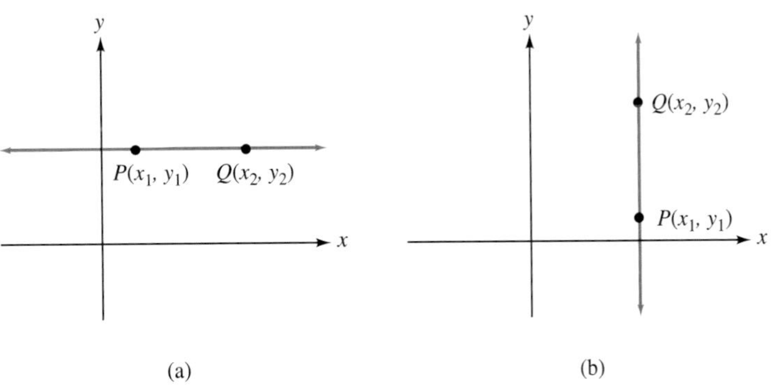

Figure 2-22

Horizontal and Vertical Lines

The slope of a horizontal line (a line with an equation of the form $y = b$) is 0.

The slope of a vertical line (a line with an equation of the form $x = a$) is not defined.

If a line rises as we follow it from left to right, as in Figure 2-23(a), its slope is positive. If a line drops as we follow it from left to right, as in Figure 2-23(b), its slope is negative.

If a line is horizontal, as in Figure 2-23(c), its slope is 0. If a line is vertical, as in Figure 2-23(d), it has no defined slope.

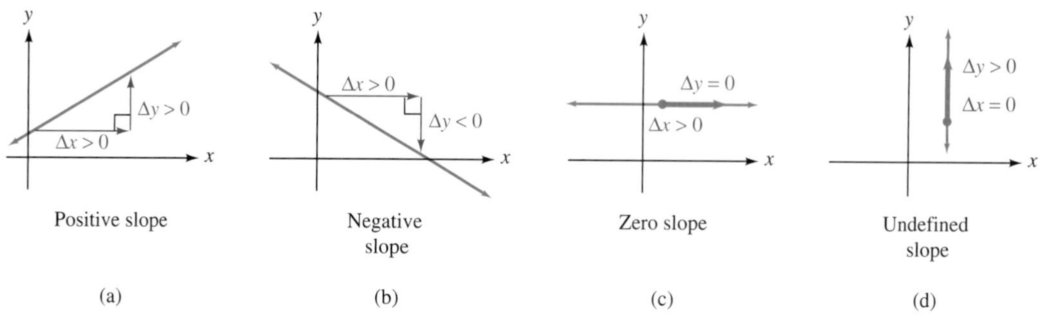

Figure 2-23

Slopes of Parallel Lines

To see a relationship between parallel lines and their slopes, we refer to the parallel lines l_1 and l_2 shown in Figure 2-24, with slopes of m_1 and m_2, respectively. Because right triangles ABC and DEF are similar, it follows that

$$\begin{aligned} m_1 &= \frac{\Delta y \text{ of } l_1}{\Delta x \text{ of } l_1} \\ &= \frac{\Delta y \text{ of } l_2}{\Delta x \text{ of } l_2} \\ &= m_2 \end{aligned}$$

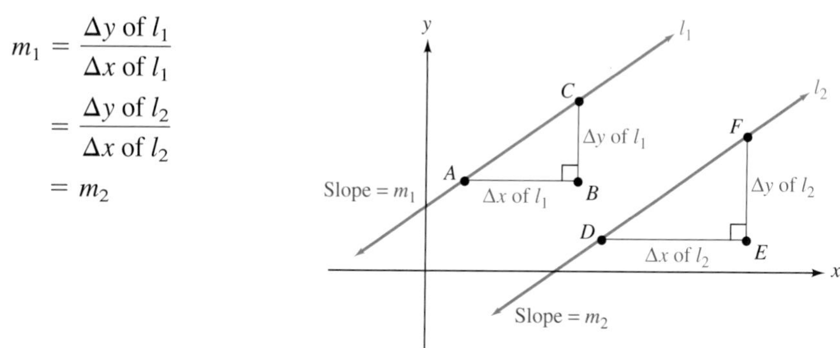

Figure 2-24

This shows that if two nonvertical lines are parallel, they have the same slope. It is also true that when two lines have the same slope, they are parallel.

Slopes of Parallel Lines

Nonvertical parallel lines have the same slope, and lines having the same slope are parallel.

Since vertical lines are parallel, lines with undefined slopes are parallel.

EXAMPLE 5 The lines in Figure 2-25 are parallel. Find y.

Solution Since the lines are parallel, their slopes are equal. To find y, we find the slope of each line, set them equal, and solve the resulting equation.

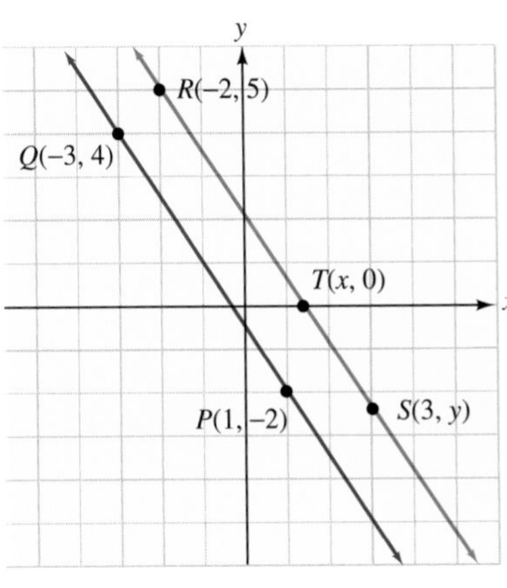

Figure 2-25

Slope of PQ **= slope of** RS

$$\frac{-2-4}{1-(-3)} = \frac{y-5}{3-(-2)}$$

$$\frac{-6}{4} = \frac{y-5}{5} \qquad \text{Simplify.}$$

$$-30 = 4(y-5) \qquad \text{Multiply both sides by 20.}$$

$$-30 = 4y - 20 \qquad \text{Remove parentheses and simplify.}$$

$$-10 = 4y \qquad \text{Add 20 to both sides.}$$

$$-\frac{5}{2} = y \qquad \text{Divide both sides by 4 and simplify.}$$

Thus, $y = -\frac{5}{2}$.

Self Check Find x in Figure 2-25. ■

■ Slopes of Perpendicular Lines

The following theorem relates perpendicular lines and their slopes.

Slopes of Perpendicular Lines

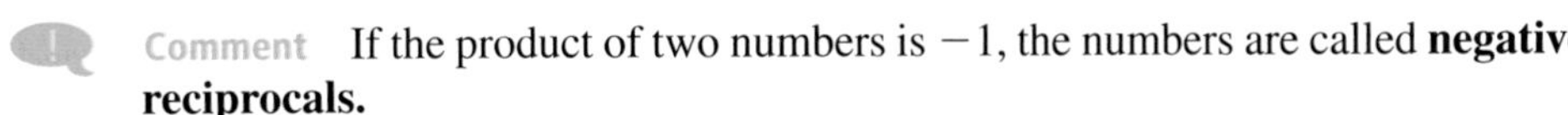
If two nonvertical lines are perpendicular, the product of their slopes is -1.

If the product of the slopes of two lines is -1, the lines are perpendicular.

Comment If the product of two numbers is -1, the numbers are called **negative reciprocals.**

Proof Suppose l_1 and l_2 are lines with slopes of m_1 and m_2 that intersect at some point. See Figure 2-26. Then superimpose a coordinate system over the lines so that the intersection point is the origin. Let $P(a, b)$ be a point on l_1, and let $Q(c, d)$ be a point on l_2. Neither point P nor point Q can be the origin.

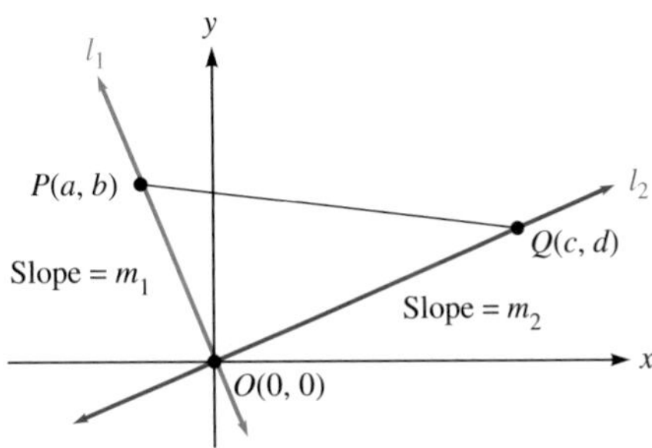

Figure 2-26

First, we suppose that l_1 and l_2 are perpendicular. Then triangle POQ is a right triangle with its right angle at O. By the Pythagorean theorem,

$$d(OP)^2 + d(OQ)^2 = d(PQ)^2$$
$$(a - 0)^2 + (b - 0)^2 + (c - 0)^2 + (d - 0)^2 = (a - c)^2 + (b - d)^2$$
$$a^2 + b^2 + c^2 + d^2 = a^2 - 2ac + c^2 + b^2 - 2bd + d^2$$
$$0 = -2ac - 2bd$$
$$bd = -ac$$
$$(1) \qquad \frac{b}{a} \cdot \frac{d}{c} = -1 \qquad \text{Divide both sides by } ac.$$

The coordinates of P are (a, b), and the coordinates of O are $(0, 0)$. Using the definition of slope, we have

$$m_1 = \frac{b - 0}{a - 0} = \frac{b}{a}$$

Similarly, we have

$$m_2 = \frac{d}{c}$$

We substitute m_1 for $\frac{b}{a}$ and m_2 for $\frac{d}{c}$ in Equation 1 to obtain

$$m_1 m_2 = -1$$

Hence, if lines l_1 and l_2 are perpendicular, the product of their slopes is -1.

Conversely, we suppose that the product of the slopes of lines l_1 and l_2 is -1. Because the steps in the previous discussion are reversible, we have $d(OP)^2 + d(OQ)^2 = d(PQ)^2$. By the Pythagorean theorem, triangle POQ is a right triangle. Thus, l_1 and l_2 are perpendicular. ■

COLLEGE **Algebra** f(x) **Now**™
Go to **http://1pass.thomson.com** or your CD to practice this example.

It is also true that a horizontal line is perpendicular to a vertical line.

ACTIVE EXAMPLE 6 Are the lines shown in Figure 2-27 perpendicular?

Solution We find the slopes of lines and see whether their product is -1.

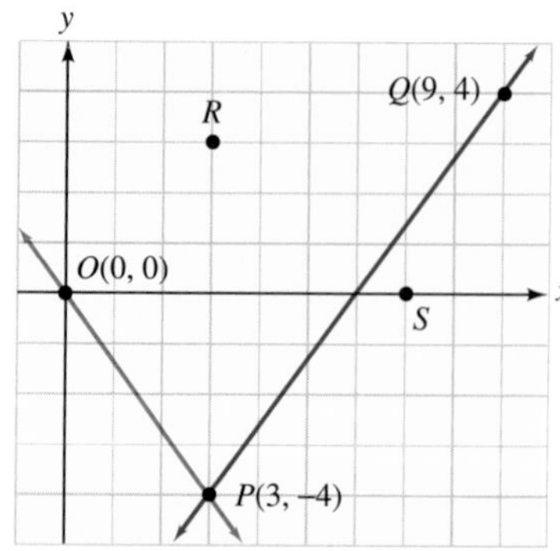

Figure 2-27

$$\text{Slope of } OP = \frac{\Delta y}{\Delta x} = \frac{y_2 - y_1}{\mathbf{x_2} - \mathbf{x_1}} = \frac{-4 - 0}{\mathbf{3} - \mathbf{0}} = -\frac{4}{3}$$

$$\text{Slope of } PQ = \frac{\Delta y}{\Delta x} = \frac{y_2 - y_1}{\mathbf{x_2} - \mathbf{x_1}} = \frac{4 - (-4)}{\mathbf{9} - \mathbf{3}} = \frac{8}{6} = \frac{4}{3}$$

Since the product of the slopes is $-\frac{16}{9}$ and not -1, the lines are not perpendicular.

Self Check Is either line in Figure 2-27 perpendicular to the line passing through R and S? ■

Self Check Answers

1. $\frac{13}{8}$ **2.** $\frac{3}{2}$ **5.** $\frac{4}{3}$ **6.** yes

2.2 Exercises

VOCABULARY AND CONCEPTS *Fill in the blanks.*

1. The slope of a nonvertical line is defined to be the change in y **divided** by the change in x.
2. The change in **y** is often called the rise.
3. The change in x is often called the **run**.
4. When computing the slope from the coordinates of two points, always subtract the y-values and the x-values in the **same order**.
5. The symbol Δy means **the change in** y.
6. The slope of a **horizontal** line is 0.
7. The slope of a **vertical** line is undefined.
8. If the slopes of two lines are equal, the lines are **parallel**.
9. If the product of the slopes of two lines is -1, the lines are **perpendicular**.
10. If two lines are **perpendicular**, the product of their slopes is **-1**.

PRACTICE *Find the slope of the line passing through each pair of points, if possible.*

11. 1

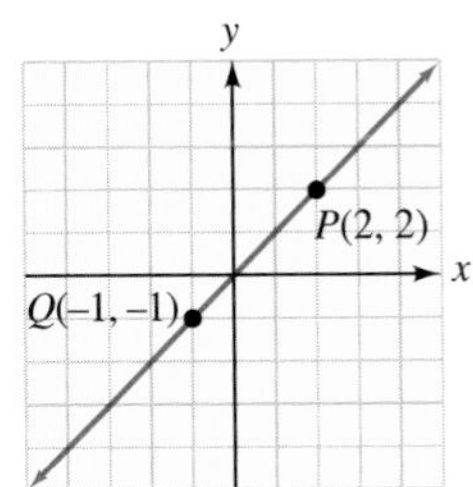

12. $-\frac{5}{12}$

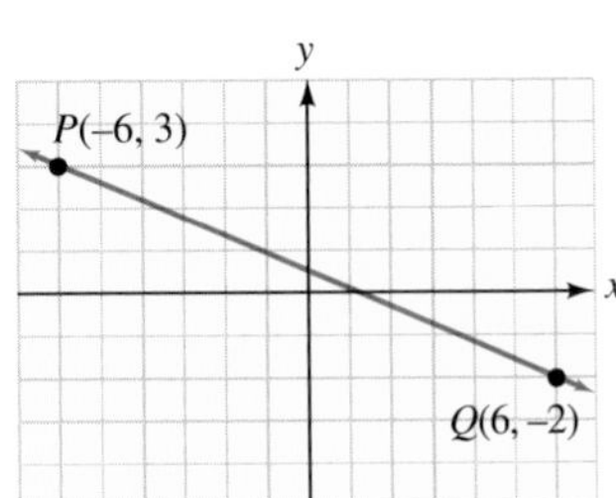

13. $P(2, 5)$; $Q(3, 10)$

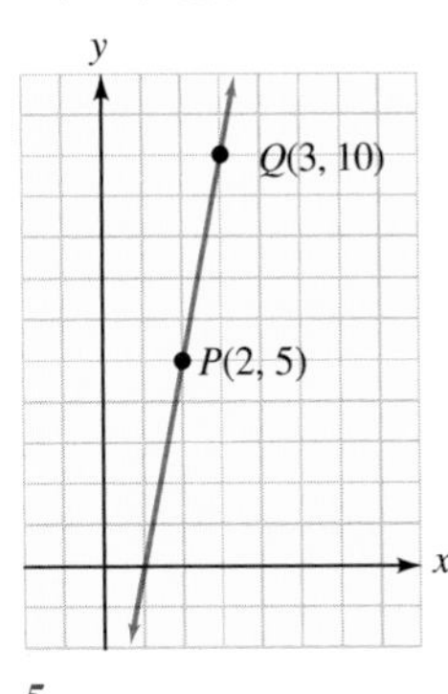

5

14. $P(3, -1)$; $Q(5, 3)$

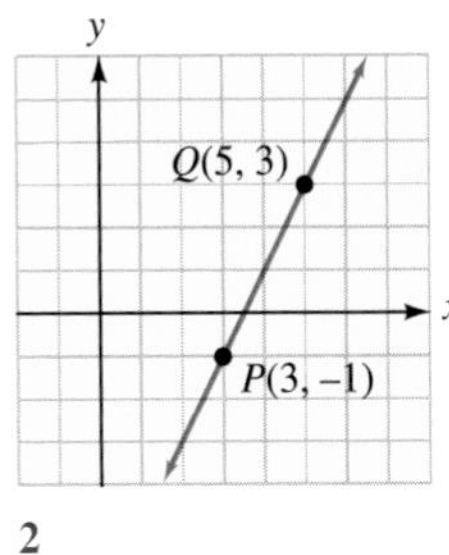

2

15. $P(3, -2)$; $Q(-1, 5)$ $-\frac{7}{4}$

16. $P(3, 7)$; $Q(6, 16)$ 3

17. $P(8, -7)$; $Q(4, 1)$ -2

18. $P(5, 17)$; $Q(17, 17)$ 0

19. $P(-4, 3)$; $Q(-4, -3)$ undefined

20. $P(2, \sqrt{7})$; $Q(\sqrt{7}, 2)$ -1

21. $P\left(\frac{3}{2}, \frac{2}{3}\right)$; $Q\left(\frac{5}{2}, \frac{7}{3}\right)$ $\frac{5}{3}$

22. $P\left(-\frac{2}{5}, \frac{1}{3}\right)$; $Q\left(\frac{3}{5}, -\frac{5}{3}\right)$ -2

23. $P(a + b, c)$; $Q(b + c, a)$ assume $c \neq a$ -1

24. $P(b, 0)$; $Q(a + b, a)$ assume $a \neq 0$ 1

Find two points on the line and find the slope of the line.

25. $y = 3x + 2$ 3

26. $y = 5x - 8$ 5

27. $5x - 10y = 3$ $\frac{1}{2}$

28. $8y + 2x = 5$ $-\frac{1}{4}$

29. $3(y + 2) = 2x - 3$ $\frac{2}{3}$

30. $4(x - 2) = 3y + 2$ $\frac{4}{3}$

31. $3(y + x) = 3(x - 1)$ 0

32. $2x + 5 = 2(y + x)$ 0

Determine whether the slope of the line is positive, negative, 0, or undefined.

33.

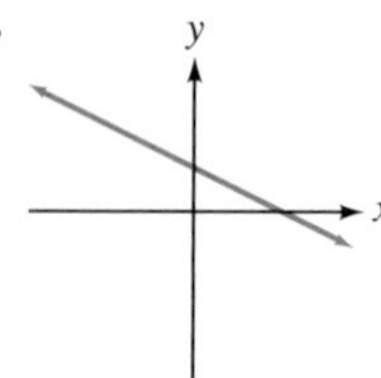

negative

34.

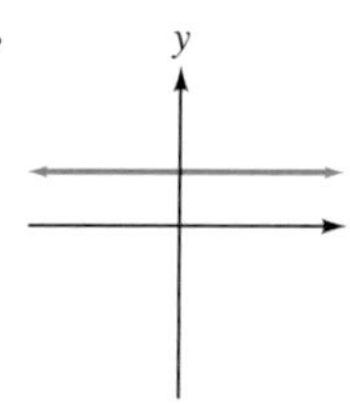

zero

35.

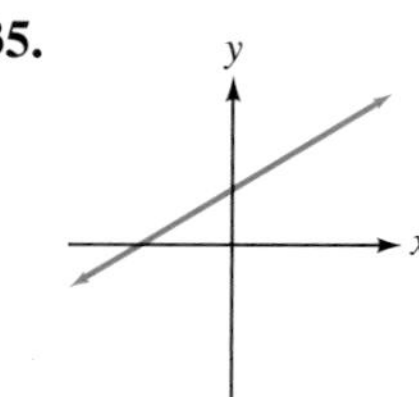

positive

36.

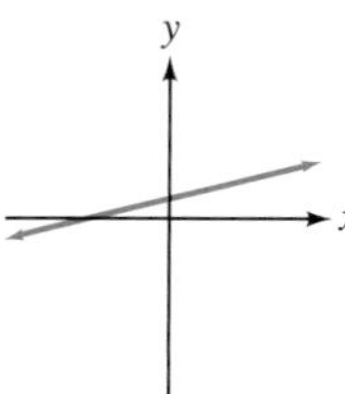

positive

37.

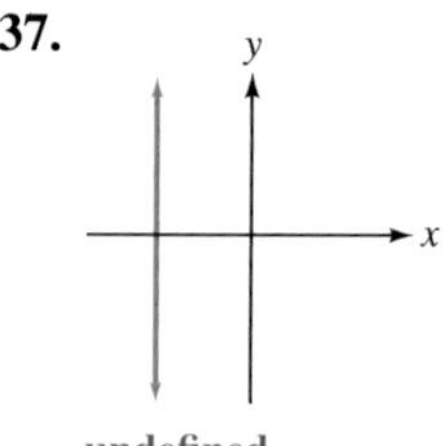

undefined

38.

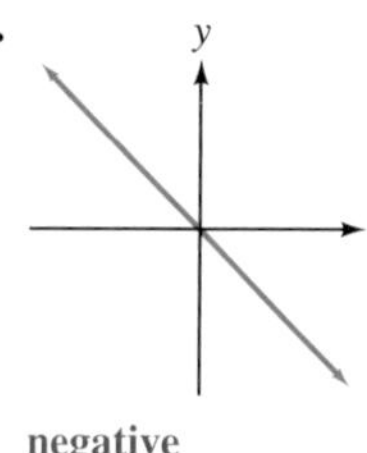

negative

Determine whether the lines with the given slopes are parallel, perpendicular, or neither.

39. $m_1 = 3; m_2 = -\frac{1}{3}$
perpendicular

40. $m_1 = \frac{2}{3}; m_2 = \frac{3}{2}$
neither

41. $m_1 = \sqrt{8}; m_2 = 2\sqrt{2}$
parallel

42. $m_1 = 1; m_2 = -1$
perpendicular

43. $m_1 = -\sqrt{2}; m_2 = \frac{\sqrt{2}}{2}$ perpendicular

44. $m_1 = 2\sqrt{7}; m_2 = \sqrt{28}$ parallel

45. $m_1 = -0.125; m_2 = 8$ perpendicular

46. $m_1 = 0.125; m_2 = \frac{1}{8}$ parallel

47. $m_1 = ab^{-1}; m_2 = -a^{-1}b \quad (a \neq 0, b \neq 0)$
perpendicular

48. $m_1 = \left(\frac{a}{b}\right)^{-1}; m_2 = -\frac{b}{a} \quad (a \neq 0, b \neq 0, a \neq b)$
neither

Determine whether the line through the given points and the line through R(−3, 5) and S(2, 7) are parallel, perpendicular, or neither.

49. $P(2, 4); Q(7, 6)$
parallel

50. $P(-3, 8); Q(-13, 4)$
parallel

51. $P(-4, 6); Q(-2, 1)$
perpendicular

52. $P(0, -9); Q(4, 1)$
neither

53. $P(a, a); Q(3a, 6a) \quad (a \neq 0)$ neither

54. $P(b, b); Q(-b, 6b) \quad (b \neq 0)$ perpendicular

Find the slopes of lines PQ and PR, and determine whether points P, Q, and R lie on the same line.

55. $P(-2, 8); Q(-6, 9); R(2, 5)$ not on same line

56. $P(1, -1); Q(3, -2); R(-3, 0)$ not on same line

57. $P(-a, a); Q(0, 0); R(a, -a)$ on same line

58. $P(a, a + b); Q(a + b, b); R(a - b, a)$
not on same line

Determine which, if any, of the three lines PQ, PR, and QR are perpendicular.

59. $P(5, 4); Q(2, -5); R(8, -3)$
No two are perpendicular.

60. $P(8, -2); Q(4, 6); R(6, 7)$
PQ and *QR* are perpendicular.

61. $P(1, 3); Q(1, 9); R(7, 3)$
PQ and *PR* are perpendicular.

62. $P(2, -3); Q(-3, 2); R(3, 8)$
PQ and *QR* are perpendicular.

63. $P(0, 0); Q(a, b); R(-b, a)$
PQ and *PR* are perpendicular.

64. $P(a, b); Q(-b, a); R(a - b, a + b)$
PR and *QR* are perpendicular.

65. Right triangles Show that the points $A(-1, -1)$, $B(-3, 4)$, and $C(4, 1)$ are the vertices of a right triangle.

66. Right triangles Show that the points $D(0, 1)$, $E(-1, 3)$, and $F(3, 5)$ are the vertices of a right triangle.

67. Squares Show that the points $A(1, -1)$, $B(3, 0)$, $C(2, 2)$, and $D(0, 1)$ are the vertices of a square.

68. Squares Show that the points $E(-1, -1)$, $F(3, 0)$, $G(2, 4)$, and $H(-2, 3)$ are the vertices of a square.

69. Parallelograms Show that the points $A(-2, -2)$, $B(3, 3)$, $C(2, 6)$, and $D(-3, 1)$ are the vertices of a parallelogram. (Show that both pairs of opposite sides are parallel.)

70. Trapezoids Show that points $E(1, -2)$, $F(5, 1)$, $G(3, 4)$, and $H(-3, 4)$ are the vertices of a trapezoid. (Show that only one pair of opposite sides is parallel.)

71. Geometry In the illustration, points M and N are midpoints of CB and BA, respectively. Show that MN is parallel to AC.

72. Geometry In the illustration, $d(AB) = d(AC)$. Show that AD is perpendicular to BC.

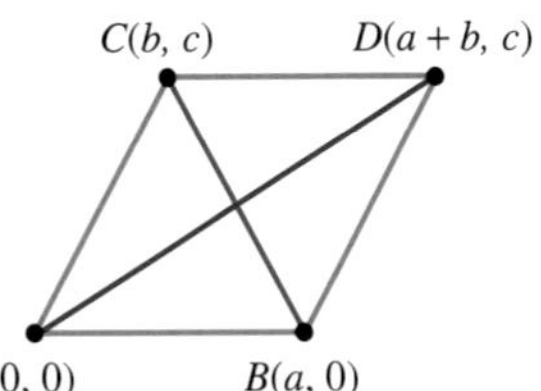

APPLICATIONS

73. Rate of growth When a college started an aviation program, the administration agreed to predict enrollments using a straight-line method. If the enrollment during the first year was 12, and the enrollment during the fifth year was 26, find the rate of growth per year (the slope of the line). See the illustration. **3.5 students per yr**

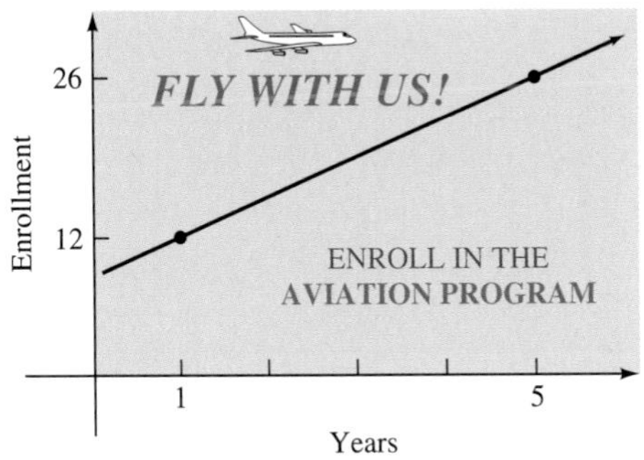

74. Rate of growth A small business predicts sales according to a straight-line method. If sales were $50,000 in the first year and $110,000 in the third year, find the rate of growth in dollars per year (the slope of the line). **$30,000 per year**

75. Rate of decrease The price of computers has been dropping steadily for the past ten years. If a desktop PC cost $6,700 ten years ago, and the same computing power cost $2,200 three years ago, find the rate of decrease per year. (Assume a straight-line model.) **$642.86 per year**

76. Hospital costs The table shows the changing mean daily cost for a hospital room. For the ten-year period, find the rate of change per year of the portion of the room cost that is absorbed by the hospital. **$26 per year**

Year	Total cost to the hospital	Amount passed on to patient
1990	$459	$214
1995	670	295
2000	812	307

77. Charting temperature changes The following Fahrenheit temperature readings were recorded over a four-hour period.

Time	12:00	1:00	2:00	3:00	4:00
Temperature	47°	53°	59°	65°	71°

Let t represent the time (in hours), with 12:00 corresponding to $t = 0$. Let T represent the temperature. Plot the points (t, T), and draw the line through those points. Explain the meaning of $\frac{\Delta T}{\Delta t}$.

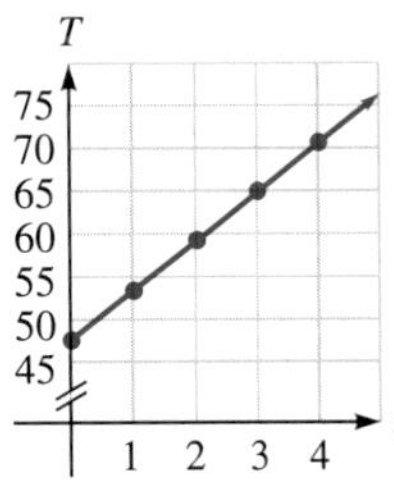

$\frac{\Delta T}{\Delta t}$ is the hourly rate of change in temperature.

78. Tracking the Dow The Dow Jones Industrial Averages at the close of trade on three consecutive days were as follows:

Day	Monday	Tuesday	Wednesday
Close	10,981	10,964	10,947

Let d represent the day, with $d = 0$ corresponding to Monday, and let D represent the Dow Jones average. Plot the points (d, D), and draw the graph. Explain the meaning of $\frac{\Delta D}{\Delta d}$.

$\frac{\Delta D}{\Delta d}$ is the daily rate of change in the Dow.

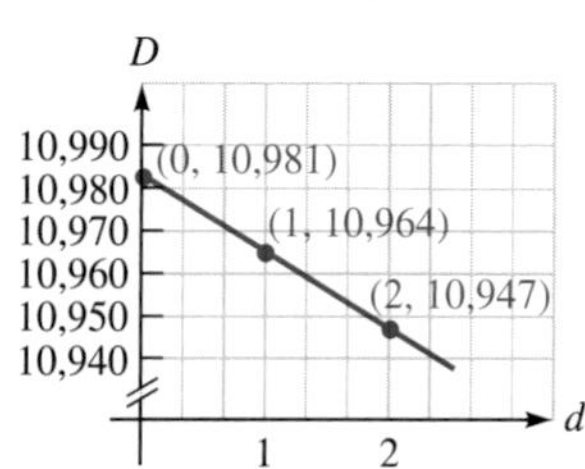

79. Speed of an airplane A pilot files a flight plan indicating her intention to fly at a constant speed of 590 mph. Write an equation that expresses the distance traveled in terms of the flying time. Then graph the equation and interpret the slope of the line. (*Hint:* $d = rt$.)

The slope is the speed of the plane.

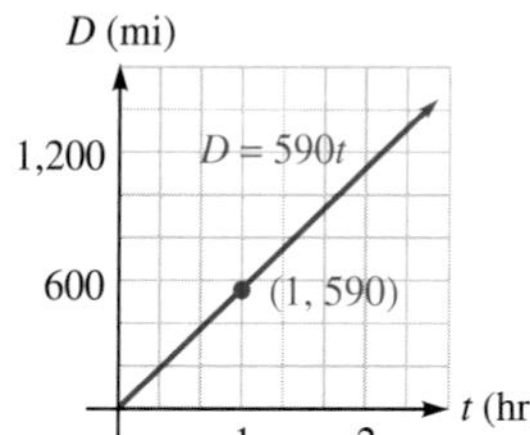

80. Growth of savings A student deposits $25 each month in a Holiday Club account at her bank. The account pays no interest. Write an equation that expresses the amount in her account in terms of the number of deposits. Then graph the line, and interpret the slope of the line.

The slope is the rate of increase, in dollars per month.

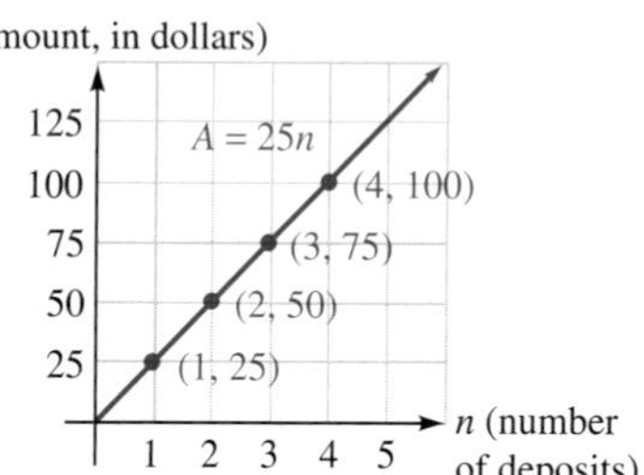

DISCOVERY AND WRITING

81. Explain why the slope of a vertical line is undefined.

82. Explain how to determine whether two lines are parallel, perpendicular, or neither.

REVIEW ***Solve each equation for y and simplify.***

83. $3x + 7y = 21$ $y = -\frac{3}{7}x + 3$

84. $y - 3 = 5(x + 2)$ $y = 5x + 13$

85. $\frac{x}{5} + \frac{y}{2} = 1$ $y = -\frac{2}{5}x + 2$

86. $x - 5y = 15$ $y = \frac{1}{5}x - 3$

Factor each expression.

87. $6p^2 + p - 12$ $(2p + 3)(3p - 4)$

88. $b^3 - 27$ $(b - 3)(b^2 + 3b + 9)$

89. $mp + mq + np + nq$ $(m + n)(p + q)$

90. $x^4 + x^2 - 2$ $(x^2 + 2)(x + 1)(x - 1)$

2.3 Writing Equations of Lines

In this section, you will learn about

- Point–Slope Form of the Equation of a Line
- Slope–Intercept Form of the Equation of a Line
- Graphing Equations Written in Slope–Intercept Form
- General Form of the Equation of a Line
- Straight-Line Depreciation
- Linear Curve Fitting

We have seen that we can find the graph of a linear equation in two variables. In this section, we will begin with a graph and find its equation.

■ Point–Slope Form of the Equation of a Line

Suppose that line l in Figure 2-28 has a slope of m and passes through the point $P(x_1, y_1)$. If $Q(x, y)$ is any other point on line l, we have

$$m = \frac{y - y_1}{x - x_1}$$

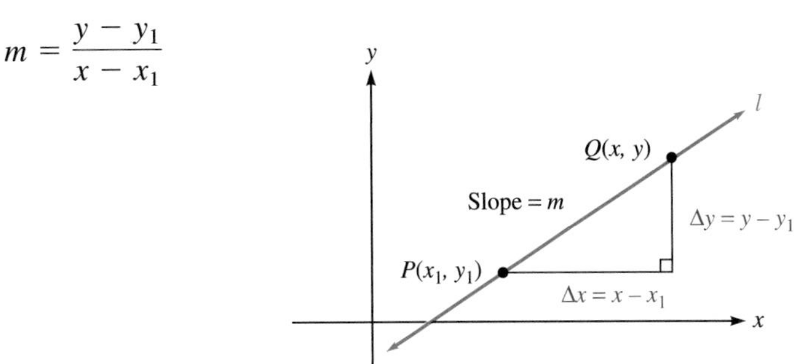

Figure 2-28

If we multiply both sides by $x - x_1$, we have

(1) $$y - y_1 = m(x - x_1)$$

Since Equation 1 displays the coordinates of the point (x_1, y_1) on the line and the slope m of the line, it is called the **point–slope form** of the equation of a line.

Point–Slope Form The equation of the line passing through $P(x_1, y_1)$ and with slope m is

$$y - y_1 = m(x - x_1)$$

EXAMPLE 1 Write the equation of the line with slope $-\frac{5}{3}$ and passing through $P(3, -1)$.

Solution We substitute $-\frac{5}{3}$ for m, 3 for x_1, and -1 for y_1 in the point–slope form and simplify.

$$y - y_1 = m(x - x_1)$$

$$y - (-1) = -\frac{5}{3}(x - 3) \quad \text{Substitute } -\tfrac{5}{3} \text{ for } m, 3 \text{ for } x_1, \text{ and } -1 \text{ for } y_1.$$

$$y + 1 = -\frac{5}{3}x + 5 \quad \text{Remove parentheses.}$$

$$y = -\frac{5}{3}x + 4 \quad \text{Subtract 1 from both sides.}$$

The equation of the line is $y = -\frac{5}{3}x + 4$.

Self Check Write the equation of the line with slope $-\frac{2}{3}$ and passing through $P(-4, 5)$. ■

ACTIVE EXAMPLE 2 Find the equation of the line passing through $P(3, 7)$ and $Q(-5, 3)$.

Solution First we find the slope of the line.

COLLEGE **Algebra f(x) Now™**
Explore this Active Example and test yourself on the steps taken here at **http://1pass.thomson.com**. You can also find this example on the Interactive Video Skillbuilder CD-ROM.

$$
\begin{aligned}
m &= \frac{y_2 - y_1}{x_2 - x_1} \\
&= \frac{3 - 7}{-5 - 3} && \text{Substitute 3 for } y_2\text{, 7 for } y_1\text{, } -5 \text{ for } x_2\text{, and 3 for } x_1. \\
&= \frac{-4}{-8} \\
&= \frac{1}{2}
\end{aligned}
$$

We can choose either point P or point Q and substitute its coordinates into the point–slope form. If we choose $P(3, 7)$, we substitute $\frac{1}{2}$ for m, 3 for x_1, and 7 for y_1.

$$
\begin{aligned}
y - y_1 &= m(x - x_1) \\
y - 7 &= \frac{1}{2}(x - 3) && \text{Substitute } \tfrac{1}{2} \text{ for } m\text{, 3 for } x_1\text{, and 7 for } y_1. \\
y &= \frac{1}{2}x - \frac{3}{2} + 7 && \text{Remove parentheses and add 7 to both sides.} \\
y &= \frac{1}{2}x + \frac{11}{2} && -\tfrac{3}{2} + 7 = -\tfrac{3}{2} + \tfrac{14}{2} = \tfrac{11}{2}.
\end{aligned}
$$

The equation of the line is $y = \frac{1}{2}x + \frac{11}{2}$.

Self Check Find the equation of the line passing through $P(-5, 4)$ and $Q(8, -6)$. ■

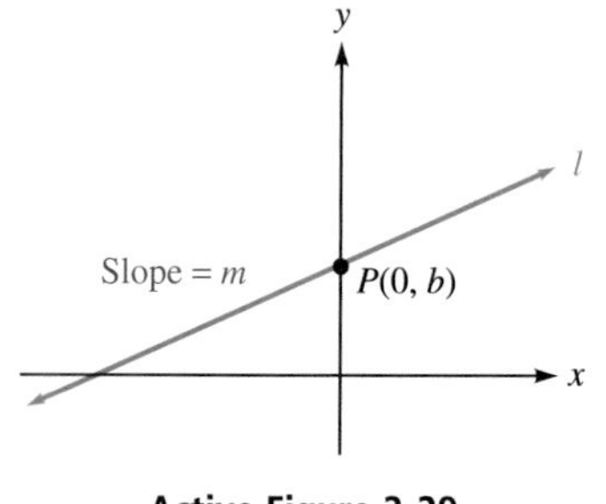

Active Figure 2-29

COLLEGE **Algebra f(x) Now™**
Explore this Active Figure and test your understanding at **http://1pass.thomson.com**.

■ Slope–Intercept Form of the Equation of a Line

Since the y-intercept of the line shown in Figure 2-29 is the point $(0, b)$, we can write the equation of the line by substituting 0 for x_1 and b for y_1 in the point–slope form and simplifying.

$$
\begin{aligned}
y - y_1 &= m(x - x_1) \\
y - b &= m(x - 0) && \text{Substitute 0 for } x_1 \text{ and } b \text{ for } y_1. \\
y - b &= mx && x - 0 = x. \\
y &= mx + b && \text{Add } b \text{ to both sides.}
\end{aligned}
\tag{2}
$$

Because Equation 2 displays the slope m and the y-coordinate b of the y-intercept, it is called the **slope–intercept form** of the equation of a line.

Slope–Intercept Form

The equation of the line with slope m and y-intercept $(0, b)$ is

$$y = mx + b$$

EXAMPLE 3 Use slope–intercept form to write the equation of the line with slope 4 that passes through $P(5, 9)$.

Solution Since we know that $m = 4$ and that the pair $(5, 9)$ satisfies the equation, we substitute 4 for m, 5 for x, and 9 for y in the equation $y = mx + b$ and solve for b.

$$
\begin{aligned}
y &= mx + b \\
9 &= 4(5) + b && \text{Substitute 4 for } m\text{, 5 for } x\text{, and 9 for } y. \\
9 &= 20 + b && \text{Simplify.} \\
-11 &= b && \text{Subtract 20 from both sides.}
\end{aligned}
$$

Because $m = 4$ and $b = -11$, the equation is $y = 4x - 11$.

Self Check Use slope–intercept form to write the equation of the line with slope $\frac{7}{3}$ and passing through (3, 1). ■

■ Graphing Equations Written in Slope–Intercept Form

It is easy to graph a linear equation when it is written in slope–intercept form. For example, to graph $y = \frac{4}{3}x - 2$, we note that $b = -2$ and that the y-intercept is $(0, b) = (0, -2)$. (See Figure 2-30.)

Because the slope is $\frac{\Delta y}{\Delta x} = \frac{4}{3}$, we can locate another point Q on the line by starting at point P and counting 3 units to the right and 4 units up. The change in x from point P to point Q is $\Delta x = 3$, and the corresponding change in y is $\Delta y = 4$. The line joining points P and Q is the graph of the equation.

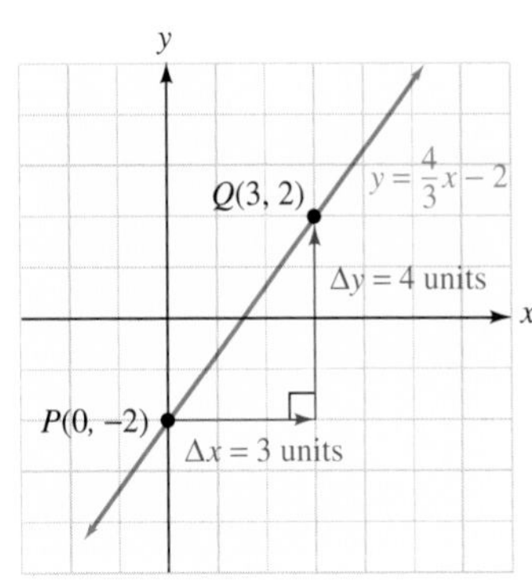

Figure 2-30

ACTIVE EXAMPLE 4 Find the slope and the y-intercept of the line with equation $3(y + 2) = 6x - 1$. Then graph it.

COLLEGE **Algebra** *f(x)* **Now**™
Go to **http://1pass.thomson.com** or your CD to practice this example.

Solution We write the equation in the form $y = mx + b$ to find the slope m and the y-intercept $(0, b)$.

$$
\begin{aligned}
3(y + 2) &= 6x - 1 \\
3y + 6 &= 6x - 1 && \text{Remove parentheses.} \\
3y &= 6x - 7 && \text{Subtract 6 from both sides.} \\
y &= 2x - \frac{7}{3} && \text{Divide both sides by 3.}
\end{aligned}
$$

The slope is 2, and the y-intercept is $\left(0, -\frac{7}{3}\right)$. We plot the y-intercept. Then we find a second point on the line by moving 1 unit to the right and 2 units up to the point $\left(1, -\frac{1}{3}\right)$. To get the graph, we draw a line through the two points, as shown in Figure 2-31.

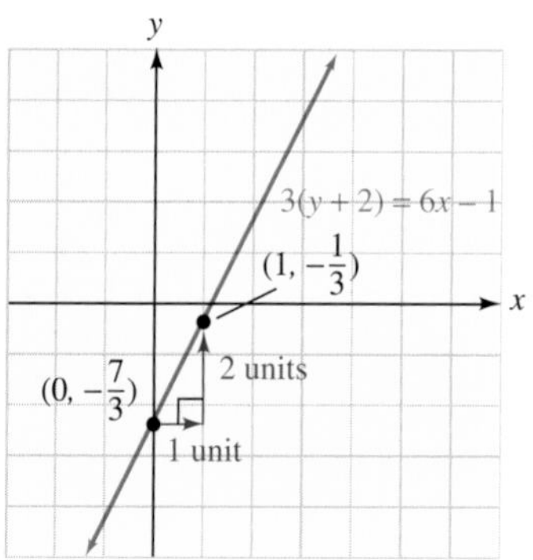

Figure 2-31

Self Check Find the slope and the y-intercept of the line with equation $2(x - 3) = -3(y + 5)$. Then graph it. ■

EXAMPLE 5 Show that the lines represented by $4x + 8y = 10$ and $2x = 12 - 4y$ are parallel.

Solution We solve each equation for y to see that the lines are distinct and that their slopes are equal.

$$\begin{aligned} 4x + 8y &= 10 \\ 8y &= -4x + 10 \\ y &= \frac{-4x}{8} + \frac{10}{8} \\ y &= -\frac{1}{2}x + \frac{5}{4} \end{aligned} \qquad \begin{aligned} 2x &= 12 - 4y \\ 4y &= -2x + 12 \\ y &= \frac{-2x}{4} + \frac{12}{4} \\ y &= -\frac{1}{2}x + 3 \end{aligned}$$

Since the values of b are different, the lines are distinct. Since each slope is $-\frac{1}{2}$, the lines are parallel.

Self Check Are the lines represented by $y = 3x + 2$ and $6x + 2y = 5$ parallel? ■

EXAMPLE 6 Show that the lines represented by $4x + 8y = 10$ and $4x - 2y = 21$ are perpendicular.

Solution We solve each equation for y to see whether the product of the slopes of their straight-line graphs is -1.

$$\begin{aligned} 4x + 8y &= 10 \\ 8y &= -4x + 10 \\ y &= \frac{-4x}{8} + \frac{10}{8} \\ y &= -\frac{1}{2}x + \frac{5}{4} \end{aligned} \qquad \begin{aligned} 4x - 2y &= 21 \\ -2y &= -4x + 21 \\ y &= \frac{-4x}{-2} + \frac{21}{-2} \\ y &= 2x - \frac{21}{2} \end{aligned}$$

Since the product of their slopes $\left(-\frac{1}{2} \text{ and } 2\right)$ is -1, the lines are perpendicular.

Self Check Are the lines represented by $3x + 2y = 7$ and $y = \frac{2}{3}x + 3$ perpendicular? ■

EXAMPLE 7 Write the equation of the line passing through $P(-2, 5)$ and parallel to the line $y = 8x - 3$.

Solution The slope of the line given by $y = 8x - 3$ is 8, the coefficient of x. Since the graph of the desired equation is to be parallel to the graph of $y = 8x - 3$, its slope must also be 8.

We substitute -2 for x_1, 5 for y_1, and 8 for m in the point–slope form and simplify.

$$y - y_1 = m(x - x_1)$$

$$y - 5 = 8[x - (-2)] \quad \text{Substitute 5 for } y_1, \text{ 8 for } m, \text{ and } -2 \text{ for } x_1.$$

$$y - 5 = 8(x + 2) \quad -(-2) = 2.$$

$$y - 5 = 8x + 16 \quad \text{Use the distributive property to remove parentheses.}$$

$$y = 8x + 21 \quad \text{Add 5 to both sides.}$$

The equation of the desired line is $y = 8x + 21$.

Self Check Write the equation of the line passing through $Q(1, 2)$ and parallel to the line $y = 8x - 3$. ■

ACTIVE EXAMPLE 8 Write the equation of the line passing through $P(-2, 5)$ and perpendicular to the line $y = 8x - 3$.

Solution Because the slope of the given line is 8, the slope of the desired perpendicular line must be $-\frac{1}{8}$.

We substitute -2 for x_1, 5 for y_1, and $-\frac{1}{8}$ for m into the point–slope form and simplify.

COLLEGE **Algebra f(x) Now™**
Go to **http://1pass.thomson.com** or your CD to practice this example.

$$y - y_1 = m(x - x_1)$$

$$y - 5 = -\frac{1}{8}[x - (-2)] \quad \text{Substitute 5 for } y_1, -\tfrac{1}{8} \text{ for } m, \text{ and } -2 \text{ for } x_1.$$

$$y - 5 = -\frac{1}{8}(x + 2) \quad -(-2) = 2.$$

$$y = -\frac{1}{8}x - \frac{1}{4} + 5 \quad \text{Remove parentheses and add 5 to both sides.}$$

$$y = -\frac{1}{8}x + \frac{19}{4} \quad -\tfrac{1}{4} + 5 = -\tfrac{1}{4} + \tfrac{20}{4} = \tfrac{19}{4}.$$

The equation of the line is $y = -\frac{1}{8}x + \frac{19}{4}$.

Self Check Write the equation of the line passing through $Q(1, 2)$ and perpendicular to $y = 8x - 3$. ■

■ General Form of the Equation of a Line

We have shown that the graph of any equation of the form $y = mx + b$ is a line with slope m and y-intercept $(0, b)$. The graph of any equation of the form

$Ax + By = C$ (where A and B are not *both* zero) is also a line. To see why, we look at three possibilities.

- If $A \neq 0$ and $B \neq 0$, the equation $Ax + By = C$ can be written in slope–intercept form.

$$\begin{aligned} Ax + By &= C \\ By &= -Ax + C && \text{Subtract } Ax \text{ from both sides.} \\ y &= -\frac{A}{B}x + \frac{C}{B} && \text{Divide both sides by } B. \end{aligned}$$

  This is the equation of a line with slope $-\frac{A}{B}$ and y-intercept $\left(0, \frac{C}{B}\right)$.

- If $A = 0$ and $B \neq 0$, the equation $Ax + By = C$ can be written in the form $y = \frac{C}{B}$. This is the equation of a horizontal line with y-intercept $\left(0, \frac{C}{B}\right)$.
- If $A \neq 0$ and $B = 0$, the equation $Ax + By = C$ can be written in the form $x = \frac{C}{A}$. This is the equation of a vertical line with x-intercept at $\left(\frac{C}{A}, 0\right)$.

Recall that $Ax + By = C$ is called the **general form of the equation of a line.**

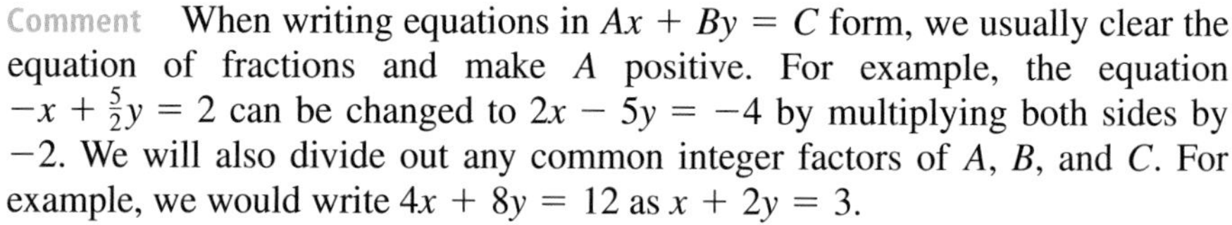

Comment When writing equations in $Ax + By = C$ form, we usually clear the equation of fractions and make A positive. For example, the equation $-x + \frac{5}{2}y = 2$ can be changed to $2x - 5y = -4$ by multiplying both sides by -2. We will also divide out any common integer factors of A, B, and C. For example, we would write $4x + 8y = 12$ as $x + 2y = 3$.

General Form of the Equation of a Line

If A, B, and C are real numbers and $B \neq 0$, the graph of

$$Ax + By = C$$

is a nonvertical line with slope of $-\frac{A}{B}$ and a y-intercept of $\left(0, \frac{C}{B}\right)$.

If $B = 0$, the graph is a vertical line with x-intercept of $\left(\frac{C}{A}, 0\right)$.

EXAMPLE 9 Find the slope and the y-intercept of the graph of $3x - 2y = 5$.

Solution The equation $3x - 2y = 5$ is in general form, with $A = 3$, $B = -2$, and $C = 5$. By the previous theorem, the slope of the graph is

$$m = -\frac{A}{B} = -\frac{3}{-2} = \frac{3}{2}$$

and the y-intercept is

$$\left(0, \frac{C}{B}\right) = \left(0, \frac{5}{-2}\right)$$

The slope is $\frac{3}{2}$, and the y-intercept is $\left(0, -\frac{5}{2}\right)$.

Self Check Find the slope and the y-intercept of the graph of $3x - 4y = 12$. ■

We summarize the various forms of the equation of a line as follows.

General form	$Ax + By = C$ A and B cannot both be 0.
Slope–intercept form	$y = mx + b$ The slope is m, and the y-intercept is $(0, b)$.
Point–slope form	$y - y_1 = m(x - x_1)$ The slope is m, and the line passes through (x_1, y_1).
A horizontal line	$y = b$ The slope is 0, and the y-intercept is $(0, b)$.
A vertical line	$x = a$ There is no defined slope, and the x-intercept is $(a, 0)$.

Straight-Line Depreciation

For tax purposes, many businesses use *straight-line depreciation* to find the declining value of aging equipment.

EXAMPLE 10 A machine shop buys a lathe for \$1,970 and expects it to last for ten years. It can then be sold as scrap for a *salvage value* of \$270. If y is the value of the lathe after x years of use, and y and x are related by the equation of a line, **a.** Find the equation of the line. **b.** Find the value of the lathe after $2\frac{1}{2}$ years. **c.** Find the economic meaning of the y-intercept of the line. **d.** Find the economic meaning of the slope of the line.

Solution **a.** We find the slope and use the point–slope form to find the equation of the line. (See Figure 2-32.)

When the lathe is new, its age x is 0, and its value y is \$1,970. When the lathe is 10 years old, $x = 10$ and $y = \$270$. Since the line passes through the points (0, 1,970) and (10, 270), the slope of the line is

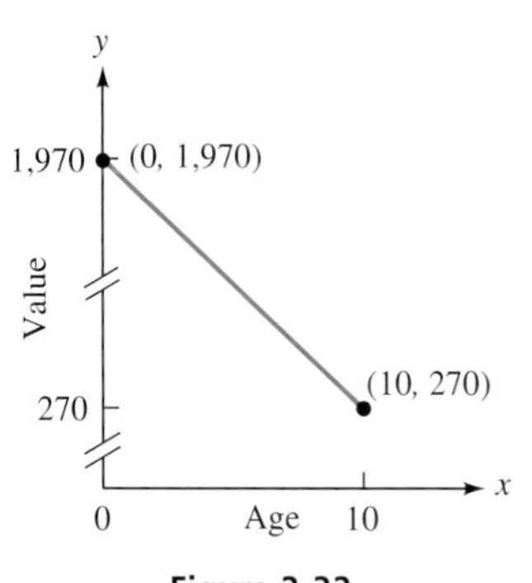

Figure 2-32

$$\begin{aligned} m &= \frac{y_2 - y_1}{x_2 - x_1} \\ &= \frac{\mathbf{270 - 1{,}970}}{\mathbf{10 - 0}} \quad \text{Substitute 270 for } y_2\text{, 1,970 for } y_1\text{, 10 for } x_2\text{, and 0 for } x_1. \\ &= \frac{-1{,}700}{10} \\ &= -170 \end{aligned}$$

To find the equation of the line, we substitute -170 for m, 0 for x_1, and 1,970 for y_1 in the point–slope form and simplify.

$$\begin{aligned} y - y_1 &= m(x - x_1) \\ y - \mathbf{1{,}970} &= \mathbf{-170}(x - \mathbf{0}) \\ y &= -170x + 1{,}970 \quad (3) \end{aligned}$$

The value y of the lathe is related to its age x by the equation $y = -170x + 1{,}970$.

b. To find the value after $2\frac{1}{2}$ years, we substitute 2.5 for x in Equation 3 and solve for y.

$$\begin{aligned} y &= -170x + 1{,}970 \\ &= -170(2.5) + 1{,}970 && \text{Substitute 2.5 for } x. \\ &= -425 + 1{,}970 \\ &= 1{,}545 \end{aligned}$$

In $2\frac{1}{2}$ years, the lathe will be worth \$1,545.

c. The y-intercept of the graph is $(0, b)$, where b is the value of y when $x = 0$.

$$\begin{aligned} y &= -170x + 1{,}970 \\ y &= -170(0) + 1{,}970 && \text{Substitute 0 for } x. \\ y &= 1{,}970 \end{aligned}$$

The y-coordinate b of the y-intercept is the value of a 0-year-old lathe, which is the lathe's original cost.

d. Each year, the value decreases by \$170, because the slope of the line is -170. The slope of the depreciation line is the *annual depreciation rate.* ■

Linear Curve Fitting

In statistics, the process of using one variable to predict another is called **regression.** For example, if we know a woman's height, we can make a good prediction about her weight, because taller women usually weigh more than shorter women.

Figure 2-33 shows the result of sampling ten women and finding their heights and weights. The graph of the ordered pairs (h, w) is called a **scattergram.**

Woman	Height (h) in inches	Weight (w) in pounds
1	60	100
2	61	105
3	62	120
4	62	130
5	63	135
6	64	120
7	64	125
8	65	155
9	67	155
10	69	160

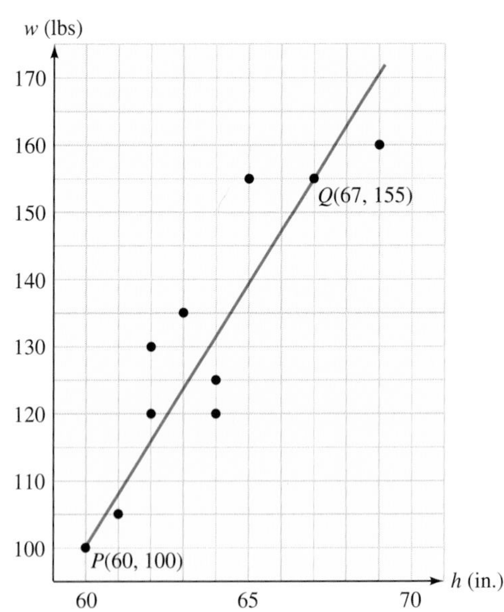

Figure 2-33

To write a **prediction equation** (sometimes called a **regression equation**), we must find the equation of the line that comes closer to all of the points in the scattergram than any other possible line. There are statistical methods to find this equation, but we can only approximate it here.

To write an approximation of the regression equation, we place a straightedge on the scattergram shown in Figure 2-33 and draw the line joining two points that seems to best fit all the points. In the figure, line PQ is drawn, where point P has coordinates of (60, 100) and point Q has coordinates of (67, 155).

Our approximation of the regression equation will be the equation of the line passing through points P and Q. To find the equation of this line, we first find its slope.

$$\begin{aligned} m &= \frac{y_2 - y_1}{x_2 - x_1} \\ &= \frac{\mathbf{155} - \mathbf{100}}{\mathbf{67} - \mathbf{60}} \\ &= \frac{55}{7} \end{aligned}$$

We can then use point–slope form to find the equation of the line.

$$y - y_1 = m(x - x_1)$$

$$y - \mathbf{100} = \frac{\mathbf{55}}{\mathbf{7}}(x - \mathbf{60}) \quad \text{Choose (60, 100) for } (x_1, y_1)$$

$$y = \frac{55}{7}x - \frac{3{,}300}{7} + 100 \quad \text{Remove parentheses and add 100 to both sides.}$$

$$(4) \qquad y = \frac{55}{7}x - \frac{2{,}600}{7} \quad \text{Simplify.}$$

Our approximation of the regression equation is $y = \frac{55}{7}x - \frac{2{,}600}{7}$.

To predict the weight of a woman who is 66 inches tall, for example, we substitute 66 for x in Equation 4 and simplify.

$$\begin{aligned} y &= \frac{55}{7}x - \frac{2{,}600}{7} \\ y &= \frac{55}{7}(66) - \frac{2{,}600}{7} \\ y &\approx 147.1428571 \end{aligned}$$

We would predict that a 66-inch-tall woman chosen at random will weigh about 147 pounds.

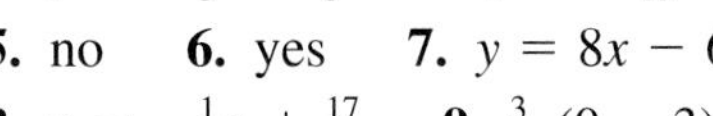

1. $y = -\frac{2}{3}x + \frac{7}{3}$ **2.** $y = -\frac{10}{13}x + \frac{2}{13}$ **3.** $y = \frac{7}{3}x - 6$ **4.** $-\frac{2}{3}$, $(0, -3)$

5. no **6.** yes **7.** $y = 8x - 6$

8. $y = -\frac{1}{8}x + \frac{17}{8}$ **9.** $\frac{3}{4}$, $(0, -3)$

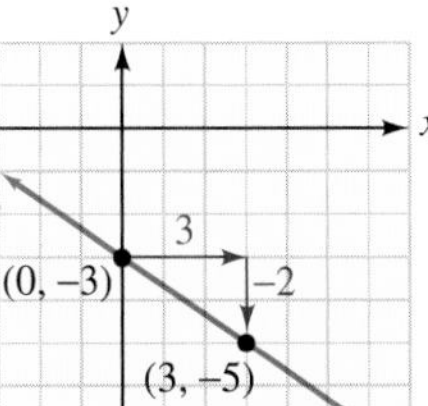

2.3 Exercises

VOCABULARY AND CONCEPTS ***Fill in the blanks.***

1. The formula for the point–slope form of a line is $\underline{y - y_1 = m(x - x_1)}$.
2. In the equation $y = mx + b$, $\underline{m}$ is the slope of the graph of the line, and $(0, b)$ is the $\underline{\text{y-intercept}}$.
3. The equation $y = mx + b$ is called the $\underline{\text{slope–intercept}}$ form of the equation of a line.
4. The general form of the equation of a line is $\underline{Ax + By = C}$.
5. The slope of the graph of $Ax + By = C$ is $\underline{-\frac{A}{B}}$.
6. The y-intercept of the graph of $Ax + By = C$ is $\underline{\left(0, \frac{C}{B}\right)}$.

PRACTICE ***Use point–slope form to write the equation of the line with the given properties. Write each equation in general form.***

7. $m = 2$, passing through $P(2, 4)$ $2x - y = 0$
8. $m = -3$, passing through $P(3, 5)$ $3x + y = 14$
9. $m = 2$, passing through $P\left(-\frac{3}{2}, \frac{1}{2}\right)$ $4x - 2y = -7$
10. $m = -6$, passing through $P\left(\frac{1}{4}, -2\right)$ $12x + 2y = -1$
11. $m = \pi$, passing through $P(\pi, 0)$ $\pi x - y = \pi^2$
12. $m = \pi$, passing through $P(0, \pi)$ $\pi x - y = -\pi$

Use point–slope form to write the equation of each line. Write the equation in general form.

13.

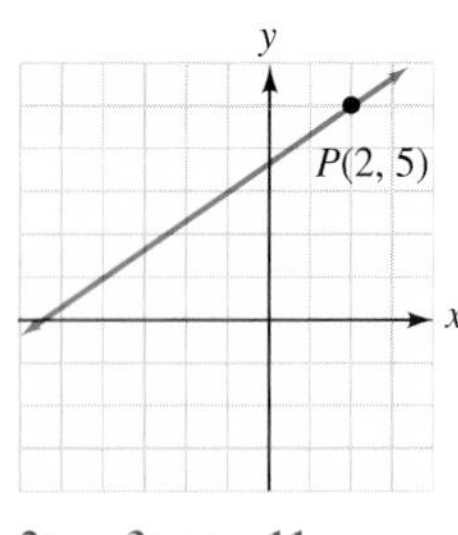

$2x - 3y = -11$

14.

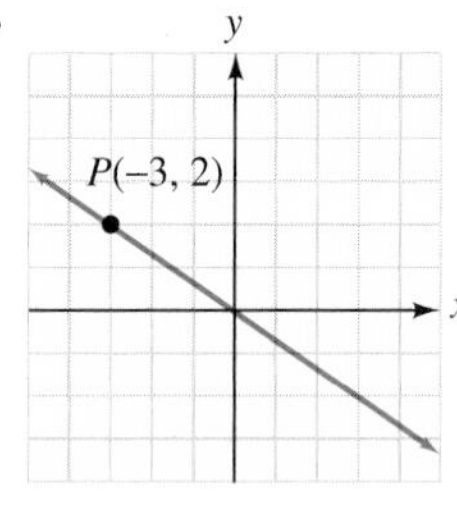

$2x + 3y = 0$

Use point–slope form to write the equation of the line passing through the two given points. Write each equation in slope–intercept form.

15. $P(0, 0)$, $Q(4, 4)$
 $y = x$
16. $P(-5, -5)$, $Q(0, 0)$
 $y = x$
17. $P(3, 4)$, $Q(0, -3)$
 $y = \frac{7}{3}x - 3$
18. $P(4, 0)$, $Q(6, -8)$
 $y = -4x + 16$

Use point–slope form to write the equation of each line. Write each answer in slope–intercept form.

19.

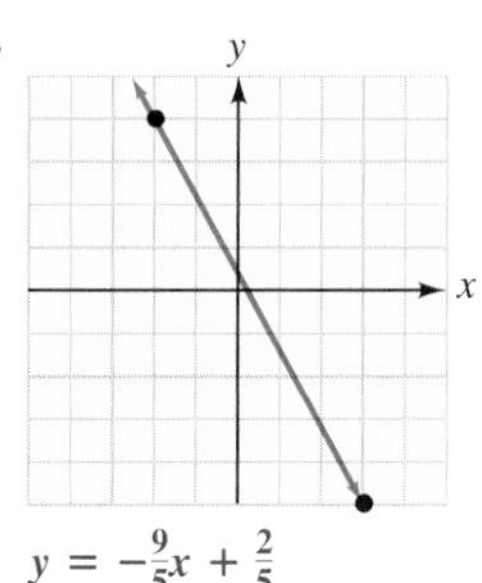

$y = -\frac{9}{5}x + \frac{2}{5}$

20.

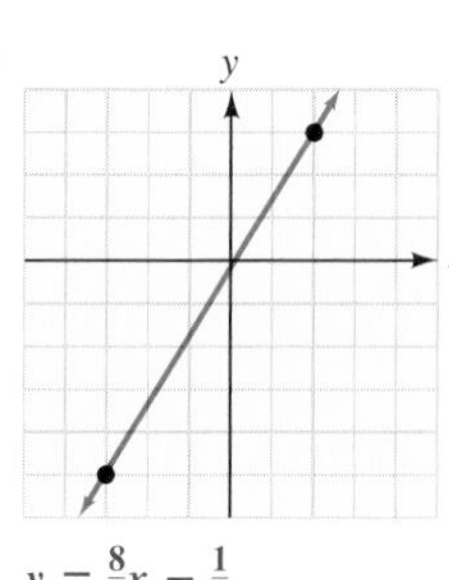

$y = \frac{8}{5}x - \frac{1}{5}$

Use slope–intercept form to write the equation of the line with the given properties. Write each equation in slope–intercept form.

21. $m = 3; b = -2$
 $y = 3x - 2$
22. $m = -\frac{1}{3}; b = \frac{2}{3}$
 $y = -\frac{1}{3}x + \frac{2}{3}$
23. $m = 5; b = -\frac{1}{5}$
 $y = 5x - \frac{1}{5}$
24. $m = \sqrt{2}; b = \sqrt{2}$
 $y = \sqrt{2}x + \sqrt{2}$
25. $m = a; b = \frac{1}{a}$
 $y = ax + \frac{1}{a}$
26. $m = a; b = 2a$
 $y = ax + 2a$
27. $m = a; b = a$
 $y = ax + a$
28. $m = \frac{1}{a}; b = a$
 $y = \frac{1}{a}x + a$

Use slope–intercept form to write the equation of a line passing through the given point and having the given slope. Express the answer in general form.

29. $P(0, 0); m = \frac{3}{2}$
 $3x - 2y = 0$
30. $P(-3, -7); m = -\frac{2}{3}$
 $2x + 3y = -27$
31. $P(-3, 5); m = -3$
 $3x + y = -4$
32. $P(-5, 1); m = 1$
 $x - y = -6$
33. $P\left(0, \sqrt{2}\right); m = \sqrt{2}$ $\sqrt{2}x - y = -\sqrt{2}$
34. $P\left(-\sqrt{3}, 0\right); m = 2\sqrt{3}$ $2\sqrt{3}x - y = -6$

Write each equation in slope–intercept form to find the slope and the y-intercept. Then use the slope and y-intercept to draw the line.

35. $y + 1 = x$
1, (0, −1)

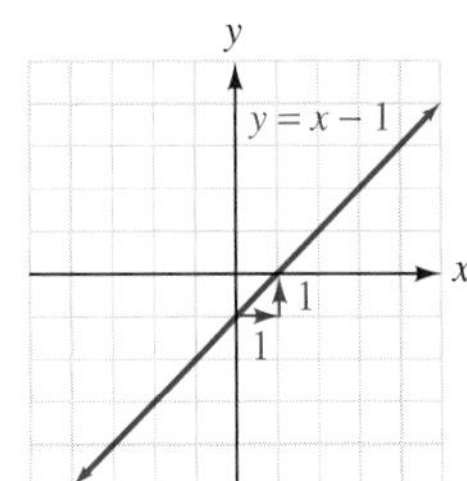

36. $x + y = 2$
−1, (0, 2)

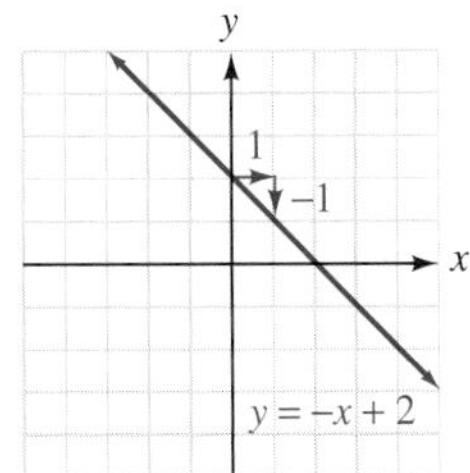

37. $x = \frac{3}{2}y - 3$
$\frac{2}{3}$, **(0, 2)**

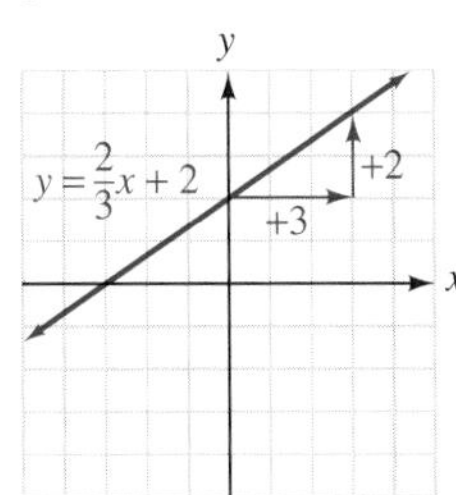

38. $x = -\frac{4}{5}y + 2$
$-\frac{5}{4}$, $\left(0, \frac{5}{2}\right)$

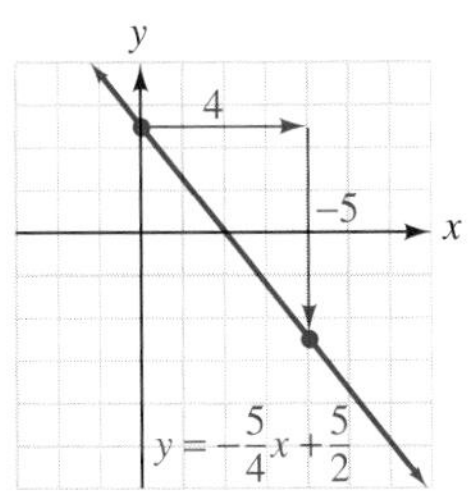

39. $3(y - 4) = -2(x - 3)$ $-\frac{2}{3}$, **(0, 6)**

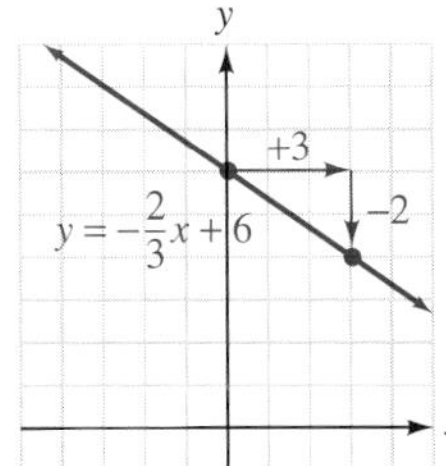

40. $-4(2x + 3) = 3(3y + 8)$ $-\frac{8}{9}$, **(0, −4)**

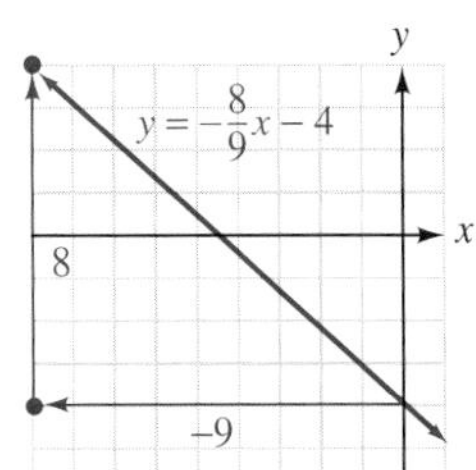

Find the slope and the y-intercept of the line determined by the given equation.

41. $3x - 2y = 8$
$\frac{3}{2}$, **(0, −4)**

42. $-2x + 4y = 12$
$\frac{1}{2}$, **(0, 3)**

43. $-2(x + 3y) = 5$
$-\frac{1}{3}$, $\left(0, -\frac{5}{6}\right)$

44. $5(2x - 3y) = 4$
$\frac{2}{3}$, $\left(0, -\frac{4}{15}\right)$

45. $x = \frac{2y - 4}{7}$
$\frac{7}{2}$, **(0, 2)**

46. $3x + 4 = -\frac{2(y - 3)}{5}$
$-\frac{15}{2}$, **(0, −7)**

Determine whether the graphs of each pair of equations are parallel, perpendicular, or neither.

47. $y = 3x + 4, y = 3x - 7$ **parallel**

48. $y = 4x - 13, y = \frac{1}{4}x + 13$ **neither**

49. $x + y = 2, y = x + 5$ **perpendicular**

50. $x = y + 2, y = x + 3$ **parallel**

51. $y = 3x + 7, 2y = 6x - 9$ **parallel**

52. $2x + 3y = 9, 3x - 2y = 5$ **perpendicular**

53. $x = 3y + 4, y = -3x + 7$ **perpendicular**

54. $3x + 6y = 1, y = \frac{1}{2}x$ **neither**

55. $y = 3, x = 4$ **perpendicular**

56. $y = -3, y = -7$ **parallel**

57. $x = \frac{y - 2}{3}, 3(y - 3) + x = 0$ **perpendicular**

58. $2y = 8, 3(2 + x) = 3(y + 2)$ **neither**

Write the equation of the line that passes through the given point and is parallel to the given line. Write the answer in slope–intercept form.

59. $P(0, 0), y = 4x - 7$ $y = 4x$

60. $P(0, 0), x = -3y - 12$ $y = -\frac{1}{3}x$

61. $P(2, 5), 4x - y = 7$ $y = 4x - 3$

62. $P(-6, 3), y + 3x = -12$ $y = -3x - 15$

63. $P(4, -2), x = \frac{5}{4}y - 2$ $y = \frac{4}{5}x - \frac{26}{5}$

64. $P(1, -5), x = -\frac{3}{4}y + 5$ $y = -\frac{4}{3}x - \frac{11}{3}$

Write the equation of the line that passes through the given point and is perpendicular to the given line. Write the answer in slope–intercept form.

65. $P(0, 0), y = 4x - 7$ $y = -\frac{1}{4}x$

66. $P(0, 0), x = -3y - 12$ $y = 3x$

67. $P(2, 5), 4x - y = 7$ $y = -\frac{1}{4}x + \frac{11}{2}$

68. $P(-6, 3), y + 3x = -12$ $y = \frac{1}{3}x + 5$

69. $P(4, -2), x = \frac{5}{4}y - 2$ $y = -\frac{5}{4}x + 3$

70. $P(1, -5), x = -\frac{3}{4}y + 5$ $y = \frac{3}{4}x - \frac{23}{4}$

Use the method of Example 9 to find the slope and the y-intercept of the graph of each equation.

71. $4x + 5y = 20$
$m = -\frac{4}{5}$; $(0, 4)$

72. $9x - 12y = 17$
$m = \frac{3}{4}$; $\left(0, -\frac{17}{12}\right)$

73. $2x + 3y = 12$
$m = -\frac{2}{3}$; $(0, 4)$

74. $5x + 6y = 30$
$m = -\frac{5}{6}$; $(0, 5)$

75. Find the equation of the line perpendicular to the line $y = 3$ and passing through the midpoint of the segment joining $(2, 4)$ and $(-6, 10)$. $x = -2$

76. Find the equation of the line parallel to the line $y = -8$ and passing through the midpoint of the segment joining $(-4, 2)$ and $(-2, 8)$. $y = 5$

77. Find the equation of the line parallel to the line $x = 3$ and passing through the midpoint of the segment joining $(2, -4)$ and $(8, 12)$. $x = 5$

78. Find the equation of the line perpendicular to the line $x = 3$ and passing through the midpoint of the segment joining $(-2, 2)$ and $(4, -8)$. $y = -3$

APPLICATIONS ***In Exercises 79–89, assume straight-line depreciation or straight-line appreciation.***

79. Depreciation A taxicab was purchased for $24,300. Its salvage value at the end of its 7-year useful life is expected to be $1,900. Find the depreciation equation. $y = -3{,}200x + 24{,}300$

80. Depreciation A small business purchases the laptop computer shown. It will be depreciated over a 4-year period, when its salvage value will be $300. Find the depreciation equation. $y = -600x + 2{,}700$

81. Appreciation An apartment building was purchased for $475,000, excluding the cost of land. The owners expect the property to double in value in 10 years. Find the appreciation equation.
$y = 47{,}500x + 475{,}000$

82. Appreciation A house purchased for $112,000 is expected to double in value in 12 years. Find its appreciation equation. $y = \frac{28{,}000}{3}x + 112{,}000$

83. Depreciation Find the depreciation equation for the TV in the following want ad. $y = -\frac{710}{3}x + 1{,}900$

For Sale: 3-year-old 54-inch TV, $1,900 new. Asking $1,190. Call 875-5555. Ask for Mike.

84. Depreciation A word processor cost $555 when new and is expected to be worth $80 after 5 years. What will it be worth after 3 years? $270

85. Salvage value A copier cost $1,050 when new and will be depreciated at the rate of $120 per year. If the useful life of the copier is 8 years, find its salvage value. $90

86. Rate of depreciation A truck that cost $27,600 when new will have no salvage value after 12 years. Find its annual rate of depreciation. $2,300

87. Value of an antique An antique table is expected to appreciate $40 each year. If the table will be worth $450 in 2 years, what will it be worth in 13 years? $890

88. Value of an antique An antique clock is expected to be worth $350 after 2 years and $530 after 5 years. What will the clock be worth after 7 years? $650

89. Purchase price of real estate A cottage that was purchased 3 years ago is now appraised at $47,700. If the property has been appreciating $3,500 per year, find its original purchase price. $37,200

90. Computer repair A computer repair company charges a fixed amount, plus an hourly rate, for a service call. Use the information in the illustration to find the hourly rate. $17.50

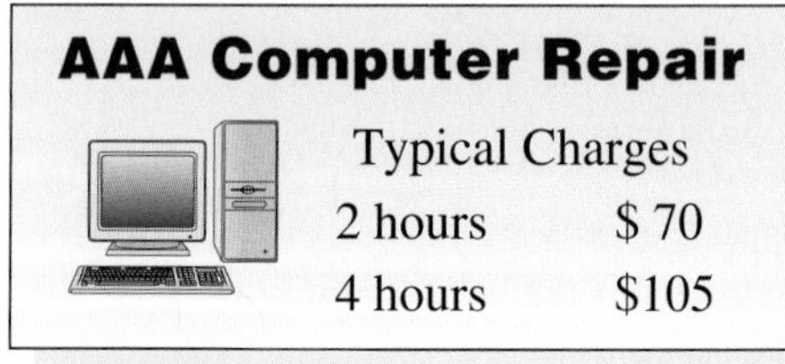

91. Automobile repair An auto repair shop charges an hourly rate, plus the cost of parts. If the cost of labor for a $1\frac{1}{2}$-hour radiator repair is $69, find the cost of labor for a 5-hour transmission overhaul. $230

92. Printer charges A printer charges a fixed setup cost, plus \$1 for every 100 copies. If 700 copies cost \$52, how much will it cost to print 1,000 copies? **\$55**

93. Predicting fires A local fire department recognizes that city growth and the number of reported fires are related by a linear equation. City records show that 300 fires were reported in a year when the local population was 57,000 persons, and 325 fires were reported in a year when the population was 59,000 persons. How many fires can be expected in the year when the population reaches 100,000 persons? **about 838**

94. Estimating the cost of rain gutter A neighbor tells you that an installer of rain gutter charges \$60, plus a dollar amount per foot. If the neighbor paid \$435 for the installation of 250 feet of gutter, how much will it cost you to have 300 feet installed? **\$510**

95. Converting temperatures Water freezes at 32° Fahrenheit, or 0° Celsius. Water boils at 212° F, or 100° C. Find a formula for converting a temperature from degrees Fahrenheit to degrees Celsius. $C = \frac{5}{9}(F - 32)$

96. Converting units A speed of 1 mile per hour is equal to 88 feet per minute, and of course, 0 miles per hour is 0 feet per minute. Find an equation for converting a speed x, in miles per hour, to the corresponding speed y, in feet per minute. $y = 88x$

97. Smoking The percent y of 18-to-25-year-old smokers in the United States has been declining at a constant rate since 1974. If about 47% of this group smoked in 1974 and about 29% smoked in 1994, find a linear equation that models this decline. If this trend continues, estimate what percent will smoke in 2014. $y = -\frac{9}{10}x + 47$; **11%**

98. Forensic science Scientists believe there is a linear relationship between the height h (in centimeters) of a male and the length f (in centimeters) of his femur bone. Use the data in the table to find a linear equation that expresses the height h in terms of f. Round all constants to the nearest thousandth. How tall would you expect a man to be if his femur measures 50 cm? Round to the nearest centimeter.

Person	Length of femur (f)	Height (h)
A	62.5 cm	200 cm
B	40.2 cm	150 cm

$h = 2.242f + 59.875$; **about 172 cm**

99. Predicting stock prices The value of the stock of ABC Corporation has been increasing by the same fixed dollar amount each year. The pattern is expected to continue. Let 2001 be the base year corresponding to $x = 0$, with $x = 1, 2, 3, \ldots$ corresponding to later years. ABC stock was selling at $\$37\frac{1}{2}$ in 2001 and at \$45 in 2003. If y represents the price of ABC stock, find the equation $y = mx + b$ that relates x and y, and predict the price in the year 2005. $y = 3.75x + 37.5$; $\$52\frac{1}{2}$

100. Estimating inventory Inventory of unsold goods showed a surplus of 375 units in January and 264 in April. Assume that the relationship between inventory and time is given by the equation of a line, and estimate the expected inventory in March. Because March lies between January and April, this estimation is called **interpolation.** **301**

101. Oil depletion When a Petroland oil well was first brought on line, it produced 1,900 barrels of crude oil per day. In each later year, owners expect its daily production to drop by 70 barrels. Find the daily production after $3\frac{1}{2}$ years. **1,655 barrels per day**

102. Waste management The corrosive waste in industrial sewage limits the useful life of the piping in a waste processing plant to 12 years. The piping system was originally worth \$137,000, and it will cost the company \$33,000 to remove it at the end of its 12-year useful life. Find the depreciation equation. $y = 137{,}000 - \frac{42{,}500}{3}x$

103. Crickets The table shows the approximate chirping rate at various temperatures for one type of cricket.

Temperature (°F)	Chirps per minute
50	20
60	80
70	115
80	150
100	250

a. Construct a scattergram shown on the next page.

b. Assume a linear relationship and write a regression equation. $y = \frac{23}{5}x - 210$ **(answers may vary)**

c. Estimate the chirping rate at a temperature of 90° F. **204 (answers may vary)**

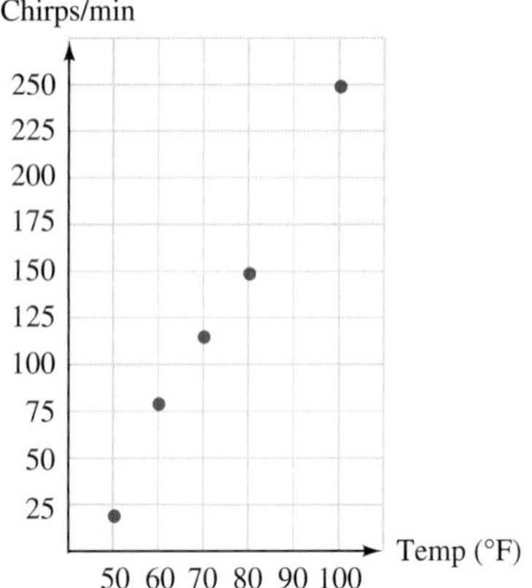

104. Fishing The table shows the lengths and weights of seven muskies captured by the Department of Natural Resourses in Catfish Lake in Eagle River, Wisconsin.

Musky	Length (in.)	Weight (lb)
1	26	5
2	27	8
3	29	9
4	33	12
5	35	14
6	36	14
7	38	19

a. Construct a scattergram for the data.
b. Assume a linear relationship and write a regression equation. $y = \frac{7}{6}x - \frac{76}{3}$ (answers may vary)
c. Estimate the weight of a musky that is 32 inches long. 12 lb (answers may vary)

DISCOVERY AND WRITING

105. Explain how to find the equation of a line passing through two given points.

106. In straight-line depreciation, explain why the slope of the line is called the *rate of depreciation.*

107. Prove that the equation of a line with x-intercept of $(a, 0)$ and y-intercept of $(0, b)$ can be written in the form

$$\frac{x}{a} + \frac{y}{b} = 1$$

108. Find the x- and y-intercepts of the line $bx + ay = ab$. $(a, 0), (0, b)$

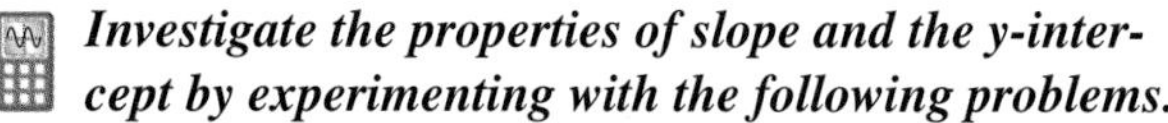
Investigate the properties of slope and the y-intercept by experimenting with the following problems.

109. Graph $y = mx + 2$ for several positive values of m. What do you notice?

110. Graph $y = mx + 2$ for several negative values of m. What do you notice?

111. Graph $y = 2x + b$ for several increasing positive values of b. What do you notice?

112. Graph $y = 2x + b$ for several decreasing negative values of b. What do you notice?

113. How will the graph of $y = \frac{1}{2}x + 5$ compare to the graph of $y = \frac{1}{2}x - 5$?

114. How will the graph of $y = \frac{1}{2}x - 5$ compare to the graph of $y = \frac{1}{2}x$?

Much of the hard work done in problems such as 103–104 can be done automatically by using a computer program, such as Excel.

115. The following table shows the length of a femur bone and the height for ten Caucasian males.

Length of femur bone (cm) x	Height (cm) y
49.3	180.8
47.4	176.7
47.1	176.1
48.5	176.5
45.2	170.6
47.8	176.8
49.4	178.6
49.6	179.6
50.9	185.2
47.8	176.2

a. Enter the data in an Excel Spreadsheet. Enter the femur values in Column A and the height values in Column B.
b. To plot the data, click and drag to highlight the two columns, select Insert on the menu bar, and then select Chart. When a window pops up, select XY (Scatter) as the chart type. When you click Finish, a **scattergram** will appear on the spreadsheet.
c. Observe that the points are approximately linear. To draw the regression line, select Chart on the menu bar and select Add Trendline. When the dialog box appears, be sure that the Linear Regression type is selected, and then click OK. The regression line will appear.

d. To find the equation of the regression line, estimate the coordinates of two points on the regression line. Then use the two points to write the equation of the line.
e. Use the equation of the regression line to predict the height of a male whose femur bone measures 46 cm.

116. The following table shows the personal income (per person, including the unemployed) and the personal outlays (per person, including the unemployed) for people living in the United States for 1994–2003.

Year	Per Capita Personal Income ($) x	Per Capita Personal Outlays ($) y
1994	5,842.5	4,902.4
1995	6,152.3	5,157.3
1996	6,520.6	5,460.0
1997	6,915.1	5,770.5
1998	7,423.0	6,119.1
1999	7,802.4	6,536.4
2000	8,429.7	7,025.6
2001	8,724.1	7,354.5
2002	8,878.9	7,668.5
2003	9,161.8	8,049.3

Source: Bureau of Economic Analysis

a. Use Excel to make a scattergram of the given data. See problem 115.
b. Use Excel to draw the regression line.
c. Write the equation of the regression line.
d. Interpret the slope of the regression line.
e. How much would you expect a person to spend if he/she had a personal income of $9,000?

REVIEW ***Simplify each expression.***

117. $x^7x^3x^{-5}$ x^5

118. $\dfrac{y^3y^{-4}}{y^{-5}}$ y^4

119. $\left(\dfrac{81}{25}\right)^{-3/2}$ $\frac{125}{729}$

120. $\sqrt[3]{27x^7}$ $3x^2\sqrt[3]{x}$

121. $\sqrt{27} - 2\sqrt{12}$ $-\sqrt{3}$

122. $\dfrac{5}{\sqrt{5}}$ $\sqrt{5}$

123. $\dfrac{5}{\sqrt{x} + 2}$ $\dfrac{5(\sqrt{x} - 2)}{x - 4}$

124. $(\sqrt{x} - 2)^2$ $x - 4\sqrt{x} + 4$

2.4 Graphs of Equations

In this section, you will learn about

- Intercepts of Graphs
- Symmetries of Graphs
- Miscellaneous Graphs
- Circles
- Graphing Equations of Circles
- Solving Equations by Graphing

The graphs of many equations are curves. If we plot several points (x, y) that satisfy such an equation, the shape of the graph will usually become evident. We can then sketch the graph by joining these points with a smooth curve.

Intercepts of Graphs

In Figure 2-34(a), the ***x*-intercepts** of the graph are $(a, 0)$ and $(b, 0)$, the points where the graph intersects the x-axis. In Figure 2-34(b), the ***y*-intercept** is $(0, c)$, the point where the graph intersects the y-axis.

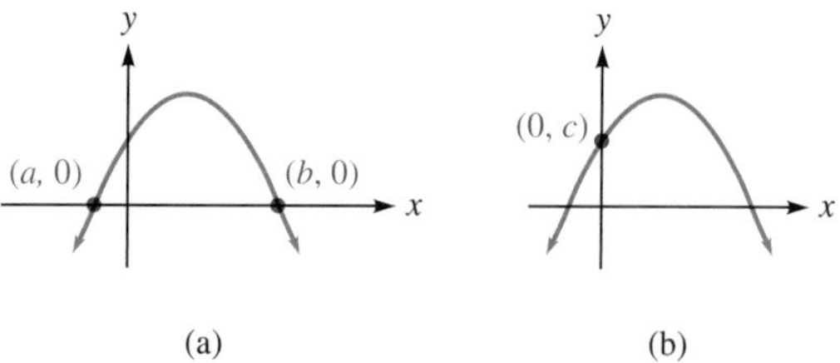

Figure 2-34

To graph the equation $y = x^2 - 4$, we first find the x- and y-intercepts. To find the x-intercepts, we let $y = 0$ and solve for x.

$$y = x^2 - 4$$

$$0 = x^2 - 4 \quad \text{Substitute 0 for } y.$$

$$0 = (x + 2)(x - 2) \quad \text{Factor } x^2 - 4.$$

$$x + 2 = 0 \quad \text{or} \quad x - 2 = 0 \quad \text{Set each factor equal to 0.}$$

$$x = -2 \quad | \quad x = 2$$

Since $y = 0$ when $x = -2$ and $x = 2$, the x-intercepts are $(-2, 0)$ and $(2, 0)$. (See Figure 2-35.)

To find the y-intercept, we let $x = 0$ and solve for y.

$$y = x^2 - 4$$

$$y = 0^2 - 4 \quad \text{Substitute 0 for } x.$$

$$y = -4$$

Since $y = -4$ when $x = 0$, the y-intercept is $(0, -4)$.

We can find other pairs (x, y) that satisfy the equation by substituting numbers for x and finding the corresponding values of y. For example, if $x = -3$, then $y = (-3)^2 - 4$, or 5, and the point $(-3, 5)$ lies on the graph.

The coordinates of the intercepts and other points appear in Figure 2-35. If we plot the points and draw a curve through them, we obtain the graph of the equation.

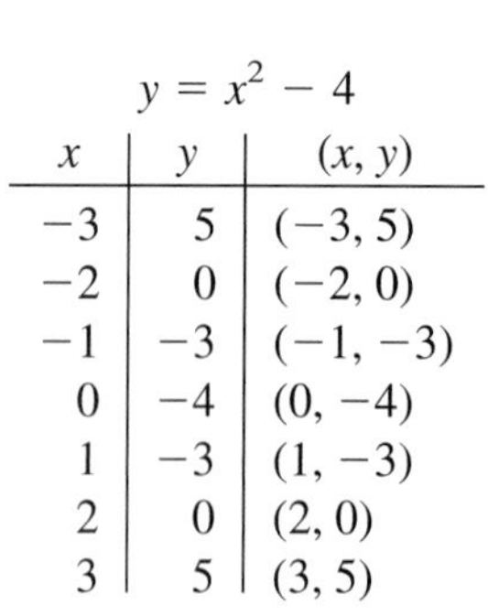

$y = x^2 - 4$

x	y	(x, y)
−3	5	(−3, 5)
−2	0	(−2, 0)
−1	−3	(−1, −3)
0	−4	(0, −4)
1	−3	(1, −3)
2	0	(2, 0)
3	5	(3, 5)

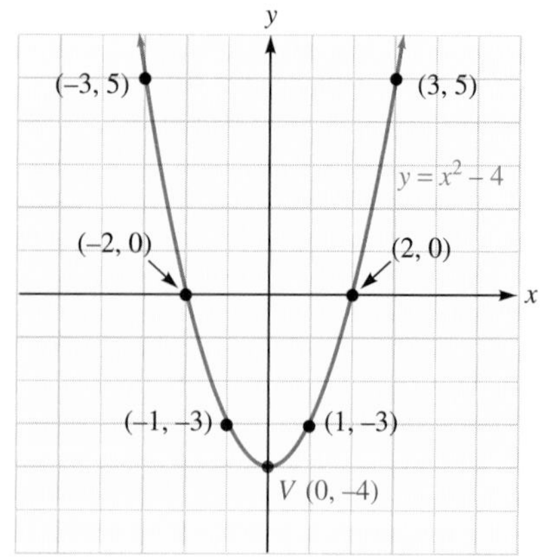

Figure 2-35

This graph is called a **parabola.** Its lowest point, $V(0, -4)$, is called the **vertex.** Because the y-axis divides the parabola into two congruent halves, it is called an **axis of symmetry.** We say that the parabola is **symmetric about the *y*-axis.**

Accent on Technology GRAPHING $y = x^2 - 4$

To use a graphing calculator to graph $y = x^2 - 4$, we enter the right-hand side of the equation after the symbol $Y_1 =$. The display will show the equation

$$Y_1 = X \wedge 2 - 4$$

If we use window settings of $[-10, 10]$ for x and $[-10, 10]$ for y and press GRAPH, we will obtain the graph shown in Figure 2-36(a). If we use settings of $[-4, 4]$ for x and $[-4, 4]$ for y, we will obtain the graph shown in Figure 2-36(b).

From the graph, we can see that the x-intercepts are $(-2, 0)$ and $(2, 0)$ and that the y-intercept is $(0, -4)$.

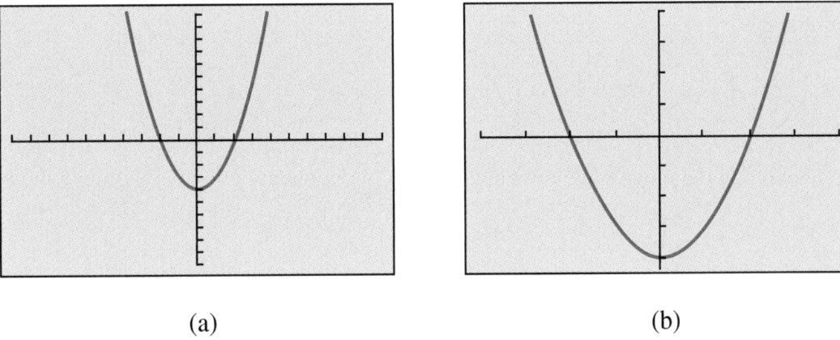

(a) (b)

Figure 2-36

Symmetries of Graphs

There are several ways in which a graph can have symmetry.

1. If the point $(-x, y)$ lies on a graph whenever the point (x, y) does, the graph is **symmetric about the y-axis.** (See Figure 2-37(a).) This implies that a graph is symmetric about the y-axis if we get the same y-coordinate when we evaluate its equation at x or at $-x$.
2. If the point $(-x, -y)$ lies on the graph whenever the point (x, y) does, the graph is **symmetric about the origin.** (See Figure 2-37(b).) This implies that the graph is symmetric about the origin if we get opposite values of y when we evaluate its equation at x or at $-x$.
3. If the point $(x, -y)$ lies on the graph whenever the point (x, y) does, the graph is **symmetric about the x-axis.** (See Figure 2-37(c).)

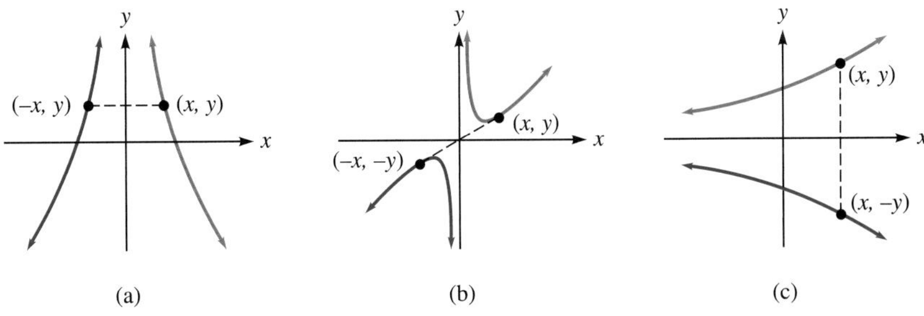

(a) (b) (c)

Figure 2-37

Tests for Symmetry for Graphs of Equations in x and y

1. To test for y-axis symmetry, replace x with $-x$. If the resulting equation is equivalent to the original, the graph is symmetric about the y-axis.
2. To test for symmetry about the origin, replace x with $-x$ and y with $-y$. If the resulting equation is equivalent to the original one, the graph is symmetric about the origin.
3. To test for x-axis symmetry, replace y with $-y$. If the resulting equation is equivalent to the original, the graph is symmetric about the x-axis.

Miscellaneous Graphs

To graph various equations, we will find the x- and y-intercepts, test for symmetries, plot points, and join the points with a smooth curve.

ACTIVE EXAMPLE 1 Graph: $y = |x|$.

Solution **x-intercepts.** To find the x-intercepts, we let $y = 0$ and solve for x.

COLLEGE **Algebra** $f(x)$ **Now**™
Explore this Active Example and test yourself on the steps taken here at **http://1pass.thomson.com**. You can also find this example on the Interactive Video Skillbuilder CD-ROM.

$$y = |x|$$
$$0 = |x| \quad \textbf{Substitute 0 for } y.$$
$$|x| = 0$$
$$x = 0$$

The x-intercept is (0, 0). (See Figure 2-38.)

y-intercepts. To find the y-intercepts, we let $x = 0$ and solve for y.

$$y = |x|$$
$$y = |0| \quad \textbf{Substitute 0 for } x.$$
$$y = 0$$

The y-intercept is (0, 0).

Symmetry. To test for y-axis symmetry, we replace x with $-x$.

(1) $y = |x|$ **The original equation.**
$y = |-x|$ **Replace x with $-x$.**
(2) $y = |x|$ $|-x| = |x|$.

Since Equations 1 and 2 are the same, the graph is symmetric about the y-axis.
To test for symmetry about the origin, we replace x with $-x$ and y with $-y$.

(1) $y = |x|$ **The original equation.**
$-y = |-x|$ **Replace x with $-x$ and y with $-y$.**
(3) $-y = |x|$ $|-x| = |x|$.

Since Equations 1 and 3 are different, the graph is not symmetric about the origin.
To test for x-axis symmetry, we replace y with $-y$.

(1) $y = |x|$ **The original equation.**
(4) $-y = |x|$ **Replace y with $-y$.**

Since Equations 1 and 4 are different, the graph is not symmetric about the x-axis.

To graph the equation, we plot the x- and y-intercepts and several other pairs (x, y) with positive values of x. We can use the property of y-axis symmetry to draw the graph for negative values of x. See Figure 2-38.

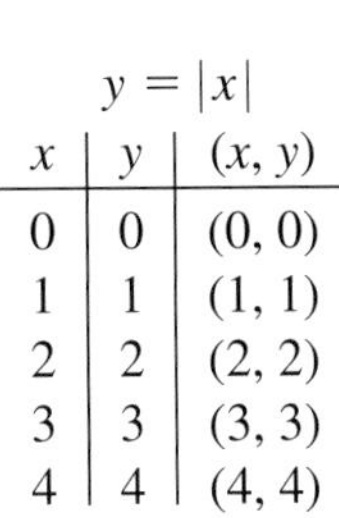
$y = |x|$

x	y	(x, y)
0	0	(0, 0)
1	1	(1, 1)
2	2	(2, 2)
3	3	(3, 3)
4	4	(4, 4)

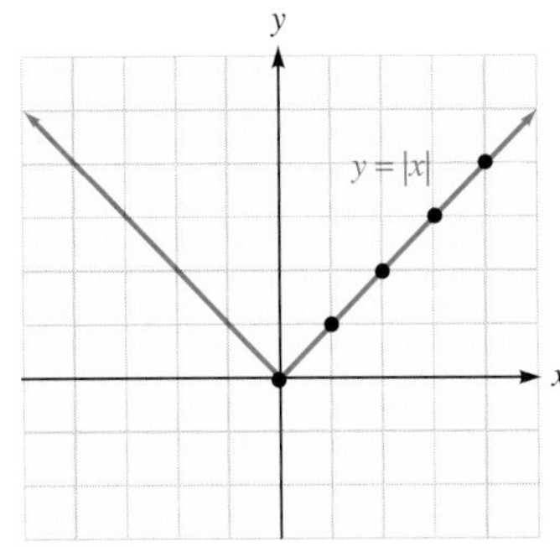

Figure 2-38

Self Check Graph: $y = -|x|$. ■

Accent on Technology GRAPHING $y = |0.5x|$

To use a graphing calculator to graph $y = |0.5x|$, we enter the right-hand side of the equation after the symbol $Y_1 =$. The display will look like

$Y_1 = \text{abs}\ (.5X)$

If we use window settings of $[-10, 10]$ for x and $[-10, 10]$ for y and press GRAPH, we will obtain the graph shown in Figure 2-39.

From the graph, we see that there is one x- and y-intercept, the point (0, 0).

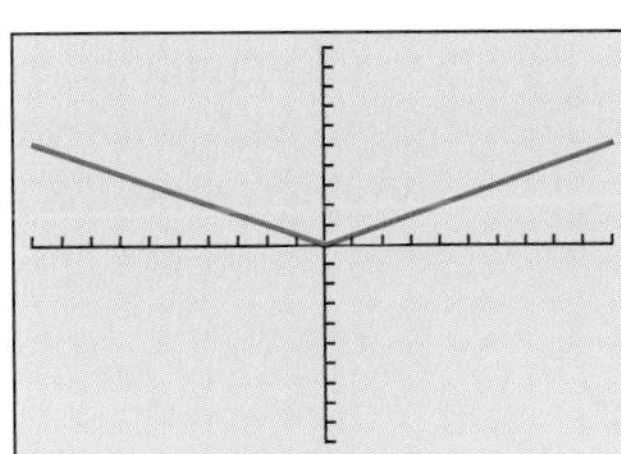

Figure 2-39

EXAMPLE 2 Graph: $y = x^3 - x$.

Solution **x-intercepts.** To find the x-intercepts, we let $y = 0$ and solve for x.

$$y = x^3 - x$$
$$0 = x^3 - x \quad \text{Substitute 0 for } y.$$
$$0 = x(x^2 - 1) \quad \text{Factor out } x.$$
$$0 = x(x + 1)(x - 1) \quad \text{Factor } x^2 - 1.$$
$$x = 0 \quad \text{or} \quad x + 1 = 0 \quad \text{or} \quad x - 1 = 0 \quad \text{Set each factor equal to 0.}$$
$$x = -1 \qquad\qquad x = 1$$

The x-intercepts are (0, 0), $(-1, 0)$, and (1, 0).

y-intercepts. To find the y-intercepts, we let $x = 0$ and solve for y.

$$y = x^3 - x$$
$$y = 0^3 - 0 \quad \text{Substitute 0 for } x.$$
$$y = 0$$

The y-intercept is $(0, 0)$.

Symmetry. To test for y-axis symmetry, we replace x with $-x$.

(1) $y = x^3 - x$ The original equation.

$y = (-x)^3 - (-x)$ Replace x with $-x$.

(2) $y = -x^3 + x$ Simplify.

Since Equations 1 and 2 are different, the graph is not symmetric about the y-axis.

To test for symmetry about the origin, we replace x with $-x$ and y with $-y$.

(1) $y = x^3 - x$ The original equation.

$-y = (-x)^3 - (-x)$ Replace x with $-x$ and y with $-y$.

$-y = -x^3 + x$ Simplify.

(3) $y = x^3 - x$ Multiply both sides by -1.

Since Equations 1 and 3 are the same, the graph is symmetric about the origin.

We test for symmetry about the x-axis by replacing y and $-y$.

(1) $y = x^3 - x$ The original equation.

$-y = x^3 - x$ Replace y with $-y$.

(4) $y = -x^3 + x$ Multiply both sides by -1.

Since Equations 1 and 4 are different, the graph is not symmetric about the x-axis.

To graph the equation, we plot the x- and y-intercepts and several other pairs (x, y) with positive values of x. We can use the property of symmetry about the origin to draw the graph for negative values of x. See Figure 2-40.

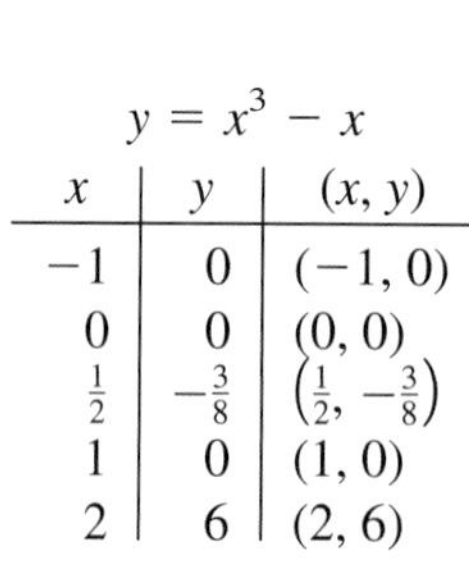

$y = x^3 - x$

x	y	(x, y)
-1	0	$(-1, 0)$
0	0	$(0, 0)$
$\frac{1}{2}$	$-\frac{3}{8}$	$\left(\frac{1}{2}, -\frac{3}{8}\right)$
1	0	$(1, 0)$
2	6	$(2, 6)$

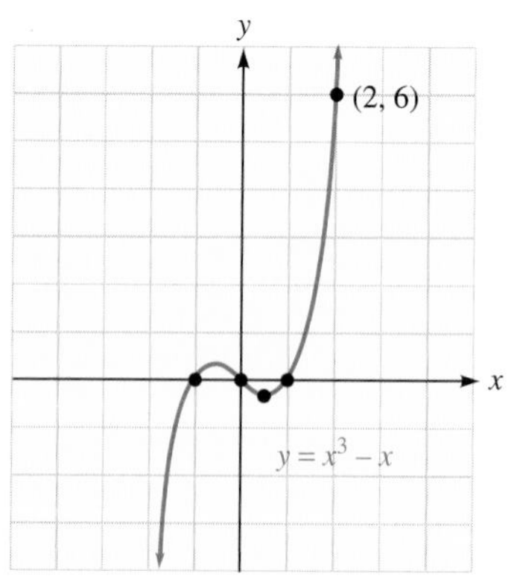

Figure 2-40

Self Check Graph: $y = x^3 - 9x$.

Accent on Technology FINDING THE INTERCEPTS OF THE GRAPH OF $y = 2x^3 - 3x$

To use a graphing calculator to graph $y = 2x^3 - 3x$, we enter the right-hand side of the equation after the symbol $Y_1 =$. The display will show the equation

$$Y_1 = 2X\text{^}3 - 3X$$

If we use window settings of $[-7, 7]$ for x and $[-10, 10]$ for y and press GRAPH, we will obtain the graph shown in Figure 2-41(a).

If we zoom in and trace to get Figures 2-41(b) and (c), we can see that the x-intercepts are (0, 0) and approximately $(-1.2, 0)$ and (1.2, 0). The only y-intercepts is (0, 0).

We can also find the intercepts by using the ZERO and VALUE commands.

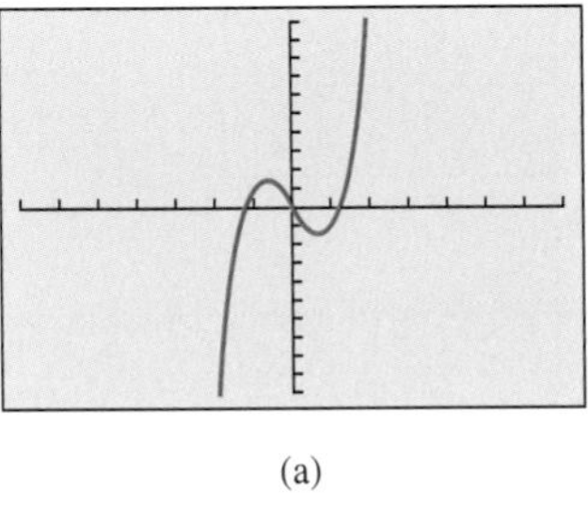

(a)

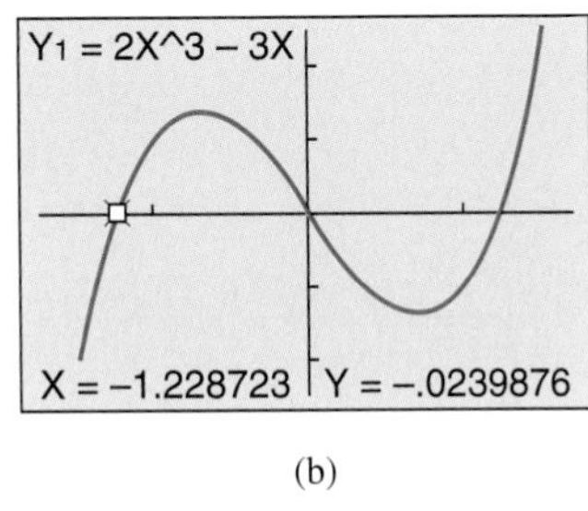

(b)

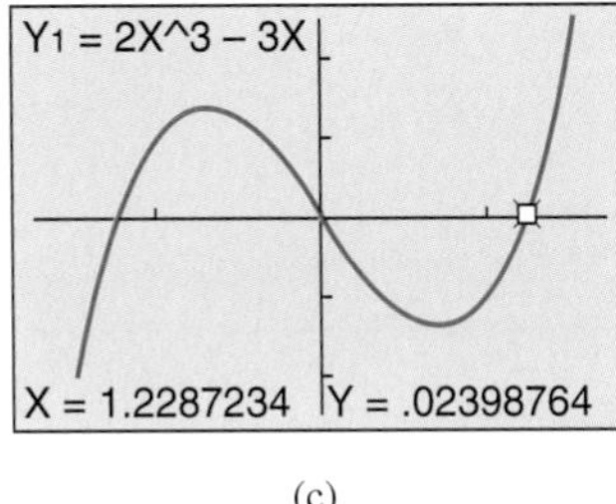

(c)

Figure 2-41

EXAMPLE 3 Graph: $y = \sqrt{x}$.

Solution Because the x- and y-intercepts are both (0, 0), the graph passes through the origin.

We now test for symmetries. Since $x \geq 0$ in the radical $\sqrt{x}$, there is no point in replacing x with $-x$. Since $y = \sqrt{x}$ and $\sqrt{x} \geq 0$, there is also no point in replacing y with $-y$. The graph of the equation has no symmetries. We plot several points to obtain the graph in Figure 2-42.

$y = \sqrt{x}$

x	y	(x, y)
0	0	(0, 0)
1	1	(1, 1)
4	2	(4, 2)
9	3	(9, 3)

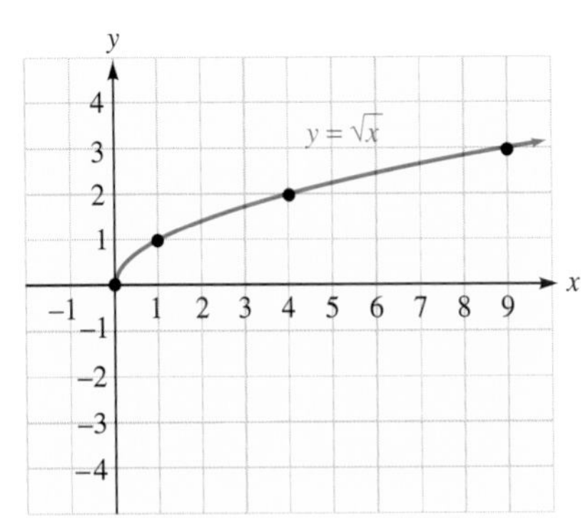

Figure 2-42

Self Check Graph: $y = -\sqrt{x}$.

■

Accent on Technology GRAPHING $y = \sqrt{3.5x}$

To use a graphing calculator to graph $y = \sqrt{3.5x}$, we enter the right-hand side of the equation after the symbol $Y_1 =$. The display will show the equation

$$Y_1 = \sqrt{}\ (3.5X)$$

If we use window settings of $[-2, 8]$ for x and $[-2, 8]$ for y and press GRAPH , we will obtain the graph shown in Figure 2-43.

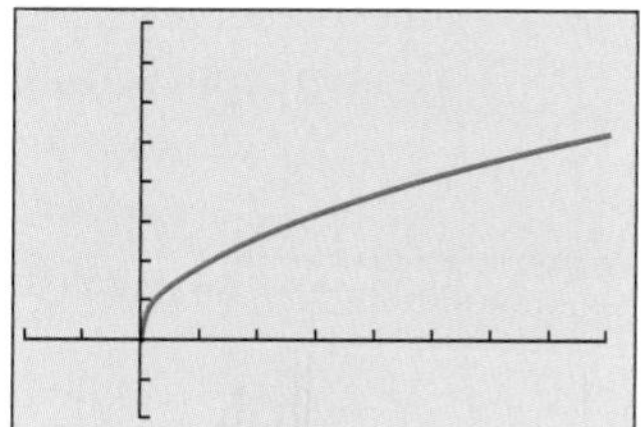

Figure 2-43

EXAMPLE 4 Graph: $y^2 = x$.

Solution Since the x- and y-intercepts are both (0, 0), the graph passes through the origin.

To test for x-axis symmetry, we replace y with $-y$.

(1) $y^2 = x$ This is the original equation.

$(-y)^2 = x$ Replace y with $-y$.

(2) $y^2 = x$ Simplify.

Since Equations 1 and 2 are the same, the graph is symmetric about the x-axis. There are no other symmetries.

To graph the equation, we plot the x- and y-intercept and several other pairs (x, y) with positive values of y. We can use the property of x-axis symmetry to draw the graph for negative values of y. (See Figure 2-44.)

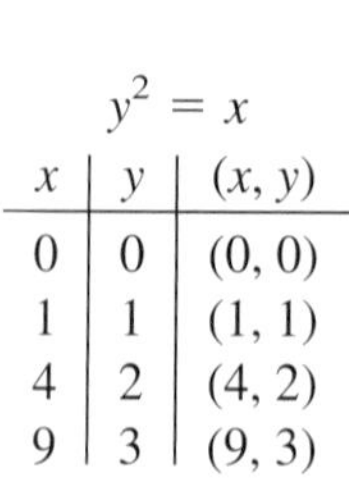

$y^2 = x$

x	y	(x, y)
0	0	(0, 0)
1	1	(1, 1)
4	2	(4, 2)
9	3	(9, 3)

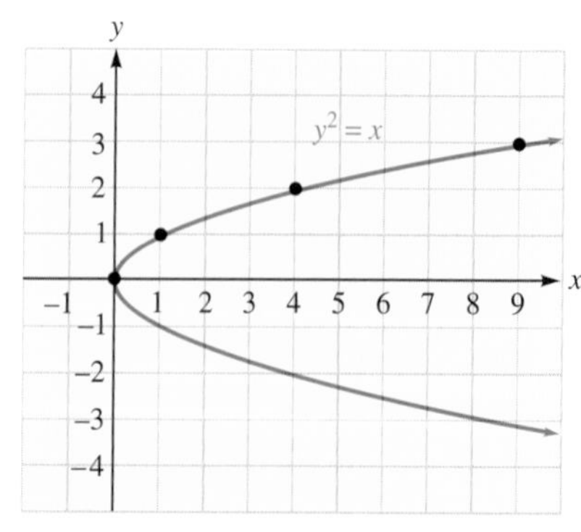

Figure 2-44

Self Check Graph: $y^2 = -x$. ■

Accent on Technology GRAPHING $y^2 = 2x$

To use a graphing calculator to graph $y^2 = 2x$, we must first solve the equation for y.

$$y^2 = 2x$$
$$y = \pm\sqrt{2x}$$
$$y = \sqrt{2x} \quad \text{or} \quad y = -\sqrt{2x}$$

We graph both equations on the same coordinate axes by entering the first equation as Y_1 and the second equation as Y_2.

$$Y_1 = \sqrt{\ }(2X) \qquad \text{and} \qquad Y_2 = -\sqrt{\ }(2X)$$

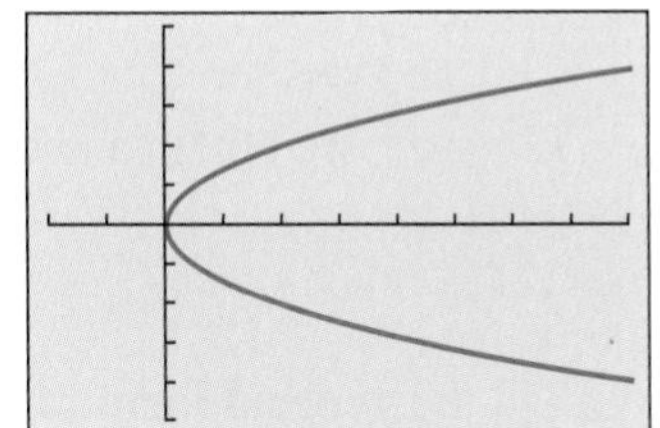

Figure 2-45

If we use window settings of $[-2, 8]$ for x and $[-5, 5]$ for y and press GRAPH, we will obtain the graph shown in Figure 2-45.

The graphs in Examples 3 and 4 are related. We have solved the equation $y^2 = x$ for y, and two equations resulted.

$$y = \sqrt{x} \qquad \text{and} \qquad y = -\sqrt{x}$$

The first equation, $y = \sqrt{x}$, was graphed in Example 3. It is the top half of the parabola shown in Example 4. The second equation, $y = -\sqrt{x}$, is the bottom half.

Circles

Circles

A **circle** is the set of all points in a plane that are a fixed distance from a point called its **center.** The fixed distance is the **radius of the circle.**

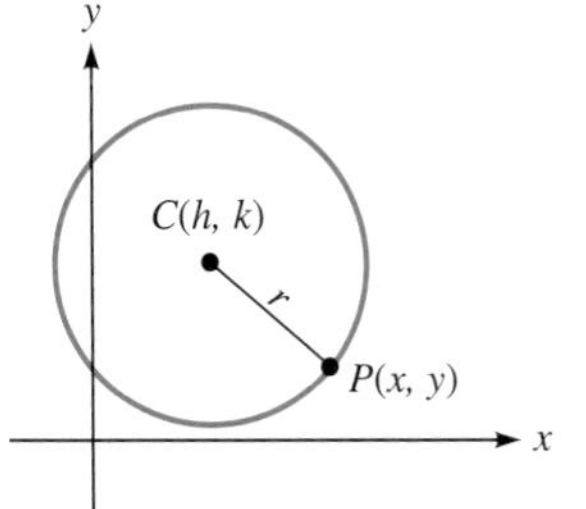

Active Figure 2-46

COLLEGE **Algebra** $f(x)$ **Now**™
Explore this Active Figure and test your understanding at **http://1pass.thomson.com**.

To find the equation of a circle with radius r and center at $C(h, k)$, we must find all points $P(x, y)$ in the xy-plane such that the length of line segment PC is r. (See Figure 2-46.) We can use the distance formula to find the length of CP, which is r:

$$r = \sqrt{(x - h)^2 + (y - k)^2}$$

After squaring both sides, we get

$$r^2 = (x - h)^2 + (y - k)^2$$

This equation is called the **standard equation of a circle.**

The Standard Equation of a Circle with Center at (h, k)

The graph of any equation that can be written in the form

$$(x - h)^2 + (y - k)^2 = r^2$$

is a circle with radius r and center at point (h, k).

If $r = 0$, the circle is a single point called a **point circle.** If the center of a circle is the origin, then $(h, k) = (0, 0)$, and we have the following result.

The Standard Equation of a Circle with Center at (0, 0)

The graph of any equation that can be written in the form

$$x^2 + y^2 = r^2$$

is a circle with radius r and center at the origin.

If we square the binomials in $(x - h)^2 + (y - k)^2 = r^2$, we obtain an equation of the form

$$x^2 + y^2 + cx + dy + e = 0$$

where c, d, and e are real numbers. This form is called the **general form of the equation of a circle.**

EXAMPLE 5 Find the general form of the equation of the circle with radius 5 and center at (3, 2).

Solution We substitute 5 for r, 3 for h, and 2 for k in the standard equation of a circle and simplify:

$$(x - h)^2 + (y - k)^2 = r^2$$
$$(x - 3)^2 + (y - 2)^2 = 5^2$$
$$x^2 - 6x + 9 + y^2 - 4y + 4 = 25 \quad \text{Remove parentheses.}$$
$$x^2 + y^2 - 6x - 4y - 12 = 0 \quad \text{Subtract 25 from both sides and simplify.}$$

The general form is $x^2 + y^2 - 6x - 4y - 12 = 0$.

Self Check Find the general form of the equation of a circle with radius 6 and center at $(-2, 5)$. ■

ACTIVE EXAMPLE 6 Find the general form of the equation of the circle with endpoints of its diameter at $(8, -3)$ and $(-4, 13)$.

COLLEGE **Algebra f(x) Now**™
Go to **http://1pass.thomson.com** or your CD to practice this example.

Solution We can find the center $O(h, k)$ of the circle by finding the midpoint of its diameter. By the midpoint formulas and $(x_1, y_1) = (8, -3)$ and $(x_2, y_2) = (-4, 13)$, we have

$$h = \frac{x_1 + x_2}{2} \qquad k = \frac{y_1 + y_2}{2}$$
$$h = \frac{8 + (-4)}{2} \qquad k = \frac{-3 + 13}{2}$$
$$= \frac{4}{2} \qquad = \frac{10}{2}$$
$$= 2 \qquad = 5$$

The center is $O(h, k) = O(2, 5)$.

To find the radius, we find the distance between the center and one endpoint of the diameter. The center is $O(2, 5)$, and one endpoint is $(8, -3)$.

$$r = \sqrt{(x_2 - x_1)^2 + (y_2 - y_1)^2}$$ This is the distance formula.

$$r = \sqrt{(2 - 8)^2 + [5 - (-3)]^2}$$ Substitute 8 for x_1, -3 for y_1, 2 for x_2, and 5 for y_2.

$$= \sqrt{(-6)^2 + (8)^2}$$

$$= \sqrt{36 + 64}$$

$$= 10$$ $\sqrt{36 + 64} = \sqrt{100} = 10.$

To find the equation of a circle with center at $(2, 5)$ and radius 10, we substitute 2 for h, 5 for k, and 10 for r in the standard equation of the circle and simplify:

$$(x - h)^2 + (y - k)^2 = r^2$$

$$(x - 2)^2 + (y - 5)^2 = 10^2$$

$$x^2 - 4x + 4 + y^2 - 10y + 25 = 100$$ Remove parentheses.

$$x^2 + y^2 - 4x - 10y - 71 = 0$$ Subtract 100 from both sides and simplify.

Self Check Find the equation of a circle with endpoints of its diameter at $(-2, 2)$ and $(6, 8)$. ■

■ Graphing Equations of Circles

ACTIVE EXAMPLE 7 Graph the circle whose equation is $2x^2 + 2y^2 - 8x + 4y = 40$.

Solution We first divide both sides of the equation by 2 to make the coefficients of x^2 and y^2 equal to 1.

COLLEGE **Algebra** $f(x)$ **Now**™
Go to **http://1pass.thomson.com** or your CD to practice this example.

$$2x^2 + 2y^2 - 8x + 4y = 40$$

$$x^2 + y^2 - 4x + 2y = 20$$

To find the coordinates of the center and the radius, we write the equation in standard form by completing the square on both x and y:

$$x^2 + y^2 - 4x + 2y = 20$$

$$x^2 - 4x + y^2 + 2y = 20$$

$$x^2 - 4x + 4 + y^2 + 2y + 1 = 20 + 4 + 1$$ Add 4 and 1 to both sides to complete the square.

$$(x - 2)^2 + (y + 1)^2 = 25$$ Factor $x^2 - 4x + 4$ and $y^2 + 2y + 1$.

$$(x - 2)^2 + [y - (-1)]^2 = 5^2$$

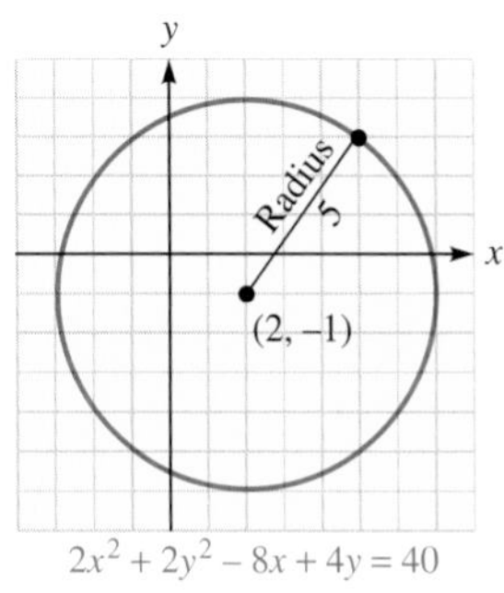

Figure 2-47

From the equation of the circle, we see that its radius is 5 and that the coordinates of its center are $h = 2$ and $k = -1$. The graph is shown in Figure 2-47.

Self Check Graph: $2x^2 + 2y^2 + 4x - 8y = -2$. ■

Accent on Technology GRAPHING CIRCLES

To use a graphing calculator to graph $(x - 2)^2 + (y + 1)^2 = 25$, we must solve the equation for y:

$$(x - 2)^2 + (y + 1)^2 = 25$$
$$(y + 1)^2 = 25 - (x - 2)^2$$
$$y + 1 = \pm\sqrt{25 - (x - 2)^2}$$
$$y = -1 \pm \sqrt{25 - (x - 2)^2}$$

This last expression represents two equations: $y = -1 + \sqrt{25 - (x - 2)^2}$ and $y = -1 - \sqrt{25 - (x - 2)^2}$. We graph both of these equations separately on the same coordinate axes by entering the first equation as Y_1 and the second as Y_2:

$$Y_1 = -1 + \sqrt{\ }(25 - (X - 2) \wedge 2)$$
$$Y_2 = -1 - \sqrt{\ }(25 - (X - 2) \wedge 2)$$

Depending on the setting of the maximum and minimum values of x and y, the graph may not appear to be a circle, and it may look like Figure 2-48(a). Most graphing calculators can be set to display equal-sized divisions on the x- and y-axes, producing a better graph—like that shown in Figure 2-48(b).

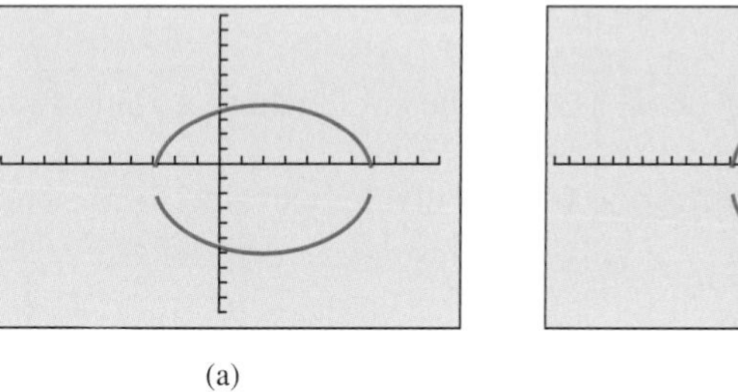

(a)

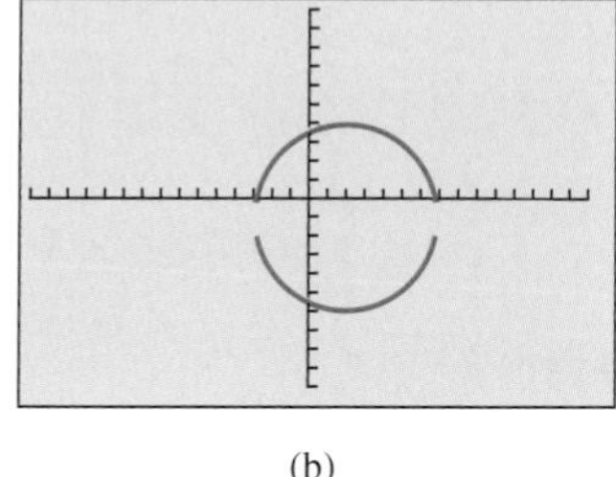

(b)

Figure 2-48

Note that in this example, it is easier to graph the circle by hand than with a calculator.

■ Solving Equations by Graphing

We can solve many equations using the graphing concepts discussed in this chapter and a graphing calculator. For example, the solutions of $x^2 - x - 3 = 0$ will be the numbers x that will make $y = 0$ in the equation $y = x^2 - x - 3$. These numbers will be the x-coordinates of the x-intercepts of the graph of $y = x^2 - x - 3$.

Accent on Technology SOLVING EQUATIONS

To use a graphing calculator to solve $x^2 - x - 3 = 0$, we graph the equation $y = x^2 - x - 3$, and find the x-intercepts. The x-coordinates of the x-intercepts will be the solutions of the equation. The graph of $y = x^2 - x - 3$ is shown in Figure 2-49(a). To find the x-coordinates of the x-intercepts, we trace and move the cursor near the negative x-intercept, as shown in Figure 2-49(b). After zooming and tracing, we can determine that $x \approx -1.302776$. We then trace and move the cursor near the positive x-intercept, as shown in Figure 2-49(c). After zooming and tracing, we can determine that $x \approx 2.3027655$. From these results, we can see that to the nearest hundredth, the solutions of the equation $x^2 - x - 3 = 0$ are $x = -1.30$ and $x = 2.30$.

We can also find the x-intercepts by using the ZERO command.

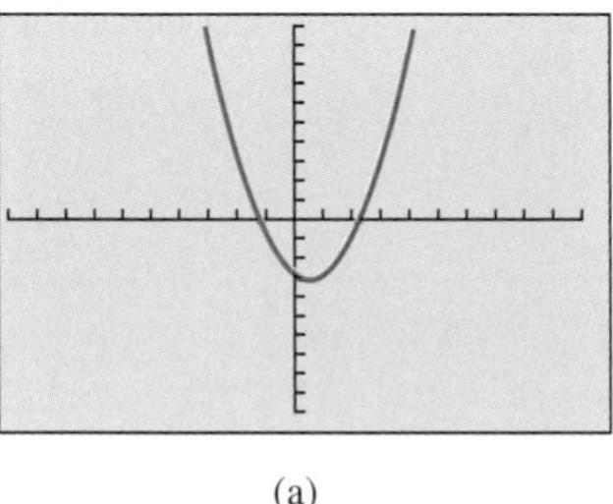

(a)

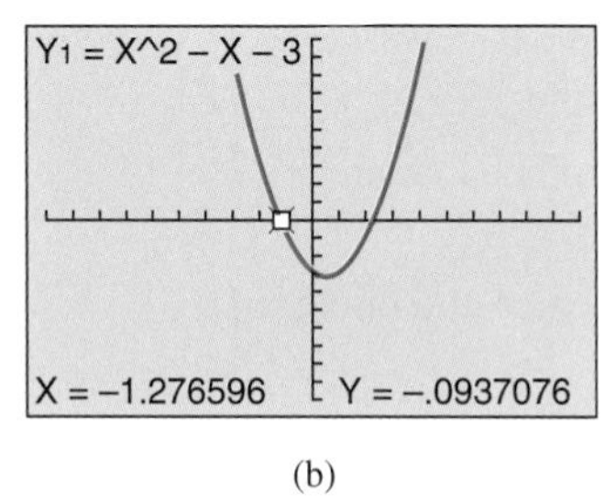

(b)

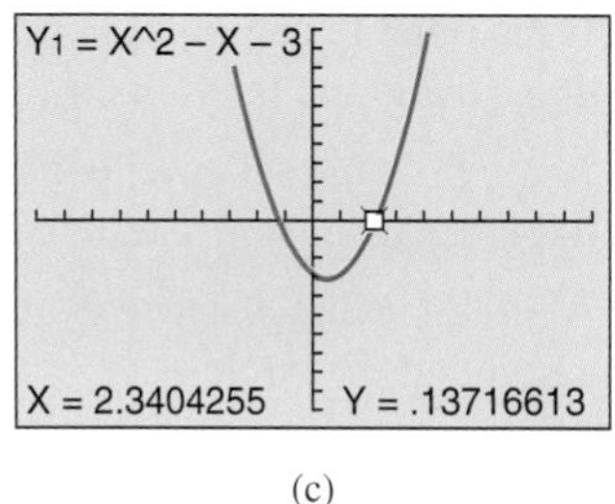

(c)

Figure 2-49

Self Check Answers

1.

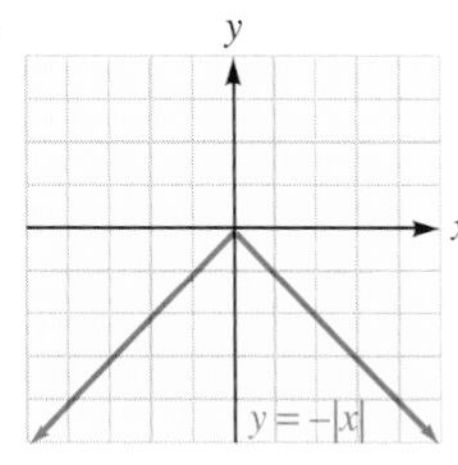

2.

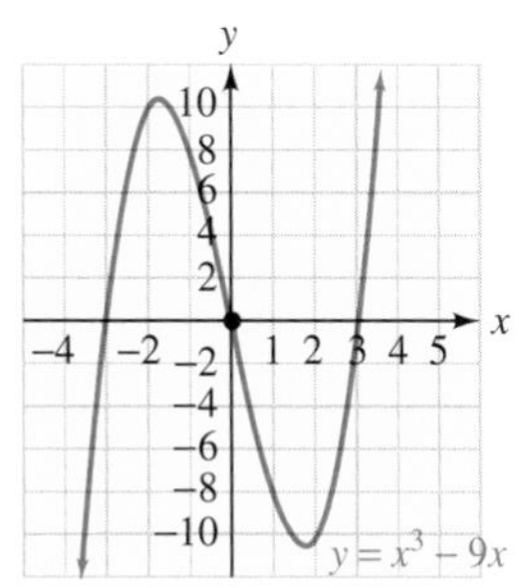

3.

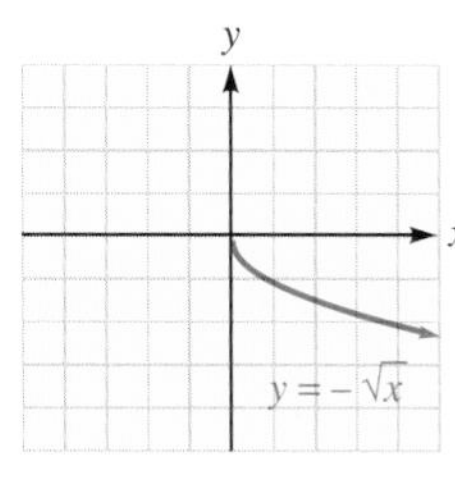

4.

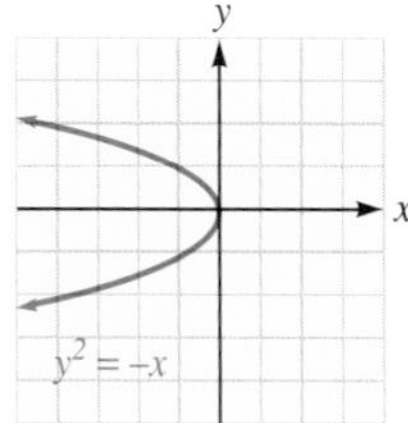

5. $x^2 + y^2 + 4x - 10y - 7 = 0$ **6.** $x^2 + y^2 - 4x - 10y + 4 = 0$ **7.**

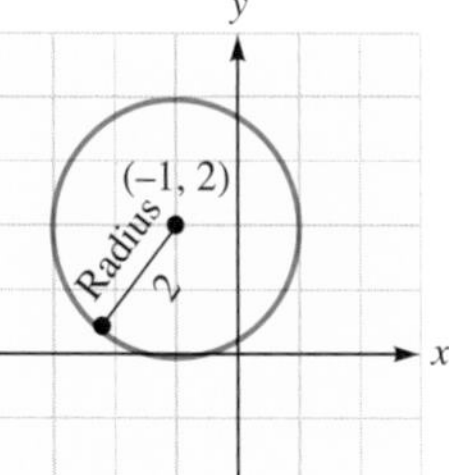

2.4 Exercises

VOCABULARY AND CONCEPTS ***Fill in the blanks.***

1. The point where a graph intersects the x-axis is called the **x-intercept**.
2. The y-intercept is the point where a graph intersects the **y-axis**.
3. If a line divides a graph into two congruent halves, we call the line an **axis of symmetry**.
4. If the point $(-x, y)$ lies on a graph whenever (x, y) does, the graph is symmetric about the **y-axis**.
5. If the point $(x, -y)$ lies on a graph whenever (x, y) does, the graph is symmetric about the **x-axis**.
6. If the point $(-x, -y)$ lies on a graph whenever (x, y) does, the graph is symmetric about the **origin**.
7. A **circle** is the set of all points in a plane that are a fixed distance from a point called its **center**.
8. A **radius** is the distance from the center of a circle to a point on the circle.
9. The standard equation of a circle with center at the origin and radius r is $x^2 + y^2 = r^2$.
10. The standard equation of a circle with center at (h, k) and radius r is $(x - h)^2 + (y - k)^2 = r^2$.

PRACTICE ***Find the x- and y-intercepts of each graph. Do not graph the equation.***

11. $y = x^2 - 4$
    $(-2, 0), (2, 0); (0, -4)$
12. $y = x^2 - 9$
    $(-3, 0), (3, 0); (0, -9)$
13. $y = 4x^2 - 2x$
    $(0, 0), (\frac{1}{2}, 0); (0, 0)$
14. $y = 2x - 4x^2$
    $(0, 0), (\frac{1}{2}, 0); (0, 0)$
15. $y = x^2 - 4x - 5$
    $(-1, 0), (5, 0); (0, -5)$
16. $y = x^2 - 10x + 21$
    $(3, 0), (7, 0); (0, 21)$
17. $y = x^2 + x - 2$
    $(1, 0), (-2, 0); (0, -2)$
18. $y = x^2 + 2x - 3$
    $(1, 0), (-3, 0); (0, -3)$
19. $y = x^3 - 9x$
    $(-3, 0), (0, 0), (3, 0); (0, 0)$
20. $y = x^3 + x$
    $(0, 0); (0, 0)$
21. $y = x^4 - 1$
    $(-1, 0), (1, 0); (0, -1)$
22. $y = x^4 - 25x^2$
    $(-5, 0), (0, 0), (5, 0); (0, 0)$

Graph each equation. Check your graph with a graphing calculator.

23. $y = x^2$

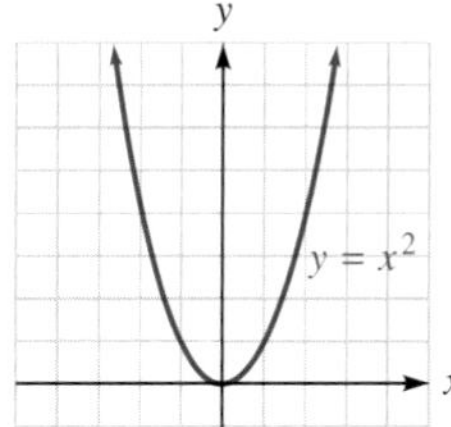

24. $y = -x^2$

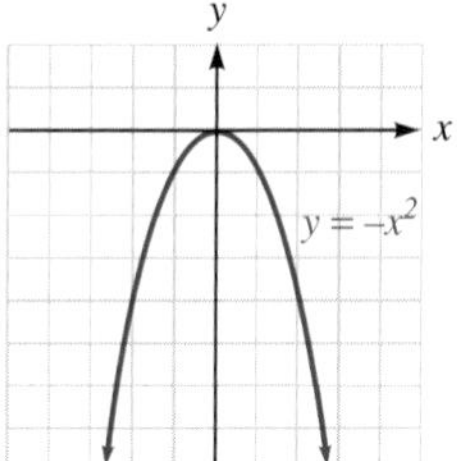

25. $y = -x^2 + 2$

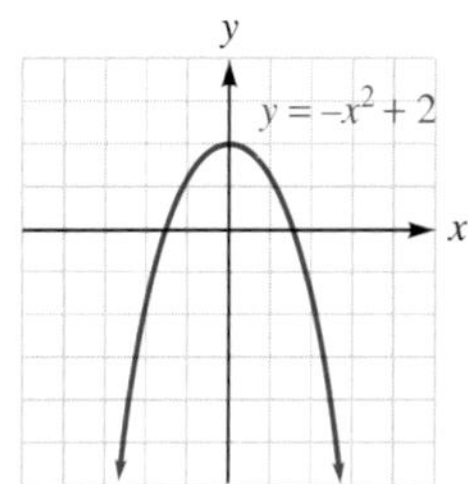

26. $y = x^2 - 1$

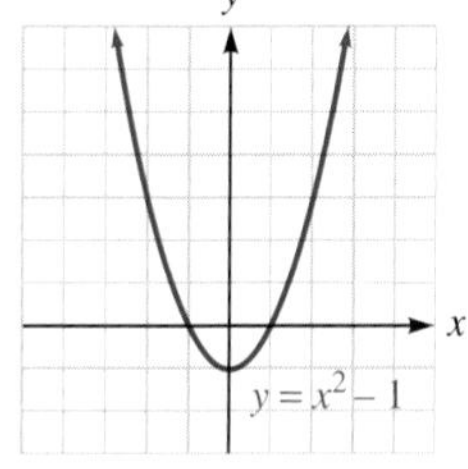

27. $y = x^2 - 4x$

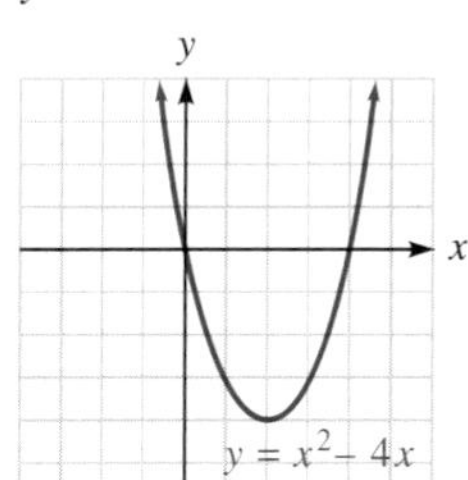

28. $y = x^2 + 2x$

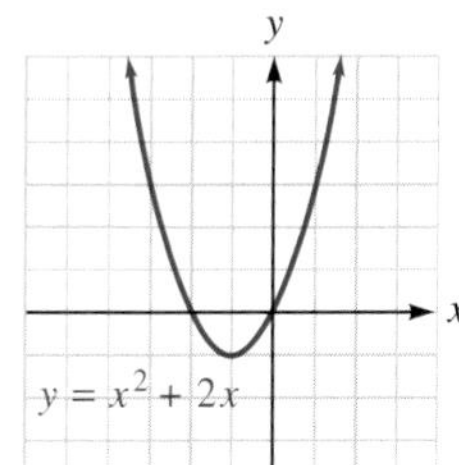

29. $y = \frac{1}{2}x^2 - 2x$

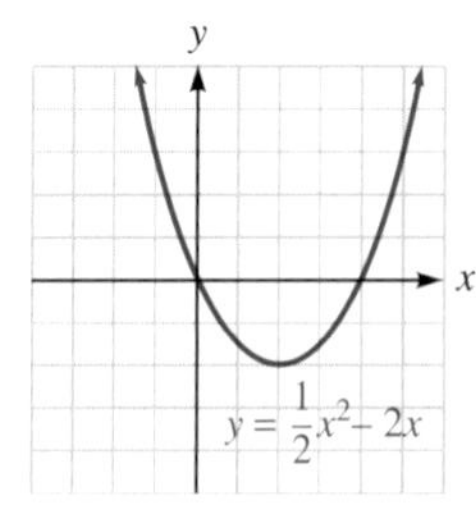

30. $y = \frac{1}{2}x^2 + 3$

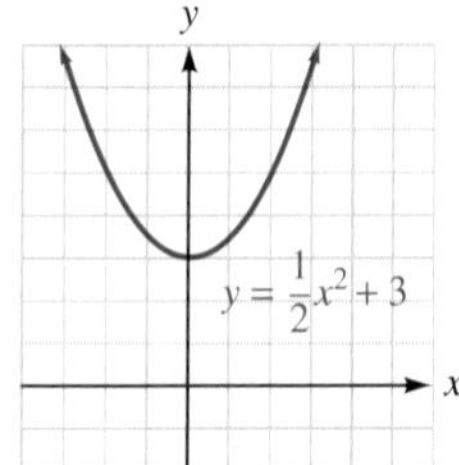

***Find the symmetries, if any, of the graph of each equation.* Do not graph the equation.**

31. $y = x^2 + 2$
about the y-axis

32. $y = 3x + 2$
none

33. $y^2 + 1 = x$
about the x-axis

34. $y^2 + y = x$
none

35. $y^2 = x^2$
about the x-axis, the y-axis, and the origin

36. $y = 3x + 7$
none

37. $y = 3x^2 + 7$
about the y-axis

38. $x^2 + y^2 = 1$
about the x-axis, the y-axis, and the origin

39. $y = 3x^3 + 7$
none

40. $y = 3x^3 + 7x$
about the origin

41. $y^2 = 3x$
about the x-axis

42. $y = 3x^4 + 7$
about the y-axis

43. $y = |x|$
about the y-axis

44. $y = |x + 1|$
none

45. $|y| = x$
about the x-axis

46. $|y| = |x|$
about the x-axis, the y-axis, and the origin

Graph each equation. Be sure to find any intercepts and symmetries. Check your graph with a graphing calculator.

47. $y = x^2 + 4x$

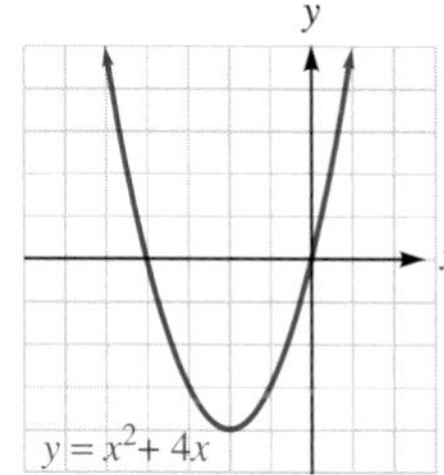

48. $y = x^2 - 6x$

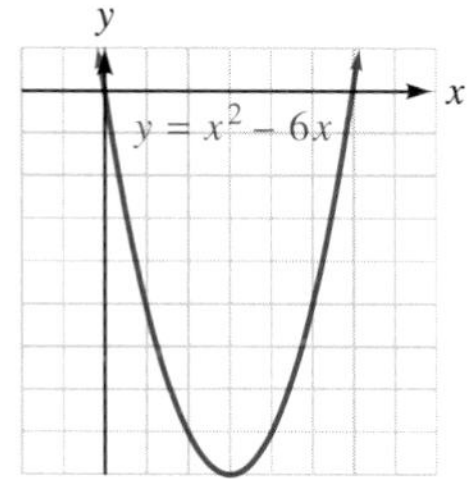

49. $y = x^3$

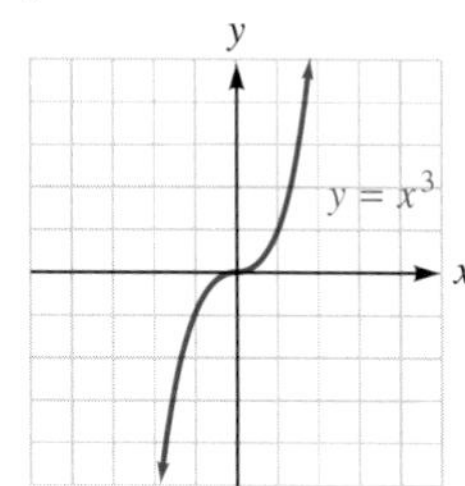

50. $y = x^3 + x$

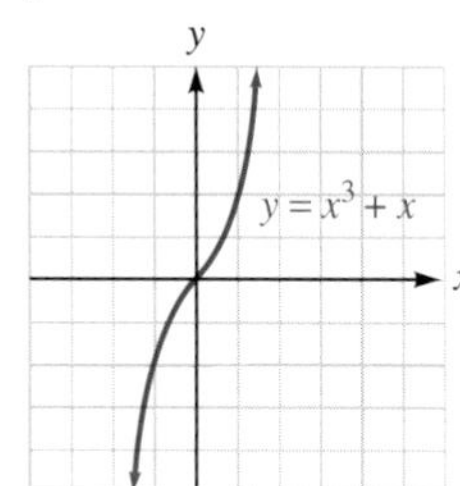

51. $y = |x - 2|$

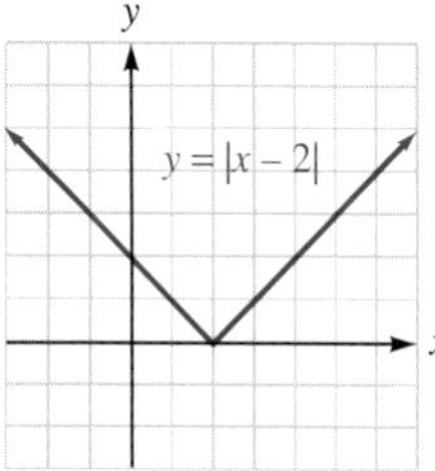

52. $y = |x| - 2$

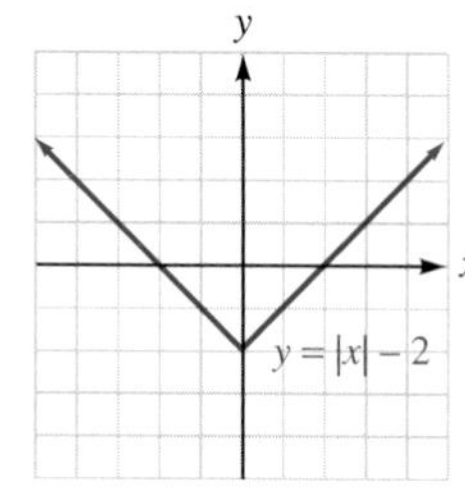

53. $y = 3 - |x|$

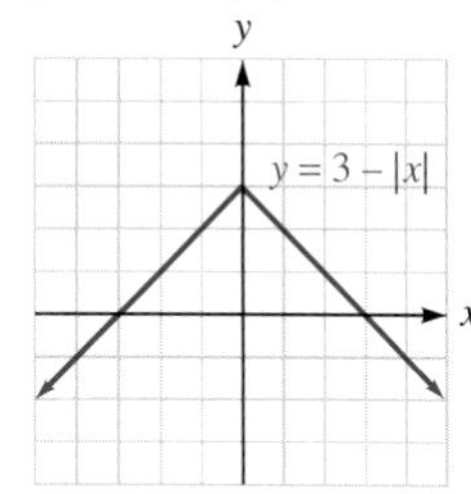

54. $y = 3|x|$

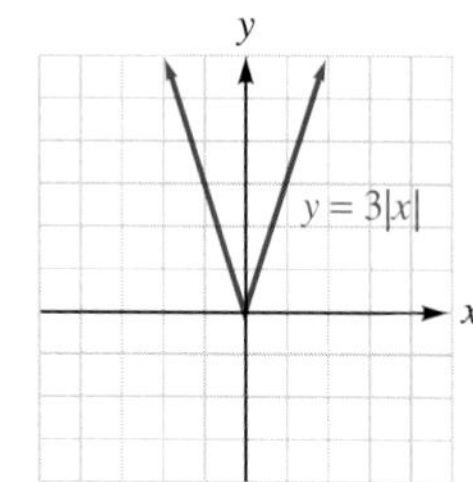

55. $y^2 = -x$

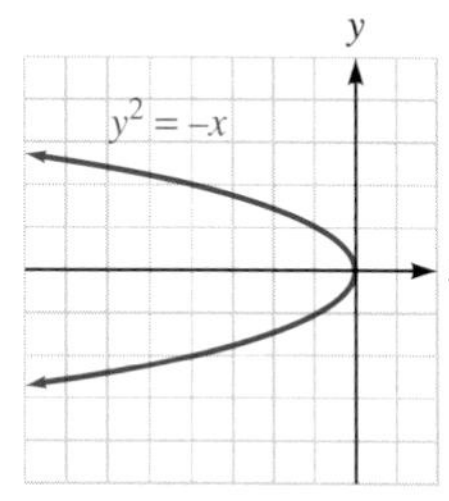

56. $y^2 = 4x$

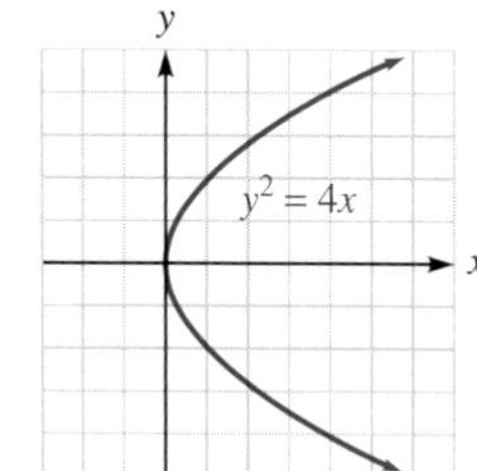

57. $y^2 = 9x$

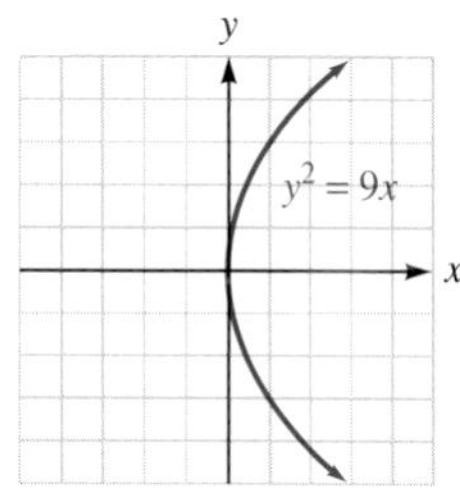

58. $y^2 = -4x$

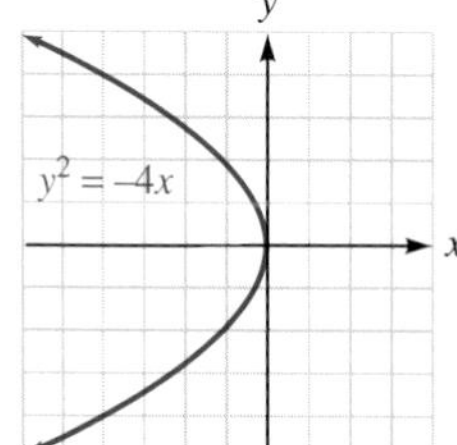

59. $y = \sqrt{x} - 1$

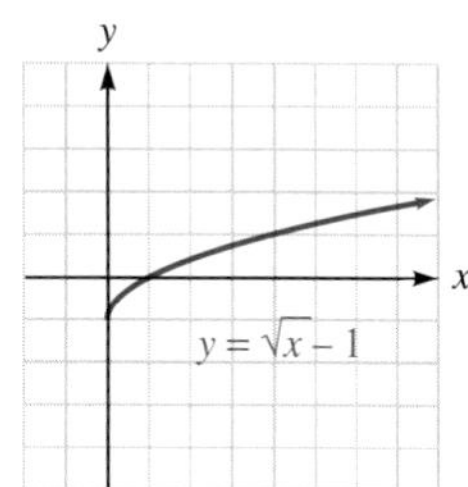

60. $y = 1 - \sqrt{x}$

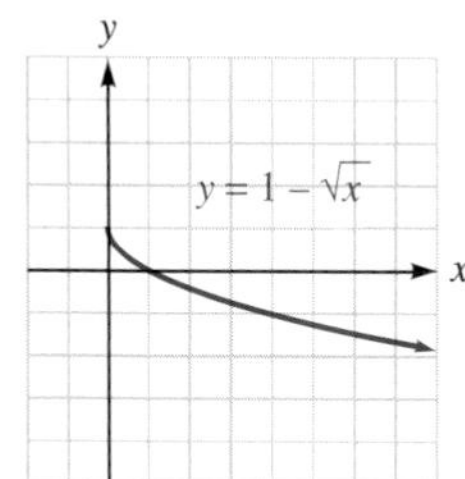

61. $xy = 4$

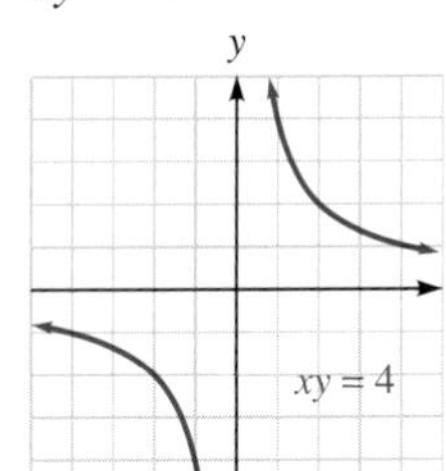

62. $xy = -9$

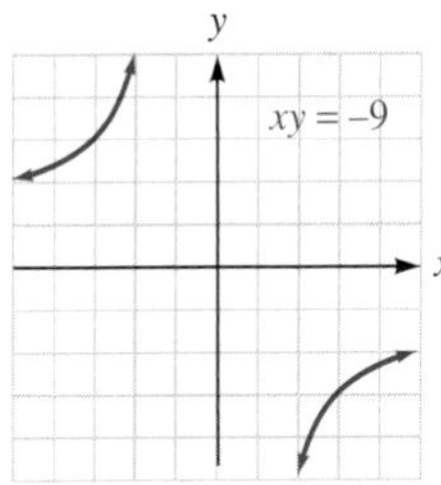

Find the equation in general form of each circle with the given properties.

63. Center at the origin; $r = 1$ $\quad$ $\mathbf{x^2 + y^2 - 1 = 0}$

64. Center at the origin; $r = 4$ $\quad$ $\mathbf{x^2 + y^2 - 16 = 0}$

65. Center at $(6, 8)$; $r = 4$ $\quad$ $\mathbf{x^2 + y^2 - 12x - 16y + 84 = 0}$

66. Center at $(5, 3)$; $r = 2$ $\quad$ $\mathbf{x^2 + y^2 - 10x - 6y + 30 = 0}$

67. Center at $(3, -4)$; $r = \sqrt{2}$
$\mathbf{x^2 + y^2 - 6x + 8y + 23 = 0}$

68. Center at $(-9, 8)$; $r = 2\sqrt{3}$
$\mathbf{x^2 + y^2 + 18x - 16y + 133 = 0}$

69. Ends of diameter at $(3, -2)$ and $(3, 8)$
$\mathbf{x^2 + y^2 - 6x - 6y - 7 = 0}$

70. Ends of diameter at $(5, 9)$ and $(-5, -9)$
$\mathbf{x^2 + y^2 - 106 = 0}$

71. Center at $(-3, 4)$ and passing through the origin
$\mathbf{x^2 + y^2 + 6x - 8y = 0}$

72. Center at $(-2, 6)$ and passing through the origin
$\mathbf{x^2 + y^2 + 4x - 12y = 0}$

Graph each equation.

73. $x^2 + y^2 - 25 = 0$

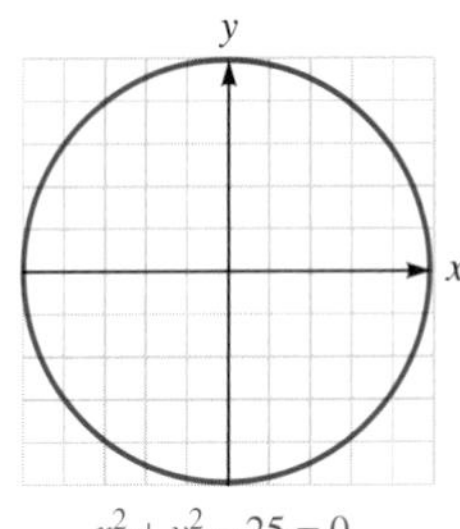

$x^2 + y^2 - 25 = 0$

74. $x^2 + y^2 - 8 = 0$

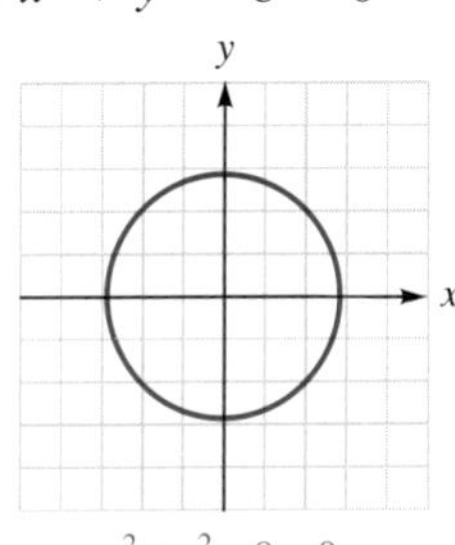

$x^2 + y^2 - 8 = 0$

75. $(x - 1)^2 + (y + 2)^2 = 4$

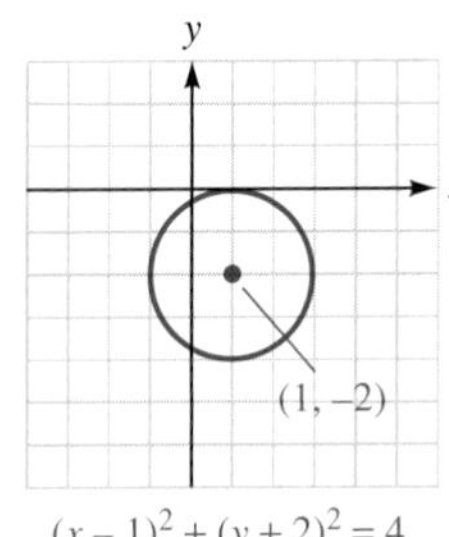

$(x - 1)^2 + (y + 2)^2 = 4$

76. $(x + 1)^2 + (y - 2)^2 = 9$

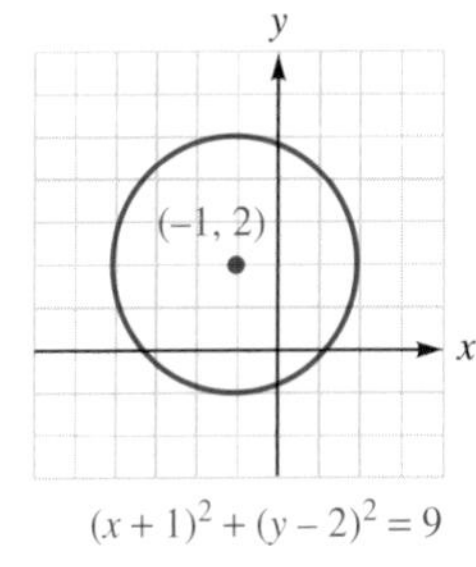

$(x + 1)^2 + (y - 2)^2 = 9$

77. $x^2 + y^2 + 2x - 24 = 0$

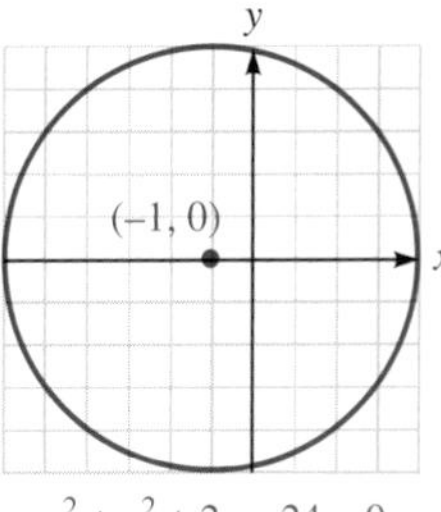

$x^2 + y^2 + 2x - 24 = 0$

78. $x^2 + y^2 - 4y = 12$

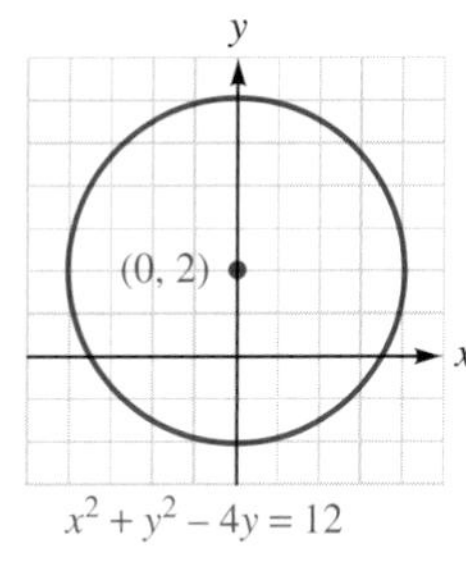

$x^2 + y^2 - 4y = 12$

79. $9x^2 + 9y^2 - 12y = 5$

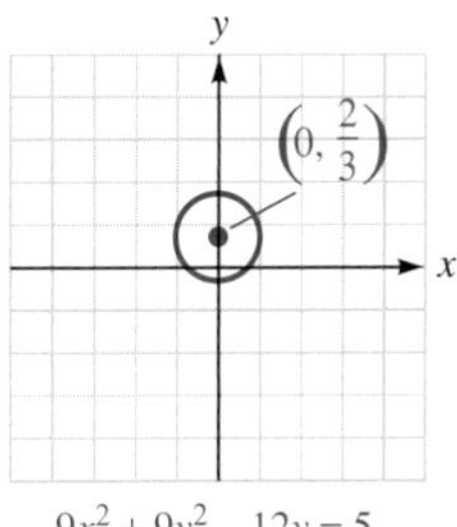

$9x^2 + 9y^2 - 12y = 5$

80. $4x^2 + 4y^2 + 4y = 15$

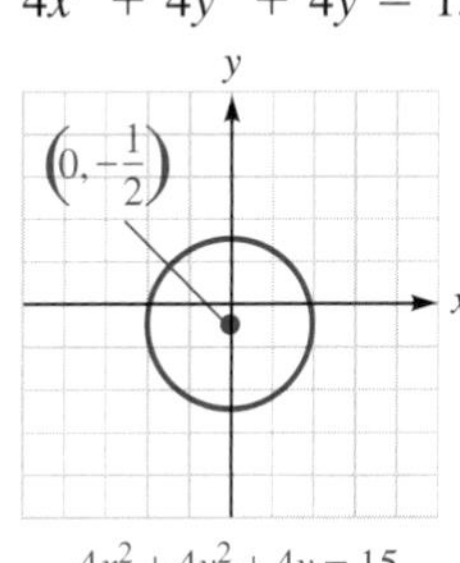

$4x^2 + 4y^2 + 4y = 15$

81. $4x^2 + 4y^2 - 4x + 8y + 1 = 0$

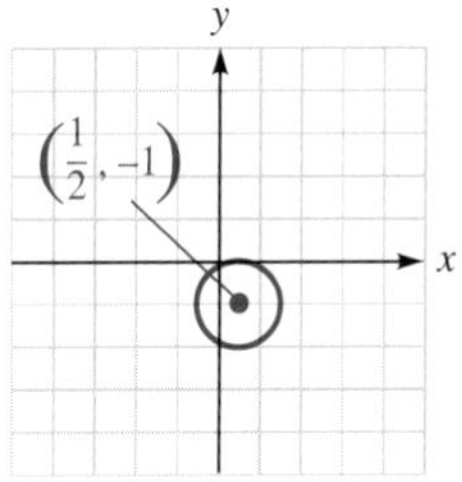

(1/2, –1)

$4x^2 + 4y^2 - 4x + 8y + 1 = 0$

82. $9x^2 + 9y^2 - 6x + 18y + 1 = 0$

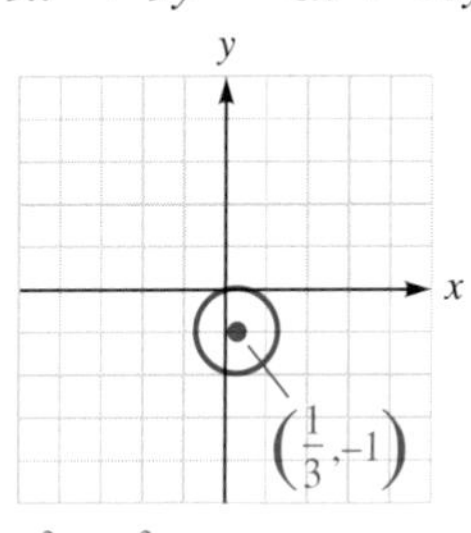

$9x^2 + 9y^2 - 6x + 18y + 1 = 0$

Use a graphing calculator to graph each equation. Then find the coordinates of the vertex to the nearest hundredth.

83. $y = 2x^2 - x + 1$
(0.25, 0.88)

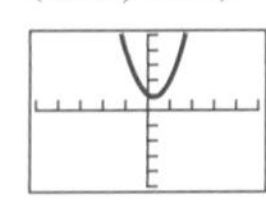

84. $y = x^2 + 5x - 6$
(−2.50, −12.25)

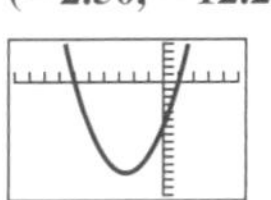

85. $y = 7 + x - x^2$

(0.50, 7.25)

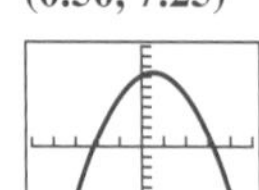

86. $y = 2x^2 - 3x + 2$

(0.75, 0.88)

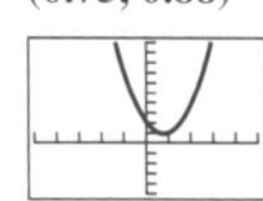

Use a graphing calculator to solve each equation. Round to the nearest hundredth.

87. $x^2 - 7 = 0$

±2.65

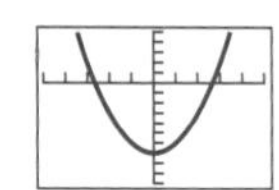

88. $x^2 - 3x + 2 = 0$

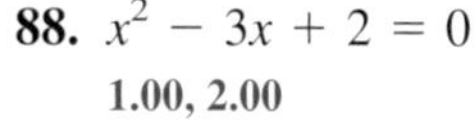

1.00, 2.00

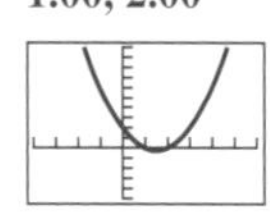

89. $x^3 - 3 = 0$

1.44

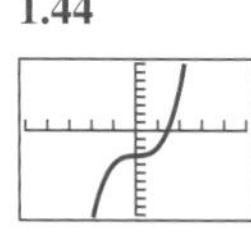

90. $3x^3 - x^2 - x = 0$

−0.44, 0, 0.77

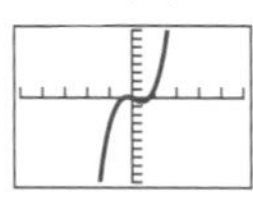

APPLICATIONS

91. Golfing A golfer's tee shot follows a path given by $y = 64t - 16t^2$, where y is the height of the ball (in feet) after t seconds of flight. How long will it take for the ball to strike the ground? **4 sec**

92. Golfing Halfway through its flight, the golf ball of Exercise 91 reaches the highest point of its trajectory. How high is that? **64 ft**

93. Stopping distances The stopping distance D (in feet) for a car moving V miles per hour is given by $D = 0.08V^2 + 0.9V$. Graph the equation for velocities between 0 and 60 mph.

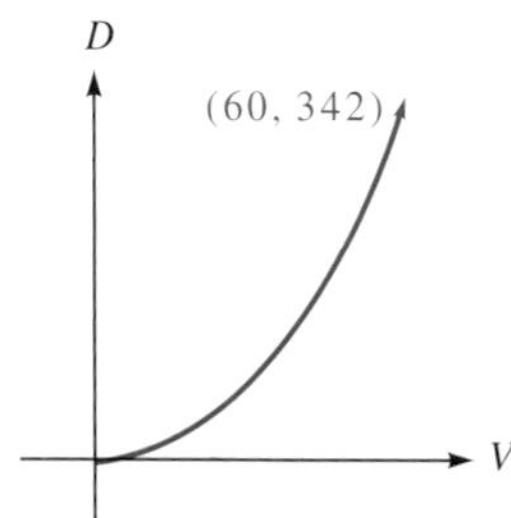

94. Stopping distances See Exercise 93. How much farther does it take to stop at 60 mph than at 30 mph? **243 ft farther**

95. CB radios The CB radio of a trucker covers the circular area shown in the illustration. Find the equation of that circle, in general form.

$\mathbf{x^2 + y^2 - 14x - 8y + 40 = 0}$

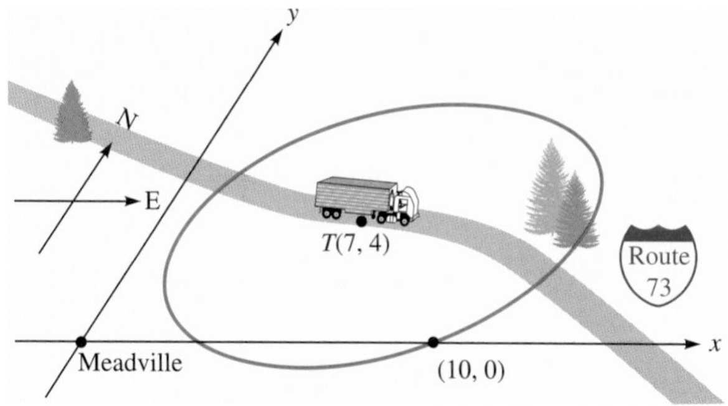

96. Tires Two 24-inch-diameter tires stand against a wall, as shown in the illustration. Find the equations of the circular boundaries of the tires.

$\mathbf{x^2 + y^2 - 24x - 24y + 144 = 0;}$

$\mathbf{x^2 + y^2 - 72x - 24y + 1{,}296 = 0}$

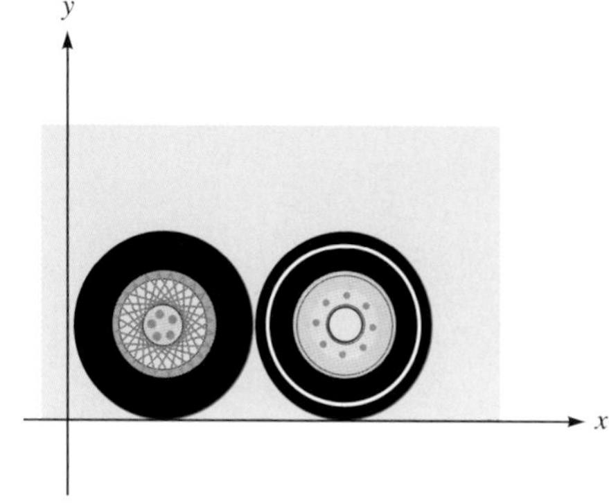

DISCOVERY AND WRITING

The solution of the inequality $P(x) < 0$ consists of those numbers x for which the graph of $y = P(x)$ lies below the x-axis. To solve $P(x) < 0$, we graph $y = P(x)$ and trace to find numbers x that produce negative values of y. Solve each inequality.

97. $x^2 + x - 6 < 0$

98. $x^2 - 3x - 10 > 0$

REVIEW

Solve each equation.

99. $3(x + 2) + x = 5x$ **6**

100. $12b + 6(3 - b) = b + 3$ **−3**

101. $\dfrac{5(2 - x)}{3} - 1 = x + 5$ **−1**

102. $\dfrac{r - 1}{3} = \dfrac{r + 2}{6} + 2$ **16**

103. Mixing an alloy In 60 ounces of alloy for watch cases, there are 20 ounces of gold. How much copper must be added to the alloy so that a watch case weighing 4 ounces, made from the new alloy, will contain exactly 1 ounce of gold? **20 oz**

104. Mixing coffee To make a mixture of 80 pounds of coffee worth \$272, a grocer mixes coffee worth \$3.25 a pound with coffee worth \$3.85 a pound. How many pounds of cheaper coffee should the grocer use? **60 lb**

2.5 Proportion and Variation

In this section, you will learn about

- Proportions
- Solving Proportions
- Direct Variation
- Inverse Variation
- Joint Variation
- Combined Variation

The quotient of two numbers is often called a **ratio.** For example, the fraction $\frac{3}{2}$ (or the expression 3:2) can be read as "the ratio of 3 to 2." Some examples are

$$\frac{3}{5}, \quad 7:9, \quad \frac{x+1}{9}, \quad \frac{a}{b}, \quad \text{and} \quad \frac{x^2-4}{x+5}$$

Proportions

An equation indicating that two ratios are equal is called a **proportion.** Some examples of proportions are

$$\frac{2}{3}=\frac{4}{6}, \quad \frac{x}{y}=\frac{3}{5}, \quad \text{and} \quad \frac{x^2+8}{2(x+3)}=\frac{17(x+3)}{2}$$

In the proportion $\frac{a}{b}=\frac{c}{d}$, the numbers a and d are called the **extremes,** and the numbers b and c are called the **means.**

To develop an important property of proportions, we suppose that

$$\frac{a}{b}=\frac{c}{d}$$

and multiply both sides by bd to get

$$bd\left(\frac{a}{b}\right)=bd\left(\frac{c}{d}\right)$$

$$\frac{bda}{b}=\frac{bdc}{d}$$

$$da=bc$$

Thus, if $\frac{a}{b}=\frac{c}{d}$, then $ad=bc$. This proves the following statement.

Property of Proportions In any proportion, the product of the extremes is equal to the product of the means.

We can use this property to solve proportions.

Solving Proportions

EXAMPLE 1 Solve the proportion: $\frac{x}{5} = \frac{2}{x+3}$.

Solution

$$\frac{x}{5} = \frac{2}{x+3}$$

$$x(x+3) = 5 \cdot 2$$ The product of the extremes equals the product of the means.

$$x^2 + 3x = 10$$ Remove parentheses and simplify.

$$x^2 + 3x - 10 = 0$$ Subtract 10 from both sides.

$$(x-2)(x+5) = 0$$ Factor the trinomial.

$$x - 2 = 0 \quad \text{or} \quad x + 5 = 0$$ Set each factor equal to 0.

$$x = 2 \quad | \quad x = -5$$

Thus, $x = 2$ or $x = -5$. Verify each solution.

Self Check Solve: $\frac{2}{5} = \frac{3}{x-4}$. ■

EXAMPLE 2 Gasoline and oil for a lawn mower are to be mixed in a 50-to-1 ratio. How many ounces of oil should be mixed with 6 gallons of gasoline?

Solution We first express 6 gallons as $6 \cdot 128$ ounces $= 768$ ounces. We then let x represent the number of ounces of oil needed, set up the proportion, and solve it.

$$\frac{50}{1} = \frac{768}{x}$$

$$50x = 768$$ The product of the extremes equals the product of the means.

$$x = \frac{768}{50}$$ Divide both sides by 50.

$$x = 15.36$$

Approximately 15 ounces of oil should be added to 6 gallons of gasoline.

Self Check How many ounces of oil should be mixed with 6 gallons of gas if the ratio is to be 40 parts of gas to 1 part of oil? ■

Direct Variation

Two variables are said to **vary directly** or be **directly proportional** if their ratio is a constant. The variables x and y vary directly when

$$\frac{y}{x} = k \qquad \text{or, equivalently,} \qquad y = kx \quad (k \text{ is a constant})$$

Direct Variation

> The words **"y varies directly with x,"** or **"y is directly proportional to x,"** mean that $y = kx$ for some real-number constant k.
>
> The number k is called the **constant of proportionality.**

ACTIVE EXAMPLE 3 Distance traveled in a given time varies directly with the speed. If a car travels 70 miles at 30 mph, how far will it travel in the same time at 45 mph?

COLLEGE **Algebra** $f(x)$ **Now**™

Explore this Active Example and test yourself on the steps taken here at **http://1pass.thomson.com**. You can also find this example on the Interactive Video Skillbuilder CD-ROM.

Solution The phrase "distance varies directly with speed" translates into the formula $d = ks$, where d represents the distance traveled and s represents the speed. The constant of proportionality k can be found by substituting 70 for d and 30 for s in the equation $d = ks$:

$$d = ks$$
$$70 = k(30)$$
$$k = \frac{7}{3}$$

To evaluate the distance d traveled at 45 mph, we substitute $\frac{7}{3}$ for k and 45 for s into the formula $d = ks$.

$$d = \frac{7}{3}s$$
$$= \frac{7}{3}(45)$$
$$= 105$$

In the time it takes to go 70 miles at 30 mph, the car could travel 105 miles at 45 mph.

Self Check How far will the car travel in the same time if its speed is 60 mph? ■

The statement "y varies directly with x" is equivalent to the equation $y = mx$, where m is the constant of proportionality. Because the equation is in the form $y = mx + b$ with $b = 0$, its graph is a line with slope m and y-intercept 0. The graphs of $y = mx$ for several values of m are shown in Figure 2-50.

The graph of the relationship of direct variation is always a line that passes through the origin.

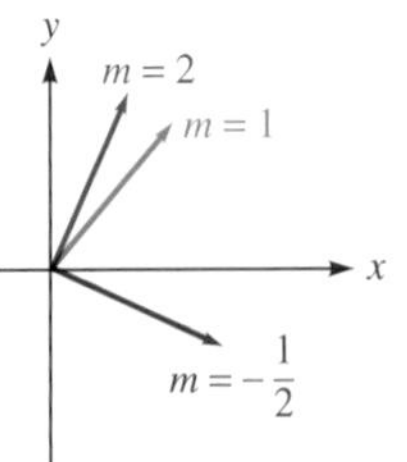

Figure 2-50

■ Inverse Variation

Two variables are said to **vary inversely** or be **inversely proportional** if their product is a constant.

$$xy = k \qquad \text{or, equivalently,} \qquad y = \frac{k}{x} \quad (k \text{ is a constant})$$

Inverse Variation

> The words **"y varies inversely with x,"** or **"y is inversely proportional to x,"** mean that $y = \frac{k}{x}$ for some real-number constant k.

ACTIVE EXAMPLE 4 Intensity of illumination from a light source varies inversely with the square of the distance from the source. If the intensity of a light source is 100 lumens at a distance of 20 feet, find the intensity at 30 feet.

Solution If I is the intensity and d is the distance from the light source, the phrase "intensity varies inversely with the square of the distance" translates into the formula

COLLEGE **Algebra** $f(x)$ **Now**™
Go to **http://1pass.thomson.com** or your CD to practice this example.

$$I = \frac{k}{d^2}$$

We can evaluate k by substituting 100 for I and 20 for d in the formula and solving for k.

$$I = \frac{k}{d^2}$$

$$100 = \frac{k}{20^2}$$

$$k = 40{,}000$$

To find the intensity at a distance of 30 feet, we substitute 40,000 for k and 30 for d in the formula

$$I = \frac{k}{d^2}$$

$$I = \frac{40{,}000}{30^2}$$

$$= \frac{400}{9}$$

At 30 feet, the intensity of light would be $\frac{400}{9}$ lumens per square centimeter.

Self Check Find the intensity at 50 feet. ■

The statement "y is inversely proportional to x" is equivalent to the equation $y = \frac{k}{x}$, where k is a constant. Figure 2-51 shows the graphs of $y = \frac{k}{x}$ $(x > 0)$ for three values of k. In each case, the equation determines one branch of a curve called a **hyperbola.** Verify these graphs with a graphing calculator.

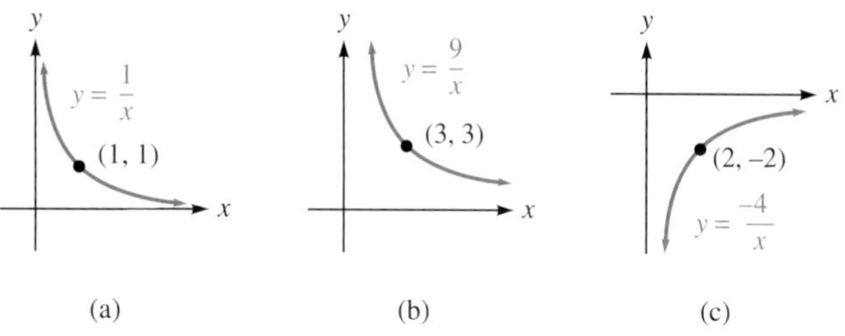

Figure 2-51

■ Joint Variation

Joint Variation The words **"y varies jointly with w and x"** mean that $y = kwx$ for some real-number constant k.

ACTIVE EXAMPLE 5 Kinetic energy of an object varies jointly with its mass and the square of its velocity. A 25-gram mass moving at the rate of 30 centimeters per second has a kinetic energy of 11,250 dyne-centimeters. Find the kinetic energy of a 10-gram mass that is moving at 40 centimeters per second.

Solution If we let E, m, and v represent the kinetic energy, mass, and velocity, respectively, the phrase "energy varies jointly with the mass and the square of its velocity" translates into the formula

COLLEGE **Algebra** f(x) **Now**™

Go to **http://1pass.thomson.com** or your CD to practice this example.

$$E = kmv^2$$

The constant k can be evaluated by substituting 11,250 for E, 25 for m, and 30 for v in the formula.

$$\begin{aligned} E &= kmv^2 \\ \mathbf{11{,}250} &= k(\mathbf{25})(\mathbf{30})^2 \\ 11{,}250 &= 22{,}500k \\ k &= \frac{1}{2} \end{aligned}$$

We can now substitute $\frac{1}{2}$ for k, 10 for m, and 40 for v in the formula and evaluate E.

$$\begin{aligned} E &= kmv^2 \\ &= \frac{\mathbf{1}}{\mathbf{2}}(\mathbf{10})(\mathbf{40})^2 \\ &= 8{,}000 \end{aligned}$$

A 10-gram mass that is moving at 40 centimeters per second has a kinetic energy of 8,000 dyne-centimeters.

Self Check Find the kinetic energy of a 25-gram mass that is moving at 100 centimeters per second. ■

■ Combined Variation

The preceding terminology can be used in various combinations. In each of the following statements, the formula on the left translates into the words on the right.

$y = \frac{kx}{z}$	y varies directly with x and inversely with z.
$y = kx^2\sqrt[3]{z}$	y varies jointly with the square of x and the cube root of z.
$y = \frac{kx\sqrt{z}}{\sqrt[3]{t}}$	y varies jointly with x and the square root of z and inversely with the cube root of t.
$y = \frac{k}{xz}$	y varies inversely with the product of x and z.

EXAMPLE 6 If it takes 100 workers 4 weeks to build 2 miles of highway, how long will it take 80 workers to build 10 miles of highway?

Solution The time it takes to build a highway varies directly with the length of the road, but inversely with the number of workers. We can let t represent the time in weeks, l represent the length in miles, and w represent the number of workers. The relationship between these variables can be expressed by the equation

$$t = \frac{kl}{w}$$

We substitute 4 for t, 100 for w, and 2 for l to find k:

$$4 = \frac{k(2)}{100}$$

$$400 = 2k \quad \text{Multiply both sides by 100.}$$

$$200 = k \quad \text{Divide both sides by 2.}$$

We now substitute 80 for w, 10 for l, and 200 for k in the equation $t = \frac{kl}{w}$ and simplify:

$$t = \frac{kl}{w}$$

$$t = \frac{200(10)}{80}$$

$$= 25$$

It will take 25 weeks for 80 workers to build 10 miles of highway.

Self Check How long will it take 100 workers to build 20 miles of highway? ■

Self Check Answers

1. $\frac{23}{2}$ **2.** 19.2 oz **3.** 140 mi **4.** 16 lumens per cm^2 **5.** 125,000 dyne-centimeters **6.** 40 weeks

2.5 Exercises

VOCABULARY AND CONCEPTS ***Fill in the blanks.***

1. A ratio is the quotient of two numbers.
2. A proportion is a statement that two ratios are equal.
3. In the proportion $\frac{a}{b} = \frac{c}{d}$, b and c are called the means.
4. In the proportion $\frac{a}{b} = \frac{c}{d}$, a and d are called the extremes.
5. In a proportion, the product of the extremes is equal to the product of the means.
6. Direct variation translates into the equation $y = kx$.
7. The equation $y = \frac{k}{x}$ indicates inverse variation.
8. In the equation $y = kx$, k is called the constant of proportionality.
9. The equation $y = kxz$ represents joint variation.
10. In the equation $y = \frac{kx^2}{z^3}$, y varies directly with x^2 and inversely with z^3.

PRACTICE ***Solve each proportion.***

11. $\frac{4}{x} = \frac{2}{7}$ 14
12. $\frac{5}{2} = \frac{x}{6}$ 15
13. $\frac{x}{2} = \frac{3}{x+1}$ 2, −3
14. $\frac{x+5}{6} = \frac{7}{8-x}$ 1, 2

Set up and solve a proportion to answer each question.

15. The ratio of women to men in a mathematics class is 3 to 5. How many women are in the class if there are 30 men? 18
16. The ratio of lime to sand in mortar is 3 to 7. How much lime must be mixed with 21 bags of sand to make mortar? 9 bags

Find the constant of proportionality.

17. y is directly proportional to x. If $x = 30$, then $y = 15$. $\frac{1}{2}$

18. z is directly proportional to t. If $t = 7$, then $z = 21$. **3**

19. I is inversely proportional to R. If $R = 20$, then $I = 50$. **1,000**

20. R is inversely proportional to the square of I. If $I = 25$, then $R = 100$. **62,500**

21. E varies jointly with I and R. If $R = 25$ and $I = 5$, then $E = 125$. **1**

22. z is directly proportional to the sum of x and y. If $x = 2$ and $y = 5$, then $z = 28$. **4**

Solve each problem.

23. y is directly proportional to x. If $y = 15$ when $x = 4$, find y when $x = \frac{7}{5}$. $\frac{21}{4}$

24. w is directly proportional to z. If $w = -6$ when $z = 2$, find w when $z = -3$. **9**

25. P varies jointly with r and s. If $P = 16$ when $r = 5$ and $s = -8$, find P when $r = 2$ and $s = 10$. **−8**

26. m varies jointly with the square of n and the square root of q. If $m = 24$ when $n = 2$ and $q = 4$, find m when $n = 5$ and $q = 9$. **225**

Decide whether the graph could represent direct variation, inverse variation, or neither.

27.

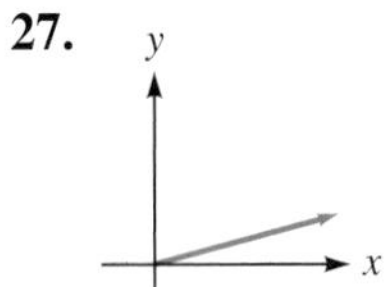

direct variation

28.

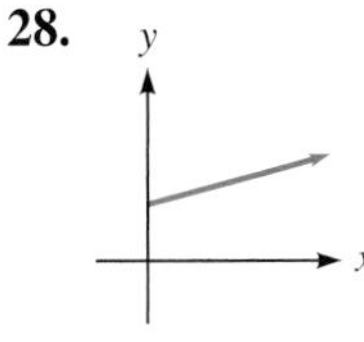

neither

29.

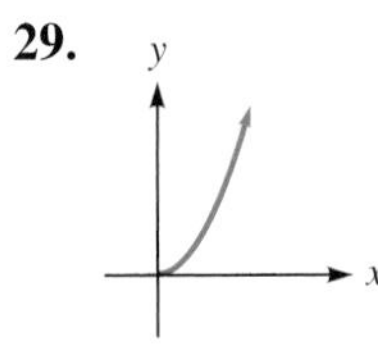

neither

30.

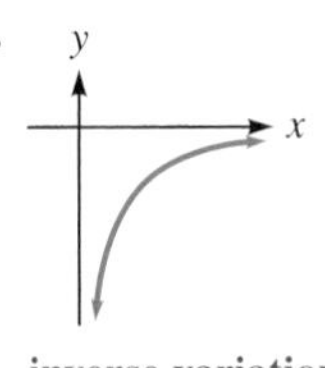

inverse variation

APPLICATIONS ***Set up and solve the required proportion.***

31. Caffeine Many convenience stores sell supersize 44-ounce soft drinks in refillable cups. For each of the products listed in the table, find the amount of caffeine contained in one of the large cups. Round to the nearest milligram. **202, 172, 136**

Soft drink, 12 oz	Caffeine (mg)
Mountain Dew	55
Coca-Cola Classic	47
Pepsi	37

Based on data from *Los Angeles Times* (November 11, 1997) p. S4

32. Telephones As of 2003, Iceland had 221 mobile cellular telephones per 250 inhabitants—the highest rate of any country in the world. If Iceland's population is about 280,000, how many mobile cellular telephones does the country have? **247,520**

33. Wallpapering Read the instructions on the label of wallpaper adhesive. Estimate the amount of adhesive needed to paper 500 square feet of kitchen walls if a heavy wallpaper will be used. **about 2 gal**

COVERAGE: One-half gallon will hang approximately 4 single rolls (140 sq ft), depending on the weight of the wall covering and the condition of the wall.

34. Recommended dosages The recommended child's dose of the sedative hydroxine is 0.006 gram per kilogram of body mass. Find the dosage for a 30-kg child in milligrams. **0.18 g or 180 mg**

35. Gas laws The volume of a gas varies directly with the temperature and inversely with the pressure. When the temperature of a certain gas is 330°, the pressure is 40 pounds per square inch and the volume is 20 cubic feet. Find the volume when the pressure increases 10 pounds per square inch and the temperature decreases to 300°. $14\frac{6}{11}$ **ft**^{3}

36. Hooke's law The force f required to stretch a spring a distance d is directly proportional to d. A force of 5 newtons stretches a spring 0.2 meter. What force will stretch the spring 0.35 meter? $8\frac{3}{4}$ **newtons**

37. Free-falling object The distance that an object will fall in t seconds varies directly with the square of t. An object falls 16 feet in 1 second. How long will it take the object to fall 144 feet? **3 sec**

38. Heat dissipation The power, in watts, dissipated as heat in a resistor varies jointly with the resistance, in ohms, and the square of the current, in amperes. A 10-ohm resistor carrying a current of 1 ampere dissipates 10 watts. How much power is dissipated in a 5-ohm resistor carrying a current of 3 amperes? **45 watts**

39. **Heat dissipation** The power, in watts, dissipated as heat in a resistor varies directly with the square of the voltage and inversely with the resistance. If 20 volts are placed across a 20-ohm resistor, it will dissipate 20 watts. What voltage across a 10-ohm resistor will dissipate 40 watts? **20 volts**

40. **Period of a pendulum** The time required for one complete swing of a pendulum is called the **period** of the pendulum. The period varies directly with the square of its length. If a 1-meter pendulum has a period of 1 second, find the length of a pendulum with a period of 2 seconds. $\sqrt{2}$ **m**

41. **Frequency of vibration** The **pitch,** or **frequency,** of a vibrating string varies directly with the square root of the tension. If a string vibrates at a frequency of 144 hertz due to a tension of 2 pounds, find the frequency when the tension is 18 pounds. **432 hertz**

42. **Kinetic energy** The kinetic energy of an object varies jointly with its mass and the square of its velocity. What happens to the energy when the mass is doubled and the velocity is tripled? **It is multiplied by 18.**

43. **Gravitational attraction** The gravitational attraction between two massive objects varies jointly with their masses and inversely with the square of the distance between them. What happens to this force if each mass is tripled and the distance between them is doubled? **The force is multiplied by $\frac{9}{4}$.**

44. **Gravitational attraction** In Problem 43, what happens to the force if one mass is doubled and the other tripled and the distance between them is halved? **It is multiplied by 24.**

45. **Plane geometry** The area of an equilateral triangle varies directly with the square of the length of a side. Find the constant of proportionality. $\frac{\sqrt{3}}{4}$

46. **Solid geometry** The diagonal of a cube varies directly with the length of a side. Find the constant of proportionality. $\sqrt{3}$

DISCOVERY AND WRITING

47. Explain the terms *extremes* and *means.*

48. Distinguish between a *ratio* and a *proportion.*

49. Explain the term *joint variation.*

50. Explain why $\frac{y}{x} = k$ indicates that y varies directly with x.

51. Explain why $xy = k$ indicates that y varies inversely with x.

52. As temperature increases on the Fahrenheit scale, it also increases on the Celsius scale. Is this direct variation? Explain.

REVIEW ***Perform each operation and simplify.***

53. $\dfrac{1}{x+2} + \dfrac{2}{x+1}$ $\quad \dfrac{3x+5}{(x+2)(x+1)}$

54. $\dfrac{x^2-1}{x+1} \cdot \dfrac{x-1}{x^2-2x+1}$ $\quad 1$

55. $\dfrac{x^2+3x-4}{x^2-5x+4} \div \dfrac{x-1}{x^2-3x-4}$ $\quad \dfrac{(x+4)(x+1)}{x-1}$

56. $\dfrac{x+2}{3x-3} \div (2x+4)$ $\quad \dfrac{1}{6(x-1)}$

57. $\dfrac{x^2+4-(x+2)^2}{4x^2}$ $\quad -\dfrac{1}{x}$

58. $\dfrac{\frac{1}{x}-\frac{1}{3}}{\frac{1}{x}-1}$ $\quad \dfrac{3-x}{3-3x}$

PROBLEMS AND PROJECTS

1. Seven years ago, the ABC Corporation bought a network server for \$10,350. Now, its salvage value is \$250. The current price of a replacement computer is \$4,500, but it will be worthless in four years. Seven years ago, XYZ Incorporated also bought a server, but paid \$17,500. They expect five more years of useful life, but then it will be worthless. Find the depreciation equation of each computer. Which company made the best purchasing choice, and why?
$y = -\frac{10,100}{7}x + 10,350, y = -1,125x + 4,500,$
$y = -\frac{4,375}{3}x + 17,500$, **ABC**

2. Two lines pass through the origin. One passes through (a, b) and the other through (c, d). The lines are also perpendicular. Show that $ac + bd = 0$.

3. The highway department plans to extend Diagonal Road to 15th Avenue, as in the illustration. Where will it cross 11th Street? **between 13th & 14th Aves.**

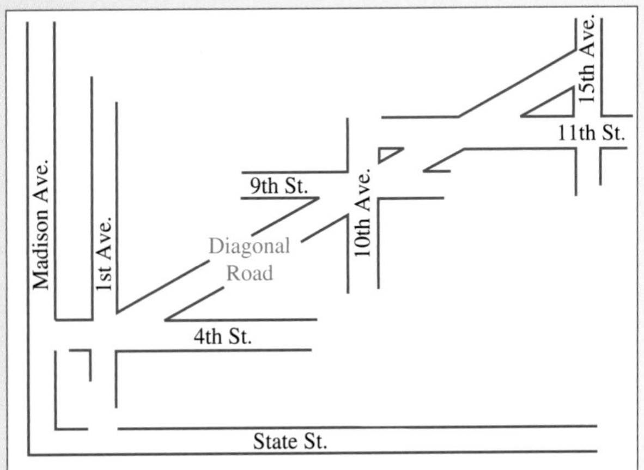

4. A photographer uses the graph shown to determine how much concentrate to use to make a batch of developer. Find the equation of the line and graph it using a graphing calculator. Use both the equation and the calculator graph to find the amount of concentrate needed to make 40 ounces of solution. Do the graph and equation give the same answer? $y = \frac{13}{160}x$**, 3.25 oz, yes**

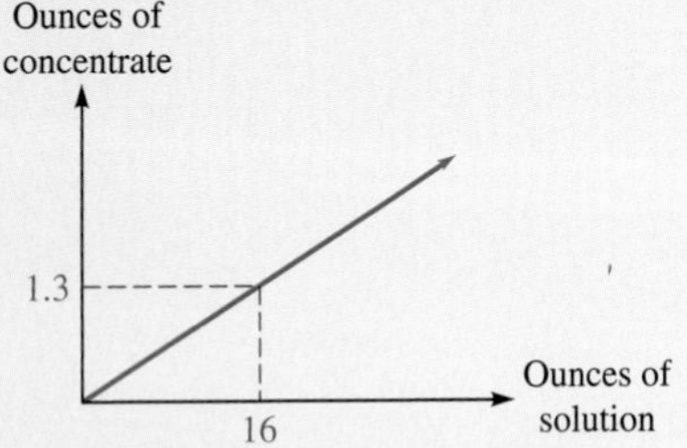

Project 1

The time required to develop film depends on temperature. A photographer would like to use graphs like those shown to determine proper development times for temperatures between 60° and 80° F. The film manufacturer supplied the information in the table. Find the equations of the three lines and graph them. For which contrast is the correct processing temperature most critical? Why?

	Processing time (minutes)		
Temperature	**High contrast**	**Average contrast**	**Low contrast**
60°	21	15	10
80°	8	$5\frac{1}{2}$	4

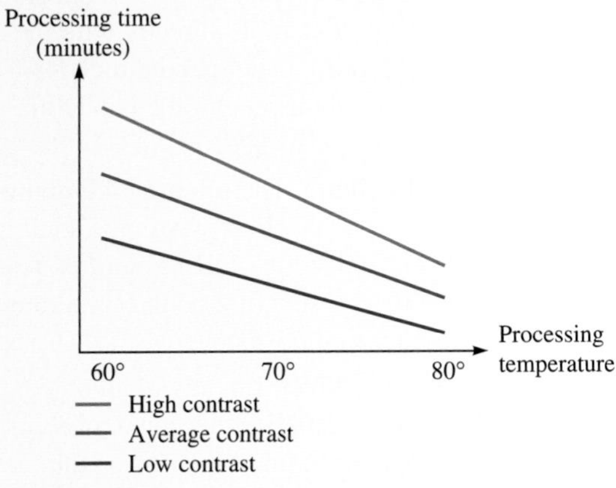

Project 2

It is the weight of the water that keeps objects from sinking. For example, the weights of the two buoys in the illustration are supported by forces exactly equal to the weight of the volume of water displaced by the submerged portions.

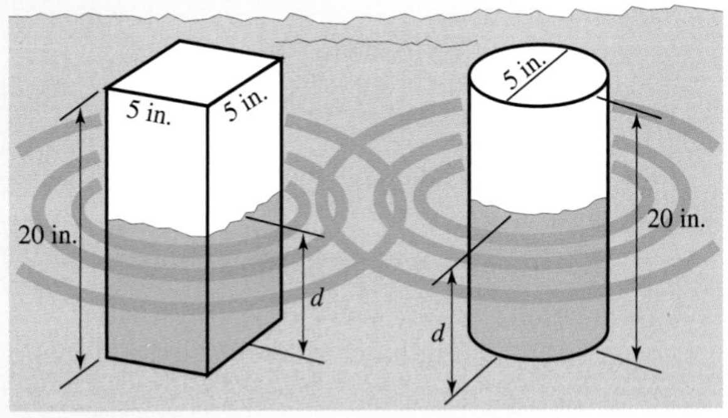

For each buoy, find an equation relating d, the depth it is submerged, to w, the weight. Graph each equation. Are there restrictions on the variable d? How can you determine, from the equations and from the graphs, the greatest floatable weights? Each cubic foot of fresh water weighs 62.4 pounds.

Project 3

Any two points lie on a line. Three or more points, picked at random, will probably not lie on the same line. There is, however, a line that comes as close as possible to passing through several points. The four points in the illustration do not lie on the same line. Find equations of the six lines determined by pairs of points: lines *AB*, *AC*, *AD*, *BC*, *BD*, and *CD*. One of these lines is drawn. Which of the six do you think is the "best fit" of all four points, and why? (You might consider the deviations of the *y*-values, marked *a* and *b* on the graph.) Is there another line that you think is a better fit than any of these six? Why?

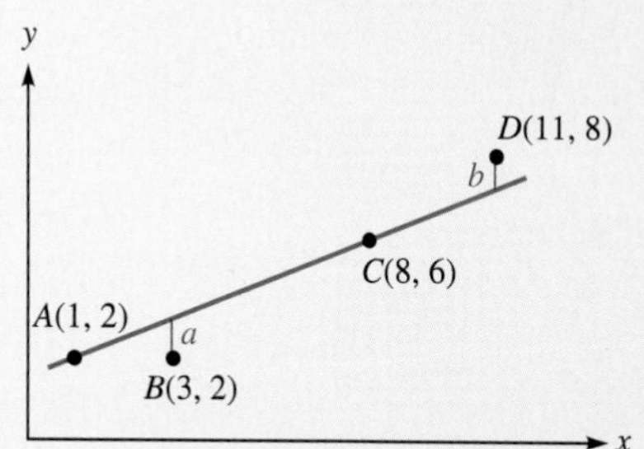

CHAPTER SUMMARY

CONCEPTS | **REVIEW EXERCISES**

2.1 The Rectangular Coordinate System

The rectangular coordinate system divides the plane into four quadrants.

Refer to the illustration and find the coordinates of each point.

1. *A* (2, 0)

2. *B* (−2, 1)

3. *C* (0, −1)

4. *D* (3, −1)

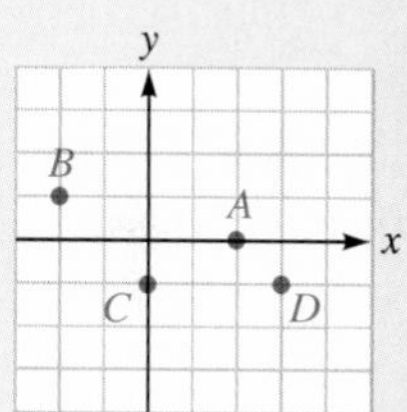

Graph each point. Indicate the quadrant in which the point lies, or the axis on which it lies.

5. (−3, 5)

6. (5, −3)

7. (0, −7)

8. $\left(-\frac{1}{2}, 0\right)$

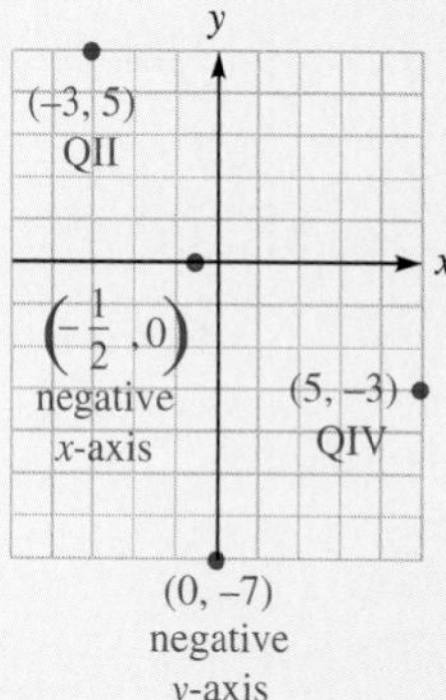

The graph of an equation in x and y is the set of all points (x, y) that satisfy the equation.

Use the x- and the y-intercepts to graph each equation.

9. $3x - 5y = 15$

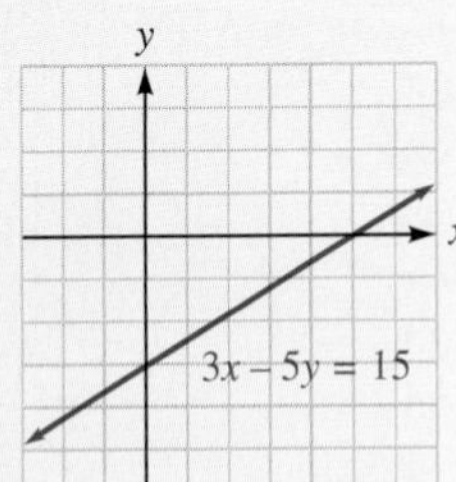

10. $x + y = 7$

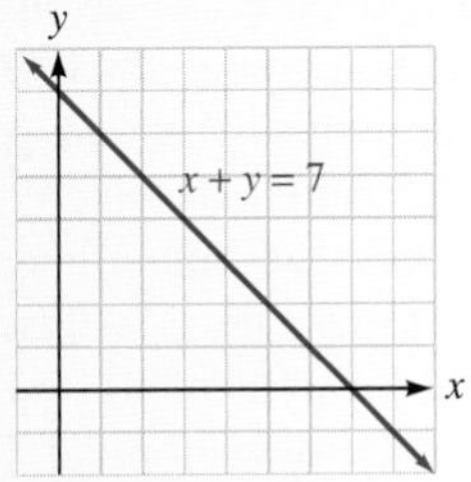

11. $x + y = -7$

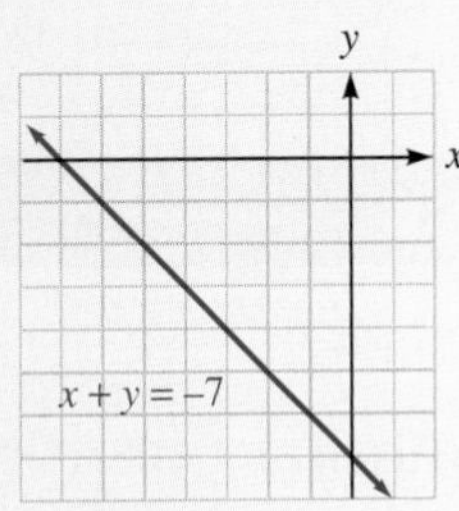

12. $x - 5y = 5$

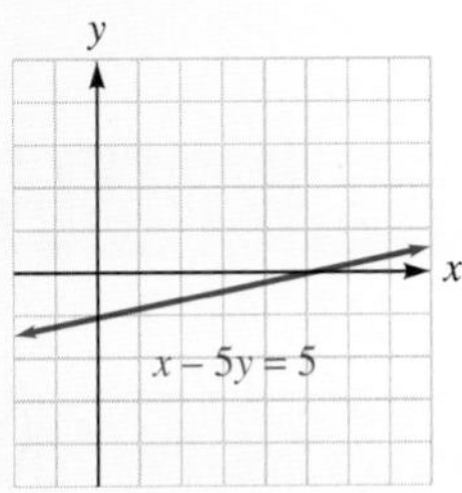

Equation of a vertical line through (a, b):

$x = a$

Equation of a horizontal line through (a, b):

$y = b$

Graph each equation.

13. $y = 4$

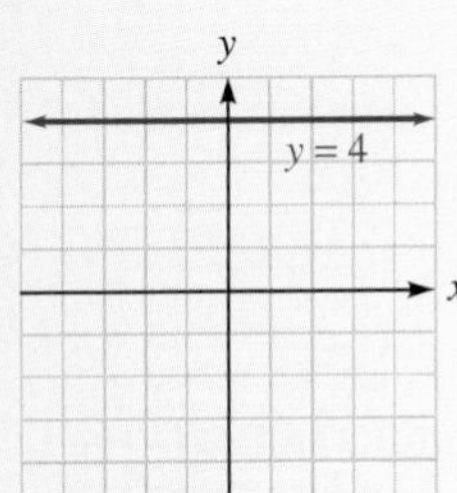

14. $x = -2$

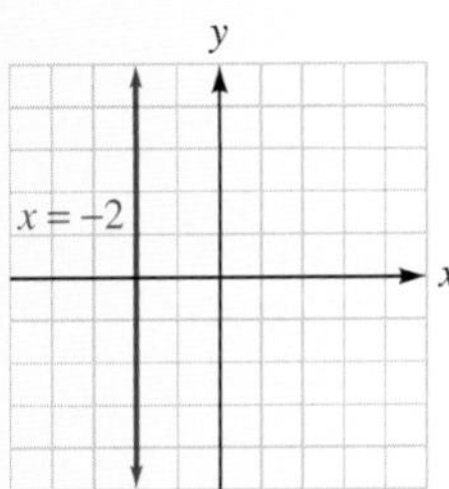

15. Depreciation A car purchased for \$18,750 is expected to depreciate according to the formula $y = -2{,}200x + 18{,}750$. Find its value after 3 years. **\$12,150**

The distance formula:

$$d = \sqrt{(x_2 - x_1)^2 + (y_2 - y_1)^2}$$

Find the length of the segment PQ.

16. $P(-3, 7)$; $Q(3, -1)$ **10**

17. $P(0, 5)$; $Q(-12, 10)$ **13**

18. $P(\sqrt{3}, 9)$; $Q(\sqrt{3}, 7)$ **2**

19. $P(a, -a)$; $Q(-a, a)$ $2\sqrt{2}|a|$

The midpoint formula:
The midpoint of the line segment joining (x_1, y_1) and (x_2, y_2) is the point M with coordinates

$$\left(\frac{x_1 + x_2}{2}, \frac{y_1 + y_2}{2}\right)$$

Find the midpoint of the segment PQ.

20. $P(-3, 7)$; $Q(3, -1)$ **(0, 3)**

21. $P(0, 5)$; $Q(-12, 10)$ $\left(-6, \frac{15}{2}\right)$

22. $P(\sqrt{3}, 9)$; $Q(\sqrt{3}, 7)$ $(\sqrt{3}, 8)$

23. $P(a, -a)$; $Q(-a, a)$ **(0, 0)**

2.2 The Slope of a Nonvertical Line

The slope m of a line passing through (x_1, y_1) and (x_2, y_2) is given by

$$m = \frac{y_2 - y_1}{x_2 - x_1} \quad (x_2 \neq x_1)$$

The slope of a vertical line is undefined.

Find the slope of the line PQ, if possible.

24. $P(3, -5)$; $Q(1, 7)$ **−6**

25. $P(2, 7)$; $Q(-5, -7)$ **2**

26. $P(b, a)$; $Q(a, b)$ **−1**

27. $P(a + b, b)$; $Q(b, b - a)$ **1**

28. Rate of descent If an airplane descends 3,000 feet in 15 minutes, what is the average rate of descent in feet per minute? **200 ft/min**

Determine whether the slope of each line is 0 or undefined.

29.

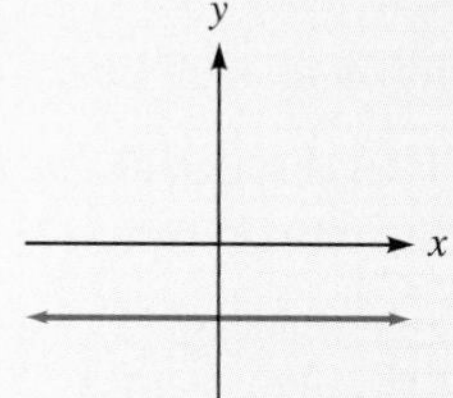

0

30.

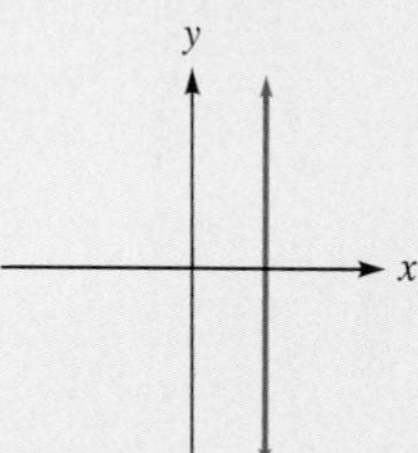

undefined

Determine whether the slope of each line is positive or negative.

31.

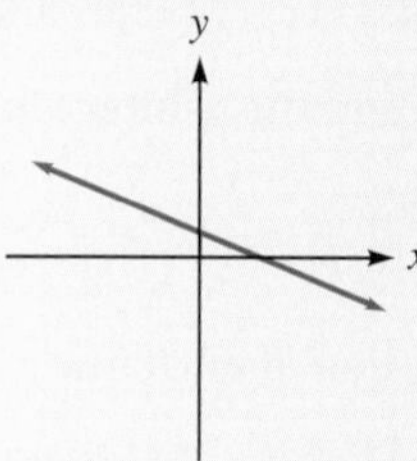

negative

32.

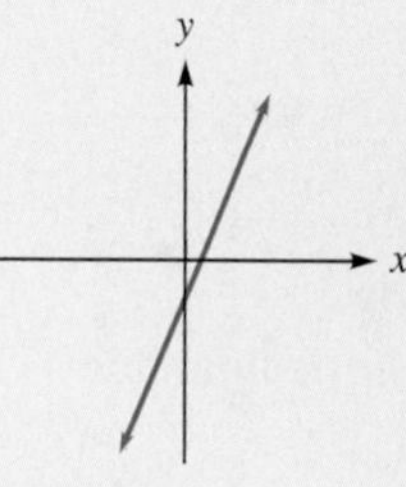

positive

Nonvertical parallel lines have the same slope.

The product of the slopes of two perpendicular lines is -1, provided neither line is vertical.

33. A line passes through $(-2, 5)$ and $(6, 10)$. A line parallel to it passes through $(2, 2)$ and $(10, y)$. Find y. $y = 7$

34. A line passes through $(-2, 5)$ and $(6, 10)$. A line perpendicular to it passes through $(-2, 5)$ and $(x, -3)$. Find x. $x = 3$

2.3 Writing Equations of Lines

Point–slope form:

$$y - y_1 = m(x - x_1)$$

Use point–slope form to write the equation of each line.

35. The line passes through the origin and the point $(-5, 7)$. $y = -\frac{7}{5}x$

36. The line passes through $(-2, 1)$ and has a slope of -4. $y - 1 = -4(x + 2)$

37. The line passes through $(7, -5)$ and $(4, 1)$.
$y + 5 = -2(x - 7)$ or $y - 1 = -2(x - 4)$

Slope–intercept form:

$$y = mx + b$$

Use slope–intercept form to write the equation of each line.

38. The line has a slope of $\frac{2}{3}$ and a y-intercept of 3. $y = \frac{2}{3}x + 3$

39. The slope is $-\frac{3}{2}$ and the line passes through $(0, -5)$. $y = -\frac{3}{2}x - 5$

Use slope–intercept form to graph each equation.

40. $y = \frac{3}{5}x - 2$

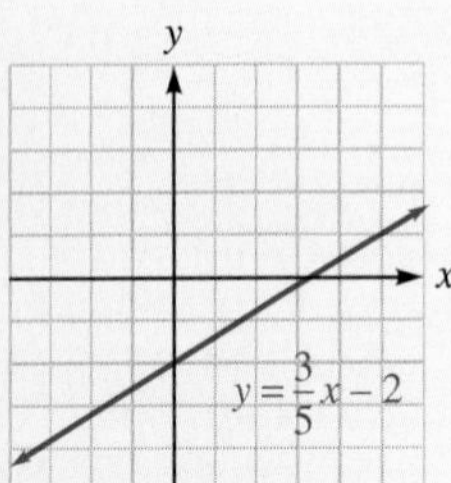

41. $y = -\frac{4}{3}x + 3$

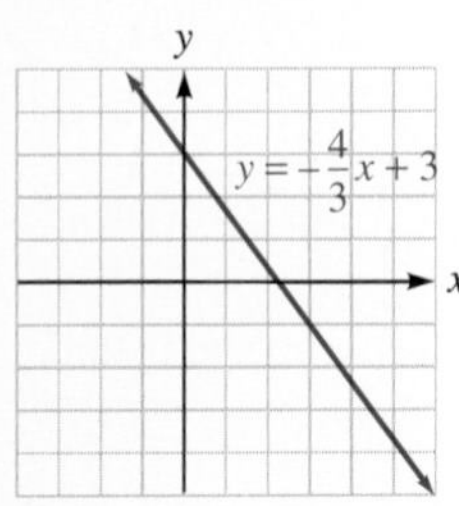

Write the equation of each line.

42. The line passes through $(7, -2)$ and is parallel to the line segment joining $(2, 4)$ and $(4, -10)$. $y = -7x + 47$

43. The line passes through $(7, -2)$ and is perpendicular to the line segment joining $(2, 4)$ and $(4, -10)$. $y = \frac{1}{7}x - 3$

44. The line is parallel to $3x - 4y = 7$ and passes through $(2, 0)$. $y = \frac{3}{4}x - \frac{3}{2}$

45. The line passes through $(0, 5)$ and is perpendicular to the line $3y + x - 4 = 0$. $y = 3x + 5$

General form of a line:

$Ax + By = C$

Find the slope and the *y*-intercept of the graph of each line.

46. $5x + 2y = 7$
slope $= -\frac{5}{2}$; $\left(0, \frac{7}{2}\right)$

47. $3x - 4y = 14$
slope $= \frac{3}{4}$; $\left(0, -\frac{7}{2}\right)$

Equation of a horizontal line through (a, b):

$y = b$

Equation of a vertical line through (a, b):

$x = a$

Write the equation of each line.

48. The line has a slope of 0 and passes through $(-5, 17)$. $y = 17$

49. The line has no defined slope and passes through $(-5, 17)$. $x = -5$

2.4 Graphs of Equations

Intercepts of a graph:

To find the ***x*-intercepts,** let $y = 0$ and solve for x.

To find the ***y*-intercepts,** let $x = 0$ and solve for y.

Tests for symmetry:
If $(-x, y)$ lies on the graph whenever (x, y) does, the graph is symmetric about the y-axis.

Graph each equation. Find all intercepts and symmetries.

50. $y = x^2 + 2$

51. $y = x^3 - 2$

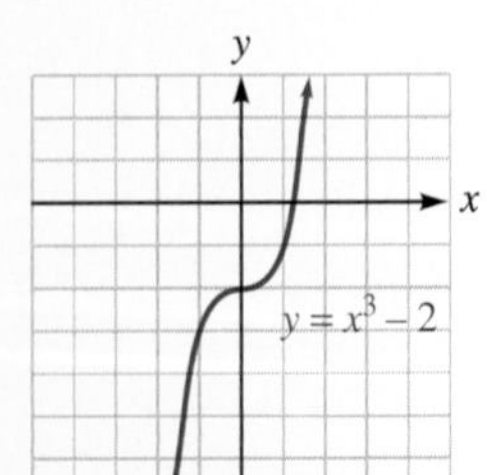

If $(x, -y)$ lies on the graph whenever (x, y) does, the graph is symmetric about the x-axis.

If $(-x, -y)$ lies on the graph whenever (x, y) does, the graph is symmetric about the origin.

52. $y = \frac{1}{2}|x|$

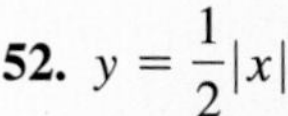

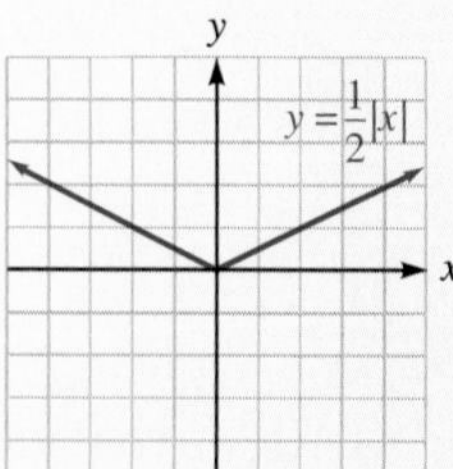

53. $y = -\sqrt{x - 4}$

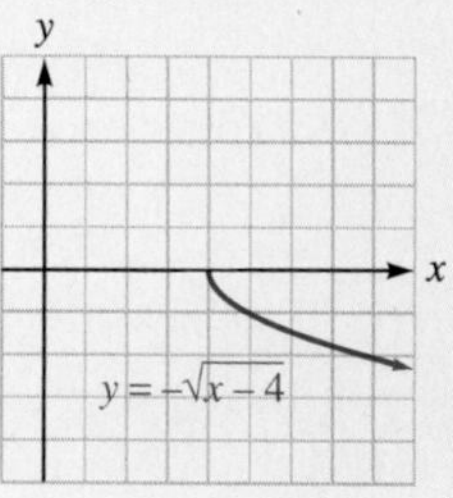

54. $y = \sqrt{x} + 2$

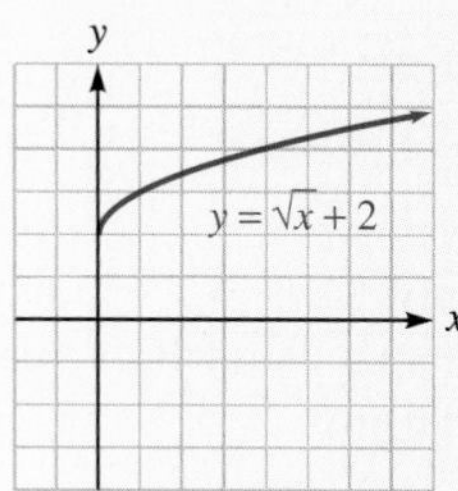

55. $y = |x + 1| + 2$

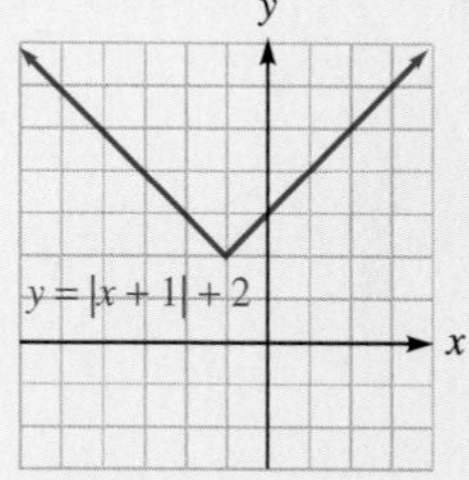

Use a graphing calculator to graph each equation.

56. $y = |x - 4| + 2$

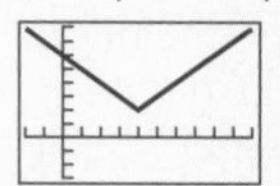

57. $y = -\sqrt{x + 2} + 3$

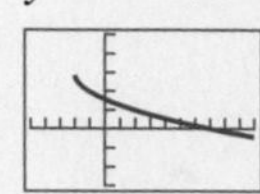

58. $y = x + 2|x|$

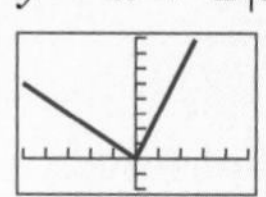

59. $y^2 = x - 3$

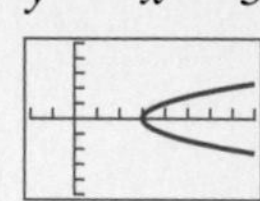

Equation of a circle:
Center at (h, k):
$$(x - h)^2 + (y - k)^2 = r^2$$
Center at $(0, 0)$:
$$x^2 + y^2 = r^2$$

Write the equation of each circle.

60. Center at $(-3, 4)$; radius 12
$(x + 3)^2 + (y - 4)^2 = 144$, **or**
$x^2 + y^2 + 6x - 8y - 119 = 0$

61. Ends of diameter at $(-6, -3)$ and $(5, 8)$
$\left(x + \frac{1}{2}\right)^2 + \left(y - \frac{5}{2}\right)^2 = \frac{121}{2}$, **or**
$x^2 + y^2 + x - 5y - 54 = 0$

Graph each equation.

62. $x^2 + y^2 - 2y = 15$

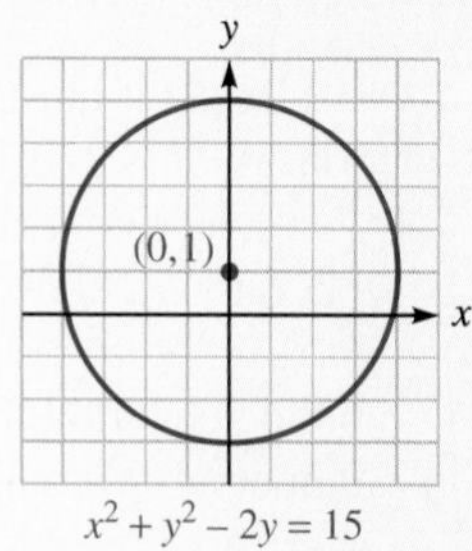

63. $x^2 + y^2 - 4x + 2y = 4$

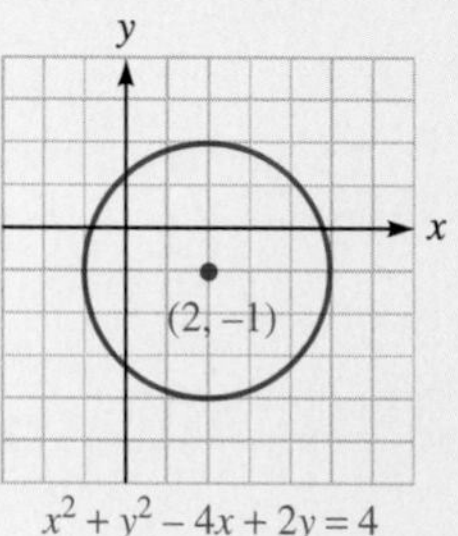

Use a graphing calculator to solve each equation. If an answer is not exact, round to the nearest hundredth.

64. $x^2 - 11 = 0$ **3.32, −3.32**

65. $x^3 - x = 0$ **−1, 0, 1**

66. $|x^2 - 2| - 1 = 0$ **1.73, −1.73, 1, −1**

67. $x^2 - 3x = 5$ **4.19, −1.19**

2.5 Proportion and Variation

In any proportion, the product of the extremes is equal to the product of the means.

Solve each proportion.

68. $\frac{x+3}{10} = \frac{x-1}{x}$ $x = 2, x = 5$

69. $\frac{x-1}{2} = \frac{12}{x+1}$ $x = -5, x = 5$

y varies directly with x: $y = kx$

70. Hooke's law The force required to stretch a spring is directly proportional to the amount of stretch. If a 3-pound force stretches a spring 5 inches, what force would stretch the spring 3 inches? **$\frac{9}{5}$ lb**

71. Kinetic energy A moving body has a kinetic energy directly proportional to the square of its velocity. By what factor does the kinetic energy of an automobile increase if its speed increases from 30 mph to 50 mph? **$\frac{25}{9}$**

y varies inversely with x: $y = \frac{k}{x}$

72. Gas laws The volume of gas in a balloon varies directly as the temperature and inversely as the pressure. If the volume is 400 cubic centimeters when the temperature is 300 K and the pressure is 25 dynes per square centimeter, find the volume when the temperature is 200 K and the pressure is 20 dynes per square centimeter. **$333\frac{1}{3}$ cc**

y varies jointly with w and x: $y = kwx$

73. The area of a rectangle varies jointly with its length and width. Find the constant of proportionality. **1**

74. Electrical resistance The resistance of a wire varies directly as the length of the wire and inversely as the square of its diameter. A 1,000-foot length of wire, 0.05 inches in diameter, has a resistance of 200 ohms. What would be the resistance of a 1,500-foot length of wire that is 0.08 inches in diameter? **about 117 ohms**

75. Billing for services Angie's Painting and Decorating Service charges a fixed amount for accepting a wallpapering job and adds a fixed dollar amount for each roll hung. If the company bills a customer \$177 to hang 11 rolls, and \$294 to hang 20 rolls, find the cost to hang 27 rolls. **\$385**

76. Paying for college Rolf must earn \$5,040 for next semester's tuition. Assume he works x hours tutoring algebra at \$14 per hour and y hours tutoring Spanish at \$18 per hour and makes his goal. Write an equation expressing the relationship between x and y, and graph the equation. If Rolf tutors algebra for 180 hours, how long must he tutor Spanish? **140 hr**

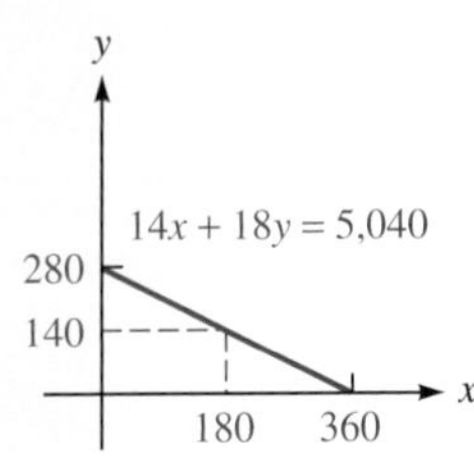

CHAPTER TEST

Indicate the quadrant in which the point lies, or the axis on which it lies.

1. $(-3, \pi)$ **QII**

2. $(0, -8)$ **negative y-axis**

Find the x- and y-intercepts and use them to graph the equation.

3. $x + 3y = 6$

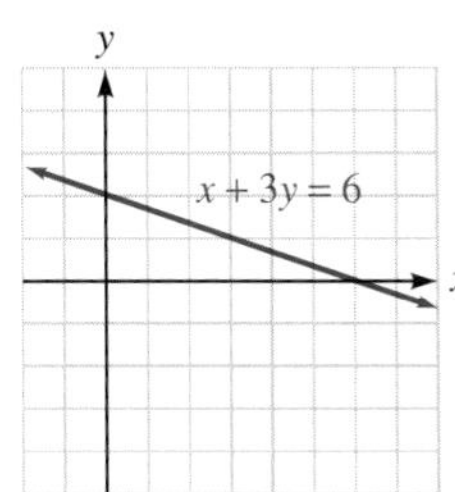

4. $2x - 5y = 10$

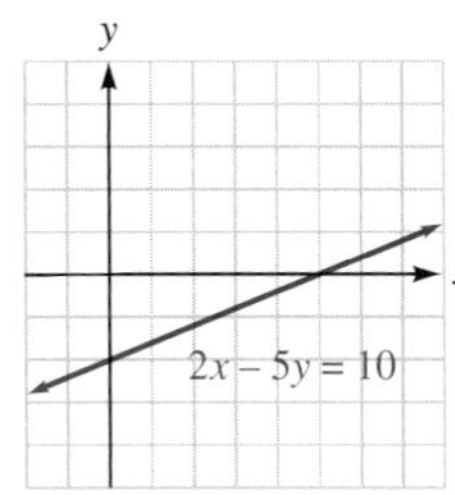

Graph each equation.

5. $2(x + y) = 3x + 5$

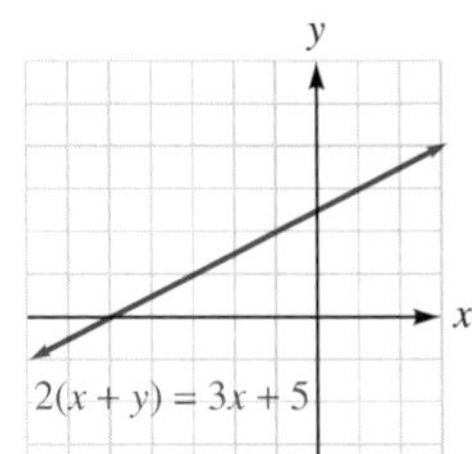

6. $3x - 5y = 3(x - 5)$

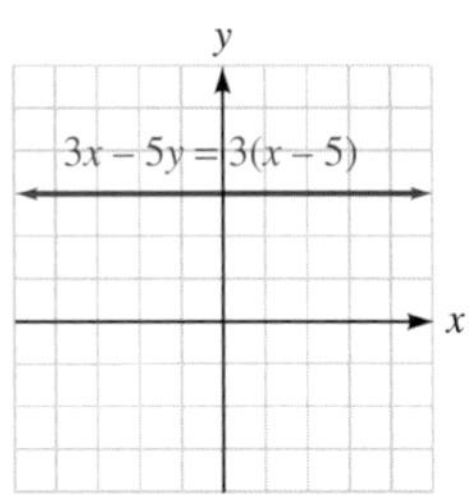

7. $\frac{1}{2}(x - 2y) = y - 1$

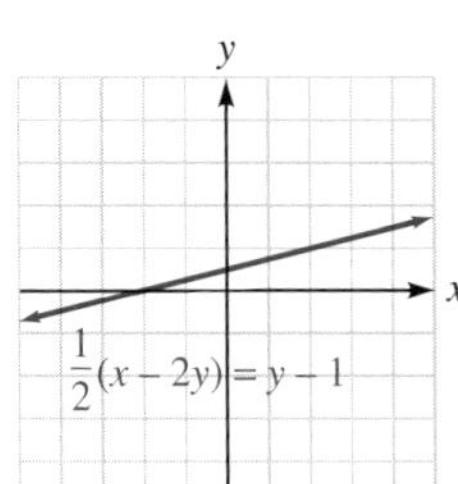

8. $\frac{x + y - 5}{7} = 3x$

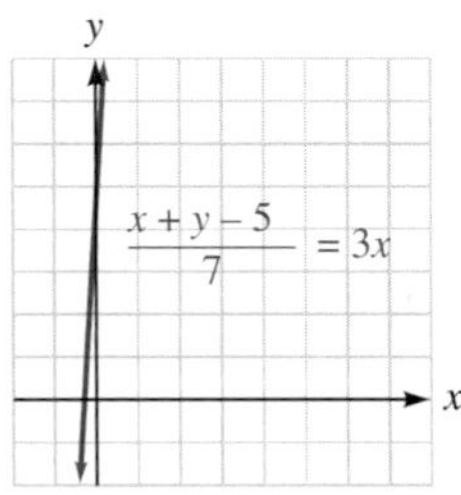

Find the distance between points P and Q.

9. $P(1, -1); Q(-3, 4)$ **$\sqrt{41}$**

10. $P(0, \pi); Q(-\pi, 0)$ **approximately 4.44**

Find the midpoint of the line segment PQ.

11. $P(3, -7); Q(-3, 7)$ **(0, 0)**

12. $P(0, \sqrt{2}); Q(\sqrt{8}, \sqrt{18})$ $(\sqrt{2}, 2\sqrt{2})$

Find the slope of the line PQ.

13. $P(3, -9); Q(-5, 1)$ $-\frac{5}{4}$

14. $P(\sqrt{3}, 3); Q(-\sqrt{12}, 0)$ $\frac{\sqrt{3}}{3}$

Determine whether the two lines are parallel, perpendicular, or neither.

15. $y = 3x - 2; y = 2x - 3$ **neither**

16. $2x - 3y = 5; 3x + 2y = 7$ **perpendicular**

Write the equation of the line with the given properties.

17. Passing through $(3, -5)$; $m = 2$ $y = 2x - 11$

18. $m = 3; b = \frac{1}{2}$ $y = 3x + \frac{1}{2}$

19. Parallel to $2x - y = 3$; $b = 5$ $y = 2x + 5$

20. Perpendicular to $2x - y = 3$; $b = 5$ $y = -\frac{1}{2}x + 5$

21. Passing through $(2, -\frac{3}{2})$ and $(3, \frac{1}{2})$ $y = 2x - \frac{11}{2}$

22. Parallel to the y-axis and passing through $(3, -4)$ $x = 3$

Find the x- and y-intercepts of each graph.

23. $y = x^3 - 16x$ **$(-4, 0), (0, 0), (4, 0); (0, 0)$**

24. $y = |x - 4|$ **$(4, 0); (0, 4)$**

Find the symmetries of each graph.

25. $y^2 = x - 1$ **about the x-axis**

26. $y = x^4 + 1$ **about the y-axis**

Graph each equation.

27. $y = x^2 - 9$

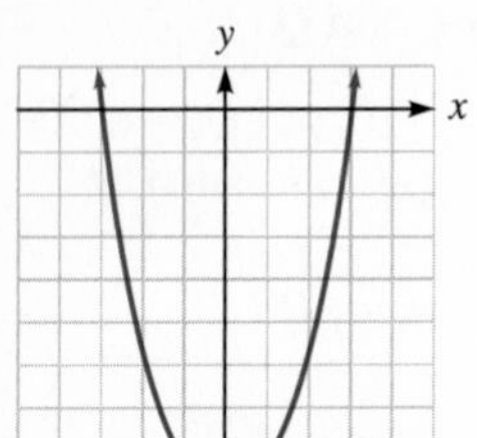

28. $x = |y|$

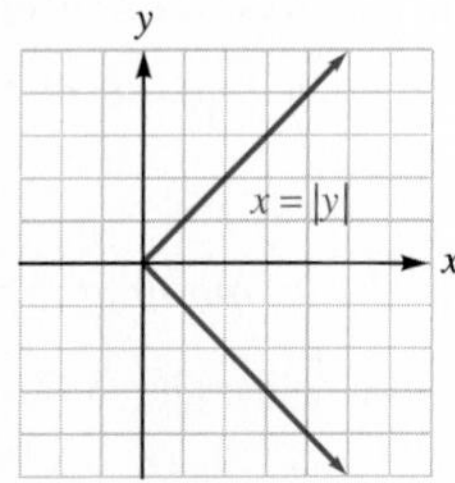

29. $y = 2\sqrt{x}$

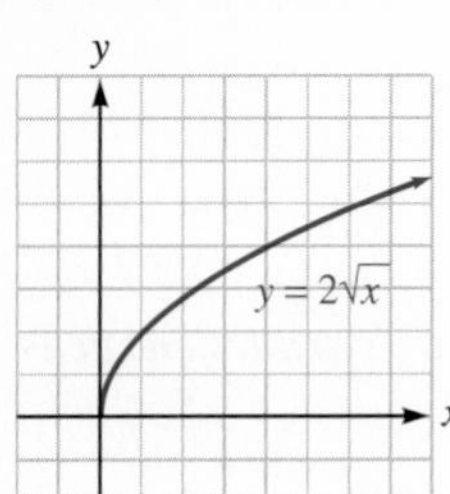

30. $x = y^3$

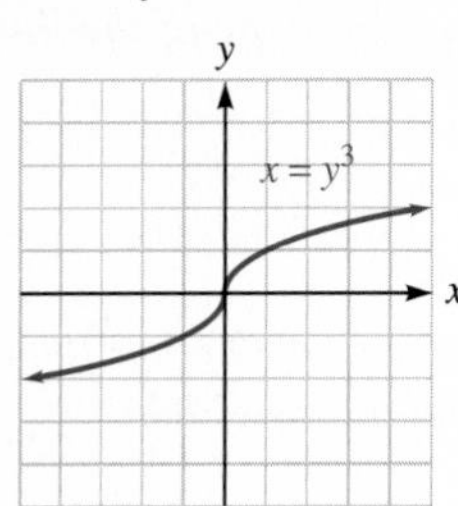

Write the equation of each circle.

31. Center at (5, 7); radius of 8
$(x - 5)^2 + (y - 7)^2 = 64$

32. Center at (2, 4); passing through (6, 8)
$(x - 2)^2 + (y - 4)^2 = 32$

Graph each equation.

33. $x^2 + y^2 = 9$

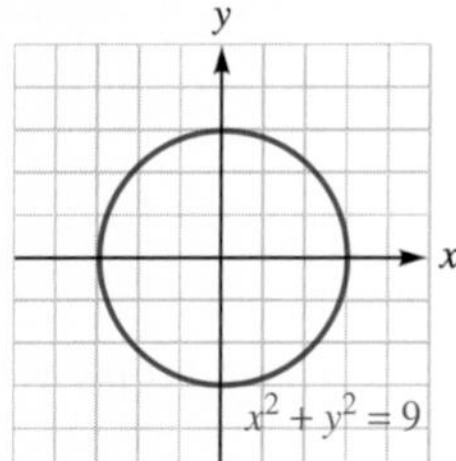

34. $x^2 - 4x + y^2 + 3 = 0$

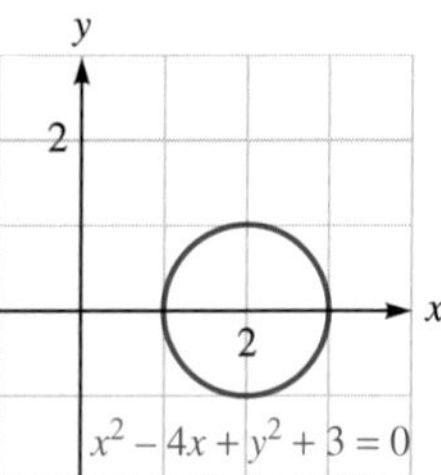

Write each statement as an equation.

35. y varies directly as the square of z. $y = kz^2$

36. w varies jointly with r and the square of s. $w = krs^2$

37. P varies directly with Q. $P = 7$ when $Q = 2$. Find P when $Q = 5$. $P = \frac{35}{2}$

38. y is directly proportional to x and inversely proportional to the square of z, and $y = 16$ when $x = 3$ and $z = 2$. Find x when $y = 2$ and $z = 3$. $x = \frac{27}{32}$

Use a graphing calculator to find the positive root of each equation.

39. $x^2 - 7 = 0$
$x = 2.65$

40. $x^2 - 5x - 5 = 0$
$x = 5.85$

3 Functions

COLLEGE **Algebra** $f(x)$ **Now**™

Throughout the chapter, this icon introduces resources on the College AlgebraNow Web Site, accessed through **http://1pass.thomson.com**, that will:

- Help you test your knowledge of the material in this chapter prior to reading it,
- Allow you to take an exam-prep quiz, and
- Provide a Personalized Learning Plan targeting areas you should study.

Careers and Mathematics

COMPUTER PROGRAMMER

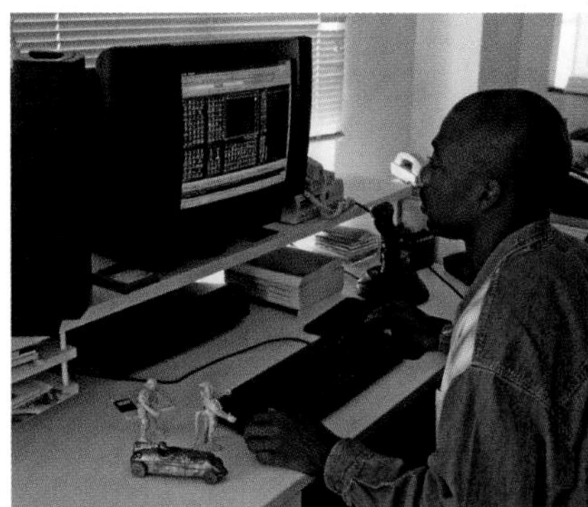
David Young-Wolff/PhotoEdit

Computer programmers write, test, and maintain the detailed instructions, called **programs,** that computers follow to perform their functions. They are grouped into two broad types—applications programmers and systems programmers. Applications programmers write programs to handle a specific job, such as tracking inventory. Systems programmers write programs to maintain and control computer systems software.

Computer programmers held about 499,000 jobs in 2002.

JOB OUTLOOK

Employment of programmers is expected to grow about as fast as the average for all occupations through 2012. Median annual earnings of programmers were $60,290 in 2002.

For a sample application, see Exercise 56 in Section 3.3.

In this chapter, we will discuss one of the most important concepts in mathematics—the concept of function.

3.1 Functions and Function Notation

In this section, you will learn about

- Functions ■ Function Notation ■ The Difference Quotient
- Graphs of Functions ■ The Vertical Line Test ■ Linear Functions

We have seen that an equation in x and y sets up a correspondence between numbers x and values y. Correspondences between the elements of two sets is a common occurrence in everyday life. For example,

- To every house, there corresponds exactly one address.
- To every car, there corresponds exactly one license plate.
- To every state, there corresponds exactly one governor.

In this chapter, we will discuss situations in which exactly one quantity corresponds to (or depends on) another quantity according to some specific rule. For example, the equation $y = x^2 - 1$ sets up a correspondence where each number x determines exactly one value y, according to the rule *square x and subtract 1.* In this case, the value y depends on x. Since the value of y depends on the number x, we call y the **dependent variable** and x the **independent variable.**

The equation $y = x^2 - 1$ determines what output value y will result from each input value x. This idea of inputs and outputs is shown in Figure 3-1(a). In the equation $y = x^2 - 1$, if the input x is 2, the output y is

$$y = 2^2 - 1 = 3$$

This is illustrated in Figure 3-1(b).

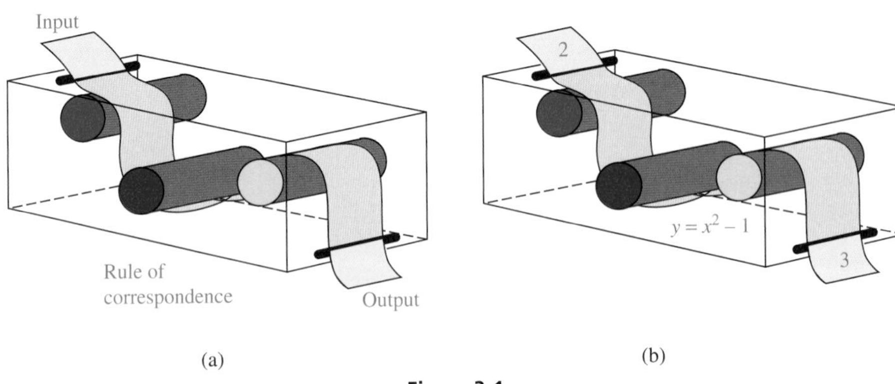

Figure 3-1

The equation $y = x^2 - 1$ also determines the table of ordered pairs and the graph shown in Figure 3-2. To see how the table determines the correspondence, we simply find an input in the x-column and then read across to find the corresponding output in the y-column. If we select $x = 2$ as an input, we get $y = 3$ for the output.

To see how the graph of $y = x^2 - 1$ determines the correspondence, we draw a vertical and horizontal line through any point (say, point P) on the graph shown in Figure 3-2. Because these lines intersect the x-axis at 2 and the y-axis at 3, the point $P(2, 3)$ associates 3 on the y-axis with 2 on the x-axis.

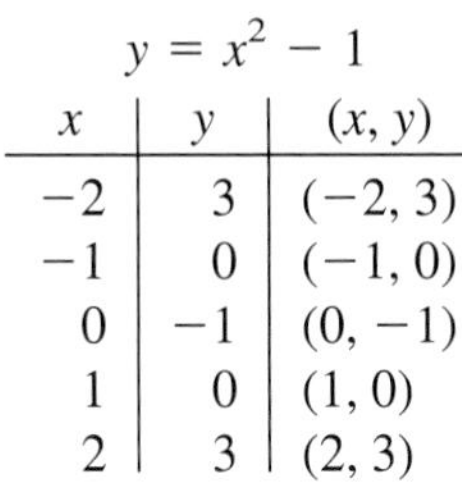

$y = x^2 - 1$

x	y	(x, y)
-2	3	$(-2, 3)$
-1	0	$(-1, 0)$
0	-1	$(0, -1)$
1	0	$(1, 0)$
2	3	$(2, 3)$

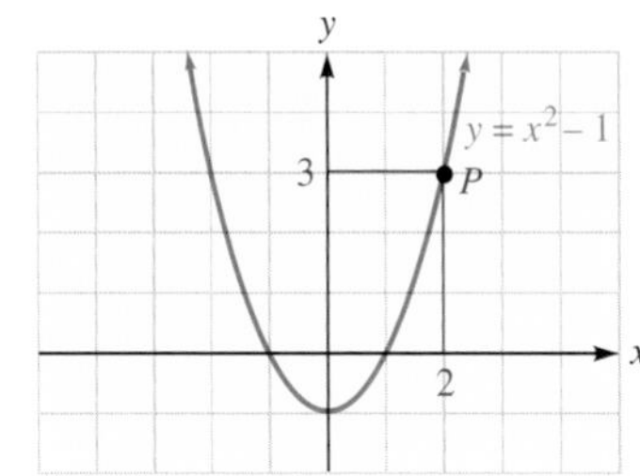

Figure 3-2

Functions

Any correspondence that assigns exactly one value of y to each number x is called a **function.**

Functions

A **function** f is a correspondence between a set of input values x (called the **domain**) and a set of output values y (called the **range**), where to each x-value in the domain there corresponds exactly one y-value in the range.

We can think of the domain of a function as the set of all possible inputs to the function. The range is the set of all possible outputs.

Unless otherwise indicated, we will assume that the domain of a function is the set of all real numbers x for which the function is defined.

EXAMPLE 1 Decide whether the following equations define y to be a function of x. If so, find the domain and range: **a.** $7x + y = 5$ and **b.** $y = \dfrac{3}{x - 2}$.

Solution **a.** First, we solve $7x + y = 5$ for y to get

$$y = 5 - 7x$$

Since there is exactly one value of y for each number we substitute for x, the equation determines y to be a function of x.

Because x can be any real number, the domain is the interval $(-\infty, \infty)$.

Because y can be any real number, the range is the interval $(-\infty, \infty)$.

b. Since each input x (except 2) gives exactly one output y, the equation $y = \frac{3}{x - 2}$ defines y to be a function of x.

The domain is $(-\infty, 2) \cup (2, \infty)$, which is the set of all real numbers except 2. We cannot substitute 2 for x, because division by 0 is undefined.

The range is $(-\infty, 0) \cup (0, \infty)$, which is the set of all real numbers except 0. Here y cannot be 0, because a fraction with a numerator of 3 cannot be 0.

Self Check Decide whether each equation defines y to be a function of x. If so, find the domain and range. **a.** $y = |x|$ and **b.** $y = \sqrt{x}$. ■

EXAMPLE 2 Find the domain and range of the function defined by each equation:

a. $y = \sqrt{3x - 2}$ and **b.** $y = \dfrac{1}{x^2 - 5x - 6}$.

Solution **a.** Since the radicand must be nonnegative, we have

$$3x - 2 \geq 0$$

$$3x \geq 2 \quad \text{Add 2 to both sides.}$$

$$x \geq \frac{2}{3} \quad \text{Divide both sides by 3.}$$

The domain is the interval $\left[\frac{2}{3}, \infty\right)$. Because the radical sign calls for the nonnegative square root, the range is the interval $[0, \infty)$.

b. We can factor the denominator to see what values of x will give 0's in the denominator. These values are not in the domain.

$$x^2 - 5x - 6 = 0$$

$$(x - 6)(x + 1) = 0$$

$$x - 6 = 0 \quad \text{or} \quad x + 1 = 0$$

$$x = 6 \quad | \quad x = -1$$

The domain is $(-\infty, -1) \cup (-1, 6) \cup (6, \infty)$. Since the numerator of the fraction is 1, y cannot be 0. The range is $(-\infty, 0) \cup (0, \infty)$.

Self Check Find the domain and range of the function defined by $y = \sqrt[3]{x + 2}$. ■

ACTIVE EXAMPLE 3 Find the domain and range of the function defined by the equation $y = \dfrac{x + 1}{x - 2}$.

Solution Here, x cannot be 2, because that would give a 0 in the denominator. So the domain is $(-\infty, 2) \cup (2, \infty)$.

One way to find the restrictions on y is to solve the equation for x.

COLLEGE **Algebra** *f(x)* **Now**™

Explore this Active Example and test yourself on the steps taken here at **http://1pass.thomson.com**. You can also find this example on the Interactive Video Skillbuilder CD-ROM.

$$y = \frac{x + 1}{x - 2}$$

$$y(x - 2) = x + 1 \quad \text{If } x \neq 2\text{, we can multiply both sides by } x - 2.$$

$$xy - 2y = x + 1 \quad \text{Use the distributive property to remove parentheses.}$$

$$xy - x = 2y + 1 \quad \text{Add } 2y \text{ and subtract } x \text{ from both sides.}$$

$$x(y - 1) = 2y + 1 \quad \text{Factor out } x.$$

$$x = \frac{2y + 1}{y - 1} \quad \text{Divide both sides by } y - 1.$$

From the last equation, we can see that y cannot be 1. So the range is $(-\infty, 1) \cup (1, \infty)$.

Self Check Find the domain and range of the function defined by $y = \dfrac{x - 2}{x + 1}$. ■

Function Notation

To indicate that y is a function of x, we often use **function notation** and write

$$y = f(x) \quad \text{Read as "}y\text{ is a function of }x.\text{"}$$

The notation $y = f(x)$ provides a way of denoting the value of y (the dependent variable) that corresponds to some number x (the independent variable). For example, if $y = f(x)$, the value of y that is determined when $x = 2$ is denoted by $f(2)$, read as "f of 2." If $f(x) = 5 - 7x$, we can evaluate $f(2)$ by substituting 2 for x.

$$\begin{aligned} f(x) &= 5 - 7x \\ f(2) &= 5 - 7(2) \quad \text{Substitute 2 for } x. \\ &= -9 \end{aligned}$$

If $x = 2$, then $y = f(2) = -9$.

To evaluate $f(-5)$, we substitute -5 for x.

$$\begin{aligned} f(x) &= 5 - 7x \\ f(-5) &= 5 - 7(-5) \quad \text{Substitute } -5 \text{ for } x. \\ &= 40 \end{aligned}$$

If $x = -5$, then $y = f(-5) = 40$.

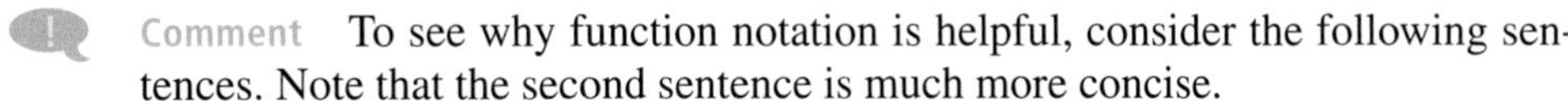

Comment To see why function notation is helpful, consider the following sentences. Note that the second sentence is much more concise.

1. In the function $y = 3x^2 + x - 4$, find the value of y when $x = -3$.
2. In the function $f(x) = 3x^2 + x - 4$, find $f(-3)$.

In this context, the notations y and $f(x)$ both represent the output of a function and can be used interchangeably.

Sometimes functions are denoted by letters other than f. The notations $y = g(x)$ and $y = h(x)$ also denote functions involving the independent variable x.

EXAMPLE 4 Let $g(x) = 3x^2 + x - 4$. Find **a.** $g(-3)$, **b.** $g(k)$, **c.** $g(-t^3)$, and **d.** $g(k + 1)$.

Solution

a.
$$\begin{aligned} g(x) &= 3x^2 + x - 4 \\ g(-3) &= 3(-3)^2 + (-3) - 4 \\ &= 3(9) - 3 - 4 \\ &= 20 \end{aligned}$$

b.
$$\begin{aligned} g(x) &= 3x^2 + x - 4 \\ g(k) &= 3k^2 + k - 4 \end{aligned}$$

c.
$$\begin{aligned} g(x) &= 3x^2 + x - 4 \\ g(-t^3) &= 3(-t^3)^2 + (-t^3) - 4 \\ &= 3t^6 - t^3 - 4 \end{aligned}$$

d.
$$\begin{aligned} g(x) &= 3x^2 + x - 4 \\ g(k + 1) &= 3(k + 1)^2 + (k + 1) - 4 \\ &= 3(k^2 + 2k + 1) + k + 1 - 4 \\ &= 3k^2 + 6k + 3 + k + 1 - 4 \\ &= 3k^2 + 7k \end{aligned}$$

Self Check Evaluate: **a.** $g(0)$, **b.** $g(2)$, **c.** $g(k - 1)$. ■

The Difference Quotient

The fraction $\dfrac{f(x + h) - f(x)}{h}$ is called the **difference quotient** and is important in calculus.

ACTIVE EXAMPLE 5 If $f(x) = 2x + 1$, evaluate $\dfrac{f(x + h) - f(x)}{h}$.

Solution First, we find $f(x + h)$.

$$f(x + h) = 2(x + h) + 1 \quad \text{Substitute } x + h \text{ for } x.$$
$$= 2x + 2h + 1$$

COLLEGE Algebra f(x) Now™
Go to **http://1pass.thomson.com** or your CD to practice this example.

Then we substitute $2x + 2h + 1$ for $f(x + h)$ and $2x + 1$ for $f(x)$ in the difference quotient.

$$\frac{f(x + h) - f(x)}{h} = \frac{2x + 2h + 1 - (2x + 1)}{h}$$
$$= \frac{2x + 2h + 1 - 2x - 1}{h}$$
$$= \frac{2h}{h}$$
$$= 2$$

The value is 2.

Self Check If $f(x) = x^2 + 2$, evaluate the difference quotient. ■

Graphs of Functions

If f is a function whose domain and range are sets of real numbers, its graph is the set of all points $(x, f(x))$ in the xy-plane. In other words, the graph of f is the graph of the equation $y = f(x)$. For example, the graph of the function $y = f(x) = -7x + 5$ is a line with slope -7 and y-intercept $(0, 5)$. (See Figure 3-3.) If the graph of a function is a nonvertical line, the function is called a **linear function.**

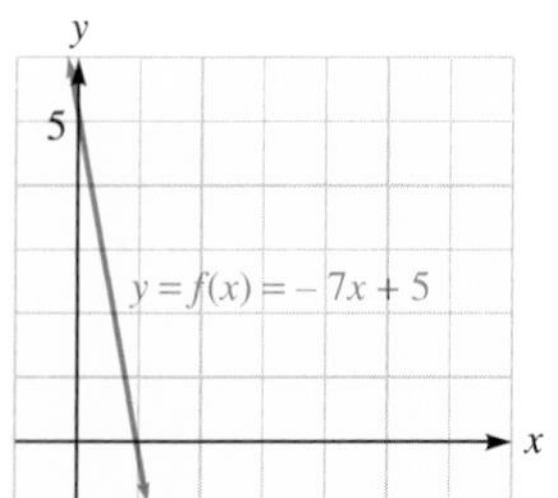

Figure 3-3

Graph of a Function

> The **graph** of a function f in the xy-plane is the set of all points (x, y) where x is in the domain of f, y is in the range of f, and $y = f(x)$.

In Figure 3-4, we see the graphs of several basic functions. Figure 3-5 illustrates how we can read the domain and range of the first three functions from their graphs.

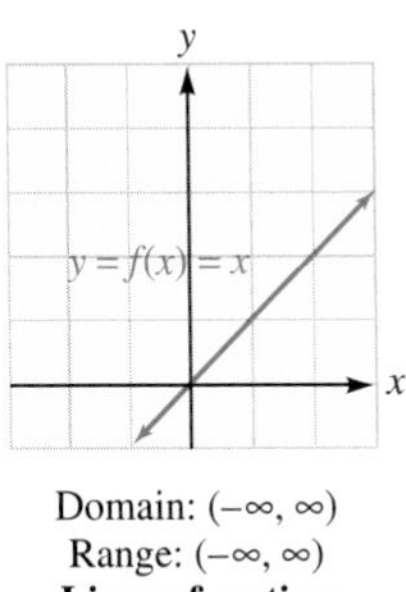

Domain: $(-\infty, \infty)$
Range: $(-\infty, \infty)$
Linear function
(a)

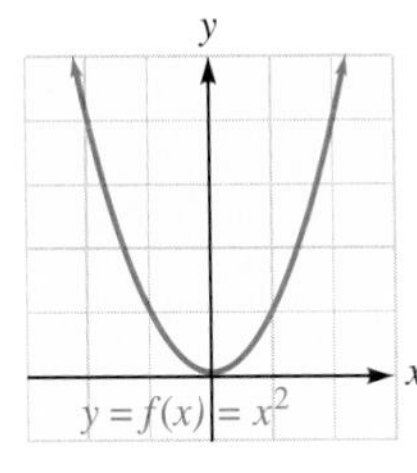

Domain: $(-\infty, \infty)$
Range: $[0, \infty)$
The squaring function
(b)

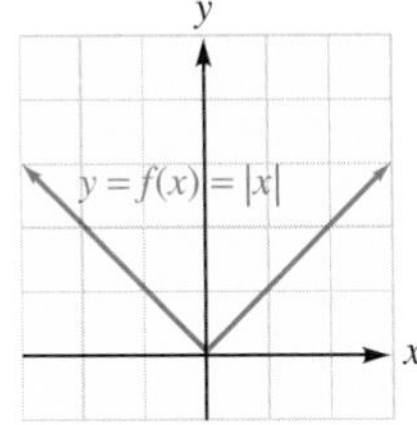

Domain: $(-\infty, \infty)$
Range: $[0, \infty)$
The absolute value function
(c)

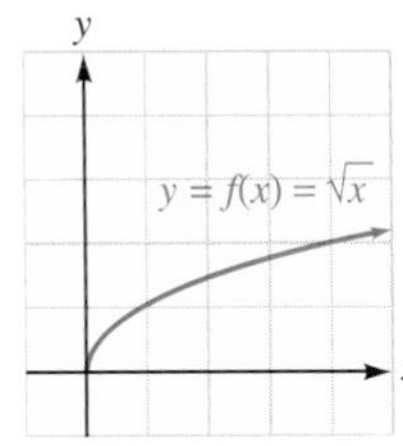

Domain: $[0, \infty)$
Range: $[0, \infty)$
The square root function
(d)

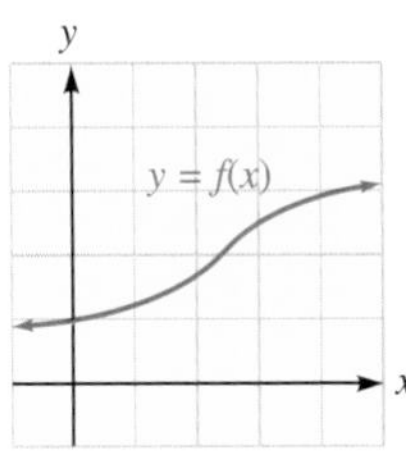

A general function
(e)

Figure 3-4

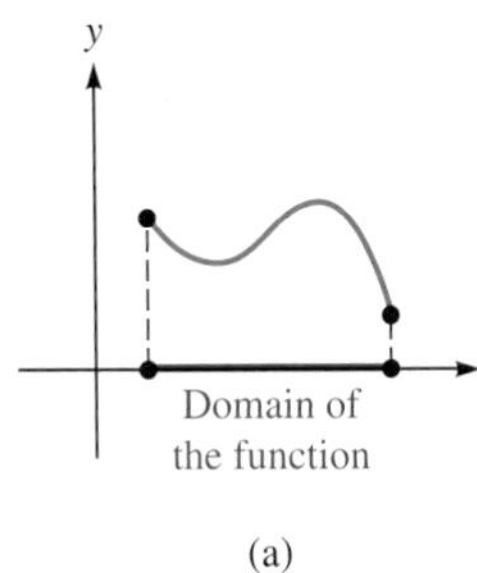

(a)

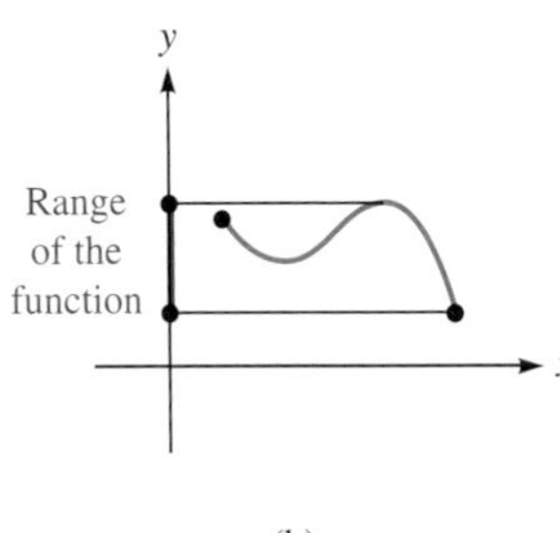

(b)

Figure 3-5

The Vertical Line Test

A **vertical line test** can be applied to a graph to determine whether it represents a function. If each vertical line that intersects a graph does so exactly once, each number x determines exactly one value of y, and the graph represents a function. (See Figure 3-6(a).)

If any vertical line intersects a graph more than once, more than one value of y corresponds to some numbers x, and the graph does not represent a function. (See Figure 3-6(b).)

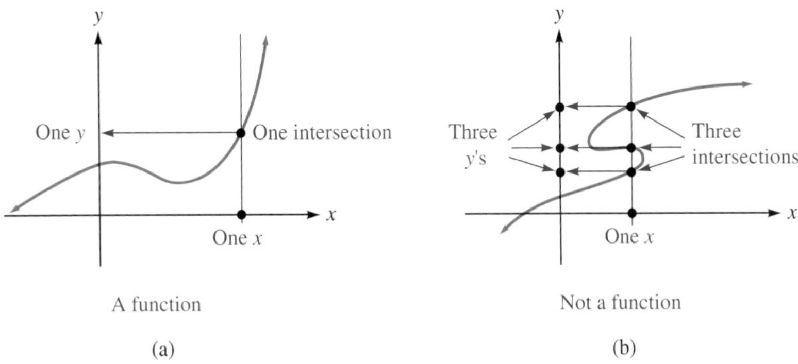

Figure 3-6

EXAMPLE 6 Determine which of the following graphs represent functions.

a. **b.** **c.**

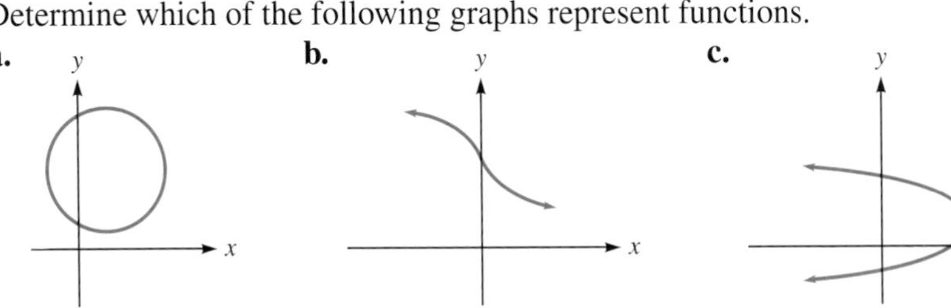

Solution

a. This graph fails the vertical line test, so it does not represent a function.

b. This graph passes the vertical line test, so it does represent a function.

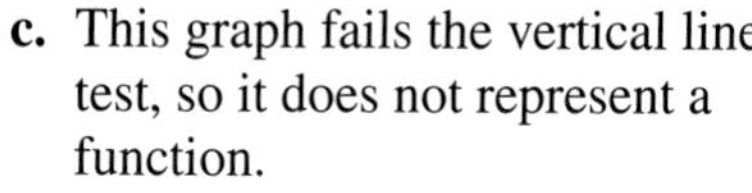

c. This graph fails the vertical line test, so it does not represent a function.

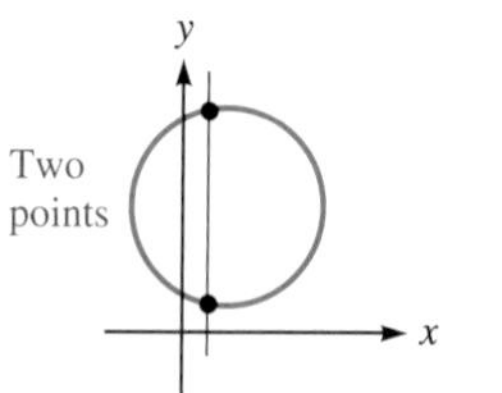

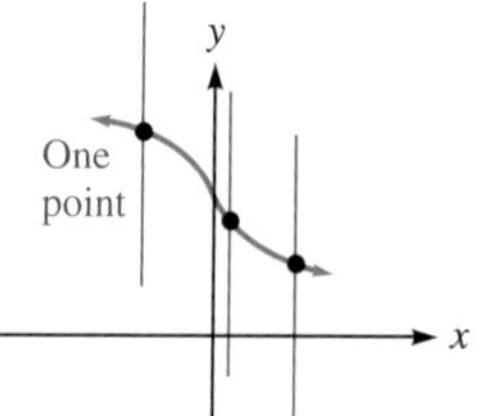

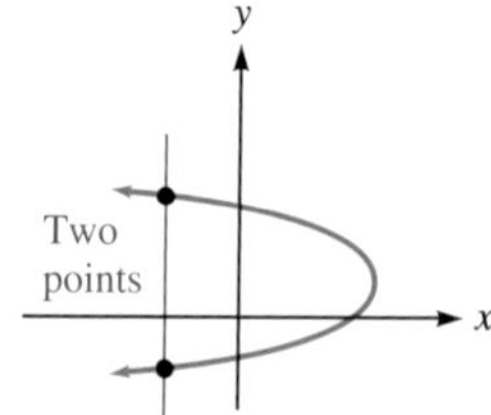

Self Check Does the graph shown in Figure 3-2 represent a function? ■

ACTIVE EXAMPLE 7 Graph the function $f(x) = \sqrt{x - 2}$ and compare the graph to Figure 3-4(d).

Solution We choose several ordered pairs that satisfy the equation, plot them, and draw the graph. (See Figure 3-7.) The graph is the same as $f(x) = \sqrt{x}$, except that it has been shifted two units to the right.

$f(x) = \sqrt{x - 2}$

x	$f(x)$	$(x, f(x))$
2	0	(2, 0)
6	2	(6, 2)
11	3	(11, 3)
18	4	(18, 4)

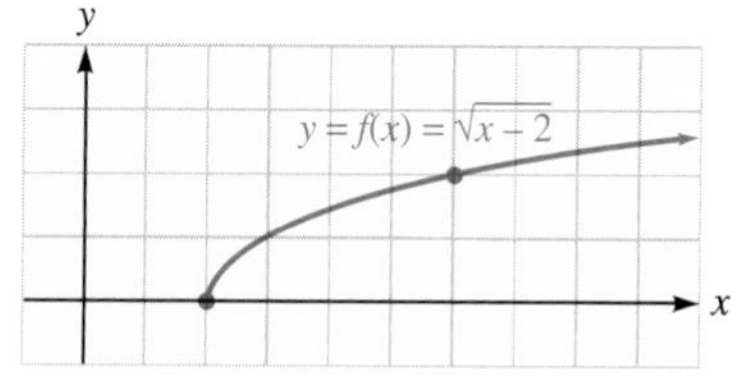

Figure 3-7

Self Check Graph $f(x) = |x + 3|$ and compare the graph to Figure 3-4(c). ■

Accent on Technology CALCULATOR GRAPHS

The calculator graphs of $f(x) = \sqrt{x - 2}$ and $f(x) = |x + 3|$ are shown in Figure 3-8. From Figure 3-8(a), we can see that the domain of $f(x) = \sqrt{x - 2}$ is $[2, \infty)$ and the range is $[0, \infty)$. From Figure 3-8(b), we can see that the domain of $f(x) = |x + 3|$ is $(-\infty, \infty)$ and the range is $[0, \infty)$.

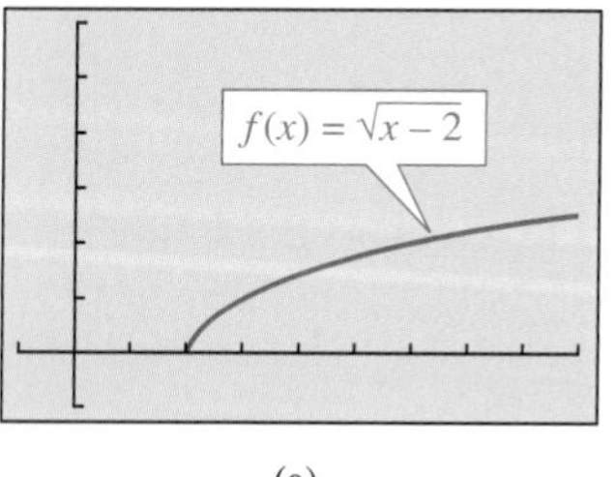

(a)

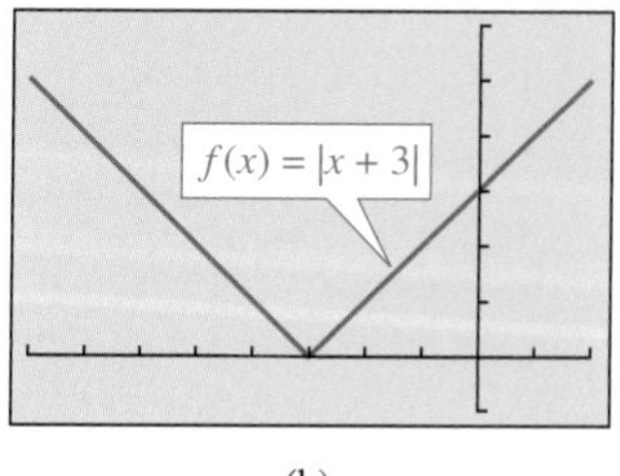

(b)

Figure 3-8

Linear Functions

We have seen that the equation of a nonvertical line defines a linear function—an important function in mathematics and its applications.

Linear Functions

A **linear function** is a function determined by an equation of the form

$$f(x) = mx + b \quad \text{or} \quad y = mx + b$$

EXAMPLE 8 The cost of electricity in Eagle River is a linear function of x, the number of kilowatt-hours (kwh) used. If the cost of 100 kwh is \$17 and the cost of 500 kwh is \$57, find an equation that expresses the function.

Solution Since c (the cost of electricity) is given to be a linear function of x, there are constants m and b such that

$$c = mx + b \qquad (1)$$

Since $c = 17$ when $x = 100$, the point $(x_1, c_1) = (100, 17)$ lies on the straight-line graph of this function. Since $c = 57$ when $x = 500$, the point $(x_2, c_2) = (500, 57)$ also lies on that line. The slope of the line is

$$\begin{aligned} m &= \frac{c_2 - c_1}{x_2 - x_1} \\ &= \frac{57 - 17}{500 - 100} \\ &= \frac{40}{400} \\ &= 0.10 \end{aligned}$$

Thus, $m = 0.10$. To determine b, we can substitute 0.10 for m and the coordinates of point $P(100, 17)$ in the equation $c = mx + b$ and solve for b.

$$\begin{aligned} c &= mx + b \\ 17 &= 0.10(100) + b \\ 17 &= 10 + b \\ 7 &= b \end{aligned}$$

Thus, $c = 0.10x + 7$. The electric company charges \$7 plus 10¢ per kwh used.

Self Check Find the cost of using 400 kwh of electricity. ■

Not all equations define functions. For example, the equation $x = |y|$ does not define a function, because two values of y can correspond to one number x. For example, if $x = 2$, then y can be either 2 or -2. The graph of the equation is shown in Figure 3-9. Since the graph does not pass the vertical line test, it does not represent a function.

Srinivasa Ramanujan

(1887–1920)

Ramanujan was one of India's most prominent mathematicians. When in high school, he read *Synopsis of Elementary Results in Pure Mathematics* and from this book taught himself mathematics. He entered college but failed several times because he would only study mathematics. He went on to teach at Cambridge University in England.

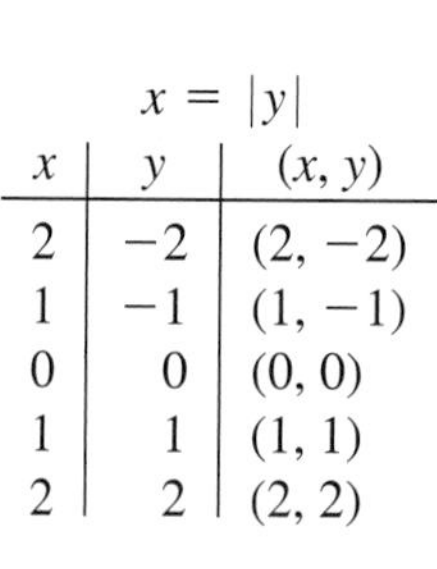

$x = |y|$

x	y	(x, y)
2	−2	(2, −2)
1	−1	(1, −1)
0	0	(0, 0)
1	1	(1, 1)
2	2	(2, 2)

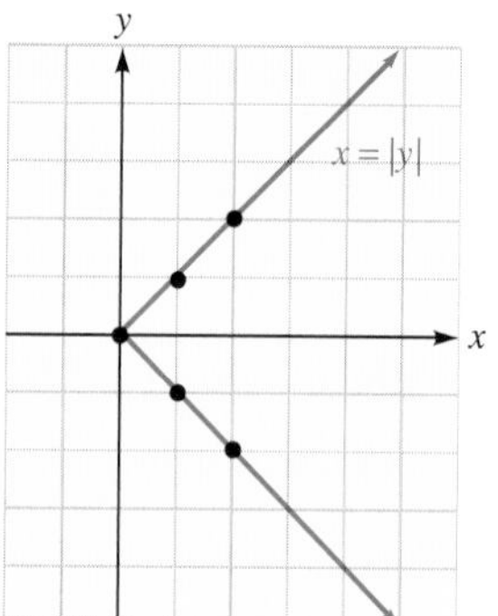

Figure 3-9

Correspondences between a set of input values x (called the domain) and a set of output values y (called the range), where to each x-value in the domain there corresponds one or more y-values in the range, are called **relations.** Although the graph in Figure 3-9 does not represent a function, it does represent a relation. We note that all functions are relations, but not all relations are functions.

Another way to visualize the definition of function is to consider the diagram shown in Figure 3-10(a). The function f that assigns the element y to the element x is represented by an arrow leaving x and pointing to y. The set of elements in **X** from which arrows originate is the domain of the function. The set of elements in **Y** to which arrows point is the range.

To constitute a function, each element of the domain must determine exactly one y-value in the range. However, the same value of y could correspond to several numbers x. In the function shown in Figure 3-10(b), the single value y corresponds to the three numbers x_1, x_2, and x_3 in the domain.

The correspondence shown in Figure 3-10(c) is not a function, because two values of y correspond to the same number x. However, it is still a relation.

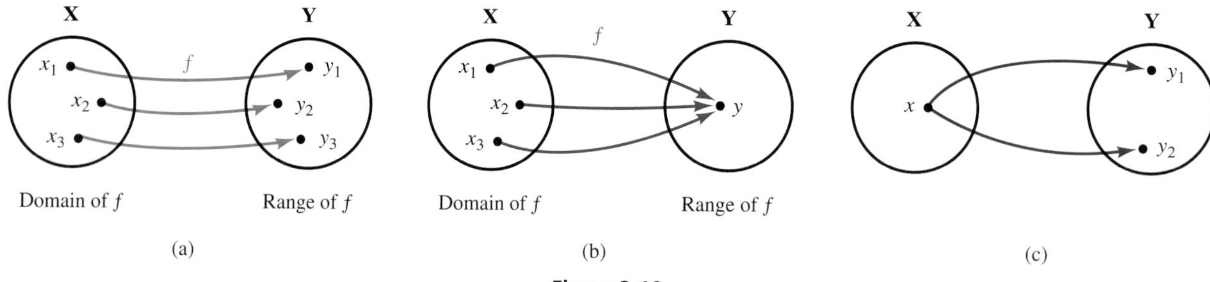

Figure 3-10

Self Check Answers

1. a. a function, $(-\infty, \infty)$, $[0, \infty)$ **b.** a function, $[0, \infty)$, $[0, \infty)$ **2.** Both are $(-\infty, \infty)$.
3. $(-\infty, -1) \cup (-1, \infty)$; $(-\infty, 1) \cup (1, \infty)$ **4. a.** -4 **b.** 10 **c.** $3k^2 - 5k - 2$ **5.** $2x + h$ **6.** yes
7. The graph is the same as $f(x) = |x|$, except that it has been shifted three units to the left.
8. \$47

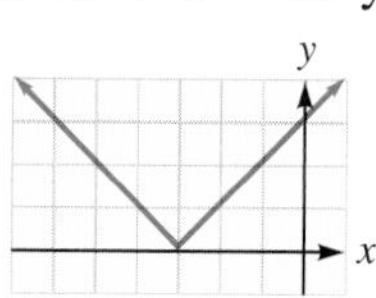

3.1 Exercises

VOCABULARY AND CONCEPTS *Fill in the blanks.*

1. A correspondence that assigns exactly one value of y to any number x is called a **function**.
2. A correspondence that assigns one or more values of y to any number x is called a **relation**.
3. The set of input numbers x in a function is called the **domain** of the function.
4. The set of all output values y in a function is called the **range** of the function.
5. The statement "y is a function of x" can be written as the equation $y = f(x)$.
6. The graph of a function $y = f(x)$ in the xy-plane is the set of all points (x, y) that satisfy the equation, where x is in the **domain** of f and y is in the **range** of f.
7. In the function of Exercise 5, x is called the independent variable.
8. In the function of Exercise 5, y is called the **dependent** variable.
9. If every **vertical** line that intersects a graph does so **once**, the graph represents a function.
10. A function that can be written in the form $y = mx + b$ is called a **linear** function.

PRACTICE ***Assume that all variables represent real numbers. Determine whether each equation determines y to be a function of x.***

11. $y = x$ **a function**
12. $y - 2x = 0$ **a function**
13. $y^2 = x$ **not a function**
14. $|y| = x$ **not a function**
15. $y = x^2$ **a function**
16. $y - 7 = 7$ **a function**
17. $y^2 - 4x = 1$ **not a function**
18. $|x - 2| = y$ **a function**
19. $|x| = |y|$ **not a function**
20. $x = 7$ **not a function**
21. $y = 7$ **a function**
22. $|x + y| = 7$ **not a function**

Let the function f be defined by the equation y = f(x), where x and f(x) are real numbers. Find the domain and range of each function.

23. $f(x) = 3x + 5$ **domain: $(-\infty, \infty)$; range: $(-\infty, \infty)$**
24. $f(x) = -5x + 2$ **domain: $(-\infty, \infty)$; range: $(-\infty, \infty)$**
25. $f(x) = x^2$ **domain: $(-\infty, \infty)$; range: $[0, \infty)$**
26. $f(x) = x^3$ **domain: $(-\infty, \infty)$; range: $(-\infty, \infty)$**
27. $f(x) = \dfrac{3}{x + 1}$ **domain: $(-\infty, -1) \cup (-1, \infty)$; range: $(-\infty, 0) \cup (0, \infty)$**
28. $f(x) = \dfrac{-7}{x + 3}$ **domain: $(-\infty, -3) \cup (-3, \infty)$; range: $(-\infty, 0) \cup (0, \infty)$**
29. $f(x) = \sqrt{x}$ **domain: $[0, \infty)$; range: $[0, \infty)$**
30. $f(x) = \sqrt{x^2 - 1}$ **domain: $(-\infty, -1] \cup [1, \infty)$; range: $[0, \infty)$**
31. $f(x) = \dfrac{x}{x + 3}$ **domain: $(-\infty, -3) \cup (-3, \infty)$; range: $(-\infty, 1) \cup (1, \infty)$**
32. $f(x) = \dfrac{x}{x - 3}$ **domain: $(-\infty, 3) \cup (3, \infty)$; range: $(-\infty, 1) \cup (1, \infty)$**
33. $f(x) = \dfrac{x - 2}{x + 3}$ **domain: $(-\infty, -3) \cup (-3, \infty)$; range: $(-\infty, 1) \cup (1, \infty)$**
34. $f(x) = \dfrac{x + 2}{x - 1}$ **domain: $(-\infty, 1) \cup (1, \infty)$; range: $(-\infty, 1) \cup (1, \infty)$**

Let the function f be defined by y = f(x), where x and f(x) are real numbers. Find f(2), f(−3), f(k), and f(k² − 1).

35. $f(x) = 3x - 2$ **$4; -11; 3k - 2; 3k^2 - 5$**
36. $f(x) = 5x + 7$ **$17; -8; 5k + 7; 5k^2 + 2$**
37. $f(x) = \dfrac{1}{2}x + 3$ **$4; \frac{3}{2}; \frac{1}{2}k + 3; \frac{1}{2}k^2 + \frac{5}{2}$**
38. $f(x) = \dfrac{2}{3}x + 5$ **$\frac{19}{3}; 3; \frac{2}{3}k + 5; \frac{2}{3}k^2 + \frac{13}{3}$**
39. $f(x) = x^2$ **$4; 9; k^2; k^4 - 2k^2 + 1$**
40. $f(x) = 3 - x^2$ **$-1; -6; 3 - k^2; 2 - k^4 + 2k^2$**
41. $f(x) = |x^2 + 1|$ **$5, 10, k^2 + 1, k^4 - 2k^2 + 2$**
42. $f(x) = |x^2 + x + 4|$ **$10, 10, k^2 + k + 4, k^4 - k^2 + 4$**
43. $f(x) = \dfrac{2}{x + 4}$ **$\frac{1}{3}; 2; \frac{2}{k + 4}; \frac{2}{k^2 + 3}$**
44. $f(x) = \dfrac{3}{x - 5}$ **$-1; -\frac{3}{8}; \frac{3}{k - 5}; \frac{3}{k^2 - 6}$**
45. $f(x) = \dfrac{1}{x^2 - 1}$ **$\frac{1}{3}; \frac{1}{8}; \frac{1}{k^2 - 1}; \frac{1}{k^4 - 2k^2}$**
46. $f(x) = \dfrac{3}{x^2 + 3}$ **$\frac{3}{7}; \frac{1}{4}; \frac{3}{k^2 + 3}; \frac{3}{k^4 - 2k^2 + 4}$**
47. $f(x) = \sqrt{x^2 + 1}$ **$\sqrt{5}; \sqrt{10}; \sqrt{k^2 + 1}; \sqrt{k^4 - 2k^2 + 2}$**

48. $f(x) = \sqrt{x^2 - 1}$
$\sqrt{3}$; $2\sqrt{2}$; $\sqrt{k^2 - 1}$; $\sqrt{k^4 - 2k^2} = |k|\sqrt{k^2 - 2}$

Evaluate the difference quotient for each function f(x).

49. $f(x) = 3x + 1$ 3

50. $f(x) = 5x - 1$ 5

51. $f(x) = x^2 + 1$ $2x + h$

52. $f(x) = x^2 - 3$ $2x + h$

Draw lines to indicate the domain and range of each function as intervals on the x- and y-axes.

53.

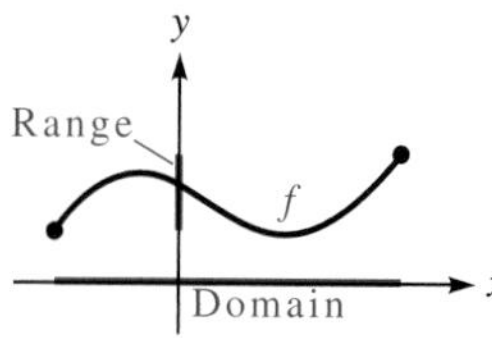

54.

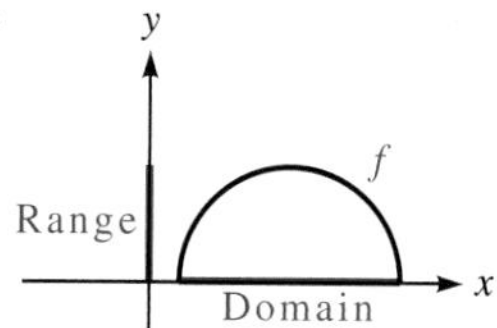

Determine whether each graph represents a function.

55.

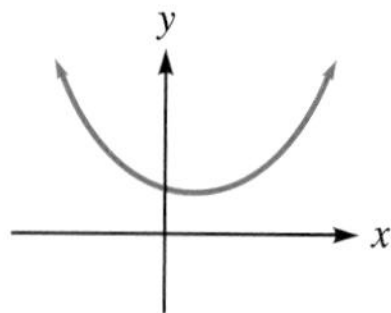

a function

56.

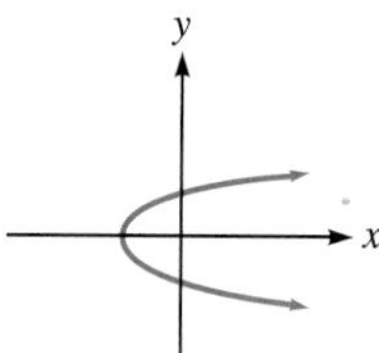

not a function

57.

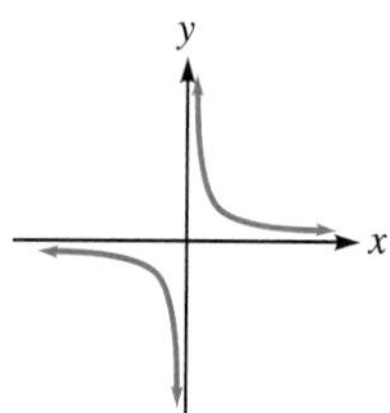

a function

58.

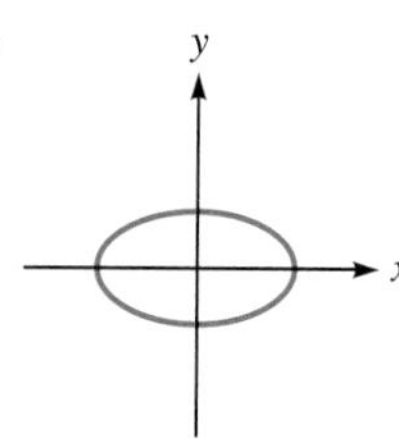

not a function

59.

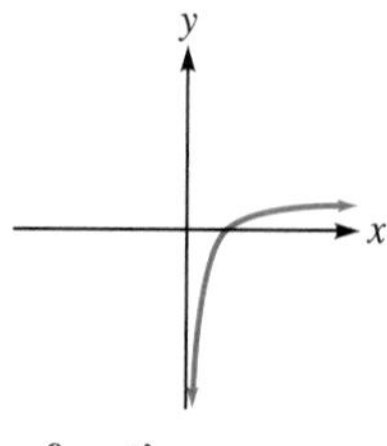

a function

60.

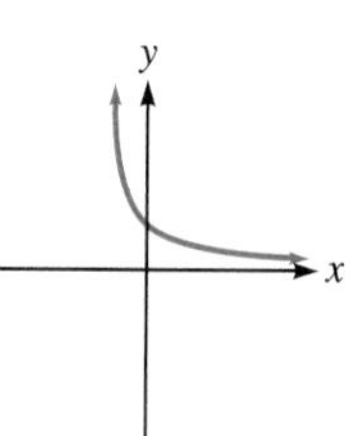

a function

Graph each function.

61. $f(x) = 2x + 3$

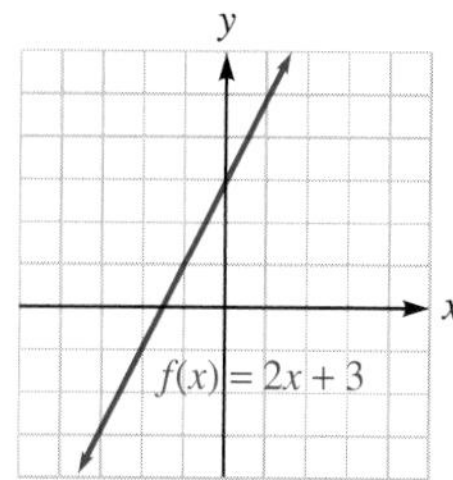

62. $f(x) = 3x + 2$

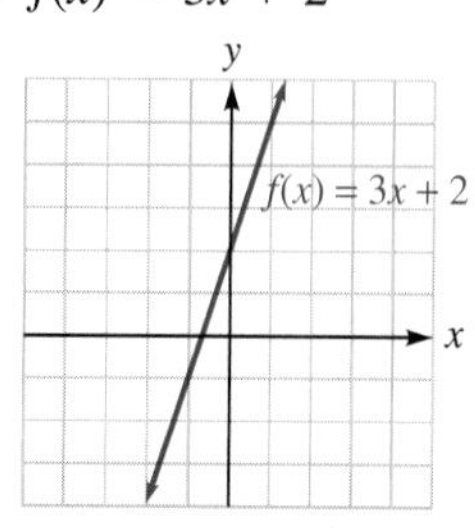

63. $f(x) = \frac{1}{2}x - 3$

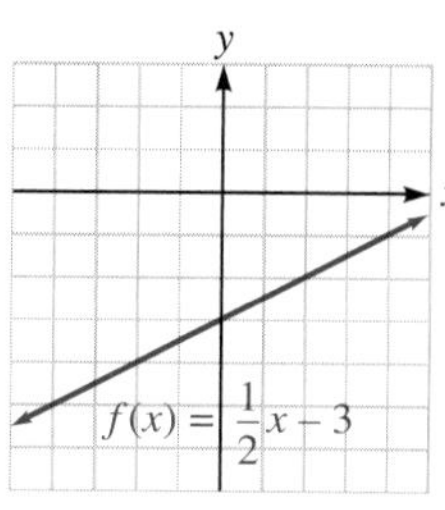

64. $f(x) = -\frac{3}{4}x + 4$

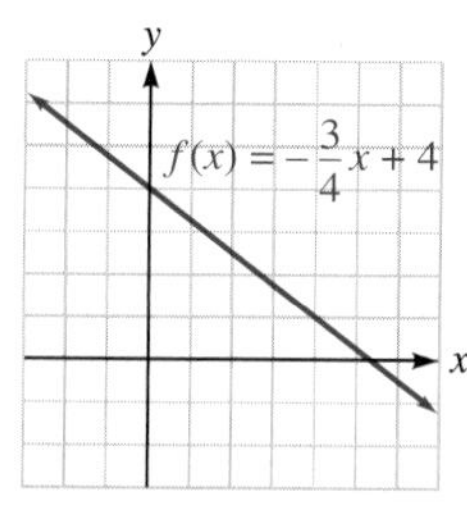

65. $2x = 3y - 3$

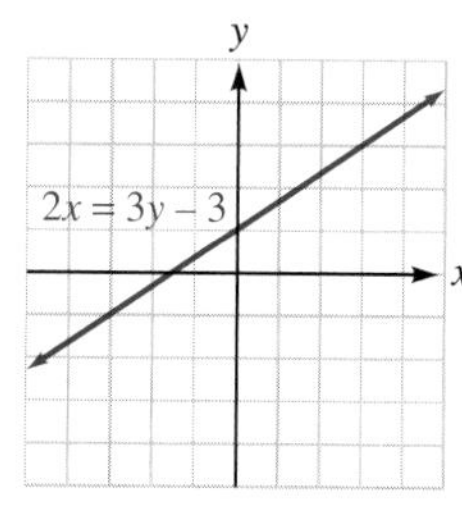

66. $3x = 2(y + 1)$

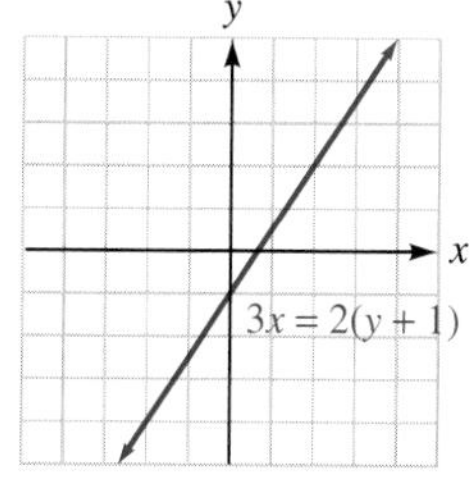

67. $f(x) = -\sqrt{x}$

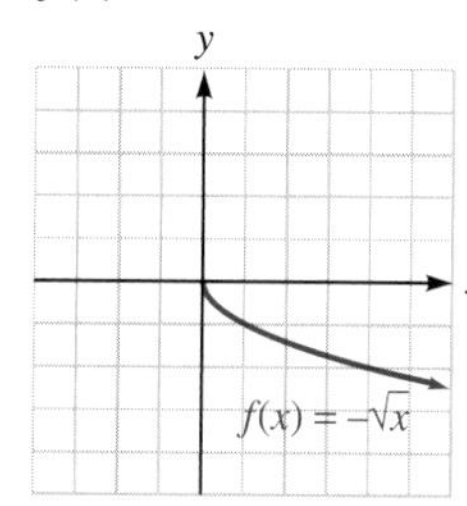

68. $f(x) = \sqrt{x + 1}$

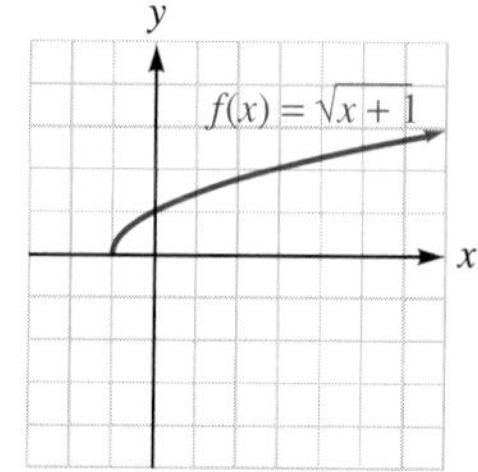

69. $f(x) = -\sqrt{x + 1}$

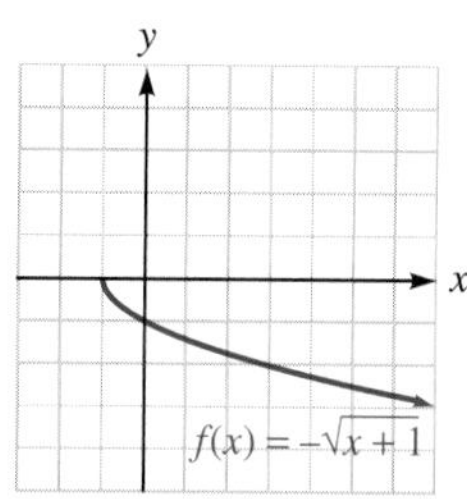

70. $f(x) = \sqrt{x} + 2$

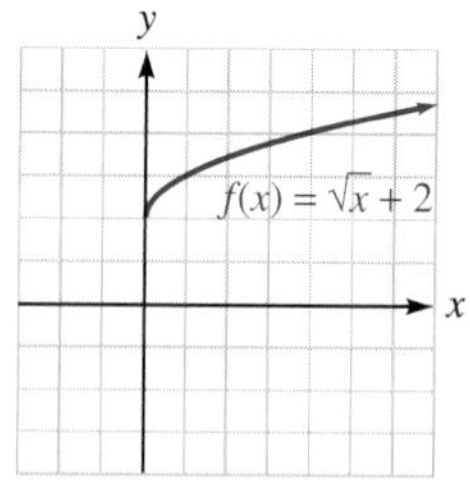

71. $f(x) = -|x|$

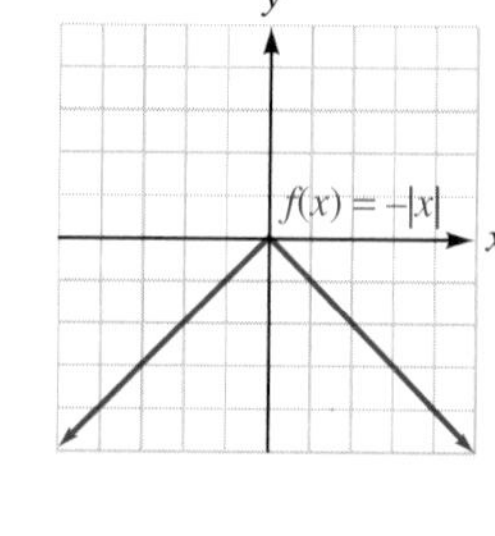

72. $f(x) = -|x| - 3$

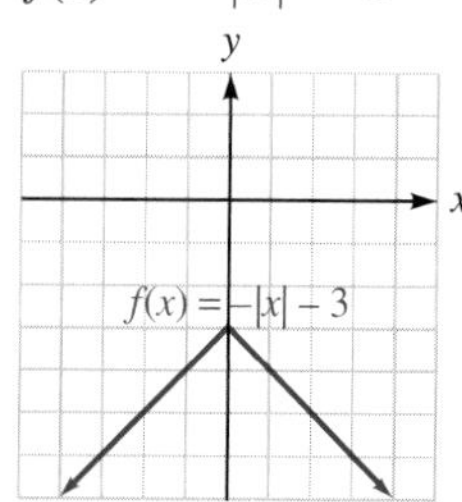

73. $f(x) = |x - 2|$

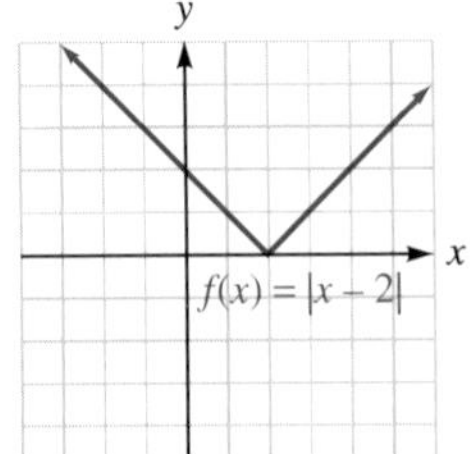

74. $f(x) = -|x - 2|$

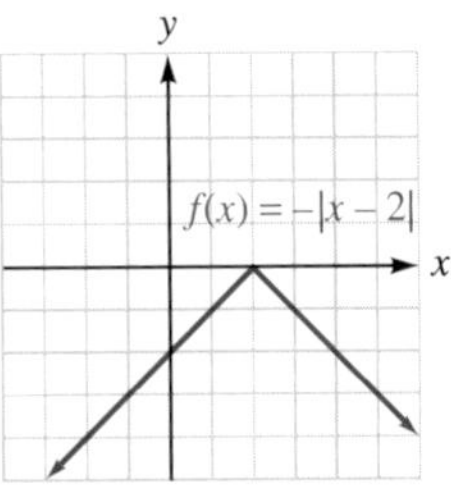

75. $f(x) = \sqrt{2x - 4}$

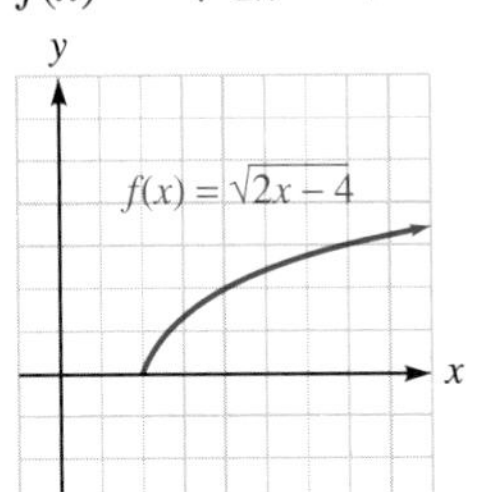

76. $f(x) = -\sqrt{2x - 4}$

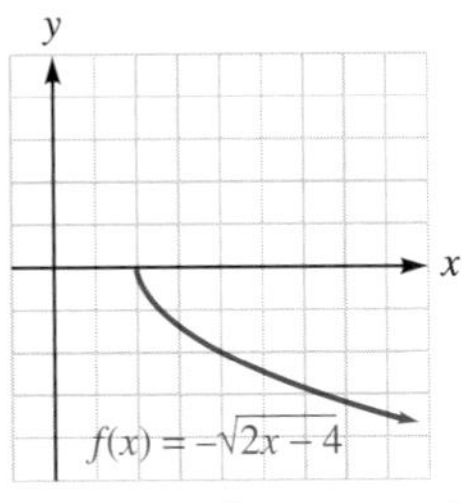

77. $f(x) = \left|\frac{1}{2}x + 3\right|$

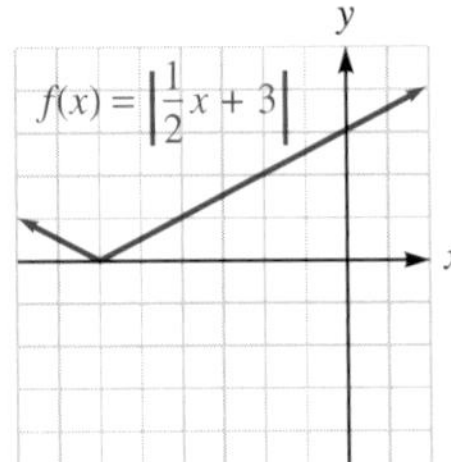

78. $f(x) = -\left|\frac{1}{2}x + 3\right|$

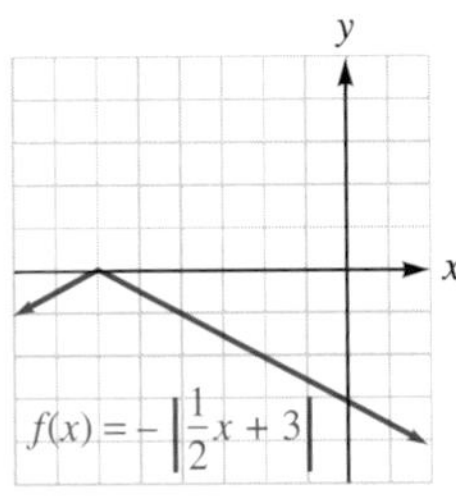

Use a graphing calculator to graph each function. Then determine the domain and range of the function.

79. $f(x) = \sqrt{2x - 5}$

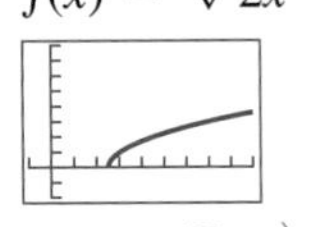

domain: $\left[\frac{5}{2}, \infty\right)$; range: $[0, \infty)$

80. $f(x) = |3x + 2|$

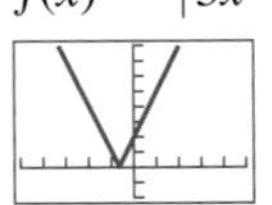

domain: $(-\infty, \infty)$; range: $[0, \infty)$

81. $f(x) = \sqrt[3]{5x - 1}$

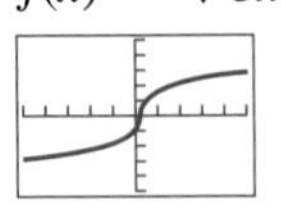

domain: $(-\infty, \infty)$; range: $(-\infty, \infty)$

82. $f(x) = -\sqrt[3]{3x + 2}$

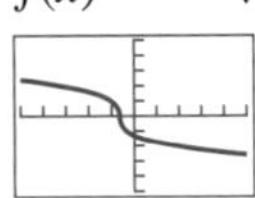

domain: $(-\infty, \infty)$; range: $(-\infty, \infty)$

APPLICATIONS

83. Temperature conversion The Fahrenheit temperature reading F is a linear function of the Celsius reading C. If $C = 0$ when $F = 32$ and the readings are the same at $-40°$, express F as a function of C. **$F = \frac{9}{5}C + 32$**

84. Free-falling objects The velocity of a falling object is a linear function of the time t it has been falling. If $v = 15$ when $t = 0$ and $v = 79$ when $t = 2$, express v as a function of t. **$v = 32t + 15$**

85. Water billing The cost c of water is a linear function of n, the number of gallons used. If 1,000 gallons cost \$4.70 and 9,000 gallons cost \$14.30, express c as a function of n. **$c = \frac{3}{2{,}500}n + \frac{7}{2}$**

86. Simple interest The amount A of money on deposit for t years in an account earning simple interest is a linear function of t. Express that function as an equation if $A = \$272$ when $t = 3$ and $A = \$320$ when $t = 5$. **$A = 24t + 200$**

87. Home construction In a proposal to prospective clients, a contractor listed the following costs:

1. Fees, permits, site preparation	\$14,000
2. Construction, per square foot	\$95

a. Write a linear function the clients can use to determine the cost c of building a house having f square feet. **$c = 95f + 14{,}000$**

b. Find the cost to build a 2,600-square-foot house. **\$261,000**

88. Concessions A concessionaire at a football game pays a vendor \$40 per game for selling hot dogs at \$2.50 each.

a. Write a linear function that describes the income I the vendor earns for the concessionaire during the game if the vendor sells h hot dogs. **$I = 2.5h - 40$**

b. Find the income if the vendor sells 175 hot dogs. **\$397.50**

DISCOVERY AND WRITING

Find all values of x that will make f(x) = 0.

89. $f(x) = 3x + 2$ **$-\frac{2}{3}$**

90. $f(x) = -2x - 5$ **$-\frac{5}{2}$**

91. Write a brief paragraph explaining how to find the domain of a function.

92. Write a brief paragraph explaining how to find the range of a function.

93. Explain why all functions are relations, but not all relations are functions.

94. Use a graphing calculator to graph the function $f(x) = \sqrt{x}$, and use TRACE and ZOOM to find $\sqrt{5}$ to 3 decimal places.

REVIEW ***Consider this set:*** $\{-3, -1, 0, 0.5, \frac{3}{4}, 1, \pi, 7, 8\}$

95. Which numbers are natural numbers? 1, 7, 8

96. Which numbers are rational numbers? $-3, -1, 0, 0.5, \frac{3}{4}, 1, 7, 8$

97. Which numbers are prime numbers? 7

98. Which numbers are even numbers? 0, 8

Write each set of numbers in interval notation.

99. 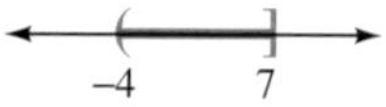

$(-4, 7]$

100. 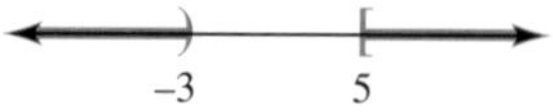

$(-\infty, -3) \cup [5, \infty)$

Graph each union of two intervals.

101. $(-3, 5) \cup [6, \infty)$

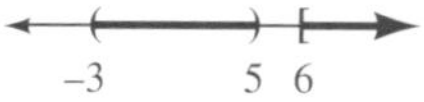

102. $(-\infty, 0) \cup (0, \infty)$

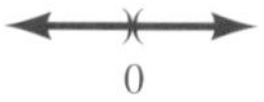

3.2 Quadratic Functions

In this section, you will learn about

- Quadratic Functions
- Finding the Vertex of a Parabola
- Maximizing Area
- Maximizing Revenue

The linear function defined by $y = f(x) = mx + b$ $(m \neq 0)$ is a first-degree polynomial function, because its right-hand side is a first-degree polynomial in the variable x. In this section, we will discuss functions of the form $y = f(x) = ax^2 + bx + c$ $(a \neq 0)$ that involve polynomials of second degree.

Quadratic Functions

A function defined by a polynomial of second-degree is called a **quadratic function.**

Quadratic Functions

A **quadratic function** is a second-degree polynomial function in one variable. It is defined by an equation of the form $f(x) = ax^2 + bx + c$ $(a \neq 0)$ or $y = ax^2 + bx + c$ $(a \neq 0)$, where a, b, and c are constants.

The most basic way to graph quadratic functions is by plotting points. For example, to graph the function $f(x) = x^2 - 2x - 3$, we plot several points with coordinates that satisfy the equation. We then join them with a smooth curve to obtain the graph shown in Figure 3-11(a). From the graph, we see that the domain is the interval $(-\infty, \infty)$ and the range is the interval $[-4, \infty)$.

$f(x) = x^2 - 2x - 3$

x	$f(x)$	$(x, f(x))$
−2	5	(−2, 5)
−1	0	(−1, 0)
0	−3	(0, −3)
1	−4	(1, −4)
2	−3	(2, −3)
3	0	(3, 0)
4	5	(4, 5)

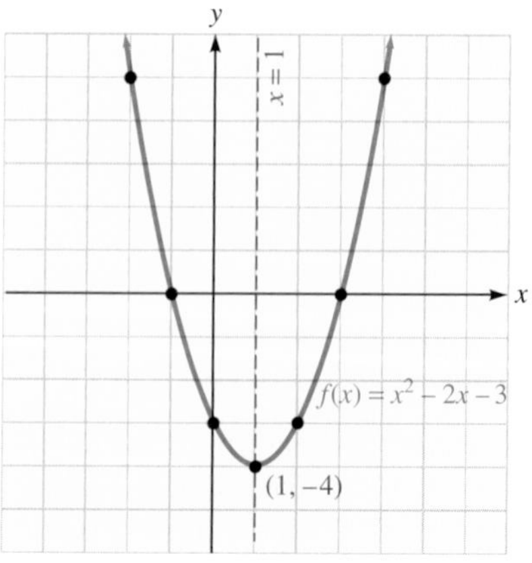

(a)

Figure 3-11

Comment In sports, the paths of many balls approximate the graph of a quadratic function.

$f(x) = -2x^2 - 8x - 3$

x	$f(x)$	$(x, f(x))$
-4	-3	$(-4, -3)$
-3	3	$(-3, 3)$
-2	5	$(-2, 5)$
-1	3	$(-1, 3)$
0	-3	$(0, -3)$

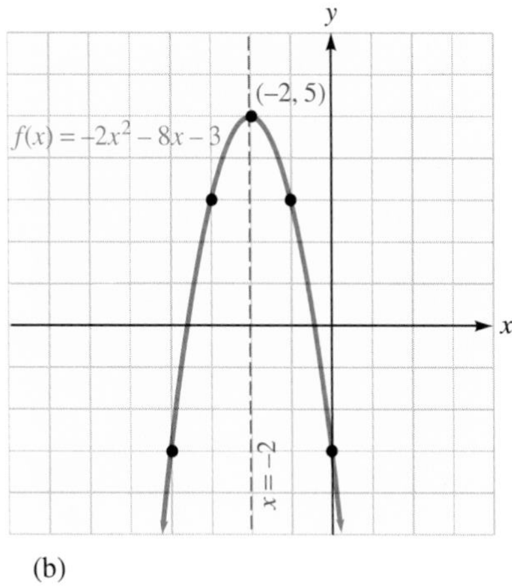

(b)

Figure 3-11 (continued)

A table of values and the graph of $f(x) = -2x^2 - 8x - 3$ is shown in Figure 3-11(b). From the graph, we see that the domain is the interval $(-\infty, \infty)$ and the range is the interval $(-\infty, 5]$.

The vertex of the parabola shown in Figure 3-11(a) is the point $(1, -4)$. The axis of symmetry is the line with equation $x = 1$, the vertical line that passes through the vertex. The vertex of the parabola in Figure 3-11(b) is the point $(-2, 5)$. Its axis of symmetry is the line whose equation is $x = -2$.

The graphs in Figure 3-11 illustrate that the graphs of $f(x) = ax^2 + bx + c$ $(a \neq 0)$ open up when $a > 0$ and down when $a < 0$.

Accent on Technology — GRAPHING QUADRATIC FUNCTIONS

We can use a graphing calculator to graph quadratic functions. If we use window settings of $[-10, 10]$ for x and $[-10, 10]$ for y, the graph of $f(x) = x^2 - 2x - 3$ will look like Figure 3-12(a). The graph of $f(x) = -2x^2 - 8x - 3$ will look like Figure 3-12(b).

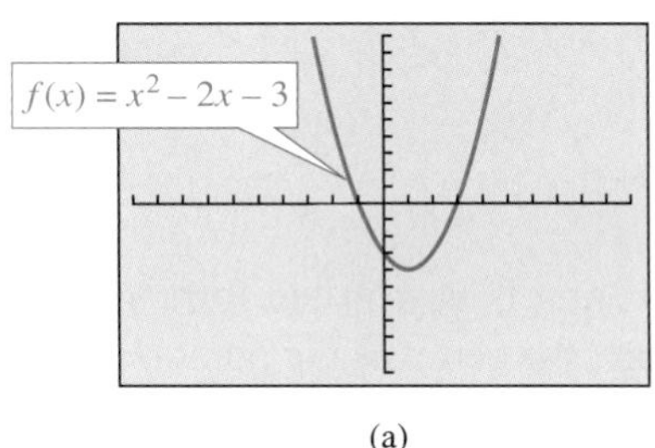

(a)

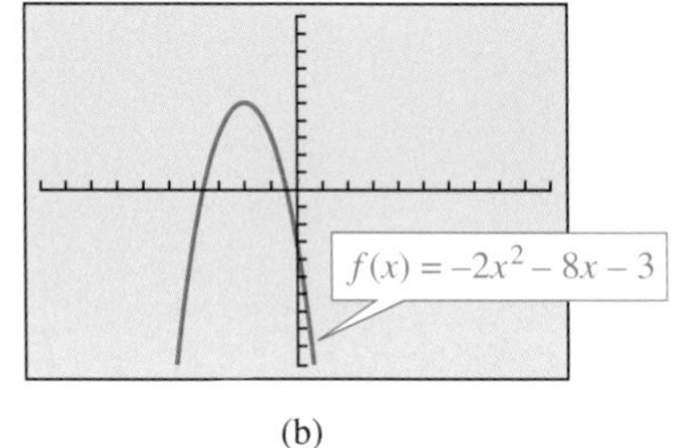

(b)

Figure 3-12

Finding the Vertex of a Parabola

If a, h, and k are constants and $a \neq 0$, the graph of

$$y = a(x - h)^2 + k \tag{1}$$

is a parabola, because if the parentheses were removed, the right-hand side would be a second-degree polynomial. If $a > 0$, the parabola would open upward, as in

Figure 3-12(a). If $a < 0$, the parabola would open downward, as in Figure 3-12(b). The values of h and k in Equation 1 determine the coordinates of the vertex of the function's parabolic graph, as the following discussion will show.

If $a > 0$, the graph of $y = a(x - h)^2 + k$ is a parabola opening upward. The vertex is the point on the graph that has the smallest possible y-coordinate. Since a is positive and $(x - h)^2$ is never negative, the minimum value of $a(x - h)^2$ is 0, and this occurs when $x = h$. If we substitute h for x in Equation 1, we see that $y = k$. Therefore, the parabola's vertex is the point (h, k).

A similar argument will show that when $a < 0$, the vertex is the highest point on the graph, which is the point (h, k).

Vertex of a Parabola

The graph of the function

$$y = f(x) = a(x - h)^2 + k \quad (a \neq 0)$$

is a parabola with its vertex at (h, k). The parabola opens upward when $a > 0$ and downward when $a < 0$.

To find the coordinates of the vertex of a parabola whose equation is in the form $y = ax^2 + bx + c$ $(a \neq 0)$, we can complete the square on $ax^2 + bx$ to change the equation into the form $y = a(x - h)^2 + k$. In this form, we can read the coordinates (h, k) of the vertex from the equation.

ACTIVE EXAMPLE 1 Find the vertex of the parabola whose equation is $y = 2x^2 - 5x - 3$.

COLLEGE **Algebra** $f(x)$ **Now**™

Explore this Active Example and test yourself on the steps taken here at **http://1pass.thomson.com**. You can also find this example on the Interactive Video Skillbuilder CD-ROM.

Solution We complete the square on $2x^2 - 5x$:

$$y = 2x^2 - 5x - 3$$

$$y = 2\left(x^2 - \frac{5}{2}x\right) - 3 \qquad \text{Factor 2 from } 2x^2 - 5x.$$

$$y = 2\left(x^2 - \frac{5}{2}x + \frac{25}{16} - \frac{25}{16}\right) - 3 \qquad \text{Add and subtract } \tfrac{25}{16} \text{ within the parentheses.}$$

$$y = 2\left(x^2 - \frac{5}{2}x + \frac{25}{16}\right) - 2\left(\frac{25}{16}\right) - 3 \qquad \text{Distribute the multiplication of 2.}$$

$$y = 2\left(x - \frac{5}{4}\right)^2 - \frac{49}{8} \qquad \text{Factor } x^2 - \tfrac{5}{2}x + \tfrac{25}{16}. \text{ Simplify: } -2\left(\tfrac{25}{16}\right) - 3 = -\tfrac{25}{8} - \tfrac{24}{8} = -\tfrac{49}{8}.$$

The equation is now in the form $y = a(x - h)^2 + k$ with $h = \frac{5}{4}$ and $k = -\frac{49}{8}$. Therefore, the vertex is the point $(h, k) = \left(\frac{5}{4}, -\frac{49}{8}\right)$.

Self Check Find the vertex of the graph of $y = 4x^2 - 16x + 19$. ■

To find formulas for the coordinates of the vertex of any parabola defined by $y = ax^2 + bx + c$ $(a \neq 0)$, we can write the equation in the form $y = a(x - h)^2 + k$ by completing the square:

$$y = ax^2 + bx + c$$

$$y = a\left(x^2 + \frac{b}{a}x\right) + c \qquad \text{Factor } a \text{ from } ax^2 + bx.$$

$$y = a\left(x^2 + \frac{b}{a}x + \frac{b^2}{4a^2} - \frac{b^2}{4a^2}\right) + c \qquad \text{Add and subtract } \frac{b^2}{4a^2} \text{ within the parentheses.}$$

$$y = a\left(x^2 + \frac{b}{a}x + \frac{b^2}{4a^2}\right) - \frac{ab^2}{4a^2} + c \qquad \text{Distribute the multiplication of } a.$$

$$y = a\left(x + \frac{b}{2a}\right)^2 + c - \frac{b^2}{4a} \qquad \text{Factor } x^2 + \frac{b}{a}x + \frac{b^2}{4a^2} \text{ and simplify } \frac{ab^2}{4a^2}.$$

(2) $$y = a\left[x - \left(-\frac{b}{2a}\right)\right]^2 + c - \frac{b^2}{4a} \qquad -\left(-\frac{b}{2a}\right) = \frac{b}{2a}.$$

If we compare Equation 2 to the form $y = a(x - h)^2 + k$, we see that $h = -\frac{b}{2a}$ and $k = c - \frac{b^2}{4a}$. This proves the following theorem.

Vertex of a Parabola

The graph of the function

$$y = f(x) = ax^2 + bx + c \quad (a \neq 0)$$

is a parabola with vertex at $\left(-\frac{b}{2a}, c - \frac{b^2}{4a}\right)$.

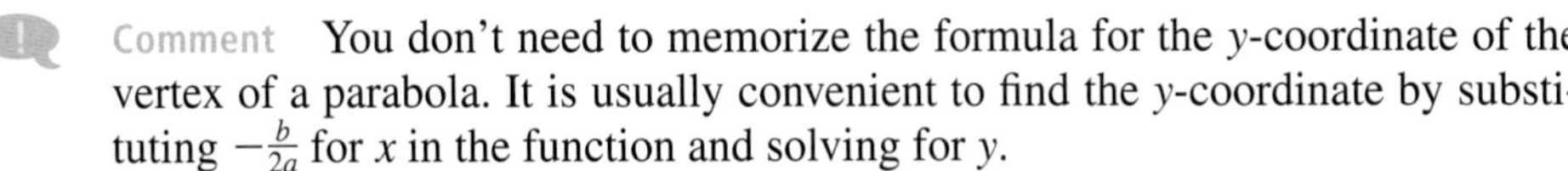

Comment You don't need to memorize the formula for the y-coordinate of the vertex of a parabola. It is usually convenient to find the y-coordinate by substituting $-\frac{b}{2a}$ for x in the function and solving for y.

It is usually easier to graph a quadratic function by finding the vertex and intercepts of its graph.

ACTIVE EXAMPLE 2 Graph the function: $y = f(x) = -2x^2 - 5x + 3$.

Solution

Determine whether the parabola opens up or down: The equation has the form $y = ax^2 + bx + c$, where $a = -2$, $b = -5$, and $c = 3$. Since $a < 0$, the parabola opens downward.

COLLEGE Algebra f(x) Now™

Go to http://1pass.thomson.com or your CD to practice this example.

Find the vertex and draw the axis of symmetry: To find the x-coordinate of the vertex, we substitute the values of a and b into the formula $x = -\frac{b}{2a}$.

$$x = -\frac{b}{2a} = -\frac{-5}{2(-2)} = -\frac{5}{4}$$

The x-coordinate of the vertex is $-\frac{5}{4}$. To find the y-coordinate, we substitute $-\frac{5}{4}$ for x in the equation and solve for y.

$$\begin{aligned} y &= -2x^2 - 5x + 3 \\ y &= -2\left(-\frac{5}{4}\right)^2 - 5\left(-\frac{5}{4}\right) + 3 \\ &= -2\left(\frac{25}{16}\right) + \frac{25}{4} + 3 \\ &= -\frac{25}{8} + \frac{50}{8} + \frac{24}{8} \\ &= \frac{49}{8} \end{aligned}$$

The vertex is the point $\left(-\frac{5}{4}, \frac{49}{8}\right)$. We plot the vertex on the coordinate system in Figure 3-13(a) and draw the axis of symmetry.

Find the x- and y-intercepts: To find the y-intercept, we let $x = 0$ and solve for y. To find the x-intercepts, we let $y = 0$ and solve for x.

To find the y-intercept, we substitute 0 for x.	***To find the x-intercepts, we substitute 0 for y.***	
$y = -2x^2 - 5x + 3$	$y = -2x^2 - 5x + 3$	
$= -2(0)^2 - 5(0) + 3$	$0 = -2x^2 - 5x + 3$	**Set y equal to 0.**
$= 0 - 0 + 3$	$0 = 2x^2 + 5x - 3$	**Divide both sides by -1 to make the lead coefficient positive.**
$= 3$		
	$0 = (2x - 1)(x + 3)$	**Factor the trinomial.**
	$2x - 1 = 0 \quad \text{or} \quad x + 3 = 0$	**Set each factor equal to 0.**
	$x = \frac{1}{2} \quad \big\vert \quad x = -3$	**Solve each linear equation.**

The y-intercept is the point (0, 3). The x-intercepts are $\left(\frac{1}{2}, 0\right)$ and $(-3, 0)$. We plot these intercepts as shown in Figure 3-13(a).

Plot one additional point: Because of symmetry, we know that the point $\left(-2\frac{1}{2}, 3\right)$ is on the graph. We plot this point on the coordinate system in Figure 3-13(a).

We can now draw the graph of the function, as shown in Figure 3-13(b).

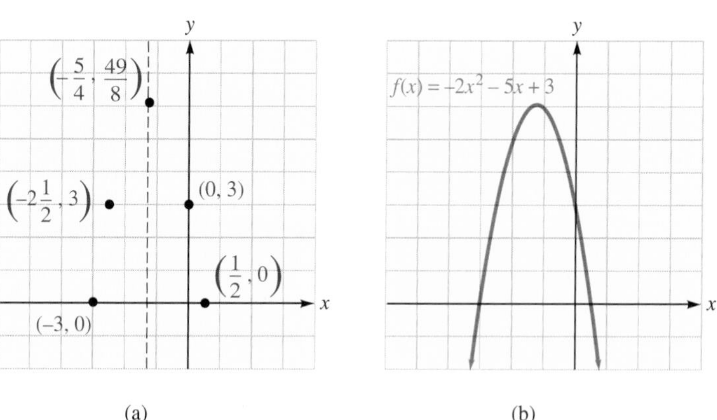

Figure 3-13

Self Check Graph the function: $f(x) = -3x^2 + 7x - 2$. ■

■ Maximizing Area

ACTIVE EXAMPLE 3 A farmer has 400 feet of fencing to enclose a rectangular corral. To save money and fencing, he intends to use the bank of a river as one boundary of the corral, as in Figure 3-14. Find the dimensions that would enclose the largest area.

Solution We let x represent the width of the fenced area. Then $400 - 2x$ represents the length. Because the area A of a rectangle is the product of the length and the width, we have

COLLEGE **Algebra** $f(x)$ **Now**™
Go to **http://1pass.thomson.com** or your CD to practice this example.

$$A = (400 - 2x)x \qquad \text{or} \qquad A = -2x^2 + 400x$$

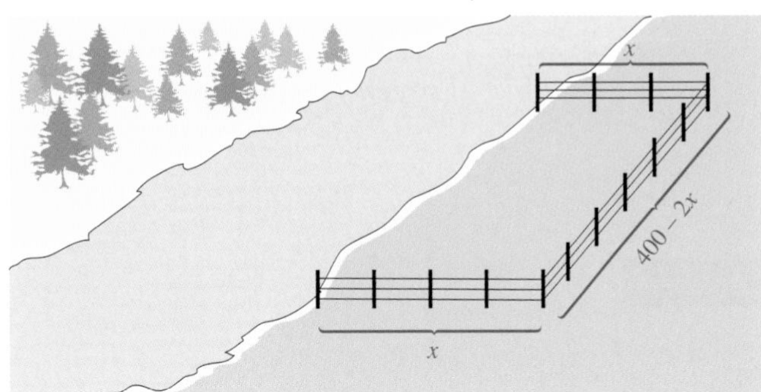

Figure 3-14

The graph of this equation is a parabola, and since the coefficient of x^2 is negative, the parabola opens downward. Since its vertex is its highest point, the A-coordinate of the vertex represents the maximum area, and the x-coordinate represents the width of the corral that will give the maximum area. We compare the equations

$$A = -2x^2 + 400x \qquad \text{and} \qquad y = ax^2 + bx + c$$

to determine that $a = -2$, $b = 400$, and $c = 0$. The vertex of the parabola is the point

$$\left(-\frac{b}{2a}, c - \frac{b^2}{4a}\right) = \left(-\frac{400}{2(-2)}, 0 - \frac{400^2}{4(-2)}\right) = (100, 20{,}000)$$

If the farmer's fence runs 100 feet out from the river, 200 feet parallel to the river, and 100 feet back to the river, it will enclose the largest possible area, which is 20,000 square feet. ■

■ Maximizing Revenue

To determine a selling price that will maximize revenue, manufacturers must consider the economic principle of supply and demand: Increasing the number of units manufactured decreases the price that can be charged for each unit.

EXAMPLE 4 A manufacturer of automobile airbags has determined that x units can be manufactured and sold each week at $\$(384 - 0.1x)$ each. Find the weekly production level that will maximize the revenue from sales, and find that revenue.

Solution We let y represent the revenue from sales. Because the revenue is the product of the number of units sold and the price charged for each unit, we have

$$y = x(384 - 0.1x) \qquad \text{or} \qquad y = -0.1x^2 + 384x$$

The graph of this equation is a parabola. Since the coefficient of x^2 is negative, it opens downward, and its vertex represents its highest point. The x-coordinate of the vertex is the production level that will maximize revenue, and the y-coordinate is that maximum revenue. We compare the equations

$$y = -0.1x^2 + 384x \quad \text{and} \quad y = ax^2 + bx + c$$

to see that $a = -0.1$, $b = 384$, and $c = 0$. Thus, the vertex of the parabola is the point

$$\left(-\frac{b}{2a}, c - \frac{b^2}{4a}\right) = \left(-\frac{384}{2(-0.1)}, 0 - \frac{384^2}{4(-0.1)}\right) = (1{,}920, 368{,}640)$$

The greatest possible revenue from sales is $368,640, which is attained at a weekly production level of 1,920 units. ■

Self Check Answers

1. (2, 3) **2.**

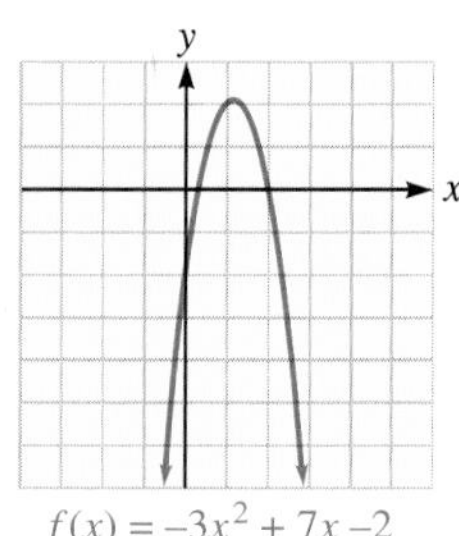

3.2 Exercises

VOCABULARY AND CONCEPTS ***Fill in the blanks.***

1. A quadratic function is defined by the equation $\underline{y = ax^2 + bx + c}$ $(a \neq 0)$.

2. The vertex of the parabolic graph of the equation $y = 2(x - 3)^2 + 5$ will be at $\underline{(3, 5)}$.

3. The x-coordinate of the vertex of the parabolic graph of $y = ax^2 + bx + c$ is $\underline{-\frac{b}{2a}}$.

4. The y-coordinate of the vertex of the parabolic graph of $y = ax^2 + bx + c$ is $\underline{c - \frac{b^2}{4a}}$.

PRACTICE ***Graph each quadratic function.***

5. $f(x) = x^2 - x$

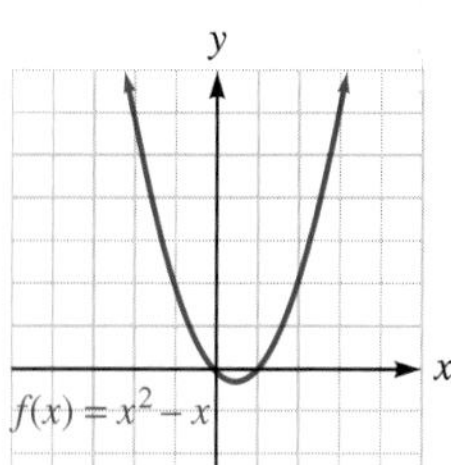

6. $f(x) = x^2 + 2x$

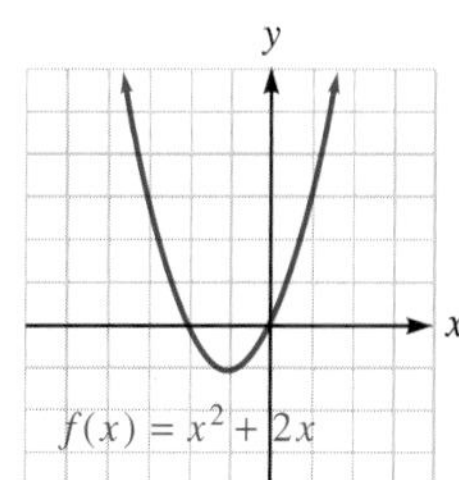

7. $f(x) = -3x^2 + 2$

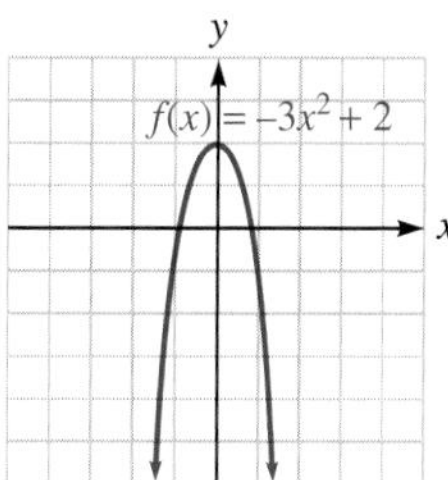

8. $f(x) = -3x^2 + 4$

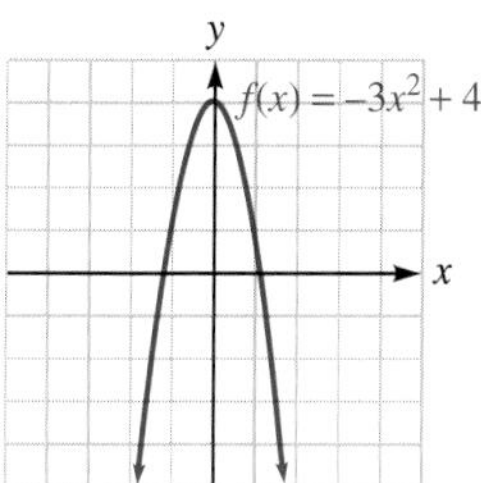

9. $f(x) = -\frac{1}{2}x^2 + 3$

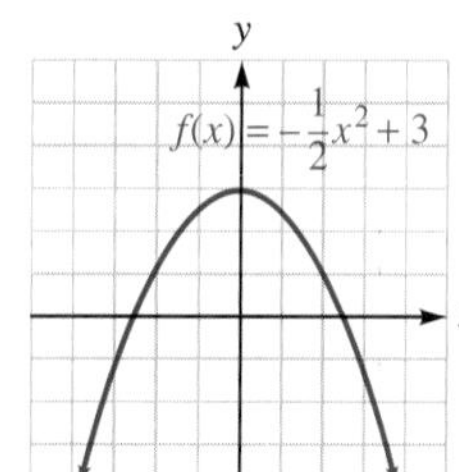

10. $f(x) = \frac{1}{2}x^2 - 2$

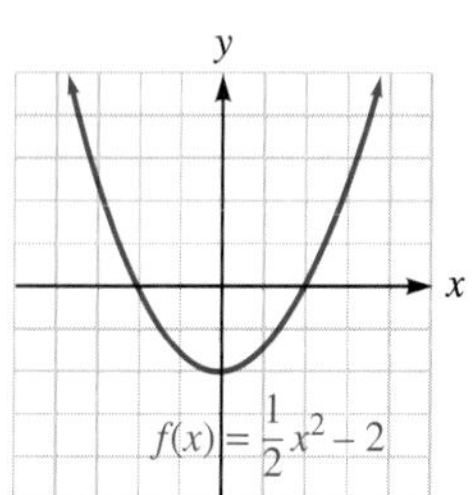

11. $f(x) = x^2 - 4x + 1$

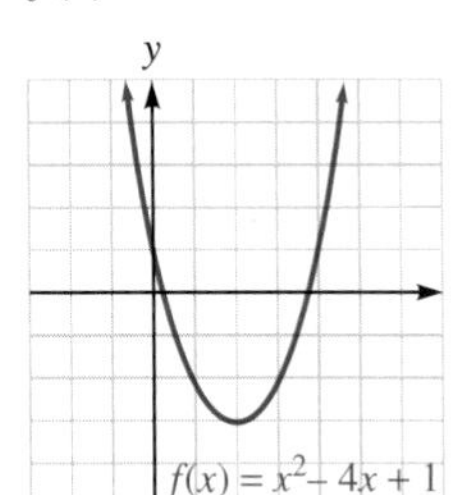

12. $f(x) = -x^2 - 4x + 1$

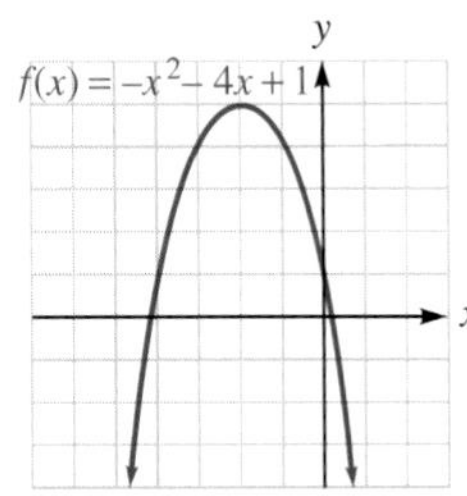

Find the vertex of each parabola.

13. $y = x^2 - 1$ **(0, −1)**

14. $y = -x^2 + 2$ **(0, 2)**

15. $y = x^2 - 4x + 4$ **(2, 0)**

16. $y = x^2 - 10x + 25$ **(5, 0)**

17. $y = x^2 + 6x - 3$ **(−3, −12)**

18. $y = -x^2 + 9x - 2$ $\left(\frac{9}{2}, \frac{73}{4}\right)$

19. $y = -2x^2 + 12x - 17$ **(3, 1)**

20. $y = 2x^2 + 16x + 33$ **(−4, 1)**

21. $y = 3x^2 - 4x + 5$ $\left(\frac{2}{3}, \frac{11}{3}\right)$

22. $y = -4x^2 + 3x + 4$ $\left(\frac{3}{8}, \frac{73}{16}\right)$

23. $y = \frac{1}{2}x^2 + 4x - 3$ **(−4, −11)**

24. $y = -\frac{2}{3}x^2 + 3x - 5$ $\left(\frac{9}{4}, -\frac{13}{8}\right)$

APPLICATIONS

25. Architecture A parabolic arch has an equation of $x^2 + 20y - 400 = 0$, where x is measured in feet. Find the maximum height of the arch. **20 ft**

26. Ballistics An object is thrown from the origin of a coordinate system with the x-axis along the ground and the y-axis vertical. Its path, or **trajectory,** is given by the equation $y = 400x - 16x^2$. Find the object's maximum height. **2,500 units**

27. Ballistics A child throws a ball up a hill that makes an angle of 45° with the horizontal. The ball lands 100 feet up the hill. Its trajectory is a parabola with equation $y = -x^2 + ax$ for some number a. Find a. $\mathbf{50\sqrt{2} + 1}$

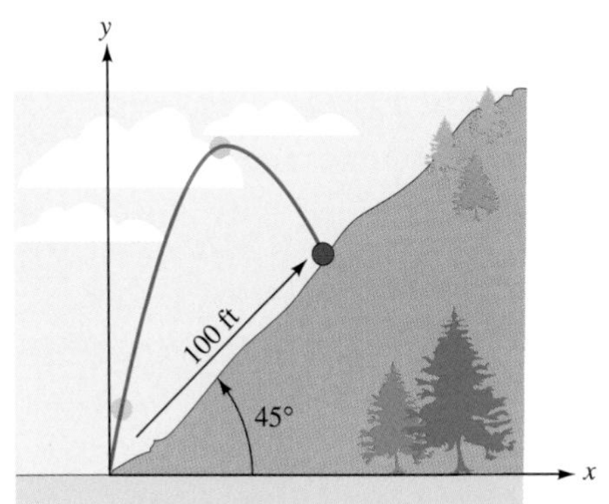

28. Maximizing height A ball is thrown straight up from the top of a building 144 ft. tall with an initial velocity of 64 ft. per second. The distance s (in feet) of the ball from the ground is given by $s = 144 + 64t - 16t^2$. Find the maximum height attained by the ball. **208 ft.**

29. Police investigations A police officer seals off the scene of an accident using a roll of yellow tape that is 300 feet long. What dimensions should be used to seal off the maximum rectangular area around the collision? Find the maximum area. **75 ft. by 75 ft., 5,625 ft.^{2}**

30. Maximizing area The rectangular garden shown has a width of x and a perimeter of 100 feet. Find x such that the area of the rectangle is maximum. **25 ft**

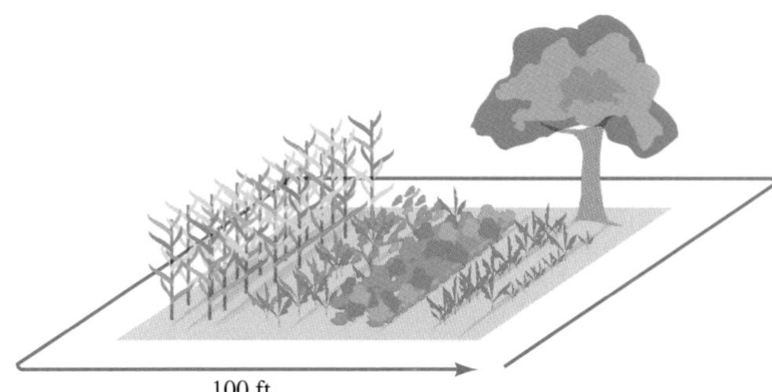

31. Maximizing storage area A farmer wants to partition a rectangular feed storage area in a corner of his barn, as shown in the illustration. The barn walls form two sides of the stall, and the farmer has 50 feet of partition for the remaining two sides. What dimensions will maximize the area? **25 ft by 25 ft**

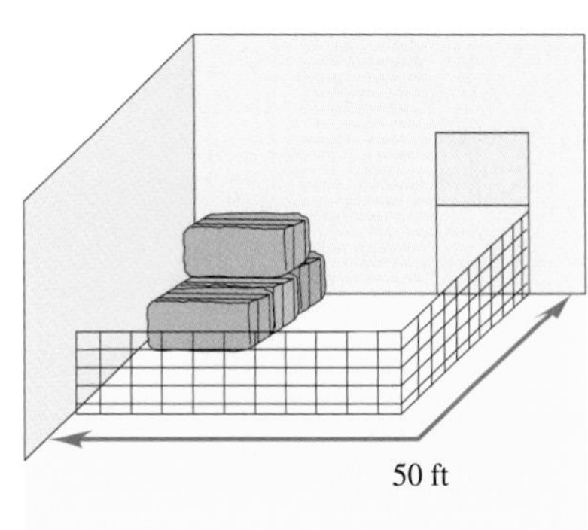

32. **Maximizing grazing area** A rancher wishes to enclose a rectangular partitioned corral with 1,800 feet of fencing. (See the illustration.) What dimensions of the corral would enclose the largest possible area? Find the maximum area.
300 ft by 450 ft; 135,000 ft^2

33. **School enrollments** The total annual enrollment (in millions) in U.S. elementary schools for the years 1975–1996 is given by the model

$$E = 0.058x^2 - 1.162x + 50.604$$

where $x = 0$ corresponds to 1975, $x = 1$ corresponds to 1976, and so on. For this period, when was enrollment the lowest? To the nearest tenth of a million, what was that enrollment? **1985, 44.8 million**

34. **U.S. Army** The function

$$N(x) = -0.0534x^2 + 0.337x + 0.969$$

gives the number of active-duty military personnel in the United States Army (in millions) for the years 1965–1972, where $x = 0$ corresponds to 1965, $x = 1$ corresponds to 1966, and so on. For this period, when was the Army's personnel strength level at its highest, and what was it? **1968, 1.5 million**

35. **Sheet metal fabrication** A 24-inch-wide sheet of metal is to be bent into a rectangular trough with the cross section shown in the illustration. Find the dimensions that will maximize the amount of water the trough can hold. That is, find the dimensions that will maximize the cross-sectional area.
$w = 12$ in.; $d = 6$ in.

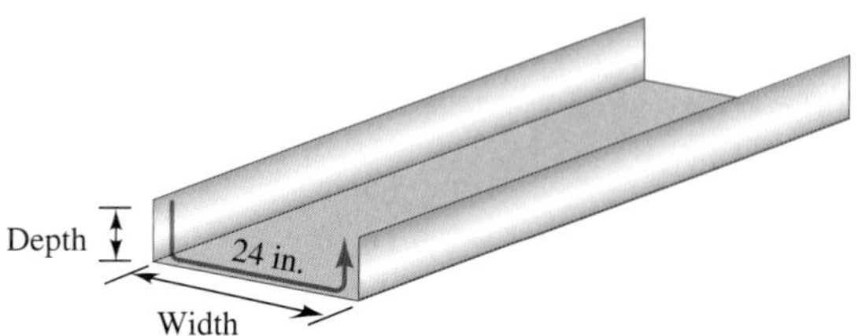

36. **Landscape design** A gardener will use D feet of edging to border a rectangular plot of ground. Show that the maximum area will be enclosed if the rectangle is a square.

37. **Selling television sets** A wholesaler of appliances finds that she can sell $(1{,}200 - p)$ television sets each week when the price is p dollars. What price will maximize revenue? **$600**

38. **Finding mass transit fares** The Municipal Transit Authority serves 150,000 commuters daily when the fare is $1.80. Market research has determined that every penny decrease in the fare will result in 1,000 new riders. What fare will maximize revenue?
$1.65

39. **Finding hotel rates** A 300-room hotel is two-thirds filled when the nightly room rate is $90. Experience has shown that each $5 increase in cost results in 10 fewer occupied rooms. Find the nightly rate that will maximize income. **$95**

40. Selling concert tickets Tickets for a concert are cheaper when purchased in quantity. The first 100 tickets are priced at \$10 each, but each additional block of 100 tickets purchased decreases the cost of each ticket by 50¢. How many blocks of tickets should be sold to maximize the revenue?
10 or 11 blocks

Use this information: At a time t seconds after an object is tossed vertically upward, it reaches a height s in feet given by the equation $s = 80t - 16t^2$.

41. In how many seconds does the object reach its maximum height? **$\frac{5}{2}$ sec**

42. In how many seconds does the object return to the point from which it was thrown? **5 sec**

43. What is the maximum height reached by the object?
100 ft

44. Show that it takes the same amount of time for the object to reach its maximum height as it does to return from that height to the point from which it was thrown.

Use a graphing calculator to determine the coordinates of the vertex of each parabola. You will have to select appropriate viewing windows.

45. $y = 2x^2 + 9x - 56$
(−2.25, −66.13)

46. $y = 14x - \dfrac{x^2}{5}$
(35, 245)

47. $y = (x - 7)(5x + 2)$
(3.3, −68.5)

48. $y = -x(0.2 + 0.1x)$
(−1, 0.1)

DISCOVERY AND WRITING ***Find all values of x that will make $f(x) = 0$.***

49. $f(x) = x^2 - 5x + 6$
2, 3

50. $f(x) = 6x^2 + x - 2$
$\frac{1}{2}, -\frac{2}{3}$

51. Find the dimensions of the largest rectangle that can be inscribed in the right triangle ABC shown in the illustration. **6 by $4\frac{1}{2}$ units**

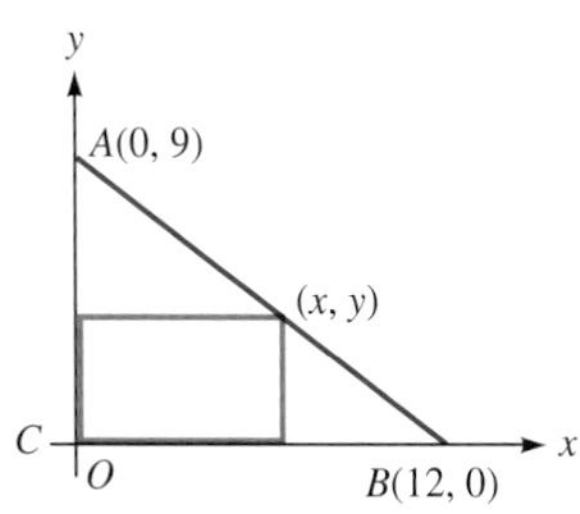

52. Point P lies in the first quadrant and on the line $x + y = 1$ in such a position that the area of triangle OPA is maximum. Find the coordinates of P. (See the illustration.) **$P\left(\frac{1}{2}, \frac{1}{2}\right)$**

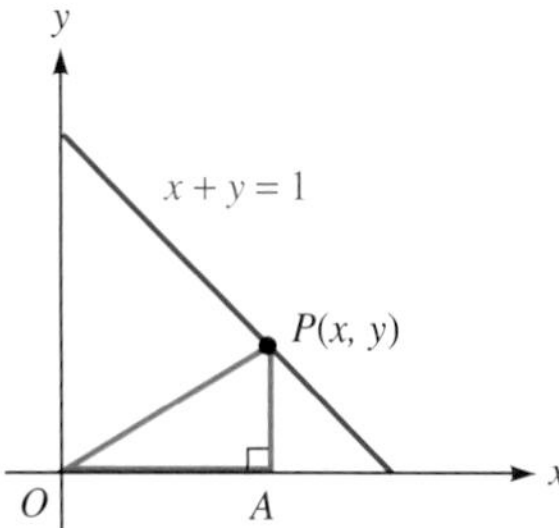

53. The sum of two numbers is 6, and the sum of the squares of those two numbers is as small as possible. What are the numbers? **Both numbers are 3.**

54. What number most exceeds its square? **$\frac{1}{2}$**

The maximum or minimum value of a quadratic function can be found automatically by using a computer program, such as Excel.

55. Find the minimum value of the function $f(x) = 2x^2 - 3x - 4$ by using the Solver in Excel. Give the value of x that minimizes the function as well as the minimum value of the function. See Problem 109 in Section 1.3.

56. Find the maximum value of the function $f(x) = -2x^2 + 3x + 4$ by using the Solver in Excel. Give the value of x that maximizes the function as well as the maximum value of the function. See Problem 109 in Section 1.3.

REVIEW ***Find $f(a)$ and $f(-a)$.***

57. $f(x) = x^2 - 3x$
$a^2 - 3a; a^2 + 3a$

58. $f(x) = x^3 - 3x$
$a^3 - 3a; -a^3 + 3a$

59. $f(x) = (5 - x)^2$
$(5 - a)^2; (5 + a)^2$

60. $f(x) = \dfrac{1}{x^2 - 4}$
$\dfrac{1}{a^2 - 4}; \dfrac{1}{a^2 - 4}$

61. $f(x) = 7$
7; 7

62. $f(x) = -|x|$
$-|a|; -|a|$

3.3 Polynomial and Other Functions

In this section, you will learn about

- Graphing Polynomial Functions
- Even and Odd Functions
- Increasing and Decreasing Functions
- Piecewise-Defined Functions
- The Greatest Integer Function

So far, we have discussed two types of polynomial functions—first-degree (or linear) functions, and second-degree (or quadratic) functions. In this section, we will discuss polynomial functions of higher degree.

Polynomial Functions

A **polynomial function in one variable (say, x)** is a function of the form

$$f(x) = a_n x^n + a_{n-1}x^{n-1} + \cdots + a_1 x + a_0$$

where $a_n, a_{n-1}, \ldots, a_1$, and a_0 are real numbers and n is a whole number.

The **degree of a polynomial function** is the largest power of x that appears in the polynomial.

Four basic polynomial functions are as follows:

Name	Function	Degree	Graph
Zero function	$f(x) = 0$	None	x-axis
Constant function	$f(x) = a_0 \quad (a_0 \neq 0)$	0	Horizontal line, y-intercept at $(0, a_0)$
Linear function	$f(x) = a_1 x + a_0 \quad (a_1 \neq 0)$	1	Nonvertical line, slope of a_1, y-intercept of $(0, a_0)$
Quadratic function	$f(x) = a_2 x^2 + a_1 x + a_0 \quad (a_2 \neq 0)$	2	Parabola, opens upward when $a_2 > 0$ and downward when $a_2 < 0$

Some examples of higher-degree polynomial functions are

$$f(x) = 3x^3 + 2x^2 - x + 4 \quad \text{(with a degree of 3)}$$

$$f(x) = \frac{1}{2}x^4 - 3x^2 + 2 \quad \text{(with a degree of 4)}$$

To graph a polynomial function, we will make use of any possible symmetries of its graph. Recall that in Section 2.4, we learned that a graph is symmetric about the y-axis if the graph of $y = f(x)$ has the same y-coordinate when the function is evaluated at x or at $-x$. Thus, a function is symmetric about the y-axis if $f(x) = f(-x)$ for all values of x that are in the domain of the function. See Figure 3-15(a).

Also recall that a graph is symmetric about the origin if the point $(-x, -f(x))$ lies on the graph whenever $(x, f(x))$ does. In this case, $f(-x) = -f(x)$. See Figure 3-15(b).

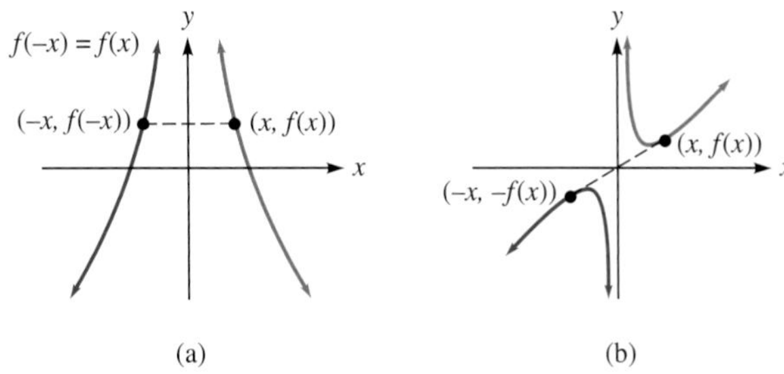

Figure 3-15

Graphing Polynomial Functions

Like the graphs of linear and quadratic functions, the graphs of higher degree polynomial functions are smooth continuous curves. Because their graphs are smooth, they have no cusps or corners. Because they are continuous, their graphs have no breaks or holes; they can be drawn without lifting the pencil from the paper. See Figure 3-16.

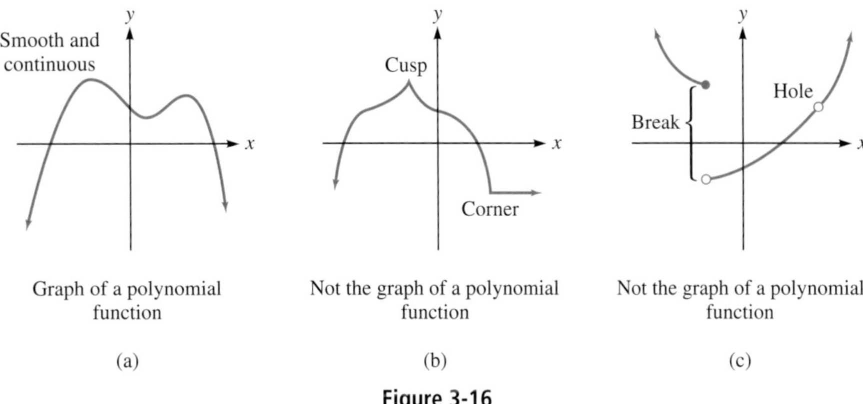

Figure 3-16

Some basic polynomial functions are functions of the form $f(x) = x^n$. Several of their graphs are shown in Figure 3-17. Note that when n is even, the graph has the same general shape as $y = x^2$. When n is odd and greater than 1, the graph has the same general shape as $y = x^3$. However, the graphs are flatter at the origin and steeper as n becomes large.

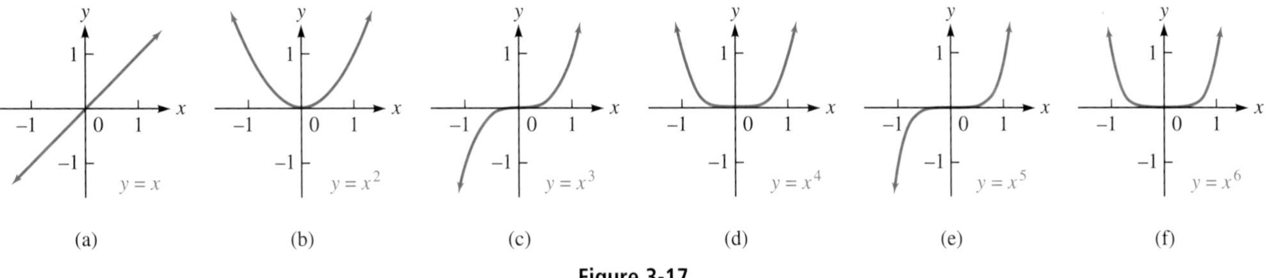

Figure 3-17

The ends of the graph of any polynomial function will be similar to the graph of its term with the highest power of x, because when n becomes large, the other terms become relatively insignificant. If the term with the highest power of x is preceded by a negative sign, the ends of the graph will turn in the opposite direction.

To graph polynomial functions, we can use the following steps.

Graphing Polynomial Functions

1. Find any symmetries of the graph.
2. Find the x- and y-intercepts of the graph.
3. Determine where the graph is above and below the x-axis.
4. If necessary, plot a few more points.
5. Draw the graph as a smooth continuous curve.

EXAMPLE 1 Graph the function f: $f(x) = x^3 - 4x$.

Solution To test for symmetry about the y-axis, we check to see whether $f(x) = f(-x)$. To test for symmetry about the origin, we check to see whether $f(-x) = -f(x)$.

$$f(x) = x^3 - 4x$$
$$f(-x) = (-x)^3 - 4(-x) \quad \text{Substitute } -x \text{ for } x.$$
$$f(-x) = -x^3 + 4x \quad \text{Simplify.}$$

Since $f(x) \neq f(-x)$, there is no symmetry about the y-axis. However, since $f(-x) = -f(x)$, there is symmetry about the origin.

To find the x-intercepts, we let $f(x) = 0$ and solve for x.

$$x^3 - 4x = 0$$
$$x(x^2 - 4) = 0 \quad \text{Factor out } x.$$
$$x(x + 2)(x - 2) = 0 \quad \text{Factor } x^2 - 4.$$
$$x = 0 \quad \text{or} \quad x + 2 = 0 \quad \text{or} \quad x - 2 = 0 \quad \text{Set each factor equal to 0.}$$
$$\qquad x = -2 \qquad\qquad x = 2$$

The x-intercepts are $(0, 0)$, $(-2, 0)$, and $(2, 0)$. If we let $x = 0$ and solve for $f(x)$, we see that the y-intercept is also $(0, 0)$.

To determine where the graph is above or below the x-axis, we plot the solutions of $x^3 - 4x = 0$ on a number line and establish the four intervals shown in Figure 3-18. (For a review of this process, see Example 7 in Section 1.7.)

Sign of $f(x) = x^3 - 4x$	$-$	$+$	$-$	$+$
	$(-\infty, -2)$	$(-2, 0)$	$(0, 2)$	$(2, \infty)$
Test point	$f(-3) = -15$	$f(-1) = 3$	$f(1) = -3$	$f(3) = 15$
Graph of $f(x)$	below the x-axis	above the x-axis	below the x-axis	above the x-axis

(Number line points: -2, 0, 2)

Figure 3-18

We can now plot the intercepts and one additional point for positive x as shown in Figure 3-19(a). We then use our knowledge of symmetry and where the graph is above and below the x-axis to draw the graph.

A calculator graph is shown in Figure 3-19(b).

$f(x) = x^3 - 4x$

x	$f(x)$	$(x, f(x))$
-2	0	$(-2, 0)$
0	0	$(0, 0)$
1	-3	$(1, -3)$
2	0	$(2, 0)$

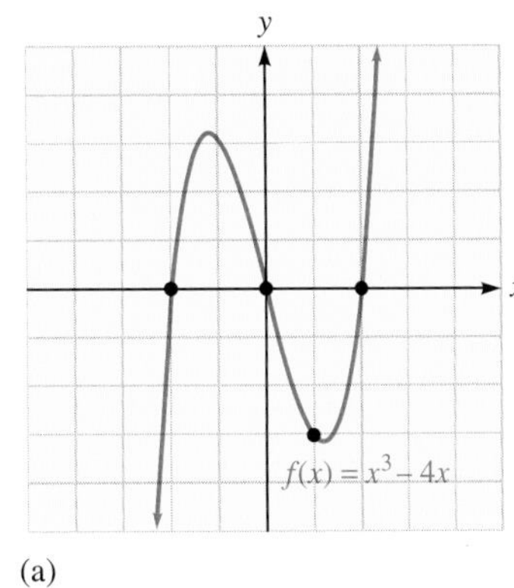

(a)

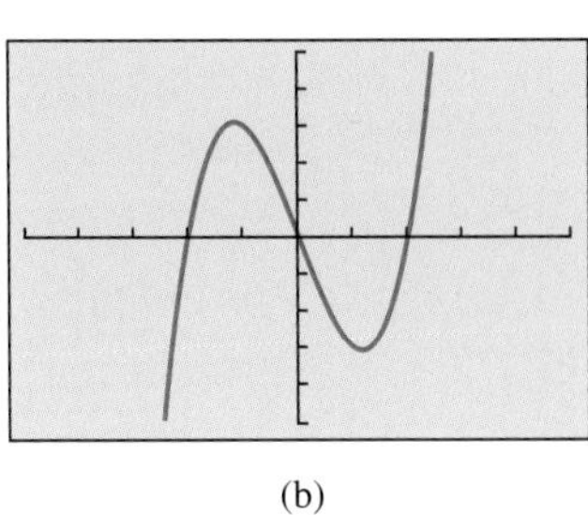

(b)

Figure 3-19

Comment Note that the far right and far left ends of the graph are similar to the ends of the graph of $f(x) = x^3$, which is the term of the function $f(x) = x^3 - 4x$ with highest degree.

Self Check Graph: $f(x) = x^3 - 9x$. ■

The peak and valley of the graph shown in Figure 3-19 are called **turning points.** In calculus, such points are called **local minima** and **local maxima.** Although we cannot find these points without using calculus, we can approximate them by plotting points or by using the TRACE feature of a graphing calculator.

Note that Figure 3-19 shows the graph of a third-degree polynomial and the graph has 2 turning points. This suggests the following result from calculus that helps us understand the shape of many polynomial graphs.

Number of Turning Points If $f(x)$ is a polynomial function of nth degree, then the graph of $f(x)$ will have $n - 1$, or fewer, turning points.

ACTIVE EXAMPLE 2 Graph the function f: $f(x) = x^4 - 5x^2 + 4$.

Solution Because x appears with only even exponents, $f(x) = f(-x)$, and the graph is symmetric about the y-axis. The graph is not symmetric about the origin.

COLLEGE **Algebra f(x) Now**™

Explore this Active Example and test yourself on the steps taken here at **http://1pass.thomson.com**. You can also find this example on the Interactive Video Skillbuilder CD-ROM.

To find the x-intercepts, we let $f(x) = 0$ and solve for x.

$$x^4 - 5x^2 + 4 = 0$$
$$(x^2 - 4)(x^2 - 1) = 0$$
$$(x + 2)(x - 2)(x + 1)(x - 1) = 0$$

$$x + 2 = 0 \quad \text{or} \quad x - 2 = 0 \quad \text{or} \quad x + 1 = 0 \quad \text{or} \quad x - 1 = 0$$
$$x = -2 \quad | \quad x = 2 \quad | \quad x = -1 \quad | \quad x = 1$$

The x-intercepts are $(-2, 0)$, $(2, 0)$, $(-1, 0)$, and $(1, 0)$. To find the y-intercept, we let $x = 0$ and see that the y-intercept is $(0, 4)$.

To determine where the graph is above or below the x-axis, we plot the x-intercepts on a number line and establish the intervals shown in Figure 3-20.

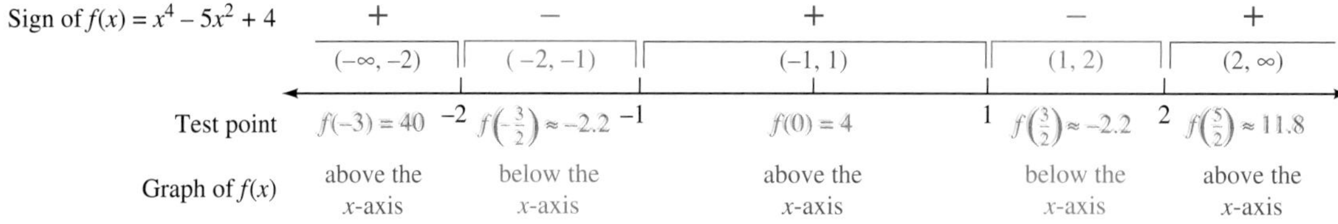

Figure 3-20

We can now plot the intercepts and use our knowledge of symmetry and where the graph is above and below the x-axis to draw the graph, as in Figure 3-21(a). A calculator graph is shown in Figure 3-21(b).

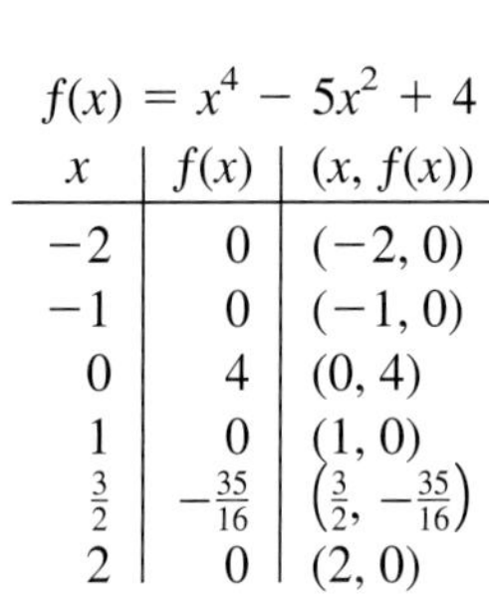
$f(x) = x^4 - 5x^2 + 4$

x	$f(x)$	$(x, f(x))$
-2	0	$(-2, 0)$
-1	0	$(-1, 0)$
0	4	$(0, 4)$
1	0	$(1, 0)$
$\frac{3}{2}$	$-\frac{35}{16}$	$\left(\frac{3}{2}, -\frac{35}{16}\right)$
2	0	$(2, 0)$

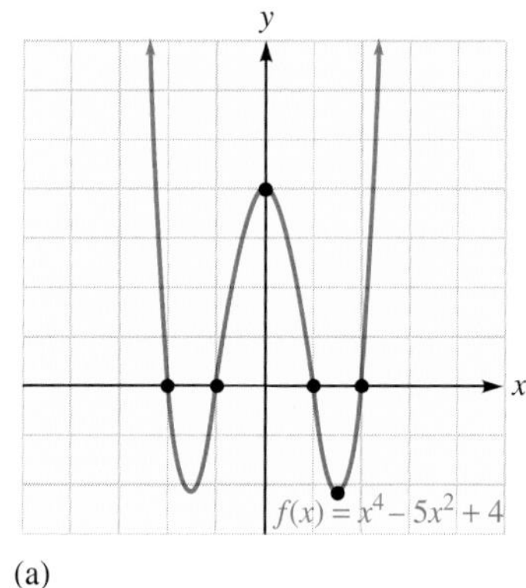

(a)

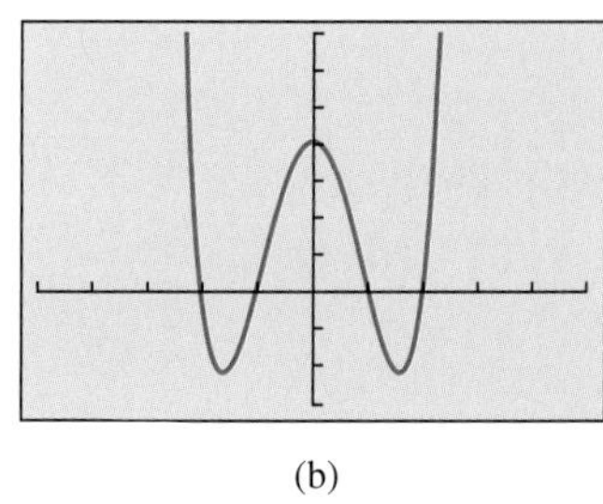

(b)

Figure 3-21

Comment Note that the far right and far left ends of the graph are similar to the ends of the graph of $f(x) = x^4$, which is the term of the function $f(x) = x^4 - 5x^2 + 4$ with highest degree. Also note that the graph has 3 turning points.

Self Check Graph: $f(x) = x^4 - 10x^2 + 9$. ■

EXAMPLE 3 Graph the function f: $f(x) = -\frac{1}{5}x^4 - x^2$.

Solution Since all powers of x are even, this graph has symmetry about the y-axis.

To find the x-intercepts, we let $f(x) = 0$ and solve for x.

$$-\frac{1}{5}x^4 - x^2 = 0$$

$$x^4 + 5x^2 = 0 \qquad \text{Multiply both sides by } -5.$$

$$x^2(x^2 + 5) = 0 \qquad \text{Factor.}$$

$$x = 0 \quad \text{or} \quad x = 0 \quad \text{or} \quad x^2 + 5 = 0$$

$$x = 0 \quad | \quad x = 0 \quad | \quad \text{This equation has no real roots.}$$

The only x-intercept is (0, 0). To find the y-intercept, we let $x = 0$ and see that the y-intercept is (0, 0).

To determine where the graph is above or below the x-axis, we plot the x-intercept on a number line and establish the intervals shown in Figure 3-22.

Sign of $f(x) = -\frac{1}{5}x^4 - x^2$	$-$		$-$
	$(-\infty, 0)$		$(0, \infty)$
Test point	$f(-1) = -1.2$	0	$f(1) = -1.2$
Graph of $f(x)$	below the x-axis		below the x-axis

Figure 3-22

We can now plot the intercepts and use our knowledge of symmetry and where the graph is above and below the x-axis to draw the graph, as in Figure 3-23(a). A calculator graph is shown in Figure 3-23(b).

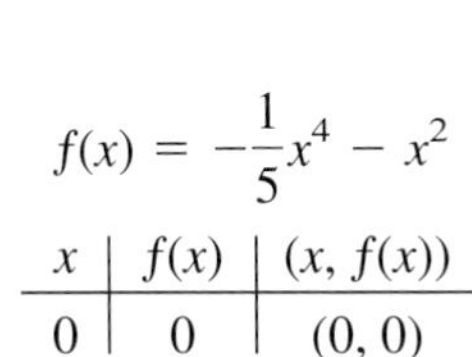

$$f(x) = -\frac{1}{5}x^4 - x^2$$

x	$f(x)$	$(x, f(x))$
0	0	(0, 0)

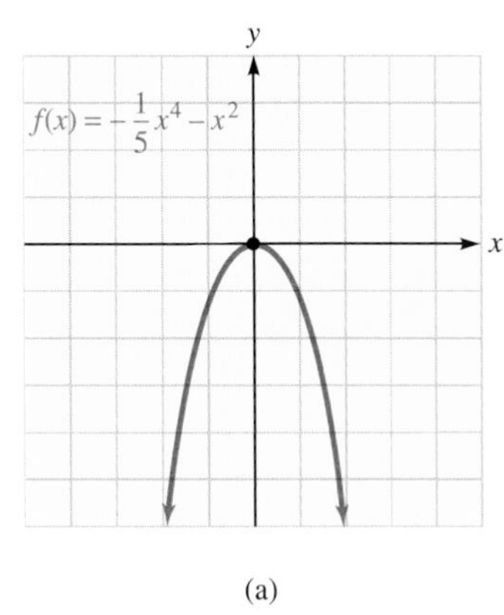

(a)

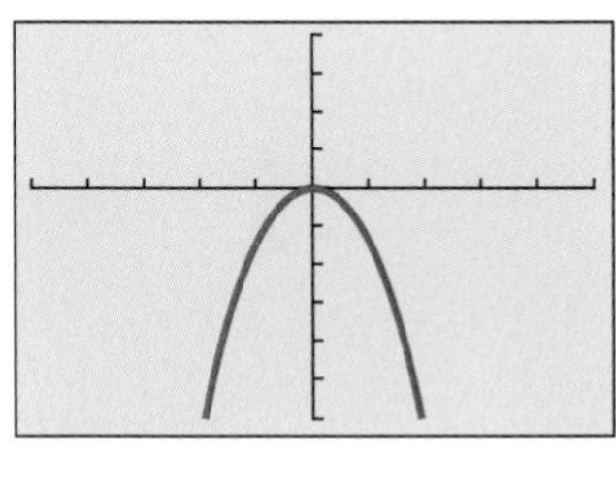

(b)

Figure 3-23

Comment Note that the far right and far left ends of the graph are similar to the ends of the graph of $f(x) = x^4$, except they turn in the opposite direction. Also note that the graph has only 1 turning point.

Self Check Graph: $f(x) = -x^3 + 1$. ■

Even and Odd Functions

If $f(-x) = f(x)$ for all x in the domain of f, the graph of the function is symmetric about the y-axis, and the function is called an **even function.** If $f(-x) = -f(x)$ for all x in the domain of f, the function is symmetric about the origin, and the function is called an **odd function.**

Since the graph in Example 1 is symmetric about the origin, it represents an odd function. Since the graphs in Examples 2 and 3 are symmetric about the y-axis, they represent even functions. If $f(x)$ does not have either of these symmetries, it is neither even nor odd.

ACTIVE EXAMPLE 4 Determine whether each function is even or odd: **a.** $f(x) = x^2$ and **b.** $f(x) = x^3$.

Solution **a.** To check whether the function is an even function, we find $f(-x)$ and see whether $f(-x) = f(x)$.

$$f(-x) = (-x)^2 = x^2 = f(x)$$

Since $f(-x) = f(x)$, the function is an even function.

COLLEGE **Algebra** $f(x)$ **Now**™
Go to **http://1pass.thomson.com** or your CD to practice this example.

b. To check whether the function is an even function, we find $f(-x)$ and see whether $f(-x) = f(x)$.

$$f(-x) = (-x)^3 = -x^3 = -f(x)$$

Since $f(-x) \neq f(x)$, the function is not an even function. However, the function is an odd function, because $f(-x) = -f(x)$.

Self Check Classify each function as even or odd: **a.** $f(x) = x^3 + x$ and **b.** $f(x) = x^2 + 4$. ■

■ Increasing and Decreasing Functions

If the values $f(x)$ increase on an interval $[a, b]$, as in Figure 3-24(a), we say that the function is **increasing on the interval.** If the values $f(x)$ decrease as x increases on $[a, b]$, as in Figure 3-24(b), we say that the function is **decreasing on the interval.** If the values $f(x)$ remain unchanged as x increases on $[a, b]$, as in Figure 3-24(c), we say that the function is **constant on the interval.**

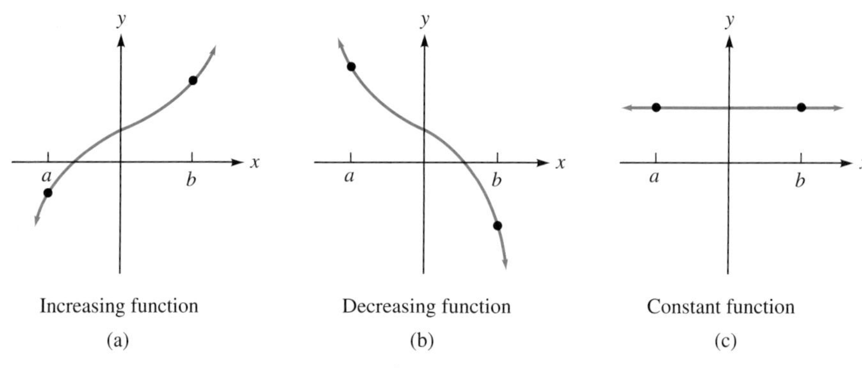

Figure 3-24

■ Piecewise-Defined Functions

Some functions, called **piecewise-defined functions,** are defined by using different equations for different intervals in their domains. To illustrate, we will graph the piecewise-defined function f given by

$$f(x) = \begin{cases} -2 & \text{if } x \leq 0 \\ x + 1 & \text{if } x > 0 \end{cases}$$

For each number x, we decide which part of the definition to use. If $x \leq 0$, the corresponding value of $f(x)$ is -2. In the interval $(-\infty, 0]$, the function is constant.

If $x > 0$, the corresponding value of $f(x)$ is $x + 1$. In the interval $(0, \infty)$, the function is increasing. The graph appears in Figure 3-25.

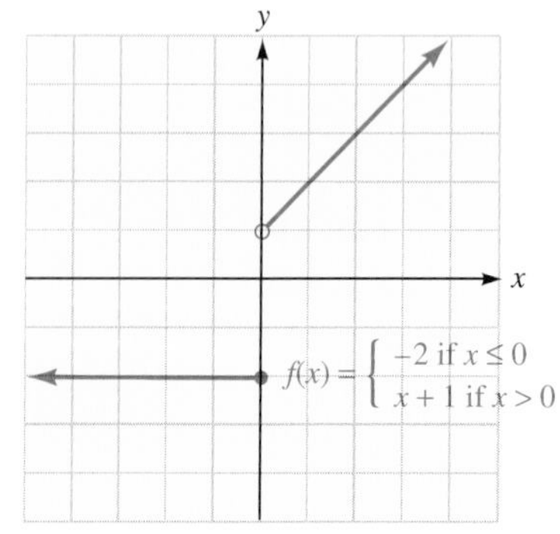

Figure 3-25

ACTIVE EXAMPLE 5 Graph the function: $f(x) = \begin{cases} -x & \text{if } x < 0 \\ x^2 & \text{if } 0 \le x \le 1. \\ 1 & \text{if } x > 1 \end{cases}$

Solution For each number x, we decide which part of the definition to use. If $x < 0$, the value of $f(x)$ is determined by the equation $f(x) = -x$. In the interval $(-\infty, 0)$, the function is decreasing.

If $0 \le x \le 1$, the value of $f(x)$ is x^2. In the interval $[0, 1]$, the function is increasing.

If $x > 1$, the value of $f(x)$ is 1. In the interval $(1, \infty)$, the function is constant. The graph appears in Figure 3-26.

COLLEGE **Algebra** $f(x)$ **Now**™
Go to **http://1pass.thomson.com** or your CD to practice this example.

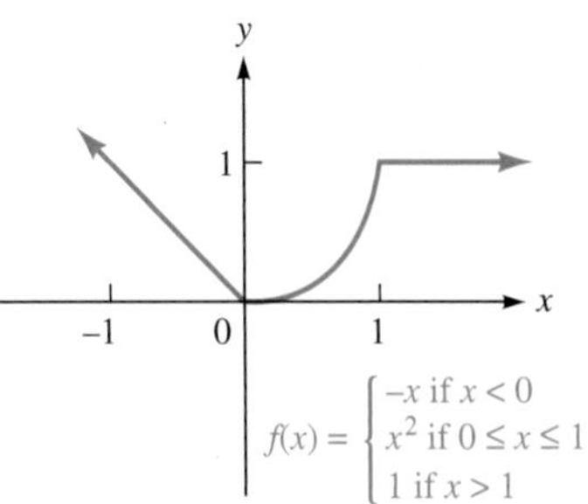

Figure 3-26

Self Check Graph: $f(x) = \begin{cases} 2x & \text{if } x \le 0 \\ x - 1 & \text{if } x > 0 \end{cases}$.

The Greatest Integer Function

The **greatest integer function** is important in computer science. This function is determined by the equation $y = [\![x]\!]$, where the value of y that corresponds to x is the greatest integer that is less than or equal to x. For example,

$$[\![2.71]\!] = 2, \quad [\![23\tfrac{1}{2}]\!] = 23, \quad [\![10]\!] = 10, \quad [\![\pi]\!] = 3, \quad [\![-2.5]\!] = -3$$

EXAMPLE 6 Graph: $y = [\![x]\!]$.

Solution We list several intervals and the corresponding values of the greatest integer function:

$[0, 1)$	$y = [\![x]\!] = 0$	For numbers from 0 to 1 (not including 1), the greatest integer in the interval is 0.
$[1, 2)$	$y = [\![x]\!] = 1$	For numbers from 1 to 2 (not including 2), the greatest integer in the interval is 1.
$[2, 3)$	$y = [\![x]\!] = 2$	For numbers from 2 to 3 (not including 3), the greatest integer in the interval is 2.

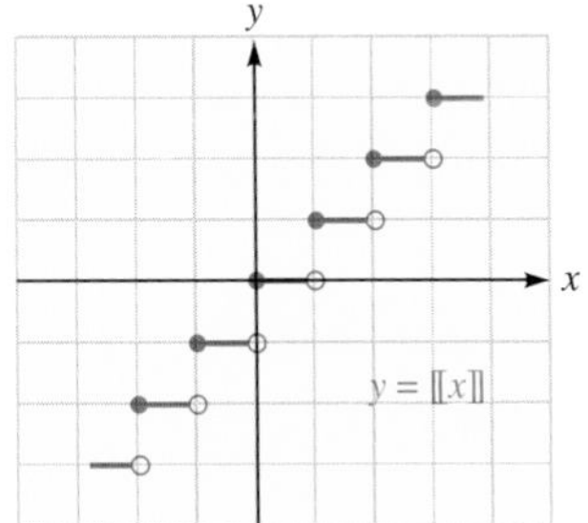

Figure 3-27

Within each interval, the values of y are constant, but they jump by 1 at integer values of x. The graph is shown in Figure 3-27. From the graph, we can see that the domain of the greatest integer function is the interval $(-\infty, \infty)$. The range is the set of integers.

Self Check Find **a.** $[\![7.61]\!]$ and **b.** $[\![-3.75]\!]$.

Since the greatest integer function is made up of a series of horizontal line segments, it is an example of a group of functions called **step functions.**

EXAMPLE 7 To print a business form, a printer charges \$10 for the order, plus \$20 for each box containing 500 forms. The printer counts any portion of a box as a full box. Graph this step function.

Solution If we order the forms and then change our minds before the forms are printed, the cost will be \$10. Thus, the ordered pair (0, 10) will be on the graph. If we purchase up to one full box, the cost will be \$10 for the order and \$20 for the printing, for a total of \$30. Thus, the ordered pair (1, 30) will be on the graph. The cost for $1\frac{1}{2}$ boxes will be the same as the cost for 2 full boxes, or \$50. Thus, the ordered pairs (1.5, 50) and (2, 50) are on the graph.

The complete graph is shown in Figure 3-28.

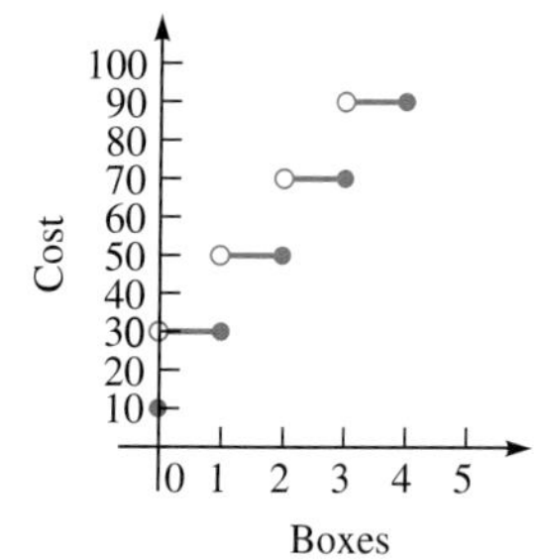

Figure 3-28

Self Check Find the cost of $4\frac{1}{2}$ boxes.

Self Check Answers

1.

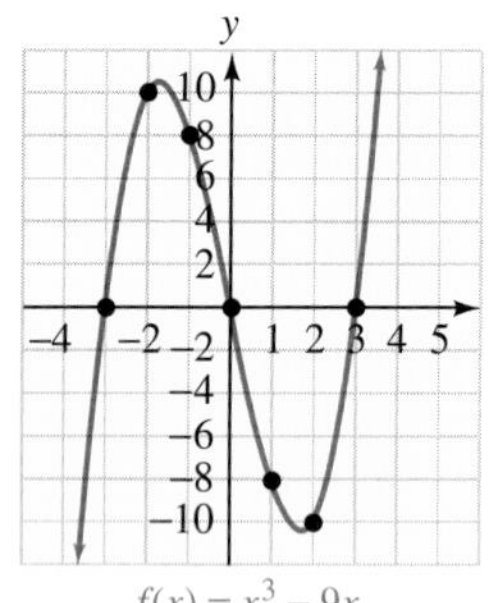

2.

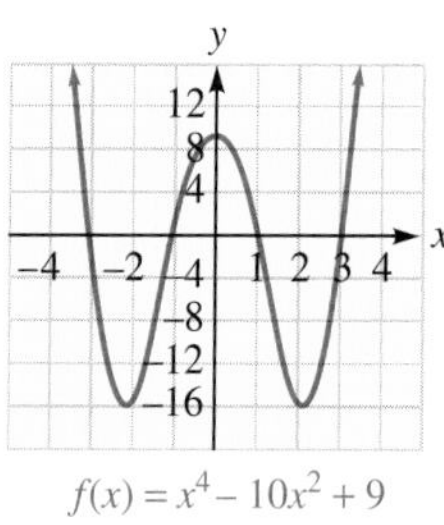

3.

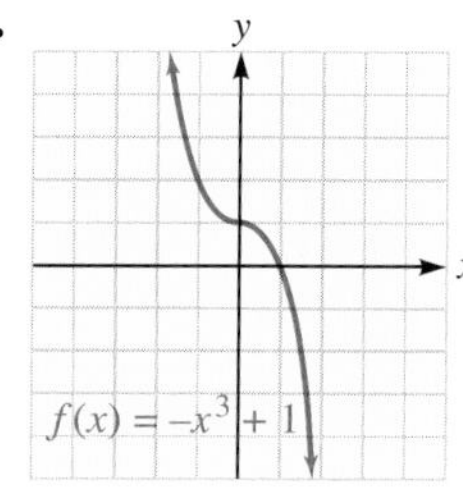

4. a. odd **b.** even

5.

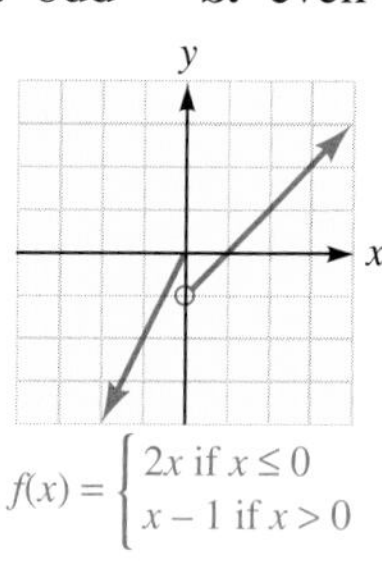

6. a. 7 **b.** −4 **7.** \$110

3.3 Exercises

VOCABULARY AND CONCEPTS ***Fill in the blanks.***

1. The degree of the function $y = f(x) = x^4 - 3$ is **4**.
2. Peaks and valleys on a polynomial graph are called **turning** points.
3. The graph of a *n*th degree polynomial function can have at most **$n - 1$** turning points.
4. If the graph of a function is symmetric about the **y-axis**, it is called an even function.
5. If the graph of a function is symmetric about the origin, it is called an **odd** function.
6. If the values of $f(x)$ get larger as x increases on an interval, we say that the function is **increasing** on the interval.
7. **Piecewise-defined** functions are defined by different equations for different intervals in their domains.
8. If the values of $f(x)$ get smaller as x increases on an interval, we say that the function is **decreasing** on the interval.
9. $[\![3.69]\!] =$ **3**.
10. If the values of $f(x)$ do not change as x increases on an interval, we say that the function is **constant** on the interval.

PRACTICE ***Graph each polynomial function.***

11. $f(x) = x^3 - x$

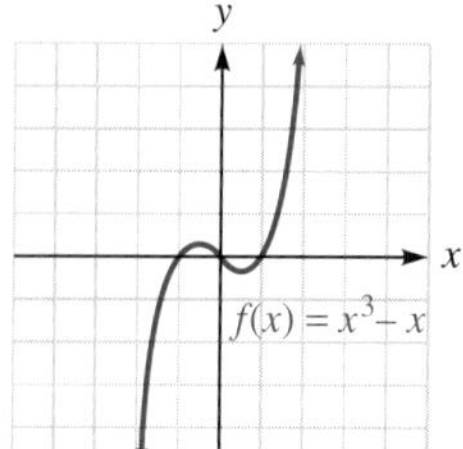

12. $f(x) = x^3 + x^2$

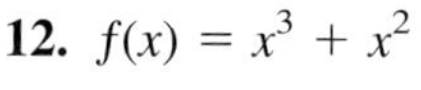

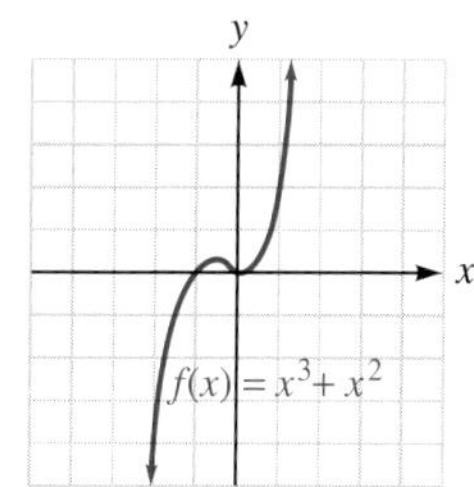

13. $f(x) = -x^3$

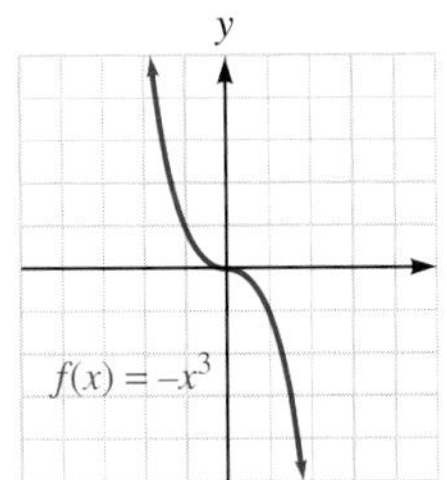

14. $f(x) = -x^3 + 1$

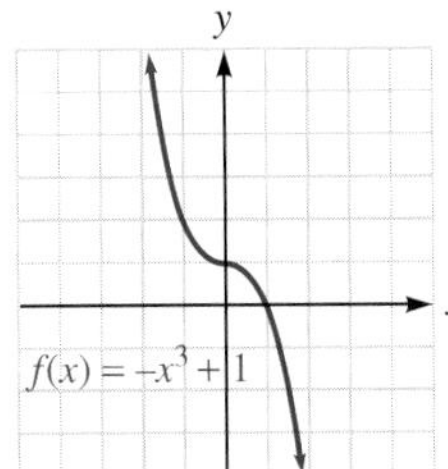

15. $f(x) = x^4 - 2x^2 + 1$

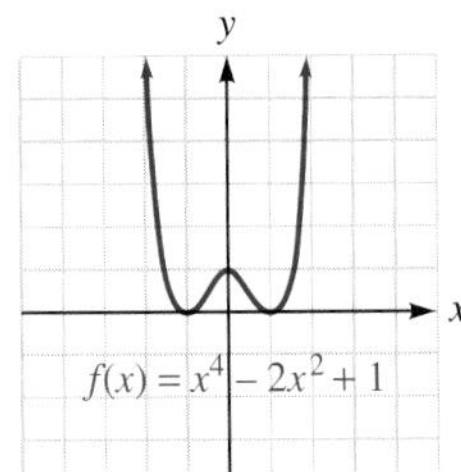

16. $f(x) = x^4 - 5x^2 + 4$

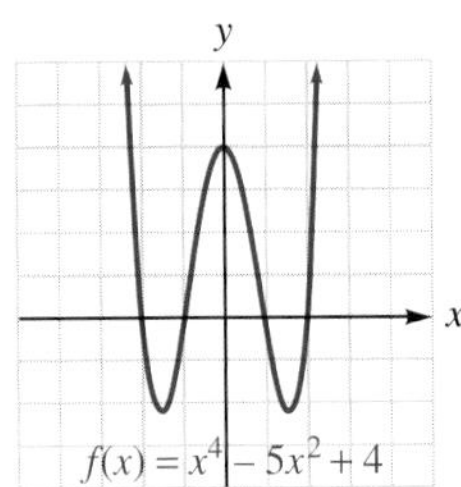

17. $f(x) = x^3 - x^2 - 4x + 4$

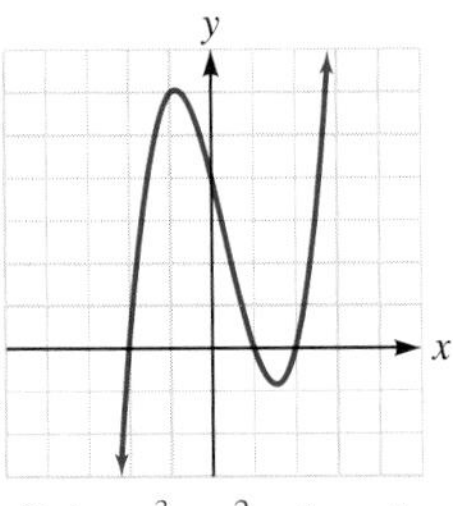

$f(x) = x^3 - x^2 - 4x + 4$

18. $f(x) = 4x^3 - 4x^2 - x + 1$

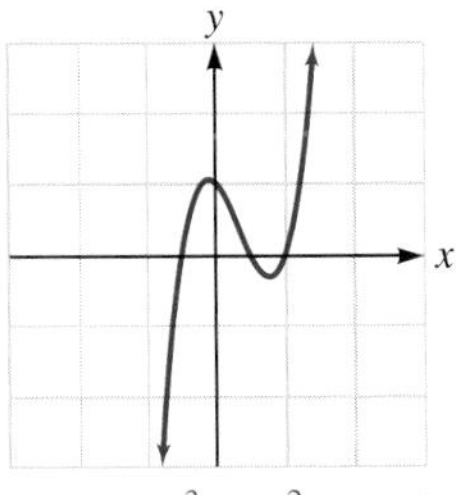

$f(x) = 4x^3 - 4x^2 - x + 1$

19. $f(x) = -x^4 + 5x^2 - 4$

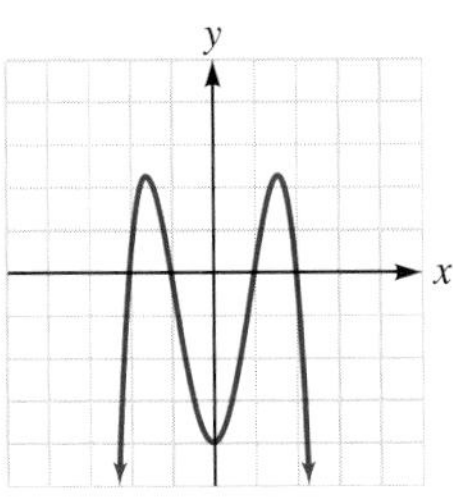

$f(x) = -x^4 + 5x^2 - 4$

20. $f(x) = x^4 - 4x^3 + x^2 + 6x$ (*Hint:* A partial factorization of $f(x)$ is $(x - 3)(x^3 - x^2 - 2x)$.)

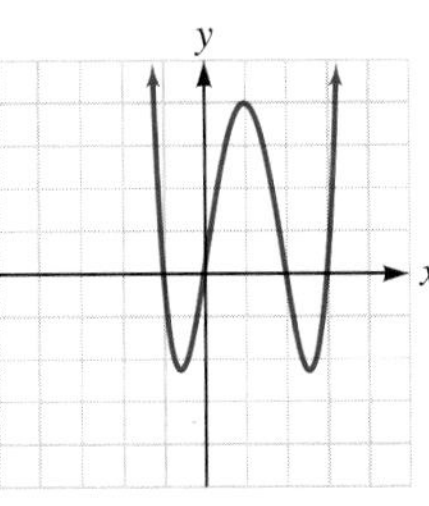

$f(x) = x^4 - 4x^3 + x^2 + 6x$

Determine whether each function is even or odd. If it is neither, so indicate.

21. $f(x) = x^4 + x^2$ even

22. $f(x) = x^3 - 2x$ odd

23. $f(x) = x^3 + x^2$ neither

24. $f(x) = x^6 - x^2$ even

25. $f(x) = x^5 + x^3$ odd

26. $f(x) = x^3 - x^2$ neither

27. $f(x) = 2x^3 - 3x$ odd

28. $f(x) = 4x^2 - 5$ even

Determine where each function is increasing, decreasing, or constant.

29.

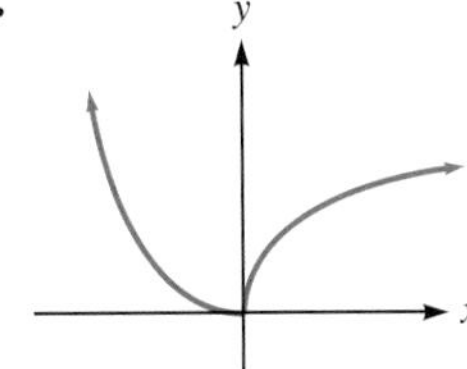

decreasing for $x < 0$;
increasing for $x > 0$

30.

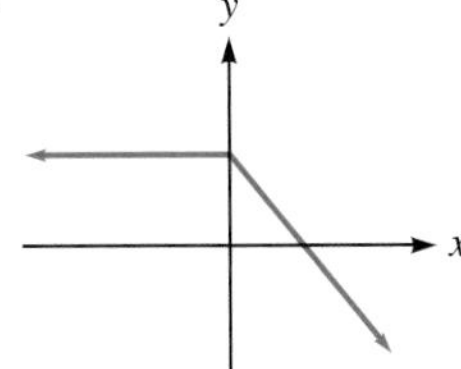

constant for $x < 0$;
decreasing for $x > 0$

31.

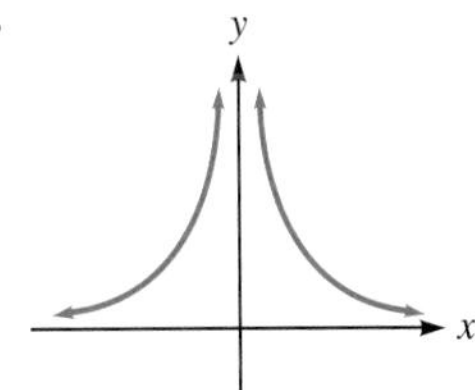

increasing for $x < 0$;
decreasing for $x > 0$

32.

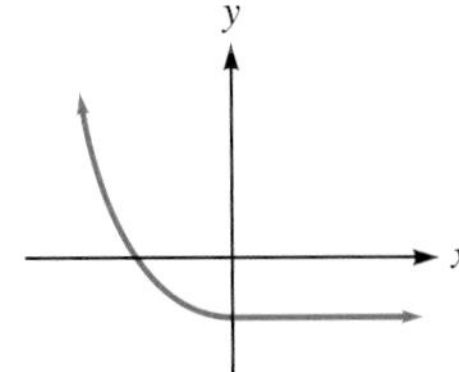

decreasing for $x < 0$;
constant for $x > 0$

Determine the interval in which each function is decreasing, increasing, or is constant.

33.

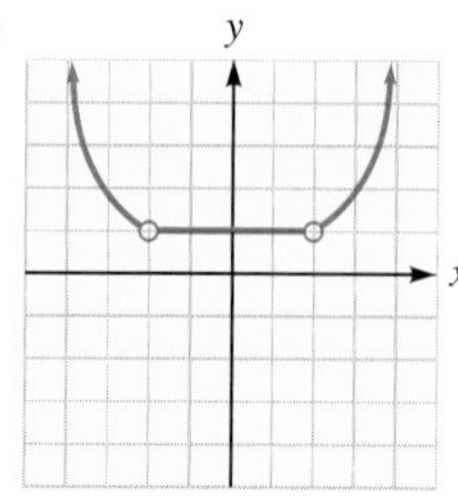

decreasing on $(-\infty, -2)$; constant on $(-2, 2)$; increasing on $(2, \infty)$

34.

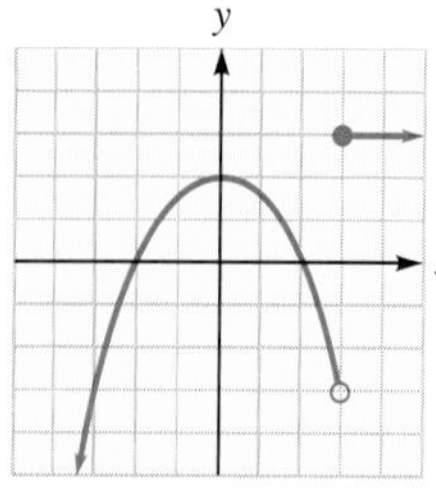

increasing on $(-\infty, 0)$; decreasing on $(0, 3)$; constant on $[3, \infty)$

35. $f(x) = x^2 - 4x + 4$

decreasing for $x < 2$; increasing for $x > 2$

36. $f(x) = 4 - x^2$

increasing for $x < 0$; decreasing for $x > 0$

Graph each piecewise-defined function.

37. $f(x) = \begin{cases} x + 2 & \text{if } x < 0 \\ 2 & \text{if } x \geq 0 \end{cases}$

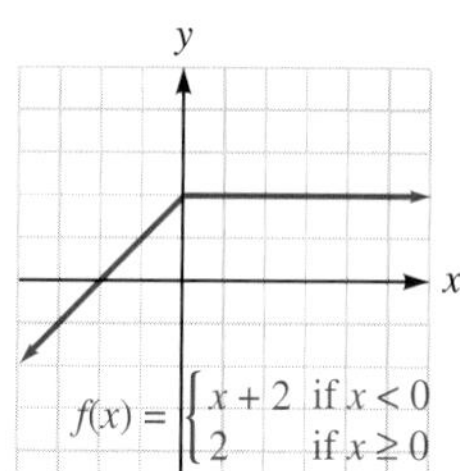

38. $f(x) = \begin{cases} 2x & \text{if } x < 0 \\ -2x & \text{if } x \geq 0 \end{cases}$

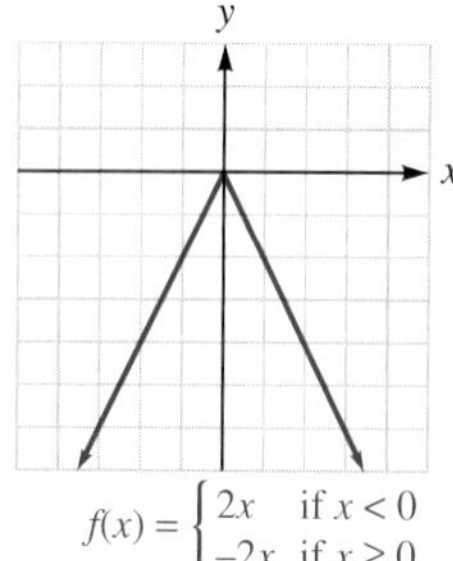

39. $f(x) = \begin{cases} -x & \text{if } x < 0 \\ x^2 & \text{if } x \geq 0 \end{cases}$

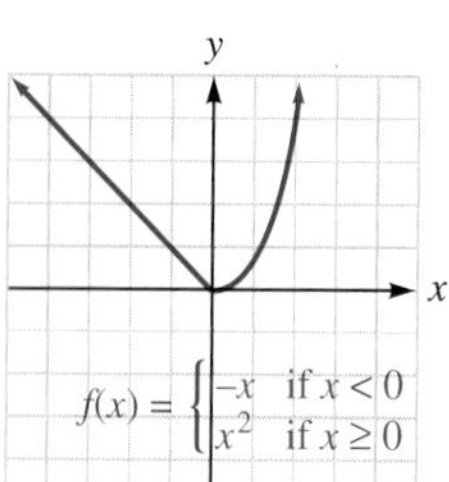

40. $f(x) = \begin{cases} |x| & \text{if } x < 0 \\ \sqrt{x} & \text{if } x \geq 0 \end{cases}$

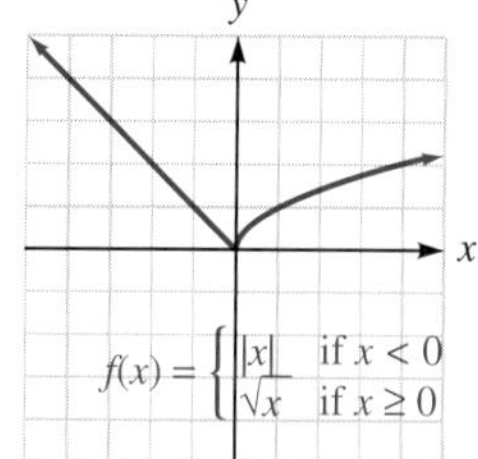

41. $f(x) = \begin{cases} x & \text{if } x \leq 0 \\ 2 & \text{if } x > 0 \end{cases}$

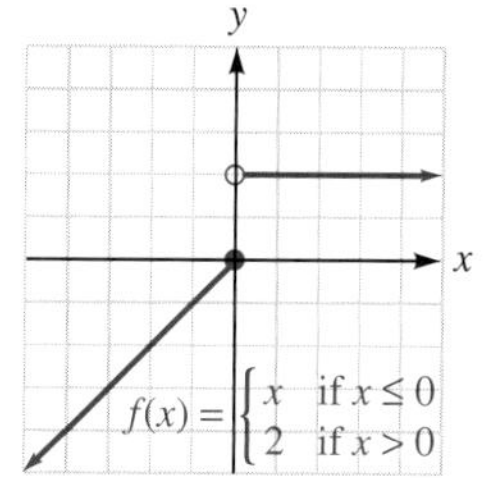

42. $f(x) = \begin{cases} -x & \text{if } x < 0 \\ \frac{1}{2}x & \text{if } x > 0 \end{cases}$

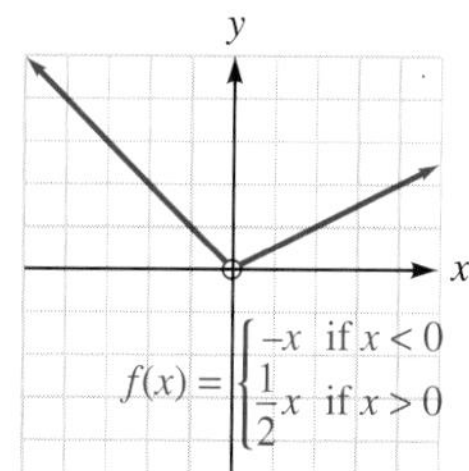

43. $f(x) = \begin{cases} 0 & \text{if } x < 0 \\ x^2 & \text{if } 0 \leq x \leq 2 \\ 4 - 2x & \text{if } x > 2 \end{cases}$

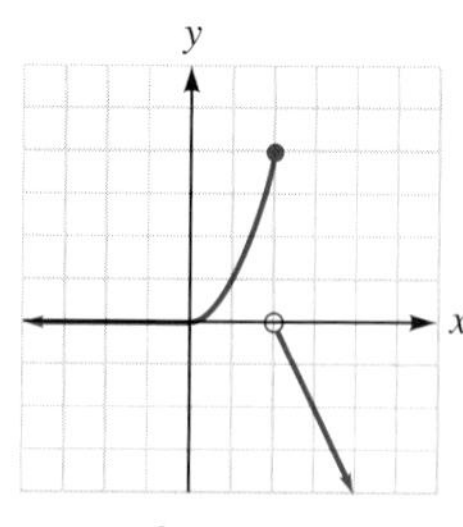

$f(x) = \begin{cases} 0 & \text{if } x < 0 \\ x^2 & \text{if } 0 \leq x \leq 2 \\ 4 - 2x & \text{if } x > 2 \end{cases}$

44. $f(x) = \begin{cases} 2 & \text{if } x < 0 \\ 2 - x & \text{if } 0 \le x < 2 \\ x & \text{if } x \ge 2 \end{cases}$

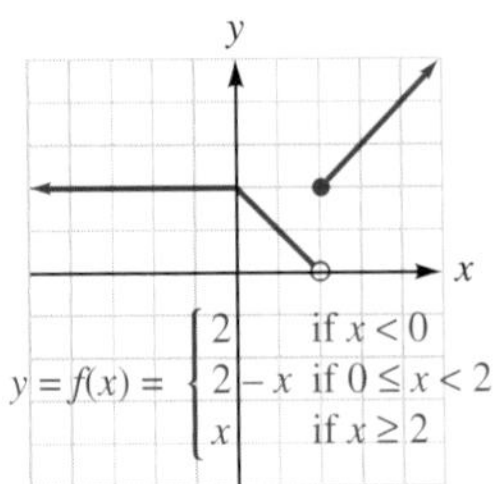

Graph each function.

45. $y = [\![2x]\!]$

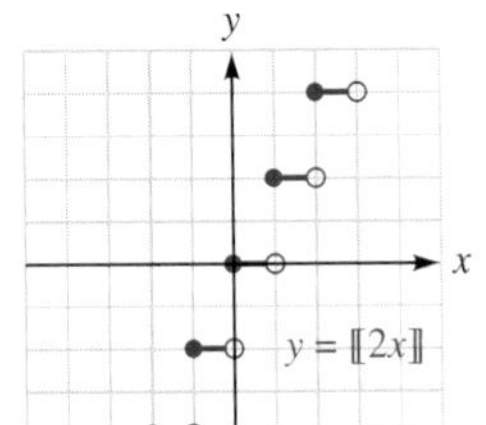

46. $y = \left[\!\left[\frac{1}{3}x + 3\right]\!\right]$

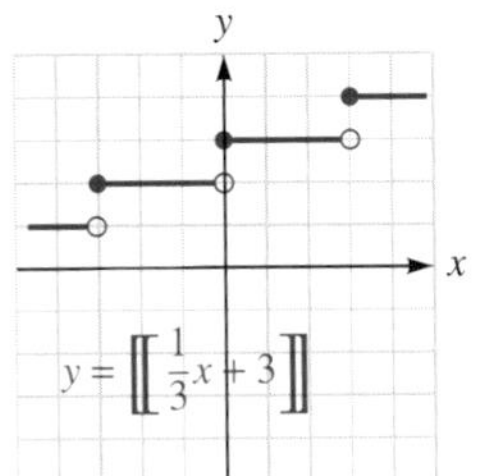

47. $y = [\![x]\!] - 1$

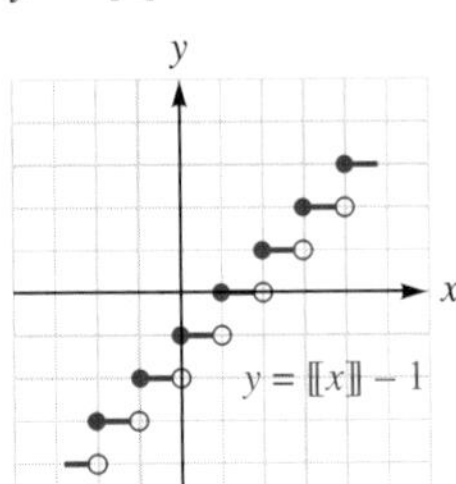

48. $y = [\![x + 2]\!]$

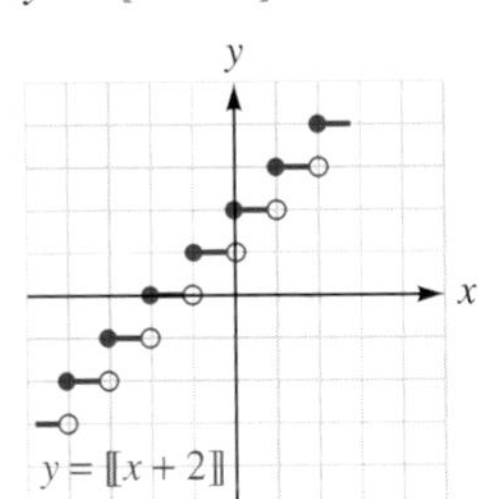

APPLICATIONS

49. **Grading scales** A mathematics instructor assigns letter grades according to the following scale.

From	Up to but less than	Grade
60%	70%	D
70%	80%	C
80%	90%	B
90%	100% (including 100%)	A

Graph the ordered pairs (p, g), where p represents the percent and g represents the grade. Find the final semester grade of a student who has test scores of 67%, 73%, 84%, 87%, and 93%. **B**

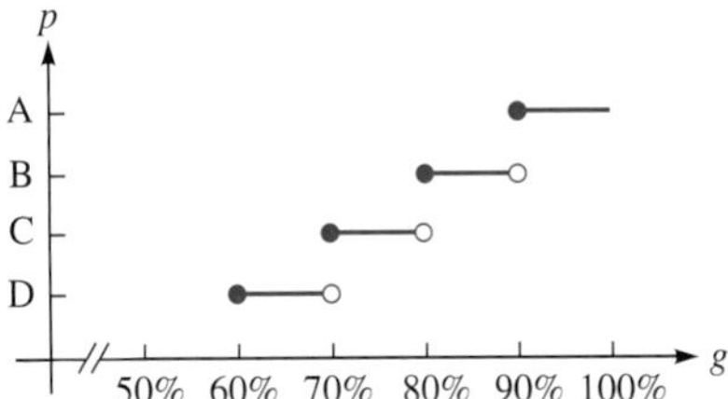

50. **Calculating grades** See Exercise 49 and find the final semester grade of a student who has test scores of 53%, 65%, 64%, 73%, 89%, and 82%. **C**

51. **Renting a car** A rental company charges \$20 to rent a car for one day, plus \$2 for every 100 miles (or portion of 100 miles) that it is driven. Graph the ordered pairs (m, c), where m represents the miles driven and c represents the cost. Find the cost if the car is driven 275 miles. **\$26**

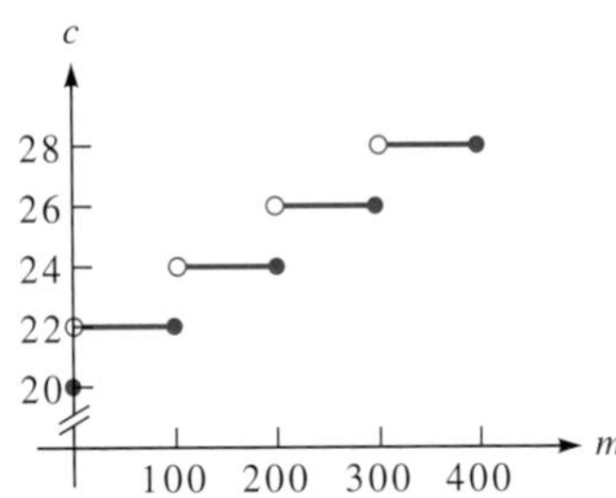

52. **Riding in a taxi** A taxicab company charges \$3 for a trip up to 1 mile, and \$2 for every extra mile (or portion of a mile). Graph the ordered pairs (m, c), where m represents the miles traveled and c represents the cost. Find the cost to ride $10\frac{1}{4}$ miles. **\$23**

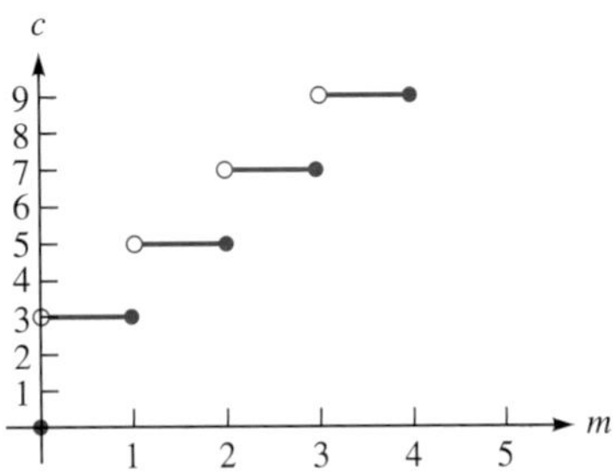

53. Computer communications An on-line information service charges for connect time at a rate of \$12 per hour, computed for every minute or fraction of a minute. Graph the points (t, c), where c is the cost of t minutes of connect time. Find the cost of $7\frac{1}{2}$ minutes. **\$1.60**

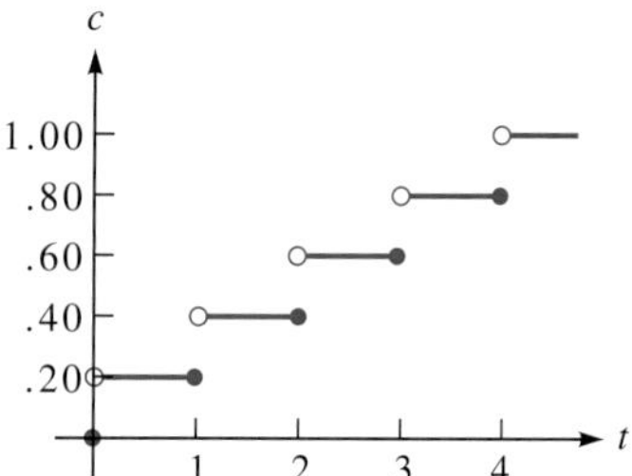

54. Plumbing repairs A plumber charges \$30, plus \$40 per hour (or fraction of an hour), to install a new bathtub. Graph the points (t, c), where t is the time it takes to do the job and c is the cost. If the job took 4 hours, how much did it cost? **\$190**

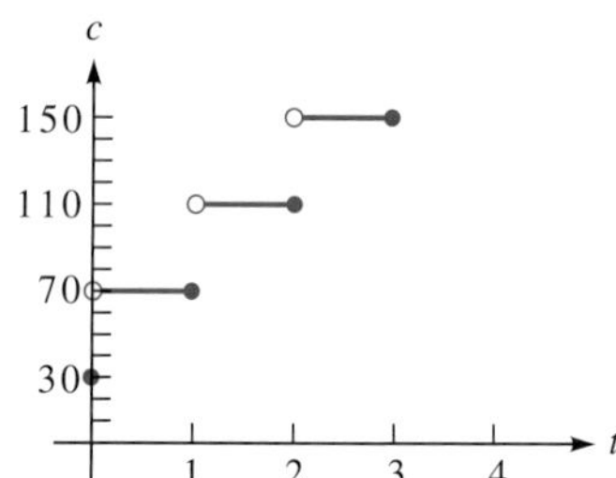

55. Rounding numbers Measurements are rarely exact; they are often *rounded* to an appropriate precision. Graph the points (x, y), where y is the result of rounding the number x to the nearest ten.

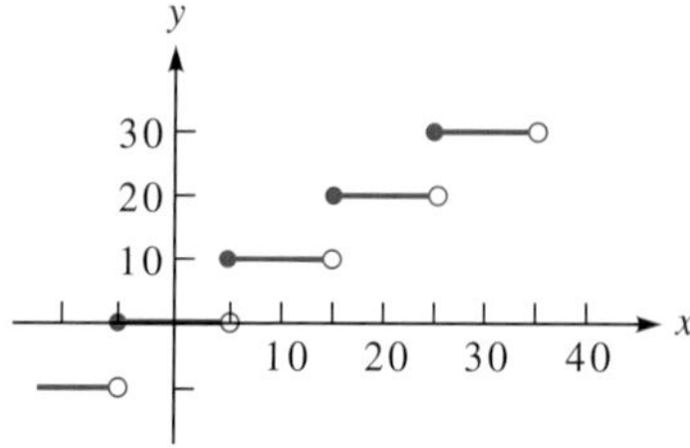

56. Signum function Computer programmers often use the following function, denoted by $y = \text{sgn } x$. Graph this function and find its domain and range.

$$y = \begin{cases} -1 & \text{if } x < 0 \\ 0 & \text{if } x = 0 \\ 1 & \text{if } x > 0 \end{cases}$$

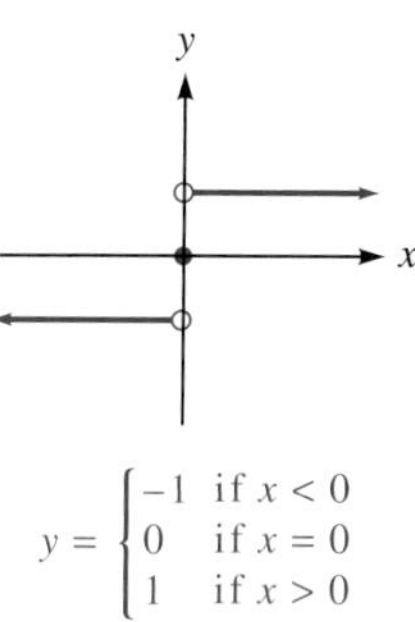

$$y = \begin{cases} -1 & \text{if } x < 0 \\ 0 & \text{if } x = 0 \\ 1 & \text{if } x > 0 \end{cases}$$

57. Graph the function defined by $y = \frac{|x|}{x}$ and compare it to the graph in Exercise 56. Are the graphs the same? **no; this is not defined at $x = 0$**

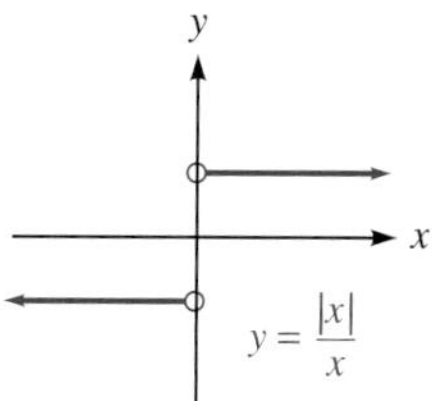

58. Graph: $y = x + |x|$.

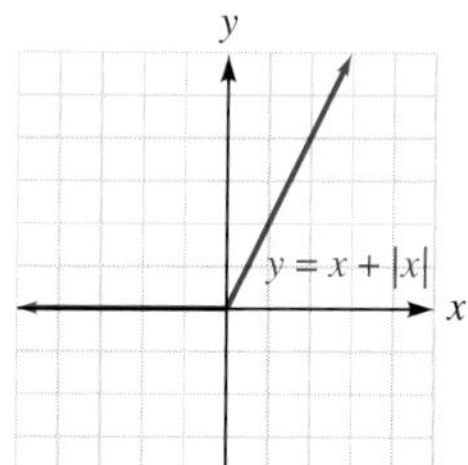

DISCOVERY AND WRITING ***Use a graphing calculator to explore the properties of graphs of polynomial functions. Write a paragraph summarizing your observations.***

59. Graph the function $y = x^2 + ax$ for several values of a. How does the graph change?

60. Graph the function $y = x^3 + ax$ for several values of a. How does the graph change?

61. Graph the function $y = (x - a)(x - b)$ for several values of a and b. What is the relationship between the x-intercepts and the equation?

62. Use the insight you gained in Exercise 61 to factor $x^3 - 3x^2 - 4x + 12$.

REVIEW

63. If $f(x) = 3x + 2$, find $f(x + 1)$ and $f(x) + 1$.
$3x + 5$; $3x + 3$

64. If $f(x) = x^2$, find $f(x - 2)$ and $f(x) - 2$.
$x^2 - 4x + 4$; $x^2 - 2$

65. If $f(x) = \frac{3x + 1}{5}$, find $f(x - 3)$ and $f(x) - 3$.
$\frac{3x - 8}{5}$; $\frac{3x + 1}{5} - 3$ or $\frac{3x - 14}{5}$

66. If $f(x) = 8$, find $f(x + 8)$ and $f(x) + 8$. 8; 16

67. Solve: $2x^2 - 3 = x$. $-1, \frac{3}{2}$

68. Solve: $4x^2 = 24x - 37$. $3 \pm \frac{1}{2}i$

3.4 Translating and Stretching Graphs

In this section, you will learn about

- Vertical Translations
- Horizontal Translations
- Shifts Involving Two Translations
- Reflection about the *x*- and *y*-Axes
- Vertical and Horizontal Stretchings

Vertical Translations

The graphs of functions can be identical except for their position in the *xy*-plane. For example, Figure 3-29 shows the graph of $y = x^2 + k$ for three values of k. If $k = 0$, we have the graph of $y = x^2$. The graph of $y = x^2 + 2$ is identical to the graph of $y = x^2$ except that it is shifted 2 units up. The graph of $y = x^2 - 3$ is identical to the graph of $y = x^2$ except that it is shifted 3 units down. These shifts are called **vertical translations.**

$y = x^2$

x	y	(x, y)
-2	4	$(-2, 4)$
-1	1	$(-1, 1)$
0	0	$(0, 0)$
1	1	$(1, 1)$
2	4	$(2, 4)$

$y = x^2 + 2$

x	y	(x, y)
-2	6	$(-2, 6)$
-1	3	$(-1, 3)$
0	2	$(0, 2)$
1	3	$(1, 3)$
2	6	$(2, 6)$

$y = x^2 - 3$

x	y	(x, y)
-2	1	$(-2, 1)$
-1	-2	$(-1, -2)$
0	-3	$(0, -3)$
1	-2	$(1, -2)$
2	1	$(2, 1)$

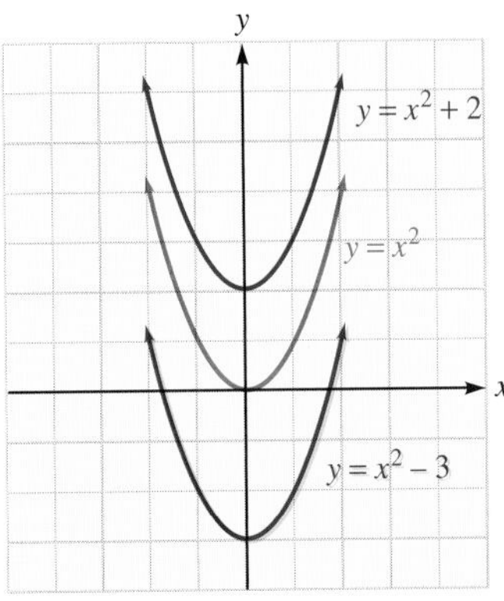

COLLEGE Algebra f(x) Now™
Explore this Active Figure and test your understanding at **http://1pass.thomson.com**.

Active Figure 3-29

In general, we can make the following observations.

Vertical Translations

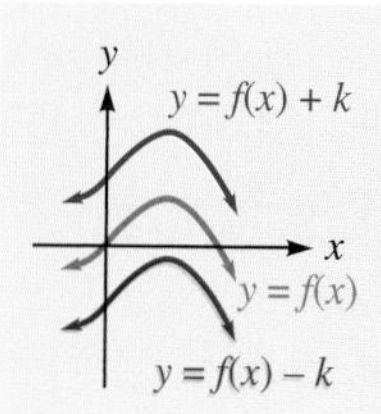

If f is a function and k is a positive number, then

- The graph of $y = f(x) + k$ is identical to the graph of $y = f(x)$ except that it is translated k units up.
- The graph of $y = f(x) - k$ is identical to the graph of $y = f(x)$ except that it is translated k units down.

ACTIVE EXAMPLE 1 Graph each function: **a.** $g(x) = |x| - 2$ and **b.** $h(x) = |x| + 3$.

Solution The graph of $g(x) = |x| - 2$ is identical to the graph of $f(x) = |x|$ except that it is translated 2 units down, as shown in Figure 3-30(a). The graph of $h(x) = |x| + 3$ is identical to the graph of $f(x) = |x|$ except that it is translated 3 units up, as shown in Figure 3-30(b).

COLLEGE **Algebra** $f(x)$ **Now**™
Explore this Active Example and test yourself on the steps taken here at **http://1pass.thomson.com**. You can also find this example on the Interactive Video Skillbuilder CD-ROM.

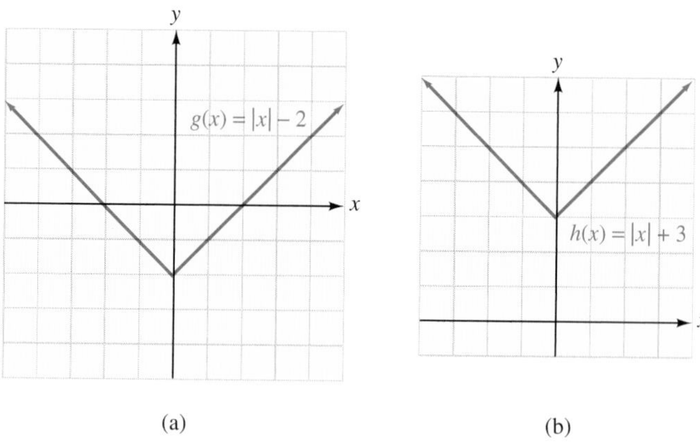

Figure 3-30

Self Check *Fill in the blanks:* The graph of $g(x) = x^2 + 3$ is identical to the graph of $f(x) = x^2$ except that it is translated <u>3</u> units <u>up</u>. The graph of $h(x) = x^2 - 4$ is identical to the graph of $f(x) = x^2$ except that it is translated <u>4</u> units <u>down</u>. ■

Horizontal Translations

Figure 3-31 shows the graph of $y = (x + h)^2$ for three values of h. If $h = 0$, we have the graph of $y = x^2$. The graph of $y = (x - 2)^2$ is identical to the graph of $y = x^2$ except that it is shifted 2 units to the right. The graph of $y = (x + 3)^2$ is identical to the graph of $y = x^2$ except that it is shifted 3 units to the left. These shifts are called **horizontal translations.**

COLLEGE **Algebra** $f(x)$ **Now**™
Explore this Active Figure and test your understanding at **http://1pass.thomson.com**.

$y = x^2$

x	y	(x, y)
−2	4	(−2, 4)
−1	1	(−1, 1)
0	0	(0, 0)
1	1	(1, 1)
2	4	(2, 4)

$y = (x - 2)^2$

x	y	(x, y)
0	4	(0, 4)
1	1	(1, 1)
2	0	(2, 0)
3	1	(3, 1)
4	4	(4, 4)

$y = (x + 3)^2$

x	y	(x, y)
−5	4	(−5, 4)
−4	1	(−4, 1)
−3	0	(−3, 0)
−2	1	(−2, 1)
−1	4	(−1, 4)

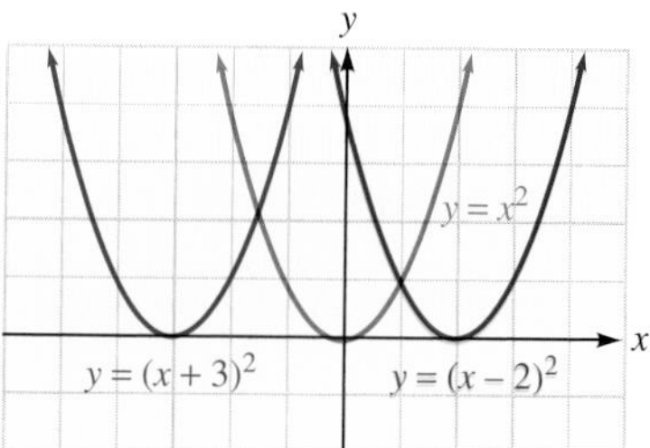

Active Figure 3-31

In general, we can make the following observations.

Horizontal Translations

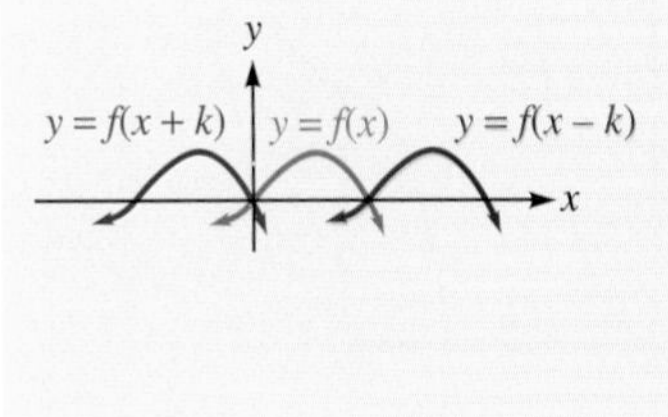

If f is a function and k is a positive number, then

- The graph of $y = f(x - k)$ is identical to the graph of $y = f(x)$ except that it is translated k units to the right.
- The graph of $y = f(x + k)$ is identical to the graph of $y = f(x)$ except that it is translated k units to the left.

ACTIVE EXAMPLE 2 Graph each function: **a.** $g(x) = |x - 4|$ and **b.** $h(x) = |x + 2|$.

COLLEGE **Algebra** $f(x)$ **Now**™
Go to **http://1pass.thomson.com** or your CD to practice this example.

Solution The graph of $g(x) = |x - 4|$ is identical to the graph of $f(x) = |x|$ except that it is translated 4 units to the right, as shown in Figure 3-32(a). The graph of $h(x) = |x + 2|$ is identical to the graph of $f(x) = |x|$ except that it is translated 2 units to the left, as shown in Figure 3-32(b).

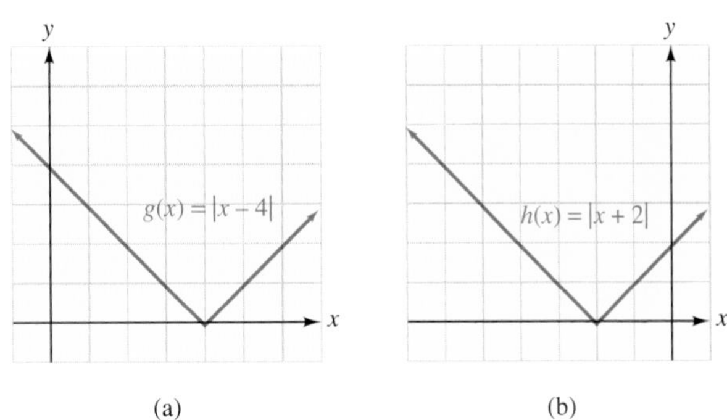

Figure 3-32

Self Check *Fill in the blanks:* The graph of $g(x) = (x - 3)^2$ is identical to the graph of $f(x) = x^2$ except that it is translated 3 units to the right. The graph of $h(x) = (x + 2)^2$ is identical to the graph of $f(x) = x^2$ except that it is translated 2 units to the left. ■

Shifts Involving Two Translations

Sometimes we can obtain a graph by using both a horizontal and a vertical translation.

ACTIVE EXAMPLE 3 Graph each function: **a.** $g(x) = (x - 5)^3 + 4$ and **b.** $h(x) = (x + 2)^2 - 2$.

Solution The graph of $g(x) = (x - 5)^3 + 4$ is identical to the graph of $f(x) = x^3$ except that it is translated 5 units to the right and 4 units up, as shown in Figure 3-33(a). The graph of $h(x) = (x + 2)^2 - 2$ is identical to the graph of $f(x) = x^2$ except that it is translated 2 units to the left and 2 units down, as shown in Figure 3-33(b).

COLLEGE **Algebra** $f(x)$ **Now**™
Go to **http://1pass.thomson.com** or your CD to practice this example.

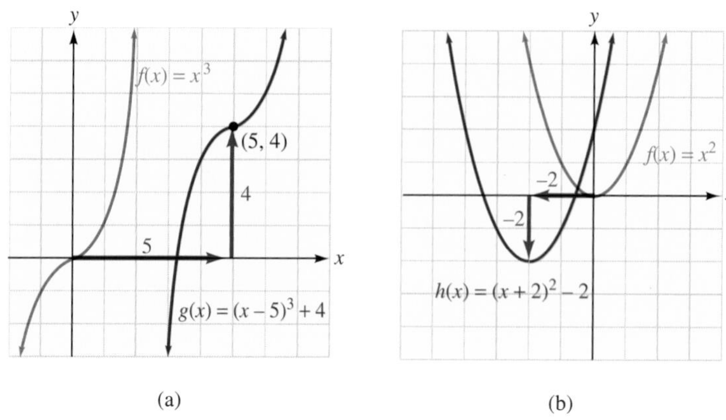

Figure 3-33

Self Check *Fill in the blanks:* The graph of $g(x) = |x - 4| + 5$ is identical to the graph of $f(x) = |x|$ except that it is translated **4** units to the **right** and **5** units **up**. ■

Accent on Technology CALCULATOR GRAPHS

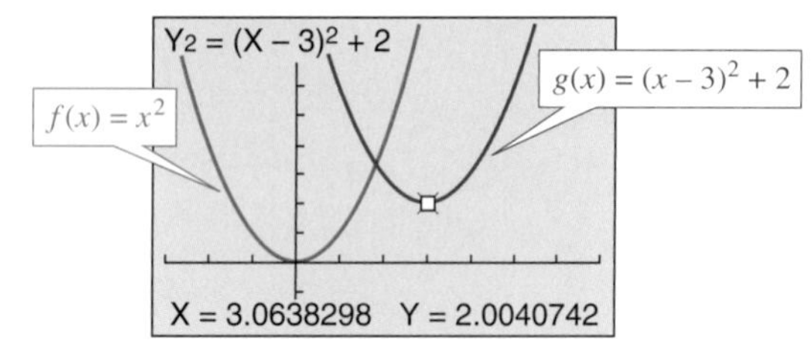

Figure 3-34

We can use a graphing calculator to show the effect of vertical and horizontal translations by graphing $f(x) = x^2$ and $g(x) = (x - 3)^2 + 2$. The graph of $f(x) = x^2$ is a parabola opening upward, with vertex at the origin. The graph of $g(x) = (x - 3)^2 + 2$ should be that same parabola translated 3 units to the right and 2 units up. Figure 3-34 shows the result of graphing these functions. After tracing, we see that the vertex of the translated graph is the point (3, 2), as expected. (More zooms and traces will give more accurate results.)

■ Reflections about the *x*- and *y*-Axes

Figure 3-35(a) shows that the graph of $y = -\sqrt{x}$ is identical to the graph of $y = \sqrt{x}$ except that it is reflected about the *x*-axis. Figure 3-35(b) shows that the graph of $y = \sqrt{-x}$ is identical to the graph of $y = \sqrt{x}$ except that it is reflected about the *y*-axis.

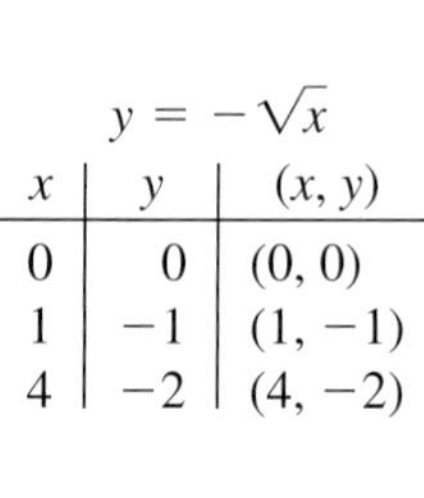

$y = -\sqrt{x}$

x	y	(x, y)
0	0	(0, 0)
1	−1	(1, −1)
4	−2	(4, −2)

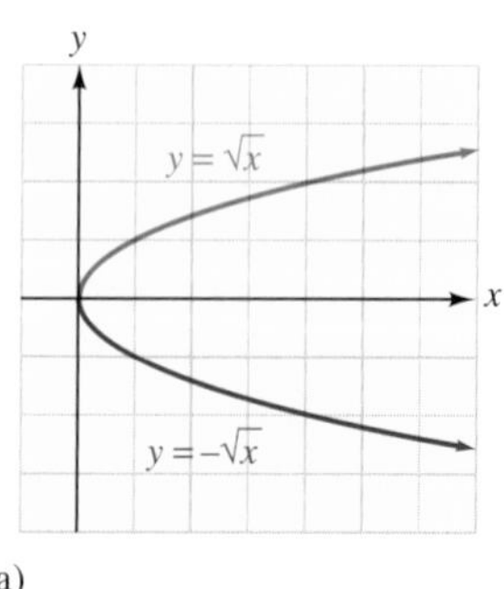

(a)

Figure 3-35

$y = \sqrt{-x}$

x	y	(x, y)
0	0	(0, 0)
−1	1	(−1, 1)
−4	2	(−4, 2)

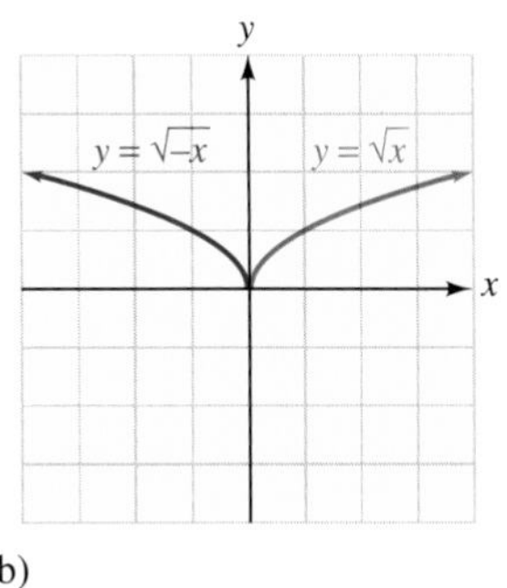

(b)

Figure 3-35 (continued)

In general, we can make the following observations.

Reflections

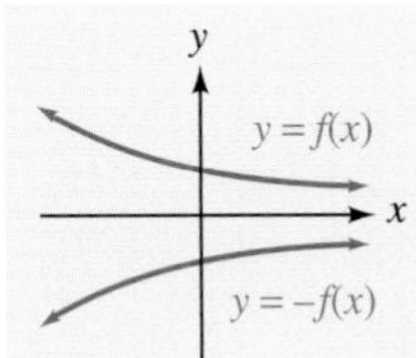

If f is a function, then

- The graph of $y = -f(x)$ is identical to the graph of $y = f(x)$ except that it is reflected about the x-axis.
- The graph of $y = f(-x)$ is identical to the graph of $y = f(x)$ except that it is reflected about the y-axis.

EXAMPLE 4 Graph each function: **a.** $g(x) = -|x + 1|$ and **b.** $h(x) = |-x + 1|$.

Solution The graph of $g(x) = -|x + 1|$ is identical to the graph of $f(x) = |x + 1|$ except that it is reflected about the x-axis, as shown in Figure 3-36(a). The graph of $h(x) = |-x + 1|$ is identical to the graph of $f(x) = |x + 1|$ except that it is reflected about the y-axis, as shown in Figure 3-36(b).

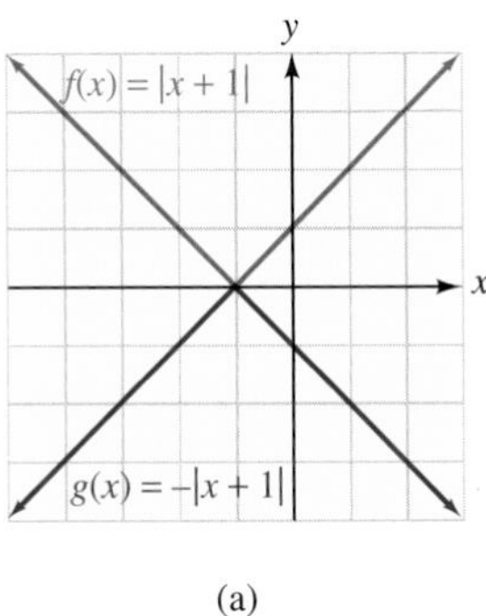

(a)

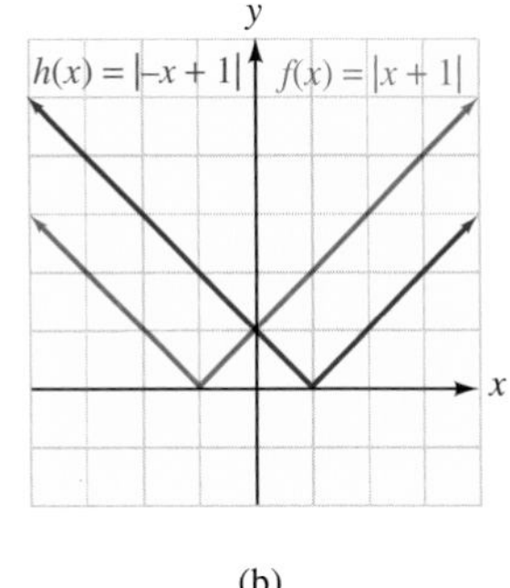

(b)

Figure 3-36

Self Check *Fill in the blanks:* The graph of $g(x) = -(x - 4)^2$ is identical to the graph of $f(x) = (x - 4)^2$ except that it is reflected about the x- axis. The graph of $h(x) = (-x - 4)^2$ is identical to the graph of $f(x) = (x - 4)^2$ except that it is reflected about the y- axis. ■

Vertical and Horizontal Stretchings

Figure 3-37 shows the graphs of $y = x^2 - 1$ and $y = 3(x^2 - 1)$. Because each value of $y = 3(x^2 - 1)$ is 3 times greater than the corresponding value of $y = x^2 - 1$, its graph is stretched vertically by a factor of 3. The x-intercepts of both graphs are $(1, 0)$ and $(-1, 0)$.

$y = x^2 - 1$

x	y	(x, y)
-2	3	$(-2, 3)$
-1	0	$(-1, 0)$
0	-1	$(0, -1)$
1	0	$(1, 0)$
2	3	$(2, 3)$

$y = 3(x^2 - 1)$

x	y	(x, y)
-2	9	$(-2, 9)$
-1	0	$(-1, 0)$
0	-3	$(0, -3)$
1	0	$(1, 0)$
2	9	$(2, 9)$

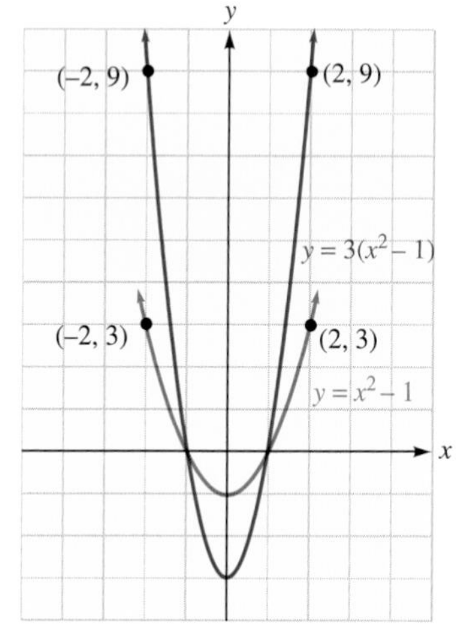

Figure 3-37

Figure 3-38 shows the graph of $y = x^2 - 1$ and $y = (3x)^2 - 1$. Because each value of x in $y = (3x)^2 - 1$ is $\frac{1}{3}$ of the corresponding value of x in $y = x^2 - 1$, the graph of $y = (3x)^2 - 1$ is stretched horizontally by a factor of $\frac{1}{3}$. The y-intercepts of both graphs are $(0, -1)$.

$y = x^2 - 1$

x	y	(x, y)
-2	3	$(-2, 3)$
-1	0	$(-1, 0)$
0	-1	$(0, -1)$
1	0	$(1, 0)$
2	3	$(2, 3)$

$y = (3x)^2 - 1$

x	y	(x, y)
$-\frac{2}{3}$	3	$\left(-\frac{2}{3}, 3\right)$
$-\frac{1}{3}$	0	$\left(-\frac{1}{3}, 0\right)$
0	-1	$(0, -1)$
$\frac{1}{3}$	0	$\left(\frac{1}{3}, 0\right)$
$\frac{2}{3}$	3	$\left(\frac{2}{3}, 3\right)$

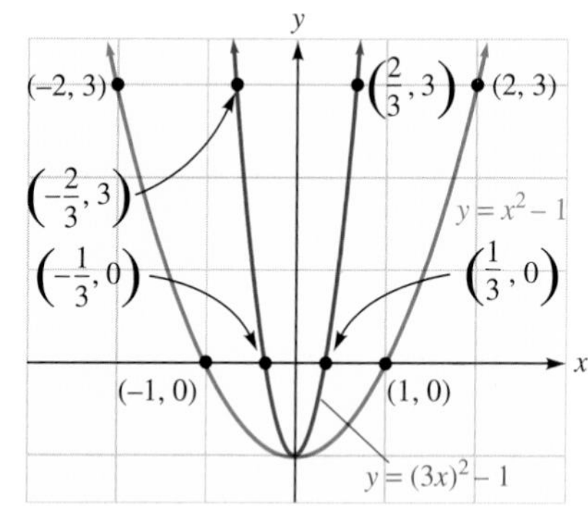

Figure 3-38

In general, we can make the following observations.

Vertical Stretching

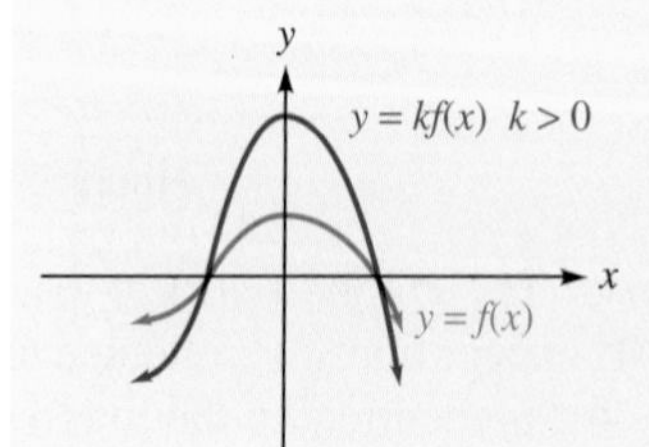

If f is a function and k is a positive number, then

- The graph of $y = kf(x)$ can be obtained by stretching the graph of $y = f(x)$ vertically by a factor of k.

Horizontal Stretching

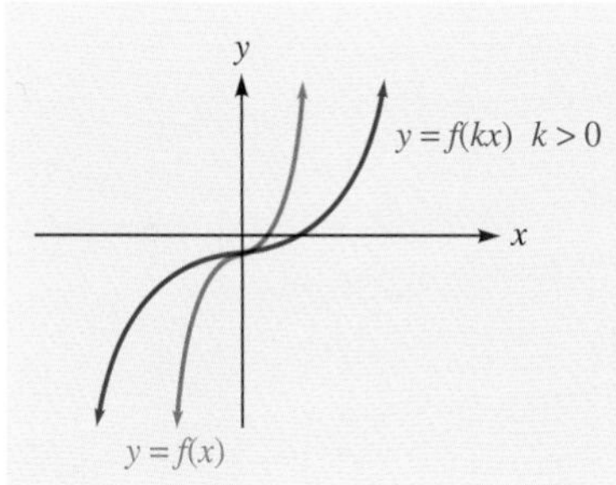

- The graph of $y = f(kx)$ can be obtained by stretching the graph of $y = f(x)$ horizontally by a factor of $\frac{1}{k}$.

Accent on Technology CALCULATOR GRAPHS

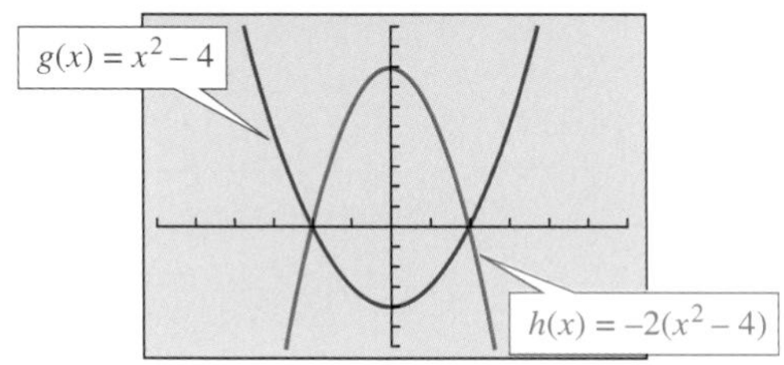

Figure 3-39

We can use a graphing calculator to show the effect of a vertical stretching and reflection by graphing $g(x) = x^2 - 4$ and $h(x) = -2(x^2 - 4)$. The graph of $g(x) = x^2 - 4$ is a parabola opening up with vertex at $(0, -4)$. If that graph is stretched vertically by a factor of 2 and then reflected about the x-axis, the graph of $h(x) = -2(x^2 - 4)$ should be the result. That is what happens, as shown in Figure 3-39. Notice that the x-intercepts of the graphs are the same.

We can summarize the ideas in the section as follows.

Translations and Reflections

If f is a function and k represents a positive number, then

The graph of	***can be obtained by graphing y = f(x) and***
$y = f(x) + k$	translating the graph k units up.
$y = f(x) - k$	translating the graph k units down.
$y = f(x + k)$	translating the graph k units to the left.
$y = f(x - k)$	translating the graph k units to the right.
$y = -f(x)$	reflecting the graph about the x-axis.
$y = f(-x)$	reflecting the graph about the y-axis.
$y = kf(x)$	stretching the graph vertically by a factor of k.
$y = f(kx)$	stretching the graph horizontally by a factor of $\frac{1}{k}$.

EXAMPLE 5 Figure 3-40 on the next page shows the graph of $y = f(x)$. Use this graph and a translation to find the graph of **a.** $y = f(x) + 2$, **b.** $y = f(x - 2)$, and **c.** $y = f(2x)$.

Solution The graph of $y = f(x) + 2$ is identical to the graph of $y = f(x)$ except that it is translated 2 units up. See Figure 3-41(a). The graph of $y = f(x - 2)$ is identical to the graph of $y = f(x)$ except that it is translated 2 units to the right. See Figure 3-41(b). The graph of $y = f(2x)$ is identical to the graph of $y = f(x)$ except that it is stretched horizontally by a factor of $\frac{1}{2}$. See Figure 3-41(c).

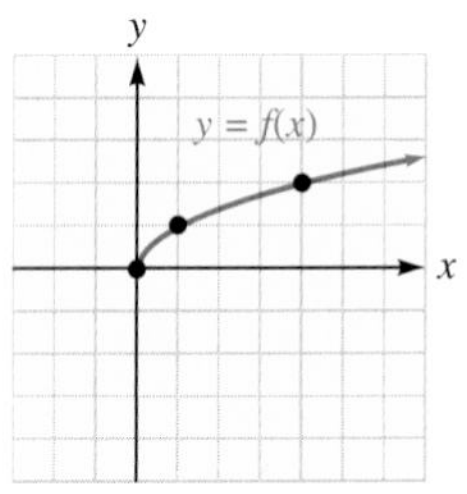

Figure 3-40

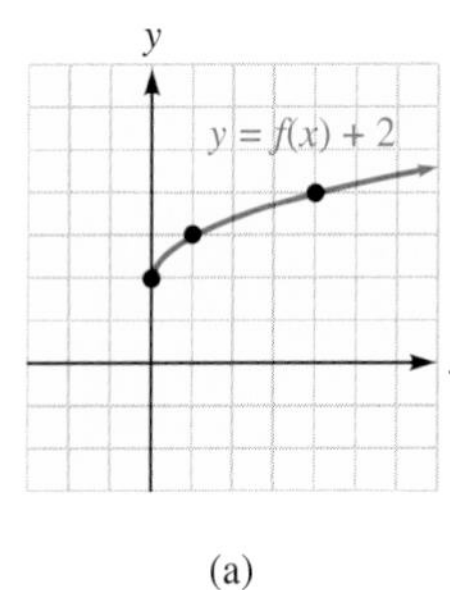

(a)

(b)

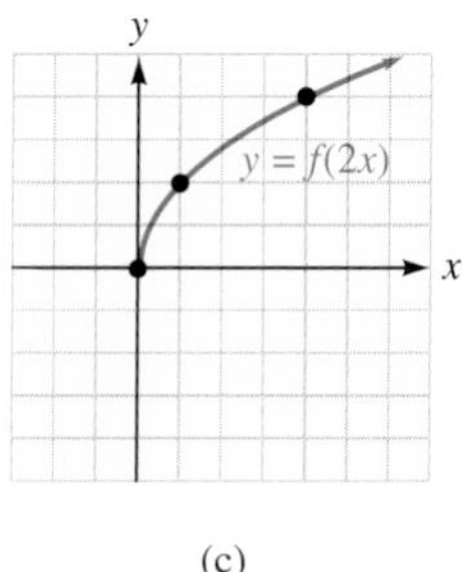

(c)

Figure 3-41

Self Check Use Figure 3-40 and a reflection to find the graph of **a.** $y = -f(x)$ and **b.** $y = f(-x)$. ■

Self Check Answers

1. 3, up; 4, down **2.** 3, right; 2, left **3.** 4, right; 5, up **4.** x; y

5. a.

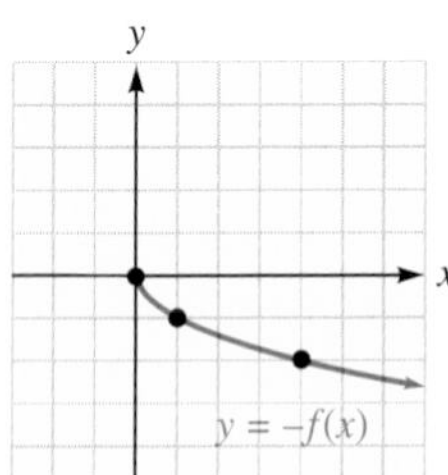

b.

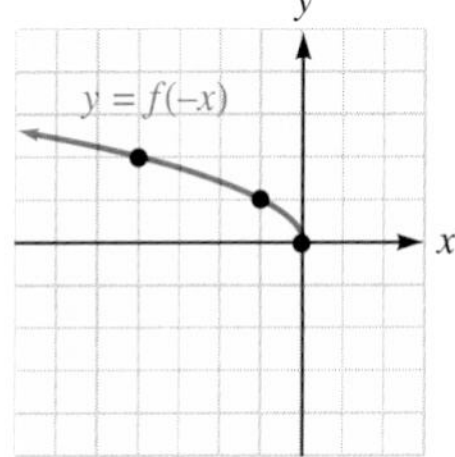

3.4 Exercises

VOCABULARY AND CONCEPTS ***Fill in the blanks.***

1. The graph of $y = f(x) + 5$ is identical to the graph of $y = f(x)$ except that it is translated 5 units **up**.
2. The graph of $y =$ $f(x) - 7$ is identical to the graph of $y = f(x)$ except that it is translated 7 units down.
3. The graph of $y = f(x - 3)$ is identical to the graph of $y = f(x)$ except that it is translated 3 units **to the right**.
4. The graph of $y = f(x + 2)$ is identical to the graph of $y = f(x)$ except that it is translated 2 units **to the left**.
5. To draw the graph of $y = (x + 2)^2 - 3$, translate the graph of $y = x^2$ **2** units to the left and 3 units **down**.
6. To draw the graph of $y = (x - 3)^3 + 1$, translate the graph of $y = x^3$ 3 units to the **right** and 1 unit **up**.
7. The graph of $y = f(-x)$ is a reflection of the graph of $y = f(x)$ about the **y-axis**.
8. The graph of $y = -f(x)$ is a reflection of the graph of $y = f(x)$ about the x-axis.
9. The graph of $y = f(4x)$ stretches the graph of $y = f(x)$ **horizontally** by a factor of $\frac{1}{4}$.
10. The graph of $y = 8f(x)$ stretches the graph of $y = f(x)$ **vertically** by a factor of 8.

The graph of each function is a translation of the graph of $f(x) = x^2$. Graph each function.

11. $g(x) = x^2 - 2$

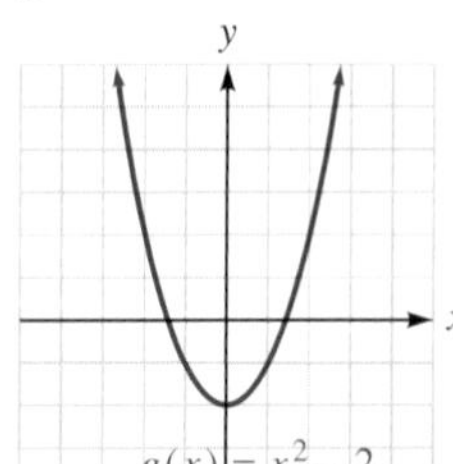

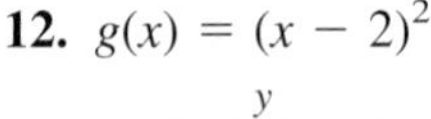

12. $g(x) = (x - 2)^2$

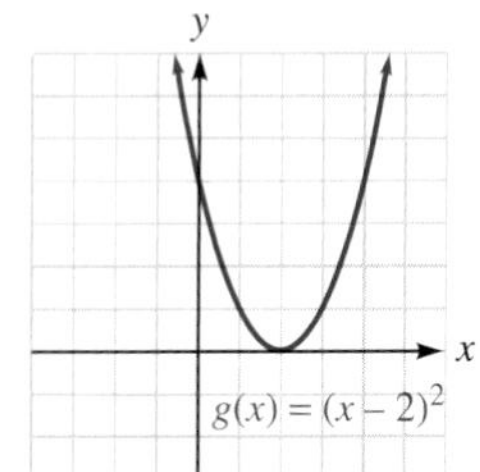

13. $g(x) = (x + 3)^2$

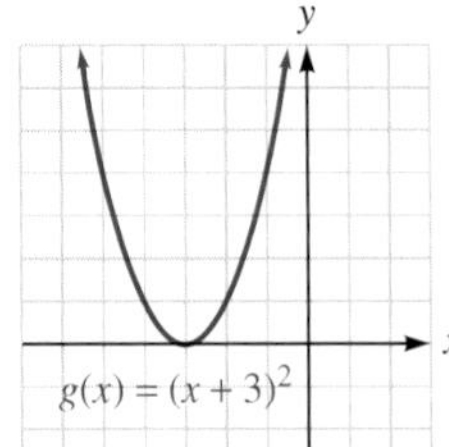

14. $g(x) = x^2 + 3$

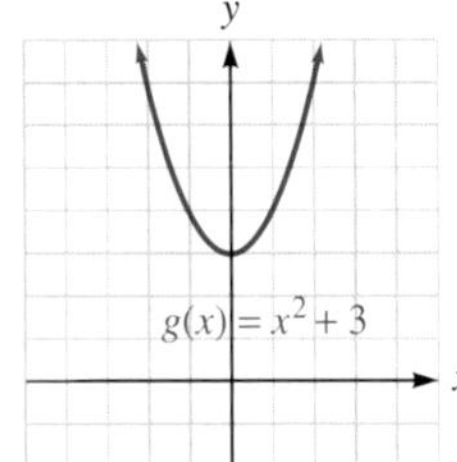

15. $h(x) = (x + 1)^2 + 2$

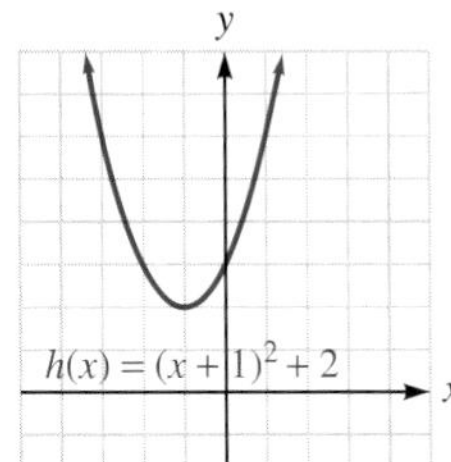

16. $h(x) = (x - 3)^2 - 1$

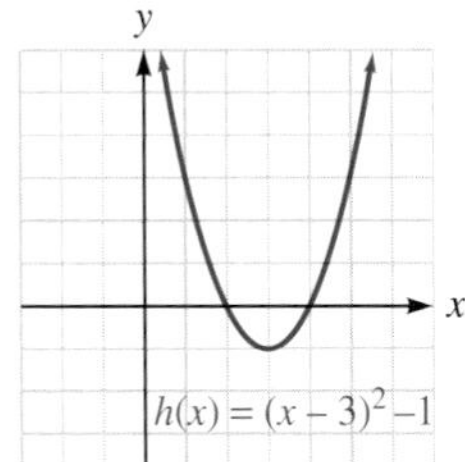

17. $h(x) = \left(x + \frac{1}{2}\right)^2 - \frac{1}{2}$

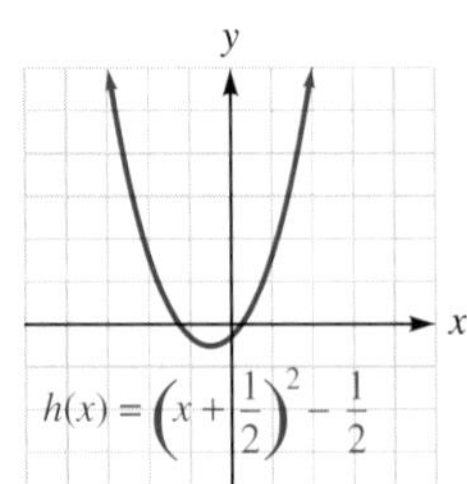

18. $h(x) = \left(x - \frac{3}{2}\right)^2 + \frac{5}{2}$

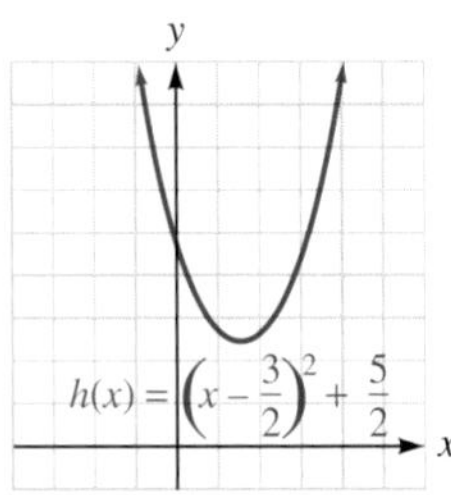

The graph of each function is a translation of the graph of $f(x) = x^3$. Graph each function.

19. $g(x) = x^3 + 1$

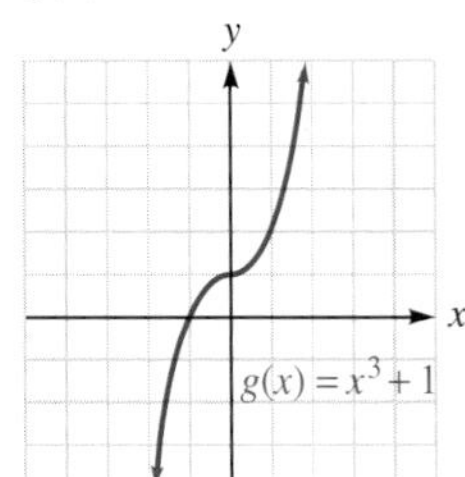

20. $g(x) = x^3 - 3$

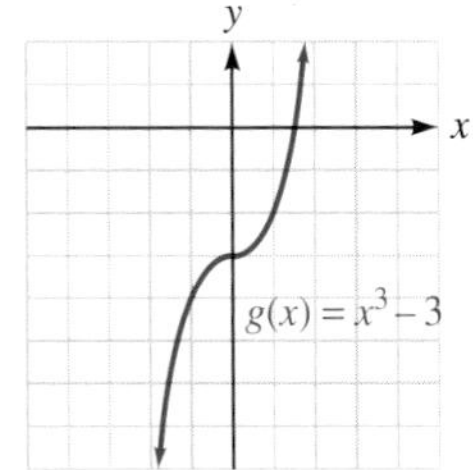

21. $g(x) = (x - 2)^3$

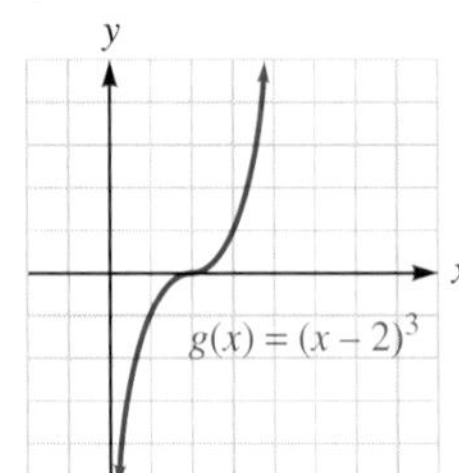

22. $g(x) = (x + 3)^3$

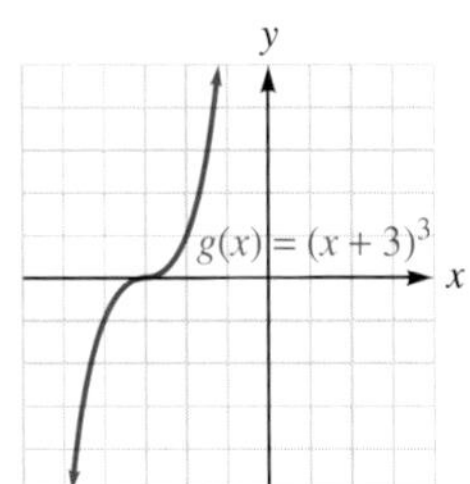

23. $h(x) = (x - 2)^3 - 3$

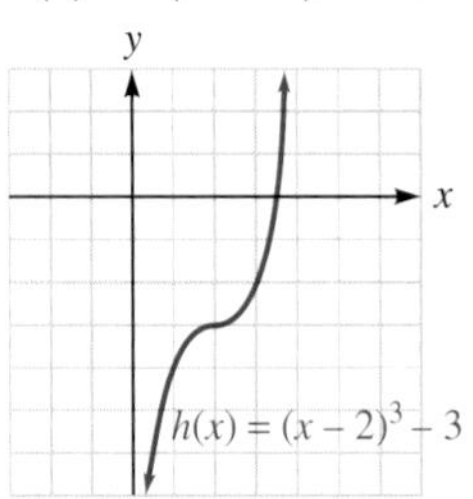

24. $h(x) = (x + 1)^3 + 4$

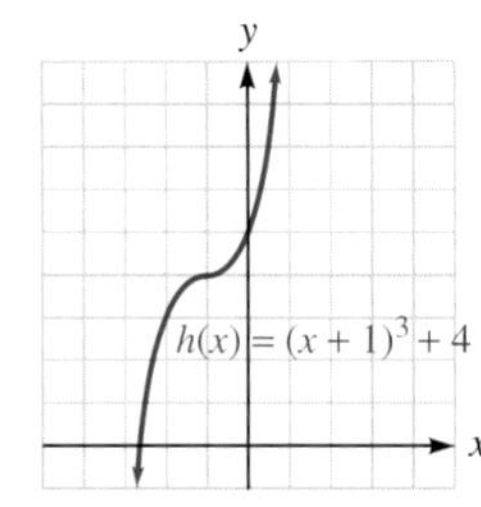

25. $y + 2 = x^3$

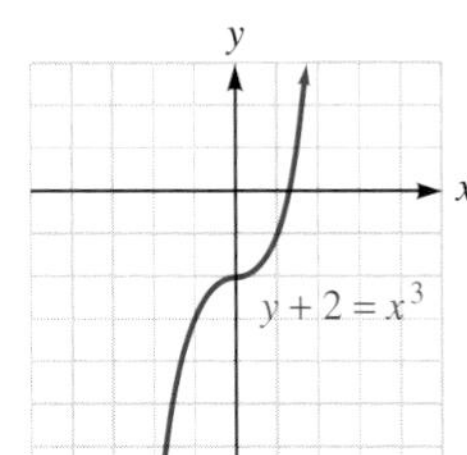

26. $y - 7 = (x - 5)^3$

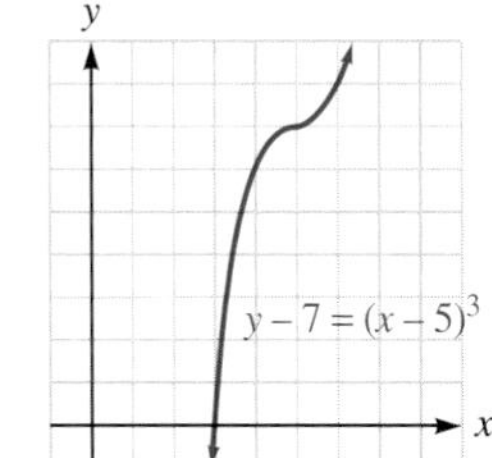

The graph of each function is a reflection of the graph of $y = x^2$, $y = x^3$, or $y = (x - 1)^2$. Graph each function.

27. $f(x) = -x^2$

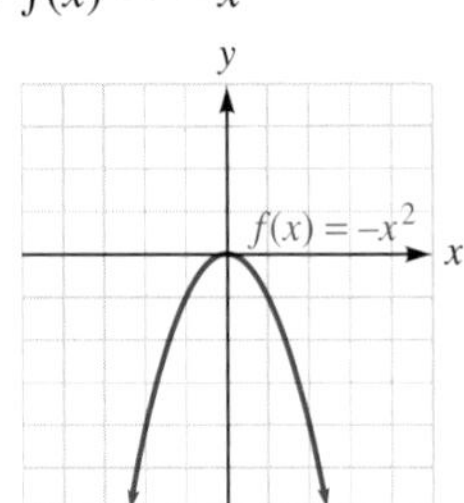

28. $g(x) = (-x)^3$

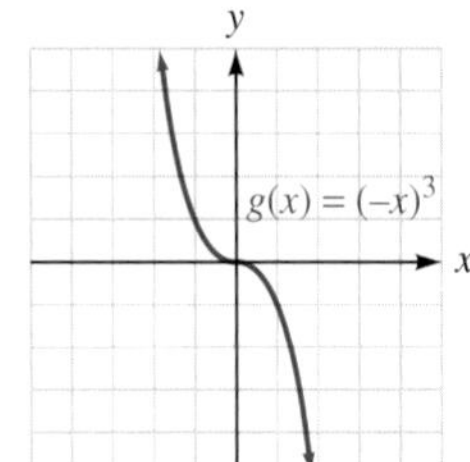

29. $h(x) = -x^3$

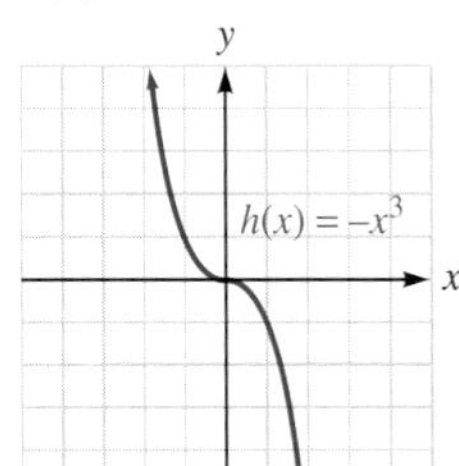

30. $f(x) = (-x - 1)^2$

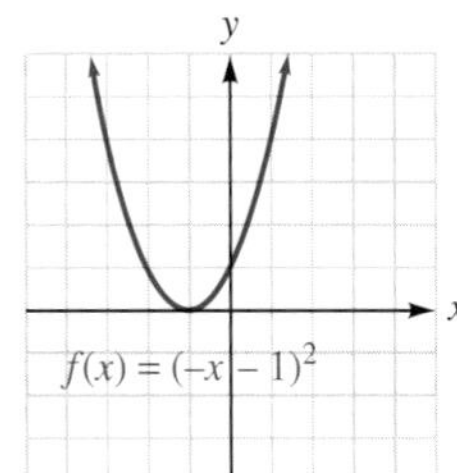

The graph of each function is a stretching of the graph of $y = x^2$. Graph each function.

31. $f(x) = 2x^2$

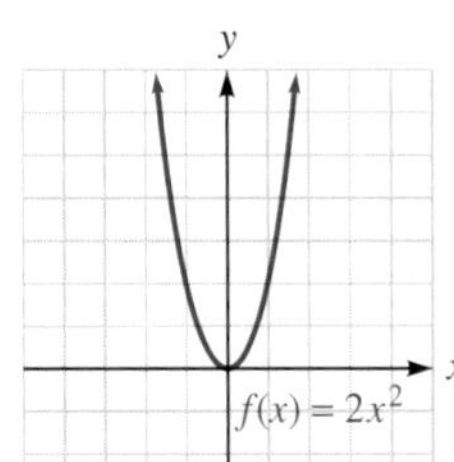

32. $g(x) = \frac{1}{2}x^2$

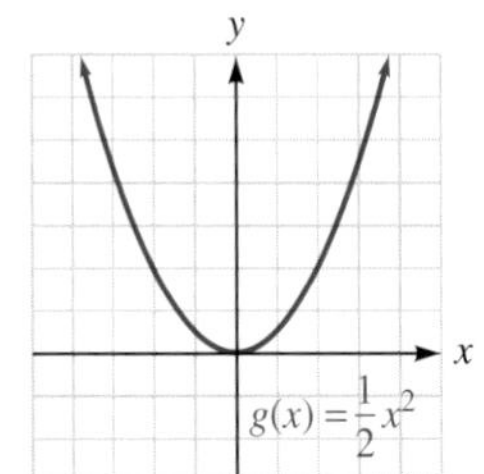

33. $h(x) = -3x^2$

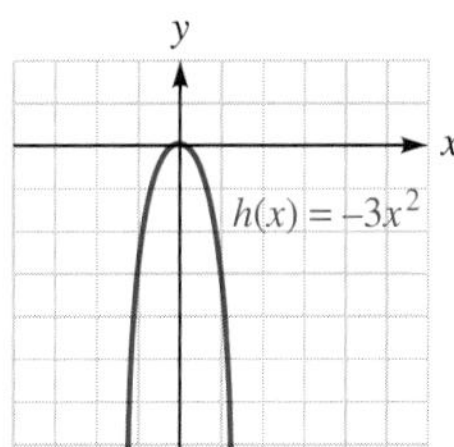

34. $f(x) = -\frac{1}{3}x^2$

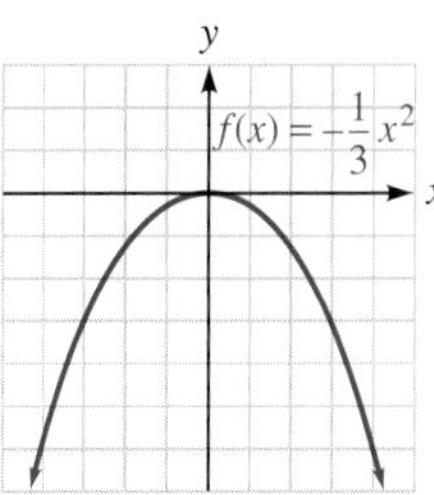

The graph of each function is a stretching of the graph of $y = x^3$. Graph each function.

35. $f(x) = \left(\frac{1}{2}x\right)^3$

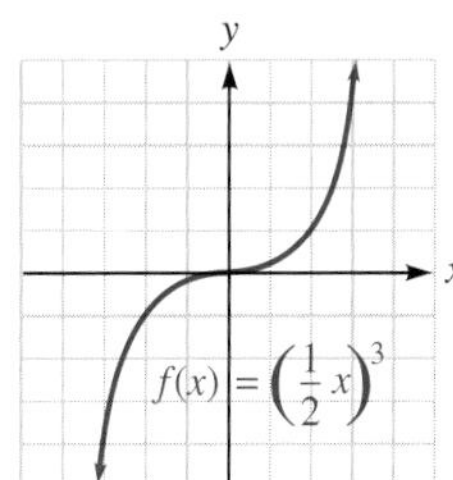

36. $f(x) = \frac{1}{8}x^3$

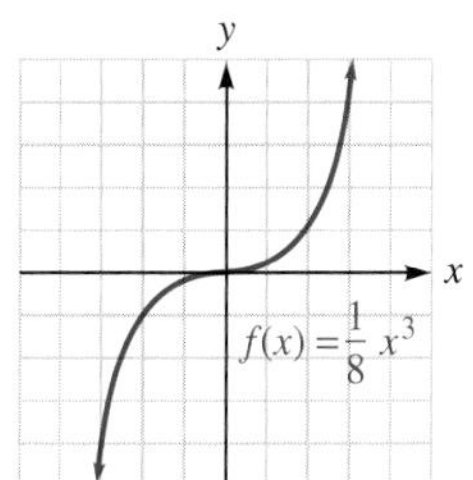

37. $f(x) = -8x^3$

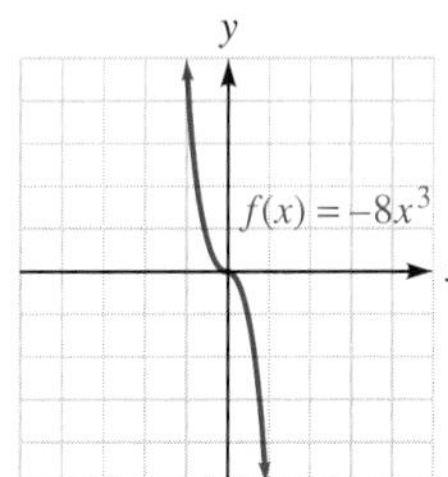

38. $f(x) = (-2x)^3$

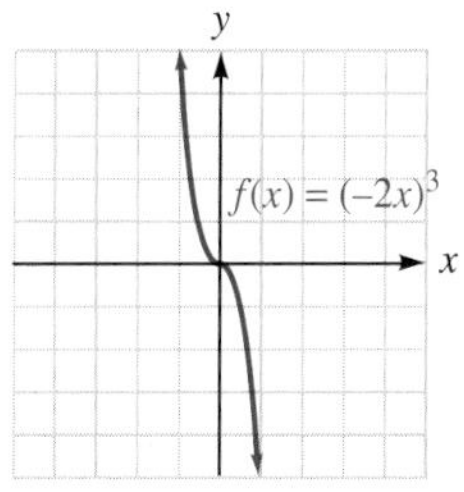

Use a translation to graph each equation.

39. $f(x) = |x - 2| + 1$

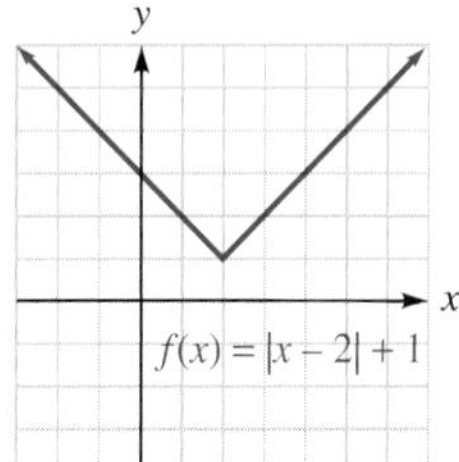

40. $f(x) = |x + 5| - 2$

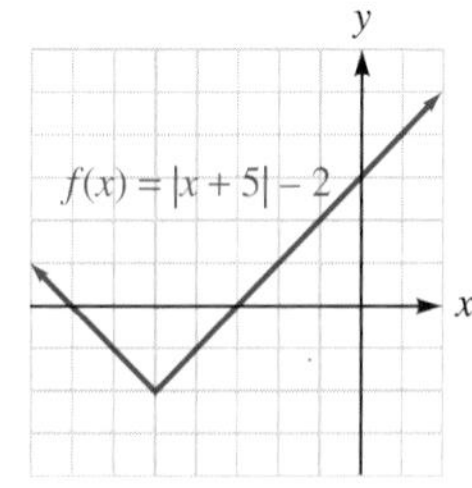

41. $g(x) = |3x|$

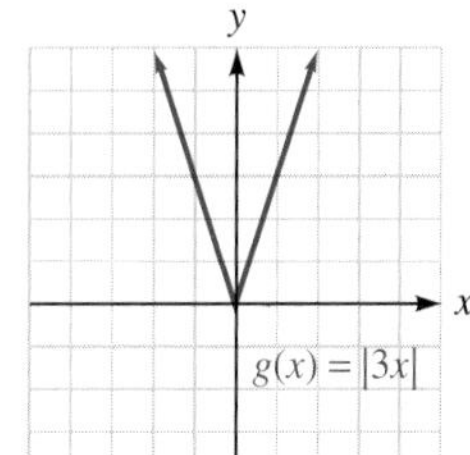

42. $g(x) = 3|x|$

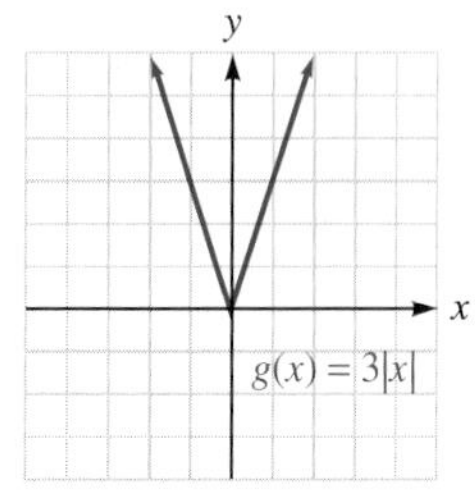

43. $h(x) = \sqrt{x - 2} + 1$ $(x \geq 2)$

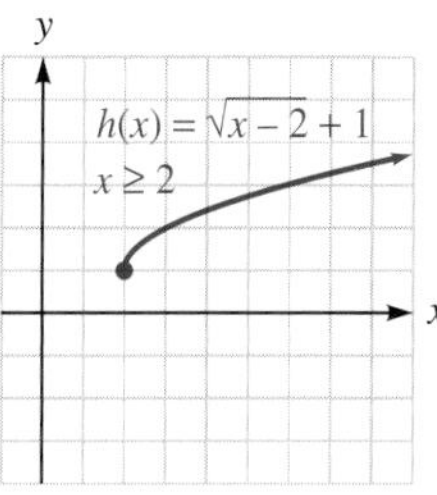

44. $h(x) = \sqrt{x + 5} - 2$ $(x \geq -5)$

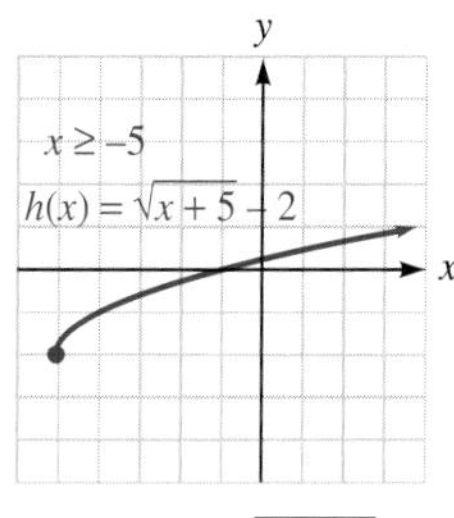

45. $f(x) = 2\sqrt{x} + 3$ $(x \geq 0)$

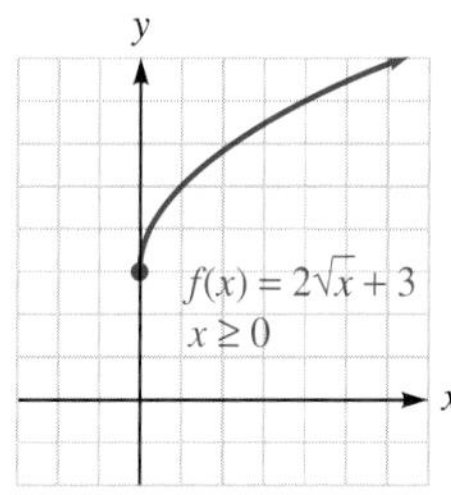

46. $g(x) = 2\sqrt{x + 3}$ $(x \geq -3)$

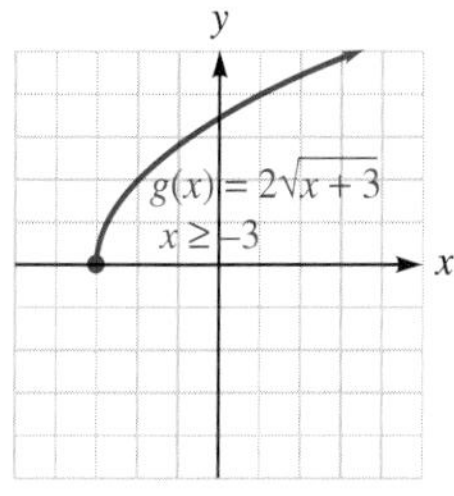

47. $h(x) = -2|x| + 3$

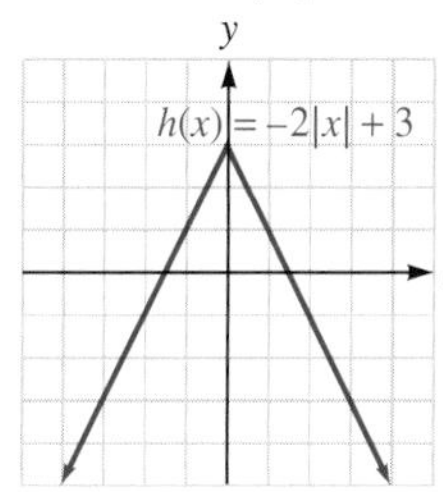

48. $f(x) = -2|x + 3|$

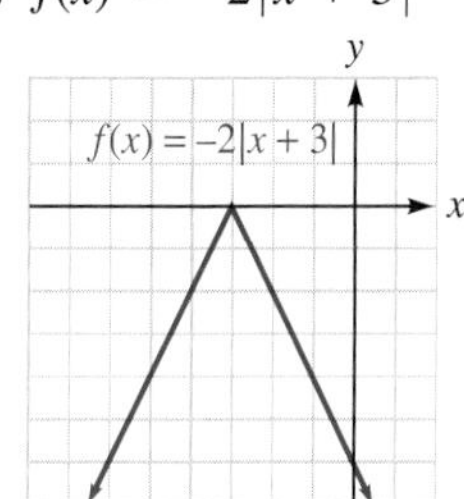

Use the following graph and a translation or reflection to find the graph of each function.

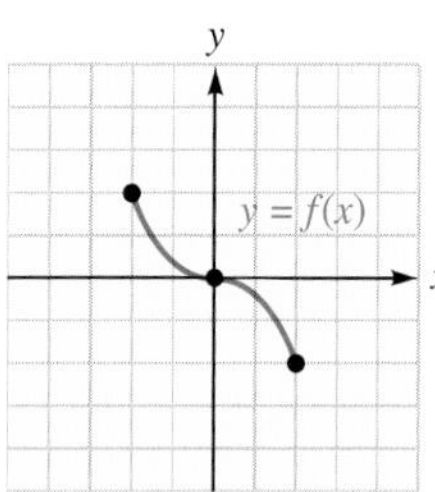

49. $y = f(x) + 1$

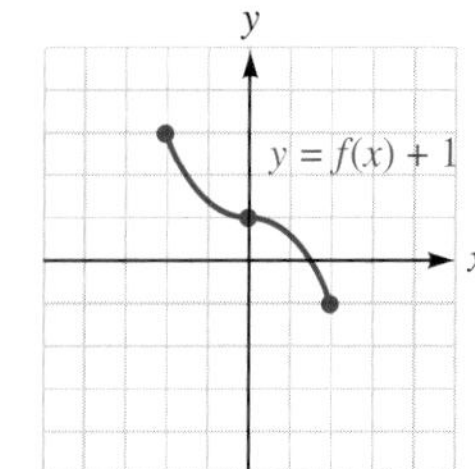

50. $y = f(x + 1)$

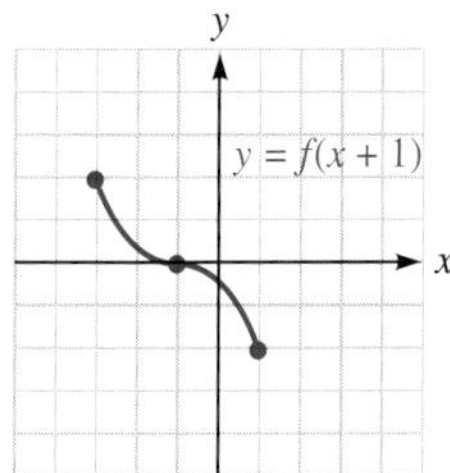

51. $y = 2f(x)$

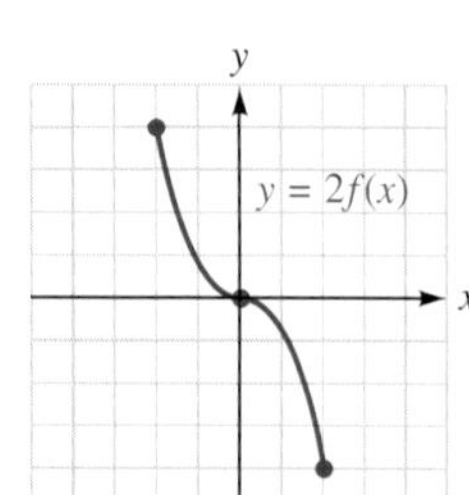

52. $y = f\left(\frac{x}{2}\right)$

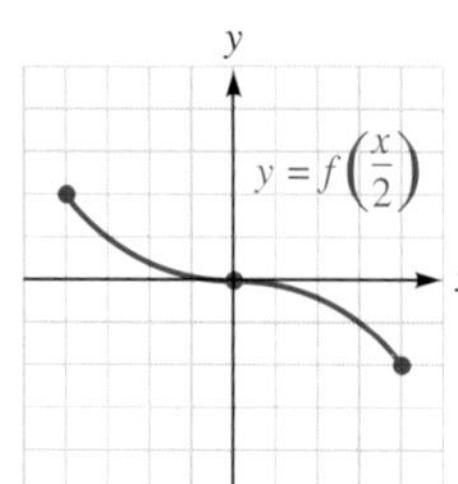

53. $y = f(x - 2) + 1$

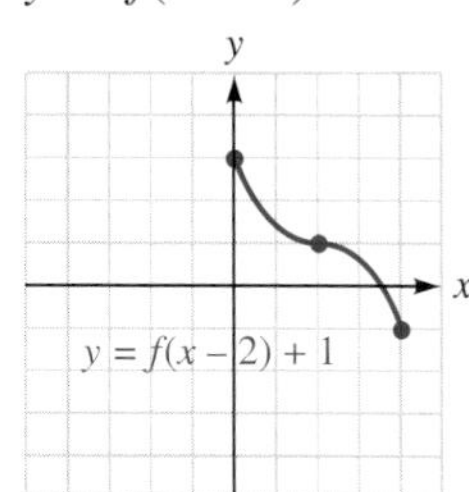

54. $y = -f(x)$

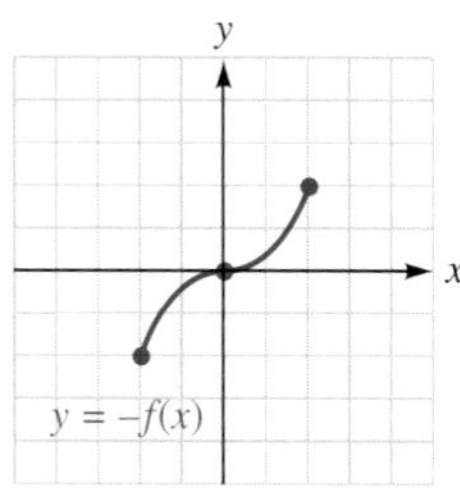

55. $y = 2f(-x)$

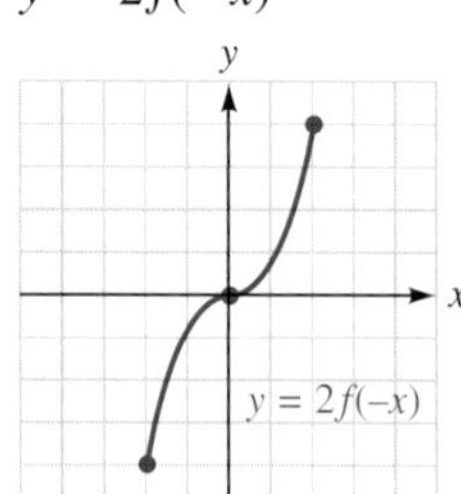

56. $y = f(x + 1) - 2$

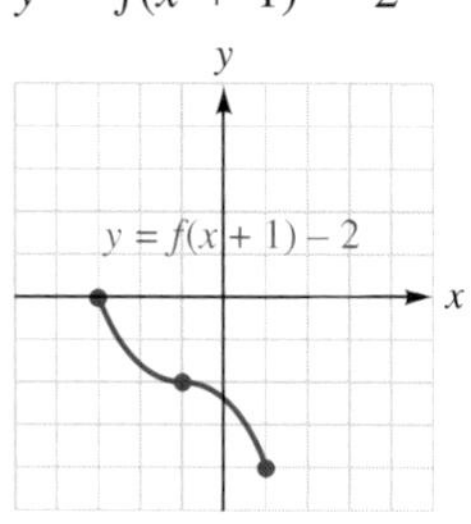

DISCOVERY AND WRITING ***Use a graphing calculator to perform each experiment. Write a brief paragraph describing your findings.***

57. Investigate the translations of the graph of a function by graphing the parabola $y = (x - k)^2 + k$ for several values of k. What do you observe about successive positions of the vertex?

58. Investigate the translations of the graph of a function by graphing the parabola $y = (x - k)^2 + k^2$ for several values of k. What do you observe about successive positions of the vertex?

59. Investigate the horizontal stretching of the graph of a function by graphing $y = \sqrt{ax}$ for several values of a. What do you observe?

60. Investigate the vertical stretching of the graph of a function by graphing $y = b\sqrt{x}$ for several values of b. What do you observe? Are these graphs different from the graphs in Exercise 59?

Write a paragraph using your own words.

61. Explain why the effect of vertically stretching a graph by a factor of -1 is to reflect the graph in the x-axis.

62. Explain why the effect of horizontally stretching a graph by a factor of -1 is to reflect the graph in the y-axis.

REVIEW ***Simplify each function.***

63. $\dfrac{x^2 + x - 6}{x^2 + 5x + 6}$ $\quad \dfrac{x - 2}{x + 2}$

64. $\dfrac{2x^2 + 3x}{2x^2 + x - 3}$ $\quad \dfrac{x}{x - 1}$

Find the domain of each function.

65. $f(x) = \dfrac{x + 7}{x - 3}$

all real numbers except 3

66. $f(x) = \dfrac{x^2 + 1}{x^2 + 3x + 2}$

all real numbers except −1 and −2

Perform each division and write the answer in quotient + $\frac{\text{remainder}}{\text{divisor}}$ form.

67. $\dfrac{x^2 + 3x}{x + 1}$

$x + 2 + \dfrac{-2}{x + 1}$

68. $\dfrac{x^2 + 3}{x + 1}$

$x - 1 + \dfrac{4}{x + 1}$

3.5 Rational Functions

In this section, you will learn about

- Average Hourly Cost
- Rational Functions
- Vertical and Horizontal Asymptotes
- Finding Asymptotes
- Graphing Rational Functions
- Graphs with Missing Points

Average Hourly Cost

Rational expressions often define functions. For example, if the cost of subscribing to an on-line research service is \$6 per month plus \$1.50 per hour of access time, the average (mean) hourly cost of the service is the total monthly cost, divided by the number of hours of access time:

$$\bar{c} = \frac{C}{n} = \frac{1.50n + 6}{n}$$ $\bar{c}$ is the mean hourly cost, C is the total monthly cost, and n is the number of hours the service is used.

The function

(1) $$\bar{c} = f(n) = \frac{1.50n + 6}{n} \quad (n > 0)$$

gives the mean hourly cost of using the service for n hours per month. Since $n > 0$, the domain of the function is the interval $(0, \infty)$.

Accent on Technology CALCULATOR GRAPHS

If we use a graphing calculator with window settings of [0, 10] for x and [0, 10] for y to graph the function $f(n) = \frac{1.50n + 6}{n}$, we will obtain the graph shown in Figure 3-42. Note that the graph of the function passes the vertical line test, as expected.

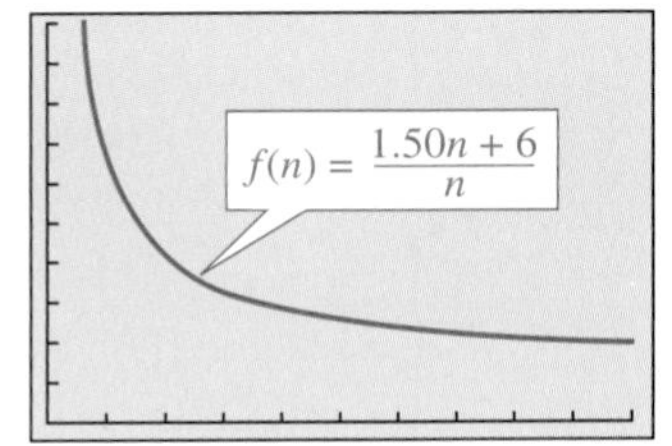

Figure 3-42

From the graph in Figure 3-42, we can see that the mean hourly cost decreases as the number of hours of access time increases. Since the cost of each extra hour of access time is \$1.50, the mean hourly cost can approach \$1.50 but never drop below it. The graph of the function approaches the line $y = 1.5$ as n increases without bound. When a graph approaches a line, we call the line an **asymptote.** The line $y = 1.5$ is a **horizontal asymptote** of the graph.

As n gets smaller and approaches 0, the graph approaches the y-axis. The y-axis is a **vertical asymptote** of the graph.

EXAMPLE 1 Find the mean hourly cost when the service described above is used for **a.** 3 hours and **b.** 10.4 hours.

Solution **a.** To find the mean cost for 3 hours of access time, we substitute 3 for n in Equation 1 and simplify:

$$\bar{c} = f(3) = \frac{1.50(3) + 6}{3} = 3.5$$

The mean cost for 3 hours of access time is \$3.50 per hour.

b. To find the mean hourly cost for 10.4 hours of access time, we substitute 10.4 for n in Equation 1 and simplify:

$$\bar{c} = f(10.4) = \frac{1.50(10.4) + 6}{10.4} = 2.076923$$

The mean cost for 10.4 hours of access time is approximately \$2.08 per hour.

Self Check Find the mean hourly cost when the service is used for $\frac{1}{2}$ hour. ■

■ Rational Functions

The function $f(n) = \frac{1.50n + 6}{n}$ $(n > 0)$ is an example of a **rational function.** Such functions are defined by equations of the form

$$y = \frac{P(x)}{Q(x)}$$

where $P(x)$ and $Q(x)$ are polynomials. Because $Q(x)$ appears in the denominator of a fraction, it cannot equal 0. Thus, the domain of a rational function must exclude all values of x for which $Q(x) = 0$.

Here are some examples of rational functions and their domains:

$$f(x) = \frac{3}{x - 2} \qquad \text{Domain: } (-\infty, 2) \cup (2, \infty)$$

$$f(x) = \frac{5x + 2}{x^2 - 4} \qquad \text{Domain: } (-\infty, -2) \cup (-2, 2) \cup (2, \infty)$$

$$f(x) = \frac{3x^2 - 2}{(2x + 1)(x - 3)} \qquad \text{Domain: } \left(-\infty, -\tfrac{1}{2}\right) \cup \left(-\tfrac{1}{2}, 3\right) \cup (3, \infty)$$

EXAMPLE 2 Find the domain of $f(x) = \dfrac{3x + 2}{x^2 - 7x + 12}$.

Solution To find the numbers x that make the denominator 0, we set $x^2 - 7x + 12$ equal to 0 and solve for x.

$$x^2 - 7x + 12 = 0$$

$$(x - 4)(x - 3) = 0 \qquad \text{Factor } x^2 - 7x + 12.$$

$$x - 4 = 0 \quad \text{or} \quad x - 3 = 0 \qquad \text{Set each factor equal to 0.}$$

$$x = 4 \quad | \quad x = 3 \qquad \text{Solve each linear equation.}$$

The domain is $(-\infty, 3) \cup (3, 4) \cup (4, \infty)$.

Self Check Find the domain of $f(x) = \dfrac{2x - 3}{x^2 - x - 2}$. ■

Accent on Technology CALCULATOR GRAPHS

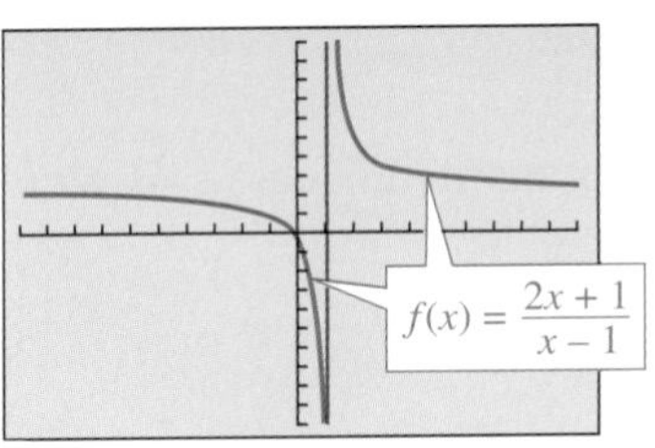

Figure 3-43

We can use graphing calculators to find domains and ranges of rational functions. If we use settings of $[-10, 10]$ for x and $[-10, 10]$ for y and graph $f(x) = \frac{2x+1}{x-1}$, we will obtain Figure 3-43.

From the graph, we can see that every real number x except 1 gives a value of y. Thus, the domain of the function is $(-\infty, 1) \cup (1, \infty)$. We can also see that y can be any value except 2. The range of the function is $(-\infty, 2) \cup (2, \infty)$.

■ Vertical and Horizontal Asymptotes

We have seen that when a graph approaches a line, we call the line an *asymptote.* Two asymptotes are apparent in Figure 3-43. From the figure, we can see that

- As x approaches 1 from the left, the values of y decrease, and the graph approaches the vertical line $x = 1$.
- As x approaches 1 from the right, the values of y increase, and the graph approaches the vertical line $x = 1$.

For this reason, the line $x = 1$ is a *vertical asymptote.* Although the vertical line in the graph appears to be the asymptote, it is not. Graphing calculators draw graphs by connecting dots whose x-coordinates are close together. When two points straddle a vertical asymptote and their y-coordinates are far apart, the calculator draws a line between them producing what appears to be the vertical asymptote in the figure. If you set your calculator to dot mode instead of connected mode, the vertical line will not appear.

From the figure, we can also see that

- As x increases to the right of 1, the values of y decrease and approach the value $y = 2$.
- As x decreases to the left of 1, the values of y increase and approach the value $y = 2$.

If we were to draw the line $y = 2$, it would be a *horizontal asymptote.* Graphing calculators do not draw lines that appear to be horizontal asymptotes.

Figure 3-44 shows a typical vertical and horizontal asymptote.

Vertical asymptote at $x = a$

$f(x)$ approaches ∞ as x approaches a from left.

$f(x)$ approaches $-\infty$ as x approaches a from right.

a x

a x

Horizontal asymptote at $y = b$

$f(x)$ approaches b as x approaches ∞ or as x approaches $-\infty$.

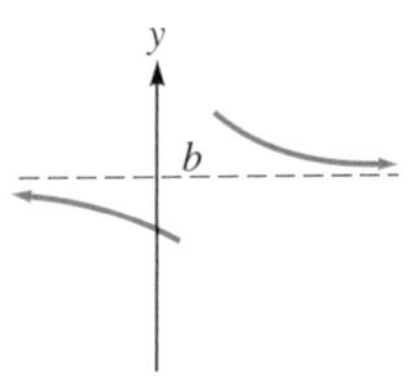

Figure 3-44

Finding Asymptotes

To find the vertical asymptotes of a rational function that is in simplest form, we must find the values of x for which the function is not defined. For example, since the denominator of $f(x) = \frac{2x-1}{x+2}$ is zero when $x = -2$, the line $x = -2$ is a vertical asymptote.

To find the horizontal asymptote of $f(x) = \frac{2x-1}{x+2}$, we can do a long division and write the result in quotient + $\frac{\text{remainder}}{\text{divisor}}$ form.

$$\begin{array}{r} 2 \\ x+2\overline{)2x-1} \\ \underline{2x+4} \\ -5 \end{array}$$

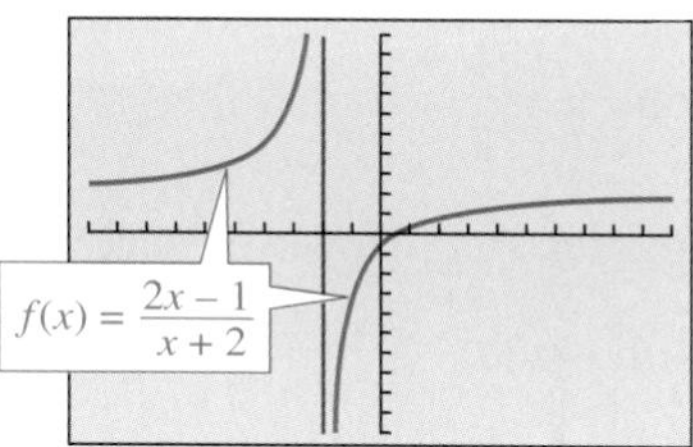

Figure 3-45

This shows that the function $f(x) = \frac{2x-1}{x+2}$ can be written in the form $f(x) = 2 + \frac{-5}{x+2}$. As $|x|$ gets very large, the fraction $\frac{-5}{x+2}$ approaches 0, and the value of $f(x)$ approaches 2. Thus, the line $y = 2$ is a horizontal asymptote. A calculator graph of the function appears in Figure 3-45.

EXAMPLE 3 Find the asymptotes of each function: **a.** $y = f(x) = \dfrac{x+2}{x^2-1}$ and

b. $y = f(x) = \dfrac{2x^2+x+2}{x^2-1}$.

Solution Each function has a vertical asymptote at $x = 1$ and $x = -1$, because at these values, their denominators are 0. We must also find the horizontal asymptotes, if any.

a. Because the degree of the numerator is less than the degree of the denominator, we can find the horizontal asymptote by dividing the numerator and the denominator by x^2, which is the largest power of x in the denominator.

$$f(x) = \frac{x+2}{x^2-1} = \frac{\frac{x}{x^2}+\frac{2}{x^2}}{\frac{x^2}{x^2}-\frac{1}{x^2}} = \frac{\frac{1}{x}+\frac{2}{x^2}}{1-\frac{1}{x^2}}$$

Since $\frac{1}{x}$, $\frac{2}{x^2}$, and $\frac{1}{x^2}$ all approach 0 as x approaches ∞, it follows that y approaches

$$\frac{0+0}{1-0} = 0$$

The horizontal asymptote is the line $y = 0$. If the degree of the numerator is less than the degree of the denominator, the horizontal asymptote is always $y = 0$.

b. To find the horizontal asymptote, we do a long division and write the result in quotient + $\frac{\text{remainder}}{\text{divisor}}$ form.

$$\begin{array}{r} 2 \\ x^2-1\overline{)2x^2+x+2} \\ \underline{2x^2\quad\;\, -2} \\ x+4 \end{array}$$

Thus,

$$y = \frac{2x^2 + x + 2}{x^2 - 1} = 2 + \frac{x + 4}{x^2 - 1}$$

The last fraction approaches 0 as x approaches ∞, for reasons discussed in part **a.** So, y approaches 2. The horizontal asymptote is the line $y = 2$. If the degrees of the numerator and denominator are the same, the horizontal asymptote is always $y =$ the lead coefficient of the numerator divided by the lead coefficient of the denominator.

Comment Note that when $x = -4$, the fraction $\frac{x + 4}{x^2 - 1} = 0$ and the graph of $y = \frac{2x^2 + x + 2}{x^2 - 1}$ touches the horizontal asymptote $y = 2$. However, this is the only place where the graph and the asymptote will touch.

Self Check Find the asymptotes of $f(x) = \frac{x^2}{x^2 - 4}$. ■

EXAMPLE 4 Find the asymptotes of each function. **a.** $y = f(x) = \frac{3x^3 + 2x^2 + 2}{x^2 - 1}$ and **b.** $y = f(x) = \frac{x^4 + x + 2}{x^2 - 1}$.

Solution Again, the vertical asymptotes are at $x = 1$ and $x = -1$.

a. To find the horizontal asymptote, we do a long division and write the result in quotient + $\frac{\text{remainder}}{\text{divisor}}$ form.

$$\begin{array}{r} 3x + 2 \\ x^2 - 1 \overline{\smash{)}3x^3 + 2x^2 + 2} \\ \underline{3x^3 - 3x } \\ 2x^2 + 3x + 2 \\ \underline{2x^2 - 2} \\ 3x + 4 \end{array}$$

Thus,

$$y = \frac{3x^3 + 2x^2 + 2}{x^2 - 1} = 3x + 2 + \frac{3x + 4}{x^2 - 1}$$

The last fraction approaches 0 as x approaches ∞, and the graph approaches the line $y = 3x + 2$. Since the line is not horizontal, we call the asymptote a **slant asymptote.** If the degree of the numerator is 1 more than the degree of the denominator, there will be a slant asymptote.

Comment Note that when $x = -\frac{4}{3}$, the fraction $\frac{3x + 4}{x^2 - 1}$ is 0. This is the only place where the graph and slant asymptote will touch.

b. To find the horizontal asymptote, we do a long division and write the result in quotient + $\frac{\text{remainder}}{\text{divisor}}$ form.

$$
\begin{array}{r}
x^2 + 1 \\
x^2 - 1 \overline{\big) x^4 + x + 2} \\
\underline{x^4 - x^2} \\
x^2 + x + 2 \\
\underline{x^2 - 1} \\
x + 3
\end{array}
$$

Thus,

$$y = \frac{x^4 + x + 2}{x^2 - 1} = x^2 + 1 + \frac{x + 3}{x^2 - 1}$$

The last fraction approaches 0 as x approaches ∞, so the curve approaches the parabola $y = x^2 + 1$. Since a parabola is not a line, the graph has no horizontal or slant asymptotes.

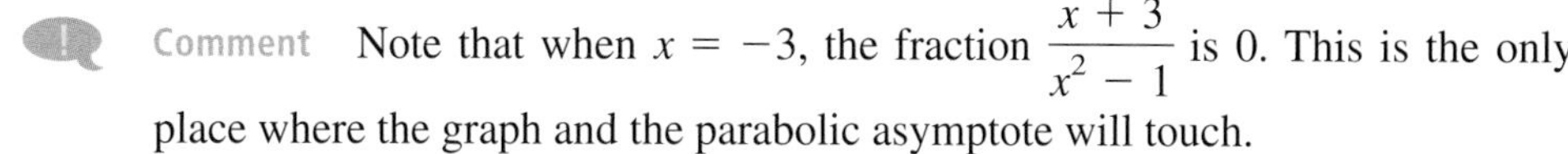

Comment Note that when $x = -3$, the fraction $\frac{x + 3}{x^2 - 1}$ is 0. This is the only place where the graph and the parabolic asymptote will touch.

Self Check Find the asymptotes of $f(x) = \frac{2x^3 - 3x + 1}{x^2 - 4}$. ■

■ Graphing Rational Functions

We follow these steps to graph the rational function $f(x) = \frac{P(x)}{Q(x)}$, where $P(x)$ and $Q(x)$ are polynomials written in descending powers of x and $\frac{P(x)}{Q(x)}$ is in simplest form.

Strategy for Graphing Rational Functions

Check for symmetry. If $P(x)$ and $Q(x)$ involve only even powers of x, or if $f(x) = f(-x)$, the graph is symmetric about the y-axis. Check for symmetry about the origin.

Look for vertical asymptotes. The real roots of $Q(x) = 0$, if any, determine the vertical asymptotes of the graph.

Look for the y- and x-intercepts. Let $x = 0$. The resulting value of y, if any, is the y-intercept of the graph. The real roots of $P(x) = 0$, if any, are the x-intercepts of the graph.

Look for horizontal asymptotes.

- If the degree of $P(x)$ is less than the degree of $Q(x)$, the line $y = 0$ is a horizontal asymptote.
- If the degrees of $P(x)$ and $Q(x)$ are equal, the line $y = \frac{p}{q}$, where p and q are the lead coefficients of $P(x)$ and $Q(x)$, is a horizontal asymptote.
- If the degree of $P(x)$ is greater than the degree of $Q(x)$, there is no horizontal asymptote.

Look for slant asymptotes. If the degree of $P(x)$ is 1 greater than the degree of $Q(x)$, there is a slant asymptote. To find it, divide $P(x)$ by $Q(x)$ and ignore the remainder.

Comment To find a horizontal or slant asymptote of $f(x) = \frac{P(x)}{Q(x)}$, we divide $P(x)$ by $Q(x)$ and ignore the remainder. If any value of x in the domain of $P(x)$ and $Q(x)$ makes the remainder equal to 0, the graph and the asymptote will cross at that point. Otherwise, the graph of the function will not cross the asymptote.

ACTIVE EXAMPLE 5 Graph: $y = f(x) = \frac{x^2 - 4}{x^2 - 1}$.

COLLEGE **Algebra f(x) Now™**
Explore this Active Example and test yourself on the steps taken here at **http://1pass.thomson.com**. You can also find this example on the Interactive Video Skillbuilder CD-ROM.

Solution **Symmetry:** Because x appears to even powers only, there is symmetry about the y-axis. There is no symmetry about the origin.

Vertical asymptotes: To find the vertical asymptotes, we set the denominator equal to 0 and solve for x.

$$x^2 - 1 = 0$$
$$x + 1 = 0 \quad \text{or} \quad x - 1 = 0$$
$$x = -1 \quad | \quad x = 1$$

There will be vertical asymptotes at $x = -1$ and $x = 1$.

***y*- and *x*-intercepts:** We can find the y-intercept by setting x equal to 0 and solving for y:

$$y = \frac{x^2 - 4}{x^2 - 1} = \frac{0^2 - 4}{0^2 - 1} = \frac{-4}{-1} = 4$$

The y-intercept is (0, 4).

We can find the x-intercepts by setting the numerator equal to 0 and solving for x:

$$x^2 - 4 = 0$$
$$(x + 2)(x - 2) = 0$$
$$x + 2 = 0 \quad \text{or} \quad x - 2 = 0$$
$$x = -2 \quad | \quad x = 2$$

The x-intercepts are (2, 0) and (−2, 0).

Horizontal asymptotes: Since the degrees of the numerator and denominator polynomials are the same, the line

$$y = \frac{1}{1}$$ 1 is the lead coefficient of the numerator. 1 is the lead coefficient of the denominator.

is a horizontal asymptote. The horizontal asymptote is the line $y = 1$. The graph is shown in Figure 3-46.

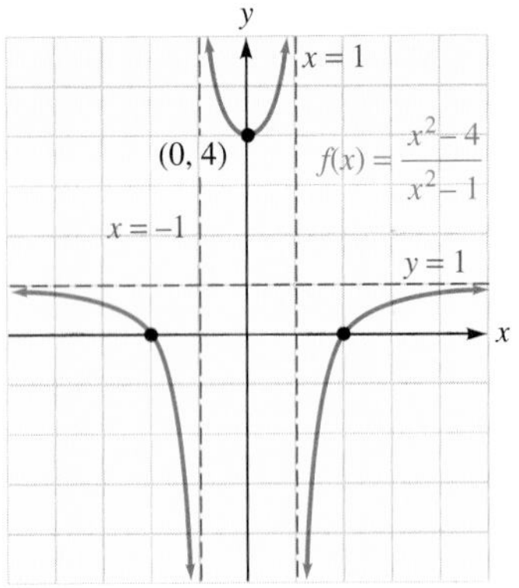

Figure 3-46

Self Check Use a graphing calculator to graph the function in Example 5. ■

EXAMPLE 6 Graph the function: $y = f(x) = \frac{3x}{x - 2}$.

Solution **Symmetry:** Because $f(-x) \neq -f(x)$, there is no symmetry about the y-axis. There is no symmetry about the origin either.

Vertical asymptotes: To find the vertical asymptotes, we set the denominator equal to 0 and solve for x. Since the solution is 2, there will be a vertical asymptote at $x = 2$.

y- and x-intercepts: We can find the y-intercept by setting x equal to 0 and solving for y. Since the solution is 0, the y-intercept is (0, 0). The graph passes through the origin.

We can find the x-intercepts by setting the numerator equal to 0 and solving for x:

$$3x = 0$$
$$x = 0$$

The only x-intercept is (0, 0).

Horizontal asymptotes: Since the degrees of the numerator and denominator polynomials are the same, the line

$$y = \frac{3}{1}$$ **3 is the lead coefficient of the numerator. 1 is the lead coefficient of the denominator.**

is a horizontal asymptote. The horizontal asymptote is the line $y = 3$.

To find what happens when x is greater than 2, we pick a value of x greater than 2 and find the corresponding value of y. If $x = 3$, then $y = 9$. After plotting the point (3, 9), we use the intercepts and asymptotes to sketch the graph shown in Figure 3-47.

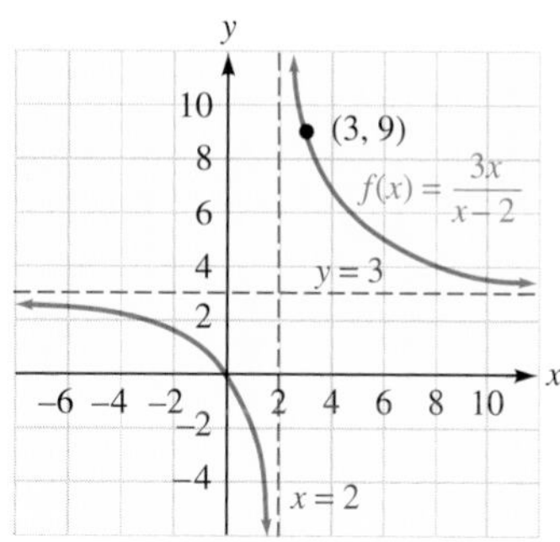

Figure 3-47

Self Check Use a graphing calculator to graph the function in Example 6. ■

Everyday Connections Why Do Women Live Longer Than Men?

"Research by exercise scientists at Liverpool John Moores University (LJMU) may have an answer to the age old question of why women live longer than men. LJMU's findings show that women's longevity may be linked to the fact that their hearts age differently to men's and do not lose their pumping power as they get older." ***Medical Research News, Published: Wednesday, 12-Jan-2005.*** http://www.news-medical.net

(Continued)

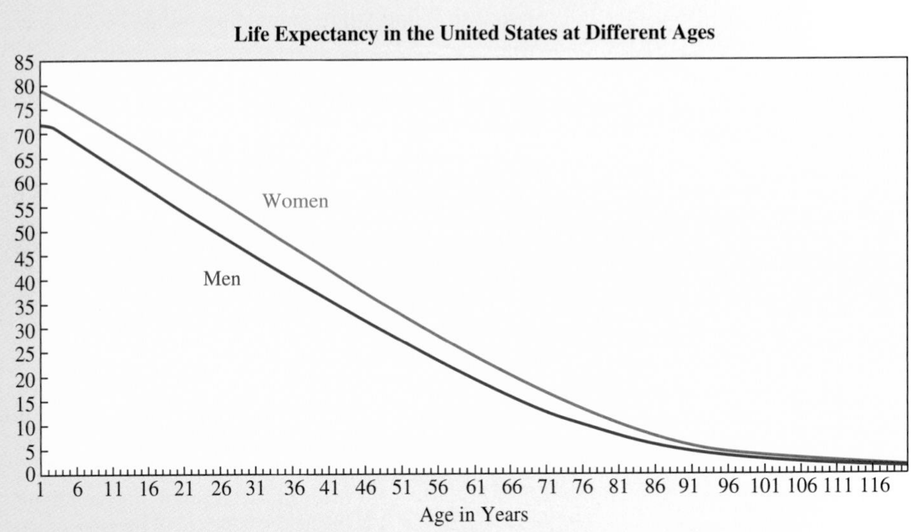

Source: http://www.annuityadvantage.com/lifeexpectancy.htm

The graph illustrates the life expectancies for men and women in the United States. Use the graph to estimate each of the following.

1. Determine the life expectancy at birth of a man and of a woman.
   man = 72 years, woman = 79 years
2. Determine the life expectancy of a 26-year-old man and of a 26-year-old woman. **man = 48 years, woman = 54 years**
3. Each curve appears to have the same horizontal asymptote. Find the equation of this asymptote and explain its significance.
   $y = 0$ is the horizontal asymptote.

EXAMPLE 7 Graph the function: $y = \dfrac{1}{x(x-1)^2}$.

Solution **Symmetry:** Because $y = f(x)$ is not equivalent to $y = f(-x)$, there is no symmetry about the y-axis. Because $y = f(x)$ and $-y = f(-x)$ are not equivalent, there is no symmetry about the origin.

Vertical asymptotes: Since 0 and 1 make the denominator 0, the vertical asymptotes are the lines $x = 0$ and $x = 1$.

y- and x-intercepts: Since x cannot be 0, the graph has no y-intercept. Since y cannot be 0, the graph has no x-intercept.

Horizontal asymptotes: As x approaches ∞, the denominator of the fraction becomes large, and the corresponding values of y approach 0. Thus, the line $y = 0$ is a horizontal asymptote.

The table in Figure 3-48 gives three points lying in different intervals separated by the asymptotes. The values of y change sign at $x = 0$: to the left of the y-axis, $y < 0$, and to the right of the y-axis, $y > 0$. This happens because x appears to an odd power as a factor in the denominator. The value of y does not change sign at $x = 1$: To the left and to the right of the asymptote $x = 1$, y is positive. The function behaves this way because $(x - 1)$ appears to an even power in the denominator. The graph of the function appears in Figure 3-48.

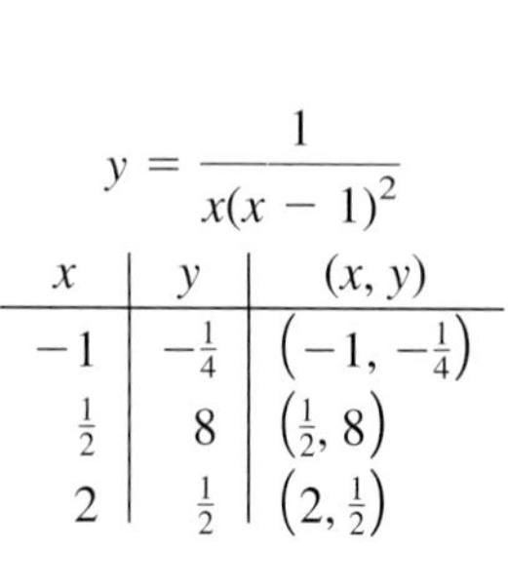

$$y = \frac{1}{x(x-1)^2}$$

x	y	(x, y)
-1	$-\frac{1}{4}$	$\left(-1, -\frac{1}{4}\right)$
$\frac{1}{2}$	8	$\left(\frac{1}{2}, 8\right)$
2	$\frac{1}{2}$	$\left(2, \frac{1}{2}\right)$

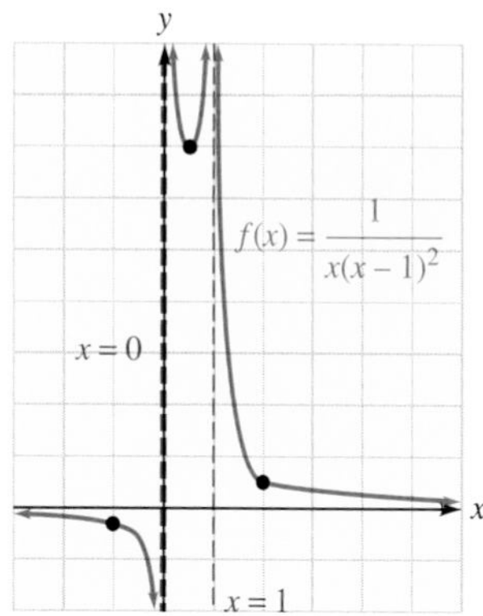

Figure 3-48

Self Check Use a graphing calculator to graph the function in Example 7. ■

EXAMPLE 8 Graph: $y = f(x) = \dfrac{1}{x^2 + 1}$.

Solution **Symmetry:** Because x appears only to an even power, the graph will be symmetric about the y-axis.

Vertical asymptotes: Since no numbers x make the denominator 0, the graph has no vertical asymptotes.

y- and x-intercepts: Since $f(0) = 1$, the y-intercept of the graph is $(0, 1)$. Because the denominator is always positive, the fraction is always positive and the graph lies entirely above the x-axis. There are no x-intercepts.

Horizontal asymptotes: Since $f(x)$ approaches 0 as x approaches ∞, the graph has $y = 0$ as a horizontal asymptote.

The graph appears in Figure 3-49.

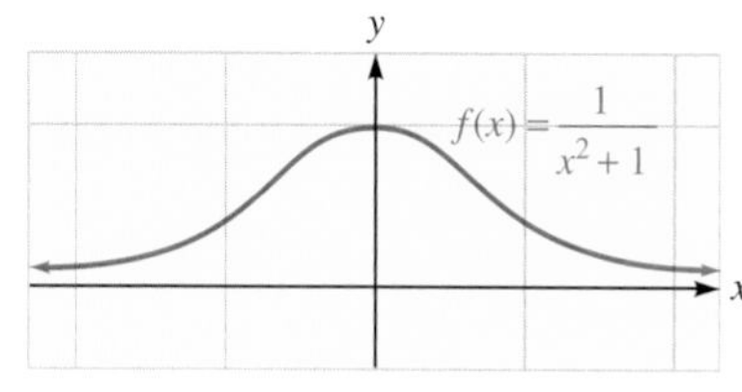

Figure 3-49

Self Check Use a graphing calculator to graph the function in Example 8. ■

ACTIVE EXAMPLE 9 Graph: $y = f(x) = \dfrac{x^2 + x - 2}{x - 3}$.

Solution We first factor the numerator of the expression

COLLEGE **Algebra** *f(x)* **Now**™
Go to **http://1pass.thomson.com** or your CD to practice this example.

$$y = \frac{(x - 1)(x + 2)}{x - 3}$$

Symmetry: The function is not symmetric about the y-axis or the origin.

Vertical asymptotes: The vertical asymptote is the line $x = 3$.

y- and x-intercepts: The y-intercept is $\left(0, \frac{2}{3}\right)$. The x-intercepts are $(1, 0)$ and $(-2, 0)$.

Slant asymptote: Because the degree of the numerator is 1 greater than the degree of the denominator, this graph will have a slant asymptote. To find it, we do a long division and write the expression as

$$y = \frac{x^2 + x - 2}{x - 3} = x + 4 + \frac{10}{x - 3}$$

The fraction $\frac{10}{x-3}$ approaches 0 as x approaches ∞, and the graph approaches a slant asymptote: the line $y = x + 4$. The graph appears in Figure 3-50.

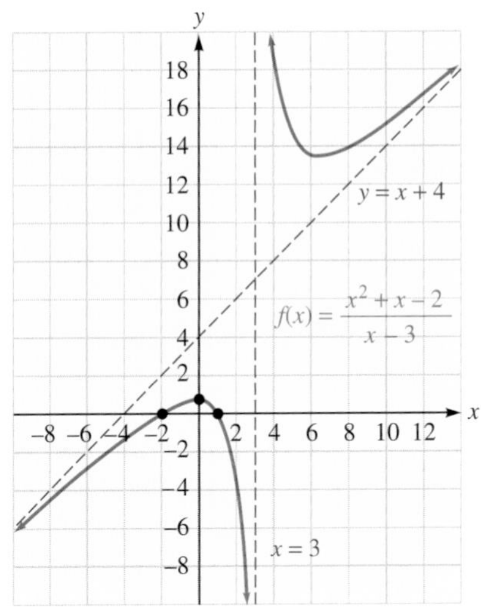

Figure 3-50

Self Check Use a graphing calculator to graph the function in Example 9. ■

■ Graphs with Missing Points

We have discussed rational functions where the fraction is in simplified form. We now consider a rational function $f(x) = \frac{P(x)}{Q(x)}$ where $P(x)$ and $Q(x)$ have a common factor. Graphs of such functions have gaps or missing points that are not the result of vertical asymptotes.

ACTIVE EXAMPLE 10 Find the domain of the function $f(x) = \dfrac{x^2 - x - 12}{x - 4}$ and graph it.

Solution Since a denominator cannot be 0, $x \neq 4$. The domain is the set of all real numbers except 4.

When we factor the numerator of the expression, we see that the numerator and denominator have a common factor of $x - 4$.

$$y = \frac{x^2 - x - 12}{x - 4}$$
$$= \frac{(x + 3)(x - 4)}{x - 4}$$

If $x \neq 4$, the common factor of $x - 4$ can be divided out. The resulting function is equivalent to the original function only when we keep the restriction that $x \neq 4$. Thus,

$$y = \frac{(x + 3)(x - 4)}{x - 4} = x + 3 \quad \text{(provided that } x \neq 4\text{)}$$

When $x = 4$, the function is not defined. The graph of the function appears in Figure 3-51. It is a line with the point with x-coordinate of 4 missing.

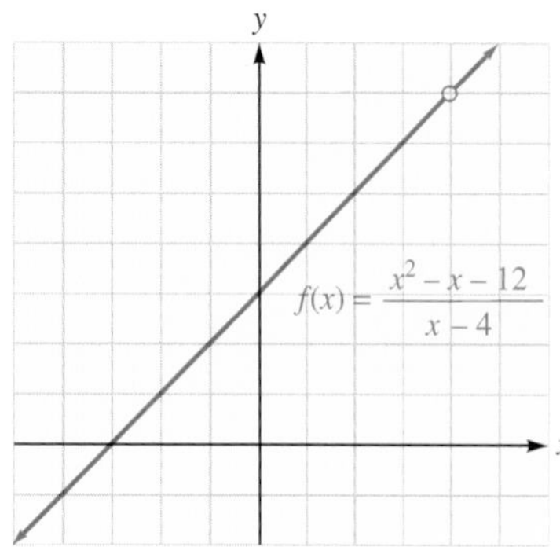

Figure 3-51

Self Check Use a graphing calculator to graph the function in Example 10. ■

Self Check Answers

1. \$13.50 per hour **2.** $(-\infty, -1) \cup (-1, 2) \cup (2, \infty)$ **3.** vertical asymptotes at $x = 2$ and $x = -2$; horizontal asymptote at $y = 1$ **4.** vertical asymptotes at $x = 2$ and $x = -2$; slant asymptote at $y = 2x$

5.
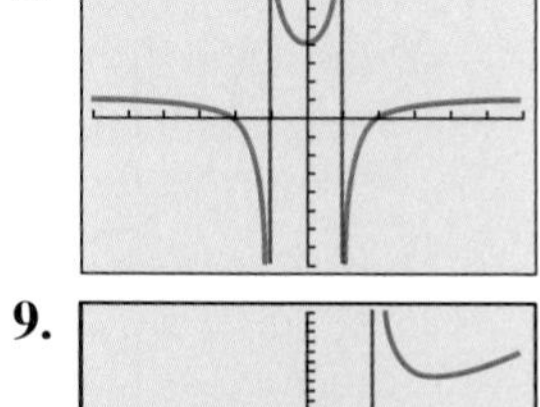

6.
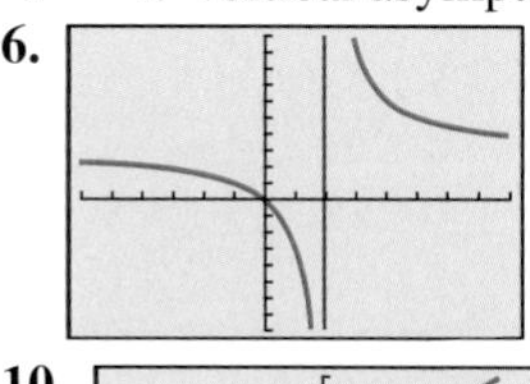

7.
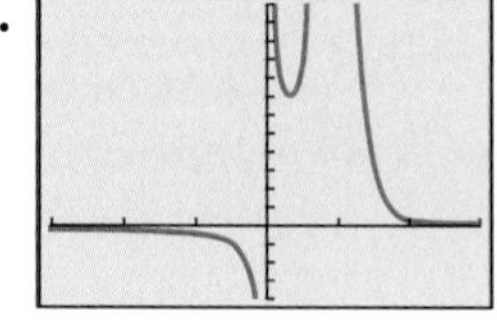

8.
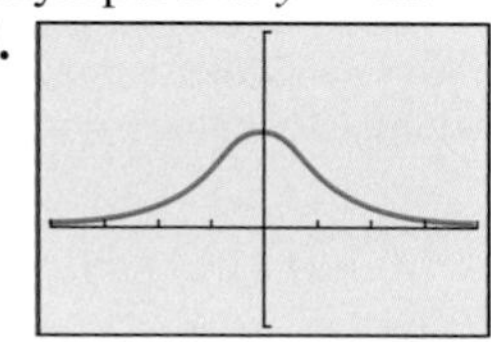

9.
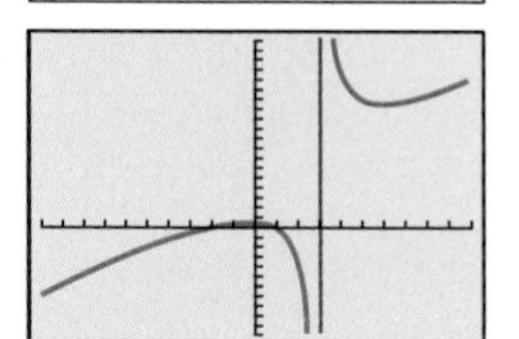

10.
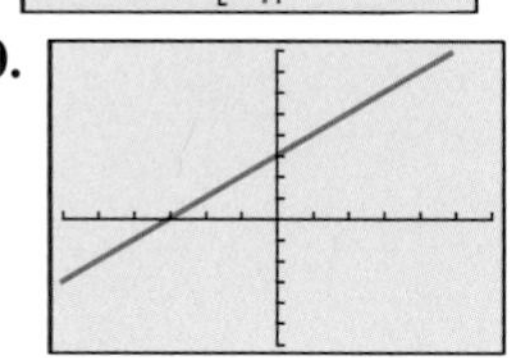

Note that the calculator graph does not indicate the excluded point.

3.5 Exercises

VOCABULARY AND CONCEPTS ***Fill in the blanks.***

1. When a graph approaches a vertical line but never touches it, we call the line an **asymptote**.
2. A rational function is a function with a polynomial numerator and a **nonzero** polynomial denominator.
3. To find a **vertical** asymptote, set the denominator polynomial equal to 0 and solve the equation.
4. To find the **y-intercept** of a rational function, let $x = 0$ and solve for y.
5. To find the **x-intercept** of a rational function, set the numerator equal to 0 and solve the equation.
6. In the function $\frac{P(x)}{Q(x)}$, if the degree of $P(x)$ is less than the degree of $Q(x)$, the horizontal asymptote is **$y = 0$**.
7. In the function $f(x) = \frac{P(x)}{Q(x)}$, if the degree of $P(x)$ and $Q(x)$ are **the same**, the horizontal asymptote is

$$y = \frac{\text{the lead coefficient of the numerator}}{\text{the lead coefficient of the denominator}}$$

8. In a rational function, if the degree of the numerator is 1 greater than the degree of the denominator, the graph will have a **slant asymptote**.
9. A graph can cross a **horizontal** asymptote but can never cross a **vertical** asymptote.
10. The graph of $f(x) = \dfrac{x^2 - 4}{x + 2}$ will have a **missing** point.

Find the equations of the vertical and horizontal asymptotes of each graph.

11.

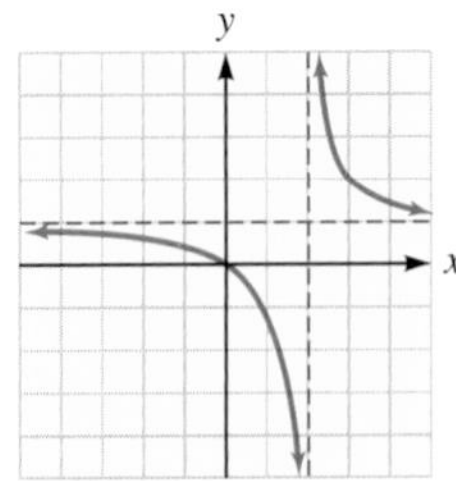

vertical asymptote: $x = 2$; horizontal asymptote: $y = 1$

12.

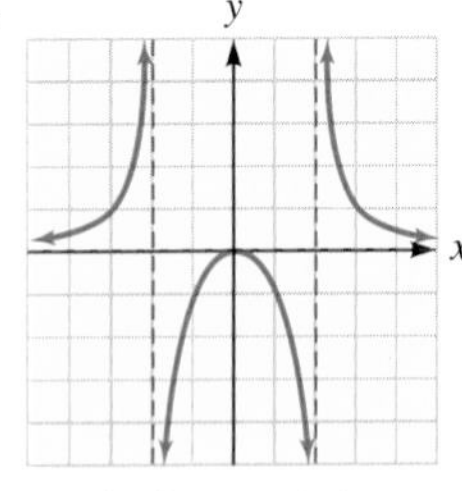

vertical asymptotes: $x = -2, x = 2$; horizontal asymptote: $y = 0$

PRACTICE ***The time t it takes to travel 600 miles is a function of the mean rate of speed r:***

$$t = f(r) = \frac{600}{r}$$

Find t for the given values of r.

13. 30 mph **20 hr**
14. 40 mph **15 hr**
15. 50 mph **12 hr**
16. 60 mph **10 hr**

Suppose the cost (in dollars) of removing p% of the pollution in a river is given by the function

$$c = f(p) = \frac{50{,}000p}{100 - p} \quad (0 \le p < 100)$$

Find the cost of removing each percent of pollution.

17. 10% **\$5,555.56**
18. 30% **\$21,428.57**
19. 50% **\$50,000**
20. 80% **\$200,000**

Find the domain of each rational function. Do not graph the function.

21. $f(x) = \dfrac{x^2}{x - 2}$
$(-\infty, 2) \cup (2, \infty)$
22. $f(x) = \dfrac{x^3 - 3x^2 + 1}{x + 3}$
$(-\infty, -3) \cup (-3, \infty)$
23. $f(x) = \dfrac{2x^2 + 7x - 2}{x^2 - 25}$
$(-\infty, -5) \cup (-5, 5) \cup (5, \infty)$
24. $f(x) = \dfrac{5x^2 + 1}{x^2 + 5}$
$(-\infty, \infty)$
25. $f(x) = \dfrac{x - 1}{x^3 - x}$
$(-\infty, -1) \cup (-1, 0) \cup (0, 1) \cup (1, \infty)$
26. $f(x) = \dfrac{x + 2}{2x^2 - 9x + 9}$
$(-\infty, \frac{3}{2}) \cup (\frac{3}{2}, 3) \cup (3, \infty)$
27. $f(x) = \dfrac{3x^2 + 5}{x^2 + 1}$
$(-\infty, \infty)$
28. $f(x) = \dfrac{7x^2 - x + 2}{x^4 + 4}$
$(-\infty, \infty)$

Find all vertical, horizontal, and slant asymptotes, x- and y-intercepts, and symmetries, and then graph each function. Check your work with a graphing calculator.

29. $y = \dfrac{1}{x - 2}$

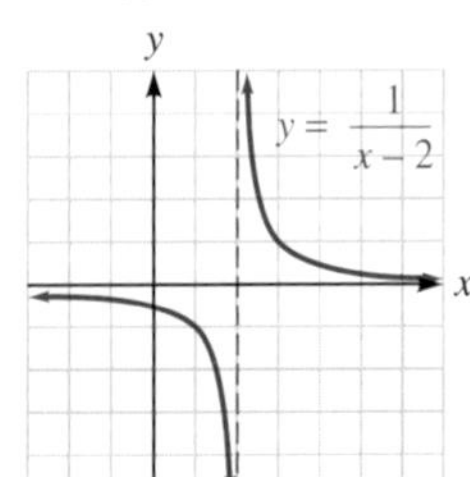

30. $y = \dfrac{3}{x + 3}$

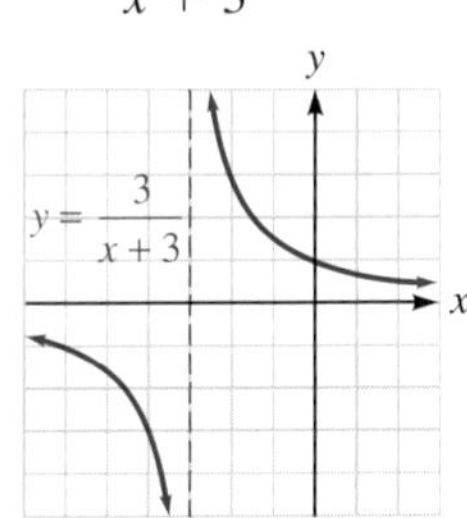

31. $y = \dfrac{x}{x - 1}$

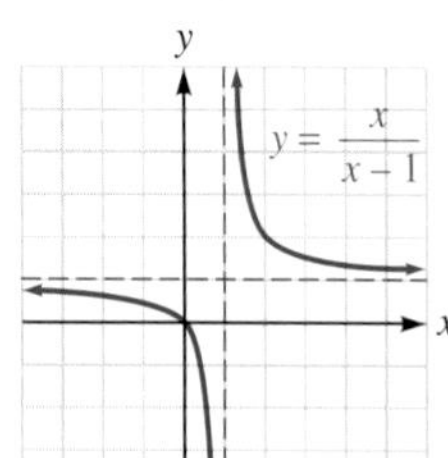

32. $y = \dfrac{x}{x + 2}$

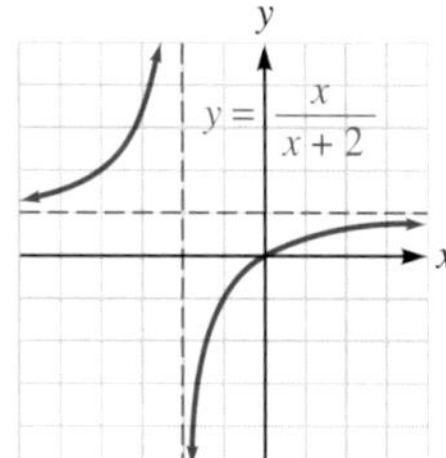

33. $f(x) = \dfrac{x + 1}{x + 2}$

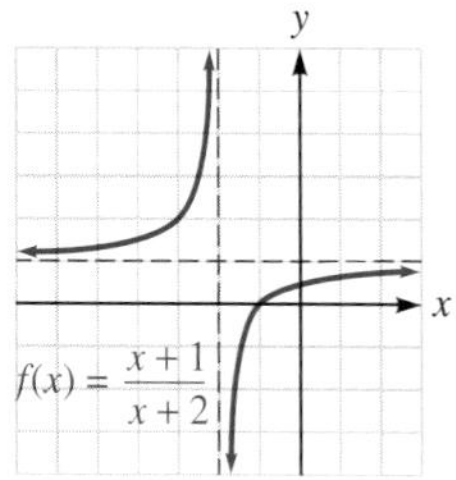

34. $f(x) = \dfrac{x - 1}{x - 2}$

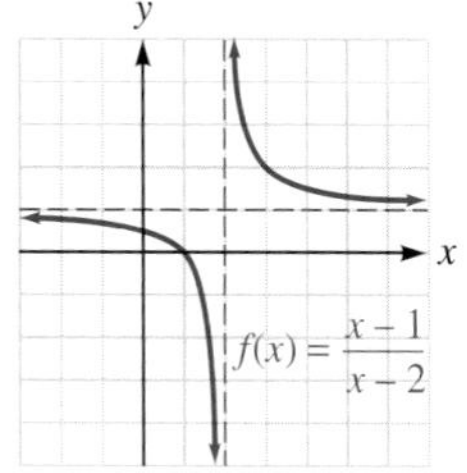

35. $f(x) = \dfrac{2x - 1}{x - 1}$

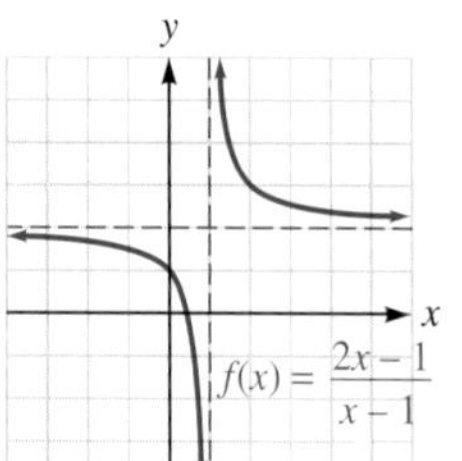

36. $f(x) = \dfrac{3x + 2}{x^2 - 4}$

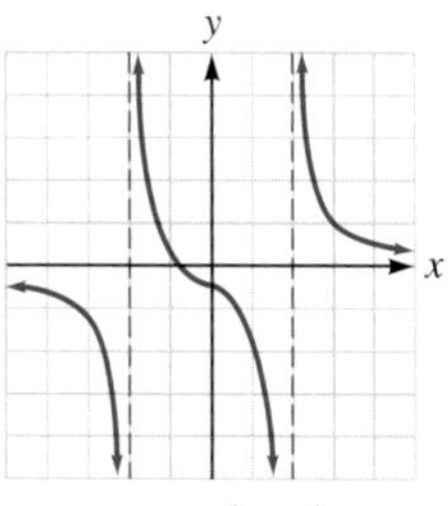

$f(x) = \dfrac{3x+2}{x^2-4}$

37. $g(x) = \dfrac{x^2 - 9}{x^2 - 4}$

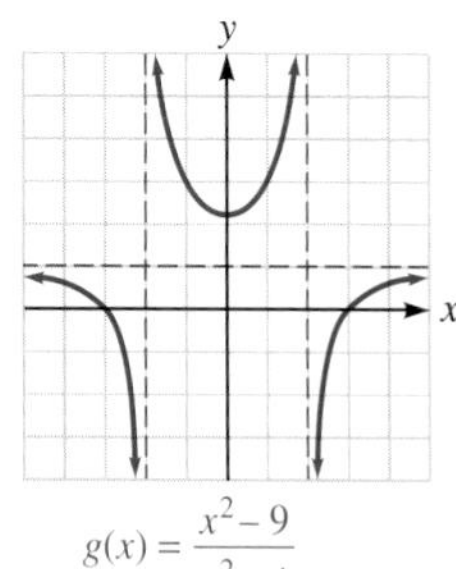

$g(x) = \dfrac{x^2-9}{x^2-4}$

38. $g(x) = \dfrac{x^2 - 4}{x^2 - 9}$

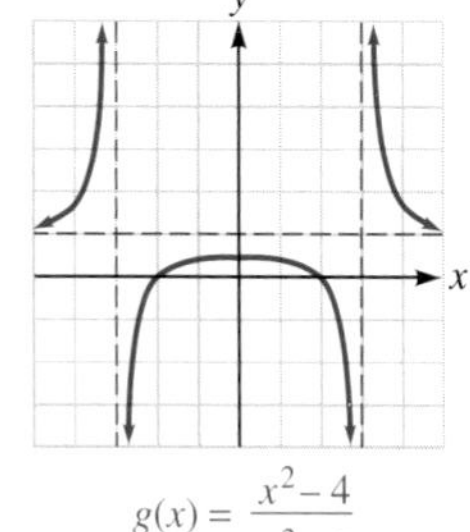

$g(x) = \dfrac{x^2-4}{x^2-9}$

39. $g(x) = \dfrac{x^2 - x - 2}{x^2 - 4x + 3}$

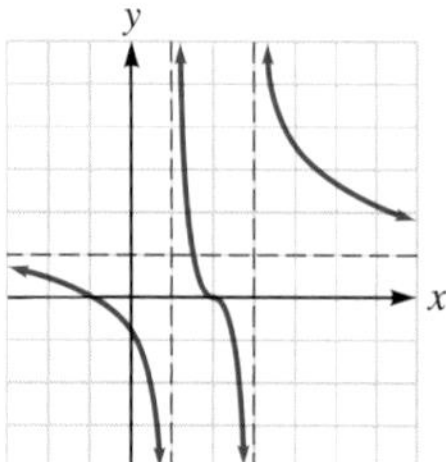

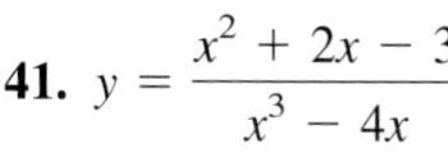

$g(x) = \dfrac{x^2-x-2}{x^2-4x+3}$

40. $g(x) = \dfrac{x^2 + 7x + 12}{x^2 - 7x + 12}$

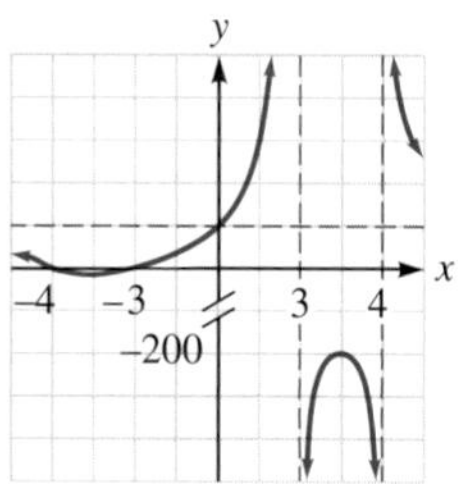

$g(x) = \dfrac{x^2+7x+12}{x^2-7x+12}$

41. $y = \dfrac{x^2 + 2x - 3}{x^3 - 4x}$

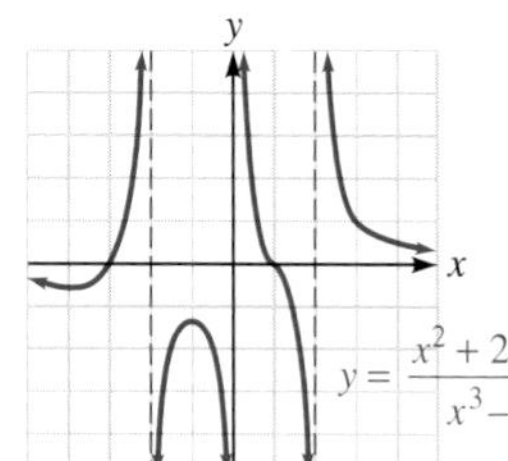

42. $y = \dfrac{3x^2 - 4x + 1}{2x^3 + 3x^2 + x}$

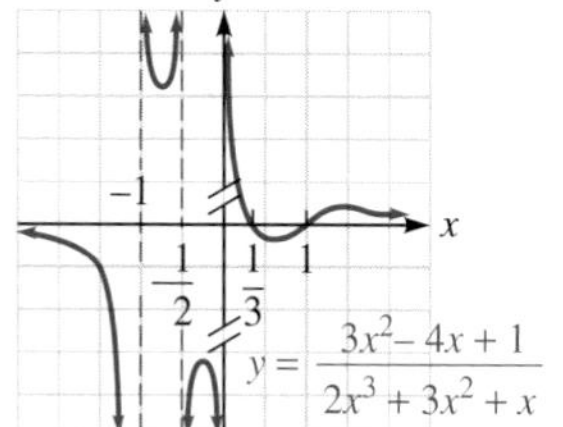

43. $y = \dfrac{x^2 - 9}{x^2}$

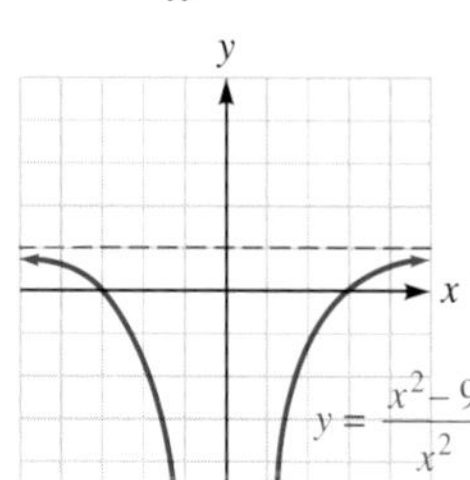

44. $y = \dfrac{3x^2 - 12}{x^2}$

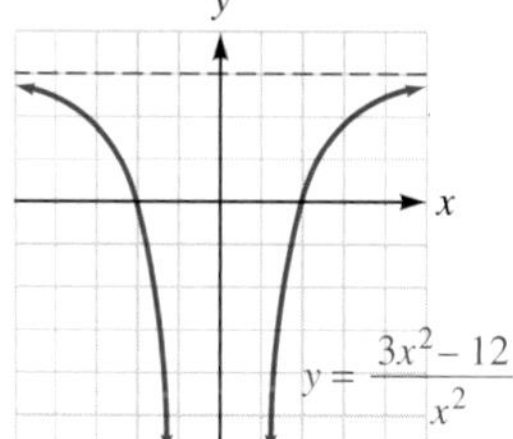

45. $f(x) = \dfrac{x}{(x + 3)^2}$

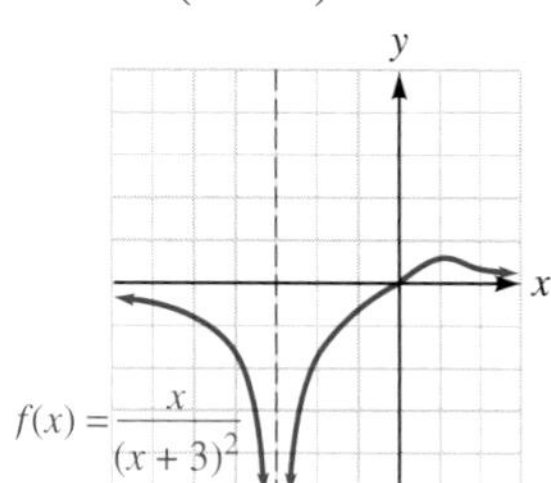

46. $f(x) = \dfrac{x}{(x - 1)^2}$

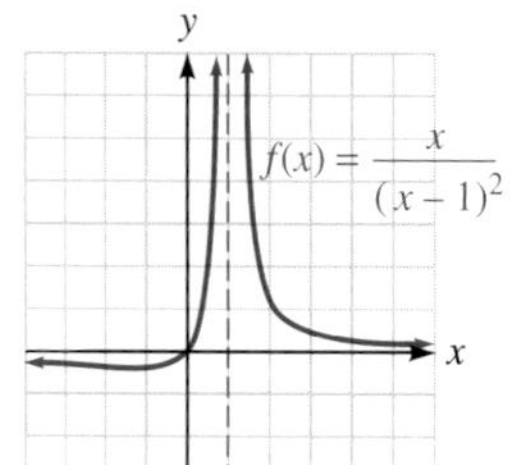

47. $f(x) = \dfrac{x + 1}{x^2(x - 2)}$

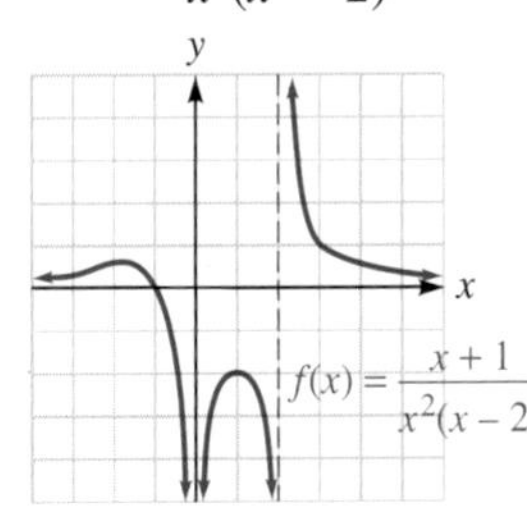

48. $f(x) = \dfrac{x - 1}{x^2(x + 2)^2}$

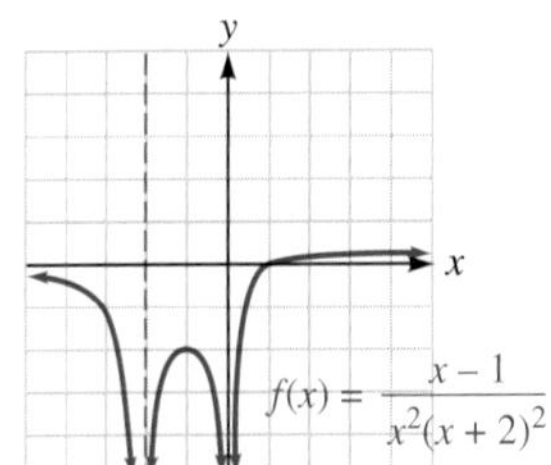

49. $y = \dfrac{x}{x^2 + 1}$

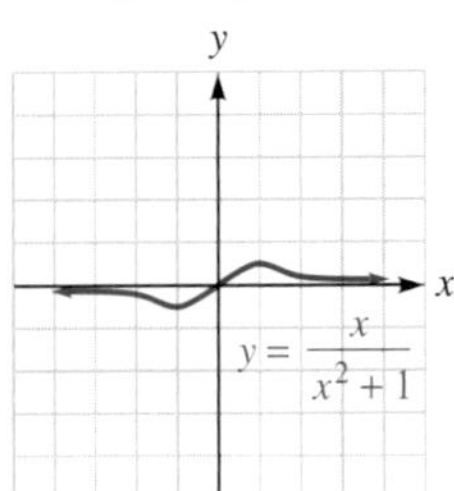

50. $y = \dfrac{x - 1}{x^2 + 2}$

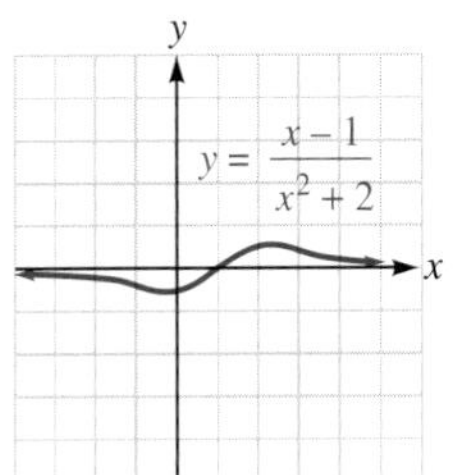

51. $y = \dfrac{3x^2}{x^2 + 1}$

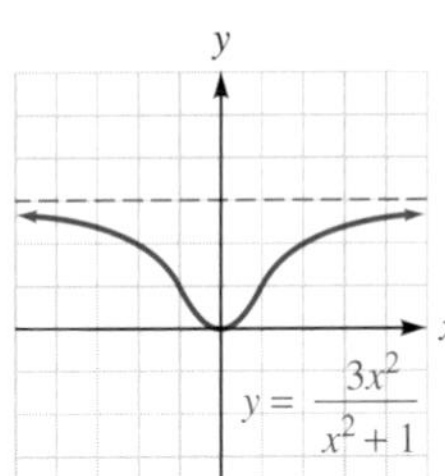

52. $y = \dfrac{x^2 - 9}{2x^2 + 1}$

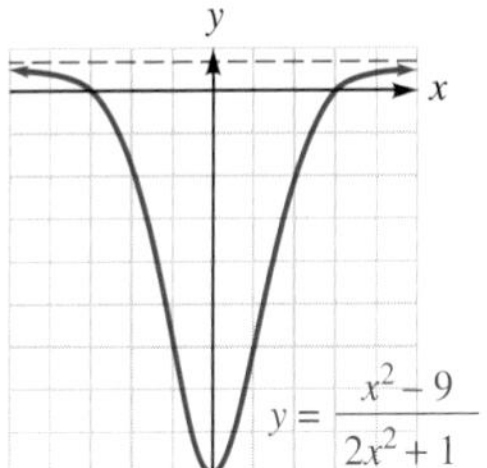

53. $h(x) = \dfrac{x^2 - 2x - 8}{x - 1}$

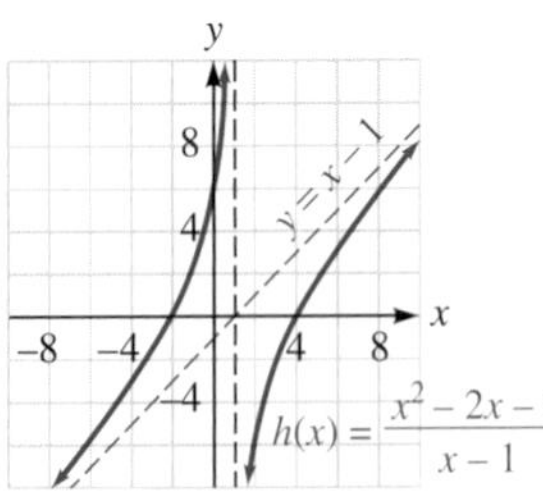

54. $h(x) = \dfrac{x^2 + x - 6}{x + 2}$

55. $f(x) = \dfrac{x^3 + x^2 + 6x}{x^2 - 1}$

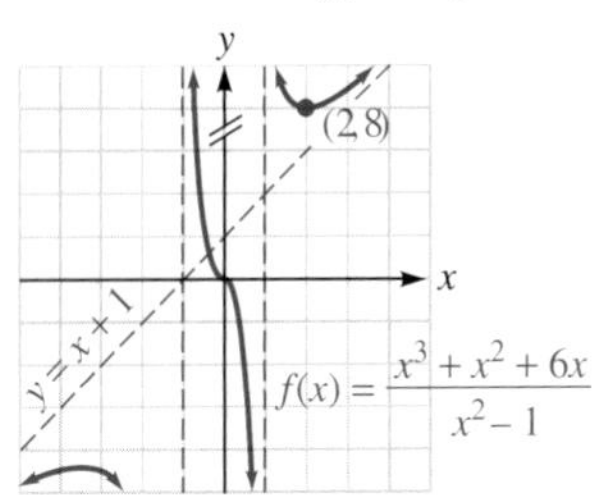

56. $f(x) = \dfrac{x^3 - 2x^2 + x}{x^2 - 4}$

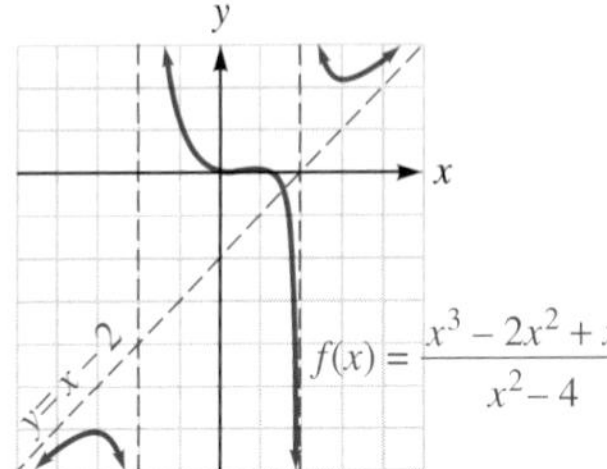

Graph each rational function. Note that the numerator and denominator of the fraction share a common factor.

57. $f(x) = \dfrac{x^2}{x}$

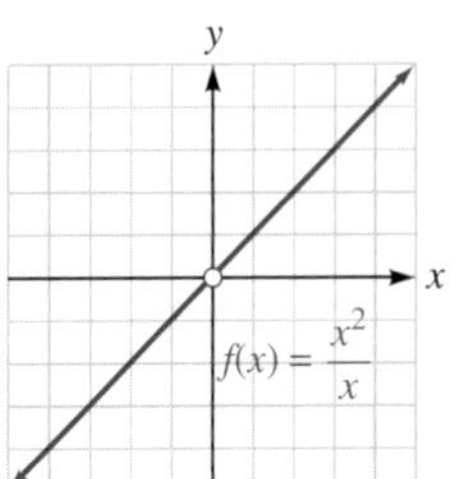

58. $f(x) = \dfrac{x^2 - 1}{x - 1}$

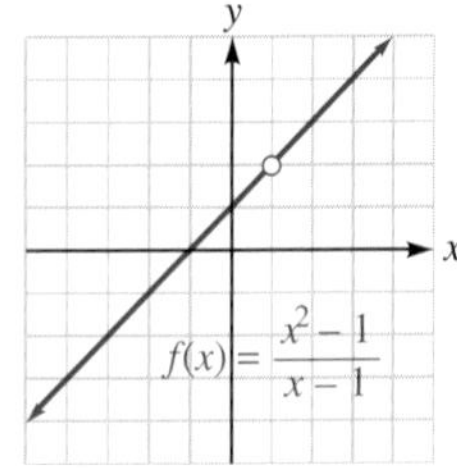

59. $f(x) = \dfrac{x^3 + x}{x}$

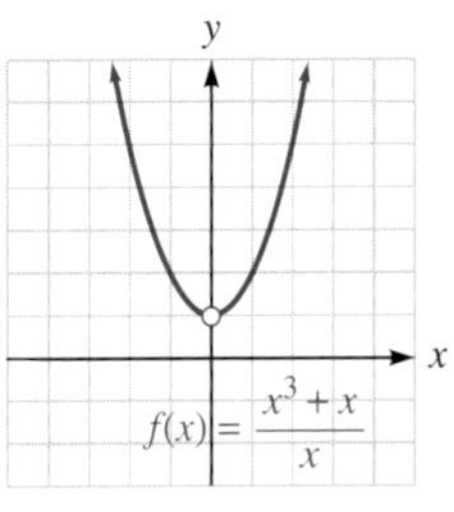

60. $f(x) = \dfrac{x^3 - x^2}{x - 1}$

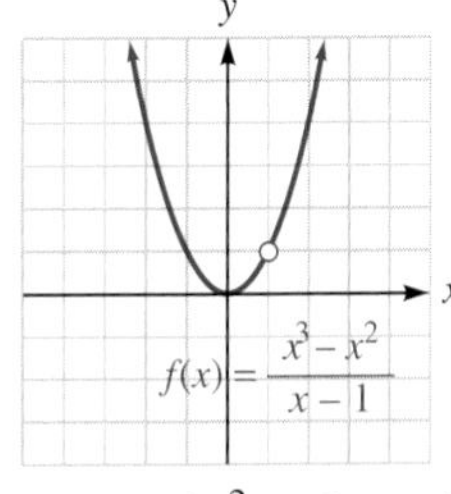

61. $f(x) = \dfrac{x^2 - 2x + 1}{x - 1}$

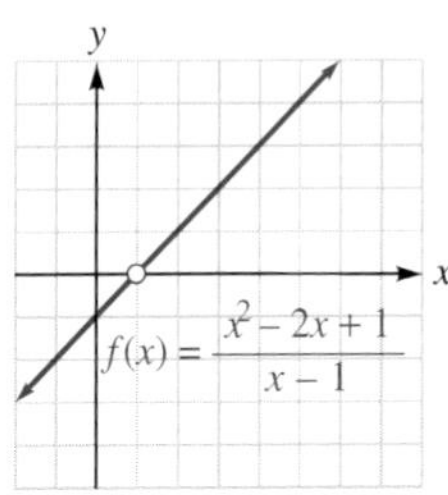

62. $f(x) = \dfrac{2x^2 + 3x - 2}{x + 2}$

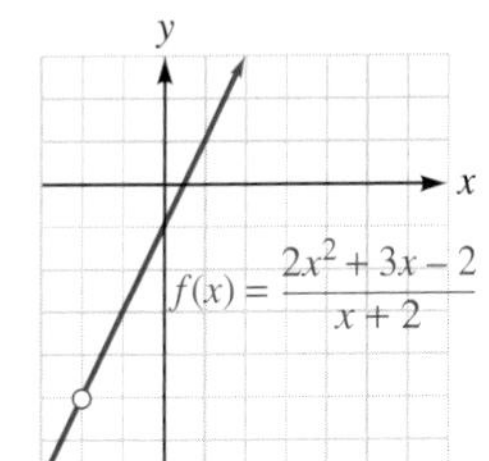

63. $f(x) = \dfrac{x^3 - 1}{x - 1}$

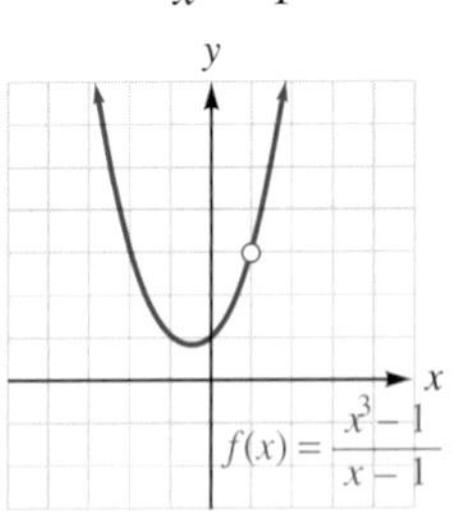

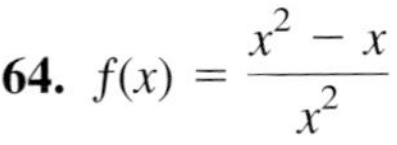

64. $f(x) = \dfrac{x^2 - x}{x^2}$

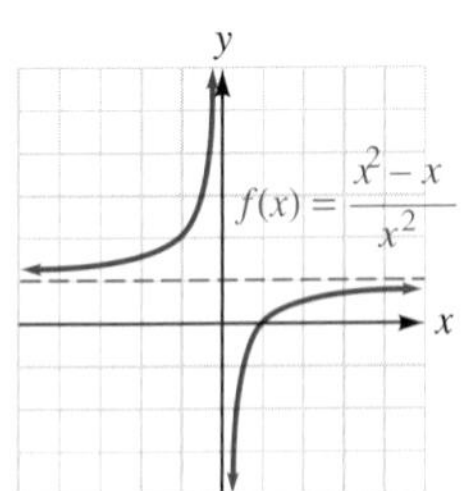

APPLICATIONS

A service club wants to publish a directory of its members. Some investigation shows that the cost of typesetting and photography will be \$700, and the cost of printing each directory will be \$1.25.

65. Find a function that gives the total cost c of printing x directories. **$c = f(x) = 1.25x + 700$**

66. Find the mean cost per directory if 500 directories are printed. **\$2.65**

67. Find a function that gives the mean cost per directory $\bar{c}$ of printing x directories. **$\bar{c} = f(x) = \frac{1.25x + 700}{x}$**

68. Find the mean cost per directory if 1,000 directories are printed. **\$1.95**

69. Find the total cost of printing 500 directories. **\$1,325**

70. Find the mean cost per directory if 2,000 directories are printed. **\$1.60**

An electric company charges \$7.50 per month plus 9¢ for each kilowatt hour (kwh) of electricity used.

71. Find a function that gives the total cost c of n kwh of electricity. **$c = f(n) = 0.09n + 7.50$**

72. Find a function that gives the mean cost per kwh, $\bar{c}$, when using n kwh. **$\bar{c} = \frac{c}{n} = \frac{0.09n + 7.50}{n}$**

73. Find the total cost for using 775 kwh. **\$77.25**

74. Find the mean cost per kwh when 775 kwh are used. **9.97¢**

75. Find the mean cost per kwh when 1,000 kwh are used. **9.75¢**

76. Find the mean cost per kwh when 1,200 kwh are used. **9.625¢**

77. Utility costs An electric company charges \$8.50 per month plus 9.5¢ for each kilowatt hour (kwh) of electricity used.

a. Find a linear function that gives the total cost of n kwh of electricity. **$c(n) = 0.095n + 8.50$**

b. Find a rational function that gives the average cost per kwh when using n kwh. **$c(n) = \frac{0.095n + 8.50}{n}$**

c. Find the average cost per kwh when 850 kwh are used. **10.5¢**

78. Scheduling work crews The rational function

$$f(t) = \frac{t^2 + 3t}{2t + 3}$$

gives the number of days it would take two construction crews, working together, to frame a house that crew 1 (working alone) could complete in t days and crew 2 (working alone) could complete in $(t + 3)$ days.

a. If crew 1 could frame a certain house in 21 days, how long would it take both crews working together? **11.2 days**

b. If crew 2 could frame a certain house in 25 days, how long would it take both crews working together? **about 11.7 days**

DISCOVERY AND WRITING

79. Can a rational function have two horizontal asymptotes? Explain.

80. Can a rational function have two slant asymptotes? Explain.

In Exercises 81–82, a, b, c, and d are nonzero constants.

81. Show that the graph of

$$y = \frac{ax + b}{cx^2 + d}$$

has the horizontal asymptote $y = 0$.

82. Show that the graph of

$$y = \frac{ax^3 + b}{cx^2 + d}$$

has the slant asymptote $y = \frac{a}{c}x$.

83. Show that the graph of

$$y = \frac{ax^2 + b}{cx^2 + d}$$

has the horizontal asymptote $y = \frac{a}{c}$.

84. Graph the rational function $y = \frac{x^3 + 1}{x}$ and explain why the curve is said to have a *parabolic asymptote*.

Use a graphing calculator to perform each experiment. Write a brief paragraph describing your findings.

85. Investigate the positioning of the vertical asymptotes of a rational function by graphing $y = \dfrac{x}{x - k}$ for several values of k. What do you observe?

86. Investigate the positioning of the vertical asymptotes of a rational function by graphing $y = \dfrac{x}{x^2 - k}$ for $k = 4, 1, -1$, and 0. What do you observe?

87. Find the range of the rational function $y = \dfrac{kx^2}{x^2 + 1}$ for several values of k. What do you observe?

88. Investigate the positioning of the x-intercepts of a rational function by graphing $y = \dfrac{x^2 - k}{x}$ for $k = 1$, -1, and 0. What do you observe?

REVIEW ***Perform each operation.***

89. $(2x^2 + 3x) + (x^2 - 2x)$ $3x^2 + x$

90. $(3x + 2) - (x^2 + 2)$ $-x^2 + 3x$

91. $(5x + 2)(2x + 5)$ $10x^2 + 29x + 10$

92. $\dfrac{2x^2 + 3x + 1}{x + 1}$ $2x + 1$

93. If $f(x) = 3x + 2$, find $f(x + 1)$. $3x + 5$

94. If $f(x) = x^2 + x$, find $f(2x + 1)$. $4x^2 + 6x + 2$

3.6 Operations on Functions

In this section, you will learn about

- Algebra of Functions
- Composition of Functions
- The Identity Function
- Problem Solving

Algebra of Functions

With the following definitions, it is possible to perform arithmetic operations on algebraic functions.

Adding, Subtracting, Multiplying, and Dividing Functions

If the ranges of the functions f and g are subsets of the real numbers, then

1. The **sum** of f and g, denoted as $\boldsymbol{f + g}$, is defined by

$$(f + g)(x) = f(x) + g(x)$$

2. The **difference** of f and g, denoted as $\boldsymbol{f - g}$, is defined by

$$(f - g)(x) = f(x) - g(x)$$

3. The **product** of f and g, denoted as $\boldsymbol{f \cdot g}$, is defined by

$$(f \cdot g)(x) = f(x)g(x)$$

4. The **quotient** of f and g, denoted as $\boldsymbol{f/g}$, is defined by

$$(f/g)(x) = \frac{f(x)}{g(x)} \quad (g(x) \neq 0)$$

The domain of each function, unless otherwise restricted, is the set of real numbers x that are in the domains of both f and g. In the case of the quotient f/g, there is the further restriction that $g(x) \neq 0$.

EXAMPLE 1 Let $f(x) = 3x + 1$ and $g(x) = 2x - 3$. Find each of the following functions and its domain: **a.** $f + g$ and **b.** $f - g$.

Solution **a.** $(f + g)(x) = f(x) + g(x)$
$= (3x + 1) + (2x - 3)$
$= 5x - 2$

Since the domain of both f and g is the set of real numbers, the domain of $f + g$ is the interval $(-\infty, \infty)$.

b. $(f - g)(x) = f(x) - g(x)$
$= (3x + 1) - (2x - 3)$
$= x + 4$

Since the domain of both f and g is the set of real numbers, the domain of $f - g$ is the interval $(-\infty, \infty)$.

Self Check Find $g - f$. ■

ACTIVE EXAMPLE 2 Let $f(x) = 3x + 1$ and $g(x) = 2x - 3$. Find each of the following functions and its domain: **a.** $f \cdot g$ and **b.** f/g.

Solution **a.** $(f \cdot g)(x) = f(x) \cdot g(x)$
$= (3x + 1)(2x - 3)$
$= 6x^2 - 7x - 3$

COLLEGE **Algebra f(x) Now**™
Explore this Active Example and test yourself on the steps taken here at **http://1pass.thomson.com**. You can also find this example on the Interactive Video Skillbuilder CD-ROM.

Since the domain of both f and g is the set of real numbers, the domain of $f \cdot g$ is the interval $(-\infty, \infty)$.

b. $(f/g)(x) = \dfrac{f(x)}{g(x)}$

$= \dfrac{3x + 1}{2x - 3} \quad (2x - 3 \neq 0)$

Since $2x - 3 \neq 0$, the domain of f/g is the set of all real numbers except $\frac{3}{2}$. This is $\left(-\infty, \frac{3}{2}\right) \cup \left(\frac{3}{2}, \infty\right)$.

Self Check Find g/f and its domain. ■

EXAMPLE 3 Let $f(x) = x^2 - 4$ and $g(x) = \sqrt{x}$. Find each function and its domain: **a.** $f + g$, **b.** $f \cdot g$, **c.** f/g, and **d.** g/f.

Solution Because all real numbers can be squared, the domain of f is the interval $(-\infty, \infty)$. Because $\sqrt{x}$ is to be a real number, the domain of g is the interval $[0, \infty)$.

a. $(f + g)(x) = f(x) + g(x)$
$= x^2 - 4 + \sqrt{x}$

The domain consists of the numbers x that are in the domain of both f and g. This is $(-\infty, \infty) \cap [0, \infty)$, which is $[0, \infty)$. The domain of $f + g$ is $[0, \infty)$.

b. $(f \cdot g)(x) = f(x)g(x)$
$= (x^2 - 4)\sqrt{x}$
$= x^2\sqrt{x} - 4\sqrt{x}$

The domain consists of the numbers x that are in the domain of both f and g. The domain of $f \cdot g$ is $[0, \infty)$.

c. $(f/g)(x) = \dfrac{f(x)}{g(x)}$

$= \dfrac{x^2 - 4}{\sqrt{x}}$

The domain consists of the numbers x that are in the domain of both f and g, except 0 (because division by 0 is undefined). The domain of f/g is $(0, \infty)$.

d. $(g/f)(x) = \dfrac{g(x)}{f(x)}$

$= \dfrac{\sqrt{x}}{x^2 - 4}$

The domain consists of the numbers x that are in the domain of both f and g, except 2 (because division by 0 is undefined). The domain of f/g is $[0, 2) \cup (2, \infty)$.

Self Check Find $g - f$ and its domain. ■

EXAMPLE 4 Find $(f + g)(3)$ when $f(x) = x^2 + 1$ and $g(x) = 2x + 1$.

Solution We first find $(f + g)(x)$.

$$\begin{aligned}(f + g)(x) &= f(x) + g(x)\\ &= x^2 + 1 + 2x + 1\\ &= x^2 + 2x + 2\end{aligned}$$

We then find $(f + g)(3)$.

$$\begin{aligned}(f + g)(x) &= x^2 + 2(x) + 2\\ (f + g)(3) &= 3^2 + 2(3) + 2 \quad \text{Substitute 3 for } x.\\ &= 9 + 6 + 2\\ &= 17\end{aligned}$$

Self Check Find $(f \cdot g)(-2)$. ■

EXAMPLE 5 Let $h(x) = x^2 + 3x + 2$. Find two functions f and g such that **a.** $f + g = h$ and **b.** $f \cdot g = h$.

Solution **a.** There are many possibilities. One is $f(x) = x^2$ and $g(x) = 3x + 2$, for then

$$\begin{aligned}(f + g)(x) &= f(x) + g(x)\\ &= (x^2) + (3x + 2)\\ &= x^2 + 3x + 2\\ &= h(x)\end{aligned}$$

Another possibility is $f(x) = x^2 + 2x$ and $g(x) = x + 2$.

b. Again, there are many possibilities. One is suggested by factoring $x^2 + 3x + 2$:

$$x^2 + 3x + 2 = (x + 1)(x + 2)$$

If we let $f(x) = x + 1$ and $g(x) = x + 2$, then

$$\begin{aligned}(f \cdot g)(x) &= f(x) \cdot g(x)\\ &= (x + 1)(x + 2)\\ &= x^2 + 3x + 2\\ &= h(x)\end{aligned}$$

Another possibility is $f(x) = 3$ and $g(x) = \frac{x^2}{3} + x + \frac{2}{3}$.

Self Check Find two functions f and g such that $f - g = h$. ■

■ Composition of Functions

Often one quantity is a function of a second quantity that depends, in turn, on a third quantity. For example, the cost of a car trip is a function of the gasoline consumed. The amount of gasoline consumed, in turn, is a function of the number of miles driven. Such chains of dependence are analyzed mathematically as *composition of functions.*

Suppose that $y = f(x)$ and $y = g(x)$ define two functions. Any number x in the domain of g will produce a corresponding value $g(x)$ in the range of g. If $g(x)$ is in the domain of function f, then $g(x)$ can be substituted into f, and a corresponding value $f(g(x))$ will be determined. This two-step process defines a new function, called a **composite function,** denoted by $f \circ g$. (See Figure 3-52.)

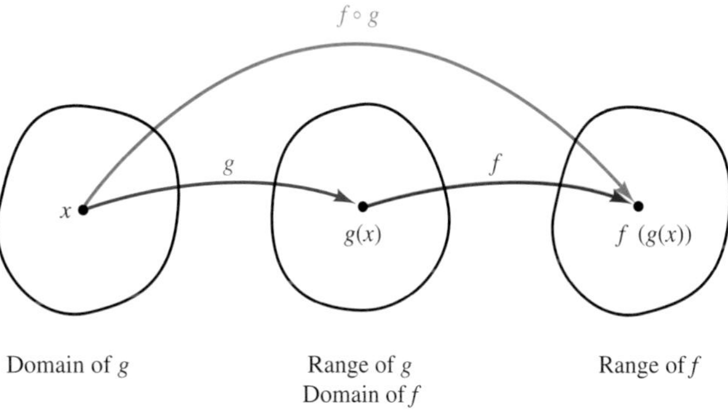

Figure 3-52

Composite Functions

The **composite function $f \circ g$** is defined by

$$(f \circ g)(x) = f(g(x))$$

The **domain** of $f \circ g$ consists of all those numbers in the domain of g for which $g(x)$ is in the domain of f.

For example, if $f(x) = 5x + 1$ and $g(x) = 4x - 3$, then

$$\begin{aligned}(f \circ g)(x) &= f(g(x))\\ &= f(4x - 3)\\ &= 5(4x - 3) + 1\\ &= 20x - 14\end{aligned}$$

$$\begin{aligned}(g \circ f)(x) &= g(f(x))\\ &= g(5x + 1)\\ &= 4(5x + 1) - 3\\ &= 20x + 1\end{aligned}$$

Comment Note that in the previous example, $(f \circ g)(x) \neq (g \circ f)(x)$. This shows that the composition of functions is not commutative.

We have seen that a function can be represented by a machine. If we put a number from the domain into the machine (the input), a number from the range comes out (the output). For example, if we put 2 into the machine shown in Figure 3-53(a), the number $f(2) = 5(2) - 2 = 8$ comes out. In general, if we put x into the machine shown in Figure 3-53(b), the value $f(x)$ comes out.

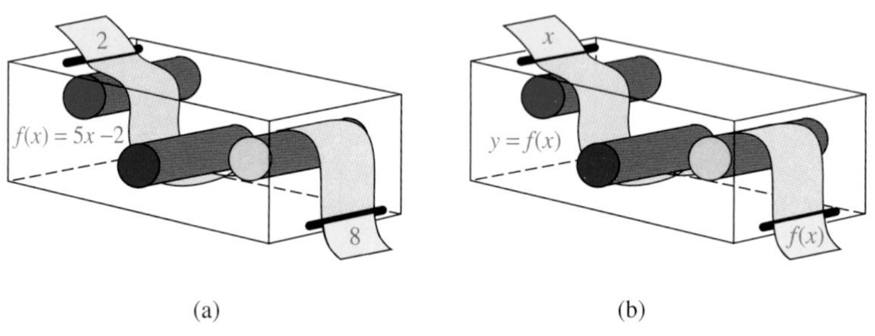

Figure 3-53

The function machines shown in Figure 3-54 illustrate the composition $f \circ g$. When we put a number x into the function g, the value $g(x)$ comes out. The value $g(x)$ then goes into function f, and $f(g(x))$ comes out.

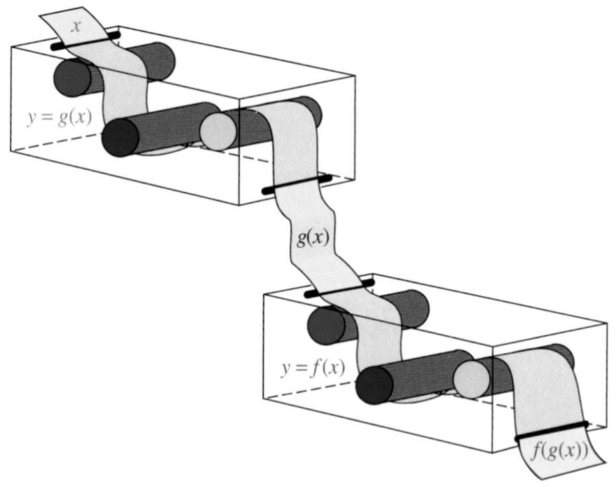

Figure 3-54

To further illustrate these ideas, suppose we let $f(x) = 2x + 1$ and $g(x) = x - 4$.

- $(f \circ g)(9)$ means $f(g(9))$. In Figure 3-55(a), function g receives the number 9 and subtracts 4, and the number $g(x) = 5$ comes out. The 5 goes into the f function, which doubles it and adds 1. The final result, 11, is the output of the composite function $f \circ g$:

$$(f \circ g)(9) = f(g(9)) = f(5) = 2(5) + 1 = 11$$

- $(f \circ g)(x)$ means $f(g(x))$. In Figure 3-55(a), function g receives the number x and subtracts 4, and the number $x - 4$ comes out. The $x - 4$ goes into the f function, which doubles it and adds 1. The final result, $2x - 7$, is the output of the composite function $f \circ g$.

$$(f \circ g)(x) = f(g(x)) = f(x - 4) = 2(x - 4) + 1 = 2x - 7$$

- $(g \circ f)(-2)$ means $g(f(-2))$. In Figure 3-55(b), function f receives the number -2, doubles it and adds 1, and releases -3 into the g function. Function g subtracts 4 from -3 and releases a final output of -7. Thus,

$$(g \circ f)(-2) = g(f(-2)) = g(-3) = -3 - 4 = -7$$

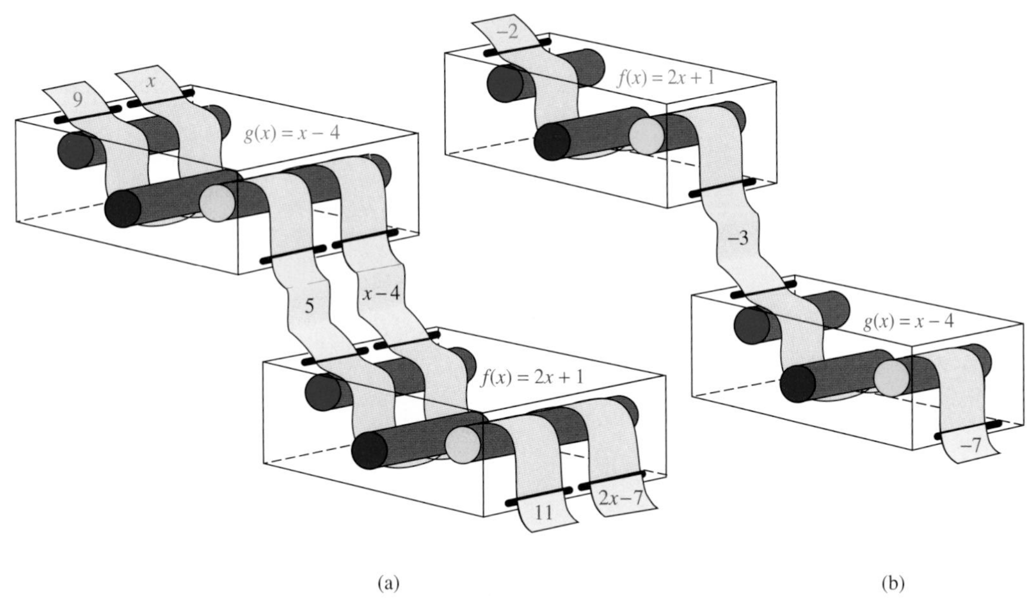

Figure 3-55

ACTIVE EXAMPLE 6 If $f(x) = 2x + 7$ and $g(x) = 4x + 1$, find **a.** $(f \circ g)(x)$ and **b.** $(g \circ f)(x)$.

Solution

a. $(f \circ g) = f(g(x)) = f(4x + 1)$
$= 2(4x + 1) + 7$
$= 8x + 9$

b. $(g \circ f)(x) = g(f(x)) = g(2x + 7)$
$= 4(2x + 7) + 1$
$= 8x + 29$

COLLEGE **Algebra** $f(x)$ **Now**™
Go to **http://1pass.thomson.com** or your CD to practice this example.

Self Check If $h(x) = x + 1$, find $(f \circ h)(x)$. ■

To be in the domain of the composite function $f \circ g$, a number x has to be in the domain of g, and the output of g must be in the domain of f. Thus, the domain of $f \circ g$ consists of those inputs x that are in the domain of g, and for which $g(x)$ is in the domain of f.

EXAMPLE 7 Let $f(x) = \sqrt{x}$ and $g(x) = x - 3$. Find the domain of **a.** $f \circ g$, and **b.** $g \circ f$.

Solution For $\sqrt{x}$ to be a real number, x must be a nonnegative real number. Thus, the domain of f is the interval $[0, \infty)$. Since any real number x can be an input into g, the domain of g is the interval $(-\infty, \infty)$.

a. The domain of $f \circ g$ is the set of real numbers x such that x is in the domain of g and $g(x)$ is in the domain of f. We have seen that all values of x are in the domain of g. However, $g(x)$ must be nonnegative, because $g(x)$ must be in the domain of f. So we have

$$g(x) \geq 0 \quad g(x) \text{ must be nonnegative.}$$
$$x - 3 \geq 0 \quad \text{Substitute } x - 3 \text{ for } g(x).$$
$$x \geq 3 \quad \text{Add 3 to both sides.}$$

Since $x \geq 3$, the domain of $f \circ g$ is the interval $[3, \infty)$.

b. The domain of $g \circ f$ is the set of real numbers x such that x is in the domain of f and $f(x)$ is in the domain of g. We have seen that only nonnegative values of x are in the domain of f. Because all values of $f(x)$ are in the domain of g, the domain of $g \circ f$ is the domain of f, which is the interval $[0, \infty)$.

Self Check Find the domain of $f \circ f$. ■

ACTIVE EXAMPLE 8 Let $f(x) = \dfrac{x + 3}{x - 2}$ and $g(x) = \dfrac{1}{x}$. **a.** Find the domain of $f \circ g$ and **b.** find $f \circ g$.

Solution For $\frac{1}{x}$ to be a real number, x cannot be 0. Thus, the domain of g is $(-\infty, 0) \cup (0, \infty)$. Since any real number except 2 can be an input into f, the domain of f is $(-\infty, 2) \cup (2, \infty)$.

COLLEGE **Algebra** $f(x)$ **Now**™
Go to **http://1pass.thomson.com** or your CD to practice this example.

a. The domain of $f \circ g$ is the set of real numbers x such that x is in the domain of g and $g(x)$ is in the domain of f. We have seen that all values of x but 0 are in the domain of g and that all values of $g(x)$ but 2 are in the domain of f. So we must exclude 0 from the domain of $f \circ g$ and all values of x where $g(x) = 2$.

$$g(x) = 2$$
$$\frac{1}{x} = 2 \quad \text{Substitute } \tfrac{1}{x} \text{ for } g(x).$$
$$1 = 2x \quad \text{Since } x \neq 0, \text{ we can multiply both sides by } x.$$
$$x = \frac{1}{2} \quad \text{Divide both sides by 2.}$$

The domain of $f \circ g$ is the set of all real numbers except 0 and $\frac{1}{2}$, which is $(-\infty, 0) \cup \left(0, \frac{1}{2}\right) \cup \left(\frac{1}{2}, \infty\right)$.

b. $(f \circ g)(x) = f(g(x))$

$$= f\left(\frac{1}{x}\right) \quad \text{Substitute } \tfrac{1}{x} \text{ for } g(x).$$

$$= \frac{\frac{1}{x} + 3}{\frac{1}{x} - 2} \quad \text{Substitute } \tfrac{1}{x} \text{ for } x \text{ in } f.$$

$$= \frac{1 + 3x}{1 - 2x} \quad \text{Multiply numerator and denominator by } x.$$

Thus, $(f \circ g)(x) = \dfrac{1 + 3x}{1 - 2x}$.

Self Check Let $f(x) = \dfrac{x}{x - 1}$ and $g(x) = \dfrac{1}{x}$. Find the domain of $f \circ g$, and then find the function. ■

■ The Identity Function

The **identity function** is defined by the equation $I(x) = x$. Under this function, the value that corresponds to any real number x is x itself. If f is any function, the composition of f with the identity function is the function f:

$$(f \circ I)(x) = (I \circ f)(x) = f(x)$$

EXAMPLE 9 Let f be a function and I be the identity function. Show that **a.** $(f \circ I)(x) = f(x)$ and **b.** $(I \circ f)(x) = f(x)$.

Solution **a.** $(f \circ I)(x)$ means $f(I(x))$. Because $I(x) = x$, we have

$$(f \circ I)(x) = f(I(x)) = f(x)$$

b. $(I \circ f)(x)$ means $I(f(x))$. Because I passes any number through unchanged, we have $I(f(x)) = f(x)$ and

$$(I \circ f)(x) = I(f(x)) = f(x)$$

Self Check Find **a.** $(f \circ I)(5)$ and **b.** $(I \circ f)(5)$. ■

■ Problem Solving

EXAMPLE 10 A laboratory sample is removed from a cooler at a temperature of 15° F. Technicians then warm the sample at the rate of 3° F per hour. Express the sample's temperature in degrees Celsius as a function of the time t (in hours) since it was removed from the cooler.

Solution The temperature of the sample is 15° F when $t = 0$. Because it warms at 3° F per hour, it warms $3t$° after t hours. Thus, the Fahrenheit temperature after t hours is given by the function

$$F(t) = 3t + 15 \quad F(t) \text{ is the Fahrenheit temperature, and } t \text{ represents the time in hours.}$$

The Celsius temperature is a function of the Fahrenheit temperature $F(t)$, given by the formula

$$C(F(t)) = \frac{5}{9}(F(t) - 32)$$

To express the sample's Celsius temperature as a function of time, we find the composition function $C \circ F$.

$$\begin{aligned}(C \circ F)(t) &= C(F(t)) \\ &= C(3t + 15) && \text{Substitute } 3t + 15 \text{ for } F(t). \\ &= \frac{5}{9}[(3t + 15) - 32] && \text{Substitute } 3t + 15 \text{ for } F(t) \text{ in } C(F(t)). \\ &= \frac{5}{9}(3t - 17) && \text{Simplify.} \\ &= \frac{15}{9}t - \frac{85}{9} \\ &= \frac{5}{3}t - \frac{85}{9}\end{aligned}$$

■

Self Check Answers

1. $(g - f)(x) = -x - 4$ **2.** $(g/f)(x) = \frac{2x-3}{3x+1}, \left(-\infty, -\frac{1}{3}\right) \cup \left(-\frac{1}{3}, \infty\right)$ **3.** $(g - f)(x) = \sqrt{x} - x^2 + 4, [0, \infty)$ **4.** -15 **5.** One possibility is $f(x) = 2x^2$ and $g(x) = x^2 - 3x - 2$. **6.** $(f \circ h)(x) = 2x + 9$ **7.** $[0, \infty)$ **8.** $(-\infty, 0) \cup (0, 1) \cup (1, \infty)$; $(f \circ g)(x) = \frac{1}{1-x}$ **9. a.** $f(5)$ **b.** $f(5)$

3.6 Exercises

VOCABULARY AND CONCEPTS ***Fill in the blanks.***

1. $(f + g)(x) =$ $\underline{f(x) + g(x)}$

2. $(f - g)(x) =$ $\underline{f(x) - g(x)}$

3. $(f \cdot g)(x) =$ $\underline{f(x)g(x)}$

4. $(f/g)(x) =$ $\underline{f(x)/g(x)}$, where $g(x) \neq 0$.

5. The set of real numbers is the interval $\underline{(-\infty, \infty)}$.

6. $(f \circ g)(x) =$ $\underline{f(g(x))}$

7. $(g \circ f)(x) =$ $\underline{g(f(x))}$

8. $(I \circ g)(x) =$ $\underline{g(x)}$

9. The function $I(x) = x$ is called the $\underline{\text{identity}}$ function.

10. $(f \circ I)(x) = (I \circ f)(x) =$ $\underline{f(x)}$

PRACTICE ***Let $f(x) = 2x + 1$ and $g(x) = 3x - 2$. Find each function and its domain.***

11. $f + g$
$(f + g)(x) = 5x - 1$; $(-\infty, \infty)$

12. $f - g$
$(f - g)(x) = -x + 3$; $(-\infty, \infty)$

13. $f \cdot g$
$(f \cdot g)(x) = 6x^2 - x - 2$; $(-\infty, \infty)$

14. f/g
$(f/g)(x) = \frac{2x+1}{3x-2}$; $\left(-\infty, \frac{2}{3}\right) \cup \left(\frac{2}{3}, \infty\right)$

Let $f(x) = x^2 + x$ and $g(x) = x^2 - 1$. Find each function and its domain.

15. $f - g$
$(f - g)(x) = x + 1$; $(-\infty, \infty)$

16. $f + g$
$(f + g)(x) = 2x^2 + x - 1$; $(-\infty, \infty)$

17. f/g
$(f/g)(x) = \frac{x^2 + x}{x^2 - 1} = \frac{x}{x - 1}$; $(-\infty, -1) \cup (-1, 1) \cup (1, \infty)$

18. $f \cdot g$
$(f \cdot g)(x) = x^4 + x^3 - x^2 - x$; $(-\infty, \infty)$

Let $f(x) = x^2 - 1$ and $g(x) = 3x - 2$. Find each value, if possible.

19. $(f + g)(2)$ 7

20. $(f + g)(-3)$ −3

21. $(f - g)(0)$ 1

22. $(f - g)(-5)$ 41

23. $(f \cdot g)(2)$ 12

24. $(f \cdot g)(-1)$ 0

25. $(f/g)\left(\frac{2}{3}\right)$ no value

26. $(f/g)(t)$ $\frac{t^2 - 1}{3t - 2}$

Find two functions f and g such that h(x) can be expressed as the function indicated. Several answers are possible.

27. $h(x) = 3x^2 + 2x$; $f + g$ $f(x) = 3x^2$; $g(x) = 2x$

28. $h(x) = 3x^2$; $f \cdot g$ $f(x) = 3$; $g(x) = x^2$

29. $h(x) = \frac{3x^2}{x^2 - 1}$; f/g $f(x) = 3x^2$; $g(x) = x^2 - 1$

30. $h(x) = 5x + x^2$; $f - g$ $f(x) = 5x$; $g(x) = -x^2$

31. $h(x) = x(3x^2 + 1)$; $f - g$ $f(x) = 3x^3$; $g(x) = -x$

32. $h(x) = (3x - 2)(3x + 2)$; $f + g$
$f(x) = 9x^2$; $g(x) = -4$

33. $h(x) = x^2 + 7x - 18$; $f \cdot g$
$f(x) = x + 9$; $g(x) = x - 2$

34. $h(x) = 5x^5$; f/g $f(x) = 5x^6$; $g(x) = x$

Let $f(x) = 2x - 5$ and $g(x) = 5x - 2$. Find each value.

35. $(f \circ g)(2)$ 11

36. $(g \circ f)(-3)$ −57

37. $(f \circ f)\left(-\frac{1}{2}\right)$ −17

38. $(g \circ g)\left(\frac{3}{5}\right)$ 3

Let $f(x) = 3x^2 - 2$ and $g(x) = 4x + 4$. Find each value.

39. $(f \circ g)(-3)$ 190

40. $(g \circ f)(3)$ 104

41. $(f \circ f)(\sqrt{3})$ 145

42. $(g \circ g)(-4)$ −44

Let $f(x) = 3x$ and $g(x) = x + 1$. Determine the domain of each composite function and then find the composite function.

43. $f \circ g$
$(-\infty, \infty)$;
$(f \circ g)(x) = 3x + 3$

44. $g \circ f$
$(-\infty, \infty)$;
$(g \circ f)(x) = 3x + 1$

45. $f \circ f$
$(-\infty, \infty)$;
$(f \circ f)(x) = 9x$

46. $g \circ g$
$(-\infty, \infty)$;
$(g \circ g)(x) = x + 2$

Let $f(x) = x^2$ and $g(x) = 2x$. Determine the domain of each composite function and then find the composite function.

47. $g \circ f$
$(-\infty, \infty)$; $(g \circ f)(x) = 2x^2$

48. $f \circ g$
$(-\infty, \infty)$; $(f \circ g)(x) = 4x^2$

49. $g \circ g$
$(-\infty, \infty)$; $(g \circ g)(x) = 4x$

50. $f \circ f$
$(-\infty, \infty)$; $(f \circ f)(x) = x^4$

Let $f(x) = \sqrt{x}$ and $g(x) = x + 1$. Determine the domain of each composite function and then find the composite function.

51. $f \circ g$
$[-1, \infty)$;
$(f \circ g)(x) = \sqrt{x + 1}$

52. $g \circ f$
$[0, \infty)$; $(g \circ f)(x) =$
$\sqrt{x} + 1$

53. $f \circ f$
$[0, \infty)$;
$(f \circ f)(x) = \sqrt[4]{x}$

54. $g \circ g$
$(-\infty, \infty)$;
$(g \circ g)(x) = x + 2$

Let $f(x) = \sqrt{x + 1}$ and $g(x) = x^2 - 1$. Determine the domain of each composite function and then find the composite function.

55. $g \circ f$
$[-1, \infty)$; $(g \circ f)(x) = x$

56. $f \circ g$
$(-\infty, \infty)$;
$(f \circ g)(x) = \sqrt{x^2} = |x|$

57. $g \circ g$
$(-\infty, \infty)$;
$(g \circ g)(x) = x^4 - 2x^2$

58. $f \circ f$
$[-1, \infty)$;
$(f \circ f)(x) =$
$\sqrt{\sqrt{x + 1} + 1}$

Let $f(x) = \frac{1}{x - 1}$ and $g(x) = \frac{1}{x - 2}$. Determine the domain of each composite function and then find the composite function.

59. $f \circ g$
$(-\infty, 2) \cup (2, 3) \cup (3, \infty)$;
$(f \circ g)(x) = \frac{x - 2}{3 - x}$

60. $g \circ f$
$(-\infty, 1) \cup \left(1, \frac{3}{2}\right) \cup \left(\frac{3}{2}, \infty\right)$;
$(g \circ f)(x) = \frac{x - 1}{3 - 2x}$

61. $f \circ f$
$(-\infty, 1) \cup (1, 2) \cup (2, \infty)$;
$(f \circ f)(x) = \frac{x - 1}{2 - x}$

62. $g \circ g$
$(-\infty, 2) \cup \left(2, \frac{5}{2}\right) \cup \left(\frac{5}{2}, \infty\right)$;
$(g \circ g)(x) = \frac{x - 2}{5 - 2x}$

Find two functions f and g such that the composition $f \circ g$ expresses the given correspondence. Several answers are possible.

63. $y = 3x - 2$
$f(x) = x - 2$; $g(x) = 3x$

64. $y = 7x - 5$
$f(x) = x - 5$; $g(x) = 7x$

65. $y = x^2 - 2$
$f(x) = x - 2$; $g(x) = x^2$

66. $y = x^3 - 3$
$f(x) = x - 3$; $g(x) = x^3$

67. $y = (x - 2)^2$
$f(x) = x^2$; $g(x) = x - 2$

68. $y = (x - 3)^3$
$f(x) = x^3$; $g(x) = x - 3$

69. $y = \sqrt{x + 2}$
$f(x) = \sqrt{x}$; $g(x) = x + 2$

70. $y = \frac{1}{x - 5}$
$f(x) = \frac{1}{x}$; $g(x) = x - 5$

71. $y = \sqrt{x} + 2$
$f(x) = x + 2$; $g(x) = \sqrt{x}$

72. $y = \frac{1}{x} - 5$
$f(x) = x - 5$; $g(x) = \frac{1}{x}$

73. $y = x$
$f(x) = x$; $g(x) = x$

74. $y = 3$
$f(x) = 3$; $g(x) = x$

APPLICATIONS

75. Picture tubes Refer to the television picture tube shown.

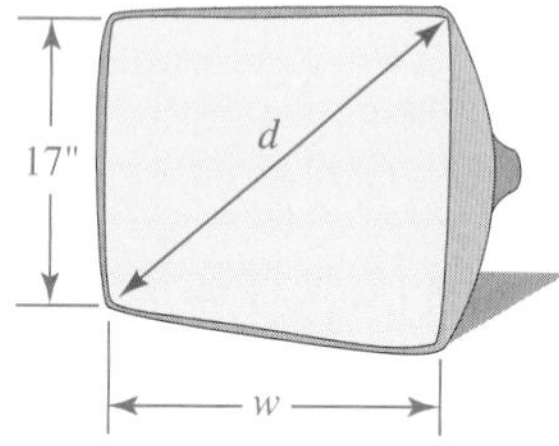

a. Write a formula to find the area of the tube.
$A = 17w$

b. Use the Pythagorean theorem to write a formula to find the width w of the tube.
$w = \sqrt{d^2 - 289}$

c. Write a formula to find the area of the tube as a function of the diagonal d. $A = 17\sqrt{d^2 - 289}$

76. Area of a square Write a formula for the area of a square in terms of its perimeter. $A = \frac{p^2}{16}$

77. Perimeter of a square Write a formula for the perimeter of a square in terms of its area.
$P = 4\sqrt{A}$

78. Ceramics When the temperature of a pot in a kiln is 1,200° F, an artist turns off the heat and leaves the pot to cool at a controlled rate of 81° F per hour. Express the temperature of the pot in degrees Celsius as a function of the time t (in hours) since the kiln was turned off. $(C \circ F)(t) = -45t + \frac{5{,}840}{9}$

DISCOVERY AND WRITING

79. Let $f(x) = 3x$. Show that $(f + f)(x) = f(x + x)$.

80. Let $g(x) = x^2$. Show that $(g + g)(x) \neq g(x + x)$.

81. Let $f(x) = \dfrac{x - 1}{x + 1}$. Find $(f \circ f)(x)$.

82. Let $g(x) = \dfrac{x}{x - 1}$. Find $(g \circ g)(x)$.

Let $f(x) = x^2 - x$, $g(x) = x - 3$, and $h(x) = 3x$. Use a graphing calculator to graph both functions on the same axis. Write a brief paragraph summarizing your observations.

83. f and $f \circ g$

84. f and $g \circ f$

85. f and $f \circ h$

86. f and $h \circ f$

REVIEW *Solve each equation for y.*

87. $x = 3y - 7$ $y = \frac{x + 7}{3}$

88. $x = \dfrac{7}{y}$ $y = \frac{7}{x}$

89. $x = \dfrac{y}{y + 3}$ $y = \frac{3x}{1 - x}$

90. $x = \dfrac{y - 1}{y}$ $y = \frac{1}{1 - x}$

3.7 Inverse Functions

In this section, you will learn about

- One-to-One Functions
- The Horizontal Line Test
- Finding the Inverse of a One-to-One Function
- The Relationship between the Graphs of f and f^{-1}

The linear function defined by $C = \frac{5}{9}(F - 32)$ gives a formula to convert degrees Fahrenheit to degrees Celsius. If we substitute a Fahrenheit reading into the formula, a Celsius reading comes out. For example, if we substitute 41° for F, we obtain a Celsius reading of 5°:

$$C = \frac{5}{9}(F - 32)$$

$$= \frac{5}{9}(41 - 32)$$

$$= \frac{5}{9}(9)$$
$$= 5$$

If we want to find a Fahrenheit reading from a Celsius reading, we need a formula into which we can substitute a Celsius reading and have a Fahrenheit reading come out. Such a formula is $F = \frac{9}{5}C + 32$, which takes the Celsius reading of 5° and turns it back into a Fahrenheit reading of 41°.

$$F = \frac{9}{5}C + 32$$
$$= \frac{9}{5}(5) + 32$$
$$= 41$$

The functions defined by these two formulas do opposite things. The first turns 41° F into 5° Celsius, and the second turns 5° Celsius back into 41° F. Such functions are called *inverse functions.*

Some functions have inverses that are functions and some do not. To guarantee that the inverse of a function will also be a function, we will only find inverses of one-to-one functions.

One-to-One Functions

Recall that each element x in the domain of a function has a single output y. For some functions, different numbers x in the domain can have the same output. (See Figure 3-56(a).) For other functions, called **one-to-one functions,** different numbers x have different outputs. (See Figure 3-56(b).)

One-to-One Functions

A function f from a set **X** to a set **Y** is called **one-to-one** if and only if different numbers in the domain of f have different outputs in the range of f.

The previous definition implies that if x_1 and x_2 are two numbers in the domain of f and $x_1 \neq x_2$, then $f(x_1) \neq f(x_2)$.

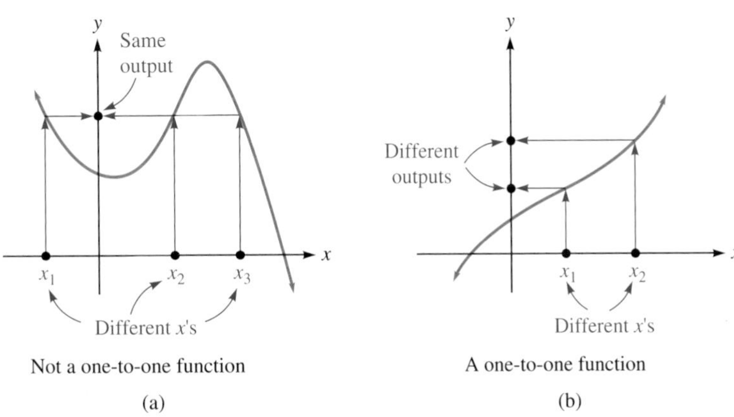

Figure 3-56

EXAMPLE 1 Determine whether the functions **a.** $f(x) = x^4 + x^2$ and **b.** $f(x) = x^3$ are one-to-one.

Solution **a.** The function $f(x) = x^4 + x^2$ is not one-to-one, because different numbers in the domain have the same output. For example, 2 and -2 have the same output: $f(2) = f(-2) = 20$.

b. The function $f(x) = x^3$ is one-to-one, because different numbers x produce different outputs $f(x)$. This is because different numbers have different cubes.

Self Check Determine whether $f(x) = \sqrt{x}$ is one-to-one. ■

■ The Horizontal Line Test

A **horizontal line test** can be used to determine whether the graph of a function represents a one-to-one function. If every horizontal line that intersects the graph of a function does so only once, the function is one-to-one. (See Figure 3-57(a).) If any horizontal line intersects the graph of a function more than once, the function is not one-to-one. (See Figure 3-57(b).)

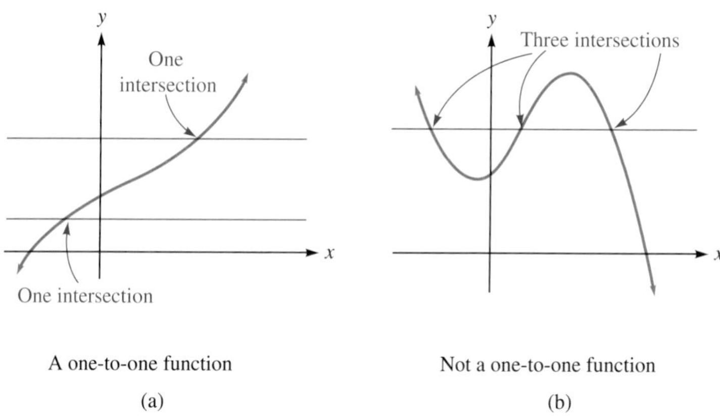

A one-to-one function

(a)

Not a one-to-one function

(b)

Figure 3-57

Figure 3-58(a) illustrates a function f from set **X** to set **Y.** Since three arrows point to a single y, the function f is not one-to-one. If the arrows in Figure 3-58(a) were reversed, the diagram would not represent a function, because three values of x in set **X** would correspond to a value y in set **Y**.

If the arrows of the one-to-one function f in Figure 3-58(b) were reversed, as in Figure 3-58(c), the diagram would represent a function. This function is called the **inverse of function *f*** and is denoted by the symbol f^{-1}.

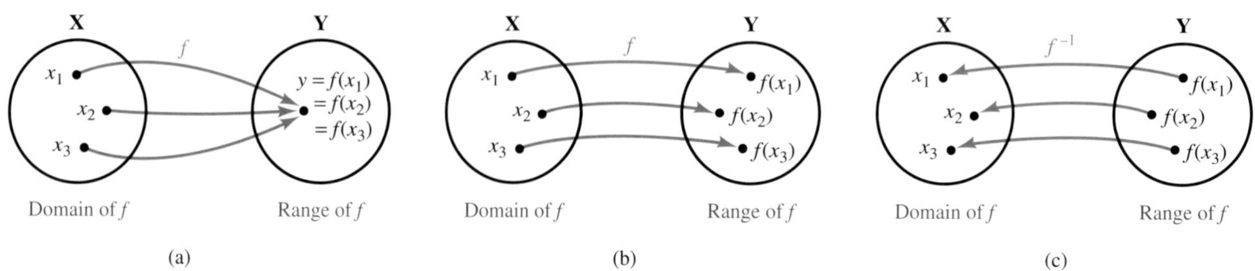

Figure 3-58

Inverse Functions

If f is a one-to-one function with domain **X** and range **Y**, then f^{-1}, called the **inverse function of f**, has domain **Y** and range **X**. The function f^{-1} defined by

$$f^{-1}(y) = x \qquad \text{is equivalent to} \qquad f(x) = y$$

for all y in **Y**.

Comment The -1 in the notation for inverse function is not an exponent. Remember that

$$f^{-1}(x) \neq \frac{1}{f(x)}$$

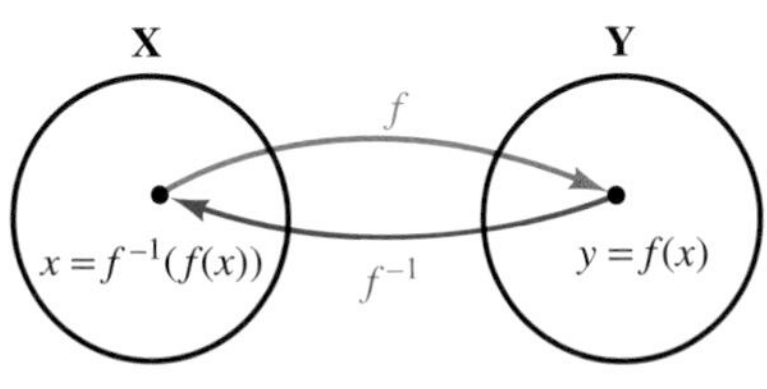

Figure 3-59

Figure 3-59 shows a one-to-one function f and its inverse f^{-1}. To the number x in the domain of f, there corresponds an output $f(x)$ in the range of f. Since $f(x)$ is in the domain of f^{-1}, the output for $f(x)$ under the function f^{-1} is $f^{-1}(f(x)) = x$. Thus,

$$(f^{-1} \circ f)(x) = f^{-1}(f(x)) = x$$

Furthermore, if y is any number in the domain of f^{-1}, then

$$(f \circ f^{-1})(y) = f(f^{-1}(y)) = y$$

Properties of a One-to-One Function

If f is a one-to-one function with domain **X** and range **Y**, there is a one-to-one function f^{-1} with domain **Y** and range **X** such that

$$(f^{-1} \circ f)(x) = x \qquad \text{and} \qquad (f \circ f^{-1})(y) = y$$

To show that one function is the inverse of another, we must show that their composition is the *identity function.*

EXAMPLE 2 Show that $f(x) = x^3$ is the inverse function of $g(x) = \sqrt[3]{x}$.

Solution To show that f is the inverse function of g, we must show that $f \circ g$ and $g \circ f$ are the identity function.

$$(f \circ g)(x) = f(g(x)) = f\left(\sqrt[3]{x}\right) = \left(\sqrt[3]{x}\right)^3 = x$$
$$(g \circ f)(x) = g(f(x)) = g(x^3) = \sqrt[3]{x^3} = x$$

Self Check If $x \geq 0$, is $f(x) = x^2$ the inverse of $g(x) = \sqrt{x}$? ■

■ Finding the Inverse of a One-to-One Function

If f is the one-to-one function $y = f(x)$, then f^{-1} reverses the correspondence of f. That is, if $f(a) = b$, then $f^{-1}(b) = a$. To determine f^{-1}, we follow these steps.

Strategy for Finding f^{-1}

1. Write the function in the form $y = f(x)$.
2. Interchange the positions of variables x and y in the equation $y = f(x)$. The resulting equation, $x = f(y)$, defines the inverse function f^{-1}.
3. Solve the equation $x = f(y)$ for y, if possible. The result is $y = f^{-1}(x)$.

ACTIVE EXAMPLE 3 Find the inverse of $y = f(x) = \frac{3}{2}x + 2$ and verify the result.

Solution Since the function is written in the form $y = f(x)$, we can find the inverse function by interchanging the positions of x and y.

COLLEGE **Algebra** $f(x)$ **Now**™
Explore this Active Example and test yourself on the steps taken here at **http://1pass.thomson.com**. You can also find this example on the Interactive Video Skillbuilder CD-ROM.

$$x = \frac{3}{2}y + 2$$

To write the inverse function in the form $y = f^{-1}(x)$, we solve the equation for y.

$$\begin{aligned} x &= \frac{3}{2}y + 2 \\ 2x &= 3y + 4 && \text{Multiply both sides by 2.} \\ 2x - 4 &= 3y && \text{Subtract 4 from both sides.} \\ y &= \frac{2x - 4}{3} && \text{Divide both sides by 3.} \end{aligned}$$

The inverse of $y = f(x) = \frac{3}{2}x + 2$ is $y = f^{-1}(x) = \frac{2x - 4}{3}$.

To verify that

$$f(x) = \frac{3}{2}x + 2 \qquad \text{and} \qquad f^{-1}(x) = \frac{2x - 4}{3}$$

are inverses, we must show that $f \circ g$ and $g \circ f$ are the identity function.

$$\begin{aligned} (f \circ f^{-1})(x) &= f(f^{-1}(x)) \\ &= f\left(\frac{2x - 4}{3}\right) \\ &= \frac{3}{2}\left(\frac{2x - 4}{3}\right) + 2 \\ &= x - 2 + 2 \\ &= x \end{aligned}$$

$$\begin{aligned} (f^{-1} \circ f)(x) &= f^{-1}(f(x)) \\ &= f^{-1}\left(\frac{3}{2}x + 2\right) \\ &= \frac{2\left(\frac{3}{2}x + 2\right) - 4}{3} \\ &= \frac{3x + 4 - 4}{3} \\ &= x \end{aligned}$$

Self Check Find $f(2)$. Then find $f^{-1}(5)$. Explain the significance of the results. ■

■ The Relationship between the Graphs of f and f^{-1}

Because we interchange the positions of x and y to find the inverse of a function, the point (b, a) lies on the graph of $y = f^{-1}(x)$ whenever the point (a, b) lies on the graph of $y = f(x)$. Thus, the graph of a function and its inverse are reflections of each other about the line $y = x$.

EXAMPLE 4 Find the inverse of $y = f(x) = x^3 + 3$. Graph the function and its inverse on the same set of coordinate axes.

Solution We find the inverse of the function $y = f(x) = x^3 + 3$ by interchanging x and y.

$$x = y^3 + 3 \qquad \text{The inverse function.}$$

To write the inverse in the form $f^{-1}(x)$, we solve the equation for y to get

$$x - 3 = y^3$$
$$y = \sqrt[3]{x - 3}$$

Thus, $f^{-1}(x) = \sqrt[3]{x - 3}$. In the graph that appears in Figure 3-60, the line $y = x$ is the axis of symmetry.

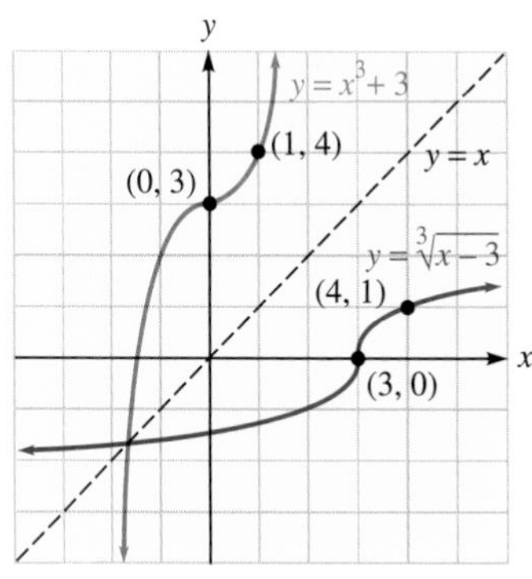

Figure 3-60

Self Check Find $f(2)$. Then find $f^{-1}(11)$. Explain the significance of the result. ■

ACTIVE EXAMPLE 5 The function $y = f(x) = x^2 + 3$ is not one-to-one. However, it becomes one-to-one when we restrict its domain to the interval $(-\infty, 0]$. Under this restriction: **a.** Find the range of f. **b.** Find the inverse of f along with its domain and range. **c.** Graph each function.

COLLEGE **Algebra** f(x) **Now**™
Go to **http://1pass.thomson.com** or your CD to practice this example.

Solution **a.** The function f is defined by $y = f(x) = x^2 + 3$, with domain $(-\infty, 0]$. If x is replaced with numbers in this interval, y ranges over the values 3 and above. (See Figure 3-61.) Thus, the range of f is the interval $[3, \infty)$.

b. To find the inverse of f, we interchange x and y in the equation that defines f in the restricted domain. Then solve for y.

$$y = x^2 + 3 \quad (x \le 0)$$
$$x = y^2 + 3 \quad (y \le 0) \quad \text{Interchange } x \text{ and } y.$$
$$x - 3 = y^2 \quad (y \le 0) \quad \text{Subtract 3 from both sides.}$$

To solve this equation for y, we take the square root of both sides. Because $y \le 0$, we have

$$-\sqrt{x - 3} = y \quad (y \le 0)$$

The inverse of f is defined by

$$y = f^{-1}(x) = -\sqrt{x - 3}$$

It has domain $[3, \infty)$ and range $(-\infty, 0]$. (See Figure 3-61.)

c. The graphs of the functions appear in Figure 3-61. Note that the line of symmetry is $y = x$.

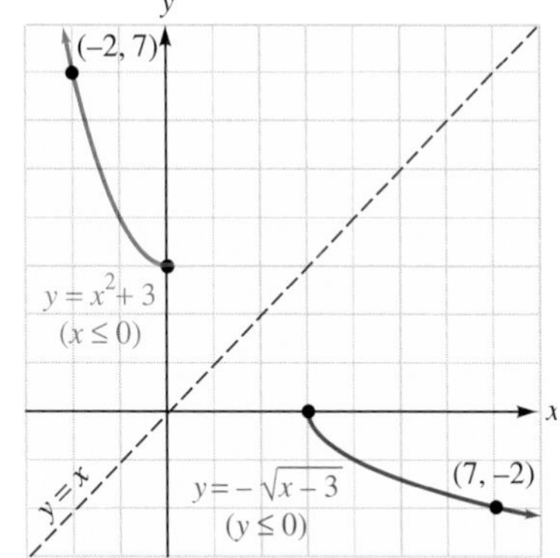

Figure 3-61

Self Check Find the inverse of f when its domain is restricted to the interval $[0, \infty)$. ■

If a function is defined by the equation $y = f(x)$, we can often find the domain of f by inspection. Finding the range can be more difficult. One way to find the range of f is to find the domain of f^{-1}.

ACTIVE EXAMPLE 6 Find the domain and range of $y = f(x) = \dfrac{2}{x} + 3$.

Solution Because x cannot be 0, the domain of f is $(-\infty, 0) \cup (0, \infty)$.

To find the range of f, we find the domain of f^{-1} as follows. (Remember that no denominators can be 0.)

COLLEGE **Algebra** $f(x)$ **Now**™
Go to **http://1pass.thomson.com** or your CD to practice this example.

$$y = \frac{2}{x} + 3$$

$$x = \frac{2}{y} + 3 \quad \text{Interchange } x \text{ and } y.$$

$$xy = 2 + 3y \quad \text{Multiply both sides by } y.$$

$$xy - 3y = 2 \quad \text{Subtract } 3y \text{ from both sides.}$$

$$y(x - 3) = 2 \quad \text{Factor out } y.$$

$$y = \frac{2}{x - 3} \quad \text{Divide both sides by } x - 3.$$

This final equation defines f^{-1} whose domain is $(-\infty, 3) \cup (3, \infty)$. Because the range of f is the domain of f^{-1}, the range of f is $(-\infty, 3) \cup (3, \infty)$.

Self Check Find the range of $y = f(x) = \frac{3}{x} - 1$. ■

Self Check Answers

1. yes **2.** yes **3.** 5, 2 **4.** 11, 2 **5.** $y = f^{-1}(x) = \sqrt{x - 3}$ **6.** $(-\infty, -1) \cup (-1, \infty)$

3.7 Exercises

VOCABULARY AND CONCEPTS *Fill in the blanks.*

1. If different numbers in the domain of a function have different outputs, the function is called a one-to-one function.
2. If every horizontal line intersects the graph of a function only once, the function is one-to-one.
3. To find the inverse of $y = f(x)$, we interchange the variables x and y.
4. The graph of a function and its inverse are reflections of each other about the line $y = x$.

PRACTICE *Determine whether each function is one-to-one.*

5. $y = 3x$ one-to-one
6. $y = \frac{1}{2}x$ one-to-one
7. $y = x^2 + 3$ not one-to-one
8. $y = x^4 - x^2$ not one-to-one
9. $y = x^3 - x$ not one-to-one
10. $y = x^2 - x$ not one-to-one
11. $y = |x|$ not one-to-one
12. $y = |x - 3|$ not one-to-one
13. $y = 5$ not one-to-one
14. $y = \sqrt{x - 5}$ one-to-one
15. $y = (x - 2)^2; x \geq 2$ one-to-one
16. $y = \frac{1}{x}$ one-to-one

Use the horizontal line test to determine whether each graph represents a one-to-one function.

17.
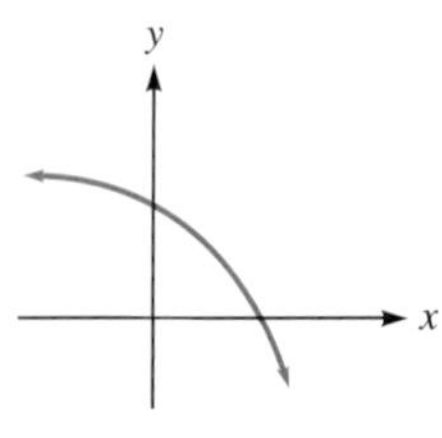

one-to-one

18.
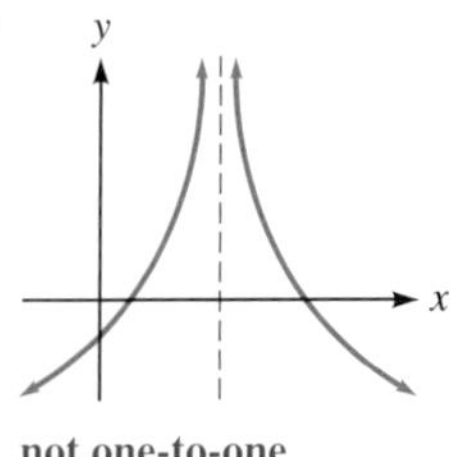

not one-to-one

19.

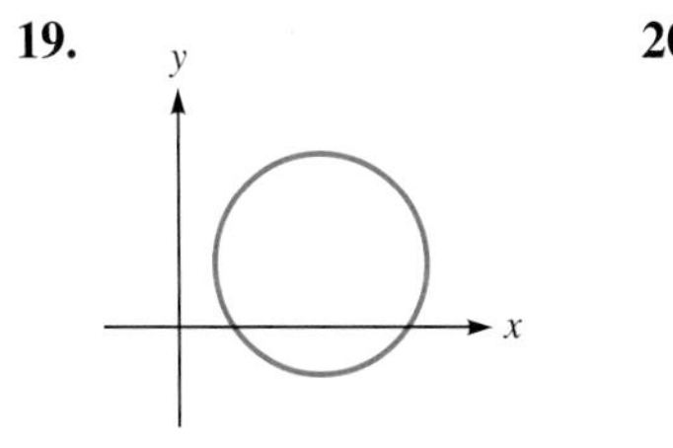

not a function

20.

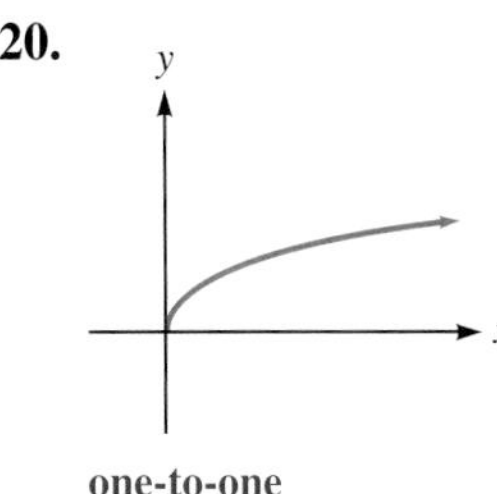

one-to-one

Verify that the functions are inverses by showing that $f \circ g$ and $g \circ f$ are identity functions.

21. $f(x) = 5x$ and $g(x) = \frac{1}{5}x$

22. $f(x) = 4x + 5$ and $g(x) = \frac{x - 5}{4}$

23. $f(x) = \frac{x + 1}{x}$ and $g(x) = \frac{1}{x - 1}$

24. $f(x) = \frac{x + 1}{x - 1}$ and $g(x) = \frac{x + 1}{x - 1}$

Each equation defines a one-to-one function f. Determine f^{-1} and verify that $f \circ f^{-1}$ and $f^{-1} \circ f$ are the identity function.

25. $y = 3x$ $f^{-1}(x) = \frac{1}{3}x$

26. $y = \frac{1}{3}x$ $f^{-1}(x) = 3x$

27. $y = 3x + 2$
$f^{-1}(x) = \frac{x - 2}{3}$

28. $y = 2x - 5$
$f^{-1}(x) = \frac{x + 5}{2}$

29. $y = \frac{1}{x + 3}$
$f^{-1}(x) = \frac{1}{x} - 3$

30. $y = \frac{1}{x - 2}$
$f^{-1}(x) = \frac{1}{x} + 2$

31. $y = \frac{1}{2x}$ $f^{-1}(x) = \frac{1}{2x}$

32. $y = \frac{1}{x^3}$ $f^{-1}(x) = \sqrt[3]{\frac{1}{x}}$

Find the inverse of each one-to-one function and graph both the function and its inverse on the same set of coordinate axes.

33. $y = 5x$

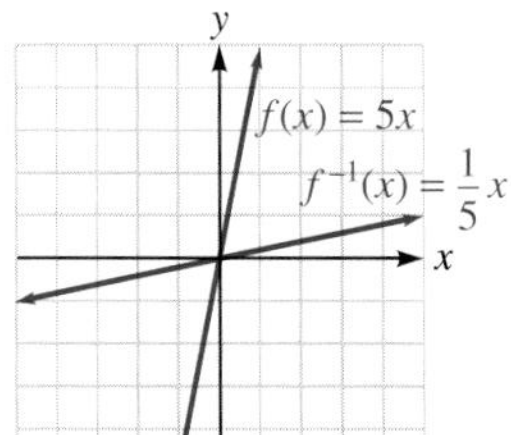

34. $y = \frac{3}{2}x$

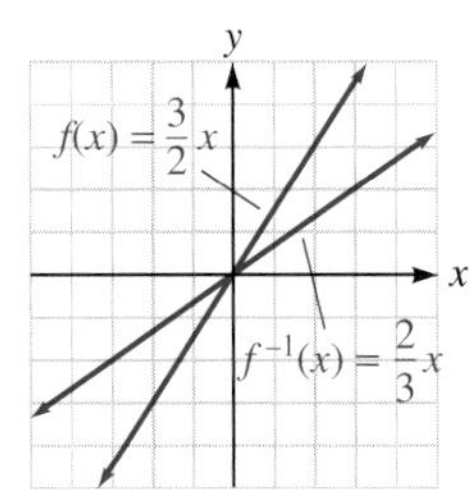

35. $y = 2x - 4$

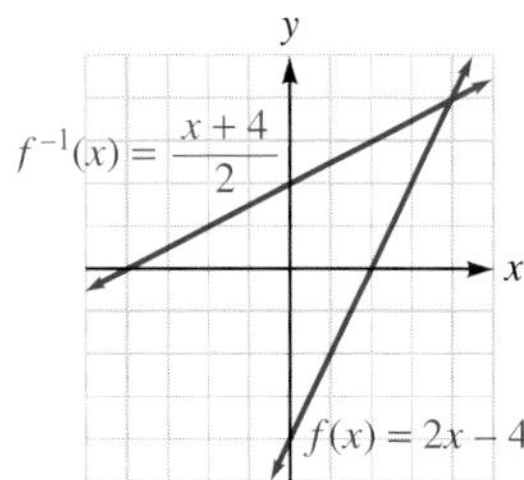

36. $y = \frac{3}{2}x - 2$

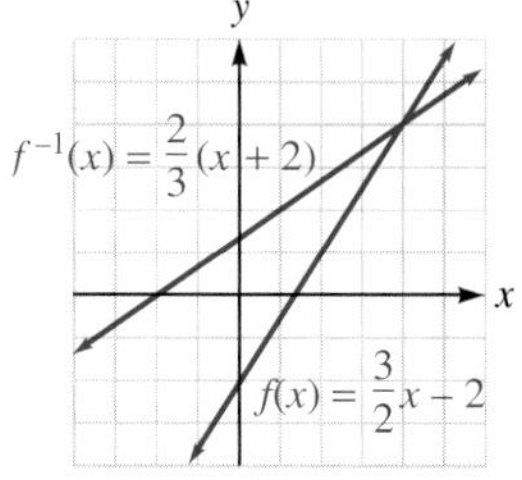

37. $x - y = 2$

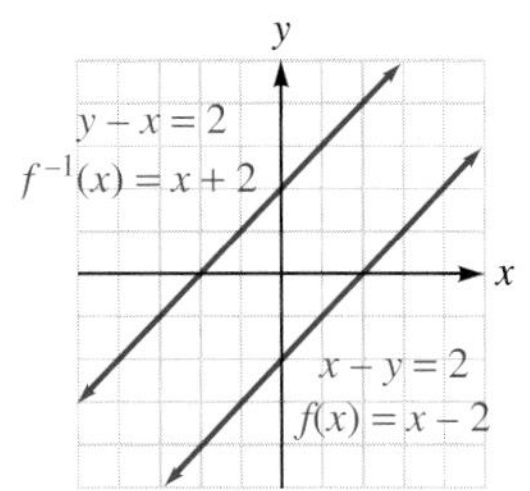

38. $x + y = 0$

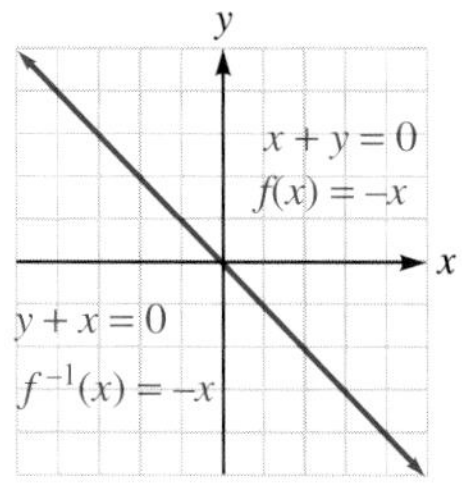

39. $2x + y = 4$

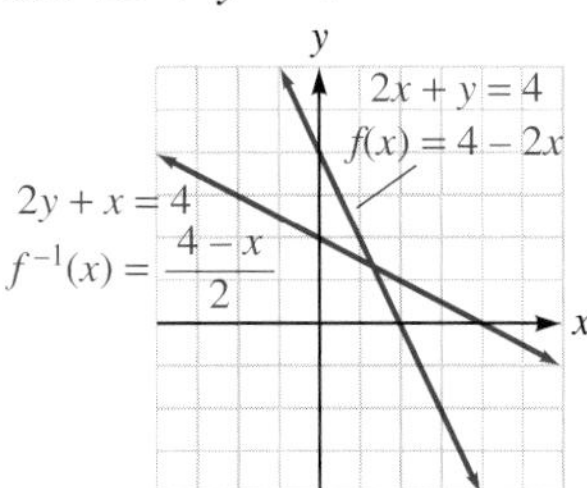

40. $3x + 2y = 6$

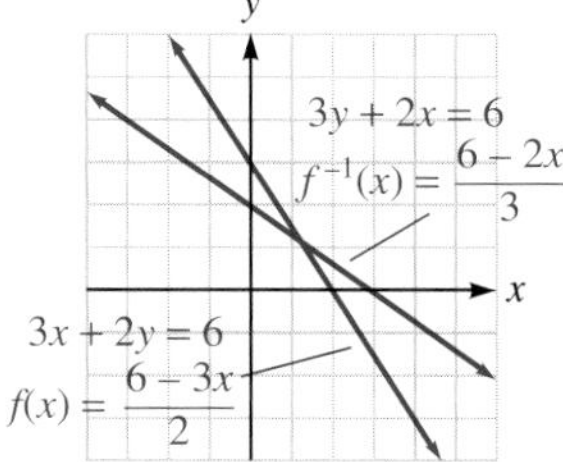

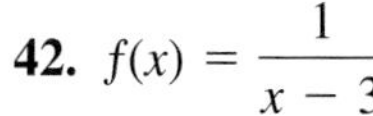

41. $f(x) = \frac{1}{2x}$

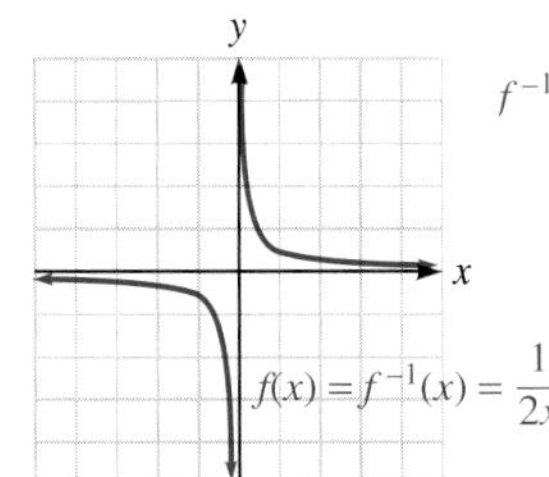

42. $f(x) = \frac{1}{x - 3}$

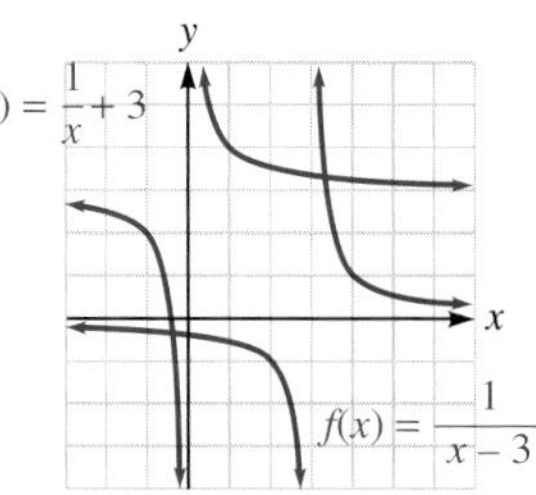

43. $f(x) = \frac{x + 1}{x - 1}$

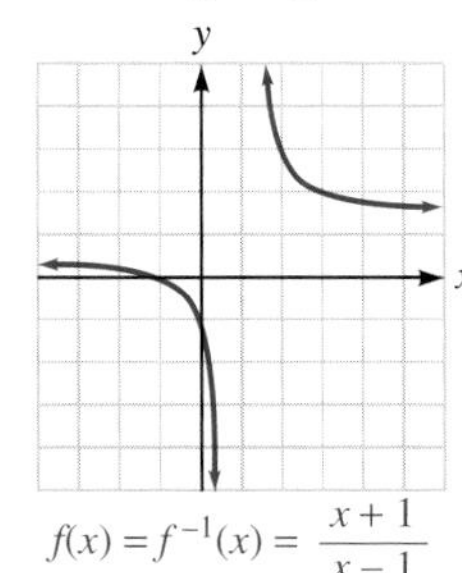

44. $f(x) = \frac{x - 1}{x}$

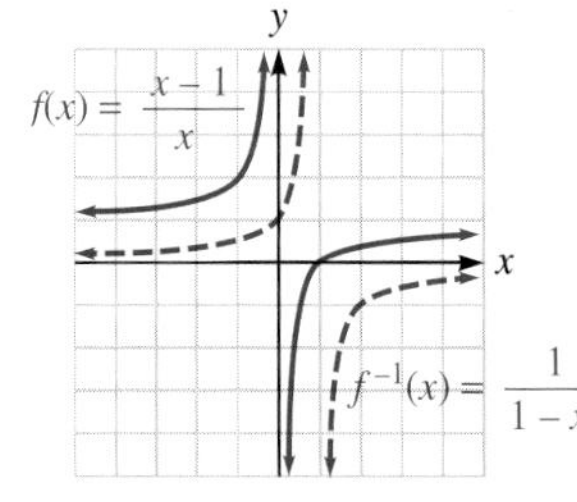

The function f defined by the given equation is one-to-one on the given domain. Find $f^{-1}(x)$.

45. $f(x) = x^2 - 3 \quad (x \le 0)$
$f^{-1}(x) = -\sqrt{x+3} \quad (x \ge -3)$

46. $f(x) = \dfrac{1}{x^2} \quad (x > 0) \qquad f^{-1}(x) = \sqrt{\frac{1}{x}} \quad (x > 0)$

47. $f(x) = x^4 - 8 \quad (x \ge 0)$
$f^{-1}(x) = \sqrt[4]{x+8} \quad (x \ge -8)$

48. $f(x) = \dfrac{-1}{x^4} \quad (x < 0) \qquad f^{-1}(x) = -\sqrt[4]{-\frac{1}{x}} \quad (x < 0)$

49. $f(x) = \sqrt{4 - x^2} \quad (0 \le x \le 2)$
$f^{-1}(x) = \sqrt{4 - x^2} \quad (0 \le x \le 2)$

50. $f(x) = \sqrt{x^2 - 1} \quad (x \le -1)$
$f^{-1}(x) = -\sqrt{x^2 + 1} \quad (x \ge 0)$

Find the domain and the range of f. Find the range by finding the domain of f^{-1}.

51. $f(x) = \dfrac{x}{x-2}$
domain: $(-\infty, 2) \cup (2, \infty)$; range: $(-\infty, 1) \cup (1, \infty)$

52. $f(x) = \dfrac{x-2}{x+3}$
domain: $(-\infty, -3) \cup (-3, \infty)$; range: $(-\infty, 1) \cup (1, \infty)$

53. $f(x) = \dfrac{1}{x} - 2$
domain: $(-\infty, 0) \cup (0, \infty)$; range: $(-\infty, -2) \cup (-2, \infty)$

54. $f(x) = \dfrac{3}{x} - \dfrac{1}{2}$
domain: $(-\infty, 0) \cup (0, \infty)$; range: $\left(-\infty, -\frac{1}{2}\right) \cup \left(-\frac{1}{2}, \infty\right)$

APPLICATIONS

55. Buying pizza A pizzeria charges \$8.50 plus 75¢ per topping for a large pizza.

a. Find a linear function that expresses the cost y of a large pizza in terms of the number of toppings x. $y = 0.75x + 8.50$

b. Find the cost of a pizza that has four toppings. \$11.50

c. Find the inverse of the function found in part a, to find a formula that gives the number of toppings y in terms of the cost x. $y = \frac{x - 8.50}{0.75}$

d. If Josh has \$10, how many toppings can he afford? 2

56. Phone bills A phone company charges \$11 per month plus a nickel per call.

a. Find a rational function that expresses the average cost y of a call in a month when x calls were made. $y = \frac{0.05x + 11}{x}$

b. To the nearest tenth of a cent, find the average cost of a call in a month when 68 calls were made. 21.2¢

c. Find the inverse of the function found in part a, to find a formula that gives the number of calls y that can be made for an average cost x. $y = \frac{11}{x - 0.05}$

d. How many calls need to be made for an average cost of 15¢ per call? 110

DISCOVERY AND WRITING

57. Write a brief paragraph to explain why the range of f is the domain of f^{-1}.

58. Write a brief paragraph to explain why the graphs of a function and its inverse are reflections in the line $y = x$.

Use a graphing calculator to graph each function for various values of a.

59. For what values of a is $f(x) = x^3 + ax$ a one-to-one function? $a \ge 0$

60. For what values of a is $f(x) = x^3 + ax^2$ a one-to-one function? $a = 0$

REVIEW *Simplify each expression.*

61. $16^{3/4}$ 8

62. $25^{-1/2}$ $\frac{1}{5}$

63. $(-8)^{2/3}$ 4

64. $-8^{2/3}$ -4

65. $\left(\dfrac{64}{125}\right)^{-1/3}$ $\frac{5}{4}$

66. $49^{3/2}$ 343

67. $49^{-1/2}$ $\frac{1}{7}$

68. $\left(\dfrac{9}{25}\right)^{-3/2}$ $\frac{125}{27}$

PROBLEMS AND PROJECTS

1. To encourage their sales staffs, managers of two jewelry stores offer bonuses that are functions of weekly sales, as shown in the table. For each plan, draw a graph, using the same set of axes. Place the earned bonus on the vertical axis and sales on the horizontal axis.

Sales	**Bonus**	
	Golden Gifts	**Glitter and Co.**
Sales < \$3,000	1% of sales	zero
\$3,000 ≤ sales < \$10,000	3% of sales	5% of sales over \$3,000
\$10,000 ≤ sales	5% of sales	\$350, plus 7% of sales over \$10,000

2. Which plan in Problem 1 do you think is more fair? For any given level of sales, is one plan consistently better than the other? Which provides the greater incentive to sell more? Write brief paragraphs to explain your choices.

3. For the **modular function** $y = \text{mod}(x)$, define y to be the remainder obtained when the integer x is divided by 5. For example, $\text{mod}(9) = 4$, because when 9 is divided by 5, the remainder is 4. Similarly, $\text{mod}(15) = 0$, and $\text{mod}(3) = 3$.

   Graph this function. (*Hint:* Your graph will consist of unconnected points. What are the domain and the range of the function?)

4. Look at your graph of the modular function in Problem 3. How would you extend the domain to negative values of x? How does your choice fit with the definition "remainder when x is divided by 5"?

Project 1

Understanding the behavior of a liquid flowing in a pipe has applications in many systems, including home plumbing, oil pipelines, veins and arteries, and hydraulic control lines in aircraft. The *velocity profile* in the illustration shows that fluid next to the pipe wall is moving slowly, but nearer the center of the pipe, flow becomes more rapid.

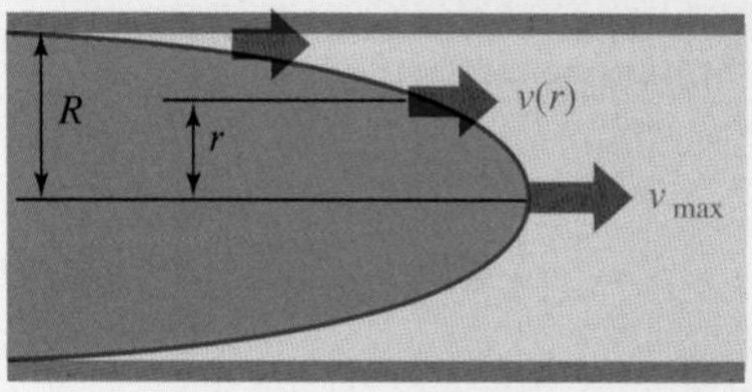

The velocity at a distance r from the pipe's center is given by

$$v(r) = v_{\max}\left(1 - \frac{r}{R}\right)^{1/k}$$

where R is the radius of the pipe, $v_{\max}$ is the greatest velocity, and k depends on properties of the fluid. For one particular fluid in a 2-inch-diameter pipe, $v_{\max} = 8$ and $k = 10$.

a. Graph the function. In the context of this application, what is the domain of the function? (*Hint:* Restrictions on r come from the pipe, not the equation.)
b. Where is the velocity greatest? Does your graph show that this greatest velocity is $v_{\max}$?
c. Graph the function again for $k = 4$. Which graph describes the flow of a thick, syrupy fluid, and which describes a fluid more like water? Why?

Project 2

The illustration shows the geometric basis of the algebraic process of completing the square. Study the figure, where each algebraic expression is related to the area of the corresponding geometric figure. What is the area of the missing piece that will "complete the square"?

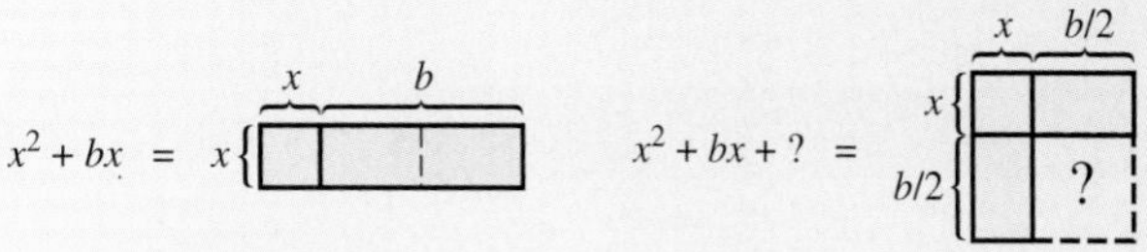

Draw a similar figure that shows geometrically how to complete the square on the expression $x^2 + 8x$.

Project 3

The surface area of a sphere is a function of the radius, given by

(1) $A = 4\pi r^2$

Its volume is also a function of the radius, given by

(2) $V = \frac{4}{3}\pi r^3$

a. Express the surface area of a sphere as a function of its volume. (*Hint:* Solve Equation 2 for r and substitute into Equation 1.)

b. Use a graphing calculator to graph the function you found in part a.

c. Insects don't have lungs; they breathe through their skin. Oxygen requirements increase with an insect's volume, but oxygen intake depends on the insect's surface area. Refer to your graph and write a paragraph explaining why insects must be limited in size.

CHAPTER SUMMARY

CONCEPTS	REVIEW EXERCISES

3.1 Functions and Function Notation

A **function** is a correspondence that assigns to each number x in some set **X** a single value y in some set **Y**.

Determine whether the equation defines y as a function of x. Assume that all variables represent real numbers.

1. $y = 3$ **a function**

2. $y + 5x^2 = 2$ **a function**

3. $y^2 - x = 5$ **not a function**

4. $y = |x| + x$ **a function**

Find the domain and the range of each function.

5. $f(x) = 3x^2 - 5$
domain: $(-\infty, \infty)$; range: $[-5, \infty)$

6. $f(x) = \frac{3x}{x - 5}$
domain: $(-\infty, 5) \cup (5, \infty)$; range: $(-\infty, 3) \cup (3, \infty)$

7. $f(x) = \sqrt{x - 1}$
domain: $[1, \infty)$; range: $[0, \infty)$

8. $f(x) = \sqrt{x^2 + 1}$
domain: $(-\infty, \infty)$; range: $[1, \infty)$

Find $f(2)$, $f(-3)$, and $f(0)$.

9. $f(x) = 5x - 2$ **8; −17; −2**

10. $f(x) = \frac{6}{x - 5}$ **$-2; -\frac{3}{4}; -\frac{6}{5}$**

11. $f(x) = |x - 2|$ **0; 5; 2**

12. $f(x) = \frac{x^2 - 3}{x^2 + 3}$ **$\frac{1}{7}; \frac{1}{2}; -1$**

3.2 Quadratic Functions

The graph of $y = a(x - h)^2 + k$ $(a \neq 0)$ is a parabola with its vertex at (h, k). It opens up if $a > 0$ and down if $a < 0$.

Graph each parabola and find its vertex.

13. $y = x^2 - x$

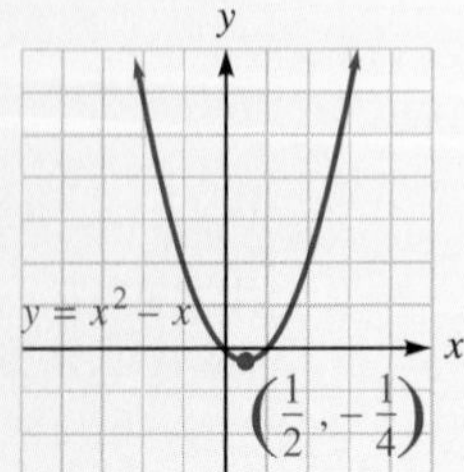

14. $y = x - x^2$

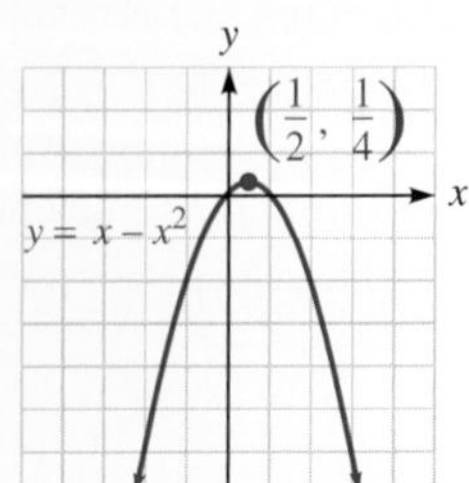

The graph of $y = ax^2 + bx + c$ $(a \neq 0)$ is a parabola with vertex
$$\left(-\frac{b}{2a}, c - \frac{b^2}{4a}\right)$$

15. $y = x^2 - 3x - 4$

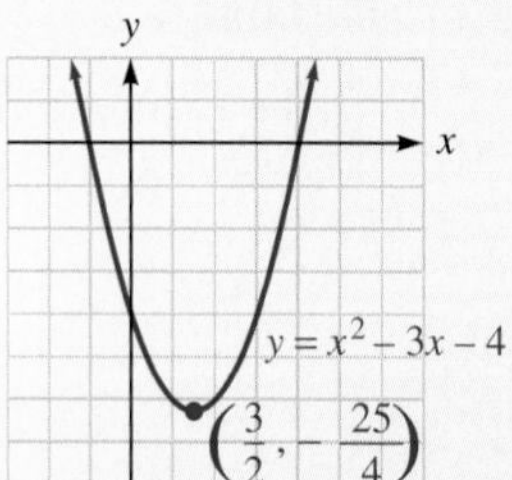

16. $y = 3x^2 - 8x - 3$

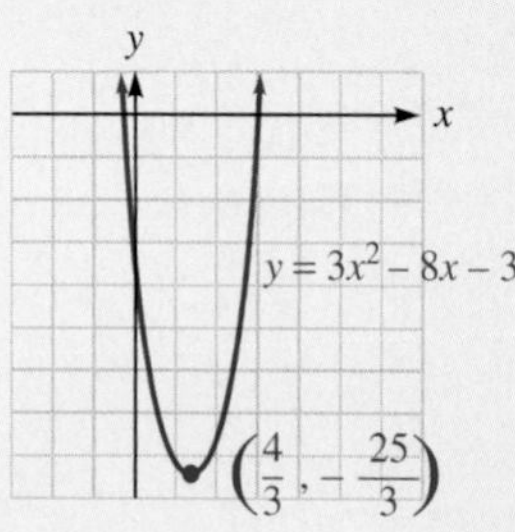

17. Architecture A parabolic arch has an equation of $3x^2 + y - 300 = 0$. Find the maximum height of the arch. **300 units**

18. Puzzle problem The sum of two numbers is 1, and their product is as small as possible. Find the numbers. **Both numbers are $\frac{1}{2}$.**

3.3 Polynomial and Other Functions

If $f(x)$ is a polynomial function of nth degree, then the graph of $f(x)$ will have $n - 1$, or fewer, turning points.

If $f(-x) = f(x)$ for all x in the domain, then f is an **even function.**

If $f(-x) = -f(x)$ for all x in the domain, then f is an **odd function.**

Graph each polynomial function and determine whether it is even, odd, or neither.

19. $y = x^3 - x$

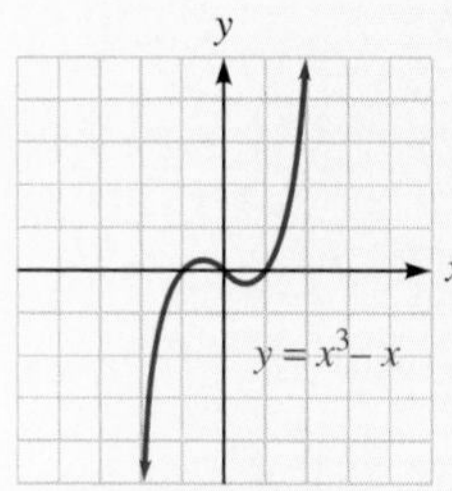

an odd function

20. $y = x^2 - 4x$

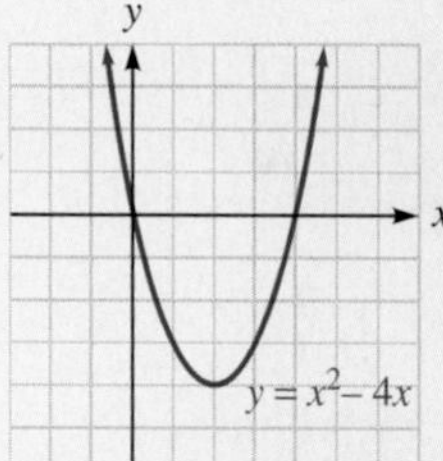

neither even nor odd

21. $y = x^3 - x^2$

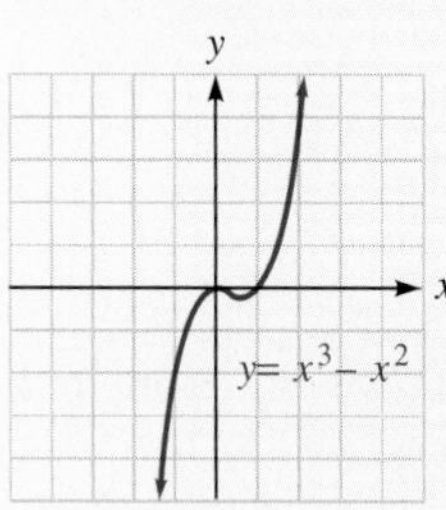

neither even nor odd

22. $y = 1 - x^4$

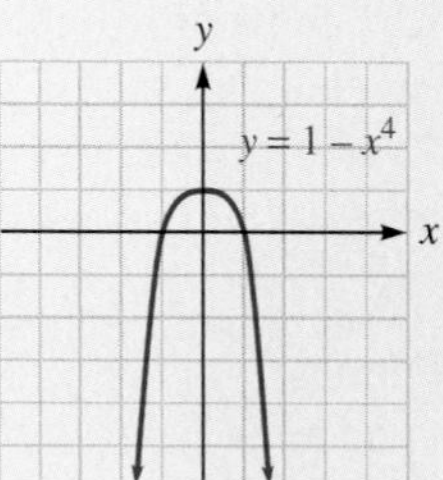

an even function

Graph each piecewise-defined function and determine when it is increasing, decreasing, or constant.

23. $y = f(x) = \begin{cases} x + 5 & \text{if } x \leq 0 \\ 5 - x & \text{if } x > 0 \end{cases}$

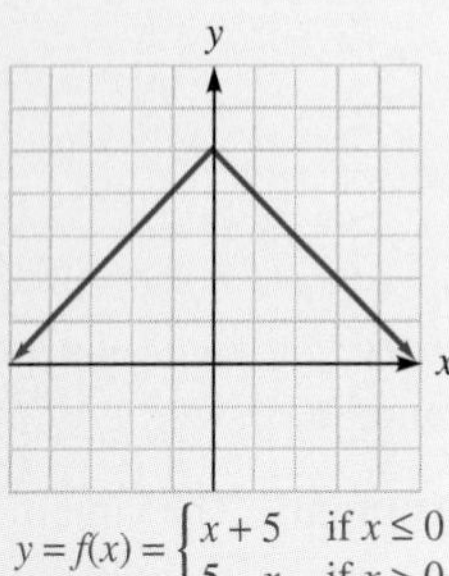

$y = f(x) = \begin{cases} x+5 & \text{if } x \leq 0 \\ 5-x & \text{if } x > 0 \end{cases}$

increasing for $x < 0$; decreasing for $x > 0$

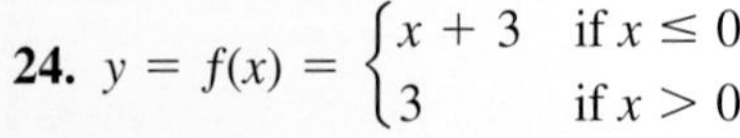

24. $y = f(x) = \begin{cases} x + 3 & \text{if } x \leq 0 \\ 3 & \text{if } x > 0 \end{cases}$

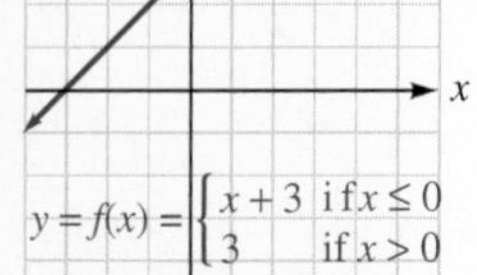

increasing for $x < 0$; constant for $x > 0$

3.4 Translating and Stretching Graphs

Vertical translations:
If $k > 0$, the graph of $\begin{cases} y = f(x) + k \\ y = f(x) - k \end{cases}$ is identical to the graph of $y = f(x)$, except that it is translated k units $\begin{cases} \text{up} \\ \text{down} \end{cases}$.

Horizontal translations:
If $k > 0$, the graph of $\begin{cases} y = f(x - k) \\ y = f(x + k) \end{cases}$ is identical to the graph of $y = f(x)$, except that it is translated k units to the $\begin{cases} \text{right} \\ \text{left} \end{cases}$.

Vertical stretchings:
If $k > 0$, the graph of $y = kf(x)$ can be obtained by stretching the graph of $y = f(x)$ vertically by a factor of k.

If $k < 0$, the graph of $y = kf(x)$ can be obtained by stretching the reflection of the graph of $y = f(x)$ in the x-axis vertically by a factor of $|k|$.

Each function is a translation of a simpler function. Graph both on one set of coordinate axes.

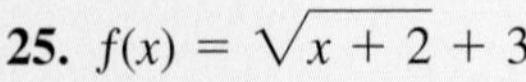

25. $f(x) = \sqrt{x + 2} + 3$

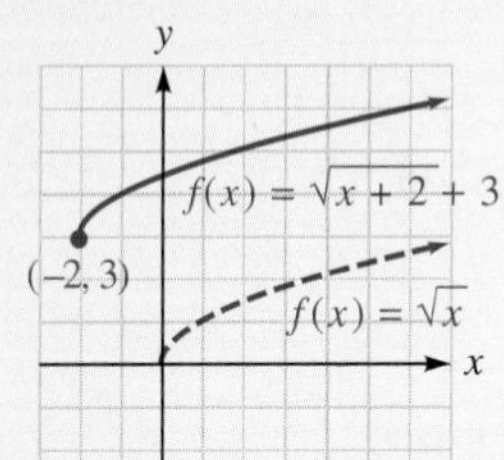

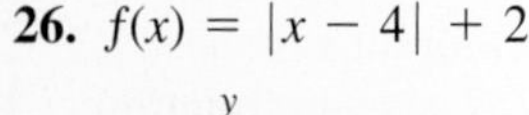

26. $f(x) = |x - 4| + 2$

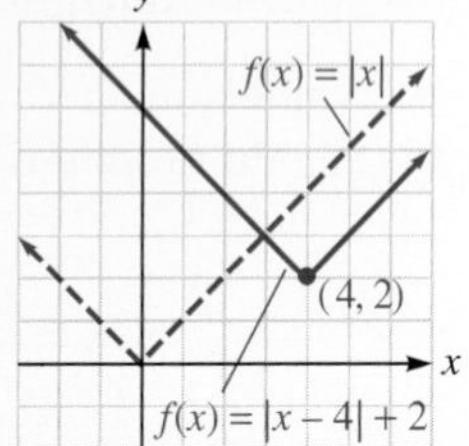

Each function is a stretching of a simpler function. Graph both on one set of coordinate axes.

27. $f(x) = \dfrac{1}{3}x^3$

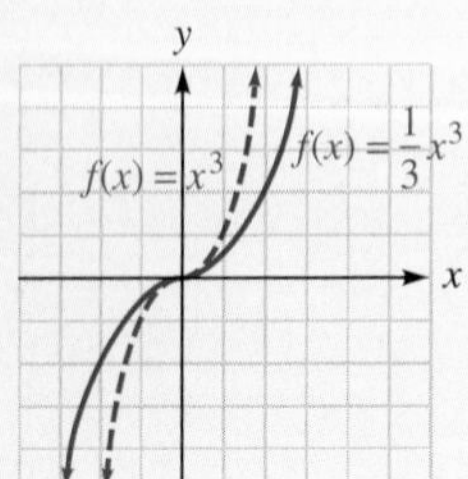

Horizontal stretchings:
If $k > 0$, the graph of $y = f(kx)$ can be obtained by stretching the graph of $y = f(x)$ horizontally by a factor of $\frac{1}{k}$.

If $k < 0$, the graph of $y = f(kx)$ can be obtained by stretching the reflection of the graph of $y = f(x)$ in the y-axis horizontally by a factor of $|\frac{1}{k}|$.

28. $f(x) = (-5x)^3$

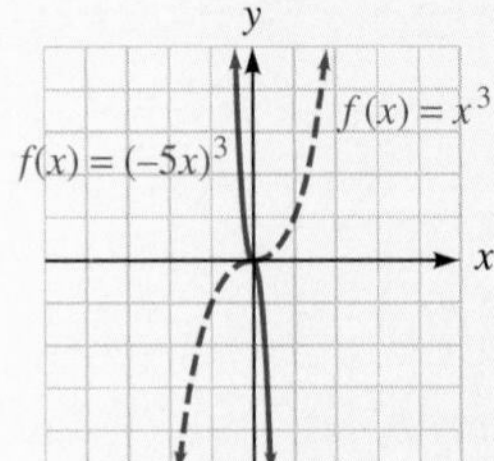

3.5 Rational Functions

To graph the rational function $y = \dfrac{P(x)}{Q(x)}$:

1. Check symmetries.
2. Look for vertical asymptotes.
3. Look for the y- and x-intercepts.
4. Look for horizontal asymptotes.
5. Look for slant asymptotes.

Graph each rational function.

29. $f(x) = \dfrac{x}{(x-1)^2}$

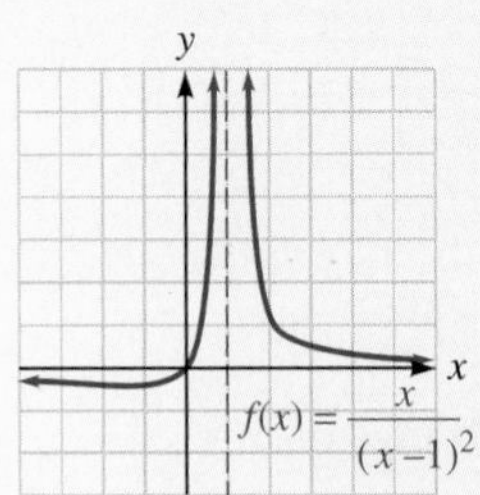

30. $f(x) = \dfrac{(x-1)^2}{x}$

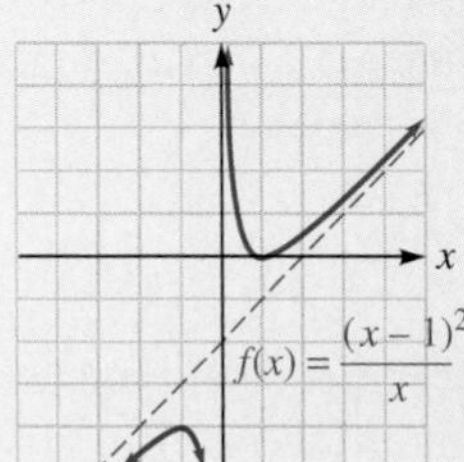

31. $f(x) = \dfrac{x^2 - x - 2}{x^2 + x - 2}$

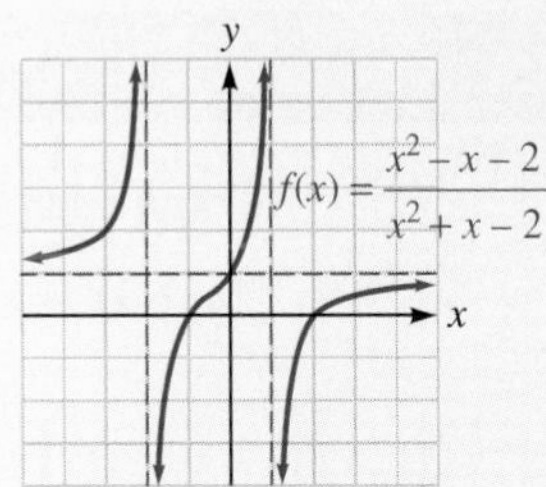

32. $f(x) = \dfrac{x^3 + x}{x^2 - 4}$

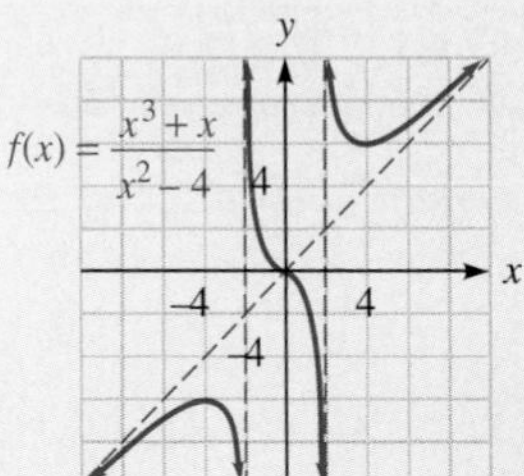

3.6 Operations on Functions

If the ranges of the functions f and g are subsets of the real numbers, then

1. $(f + g)(x) = f(x) + g(x)$
2. $(f - g)(x) = f(x) - g(x)$
3. $(f \cdot g)(x) = f(x) \cdot g(x)$
4. $(f/g)(x) = \dfrac{f(x)}{g(x)}$

Let $f(x) = x^2 - 1$ and $g(x) = 2x + 1$. Find each function.

33. $f + g$
$(f + g)(x) = x^2 + 2x$

34. $f \cdot g$
$(f \cdot g)(x) = 2x^3 + x^2 - 2x - 1$

The domain of each function, unless otherwise restricted, is the set of real numbers x that are in the domain of both f and g. In the case of the quotient f/g, there is the further restriction that $g(x) \neq 0$.

Composition of functions:

$$(f \circ g)(x) = f(g(x))$$

The domain of $f \circ g$ is the set of all x in the domain of g for which $g(x)$ is in the domain of f.

35. $f - g$
$(f - g)(x) = x^2 - 2x - 2$

36. f/g
$(f/g)(x) = \dfrac{f(x)}{g(x)} = \dfrac{x^2 - 1}{2x + 1}$

37. $f \circ g$
$(f \circ g)(x) = f(g(x)) = 4x^2 + 4x$

38. $g \circ f$
$(g \circ f)(x) = g(f(x)) = 2x^2 - 1$

3.7 Inverse Functions

If f is a one-to-one function with domain **X** and range **Y**, there is a one-to-one function f^{-1} with domain **Y** and range **X** such that

$$(f^{-1} \circ f)(x) = x \quad \text{and}$$
$$(f \circ f^{-1})(y) = y$$

The graph of a function is symmetric to the graph of its inverse. The axis of symmetry is the line $y = x$.

Each equation defines a one-to-one function. Find f^{-1}.

39. $y = 7x - 1$ $\quad f^{-1}(x) = \frac{x+1}{7}$

40. $y = \dfrac{1}{2 - x}$ $\quad f^{-1}(x) = 2 - \frac{1}{x}$

41. $y = \dfrac{x}{1 - x}$ $\quad f^{-1}(x) = \frac{x}{x+1}$

42. $y = \dfrac{3}{x^3}$ $\quad f^{-1}(x) = \sqrt[3]{\frac{3}{x}}$

43. Find the range of $y = \dfrac{2x + 3}{5x - 10}$ by finding the domain of f^{-1}.
$\left(-\infty, \frac{2}{5}\right) \cup \left(\frac{2}{5}, \infty\right)$

CHAPTER TEST

Find the domain and range of each function.

1. $f(x) = \dfrac{3}{x - 5}$
domain: $(-\infty, 5) \cup (5, \infty)$; range: $(-\infty, 0) \cup (0, \infty)$

2. $f(x) = \sqrt{x + 3}$
domain: $[-3, \infty)$; range: $[0, \infty)$

Find $f(-1)$ and $f(2)$.

3. $f(x) = \dfrac{x}{x - 1}$
$\frac{1}{2}, 2$

4. $f(x) = \sqrt{x + 7}$
$\sqrt{6}, 3$

Find the vertex of each parabola.

5. $y = 3(x - 7)^2 - 3$
(7, −3)

6. $y = x^2 - 2x - 3$
(1, −4)

7. $f(x) = 3x^2 - 24x + 38$
(4, −10)

8. $f(x) = 5 - 4x - x^2$
(−2, 9)

Graph each function.

9. $f(x) = x^4 - x^2$

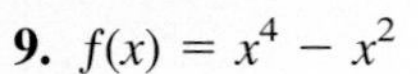
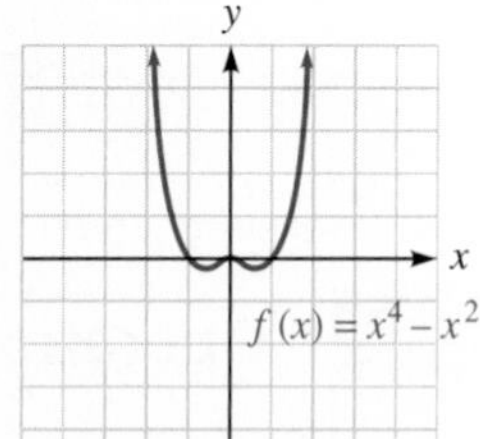

10. $f(x) = x^5 - x^3$

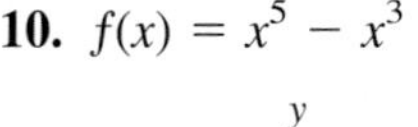
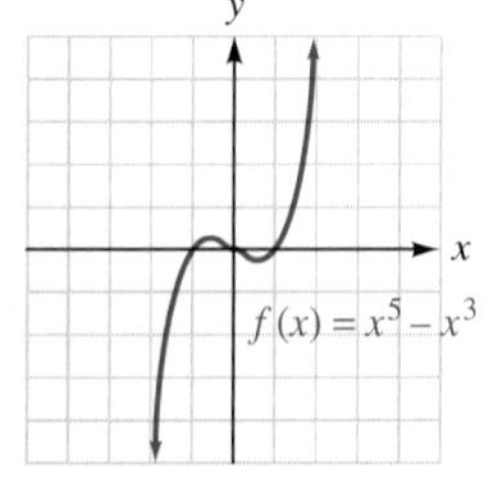

Assume that an object tossed vertically upward reaches a height of h feet after t seconds, where h = 100t − 16t².

11. In how many seconds does the object reach its maximum height? $\frac{25}{8}$ **sec**

12. What is that maximum height? $\frac{625}{4}$ **ft**

13. Suspension bridges The cable of a suspension bridge is in the shape of the parabola $x^2 - 2{,}500y + 25{,}000 = 0$ in the coordinate system shown in the illustration. (Distances are in feet.) How far above the roadway is the cable's lowest point?
10 ft

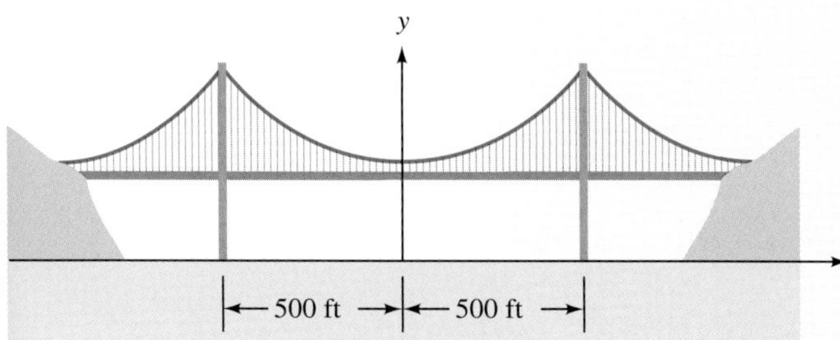

14. Refer to Question 13. How far above the roadway does the cable attach to the vertical pillars? **110 ft**

Graph each function.

15. $f(x) = (x - 3)^2 + 1$

16. $f(x) = \sqrt{x - 1} + 5$

y
x
$f(x) = \sqrt{x - 1} + 5$
(1, 5)

Find all asymptotes of the graph of each rational function. Do not graph the function.

17. $y = \dfrac{x - 1}{x^2 - 9}$ **vertical asymptotes: $x = -3$ and $x = 3$; horizontal asymptote: $y = 0$**

18. $y = \dfrac{x^2 - 5x - 14}{x - 3}$ **vertical asymptote: $x = 3$; horizontal asymptote: none; slant asymptote: $y = x - 2$**

Graph each rational function. Check for asymptotes, intercepts, and symmetry.

19. $y = \dfrac{x^2}{x^2 - 9}$

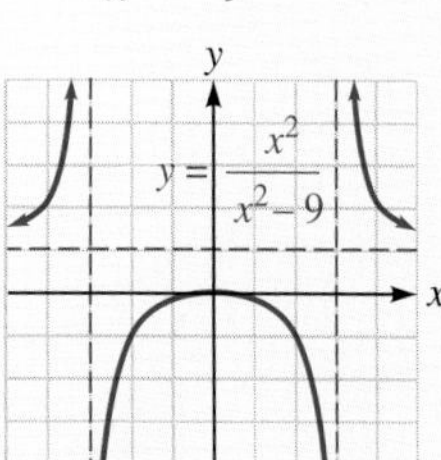

20. $y = \dfrac{x}{x^2 + 1}$

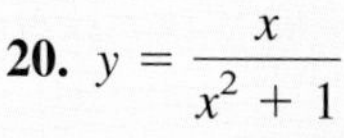

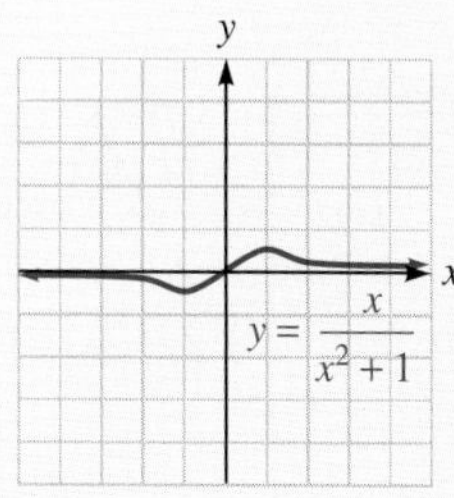

Graph each rational function. The numerator and denominator share a common factor.

21. $y = \dfrac{2x^2 - 3x - 2}{x - 2}$

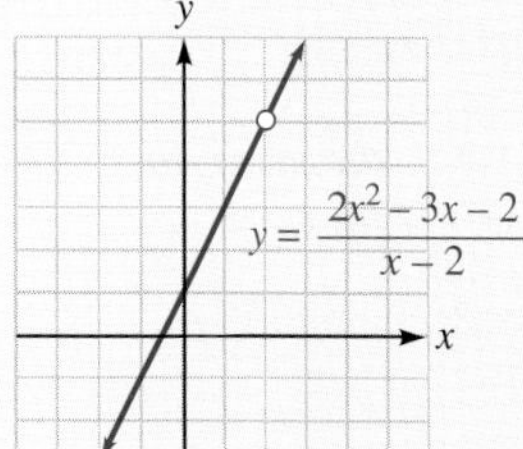

22. $y = \dfrac{x}{x^2 - x}$

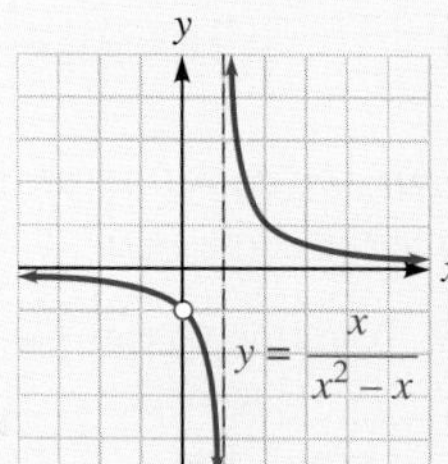

Let $f(x) = 3x$ and $g(x) = x^2 + 2$. Find each function.

23. $f + g$ $(f + g)(x) = f(x) + g(x) = x^2 + 3x + 2$

24. $g \circ f$ $(g \circ f)(x) = g(f(x)) = 9x^2 + 2$

25. f/g $(f/g)(x) = \dfrac{f(x)}{g(x)} = \dfrac{3x}{x^2 + 2}$

26. $f \circ g$ $(f \circ g)(x) = f(g(x)) = 3x^2 + 6$

Assume that $f(x)$ is one-to-one. Find f^{-1}.

27. $f(x) = \dfrac{x + 1}{x - 1}$
$f^{-1}(x) = \dfrac{x + 1}{x - 1}$

28. $f(x) = x^3 - 3$
$f^{-1}(x) = \sqrt[3]{x + 3}$

Find the range of f by finding the domain of f^{-1}.

29. $y = \dfrac{3}{x} - 2$
range: $(-\infty, -2) \cup (-2, \infty)$

30. $y = \dfrac{3x - 1}{x - 3}$
range: $(-\infty, 3) \cup (3, \infty)$

CUMULATIVE REVIEW EXERCISES

Use the x- and y-intercepts to graph each equation.

1. $5x - 3y = 15$

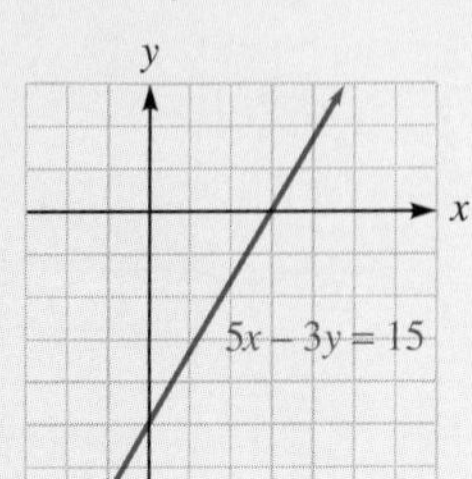

2. $3x + 2y = 12$

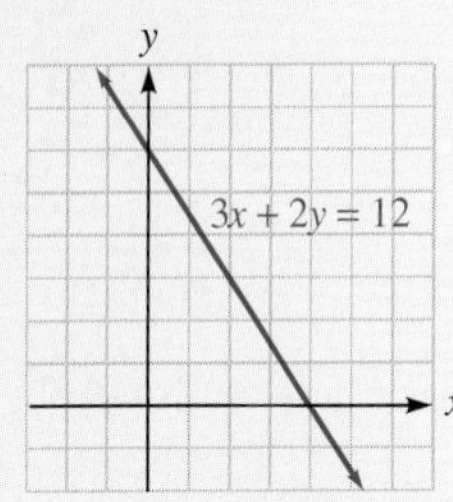

Find the length, the midpoint, and the slope of the line segment PQ.

3. $P\left(-2, \frac{7}{2}\right); Q\left(3, -\frac{1}{2}\right)$ **$\sqrt{41}$; $\left(\frac{1}{2}, \frac{3}{2}\right)$; $-\frac{4}{5}$**

4. $P(3, 7); Q(-7, 3)$ **$2\sqrt{29}$; $(-2, 5)$; $\frac{2}{5}$**

Write the equation of the line with the given properties. Give the answer in slope–intercept form.

5. The line passes through $(-3, 5)$ and $(3, -7)$.
$y = -2x - 1$

6. The line passes through $\left(\frac{3}{2}, \frac{5}{2}\right)$ and has a slope of $\frac{7}{2}$.
$y = \frac{7}{2}x - \frac{11}{4}$

7. The line is parallel to $3x - 5y = 7$ and passes through $(-5, 3)$. **$y = \frac{3}{5}x + 6$**

8. The line is perpendicular to $x - 4y = 12$ and passes through the origin. **$y = -4x$**

Graph each equation. Make use of intercepts and symmetries.

9. $x^2 = y - 2$

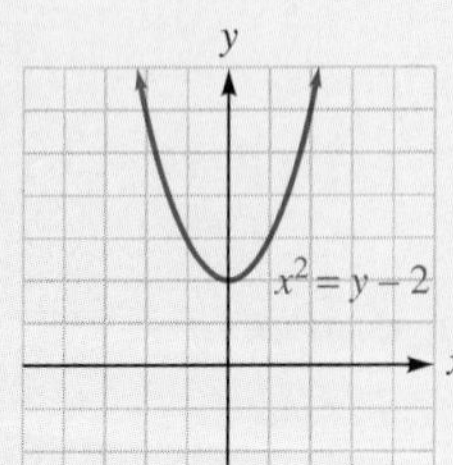

10. $y^2 = x - 2$

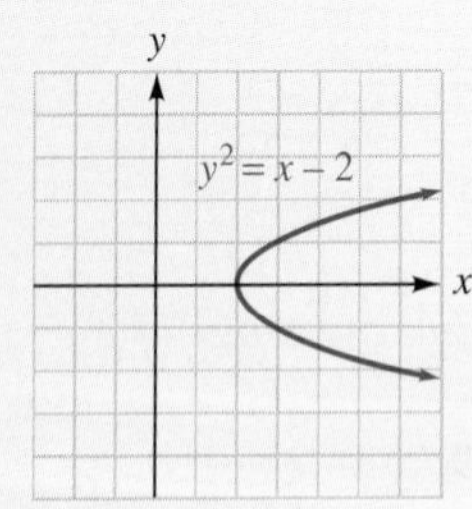

11. $x^2 + y^2 = 100$

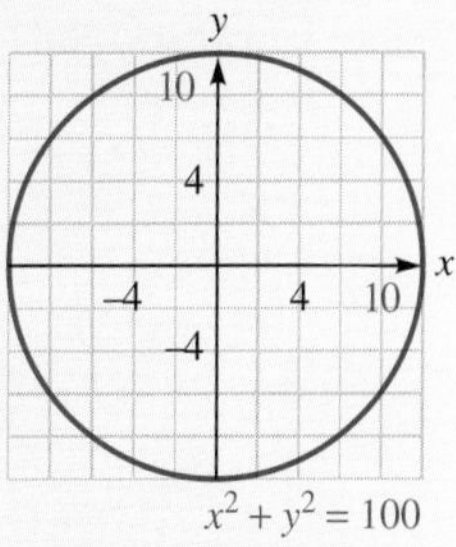

12. $x^2 - 2x + y^2 = 8$

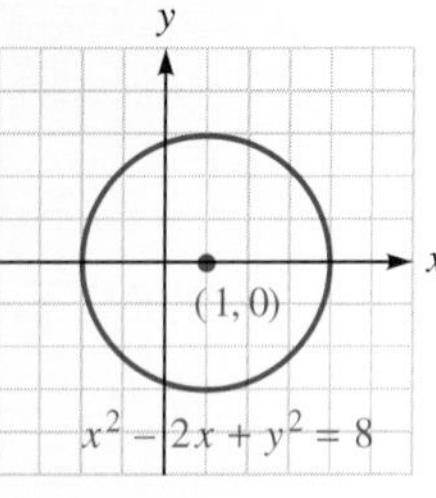

Solve each proportion.

13. $\dfrac{x - 2}{x} = \dfrac{x - 6}{5}$
$x = 1, x = 10$

14. $\dfrac{x + 2}{x - 6} = \dfrac{3x + 1}{2x - 11}$
$x = 2, x = 8$

15. Dental billing The billing schedule for dental X rays specifies a fixed amount for the office visit plus a fixed amount for each X ray exposure. If 2 X rays cost \$37 and 4 cost \$54, find the cost of 5 exposures. **\$62.50**

16. Automobile collisions The energy dissipated in an automobile collision varies directly with the square of the speed. By what factor does the energy increase in a 50-mph collision compared with a 20-mph collision? **$\frac{25}{4}$**

Determine whether each equation defines a function.

17. $y = 3x - 1$ **a function**

18. $y = x^2 + 3$ **a function**

19. $y = \dfrac{1}{x - 2}$ **a function**

20. $y^2 = 4x$ **not a function**

Find the domain and range of each function.

21. $f(x) = x^2 + 5$
domain: $(-\infty, \infty)$; range: $[5, \infty)$

22. $f(x) = \dfrac{7}{x + 2}$
domain: $(-\infty, -2) \cup (-2, \infty)$; range: $(-\infty, 0) \cup (0, \infty)$

23. $f(x) = -\sqrt{x - 2}$
domain: $[2, \infty)$; range: $(-\infty, 0]$

24. $f(x) = \sqrt{x + 4}$
domain: $[-4, \infty)$; range: $[0, \infty)$

Find the vertex of the parabolic graph of each equation.

25. $y = x^2 + 5x - 6$
$\left(-\frac{5}{2}, -\frac{49}{4}\right)$

26. $f(x) = -x^2 + 5x + 6$
$\left(\frac{5}{2}, \frac{49}{4}\right)$

Graph each function.

27. $f(x) = x^2 - 4$

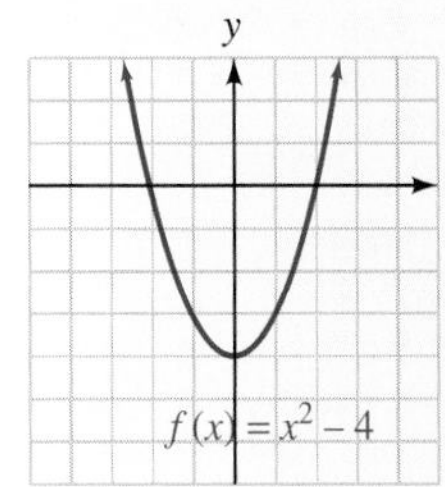

28. $f(x) = -x^2 + 4$

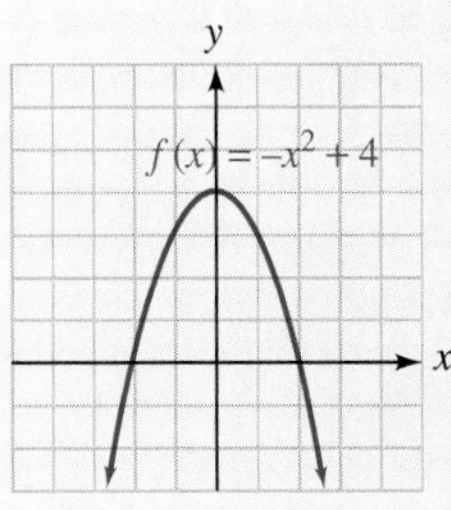

29. $f(x) = x^3 + x$

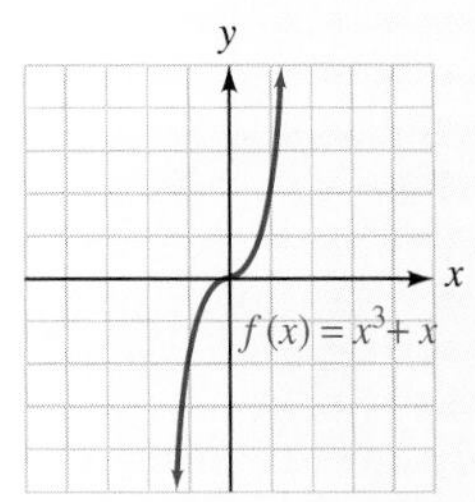

30. $f(x) = -x^4 + 2x^2 + 1$

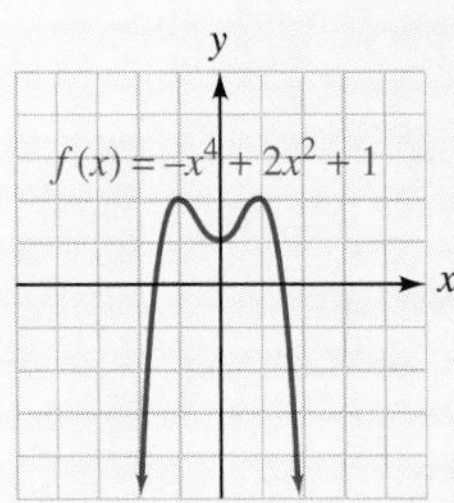

Graph each function. Show all asymptotes.

31. $f(x) = \dfrac{x}{x - 3}$

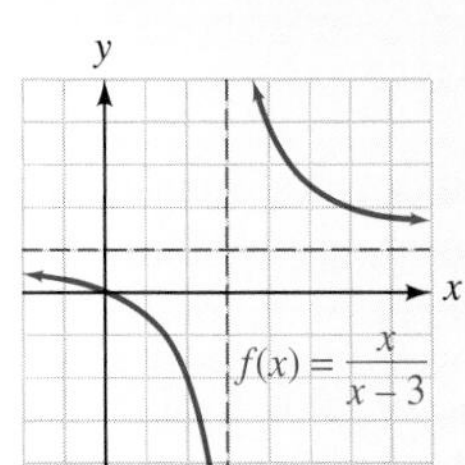

32. $f(x) = \dfrac{x^2 - 1}{x^2 - 9}$

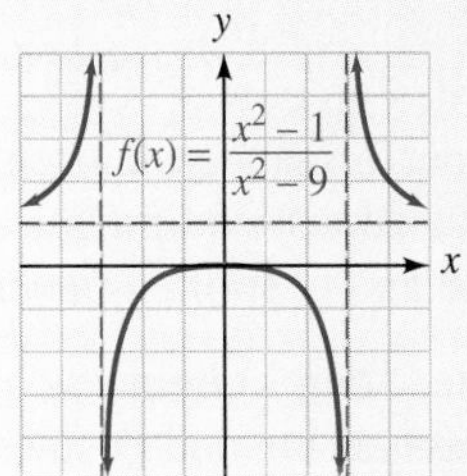

Let $f(x) = 3x - 4$ and $g(x) = x^2 + 1$. Find each function and its domain.

33. $(f + g)(x)$ $\quad (f + g)(x) = f(x) + g(x) = x^2 + 3x - 3$
domain: $(-\infty, \infty)$

34. $(f - g)(x)$ $\quad (f - g)(x) = f(x) - g(x) = -x^2 + 3x - 5$
domain: $(-\infty, \infty)$

35. $(f \cdot g)(x)$ $\quad (f \cdot g)(x) = f(x) \cdot g(x) = 3x^3 - 4x^2 + 3x - 4$
domain: $(-\infty, \infty)$

36. $(f/g)(x)$ $\quad (f/g)(x) = \dfrac{f(x)}{g(x)} = \dfrac{3x - 4}{x^2 + 1}$
domain: $(-\infty, \infty)$

Let $f(x) = 3x - 4$ and $g(x) = x^2 + 1$. Find each value.

37. $(f \circ g)(2)$
$(f \circ g)(2) = 11$

38. $(g \circ f)(2)$
$(g \circ f)(2) = 5$

39. $(f \circ g)(x)$ $\quad (f \circ g)(x) = 3x^2 - 1$

40. $(g \circ f)(x)$ $\quad (g \circ f)(x) = 9x^2 - 24x + 17$

Find the inverse of the function defined by each equation.

41. $y = 3x + 2$
$f^{-1}(x) = \frac{x - 2}{3}$

42. $y = \dfrac{1}{x - 3}$
$f^{-1}(x) = \frac{1}{x} + 3$

43. $y = x^2 + 5 \ (x \geq 0)$
$f^{-1}(x) = \sqrt{x - 5}$

44. $3x - y = 1$
$f^{-1}(x) = \frac{x + 1}{3}$

Write each sentence as an equation.

45. y varies directly with the product of w and z.
$y = kwz$

46. y varies directly with x and inversely with the square of t. $\quad y = \dfrac{kx}{t^2}$

4 Exponential and Logarithmic Functions

COLLEGE **Algebra** $f(x)$ **Now**™

Throughout the chapter, this icon introduces resources on the College AlgebraNow Web Site, accessed through **http://1pass.thomson.com**, that will:

- Help you test your knowledge of the material in this chapter prior to reading it,
- Allow you to take an exam-prep quiz, and
- Provide a Personalized Learning Plan targeting areas you should study.

Careers and Mathematics

ELECTRICAL AND ELECTRONICS ENGINEERS

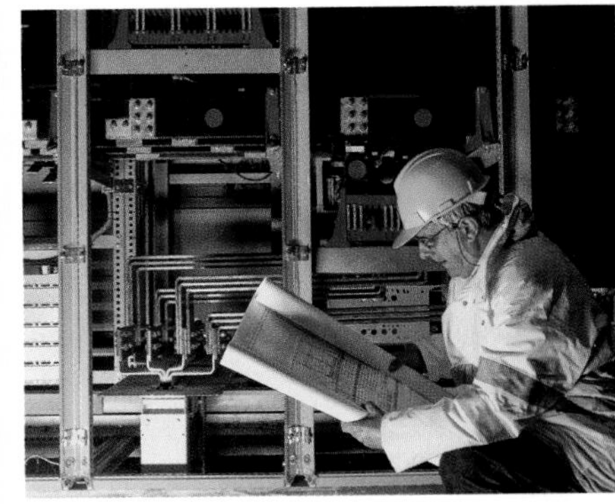

age fotostock/SuperStock

Electrical and electronics engineers design, develop, test, and supervise the manufacture of electrical and electronic equipment. They specialize in different areas such as power generation, transmission, and distribution; communications; and electrical equipment manufacturing. They design new products, write performance requirements, and develop maintenance schedules. They also test equipment, solve operating problems, and estimate the time and cost of engineering projects.

Electrical and electronics engineers held about 292,000 jobs in 2002. Most were in engineering and business consulting firms, government agencies, and manufacturers of electrical and electronics equipment, industrial machinery, and scientific instruments.

JOB OUTLOOK

Employment of electrical and electronics engineers is expected to grow more slowly than the average for all occupations through 2012. Median annual earnings of electrical engineers were $68,180 in 2002.

Median annual earnings of electronics engineers were $69,930 in 2002.

For a sample application, see Example 1 in Section 4.4.

In this chapter, we will discuss exponential functions, which are often used in banking, ecology, and science. We will also discuss logarithmic functions, which are applied in chemistry, geology, and environmental science.

4.1 Exponential Functions and Their Graphs

In this section, you will learn about

- **Irrational Exponents**
- **Exponential Functions and Their Graphs**
- **Compound Interest**
- **Base-*e* Exponential Functions and Their Graphs**
- **Vertical and Horizontal Translations**
- **Exponential Functions of the Form $f(x) = kb^x$ and $f(x) = b^{kx}$**

The graph in Figure 4-1 shows the balance in a bank account in which \$5,000 was invested in 1990 at 8%, compounded monthly. The graph shows that in the year 2015, the value of the account will be approximately \$38,000, and in the year 2030, the value will be approximately \$121,000.

Sir Isaac Newton

(1642–1727)

Newton was an English scientist and mathematician. Because he was not a good farmer, he went to Cambridge University to become a preacher. When he had to leave Cambridge because of the plague, he made some of his most important discoveries. He is best known in mathematics for developing calculus and in physics for discovering the laws of motion. Newton probably contributed more to science and mathematics than anyone else in history.

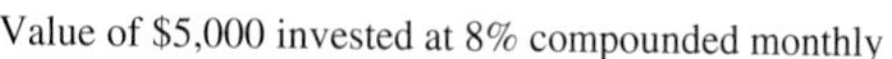

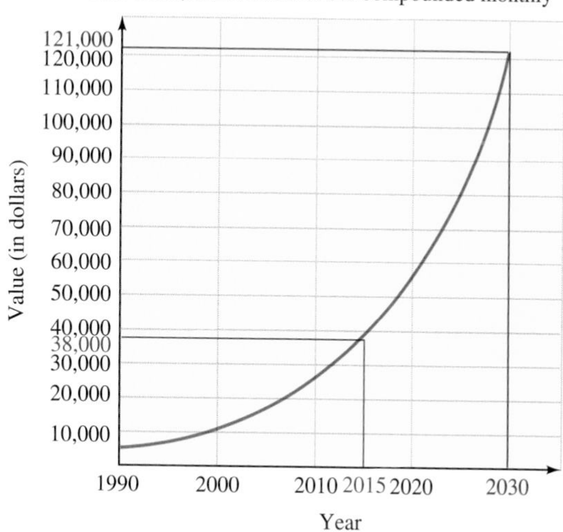

Figure 4-1

The curve in Figure 4-1 is the graph of a function called an *exponential function.* From the graph, we can see that the longer the money is kept on deposit, the more rapidly it will grow.

Before we can discuss exponential functions, we must define irrational exponents.

Irrational Exponents

We have discussed expressions of the form b^x, where x is a rational number.

5^2 means "the square of 5."

$4^{1/3}$ means "the cube root of 4."

$6^{-2/5} = \frac{1}{6^{2/5}}$ means "the reciprocal of the fifth root of 6^2."

To give meaning to b^x when x is an irrational number, we consider the expression

$3^{\sqrt{2}}$ where $\sqrt{2}$ is the irrational number 1.414213562. . .

Since the exponents in the following list of expressions are rational numbers, each one is defined, and we can use a calculator to approximate the value of each expression.

	Scientific calculator	*Graphing calculator*
$3^{1.4} \approx 4.655536722$	Press 3 y^x 1.4 =	or 3 ^ 1.4 ENTER.
$3^{1.41} \approx 4.706965002$	Press 3 y^x 1.41 =	or 3 ^ 1.41 ENTER.
$3^{1.414} \approx 4.727695035$	Press 3 y^x 1.414 =	or 3 ^ 1.414 ENTER.
$3^{1.4142} \approx 4.72873393$	Press 3 y^x 1.4142 =	or 3 ^ 1.4142 ENTER.

Since the exponents of the expressions in the list are getting closer to $\sqrt{2}$, the values of the expressions are getting closer to the value of $3^{\sqrt{2}}$. To find an even better approximation of $3^{\sqrt{2}}$, we use a calculator.

	Scientific calculator	*Graphing calculator*
$3^{\sqrt{2}} \approx 4.728804388$	Press 3 y^x 2 $\sqrt{\ }$ =	or 3 ^ $\sqrt{\ }$ 2 ENTER.

If b is a positive number and x is a real number, the expression b^x always represents a positive number. It is also true that the familiar properties of exponents hold for irrational exponents.

EXAMPLE 1 Use properties of exponents to simplify: **a.** $(3^{\sqrt{2}})^{\sqrt{2}}$ and **b.** $a^{\sqrt{8}} \cdot a^{\sqrt{2}}$.

Solution **a.** $(3^{\sqrt{2}})^{\sqrt{2}} = 3^{\sqrt{2}\sqrt{2}}$ Keep the base and multiply the exponents.

$= 3^2$ $\sqrt{2}\sqrt{2} = \sqrt{4} = 2.$

$= 9$

b. $a^{\sqrt{8}} \cdot a^{\sqrt{2}} = a^{\sqrt{8}+\sqrt{2}}$ Keep the base and add the exponents.

$= a^{2\sqrt{2}+\sqrt{2}}$ $\sqrt{8} = \sqrt{4}\sqrt{2} = 2\sqrt{2}.$

$= a^{3\sqrt{2}}$ $2\sqrt{2} + \sqrt{2} = 3\sqrt{2}.$

Self Check Simplify: **a.** $(2^{\sqrt{3}})^{\sqrt{12}}$ and **b.** $x^{\sqrt{20}} \cdot x^{\sqrt{5}}$. ■

Exponential Functions and Their Graphs

If $b > 0$ and $b \neq 1$, the function $y = b^x$ defines a function, because for each input x, there is exactly one output y. Since x can be any real number, the domain of the function is the set of real numbers. Since the base b of the expression b^x is positive, y is always positive, and the range is the set of positive numbers. Since b^x is an exponential expression, the function is called an **exponential function.**

We make the restriction that $b > 0$ to exclude any imaginary numbers that might result from taking even roots of negative numbers. The restriction that $b \neq 1$ excludes the constant function $f(x) = 1^x$, in which $f(x) = 1$ for every real number x.

Exponential Functions

An **exponential function with base b** is defined by the equation

$$f(x) = b^x \qquad \text{or} \qquad y = b^x \quad (b > 0, b \neq 1, \text{and } x \text{ is a real number})$$

The **domain of any exponential function** is the interval $(-\infty, \infty)$. The **range** is the interval $(0, \infty)$.

Since the domain and range of $f(x) = b^x$ are sets of real numbers, we can graph exponential functions. For example, to graph

$$f(x) = 2^x$$

we find several points $(x, f(x))$ whose coordinates satisfy the equation, plot the points, and join them with a smooth curve, as in Figure 4-2(a). To graph the function

$$f(x) = \left(\frac{1}{2}\right)^x$$

we find several points $(x, f(x))$ whose coordinates satisfy the equation, plot the points, and join them with a smooth curve, as shown in Figure 4-2(b).

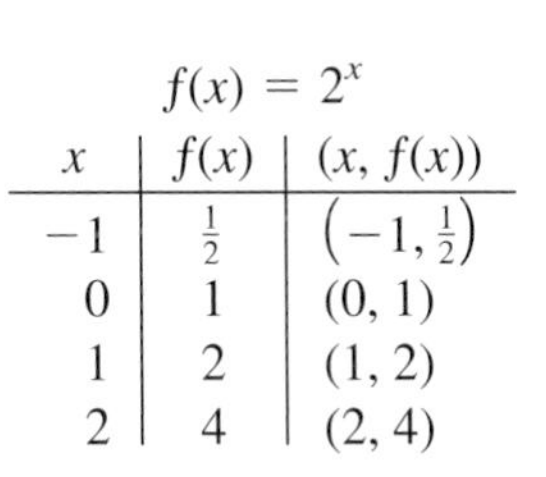

$f(x) = 2^x$

x	$f(x)$	$(x, f(x))$
-1	$\frac{1}{2}$	$\left(-1, \frac{1}{2}\right)$
0	1	(0, 1)
1	2	(1, 2)
2	4	(2, 4)

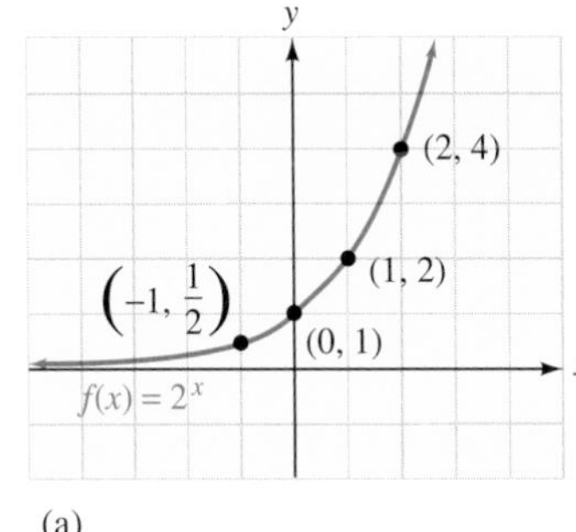

(a)

$f(x) = \left(\frac{1}{2}\right)^x$

x	$f(x)$	$(x, f(x))$
-2	4	$(-2, 4)$
-1	2	$(-1, 2)$
0	1	(0, 1)
1	$\frac{1}{2}$	$\left(1, \frac{1}{2}\right)$

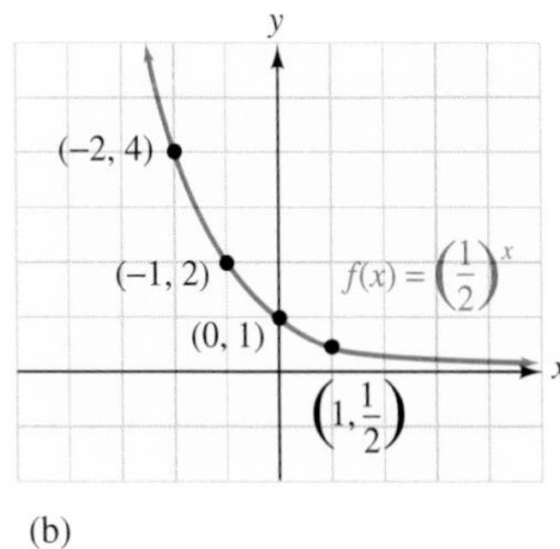

(b)

Figure 4-2

By looking at the graphs in Figure 4-2, we can see that the domain of each function is the interval $(-\infty, \infty)$ and that the range is the interval $(0, \infty)$.

ACTIVE EXAMPLE 2 Graph: $f(x) = 4^x$.

Solution We find several points (x, y) that satisfy the equation, plot the points, and join them with a smooth curve, as in Figure 4-3.

$f(x) = 4^x$

x	$f(x)$	$(x, f(x))$
-1	$\frac{1}{4}$	$\left(-1, \frac{1}{4}\right)$
0	1	(0, 1)
1	4	(1, 4)

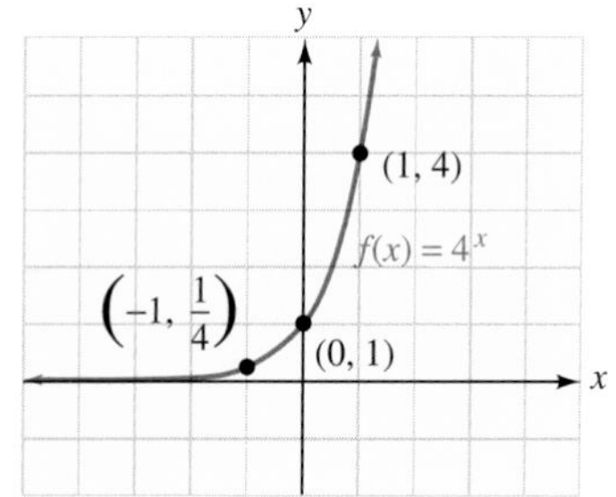

Figure 4-3

Self Check Graph: $f(x) = \left(\frac{1}{4}\right)^x$. ■

The graph of $f(x) = 4^x$ in Example 2 has the following properties:

1. It passes through the point (0, 1).
2. It passes through the point (1, 4).
3. It approaches the x-axis. The x-axis is an asymptote.
4. The domain is the interval $(-\infty, \infty)$, and the range is the interval $(0, \infty)$.

This example illustrates the following properties of exponential functions.

Properties of Exponential Functions

The **domain of the exponential function** $f(x) = b^x$ is $(-\infty, \infty)$, the set of real numbers.

The **range** is $(0, \infty)$, the set of positive real numbers.

The graph has a y-intercept at (0, 1).

The x-axis is an asymptote of the graph.

The graph of $f(x) = \boldsymbol{b}^x$ passes through the point $(1, \boldsymbol{b})$.

EXAMPLE 3 From the graph of $f(x) = b^x$ (shown in Figure 4-4), find the value of b.

Solution Since the graph passes through the point (0, 1), the graph could be the graph of $f(x) = b^x$. If it is, we can find the base by substituting 2 for x and 25 for $f(x)$ in the equation $f(x) = b^x$ to get

$$f(x) = b^x$$
$$f(2) = b^2$$
$$25 = b^2$$
$$5 = b \quad \textbf{\textit{b} must be positive.}$$

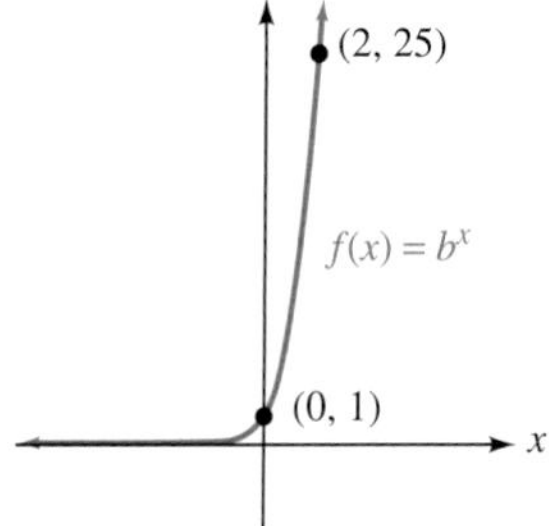

Figure 4-4

The base b is 5. Note that the points (0, 1) and (2, 25) satisfy the equation $f(x) = 5^x$.

Self Check Can a graph passing through (0, 2) and $\left(1, \frac{3}{2}\right)$ be the graph of $f(x) = b^x$? ■

In Figure 4-2(a) (where $b = 2$ and $2 > 1$), the values of y increase as the values of x increase. Since the graph rises as we move to the right, the function is an increasing function. Such a function is said to model *exponential growth.*

In Figure 4-2(b) $\left(\text{where } b = \frac{1}{2} \text{ and } 0 < \frac{1}{2} < 1\right)$, the values of y decrease as the values of x increase. Since the graph drops as we move to the right, the function is a decreasing function. Such a function is said to model *exponential decay.*

Increasing and Decreasing Functions

If $b > 1$, then $f(x) = b^x$ is an **increasing function.** This function models **exponential growth.**

If $0 < b < 1$, then $f(x) = b^x$ is a **decreasing function.** This function models **exponential decay.**

Increasing function

Decreasing function

Comment Recall that $b^{-x} = \frac{1}{b^x} = \left(\frac{1}{b}\right)^x$. If $b > 1$, any function of the form $f(x) = b^{-x}$ models exponential decay, because $0 < \frac{1}{b} < 1$.

An exponential function $f(x) = b^x$ is either increasing (for $b > 1$) or decreasing (for $0 < b < 1$). Because different real numbers x always determine different values of b^x, an exponential function is one-to-one.

An exponential function defined by

$$f(x) = b^x \qquad \text{or} \qquad y = b^x, \quad \text{where } b > 0 \text{ and } b \neq 1$$

is one-to-one. This implies that

1. If $b^r = b^s$, then $r = s$.

2. If $r \neq s$, then $b^r \neq b^s$.

Compound Interest

Banks pay **interest** for using their customers' money. Interest is calculated as a percent of the amount on deposit in an account and is paid annually (once a year), quarterly (four times per year), monthly, or daily. Interest left on deposit in a bank account will also earn interest. Such accounts are said to earn **compound interest.**

Compound Interest Formula

If P dollars are deposited in an account earning interest at an annual rate r, compounded k times each year, the amount A in the account after t years is given by

$$A = P\left(1 + \frac{r}{k}\right)^{kt}$$

ACTIVE EXAMPLE 4 The parents of a newborn child invest \$8,000 in a plan that earns 9% interest, compounded quarterly. If the money is left untouched, how much will the child have in the account in 55 years?

Solution We substitute 8,000 for P, 0.09 for r, and 55 for t in the formula for compound interest. Because quarterly compounding means four times per year, $k = 4$.

COLLEGE **Algebra** $f(x)$ **Now**™
Go to **http://1pass.thomson.com** or your CD to practice this example.

$$A = P\left(1 + \frac{r}{k}\right)^{kt}$$

$$A = \mathbf{8{,}000}\left(1 + \frac{\mathbf{0.09}}{\mathbf{4}}\right)^{\mathbf{4 \cdot 55}}$$

$$= 8{,}000(1.0225)^{220}$$

$$\approx 1{,}069{,}103.266 \qquad \text{Use a calculator.}$$

In 55 years, the account will be worth \$1,069,103.27.

Self Check Would \$20,000 invested at 7% interest, compounded monthly, have provided more income at age 55? ■

In financial calculations, the initial amount deposited is often called the **present value,** denoted by PV. The amount to which the account will grow is called the **future value,** denoted by FV. The interest rate for each compounding period is called the **periodic interest rate,** i, and the number of times interest is compounded is the **number of compounding periods,** n. Using these definitions, an alternate formula for compound interest is as follows.

$$FV = PV(1 + i)^n$$

To use this formula to solve Example 4, we proceed as follows:

$$FV = PV(1 + i)^n$$

$$FV = \mathbf{8{,}000}(1 + \mathbf{0.0225})^{\mathbf{220}} \qquad i = \tfrac{0.09}{4} = 0.0225, \text{ and } n = 4(55) = 220.$$

$$= 8{,}000(1.0225)^{220}$$

$$\approx 1{,}069{,}103.266 \qquad \text{Use a calculator.}$$

■ Base-*e* Exponential Functions and Their Graphs

In mathematical models of natural events, the number $e = \mathbf{2.71828182845904...}$ often appears as the base of an exponential function. We can introduce this number by considering the compound interest formula

$$A = P\left(1 + \frac{r}{k}\right)^{kt}$$

A is the amount, P is the initial deposit, r is the annual rate, k is the number of compoundings per year, and t is the time in years.

and allowing k to become very large. To see what happens, we let $k = rx$, where x is another variable.

$$A = P\left(1 + \frac{r}{k}\right)^{kt}$$

$$A = P\left(1 + \frac{r}{rx}\right)^{rxt} \qquad \text{Substitute } rx \text{ for } k.$$

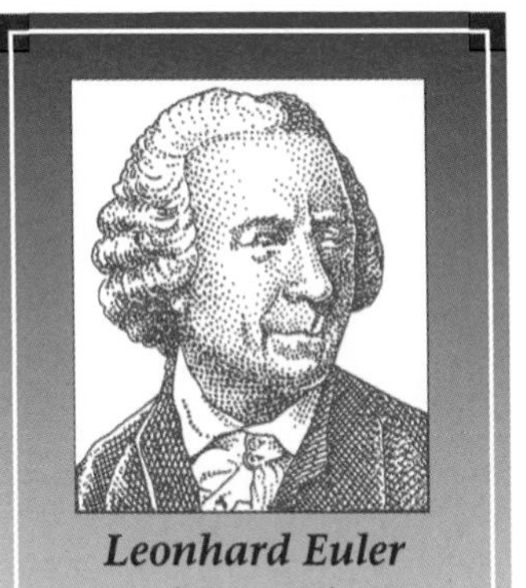

Leonhard Euler
(1707–1783)
Euler first used the letter i to represent $\sqrt{-1}$, the letter e for the base of natural logarithms, and the symbol Σ for summation. Euler was one of the most prolific mathematicians of all time, contributing to almost all areas of mathematics. Much of his work was accomplished after he became blind.

$$A = P\left(1 + \frac{1}{x}\right)^{rxt} \qquad \text{Simplify } \frac{r}{rx}.$$

$$A = P\left[\left(1 + \frac{1}{x}\right)^{x}\right]^{rt} \qquad \text{Remember that } (a^m)^n = a^{mn}.$$

Since all variables in this formula are positive, r is a constant rate, and $k = rx$, it follows that as k becomes large, so does x. What happens to the value of A as k becomes large will depend on the value of $\left(1 + \frac{1}{x}\right)^x$ as x becomes large. Some results calculated for increasing values of x appear in Table 4-1.

x	$\left(1 + \frac{1}{x}\right)^x$
1	2
10	2.5937425
100	2.7048138
1,000	2.7169239
1,000,000	2.7182805
1,000,000,000	2.7182818

Table 4-1

From the table, we can see that as x increases, the value of $\left(1 + \frac{1}{x}\right)^x$ approaches the value of e, and the formula

$$A = P\left[\left(1 + \frac{1}{x}\right)^{x}\right]^{rt}$$

becomes

$$A = Pe^{rt} \qquad \text{Substitute } e \text{ for } \left(1 + \tfrac{1}{x}\right)^x.$$

When the amount invested grows exponentially according to the formula $A = Pe^{rt}$, we say that interest is **compounded continuously.**

Continuous Compound Interest Formula

If P dollars are deposited in an account earning interest at an annual rate r, compounded continuously, the amount A after t years is given by the formula

$$A = Pe^{rt}$$

EXAMPLE 5 If the parents of the newborn child in Example 4 had invested \$8,000 at an annual rate of 9%, compounded continuously, how much would the child have in the account in 55 years?

Solution

$$A = Pe^{rt}$$
$$A = 8{,}000e^{(0.09)(55)}$$
$$= 8{,}000e^{4.95}$$
$$\approx 1{,}129{,}399.711 \qquad \text{Use a calculator.}$$

In 55 years, the balance will be \$1,129,399.71, which is \$60,296.44 more than the amount earned with quarterly compounding.

Self Check Find the balance in 60 years. ■

To graph the exponential function $f(x) = e^x$, we plot several points and join them with a smooth curve (as in Figure 4-5(a)) or use a graphing calculator (as in Figure 4-5(b)).

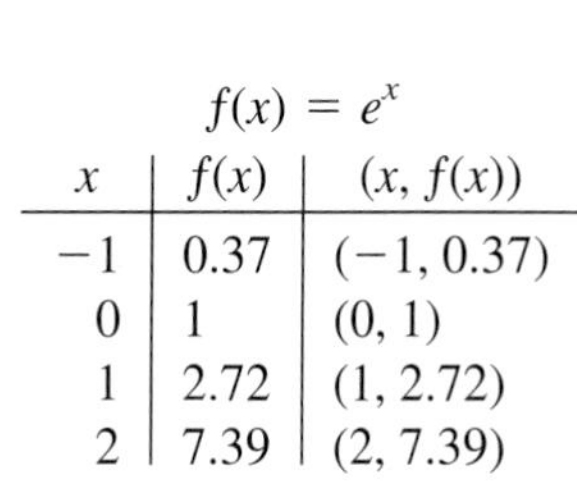

$f(x) = e^x$

x	$f(x)$	$(x, f(x))$
-1	0.37	$(-1, 0.37)$
0	1	$(0, 1)$
1	2.72	$(1, 2.72)$
2	7.39	$(2, 7.39)$

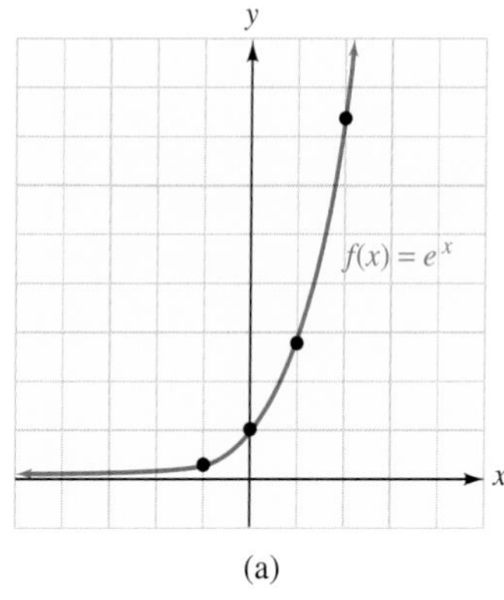

(a)

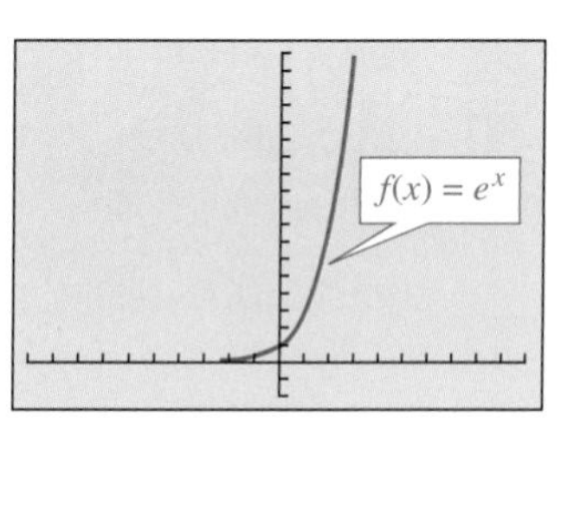

(b)

Figure 4-5

■ Vertical and Horizontal Translations

We have seen that when $k > 0$ the graph of

1. $y = f(x) + k$ **2.** $y = f(x) - k$ **3.** $y = f(x - k)$ **4.** $y = f(x + k)$

is identical to the graph of $y = f(x)$, except that it is translated k units

1. up **2.** down **3.** to the right **4.** to the left

ACTIVE EXAMPLE 6 On one set of axes, graph $f(x) = 2^x$ and $f(x) = 2^x + 3$.

Solution The graph of $f(x) = 2^x + 3$ is identical to the graph of $f(x) = 2^x$, except that it is translated 3 units up. (See Figure 4-6.)

COLLEGE **Algebra** f(x) **Now**™
Go to **http://1pass.thomson.com** or your CD to practice this example.

$f(x) = 2^x$

x	$f(x)$	$(x, f(x))$
-4	$\frac{1}{16}$	$\left(-4, \frac{1}{16}\right)$
0	1	$(0, 1)$
2	4	$(2, 4)$

$f(x) = 2^x + 3$

x	$f(x)$	$(x, f(x))$
-4	$3\frac{1}{16}$	$\left(-4, 3\frac{1}{16}\right)$
0	4	$(0, 4)$
2	7	$(2, 7)$

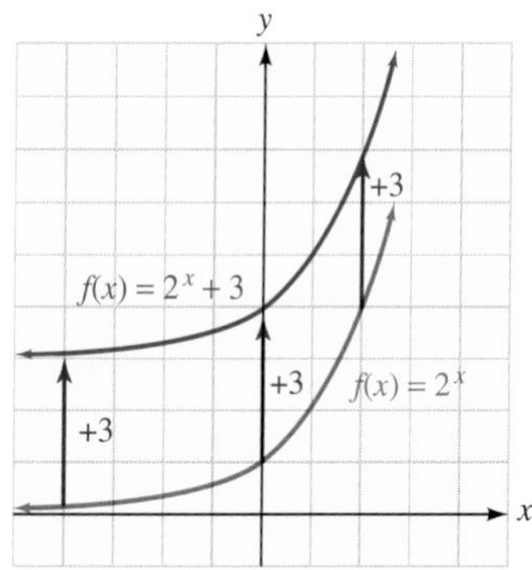

Figure 4-6

Self Check On one set of axes, graph $f(x) = 2^x$ and $f(x) = 2^x - 2$. ■

EXAMPLE 7 On one set of axes, graph $f(x) = e^x$ and $f(x) = e^{x-3}$.

Solution The graph of $f(x) = e^{x-3}$ is identical to the graph of $f(x) = e^x$, except that it is translated 3 units to the right. (See Figure 4-7.)

$f(x) = e^x$

x	$f(x)$	$(x, f(x))$
-1	0.37	$(-1, 0.37)$
0	1	$(0, 1)$
1	2.72	$(1, 2.72)$
2	7.39	$(2, 7.39)$

$f(x) = e^{x-3}$

x	$f(x)$	$(x, f(x))$
2	0.37	$(2, 0.37)$
3	1	$(3, 1)$
4	2.72	$(4, 2.72)$
5	7.39	$(5, 7.39)$

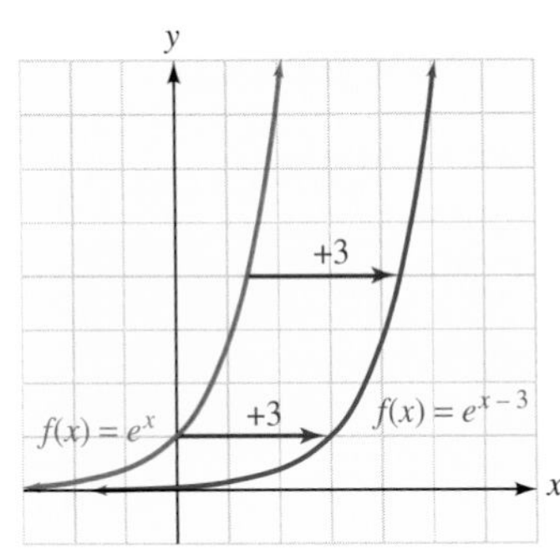

Figure 4-7

Self Check On one set of axes, graph $f(x) = e^x$ and $f(x) = e^{x+2}$. ■

■ Exponential Functions of the Form $f(x) = kb^x$ and $f(x) = b^{kx}$

The graphs of $f(x) = kb^x$ and $f(x) = b^{kx}$ are vertical and horizontal stretchings of the graph of $f(x) = b^x$. To graph these functions, we can use a graphing calculator.

Accent on Technology — GRAPHING EXPONENTIAL FUNCTIONS

To use a graphing calculator to graph the exponential function $f(x) = 2(3^{x/2})$, we enter the right-hand side of the equation after the symbol Y_1 =. The display will show the equation

$$Y_1 = 2(3 \wedge (X/2))$$

If we use window settings of $[-10, 10]$ for x and $[-2, 18]$ for y and press GRAPH, we will obtain the graph shown in Figure 4-8(a).

To graph the exponential function $f(x) = 3e^{-x/2}$, we enter the right-hand side of the equation after the symbol Y_1 =. The display will show the equation

$$Y_1 = 3(e \wedge (-x/2))$$

If we use window settings of $[-10, 10]$ for x and $[-2, 18]$ for y and press GRAPH, we will obtain the graph shown in Figure 4-8(b).

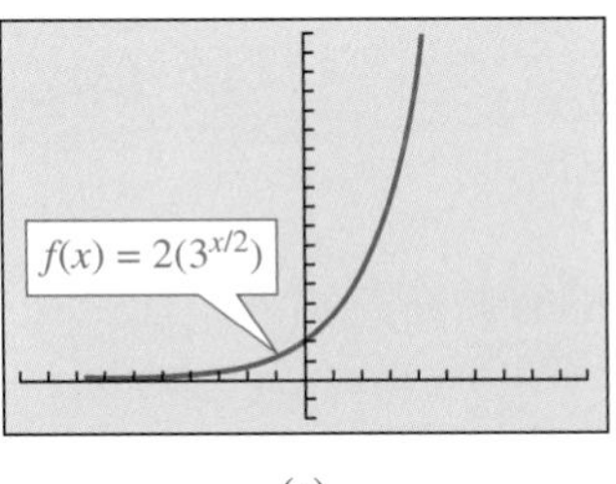

(a)

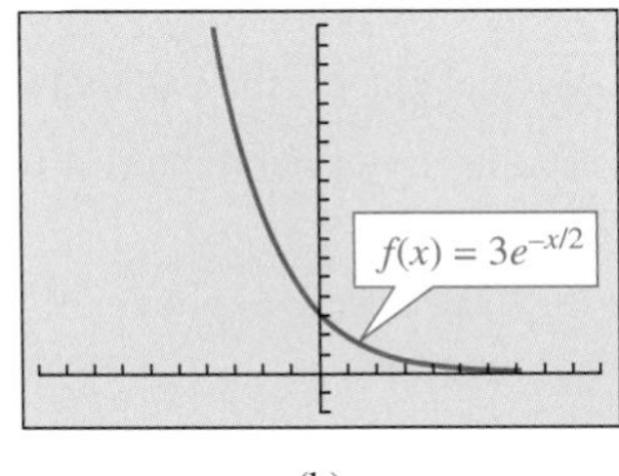

(b)

Figure 4-8

Self Check Answers

1. a. 64 **b.** $x^{3\sqrt{5}}$

2.

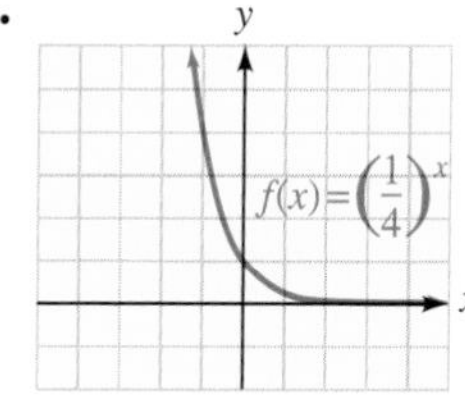

3. no
4. no
5. $1,771,251.33

6.

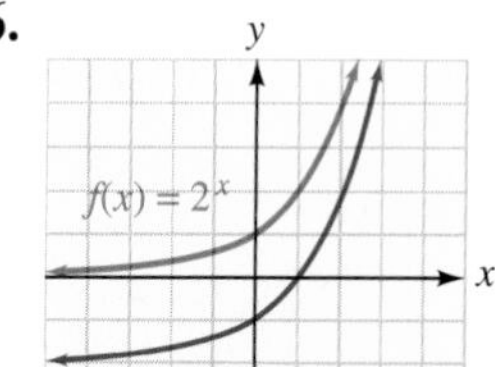

7.

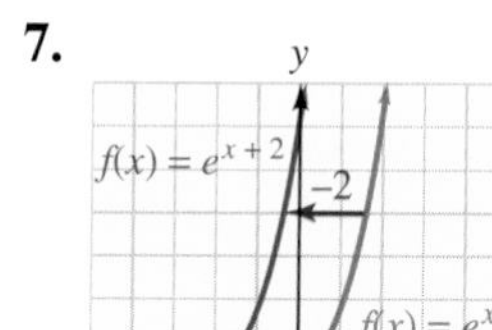

4.1 Exercises

VOCABULARY AND CONCEPTS ***Fill in the blanks.***

1. If $b > 0$ and $b \neq 1$, $y = b^x$ represents an **exponential** function.
2. If $f(x) = b^x$ represents an increasing function, then $b >$ **1**.
3. In interval notation, the domain of the exponential function $f(x) = b^x$ is $(-\infty, \infty)$.
4. The number b is called the **base** of the exponential function $y = b^x$.
5. The range of the exponential function $f(x) = b^x$ is $(0, \infty)$.
6. The graphs of all exponential functions $y = b^x$ have the same *y*-intercept, the point **(0, 1)**.
7. If $b > 0$ and $b \neq 1$, the graph of $y = b^x$ approaches the *x*-axis, which is called an **asymptote** of the curve.
8. If $f(x) = b^x$ represents a decreasing function, then **0** $< b <$ **1**.
9. If $b > 1$, then $y = b^x$ defines a (an) **increasing** function.
10. The graph of an exponential function $y = b^x$ always passes through the points (0, 1) and $(1, b)$.
11. To two decimal places, the value of e is **2.72**.
12. The continuous compound interest formula is $A = Pe^{rt}$.
13. Since $e > 1$, the base-e exponential function is a (an) **increasing** function.
14. The graph of the exponential function $y = e^x$ passes through the points (0, 1) and $(1, e)$.

PRACTICE ***Use a calculator to find each value to four decimal places.***

15. $4^{\sqrt{3}}$ **11.0357**
16. $5^{\sqrt{2}}$ **9.7385**
17. 7^{π} **451.8079**
18. $3^{-\pi}$ **0.0317**

Simplify each expression.

19. $5^{\sqrt{2}}5^{\sqrt{2}}$
$5^{2\sqrt{2}} = 25^{\sqrt{2}}$
20. $\left(5^{\sqrt{2}}\right)^{\sqrt{2}}$
25
21. $\left(a^{\sqrt{8}}\right)^{\sqrt{2}}$
a^4
22. $a^{\sqrt{12}}a^{\sqrt{3}}$
$a^{3\sqrt{3}}$

Graph each exponential function.

23. $f(x) = 3^x$

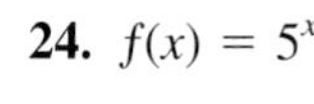

24. $f(x) = 5^x$

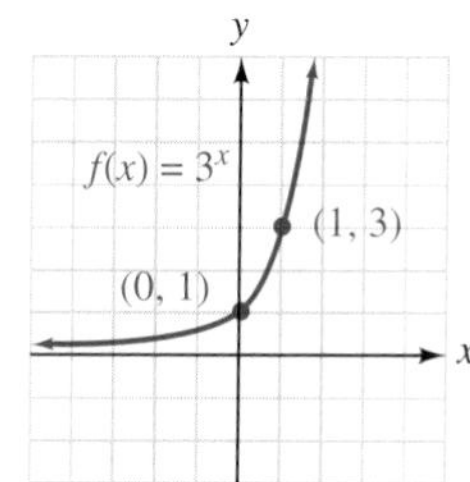

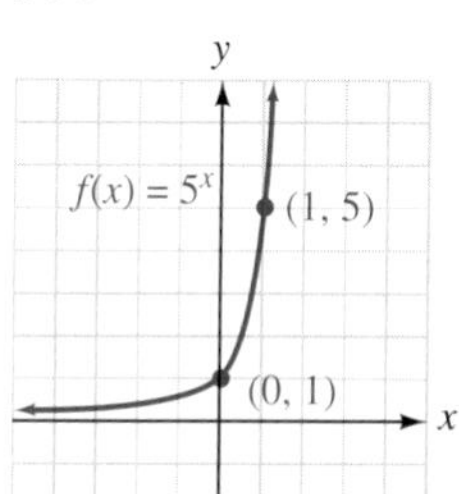

25. $f(x) = \left(\frac{1}{5}\right)^x$

26. $f(x) = \left(\frac{1}{3}\right)^x$

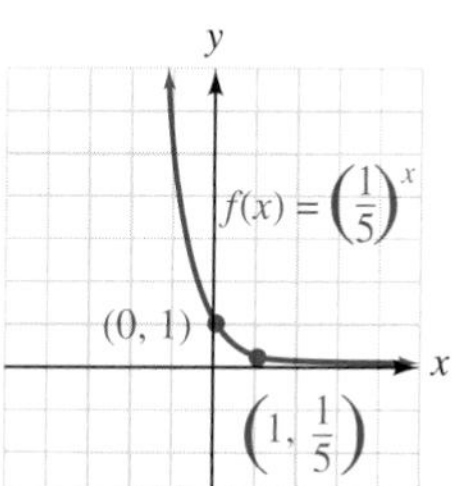

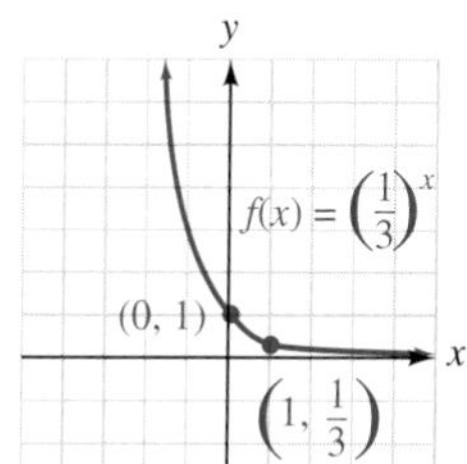

27. $f(x) = \left(\frac{3}{4}\right)^x$

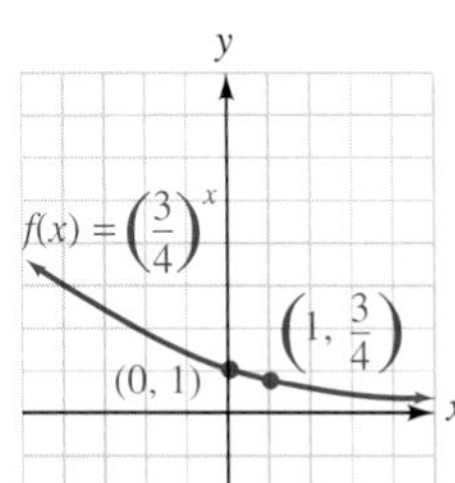

28. $f(x) = \left(\frac{4}{3}\right)^x$

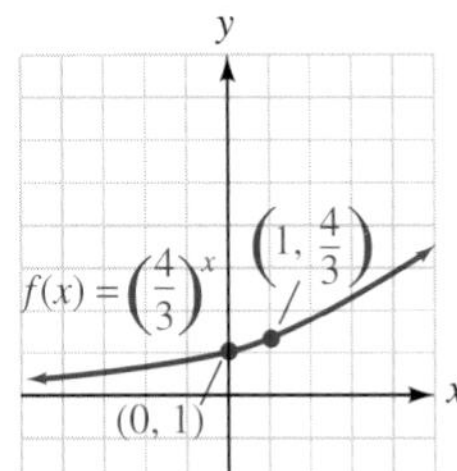

29. $f(x) = (1.5)^x$

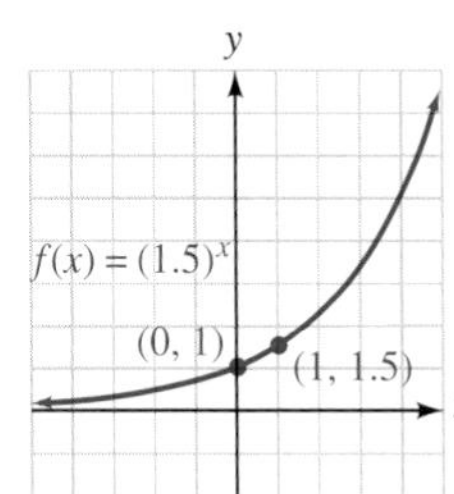

30. $f(x) = (0.3)^x$

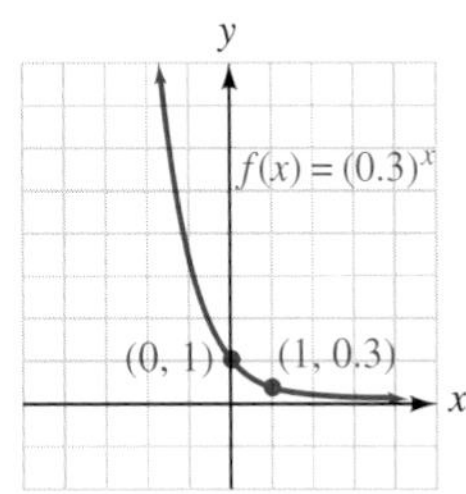

Determine whether the graph could represent an exponential function of the form $f(x) = b^x$.

31.

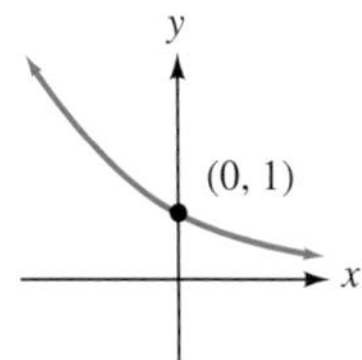

yes

32.

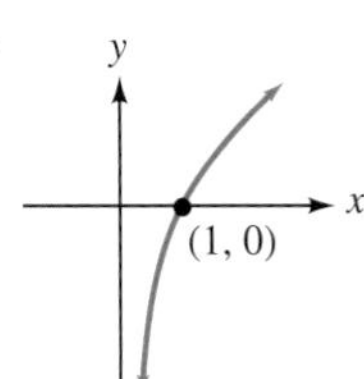

no

33.

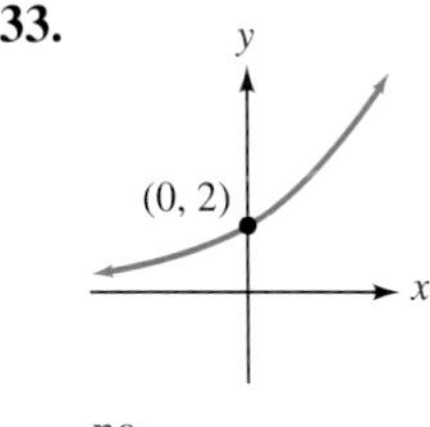

no

34.

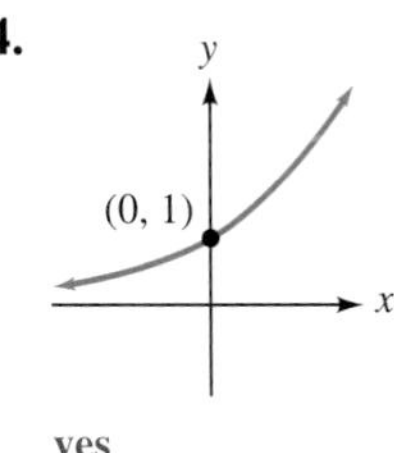

yes

Find the value of b, if any, that would cause the graph of $y = b^x$ to look like the graph indicated.

35.

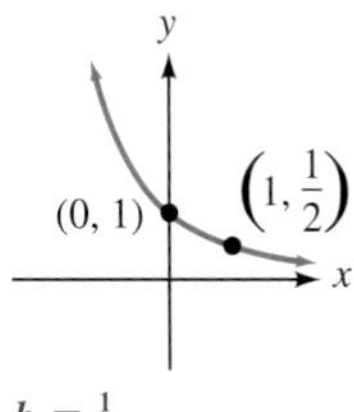

$b = \frac{1}{2}$

36.

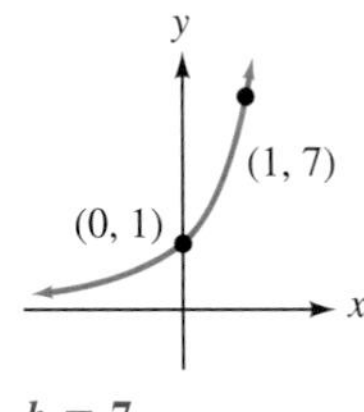

$b = 7$

37.

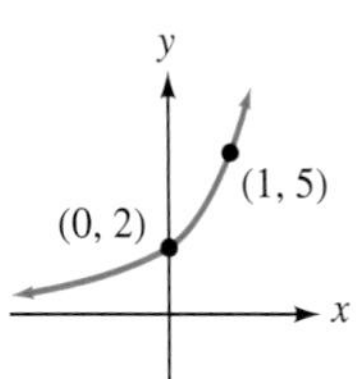

no value of b

38.

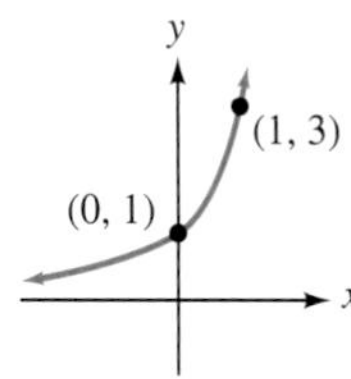

$b = 3$

39.

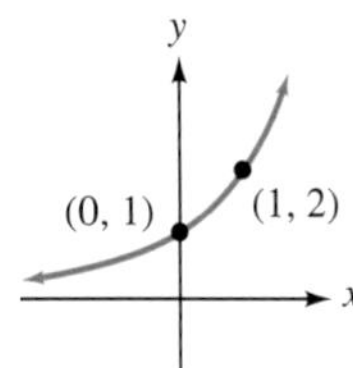

$b = 2$

40.

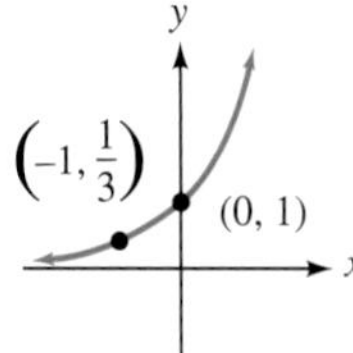

$b = 3$

41.

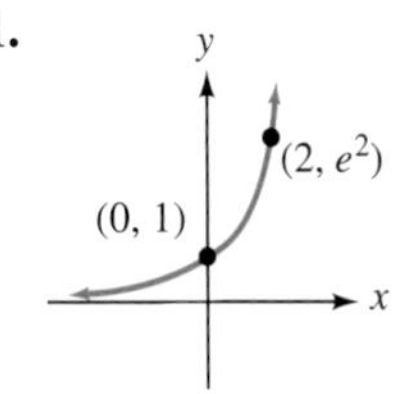

$b = e$

42.

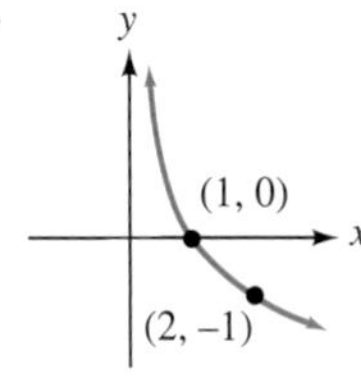

no value of b

***Graph each function using translations.* Do not use a graphing calculator.**

43. $f(x) = 3^x - 1$

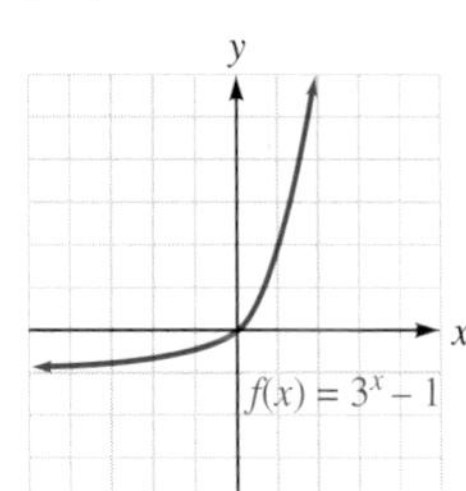

44. $f(x) = 2^x + 3$

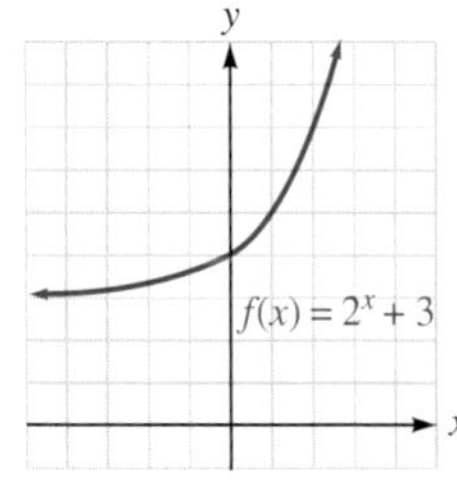

45. $f(x) = 2^x + 1$

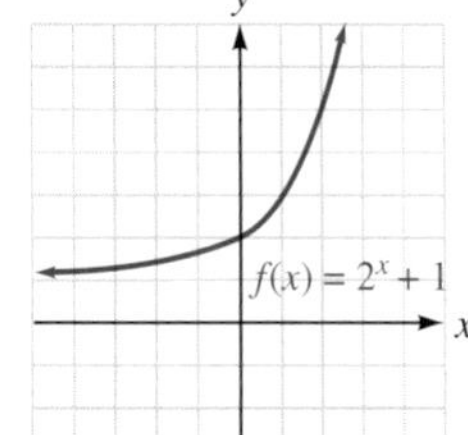

46. $f(x) = 4^x - 4$

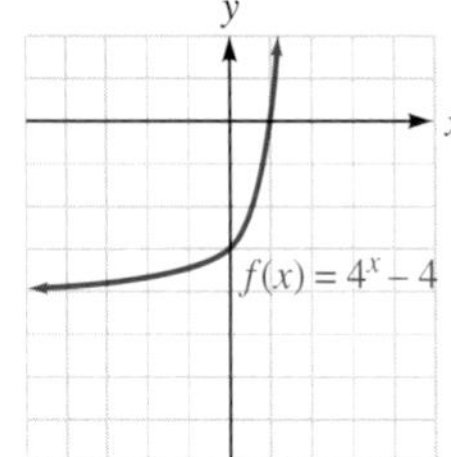

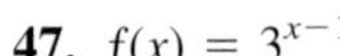

47. $f(x) = 3^{x-1}$

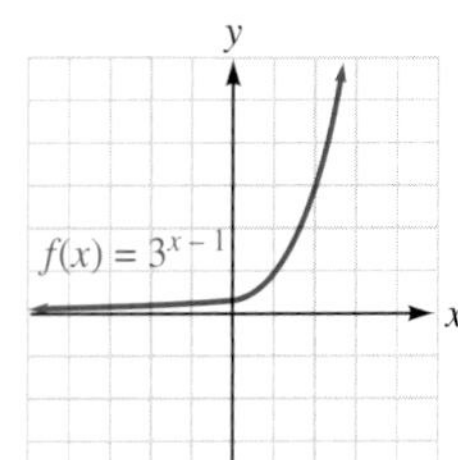

48. $f(x) = 2^{x+3}$

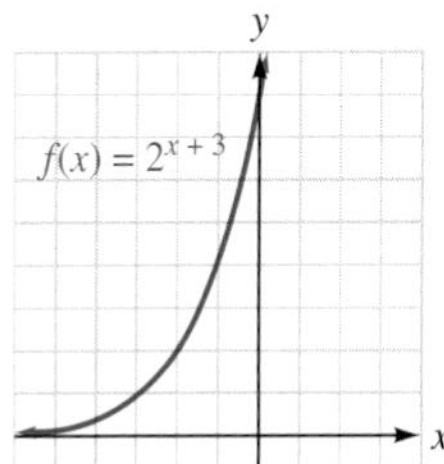

49. $f(x) = 3^{x+1}$

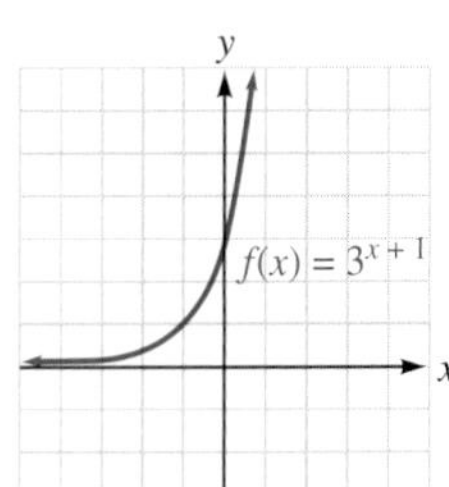

50. $f(x) = 2^{x-3}$

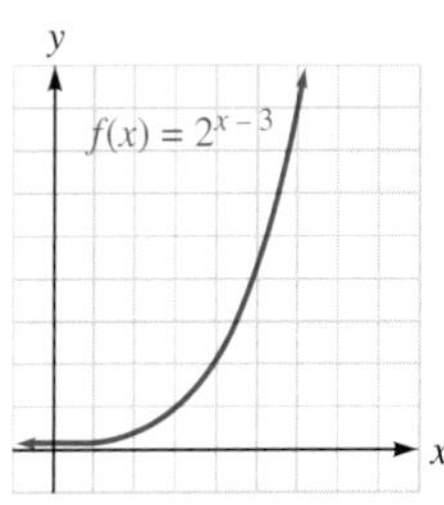

51. $f(x) = e^x - 4$

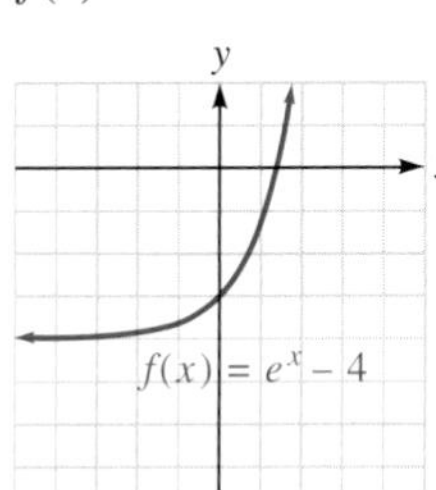

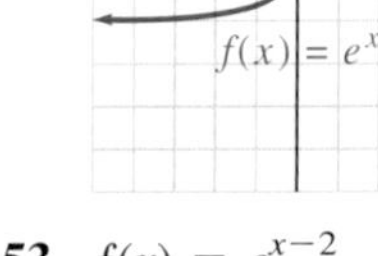

52. $f(x) = e^x + 2$

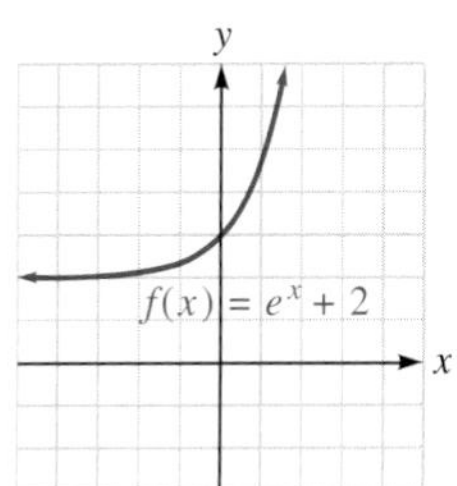
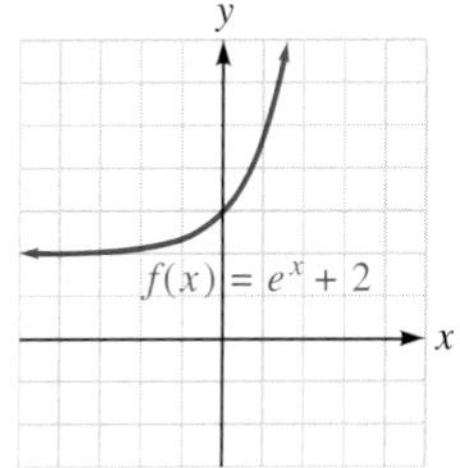

53. $f(x) = e^{x-2}$

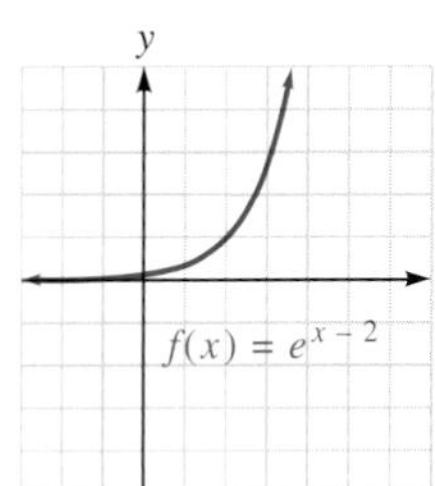

54. $f(x) = e^{x+3}$

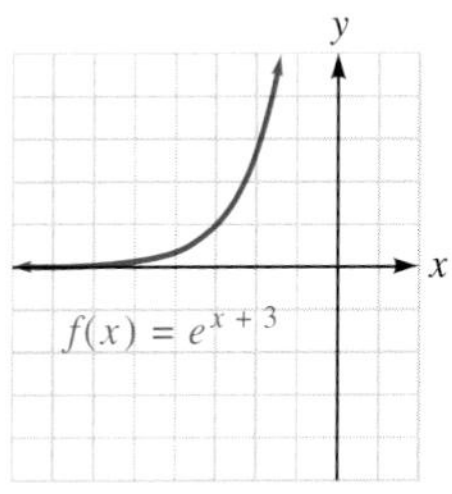

55. $f(x) = 2^{x+1} - 2$

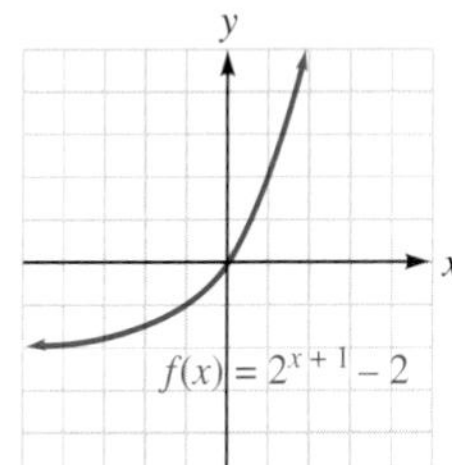

56. $f(x) = 3^{x-1} + 2$

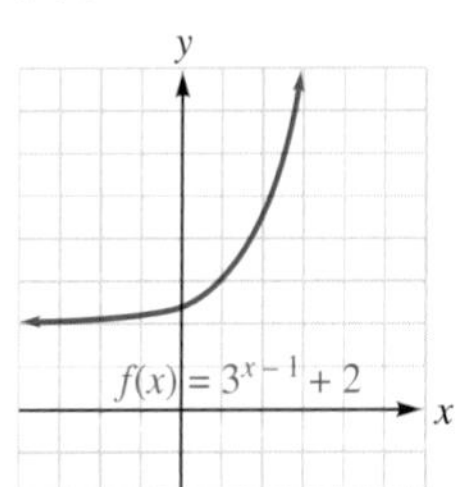

57. $y = 3^{x-2} + 1$

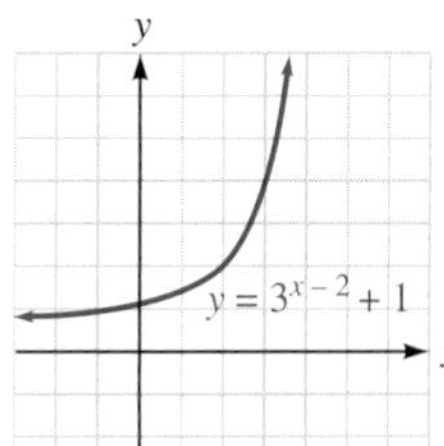

58. $y = 3^{x+2} - 1$

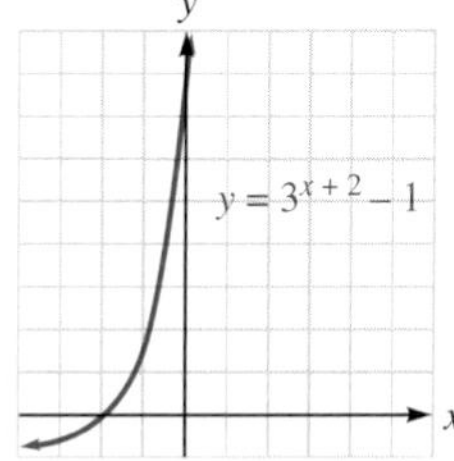

Use a graphing calculator to graph each function.

59. $f(x) = 5(2^x)$

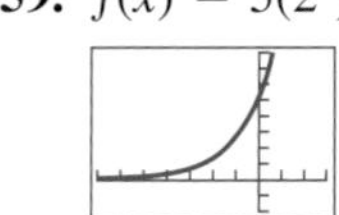

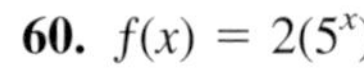

60. $f(x) = 2(5^x)$

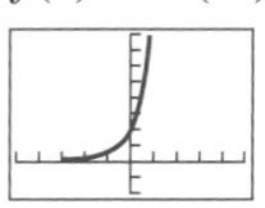

61. $f(x) = 3^{-x}$

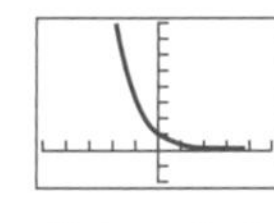

62. $f(x) = 2^{-x}$

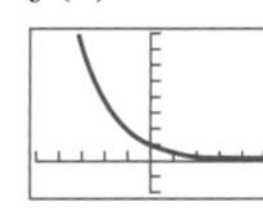

63. $f(x) = 2e^x$

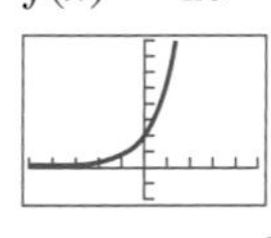

64. $f(x) = 3e^{-x}$

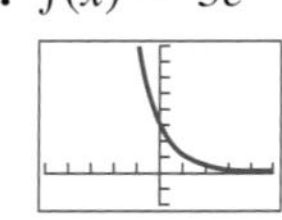

65. $f(x) = 5e^{-0.5x}$

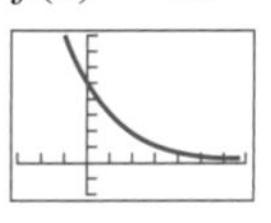

66. $f(x) = -3e^{2x}$

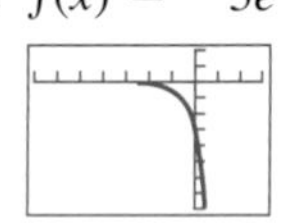

APPLICATIONS ***In Exercises 67–70, assume that there are no deposits or withdrawals.***

67. Compound interest An initial deposit of \$10,000 earns 8% interest, compounded quarterly. How much will be in the account in 10 years? **\$22,080.40**

68. Compound interest An initial deposit of \$1,000 earns 9% interest, compounded monthly. How much will be in the account in $4\frac{1}{2}$ years? **\$1,497.04**

69. Comparing interest rates How much more interest could \$500 earn in 5 years, compounded semiannually (two times a year), if the annual interest rate were $5\frac{1}{2}$% instead of 5%? **\$15.79**

70. Comparing savings plans Which institution in the ads provides the better investment? **Fidelity**

71. **Compound interest** If \$1 had been invested on July 4, 1776, at 5% interest, compounded annually, what would it be worth on July 4, 2076? **\$2,273,996.13**

72. **360/365 method** Some financial institutions pay daily interest, compounded by the 360/365 method, using the formula

$$A = A_0\left(1 + \frac{r}{360}\right)^{365t}$$

(t is in years). Using this method, what will an initial investment of \$1,000 be worth in 5 years, assuming a 7% annual interest rate? **\$1,425.93**

73. **Carrying charges** A college student takes advantage of the ad shown and buys a bedroom set for \$1,100. He plans to pay the \$1,100 plus interest when his income tax refund comes in 8 months. At that time, what will he need to pay? **\$1,263.77**

74. **Credit card interest** A bank credit card charges interest at the rate of 21% per year, compounded monthly. If a senior in college charges her last tuition bill of \$1,500 and intends to pay it in one year, what will she have to pay? **\$1,847.16**

75. **Continuous compound interest** An initial investment of \$5,000 earns 8.2% interest, compounded continuously. What will the investment be worth in 12 years? **\$13,375.68**

76. **Continuous compound interest** An initial investment of \$2,000 earns 8% interest, compounded continuously. What will the investment be worth in 15 years? **\$6,640.23**

77. **Comparison of compounding methods** An initial deposit of \$5,000 grows at an annual rate of 8.5% for 5 years. Compare the final balances resulting from continuous compounding and annual compounding. **\$7,647.95 from continuous compounding, \$7,518.28 from annual compounding**

78. **Comparison of compounding methods** An initial deposit of \$30,000 grows at an annual rate of 8% for 20 years. Compare the final balances resulting from continuous compounding and annual compounding. **\$148,590.97 from continuous compounding, \$139,828.71 from annual compounding**

79. **Frequency of compounding** \$10,000 is invested in each of two accounts, both paying 6% annual interest. In the first account, interest compounds quarterly, and in the second account, interest compounds daily. Find the difference between the accounts after 20 years. **\$291.27**

80. **Determining an initial deposit** An account now contains \$11,180 and has been accumulating interest at a 7% annual rate, compounded continuously, for 7 years. Find the initial deposit. **\$6,849.16**

81. **Saving for college** In 20 years, a father wants to accumulate \$40,000 to pay for his daughter's college expenses. If he can get 6% interest, compounded quarterly, how much must he invest now to achieve his goal? **\$12,155.61**

82. **Saving for college** In Problem 81, how much should he invest to achieve his goal if he can get 6% interest, compounded continuously? **\$12,047.77**

DISCOVERY AND WRITING

83. **Financial planning** To have \$$P$ available in n years, \$$A$ can be invested now in an account paying interest at an annual rate r, compounded annually. Show that

$$A = P(1 + r)^{-n}$$

84. If $2^{t+4} = k2^t$, find k. **16**

85. If $5^{3t} = k^t$, find k. **125**

86. **a.** If $e^{t+3} = ke^t$, find k. e^3

b. If $e^{3t} = k^t$, find k. e^3

REVIEW *Factor each expression completely.*

87. $x^2 + 9x^4$
$x^2(1 + 9x^2)$

88. $x^2 - 9x^4$
$x^2(1 + 3x)(1 - 3x)$

89. $x^2 + x - 12$
$(x + 4)(x - 3)$

90. $x^3 + 27$
$(x + 3)(x^2 - 3x + 9)$

4.2 Applications of Exponential Functions

In this section, you will learn about

- Radioactive Decay
- Oceanography
- Malthusian Population Growth
- Epidemiology
- The Malthusian Theory

A mathematical description of an observed event is called a **model** of that event. Many events that change with time can be modeled by exponential functions of the form

$$y = f(t) = ab^{kt}$$ Remember that ab^{kt} means $a(b^{kt})$.

where a, b, and k are constants and t represents time. If f is an increasing function, we say that *y grows exponentially.* If f is a decreasing function, we say that *y decays exponentially.*

Radioactive Decay

The atomic structure of a radioactive material changes as the material emits radiation. Uranium, for example, changes (decays) into thorium, then into radium, and eventually into lead.

Experiments have determined the time it takes for one-half of a sample of a given radioactive element to decompose. That time is a constant, called the element's **half-life.** The amount present decays exponentially according to this formula.

Radioactive Decay Formula

The amount A of radioactive material present at time t is given by

$$A = A_0 2^{-t/h}$$

where A_0 is the amount that was present initially (at $t = 0$) and h is the material's half-life.

ACTIVE EXAMPLE 1 The half-life of radium is approximately 1,600 years. How much of a 1-gram sample will remain after 1,000 years?

Solution In this example, $A_0 = 1$, $h = 1{,}600$, and $t = 1{,}000$. We substitute these values into the formula for radioactive decay and simplify.

$$A = A_0 2^{-t/h}$$

$$A = 1 \cdot 2^{-1{,}000/1{,}600}$$

$$\approx 0.648419777$$ Use a calculator.

After 1,000 years, approximately 0.65 gram of radium will remain.

COLLEGE **Algebra** $f(x)$ **Now**™

Explore this Active Example and test yourself on the steps taken here at **http://1pass.thomson.com**. You can also find this example on the Interactive Video Skillbuilder CD-ROM.

Self Check After 800 years, how much radium will remain. ■

Oceanography

ACTIVE EXAMPLE 2 The intensity I of light (in lumens) at a distance x meters below the surface of a body of water decreases exponentially according to the formula

COLLEGE **Algebra** $f(x)$ **Now**™
Go to **http://1pass.thomson.com** or your CD to practice this example.

$$I = I_0 k^x$$

where I_0 is the intensity of light above the water and k is a constant that depends on the clarity of the water. For a certain area of the Atlantic Ocean, $I_0 = 12$ and $k = 0.6$. Find the intensity of light at a depth of 5 meters.

Solution After substituting 12 for I_0 and 0.6 for k, we have the formula

$$I = 12(0.6)^x$$

We substitute 5 for x and calculate I.

$$I = 12(0.6)^5$$
$$I \approx 0.93312$$

At a depth of 5 meters, the intensity of the light is slightly less than 1 lumen.

Self Check Find the intensity at a depth of 10 meters. ■

Malthusian Population Growth

An equation based on the exponential function provides a model for **population growth.** One such model, called the **Malthusian model of population growth,** assumes a constant birth rate and a constant death rate. In this model, the population P grows exponentially according to the following formula.

Malthusian Model of Population Growth

If b is the annual birth rate, d is the annual death rate, t is the time (in years), P_0 is the initial population at $t = 0$, and P is the current population, then

$$P = P_0 e^{kt}$$

where $k = b - d$ is the **annual growth rate,** the difference between the annual birth rate and death rate.

ACTIVE EXAMPLE 3 The population of the United States is approximately 300 million people. Assuming that the annual birth rate is 19 per 1,000 and the annual death rate is 7 per 1,000, what does the Malthusian model predict the U.S. population will be in 50 years?

Solution Since k is the difference between the birth and death rates, we have

COLLEGE **Algebra** $f(x)$ **Now**™
Go to **http://1pass.thomson.com** or your CD to practice this example.

$$k = b - d$$

$$k = \frac{19}{1,000} - \frac{7}{1,000} \quad \text{Substitute } \tfrac{19}{1,000} \text{ for } b \text{ and } \tfrac{7}{1,000} \text{ for } d.$$

$$k = 0.019 - 0.007$$
$$= 0.012$$

We can now substitute 300,000,000 for P_0, 50 for t, and 0.012 for k in the formula for the Malthusian model of population growth and simplify.

$$P = P_0e^{kt}$$
$$P = (300{,}000{,}000)e^{(0.012)(50)}$$
$$= (300{,}000{,}000)e^{0.6}$$
$$\approx 546{,}635{,}640.1 \qquad \text{Use a calculator.}$$

After 50 years, the U.S. population will exceed 546 million people.

Self Check Find the population in 100 years. ■

■ Epidemiology

Many infectious diseases, including some caused by viruses, spread most rapidly when they first infect a population, but then more slowly as the number of uninfected individuals decreases. These situations are often modeled by a function, called a **logistic function,** of the form

$$P = \frac{M}{1 + \left(\frac{M}{P_0} - 1\right)e^{-kt}}$$

where P is the size of the infected population at any time t, P_0 is the infected population size at $t = 0$, and k is a constant determined by how contagious the virus is in a given environment. M is the theoretical maximum size of the population P.

EXAMPLE 4 In a city with a population of 1,200,000, there are currently 1,000 cases of infection with the HIV virus. If the spread of the disease is projected by the formula

$$P = \frac{1{,}200{,}000}{1 + (1{,}200 - 1)e^{-0.4t}}$$

how many people will be infected in 3 years?

Solution We can substitute 3 for t in the given formula and calculate P.

$$P = \frac{1{,}200{,}000}{1 + (1{,}200 - 1)e^{-0.4t}}$$
$$P = \frac{1{,}200{,}000}{1 + (1{,}199)e^{-0.4(3)}}$$
$$\approx 3{,}313.710094$$

In 3 years, approximately 3,300 people are expected to be infected.

Self Check How many will be infected in 10 years? ■

■ The Malthusian Theory

The English economist Thomas Robert Malthus (1766–1834) pioneered in population study. He believed that poverty and starvation were unavoidable, because the human population tends to grow exponentially, whereas the food supply tends to grow linearly.

EXAMPLE 5 Suppose that a country with a population of 1,000 people is growing exponentially according to the formula

$$P = 1{,}000e^{0.02t}$$

where t is in years. Furthermore, assume that the food supply, measured in adequate food per day per person, is growing linearly according to the formula

$$y = 30.625x + 2{,}000$$

In how many years will the population outstrip the food supply?

Solution We can use a graphing calculator with window settings of [0, 100] for x and [0, 10,000] for y. After graphing the functions as shown in Figure 4-9, we trace and zoom to find the point where the two graphs intersect. From the graph, we can see that the food supply will be adequate for about 71 years. At that time, the population of approximately 4,200 people will have problems.

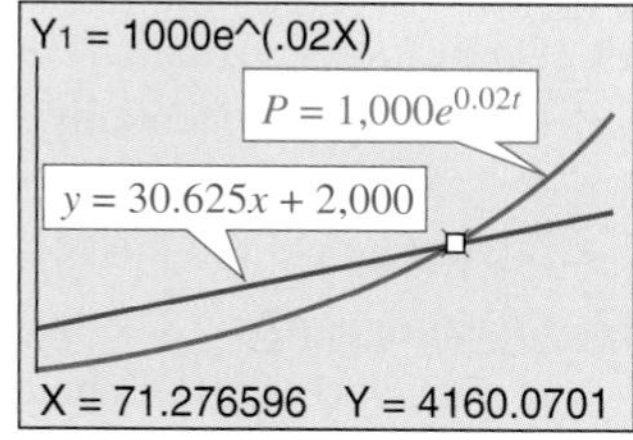

Figure 4-9

■

Self Check Answers

1. about 0.71 g **2.** 0.073 lumen **3.** over 996 million **4.** about 52,000

4.2 Exercises

VOCABULARY AND CONCEPTS ***Fill in the blanks.***

1. The Malthusian model assumes a constant **birth** rate and a constant **death** rate.
2. The Malthusian prediction is pessimistic, because a **population** grows exponentially, but food supplies grow **linearly**.

APPLICATIONS ***Use a calculator to help solve each problem.***

3. **Tritium decay** Tritium, a radioactive isotope of hydrogen, has a half-life of 12.4 years. Of an initial sample of 50 grams, how much will remain after 100 years? **0.1868 g**
4. **Chernobyl** In April 1986, the world's worst nuclear power disaster occurred at Chernobyl in the former USSR. An explosion released about 1,000 kilograms of radioactive cesium-137 (^{137}Cs) into the atmosphere. If the half-life of ^{137}Cs is 30.17 years, how much will remain in the atmosphere in 100 years? **about 101 kg**
5. **Chernobyl** Refer to Exercise 4. How much ^{137}Cs will remain in 200 years? **about 10 kg**
6. **Carbon-14 decay** The half-life of radioactive carbon-14 is 5,700 years. How much of an initial sample will remain after 3,000 years? **about 69.4%**
7. **Plutonium decay** One of the isotopes of plutonium, ^{237}Pu, decays with a half-life of 40 days. How much of an initial sample will remain after 60 days? **about 35.4%**
8. **Comparing radioactive decay** One isotope of holmium, ^{162}Ho, has a half-life of 22 minutes. The half-life of a second isotope, ^{164}Ho, is 37 minutes. Starting with a sample containing equal amounts, find the ratio of the amounts of ^{162}Ho to ^{164}Ho after one hour. **0.465**
9. **Drug absorption in smokers** The biological half-life of the asthma medication theophylline is 4.5 hours for smokers. Find the amount of the drug retained in a smoker's system 12 hours after a dose of 1 unit is taken. **0.1575 unit**

10. Drug absorption in nonsmokers For a nonsmoker, the biological half-life of theophylline is 8 hours. Find the amount of the drug retained in a nonsmoker's system 12 hours after taking a one-unit dose. **0.3536 unit**

11. Bluegill population A Wisconsin lake is stocked with 10,000 bluegill. The population is expected to grow exponentially according to the model $P = P_0 2^{t/2}$. How many bluegill will be in the lake in 5 years? **about 56,570**

12. Community growth The population of Eagle River is growing exponentially according to the model $P = 375(1.3)^t$, where t is measured in years from the present date. Find the population in 3 years. **824**

13. Oceanography The intensity I of light (in lumens) at a distance x meters below the surface is given by $I = I_0 k^x$, where I_0 is the intensity at the surface and k depends on the clarity of the water.

At one location in the Arctic Ocean, $I_0 = 8$ and $k = 0.5$. Find the intensity at a depth of 2 meters. **2 lumens**

14. Oceanography At one location in the Atlantic Ocean, $I_0 = 14$ and $k = 0.7$. Find the intensity of light at a depth of 12 meters. (See Exercise 13.) **0.194 lumen**

15. Newton's law of cooling Some hot water, initially at 100°C, is placed in a room with a temperature of 40°C. The temperature T of the water after t hours is given by $T = 40 + 60(0.75)^t$. Find the temperature in $3\frac{1}{2}$ hours. **61.9°C**

16. Bacterial cultures A colony of 6 million bacteria is growing in a culture medium. The population P after t hours is given by the formula $P = (6 \times 10^6)(2.3)^t$. Find the population after 4 hours. **167,904,600**

17. Population growth The growth of a town's population is modeled by $P = 173e^{0.03t}$. How large will the population be when $t = 20$? **315**

18. Population decline The decline of a city's population is modeled by $P = 1.2 \times 10^6 e^{-0.008t}$. How large will the population be when $t = 30$? **9.44×10^5**

19. Epidemics The spread of hoof and mouth disease through a herd of cattle can be modeled by the formula $P = P_0 e^{0.27t}$, where P is the size of the infected population, P_0 is the infected population size at $t = 0$, and t is in days. If a rancher does not act quickly to treat two cases, how many cattle will have the disease in one week? **13**

20. Alcohol absorption In one individual, the percent of alcohol absorbed into the bloodstream after drinking two shots of whiskey is given by

$$P = 0.3(1 - e^{-0.05t})$$

where t is in minutes. Find the percent of alcohol absorbed into the blood after $\frac{1}{2}$ hour. **23.3%**

21. World population growth The population of the Earth is approximately 6 billion people and is growing at an annual rate of 1.9%. Assuming a Malthusian growth model, find the world population in 30 years. **10.6 billion**

22. World population growth See Exercise 21. Assuming a Malthusian growth model, find the world population in 40 years. **12.8 billion**

23. World population growth See Exercise 21. By what factor will the current population of the Earth increase in 50 years? **2.6**

24. World population growth See Exercise 21. By what factor will the current population of the Earth increase in 100 years? **6.7**

25. Drug absorption The percent P of the drug triazolam (a drug for treating insomnia) remaining in a person's bloodstream after t hours is given by $P = e^{-0.3t}$. What percent will remain in the bloodstream after 24 hours? **about 0.07%**

26. Medicine The concentration x of a certain drug in an organ after t minutes is given by $x = 0.08(1 - e^{-0.1t})$. Find the concentration of the drug in $\frac{1}{2}$ hour. **0.076**

27. Medicine Refer to Exercise 26. Find the initial concentration of the drug (*Hint:* when $t = 0$). **0**

28. Spreading the news Suppose the function

$$N = P(1 - e^{-0.1t})$$

is used to model the length of time t (in hours) it takes for N people living in a town with population P to hear a news flash. How many people in a town of 50,000 will hear the news between 1 and 2 hours after it happened? **4,305**

29. Spreading the news How many people in the town described in Problem 28 will not have heard the news after 10 hours? **18,394**

30. Epidemics Refer to Example 4. How many people will have the HIV virus in 5 years? **about 7,350**

31. Epidemics Refer to Example 4. How many people will have the HIV virus in 8 years? **about 24,060**

32. Life expectancy The life expectancy of white females can be estimated by using the function $l = 78.5(1.001)^x$, where x is the current age. Find the life expectancy of a white female who is currently 50 years old. Give the answer to the nearest tenth. **82.5 yr**

33. Oceanography The width w (in millimeters) of successive growth spirals of the sea shell *Catapulus voluto,* shown in the illustration, is given by the function $w = 1.54e^{0.503n}$, where n is the spiral number. To the nearest tenth of a millimeter, find the width of the fifth spiral. **19.0 mm**

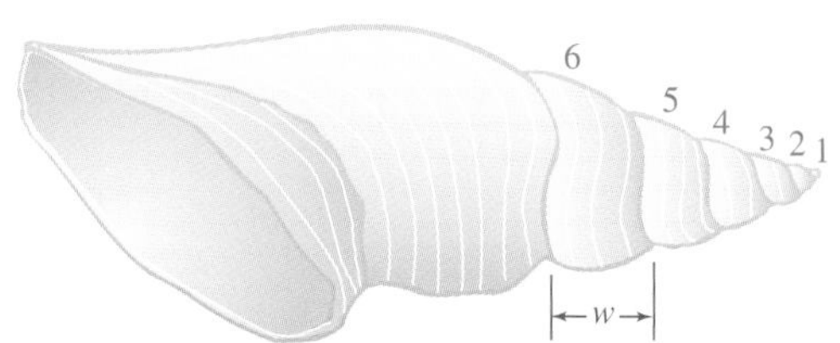

34. Skydiving Before the parachute opens, the velocity v (in meters per second) of a skydiver is given by $v = 50(1 - e^{-0.2t})$. Find the initial velocity. **0 mps**

35. Skydiving Refer to Exercise 34 and find the velocity after 20 seconds. **49 mps**

36. Free-falling objects After t seconds, a certain falling object has a velocity v given by $v = 50(1 - e^{-0.3t})$. Which is falling faster after 2 seconds, this object or the skydiver in Exercise 34? **this object**

37. Population growth In 1999, the male population of the United States was about 133 million, and the female population was about 139 million. Assuming a Malthusian growth model with a 1% annual growth rate, how many more females than males will there be in 20 years? **about 7 million**

38. Population growth See Exercise 37. How many more females than males will there be in 50 years? **about 10 million**

Use a graphing calculator to solve each problem.

39. In Example 5, suppose that better farming methods change the formula for food growth to $y = 31x + 2{,}000$. How long will the food supply be adequate? **about 72.2 years**

40. In Example 5, suppose that a birth control program changed the formula for population growth to $P = 1{,}000e^{0.01t}$. How long will the food supply be adequate? **about 215 years**

DISCOVERY AND WRITING

41. The value of e can be calculated to any degree of accuracy by adding the first several terms of the following list:

$$1,\ 1,\ \frac{1}{2},\ \frac{1}{2 \cdot 3},\ \frac{1}{2 \cdot 3 \cdot 4},\ \frac{1}{2 \cdot 3 \cdot 4 \cdot 5},\ \ldots$$

The more terms that are added, the closer the sum will be to e. Add the first six numbers in the preceding list. To how many decimal places is the sum accurate?

42. Graph the function defined by the equation $f(x) = \frac{e^x + e^{-x}}{2}$ from $x = -2$ to $x = 2$. The graph will look like a parabola, but it is not. The graph, called a **catenary,** is important in the design of power distribution networks, because it represents the shape of a uniform flexible cable whose ends are suspended from the same height. The function is called the **hyperbolic cosine function.**

43. Graph the logistic function in Example 4:

$$P = \frac{1{,}200{,}000}{1 + (1{,}199)e^{-0.4t}}$$

Use window settings of [0, 20] for x and [0, 1,500,000] for y.

44. Use the trace capabilities of your graphing calculator to explore the logistic function of Example 4 and Exercise 43. As time passes, what value does P approach? How many years does it take for 20% of the population to become infected? For 80%?

REVIEW *Find the value of x that makes each statement true.*

45. $2^3 = x$ 8

46. $3^x = 9$ 2

47. $x^3 = 27$ 3

48. $3^{-2} = x$ $\frac{1}{9}$

49. $x^{-3} = \frac{1}{8}$ 2

50. $3^x = \frac{1}{3}$ -1

51. $9^{1/2} = x$ 3

52. $x^{1/3} = 3$ 27

4.3 Logarithmic Functions and Their Graphs

In this section, you will learn about

- Logarithms ■ Base-10 Logarithms ■ Base-*e* Logarithms
- Graphs of Logarithmic Functions
- Vertical and Horizontal Translations

Since exponential functions are one-to-one functions, each one has an inverse. For example, to find the inverse of the function $y = 3^x$, we interchange the positions of x and y to obtain $x = 3^y$. The graphs of these two functions are shown in Figure 4-10(a).

To find the inverse of the function $y = \left(\frac{1}{3}\right)^x$, we again interchange the positions of x and y to obtain $x = \left(\frac{1}{3}\right)^y$. The graphs of these two functions are shown in Figure 4-10(b).

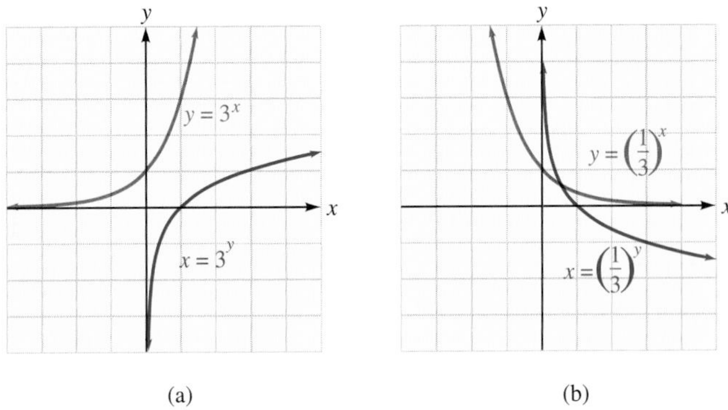

Figure 4-10

In general, the inverse of the function $y = b^x$ is $x = b^y$. When $b > 0$, their graphs appear as shown in Figure 4-11(a). When $0 < b < 1$, their graphs appear as shown in Figure 4-11(b).

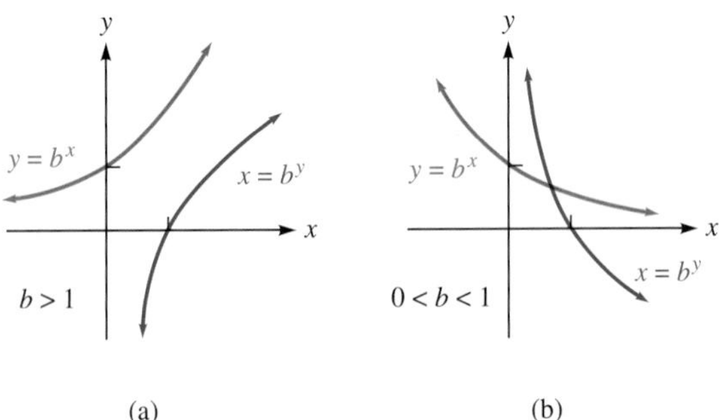

Figure 4-11

■ Logarithms

Since an exponential function defined by $y = b^x$ is one-to-one, it has an inverse function that is defined by the equation $x = b^y$. To express this inverse function in the form $y = f^{-1}(x)$, we must solve the equation $x = b^y$ for y. To do this, we need the following definition.

Logarithmic Functions

If $b > 0$ and $b \neq 1$, the **logarithmic function with base *b*** is defined by

$$y = \log_b x \qquad \text{if and only if} \qquad x = b^y$$

The **domain of the logarithmic function** is the interval $(0, \infty)$. The **range** is the interval $(-\infty, \infty)$. The logarithmic function is also denoted as $f(x) = \log_b x$.

The range of the logarithmic function is the set of real numbers, because the value of y in the equation $x = b^y$ can be any real number. The domain is the set of positive numbers, because the value of x in the equation $x = b^y$ $(b > 0)$ is always positive.

Since the function $y = \log_b x$ is the inverse of the one-to-one exponential function $y = b^x$, the logarithmic function is also one-to-one.

The expression $x = b^y$ is said to be written in *exponential form.* The equivalent expression $y = \log_b x$ is said to be written in *logarithmic form.* To translate from one form to the other, it is helpful to keep track of the base and the exponent.

Exponential form	***Logarithmic form***
$x = b^y$	$y = \log_b x$
↑↑	↑ ↑
Base Exponent	**Exponent Base**

Comment Since the domain of the logarithmic function is the set of positive numbers, the logarithm of 0 and the logarithm of a negative number are undefined in the set of real numbers.

The definition of logarithm guarantees that any pair (x, y) that satisfies the equation $y = \log_b x$ also satisfies the equation $x = b^y$.

$\log_b x = y$	because	$x = b^y$
$\log_5 25 = 2$	because	$25 = 5^2$
$\log_7 1 = 0$	because	$1 = 7^0$
$\log_{16} 4 = \frac{1}{2}$	because	$4 = 16^{1/2}$
$\log_2 \frac{1}{8} = -3$	because	$\frac{1}{8} = 2^{-3}$

In each of these examples, the logarithm of a number is an exponent. In fact,

$\log_b x$ is the exponent to which b is raised to get x.

To express this as an equation, we write

$$b^{\log_b x} = x$$

COLLEGE **Algebra** f(x) **Now**™

Explore this Active Example and test yourself on the steps taken here at **http://1pass.thomson.com**. You can also find this example on the Interactive Video Skillbuilder CD-ROM.

ACTIVE EXAMPLE 1 Find y in each equation: **a.** $\log_2 8 = y$, **b.** $\log_5 1 = y$, and **c.** $\log_7 \frac{1}{49} = y$.

Solution

a. $\log_2 8 = y$ is equivalent to $8 = 2^y$. Since $8 = 2^3$, we have $2^y = 2^3$ and $y = 3$.

b. $\log_5 1 = y$ is equivalent to $1 = 5^y$. Since $1 = 5^0$, we have $5^y = 5^0$ and $y = 0$.

c. $\log_7 \frac{1}{49} = y$ is equivalent to $\frac{1}{49} = 7^y$. Since $\frac{1}{49} = 7^{-2}$, we have $7^y = 7^{-2}$ and $y = -2$.

Self Check Find y in each equation: **a.** $\log_3 9 = y$, **b.** $\log_2 16 = y$, and **c.** $\log_5 \dfrac{1}{25} = y$. ■

ACTIVE EXAMPLE 2 Find a in each equation: **a.** $\log_a 32 = 5$, **b.** $\log_9 a = -\dfrac{1}{2}$, and **c.** $\log_9 3 = a$.

Solution

a. $\log_a 32 = 5$ is equivalent to $a^5 = 32$. Since $2^5 = 32$, we have $a^5 = 2^5$ and $a = 2$.

b. $\log_9 a = -\frac{1}{2}$ is equivalent to $9^{-1/2} = a$. Since $9^{-1/2} = \frac{1}{3}$, it follows that $a = \frac{1}{3}$.

c. $\log_9 3 = a$ is equivalent to $3 = 9^a$. Since $3 = 9^{1/2}$, we have $9^a = 9^{1/2}$ and $a = \frac{1}{2}$.

COLLEGE **Algebra** $f(x)$ **Now**™
Go to **http://1pass.thomson.com** or your CD to practice this example.

Self Check Find d in each equation: **a.** $\log_4 \frac{1}{16} = d$, **b.** $\log_d 36 = 2$, and **c.** $\log_8 d = -\frac{1}{3}$. ■

■ Base-10 Logarithms

Many applications use base-10 logarithms (also called **common logarithms**). When the base b is not indicated in the notation $\log x$, we assume that $b = 10$:

$$\log x \quad \text{means} \quad \log_{10} x$$

Because base-10 logarithms appear so often, you should become familiar with the following base-10 logarithms:

$$\log_{10} \frac{1}{100} = -2 \quad \text{because} \quad 10^{-2} = \frac{1}{100}$$

$$\log_{10} \frac{1}{10} = -1 \quad \text{because} \quad 10^{-1} = \frac{1}{10}$$

$$\log_{10} 1 = 0 \quad \text{because} \quad 10^0 = 1$$

$$\log_{10} 10 = 1 \quad \text{because} \quad 10^1 = 10$$

$$\log_{10} 100 = 2 \quad \text{because} \quad 10^2 = 100$$

$$\log_{10} 1{,}000 = 3 \quad \text{because} \quad 10^3 = 1{,}000$$

In general, we have

$$\log_{10} 10^x = x$$

Accent on Technology USING CALCULATORS TO FIND LOGARITHMS

Before calculators, extensive tables were used to provide logarithms of numbers. Today, logarithms are easy to find with a calculator. For example, to find log 2.34 with a scientific calculator, we enter these numbers and press these keys:

Continued

2.34 LOG

The display will read .369215857. To four decimal places, $\log 2.34 = 0.3692$.

To use a graphing calculator, we enter these numbers and press these keys:

LOG 2.34 ENTER

The display will read log 2.34
.3692158574

EXAMPLE 3 Find x in the equation $\log x = 0.7482$ to four decimal places.

Solution The equation $\log x = 0.7482$ is equivalent to $10^{0.7482} = x$. To find x with a calculator, we enter these numbers and press these keys:

Scientific calculator	***Graphing calculator***
10 y^x .7482 =	10 ^ .7482 ENTER

or

.7482 10^x	10^x .7482 ENTER

Either way, the result is 5.600154388. To four decimal places, $x = 5.6002$.

Self Check Solve: $\log x = 1.87737$. Give the result to four decimal places. ■

■ Base-*e* Logarithms

We have seen the importance of the number e in mathematical models of events in nature. Base-e logarithms are just as important. They are called **natural logarithms** or **Napierian logarithms** after John Napier (1550–1617). They are usually written as $\ln x$, rather than $\log_e x$:

$$\ln x \quad \text{means} \quad \log_e x$$

Like all logarithmic functions, the domain of $f(x) = \ln x$ is the interval $(0, \infty)$, and the range is the interval $(-\infty, \infty)$.

To estimate the base-e logarithms of numbers, we can use a calculator.

Accent on Technology — USING CALCULATORS TO FIND LOGARITHMS

To use a calculator to estimate the value of $\ln 2.34$, we enter these numbers and press these keys:

Scientific calculator	***Graphing calculator***
2.34 LN	LN 2.34 ENTER

Either way, the result is .8501509294. To four decimal places, $\ln 2.34 = 0.8502$.

EXAMPLE 4 Use a calculator to find **a.** $\ln 17.32$ and **b.** $\ln(\log 0.05)$.

Solution **a.** Enter these numbers and press these keys:

Scientific calculator	*Graphing calculator*
17.32 LN	LN 17.32 ENTER

Either way, the result is 2.851861903.

b. Enter these numbers and press these keys:

Scientific calculator	*Graphing calculator*
0.05 LOG LN	LN LOG 0.05) ENTER

Either way, we obtain an error, because log 0.05 is a negative number, and we cannot take the logarithm of a negative number.

Self Check Find each value to four decimal places: **a.** $\ln \pi$ and **b.** $\ln\left(\log \frac{1}{3}\right)$. ■

EXAMPLE 5 Solve each equation: **a.** $\ln x = 1.335$ and **b.** $\ln x = \log 5.5$. Give each result to four decimal places.

Solution **a.** The equation $\ln x = 1.335$ is equivalent to $e^{1.335} = x$. To use a calculator to find x, we enter these numbers and press these keys:

Scientific calculator	*Graphing calculator*
1.335 e^x	e^x 1.335 ENTER

Either way, the result is 3.799995946. To four decimal places, $x = 3.8000$.

b. The equation $\ln x = \log 5.5$ is equivalent to $e^{\log 5.5} = x$. To use a calculator to find x, we enter these numbers and press these keys:

Scientific calculator	*Graphing calculator*
5.5 LOG e^x	e^x LOG 5.5 ENTER

Either way, the result is 2.096695826. To four decimal places, $x = 2.0967$.

Self Check Solve: **a.** $\ln x = 1.9344$ and **b.** $\log x = \ln 3.2$. Give each result to four decimal places. ■

■ Graphs of Logarithmic Functions

To graph the logarithmic function $y = f(x) = \log_2 x$, we calculate and plot several points with coordinates (x, y) that satisfy the equivalent equation $x = 2^y$. After joining these points with a smooth curve, we have the graph shown in Figure 4-12(a).

To graph $y = f(x) = \log_{1/2} x$, we calculate and plot several points with coordinates (x, y) that satisfy the equation $x = \left(\frac{1}{2}\right)^y$. After joining these points with a smooth curve, we have the graph shown in Figure 4-12(b).

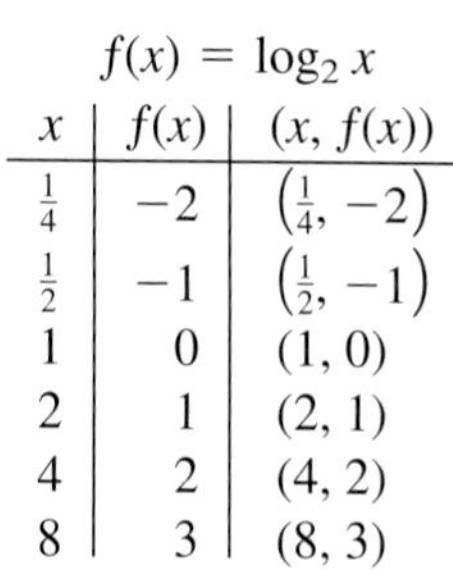

$f(x) = \log_2 x$

x	$f(x)$	$(x, f(x))$
$\frac{1}{4}$	-2	$\left(\frac{1}{4}, -2\right)$
$\frac{1}{2}$	-1	$\left(\frac{1}{2}, -1\right)$
1	0	(1, 0)
2	1	(2, 1)
4	2	(4, 2)
8	3	(8, 3)

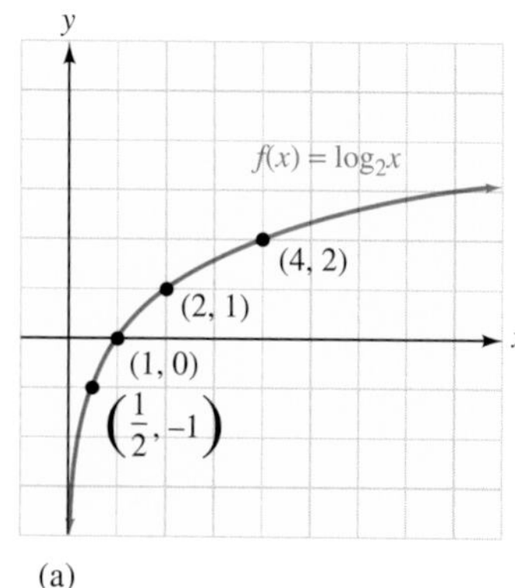

(a)

$f(x) = \log_{1/2} x$

x	$f(x)$	$(x, f(x))$
$\frac{1}{4}$	2	$\left(\frac{1}{4}, 2\right)$
$\frac{1}{2}$	1	$\left(\frac{1}{2}, 1\right)$
1	0	(1, 0)
2	-1	$(2, -1)$
4	-2	$(4, -2)$
8	-3	$(8, -3)$

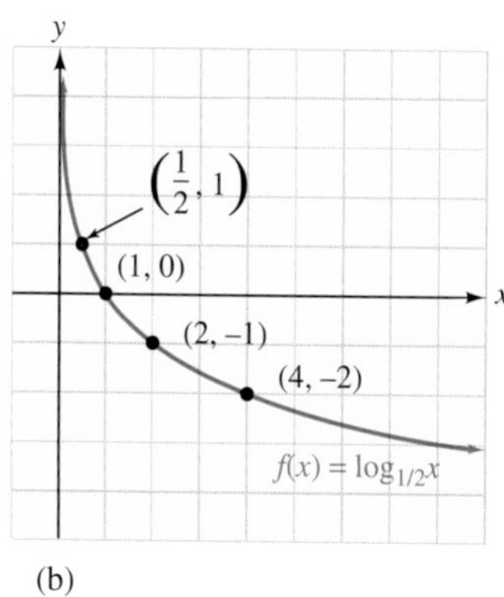

(b)

Figure 4-12

A table of values and the graph of $f(x) = \log_{10} x$ appears in Figure 4-13.

The graphs of all logarithmic functions are similar to those in Figure 4-14. If $b > 1$, the logarithmic function is an increasing function, as in Figure 4-14(a). If $0 < b < 1$, the logarithmic function is a decreasing function, as in Figure 4-14(b).

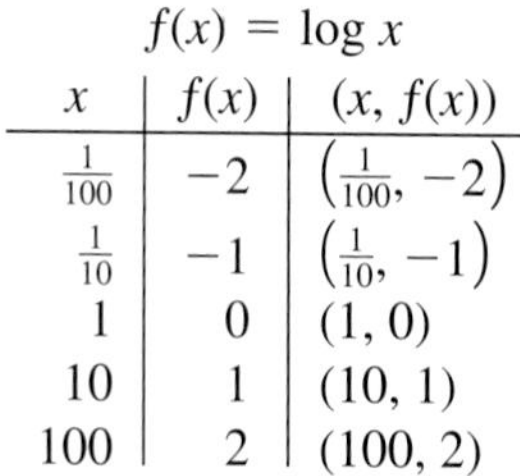

$f(x) = \log x$

x	$f(x)$	$(x, f(x))$
$\frac{1}{100}$	-2	$\left(\frac{1}{100}, -2\right)$
$\frac{1}{10}$	-1	$\left(\frac{1}{10}, -1\right)$
1	0	(1, 0)
10	1	(10, 1)
100	2	(100, 2)

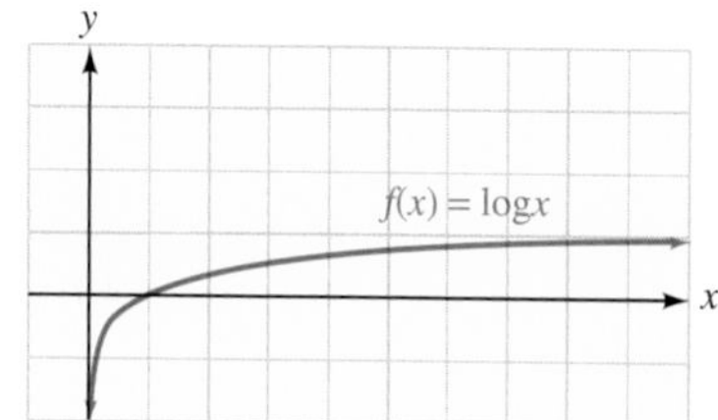

Figure 4-13

As Figures 4-14 (a) and (b) show, the graph of $f(x) = \log_b x$ has these properties:

1. It passes through the point (1, 0).
2. It passes through the point $(b, 1)$.
3. The y-axis is an asymptote.
4. The domain is $(0, \infty)$, and the range is $(-\infty, \infty)$.

Figures 4-14(c) and (d) show that the exponential and logarithmic functions are inverses of each other and have symmetry about the line $y = x$.

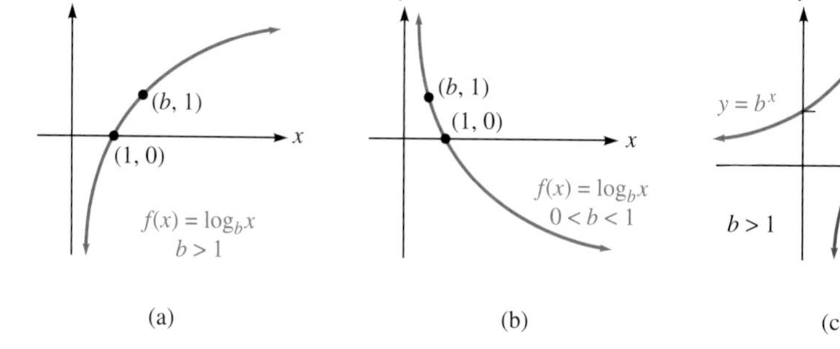

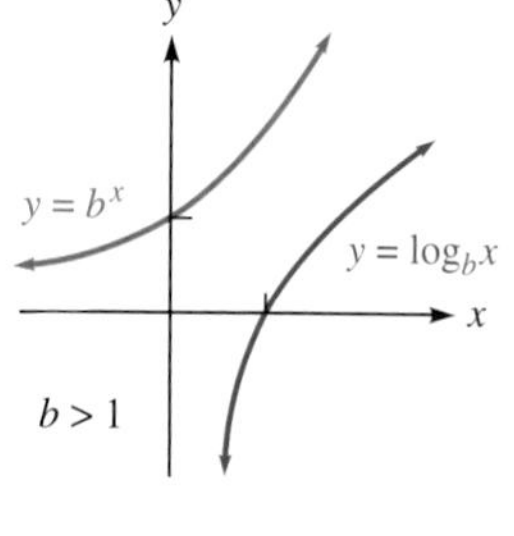

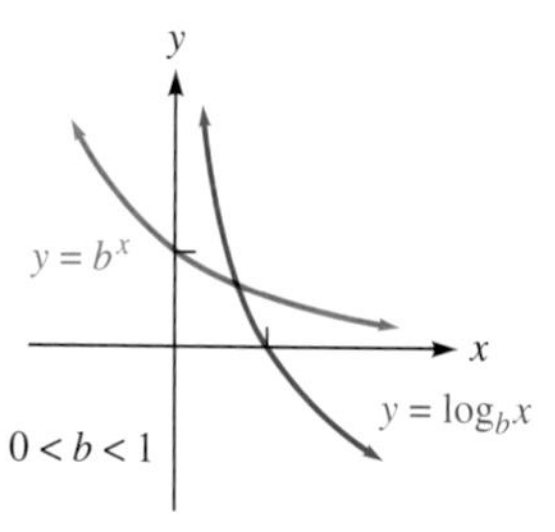

Figure 4-14

To graph $f(x) = \ln x$, we can plot points that satisfy the equation $x = e^y$ and join them with a smooth curve, as shown in Figure 4-15.

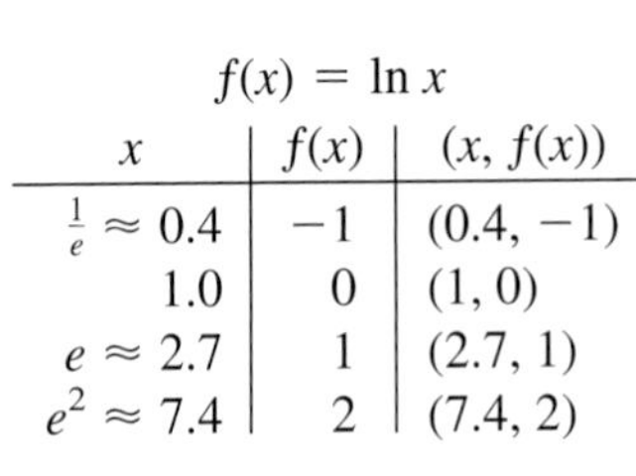

$f(x) = \ln x$

x	$f(x)$	$(x, f(x))$
$\frac{1}{e} \approx 0.4$	-1	$(0.4, -1)$
1.0	0	$(1, 0)$
$e \approx 2.7$	1	$(2.7, 1)$
$e^2 \approx 7.4$	2	$(7.4, 2)$

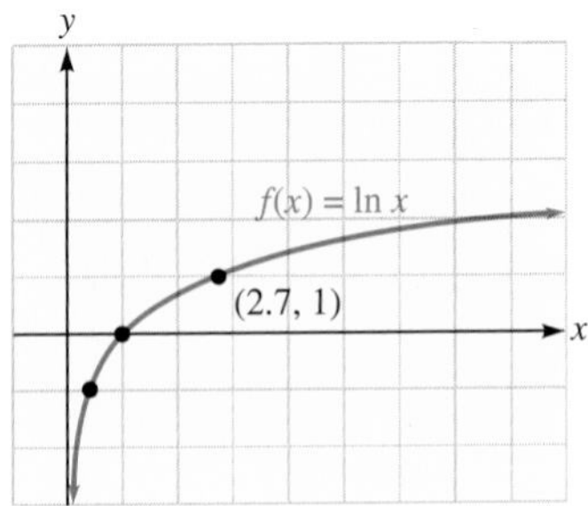

Figure 4-15

Vertical and Horizontal Translations

The graphs of many functions involving logarithms are translations of the basic logarithmic graphs.

EXAMPLE 6 Graph: $f(x) = 3 + \log_2 x$.

Solution See Figure 4-16. The graph of $f(x) = 3 + \log_2 x$ is identical to the graph of $f(x) = \log_2 x$, except that it is translated 3 units up.

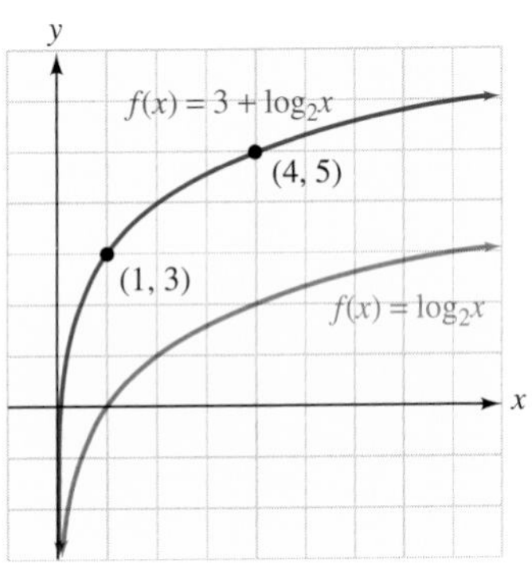

Figure 4-16

Self Check Graph: $f(x) = \log_3 x - 2$. ■

ACTIVE EXAMPLE 7 Graph: $f(x) = \log_{1/2} (x - 1)$.

Solution See Figure 4-17. The graph of $f(x) = \log_{1/2} (x - 1)$ is identical to the graph of $f(x) = \log_{1/2} x$, except that it is translated 1 unit to the right.

COLLEGE **Algebra** f(x) **Now**™
Go to **http://1pass.thomson.com** or your CD to practice this example.

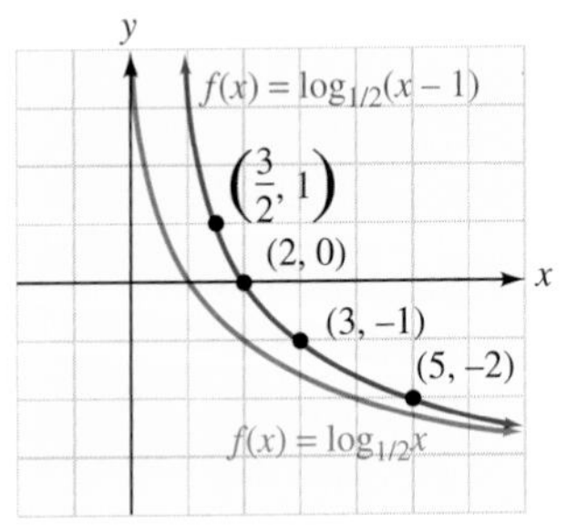

Figure 4-17

Self Check Graph: $f(x) = \log_{1/3} (x + 2)$. ■

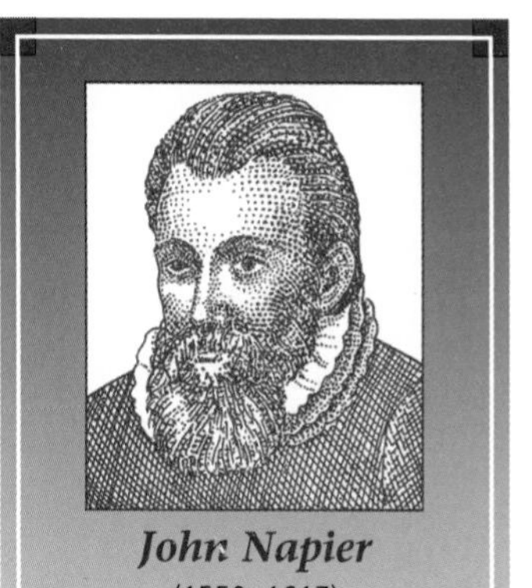

John Napier
(1550–1617)

Napier is famous for his work with natural logarithms. In fact, natural logarithms are often called Napierian logarithms. He also invented a device called Napier's rods, which did multiplications mechanically. This was a forerunner of modern-day computers.

Many graphs of logarithmic functions involve translations of the graph of $f(x) = \ln x$. For example, Figure 4-18 shows a calculator graph of the functions $f(x) = \ln x$, $f(x) = \ln x + 2$, and $f(x) = \ln x - 3$.

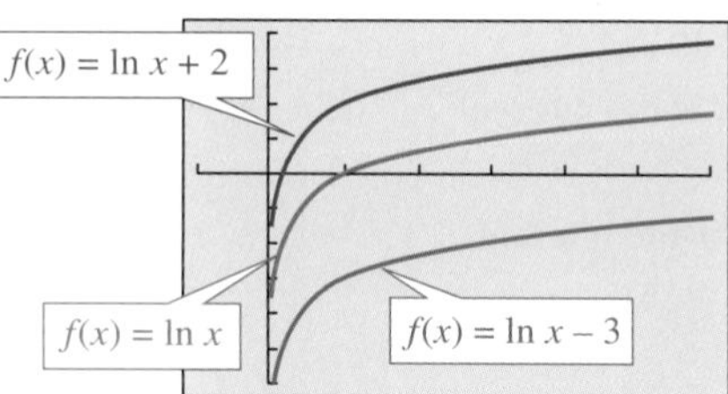

The graph of $f(x) = \ln x + 2$ is 2 units above the graph of $f(x) = \ln x$.

The graph of $f(x) = \ln x - 3$ is 3 units below the graph of $f(x) = \ln x$.

Figure 4-18

Figure 4-19 shows a calculator graph of the functions $f(x) = \ln x$, $f(x) = \ln(x - 2)$, and $f(x) = \ln(x + 3)$.

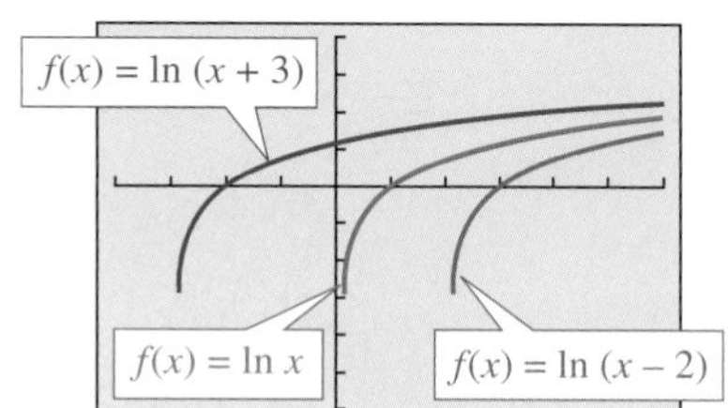

The graph of $y = \ln(x + 3)$ is 3 units to the left of the graph of $f(x) = \ln x$.

The graph of $f(x) = \ln(x - 2)$ is 2 units to the right of the graph of $f(x) = \ln x$.

Figure 4-19

Accent on Technology — USING CALCULATORS TO GRAPH LOGARITHMIC FUNCTIONS

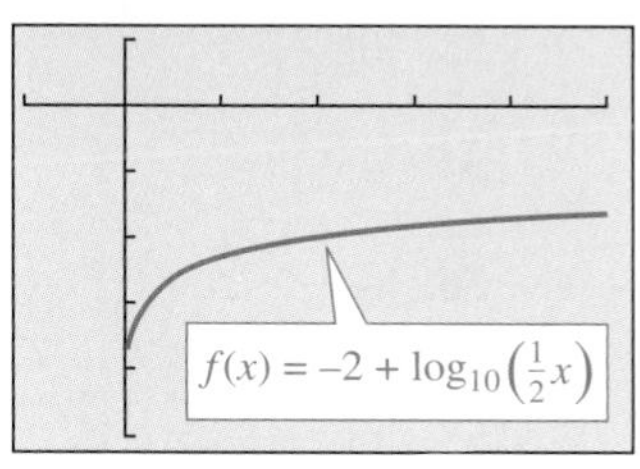

Figure 4-20

Graphing calculators can draw graphs of logarithmic functions when the base of the logarithmic function is 10 or e. To use a calculator to graph $f(x) = -2 + \log_{10}\left(\frac{1}{2}x\right)$, we enter the right-hand side of the equation after the symbol Y_1 =. The display will show the equation

$$Y_1 = -2 + \log(1/2*x)$$

If we use window settings of $[-1, 5]$ for x and $[-5, 1]$ for y and press GRAPH, we will obtain the graph shown in Figure 4-20.

Self Check Answers

1. a. 2 **b.** 4 **c.** −2 **2. a.** −2 **b.** 6 **c.** $\frac{1}{2}$ **3.** 75.3998 **4. a.** 1.1447 **b.** no value
5. a. 6.9199 **b.** 14.5596 **6.**

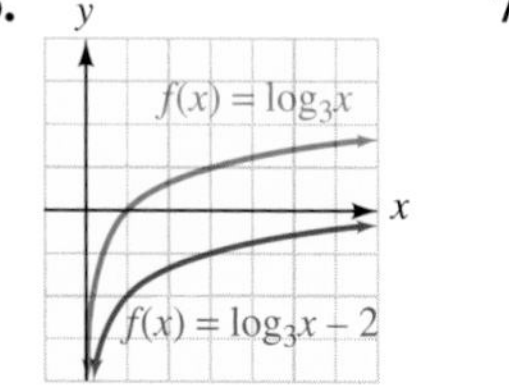

7.

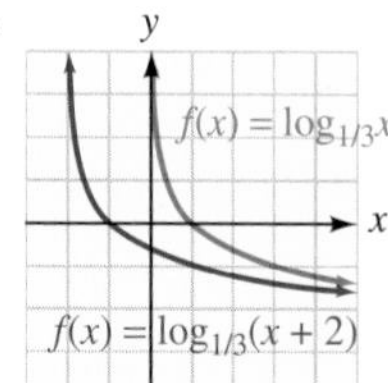

4.3 Exercises

VOCABULARY AND CONCEPTS *Fill in the blanks.*

1. The equation $y = \log_b x$ is equivalent to $x = b^y$.
2. The domain of a logarithmic function is the interval $(0, \infty)$.
3. The range of a logarithmic function is the interval $(-\infty, \infty)$.
4. $b^{\log_b x} = x$
5. Because the exponential function is one-to-one, it has an inverse function.
6. The inverse of an exponential function is called a logarithmic function.
7. $\text{Log}_b\, x$ is the exponent to which b is raised to get x.
8. The y-axis is an asymptote of the graph of $f(x) = \log_b x$.
9. The graph of $f(x) = \log_b x$ passes through the points $(b, 1)$ and $(1, 0)$.
10. $\log_{10} 10^x = x$
11. $\ln x$ means $\log_e x$.
12. The domain of the function $f(x) = \ln x$ is the interval $(0, \infty)$.
13. The range of the function $f(x) = \ln x$ is the interval $(-\infty, \infty)$.
14. The graph of $f(x) = \ln x$ has the y-axis as an asymptote.
15. In the expression $\log x$, the base is understood to be 10.
16. In the expression $\ln x$, the base is understood to be e.

PRACTICE *Write each equation in exponential form.*

17. $\log_3 81 = 4$ $\quad 3^4 = 81$
18. $\log_7 7 = 1$ $\quad 7^1 = 7$
19. $\log_{1/2} \frac{1}{8} = 3$ $\quad \left(\frac{1}{2}\right)^3 = \frac{1}{8}$
20. $\log_{1/5} 1 = 0$ $\quad \left(\frac{1}{5}\right)^0 = 1$
21. $\log_4 \frac{1}{64} = -3$ $\quad 4^{-3} = \frac{1}{64}$
22. $\log_6 \frac{1}{36} = -2$ $\quad 6^{-2} = \frac{1}{36}$
23. $\log_\pi \pi = 1$ $\quad \pi^1 = \pi$
24. $\log_7 \frac{1}{49} = -2$ $\quad 7^{-2} = \frac{1}{49}$

Write each equation in logarithmic form.

25. $8^2 = 64$ $\quad \log_8 64 = 2$
26. $10^3 = 1{,}000$ $\quad \log_{10} 1{,}000 = 3$
27. $4^{-2} = \frac{1}{16}$ $\quad \log_4 \frac{1}{16} = -2$
28. $3^{-4} = \frac{1}{81}$ $\quad \log_3 \frac{1}{81} = -4$
29. $\left(\frac{1}{2}\right)^{-5} = 32$ $\quad \log_{1/2} 32 = -5$
30. $\left(\frac{1}{3}\right)^{-3} = 27$ $\quad \log_{1/3} 27 = -3$
31. $x^y = z$ $\quad \log_x z = y$
32. $m^n = p$ $\quad \log_m p = n$

Find each value of x.

33. $\log_2 8 = x$ $\quad 3$
34. $\log_3 9 = x$ $\quad 2$
35. $\log_4 64 = x$ $\quad 3$
36. $\log_6 216 = x$ $\quad 3$
37. $\log_{1/2} \frac{1}{8} = x$ $\quad 3$
38. $\log_{1/3} \frac{1}{81} = x$ $\quad 4$
39. $\log_9 3 = x$ $\quad \frac{1}{2}$
40. $\log_{125} 5 = x$ $\quad \frac{1}{3}$
41. $\log_{1/2} 8 = x$ $\quad -3$
42. $\log_{1/2} 16 = x$ $\quad -4$
43. $\log_8 x = 2$ $\quad 64$
44. $\log_7 x = 0$ $\quad 1$
45. $\log_7 x = 1$ $\quad 7$
46. $\log_2 x = 8$ $\quad 256$
47. $\log_{25} x = \frac{1}{2}$ $\quad 5$
48. $\log_4 x = \frac{1}{2}$ $\quad 2$
49. $\log_5 x = -2$ $\quad \frac{1}{25}$
50. $\log_3 x = -4$ $\quad \frac{1}{81}$
51. $\log_{36} x = -\frac{1}{2}$ $\quad \frac{1}{6}$
52. $\log_{27} x = -\frac{1}{3}$ $\quad \frac{1}{3}$
53. $\log_x 5^3 = 3$ $\quad 5$
54. $\log_x 5 = 1$ $\quad 5$
55. $\log_x \frac{9}{4} = 2$ $\quad \frac{3}{2}$
56. $\log_x \frac{\sqrt{3}}{3} = \frac{1}{2}$ $\quad \frac{1}{3}$
57. $\log_x \frac{1}{64} = -3$ $\quad 4$
58. $\log_x \frac{1}{100} = -2$ $\quad 10$
59. $\log_x \frac{9}{4} = -2$ $\quad \frac{2}{3}$
60. $\log_x \frac{\sqrt{3}}{3} = -\frac{1}{2}$ $\quad 3$
61. $2^{\log_2 5} = x$ $\quad 5$
62. $3^{\log_3 4} = x$ $\quad 4$
63. $x^{\log_4 6} = 6$ $\quad 4$
64. $x^{\log_3 8} = 8$ $\quad 3$

Use a calculator to find each value to four decimal places.

65. $\log 3.25$ $\quad 0.5119$
66. $\log 0.57$ $\quad -0.2441$
67. $\log 0.00467$ $\quad -2.3307$
68. $\log 375.876$ $\quad 2.5750$
69. $\ln 45.7$ $\quad 3.8221$
70. $\ln 0.005$ $\quad -5.2983$

71. $\ln \frac{2}{3}$ **−0.4055**

72. $\ln \frac{12}{7}$ **0.5390**

73. $\ln 35.15$ **3.5596**

74. $\ln 0.675$ **−0.3930**

75. $\ln 7.896$ **2.0664**

76. $\ln 0.00465$ **−5.3709**

77. $\log (\ln 1.7)$ **−0.2752**

78. $\ln (\log 9.8)$ **−0.0088**

79. $\ln (\log 0.1)$ **undefined**

80. $\log (\ln 0.01)$ **undefined**

Use a calculator to find y to four decimal places, if possible.

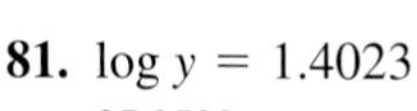

81. $\log y = 1.4023$ **25.2522**

82. $\log y = 0.926$ **8.4333**

83. $\log y = -3.71$ **1.9498×10^{-4}**

84. $\log y = \log \pi$ **3.1416**

85. $\ln y = 1.4023$ **4.0645**

86. $\ln y = 2.6490$ **14.1399**

87. $\ln y = 4.24$ **69.4079**

88. $\ln y = 0.926$ **2.5244**

89. $\ln y = -3.71$ **0.0245**

90. $\ln y = -0.28$ **0.7558**

91. $\log y = \ln 8$ **120.0719**

92. $\ln y = \log 7$ **2.3282**

Find the value of b, if any, that would cause the graph of $y = \log_b x$ to look like the graph shown.

93.

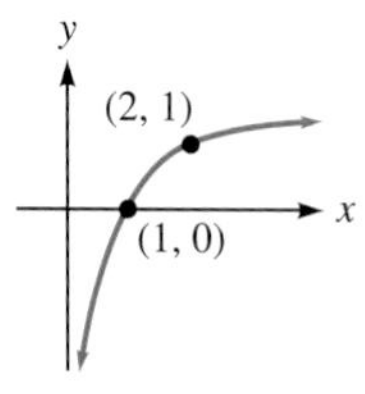

$b = 2$

94.

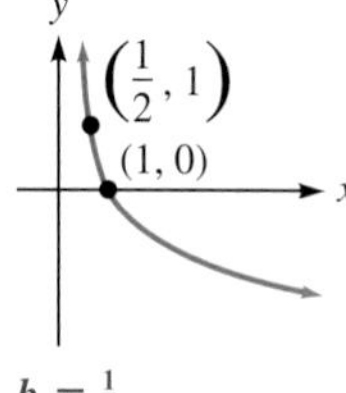

$b = \frac{1}{2}$

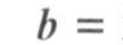

95.

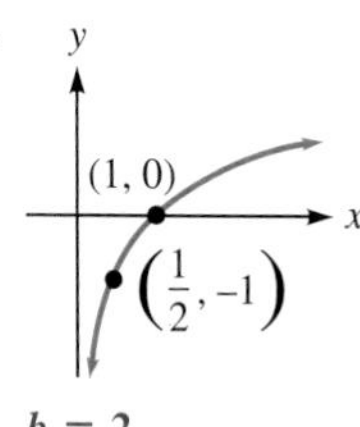

$b = 2$

96.

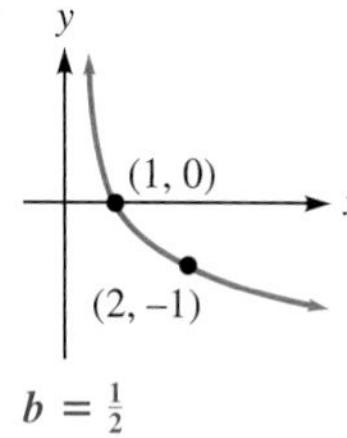

$b = \frac{1}{2}$

Graph each function.

97. $f(x) = \log_3 x$

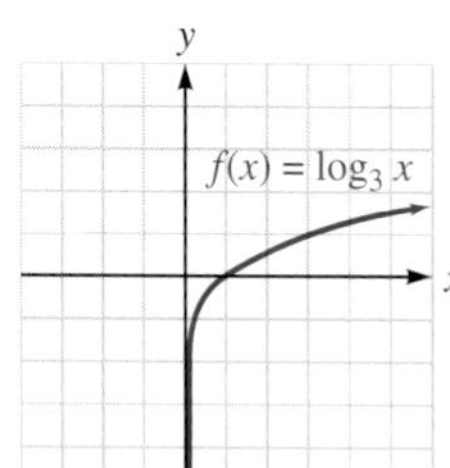

98. $f(x) = \log_4 x$

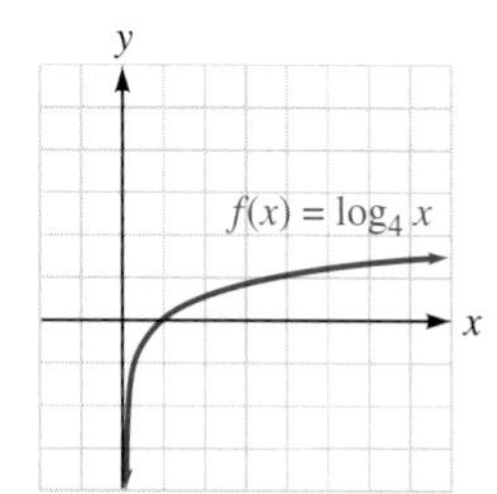

99. $f(x) = \log_{1/3} x$

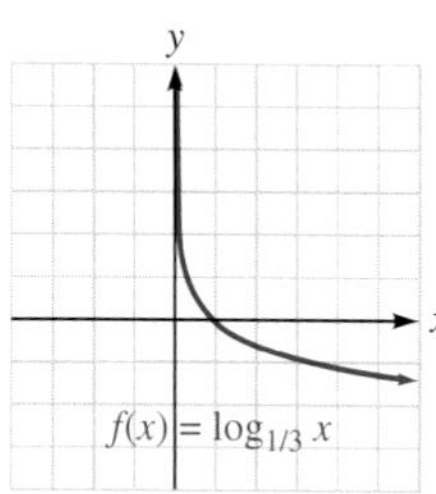

100. $f(x) = \log_{1/4} x$

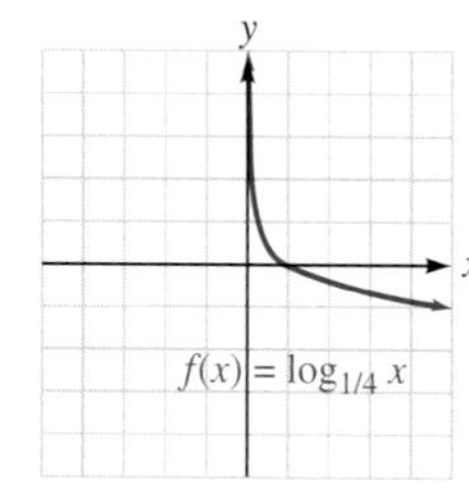

101. $f(x) = 2 + \log_2 x$

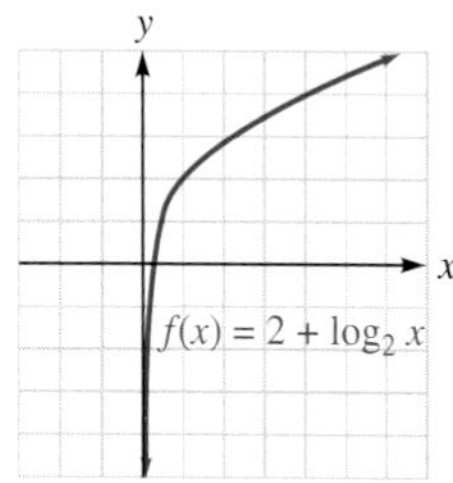

102. $f(x) = \log_2 (x - 1)$

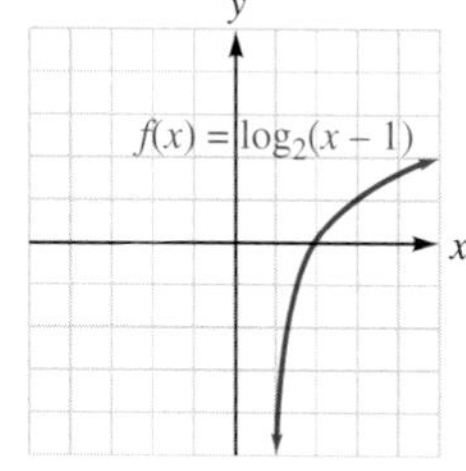

103. $f(x) = \log_3 (x + 2)$

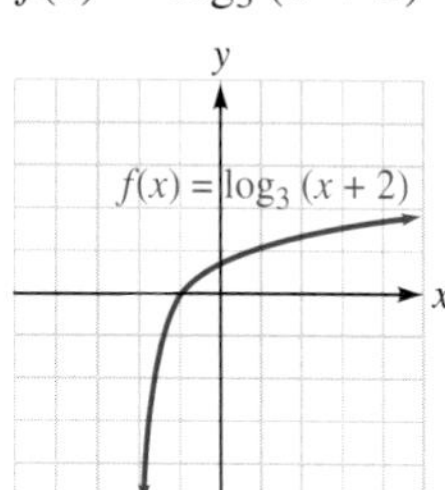

104. $f(x) = -3 + \log_3 x$

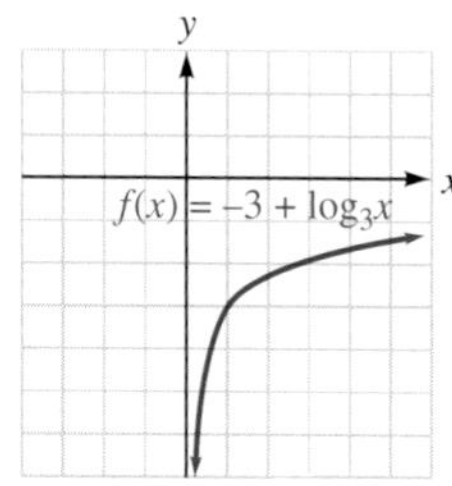

Use a graphing calculator to graph each function.

105. $f(x) = \log (3x)$

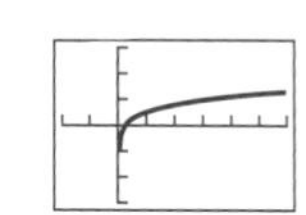

106. $f(x) = \log \left(\frac{x}{3}\right)$

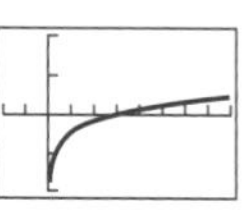

107. $f(x) = \log (-x)$

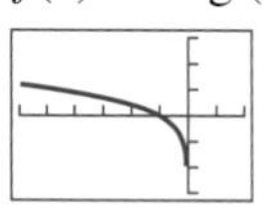

108. $f(x) = -\log x$

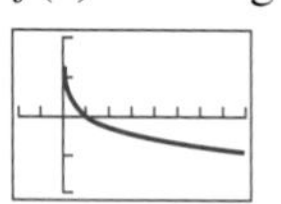

109. $f(x) = \ln \left(\frac{1}{2}x\right)$

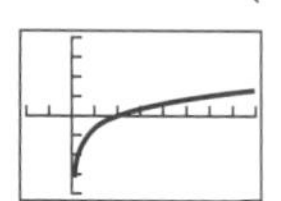

110. $f(x) = \ln x^2$

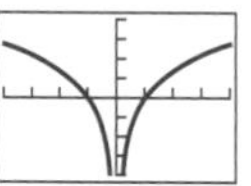

111. $f(x) = \ln (-x)$

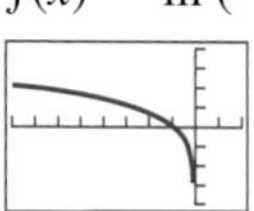

112. $f(x) = \ln (3x)$

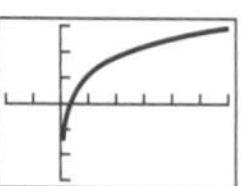

DISCOVERY AND WRITING

113. Consider the following graphs. Which is larger, a or b, and why? **b is larger.**

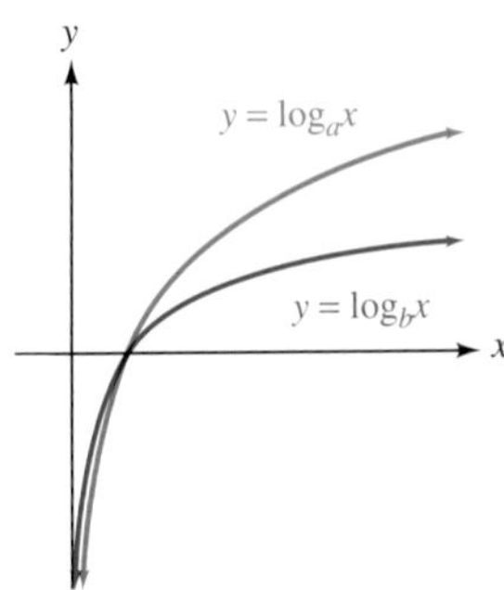

114. Consider the following graphs. Which is larger, a or b, and why? **a is larger.**

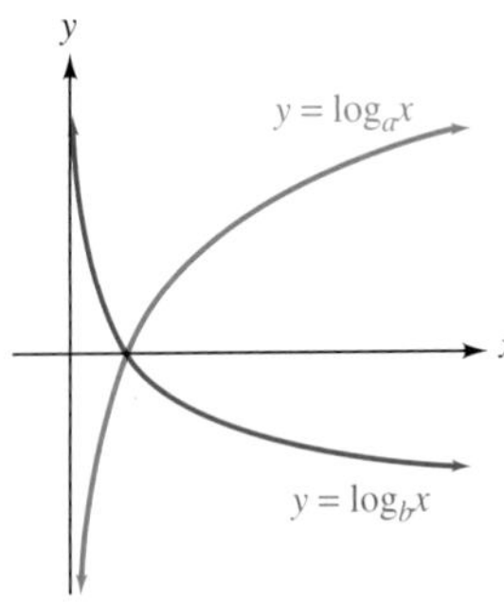

115. Pick two numbers and add their logarithms. Then find the logarithm of the product of those two numbers. What do you observe? Does it work for three numbers?

116. If $\log_a b = 7$, find $\log_b a$. $\frac{1}{7}$

REVIEW ***Find the vertex of each parabola.***

117. $y = x^2 + 7x + 3$ $\left(-\frac{7}{2}, -\frac{37}{4}\right)$

118. $y = 3x^2 - 8x - 1$ $\left(\frac{4}{3}, -\frac{19}{3}\right)$

119. Fencing a pasture A farmer will use 3,400 feet of fencing to enclose and divide the pasture shown in the illustration. What dimensions will enclose the greatest area? **425 ft by 850 ft**

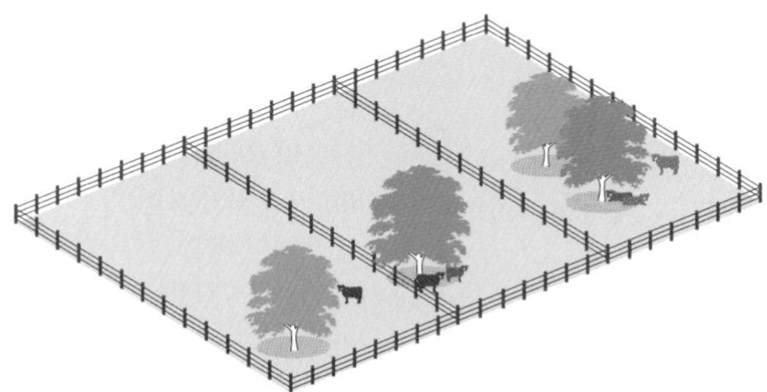

120. Selling appliances When the price is p dollars, an appliance dealer can sell $(2{,}200 - p)$ refrigerators. What price will maximize his revenue? **$1,100**

121. Write the equation of a line passing through the origin and parallel to the line $y = 5x - 8$. $y = 5x$

122. Write the equation of a line passing through $(3, 2)$ and perpendicular to the line $y = \frac{2}{3}x - 12$. $y = -\frac{3}{2}x + \frac{13}{2}$

4.4 Applications of Logarithmic Functions

In this section, you will learn about

- Electrical Engineering
- Geology
- Charging Batteries
- Population Growth
- Isothermal Expansion

Logarithmic functions have applications in fields such as engineering, geology, social science, and physics.

Electrical Engineering

Electronic engineers use common logarithms to measure the voltage gain of devices such as amplifiers or the length of a transmission line. The unit of gain, called the **decibel,** is defined by a logarithmic function.

Decibel Voltage Gain

If E_O is the output voltage of a device and E_I is the input voltage, the **decibel voltage gain** is given by

$$\text{db gain} = 20 \log \frac{E_O}{E_I}$$

ACTIVE EXAMPLE 1 Find the db gain of an amplifier if its input is 0.5 volt and its output is 40 volts.

Solution The decibel voltage gain is found by substituting these values into the formula for db gain:

COLLEGE Algebra f(x) Now™
Explore this Active Example and test yourself on the steps taken here at **http://1pass.thomson.com**. You can also find this example on the Interactive Video Skillbuilder CD-ROM.

$$\text{db voltage gain} = 20 \log \frac{E_O}{E_I}$$

$$\text{db voltage gain} = 20 \log \frac{\mathbf{40}}{\mathbf{0.5}} \quad \text{Substitute 40 for } E_O \text{ and 0.5 for } E_I.$$

$$= 20 \log 80$$

$$\approx 38.06179974 \quad \text{Use a calculator.}$$

To the nearest decibel, the db gain is 38 decibels.

Self Check Find the db gain if the input is 0.7 volt. ■

■ Geology

Seismologists measure the intensity of earthquakes on the **Richter scale,** which is based on a logarithmic function.

Richter Scale

If R is the intensity of an earthquake, A is the amplitude (measured in micrometers), and P is the period (the time of one oscillation of the Earth's surface, measured in seconds), then

$$R = \log \frac{A}{P}$$

ACTIVE EXAMPLE 2 Find the intensity of an earthquake with amplitude of 5,000 micrometers $\left(\frac{1}{2}\text{ centimeter}\right)$ and a period of 0.07 second.

Solution We substitute 5,000 for A and 0.07 for P in the Richter scale formula and simplify.

COLLEGE Algebra f(x) Now™
Go to **http://1pass.thomson.com** or your CD to practice this example.

$$R = \log \frac{A}{P}$$

$$R = \log \frac{\mathbf{5{,}000}}{\mathbf{0.07}}$$

$$\approx \log 71{,}428.57143 \quad \text{Use a calculator.}$$

$$\approx 4.853871964$$

To the nearest tenth, the earthquake measures 4.9 on the Richter scale.

Self Check Find the intensity of an aftershock with the same period but one-half of the amplitude. ■

Charging Batteries

A battery charges at a rate that depends on how close it is to being fully charged—it charges fastest when it is most discharged. The formula that determines the time required to charge a battery to a certain level is based on a natural logarithmic function.

Charging Batteries

If M is the theoretical maximum charge that a battery can hold and k is a positive constant that depends on the battery and the charger, the length of time (in minutes) required to charge the battery to a given level C is given by

$$t = -\frac{1}{k}\ln\left(1 - \frac{C}{M}\right)$$

EXAMPLE 3 How long will it take to bring a fully discharged battery to 90% of full charge? Assume that $k = 0.025$ and that time is measured in minutes.

Solution 90% of full charge means 90% of M. We substitute $0.90M$ for C and 0.025 for k in the formula for charging batteries.

$$\begin{aligned} t &= -\frac{1}{k}\ln\left(1 - \frac{C}{M}\right) \\ t &= -\frac{1}{\mathbf{0.025}}\ln\left(1 - \frac{\mathbf{0.90}M}{M}\right) \\ &= -40\ln(1 - 0.9) \\ &= -40\ln(0.1) \\ &\approx 92.10340372 && \text{Use a calculator.} \end{aligned}$$

The battery will reach 90% charge in about 92 minutes.

Self Check How long will it take this battery to reach 80% of full charge? ■

Population Growth

If a population grows exponentially at a certain annual rate, the time required for the population to double is called the **doubling time** and is given by the following formula. You will be asked to prove this formula in Exercise 82 in Section 4.6.

Population Doubling Time

If r is the annual growth rate and t is the time (in years) required for a population to double, then

$$t = \frac{\ln 2}{r}$$

EXAMPLE 4 The population of the Earth is growing at the approximate rate of 2% per year. If this rate continues, how long will it take the population to double?

Solution Because the population is growing at the rate of 2% per year, we substitute 0.02 for r in the formula for doubling time and simplify.

$$t = \frac{\ln 2}{r}$$

$$t = \frac{\ln 2}{0.02}$$

$$\approx 34.65735903$$

It will take about 35 years for the Earth's population to double.

Self Check If the world population's annual growth rate could be reduced to 1.5% per year, what would be the doubling time? ■

Isothermal Expansion

When energy is added to a gas, its temperature and volume could increase. In **isothermal expansion,** the temperature remains constant—only the volume changes. The energy required is calculated as follows.

Isothermal Expansion

If the temperature T is constant, the energy E required to increase the volume of 1 mole of gas from an initial volume V_i to a final volume V_f is given by

$$E = RT \ln \left(\frac{V_f}{V_i}\right)$$

E is measured in joules and T in Kelvins. R is the universal gas constant, which is 8.314 joules/mole/K.

ACTIVE EXAMPLE 5 Find the amount of energy that must be supplied to triple the volume of 1 mole of gas at a constant temperature of 300 K.

Solution We substitute 8.314 for R and 300 for T in the formula. Since the final volume is to be three times the initial volume, we also substitute $3V_i$ for V_f.

COLLEGE **Algebra** $f(x)$ **Now**™
Go to **http://1pass.thomson.com** or your CD to practice this example.

$$E = RT \ln \left(\frac{V_f}{V_i}\right)$$

$$E = (8.314)(300) \ln \left(\frac{3V_i}{V_i}\right)$$

$$= 2{,}494.2 \ln 3$$

$$\approx 2{,}740.15877$$

Approximately 2,740 joules of energy must be added to triple the volume.

Self Check What energy is required to double the volume? ■

Self Check Answers

1. about 35 decibels **2.** about 4.6 **3.** about 64 min **4.** about 46 years **5.** 1,729 joules

4.4 Exercises

VOCABULARY AND CONCEPTS ***Fill in the blanks.***

1. db gain = $\underline{20 \log \frac{E_O}{E_I}}$
2. The intensity of an earthquake is measured by the formula $R = \underline{\log \frac{A}{P}}$.
3. The formula for charging batteries is $\underline{t = -\frac{1}{k} \ln\left(1 - \frac{C}{M}\right)}$.
4. If a population grows exponentially at a rate r, the time it will take for the population to double is given by the formula $t = \underline{\frac{\ln 2}{r}}$.
5. The formula for isothermal expansion is $\underline{E = RT \ln\left(\frac{V_f}{V_i}\right)}$.
6. The logarithm of a negative number is undefined.

APPLICATIONS ***Use a calculator to solve each problem.***

7. **Gain of an amplifier** An amplifier produces an output of 17 volts when the input signal is 0.03 volt. Find the decibel voltage gain. **55 db**
8. **Transmission lines** A 4.9-volt input to a long transmission line decreases to 4.7 volts at the other end. Find the decibel voltage loss. **0.36 db**
9. **Gain of an amplifier** Find the db gain of an amplifier whose input voltage is 0.71 volt and whose output voltage is 20 volts. **29 db**
10. **Gain of an amplifier** Find the db gain of an amplifier whose output voltage is 2.8 volts and whose input voltage is 0.05 volt. **35 db**
11. **db gain** Find the db gain of the amplifier shown below. **49.5 db**

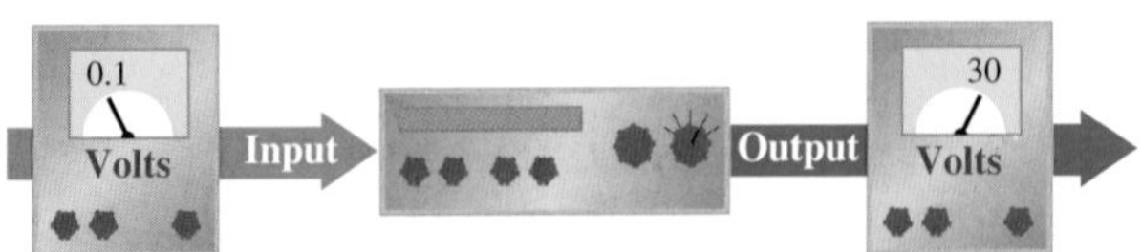

12. **db gain** Find the db gain of the amplifier shown below. **56.5 db**

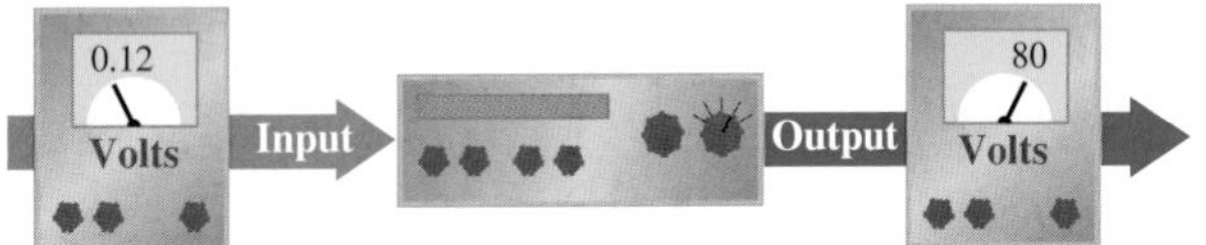

13. **Earthquakes** An earthquake has an amplitude of 5,000 micrometers and a period of 0.2 second. Find its measure on the Richter scale. **4.4**
14. **Earthquakes** An earthquake has an amplitude of 8,000 micrometers and a period of 0.008 second. Find its measure on the Richter scale. **6**
15. **Earthquakes** An earthquake with a period of $\frac{1}{4}$ second has an amplitude of 2,500 micrometers. Find its measure on the Richter scale. **4**
16. **Earthquakes** An earthquake has a period of $\frac{1}{2}$ second and an amplitude of 5 cm. Find its measure on the Richter scale. (*Hint:* 1 cm = 10,000 micrometers) **5**
17. **Battery charge** If $k = 0.116$, how long will it take a battery to reach a 90% charge? Assume that the battery was fully discharged when it began charging. **19.8 min**
18. **Battery charge** If $k = 0.201$, how long will it take a battery to reach a 40% charge? Assume that the battery was fully discharged when it began charging. **about 2.5 min**
19. **Population growth** A town's population grows at the rate of 12% per year. If this growth rate remains constant, how long will it take the population to double? **about 5.8 yr**
20. **Fish population growth** One thousand bass were stocked in Catfish Lake in Eagle River, Wisconsin, a lake with no bass population. If the population of bass is expected to grow at a rate of 25% per year, how long will it take the population to double? **about 2.8 years**
21. **Population growth** A population growing at an annual rate r will triple in a time t given by the formula $t = \frac{\ln 3}{r}$. How long will it take the population of the town in Exercise 19 to triple? **about 9.2 yr**
22. **Fish population growth** How long would it take the fish population in Exercise 20 to triple? **about 4.4 years**
23. **Isothermal expansion** One mole of gas expands isothermically to triple its volume. If the gas temperature is 400 K, what energy is absorbed? **about 3,654 joules**
24. **Isothermal expansion** One mole of gas expands isothermically to double its volume. If the gas temperature is 300K, what energy is absorbed? **about 1,729 joules**

If an investment is growing continuously for t years, its annual growth rate r is given by the formula

$$r = \frac{1}{t} \ln \frac{P}{P_0}$$

where P is the current value and P_0 is the amount originally invested.

25. **Investing** An investment of \$10,400 in America Online in 1992 was worth \$10,400,000 in 1999. Find AOL's average annual growth rate during this period. **about 99% per year**

26. **Investing** A \$5,000 investment in Dell Computer in 1995 was worth \$237,000 in 1999. Find the average annual growth rate of the stock. **about 96% per year**

27. **Depreciation** In business, equipment is often depreciated using the double declining-balance method. In this method, a piece of equipment with a life expectancy of N years, costing \$$C$, will depreciate to a value of \$$V$ in n years, where n is given by the formula

$$n = \frac{\log V - \log C}{\log\left(1 - \frac{2}{N}\right)}$$

If a computer that cost \$37,000 has a life expectancy of 5 years and has depreciated to a value of \$8,000, how old is it? **3 yr old**

28. **Depreciation** A word processor worth \$470 when new had a life expectancy of 12 years. If it is now worth \$189, how old is it? (See Exercise 27.) **about 5 yr old**

29. **Annuities** If \$$P$ is invested at the end of each year in an annuity earning interest at an annual rate r, the amount in the account will be \$$A$ after n years, where

$$n = \frac{\log\left[\frac{Ar}{P} + 1\right]}{\log(1 + r)}$$

If \$1,000 is invested each year in an annuity earning 12% annual interest, when will the account be worth \$20,000? **about 10.8 yr**

30. **Annuities** If \$5,000 is invested each year in an annuity earning 8% annual interest, when will the account be worth \$50,000? (See Exercise 29.) **about 7.6 years**

31. **Breakdown voltage** The coaxial power cable shown has a central wire with radius $R_1 = 0.25$ centimeters. It is insulated from a surrounding shield with inside radius $R_2 = 2$ centimeters. The maximum voltage the cable can withstand is called the **breakdown voltage** V of the insulation. V is given by the formula

$$V = ER_1 \ln \frac{R_2}{R_1}$$

where E is the **dielectric strength** of the insulation. If $E = 400{,}000$ volts/centimeter, find V. **about 208,000 V**

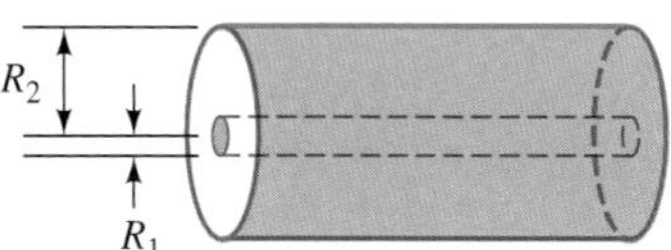

32. **Breakdown voltage** In Exercise 31, if the inside diameter of the shield were doubled, what voltage could the cable withstand? **277,000 V**

33. Suppose you graph the function $f(x) = \ln x$ on a coordinate grid with a unit distance of 1 centimeter on the x- and y-axes. How far out must you go on the x-axis so that $f(x) = 12$? Give your result to the nearest mile. **1 mi**

34. Suppose you graph the function $f(x) = \log x$ on a coordinate grid with a unit distance of 1 centimeter on the x- and y-axes. How far out must you go on the x-axis so that $f(x) = 12$? Give the result to the nearest mile. Why is this result so much larger than the result in Exercise 33? **6,214,000 mi**

DISCOVERY AND WRITING

35. One form of the logistic function is given by the equation

$$y = \frac{1}{1 + e^{-2x}}$$

Explain how you would find the y-intercept of its graph.

36. Graph the function $y = \ln |x|$. Explain why the graph looks the way it does.

REVIEW *Write the equation of the required line.*

37. Having a slope of 7 and a y-intercept of 3 $y = 7x + 3$

38. Parallel to the line $3x + 2y = 9$ and passing through the point $(-3, 5)$ **$3x + 2y = 1$**

39. A vertical line passing through (2, 3) $x = 2$

40. A horizontal line passing through (2, 3) $y = 3$

Simplify each expression.

41. $\dfrac{2(x+2)-1}{4x^2-9}$ $\quad \dfrac{1}{2x-3}$

42. $\dfrac{x+1}{x} + \dfrac{x-1}{x+1}$ $\quad \dfrac{2x^2+x+1}{x(x+1)}$

43. $\dfrac{x^2+3x+2}{3x+9} \cdot \dfrac{x+3}{x^2-4}$ $\quad \dfrac{x+1}{3(x-2)}$

44. $\dfrac{1+\dfrac{y}{x}}{\dfrac{y}{x}-1}$ $\quad \dfrac{x+y}{y-x}$

4.5 Properties of Logarithms

In this section, you will learn about

- Properties of Logarithms
- The Change-of-Base Formula
- pH Scale
- Electronics
- Physiology

Properties of Logarithms

Since logarithms are exponents, the properties of exponents have counterparts in the theory of logarithms. We begin with four basic properties.

Properties of Logarithms

If b is a positive number and $b \neq 1$, then

1. $\log_b 1 = 0$ **2.** $\log_b b = 1$

3. $\log_b b^x = x$ **4.** $b^{\log_b x} = x \quad (x > 0)$

Properties 1 through 4 follow directly from the definition of logarithm.

1. $\log_b 1 = 0$, because $b^0 = 1$.

2. $\log_b b = 1$, because $b^1 = b$.

3. $\log_b b^x = x$, because $b^x = b^x$.

4. $b^{\log_b x} = x$, because $\log_b x$ is the exponent to which b is raised to get x.

Properties 3 and 4 also indicate that the composition of the exponential and logarithmic functions (in both directions) is the identity function. This is expected, because the exponential and logarithmic functions with the same base are inverse functions.

EXAMPLE 1 Simplify each expression: **a.** $\log_3 1$, **b.** $\log_4 4$, **c.** $\log_7 7^3$, and **d.** $b^{\log_b 3}$.

Solution **a.** By Property 1, $\log_3 1 = 0$, because $3^0 = 1$.

b. By Property 2, $\log_4 4 = 1$, because $4^1 = 4$.

c. By Property 3, $\log_7 7^3 = 3$, because $7^3 = 7^3$.

d. By Property 4, $b^{\log_b 3} = 3$, because $\log_b 3$ is the power to which b is raised to get 3.

Self Check Simplify: **a.** $\log_4 1$, **b.** $\log_3 3$, **c.** $\log_2 2^4$, and **d.** $5^{\log_5 2}$. ■

The next two properties state that

The logarithm of a product is the sum of the logarithms.

The logarithm of a quotient is the difference of the logarithms.

Properties of Logarithms

If M, N, and b are positive numbers and $b \neq 1$, then

5. $\log_b MN = \log_b M + \log_b N$ **6.** $\log_b \frac{M}{N} = \log_b M - \log_b N$

Proof To prove Property 5, we let $x = \log_b M$ and $y = \log_b N$ and use the definition of logarithm to write each equation in exponential form.

$$M = b^x \quad \text{and} \quad N = b^y$$

Then $MN = b^x b^y$ and a property of exponents gives

$$MN = b^{x+y} \quad b^x b^y = b^{x+y}\text{; keep the base and add the exponents.}$$

We write this exponential equation in logarithmic form as

$$\log_b MN = x + y$$

Substituting the values of x and y completes the proof.

$$\log_b MN = \log_b M + \log_b N$$

■

The proof of Property 6 is similar. You will be asked to do it in an exercise.

Comment By Property 5 of logarithms, the logarithm of a *product* is equal to the *sum* of the logarithms. The logarithm of a sum or a difference usually does not simplify. In general,

$$\log_b (M + N) \neq \log_b M + \log_b N \quad \text{and} \quad \log_b (M - N) \neq \log_b M - \log_b N$$

By Property 6, the logarithm of a *quotient* is equal to the *difference* of the logarithms. The logarithm of a quotient is not the quotient of the logarithms:

$$\log_b \frac{M}{N} \neq \frac{\log_b M}{\log_b N}$$

Accent on Technology

We can use a calculator to illustrate Property 5 of logarithms by showing that

$$\ln [(3.7)(15.9)] = \ln 3.7 + \ln 15.9$$

We calculate the left- and right-hand sides of the equation separately and compare the results. To use a calculator to find ln [(3.7)(15.9)], we enter these numbers and press these keys:

Scientific calculator	***Graphing calculator***
3.7 × 15.9 = LN	LN 3.7 × 15.9) ENTER

The display will read 4.074651929.

Continued

To find $\ln 3.7 + \ln 15.9$, we enter these numbers and press these keys:

Scientific calculator	*Graphing calculator*
3.7 LN + 15.9 LN =	LN 3.7) + LN 15.9) ENTER

The display will read 4.074651929. Since the left- and right-hand sides are equal, the equation is true.

Two more properties state that

The logarithm of a power is the power times the logarithm.

If the logarithms of two numbers are equal, the numbers are equal.

Properties of Logarithms

If M and b are positive numbers and $b \neq 1$, then

7. $\log_b M^p = p \log_b M$ **8.** If $\log_b x = \log_b y$, then $x = y$.

Proof To prove Property 7, we let $x = \log_b M$, write the expression in exponential form, and raise both sides to the pth power:

$$M = b^x$$

$$(M)^p = (b^x)^p \quad \text{Raise both sides to the } p\text{th power.}$$

$$M^p = b^{px} \quad \text{Keep the base and multiply the exponents.}$$

Using the definition of logarithms gives

$$\log_b M^p = px$$

Substituting the value for x completes the proof.

$$\log_b M^p = p \log_b M$$

■

Property 8 follows from the fact that the logarithmic function is a one-to-one function. Property 8 will be important in the next section when we solve logarithmic equations.

We can use the properties of logarithms to write a logarithm as the sum or difference of several logarithms.

EXAMPLE 2 Assume that x, y, and z are positive numbers. Write each expression in terms of the logarithms of x, y, and z: **a.** $\log_b xyz$ and **b.** $\log_b \frac{x}{yz}$.

Solution **a.**

$$\log_b xyz = \log_b (xy)z$$

$$= \log_b (xy) + \log_b z \quad \text{The log of a product is the sum of the logs.}$$

$$= \log_b x + \log_b y + \log_b z \quad \text{The log of a product is the sum of the logs.}$$

b.

$$\log_b \frac{x}{yz} = \log_b x - \log_b (yz) \quad \text{The log of a quotient is the difference of the logs.}$$

$$= \log_b x - (\log_b y + \log_b z) \quad \text{The log of a product is the sum of the logs.}$$

$$= \log_b x - \log_b y - \log_b z \quad \text{Remove parentheses.}$$

Self Check Write the expression in terms of the logarithms of x, y, and z: $\log_b \frac{xy}{z}$. ■

ACTIVE EXAMPLE 3 Assume that x, y, z, and b are positive numbers and $b \neq 1$. Write each expression in terms of the logarithms of x, y, and z: **a.** $\log_b (x^3y^2z)$ and **b.** $\log_b \frac{y^2\sqrt{z}}{x}$.

Solution

a. $\log_b (x^3y^2z) = \log_b x^3 + \log_b y^2 + \log_b z$ — The log of a product is the sum of the logs.

$= 3 \log_b x + 2 \log_b y + \log_b z$ — The log of a power is the power times the log.

COLLEGE **Algebra** $f(x)$ **Now**™
Explore this Active Example and test yourself on the steps taken here at **http://1pass.thomson.com**. You can also find this example on the Interactive Video Skillbuilder CD-ROM.

b. $\log_b \frac{y^2\sqrt{z}}{x} = \log_b \left(y^2\sqrt{z}\right) - \log_b x$ — The log of a quotient is the difference of the logs.

$= \log_b y^2 + \log_b z^{1/2} - \log_b x$ — The log of a product is the sum of the logs; $\sqrt{z} = z^{1/2}$.

$= 2 \log_b y + \frac{1}{2} \log_b z - \log_b x$ — The log of a power is the power times the log.

Self Check Write the expression in terms of the logarithms of x, y, and z: $\log_b \sqrt[3]{\frac{x^2y}{z}}$. ■

We can use the properties of logarithms to combine several logarithms into one logarithm.

ACTIVE EXAMPLE 4 Assume that x, y, z, and b are positive numbers and $b \neq 1$. Write each expression as one logarithm: **a.** $2 \log_b x + \frac{1}{3} \log_b y$ **b.** $\frac{1}{2} \log_b (x - 2) - \log_b y + 3 \log_b z$.

Solution

a. $2 \log_b x + \frac{1}{3} \log_b y = \log_b x^2 + \log_b y^{1/3}$ — A power times a log is the log of the power.

$= \log_b (x^2y^{1/3})$ — The sum of two logs is the log of the product.

COLLEGE **Algebra** $f(x)$ **Now**™
Go to **http://1pass.thomson.com** or your CD to practice this example.

b. $\frac{1}{2} \log_b (x - 2) - \log_b y + 3 \log_b z$

$= \log_b (x - 2)^{1/2} - \log_b y + \log_b z^3$ — A power times a log is the log of the power.

$= \log_b \frac{(x - 2)^{1/2}}{y} + \log_b z^3$ — The difference of two logs is the log of the quotient.

$= \log_b \frac{z^3\sqrt{x - 2}}{y}$ — The sum of two logs is the log of the product.

Self Check Write as one logarithm: $2 \log_b x + \frac{1}{2} \log_b y - 3 \log_b (x - y)$. ■

We summarize the eight properties of logarithms as follows.

Properties of Logarithms

If b, M, and N are positive numbers and $b \neq 1$, then

1. $\log_b 1 = 0$ **2.** $\log_b b = 1$
3. $\log_b b^x = x$ **4.** $b^{\log_b x} = x$
5. $\log_b MN = \log_b M + \log_b N$ **6.** $\log_b \frac{M}{N} = \log_b M - \log_b N$
7. $\log_b M^p = p \log_b M$ **8.** If $\log_b x = \log_b y$, then $x = y$.

EXAMPLE 5 Given that $\log_{10} 2 \approx 0.3010$ and $\log_{10} 3 \approx 0.4771$, find approximations for **a.** $\log_{10} 18$ and **b.** $\log_{10} 2.5$.

Solution **a.**

$$\begin{aligned} \log_{10} 18 &= \log_{10} (2 \cdot 3^2) \\ &= \log_{10} 2 + \log_{10} 3^2 && \text{The log of a product is the sum of the logs.} \\ &= \log_{10} 2 + 2\log_{10} 3 && \text{The log of a power is the power times the log.} \\ &\approx 0.3010 + 2(0.4771) \\ &\approx 1.2552 \end{aligned}$$

b.

$$\begin{aligned} \log_{10} 2.5 &= \log_{10}\left(\frac{5}{2}\right) \\ &= \log_{10} 5 - \log_{10} 2 && \text{The log of a quotient is the difference of the logs.} \\ &= \log_{10}\frac{10}{2} - \log_{10} 2 && \text{Write 5 as } \tfrac{10}{2}. \\ &= \log_{10} 10 - \log_{10} 2 - \log_{10} 2 && \text{The log of a quotient is the difference of the logs.} \\ &= 1 - 2\log_{10} 2 && \log_{10} 10 = 1. \\ &\approx 1 - 2(0.3010) \\ &\approx 0.3980 \end{aligned}$$

Self Check Use the information given in Example 5 to find an approximation for $\log_{10} 0.75$. ■

■ The Change-of-Base Formula

We have seen how to use a calculator to find base-10 and base-*e* logarithms. To use a calculator to find logarithms with different bases, such as $\log_7 63$, we can divide the base-10 (or base-*e*) logarithm of 63 by the base-10 (or base-*e*) logarithm of 7.

$$\log_7 63 = \frac{\log 63}{\log 7} \approx 2.129150063 \qquad \log_7 63 = \frac{\ln 63}{\ln 7} \approx 2.129150063$$

To check the result, we verify that $7^{2.129150063} \approx 63$. This example suggests that if we know the base-*a* logarithm of a number, we can find its logarithm to some other base *b*.

Change-of-Base Formula

If *a*, *b*, and *x* are positive numbers and $a \neq 1$ and $b \neq 1$, then

$$\log_b x = \frac{\log_a x}{\log_a b}$$

To prove this formula, we begin with the equation $\log_b x = y$.

$$\begin{aligned} y &= \log_b x \\ x &= b^y && \text{Change the equation from logarithmic to exponential form.} \\ \log_a x &= \log_a b^y && \text{Take the base-}a\text{ logarithm of both sides.} \end{aligned}$$

$$\log_a x = y \log_a b \quad \text{The log of a power is the power times the log.}$$

$$y = \frac{\log_a x}{\log_a b} \quad \text{Divide both sides by } \log_a b.$$

$$\log_b x = \frac{\log_a x}{\log_a b} \quad \text{Refer to the first equation and substitute } \log_b x \text{ for } y.$$

If we know logarithms to base a (for example, $a = 10$), we can find the logarithm of x to a new base b by dividing the base-a logarithm of x by the base-a logarithm of b.

Comment $\dfrac{\log_a x}{\log_a b}$ means that one logarithm is to be divided by the other. They are not to be subtracted.

ACTIVE EXAMPLE 6 Find $\log_3 5$.

Solution We can substitute 3 for b, 10 for a, and 5 for x into the change-of-base formula:

COLLEGE Algebra f(x) Now™
Go to **http://1pass.thomson.com** or your CD to practice this example.

$$\log_b x = \frac{\log_a x}{\log_a b}$$

$$\log_3 5 = \frac{\log_{10} 5}{\log_{10} 3} \quad \text{Divide the base-10 logarithm of 5 by the base-10 logarithm of 3.}$$

$$\approx 1.464973521$$

To four decimal places, $\log_3 5 = 1.4650$.

Self Check Find $\log_5 3$ to four decimal places. ■

We must use properties of logarithms to solve many problems.

■ pH Scale

The more acidic a chemical solution, the greater the concentration of hydrogen ions. Chemists measure this concentration indirectly by the **pH scale,** or the **hydrogen ion index.**

pH of a Solution

If $[H^+]$ is the hydrogen ion concentration in gram-ions per liter, then

$$\text{pH} = -\log\,[H^+]$$

Since pure water has approximately 10^{-7} gram-ions per liter, its pH is

$$\text{pH} = -\log\,[H^+]$$

$$\text{pH} = -\log 10^{-7}$$

$$= -(-7)\log 10 \quad \text{The log of a power is the power times the log.}$$

$$= -(-7)\cdot 1 \quad \text{Use Property 2 of logarithms: } \log_b b = 1.$$

$$= 7$$

EXAMPLE 7 Seawater has a pH of approximately 8.5. Find its hydrogen ion concentration.

Solution We can substitute 8.5 for pH and solve the equation $\text{pH} = -\log[\text{H}^+]$ for $[\text{H}^+]$.

$$8.5 = -\log\,[\text{H}^+]$$
$$-8.5 = \log\,[\text{H}^+]$$
$$[\text{H}^+] = 10^{-8.5}$$
Change the equation from logarithmic form to exponential form.

We use a calculator to find that $[\text{H}^+] \approx 3.2 \times 10^{-9}$ gram-ions per liter.

Self Check The pH of a solution is 5.7. Find the hydrogen ion concentration. ■

Electronics

Recall that if E_O is the output voltage of a device and E_I is the input voltage, the decibel voltage gain is given by

$$\textbf{db gain} = 20 \log \frac{E_O}{E_I}$$

If input and output are measured in watts instead of volts, a different formula is needed.

EXAMPLE 8 Show that an alternate formula for db voltage gain is

$$\text{db gain} = 10 \log \frac{P_O}{P_I}$$

where P_I is the power input and P_O is the power output.

Solution Power is directly proportional to the square of the voltage. So for some constant k,

$$P_I = k(E_I)^2 \qquad \text{and} \qquad P_O = k(E_O)^2$$

and

$$\frac{P_O}{P_I} = \frac{k(E_O)^2}{k(E_I)^2} = \left(\frac{E_O}{E_I}\right)^2$$

We raise both sides to the $\frac{1}{2}$ power to get

$$\frac{E_O}{E_I} = \left(\frac{P_O}{P_I}\right)^{1/2}$$

which we substitute into the formula for db gain.

$$\text{db gain} = 20 \log \frac{E_O}{E_I}$$
$$= 20 \log \left(\frac{P_O}{P_I}\right)^{1/2}$$
$$= 20 \cdot \frac{1}{2} \log \frac{P_O}{P_I}$$
The log of a power is the power times the log.

$$\text{db gain} = 10 \log \frac{P_O}{P_I}$$
Simplify.

Self Check Find the db gain of a device to the nearest hundredth when $P_O = 30$ watts and $P_I = 2$ watts. ■

■ Physiology

In physiology, experiments suggest that the relationship between the loudness and the intensity of sound is a logarithmic one known as the Weber–Fechner law.

Weber–Fechner Law If L is the apparent loudness of a sound and I is the intensity, then

$$L = k \ln I$$

EXAMPLE 9 What increase in the intensity of a sound is necessary to cause a doubling of the apparent loudness?

Solution We use the formula $L = k \ln I$. To double the apparent loudness, we multiply both sides of the equation by 2 and use Property 7 of logarithms.

$$\begin{aligned} L &= k \ln I \\ 2L &= 2k \ln I \\ &= k \ln I^2 \end{aligned}$$

To double the apparent loudness, we must square the intensity.

Self Check What increase is necessary to triple the apparent loudness? ■

Self Check Answers

1. a. 0 **b.** 1 **c.** 4 **d.** 2 **2.** $\log_b x + \log_b y - \log_b z$ **3.** $\frac{1}{3}(2 \log_b x + \log_b y - \log_b z)$

4. $\log_b \frac{x^2\sqrt{y}}{(x-y)^3}$ **5.** -0.1249 **6.** 0.6826 **7.** 2×10^{-6} **8.** 11.76 **9.** Cube the intensity.

4.5 Exercises

VOCABULARY AND CONCEPTS ***Fill in the blanks.***

1. $\log_b 1 = \underline{0}$

2. $\log_b b = \underline{1}$

3. $\log_b MN = \log_b \underline{M} + \log_b \underline{N}$

4. $b^{\log_b x} = \underline{x}$

5. If $\log_b x = \log_b y$, then $\underline{x} = \underline{y}$.

6. $\log_b \frac{M}{N} = \log_b M \underline{-} \log_b N$

7. $\log_b x^p = p \cdot \log_b \underline{x}$

8. $\log_b b^x = \underline{x}$

9. $\log_b (A + B) \underline{\neq} \log_b A + \log_b B$

10. $\log_b A + \log_b B \underline{=} \log_b AB$

Simplify each expression.

11. $\log_4 1 = \underline{0}$

12. $\log_4 4 = \underline{1}$

13. $\log_4 4^7 = \underline{7}$

14. $4^{\log_4 8} = \underline{8}$

15. $5^{\log_5 10} = \underline{10}$

16. $\log_5 5^2 = \underline{2}$

17. $\log_5 5 = \underline{1}$

18. $\log_5 1 = \underline{0}$

PRACTICE ***Use a calculator to verify each equation.***

19. $\log [(3.7)(2.9)] = \log 3.7 + \log 2.9$

20. $\ln \frac{9.3}{2.1} = \ln 9.3 - \ln 2.1$

21. $\ln (3.7)^3 = 3 \ln 3.7$

22. $\log\sqrt{14.1} = \dfrac{1}{2}\log 14.1$

23. $\log 3.2 = \dfrac{\ln 3.2}{\ln 10}$

24. $\ln 9.7 = \dfrac{\log 9.7}{\log e}$

Assume that x, y, z, and b are positive numbers. Use the properties of logarithms to write each expression in terms of the logarithms of x, y, and z.

25. $\log_b 2xy$ $\quad \log_b 2 + \log_b x + \log_b y$

26. $\log_b 3xz$ $\quad \log_b 3 + \log_b x + \log_b z$

27. $\log_b \dfrac{2x}{y}$ $\quad \log_b 2 + \log_b x - \log_b y$

28. $\log_b \dfrac{x}{yz}$ $\quad \log_b x - \log_b y - \log_b z$

29. $\log_b x^2y^3$ $\quad 2\log_b x + 3\log_b y$

30. $\log_b x^3y^2z$ $\quad 3\log_b x + 2\log_b y + \log_b z$

31. $\log_b (xy)^{1/3}$ $\quad \frac{1}{3}(\log_b x + \log_b y)$

32. $\log_b x^{1/2}y^3$ $\quad \frac{1}{2}\log_b x + 3\log_b y$

33. $\log_b x\sqrt{z}$ $\quad \log_b x + \frac{1}{2}\log_b z$

34. $\log_b \sqrt{xy}$ $\quad \frac{1}{2}(\log_b x + \log_b y)$

35. $\log_b \dfrac{\sqrt[3]{x}}{\sqrt[3]{yz}}$ $\quad \frac{1}{3}\log_b x - \frac{1}{3}\log_b y - \frac{1}{3}\log_b z$

36. $\log_b \sqrt[4]{\dfrac{x^3y^2}{z^4}}$ $\quad \frac{3}{4}\log_b x + \frac{1}{2}\log_b y - \log_b z$

Assume that x, y, and z are positive numbers. Use the properties of logarithms to write each expression as the logarithm of one quantity.

37. $\log_b (x + 1) - \log_b x$ $\quad \log_b \dfrac{x+1}{x}$

38. $\log_b x + \log_b (x + 2) - \log_b 8$ $\quad \log_b \dfrac{x(x+2)}{8}$

39. $2\log_b x + \dfrac{1}{3}\log_b y$ $\quad \log_b x^2\sqrt[3]{y}$

40. $-2\log_b x - 3\log_b y + \log_b z$ $\quad \log_b \dfrac{z}{x^2y^3}$

41. $-3\log_b x - 2\log_b y + \dfrac{1}{2}\log_b z$ $\quad \log_b \dfrac{\sqrt{z}}{x^3y^2}$

42. $3\log_b (x + 1) - 2\log_b (x + 2) + \log_b x$

$\log_b \dfrac{(x+1)^3x}{(x+2)^2}$

43. $\log_b \left(\dfrac{x}{z} + x\right) - \log_b \left(\dfrac{y}{z} + y\right)$

$\log_b \dfrac{\dfrac{x}{z} + x}{\dfrac{y}{z} + y} = \log_b \dfrac{x}{y}$

44. $\log_b (xy + y^2) - \log_b (xz + yz) + \log_b z$ $\quad \log_b y$

Determine whether each statement is true or false.

45. $\log_b ab = \log_b a + 1$ **true**

46. $\log_b \dfrac{1}{a} = -\log_b a$ **true**

47. $\log_b 0 = 1$ **false**

48. $\log_b 2 = \log_2 b$ **false**

49. $\log_b (x + y) \neq \log_b x + \log_b y$ **true**

50. $\log_b xy = (\log_b x)(\log_b y)$ **false**

51. If $\log_a b = c$, then $\log_b a = c$. **false**

52. If $\log_a b = c$, then $\log_b a = \dfrac{1}{c}$. **true**

53. $\log_7 7^7 = 7$ **true**

54. $7^{\log_7 7} = 7$ **true**

55. $\log_b (-x) = -\log_b x$ **false**

56. If $\log_b a = c$, then $\log_b a^p = pc$. **true**

57. $\dfrac{\log_b A}{\log_b B} = \log_b A - \log_b B$ **false**

58. $\log_b (A - B) = \dfrac{\log_b A}{\log_b B}$ **false**

59. $\log_b \dfrac{1}{5} = -\log_b 5$ **true**

60. $3\log_b \sqrt[3]{a} = \log_b a$ **true**

61. $\dfrac{1}{3}\log_b a^3 = \log_b a$ **true**

62. $\log_{4/3} y = -\log_{3/4} y$ **true**

63. $\log_b y + \log_{1/b} y = 0$ **true**

64. $\log_{10} 10^3 = 3(10^{\log_{10} 3})$ **false**

***Assume that $\log_{10} 4 = 0.6021$, $\log_{10} 7 = 0.8451$, and $\log_{10} 9 = 0.9542$. Use these values and the properties of logarithms to find each value.* Do not use a calculator.**

65. $\log_{10} 28$ **1.4472**

66. $\log_{10} \dfrac{7}{4}$ **0.2430**

67. $\log_{10} 2.25$ **0.3521**

68. $\log_{10} 36$ **1.5563**

69. $\log_{10} \frac{63}{4}$ **1.1972**

70. $\log_{10} \frac{4}{63}$ **−1.1972**

71. $\log_{10} 252$ **2.4014**

72. $\log_{10} 49$ **1.6902**

73. $\log_{10} 112$ **2.0493**

74. $\log_{10} 324$ **2.5105**

75. $\log_{10} \frac{144}{49}$ **0.4682**

76. $\log_{10} \frac{324}{63}$ **0.7112**

Use a calculator and the change-of-base formula to find each logarithm.

77. $\log_3 7$ **1.7712**

78. $\log_7 3$ **0.5646**

79. $\log_\pi 3$ **0.9597**

80. $\log_3 \pi$ **1.0420**

81. $\log_3 8$ **1.8928**

82. $\log_5 10$ **1.4307**

83. $\log_{\sqrt{2}} \sqrt{5}$ **2.3219**

84. $\log_\pi e$ **0.8736**

APPLICATIONS

85. **pH of a solution** Find the pH of a solution with a hydrogen ion concentration of 1.7×10^{-5} gram-ions per liter. **4.77**

86. **pH of calcium hydroxide** Find the hydrogen ion concentration of a saturated solution of calcium hydroxide whose pH is 13.2. **6.3×10^{-14}**

87. **pH of apples** The pH of apples can range from 2.9 to 3.3. Find the range in the hydrogen ion concentration. **from 5.01×10^{-4} to 1.26×10^{-3}**

88. **pH of sour pickles** The hydrogen ion concentration of sour pickles is 6.31×10^{-4}. Find the pH. **3.2**

89. **db gain** An amplifier produces a 40-watt output with a $\frac{1}{2}$-watt input. Find the db gain. **19 db**

90. **db loss** Losses in a long telephone line reduce a 12-watt input signal to an output of 3 watts. Find the db gain. (Because it is a loss, the "gain" will be negative.) **−6 db**

91. **Weber–Fechner law** What increase in intensity is necessary to quadruple the loudness?
The original intensity must be raised to the 4th power.

92. **Weber–Fechner law** What decrease in intensity is necessary to make a sound half as loud?
The original intensity must be raised to the $\frac{1}{2}$ power.

93. **Isothermal expansion** If a certain amount E of energy is added to one mole of a gas, it expands from an initial volume of 1 liter to a final volume V without changing its temperature according to the formula

$$E = 8{,}300 \ln V$$

Find the volume if twice that energy is added to the gas. **The volume V is squared.**

94. **Richter scale** By what factor must the amplitude of an earthquake change to increase its severity by 1 point on the Richter scale? Assume that the period remains constant. The Richter scale is given by

$$R = \log \frac{A}{P}$$

where A is the amplitude and P the period of the tremor. **by a factor of 10**

DISCOVERY AND WRITING

95. Simplify: $3^{4\log_3 2} + 5^{\frac{1}{2}\log_5 25}$

96. Find the value of $a - b$:

$$5 \log x + \frac{1}{3}\log y - \frac{1}{2}\log x - \frac{5}{6}\log y = \log (x^a y^b)$$

97. Prove Property 6 of logarithms:

$$\log_b \frac{M}{N} = \log_b M - \log_b N$$

98. Show that $-\log_b x = \log_{1/b} x$.

99. Show that $e^{x \ln a} = a^x$.

100. Show that $e^{\ln x} = x$.

101. Show that $\ln (e^x) = x$.

102. If $\log_b 3x = 1 + \log_b x$, find b. **$b = 3$**

103. Explain why $\ln (\log 0.9)$ is undefined.

104. Explain why $\log_b (\ln 1)$ is undefined.

In Exercises 105–106, A and B are both negative. Thus, AB and $\frac{A}{B}$ are positive, and log AB and log $\frac{A}{B}$ are defined.

105. Is it still true that $\log AB = \log A + \log B$? Explain.

106. Is it still true that $\log \frac{A}{B} = \log A - \log B$? Explain.

REVIEW ***Determine whether each equation defines a function.***

107. $y = 3x - 1$ **yes**

108. $y = \frac{x+3}{x-1}$ **yes**

109. $y^2 = 4x$ **no**

110. $y = 4x^2$ **yes**

Find the domain of each function.

111. $f(x) = x^2 - 4$ **$(-\infty, \infty)$**

112. $f(x) = \frac{1}{x^2 - 4}$ **$(-\infty, -2) \cup (-2, 2) \cup (2, \infty)$**

113. $f(x) = \sqrt{x^2 + 4}$ **$(-\infty, \infty)$**

114. $f(x) = \sqrt{x^2 - 4}$ **$(-\infty, -2] \cup [2, \infty)$**

4.6 Exponential and Logarithmic Equations

In this section, you will learn about

- Solving Exponential Equations
- Solving Logarithmic Equations
- Carbon-14 Dating
- Population Growth

An **exponential equation** is an equation with a variable in one of its exponents. Some examples of exponential equations are

$$3^x = 5, \qquad 6^{x-3} = 2^x, \qquad \text{and} \qquad 3^{2x+1} - 10(3^x) + 3 = 0$$

A **logarithmic equation** is an equation with logarithmic expressions that contain a variable. Some examples of logarithmic equations are

$$\log 2x = 25, \qquad \ln x - \ln(x - 12) = 24, \qquad \text{and} \qquad \log x = \log \frac{1}{x} + 4$$

In this section, we will learn how to solve many of these equations.

Solving Exponential Equations

EXAMPLE 1 Solve the exponential equation: $3^x = 5$.

Solution Since logarithms of equal numbers are equal, we can take the common logarithm of each side of the equation. We can then move the variable x from its position as an exponent to a position as a coefficient.

$$3^x = 5$$

$$\log 3^x = \log 5 \qquad \text{Take the common logarithm of each side.}$$

$$x \log 3 = \log 5 \qquad \text{The log of a power is the power times the log.}$$

(1) $$x = \frac{\log 5}{\log 3} \qquad \text{Divide both sides by log 3.}$$

$$\approx 1.464973521 \qquad \text{Use a calculator.}$$

To four decimal places, $x = 1.4650$.

Self Check Solve: $5^x = 3$ to four decimal places. ■

Comment A careless reading of Equation 1 leads to a common error. The right-hand side of Equation 1 calls for a division, not a subtraction.

$$\frac{\log 5}{\log 3} \qquad \text{means} \qquad (\log 5) \div (\log 3)$$

It is the expression $\log \frac{5}{3}$ that means $\log 5 - \log 3$.

ACTIVE EXAMPLE 2 Solve the exponential equation: $6^{x-3} = 2^x$.

Solution

COLLEGE **Algebra** $f(x)$ **Now**™
Explore this Active Example and test yourself on the steps taken here at **http://1pass.thomson.com**. You can also find this example on the Interactive Video Skillbuilder CD-ROM.

$$6^{x-3} = 2^x$$

$$\log 6^{x-3} = \log 2^x \quad \text{Take the common logarithm of each side.}$$

$$(x-3)\log 6 = x\log 2 \quad \text{The log of a power is the power times the log.}$$

$$x\log 6 - 3\log 6 = x\log 2 \quad \text{Use the distributive property.}$$

$$x\log 6 - x\log 2 = 3\log 6 \quad \text{Add 3 log 6 and subtract } x \log 2 \text{ from both sides.}$$

$$x(\log 6 - \log 2) = 3\log 6 \quad \text{Factor out } x \text{ on the left-hand side.}$$

$$x = \frac{3\log 6}{\log 6 - \log 2} \quad \text{Divide both sides by } \log 6 - \log 2.$$

$$x \approx 4.892789261 \quad \text{Use a calculator.}$$

Self Check Solve: $5^{x+3} = 3^x$. ■

In Examples 1 and 2, we took the common logarithm of both sides of an equation to solve the equation. We could just as well have taken the natural logarithm of both sides. We would obtain the same result. In the next example, we will take the natural logarithm of both sides.

EXAMPLE 3 Solve: $4^{x+3} = 8^{2x}$.

Solution

$$4^{x+3} = 8^{2x}$$

$$\ln 4^{x+3} = \ln 8^{2x} \quad \text{Take the natural logarithm of both sides.}$$

$$(x+3)\ln 4 = (2x)\ln 8 \quad \text{The log of a power is the power times the log.}$$

$$x\ln 4 + 3\ln 4 = 2x\ln 8 \quad \text{Use the distributive property on the left-hand side.}$$

$$x\ln 4 - 2x\ln 8 = -3\ln 4 \quad \text{Subtract } 2x \ln 8 \text{ and } 3 \ln 4 \text{ from both sides.}$$

$$x(\ln 4 - 2\ln 8) = -3\ln 4 \quad \text{Factor out } x \text{ on the left-hand side.}$$

$$x = \frac{-3\ln 4}{\ln 4 - 2\ln 8} \quad \text{Divide both sides by } \ln 4 - 2\ln 8.$$

$$x = 1.5 \quad \text{Use a calculator.}$$

Self Check Solve: $8^{x+1} = 4^{2x}$. ■

Example 3 can also be solved by using the rules of exponents.

$$4^{x+3} = 8^{2x}$$

$$(2^2)^{x+3} = (2^3)^{2x} \quad \text{Write 4 as } 2^2 \text{ and 8 as } 2^3.$$

$$2^{2(x+3)} = 2^{6x} \quad \text{Multiply exponents.}$$

$$2(x+3) = 6x \quad \text{Equal quantities with equal bases have equal exponents.}$$

$$2x + 6 = 6x \quad \text{Use the distributive property.}$$

$$-4x = -6 \quad \text{Subtract 6 and } 6x \text{ from both sides.}$$

$$x = \frac{3}{2} \quad \text{Divide both sides by } -4 \text{ and simplify.}$$

The answer is the same.

EXAMPLE 4 Solve: $2^{x^2+2x} = \frac{1}{2}$.

Solution Since $\frac{1}{2} = 2^{-1}$, we can write the equation in the form

$$2^{x^2+2x} = 2^{-1}$$

Since equal quantities with equal bases have equal exponents, we have

$$x^2 + 2x = -1$$

$$x^2 + 2x + 1 = 0 \quad \text{Add 1 to both sides.}$$

$$(x + 1)(x + 1) = 0 \quad \text{Factor the trinomial.}$$

$$x + 1 = 0 \quad \text{or} \quad x + 1 = 0 \quad \text{Set each factor equal to 0.}$$

$$x = -1 \quad | \quad x = -1$$

Verify that -1 satisfies the equation.

Self Check Solve: $3^{x^2+2x} = 27$. ■

■ Solving Logarithmic Equations

In each of the following examples, we use the properties of logarithms to change a logarithmic equation into an algebraic equation.

EXAMPLE 5 Solve: $\log_b (3x + 2) - \log_b (2x - 3) = 0$.

Solution

$$\log_b (3x + 2) - \log_b (2x - 3) = 0$$

$$\log_b (3x + 2) = \log_b (2x - 3) \quad \text{Add } \log_b (2x - 3) \text{ to both sides.}$$

$$3x + 2 = 2x - 3 \quad \text{If the logs of two numbers are equal, the numbers are equal.}$$

$$x = -5 \quad \text{Subtract } 2x \text{ and 2 from both sides.}$$

Check:

$$\log_b (3x + 2) - \log_b (2x - 3) = 0$$

$$\log_b [3(-5) + 2] - \log_b [2(-5) - 3] \stackrel{?}{=} 0$$

$$\log_b (-13) - \log_b (-13) \stackrel{?}{=} 0$$

Since the logarithm of a negative number does not exist, -5 is extraneous and must be discarded. The equation has no roots.

Self Check Solve: $\log_b (5x + 2) - \log_b (6x + 1) = 0$. ■

Comment Example 5 illustrates that you *must* check the solutions of a logarithmic equation.

ACTIVE EXAMPLE 6 Solve: $\log x + \log(x - 3) = 1$.

Solution

COLLEGE **Algebra** $f(x)$ **Now**™
Go to **http://1pass.thomson.com** or your CD to practice this example.

$$\log x + \log(x - 3) = 1$$

$$\log x(x - 3) = 1 \quad \text{The sum of two logs is the log of a product.}$$

$$x(x - 3) = 10^1 \quad \text{Use the definition of logarithms to change the equation to exponential form.}$$

$$x^2 - 3x - 10 = 0 \quad \text{Remove parentheses and subtract 10 from both sides.}$$

$$(x + 2)(x - 5) = 0 \quad \text{Factor the trinomial.}$$

$$x + 2 = 0 \quad \text{or} \quad x - 5 = 0$$

$$x = -2 \quad | \quad x = 5$$

Check: The number -2 is not a solution, because it does not satisfy the equation (a negative number does not have a logarithm). We check the remaining number, 5.

$$\log x + \log(x - 3) = 1$$

$$\log 5 + \log(5 - 3) \stackrel{?}{=} 1 \quad \text{Substitute 5 for } x.$$

$$\log 5 + \log 2 \stackrel{?}{=} 1$$

$$\log 10 \stackrel{?}{=} 1 \quad \text{The sum of two logs is the log of a product.}$$

$$1 = 1 \quad \log_b b = 1.$$

Since 5 does check, it is a root.

Self Check Solve: $\log x + \log(x - 15) = 2$. ■

ACTIVE EXAMPLE 7 Solve: $\dfrac{\log(5x - 6)}{\log x} = 2$.

Solution We can multiply both sides by $\log x$ to get

COLLEGE **Algebra** $f(x)$ **Now**™
Go to **http://1pass.thomson.com** or your CD to practice this example.

$$\log(5x - 6) = 2 \log x$$

and apply Property 7 of logarithms to get

$$\log(5x - 6) = \log x^2$$

By Property 8 of logarithms, $5x - 6 = x^2$, because they have equal logarithms. So

$$x^2 = 5x - 6$$

$$x^2 - 5x + 6 = 0$$

$$(x - 3)(x - 2) = 0$$

$$x - 3 = 0 \quad \text{or} \quad x - 2 = 0$$

$$x = 3 \quad | \quad x = 2$$

Verify that both 2 and 3 satisfy the equation.

Self Check Solve: $\dfrac{\log(8x - 15)}{\log x} = 2$. ■

Accent on Technology

We can use a graphing calculator to solve logarithmic equations. For example, to solve $\log x + \log(x - 3) = 1$, we subtract 1 from both sides to get

$$\log x + \log(x - 3) - 1 = 0$$

and then graph the corresponding function

$$y = \log x + \log(x - 3) - 1$$

with settings of [0, 10] for x and [−2, 2] for y, to obtain the graph in Figure 4-21.

Since the root of the equation is the x-intercept, we can find the root by tracing to find the value of the x-intercept. The root is $x = 5$. (See Figure 4-22.)

We can also find the x-intercept using the ZERO command in the CALC menu.

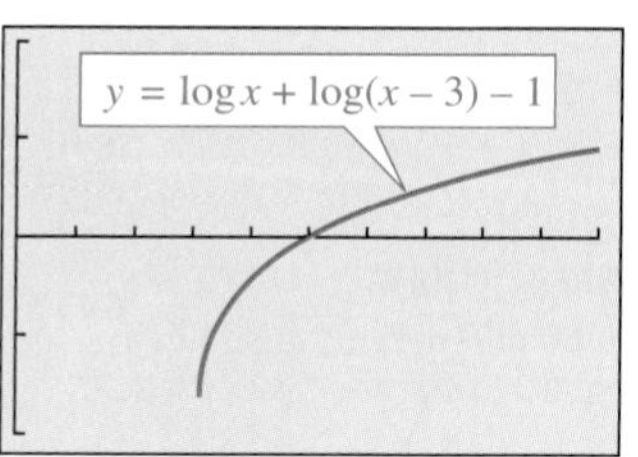

Figure 4-21

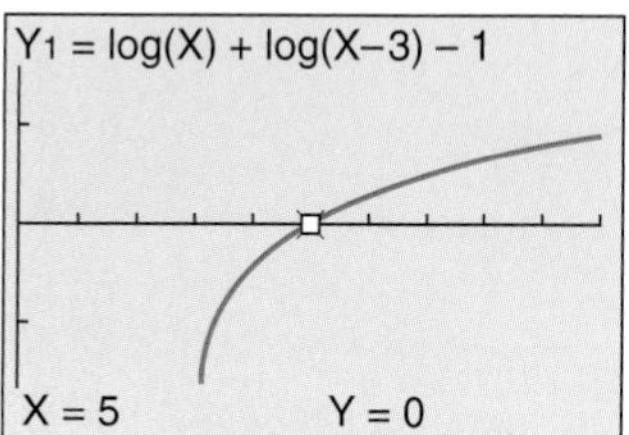

Figure 4-22

Carbon-14 Dating

When a living organism dies, the oxygen/carbon dioxide cycle common to all living things ceases; then carbon-14, a radioactive isotope with a half-life of 5,700 years, is no longer absorbed. By measuring the amount of carbon-14 present in ancient objects, archaeologists can estimate the object's age.

The amount A of radioactive material present at time t is given by the model

$$A = A_0 2^{-t/h}$$

where A_0 is the amount present initially and h is the half-life of the material.

EXAMPLE 8 How old is a wooden statue that has two-thirds of its original carbon-14 content?

Solution To find the time t when $A = \frac{2}{3}A_0$, we substitute $\frac{2A_0}{3}$ for A and 5,700 for h in the radioactive decay formula and solve for t:

$$A = A_0 2^{-t/h}$$

$$\frac{2A_0}{3} = A_0 2^{-t/5,700}$$

$$1 = \frac{3}{2}(2^{-t/5,700})$$ Divide both sides by A_0 and multiply both sides by $\frac{3}{2}$.

$$\log 1 = \log \frac{3}{2}(2^{-t/5,700})$$ Take the common logarithm of each side.

$$0 = \log \frac{3}{2} + \log 2^{-t/5,700}$$ The log of a product is the sum of the logs.

$$-\log \frac{3}{2} = -\frac{t}{5,700} \log 2$$ Subtract $\log \frac{3}{2}$ from both sides and use Property 7 of logarithms.

$$5{,}700\left(\frac{\log \frac{3}{2}}{\log 2}\right) = t \qquad \text{Multiply both sides by } -\frac{5{,}700}{\log 2}.$$

$$t \approx 3{,}334.286254 \qquad \text{Use a calculator.}$$

The wooden statue is approximately 3,300 years old.

Self Check How old is an artifact that has 60% of its original carbon-14 content? ■

Everyday Connections — Carbon-14 Dating

"With these things [cleanliness, care, seriousness, and practice], it is possible to obtain radiocarbon dates which are consistent and which may indeed help roll back the pages of history and reveal to mankind something more about his ancestors, and in this way perhaps about his future." Willard F. Libby, Physical Chemist, pioneered radiocarbon (Carbon-14) dating method, awarded Nobel Prize for Chemistry in 1960.

Turin Shroud Older Than Thought

The Shroud of Turin, the piece of linen long believed to have been wrapped around Jesus's body after the crucifixion, is much older than the date suggested by radiocarbon tests, according to new microchemical research.

Published in the current issue of *Thermochimica Acta,* a chemistry peer-reviewed scientific journal, the study dismisses the results of the 1988 Carbon-14 dating.

At that time, three reputable laboratories in Oxford, Zurich, and Tucson, Ariz., concluded that the cloth on which the smudged outline of the body of a man is indelibly impressed, was a medieval fake dating from 1260 to 1390, and not the burial cloth wrapped around the body of Christ.

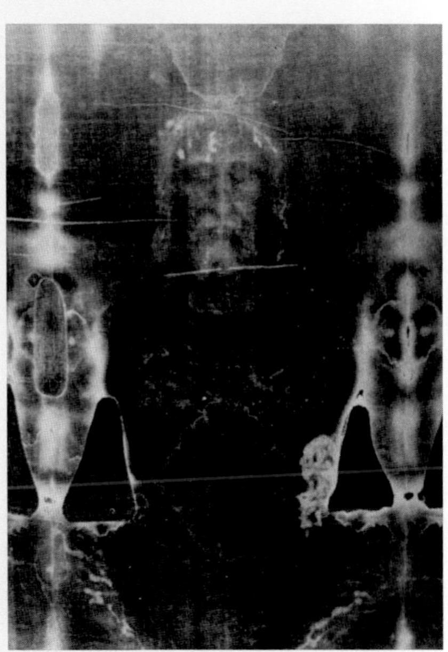

Barrie M. Schwortz/AP/Wide World Photos

Source: Rossella Lorenzi, Discovery News, http://dsc.discovery.com/news/briefs/20050124/shroud.html

Suppose a fossil sample contains 20% of the amount of Carbon-14 that was present originally. Given that the half-life of Carbon-14 is 5,700 years, answer the following questions.

1. Approximate the age of the fossil sample. **13,235 years old**
2. Approximate the amount of Carbon-14 remaining when the fossil sample is 35,000 years old. **1.42% remains**

Population Growth

When there is sufficient food and space, populations of living organisms tend to increase exponentially according to the Malthusian growth model

$$P = P_0e^{kt}$$

where P_0 is the initial population at $t = 0$ and k depends on the rate of growth.

EXAMPLE 9 The bacteria in a laboratory culture increased from an initial population of 500 to 1,500 in 3 hours. Find the time it will take for the population to reach 10,000.

Solution

$$P = P_0e^{kt}$$

$$1{,}500 = 500(e^{k3}) \quad \text{Substitute 1,500 for } P\text{, 500 for } P_0\text{, and 3 for } t.$$

$$3 = e^{3k} \quad \text{Divide both sides by 500.}$$

$$3k = \ln 3 \quad \text{Change the equation from exponential to logarithmic form.}$$

$$k = \frac{\ln 3}{3} \quad \text{Divide both sides by 3.}$$

To find out when the population will reach 10,000, we substitute 10,000 for P, 500 for P_0, and $\frac{\ln 3}{3}$ for k in the equation $P = P_0e^{kt}$ and solve for t:

$$P = P_0e^{kt}$$

$$10{,}000 = 500e^{\left(\frac{\ln 3}{3}\right)t}$$

$$20 = e^{\left(\frac{\ln 3}{3}\right)t} \quad \text{Divide both sides by 500.}$$

$$\left(\frac{\ln 3}{3}\right)t = \ln 20 \quad \text{Change the equation to logarithmic form.}$$

$$t = \frac{3 \ln 20}{\ln 3} \quad \text{Multiply both sides by } \frac{3}{\ln 3}.$$

$$\approx 8.180499084 \quad \text{Use a calculator.}$$

The culture will reach 10,000 bacteria in a little more than 8 hours.

Self Check If the population increases from 1,000 to 3,000 in 3 hours, how long will it take to reach 20,000? ■

Self Check Answers

1. $x = 0.6826$ **2.** -9.451980309 **3.** 3 **4.** 1, -3 **5.** 1 **6.** 20; -5 is extraneous **7.** 3, 5
8. about 4,200 years **9.** about 8 hr

4.6 Exercises

VOCABULARY AND CONCEPTS *Fill in the blanks.*

1. An equation with a variable in its exponent is called a(n) exponential equation.
2. An equation with a logarithmic expression that contains a variable is a(n) logarithmic equation.
3. The formula for carbon dating is $A = $ $A_0 2^{-t/h}$.
4. The formula for population growth is $P = $ P_0e^{kt}.

PRACTICE ***Solve each equation. If an answer is not exact, give the answer to four decimal places.***

5. $4^x = 5$ **1.1610**

6. $7^x = 12$ **1.2770**

7. $13^{x-1} = 2$ **1.2702**

8. $5^{x+1} = 3$ **−0.3174**

9. $2^{x+1} = 3^x$ **1.7095**

10. $5^{x-3} = 3^{2x}$ **−8.2144**

11. $2^x = 3^x$ **0**

12. $3^{2x} = 4^x$ **0**

13. $4^{x+2} = 8^x$ **4**

14. $27^{x+1} = 3^{2x+1}$ **−2**

15. $3^{x-1} = 9^{2x}$ $-\frac{1}{3}$

16. $5^{2x+1} = 125^x$ **1**

17. $7^{x^2} = 10$ **±1.0878**

18. $8^{x^2} = 11$ **±1.0738**

19. $8^{x^2} = 9^x$ **0, 1.0566**

20. $5^{x^2} = 2^{5x}$ **0, 2.1534**

21. $2^{x^2-2x} = 8$ **3, −1**

22. $5^{x^2-3x} = 625$ **4, −1**

23. $3^{x^2+4x} = \frac{1}{81}$ **−2, −2**

24. $7^{x^2+3x} = \frac{1}{49}$ **−2, −1**

25. $4^{x+2} - 4^x = 15$ (*Hint:* $4^{x+2} = 4^x4^2$.) **0**

26. $3^{x+3} + 3^x = 84$ (*Hint:* $3^{x+3} = 3^x3^3$.) **1**

27. $2(3^x) = 6^{2x}$ **0.2789**

28. $2(3^{x+1}) = 3(2^{x-1})$ **−3.4190**

29. $2^{2x} - 10(2^x) + 16 = 0$ (*Hint:* Let $y = 2^x$.) **1, 3**

30. $3^{2x} - 10(3^x) + 9 = 0$ (*Hint:* Let $y = 3^x$.) **0, 2**

31. $2^{2x+1} - 2^x = 1$ (*Hint:* $2^{a+b} = 2^a2^b$.) **0**

32. $3^{2x+1} - 10(3^x) + 3 = 0$ (*Hint:* $3^{a+b} = 3^a3^b$.) **−1, 1**

Solve each equation.

33. $\log (2x - 3) = \log (x + 4)$ **7**

34. $\log (3x + 5) - \log (2x + 6) = 0$ **1**

35. $\log \frac{4x + 1}{2x + 9} = 0$ **4**

36. $\log \frac{5x + 2}{2(x + 7)} = 0$ **4**

37. $\log x^2 = 2$ **10, −10**

38. $\log x^3 = 3$ **10**

39. $\log x + \log (x - 48) = 2$ **50**

40. $\log x + \log (x + 9) = 1$ **1**

41. $\log x + \log (x - 15) = 2$ **20**

42. $\log x + \log (x + 21) = 2$ **4**

43. $\log (x + 90) = 3 - \log x$ **10**

44. $\log (x - 6) - \log (x - 2) = \log \frac{5}{x}$ **10**

45. $\log (x - 1) - \log 6 = \log (x - 2) - \log x$ **3, 4**

46. $\log (2x - 3) - \log (x - 1) = 0$ **2**

47. $\log x^2 = (\log x)^2$ **1, 100**

48. $\log (\log x) = 1$ 10^{10}

49. $\frac{\log (3x - 4)}{\log x} = 2$ **no solution**

50. $\frac{\log (8x - 7)}{\log x} = 2$ **7**

51. $\frac{\log (5x + 6)}{2} = \log x$ **6**

52. $\frac{1}{2} \log (4x + 5) = \log x$ **5**

53. $\log_3 x = \log_3 \left(\frac{1}{x}\right) + 4$ **9**

54. $\log_5 (7 + x) + \log_5 (8 - x) - \log_5 2 = 2$ **3, −2**

55. $2 \log_2 x = 3 + \log_2 (x - 2)$ **4**

56. $2 \log_3 x - \log_3 (x - 4) = 2 + \log_3 2$ **6, 12**

57. $\log (7y + 1) = 2 \log (y + 3) - \log 2$ **7, 1**

58. $2 \log (y + 2) = \log (y + 2) - \log 12$ $-\frac{23}{12}$

Use a graphing calculator to solve each equation. If an answer is not exact, give the result to the nearest hundredth.

59. $\log x + \log (x - 15) = 2$ **20**

60. $\log x + \log (x + 3) = 1$ **2**

61. $2^{x+1} = 7$ **1.81**

62. $\ln (2x + 5) - \ln 3 = \ln (x - 1)$ **8**

APPLICATIONS ***Use a calculator to help solve each problem.***

63. Tritium decay The half-life of tritium is 12.4 years. How long will it take for 25% of a sample of tritium to decompose? **about 5.1 yr**

64. Radioactive decay In 2 years, 20% of a radioactive element decays. Find its half-life. **about 6.2 yr**

65. Thorium decay An isotope of thorium, ^{227}Th, has a half-life of 18.4 days. How long will it take 80% of the sample to decompose? **about 42.7 days**

66. Lead decay An isotope of lead, ^{201}Pb, has a half-life of 8.4 hours. How many hours ago was there 30% more of the substance? **about 3.2 hr**

67. Carbon-14 dating A cloth fragment is found in an ancient tomb. It contains 70% of the carbon-14 that it is assumed to have had initially. How old is the cloth? **about 2,900 yr**

68. Carbon-14 dating Only 25% of the carbon-14 in a wooden bowl remains. How old is the bowl? **about 11,000 yr**

69. Compound interest If $500 is deposited in an account paying 8.5% annual interest, compounded semiannually, how long will it take for the account to increase to $800? **about 5.6 yr**

70. Continuous compound interest In Exercise 69, how long will it take if the interest is compounded continuously? **about 5.5 yr**

71. Compound interest If \$1,300 is deposited in a savings account paying 9% interest, compounded quarterly, how long will it take the account to increase to \$2,100? **about 5.4 yr**

72. Compound interest A sum of \$5,000 deposited in an account grows to \$7,000 in 5 years. Assuming annual compounding, what interest rate is being paid? **about 6.96%**

73. Rule of seventy A rule of thumb for finding how long it takes an investment to double is called the **rule of seventy.** To apply the rule, divide 70 by the interest rate (expressed as a percent). At 5%, it takes $\frac{70}{5} = 14$ years to double the investment. At 7%, it takes $\frac{70}{7} = 10$ years. Explain why this formula works. **because $\ln 2 \approx 0.70$**

74. Bacterial growth A bacterial culture grows according to the formula $P = P_0 a^t$. If it takes 5 days for the culture to triple in size, how long will it take to double in size? **about 3.2 days**

75. Oceanography The intensity I of a light a distance x meters beneath the surface of a lake decreases exponentially. If the light intensity at 6 meters is 70% of the intensity at the surface, at what depth will the intensity be 20%? **about 27 m**

76. Rodent control The rodent population in a city is currently estimated at 30,000. If it is expected to double every 5 years, when will the population reach 1 million? **in about 25.3 yr**

77. Newton's law of cooling Water whose temperature is at 100°C is left to cool in a room where the temperature is 60°C. After 3 minutes, the water temperature is 90°. If the water temperature T is a function of time t given by $T = 60 + 40e^{kt}$, find k.

$$k = \frac{\ln 0.75}{3}$$

78. Newton's law of cooling Refer to Exercise 77 and find the time for the water temperature to reach 70°C. **about 14.5 min**

79. Newton's law of cooling A block of steel, initially at 0°C, is placed in an oven heated to 300°C. After 5 minutes, the temperature of the steel is 100°C. If the steel temperature T is a function of time t given by $T = 300 - 300e^{kt}$, find the value of k.

$$\frac{\ln (2/3)}{5}$$

80. Newton's law of cooling Refer to Exercise 79 and find the time for the steel temperature to reach 200°C. **about 13.5 min**

DISCOVERY AND WRITING

81. Explain why it is necessary to check the solutions of a logarithmic equation.

82. Use the population growth formula to show that the doubling time for population growth is given by

$$T = \frac{\ln 2}{r}.$$

83. Use the population growth formula to show that the tripling time for population growth is given by

$$T = \frac{\ln 3}{r}.$$

84. Can you solve $x = \log x$ algebraically? Can you find an approximate solution?

Find x.

85. $\log_2 (\log_5 (\log_7 x)) = 2$

86. $\log_8 \left[16\sqrt[3]{4{,}096}\right]^{\frac{1}{6}} = x$

REVIEW *Find the inverse of the function defined by each equation.*

87. $y = 3x + 2$
$y = \frac{x-2}{3}$

88. $y = \dfrac{1}{x - 3}$
$y = \frac{1}{x} + 3$

Let $f(x) = 5x - 1$ and $g(x) = x^2$. Find each value.

89. $(f \circ g)(2)$ **19**

90. $(g \circ f)(2)$ **81**

91. $(f \circ g)(x)$ **$5x^2 - 1$**

92. $(g \circ f)(x)$ **$(5x - 1)^2$**

PROBLEMS AND PROJECTS

1. Match each equation with its graph.

x a. $y = 3^x$ u b. $y = 3^x - 1$

v c. $y = 3^{x-1}$ s d. $y = \log_3 x$

y e. $y = \log_3 (x - 1)$ t f. $y = \log_{1/3} x$

s.

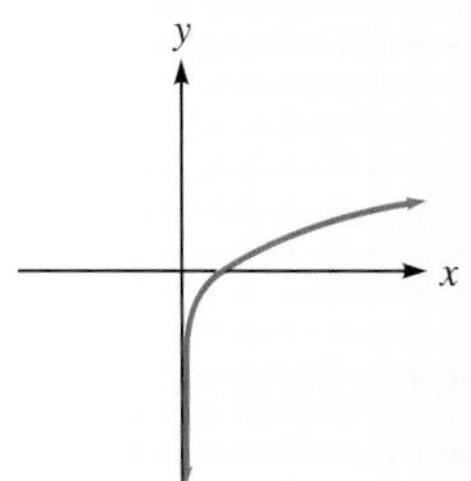

t.

u.

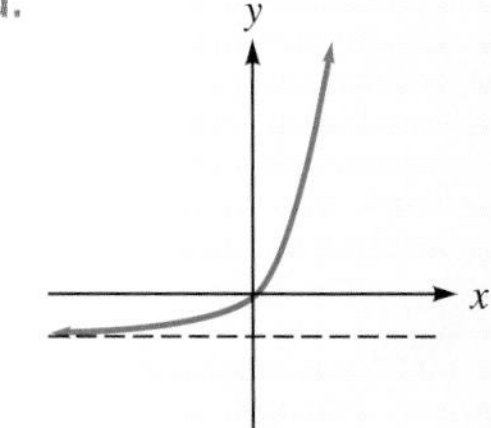

v.

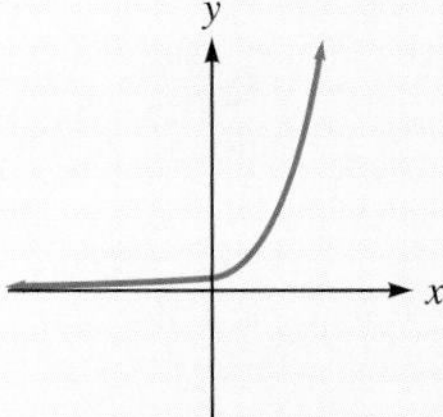

w.

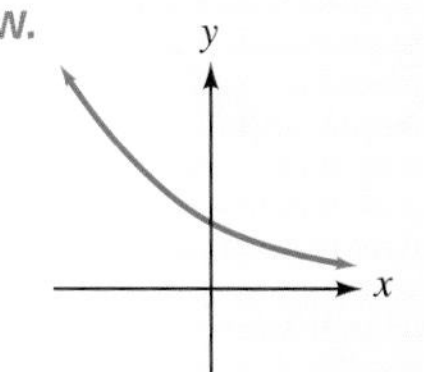

x.

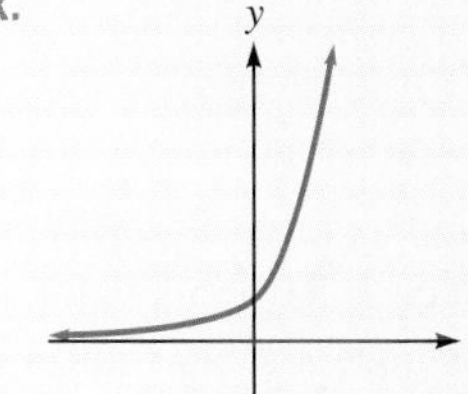

y.

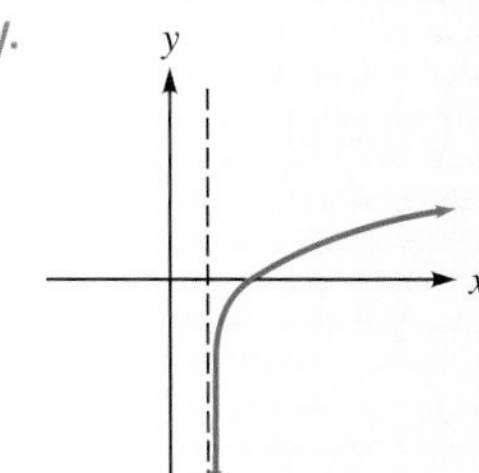

z.

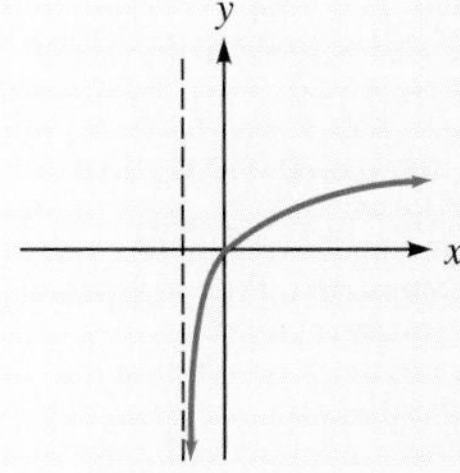

2. Use a graphing calculator to graph the function $y = \ln (e^x)$. Explain why the graph is a line. What is a simpler form of the equation of that line?

3. Use a graphing calculator to graph the function $y = e^{\ln x}$. Is its graph a line, or just part of a line? Explain.

4. The tank in the illustration initially contains 20 gallons of pure water. A brine solution containing 0.5 pounds of salt per gallon is pumped into the tank, and the well-stirred mixture leaves at the same rate. The amount A of salt in the tank after t minutes is given by

$$A = 10(1 - e^{-0.03t})$$

a. Graph this function.
b. What is A when $t = 0$? Explain why that value is expected.
c. What is A after 2 minutes? After 10 minutes?
d. What value does A approach after a long time (as t becomes large)? Explain why this is the value you would expect.

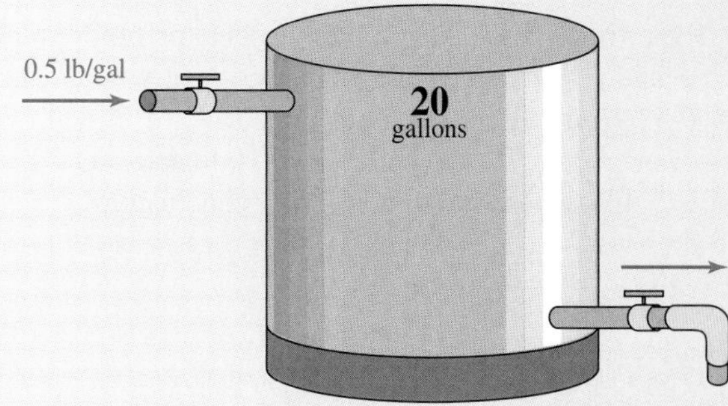

Project 1

Graphing calculators graph base-e and base-10 logarithmic functions easily, because logarithms to these bases are built-in. Find a way of graphing the function $y = \log_3 x$, even though there is no $\log_3$ key.

Project 2

For positive values of x, the graphs of the polynomial function $y = x^3$ and the exponential function $y = e^x$ are both increasing, as shown in Illustration 1. Which is increasing faster? At $x = 2$, 3, and 4, the graph of the polynomial is winning, but when $x = 5$, the graph of the exponential has caught up to and passed the polynomial graph.

The higher a polynomial's degree, the faster its graph rises. Illustration 2 shows the graphs of $y = x^5$ and $y = e^x$. It looks as if the graph of the polynomial is rising more rapidly. Will the exponential graph ever catch up? In a race between a polynomial and an exponential function, which function will eventually win?

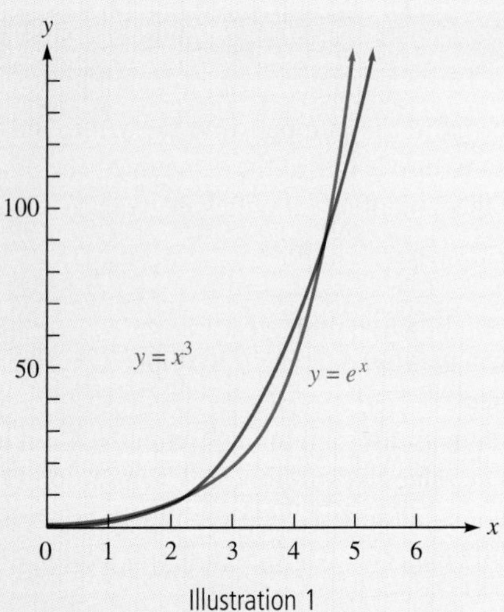

Illustration 1

Experiment with a graphing calculator, and graph $y = x^5$ and $y = e^x$. Then try a race between $y = x^{20}$ and $y = e^x$. (*Hint:* Useful viewing windows are difficult to find. For $y = x^{20}$, try $80 \leq x \leq 100$ and $0 \leq Y \leq 5 \times 10^{39}$.) Write a brief report of your conclusions.

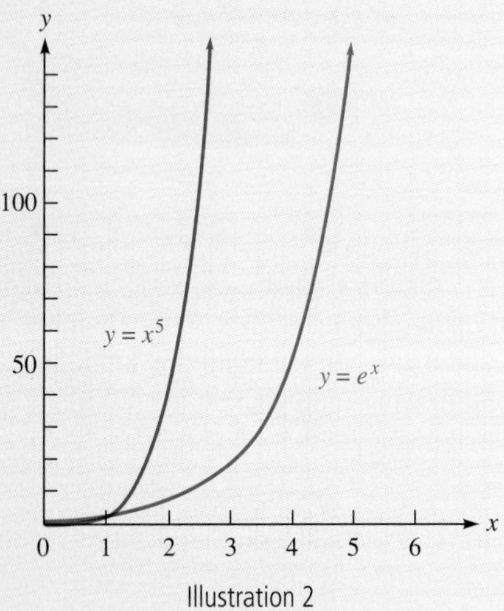

Illustration 2

Project 3

When graphing a function $y = f(x)$, various numbers x are used to produce corresponding values of y, and many pairs (x, y) are plotted to produce the graph. In **parametric equations,** various values of a third variable, called a **parameter,** are used to generate both x- and y-values. Set your graphing calculator for parametric equations (try the MODE key, or consult the owner's manual). Set the range of the parameter, t:

Tmin = 0

Tmax = 10

Tstep = 0.1

Then graph the parametric equations

$$X_{1_T} = \ln t$$
$$Y_{1_T} = t$$

The resulting graph is that of $y = e^x$, which is the inverse of $y = \ln x$. Explain why.

Project 4

The 19 frets on the neck of the classical guitar shown in the illustration are positioned according to the exponential function

$$d = L\left(\frac{1}{2}\right)^{n/12}$$

where L is the distance between the nut and the bridge and d is the distance from the bridge to fret number n.

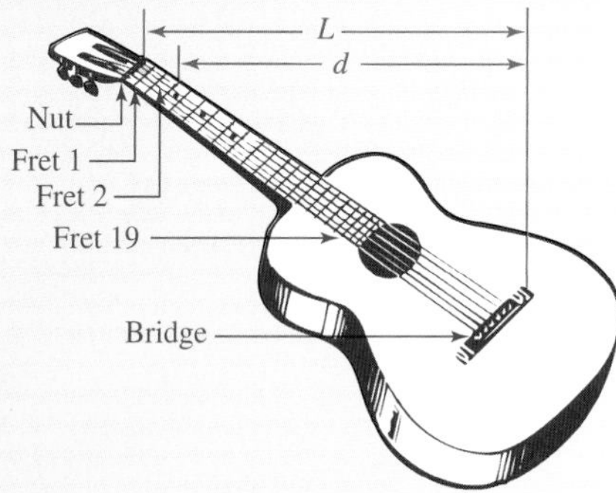

- Find a classical guitar and measure its nut-to-bridge distance L. Use this value in the equation and calculate the distance to frets 1 and 19. Do the calculated values agree with the measured distances?
- Calculate and measure the distance to fret 12, which produces a tone one octave above that of the open string. What fractional part of L is this distance?
- Calculate the distance to fret zero. From the value calculated, where is fret zero?
- Does this formula work for an electric guitar?

Project 5

Determine the current interest rate for Government EE savings bonds by calling a bank or visiting the Web site www.savingsbonds.gov. Then calculate how long it will take an EE bond to double in value.

Determine the current interest rate for Government I savings bonds by calling a bank or visiting www.savingsbonds.gov. Then calculate how long it will take an I bond to double in value.

Call a bank and determine the interest rate on a 5-year certificate of deposit. Then calculate how long it will take the value of the CD to double.

Write a paper comparing these three investments. Tell which one you think is the better investment for you at this time.

CHAPTER SUMMARY

CONCEPTS	REVIEW EXERCISES

4.1 Exponential Functions and Their Graphs

An exponential function with base b is defined by the equation

$y = f(x) = b^x$

$(b > 0, b \neq 1)$

Use properties of exponents to simplify.

1. $5^{\sqrt{2}} \cdot 5^{\sqrt{2}}$ $5^{2\sqrt{2}}$

2. $(2^{\sqrt{5}})^{\sqrt{2}}$ $2^{\sqrt{10}}$

Graph the function defined by each equation.

3. $f(x) = 3^x$

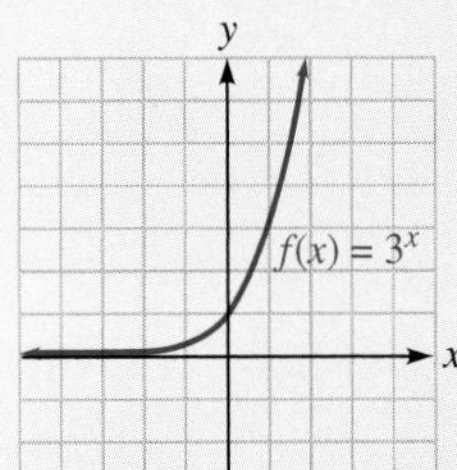

4. $f(x) = \left(\frac{1}{3}\right)^x$

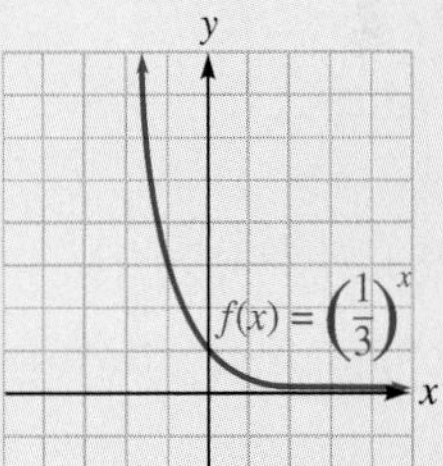

5. The graph of $f(x) = 6^x$ will pass through the points $(0, p)$ and $(1, q)$. Find p and q. $p = 1, q = 6$

6. Give the domain and range of the function $f(x) = b^x$, with $b > 0$ and $b \neq 1$. **domain:** $(-\infty, \infty)$; **range:** $(0, \infty)$

Graph each function by using a translation.

7. $f(x) = \left(\frac{1}{2}\right)^x - 2$

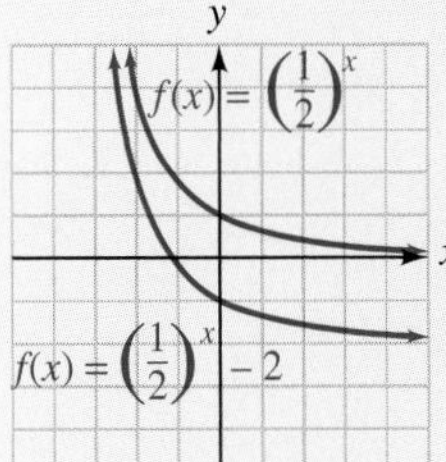

8. $f(x) = \left(\frac{1}{2}\right)^{x+2}$

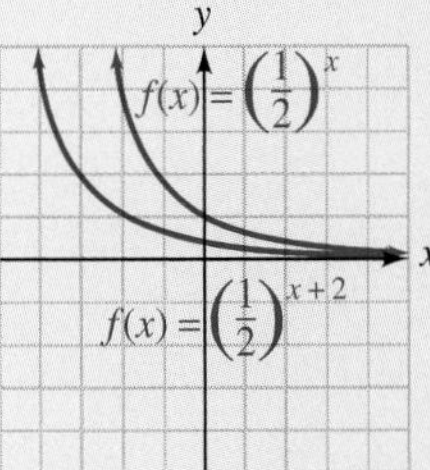

Graph each function.

9. $f(x) = e^x + 1$

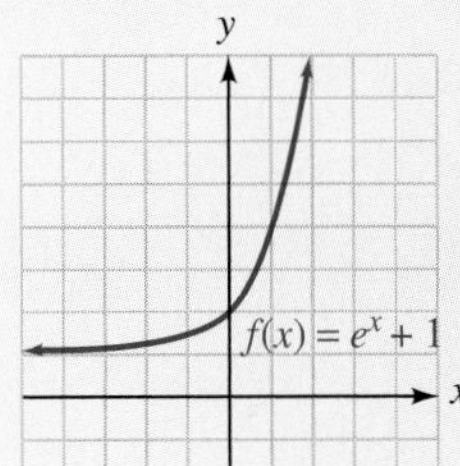

10. $f(x) = e^{x-3}$

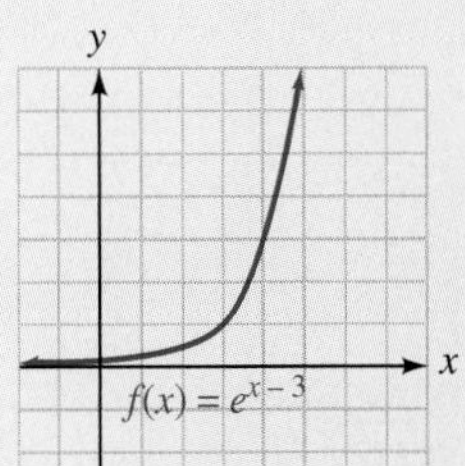

Compound interest:

$$A = P\left(1 + \frac{r}{k}\right)^{kt}$$

Continuous compound interest:

$$A = Pe^{rt}$$

11. How much will \$10,500 become if it earns 9% per year for 60 years, compounded quarterly? **\$2,189,703.45**

12. If \$10,500 accumulates interest at an annual rate of 9%, compounded continuously, how much will be in the account in 60 years? **\$2,324,767.37**

4.2 Applications of Exponential Functions

Radioactive decay:

$$A = A_0 2^{-t/h}$$

Intensity of light:

$$I = I_0 k^x$$

$$e = 2.71828182845904\ldots$$

Malthusian population growth:

$$P = P_0 e^{kt}$$

13. The half-life of a radioactive material is about 34.2 years. How much of the material is left after 20 years? $\frac{2}{3}$

14. Find the intensity of light at a depth of 12 meters if $I_0 = 14$ and $k = 0.7$. **0.19 lumen**

15. The population of the United States is approximately 300,000,000 people. Find the population in 50 years if $k = 0.015$. **about 635,000,000**

4.3 Logarithmic Functions and Their Graphs

If $b > 0$ and $b \neq 1$, then

$y = \log_b x$ means $x = b^y$

16. Give the domain and range of the logarithmic function. $(0, \infty)$; $(-\infty, \infty)$

Find each value.

17. $\log_3 9$ 2

18. $\log_9 \frac{1}{3}$ $-\frac{1}{2}$

19. $\log_x 1$ 0

20. $\log_5 0.04$ -2

21. $\log_a \sqrt{a}$ $\frac{1}{2}$

22. $\log_a \sqrt[3]{a}$ $\frac{1}{3}$

Find *x*.

23. $\log_2 x = 5$ 32

24. $\log_{\sqrt{3}} x = 4$ 9

25. $\log_{\sqrt{2}} x = 6$ 8

26. $\log_{0.1} 10 = x$ -1

27. $\log_x 2 = -\frac{1}{3}$ $\frac{1}{8}$

28. $\log_x 32 = 5$ 2

29. $\log_{0.25} x = -1$ 4

30. $\log_{0.125} x = -\frac{1}{3}$ 2

31. $\log_{\sqrt{2}} 32 = x$ 10

32. $\log_{\sqrt{5}} x = -4$ $\frac{1}{25}$

33. $\log_{\sqrt{3}} 9\sqrt{3} = x$ 5

34. $\log_{\sqrt{5}} 5\sqrt{5} = x$ 3

Graph each function.

35. $f(x) = \log(x - 2)$

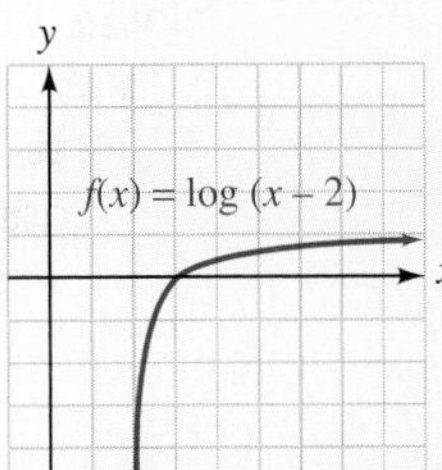

36. $f(x) = 3 + \log x$

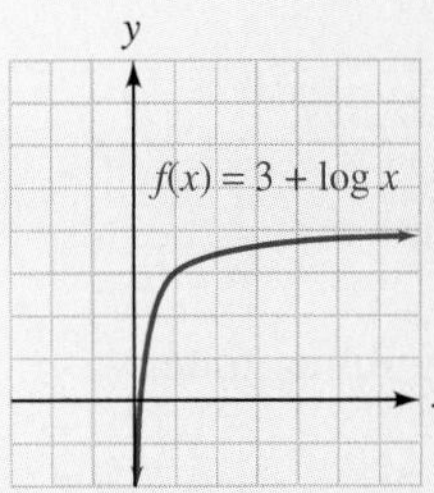

Graph each pair of equations on one set of coordinate axes.

37. $y = 4^x$ and $y = \log_4 x$

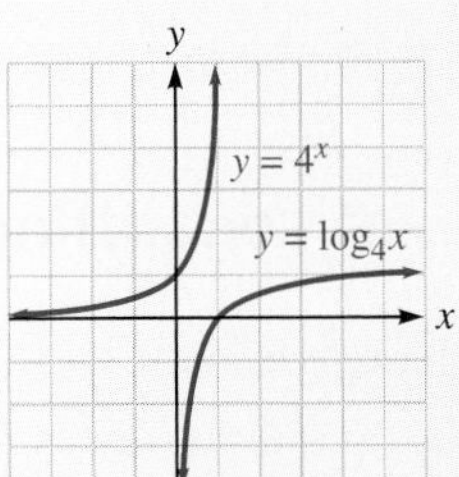

38. $y = \left(\frac{1}{3}\right)^x$ and $y = \log_{1/3} x$

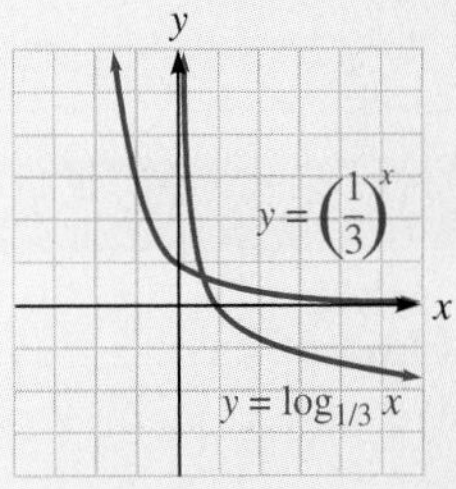

ln x means $\log_e x$.

 **Use a calculator to find each value to four decimal places.**

39. ln 452 **6.1137**

40. ln (log 7.85) **−0.1111**

Use a calculator to solve each equation. Round each answer to four decimal places.

41. $\ln x = 2.336$ **10.3398**

42. $\ln x = \log 8.8$ **2.5715**

Graph each function.

43. $y = f(x) = 1 + \ln x$

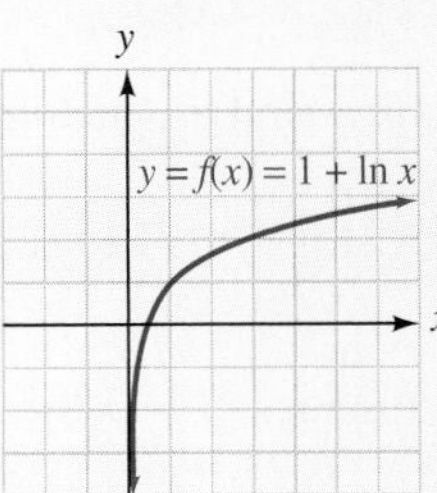

44. $y = f(x) = \ln(x + 1)$

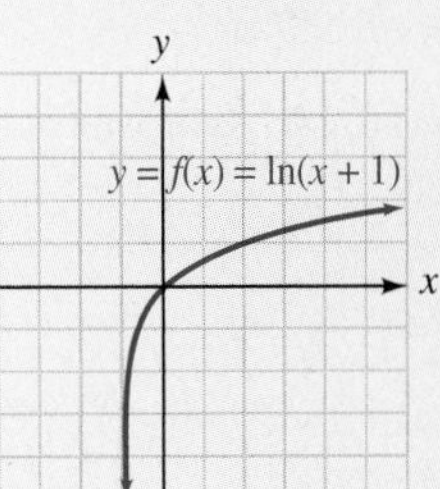

4.4 Applications of Logarithmic Functions

Decibel voltage gain:

$$\text{db gain} = 20 \log \frac{E_O}{E_I}$$

45. An amplifier has an output of 18 volts when the input is 0.04 volt. Find the db gain. **53 db**

Richter scale:

$$R = \log \frac{A}{P}$$

46. An earthquake had a period of 0.3 second and an amplitude of 7,500 micrometers. Find its measure on the Richter scale. **4.4**

Charging batteries:

$$t = -\frac{1}{k} \ln\left(1 - \frac{C}{M}\right)$$

47. How long will it take a dead battery to reach an 80% charge? (Assume $k = 0.17$.) **$9\frac{1}{2}$ min**

Population doubling time:

$$t = \frac{\ln 2}{r}$$

48. How long will it take the population of the United States to double if the growth rate is 3% per year? **23 yr**

Isothermal expansion:

$$E = RT \ln\left(\frac{V_f}{V_i}\right)$$

49. Find the amount of energy that must be supplied to double the volume of 1 mole of gas at a constant temperature of 350K. (*Hint:* $R = 8.314$.) **2,017 joules**

4.5 Properties of Logarithms

Properties of logarithms:
If b is a positive number and $b \neq 1$,

1. $\log_b 1 = 0$
2. $\log_b b = 1$
3. $\log_b b^x = x$
4. $b^{\log_b x} = x$
5. $\log_b MN = \log_b M + \log_b N$
6. $\log_b \frac{M}{N} = \log_b M - \log_b N$
7. $\log_b M^p = p \log_b M$
8. If $\log_b x = \log_b y$, then $x = y$.

Simplify each expression.

50. $\log_7 1$ **0**

51. $\log_7 7$ **1**

52. $\log_7 7^3$ **3**

53. $7^{\log_7 4}$ **4**

54. $\ln e^4$ **4**

55. $\ln 1$ **0**

56. $10^{\log_{10} 7}$ **7**

57. $e^{\ln 3}$ **3**

58. $\log_b b^4$ **4**

59. $\ln e^9$ **9**

Write each expression in terms of the logarithms of x, y, and z.

60. $\log_b \frac{x^2y^3}{z^4}$

$2 \log_b x + 3 \log_b y - 4 \log_b z$

61. $\log_b \sqrt{\frac{x}{yz^2}}$

$\frac{1}{2}(\log_b x - \log_b y - 2 \log_b z)$

Write each expression as the logarithm of one quantity.

62. $3 \log_b x - 5 \log_b y + 7 \log_b z$ $\quad \log_b \frac{x^3z^7}{y^5}$

63. $\frac{1}{2}(\log_b x + 3 \log_b y) - 7 \log_b z$ $\quad \log_b \frac{\sqrt{xy^3}}{z^7}$

Assume that log a = 0.6, log b = 0.36, and log c = 2.4. Find the value of each expression.

64. $\log abc$ **3.36**

65. $\log a^2b$ **1.56**

66. $\log \frac{ac}{b}$ **2.64**

67. $\log \frac{a^2}{c^3b^2}$ **−6.72**

Change-of-base formula:

$$\log_b y = \frac{\log_a y}{\log_a b}$$

68. To four decimal places, find $\log_5 17$. **1.7604**

pH scale:

$\text{pH} = -\log [\text{H}^+]$

Weber–Fechner law:

$L = k \ln I$

69. pH of grapefruit The pH of grapefruit juice is about 3.1. Find its hydrogen ion concentration. **about 7.94×10^{-4} gram-ions per liter**

70. Find the decrease in loudness if the intensity is cut in half. **$k \ln 2$ less**

4.6 Exponential and Logarithmic Equations

Solve each equation for x.

71. $3^x = 7$ $\quad \frac{\log 7}{\log 3} \approx 1.7712$

72. $5^{x+2} = 625$ $\quad$ **2**

73. $2^x = 3^{x-1}$ $\quad \frac{\log 3}{\log 3 - \log 2} \approx 2.7095$

74. $2^{x^2+4x} = \frac{1}{8}$ $\quad$ **−1, −3**

Solve each equation for x.

75. $\log x + \log (29 - x) = 2$ $\quad$ **25, 4**

76. $\log_2 x + \log_2 (x - 2) = 3$ $\quad$ **4**

77. $\log_2 (x + 2) + \log_2 (x - 1) = 2$ $\quad$ **2**

78. $\frac{\log (7x - 12)}{\log x} = 2$ $\quad$ **4, 3**

79. $\log x + \log (x - 5) = \log 6$ $\quad$ **6**

80. $\log 3 - \log (x - 1) = -1$ $\quad$ **31**

81. $e^{x \ln 2} = 9$ $\quad \frac{\ln 9}{\ln 2} \approx 3.1699$

82. $\ln x = \ln (x - 1)$ $\quad$ **no solution**

83. $\ln x = \ln (x - 1) + 1$ $\quad \frac{e}{e - 1} \approx 1.5820$

84. $\ln x = \log_{10} x$ (*Hint:* Use the change-of-base formula.) **1**

Carbon dating:

$A = A_0 2^{-t/h}$

85. Carbon-14 dating A wooden statue found in Egypt has a carbon-14 content that is two-thirds of that found in living wood. If the half-life of carbon-14 is 5,700 years, how old is the statue? **about 3,300 yr**

CHAPTER TEST

Graph each function.

1. $f(x) = 2^x + 1$

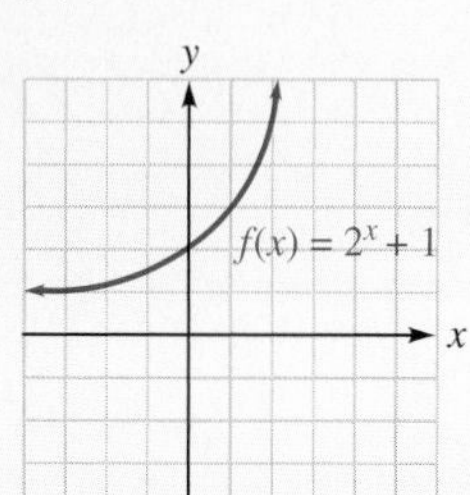

2. $f(x) = e^{x-2}$

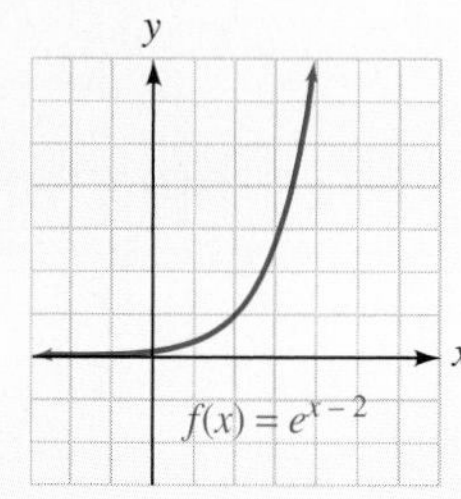

Solve each problem.

3. A radioactive material decays according to the formula $A = A_0(2)^{-t}$. How much of a 3-gram sample will be left in 6 years? $\frac{3}{64}$ **g**

4. An initial deposit of \$1,000 earns 6% interest, compounded twice a year. How much will be in the account in one year? **\$1,060.90**

5. An account contains \$2,000 and has been earning 8% interest, compounded continuously. How much will be in the account in 10 years? **\$4,451.08**

Find each value.

6. $\log_7 343$ **3**

7. $\log_3 \frac{1}{27}$ **−3**

8. $\log_{10} 10^{12} + 10^{\log_{10} 5}$ **17**

9. $\log_{3/2} \frac{9}{4}$ **2**

10. $\log_{2/3} \frac{27}{8}$ **−3**

Graph each function.

11. $f(x) = \log (x - 1)$

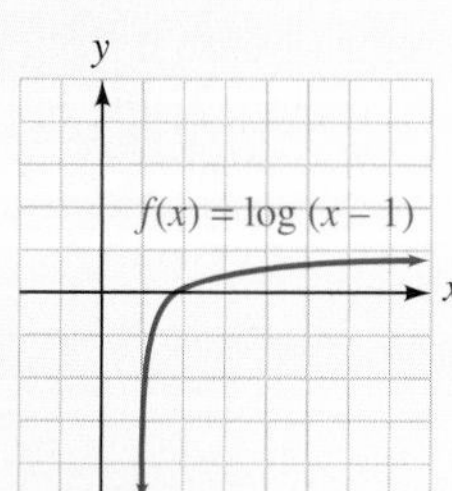

12. $f(x) = 2 + \ln x$

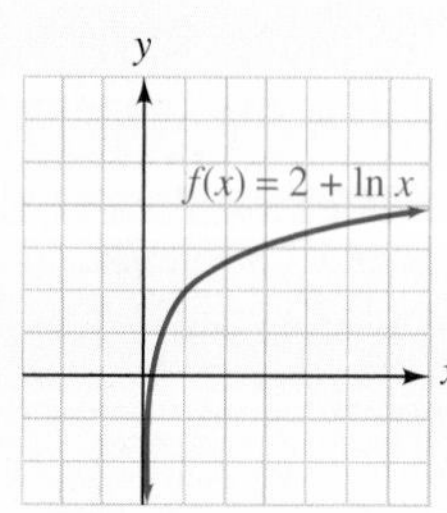

Write each expression in terms of the logarithms of a, b, and c.

13. $\log a^2bc^3$
$2 \log a + \log b + 3 \log c$

14. $\ln \sqrt{\frac{a}{b^2c}}$
$\frac{1}{2}(\ln a - 2 \ln b - \ln c)$

Write each expression as a logarithm of a single quantity.

15. $\frac{1}{2} \log (a + 2) + \log b - 2 \log c$ **$\log \frac{b\sqrt{a+2}}{c^2}$**

16. $\frac{1}{3}(\log a - 2 \log b) - \log c$ **$\log \frac{\sqrt[3]{\frac{a}{b^2}}}{c}$**

Assume that $\log 2 = 0.3010$ *and* $\log 3 = 0.4771$. *Find each value.* **Do not use a calculator.**

17. $\log 24$ **1.3801**

18. $\log \frac{8}{3}$ **0.4259**

Use the change-of-base formula to find each logarithm. **Do not attempt to simplify the answer.**

19. $\log_7 3$ **$\frac{\log 3}{\log 7}$ or $\frac{\ln 3}{\ln 7}$**

20. $\log_\pi e$ **$\frac{\log e}{\log \pi}$ or $\frac{\ln e}{\ln \pi}$**

Determine whether each statement is true or false.

21. $\log_a ab = 1 + \log_a b$
true

22. $\frac{\log a}{\log b} = \log a - \log b$
false

23. Find the pH of a solution with a hydrogen ion concentration of 3.7×10^{-7}. (*Hint:* $\text{pH} = -\log [\text{H}^+]$.)
6.4

24. Find the db gain of an amplifier when $E_O = 60$ volts and $E_I = 0.3$ volt. (*Hint:* db gain $= 20 \log (E_O/E_I)$.)
46 db

Solve each equation.

25. $3^{x-1} = 100^x$ **$\frac{\log 3}{\log 3 - 2} \approx -0.3133$**

26. $3^{x^2-2x} = 27$ **−1, 3**

27. $\log (5x + 2) = \log (2x + 5)$ **1**

28. $\log x + \log (x - 9) = 1$ **10**

5 Solving Polynomial Equations

COLLEGE **Algebra** $f(x)$ **Now**™

Throughout the chapter, this icon introduces resources on the College AlgebraNow Web Site, accessed through **http://1pass.thomson.com**, that will:

- Help you test your knowledge of the material in this chapter prior to reading it,
- Allow you to take an exam-prep quiz, and
- Provide a Personalized Learning Plan targeting areas you should study.

Careers and Mathematics

AGRICULTURAL ENGINEER

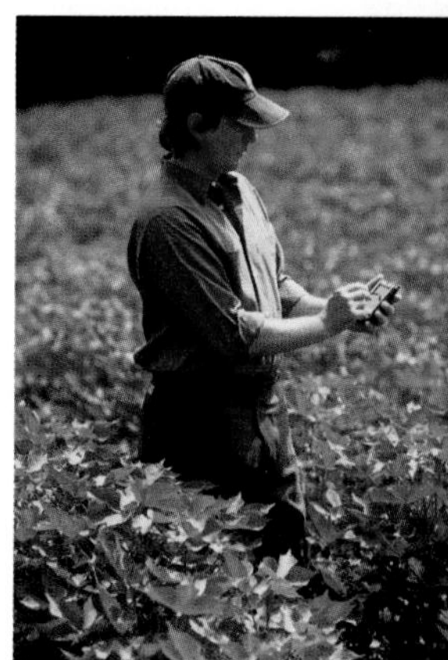
Creatas/PictureQuest

Agricultural engineers apply knowledge of engineering technology and biological science to agriculture. They design agricultural machinery and equipment and agricultural structures. They also develop ways to conserve soil and water and to improve the processing of agricultural products. Agricultural engineers work in research and development, production, sales, or management. More than one-third of the 2,400 agricultural engineers employed in 2000 worked for engineering and management services, supplying consultant services to farmers and farm-related industries. Others worked in a wide variety of industries, including crops and livestock as well as manufacturing and government.

JOB OUTLOOK

Employment of agricultural engineers is expected to increase about as fast as the average for all occupations through 2011. Increasing demand for agricultural products, continued efforts for more efficient agricultural production, and increasing emphasis on the conservation of resources should result in job opportunities for agricultural engineers. However, most openings will be created by the need to replace agricultural engineers who transfer to other occupations or leave the labor force.

Median annual earnings of agricultural engineers were $50,700 in 2002.

For a sample application, see Example 3 in Section 5.3.

A goal throughout all of algebra has been to solve equations. In this chapter, we will develop methods to solve polynomial equations of any degree.

5.1 The Remainder and Factor Theorems; Synthetic Division

In this section, you will learn about

- Zeros of a Polynomial
- The Remainder Theorem
- The Factor Theorem
- Synthetic Division
- Using Synthetic Division to Evaluate Polynomials

We have seen that many polynomial equations can be solved by factoring. For example, to solve $x^3 - 3x^2 + 2x = 0$, we proceed as follows:

$$x^3 - 3x^2 + 2x = 0$$

$$x(x^2 - 3x + 2) = 0 \qquad \text{Factor out } x.$$

$$x(x - 1)(x - 2) = 0 \qquad \text{Factor } x^2 - 3x + 2.$$

$$x = 0 \quad \text{or} \quad x - 1 = 0 \quad \text{or} \quad x - 2 = 0 \qquad \text{Set each factor equal to 0.}$$

$$x = 1 \qquad\qquad x = 2$$

The solution set of this equation is $\{0, 1, 2\}$.

Zeros of a Polynomial

In general, a **polynomial equation** is an equation that can be written in the form $P(x) = 0$, where

$$P(x) = a_n x^n + a_{n-1}x^{n-1} + a_{n-2}x^{n-2} + \cdots + a_1 x + a_0$$

where n is a natural number and the polynomial is of degree n. A **zero of the polynomial** $P(x)$ is any number r for which $P(r) = 0$. It follows that a zero of $P(x)$ is a root of the polynomial equation $P(x) = 0$. For example, 0, 1, and 2 are zeros of the polynomial

$$P(x) = x^3 - 3x^2 + 2x$$

because

$$P(0) = 0^3 - 3(0)^2 + 2(0) = 0 - 0 + 0 = 0$$

$$P(1) = 1^3 - 3(1)^2 + 2(1) = 1 - 3 + 2 = 0$$

$$P(2) = 2^3 - 3(2)^2 + 2(2) = 8 - 12 + 4 = 0$$

Comment We began this section by showing that 0, 1, and 2 are the roots of the polynomial equation

$$x^3 - 3x^2 + 2x = 0$$

We then showed that 0, 1, and 2 are the zeros of the polynomial

$$P(x) = x^3 - 3x^2 + 2x$$

In general, the real roots of the polynomial equation $P(x) = 0$ are zeros of the polynomial $P(x)$.

The Remainder Theorem

There is a relationship between a zero r of a polynomial $P(x)$ and the results of a long division of $P(x)$ by $x - r$. This relationship is best shown with an example.

EXAMPLE 1 Let $P(x) = 3x^3 - 5x^2 + 3x - 10$. Find **a.** $P(1)$ and **b.** divide $P(x)$ by $x - 1$.

Solution **a.**

$$\begin{aligned} P(1) &= 3(1)^3 - 5(1)^2 + 3(1) - 10 \\ &= 3 - 5 + 3 - 10 \\ &= \mathbf{-9} \end{aligned}$$

b.

$$\begin{array}{r} 3x^2 - 2x + 1 \\ x - 1\overline{)3x^3 - 5x^2 + 3x - 10} \\ \underline{3x^3 - 3x^2} \\ -2x^2 + 3x \\ \underline{-2x^2 + 2x} \\ +x - 10 \\ \underline{x - 1} \\ -\mathbf{9} \end{array}$$

Note that the remainder is equal to $P(1)$.

Self Check Let $P(x) = 2x^2 - 3x + 5$. Find $P(2)$ and divide $P(x)$ by $x - 2$. What do you notice about the results? ■

ACTIVE EXAMPLE 2 Let $P(x) = 3x^3 - 5x^2 + 3x - 10$. **a.** Find $P(-2)$ and **b.** divide $P(x)$ by $x + 2$.

Solution **a.**

$$\begin{aligned} P(-2) &= 3(-2)^3 - 5(-2)^2 + 3(-2) - 10 \\ &= 3(-8) - 5(4) + 3(-2) - 10 \\ &= -24 - 20 - 6 - 10 \\ &= \mathbf{-60} \end{aligned}$$

b.

$$\begin{array}{r} 3x^2 - 11x + 25 \\ x + 2\overline{)3x^3 - 5x^2 + 3x - 10} \\ \underline{3x^3 + 6x^2} \\ -11x^2 + 3x \\ \underline{-11x^2 - 22x} \\ 25x - 10 \\ \underline{25x + 50} \\ -\mathbf{60} \end{array}$$

Note that the remainder is equal to $P(-2)$.

COLLEGE **Algebra** $f(x)$ **Now**™

Explore this Active Example and test yourself on the steps taken here at **http://1pass.thomson.com**. You can also find this example on the Interactive Video Skillbuilder CD-ROM.

Self Check Let $P(x) = 2x^2 - 3x + 5$. Find $P(-3)$ and divide $P(x)$ by $x + 3$. What do you notice about the results? ■

When $P(x)$ was divided by $x - 1$ in Example 1, the remainder was $P(1) = -9$. When $P(x)$ was divided by $x + 2$, or $x - (-2)$, in Example 2, the remainder was $P(-2) = -60$. These results are not coincidental. The **remainder theorem** states that a division of any polynomial $P(x)$ by $x - r$ gives $P(r)$ as the remainder.

The Remainder Theorem

If $P(x)$ is a polynomial, r is any number, and $P(x)$ is divided by $x - r$, the remainder is $P(r)$.

Proof To divide $P(x)$ by $x - r$, we must find a quotient $Q(x)$ and a remainder $R(x)$ such that

$$\begin{array}{ccccccc} \text{Dividend} & = & \text{divisor} & \cdot & \text{quotient} & + & \text{remainder} \\ \downarrow & & \downarrow & & \downarrow & & \downarrow \\ P(x) & = & (x - r) & \cdot & Q(x) & + & R(x) \end{array}$$

Since the degree of the remainder $R(x)$ must be less than the degree of the divisor $x - r$, and the degree of $x - r$ is 1, $R(x)$ must be a constant R.

In the equation

$$P(x) = (x - r)Q(x) + R$$

the polynomial on the left-hand side is the same as the polynomial on the right-hand side, and the values that they assume for any number x are equal. If we replace x with r, we have

$$\begin{aligned} P(r) &= (r - r)Q(r) + R \\ &= (0)Q(r) + R \\ &= R \end{aligned}$$

Thus, $P(r) = R$. ■

EXAMPLE 3 Find the remainder that will occur when $P(x) = 2x^4 - 10x^3 + 17x^2 - 14x - 3$ is divided by $x - 3$.

Solution By the remainder theorem, the remainder will be $P(3)$.

$$\begin{aligned} P(x) &= 2x^4 - 10x^3 + 17x^2 - 14x - 3 \\ P(3) &= 2(3)^4 - 10(3)^3 + 17(3)^2 - 14(3) - 3 && \text{Substitute 3 for } x. \\ &= 0 \end{aligned}$$

The remainder will be 0. Although this calculation is tedious, it is easy to do with a calculator.

Self Check Find the remainder when $P(x)$ is divided by $x - 2$. ■

■ The Factor Theorem

If $R = P(r) = 0$ in the equation $P(x) = (x - r)Q(x) + R$, then $P(x)$ factors as $(x - r)Q(x)$. This fact can help us factor polynomials.

The Factor Theorem

If $P(x)$ is a polynomial and r is any number, then

If $P(r) = 0$, then $x - r$ is a factor of $P(x)$.

If $x - r$ is a factor of $P(x)$, then $P(r) = 0$.

Proof **Part 1:** First, we assume that $P(r) = 0$ and prove that $x - r$ is a factor of $P(x)$. If $P(r) = 0$, then $R = 0$, and the equation $P(x) = (x - r)Q(x) + R$ becomes

$$P(x) = (x - r)Q(x) + 0$$
$$P(x) = (x - r)Q(x)$$

Therefore, $x - r$ divides $P(x)$ evenly, and $x - r$ is a factor of $P(x)$.

Part 2: Conversely, we assume that $x - r$ is a factor of $P(x)$ and prove that $P(r) = 0$. Because, by assumption, $x - r$ is a factor of $P(x)$, $x - r$ divides $P(x)$ evenly, and the division has a remainder of 0. By the remainder theorem, this remainder is $P(r)$. Hence, $P(r) = 0$. ■

We can state the factor theorem in a slightly different way:

If r is a zero of the polynomial $P(x)$, then $x - r$ is a factor of $P(x)$.

If $x - r$ is a factor of $P(x)$, then r is a zero of the polynomial.

ACTIVE EXAMPLE 4 Determine whether $x + 2$ is a factor of $P(x) = x^4 - 7x^2 - 6x$. If so, find the other factor by long division.

Solution By the factor theorem, $x + 2$, or $x - (-2)$, is a factor of $P(x)$ if $P(-2) = 0$. So we find the value of $P(-2)$.

COLLEGE **Algebra f(x) Now™**
Go to **http://1pass.thomson.com** or your CD to practice this example.

$$\begin{aligned} P(x) &= x^4 - 7x^2 - 6x \\ P(-2) &= (-2)^4 - 7(-2)^2 - 6(-2) \quad \text{Substitute } -2 \text{ for } x. \\ &= 16 - 28 + 12 \\ &= 0 \end{aligned}$$

Since $P(-2) = 0$, we know that $x - (-2)$, or $x + 2$, is a factor of $P(x)$.

To find the other factor, we divide $x^4 - 7x^2 - 6x$ by $x + 2$.

$$\begin{array}{r} x^3 - 2x^2 - 3x \\ x + 2 \overline{\smash{)} x^4 - 7x^2 - 6x} \\ \underline{x^4 + 2x^3} \\ - 2x^3 - 7x^2 \\ \underline{- 2x^3 - 4x^2} \\ - 3x^2 - 6x \\ \underline{- 3x^2 - 6x} \\ 0 \end{array}$$

$x^4 - 7x^2 - 6x$ factors as $(x + 2)(x^3 - 2x^2 - 3x)$.

Self Check Determine whether $x + 3$ a factor of $P(x)$. ■

EXAMPLE 5 Let $P(x) = 3x^3 - 5x^2 + 3x - 10$. Show that $P(2) = 0$, and use this fact to help factor $P(x)$.

Solution We find $P(2)$.

$$\begin{aligned} P(x) &= 3x^3 - 5x^2 + 3x - 10 \\ P(2) &= 3(2)^3 - 5(2)^2 + 3(2) - 10 \quad \text{Substitute 2 for } x. \\ &= 3(8) - 5(4) + 6 - 10 \\ &= 0 \end{aligned}$$

Since $P(2) = 0$, it follows that $x - 2$ is a factor of $P(x)$. To find the other factor, we divide $P(x)$ by $x - 2$.

$$\begin{array}{r} 3x^2 + x + 5 \\ x - 2\overline{)3x^3 - 5x^2 + 3x - 10} \\ \underline{3x^3 - 6x^2} \\ x^2 + 3x \\ \underline{x^2 - 2x} \\ 5x - 10 \\ \underline{5x - 10} \\ 0 \end{array}$$

$3x^3 - 5x^2 + 3x - 10$ factors as $(x - 2)(3x^2 + x + 5)$.

Self Check Let $P(x) = x^3 - 1$. Show that $P(1) = 0$, and use this fact to help factor $x^3 - 1$. ■

EXAMPLE 6 Solve: $3x^3 - 5x^2 + 3x - 10 = 0$.

Solution From Example 5, we have

$$\begin{aligned} 3x^3 - 5x^2 + 3x - 10 &= 0 \\ (x - 2)(3x^2 + x + 5) &= 0 \end{aligned}$$

To solve for x, we set each factor equal to 0 and apply the quadratic formula to the equation $3x^2 + x + 5 = 0$.

$$x - 2 = 0 \quad \text{or} \quad 3x^2 + x + 5 = 0$$

$$x = 2 \qquad\qquad x = \frac{-1 \pm \sqrt{1^2 - 4(3)(5)}}{2(3)}$$

$$x = \frac{-1 \pm i\sqrt{59}}{6}$$

The solution set is $\left\{2,\ -\frac{1}{6} + \frac{\sqrt{59}}{6}i,\ -\frac{1}{6} - \frac{\sqrt{59}}{6}i\right\}$.

Self Check Solve: $x^3 - 1 = 0$. ■

EXAMPLE 7 Find a polynomial $P(x)$ with zeros of 3, 3, and -5.

Solution We have proved that if r is a zero of a polynomial, then $x - r$ is a factor of the polynomial. In particular, if 3, 3, and -5 are zeros of $P(x)$, then $x - 3$, $x - 3$, and $x - (-5)$ are factors of $P(x)$.

$$P(x) = (x - 3)(x - 3)(x + 5)$$
$$P(x) = (x^2 - 6x + 9)(x + 5) \qquad \textbf{Multiply } x - 3 \textbf{ and } x - 3.$$
$$P(x) = x^3 - x^2 - 21x + 45$$

The polynomial $P(x) = x^3 - x^2 - 21x + 45$ has zeros of 3, 3, and -5.

Because 3 occurs twice as a zero, we say that 3 is a **zero of multiplicity 2.**

Self Check Find a polynomial $P(x)$ with zeros of -2, 2, and 3. ■

■ Synthetic Division

Synthetic division is an easy way to divide higher-degree polynomials by binomials of the form $x - r$. To see how it works, we consider the following long division. On the left is a complete division. On the right is a modified version in which the variables have been removed.

$$\begin{array}{r} 2x^2 + 10x + 27 \\ x - 3\overline{)2x^3 + 4x^2 - 3x + 10} \\ \underline{2x^3 - 6x^2} \qquad\qquad\quad \\ 10x^2 - 3x \qquad\quad \\ \underline{10x^2 - 30x} \qquad\quad \\ 27x + 10 \\ \underline{27x - 81} \\ \text{(remainder)}\ \ 91 \end{array} \qquad\qquad \begin{array}{r} 2 + 10 + 27 \\ 1 - 3\overline{)2 + 4 - 3 + 10} \\ \underline{2 - 6 \qquad\qquad} \\ 10 - 3 \qquad\quad \\ \underline{10 - 30} \qquad\quad \\ 27 + 10 \\ \underline{27 - 81} \\ \text{(remainder)}\ \ 91 \end{array}$$

We can shorten the work even more by omitting the numbers printed in color.

$$\begin{array}{r} 2 + 10 + 27 \\ -3\overline{)2 + 4 - 3 + 10} \\ \underline{- 6} \qquad\qquad \\ 10 \qquad\qquad \\ \underline{- 30} \qquad \\ 27 \qquad \\ \underline{- 81} \\ \text{(remainder)}\ \ 91 \end{array}$$

We can then compress the work vertically to get

$$\begin{array}{r} 2 + 10 + 27 \\ -3\overline{)2 + 4 - 3 + 10} \\ \underline{- 6 - 30 - 81} \\ 10 \quad 27 \quad 91 \end{array}$$

If we write the 2 in the quotient on the bottom line, the bottom line gives both the coefficients of the quotient and the remainder. The top line can now be eliminated, and the division appears as

$$\begin{array}{r|rrrr} \underline{-3} & 2 + & 4 - & 3 + & 10 \\ & - & 6 - & 30 - & 81 \\ \hline & 2 & 10 & 27 & 91 \end{array}$$

The bottom line was obtained by subtracting the middle line from the top line. If we replace the -3 in the divisor with $+3$, the signs of each number in the middle line will be reversed in the division process. Then the bottom line can be obtained by addition, and we have the final form of the synthetic division.

$$\begin{array}{r|rrrr} +3 & 2\ + & 4\ - & 3\ + & 10 \\ & + & 6\ + & 30\ + & 81 \\ \hline & 2 & 10 & 27 & 91 \end{array}$$

The coefficients of the dividend.

The coefficients of the quotient and the remainder.

Thus,

$$\frac{2x^3 + 4x^2 - 3x + 10}{x - 3} = 2x^2 + 10x + 27 + \frac{91}{x - 3}$$

EXAMPLE 8 Use synthetic division to divide $10x + 3x^4 - 8x^3 + 3$ by $x - 2$.

Solution We first write the terms of $P(x)$ in descending powers of x:

$$3x^4 - 8x^3 + 10x + 3$$

We then write the coefficients of the dividend, with its terms in descending powers of x, and the 2 from the divisor in the following form:

$$\begin{array}{r|rrrrr} 2 & 3 & -8 & 0 & 10 & 3 \\ \hline & & & & & \end{array}$$

Write 0 for the coefficient of the missing x^2 term.

Then we follow these steps:

$$\begin{array}{r|rrrrr} 2 & 3 & -8 & 0 & 10 & 3 \\ & \downarrow & & & & \\ \hline & 3 & & & & \end{array}$$

Bring down the 3.

$$\begin{array}{r|rrrrr} 2 & 3 & -8 & 0 & 10 & 3 \\ & & 6 & & & \\ \hline & 3 & -2 & & & \end{array}$$

Multiply 2 and 3 together to get 6, and add 6 and -8 to get -2.

$$\begin{array}{r|rrrrr} 2 & 3 & -8 & 0 & 10 & 3 \\ & & 6 & -4 & & \\ \hline & 3 & -2 & -4 & & \end{array}$$

Multiply 2 and -2 together to get -4, and add -4 and 0 to get -4.

$$\begin{array}{r|rrrrr} 2 & 3 & -8 & 0 & 10 & 3 \\ & & 6 & -4 & -8 & \\ \hline & 3 & -2 & -4 & 2 & \end{array}$$

Multiply 2 and -4 together to get -8, and add -8 and 10 to get 2.

$$\begin{array}{r|rrrrr} 2 & 3 & -8 & 0 & 10 & 3 \\ & & 6 & -4 & -8 & 4 \\ \hline & 3 & -2 & -4 & 2 & 7 \end{array}$$

Multiply 2 and 2 together to get 4, and add 4 and 3 to get 7.

The coefficients of the quotient and the remainder.

Thus,

$$\frac{10x + 3x^4 - 8x^3 + 3}{x - 2} = 3x^3 - 2x^2 - 4x + 2 + \frac{7}{x - 2}$$

Self Check Divide $2x^3 - 4x^2 + 5x - 7$ by $x - 3$. ■

Using Synthetic Division to Evaluate Polynomials

ACTIVE EXAMPLE 9 Use synthetic division to find $P(-2)$ when $P(x) = 5x^3 + 3x^2 - 21x - 1$.

Solution Because of the remainder theorem, $P(-2)$ is the remainder when $P(x)$ is divided by $x - (-2)$. We use synthetic division to find the remainder.

COLLEGE **Algebra** f(x) **Now**™
Go to **http://1pass.thomson.com** or your CD to practice this example.

$$\begin{array}{r|rrrr} -2 & 5 & 3 & -21 & -1 \\ & & -10 & & \\ \hline & 5 & -7 & & \end{array}$$

$-2(5) = -10; 3 + (-10) = -7.$

$$\begin{array}{r|rrrr} -2 & 5 & 3 & -21 & -1 \\ & & -10 & 14 & \\ \hline & 5 & -7 & -7 & \end{array}$$

$-2(-7) = 14; -21 + 14 = -7.$

$$\begin{array}{r|rrrr} -2 & 5 & 3 & -21 & -1 \\ & & -10 & 14 & 14 \\ \hline & 5 & -7 & -7 & 13 \end{array}$$

$-2(-7) = 14; -1 + 14 = 13.$

Because the remainder is 13, $P(-2) = 13$.

Self Check Find $P(3)$. ■

EXAMPLE 10 If $P(x) = x^3 - x^2 + x - 1$, find $P(i)$, where $i = \sqrt{-1}$.

Solution We can use synthetic division.

$$\begin{array}{r|rrrr} i & 1 & -1 & +1 & -1 \\ & & i & -1 - i & +1 \\ \hline & 1 & i - 1 & -i & 0 \end{array}$$

Since the remainder is 0, $P(i) = 0$, and i is a zero of $P(x)$.

Self Check Find $P(-i)$. ■

Self Check Answers

1. $P(2)$ is the remainder. **2.** $P(-3)$ is the remainder. **3.** -11 **4.** no **5.** $(x - 1)(x^2 + x + 1)$ **6.** $1, \dfrac{-1 \pm i\sqrt{3}}{2}$ **7.** $P(x) = x^3 - 3x^2 - 4x + 12$ **8.** $2x^2 + 2x + 11 + \dfrac{26}{x - 3}$ **9.** 98 **10.** 0

5.1 Exercises

VOCABULARY AND CONCEPTS ***Fill in the blanks.***

1. The variables in a polynomial have <u>whole</u>-number exponents.

2. A zero of $P(x)$ is any number r for which <u>$P(r) = 0$</u>.

3. The remainder theorem holds when r is <u>any</u> number.

4. If $P(x)$ is a polynomial and $P(x)$ is divided by <u>$x - r$</u>, the remainder will be $P(r)$.

5. If $P(x)$ is a polynomial, then $P(r) = 0$ if and only if $x - r$ is a factor of $P(x)$.

6. A shortcut method for dividing a polynomial by a binomial of the form $x - r$ is called synthetic division.

PRACTICE ***Let $P(x) = 2x^4 - 2x^3 + 5x^2 - 1$. Find each value by substituting the given value of x into the polynomial and simplifying. Then find the value by doing a long division and finding the remainder.***

7. $P(1)$ 4
8. $P(2)$ 35
9. $P(-2)$ 67
10. $P(-1)$ 8

Use the remainder theorem to find the remainder that occurs when $P(x) = 3x^4 + 5x^3 - 4x^2 - 2x + 1$ is divided by each binomial.

11. $x - 12$ 70,249
12. $x + 15$ 134,131
13. $x + 3.25$ 128.3085938
14. $x - 7.12$ 9,298.469614

Use the factor theorem to decide whether each statement is true. If the statement is not true, so indicate.

15. $x - 1$ is a factor of $x^7 - 1$. true
16. $x - 2$ is a factor of $x^3 - x^2 + 2x - 8$. true
17. $x - 1$ is a factor of $3x^5 + 4x^2 - 7$. true
18. $x + 1$ is a factor of $3x^5 + 4x^2 - 7$. false
19. $x + 3$ is a factor of $2x^3 - 2x^2 + 1$. false
20. $x - 3$ is a factor of $3x^5 - 3x^4 + 5x^2 - 13x - 6$. false
21. $x - 1$ is a factor of $x^{1,984} - x^{1,776} + x^{1,492} - x^{1,066}$. true
22. $x + 1$ is a factor of $x^{1,984} + x^{1,776} - x^{1,492} - x^{1,066}$. true

A partial solution set is given for each equation. Find the complete solution set.

23. $x^3 + 3x^2 - 13x - 15 = 0; \{-1\}$ $\{-1, -5, 3\}$
24. $x^3 + 6x^2 + 5x - 12 = 0; \{1\}$ $\{1, -3, -4\}$
25. $x^4 - 2x^3 - 2x^2 + 6x - 3 = 0; \{1, 1\}$ $\{1, 1, \sqrt{3}, -\sqrt{3}\}$
26. $x^5 + 4x^4 + 4x^3 - x^2 - 4x - 4 = 0; \{1, -2, -2\}$ $\left\{-2, -2, 1, -\frac{1}{2} \pm \frac{\sqrt{3}}{2}i\right\}$
27. $x^4 - 5x^3 + 7x^2 - 5x + 6 = 0; \{2, 3\}$ $\{2, 3, i, -i\}$
28. $x^4 + 2x^3 - 3x^2 - 4x + 4 = 0; \{1, -2\}$ $\{1, -2, 1, -2\}$

Find a polynomial with the given zeros.

29. 4, 5 $x^2 - 9x + 20$
30. -3, 5 $x^2 - 2x - 15$
31. 1, 1, 1 $x^3 - 3x^2 + 3x - 1$
32. 1, 0, -1 $x^3 - x$
33. 2, 4, 5 $x^3 - 11x^2 + 38x - 40$
34. 7, 6, 3 $x^3 - 16x^2 + 81x - 126$
35. $1, -1, \sqrt{2}, -\sqrt{2}$ $x^4 - 3x^2 + 2$
36. $0, 0, 0, \sqrt{3}, -\sqrt{3}$ $x^5 - 3x^3$
37. $\sqrt{2}, i, -i$ $x^3 - \sqrt{2}x^2 + x - \sqrt{2}$
38. $i, i, 2$ $x^3 - (2 + 2i)x^2 - (1 - 4i)x + 2$
39. $0, 1 + i, 1 - i$ $x^3 - 2x^2 + 2x$
40. $i, 2 + i, 2 - i$ $x^3 - (4 + i)x^2 + (5 + 4i)x - 5i$

Use synthetic division to express $P(x) = 3x^3 - 2x^2 - 6x - 4$ in the form (divisor)(quotient) + remainder for each divisor.

41. $x - 1$ $(x - 1)(3x^2 + x - 5) - 9$
42. $x - 2$ $(x - 2)(3x^2 + 4x + 2) + 0$
43. $x - 3$ $(x - 3)(3x^2 + 7x + 15) + 41$
44. $x - 4$ $(x - 4)(3x^2 + 10x + 34) + 132$
45. $x + 1$ $(x + 1)(3x^2 - 5x - 1) - 3$
46. $x + 2$ $(x + 2)(3x^2 - 8x + 10) - 24$
47. $x + 3$ $(x + 3)(3x^2 - 11x + 27) - 85$
48. $x + 4$ $(x + 4)(3x^2 - 14x + 50) - 204$

Use synthetic division to perform each division.

49. $\dfrac{x^3 + x^2 + x - 3}{x - 1}$ $x^2 + 2x + 3$
50. $\dfrac{x^3 - x^2 - 5x + 6}{x - 2}$ $x^2 + x - 3$
51. $\dfrac{7x^3 - 3x^2 - 5x + 1}{x + 1}$ $7x^2 - 10x + 5 + \frac{-4}{x + 1}$
52. $\dfrac{2x^3 + 4x^2 - 3x + 8}{x - 3}$ $2x^2 + 10x + 27 + \frac{89}{x - 3}$
53. $\dfrac{4x^4 - 3x^3 - x + 5}{x - 3}$ $4x^3 + 9x^2 + 27x + 80 + \frac{245}{x - 3}$
54. $\dfrac{x^4 + 5x^3 - 2x^2 + x - 1}{x + 1}$ $x^3 + 4x^2 - 6x + 7 + \frac{-8}{x + 1}$
55. $\dfrac{3x^5 - 768x}{x - 4}$ $3x^4 + 12x^3 + 48x^2 + 192x$
56. $\dfrac{x^5 - 4x^2 + 4x + 4}{x + 3}$ $x^4 - 3x^3 + 9x^2 - 31x + 97 + \frac{-287}{x + 3}$

Let $P(x) = 5x^3 + 2x^2 - x + 1$. Use synthetic division to find each value.

57. $P(2)$ 47

58. $P(-2)$ -29

59. $P(-5)$ -569

60. $P(3)$ 151

Let $P(x) = 2x^4 - x^2 + 2$. Use synthetic division to find each value.

61. $P\left(\frac{1}{2}\right)$ $\frac{15}{8}$

62. $P\left(\frac{1}{3}\right)$ $\frac{155}{81}$

63. $P(i)$ 5

64. $P(-i)$ 5

Let $P(x) = x^4 - 8x^3 + 8x + 14x^2 - 15$. Write the terms of $P(x)$ in descending powers of x and use synthetic division to find each value.

65. $P(1)$ 0

66. $P(0)$ -15

67. $P(-3)$ 384

68. $P(-1)$ 0

Let $P(x) = 8 - 8x^2 + x^5 - x^3$. Write the terms of $P(x)$ in descending powers of x and use synthetic division to find each value.

69. $P(i)$ $16 + 2i$

70. $P(-i)$ $16 - 2i$

71. $P(-2i)$ $40 - 40i$

72. $P(2i)$ $40 + 40i$

DISCOVERY AND WRITING

73. If 0 is a zero of $P(x) = a_nx^n + a_{n-1}x^{n-1} + \cdots a_1x + a_0$, find a_0. 0

74. If 0 occurs twice as a zero of $P(x) = a_nx^n + a_{n-1}x^{n-1} + \cdots + a_1x + a_0$, find a_1. 0

75. If $P(2) = 0$ and $P(-2) = 0$, explain why $x^2 - 4$ is a factor of $P(x)$.

76. If $P(x) = x^4 - 3x^3 + kx^2 + 4x - 1$ and $P(2) = 11$, find k. 3

REVIEW ***Find the quadrant in which each point lies.***

77. $P(3, -2)$ QIV

78. $Q(-2, -5)$ QIII

79. $R(8, \pi)$ QI

80. $S(-9, 9)$ QII

Find the distance between each pair of points.

81. $A(3, -3), B(-5, 3)$ 10

82. $C(-8, 2), D(2, -22)$ 26

Find the slope of the line passing through each pair of points.

83. $E(3, 5), F(-5, -3)$ 1

84. $G\left(3, \frac{3}{5}\right), H\left(-\frac{3}{5}, 1\right)$ $-\frac{1}{9}$

5.2 Descartes' Rule of Signs and Bounds on Roots

In this section, you will learn about

- The Fundamental Theorem of Algebra
- The Conjugate Pairs Theorem
- Descartes' Rule of Signs
- Bounds on Roots

The remainder theorem and synthetic division provide a way of verifying that a particular number is a root of a polynomial equation, but they do not provide the roots. We need some guidelines to indicate how many roots to expect, what kind of roots to expect, and where they are located. This section develops several theorems that provide such guidelines.

The Fundamental Theorem of Algebra

Before attempting to find the roots of a polynomial equation, it would be useful to know whether any roots exist. This question was answered by Carl Friedrich Gauss (1777–1855) when he proved the **fundamental theorem of algebra.**

The Fundamental Theorem of Algebra

If $P(x)$ is a polynomial with positive degree, then $P(x)$ has at least one zero.

The fundamental theorem of algebra guarantees that polynomials such as

$$2x^3 + 3 \quad \text{and} \quad 32.75x^{1,984} + ix^3 - (2 + i)x - 5$$

all have zeros. Since all polynomials with positive degree have zeros, their corresponding polynomial equations all have roots. They are the zeros of the polynomial.

The next theorem will help us show that every nth-degree polynomial equation has exactly n roots.

The Polynomial Factorization Theorem

If $n > 0$ and $P(x)$ is an nth-degree polynomial, then $P(x)$ has exactly n linear factors:

$$P(x) = a_n(x - r_1)(x - r_2)(x - r_3) \cdots (x - r_n)$$

where $r_1, r_2, r_3, \ldots, r_n$ are numbers and a_n is the lead coefficient of $P(x)$.

Proof Let $P(x)$ be a polynomial of degree n $(n > 0)$. Because of the fundamental theorem of algebra, we know that $P(x)$ has a zero r_1 and that the equation $P(x) = 0$ has r_1 for a root. By the factor theorem, we know that $x - r_1$ is a factor of $P(x)$. Thus,

$$P(x) = (x - r_1)Q_1(x)$$

If the lead coefficient of the nth-degree polynomial $P(x)$ is a_n, then $Q_1(x)$ is a polynomial of degree $n - 1$ whose lead coefficient is also a_n.

By the fundamental theorem of algebra, we know that $Q_1(x)$ also has a zero, r_2. By the factor theorem, $x - r_2$ is a factor of $Q_1(x)$, and

$$P(x) = (x - r_1)(x - r_2)Q_2(x)$$

where $Q_2(x)$ is a polynomial of degree $n - 2$ with lead coefficient a_n.

This process can continue only to n factors of the form $x - r_i$ until the final quotient $Q_n(x)$ is a polynomial of degree $n - n$, or degree 0. Thus, the polynomial $P(x)$ factors completely as

(1) $$P(x) = a_n(x - r_1)(x - r_2)(x - r_3) \cdots (x - r_n)$$ ■

If we substitute any one of the numbers $r_1, r_2, r_3, \ldots, r_n$ for x in Equation 1, $P(x)$ will equal 0. Thus, each value of r is a zero of $P(x)$ and a root of the equation $P(x) = 0$. There can be no other roots, because no single factor in Equation 1 is 0 for any value of x not included in the list $r_1, r_2, r_3, \ldots, r_n$.

The values of r in the previous list need not be distinct. Any number r_i that occurs k times as a root of a polynomial equation is called a **root of multiplicity *k*.**

The following theorem summarizes the previous discussion.

Theorem

If multiple roots are counted individually, the polynomial equation $P(x) = 0$ with degree n $(n > 0)$ has exactly n roots among the complex numbers.

The Conjugate Pairs Theorem

Recall that the complex numbers $a + bi$ and $a - bi$ are called **complex conjugates** of each other. The next theorem points out that *complex roots of polynomial equations with real coefficients occur in complex conjugate pairs.*

The Conjugate Pairs Theorem

If a polynomial equation $P(x) = 0$ with real-number coefficients has a complex root $a + bi$ with $b \neq 0$, then its conjugate $a - bi$ is also a root.

ACTIVE EXAMPLE 1 Find a second-degree polynomial equation with real coefficients that has a root of $2 + i$.

Solution Because $2 + i$ is a root, its complex conjugate $2 - i$ is also a root. The equation is

$$[x - (2 + i)][x - (2 - i)] = 0$$
$$(x - 2 - i)(x - 2 + i) = 0$$
$$x^2 - 4x + 5 = 0 \quad \text{Multiply.}$$

The equation $x^2 - 4x + 5 = 0$ will have roots of $2 + i$ and $2 - i$.

COLLEGE **Algebra** f(x) **Now**™
Explore this Active Example and test yourself on the steps taken here at **http://1pass.thomson.com**. You can also find this example on the Interactive Video Skillbuilder CD-ROM.

Self Check Find a second-degree polynomial equation with real coefficients that has a root of $1 - i$. ■

EXAMPLE 2 Find a fourth-degree polynomial equation with real coefficients and i as a root of multiplicity 2.

Solution Because i is a root twice and a fourth-degree polynomial equation has four roots, we must find the other two roots. According to the conjugate pairs theorem, the missing roots are the conjugates of the given roots. Thus, the complete solution set is

$$\{i, i, -i, -i\}$$

The equation is

$$(x - i)(x - i)[x - (-i)][x - (-i)] = 0$$
$$(x - i)(x + i)(x - i)(x + i) = 0$$
$$(x^2 + 1)(x^2 + 1) = 0$$
$$x^4 + 2x^2 + 1 = 0$$

Self Check Find a fourth-degree polynomial equation with real coefficients and $-i$ as a root of multiplicity 2. ■

EXAMPLE 3 Find a quadratic equation with a double root of i.

Solution If i is a root twice in a quadratic equation, the equation is

$$(x - i)(x - i) = 0$$
$$x^2 - 2ix - 1 = 0$$

In this equation, the coefficient of x is $-2i$, which is not a real number. Therefore, it is not surprising that the roots are not in complex conjugate pairs.

Self Check Find a quadratic equation with a double root of $-i$. ■

Comment The theorem "Complex roots of real polynomial equations occur in complex conjugate pairs" applies only to polynomial equations with real coefficients.

Descartes' Rule of Signs

René Descartes (1596–1650) is credited with a theorem known as **Descartes' rule of signs,** which enables us to estimate the number of positive, negative, and nonreal roots of a polynomial equation.

If a polynomial is written in descending powers of x and we scan it from left to right, we say that a variation in sign occurs whenever successive terms have opposite signs. For example, the polynomial

$$P(x) = \overbrace{3x^5 - 2x^4}^{+\text{ to }-} \overbrace{- 5x^3 + x^2}^{-\text{ to }+}\!\!\!\!\!\!\overbrace{ - x}^{+\text{ to }-} - 9$$

has three variations in sign, and the polynomial

$$\begin{aligned} P(-x) &= 3(-x)^5 - 2(-x)^4 - 5(-x)^3 + (-x)^2 - (-x) - 9 \\ &= -3x^5 \overbrace{- 2x^4 + 5x^3}^{-\text{ to }+} + \overbrace{x^2 + x - 9}^{+\text{ to }-} \end{aligned}$$

has two variations in sign.

Descartes' Rule of Signs

If $P(x)$ is a polynomial with real coefficients, the number of positive roots of $P(x) = 0$ is either equal to the number of variations in sign of $P(x)$ or less than that by an even number.

The number of negative roots of $P(x) = 0$ is either equal to the number of variations in sign of $P(-x)$ or less than that by an even number.

ACTIVE EXAMPLE 4 Discuss the possibilities for the roots of $P(x) = 3x^3 - 2x^2 + x - 5 = 0$.

Solution Since there are three variations of sign in $P(x) = 3x^3 - 2x^2 + x - 5 = 0$, there can be either 3 positive roots or only 1 (1 is less than 3 by the even number 2). Because

COLLEGE **Algebra** $f(x)$ **Now**™
Go to **http://1pass.thomson.com** or your CD to practice this example.

$$\begin{aligned} P(-x) &= 3(-x)^3 - 2(-x)^2 + (-x) - 5 \\ &= -3x^3 - 2x^2 - x - 5 \end{aligned}$$

has no variations in sign, there are 0 negative roots. Furthermore, 0 is not a root, because the terms of the polynomial do not have a common factor of x.

If there are 3 positive roots, then all of the roots are accounted for. If there is 1 positive root, the 2 remaining roots must be nonreal complex numbers. The following chart shows these possibilities.

Number of positive roots	Number of negative roots	Number of nonreal roots
3	0	0
1	0	2

The number of nonreal complex roots is the number needed to bring the total number of roots up to 3.

Self Check Discuss the possibilities for the roots of $5x^3 + 2x^2 - x + 3 = 0$. ■

EXAMPLE 5 Discuss the possibilities for the roots of $P(x) = 5x^5 - 3x^3 - 2x^2 + x - 1 = 0$.

Solution Since there are three variations of sign in $P(x)$, there are either 3 or 1 positive roots. Because $P(-x) = -5x^5 + 3x^3 - 2x^2 - x - 1$ has two variations in sign, there are 2 or 0 negative roots. The possibilities are shown as follows:

Number of positive roots	Number of negative roots	Number of nonreal roots
1	0	4
3	0	2
1	2	2
3	2	0

In each case, the number of nonreal complex roots is an even number. This is expected, because this polynomial has real coefficients, and its nonreal complex roots will occur in conjugate pairs.

Self Check Discuss the possibilities for the roots of $5x^5 - 2x^2 - x - 1 = 0$. ■

Bounds on Roots

A final theorem provides a way to find **bounds** on the roots of a polynomial equation, enabling us to look for roots where they can be found.

Upper and Lower Bounds on Roots

Let the lead coefficient of a polynomial $P(x)$ with real coefficients be positive, and do a synthetic division of the coefficients by a positive number c. If each term in the last row of the division is nonnegative, no number greater than c can be a root of $P(x) = 0$. (c is called an **upper bound** of the real roots.)

If $P(x)$ is synthetically divided by a negative number d and the signs in the last row alternate,* no value less than d can be a root of $P(x) = 0$. (d is called a **lower bound** of the real roots.)

*If 0 appears in the third row, that 0 can be assigned either a + or a − sign to help the signs alternate.

ACTIVE EXAMPLE 6 Establish integer bounds for the roots of $2x^3 + 3x^2 - 5x - 7 = 0$.

Solution We will do several synthetic divisions by positive integers, looking for nonnegative values in the last row. Then we will divide by several negative integers, looking for alternating signs in the last row. Trying 1 first gives

COLLEGE **Algebra f(x) Now**™
Go to **http://1pass.thomson.com** or your CD to practice this example.

$$\begin{array}{r|rrrr} 1 & 2 & 3 & -5 & -7 \\ & & 2 & 5 & 0 \\ \hline & +2 & +5 & 0 & -7 \end{array}$$

Because one of the signs in the last row is negative, we cannot claim that 1 is an upper bound. We now try 2.

2 |	2	3	−5	−7
		4	14	18
	+2	+7	+9	+11

Because the last row is entirely nonnegative, we can claim that 2 is an upper bound. That is, no number greater than 2 can be a root of the equation.

Now we try some negative divisors, beginning with -3.

−3 |	2	3	−5	−7
		−6	9	−12
	+2	−3	+4	−19

Since the signs in the last row alternate, -3 is a lower bound. That is, no number less than -3 can be a root. To see whether there is a greater lower bound, we try -2.

−2 |	2	3	−5	−7
		−4	2	6
	+2	−1	−3	−1

Since the signs in the last row do not alternate, we cannot claim that -2 is a lower bound.

Since -3 is a lower bound and 2 is an upper bound, then all of the real roots must be in the interval $(-3, 2)$.

Self Check Establish integer bounds for the roots of $2x^3 + 3x^2 - 11x - 7 = 0$. ■

It is important to understand what the theorem on the bounds of roots says and what it doesn't say. If we divide synthetically by a positive number c and the last row of the synthetic division is entirely nonnegative, the theorem guarantees that c is an upper bound of the roots. However, if the last row contains some negative values, c could still be an upper bound.

If we divide by a negative number d and the signs in the last row alternate, the theorem guarantees that d is a lower bound of the roots. However, if the signs in the last row do not alternate, d could still be a lower bound. This is illustrated in Example 6. It can be shown that the smallest negative root of the equation is approximately -1.81. Thus, -2 is a lower bound for the roots of the equation. However, when we checked -2, the last row of the synthetic division did not have alternating signs. Unfortunately, the theorem does not always determine the best bounds for the roots of the equation.

Self Check Answers

1. $x^2 - 2x + 2 = 0$ **2.** $x^4 + 2x^2 + 1 = 0$ **3.** $x^2 + 2ix - 1 = 0$

4.

Num. of pos. roots	Num. of neg. roots	Num. of nonreal roots
2	1	0
0	1	2

5.

Num. of pos. roots	Num. of neg. roots	Num. of nonreal roots
1	2	2
1	0	4

6. Roots are in $(-4, 3)$.

5.2 Exercises

VOCABULARY AND CONCEPTS *Fill in the blanks.*

1. If $P(x)$ is a polynomial with positive degree, then $P(x)$ has at least one **zero**.
2. The statement in Exercise 1 is called the **fundamental theorem of algebra**.
3. The **conjugate** of $a + bi$ is $a - bi$.
4. The polynomial $6x^4 + 5x^3 - 2x^2 + 3$ has **2** variations in sign.
5. The polynomial $(-x)^3 - (-x)^2 - 4$ has **0** variations in sign.
6. The equation $7x^4 + 5x^3 - 2x + 1 = 0$ can have at most **2** positive roots.
7. The equation $7x^4 + 5x^3 - 2x + 1 = 0$ can have at most **2** negative roots.
8. Complex roots occur in complex **conjugate** pairs. (Assume that the equation has real coefficients.)
9. If no number less than d can be a root of $P(x) = 0$, then d is called a(n) **lower bound**.
10. If no number greater than c can be a root of $P(x) = 0$, then c is called a(n) **upper bound**.

PRACTICE *Determine how many roots each equation has.*

11. $x^{10} = 1$ **10**
12. $x^{40} = 1$ **40**
13. $3x^4 - 4x^2 - 2x = -7$ **4**
14. $-32x^{111} - x^5 = 1$ **111**
15. One root of $x(3x^4 - 2) = 12x$ is 0. How many other roots are there? **4**
16. Two roots of $3x^2(x^7 - 14x + 3) = 0$ are 0. How many other roots are there? **7**

Write a third-degree polynomial equation with real coefficients and the given roots.

17. $3, -i$ **$x^3 - 3x^2 + x - 3 = 0$**
18. $1, i$ **$x^3 - x^2 + x - 1 = 0$**
19. $2, 2 + i$ **$x^3 - 6x^2 + 13x - 10 = 0$**
20. $-2, 3 - i$ **$x^3 - 4x^2 - 2x + 20 = 0$**

Write a fourth-degree polynomial equation with real coefficients and the given roots.

21. $3, 2, i$ **$x^4 - 5x^3 + 7x^2 - 5x + 6 = 0$**
22. $1, 2, 1 + i$ **$x^4 - 5x^3 + 10x^2 - 10x + 4 = 0$**
23. $i, 1 - i$ **$x^4 - 2x^3 + 3x^2 - 2x + 2 = 0$**
24. $i, 2 - i$ **$x^4 - 4x^3 + 6x^2 - 4x + 5 = 0$**

Use Descartes' rule of signs to find the number of possible positive, negative, and nonreal roots of each equation.

25. $3x^3 + 5x^2 - 4x + 3 = 0$
0 or 2 positive; 1 negative; 0 or 2 nonreal
26. $3x^3 - 5x^2 - 4x - 3 = 0$
1 positive; 0 or 2 negative; 0 or 2 nonreal
27. $2x^3 + 7x^2 + 5x + 5 = 0$
0 positive; 1 or 3 negative; 0 or 2 nonreal
28. $-2x^3 - 7x^2 - 5x - 4 = 0$
0 positive; 1 or 3 negative; 0 or 2 nonreal
29. $8x^4 = -5$ **0 positive; 0 negative; 4 nonreal**
30. $-3x^3 = -5$ **1 positive; 0 negative; 2 nonreal**
31. $x^4 + 8x^2 - 5x - 10 = 0$
1 positive; 1 negative; 2 nonreal
32. $5x^7 + 3x^6 - 2x^5 + 3x^4 + 9x^3 + x^2 + 1 = 0$
0 or 2 positive; 1 or 3 negative; 2, 4, or 6 nonreal
33. $-x^{10} - x^8 - x^6 - x^4 - x^2 - 1 = 0$
0 positive; 0 negative; 10 nonreal
34. $x^{10} + x^8 + x^6 + x^4 + x^2 + 1 = 0$
0 positive; 0 negative; 10 nonreal
35. $x^9 + x^7 + x^5 + x^3 + x = 0$ (Is 0 a root?)
0 positive; 0 negative; 8 nonreal; yes
36. $-x^9 - x^7 - x^5 - x^3 - x = 0$ (Is 0 a root?)
0 positive; 0 negative; 8 nonreal; yes
37. $-2x^4 - 3x^2 + 2x + 3 = 0$
1 positive; 1 negative; 2 nonreal
38. $-7x^5 - 6x^4 + 3x^3 - 2x^2 + 7x - 4 = 0$
0, 2, or 4 positive; 1 negative; 0, 2, or 4 nonreal

Find integer bounds for the roots of each equation.

39. $x^2 - 2x - 4 = 0$
−2, 4
40. $9x^2 - 6x - 1 = 0$
−1, 1
41. $18x^2 - 6x - 1 = 0$
−1, 1
42. $2x^2 - 10x - 9 = 0$
−1, 6
43. $6x^3 - 13x^2 - 110x = 0$ **−4, 6**
44. $12x^3 + 20x^2 - x - 6 = 0$ **−2, 1**
45. $x^5 + x^4 - 8x^3 - 8x^2 + 15x + 15 = 0$ **−4, 3**
46. $3x^4 - 5x^3 - 9x^2 + 15x = 0$ **−2, 3**
47. $3x^5 - 11x^4 - 2x^3 + 38x^2 - 21x - 15 = 0$ **−2, 4**
48. $3x^6 - 4x^5 - 21x^4 + 4x^3 + 8x^2 + 8x + 32 = 0$
−3, 4

DISCOVERY AND WRITING

49. Explain why the fundamental theorem of algebra guarantees that every polynomial equation of positive degree has at least one root.

50. Explain why the fundamental theorem of algebra and the factor theorem guarantee that an nth-degree polynomial equation has n roots.

51. Prove that any odd-degree polynomial equation with real coefficients must have at least one real root.

52. If a, b, c, and d are positive numbers, prove that $ax^4 + bx^2 + cx - d = 0$ has exactly two nonreal roots.

REVIEW *Assume that k represents a positive real number, and complete each sentence.*

53. The graph of $y = f(x - k)$ looks like the graph of $y = f(x)$, except that it has been translated **k units to the right**.

54. The graph of $y = f(x) - k$ looks like the graph of $y = f(x)$, except that it has been translated **k units down**.

55. The graph of $y = f(-x)$ looks like the graph of $y = f(x)$, except that it has been **reflected about the y-axis**.

56. The graph of $y = -f(x)$ looks like the graph of $y = f(x)$, except that it has been **reflected about the x-axis**.

57. If $k > 0$, the graph of $y = kf(x)$ looks like the graph of $y = f(x)$, except that it has been **stretched vertically by a factor of k**.

58. If $k > 0$, the graph of $y = f(kx)$ looks like the graph of $y = f(x)$, except that it has been **stretched horizontally by a factor of $\frac{1}{k}$**.

5.3 Rational Roots of Polynomial Equations

In this section, you will learn about

- Finding Possible Rational Roots
- Finding Rational Roots
- An Application

Recall that a rational number is any number that can be written in the form $\frac{p}{q}$, where p and q are integers and $q \neq 0$. In this section, we will find rational roots of polynomial equations with integer coefficients. The following theorem enables us to list the possible rational roots of such equations.

Rational Root Theorem

Let the polynomial equation

$$P(x) = a_nx^n + a_{n-1}x^{n-1} + a_{n-2}x^{n-2} + \cdots + a_1x + a_0 = 0$$

have integer coefficients. If the rational number $\frac{p}{q}$ (written in lowest terms) is a root of $P(x) = 0$, then p is a factor of the constant a_0, and q is a factor of the lead coefficient a_n.

Proof Let $\frac{p}{q}$ (written in lowest terms) be a rational root of $P(x) = 0$. Then the equation is satisfied by $\frac{p}{q}$:

$$a_n\left(\frac{p}{q}\right)^n + a_{n-1}\left(\frac{p}{q}\right)^{n-1} + a_{n-2}\left(\frac{p}{q}\right)^{n-2} + \cdots + a_1\left(\frac{p}{q}\right) + a_0 = 0 \tag{1}$$

We can clear Equation 1 of fractions by multiplying both sides by q^n.

$$a_np^n + a_{n-1}p^{n-1}q + a_{n-2}p^{n-2}q^2 + \cdots + a_1pq^{n-1} + a_0q^n = 0 \tag{2}$$

We can factor p from all but the last term and subtract a_0q^n from both sides to get

$$p(a_np^{n-1} + a_{n-1}p^{n-2}q + a_{n-2}p^{n-3}q^2 + \cdots + a_1q^{n-1}) = -a_0q^n$$

Since p is a factor of the left-hand side, it is also a factor of the right-hand side. So p is a factor of $-a_0q^n$, but because $\frac{p}{q}$ is written in lowest terms, p cannot be a factor of q^n. Therefore, p is a factor of a_0.

We can factor q from all but the first term of Equation 2 and subtract a_np^n from both sides to get

$$q(a_{n-1}p^{n-1} + a_{n-2}p^{n-2}q + a_{n-3}p^{n-3}q^2 + \cdots + a_0q^{n-1}) = -a_np^n$$

Since q is a factor of the left-hand side, it is also a factor of the right-hand side. Because q is not a factor of p^n, it must be a factor of a_n. ■

Everyday Connections Gasoline Prices

"In the next 20 years, oil production from conventional reservoirs may begin to decline, creating a gap between supply and demand."

Stephen Holditch, President, Society of Petroleum Engineers

Gas prices seen hitting record high this year.

NEW YORK–It could be a tough time for American motorists when they hit the road for the vacation season with gasoline prices poised to hit record highs.

The average pump price for self-serve gas is currently \$1.92 a gallon [March 2005], about 13 cents below the all-time record hit last May, and 22 cents higher than a year ago, according to the American Automobile Association. Furthermore, oil prices are at their highest level in months and still rising. The head of the Organization for Petroleum Exporting Countries has said crude could get as high as \$80 per barrel, up from about \$54 currently.

"All of the dynamics are in place for U.S. motorists to pay new record high prices again this year," said AAA spokesman Geoff Sundstrom.

Source: **MSNBC staff and news service reports, March 6, 2005** ***www.msnbc.msn.com***

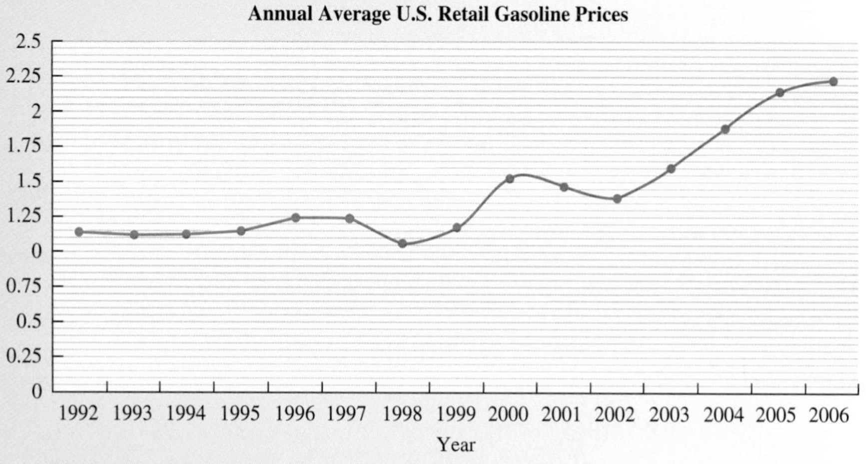

Source: ***http://www.eia.doe.gov/emeu/steo/pub/contents.html***

Continued

$$C(x) = 1.14 - 17.5799x + 56.1669x^2 - 75.8853x^3 + 58.2841x^4 - 28.6823x^5 + 9.6162x^6$$

The polynomial function shown above models the average U.S. retail gasoline prices in dollars per gallon during the period 1992–2006, where x equals the number of years since 1991. Use the graph of the function to answer the following questions.

1. How many times was the average price per gallon \$1.50 between 1999–2005? **3**
2. When was the average price per gallon the lowest between 1992–2006? **1998**
3. What is the predicted average price per gallon for 2006? **\$2.25**

Finding Possible Rational Roots

To illustrate the previous theorem, we consider the equation

$$\frac{1}{2}x^4 + \frac{2}{3}x^3 + 3x^2 - \frac{3}{2}x + 3 = 0$$

Because the theorem requires integer coefficients, we multiply both sides of the equation by 6 to clear it of fractions.

$$3x^4 + 4x^3 + 18x^2 - 9x + 18 = 0$$

By the previous theorem, the only possible numerators for the rational roots of the equation are the factors of the constant term 18:

$$\pm 1,\ \pm 2,\ \pm 3,\ \pm 6,\ \pm 9, \text{ and } \pm 18$$

The only possible denominators are the factors of the lead coefficient 3:

$$\pm 1 \quad \text{and} \quad \pm 3$$

We can form a list of all possible rational solutions by listing the combinations of possible numerators and denominators:

$$\pm\frac{1}{1}, \pm\frac{2}{1}, \pm\frac{3}{1}, \pm\frac{6}{1}, \pm\frac{9}{1}, \pm\frac{18}{1}, \pm\frac{1}{3}, \pm\frac{2}{3}, \pm\frac{3}{3}, \pm\frac{6}{3}, \pm\frac{9}{3}, \pm\frac{18}{3}$$

Since several of these possibilities are duplicates, we can condense the list to get

Possible rational roots

$$\pm 1, \quad \pm 2, \quad \pm 3, \quad \pm 6, \quad \pm 9, \quad \pm 18, \quad \pm\frac{1}{3}, \quad \pm\frac{2}{3}$$

Finding Rational Roots

EXAMPLE 1 Solve: $P(x) = 2x^3 + 3x^2 - 8x + 3 = 0$.

Solution Since the equation is of third degree, it has 3 roots. By Descartes' rule of signs, there are two possible combinations of positive, negative, and nonreal roots. They are summarized as follows:

Number of positive roots	Number of negative roots	Number of nonreal roots
2	1	0
0	1	2

Since any rational roots will have the form $\frac{\text{factor of the constant 3}}{\text{factor of the lead coefficient 2}}$, the possible rational roots are

$$\pm\frac{3}{1}, \quad \pm\frac{1}{1}, \quad \pm\frac{3}{2}, \quad \pm\frac{1}{2}$$

or, written in order of increasing size,

$$-3, \quad -\frac{3}{2}, \quad -1, \quad -\frac{1}{2}, \quad \frac{1}{2}, \quad 1, \quad \frac{3}{2}, \quad 3$$

We check each possibility to see whether it is a root. We can start with $\frac{3}{2}$.

$$\begin{array}{r|rrrr} \frac{3}{2} & 2 & 3 & -8 & 3 \\ & & 3 & 9 & \frac{3}{2} \\ \hline & 2 & 6 & 1 & \frac{9}{2} \end{array}$$

Since the remainder is not 0, $\frac{3}{2}$ is not a root and can be crossed off the list. Since every number in the last row of the synthetic division is positive, $\frac{3}{2}$ is an upper bound. So 3 cannot be a root and can be crossed off the list as well.

$$-3, \quad -\frac{3}{2}, \quad -1, \quad -\frac{1}{2}, \quad \frac{1}{2}, \quad 1, \quad \cancel{\frac{3}{2}}, \quad \cancel{3}$$

We now try $\frac{1}{2}$:

$$\begin{array}{r|rrrr} \frac{1}{2} & 2 & 3 & -8 & 3 \\ & & 1 & 2 & -3 \\ \hline & 2 & 4 & -6 & 0 \end{array}$$

Since the remainder is 0, $\frac{1}{2}$ is a root. The binomial $x - \frac{1}{2}$ is a factor of $P(x)$, and any remaining roots must be supplied by the remaining factor, which is the quotient $2x^2 + 4x - 6$. We can find the other roots by solving the equation $2x^2 + 4x - 6 = 0$, called the **depressed equation.**

$$\begin{aligned} 2x^2 + 4x - 6 &= 0 \\ x^2 + 2x - 3 &= 0 && \text{Divide both sides by 2.} \\ (x - 1)(x + 3) &= 0 && \text{Factor } x^2 + 2x - 3. \end{aligned}$$

$$x - 1 = 0 \quad \text{or} \quad x + 3 = 0$$
$$x = 1 \quad | \quad x = -3$$

The solution set of the equation is $\{\frac{1}{2}, 1, -3\}$. Note that two positive roots and one negative root is a predicted possibility.

Self Check Solve: $3x^3 - 10x^2 + 9x - 2 = 0$. ■

Accent on Technology CONFIRMING ROOTS OF AN EQUATION

We can confirm that the roots found in Example 1 are correct by graphing the function $P(x) = 2x^3 + 3x^2 - 8x + 3$ and locating the resulting x-intercepts of the graph. If we use a graphing window of $x = [-4, 4]$ and $y = [-20, 20]$ and graph the function, we will obtain the graph shown in Figure 5-1. From the graph, we can see that the x-intercepts are at $x = -3$, $x = \frac{1}{2}$, and $x = 1$. These are the roots of the equation.

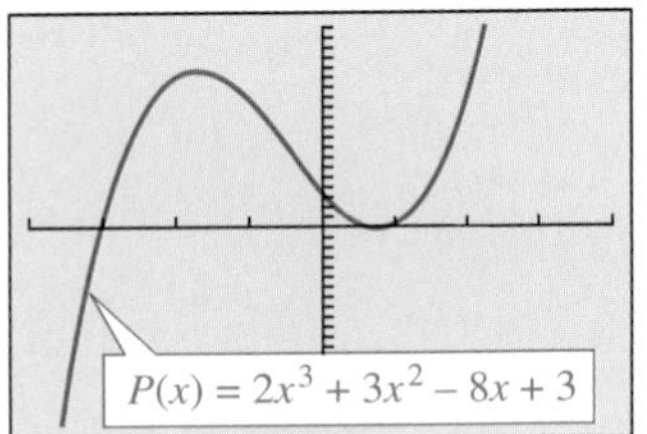

Figure 5-1

ACTIVE EXAMPLE 2 Solve: $P(x) = x^7 - 2x^6 - 5x^5 + 6x^4 - x^3 + 2x^2 + 5x - 6 = 0.$

Solution Because the equation is of seventh degree, it has 7 roots. By Descartes' rule of signs, there are six possible combinations of positive, negative, and nonreal roots.

COLLEGE **Algebra** $f(x)$ **Now**™

Explore this Active Example and test yourself on the steps taken here at **http://1pass.thomson.com**. You can also find this example on the Interactive Video Skillbuilder CD-ROM.

Number of positive roots	Number of negative roots	Number of nonreal roots
5	2	0
3	2	2
1	2	4
5	0	2
3	0	4
1	0	6

The possible rational roots are

$$-6, \quad -3, \quad -2, \quad -1, \quad 1, \quad 2, \quad 3, \quad 6$$

We check each one to eliminate those that don't satisfy the equation, beginning with -3.

$$\begin{array}{r|rrrrrrrr} -3 & 1 & -2 & -5 & 6 & -1 & 2 & 5 & -6 \\ & & -3 & 15 & -30 & 72 & -213 & 633 & -1{,}914 \\ \hline & 1 & -5 & 10 & -24 & 71 & -211 & 638 & -1{,}920 \end{array}$$

Since the last number in the synthetic division is not 0, -3 is not a root and can be crossed off the list. Since the signs in the last row alternate, -3 is a lower bound, and we can cross off -6 as well.

$$\not{-6}, \quad \not{-3}, \quad -2, \quad -1, \quad 1, \quad 2, \quad 3, \quad 6$$

We now try -2:

$$\begin{array}{r|rrrrrrrr} -2 & 1 & -2 & -5 & 6 & -1 & 2 & 5 & -6 \\ & & -2 & 8 & -6 & 0 & 2 & -8 & 6 \\ \hline & 1 & -4 & 3 & 0 & -1 & 4 & -3 & 0 \end{array}$$

Since the remainder is 0, -2 is a root.

Because this root is negative, we can revise the chart of possibilities to eliminate the possibility that there are 0 negative roots.

Number of positive roots	**Number of negative roots**	**Number of nonreal roots**
5	2	0
3	2	2
1	2	4

Because -2 is a root, the factor theorem states that $x - (-2)$, or $x + 2$, is a factor of $P(x)$. Any remaining roots can be found by solving the depressed equation

$$x^6 - 4x^5 + 3x^4 - x^2 + 4x - 3 = 0$$

Because the constant term of this equation is different from the constant term of the original equation, we can cross off other possible rational roots. The number -2 cannot be a root a second time, because it is not a factor of -3. The numbers 2 and 6 are no longer possible roots, because neither is a factor of -3.

The list of possible roots is now

$$-\not{6},\quad -\not{3},\quad -\not{2},\quad -1,\quad 1,\quad \not{2},\quad 3,\quad \not{6}$$

Since we know that there is one more negative root, we will synthetically divide the coefficients of the depressed equation by -1.

$$\begin{array}{r|rrrrrrr} -1 & 1 & -4 & 3 & 0 & -1 & 4 & -3 \\ & & -1 & 5 & -8 & 8 & -7 & 3 \\ \hline & 1 & -5 & 8 & -8 & 7 & -3 & 0 \end{array}$$

Since the remainder is $0, -1$ is a root, and the solution set so far is $\{-2, -1, \ldots\}$. The number -1 cannot be a root again, because we have found both negative roots. When we cross off -1, we have only two possibilities left.

$$-\not{6},\quad -\not{3},\quad -\not{2},\quad -\not{1},\quad 1,\quad \not{2},\quad 3,\quad \not{6}$$

The depressed equation is now $x^5 - 5x^4 + 8x^3 - 8x^2 + 7x - 3 = 0$. We can synthetically divide the coefficients of this equation by 1 to get

$$\begin{array}{r|rrrrrr} 1 & 1 & -5 & 8 & -8 & 7 & -3 \\ & & 1 & -4 & 4 & -4 & 3 \\ \hline & 1 & -4 & 4 & -4 & 3 & 0 \end{array}$$

The depressed equation is now $x^4 - 4x^3 + 4x^2 - 4x + 3 = 0$.

Thus, 1 joins the solution set $\{-2, -1, 1, \ldots\}$. To see whether 1 is a root a second time, we synthetically divide the coefficients of the new depressed equation by 1.

$$\begin{array}{r|rrrrr} 1 & 1 & -4 & 4 & -4 & 3 \\ & & 1 & -3 & 1 & -3 \\ \hline & 1 & -3 & 1 & -3 & 0 \end{array}$$

The depressed equation is now $x^3 - 3x^2 + x - 3 = 0$.

Again, 1 is a root, and the solution set is now $\{-2, -1, 1, 1, \ldots\}$. To see whether 1 is a root a third time, we synthetically divide the coefficients of the new depressed equation by 1.

$$\begin{array}{r|rrrr} 1 & 1 & -3 & 1 & -3 \\ & & 1 & -2 & -1 \\ \hline & 1 & -2 & -1 & -4 \end{array}$$

Since the remainder is not 0, the number 1 is not a root for a third time, and we can cross 1 off the list of possibilities, leaving only 3.

$-\not{6}, \quad -\not{3}, \quad -\not{2}, \quad -\not{1}, \quad \not{1}, \quad \not{2}, \quad 3, \quad \not{6}$

To see whether 3 is a root, we synthetically divide the coefficients of $x^3 - 3x^2 + x - 3 = 0$ by 3.

$$\begin{array}{r|rrrr} 3 & 1 & -3 & 1 & -3 \\ & & 3 & 0 & 3 \\ \hline & 1 & 0 & 1 & 0 \end{array}$$

Since the remainder is 0, 3 joins the solution set, which is now $\{-2, -1, 1, 1, 3 \ldots\}$.

The depressed equation is now $x^2 + 1 = 0$, which can be solved as a quadratic equation.

$$x^2 + 1 = 0$$
$$x^2 = -1$$
$$x = i \quad \text{or} \quad x = -i$$

The complete solution set is $\{-2, -1, 1, 1, 3, i, -i\}$. The solution set contains 3 positive roots, 2 negative roots, and 2 nonreal roots that are complex conjugates. This combination was one of the predicted possibilities. ■

Accent on Technology CONFIRMING ROOTS OF AN EQUATION

We can confirm that the roots found in Example 2 are correct by graphing the function $P(x) = x^7 - 2x^6 - 5x^5 + 6x^4 - x^3 + 2x^2 + 5x - 6$ and locating the resulting x-intercepts of the graph. If we use a graphing window of $x = [-4, 4]$ and $y = [-130, 50]$ and graph the function, we will obtain the graph shown in Figure 5-2. From the graph, we can see that the x-intercepts are at $x = -2$, $x = -1$, $x = 1$, and $x = 3$. We cannot detect the complex roots from the graph.

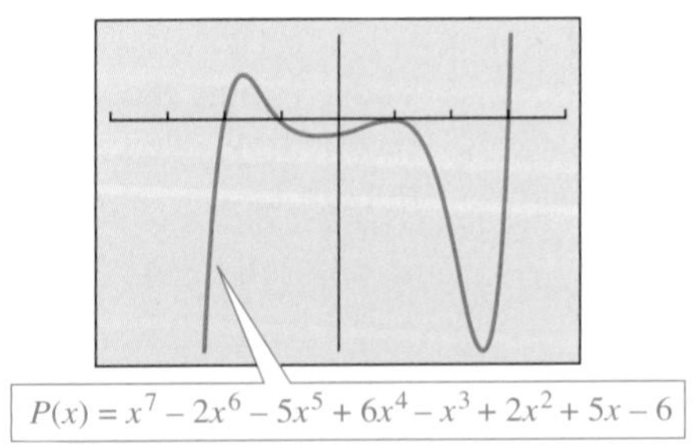

Figure 5-2

An Application

ACTIVE EXAMPLE 3 To protect cranberry crops from the damage of early freezes, growers flood the cranberry bogs. Three irrigation sources, used together, can flood a cranberry bog in one day. If the sources are used one at a time, the second source requires one day longer to flood the bog than the first, and the third requires four days longer than the first. If the bog must be flooded before a freeze that is predicted in three days, can the water in the last two sources be diverted to other bogs?

COLLEGE **Algebra** $f(x)$ **Now**™
Go to **http://1pass.thomson.com** or your CD to practice this example.

Solution Let x represent the number of days it would take the first irrigation source to flood the bog. Then $x + 1$ and $x + 4$ represent the number of days it would take the second and third sources, respectively, to flood the bog.

Because the first source, alone, requires x days to flood the bog, that source could fill $\frac{1}{x}$ of the bog in one day. In one day's time, the remaining sources could flood $\frac{1}{x+1}$ and $\frac{1}{x+4}$ of the bog. This gives the equation

The part of the bog the first source can flood in one day	plus	the part the second source can flood in one day	plus	the part the third source can flood in one day	equals	one bog.
$\frac{1}{x}$	$+$	$\frac{1}{x+1}$	$+$	$\frac{1}{x+4}$	$=$	1

We multiply both sides of the equation by $x(x + 1)(x + 4)$ to clear it of fractions and then simplify to get

$$x(x+1)(x+4)\left(\frac{1}{x} + \frac{1}{x+1} + \frac{1}{x+4}\right) = 1 \cdot x(x+1)(x+4)$$

$$(x+1)(x+4) + x(x+4) + x(x+1) = x(x+1)(x+4)$$

$$x^2 + 5x + 4 + x^2 + 4x + x^2 + x = x^3 + 5x^2 + 4x$$

$$0 = x^3 + 2x^2 - 6x - 4$$

To solve the equation $x^3 + 2x^2 - 6x - 4 = 0$, we first list its possible rational roots, which are the factors of the constant term, -4.

$$-4, \quad -2, \quad -1, \quad 1, \quad 2, \quad \text{and} \quad 4$$

One solution of this equation is $x = 2$, because when we synthetically divide by 2, the remaining is 0:

$$\begin{array}{r|rrrr} 2 & 1 & 2 & -6 & -4 \\ & & 2 & 8 & 4 \\ \hline & 1 & 4 & 2 & 0 \end{array}$$

We find the remaining solutions by using the quadratic formula to solve the depressed equation, which is $x^2 + 4x + 2 = 0$. The two solutions are $-2 + \sqrt{2}$ and $-2 - \sqrt{2}$. Since these numbers are negative and the time it takes to flood the bog cannot be negative, these roots must be discarded. The only meaningful solution is 2.

Since the first source, alone, can flood the bog in two days, and it is three days until the freeze, the other two water sources can be diverted to flood other bogs. ■

Self Check Answers

1. $1, 2, \frac{1}{3}$

5.3 Exercises

VOCABULARY AND CONCEPTS *Fill in the blanks.*

1. The rational roots of the equation $3x^3 + 4x - 7 = 0$ will have the form $\frac{p}{q}$, where p is a factor of $\underline{-7}$ and q is a factor of 3.
2. The rational roots of the equation $5x^3 + 3x^2 - 4 = 0$ will have the form $\frac{p}{q}$, where p is a factor of -4 and q is a factor of $\underline{5}$.
3. Consider the synthetic division of $5x^3 - 7x^2 - 3x - 63 = 0$ by 3.

$$\begin{array}{r|rrrr} 3 & 5 & -7 & -3 & -63 \\ & & 15 & 24 & 63 \\ \hline & 5 & 8 & 21 & 0 \end{array}$$

Since the remainder is 0, 3 is a $\underline{\text{root}}$ of the equation.

4. In Problem 3, the depressed equation is $\underline{5x^2 + 8x + 21 = 0}$.

PRACTICE *Find all rational roots of each equation.*

5. $x^3 - 5x^2 - x + 5 = 0$ $1, -1, 5$
6. $x^3 + 7x^2 - x - 7 = 0$ $1, -1, -7$
7. $x^3 - 2x^2 - x + 2 = 0$ $1, 2, -1$
8. $x^3 + x^2 - 4x - 4 = 0$ $2, -2, -1$
9. $x^3 - x^2 - 4x + 4 = 0$ $1, 2, -2$
10. $x^3 + 2x^2 - x - 2 = 0$ $1, -1, -2$
11. $x^3 - 2x^2 - 9x + 18 = 0$ $3, -3, 2$
12. $x^3 + 3x^2 - 4x - 12 = 0$ $2, -2, -3$
13. $2x^3 - x^2 - 2x + 1 = 0$ $1, -1, \frac{1}{2}$
14. $3x^3 + x^2 - 3x - 1 = 0$ $1, -1, -\frac{1}{3}$
15. $3x^3 + 5x^2 + x - 1 = 0$ $-1, -1, \frac{1}{3}$
16. $2x^3 - 3x^2 + 1 = 0$ $1, 1, -\frac{1}{2}$
17. $x^4 - 10x^3 + 35x^2 - 50x + 24 = 0$ $1, 2, 3, 4$
18. $x^4 + 4x^3 + 6x^2 + 4x + 1 = 0$ $-1, -1, -1, -1$
19. $x^4 + 3x^3 - 13x^2 - 9x + 30 = 0$ $2, -5$
20. $x^4 - 8x^3 + 14x^2 + 8x - 15 = 0$ $1, -1, 3, 5$
21. $x^5 + 3x^4 - 5x^3 - 15x^2 + 4x + 12 = 0$ $1, -1, 2, -2, -3$
22. $x^5 - 3x^4 - 5x^3 + 15x^2 + 4x - 12 = 0$ $1, -1, 2, -2, 3$
23. $x^7 - 12x^5 + 48x^3 - 64x = 0$ $0, 2, 2, 2, -2, -2, -2$
24. $x^7 + 7x^6 + 21x^5 + 35x^4 + 35x^3 + 21x^2 + 7x + 1 = 0$ $-1, -1, -1, -1, -1, -1, -1$
25. $3x^3 - 2x^2 + 12x - 8 = 0$ $\frac{2}{3}$
26. $4x^4 - 8x^3 - x^2 + 8x - 3 = 0$ $1, -1, \frac{1}{2}, \frac{3}{2}$
27. $3x^4 - 14x^3 + 11x^2 + 16x - 12 = 0$ $-1, 2, 3, \frac{2}{3}$
28. $2x^4 - x^3 - 2x^2 - 4x - 40 = 0$ $-2, \frac{5}{2}$
29. $12x^4 + 20x^3 - 41x^2 + 20x - 3 = 0$ $-3, \frac{1}{3}, \frac{1}{2}, \frac{1}{2}$
30. $4x^5 - 12x^4 + 15x^3 - 45x^2 - 4x + 12 = 0$ $3, \frac{1}{2}, -\frac{1}{2}$
31. $6x^5 - 7x^4 - 48x^3 + 81x^2 - 4x - 12 = 0$ $2, 2, -3, -\frac{1}{3}, \frac{1}{2}$
32. $36x^4 - x^2 + 2x - 1 = 0$ $\frac{1}{3}, -\frac{1}{2}$
33. $30x^3 - 47x^2 - 9x + 18 = 0$ $\frac{3}{2}, \frac{2}{3}, -\frac{3}{5}$
34. $20x^3 - 53x^2 - 27x + 18 = 0$ $3, \frac{2}{5}, -\frac{3}{4}$
35. $15x^3 - 61x^2 - 2x + 24 = 0$ $\frac{2}{3}, -\frac{3}{5}, 4$
36. $12x^4 + x^3 + 42x^2 + 4x - 24 = 0$ $\frac{2}{3}, -\frac{3}{4}$
37. $20x^3 - 44x^2 + 9x + 18 = 0$ $-\frac{1}{2}, \frac{3}{2}, \frac{6}{5}$
38. $24x^3 - 82x^2 + 89x - 30 = 0$ $\frac{3}{2}, \frac{2}{3}, \frac{5}{4}$

In Exercises 39–42, $1 + i$ is a root of each equation. Find the other roots.

39. $x^3 - 5x^2 + 8x - 6 = 0$ $3, 1 - i$
40. $x^3 - 2x + 4 = 0$ $-2, 1 - i$
41. $x^4 - 2x^3 - 7x^2 + 18x - 18 = 0$ $3, -3, 1 - i$
42. $x^4 - 2x^3 - 2x^2 + 8x - 8 = 0$ $2, -2, 1 - i$

Solve each equation.

43. $x^3 - \frac{4}{3}x^2 - \frac{13}{3}x - 2 = 0$ $-\frac{2}{3}, 3, -1$
44. $x^3 - \frac{19}{6}x^2 + \frac{1}{6}x + 1 = 0$ $-\frac{1}{2}, \frac{2}{3}, 3$
45. $x^{-5} - 8x^{-4} + 25x^{-3} - 38x^{-2} + 28x^{-1} - 8 = 0$ $\frac{1}{2}, \frac{1}{2}, \frac{1}{2}, 1, 1$
46. $1 - x^{-1} - x^{-2} - 2x^{-3} = 0$ $2, -\frac{1}{2} + \frac{\sqrt{3}}{2}i, -\frac{1}{2} - \frac{\sqrt{3}}{2}i$

APPLICATIONS

47. Parallel resistance If three resistors with resistances of R_1, R_2, and R_3 are wired in parallel, their combined resistance R is given by the formula

$$\frac{1}{R} = \frac{1}{R_1} + \frac{1}{R_2} + \frac{1}{R_3}$$

The design of a voltmeter requires that the resistance R_2 be 10 ohms greater than the resistance R_1, that the resistance R_3 be 50 ohms greater than R_1, and that their combined resistance be 6 ohms. Find the value of each resistance. **10, 20, 60 ohms**

48. Fabricating sheet metal The open tray shown in the illustration is to be manufactured from a 12-by-14-inch rectangular sheet of metal by cutting squares from each corner and folding up the sides. If the volume of the tray is to be 160 cubic inches and x is to be an integer, what size squares should be cut from each corner? **2 in. by 2 in.**

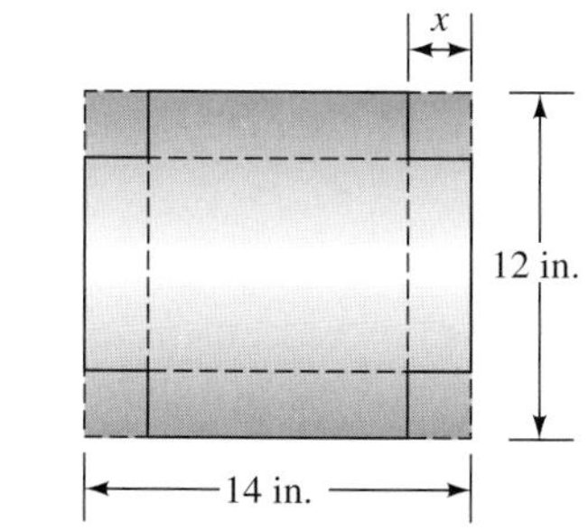

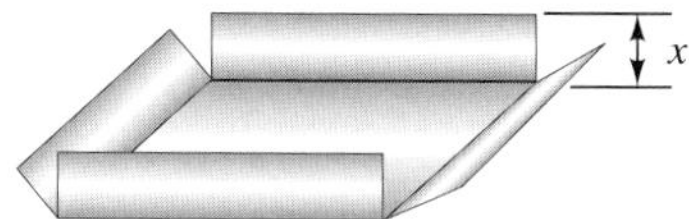

DISCOVERY AND WRITING

49. If n is an even integer and c is a positive constant, show that $x^n + c = 0$ has no real roots.

50. If n is an even positive integer and c is a positive constant, show that $x^n - c = 0$ has two real roots.

51. Precalculus A rectangle is inscribed in the parabola $y = 16 - x^2$, as shown in the illustration. Find the point (x, y) if the area of the rectangle is 42 square units. **(3, 7) or approx. (1.54, 13.63)**

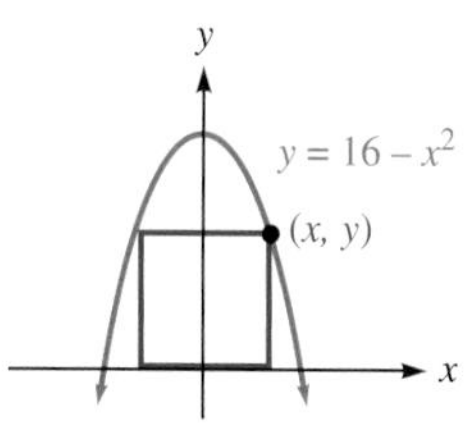

52. Precalculus One corner of the rectangle shown is at the origin, and the opposite corner (x, y) lies in the first quadrant on the curve $y = x^3 - 2x^2$. Find the point (x, y) if the area of the rectangle is 27 square units. **(3, 9)**

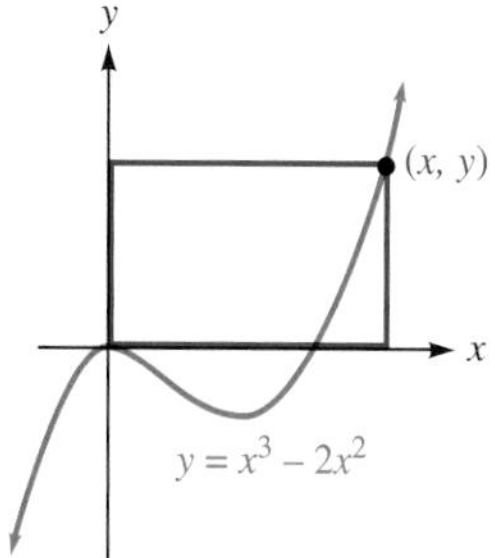

REVIEW ***Simplify each radical expression. Assume that all variables represent positive numbers.***

53. $\sqrt{72a^3b^5c}$ **$6ab^2\sqrt{2abc}$**

54. $\dfrac{5a}{\sqrt{5a}}$ **$\sqrt{5a}$**

55. $\sqrt{18a^2b} + a\sqrt{50b}$ **$8a\sqrt{2b}$**

56. $(\sqrt{3} + \sqrt{5b})(\sqrt{12} - \sqrt{45b})$ **$6 - \sqrt{15b} - 15b$**

57. $\dfrac{2}{\sqrt{3} - 1}$ **$\sqrt{3} + 1$**

58. $\dfrac{\sqrt{11} + \sqrt{x}}{\sqrt{11} - \sqrt{x}}$ **$\dfrac{11 + 2\sqrt{11x} + x}{11 - x}$**

5.4 Irrational Roots of Polynomial Equations

In this section, you will learn about

- The Intermediate Value Theorem
- The Bisection Method
- Finding Real Roots with a Graphing Calculator

First-degree equations are easy to solve, and all quadratic equations can be solved with the quadratic formula. There are formulas for solving third- and fourth-degree polynomial equations, although they are complicated. However, there are no formulas for solving polynomial equations of degree 5 or greater. This fact was proved by the Norwegian mathematician Niels Henrik Abel (1802–1829) and (for equations of degree greater than 5) by the French mathematician Evariste Galois (1811–1832).

To solve a higher-degree polynomial equation with integer coefficients, we can use the methods of the previous section to find its rational roots. Once we find them, the final depressed equation would have to be a first- or second-degree equation to enable us to complete the solution. The purpose of this section is to discuss ways of approximating irrational roots of these higher-degree polynomial equations.

The first example shows that polynomial equations can have irrational roots.

ACTIVE EXAMPLE 1 Prove that $\sqrt{2}$ is an irrational number.

Solution We know that $\sqrt{2}$ is a real root of the equation $x^2 - 2 = 0$. Since any rational root of this equation will have the form $\frac{\text{factor of the constant 2}}{\text{factor of the lead coefficient 1}}$, the possible rational roots are

$$\pm\frac{1}{1} \quad \text{and} \quad \pm\frac{2}{1}$$

Since none of the numbers 1, -1, 2, or -2 satisfies the equation, $\sqrt{2}$ must be irrational.

COLLEGE **Algebra** $f(x)$ **Now**™

Explore this Active Example and test yourself on the steps taken here at **http://1pass.thomson.com**. You can also find this example on the Interactive Video Skillbuilder CD-ROM.

Self Check Prove that $\sqrt{3}$ is irrational. ■

■ The Intermediate Value Theorem

The following theorem, called the **intermediate value theorem,** will lead to a way of locating an interval that contains a root.

The Intermediate Value Theorem

Let $P(x)$ be a polynomial with real coefficients. If $P(a) \neq P(b)$ for $a < b$, then $P(x)$ takes on all values between $P(a)$ and $P(b)$ on the closed interval $[a, b]$.

Justification This theorem becomes clear when we consider the graph of the polynomial $y = P(x)$, shown in Figure 5-3. We have seen that graphs of polynomials are *continuous* curves, a technical term that means, roughly, that they can be drawn without lifting the pencil from the paper. If $P(a) \neq P(b)$, then the continuous curve joining the points $A(a, P(a))$ and $B(b, P(b))$ must take on all values between $P(a)$ and $P(b)$ in the interval $[a, b]$, because the curve has no gaps in it.

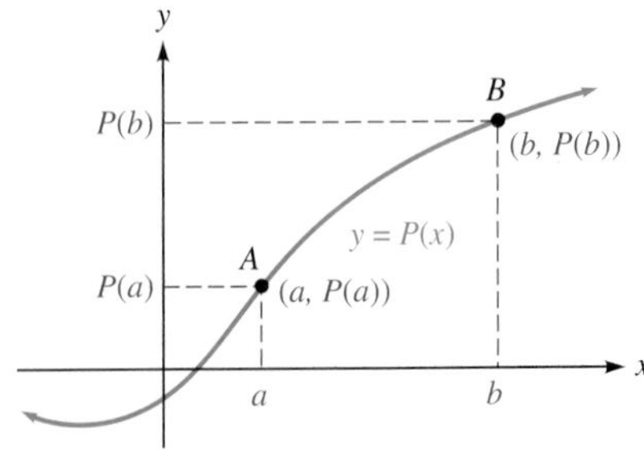

Figure 5-3

The next theorem follows from the intermediate value theorem. ■

The Location Theorem Let $P(x)$ be a polynomial with real coefficients. If $P(a)$ and $P(b)$ have opposite signs, there is at least one number r in the interval (a, b) for which $P(r) = 0$.

Proof See Figure 5-4. By the intermediate value theorem, $P(x)$ takes on all values between $P(a)$ and $P(b)$. Since $P(a)$ and $P(b)$ have opposite signs, the number 0 lies between them. Thus, there is a number r between a and b for which $P(r) = 0$. This number r is a zero of $P(x)$, and a root of the equation $P(x) = 0$.

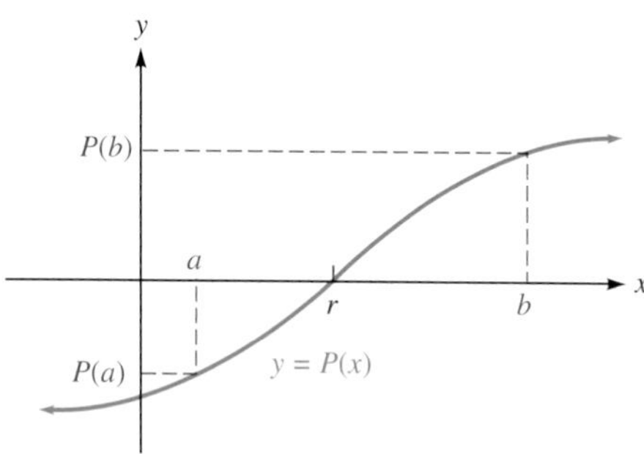

Figure 5-4

■

The Bisection Method

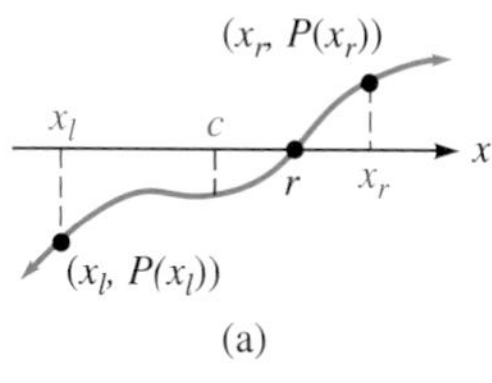

(a)

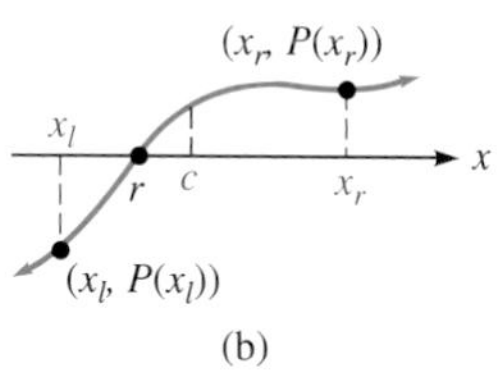

(b)

Figure 5-5

The location theorem provides a method (called the **bisection method**) for finding the roots of $P(x) = 0$ to any desired degree of accuracy.

Suppose we find, by trial and error, that the numbers x_l and x_r (for left and right) straddle a root—that is, $x_l < x_r$ and $P(x_l)$ and $P(x_r)$ have opposite signs. (See Figure 5-5.) Further suppose that $P(x_l) < 0$ and $P(x_r) > 0$. We can compute the number c that is halfway between x_l and x_r and then compute $P(c)$. If $P(c) = 0$, we've found a root. However, if $P(c)$ is not 0, we proceed in one of two ways:

1. If $P(c) < 0$, the root r lies between c and x_r, as shown in Figure 5-5(a). In this case, we let c become a new x_l and repeat the procedure.
2. If $P(c) > 0$, the root r lies between x_l and c, as shown in Figure 5-5(b). In this case, we let c become a new x_r and repeat the procedure.

At any stage in this procedure, the root is contained between the current values of x_l and x_r. If the original bounds were 1 unit apart, after 10 repetitions of this procedure, the root would be between bounds that were 2^{-10} units apart. This means that the zero of $P(x)$ would be within 0.0001 units of either x_l or x_r.

After 20 repetitions, the root would be between bounds that were 2^{-20} units apart. This means that the zero of $P(x)$ would be within 0.000001 of either x_l or x_r.

ACTIVE EXAMPLE 2 Solve $x^2 - 2 = 0$ and express the positive root to the nearest tenth.

Solution Since $P(1) = -1$ and $P(2) = 2$ have opposite signs, there is a root between 1 and 2. We can let $x_l = 1$ and $x_r = 2$ and compute the midpoint c:

$$c = \frac{x_l + x_r}{2} = \frac{1 + 2}{2} = 1.5$$

COLLEGE **Algebra f(x) Now™**
Go to **http://1pass.thomson.com** or your CD to practice this example.

Because $P(c) = P(1.5) = 0.25$ is a positive number, we let c become a new x_r and calculate a new midpoint, which we will call c_1.

$$c_1 = \frac{x_l + x_r}{2} = \frac{1 + 1.5}{2} = 1.25$$

Because $P(c_1) = P(1.25) = -0.4375$ is a negative number, we let c_1 become a new x_l and calculate a new midpoint, which we will call c_2.

$$c_2 = \frac{x_l + x_r}{2} = \frac{1.25 + 1.5}{2} = 1.375$$

Because $P(c_2) = P(1.375) = -0.109375$ is a negative number, we let c_2 become a new x_l and calculate a new midpoint, which we will call c_3.

$$c_3 = \frac{x_l + x_r}{2} = \frac{1.375 + 1.5}{2} = 1.4375$$

Because $P(c_3) = P(1.4375) = 0.066406$ is a positive number, we let c_3 become a new x_r and calculate a new midpoint, which we will call c_4.

$$c_4 = \frac{x_l + x_r}{2} = \frac{1.375 + 1.4375}{2} = 1.40625$$

From here on, the first two digits of the midpoints will remain 1.4. The root r of the equation (to the nearest tenth) is 1.4.

Self Check Find the negative root of $x^2 - 2 = 0$ to the nearest tenth. ■

Finding Real Roots with a Graphing Calculator

We can approximate the real roots of an equation either by using the TRACE and ZOOM capabilities of a graphing calculator or by using the ZERO feature.

Accent on Technology — SOLVING EQUATIONS

To use the TRACE and ZOOM capabilities of a graphing calculator to solve the equation $x^4 - 6x^2 + 9 = 0$, we graph the function $y = x^4 - 6x^2 + 9$ and trace and zoom to find the x-intercepts. If we use window settings of $[-6, 6]$ for x and $[-2, 10]$ for y and graph the function, we will obtain Figure 5-6(a). If we trace and move the cursor near the positive x-intercept, we will obtain Figure 5-6(b). If we zoom in and trace, we will obtain Figure 5-6(c). To get better results, we can zoom in and trace again to obtain Figure 5-6(d), which shows that the x-coordinate of the x-intercept is approximately 1.731383. After zooming in and tracing a few more times, we will see that to three decimal places, $x = 1.732$. By symmetry, we know that the negative solution is -1.732. Notice that we cannot find complex roots from the graph.

Continued

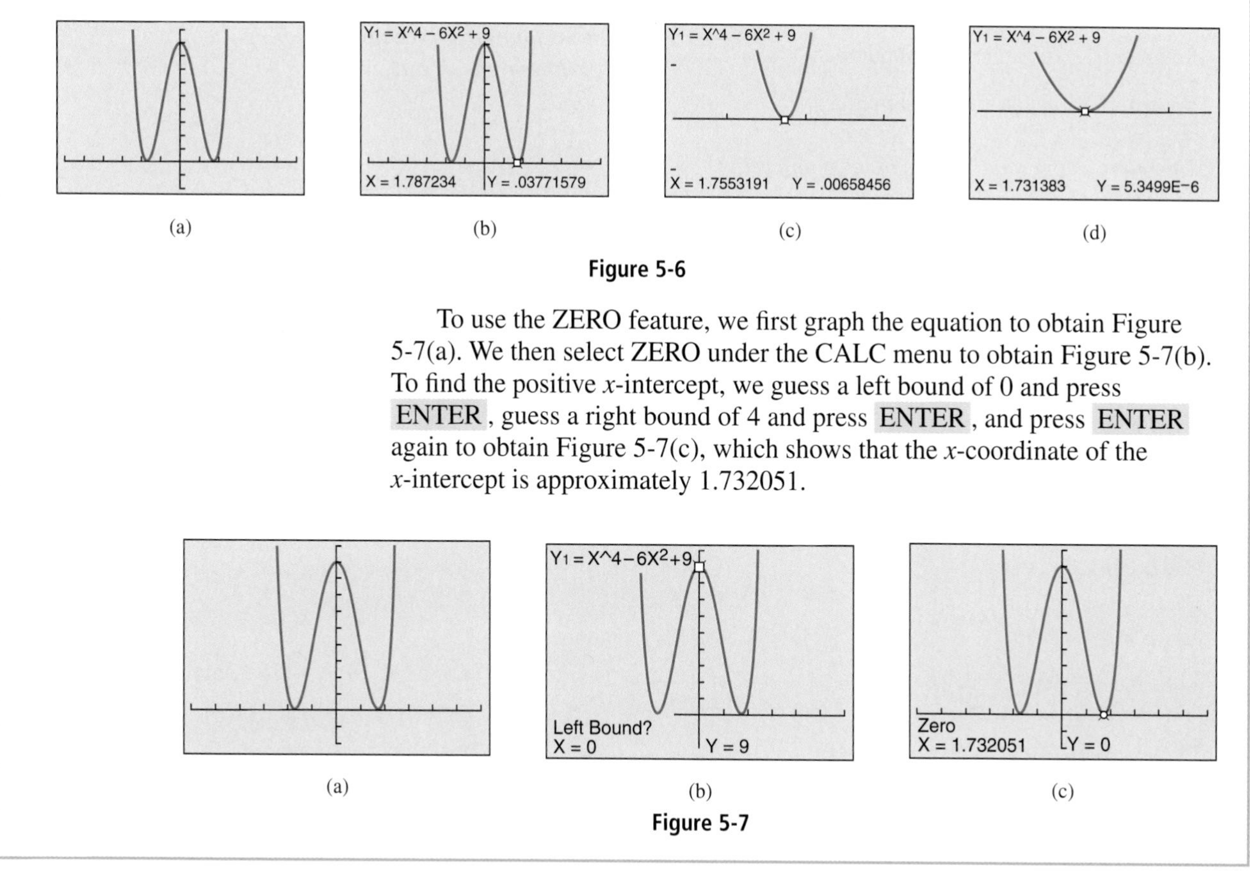

Figure 5-6

To use the ZERO feature, we first graph the equation to obtain Figure 5-7(a). We then select ZERO under the CALC menu to obtain Figure 5-7(b). To find the positive x-intercept, we guess a left bound of 0 and press ENTER, guess a right bound of 4 and press ENTER, and press ENTER again to obtain Figure 5-7(c), which shows that the x-coordinate of the x-intercept is approximately 1.732051.

Figure 5-7

The graph shown in Figure 5-7(a) illustrates a situation in which the bisection method does not work. Since the graph doesn't cross the x-axis, the values of the function on opposite sides of the intercepts do not have opposite signs. In this case, the bisection method would be useless.

Self Check Answers

1. $\pm\frac{1}{1}$ or $\pm\frac{3}{1}$ do not satisfy $x^2 - 3 = 0$. **2.** -1.4

5.4 Exercises

VOCABULARY AND CONCEPTS ***Fill in the blanks.***

1. If $P(x)$ is a polynomial with real coefficients and $P(a) \neq P(b)$ for $a < b$, then $P(x)$ takes on all values between <u>$P(a)$ and $P(b)$</u> in the interval $[a, b]$.

2. If $P(x)$ has real coefficients and $P(a)$ and $P(b)$ have opposite signs, there is at least one number r in (a, b) for which <u>$P(r) = 0$</u>.

3. In the bisection method, we find two numbers x_l and x_r that straddle a root. We then compute a new guess for the root (a number c) by finding the average of <u>x_l and x_r</u>.

4. A <u>continuous</u> curve has no gaps in it.

PRACTICE

5. Prove that $\sqrt{7}$ is an irrational number.

6. Prove that $\sqrt{5}$ is an irrational number.

Show that each equation has at least one real root between the specified numbers.

7. $2x^2 + x - 3 = 0$; -2 and -1
$P(-2) = 3; P(-1) = -2$

8. $2x^3 + 17x^2 + 31x - 20 = 0$; -1 and 2
$P(-1) = -36; P(2) = 126$

9. $3x^3 - 11x^2 - 14x = 0$; 4 and 5
$P(4) = -40; P(5) = 30$

10. $2x^3 - 3x^2 + 2x - 3 = 0$; 1 and 2
$P(1) = -2; P(2) = 5$

11. $x^4 - 8x^2 + 15 = 0$; 1 and 2
$P(1) = 8; P(2) = -1$

12. $x^4 - 8x^2 + 15 = 0$; 2 and 3
$P(2) = -1; P(3) = 24$

13. $30x^3 + 10 = 61x^2 + 39x$; 2 and 3
$P(2) = -72; P(3) = 154$

14. $30x^3 + 10 = 61x^2 + 39x$; -1 and 0
$P(-1) = -42; P(0) = 10$

15. $30x^3 + 10 = 61x^2 + 39x$; 0 and 1
$P(0) = 10; P(1) = -60$

16. $5x^3 - 9x^2 - 4x + 9 = 0$; -1 and 2
$P(-1) = -1; P(2) = 5$

Use the bisection method to find the following values to the nearest tenth.

17. The positive root of $x^2 - 3 = 0$. **1.7**

18. The negative root of $x^2 - 3 = 0$. **−1.7**

19. The negative root of $x^2 - 5 = 0$. **−2.2**

20. The positive root of $x^2 - 5 = 0$. **2.2**

21. The positive root of $x^3 - x^2 - 2 = 0$. **1.7**

22. The negative root of $x^3 - x + 2 = 0$. **−1.5**

23. The negative root of $3x^4 + 3x^3 - x^2 - 4x - 4 = 0$.
−1.2

24. The positive root of
$x^5 + x^4 - 4x^3 - 4x^2 - 5x - 5 = 0$. **2.2**

Use a graphing calculator to find the distinct real solutions of each equation to the nearest tenth. Which roots, if any, would the bisection method fail to find?

25. $x^2 - 5 = 0$ **−2.2, 2.2**

26. $x^2 - 10x + 25 = 0$
5; the bisection method fails to find the solution

27. $x^3 - 5x^2 + 8x - 4 = 0$
1, 2; the bisection method fails to find the solution 2

28. $x^3 - 5x^2 - 2x + 10 = 0$ **−1.4, 1.4, 5**

APPLICATIONS

29. Precalculus Use the bisection method or a graphing calculator to find the coordinates of the two points on the graph of $y = x^3$ that lie 1 unit from the origin. (See the illustration.) Give the result to the nearest hundredth. **(0.83, 0.56), (−0.83, −0.56)**

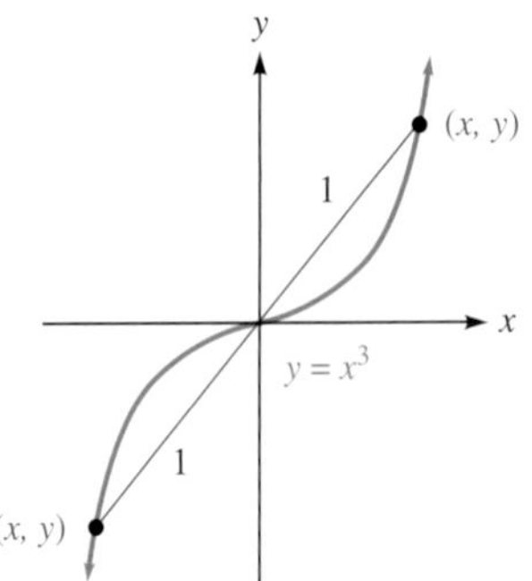

30. Building a crate The width of the shipping crate shown is to be 2 feet greater than its height and the length is to be 5 feet greater than its width, and its volume is to be 170 cubic feet. Use the bisection method or a graphing calculator to find the height of the crate to the nearest tenth of a foot. **3.2 ft**

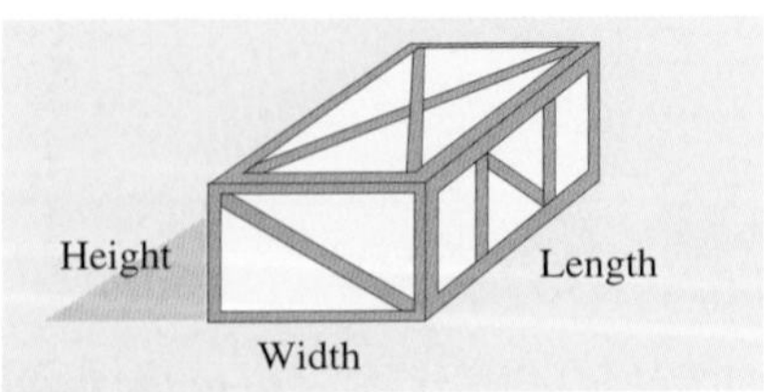

DISCOVERY AND WRITING

31. Explain when the bisection method will not work.

32. Describe how you could tell from a graph that a polynomial equation has no real roots.

REVIEW *Give the equations of all of the asymptotes of each rational function.* **Don't graph the functions.**

33. $y = \dfrac{5x - 3}{x^2 - 4}$

vertical: $x = 2, x = -2$; horizontal: $y = 0$; slant: none

34. $y = \dfrac{5x^2 - 3}{x^2 - 25}$

vertical: $x = 5, x = -5$; horizontal: $y = 5$; slant: none

35. $y = \dfrac{x^2}{x - 2}$

vertical: $x = 2$; horizontal: none; slant: $y = x + 2$

36. $y = \dfrac{x + 1}{x^2 + 1}$

vertical: none; horizontal: $y = 0$; slant: none

PROBLEMS AND PROJECTS

1. If $x^2 - 8x + 15$ is a factor of $P(x)$, find $P(3) + P(5)$. **0**

2. $x^2 - 6kx + 15k = 0$ has two positive solutions, one 5 times the other. Find k and solve the equation. **$k = 3; x = 3, 15$**

3. Using Descartes' rule of signs, construct a. a fifth-degree equation with only one positive real root and b. a fourth-degree equation with no real roots.

4. For what real value(s) of k does

$$x^3 + k^2x^2 + 2kx + 1 = 0$$

have rational solutions? **$k = 0, 2$**

Project 1

Find the dimensions of the right-triangular brace in the illustration. **$x = 4$; 12 by 16 by 20**

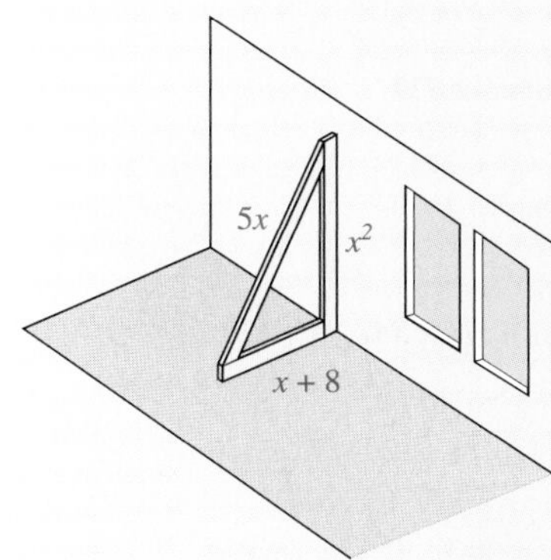

Project 2

The bisection method is one of several techniques for finding approximate solutions of equations. Another is **Newton's method,** discovered by Sir Isaac Newton (1642–1727).

The process begins with a guess of a solution and then transforms that guess into a better estimate of the solution. When Newton's method is applied to that estimate, a third solution results—one more accurate than any previous. As the method is applied repeatedly, the results converge on the actual solution.

To solve the quadratic equation $x^2 + x - 3 = 0$, for example, Newton's method uses the fraction

$$\frac{x^2 + 3}{2x + 1}$$

to generate better solutions. We begin with a guess (3, for example) and substitute 3 for x in the fraction.

$$\frac{x^2 + 3}{2x + 1} = \frac{3^2 + 3}{2(3) + 1} \approx 1.714286$$

The number 1.714286 is a better estimate of the solution than the original guess, 3. We apply Newton's method again and substitute 1.714286 for x.

$$\frac{x^2 + 3}{2x + 1} = \frac{1.714286^2 + 3}{2(1.714286) + 1} \approx 1.341014$$

The estimate 1.341014 is better yet. More passes produce even better accuracy.

- Continue with Newton's method until your results remain unchanged to six decimal places. **1.302776**
- Solve the equation using the quadratic formula. Is Newton's method correct?
- To solve the equation $ax^2 + bx + c = 0$, Newton's method uses the fraction

$$\frac{ax^2 - c}{2ax + b}$$

Solve $3x^2 - 5x - 2 = 0$ using Newton's method. Find two solutions, accurate to six decimal places.

- To solve a third-degree equation $ax^3 + bx^2 + cx + d = 0$, Newton's method uses the fraction

$$\frac{2ax^3 + bx^2 - d}{3ax^2 + 2bx + c}$$

Use this method to find three solutions of $x^3 - x^2 - 5x + 5 = 0$, accurate to six decimal places.

CHAPTER SUMMARY

CONCEPTS

REVIEW EXERCISES

5.1 The Remainder and Factor Theorems; Synthetic Division

The remainder theorem:
If $P(x)$ is a polynomial, r is any number, and $P(x)$ is divided by $x - r$, the remainder is $P(r)$.

The factor theorem:
Let $P(x)$ be any polynomial and r any number. Then $P(r) = 0$ if and only if $x - r$ is a factor of $P(x)$.

Let $P(x) = 4x^4 + 2x^3 - 3x^2 - 2$. Find the remainder when $P(x)$ is divided by each binomial.

1. $x - 1$ **1** **2.** $x - 2$ **66** **3.** $x + 3$ **241** **4.** $x + 2$ **34**

Use the factor theorem to decide whether each statement is true.

5. $x - 2$ is a factor of $x^3 + 4x^2 - 2x + 4$. **false**

6. $x + 3$ is a factor of $2x^4 + 10x^3 + 4x^2 + 7x + 21$. **false**

7. $x - 5$ is a factor of $x^5 - 3{,}125$. **true**

8. $x - 6$ is a factor of $x^5 - 6x^4 - 4x + 24$. **true**

Find the polynomial of lowest degree with integer coefficients and the given zeros.

9. $-1, 2$, and $\frac{3}{2}$
$2x^3 - 5x^2 - x + 6$

10. $1, -3$, and $\frac{1}{2}$
$2x^3 + 3x^2 - 8x + 3$

11. $2, -5, i$, and $-i$
$x^4 + 3x^3 - 9x^2 + 3x - 10$

12. $-3, 2, i$, and $-i$
$x^4 + x^3 - 5x^2 + x - 6$

Use synthetic division to find the quotient and remainder when each polynomial is divided by the given divisor.

13. $3x^4 + 2x^2 + 3x + 7; x - 3$
$3x^3 + 9x^2 + 29x + 90$ with remainder of 277

14. $2x^4 - 3x^2 + 3x - 1; x - 2$
$2x^3 + 4x^2 + 5x + 13$ with remainder of 25

15. $5x^5 - 4x^4 + 3x^3 - 2x^2 + x - 1; x + 2$
$5x^4 - 14x^3 + 31x^2 - 64x + 129$ with remainder of -259

16. $4x^5 + 2x^4 - x^3 + 3x^2 + 2x + 1; x + 1$
$4x^4 - 2x^3 + x^2 + 2x$ with remainder of 1

5.2 Descartes' Rule of Signs and Bounds on Roots

The fundamental theorem of algebra:
If $P(x)$ is a polynomial with positive degree, then $P(x)$ has at least one zero.

The polynomial equation $P(x) = 0$ with degree n ($n > 0$) has exactly n roots among the complex numbers.

How many roots does each equation have?

17. $3x^6 - 4x^5 + 3x + 2 = 0$ **6**

18. $2x^6 - 5x^4 + 5x^3 - 4x^2 + x - 12 = 0$ **6**

19. $3x^{65} - 4x^{50} + 3x^{17} + 2x = 0$ **65**

20. $x^{1{,}984} - 12 = 0$ **1,984**

Complex roots of polynomial equations with real coefficients occur in complex conjugate pairs.

Descartes' rule of signs:
If $P(x)$ has real coefficients, the number of positive roots of $P(x) = 0$ is equal to the number of variations in sign of $P(x)$, or is less than that by an even number.

The number of negative roots is equal to the number of variations in sign of $P(-x)$, or is less than that by an even number.

Upper bounds:
Let the lead coefficient of the polynomial $P(x)$ with real coefficients be positive, and do a synthetic division of the coefficients of $P(x)$ by the positive number c. If none of the terms in the bottom row is negative, then c is an upper bound for the real roots of $P(x) = 0$.

Lower bounds:
Let the lead coefficient of the polynomial $P(x)$ with real coefficients be positive, and do a synthetic division of the coefficients of $P(x)$ by the negative number d. If the signs in the bottom row alternate, then d is a lower bound for the real roots of $P(x) = 0$.

Find another root of a polynomial equation with real coefficients if the given quantity is one root.

21. $2 + i$ $2 - i$

22. $-i$ i

Find the number of possible positive, negative, and nonreal roots for each equation. Do not attempt to solve the equation.

23. $3x^4 + 2x^3 - 4x + 2 = 0$
0 or 2 positive; 0 or 2 negative; 0, 2, or 4 nonreal

24. $2x^4 - 3x^3 + 5x^2 + x - 5 = 0$
1 or 3 positive; 1 negative; 0 or 2 nonreal

25. $4x^5 + 3x^4 + 2x^3 + x^2 + x = 7$
1 positive; 0, 2, or 4 negative; 0, 2, or 4 nonreal

26. $3x^7 - 4x^5 + 3x^3 + x - 4 = 0$
1 or 3 positive; 0 or 2 negative; 2, 4, or 6 nonreal

27. $x^4 + x^2 + 24{,}567 = 0$ **0 positive; 0 negative; 4 nonreal**

28. $-x^7 - 5 = 0$ **0 positive; 1 negative; 6 nonreal**

Find integer bounds for the roots of each equation.

29. $5x^3 - 4x^2 - 2x + 4 = 0$ **−1, 2**

30. $x^4 + 3x^3 - 5x^2 - 9x + 1 = 0$ **−5, 2**

5.3 Rational Roots of Polynomial Equations

If $P(x) = 0$ has integer coefficients and $\frac{p}{q}$ (written in lowest terms) is a root, then p is a factor of the constant, and q is a factor of the lead coefficient.

Find all rational roots of each equation.

31. $2x^3 + 17x^2 + 41x + 30 = 0$ $-5, -\frac{3}{2}, -2$

32. $3x^3 + 2x^2 + 2x - 1 = 0$ $\frac{1}{3}$

33. $4x^4 - 25x^2 + 36 = 0$ $2, -2, \frac{3}{2}, -\frac{3}{2}$

34. $2x^4 - 11x^3 - 6x^2 + 64x + 32 = 0$ $4, 4, -2, -\frac{1}{2}$

5.4 Irrational Roots of Polynomial Equations

The intermediate value theorem: Let $P(x)$ be a polynomial with real coefficients. If $P(a) \neq P(b)$ for $a < b$, then $P(x)$ takes on all values between $P(a)$ and $P(b)$ on the closed interval $[a, b]$.

Let $P(x)$ be a polynomial with real coefficients. If $P(a)$ and $P(b)$ have opposite signs, then there is at least one number r between a and b for which $P(r) = 0$.

Show that each polynomial has a zero between the two given numbers.

35. $5x^3 + 37x^2 + 59x + 18 = 0$; -1 and 0 **$P(-1) = -9$; $P(0) = 18$**

36. $6x^3 - x^2 - 10x - 3 = 0$; 1 and 2 **$P(1) = -8$; $P(2) = 21$**

Use the bisection method to find the positive root of each equation to the nearest tenth.

37. $x^3 - 2x^2 - 9x - 2 = 0$ **4.2**

38. $6x^2 - 13x - 5 = 0$ **2.5**

Use a graphing calculator to find the positive root of each equation to the nearest hundredth.

39. $6x^2 - 7x - 5 = 0$ **1.67**

40. $3x^2 + x - 2 = 0$ **0.67**

41. Designing solar collectors The space available for the installation of three solar collecting panels requires that their lengths differ by the amounts shown in the illustration, and that the total of their widths be 15 meters. To be equally effective, each panel must measure exactly 60 square meters. Find the dimensions of each panel. **10 m by 6 m; 12 m by 5 m; 15 m by 4 m**

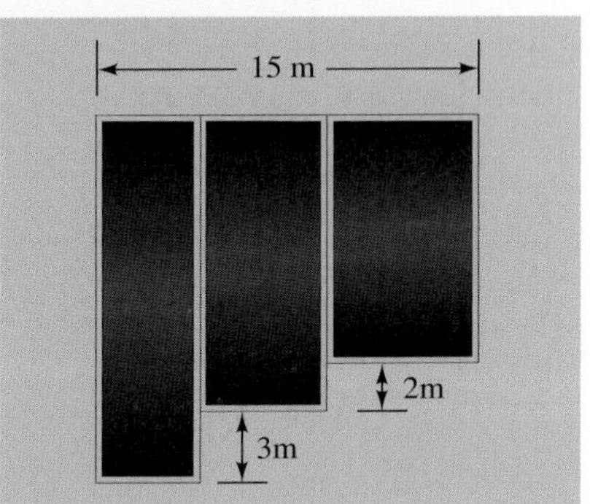

42. Designing a storage tank The design specifications for the cylindrical storage tank shown require that its height be 3 feet greater than the radius of its circular base and that the volume of the tank be 19,000 cubic feet. Use the bisection method or a graphing calculator to find the radius of the tank to the nearest hundredth of a foot. **17.27 ft**

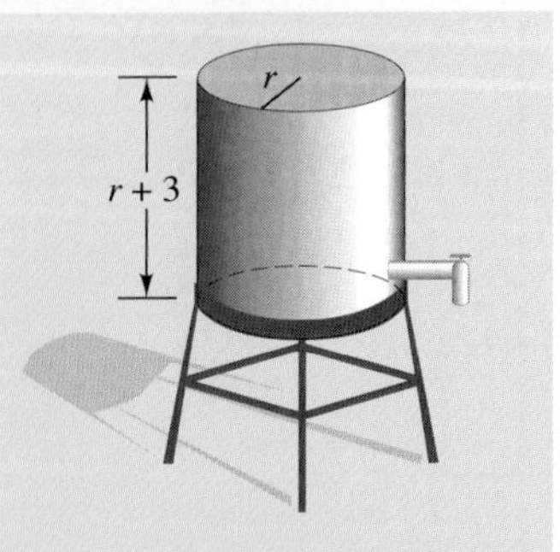

CHAPTER TEST

Use the remainder theorem to find each value.

1. $P(x) = 3x^3 - 9x - 5$; $P(2)$ 1

2. $P(x) = x^5 + 2$; $P(-2)$ −30

Find a polynomial with the given zeros.

3. $5, -1, 0$
$x^3 - 4x^2 - 5x$

4. $i, -i, \sqrt{3}, -\sqrt{3}$
$x^4 - 2x^2 - 3$

Let $P(x) = 3x^3 - 2x^2 + 4$. *Use synthetic division to find each value.*

5. $P(1)$ 5

6. $P(-2)$ −28

7. $P\left(-\frac{1}{3}\right)$ $\frac{11}{3}$

8. $P(i)$ $6 - 3i$

Use synthetic division to express $P(x) = 2x^3 - 3x^2 - 4x - 1$ *in the form* **(divisor)(quotient) + remainder** *for each divisor.*

9. $x - 2$ $(x - 2)(2x^2 + x - 2) - 5$

10. $x + 1$ $(x + 1)(2x^2 - 5x + 1) - 2$

Use synthetic division to perform each division.

11. $\dfrac{2x^2 - 7x - 15}{x - 5}$
$2x + 3$

12. $\dfrac{3x^3 + 7x^2 + 2x}{x + 2}$
$3x^2 + x$

Write a third-degree polynomial equation with real coefficients and the given roots.

13. $2, i$
$x^3 - 2x^2 + x - 2 = 0$

14. $1, 2 + i$
$x^3 - 5x^2 + 9x - 5 = 0$

Use Descartes' rule of signs to find the number of possible positive, negative, and nonreal roots of each equation.

15. $3x^5 - 2x^4 + 2x^2 - x - 3 = 0$
1 or 3 positive; 0 or 2 negative; 0, 2, or 4 nonreal

16. $2x^3 - 5x^2 - 2x - 1 = 0$
1 positive; 0 or 2 negative; 0 or 2 nonreal

Find integer bounds for the roots of each equation.

17. $x^5 - x^4 - 5x^3 + 5x^2 + 4x - 5 = 0$ −3, 3

18. $2x^3 - 11x^2 + 10x + 3 = 0$ −1, 6

Find all roots of the equation.

19. $2x^3 + 3x^2 - 11x - 6 = 0$ $2, -3, -\frac{1}{2}$

Use the bisection method to find the positive root of the equation to the nearest tenth.

20. $x^2 - 11 = 0$ 3.3

CUMULATIVE REVIEW EXERCISES

Graph the function defined by each equation.

1. $f(x) = 3^x - 2$

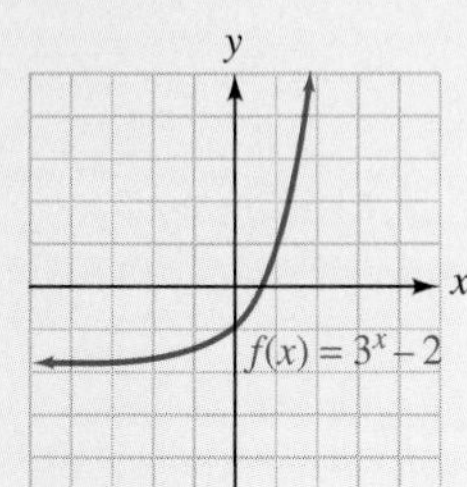

2. $f(x) = 2e^x$

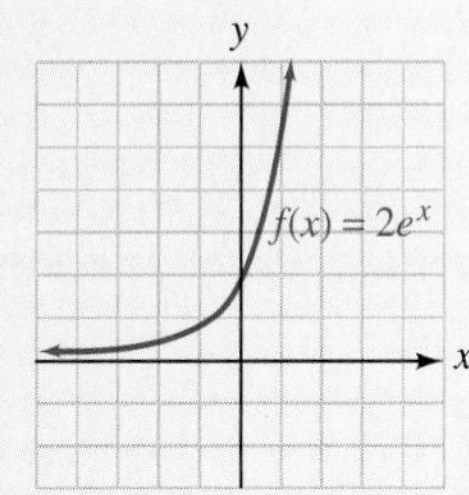

3. $f(x) = \log_3 x$

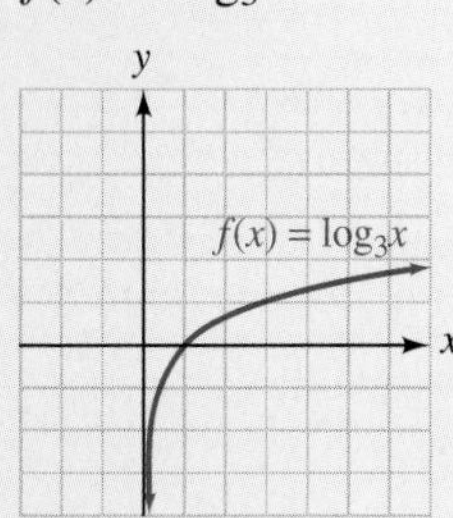

4. $f(x) = \ln(x - 2)$

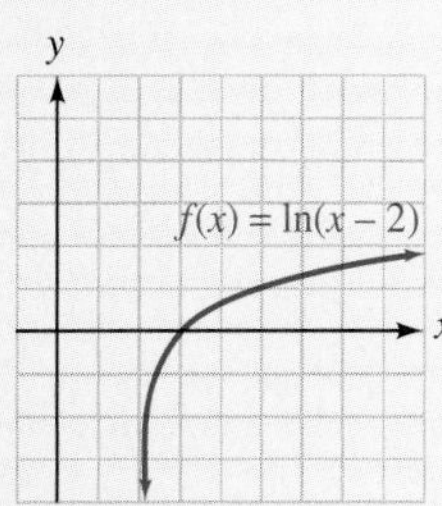

Find each value.

5. $\log_2 64$ **6**

6. $\log_{1/2} 8$ **−3**

7. $\ln e^3$ **3**

8. $2^{\log_2 2}$ **2**

Write each expression in terms of the logarithms of a, b, and c.

9. $\log abc$
$\log a + \log b + \log c$

10. $\log \frac{a^2b}{c}$
$2\log a + \log b - \log c$

11. $\log \sqrt{\frac{ab}{c^3}}$
$\frac{1}{2}(\log a + \log b - 3\log c)$

12. $\log \frac{\sqrt{ab^2}}{c}$
$\frac{1}{2}\log a + \log b - \log c$

Write each expression as the logarithm of a single quantity.

13. $3\log a - 3\log b$ **$\log \frac{a^3}{b^3}$**

14. $\frac{1}{2}\log a + 3\log b - \frac{2}{3}\log c$ **$\log \frac{\sqrt{a}\,b^3}{\sqrt[3]{c^2}}$**

Solve each equation.

15. $3^{x+1} = 8$
$x = \frac{\log 8}{\log 3} - 1$

16. $3^{x-1} = 3^{2x}$
$x = -1$

17. $\log x + \log 2 = 3$ **$x = 500$**

18. $\log(x + 1) + \log(x - 1) = 1$ **$x = \sqrt{11}$**

Let $P(x) = 4x^3 + 3x + 2$. Use synthetic division to find each value.

19. $P(1)$ **9**

20. $P(-2)$ **−36**

21. $P\left(\frac{1}{2}\right)$ **4**

22. $P(i)$ **$2 - i$**

Determine whether each binomial is a factor of $P(x) = x^3 + 2x^2 - x - 2$. Use synthetic division.

23. $x + 1$ **a factor**

24. $x - 2$ **not a factor**

25. $x - 1$ **a factor**

26. $x + 2$ **a factor**

Determine how many roots each equation has.

27. $x^{12} - 4x^8 + 2x^4 + 12 = 0$ **12**

28. $x^{2,000} - 1 = 0$ **2,000**

Determine the number of possible positive, negative, and nonreal roots of each equation.

29. $x^4 + 2x^3 - 3x^2 + x + 2 = 0$
2 or 0 positive; 2 or 0 negative; 4, 2, or 0 nonreal

30. $x^4 - 3x^3 - 2x^2 - 3x - 5 = 0$
1 positive; 3 or 1 negative; 2 or 0 nonreal

Solve each equation.

31. $x^3 + x^2 - 9x - 9 = 0$ **−1, −3, 3**

32. $x^3 - 2x^2 - x + 2 = 0$ **−1, 1, 2**

6 Linear Systems

COLLEGE **Algebra** $f(x)$ **Now**™

Throughout the chapter, this icon introduces resources on the College AlgebraNow Web Site, accessed through **http://1pass.thomson.com**, that will:

- Help you test your knowledge of the material in this chapter prior to reading it,
- Allow you to take an exam-prep quiz, and
- Provide a Personalized Learning Plan targeting areas you should study.

Careers and Mathematics

APPLIED MATHEMATICIAN

ImageState/PictureQuest

Applied mathematicians use mathematical techniques, such as modeling and computational methods, to solve practical problems in business, government, engineering, and in the physical, life, and social sciences. For example, they may analyze the most efficient way to schedule airline routes between cities, the effect and safety of new drugs, or the cost-effectiveness of alternate manufacturing processes. Some mathematicians, called **cryptanalysts,** analyze and decipher encryption systems designed to transmit military, political, financial, or law enforcement-related information in code.

Applied mathematicians held about 2,900 jobs in 2002. In addition, about 20,000 persons held full-time mathematics faculty positions in colleges and universities.

JOB OUTLOOK

Since many mathematicians will have job titles that reflect their occupation, rather than the title "mathematician," employment of applied mathematicians is expected to decline through 2012. The median annual earnings of mathematicians were \$76,470 in 2002.

For sample applications, see Section 6.8.

In this chapter, we learn to solve systems of equations—sets of several equations, each with more than one variable. One method of solution uses matrices, an important tool in mathematics and its applications.

6.1 Systems of Linear Equations

In this section, you will learn about

- The Graphing Method
- The Substitution Method
- The Addition Method
- A System with Infinitely Many Solutions
- An Inconsistent System
- Three Equations in Three Variables
- Applications

Equations with two variables have infinitely many solutions. For example, the following tables give a few of the solutions of $x + y = 5$ and $x - y = 1$.

$x + y = 5$

x	y
1	4
2	3
3	2
5	0

$x - y = 1$

x	y
8	7
3	2
1	0
0	−1

Only the pair $x = 3$ and $y = 2$ satisfies both equations. The pair of equations

$$\begin{cases} x + y = 5 \\ x - y = 1 \end{cases}$$

is called a **system of equations,** and the solution $x = 3$, $y = 2$ is called its **simultaneous solution,** or just its **solution.** The process of finding the solution of a system of equations is called **solving the system.**

The graph of an equation in two variables displays the equation's infinitely many solutions. For example, the graph of one of the lines in Figure 6-1 represents the infinitely many solutions of the equation $5x - 2y = 1$. The other line in the figure is the graph of the infinitely many solutions of $2x + 3y = 8$.

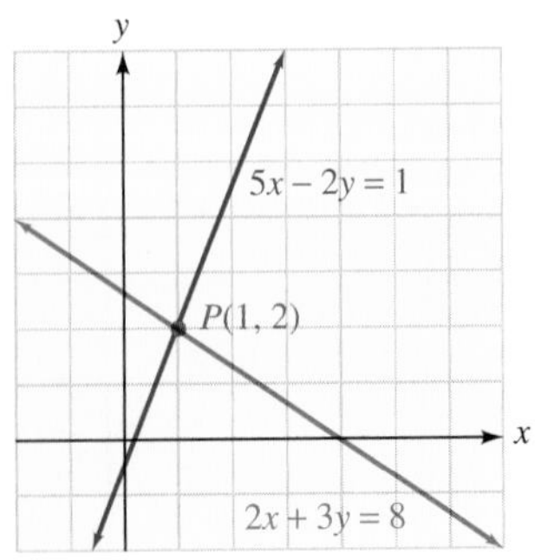

Figure 6-1

Because only point $P(1, 2)$ lies on both lines, only its coordinates satisfy both equations. Thus, the simultaneous solution of the system of equations

$$\begin{cases} 5x - 2y = 1 \\ 2x + 3y = 8 \end{cases}$$

is the pair of numbers $x = 1$ and $y = 2$, or simply the pair $(1, 2)$.

This discussion suggests a graphical method of solving systems of equations in two variables.

■ The Graphing Method

We can use the following steps to solve a system of two equations in two variables.

Strategy for Using the Graphing Method

1. On one coordinate grid, graph each equation.
2. Find the coordinates of the point or points where all of the graphs intersect. These coordinates give the solutions of the system.
3. If the graphs have no point in common, the system has no solution.

EXAMPLE 1 Use the graphing method to solve each system: **a.** $\begin{cases} 3x + y = 1 \\ -x + 2y = 9 \end{cases}$,

b. $\begin{cases} 2x - 3y = 4 \\ 4x = -4 + 6y \end{cases}$, and **c.** $\begin{cases} y = 4 - x \\ 2x + 2y = 8 \end{cases}$.

Solution **a.** The graphs of the equations are the lines shown in Figure 6-2(a). The solution of this system is given by the coordinates of the point $(-1, 4)$, where the lines intersect. By checking both values in both equations, we can verify that the solution is $x = -1$ and $y = 4$.

b. The graphs of the equations are the parallel lines shown in Figure 6-2(b). Since parallel lines do not intersect, the system has no solutions.

c. The graphs of the equations are the lines shown in Figure 6-2(c). Since the lines are the same, they have infinitely many points in common, and the system has infinitely many solutions. All ordered pairs whose coordinates satisfy one of the equations satisfy the other also.

To find some solutions, we substitute numbers for x in the first equation and solve for y. If $x = 3$, for example, then $y = 1$. One solution is the pair $(3, 1)$. Other solutions are $(0, 4)$ and $(5, -1)$.

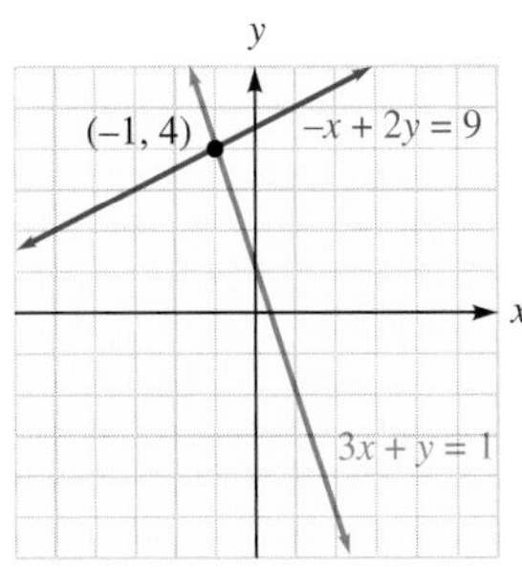

(a)

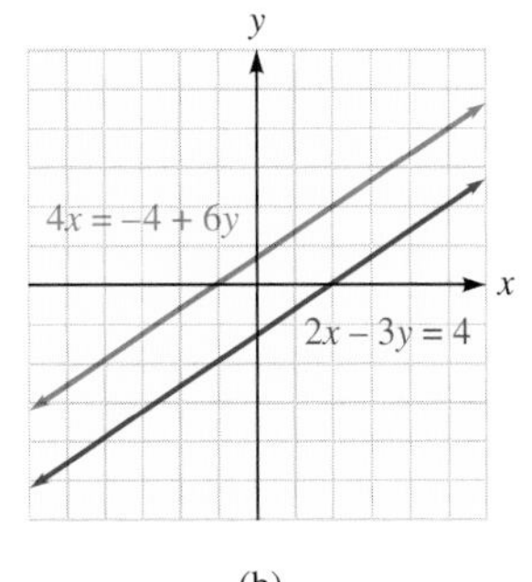

(b)

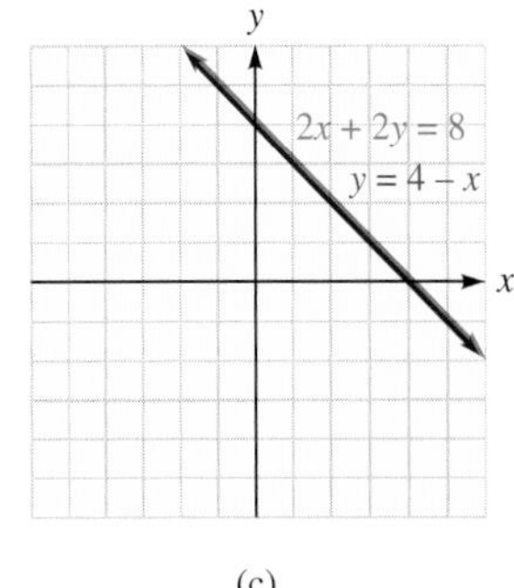

(c)

Figure 6-2

Self Check Solve: $\begin{cases} y = 2x \\ x + y = 3 \end{cases}$. ■

Example 1 illustrates three possibilities that can occur when we solve systems of equations. If a system of equations has at least one solution, as in parts a and c of Example 1, the system is called **consistent.** If it has no solutions, as in part b, it is called **inconsistent.**

If a system of two equations in two variables has exactly one solution as in part a, or no solution as in part b, the equations in the system are called **independent.** If a system of linear equations has infinitely many solutions, as in part c, the equations of the system are called **dependent.**

The following table shows the three possibilities that can occur when two equations, each with two variables, are graphed.

Possible figure	If the	Then
y, x	lines are distinct and intersect,	the equations are independent, and the system is consistent. The system has one solution.
y, x	lines are distinct and parallel,	the equations are independent, and the system is inconsistent. The system has no solutions.
y, x	lines coincide,	the equations are dependent, and the system is consistent. The system has infinitely many solutions.

Accent on Technology SOLVING SYSTEMS WITH A GRAPHING CALCULATOR

To use a graphing calculator to solve the system $\begin{cases} x - 4y = 7 \\ x + 2y = 4 \end{cases}$, we must solve both equations for y. We do so to get the equivalent system $\begin{cases} y = \frac{x - 7}{4} \\ y = \frac{4 - x}{2} \end{cases}$. Depending on the viewing window, the two graphs will be similar to those in Figure 6-3(a). We zoom in on the intersection, as in Figure 6-3(b), and trace to find the solution: $x = 5, y = -\frac{1}{2}$.

We can also find the intersection of the two lines by using the INTERSECT feature found on most graphing calculators. On a TI-83 Plus calculator, INTERSECT is found in the CALC menu. To learn how to use this feature, please consult the owner's manual. After graphing the lines and using INTERSECT, we obtain a graph similar to Figure 6-3(c). The display shows the coordinates of the point of intersection.

Continued

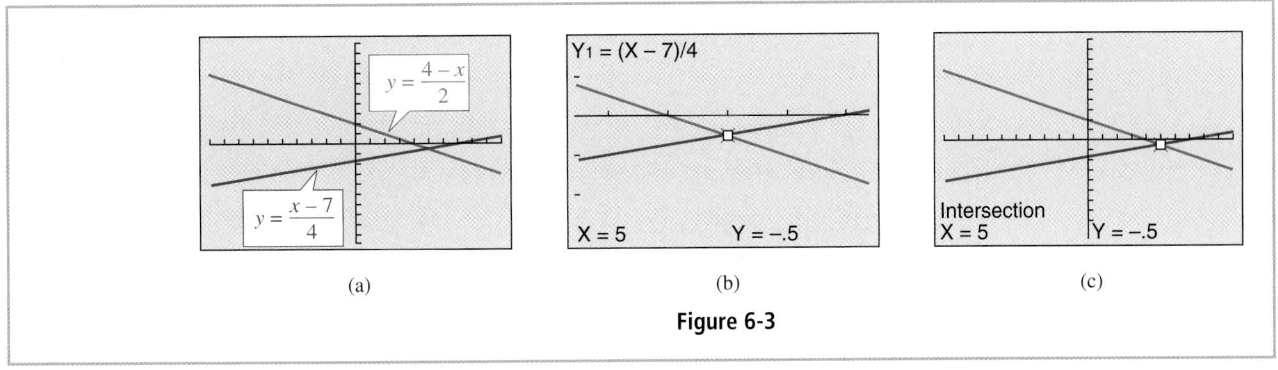

(a) (b) (c)

Figure 6-3

We now consider two algebraic methods to find solutions of systems containing any number of variables.

The Substitution Method

We use the following steps to solve a system of two equations in two variables by substitution.

Strategy for Using the Substitution Method

1. Solve one equation for a variable—say, y.
2. Substitute the expression obtained for y for every y in the second equation.
3. Solve the equation that results.
4. Substitute the solution found in Step 3 into the equation found in Step 1 and solve for y.

ACTIVE EXAMPLE 2 Use the substitution method to solve $\begin{cases} 3x + y = 1 \\ -x + 2y = 9 \end{cases}$.

Solution We can solve the first equation for y and then substitute that result for y in the second equation.

$$\begin{cases} 3x + y = 1 \rightarrow y = (1 - 3x) \\ -x + 2y = 9 \end{cases}$$

The substitution gives one linear equation with one variable, which we can solve for x:

$$-x + 2(1 - 3x) = 9$$
$$-x + 2 - 6x = 9$$
$$-7x = 7 \quad \text{Combine like terms and subtract 2 from both sides.}$$
$$x = -1 \quad \text{Divide both sides by } -7.$$

COLLEGE **Algebra** $f(x)$ **Now**™
Explore this Active Example and test yourself on the steps followed here at **http://1pass.thomson.com**. You can also find this example on the Interactive Video Skillbuilder CD-ROM.

To find y, we substitute -1 for x in the equation $y = 1 - 3x$ and simplify:

$$\begin{aligned} y &= 1 - 3x \\ &= 1 - 3(-1) \\ &= 1 + 3 \\ &= 4 \end{aligned}$$

The solution is the pair $(-1, 4)$.

Self Check Solve: $\begin{cases} 2x + 3y = 1 \\ x - y = 3 \end{cases}$. ■

■ The Addition Method

As with substitution, the addition method combines the equations of a system to eliminate terms involving one of the variables.

Strategy for Using the Addition Method

1. Write the equations of the system in general form so that terms with the same variable are aligned vertically.
2. Multiply all the terms of one or both of the equations by constants chosen to make the coefficients of x (or y) differ only in sign.
3. Add the equations. Solve the equation that results, if possible.
4. Substitute the value obtained in Step 3 into either of the original equations, and solve for the remaining variable.
5. The results obtained in Steps 3 and 4 are the solution of the system.

ACTIVE EXAMPLE 3 Use the addition method to solve $\begin{cases} 3x + 2y = 8 \\ 2x - 5y = 18 \end{cases}$.

COLLEGE **Algebra** $f(x)$ **Now**™

Go to **http://1pass.thomson.com** or your CD to practice this example.

Solution To eliminate y, we multiply the terms of the first equation by 5 and the terms of the second equation by 2, to obtain an equivalent system in which the coefficients of y differ only in sign. Then we add the equations to eliminate y.

$$\begin{cases} 3x + 2y = 8 \rightarrow 15x + 10y = 40 & \text{Multiply by 5.} \\ 2x - 5y = 18 \rightarrow \underline{4x - 10y = 36} & \text{Multiply by 2.} \end{cases}$$

$$19x = 76 \quad \text{Add the equations.}$$

$$x = 4 \quad \text{Divide both sides by 19.}$$

We now substitute 4 for x into either of the original equations and solve for y. If we use the first equation,

$$\begin{aligned} 3x + 2y &= 8 & & \\ 3(4) + 2y &= 8 & & \text{Substitute 4 for } x. \\ 12 + 2y &= 8 & & \text{Simplify.} \\ 2y &= -4 & & \text{Subtract 12 from both sides.} \\ y &= -2 & & \text{Divide both sides by 2.} \end{aligned}$$

The solution is $(4, -2)$.

Self Check Solve: $\begin{cases} 2x + 5y = 8 \\ 2x - 4y = -10 \end{cases}$. ■

■ A System with Infinitely Many Solutions

EXAMPLE 4 Use the addition method to solve $\begin{cases} x + 2y = 3 \\ 2x + 4y = 6 \end{cases}$.

Solution To eliminate the variable y, we can multiply both sides of the first equation by -2 and add the result to the second equation to get

$$\begin{array}{r} -2x - 4y = -6 \\ 2x + 4y = 6 \\ \hline 0 = 0 \end{array}$$

Although the result $0 = 0$ is true, it does not give the value of y.

Since the second equation in the original system is twice the first, the equations are equivalent. If we were to solve this system by graphing, the two lines would coincide. The (x, y) coordinates of points on that one line form the infinite set of solutions to the given system. The system is consistent, but the equations are dependent.

To find a general solution, we can solve either equation of the system for y. If we solve the first equation for y, we obtain

$$x + 2y = 3$$
$$2y = 3 - x$$
$$y = \frac{3 - x}{2}$$

All ordered pairs that are solutions will have the form $\left(x, \frac{3 - x}{2}\right)$. For example:

- If $x = 3$, then $y = 0$, and $(3, 0)$ is a solution.
- If $x = 0$, then $y = \frac{3}{2}$, and $\left(0, \frac{3}{2}\right)$ is a solution.
- If $x = -7$, then $y = 5$, and $(-7, 5)$ is a solution.

Self Check Solve: $\begin{cases} 3x - y = 2 \\ 6x - 2y = 4 \end{cases}$. ■

■ An Inconsistent System

EXAMPLE 5 Solve: $\begin{cases} x + y = 3 \\ x + y = 2 \end{cases}$.

Solution We can multiply both sides of the second equation by -1 and add the results to the first equation to get

$$\begin{array}{r} x + y = 3 \\ -x - y = -2 \\ \hline 0 = 1 \end{array}$$

Because $0 \neq 1$, the system has no solutions. If we graph each equation in this system, the graphs will be parallel lines. This system is inconsistent, and the equations are independent.

Self Check Solve: $\begin{cases} 3x + 2y = 2 \\ 3x + 2y = 3 \end{cases}$. ■

EXAMPLE 6 Solve: $\begin{cases} \dfrac{x + 2y}{4} + \dfrac{x - y}{5} = \dfrac{6}{5} \\ \dfrac{x + y}{7} - \dfrac{x - y}{3} = -\dfrac{12}{7} \end{cases}$.

Solution To clear the first equation of fractions, we multiply both sides by the lowest common denominator, 20. To clear the second equation of fractions, we multiply both sides of the second equation by the LCD, which is 21.

$$\begin{cases} 20\left(\dfrac{x + 2y}{4} + \dfrac{x - y}{5}\right) = 20\left(\dfrac{6}{5}\right) \\ 21\left(\dfrac{x + y}{7} - \dfrac{x - y}{3}\right) = 21\left(-\dfrac{12}{7}\right) \end{cases}$$

$$\begin{cases} 5(x + 2y) + 4(x - y) = 24 \\ 3(x + y) - 7(x - y) = -36 \end{cases}$$

$$\begin{cases} 9x + 6y = 24 \\ -4x + 10y = -36 \end{cases}$$ Remove the parentheses and combine like terms.

$$\begin{cases} 3x + 2y = 8 & \text{Divide both sides by 3.} \\ 2x - 5y = 18 & \text{Divide both sides by } -2. \end{cases}$$

In Example 3, we saw that the solution to this final system is $(4, -2)$.

Self Check Solve: $\begin{cases} \dfrac{x + y}{3} + \dfrac{x - y}{4} = \dfrac{3}{4} \\ \dfrac{x - y}{3} - \dfrac{2x - y}{2} = -\dfrac{1}{3} \end{cases}$. ■

■ Three Equations in Three Variables

To solve three equations in three variables, we use addition to eliminate one variable. This will produce a system of two equations in two variables, which we can solve using the methods previously discussed.

ACTIVE EXAMPLE 7 Solve the system:

COLLEGE **Algebra** *f(x)* **Now**™
Go to **http://1pass.thomson.com** or your CD to practice this example.

$$\begin{cases} x + 2y + z = 8 & (1) \\ 2x + y - z = 1 & (2) \\ x + y - 2z = -3 & (3) \end{cases}$$

Solution We can add Equations 1 and 2 to eliminate z

$$\begin{array}{ll} (1) & x + 2y + z = 8 \\ (2) & \underline{2x + \ \ y - z = 1} \\ (4) & 3x + 3y \qquad = 9 \end{array}$$

and divide both sides of Equation 4 by 3 to get

$$(5) \qquad x + y = 3$$

We now choose a different pair of equations—say, Equations 1 and 3—and eliminate z again. If we multiply both sides of Equation 1 by 2 and add the result to Equation 3, we get

$$\begin{array}{ll} & 2x + 4y + 2z = \ \ 16 \\ (3) & \underline{\ x + \ \ y - 2z = -3} \\ (6) & 3x + 5y \qquad = \ \ 13 \end{array}$$

Equations 5 and 6 form the system $\begin{cases} x + y = 3 \\ 3x + 5y = 13 \end{cases}$, which we can solve by substitution.

$$\begin{array}{ll} (5) & x + y = \ \ 3 \rightarrow y = 3 - x \\ (6) & 3x + 5y = 13 \end{array}$$

$3x + 5(3 - x) = 13$ — Substitute $3 - x$ for y.

$3x + 15 - 5x = 13$ — Remove parentheses.

$-2x = -2$ — Combine like terms and subtract 15 from both sides.

$x = 1$ — Divide both sides by -2.

To find y, we substitute 1 for x in the equation $y = 3 - x$.

$$y = 3 - 1$$
$$y = 2$$

To find z, we substitute 1 for x and 2 for y in any one of the original equations that includes z, and we find that $z = 3$. The solution is the triple

$$(x, y, z) = (1, 2, 3)$$

Because there is one solution, the system is consistent, and its equations are independent. ■

ACTIVE EXAMPLE 8 Solve the system:

COLLEGE **Algebra** $f(x)$ **Now**™
Go to **http://1pass.thomson.com** or your CD to practice this example.

$$\begin{array}{ll} (1) & \begin{cases} x + 2y + z = 8 \\ 2x + y - z = 1 \\ x - y - 2z = -7 \end{cases} \\ (2) & \\ (3) & \end{array}$$

Solution We can add Equations 1 and 2 to eliminate z.

$$\begin{array}{ll} & x + 2y + z = 8 \\ & \underline{2x + \ \ y - z = 1} \\ (4) & 3x + 3y \qquad = 9 \end{array}$$

We can now multiply Equation 1 by 2 and add it to Equation 3 to eliminate z again.

$$\begin{array}{rl} & 2x + 4y + 2z = 16 \\ & \underline{x - y - 2z = -7} \\ (5) & 3x + 3y \qquad = 9 \end{array}$$

Since Equations 4 and 5 are the same, the system is consistent, but the equations are dependent. There will be infinitely many solutions. To find a general solution, we can solve either Equation 4 or 5 for y to get

$$(6) \qquad y = 3 - x$$

We can find the value of z in terms of x by substituting the right-hand side of Equation 6 into any of the first three equations—say Equation 1.

$$x + 2y + z = 8$$

$$x + 2(3 - x) + z = 8 \qquad \text{Substitute } 3 - x \text{ for } y.$$

$$x + 6 - 2x + z = 8 \qquad \text{Use the distributive property to remove parentheses.}$$

$$-x + 6 + z = 8 \qquad \text{Combine terms.}$$

$$z = x + 2 \qquad \text{Solve for } z.$$

A general solution to this system is $(x, y, z) = (x, 3 - x, x + 2)$. To find some specific solutions, we can substitute numbers for x and compute y and z. For example,

If $x = 1$, then $y = 2$ and $z = 3$. One possible solution is $(1, 2, 3)$.

If $x = 0$, then $y = 3$ and $z = 2$. Another possible solution is $(0, 3, 2)$. ■

Everyday Connections — Child Care in the Workplace

"It makes good business sense to create a work environment that supports the needs of each individual, such as by providing access to child care. It not only benefits the individual, but it also benefits the company by enabling it to attract and retain the best people. With the changing nature of the workforce and the growing economy, this is more important to individual businesses now than ever before. And child care is also critically important to all businesses and our economy because today's children are tomorrow's workers." Robert E. Rubin, former U.S. Secretary of the Treasury

Employer-Sponsored Daycare Can Be Profitable, New Study Shows

When applied micro-economist Rachel Connelly began analyzing data on the economics of childcare in 1988, her peers doubted the subject would yield anything of significance. The rise in women's labor force participation that began in the 1960s had caused economists to take note, but there was no research on who was watching children so their mothers could work, or a breakdown of factors influencing their employment decisions.

Continued

The study suggests that on-site daycare is not only affordable, it is in many cases profitable. Further, the study found that a majority of workers would be willing to contribute to the cost of employer-sponsored daycare, whether or not they used the benefit.

Source: ***www.bowdoin.edu,* January 07, 2005**

Suppose a survey of 9,000 companies reports the following statistics:

- The number of companies offering company-supported child care is three times the number of companies offering on-site child care.
- The number of companies offering on-site child care is twice the number of companies offering subsidized child care.

Determine how many companies offer each of the three types of child care described by the survey.

6,000 companies offer company-supported child care, 2,000 companies offer on-site child care, 1,000 companies offer subsidized child care

■ APPLICATIONS

EXAMPLE 9 An airplane flies 600 miles with the wind for 2 hours and returns against the wind in 3 hours. Find the speed of the wind and the air speed of the plane.

Solution If a represents the air speed and w represents wind speed, the ground speed of the plane with the wind is the combined speed $a + w$. On the return trip, against the wind, the ground speed is $a - w$. The information in this problem is organized in Figure 6-4 and can be used to give a system of two equations in the variables a and w.

	d =	r	· t
Outbound trip	600	$a + w$	2
Return trip	600	$a - w$	3

Figure 6-4

Since $d = rt$, we have $\begin{cases} 600 = 2(a + w) \\ 600 = 3(a - w) \end{cases}$, which can be written as

$$\begin{cases} 300 = a + w \\ 200 = a - w \end{cases} \quad (7)$$

We can add these equations to get

$$500 = 2a$$

$$a = 250 \quad \text{Divide both sides by 2.}$$

To find w, we substitute 250 for a into either of the previous equations (we'll use Equation 7) and solve for w:

$$
\begin{aligned}
(7)\qquad 300 &= a + w \\
300 &= 250 + w \\
w &= 50 \qquad \text{Subtract 250 from both sides.}
\end{aligned}
$$

The air speed of the plane is 250 mph. With a 50-mph tailwind, the ground speed is 250 + 50, or 300 mph. At 300 mph, the 600-mile trip will take 2 hours.

With a 50-mph headwind, the ground speed is 250 − 50, or 200 mph. At 200 mph, the 600-mile trip will take 3 hours. The answers check. ■

Self Check Answers

1. (1, 2) 2. (2, −1) 3. (−1, 2) 4. ordered pairs of the form $(x, 3x - 2)$ 5. The system is inconsistent; no solutions 6. (1, 2)

6.1 Exercises

VOCABULARY AND CONCEPTS ***Fill in the blanks.***

1. A set of several equations with several variables is called a <u>system</u> of equations.
2. Any set of numbers that satisfies each equation of a system is called a <u>solution</u> of the system.
3. If a system of equations has a solution, the system is <u>consistent</u>.
4. If a system of equations has no solution, the system is <u>inconsistent</u>.
5. If a system of equations has only one solution, the equations of the system are <u>independent</u>.
6. If a system of equations has infinitely many solutions, the equations of the system are <u>dependent</u>.
7. The system $\begin{cases} x + y = 5 \\ x - y = 1 \end{cases}$ is <u>consistent</u> (consistent, inconsistent).
8. The system $\begin{cases} x + y = 5 \\ x + y = 1 \end{cases}$ is <u>inconsistent</u> (consistent, inconsistent).
9. The equations of the system $\begin{cases} x + y = 5 \\ 2x + 2y = 10 \end{cases}$ are <u>dependent</u> (dependent, independent).
10. The equations of the system $\begin{cases} x + y = 5 \\ x - y = 1 \end{cases}$ are <u>independent</u> (dependent, independent).
11. The pair (1, 3) <u>is</u> (is, is not) a solution of the system $\begin{cases} x + 2y = 7 \\ 2x - y = -1 \end{cases}$.
12. The pair (1, 3) <u>is</u> (is, is not) a solution of the system $\begin{cases} 3x + y = 6 \\ x - 3y = -8 \end{cases}$.

PRACTICE ***Solve each system of equations by graphing.***

13. $\begin{cases} y = -3x + 5 \\ x - 2y = -3 \end{cases}$

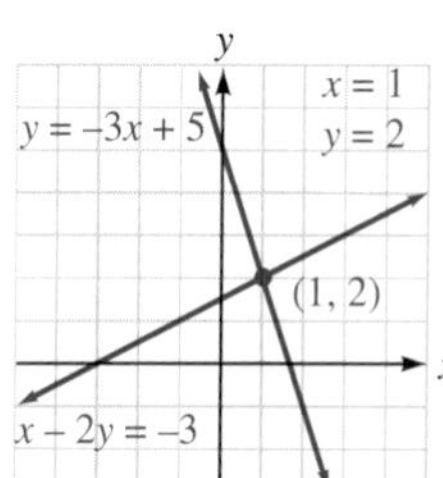

14. $\begin{cases} x - 2y = -3 \\ 3x + y = -9 \end{cases}$

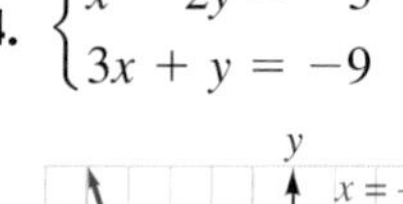
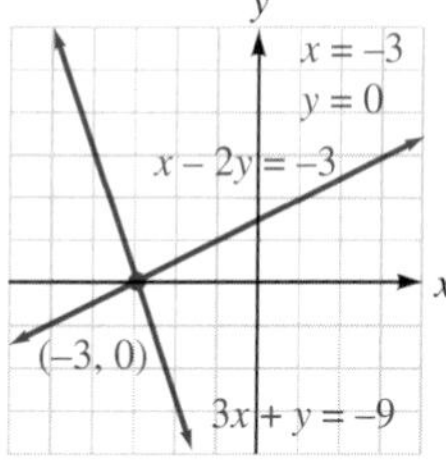

15. $\begin{cases} 3x + 2y = 2 \\ -2x + 3y = 16 \end{cases}$

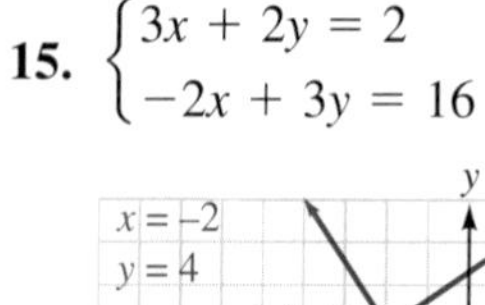
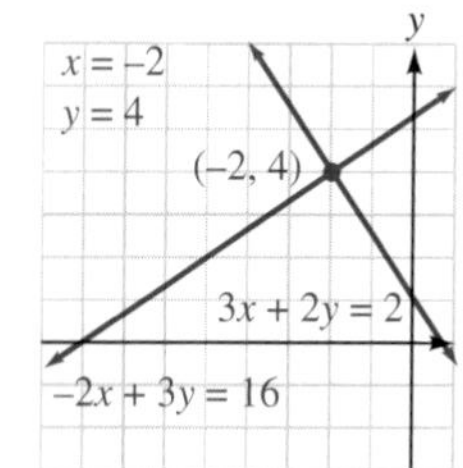

16. $\begin{cases} x + y = 0 \\ 5x - 2y = 14 \end{cases}$

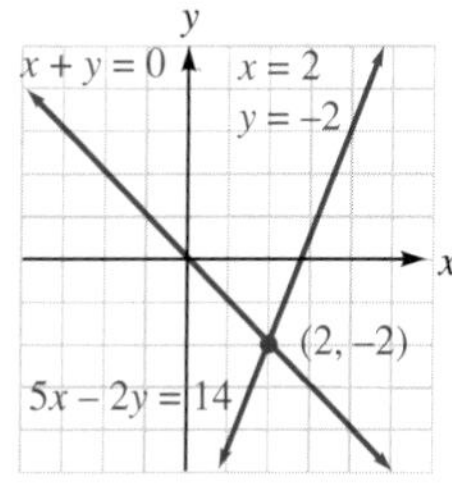

Use a graphing calculator to approximate the solutions of each system. Give answers to the nearest tenth.

17. $\begin{cases} y = -5.7x + 7.8 \\ y = 37.2 - 19.1x \end{cases}$
(2.2, −4.7)

18. $\begin{cases} y = 3.4x - 1 \\ y = -7.1x + 3.1 \end{cases}$
(0.4, 0.3)

19. $\begin{cases} y = \dfrac{5.5 - 2.7x}{3.5} \\ 5.3x - 9.2y = 6.0 \end{cases}$
(1.7, 0.3)

20. $\begin{cases} 29x + 17y = 7 \\ -17x + 23y = 19 \end{cases}$
(−0.2, 0.7)

Solve each system by substitution, if possible.

21. $\begin{cases} y = x - 1 \\ y = 2x \end{cases}$
(−1, −2)

22. $\begin{cases} y = 2x - 1 \\ x + y = 5 \end{cases}$
(2, 3)

23. $\begin{cases} 2x + 3y = 0 \\ y = 3x - 11 \end{cases}$
(3, −2)

24. $\begin{cases} 2x + y = 3 \\ y = 5x - 11 \end{cases}$
(2, −1)

25. $\begin{cases} 4x + 3y = 3 \\ 2x - 6y = -1 \end{cases}$
$\left(\frac{1}{2}, \frac{1}{3}\right)$

26. $\begin{cases} 4x + 5y = 4 \\ 8x - 15y = 3 \end{cases}$
$\left(\frac{3}{4}, \frac{1}{5}\right)$

27. $\begin{cases} x + 3y = 1 \\ 2x + 6y = 3 \end{cases}$
no solution; inconsistent system

28. $\begin{cases} x - 3y = 14 \\ 3(x - 12) = 9y \end{cases}$
no solution; inconsistent system

29. $\begin{cases} y = 3x - 6 \\ x = \dfrac{1}{3}y + 2 \end{cases}$
dependent equations; a general solution is (x, 3x − 6)

30. $\begin{cases} 3x - y = 12 \\ y = 3x - 12 \end{cases}$
dependent equations; a general solution is (x, 3x − 12)

Solve each system by the addition method, if possible.

31. $\begin{cases} 5x - 3y = 12 \\ 2x - 3y = 3 \end{cases}$
(3, 1)

32. $\begin{cases} 2x + 3y = 8 \\ -5x + y = -3 \end{cases}$
(1, 2)

33. $\begin{cases} x - 7y = -11 \\ 8x + 2y = 28 \end{cases}$
(3, 2)

34. $\begin{cases} 3x + 9y = 9 \\ -x + 5y = -3 \end{cases}$
(3, 0)

35. $\begin{cases} 3(x - y) = y - 9 \\ 5(x + y) = -15 \end{cases}$
(−3, 0)

36. $\begin{cases} 2(x + y) = y + 1 \\ 3(x + 1) = y - 3 \end{cases}$
(−1, 3)

37. $\begin{cases} 2 = \dfrac{1}{x + y} \\ 2 = \dfrac{3}{x - y} \end{cases}$
$\left(1, -\frac{1}{2}\right)$

38. $\begin{cases} \dfrac{1}{x + y} = 12 \\ \dfrac{3x}{y} = -4 \end{cases}$
$\left(\frac{1}{3}, -\frac{1}{4}\right)$

39. $\begin{cases} y + 2x = 5 \\ 0.5y = 2.5 - x \end{cases}$
dependent equations; a general solution is (x, 5 − 2x)

40. $\begin{cases} -0.3x + 0.1y = -0.1 \\ 6x - 2y = 2 \end{cases}$
dependent equations; a general solution is (x, 3x − 1)

41. $\begin{cases} x + 2(x - y) = 2 \\ 3(y - x) - y = 5 \end{cases}$
no solution; inconsistent system

42. $\begin{cases} 3x = 4(2 - y) \\ 3(x - 2) + 4y = 0 \end{cases}$
no solution; inconsistent system

43. $\begin{cases} x + \dfrac{y}{3} = \dfrac{5}{3} \\ \dfrac{x + y}{3} = 3 - x \end{cases}$
(4, −7)

44. $\begin{cases} 3x - y = 0.25 \\ x + \dfrac{3}{2}y = 2.375 \end{cases}$
(0.5, 1.25)

45. $\begin{cases} \dfrac{3}{2}x + \dfrac{1}{3}y = 2 \\ \dfrac{2}{3}x + \dfrac{1}{9}y = 1 \end{cases}$
(2, −3)

46. $\begin{cases} \dfrac{x + y}{2} + \dfrac{x - y}{5} = 2 \\ x = \dfrac{y}{2} + 1 \end{cases}$
(2, 2)

47. $\begin{cases} \dfrac{x - y}{5} + \dfrac{x + y}{2} = 6 \\ \dfrac{x - y}{2} - \dfrac{x + y}{4} = 3 \end{cases}$
(9, −1)

48. $\begin{cases} \dfrac{x - 2}{5} + \dfrac{y + 3}{2} = 5 \\ \dfrac{x + 3}{2} + \dfrac{y - 2}{3} = 6 \end{cases}$
(7, 5)

Solve each system, if possible.

49. $\begin{cases} x + y + z = 3 \\ 2x + y + z = 4 \\ 3x + y - z = 5 \end{cases}$
(1, 2, 0)

50. $\begin{cases} x - y - z = 0 \\ x + y - z = 0 \\ x - y + z = 2 \end{cases}$
(1, 0, 1)

51. $\begin{cases} x - y + z = 0 \\ x + y + 2z = -1 \\ -x - y + z = 0 \end{cases}$
$\left(0, -\frac{1}{3}, -\frac{1}{3}\right)$

52. $\begin{cases} 2x + y - z = 7 \\ x - y + z = 2 \\ x + y - 3z = 2 \end{cases}$
(3, 2, 1)

53. $\begin{cases} 2x + y = 4 \\ x - z = 2 \\ y + z = 1 \end{cases}$
(1, 2, −1)

54. $\begin{cases} 3x + y + z = 0 \\ 2x - y + z = 0 \\ 2x + y + z = 0 \end{cases}$
(0, 0, 0)

55. $\begin{cases} x + y + z = 6 \\ 2x + y + 3z = 17 \\ x + y + 2z = 11 \end{cases}$
(1, 0, 5)

56. $\begin{cases} x + y + z = 3 \\ 2x + y + z = 6 \\ x + 2y + 3z = 2 \end{cases}$
(3, 1, −1)

57. $\begin{cases} x + y + z = 3 \\ x + z = 2 \\ 2x + 2y + 2z = 3 \end{cases}$
no solution; inconsistent system

58. $\begin{cases} x + y + z = 3 \\ x + z = 2 \\ 2x + y + 2z = 5 \end{cases}$
dependent equations; a general solution is $(x, 1, 2 - x)$

59. $\begin{cases} x + 2y - z = 2 \\ 2x - y = -1 \\ 3x + y + z = 1 \end{cases}$
$(0, 1, 0)$

60. $\begin{cases} x + y = 2 \\ y + z = 2 \\ 3x + 3y = 2 \end{cases}$
no solution; inconsistent system

61. $\begin{cases} 3x + 4y + 2z = 4 \\ 6x - 2y + z = 4 \\ 3x - 8y - 6z = -3 \end{cases}$
$\left(\frac{2}{3}, \frac{1}{4}, \frac{1}{2}\right)$

62. $\begin{cases} x + y = 2 \\ y + z = 2 \\ x - z = 0 \end{cases}$
dependent equations; a general solution is $(x, 2 - x, x)$

63. $\begin{cases} 2x - y - z = 0 \\ x - 2y - z = -1 \\ x - y - 2z = -1 \end{cases}$
$\left(\frac{1}{2}, \frac{1}{2}, \frac{1}{2}\right)$

64. $\begin{cases} x + 3y - z = 5 \\ 3x - y + z = 2 \\ 2x + y = 1 \end{cases}$
no solution; inconsistent system

65. $\begin{cases} (x + y) + (y + z) + (z + x) = 6 \\ (x - y) + (y - z) + (z - x) = 0 \\ x + y + 2z = 4 \end{cases}$
dependent equations; a general solution is $(x, 2 - x, 1)$

66. $\begin{cases} (x + y) + (y + z) = 1 \\ (x + z) + (x + z) = 3 \\ (x - y) - (x - z) = -1 \end{cases}$ **$\left(\frac{11}{4}, -\frac{1}{4}, -\frac{5}{4}\right)$**

APPLICATIONS *Use systems of equations to solve each problem.*

67. **Planning for harvest** A farmer raises corn and soybeans on 350 acres of land. Because of expected prices at harvest time, he thinks it would be wise to plant 100 more acres of corn than of soybeans. How many acres of each does he plant?
225 acres of corn; 125 acres of soybeans

68. **Club memberships** There is an initiation fee to join the Pine River Country Club, as well as monthly dues. The total cost after 7 months' membership will be \$3,025, and after $1\frac{1}{2}$ years, \$3,850. Find both the initiation fee and the monthly dues.
\$2,500 initiation fee; \$75 per month dues

69. **Framing pictures** A rectangular picture frame has a perimeter of 1,900 centimeters and a width that is 250 centimeters less than its length. Find the area of the picture. **210,000 cm^2**

70. **Boating** A Mississippi riverboat can travel 30 kilometers downstream in 3 hours and can make the return trip in 5 hours. Find the speed of the boat in still water. **8 kph**

71. **Making an alloy** A metallurgist wants to make 60 grams of an alloy that is to be 34% copper. She has samples that are 9% copper and 84% copper. How many grams of each must she use?
40 g and 20 g

72. **Archimedes' law of the lever** The two weights shown will be in balance if the product of one weight and its distance from the fulcrum is equal to the product of the other weight and its distance from the fulcrum. Two weights are in balance when one is 2 meters and the other 3 meters from the fulcrum. If the fulcrum remained in the same spot and the weights were interchanged, the closer weight would need to be increased by 5 pounds to maintain balance. Find the weights. **6 lb and 4 lb**

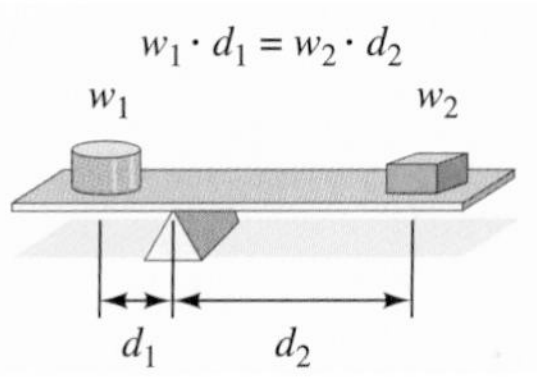

73. **Lifting weights** A 112-pound force can lift the 448-pound load shown. If the fulcrum is moved 1 additional foot away from the load, a 192-pound force is required. Find the length of the lever. **10 ft**

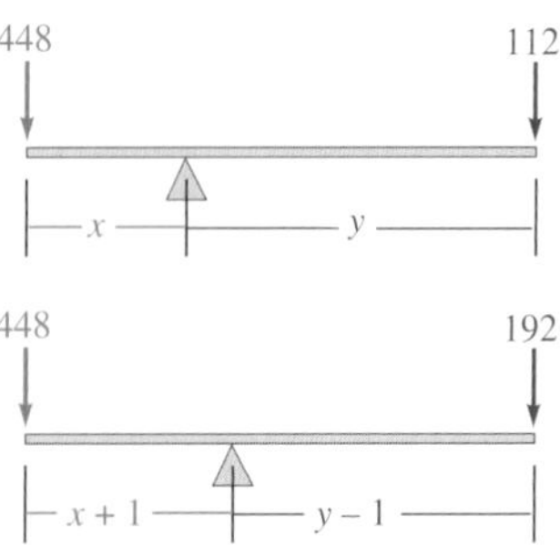

74. **Writing test questions** For a test question, a mathematics teacher wants to find two constants a and b such that the test item "Simplify $a(x + 2y) - b(2x - y)$" will have an answer of $-3x + 9y$. What constants a and b should the teacher use? **$a = 3, b = 3$**

75. Break-even point Rollowheel, Inc. can manufacture a pair of in-line skates for \$43.53. Daily fixed costs of manufacturing in-line skates amount to \$742.72. A pair of in-line skates can be sold for \$89.95. Find equations expressing the expenses E and the revenue R as functions of x, the number of pairs manufactured and sold. At what production level will expenses equal revenues?

$E(x) = 43.53x + 742.72$, $R(x) = 89.95x$; 16 pairs per day

76. Choosing salary options For its sales staff, a company offers two salary options. One is \$326 per week plus a commission of $3\frac{1}{2}\%$ of sales. The other is \$200 per week plus $4\frac{1}{4}\%$ of sales. Find equations that express incomes as functions of sales, and find the weekly sales level that produces equal salaries.

x = sales; $S_1(x) = 326 + 0.035x$, $S_2(x) = 200 + 0.0425x$; \$16,800

Use systems of three equations in three variables to solve each problem.

77. Work schedules A college student earns \$198.50 per week working three part-time jobs. Half of his 30-hour work week is spent cooking hamburgers at a fast-food chain, earning \$5.70 per hour. In addition, the student earns \$6.30 per hour working at a gas station and \$10 per hour doing janitorial work. How many hours per week does the student work at each job?

15 hr cooking hamburgers, 10 hr pumping gas, 5 hr janitorial

78. Investment income A woman invested a \$22,000 rollover IRA account in three banks paying 5%, 6%, and 7% annual interest. She invested \$2,000 more at 6% than at 5%. The total annual interest she earned was \$1,370. How much did she invest at each rate?

\$5,000 at 5%, \$7,000 at 6%, \$10,000 at 7%

79. Age distribution Approximately 3 million people live in Costa Rica. 2.61 million are less than 50 years old, and 1.95 million are older than 14. How many people are in each of the categories 0–14 years, 15–49 years, and 50 years and older?

1.05 million in 0–14 group, 1.56 million in 15–49 group, 0.39 million in 50-and-older group

80. Designing arches The engineer designing a parabolic arch knows that its equation has the form $y = ax^2 + bx + c$. Use the information in the illustration to find a, b, and c. Assume that the distances are given in feet. (*Hint:* The coordinates of points on the parabola satisfy its equation.)

$a = -\frac{3}{40}$, $b = 3$, $c = 0$

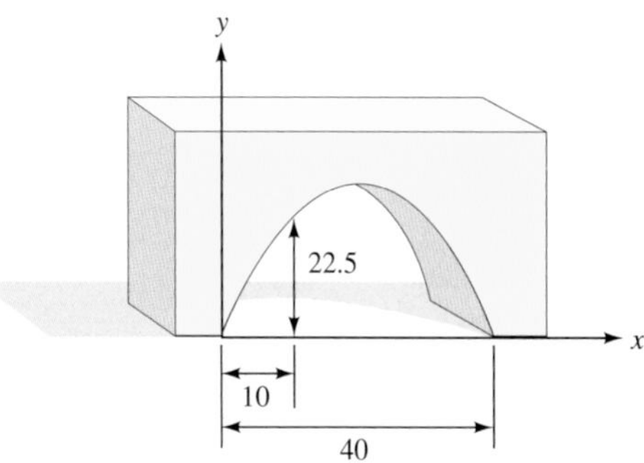

81. Geometry The sum of the angles of a triangle is 180°. In a certain triangle, the largest angle is 20° greater than the sum of the other two and is 10° greater than 3 times the smallest. How large is each angle? **30°, 50°, 100°**

82. Ballistics The path of a thrown object is a parabola with the equation $y = ax^2 + bx + c$. Use the information in the illustration to find a, b, and c. (Distances are in feet.) **$a = -\frac{1}{16}$, $b = 2$, $c = 0$**

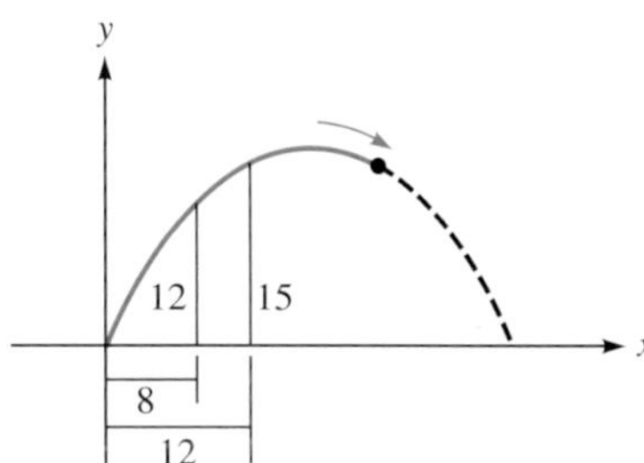

DISCOVERY AND WRITING

83. Use a graphing calculator to attempt to find the solution of the system $\begin{cases} x - 8y = -51 \\ 3x - 25y = -160 \end{cases}$.

84. Solve the system of Exercise 83 algebraically. Which method is easier, and why?

85. Use a graphing calculator to attempt to find the solution of the system $\begin{cases} 17x - 23y = -76 \\ 29x + 19y = -278 \end{cases}$.

86. Solve the system of Exercise 85 algebraically. Which method is easier, and why?

87. Invent a system of two equations with the solution $x = 1$, $y = 2$, which is difficult to solve with a graphing calculator (as in Exercise 85).

88. Invent a system of two equations with the solution $x = 1$, $y = 2$, that is easier to solve graphically than algebraically.

REVIEW *Graph each function.*

89. $f(x) = 3^x$

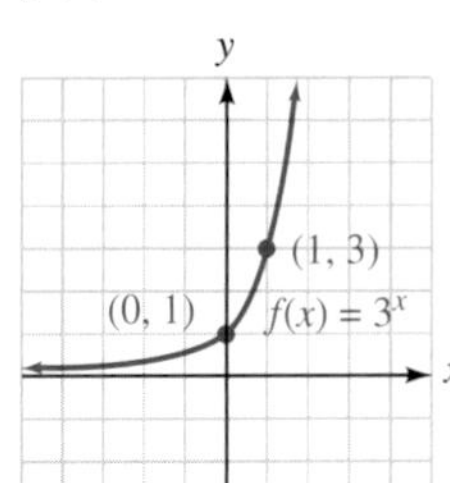

90. $f(x) = \log_2 x$

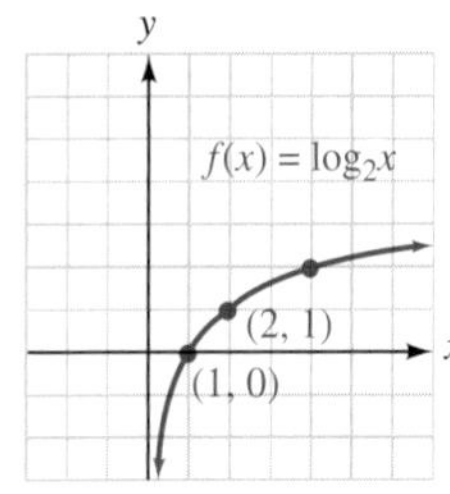

Solve for x:

91. $3 = \log_2 x$ 8

92. $x = \log_5 \frac{1}{25}$ −2

Use the properties of logarithms to write each expression in terms of the logarithms of x, y, and z.

93. $\log \frac{x}{y^2 z}$ $\log x - 2 \log y - \log z$

94. $\log x(y^z)$ $\log x + z \log y$

Use the properties of logarithms to write each expression as a single logarithm.

95. $\log x + 3 \log y - \frac{1}{2} \log z$ $\log \frac{xy^3}{\sqrt{z}}$

96. $\frac{1}{2}(\log A - 3 \log B)$ $\log \sqrt{\frac{A}{B^3}}$

6.2 Gaussian Elimination and Matrix Methods

In this section, you will learn about

- Elementary Row Operations
- Row Echelon Form of a Matrix
- An Inconsistent System
- A System of Dependent Equations
- Gauss–Jordan Elimination

In this section, we introduce a method for solving systems of equations called **Gaussian elimination,** after the German mathematician Carl Friedrich Gauss (1777–1855). In this method, we transform a system of equations into an equivalent system that can be solved by a process called **back substitution.**

Using Gaussian elimination, we solve systems of equations by working only with the coefficients of the variables. From the system of equations

$$\begin{cases} x + 2y + z = 8 \\ 2x + y - z = 1 \\ x + y - 2z = -3 \end{cases}$$

for example, we can form three rectangular arrays of numbers. Each is called a **matrix.** The first is the **coefficient matrix,** which contains the coefficients of the variables of the system. The second matrix contains the constants from the right-hand side of the system. We can write the third matrix, called the **augmented matrix** or the **system matrix,** by joining the two.

Carl Friedrich Gauss
(1777–1855)
Many people consider Gauss to be the greatest mathematician of all time. He made contributions in the areas of number theory, solutions of equations, the geometry of curved surfaces, and statistics. For his efforts, he earned the title "Prince of the Mathematicians."

Coefficient matrix *Constants* *System matrix*

$$\begin{bmatrix} 1 & 2 & 1 \\ 2 & 1 & -1 \\ 1 & 1 & -2 \end{bmatrix} \qquad \begin{bmatrix} 8 \\ 1 \\ -3 \end{bmatrix} \qquad \begin{bmatrix} 1 & 2 & 1 & 8 \\ 2 & 1 & -1 & 1 \\ 1 & 1 & -2 & -3 \end{bmatrix}$$

Each row of the system matrix represents one equation of the system.

- The first row represents the equation $x + 2y + z = 8$.
- The second row represents the equation $2x + y - z = 1$.
- The third row represents the equation $x + y - 2z = -3$.

Because the system matrix has three rows and four columns, it is called a 3×4 matrix (read as "3 by 4"). The coefficient matrix is 3×3, and the constants form a 3×1 matrix.

To illustrate Gaussian elimination, we solve this system by the addition method of the previous section. In a second column, we keep track of the changes to the system matrix.

ACTIVE EXAMPLE 1 Use Gaussian elimination to solve

$$\begin{cases} x + 2y + z = 8 & (1) \\ 2x + y - z = 1 & (2) \\ x + y - 2z = -3 & (3) \end{cases}$$

Solution We multiply each term in Equation 1 by -2 and add the result to Equation 2 to obtain Equation 4. Then we multiply each term of Equation 1 by -1 and add the result to Equation 3 to obtain Equation 5. This gives the equivalent system

COLLEGE **Algebra** $f(x)$ **Now**™

Explore this Active Example and test yourself on the steps followed here at **http://1pass.thomson.com**. You can also find this example on the Interactive Video Skillbuilder CD-ROM.

$$\begin{cases} x + 2y + z = 8 & (1) \\ -3y - 3z = -15 & (4) \\ -y - 3z = -11 & (5) \end{cases} \qquad \begin{bmatrix} 1 & 2 & 1 & 8 \\ 0 & -3 & -3 & -15 \\ 0 & -1 & -3 & -11 \end{bmatrix}$$

Now we divide both sides of Equation 4 by -3 to obtain Equation 6.

$$\begin{cases} x + 2y + z = 8 & (1) \\ y + z = 5 & (6) \\ -y - 3z = -11 & (5) \end{cases} \qquad \begin{bmatrix} 1 & 2 & 1 & 8 \\ 0 & 1 & 1 & 5 \\ 0 & -1 & -3 & -11 \end{bmatrix}$$

and add Equation 6 to Equation 5 to obtain Equation 7.

$$\begin{cases} x + 2y + z = 8 & (1) \\ y + z = 5 & (6) \\ -2z = -6 & (7) \end{cases} \qquad \begin{bmatrix} 1 & 2 & 1 & 8 \\ 0 & 1 & 1 & 5 \\ 0 & 0 & -2 & -6 \end{bmatrix}$$

Finally, we divide both sides of Equation 7 by -2 to obtain the system

$$\begin{cases} x + 2y + z = 8 & (1) \\ y + z = 5 & (6) \\ z = 3 \end{cases} \qquad \begin{bmatrix} 1 & 2 & 1 & 8 \\ 0 & 1 & 1 & 5 \\ 0 & 0 & 1 & 3 \end{bmatrix}$$

The system can now be solved by back substitution. Because $z = 3$, we can substitute 3 for z in Equation 6 and solve for y:

$$\begin{aligned} (6) \qquad y + z &= 5 \\ y + \mathbf{3} &= 5 \\ y &= 2 \end{aligned}$$

We can now substitute 2 for y and 3 for z in Equation 1 and solve for x:

$$\begin{aligned} (1) \qquad x + 2y + z &= 8 \\ x + 2(\mathbf{2}) + \mathbf{3} &= 8 \\ x + 7 &= 8 \\ x &= 1 \end{aligned}$$

The solution is the ordered triple (1, 2, 3).

Self Check Solve: $\begin{cases} x - 3y = 1 \\ 2x + y = 9 \end{cases}$. ■

■ Elementary Row Operations

In Example 1, we performed operations on the equations as well as on the corresponding matrices. These matrix operations are called **elementary row operations.**

Elementary Row Operations

If any of the following operations are performed on the rows of a system matrix, the matrix of an equivalent system results:

Type 1 row operation: Two rows of a matrix can be interchanged.

Type 2 row operation: The elements of a row of a matrix can be multiplied by a nonzero constant.

Type 3 row operation: Any row can be changed by adding a multiple of another row to it.

- A type 1 row operation is equivalent to writing the equations of a system in a different order.
- A type 2 row operation is equivalent to multiplying both sides of an equation by a nonzero constant.
- A type 3 row operation is equivalent to adding a multiple of one equation to another.

Any matrix that can be obtained from another matrix by a sequence of elementary row operations is **row equivalent** to the original matrix. This means that row equivalent matrices represent equivalent systems of equations.

■ Row Echelon Form of a Matrix

The process of Example 1 changed a system matrix into a special row equivalent form, called **row echelon form.**

Row Echelon Form of a Matrix

A matrix is in **row echelon form** if it has these three properties:

1. The first nonzero entry in each row (called the **lead entry**) is 1.
2. Lead entries appear farther to the right as we move down the rows of the matrix.
3. Any rows containing only 0's are at the bottom of the matrix.

The matrices

$$\begin{bmatrix} 1 & 3 & 0 & 7 \\ 0 & 0 & 1 & 8 \\ 0 & 0 & 0 & 0 \end{bmatrix}, \quad \begin{bmatrix} 0 & 1 & 2 \\ 0 & 0 & 1 \end{bmatrix}, \quad \text{and} \quad \begin{bmatrix} 1 & 1 \\ 0 & 0 \\ 0 & 0 \\ 0 & 0 \end{bmatrix}$$

are in row echelon form, because the first nonzero entry in each row is 1, the lead entries appear farther to the right as we move down the rows, and any rows that consist entirely of 0's are at the bottom of the matrix.

The next three matrices are not in row echelon form, for the reasons given:

$$\begin{bmatrix} 1 & 3 & 0 & 7 \\ 0 & 0 & 1 & 8 \\ 0 & 0 & 1 & 0 \end{bmatrix}$$ The lead entry in the last row is not to the right of the lead entry of the middle row.

$$\begin{bmatrix} 0 & 3 & 0 \\ 0 & 0 & 1 \end{bmatrix}$$ The lead entry of the first row is not 1.

$$\begin{bmatrix} 0 & 0 \\ 1 & 0 \\ 0 & 1 \end{bmatrix}$$ The row of 0's is not last.

ACTIVE EXAMPLE 2 Solve the following system by transforming its system matrix into row echelon form and back substituting.

COLLEGE **Algebra** $f(x)$ **Now**™
Go to **http://1pass.thomson.com** or your CD to practice this example.

$$\begin{cases} x + 2y + 3z = 4 \\ 2x - y - 2z = 0 \\ x - 3y - 3z = -2 \end{cases}$$

Solution This system is represented by the system matrix

$$\begin{bmatrix} 1 & 2 & 3 & 4 \\ 2 & -1 & -2 & 0 \\ 1 & -3 & -3 & -2 \end{bmatrix}$$

We will reduce the system matrix to row echelon form. We can use the 1 in the upper left-hand corner to zero out the rest of the first column. To do so, we multiply the first row by -2 and add the result to the second row. We indicate this operation with the notation $(-2)R1 + R2$.

$$\begin{bmatrix} 1 & 2 & 3 & 4 \\ 2 & -1 & -2 & 0 \\ 1 & -3 & -3 & -2 \end{bmatrix} (-2)R1 + R2 \rightarrow \begin{bmatrix} 1 & 2 & 3 & 4 \\ 0 & -5 & -8 & -8 \\ 1 & -3 & -3 & -2 \end{bmatrix}$$

Next, we multiply row one by -1 and add the result to row three to get a new row three. This fills the rest of the first column with 0's.

$$\begin{matrix} \\ \\ (-1)R1 + R3 \rightarrow \end{matrix} \begin{bmatrix} 1 & 2 & 3 & 4 \\ 0 & -5 & -8 & -8 \\ 0 & -5 & -6 & -6 \end{bmatrix}$$

We multiply row two by -1 and add the result to row three to get

$$\begin{matrix} \\ \\ (-1)R2 + R3 \rightarrow \end{matrix} \begin{bmatrix} 1 & 2 & 3 & 4 \\ 0 & -5 & -8 & -8 \\ 0 & 0 & 2 & 2 \end{bmatrix}$$

Finally, we multiply row two by $-\frac{1}{5}$ and row three by $\frac{1}{2}$.

$$\begin{matrix} \\ \left(-\frac{1}{5}\right)R2 \rightarrow \\ \left(\frac{1}{2}\right)R3 \rightarrow \end{matrix} \begin{bmatrix} 1 & 2 & 3 & 4 \\ 0 & 1 & \frac{8}{5} & \frac{8}{5} \\ 0 & 0 & 1 & 1 \end{bmatrix}$$

The final matrix, now in row echelon form, represents the equivalent system

$$\begin{cases} x + 2y + 3z = 4 \\ y + \frac{8}{5}z = \frac{8}{5} \\ z = 1 \end{cases}$$

We can now use back substitution to solve the system. To find y, we substitute 1 for z in the second equation.

$$\begin{aligned} y + \frac{8}{5}z &= \frac{8}{5} \\ y + \frac{8}{5}(1) &= \frac{8}{5} \\ y &= 0 \end{aligned}$$

To solve for x, we substitute 0 for y and 1 for z in the first equation.

$$\begin{aligned} x + 2y + 3z &= 4 \\ x + 2(0) + 3(1) &= 4 \\ x + 3 &= 4 \\ x &= 1 \end{aligned}$$

The solution of the original system is the triple $(x, y, z) = (1, 0, 1)$.

Self Check Solve: $\begin{cases} x + y + 2z = 5 \\ y - 2z = 0 \\ x - z = 0 \end{cases}$. ■

Accent on Technology REDUCING A MATRIX TO ROW ECHELON FORM

We can use a TI-83 Plus graphing calculator to reduce a matrix to row echelon form. For example, to reduce the matrix of Example 2

$$\begin{bmatrix} 1 & 2 & 3 & 4 \\ 2 & -1 & -2 & 0 \\ 1 & -3 & -3 & -2 \end{bmatrix}$$

to row echelon form, we press 2nd MATRIX, select EDIT, and enter the size and entries of the matrix, as shown in Figure 6-5(a). We then press 2nd QUIT to clear the screen. After pressing 2nd MATRIX again, we select MATH, scroll down to A: ref (, and press ENTER. Finally, we press 2nd MATRIX and select 1 : [A] by pressing ENTER to obtain Figure 6-5(b). Pressing ENTER one last time gives the ref (row echelon form) of the matrix, shown in Figure 6-5(c).

Continued

MATRIX[A] 3 ×4
[1 2 3 –
[2 –1 –2 –
[1 –3 –3 –
3 , 1 = 1

(a)

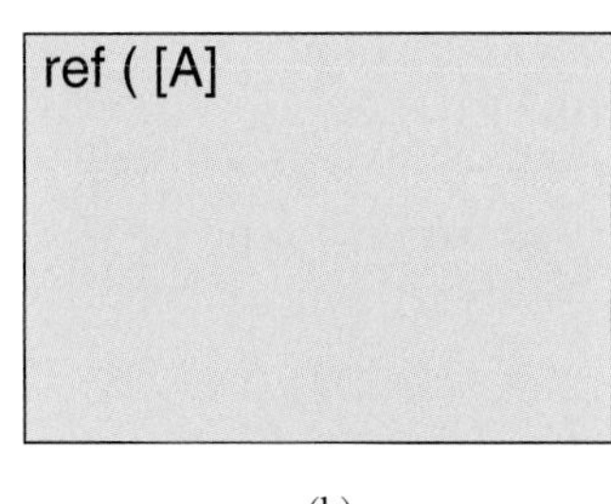

(b)

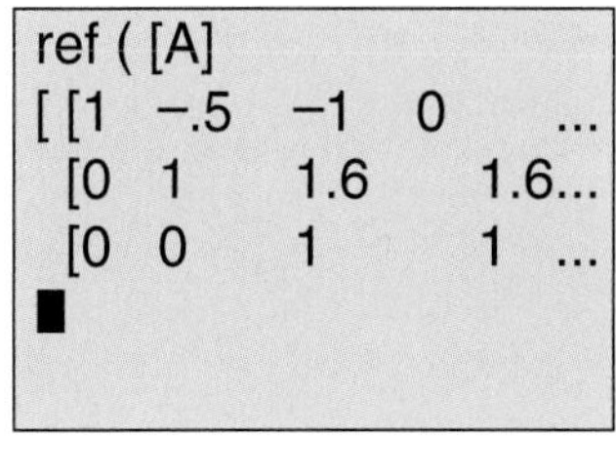

(c)

Figure 6-5

Comment The row echelon form shown in Figure 6-5(c) is not the same as the row echelon form found in Example 2. However, if we multiply row two of the matrix shown in Figure 6-5(c) by $\frac{5}{2}$ and add the results to row one, we obtain the equivalent row echelon form found in Example 2.

$$\begin{bmatrix} 1 & -.5 & -1 & 0 \\ 0 & 1 & 1.6 & 1.6 \\ 0 & 0 & 1 & 1 \end{bmatrix} \quad \left(\tfrac{5}{2}\right)R2 + R1 \rightarrow \begin{bmatrix} 1 & 2 & 3 & 4 \\ 0 & 1 & \frac{8}{5} & \frac{8}{5} \\ 0 & 0 & 1 & 1 \end{bmatrix}$$

If we back substitute into the equations represented by either row echelon form, we will obtain the same values for x, y, and z.

Since the systems of Examples 1 and 2 have solutions, each system is consistent. The next two examples illustrate a system that is inconsistent and a system whose equations are dependent.

An Inconsistent System

EXAMPLE 3 Use matrix methods to solve $\begin{cases} x + y + z = 3 \\ 2x - y + z = 2. \\ 3y + z = 1 \end{cases}$

Solution We form the system matrix and use row operations to write it in row echelon form. We use the 1 in the top left position to zero out the rest of the first column.

$$\begin{bmatrix} 1 & 1 & 1 & 3 \\ 2 & -1 & 1 & 2 \\ 0 & 3 & 1 & 1 \end{bmatrix} (-2)R1 + R2 \rightarrow \begin{bmatrix} 1 & 1 & 1 & 3 \\ 0 & -3 & -1 & -4 \\ 0 & 3 & 1 & 1 \end{bmatrix}$$

$$(1)R2 + R3 \rightarrow \begin{bmatrix} 1 & 1 & 1 & 3 \\ 0 & -3 & -1 & -4 \\ 0 & 0 & 0 & -3 \end{bmatrix}$$

Since the last row of the matrix represents the equation

$$0x + 0y + 0z = -3$$

and no values of x, y, and z could make $0 = -3$, there is no point in continuing. The given system has no solution and is inconsistent.

Self Check Solve: $\begin{cases} x + y + z = 5 \\ y - z = 1 \\ x + 2y = 3 \end{cases}$. ■

A System of Dependent Equations

EXAMPLE 4 Use matrices to solve $\begin{cases} x + 2y + z = 8 \\ 2x + y - z = 1 \\ x - y - 2z = -7 \end{cases}$.

Solution We can set up the system matrix and use row operations to reduce it to row echelon form:

$$\begin{bmatrix} 1 & 2 & 1 & 8 \\ 2 & 1 & -1 & 1 \\ 1 & -1 & -2 & -7 \end{bmatrix} \begin{matrix} \\ (-2)R1 + R2 \rightarrow \\ (-1)R1 + R3 \rightarrow \end{matrix} \begin{bmatrix} 1 & 2 & 1 & 8 \\ 0 & -3 & -3 & -15 \\ 0 & -3 & -3 & -15 \end{bmatrix}$$

$$\begin{matrix} \\ \left(-\frac{1}{3}\right)R2 \rightarrow \\ \left(-\frac{1}{3}\right)R3 \rightarrow \end{matrix} \begin{bmatrix} 1 & 2 & 1 & 8 \\ 0 & 1 & 1 & 5 \\ 0 & 1 & 1 & 5 \end{bmatrix}$$

$$\begin{matrix} \\ \\ (-1)R2 + R3 \rightarrow \end{matrix} \begin{bmatrix} 1 & 2 & 1 & 8 \\ 0 & 1 & 1 & 5 \\ 0 & 0 & 0 & 0 \end{bmatrix}$$

The final matrix is in echelon form and represents the system

$$\begin{cases} x + 2y + z = 8 & (1) \\ y + z = 5 & (2) \\ 0x + 0y + 0z = 0 & (3) \end{cases}$$

Since all coefficients in Equation 3 are 0, it can be ignored. To solve this system by back substitution, we solve Equation 2 for y in terms of z:

$$y = 5 - z$$

and then substitute $5 - z$ for y in Equation 1 and solve for x in terms of z.

$$\begin{aligned} (1) \qquad x + 2y + z &= 8 \\ x + 2(5 - z) + z &= 8 \\ x + 10 - 2z + z &= 8 && \text{Remove parentheses.} \\ x + 10 - z &= 8 && \text{Combine like terms.} \\ x &= -2 + z && \text{Solve for } x. \end{aligned}$$

The solution of this system is $(x, y, z) = (-2 + z, 5 - z, z)$.

There are infinitely many solutions to this system. We can choose any real number for z, but once it is chosen, x and y are determined. For example,

- If $z = 3$, then $x = 1$ and $y = 2$. One possible solution of this system is $x = 1$, $y = 2$, and $z = 3$.
- If $z = 2$, then $x = 0$ and $y = 3$. Another solution is $x = 0$, $y = 3$, and $z = 2$.

Because this system has solutions, it is consistent. Because there are infinitely many solutions, the equations are dependent.

Self Check Solve: $\begin{cases} x + 2y + z = 2 \\ x - z = 2 \\ 2x + 2y = 4 \end{cases}$. ■

Gauss–Jordan Elimination

In Gaussian elimination, we perform row operations until the system matrix is in row echelon form. Then we convert the matrix back into equation form and solve by back substitution. A modification of this method, called **Gauss–Jordan elimination,** uses row operations to produce a matrix in ***reduced* row echelon** form. With this method, we can obtain the solution of the system from that matrix directly, and back substitution is not needed.

Reduced Row Echelon Form of a Matrix

A matrix is in **reduced row echelon form** if

1. It is in row echelon form.
2. Entries above each lead entry are also zero.

In the next example, we solve a system by Gauss–Jordan elimination.

ACTIVE EXAMPLE 5 Use Gauss–Jordan elimination to solve $\begin{cases} w + 2x + 3y + z = 4 \\ x + 4y - z = 0 \\ w - x - y + 2z = 2 \end{cases}$.

Solution We row reduce the system matrix as follows. We first use the 1 in the upper left corner to zero out the rest of the first column:

COLLEGE **Algebra** $f(x)$ **Now**™
Go to **http://1pass.thomson.com** or your CD to practice this example.

$$\begin{bmatrix} 1 & 2 & 3 & 1 & 4 \\ 0 & 1 & 4 & -1 & 0 \\ 1 & -1 & -1 & 2 & 2 \end{bmatrix} \quad \begin{matrix} \\ \\ (-1)R1 + R3 \rightarrow \end{matrix} \begin{bmatrix} 1 & 2 & 3 & 1 & 4 \\ 0 & 1 & 4 & -1 & 0 \\ 0 & -3 & -4 & 1 & -2 \end{bmatrix}$$

The lead entry in the second row is already 1. We use it to zero out the rest of the second column.

$$\begin{matrix} (-2)R2 + R1 \rightarrow \\ \\ (3)R2 + R3 \rightarrow \end{matrix} \begin{bmatrix} 1 & 0 & -5 & 3 & 4 \\ 0 & 1 & 4 & -1 & 0 \\ 0 & 0 & 8 & -2 & -2 \end{bmatrix}$$

To make the lead entry in the third row equal to 1, we multiply the third row by $\frac{1}{8}$ and use it to zero out the rest of the third column.

$$\begin{matrix} \\ \\ \left(\frac{1}{8}\right)R3 \rightarrow \end{matrix} \begin{bmatrix} 1 & 0 & -5 & 3 & 4 \\ 0 & 1 & 4 & -1 & 0 \\ 0 & 0 & 1 & -\frac{1}{4} & -\frac{1}{4} \end{bmatrix}$$

$$\begin{matrix} (5)R3 + R1 \rightarrow \\ (-4)R3 + R2 \rightarrow \\ \\ \end{matrix} \begin{bmatrix} 1 & 0 & 0 & \frac{7}{4} & \frac{11}{4} \\ 0 & 1 & 0 & 0 & 1 \\ 0 & 0 & 1 & -\frac{1}{4} & -\frac{1}{4} \end{bmatrix}$$

The final matrix is in reduced row echelon form. Note that each lead entry is 1, and each is alone in its column. The matrix represents the system of equations

$$\begin{cases} w + \dfrac{7}{4}z = \dfrac{11}{4} \\ x = 1 \\ y - \dfrac{1}{4}z = -\dfrac{1}{4} \end{cases} \quad \text{or} \quad \begin{cases} w = \dfrac{11}{4} - \dfrac{7}{4}z \\ x = 1 \\ y = -\dfrac{1}{4} + \dfrac{1}{4}z \end{cases}$$

A general solution is $(w, x, y, z) = \left(\frac{11}{4} - \frac{7}{4}z, 1, -\frac{1}{4} + \frac{1}{4}z, z\right)$. To find some specific solutions, we choose values for z, and the corresponding values of w, x, and y will be determined. For example, if $z = 1$, then $w = 1$, $x = 1$, and $y = 0$. Thus, $(w, x, y, z) = (1, 1, 0, 1)$ is a solution. If $z = -1$, another solution is $(w, x, y, z) = \left(\frac{9}{2}, 1, -\frac{1}{2}, -1\right)$. ■

To find the row reduced form of a matrix with a TI-83 Plus calculator, enter the matrix and use the rref (row reduced echelon form) feature.

In Example 5, there were more variables than equations. In the next example, there are more equations than variables.

EXAMPLE 6 Use Gauss–Jordan elimination to solve $\begin{cases} 2x + y = 4 \\ x - 3y = 9 \\ x + 4y = -5 \end{cases}$.

Solution To get a 1 in the top left corner, we could multiply the first row by $\frac{1}{2}$, but that would introduce fractions. Instead, we can exchange the first two rows and proceed as follows:

$$\begin{bmatrix} 2 & 1 & 4 \\ 1 & -3 & 9 \\ 1 & 4 & -5 \end{bmatrix} \xrightarrow{R1 \leftrightarrow R2} \begin{bmatrix} 1 & -3 & 9 \\ 2 & 1 & 4 \\ 1 & 4 & -5 \end{bmatrix}$$

$$\begin{matrix} \\ (-2)R1 + R2 \rightarrow \\ (-1)R1 + R3 \rightarrow \end{matrix} \begin{bmatrix} 1 & -3 & 9 \\ 0 & 7 & -14 \\ 0 & 7 & -14 \end{bmatrix}$$

$$\begin{matrix} \\ \left(\frac{1}{7}\right)R2 \rightarrow \\ \left(\frac{1}{7}\right)R3 \rightarrow \end{matrix} \begin{bmatrix} 1 & -3 & 9 \\ 0 & 1 & -2 \\ 0 & 1 & -2 \end{bmatrix}$$

$$\begin{matrix} (3)R2 + R1 \rightarrow \\ \\ (-1)R2 + R3 \rightarrow \end{matrix} \begin{bmatrix} 1 & 0 & 3 \\ 0 & 1 & -2 \\ 0 & 0 & 0 \end{bmatrix}$$

This final matrix is in reduced echelon form. It represents the system

$$\begin{cases} x = 3 \\ y = -2 \end{cases}$$

Verify that $x = 3$ and $y = -2$ satisfy the equations of the original system.

Self Check Solve: $\begin{cases} x + y = 5 \\ x - y = 1 \\ x + 2y = 7 \end{cases}$. ■

Self Check Answers

1. (4, 1) **2.** (1, 2, 1) **3.** The system is inconsistent. **4.** $(x, y, z) = (2 + z, -z, z)$ **6.** $x = 3, y = 2$

6.2 Exercises

VOCABULARY AND CONCEPTS *Fill in the blanks.*

1. A rectangular array of numbers is called a **matrix**.

2. A 3×5 matrix has **3** rows and **5** columns.

3. The matrix containing the coefficients of the variables is called the **coefficient** matrix.

4. The coefficient matrix joined to the column of constants is called the **system** matrix or the **augmented** matrix.

5. Each row of a system matrix represents one **equation**.

6. The rows of the system matrix are changed using elementary **row operations**.

7. If one system matrix is changed to another using row operations, the matrices are **row equivalent**.

8. If two system matrices are row equivalent, then the systems have the **same solutions**.

9. In a type 1 row operation, two rows of a matrix can be **interchanged**.

10. In a type 2 row operation, one entire row can be **multiplied** by a nonzero constant.

11. In a type 3 row operation, any row can be changed by **adding** to it any **multiple** of another row.

12. The first nonzero entry in a row is called that row's **lead entry**.

PRACTICE *Use Gaussian elimination to solve each system.*

13. $\begin{cases} x + y = 7 \\ x - 2y = -1 \end{cases}$
$\left(\frac{13}{3}, \frac{8}{3}\right)$

14. $\begin{cases} x + 3y = 8 \\ 2x - 5y = 5 \end{cases}$
(5, 1)

15. $\begin{cases} x - y = 1 \\ 2x - y = 8 \end{cases}$
(7, 6)

16. $\begin{cases} x - 5y = 4 \\ 2x + 3y = 21 \end{cases}$
(9, 1)

17. $\begin{cases} x + 2y - z = 2 \\ x - 3y + 2z = 1 \\ x + y - 3z = -6 \end{cases}$
(1, 2, 3)

18. $\begin{cases} x + 5y - z = 2 \\ x + 2y + z = 3 \\ x + y + z = 2 \end{cases}$
(−1, 1, 2)

19. $\begin{cases} x - y - z = -3 \\ 5x + y = 6 \\ y + z = 4 \end{cases}$
(1, 1, 3)

20. $\begin{cases} x + y = 1 \\ x + z = 3 \\ y + z = 2 \end{cases}$
(1, 0, 2)

Determine whether each matrix is in row echelon form, reduced row echelon form, or neither.

21. $\begin{bmatrix} 1 & 3 & 0 & 5 \\ 0 & 1 & 2 & 7 \\ 0 & 0 & 1 & 0 \end{bmatrix}$
row echelon form

22. $\begin{bmatrix} 1 & 3 & 0 & 5 \\ 0 & 1 & 2 & 7 \\ 0 & 0 & 0 & 0 \end{bmatrix}$
row echelon form

23. $\begin{bmatrix} 1 & 0 & 1 \\ 0 & 1 & 5 \\ 0 & 0 & 0 \\ 0 & 0 & 0 \end{bmatrix}$
reduced row echelon form

24. $\begin{bmatrix} 1 & 0 & 1 \\ 0 & 1 & 5 \\ 0 & 0 & 1 \\ 0 & 0 & 0 \end{bmatrix}$
reduced row echelon form

Write each system as a matrix and solve it by Gaussian elimination.

25. $\begin{cases} 2x + y = 3 \\ x - 3y = 5 \end{cases}$
(2, −1)

26. $\begin{cases} x + 2y = -1 \\ 3x - 5y = 19 \end{cases}$
(3, −2)

27. $\begin{cases} x - 7y = -2 \\ 5x - 2y = -10 \end{cases}$
(−2, 0)

28. $\begin{cases} 3x - y = 3 \\ 2x + y = -3 \end{cases}$
(0, −3)

29. $\begin{cases} 2x - y = 5 \\ x + 3y = 6 \end{cases}$
(3, 1)

30. $\begin{cases} 3x - 5y = -25 \\ 2x + y = 5 \end{cases}$
(0, 5)

31. $\begin{cases} x - 2y = 3 \\ -2x + 4y = 6 \end{cases}$
no solution; inconsistent system

32. $\begin{cases} 2(2y - x) = 6 \\ 4y = 2(x + 3) \end{cases}$
dependent equations, a general solution is $\left(x, \frac{1}{2}x + \frac{3}{2}\right)$

33. $\begin{cases} 2x - y = 7 \\ -x + \dfrac{1}{3}y = -\dfrac{7}{3} \end{cases}$
(0, −7)

34. $\begin{cases} 45x - 6y = 60 \\ 30x + 15y = 63.75 \end{cases}$
(1.5, 1.25)

35. $\begin{cases} x - y + z = 3 \\ 2x - y + z = 4 \\ x + 2y - z = -1 \end{cases}$

(1, 0, 2)

36. $\begin{cases} 2x + y - z = 1 \\ x + y - z = 0 \\ 3x + y + 2z = 2 \end{cases}$

(1, −1, 0)

37. $\begin{cases} x + y - z = -1 \\ 3x + y = 4 \\ y - 2z = -4 \end{cases}$

(2, −2, 1)

38. $\begin{cases} 3x + y = 7 \\ x - z = 0 \\ y - 2z = -8 \end{cases}$

(3, −2, 3)

39. $\begin{cases} x - y + z = 2 \\ 2x + y + z = 5 \\ 3x - 4z = -5 \end{cases}$

(1, 1, 2)

40. $\begin{cases} x + z = -1 \\ 3x + y = 2 \\ 2x + y + 5z = 3 \end{cases}$

(−1, 5, 0)

41. $\begin{cases} x + y + 2z = 4 \\ -x - y - 3z = -5 \\ 2x + y + z = 2 \end{cases}$

(−1, 3, 1)

42. $\begin{cases} x + y - z = 5 \\ x + y + z = 2 \\ 3x + 3y - z = 12 \end{cases}$

dependent equations; a general solution is $\left(x, -x + \frac{7}{2}, -\frac{3}{2}\right)$

43. $\begin{cases} -x + 3y + 2z = -10 \\ 3x - 2y - 2z = 7 \\ -2x + y - z = -10 \end{cases}$

(1, −5, 3)

44. $\begin{cases} 2x - y + z = 6 \\ 3x + y - z = 2 \\ -x + 3y - 3z = 8 \end{cases}$

no solution; inconsistent system

Write each system as a matrix and solve it by Gauss–Jordan elimination. If a system has infinitely many solutions, show a general solution.

45. $\begin{cases} x - 2y = 7 \\ y = 3 \end{cases}$

(13, 3)

46. $\begin{cases} x - 2y = 7 \\ y = 8 \end{cases}$

(23, 8)

47. $\begin{cases} x + 2y - z = 3 \\ y + 3z = 1 \\ z = -2 \end{cases}$

(−13, 7, −2)

48. $\begin{cases} x - 3y + 2z = -1 \\ y - 2z = 3 \\ z = 5 \end{cases}$

(28, 13, 5)

49. $\begin{cases} x - y = 7 \\ x + y = 13 \end{cases}$

(10, 3)

50. $\begin{cases} x + 2y = 7 \\ 2x - y = -1 \end{cases}$

(1, 3)

51. $\begin{cases} x - \frac{1}{2}y = 0 \\ x + 2y = 0 \end{cases}$

(0, 0)

52. $\begin{cases} x - y = 5 \\ -x + \frac{1}{5}y = -9 \end{cases}$

(10, 5)

53. $\begin{cases} x + y + 2z = 0 \\ x + y + z = 2 \\ x + z = 1 \end{cases}$

(3, 1, −2)

54. $\begin{cases} x + 2y = -3 \\ x + 4y = -2 \\ 2x + z = -8 \end{cases}$

$\left(-4, \frac{1}{2}, 0\right)$

55. $\begin{cases} 2x + y - 2z = 1 \\ -x + y - 3z = 0 \\ 4x + 3y = 4 \end{cases}$

$\left(\frac{1}{4}, 1, \frac{1}{4}\right)$

56. $\begin{cases} 3x + y = 3 \\ 3x + y - z = 2 \\ 6x + z = 5 \end{cases}$

$\left(\frac{2}{3}, 1, 1\right)$

57. $\begin{cases} 2x - 2y + 3z + t = 2 \\ x + y + z + t = 5 \\ -x + 2y - 3z + 2t = 2 \\ x + y + 2z - t = 4 \end{cases}$

(1, 2, 1, 1)

58. $\begin{cases} x + y + 2z + t = 1 \\ x + 2y + z + t = 2 \\ 2x + y + z + t = 4 \\ x + y + z + 2t = 3 \end{cases}$

(2, 0, −1, 1)

59. $\begin{cases} x + y + t = 4 \\ x + z + t = 2 \\ 2x + 2y + z + 2t = 8 \\ x - y + z - t = -2 \end{cases}$

(1, 2, 0, 1)

60. $\begin{cases} x - y + 2z + t = 3 \\ 3x - 2y - z - t = 4 \\ 2x + y + 2z - t = 10 \\ x + 2y + z - 3t = 8 \end{cases}$

(3, 2, 1, 0)

61. $\begin{cases} \frac{1}{3}x + \frac{3}{4}y - \frac{2}{3}z = -2 \\ x + \frac{1}{2}y + \frac{1}{3}z = 1 \\ \frac{1}{6}x - \frac{1}{8}y - z = 0 \end{cases}$

$\left(\frac{9}{4}, -3, \frac{3}{4}\right)$

62. $\begin{cases} \frac{1}{4}x + y + 3z = 1 \\ \frac{1}{2}x - 4y + 6z = -1 \\ \frac{1}{3}x - 2y - 2z = -1 \end{cases}$

$\left(\frac{2}{3}, \frac{1}{2}, \frac{1}{9}\right)$

63. $\begin{cases} \frac{1}{2}x + \frac{1}{4}y - z = 2 \\ \frac{2}{3}x + \frac{1}{4}y + \frac{1}{2}z = \frac{3}{2} \\ \frac{2}{3}x + z = -\frac{1}{3} \end{cases}$

$\left(0, \frac{20}{3}, -\frac{1}{3}\right)$

64. $\begin{cases} \frac{5}{7}x - \frac{1}{3}y + z = 0 \\ \frac{2}{7}x + y + \frac{1}{8}z = 9 \\ 6x + 4y - \frac{27}{4}z = 20 \end{cases}$

$\left(\frac{427}{918}, \frac{2{,}617}{306}, \frac{68}{27}\right)$

65. $\begin{cases} 3x - 6y + 9z = 18 \\ 2x - 4y + 3z = 12 \\ x - 2y + 3z = 6 \end{cases}$

dependent equations; a general solution is $\left(x, \frac{1}{2}x - 3, 0\right)$

66. $\begin{cases} x + 2y - z = 7 \\ 2x - y + z = 2 \\ 3x - 4y + 3z = -3 \end{cases}$

dependent equations; a general solution is $\left(-\frac{1}{5}z + \frac{11}{5}, \frac{3}{5}z + \frac{12}{5}, z\right)$

Each system contains a different number of equations as variables. Solve each system using Gauss–Jordan elimination. If a system has infinitely many solutions, show a general solution.

67. $\begin{cases} x + y = -2 \\ 3x - y = 6 \\ 2x + 2y = -4 \\ x - y = 4 \end{cases}$

(1, −3)

68. $\begin{cases} x - y = -3 \\ 2x + y = -3 \\ 3x - y = -7 \\ 4x + y = -7 \end{cases}$

(−2, 1)

69. $\begin{cases} x + 2y + z = 4 \\ 3x - y - z = 2 \end{cases}$
dependent equations; general solution is $\left(\frac{8}{7} + \frac{1}{7}z, \frac{10}{7} - \frac{4}{7}z, z\right)$

70. $\begin{cases} x + 2y - 3z = -5 \\ 5x + y - z = -11 \end{cases}$
dependent equations; general solution is $\left(-\frac{1}{9}z - \frac{17}{9}, \frac{14}{9}z - \frac{14}{9}, z\right)$

71. $\begin{cases} w + x = 1 \\ w + y = 0 \\ x + z = 0 \end{cases}$
dependent equations; general solution is $(1 + z, -z, -1 - z, z)$

72. $\begin{cases} w + x - y + z = 2 \\ 2w - x - 2y + z = 0 \\ w - 2x - y + z = -1 \end{cases}$
dependent equations; general solution is $(y, 1, y, 1)$

73. $\begin{cases} x + y = 3 \\ 2x + y = 1 \\ 3x + 2y = 2 \end{cases}$
no solution; inconsistent system

74. $\begin{cases} x + 2y + z = 4 \\ x - y + z = 1 \\ 2x + y + 2z = 2 \\ 3x + 3z = 6 \end{cases}$
no solution; inconsistent system

APPLICATIONS ***Use matrix methods to solve each problem.***

75. **Flight range** The speed of an airplane with a tailwind is 300 miles per hour, and with a headwind 220 miles per hour. On a day with no wind, how far could the plane travel on a 5-hour fuel supply? **1,300 mi**

76. **Resource allocation** 120,000 gallons of fuel are to be divided between two airlines. Triple A Airways requires twice as much as UnityAir. How much fuel should be allocated to Triple A? **80,000 gal**

77. **Library shelving** To use space effectively, librarians like to fill shelves completely. One 35-inch shelf can hold 3 dictionaries, 5 atlases, and 1 thesaurus; or 6 dictionaries and 2 thesauruses; or 2 dictionaries, 4 atlases, and 3 thesauruses. How wide is one copy of each book?
Dictionaries are 4.5 in. wide; atlases are 3.5 in. wide; thesauruses are 4 in. wide.

78. **Copying machine productivity** When both copying machines A and B are working, secretaries can make 100 copies in one minute. In one minute's time, copiers A and C together produce 140 copies, and all three working together produce 180 copies. How many copies per minute can each machine produce separately?
machine A, 60 copies; machine B, 40 copies; machine C, 80 copies

79. **Nutritional planning** One ounce of each of three foods has the vitamin and mineral content shown in the table. How many ounces of each must be used to provide exactly 22 milligrams (mg) of niacin, 12 mg of zinc, and 20 mg of vitamin C? **2, 4, 6**

Food	Niacin	Zinc	Vitamin C
A	1 mg	1 mg	2 mg
B	2 mg	1 mg	1 mg
C	2 mg	1 mg	2 mg

80. **Chainsaw sculpting** A wood sculptor carves three types of statues with a chainsaw. The number of hours required for carving, sanding, and painting a totem pole, a bear, and a deer are shown in the table. How many of each should be produced to use all available labor hours? **3 poles, 2 bears, 4 deer**

	Totem pole	Bear	Deer	Time available
Carving	2 hr	2 hr	1 hr	14 hr
Sanding	1 hr	2 hr	2 hr	15 hr
Painting	3 hr	2 hr	2 hr	21 hr

DISCOVERY AND WRITING

81. Explain the difference between the row echelon form and the reduced row echelon form of a matrix.

82. If the upper-left corner entry of a matrix is zero, what row operation might you do first?

83. What characteristics of a row reduced matrix would let you conclude that the system is inconsistent?

84. Explain the differences between Gaussian elimination and Gauss–Jordan elimination.

Use matrix methods to solve each system.

85. $\begin{cases} x^2 + y^2 + z^2 = 14 \\ 2x^2 + 3y^2 - 2z^2 = -7 \\ x^2 - 5y^2 + z^2 = 8 \end{cases}$

(*Hint:* Solve first as a system in x^2, y^2, and z^2.)
$x = \pm 2, y = \pm 1, z = \pm 3$

86. $\begin{cases} 5\sqrt{x} + 2\sqrt{x} + \sqrt{z} = 22 \\ \sqrt{x} + \sqrt{y} - \sqrt{z} = 5 \\ 3\sqrt{x} - 2\sqrt{y} - 3\sqrt{z} = 10 \end{cases}$ **$x = 16, y = 1, z = 0$**

REVIEW *Fill in the blank.*

87. The slope–intercept form of the equation of a line is $\underline{y = mx + b}$.

88. The point–slope form of the equation of a line is $\underline{y - y_1 = m(x - x_1)}$.

89. The slopes of parallel lines are $\underline{\text{equal}}$.

90. The slopes of $\underline{\text{perpendicular}}$ lines are negative reciprocals.

Write the equation of a line with the given properties.

91. The line has a slope of 2 and a y-intercept of 7. $y = 2x + 7$

92. The line has a slope of -3 and passes through the point $(2, -3)$. $y = -3x + 3$

93. The line is vertical and passes through $(2, -3)$. $x = 2$

94. The line is horizontal and passes through $(2, -3)$. $y = -3$

6.3 Matrix Algebra

In this section, you will learn about

- Multiplying a Matrix by a Constant
- Adding and Subtracting Matrices
- Multiplying Matrices
- An Application of Matrices
- The Identity Matrix

Suppose there are 66 security officers employed at two locations:

Downtown Office	Male	Female
Day shift	12	18
Night shift	3	0

Suburban Office	Male	Female
Day shift	14	12
Night shift	5	2

The information about the security force is contained in the following matrices.

$$D = \begin{bmatrix} 12 & 18 \\ 3 & 0 \end{bmatrix} \quad \text{and} \quad S = \begin{bmatrix} 14 & 12 \\ 5 & 2 \end{bmatrix}$$

The entry 12 in matrix D gives the information that 12 males work the day shift at the downtown office. Company management can add the corresponding entries of matrices D and S to find corporate-wide totals:

$$D + S = \begin{bmatrix} 12 & 18 \\ 3 & 0 \end{bmatrix} + \begin{bmatrix} 14 & 12 \\ 5 & 2 \end{bmatrix} = \begin{bmatrix} 26 & 30 \\ 8 & 2 \end{bmatrix}$$

We interpret the total to mean:

	Male	Female
Day shift	26	30
Night shift	8	2

If one-third of the force at the downtown location retires, the downtown staff would be reduced to $\frac{2}{3}D$ people. We can compute $\frac{2}{3}D$ by multiplying each entry by $\frac{2}{3}$.

$$\frac{2}{3}D = \frac{2}{3}\begin{bmatrix} 12 & 18 \\ 3 & 0 \end{bmatrix} = \begin{bmatrix} 8 & 12 \\ 2 & 0 \end{bmatrix}$$

After retirements, downtown staff would be

	Male	Female
Day shift	8	12
Night shift	2	0

These examples illustrate two calculations used in the algebra of matrices, which is the topic of this section.

Matrices

An $\boldsymbol{m \times n}$ ***matrix*** is a rectangular array of mn numbers arranged in m rows and n columns. We say that the matrix is of **size** (or **order**) $m \times n$.

Matrices are often denoted by letters such as A, B, and C. To denote the entries in an $m \times n$ matrix A, we use double-subscript notation: The entry in the first row, third column is a_{13}, and the entry in the ith row, jth column is a_{ij}. We can use any of the following notations to denote the $m \times n$ matrix A:

$$A, \quad [a_{ij}], \quad \underbrace{\begin{bmatrix} a_{11} & a_{12} & a_{13} & \cdots & a_{1n} \\ a_{21} & a_{22} & a_{23} & \cdots & a_{2n} \\ \vdots & \vdots & \vdots & \ddots & \vdots \\ a_{m1} & a_{m2} & a_{m3} & \cdots & a_{mn} \end{bmatrix}}_{n \text{ columns}} \Bigg\} \; m \text{ rows}$$

Two matrices are equal if they are the same size, with the same entries in corresponding positions.

Equality of Matrices

If $A = [a_{ij}]$ and $B = [b_{ij}]$ are both $m \times n$ matrices, then

$$A = B$$

provided that each entry a_{ij} in A is equal to the corresponding entry b_{ij} in B.

The following matrices are equal, because they are the same size and corresponding entries are equal.

$$\begin{bmatrix} \sqrt{9} & 0.5 \\ 1 & 4 \end{bmatrix} = \begin{bmatrix} 3 & \frac{1}{2} \\ 1 & 2^2 \end{bmatrix}$$

The following matrices are not equal, because they are not the same size.

$$\begin{bmatrix} 1 & 2 & 3 \\ 1 & 2 & 3 \end{bmatrix} \neq \begin{bmatrix} 1 & 2 & 3 \\ 1 & 2 & 3 \\ 1 & 2 & 3 \end{bmatrix}$$

The first matrix is 2×3, and the second is 3×3.

Multiplying a Matrix by a Constant

We can multiply a matrix by a constant by multiplying each of its entries by that constant. If

$$A = \begin{bmatrix} 1 & 2 \\ 3 & 4 \end{bmatrix}, \qquad \text{then} \quad 5A = 5\begin{bmatrix} 1 & 2 \\ 3 & 4 \end{bmatrix} = \begin{bmatrix} 5 & 10 \\ 15 & 20 \end{bmatrix}$$

If A is a matrix, the real number k in the product kA is called a **scalar.**

Multiplying a Matrix by a Scalar

If A and B are two $m \times n$ matrices and k is a scalar, then $kA = B$ where each entry b_{ij} in B is equal to k times the corresponding entry a_{ij} in A.

EXAMPLE 1 Let $\begin{bmatrix} 5 & y \\ 15 & z \end{bmatrix} = 5\begin{bmatrix} x & 3 \\ 3 & y \end{bmatrix}$. Find y and z.

Solution We simplify the right-hand side of the expression by multiplying each entry of the matrix by 5.

$$\begin{bmatrix} 5 & y \\ 15 & z \end{bmatrix} = \begin{bmatrix} 5x & 15 \\ 15 & 5y \end{bmatrix}$$

Because the matrices are equal, their corresponding entries are equal. So $y = 15$, and $z = 5y$. We conclude that $y = 15$ and $z = 75$.

Self Check Find x. ■

Adding and Subtracting Matrices

We can add matrices of the same size by adding the entries in corresponding positions.

Sum of Two Matrices

Let A and B be two $m \times n$ matrices. The sum, $A + B$, is the $m \times n$ matrix C found by adding the corresponding entries of matrices A and B:

$$A + B = C$$

where each entry c_{ij} in C is equal to the sum of a_{ij} in A and b_{ij} in B.

ACTIVE EXAMPLE 2 Add: $\begin{bmatrix} 2 & 1 & 3 \\ 1 & -1 & 0 \end{bmatrix} + \begin{bmatrix} 1 & -1 & 2 \\ -1 & 1 & 5 \end{bmatrix}$.

Solution Since each matrix is 2×3, we find their sum by adding their corresponding entries.

COLLEGE **Algebra** $f(x)$ **Now**™
Explore this Active Example and test yourself on the steps followed here at **http://1pass.thomson.com**. You can also find this example on the Interactive Video Skillbuilder CD-ROM.

$$\begin{bmatrix} 2 & 1 & 3 \\ 1 & -1 & 0 \end{bmatrix} + \begin{bmatrix} 1 & -1 & 2 \\ -1 & 1 & 5 \end{bmatrix} = \begin{bmatrix} 2+1 & 1-1 & 3+2 \\ 1-1 & -1+1 & 0+5 \end{bmatrix}$$
$$= \begin{bmatrix} 3 & 0 & 5 \\ 0 & 0 & 5 \end{bmatrix}$$

Self Check Add: $\begin{bmatrix} 3 & -5 \\ 2 & 0 \\ -6 & 5 \end{bmatrix} + \begin{bmatrix} -3 & 4 \\ -1 & -3 \\ 7 & 0 \end{bmatrix}$. ■

Comment Matrices that are not the same size cannot be added.

In arithmetic, 0 is the **additive identity,** because $a + 0 = 0 + a = a$ for any real number a. In matrix algebra, the matrix $\mathbf{0} = \begin{bmatrix} 0 & 0 \\ 0 & 0 \end{bmatrix}$ is called an additive identity, because $A + \mathbf{0} = \mathbf{0} + A = A$. For example,

$$\begin{bmatrix} 1 & 2 \\ 3 & 4 \end{bmatrix} + \begin{bmatrix} 0 & 0 \\ 0 & 0 \end{bmatrix} = \begin{bmatrix} 0 & 0 \\ 0 & 0 \end{bmatrix} + \begin{bmatrix} 1 & 2 \\ 3 & 4 \end{bmatrix} = \begin{bmatrix} 1 & 2 \\ 3 & 4 \end{bmatrix}$$

The Additive Identity Matrix

Let A be any $m \times n$ matrix. There is an $m \times n$ matrix $\mathbf{0}$, called the **zero matrix** or the **additive identity matrix,** for which

$$A + \mathbf{0} = \mathbf{0} + A = A$$

The matrix $\mathbf{0}$ consists of m rows and n columns of 0's.

Every matrix also has an additive inverse.

The Additive Inverse of a Matrix

Any $m \times n$ matrix A has an **additive inverse,** an $m \times n$ matrix $-A$ with the property that the sum of A and $-A$ is the zero matrix:

$$A + (-A) = (-A) + A = \mathbf{0}$$

The entries of $-A$ are the negatives of the corresponding entries of A:

$$-A = (-1)A$$

The additive inverse of $A = \begin{bmatrix} 1 & -3 & 2 \\ 0 & 1 & -5 \end{bmatrix}$ is the matrix

$$-A = (-1)A = \begin{bmatrix} -1 & 3 & -2 \\ 0 & -1 & 5 \end{bmatrix}$$

because their sum is the zero matrix:

$$\begin{aligned} A + (-A) &= \begin{bmatrix} 1 & -3 & 2 \\ 0 & 1 & -5 \end{bmatrix} + \begin{bmatrix} -1 & 3 & -2 \\ 0 & -1 & 5 \end{bmatrix} \\ &= \begin{bmatrix} 1 - 1 & -3 + 3 & 2 - 2 \\ 0 + 0 & 1 - 1 & -5 + 5 \end{bmatrix} \\ &= \begin{bmatrix} 0 & 0 & 0 \\ 0 & 0 & 0 \end{bmatrix} \end{aligned}$$

Subtraction of matrices is similar to the subtraction of real numbers.

Difference of Two Matrices

If A and B are $m \times n$ matrices, their difference, $A - B$, is the sum of A and the additive inverse of B:

$$A - B = A + (-B)$$

For example,

$$\begin{bmatrix} 3 & 7 \\ -4 & 0 \end{bmatrix} - \begin{bmatrix} -1 & 4 \\ -5 & 1 \end{bmatrix} = \begin{bmatrix} 3 & 7 \\ -4 & 0 \end{bmatrix} + \begin{bmatrix} 1 & -4 \\ 5 & -1 \end{bmatrix} = \begin{bmatrix} 4 & 3 \\ 1 & -1 \end{bmatrix}$$

Multiplying Matrices

We will consider how to find the product of two matrices by finding the product of a 2×3 matrix A and a 3×3 matrix B. The result will be matrix C.

$$AB = \begin{bmatrix} 1 & 2 & 3 \\ 4 & 5 & 6 \end{bmatrix} \begin{bmatrix} a & b & c \\ d & e & f \\ g & h & i \end{bmatrix} = C$$

Arthur Cayley
(1821–1895)
Cayley taught mathematics at Cambridge University. When he refused to take religious vows, he was fired and became a lawyer. After 14 years, he returned to mathematics and to Cambridge. Cayley was a major force in developing the theory of matrices.

Each entry of matrix C is the result of multiplying the entries in a row of A and the corresponding entries in a column of B and adding the results. For example, the first-row, third-column entry of matrix C is the sum of the products of corresponding entries of the first row of A and the third column of B:

$$\begin{bmatrix} 1 & 2 & 3 \\ 4 & 5 & 6 \end{bmatrix}\begin{bmatrix} a & b & c \\ d & e & f \\ g & h & i \end{bmatrix} = \begin{bmatrix} ? & ? & 1c + 2f + 3i \\ ? & ? & ? \end{bmatrix}$$

The second-row, second-column entry of matrix C is the sum of the products of the second row of A and the second column of B.

$$\begin{bmatrix} 1 & 2 & 3 \\ 4 & 5 & 6 \end{bmatrix}\begin{bmatrix} a & b & c \\ d & e & f \\ g & h & i \end{bmatrix} = \begin{bmatrix} ? & ? & 1c + 2f + 3i \\ ? & 4b + 5e + 6h & ? \end{bmatrix}$$

The other entries of the product are computed similarly.

For the product AB to exist, the number of columns of A must equal the number of rows of B. If the product exists, it will have as many rows as A and as many columns as B:

$$\underset{m \times n}{A} \cdot \underset{n \times p}{B} = \underset{m \times p}{C}$$

These must agree.

The product is of size $m \times p$.

Product of Two Matrices

Let A be an $m \times n$ matrix and B be an $n \times p$ matrix. The product, AB, is the $m \times p$ matrix C

$$AB = C$$

where each entry c_{ij} in C is the sum of the products of the corresponding entries in the ith row of A and the jth column of B, where $i = 1, 2, 3, \ldots, m$ and $j = 1, 2, 3, \ldots, p$.

ACTIVE EXAMPLE 3 Find $C = AB$ if $A = \begin{bmatrix} 1 & 2 & 4 \\ -2 & 1 & -1 \end{bmatrix}$ and $B = \begin{bmatrix} 1 & 5 \\ -2 & 4 \\ 1 & -3 \end{bmatrix}$.

Solution Because matrix A is 2×3 and B is 3×2, the product C is defined and is 2×2. To find entry c_{11} of C, we find the total of the products of the entries in the first row of A and the first column of B:

COLLEGE **Algebra f(x) Now**™
Go to **http://1pass.thomson.com** or your CD to practice this example.

$$c_{11} = 1 \cdot 1 + 2 \cdot (-2) + 4 \cdot (1) = 1$$

$$\begin{bmatrix} 1 & 2 & 4 \\ -2 & 1 & -1 \end{bmatrix}\begin{bmatrix} 1 & 5 \\ -2 & 4 \\ 1 & -3 \end{bmatrix} = \begin{bmatrix} 1 & ? \\ ? & ? \end{bmatrix}$$

To find entry c_{12}, we move across the first row of A and down the second column of B:

$$c_{12} = 1 \cdot 5 + 2 \cdot 4 + 4 \cdot (-3) = 1$$

$$\begin{bmatrix} 1 & 2 & 4 \\ -2 & 1 & -1 \end{bmatrix} \begin{bmatrix} 1 & 5 \\ -2 & 4 \\ 1 & -3 \end{bmatrix} = \begin{bmatrix} 1 & 1 \\ ? & ? \end{bmatrix}$$

To find entry c_{21}, we move across the second row of A and down the first column of B:

$$c_{21} = (-2) \cdot 1 + 1 \cdot (-2) + (-1) \cdot 1 = -5$$

$$\begin{bmatrix} 1 & 2 & 4 \\ -2 & 1 & -1 \end{bmatrix} \begin{bmatrix} 1 & 5 \\ -2 & 4 \\ 1 & -3 \end{bmatrix} = \begin{bmatrix} 1 & 1 \\ -5 & ? \end{bmatrix}$$

Finally, we find entry c_{22}:

$$c_{22} = (-2) \cdot 5 + 1 \cdot 4 + (-1) \cdot (-3) = -3$$

$$\begin{bmatrix} 1 & 2 & 4 \\ -2 & 1 & -1 \end{bmatrix} \begin{bmatrix} 1 & 5 \\ -2 & 4 \\ 1 & -3 \end{bmatrix} = \begin{bmatrix} 1 & 1 \\ -5 & -3 \end{bmatrix}$$

Self Check Find $D = EF$ if $E = \begin{bmatrix} 1 & -2 \\ 2 & 0 \end{bmatrix}$ and $F = \begin{bmatrix} 2 & 1 \\ 3 & 5 \end{bmatrix}$. ■

ACTIVE EXAMPLE 4 Find the product: $\begin{bmatrix} 1 & -1 & 2 \\ 1 & 3 & 0 \\ 0 & 1 & 1 \end{bmatrix} \begin{bmatrix} 2 & 1 \\ 1 & 3 \\ 0 & 1 \end{bmatrix}$.

Solution Because the matrices are 3×3 and 3×2, the product is a 3×2 matrix.

COLLEGE **Algebra** $f(x)$ **Now**™
Go to **http://1pass.thomson.com** or your CD to practice this example.

$$\begin{bmatrix} 1 & -1 & 2 \\ 1 & 3 & 0 \\ 0 & 1 & 1 \end{bmatrix} \begin{bmatrix} 2 & 1 \\ 1 & 3 \\ 0 & 1 \end{bmatrix} = \begin{bmatrix} 1 \cdot 2 + (-1) \cdot 1 + 2 \cdot 0 & 1 \cdot 1 + (-1) \cdot 3 + 2 \cdot 1 \\ 1 \cdot 2 + 3 \cdot 1 + 0 \cdot 0 & 1 \cdot 1 + 3 \cdot 3 + 0 \cdot 1 \\ 0 \cdot 2 + 1 \cdot 1 + 1 \cdot 0 & 0 \cdot 1 + 1 \cdot 3 + 1 \cdot 1 \end{bmatrix}$$

$$= \begin{bmatrix} 1 & 0 \\ 5 & 10 \\ 1 & 4 \end{bmatrix}$$

Self Check Find the product: $\begin{bmatrix} 1 & -2 \\ 3 & 1 \\ -4 & 0 \end{bmatrix} \begin{bmatrix} 4 & -3 \\ 1 & -1 \end{bmatrix}$. ■

EXAMPLE 5 Find each product: **a.** $\begin{bmatrix} 1 & 2 & 3 \end{bmatrix} \begin{bmatrix} 4 \\ 5 \\ 6 \end{bmatrix}$ and **b.** $\begin{bmatrix} 1 \\ 2 \\ 3 \end{bmatrix} \begin{bmatrix} 4 & 5 & 6 \end{bmatrix}$.

Solution **a.** Since the first matrix is 1×3 and the second matrix is 3×1, the product is a 1×1 matrix:

$$[1 \quad 2 \quad 3]\begin{bmatrix} 4 \\ 5 \\ 6 \end{bmatrix} = [1 \cdot 4 + 2 \cdot 5 + 3 \cdot 6] = [32]$$

b. Since the first matrix is 3×1 and the second matrix is 1×3, the product is a 3×3 matrix:

$$\begin{bmatrix} 1 \\ 2 \\ 3 \end{bmatrix}[4 \quad 5 \quad 6] = \begin{bmatrix} 1 \cdot 4 & 1 \cdot 5 & 1 \cdot 6 \\ 2 \cdot 4 & 2 \cdot 5 & 2 \cdot 6 \\ 3 \cdot 4 & 3 \cdot 5 & 3 \cdot 6 \end{bmatrix} = \begin{bmatrix} 4 & 5 & 6 \\ 8 & 10 & 12 \\ 12 & 15 & 18 \end{bmatrix}$$

Self Check Find each product: **a.** $[1 \quad 3 \quad 5]\begin{bmatrix} 1 \\ 0 \\ 1 \end{bmatrix}$ and **b.** $\begin{bmatrix} 1 \\ 2 \end{bmatrix}[3 \quad 4]$. ■

The multiplication of matrices is not commutative. To show this, we compute AB and BA, where $A = \begin{bmatrix} 1 & 1 \\ 0 & 0 \end{bmatrix}$ and $B = \begin{bmatrix} 0 & 1 \\ 0 & 1 \end{bmatrix}$.

$$AB = \begin{bmatrix} 1 & 1 \\ 0 & 0 \end{bmatrix}\begin{bmatrix} 0 & 1 \\ 0 & 1 \end{bmatrix} = \begin{bmatrix} 0 & 2 \\ 0 & 0 \end{bmatrix}$$

$$BA = \begin{bmatrix} 0 & 1 \\ 0 & 1 \end{bmatrix}\begin{bmatrix} 1 & 1 \\ 0 & 0 \end{bmatrix} = \begin{bmatrix} 0 & 0 \\ 0 & 0 \end{bmatrix}$$

Since the products are not equal, matrix multiplication is not commutative.

Accent on Technology

Several models of graphing calculators are able to do matrix arithmetic. For example, to find the sum and the product of the matrices

$$A = \begin{bmatrix} 2 & 3.7 \\ -2.1 & 3 \end{bmatrix} \quad \text{and} \quad B = \begin{bmatrix} 2 & -1 \\ 0 & 0.3 \end{bmatrix}$$

on a TI-83 Plus model, we press MATRIX, select EDIT, and enter the size and entries of the matrix A, as shown in Figure 6-6(a). Similarly, we enter matrix B as in Figure 6-6(b). Figure 6-6(c) displays the result $A + B$, as well as the product AB.

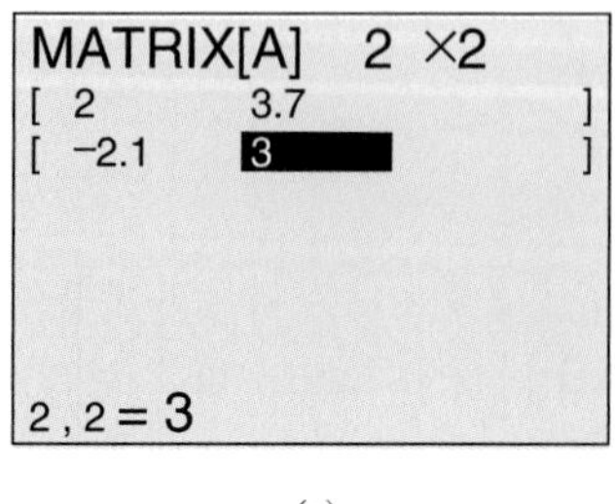

(a)

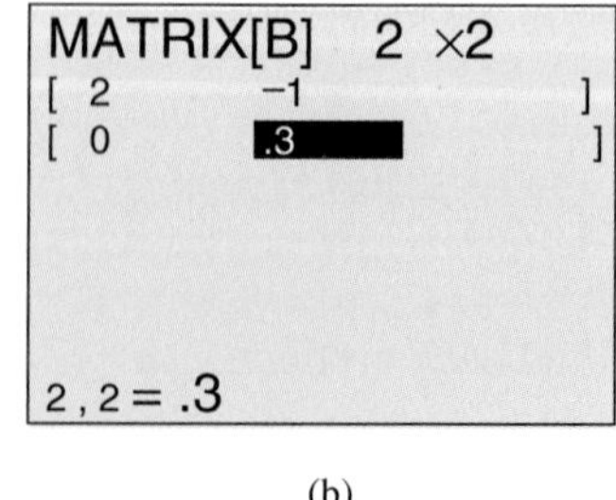

(b)

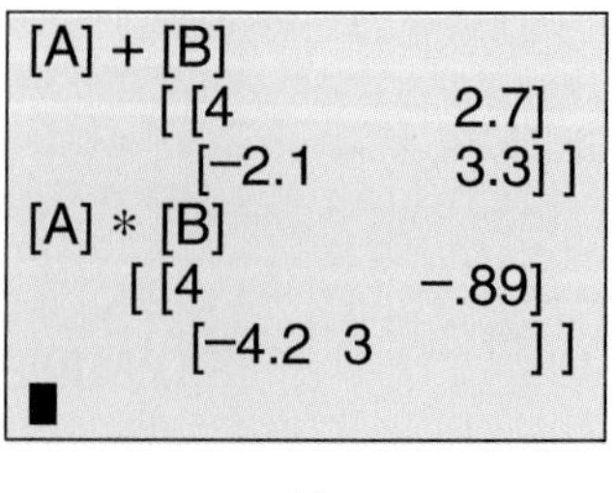

(c)

Figure 6-6

An Application of Matrices

EXAMPLE 6 Suppose that supplies must be purchased for the security officers discussed at the beginning of the section. The quantities and prices of each item required for each shift are as follows:

	Quantities			Unit Prices (in $)	
	Uniforms	**Badges**	**Whistles**	**Uniforms**	47
Day shift	17	13	19	**Badges**	7
Night shift	14	24	27	**Whistles**	5

Find the cost of supplies of each shift.

Solution We can write the quantities and prices in matrix form and multiply them to get a cost matrix.

$$C = QP$$

$$= \begin{bmatrix} 17 & 13 & 19 \\ 14 & 24 & 27 \end{bmatrix} \begin{bmatrix} 47 \\ 7 \\ 5 \end{bmatrix}$$

$$= \begin{bmatrix} 17 \cdot 47 + 13 \cdot 7 + 19 \cdot 5 \\ 14 \cdot 47 + 24 \cdot 7 + 27 \cdot 5 \end{bmatrix}$$

(17 uniforms)($47) + (13 badges)($7) + (19 whistles)($5). (14 uniforms)($47) + (24 badges)($7) + (27 whistles)($5).

$$= \begin{bmatrix} 985 \\ 961 \end{bmatrix}$$

It will cost $985 to buy supplies for the day shift and $961 to buy supplies for the night shift. ■

The Identity Matrix

The number 1 is called the **identity for multiplication,** because multiplying a number by 1 does not change the number: $a \cdot 1 = 1 \cdot a = a$. There is a **multiplicative identity matrix** with a similar property.

The Identity Matrix

Let A be an $n \times n$ matrix. There is an $\boldsymbol{n \times n}$ **identity matrix** $\boldsymbol{I}$ for which

$$AI = IA = A$$

It is the matrix I consisting of 1's on its diagonal and 0's elsewhere.

$$I = \begin{bmatrix} 1 & 0 & 0 & \cdots & 0 \\ 0 & 1 & 0 & \cdots & 0 \\ 0 & 0 & 1 & \cdots & 0 \\ \vdots & \vdots & \vdots & \ddots & \vdots \\ 0 & 0 & 0 & \cdots & 1 \end{bmatrix}$$

Comment An identity matrix is always a square matrix—a matrix that has the same number of rows and columns.

We illustrate the previous definition for the 3×3 identity matrix.

$$\begin{bmatrix} 1 & 0 & 0 \\ 0 & 1 & 0 \\ 0 & 0 & 1 \end{bmatrix}\begin{bmatrix} 1 & 2 & 3 \\ 4 & 5 & 6 \\ 7 & 8 & 9 \end{bmatrix} = \begin{bmatrix} 1\cdot 1 + 0\cdot 4 + 0\cdot 7 & 1\cdot 2 + 0\cdot 5 + 0\cdot 8 & 1\cdot 3 + 0\cdot 6 + 0\cdot 9 \\ 0\cdot 1 + 1\cdot 4 + 0\cdot 7 & 0\cdot 2 + 1\cdot 5 + 0\cdot 8 & 0\cdot 3 + 1\cdot 6 + 0\cdot 9 \\ 0\cdot 1 + 0\cdot 4 + 1\cdot 7 & 0\cdot 2 + 0\cdot 5 + 1\cdot 8 & 0\cdot 3 + 0\cdot 6 + 1\cdot 9 \end{bmatrix}$$

$$= \begin{bmatrix} 1 & 2 & 3 \\ 4 & 5 & 6 \\ 7 & 8 & 9 \end{bmatrix}$$

$$\begin{bmatrix} 1 & 2 & 3 \\ 4 & 5 & 6 \\ 7 & 8 & 9 \end{bmatrix}\begin{bmatrix} 1 & 0 & 0 \\ 0 & 1 & 0 \\ 0 & 0 & 1 \end{bmatrix} = \begin{bmatrix} 1 & 2 & 3 \\ 4 & 5 & 6 \\ 7 & 8 & 9 \end{bmatrix}$$

The following properties of real numbers carry over to matrices. In the exercises, you will be asked to illustrate many of these properties.

Properties of Matrices

Let A, B, and C be matrices and a and b be scalars.

The commutative property of addition:	$A + B = B + A$
The associative property of addition:	$A + (B + C) = (A + B) + C$
The associative properties of scalar multiplication:	$\begin{cases} a(bA) = (ab)A \\ a(AB) = (aA)B \end{cases}$
Distributive properties of scalar multiplication:	$(a + b)A = aA + bA$ $a(A + B) = aA + aB$
The associative property of multiplication:	$A(BC) = (AB)C$
The distributive properties of matrix multiplication:	$\begin{cases} A(B + C) = AB + AC \\ (A + B)C = AC + BC \end{cases}$

EXAMPLE 7 Verify the distributive property $A(B + C) = AB + AC$ using the matrices

$$A = \begin{bmatrix} 2 & 1 \\ 3 & 1 \end{bmatrix}, \quad B = \begin{bmatrix} 3 & -4 \\ 1 & -1 \end{bmatrix}, \quad \text{and} \quad C = \begin{bmatrix} -1 & 3 \\ 0 & 1 \end{bmatrix}$$

Solution We do the operations on the left-hand side and the right-hand side separately and compare the results.

$$\begin{bmatrix} 2 & 1 \\ 3 & 1 \end{bmatrix}\left(\begin{bmatrix} 3 & -4 \\ 1 & -1 \end{bmatrix} + \begin{bmatrix} -1 & 3 \\ 0 & 1 \end{bmatrix}\right) = \begin{bmatrix} 2 & 1 \\ 3 & 1 \end{bmatrix}\left(\begin{bmatrix} 2 & -1 \\ 1 & 0 \end{bmatrix}\right) \qquad \text{Do the addition within the parentheses.}$$

$$= \begin{bmatrix} 5 & -2 \\ 7 & -3 \end{bmatrix} \qquad \text{Multiply.}$$

$$\begin{bmatrix} 2 & 1 \\ 3 & 1 \end{bmatrix}\begin{bmatrix} 3 & -4 \\ 1 & -1 \end{bmatrix} + \begin{bmatrix} 2 & 1 \\ 3 & 1 \end{bmatrix}\begin{bmatrix} -1 & 3 \\ 0 & 1 \end{bmatrix} = \begin{bmatrix} 7 & -9 \\ 10 & -13 \end{bmatrix} + \begin{bmatrix} -2 & 7 \\ -3 & 10 \end{bmatrix} \qquad \text{Do the multiplications.}$$

$$= \begin{bmatrix} 5 & -2 \\ 7 & -3 \end{bmatrix} \qquad \text{Add.}$$

Because the left- and right-hand sides agree, this example illustrates the distributive property. ■

Self Check Answers

1. 1 **2.** $\begin{bmatrix} 0 & -1 \\ 1 & -3 \\ 1 & 5 \end{bmatrix}$ **3.** $\begin{bmatrix} -4 & -9 \\ 4 & 2 \end{bmatrix}$ **4.** $\begin{bmatrix} 2 & -1 \\ 13 & -10 \\ -16 & 12 \end{bmatrix}$ **5. a.** [6] **b.** $\begin{bmatrix} 3 & 4 \\ 6 & 8 \end{bmatrix}$

6.3 Exercises

VOCABULARY AND CONCEPTS *Fill in the blanks.*

1. In a matrix A, the symbol a_{ij} is the entry in row $\underline{i}$ and column $\underline{j}$.

2. For matrices A and B to be equal, they must be the same **size**, and corresponding entries must be **equal**.

3. To multiply a matrix by a scalar, we multiply **every element** by that scalar.

4. To find the sum of matrices A and B, we add the **corresponding** entries.

5. Among 2×2 matrices, $\begin{bmatrix} 0 & 0 \\ 0 & 0 \end{bmatrix}$ is the **additive identity** matrix.

6. Among 2×2 matrices, $\begin{bmatrix} 1 & 0 \\ 0 & 1 \end{bmatrix}$ is the **multiplicative identity** matrix.

PRACTICE *Find values of x and y, if any, that will make the matrices equal.*

7. $\begin{bmatrix} x & y \\ 1 & 3 \end{bmatrix} = \begin{bmatrix} 2 & 5 \\ 1 & 3 \end{bmatrix}$ $x = 2, y = 5$

8. $\begin{bmatrix} x & 5 \\ 3 & y \end{bmatrix} = \begin{bmatrix} 0 & 5 \\ 3 & 2 \end{bmatrix}$ $x = 0, y = 2$

9. $\begin{bmatrix} x + y & 3 + x \\ -2 & 5y \end{bmatrix} = \begin{bmatrix} 3 & 4 \\ -2 & 10 \end{bmatrix}$ $x = 1, y = 2$

10. $\begin{bmatrix} x + y & x - y \\ 2x & 3y \end{bmatrix} = \begin{bmatrix} -x & x - 2 \\ -y & 8 - y \end{bmatrix}$ $x = -1, y = 2$

Find 5A.

11. $A = \begin{bmatrix} 3 & -3 \\ 0 & -2 \end{bmatrix}$ $\begin{bmatrix} 15 & -15 \\ 0 & -10 \end{bmatrix}$

12. $A = \begin{bmatrix} 3 & \frac{3}{5} \\ 0 & -1 \end{bmatrix}$ $\begin{bmatrix} 15 & 3 \\ 0 & -5 \end{bmatrix}$

13. $A = \begin{bmatrix} 5 & 15 & -2 \\ -2 & -5 & 1 \end{bmatrix}$ $\begin{bmatrix} 25 & 75 & -10 \\ -10 & -25 & 5 \end{bmatrix}$

14. $A = \begin{bmatrix} -3 & 1 & 2 \\ -8 & -2 & -5 \end{bmatrix}$ $\begin{bmatrix} -15 & 5 & 10 \\ -40 & -10 & -25 \end{bmatrix}$

Find A + B.

15. $A = \begin{bmatrix} 2 & 1 & -1 \\ -3 & 2 & 5 \end{bmatrix}, B = \begin{bmatrix} -3 & 1 & 2 \\ -3 & -2 & -5 \end{bmatrix}$

$\begin{bmatrix} -1 & 2 & 1 \\ -6 & 0 & 0 \end{bmatrix}$

16. $A = \begin{bmatrix} 3 & 2 & 1 \\ -2 & 3 & -3 \\ -4 & -2 & -1 \end{bmatrix}, B = \begin{bmatrix} -2 & 6 & -2 \\ 5 & 7 & -1 \\ -4 & -6 & 7 \end{bmatrix}$

$\begin{bmatrix} 1 & 8 & -1 \\ 3 & 10 & -4 \\ -8 & -8 & 6 \end{bmatrix}$

Find A − B.

17. $A = \begin{bmatrix} -3 & 2 & -2 \\ -1 & 4 & -5 \end{bmatrix}, B = \begin{bmatrix} 3 & -3 & -2 \\ -2 & 5 & -5 \end{bmatrix}$

$\begin{bmatrix} -6 & 5 & 0 \\ 1 & -1 & 0 \end{bmatrix}$

18. $A = \begin{bmatrix} 2 & 2 & 0 \\ -2 & 8 & 1 \\ 3 & -3 & -8 \end{bmatrix}, B = \begin{bmatrix} -4 & 3 & 7 \\ -1 & 2 & 0 \\ 1 & 4 & -1 \end{bmatrix}$

$\begin{bmatrix} 6 & -1 & -7 \\ -1 & 6 & 1 \\ 2 & -7 & -7 \end{bmatrix}$

Find 5A + 3B.

19. $A = \begin{bmatrix} 3 & 1 & -2 \\ -4 & 3 & -2 \end{bmatrix}, B = \begin{bmatrix} 1 & -2 & 2 \\ -5 & -5 & 3 \end{bmatrix}$

$\begin{bmatrix} 18 & -1 & -4 \\ -35 & 0 & -1 \end{bmatrix}$

20. $A = \begin{bmatrix} 2 & -5 \\ -5 & 2 \end{bmatrix}, B = \begin{bmatrix} 5 & -2 \\ 2 & -5 \end{bmatrix}$

$\begin{bmatrix} 25 & -31 \\ -19 & -5 \end{bmatrix}$

Find the additive inverse of each matrix.

21. $A = \begin{bmatrix} 5 & -2 & 7 \\ -5 & 0 & 3 \\ -2 & 3 & -5 \end{bmatrix}$ $\begin{bmatrix} -5 & 2 & -7 \\ 5 & 0 & -3 \\ 2 & -3 & 5 \end{bmatrix}$

22. $A = \begin{bmatrix} 3 & -\frac{2}{3} & -5 & \frac{1}{2} \end{bmatrix}$ $\begin{bmatrix} -3 & \frac{2}{3} & 5 & -\frac{1}{2} \end{bmatrix}$

Find each product, if possible.

23. $\begin{bmatrix} 2 & 3 \\ 3 & -2 \end{bmatrix}\begin{bmatrix} 1 & 2 \\ 0 & -2 \end{bmatrix}$ $\begin{bmatrix} 2 & -2 \\ 3 & 10 \end{bmatrix}$

24. $\begin{bmatrix} -2 & 3 \\ 3 & -2 \end{bmatrix}\begin{bmatrix} 2 & 4 \\ -5 & 7 \end{bmatrix}$ $\begin{bmatrix} -19 & 13 \\ 16 & -2 \end{bmatrix}$

25. $\begin{bmatrix} -4 & -2 \\ 21 & 0 \end{bmatrix}\begin{bmatrix} -5 & 6 \\ 21 & -1 \end{bmatrix}$ $\begin{bmatrix} -22 & -22 \\ -105 & 126 \end{bmatrix}$

26. $\begin{bmatrix} -5 & 4 \\ 4 & -5 \end{bmatrix}\begin{bmatrix} 6 & -2 \\ 1 & 3 \end{bmatrix}$ $\begin{bmatrix} -26 & 22 \\ 19 & -23 \end{bmatrix}$

27. $\begin{bmatrix} 2 & 1 & 3 \\ 1 & 2 & -1 \\ 0 & 1 & 0 \end{bmatrix}\begin{bmatrix} 1 & 2 & 3 \\ 2 & -2 & 1 \\ 0 & 0 & 1 \end{bmatrix}$ $\begin{bmatrix} 4 & 2 & 10 \\ 5 & -2 & 4 \\ 2 & -2 & 1 \end{bmatrix}$

28. $\begin{bmatrix} 2 & 1 & 1 \\ 1 & 1 & 2 \\ 1 & -2 & -1 \end{bmatrix}\begin{bmatrix} 1 & 2 & 3 \\ 1 & 2 & -3 \\ -1 & -1 & 3 \end{bmatrix}$ $\begin{bmatrix} 2 & 5 & 6 \\ 0 & 2 & 6 \\ 0 & -1 & 6 \end{bmatrix}$

29. $\begin{bmatrix} 1 & -2 & -3 \\ 2 & 0 & 1 \end{bmatrix}\begin{bmatrix} 4 \\ -5 \\ -6 \end{bmatrix}$ $\begin{bmatrix} 32 \\ 2 \end{bmatrix}$

30. $\begin{bmatrix} 1 \\ -2 \\ -3 \end{bmatrix}\begin{bmatrix} 4 & -5 & -6 \end{bmatrix}$ $\begin{bmatrix} 4 & -5 & -6 \\ -8 & 10 & 12 \\ -12 & 15 & 18 \end{bmatrix}$

31. $\begin{bmatrix} 1 & 2 & 3 \end{bmatrix}\begin{bmatrix} 4 & 5 & 6 \\ 7 & 8 & 9 \end{bmatrix}$ not possible

32. $\begin{bmatrix} 2 & 3 & 4 \\ 1 & 2 & 3 \\ -2 & 2 & 2 \end{bmatrix}\begin{bmatrix} -1 \\ 2 \\ 3 \end{bmatrix}$ $\begin{bmatrix} 16 \\ 12 \\ 12 \end{bmatrix}$

33. $\begin{bmatrix} 2 & 5 \\ -3 & 1 \\ 0 & -2 \\ 1 & -5 \end{bmatrix}\begin{bmatrix} 3 & -2 & 4 \\ -2 & -3 & 1 \end{bmatrix}$ $\begin{bmatrix} -4 & -19 & 13 \\ -11 & 3 & -11 \\ 4 & 6 & -2 \\ 13 & 13 & -1 \end{bmatrix}$

34. $\begin{bmatrix} 1 & 4 & 0 & 0 \\ -4 & 1 & 0 & -2 \\ 0 & 0 & 1 & 0 \\ 0 & 2 & 0 & 1 \end{bmatrix}\begin{bmatrix} 1 \\ 2 \\ -2 \\ -1 \end{bmatrix}$ $\begin{bmatrix} 9 \\ 0 \\ -2 \\ 3 \end{bmatrix}$

Let $A = \begin{bmatrix} 2.3 & -1.7 & 3.1 \\ -2 & 3.5 & 1 \\ -8 & 4.7 & 9.1 \end{bmatrix}$, $B = \begin{bmatrix} -2.5 \\ 5.2 \\ -7 \end{bmatrix}$, ***and*** $C = \begin{bmatrix} -5.8 \\ 2.9 \\ 4.1 \end{bmatrix}$. ***Use a graphing calculator to find each result.***

35. AB $\begin{bmatrix} -36.29 \\ 16.2 \\ -19.26 \end{bmatrix}$

36. $B + C$ $\begin{bmatrix} -8.3 \\ 8.1 \\ -2.9 \end{bmatrix}$

37. A^2 $\begin{bmatrix} -16.11 & 4.71 & 33.64 \\ -19.6 & 20.35 & 6.4 \\ -100.6 & 72.82 & 62.71 \end{bmatrix}$

38. $AB + C$ $\begin{bmatrix} -42.09 \\ 19.1 \\ -15.16 \end{bmatrix}$

Let $A = \begin{bmatrix} 2 & 3 \\ 1 & 3 \end{bmatrix}$, $B = \begin{bmatrix} 2 & 1 & -5 \\ 1 & 1 & 2 \end{bmatrix}$, $C = \begin{bmatrix} -2 & -1 & 6 \\ 0 & -1 & -1 \end{bmatrix}$, $D = \begin{bmatrix} 1 & 2 \\ 1 & 3 \end{bmatrix}$, ***and*** $E = \begin{bmatrix} 1 & -2 \\ 2 & 3 \end{bmatrix}$. ***Verify each property by doing the operations on each side of the equation and comparing the results.***

39. Distributive property:

$$A(B + C) = AB + AC$$

40. Associative property of scalar multiplication:

$$5(6A) = (5 \cdot 6)A$$

41. Associative property of scalar multiplication:

$$3(AB) = (3A)B$$

42. Associative property of multiplication:

$$A(DE) = (AD)E$$

Let $A = \begin{bmatrix} 1 & 3 \\ 2 & 5 \end{bmatrix}$, $B = \begin{bmatrix} -1 \\ 3 \end{bmatrix}$, ***and*** $C = \begin{bmatrix} 3 & 2 \end{bmatrix}$. ***Perform the operations, if possible.***

43. $A - BC$ $\begin{bmatrix} 4 & 5 \\ -7 & -1 \end{bmatrix}$

44. $AB + B$ $\begin{bmatrix} 7 \\ 16 \end{bmatrix}$

45. $CB - AB$ not possible

46. CAB $[50]$

47. ABC $\begin{bmatrix} 24 & 16 \\ 39 & 26 \end{bmatrix}$

48. $CA + C$ $[10 \quad 21]$

49. A^2B $\begin{bmatrix} 47 \\ 81 \end{bmatrix}$

50. $(BC)^2$ $\begin{bmatrix} -9 & -6 \\ 27 & 18 \end{bmatrix}$

APPLICATIONS ***Use a graphing calculator to help solve each problem.***

51. Sporting goods Two suppliers manufactured footballs, baseballs, and basketballs in the quantities and costs given in the tables. Find matrices Q and C that represent the quantities and costs, find the product QC, and interpret the result.

	Quantities		
	Footballs	**Baseballs**	**Basketballs**
Supplier 1	200	300	100
Supplier 2	100	200	200

Unit costs	(in $)
Footballs	5
Baseballs	2
Basketballs	4

$QC = \begin{bmatrix} 2{,}000 \\ 1{,}700 \end{bmatrix}$ The cost to Supplier 1 is $2,000. The cost to Supplier 2 is $1,700.

52. Retailing Three ice cream stores sold cones, sundaes, and milkshakes in the quantities and prices given in the tables. Find matrices Q and P that represent the quantities and prices, find the product QP, and interpret the results.

	Quantities		
	Cones	**Sundaes**	**Shakes**
Store 1	75	75	32
Store 2	80	69	27
Store 3	62	40	30

Unit price	
Cones	1.50
Sundaes	1.75
Shakes	3.00

$QP = \begin{bmatrix} 339.75 \\ 321.75 \\ 253.00 \end{bmatrix}$ Store 1 made $339.75. Store 2 made $321.75. Store 3 made $253.00.

53. Beverage sales Beverages were sold to parents and children at a school basketball game in the quantities and prices given in the tables. Find matrices Q and P that represent the quantities and prices, find the product QP, and interpret the result.

$QP = \begin{bmatrix} 584.50 \\ 709.25 \\ 1{,}036.75 \end{bmatrix}$ adult males spent $584.50 adult females spent $709.25 children spent $1,036.75

	Quantities		
	Coffee	**Milk**	**Cola**
Adult males	217	23	319
Adult females	347	24	340
Children	3	97	750

Price	
Coffee	$ .75
Milk	$1.00
Cola	$1.25

54. Production costs Each of four factories manufactures three products in the daily quantities and unit costs given in the tables. Find a suitable matrix product to represent production costs.

	Production quantities		
Factory	**Product A**	**Product B**	**Product C**
Ashtabula	19	23	27
Boston	17	21	22
Chicago	21	18	20
Denver	27	25	22

	Unit Production Costs	
	Day shift	**Night shift**
Product A	$1.20	$1.35
Product B	$ .75	$ .85
Product C	$3.50	$3.70

$\begin{bmatrix} 134.55 & 145.10 \\ 113.15 & 122.20 \\ 108.70 & 117.65 \\ 128.15 & 139.10 \end{bmatrix}$

55. Connectivity matrix An entry of 1 in the following **connectivity matrix** A indicates that the person associated with that row knows the address of the person associated with that column. For example, the 1 in Bill's row and Al's column indicates that Bill can write to Al. The 0 in Bill's row and Carl's column indicates that Bill cannot write to Carl. However, Bill could ask Al to forward his letter to Carl. The matrix A^2 indicates the number of ways that one person can write to another with a letter that is forwarded exactly once. Find A^2.

	Al	**Bill**	**Carl**
Al	0	1	1
Bill	1	0	0
Carl	0	1	0

$= A$

$$\begin{bmatrix} 1 & 1 & 0 \\ 0 & 1 & 1 \\ 1 & 0 & 0 \end{bmatrix}$$

56. Communication routing Refer to Exercise 55. Find and interpret the matrix $A + A^2$. Can everyone receive a letter from everyone else with at most one forwarding? $\begin{bmatrix} 1 & 2 & 1 \\ 1 & 1 & 1 \\ 1 & 1 & 0 \end{bmatrix}$

DISCOVERY AND WRITING

57. Routing telephone calls A long-distance telephone carrier has established several direct microwave links among four cities. In the following connectivity matrix, entries a_{ij} and a_{ji} indicate the number of direct links between cities i and j. For example, cities 2 and 4 are not linked directly but could be connected through city 3. Find and interpret matrix A^2.

$$A = \begin{bmatrix} 0 & 2 & 1 & 0 \\ 2 & 0 & 1 & 0 \\ 1 & 1 & 0 & 2 \\ 0 & 0 & 2 & 0 \end{bmatrix}$$

$A^2 = \begin{bmatrix} 5 & 1 & 2 & 2 \\ 1 & 5 & 2 & 2 \\ 2 & 2 & 6 & 0 \\ 2 & 2 & 0 & 4 \end{bmatrix}$ **indicates the number of ways two cities can be linked with exactly one intermediate city to relay messages.**

58. Communication on one-way channels Three communication centers are linked as indicated in the illustration, with communication only in the direction of the arrows. Thus, location 1 can send a message directly to location 2 along two paths, but location 2 can return a message directly on only one path. Entry c_{ij} of matrix C indicates the number of channels from i to j. Find and interpret C^2.

$$C = \begin{bmatrix} 0 & 2 & 2 \\ 1 & 0 & 1 \\ 1 & 0 & 0 \end{bmatrix}$$

$C^2 = \begin{bmatrix} 4 & 0 & 2 \\ 1 & 2 & 2 \\ 0 & 2 & 2 \end{bmatrix}$**; entry i, j indicates the number of paths from i to j, with exactly one intermediate link in between.**

59. If A and B are 2×2 matrices, is $(AB)^2$ equal to A^2B^2? Support your answer.

No; if $A = \begin{bmatrix} 1 & 1 \\ 1 & 1 \end{bmatrix}$ and $B = \begin{bmatrix} 1 & 0 \\ 0 & 0 \end{bmatrix}$, then $(AB)^2 \neq A^2B^2$.

60. Let a, b, and c be real numbers. If $ab = ac$ and $a \neq 0$, then $b = c$. Find 2×2 matrices A, B, and C, where $A \neq 0$, to show that such a law does not hold for all matrices.

Let $A = \begin{bmatrix} 1 & 0 \\ 0 & 0 \end{bmatrix}$, $B = \begin{bmatrix} 0 & 1 \\ 1 & 1 \end{bmatrix}$, and $C = \begin{bmatrix} 0 & 1 \\ 1 & 2 \end{bmatrix}$. Then $AB = AC$, but $B \neq C$.

61. Another property of the real numbers is that if $ab = 0$, then either $a = 0$ or $b = 0$. To show that this property is not true for matrices, find two nonzero 2×2 matrices A and B, such that $AB = 0$.

Let $A = \begin{bmatrix} 1 & 2 \\ 1 & 2 \end{bmatrix}$ and $B = \begin{bmatrix} 2 & 2 \\ -1 & -1 \end{bmatrix}$. Neither is the zero matrix, yet $AB = 0$.

62. Find 2×2 matrices to show that $(A + B)(A - B) \neq A^2 - B^2$.

Let $A = \begin{bmatrix} 1 & 0 \\ 0 & 0 \end{bmatrix}$ and $B = \begin{bmatrix} 0 & 1 \\ 0 & 0 \end{bmatrix}$.

REVIEW ***Perform the operations and simplify.***

63. $(3x + 2)(2x - 3) - (2 - x)$ $6x^2 - 4x - 8$

64. $\dfrac{x^2 + 3x - 4}{2x + 5 - (x + 1)}$ $x - 1$

65. $\dfrac{1 + \frac{1}{x}}{1 - \frac{1}{x}}$ $\dfrac{x + 1}{x - 1}$

66. $\dfrac{1 - x^{-1}}{1 + x^{-1}}$ $\dfrac{x - 1}{x + 1}$

67. Solve the formula for a.

$$s = \frac{n(a + l)}{2} \qquad a = \frac{2s}{n} - l$$

68. Solve the formula for x_1.

$$y - y_1 = m(x - x_1) \qquad x_1 = x - \frac{y - y_1}{m}$$

6.4 Matrix Inversion

In this section, you will learn about

- Finding an Inverse by Row Operations
- Solving a System of Equations
- An Application of Matrix Inversion

Two real numbers are called **multiplicative inverses** if their product is the multiplicative identity 1. Some matrices have multiplicative inverses also.

Inverse of a Matrix

If A and B are $n \times n$ matrices, I is the $n \times n$ identity matrix, and

$$AB = BA = I$$

then A and B are called **multiplicative inverses.** Matrix A is the **inverse** of B, and B is the **inverse** of A.

It can be shown that if a matrix A has an inverse, it only has one inverse. The inverse of A is written as A^{-1}.

$$AA^{-1} = A^{-1}A = I$$

EXAMPLE 1 Show that A and B are inverses.

$$A = \begin{bmatrix} 1 & 1 & 0 \\ 4 & 3 & 0 \\ 2 & 1 & -1 \end{bmatrix} \qquad B = \begin{bmatrix} -3 & 1 & 0 \\ 4 & -1 & 0 \\ -2 & 1 & -1 \end{bmatrix}$$

Solution We must show that both AB and BA are equal to I.

$$AB = \begin{bmatrix} 1 & 1 & 0 \\ 4 & 3 & 0 \\ 2 & 1 & -1 \end{bmatrix}\begin{bmatrix} -3 & 1 & 0 \\ 4 & -1 & 0 \\ -2 & 1 & -1 \end{bmatrix}$$

$$= \begin{bmatrix} -3+4+0 & 1-1+0 & 0+0+0 \\ -12+12+0 & 4-3+0 & 0+0+0 \\ -6+4+2 & 2-1-1 & 0+0+1 \end{bmatrix} = \begin{bmatrix} 1 & 0 & 0 \\ 0 & 1 & 0 \\ 0 & 0 & 1 \end{bmatrix}$$

$$BA = \begin{bmatrix} -3 & 1 & 0 \\ 4 & -1 & 0 \\ -2 & 1 & -1 \end{bmatrix}\begin{bmatrix} 1 & 1 & 0 \\ 4 & 3 & 0 \\ 2 & 1 & -1 \end{bmatrix} = \begin{bmatrix} 1 & 0 & 0 \\ 0 & 1 & 0 \\ 0 & 0 & 1 \end{bmatrix}$$

Self Check Are C and D inverses? $C = \begin{bmatrix} 2 & 5 \\ 1 & 3 \end{bmatrix}$, $D = \begin{bmatrix} 3 & -5 \\ -1 & 2 \end{bmatrix}$ ■

■ Finding an Inverse by Row Operations

A matrix that has an inverse is called a **nonsingular matrix** and is said to be **invertible.** If it does not have an inverse, it is a **singular matrix** and is not invertible. The following method provides a way to find the inverse of an invertible matrix.

Finding a Matrix Inverse

> If a sequence of row operations performed on the $n \times n$ matrix A reduces A to the $n \times n$ identity matrix I, then those same row operations, performed in the same order on I, will transform I into A^{-1}.
>
> If no sequence of row operations will reduce A to I, then A is not invertible.

To use this method to find the inverse of an invertible matrix, we perform row operations on matrix A to change it to the identity matrix I. At the same time, we perform the same row operations on I. This changes I into A^{-1}.

A notation for this process uses an n-row-by-$2n$-column matrix, with matrix A as the left half and matrix I as the right half. If A is invertible, the proper row operations performed on $[A \mid I]$ will transform it into $[I \mid A^{-1}]$.

EXAMPLE 2 Find the inverse of matrix A if $A = \begin{bmatrix} 2 & -4 \\ 4 & -7 \end{bmatrix}$.

Solution We can set up a 2×4 matrix with A on the left and I on the right of the broken line:

$$[A \mid I] = \left[\begin{array}{cc:cc} 2 & -4 & 1 & 0 \\ 4 & -7 & 0 & 1 \end{array}\right]$$

We perform row operations on the entire matrix to transform the left half into I:

$$\left[\begin{array}{cc:cc} 2 & -4 & 1 & 0 \\ 4 & -7 & 0 & 1 \end{array}\right] \begin{array}{r} \left(\frac{1}{2}\right)R1 \rightarrow \\ (-2)R1 + R2 \rightarrow \end{array} \left[\begin{array}{cc:cc} 1 & -2 & \frac{1}{2} & 0 \\ 0 & 1 & -2 & 1 \end{array}\right]$$

$$(2)R2 + R1 \rightarrow \left[\begin{array}{cc:cc} 1 & 0 & -\frac{7}{2} & 2 \\ 0 & 1 & -2 & 1 \end{array}\right]$$

Since matrix A has been transformed into I, the right side of the previous matrix is A^{-1}. We can verify this by finding AA^{-1} and $A^{-1}A$ and showing that each product is I:

$$AA^{-1} = \begin{bmatrix} 2 & -4 \\ 4 & -7 \end{bmatrix} \begin{bmatrix} -\frac{7}{2} & 2 \\ -2 & 1 \end{bmatrix} = \begin{bmatrix} 1 & 0 \\ 0 & 1 \end{bmatrix}$$

$$A^{-1}A = \begin{bmatrix} -\frac{7}{2} & 2 \\ -2 & 1 \end{bmatrix} \begin{bmatrix} 2 & -4 \\ 4 & -7 \end{bmatrix} = \begin{bmatrix} 1 & 0 \\ 0 & 1 \end{bmatrix}$$

■

Self Check Find the inverse of $A = \begin{bmatrix} 3 & 2 \\ 4 & 3 \end{bmatrix}$.

ACTIVE EXAMPLE 3 Find the inverse of matrix A if $A = \begin{bmatrix} 1 & 1 & 0 \\ 1 & 2 & 1 \\ 2 & 3 & 2 \end{bmatrix}$.

Solution We set up a 3×6 matrix with A on the left and I on the right of the broken line.

$$[A \mid I] = \left[\begin{array}{ccc:ccc} 1 & 1 & 0 & 1 & 0 & 0 \\ 1 & 2 & 1 & 0 & 1 & 0 \\ 2 & 3 & 2 & 0 & 0 & 1 \end{array}\right]$$

COLLEGE **Algebra** $f(x)$ **Now**™
Explore this Active Example and test yourself on the steps followed here at **http://1pass.thomson.com**. You can also find this example on the Interactive Video Skillbuilder CD-ROM.

We then perform row operations on the matrix to transform the left half into I.

$$\left[\begin{array}{ccc|ccc} 1 & 1 & 0 & 1 & 0 & 0 \\ 1 & 2 & 1 & 0 & 1 & 0 \\ 2 & 3 & 2 & 0 & 0 & 1 \end{array}\right] \begin{matrix} \\ (-1)R1 + R2 \to \\ (-2)R1 + R3 \to \end{matrix} \left[\begin{array}{ccc|ccc} 1 & 1 & 0 & 1 & 0 & 0 \\ 0 & 1 & 1 & -1 & 1 & 0 \\ 0 & 1 & 2 & -2 & 0 & 1 \end{array}\right]$$

$$\begin{matrix} (-1)R2 + R1 \to \\ \\ (-1)R2 + R3 \to \end{matrix} \left[\begin{array}{ccc|ccc} 1 & 0 & -1 & 2 & -1 & 0 \\ 0 & 1 & 1 & -1 & 1 & 0 \\ 0 & 0 & 1 & -1 & -1 & 1 \end{array}\right]$$

$$\begin{matrix} R3 + R1 \to \\ (-1)R3 + R2 \to \\ \\ \end{matrix} \left[\begin{array}{ccc|ccc} 1 & 0 & 0 & 1 & -2 & 1 \\ 0 & 1 & 0 & 0 & 2 & -1 \\ 0 & 0 & 1 & -1 & -1 & 1 \end{array}\right]$$

Since the left half has been transformed into the identity matrix, the right half has become A^{-1}, and

$$A^{-1} = \begin{bmatrix} 1 & -2 & 1 \\ 0 & 2 & -1 \\ -1 & -1 & 1 \end{bmatrix}$$

Self Check Find the inverse of $B = \begin{bmatrix} 1 & 1 & 1 \\ 2 & 1 & 4 \\ 2 & 2 & 3 \end{bmatrix}$. ■

EXAMPLE 4 Find the inverse of $A = \begin{bmatrix} 1 & 2 \\ 2 & 4 \end{bmatrix}$, if possible.

Solution We form the 2×4 matrix

$$[A \mid I] = \left[\begin{array}{cc|cc} 1 & 2 & 1 & 0 \\ 2 & 4 & 0 & 1 \end{array}\right]$$

and begin to transform the left side of the matrix into the identity matrix I:

$$\left[\begin{array}{cc|cc} 1 & 2 & 1 & 0 \\ 2 & 4 & 0 & 1 \end{array}\right] \begin{matrix} \\ (-2)R1 + R2 \to \end{matrix} \left[\begin{array}{cc|cc} 1 & 2 & 1 & 0 \\ 0 & 0 & -2 & 1 \end{array}\right]$$

In obtaining the second-row, first-column position of A, the entire second row of A is zeroed out. Since we cannot transform matrix A to the identity, A is not invertible.

Self Check Find the inverse, if possible: $B = \begin{bmatrix} 1 & -2 \\ -3 & 6 \end{bmatrix}$. ■

Accent on Technology — FINDING THE INVERSE OF A MATRIX

We can use a graphing calculator to find the inverse of a matrix. For example, to find the inverse of

Continued

$$A = \begin{bmatrix} 2 & 2 & 3 \\ 1 & 2 & 3 \\ 1 & 0 & 1 \end{bmatrix}$$

on a TI-83 Plus, we press MATRIX, select EDIT, and enter matrix A as shown in Figure 6-7(a). We exit entry mode by pressing 2nd QUIT. Then we display A^{-1} by pressing MATRIX 1 x^{-1} ENTER. The display appears in Figure 6-7(b). To verify that $AA^{-1} = I$, we press CLEAR MATRIX 1 × MATRIX 1 x^{-1} ENTER to obtain the result shown in Figure 6-7(c).

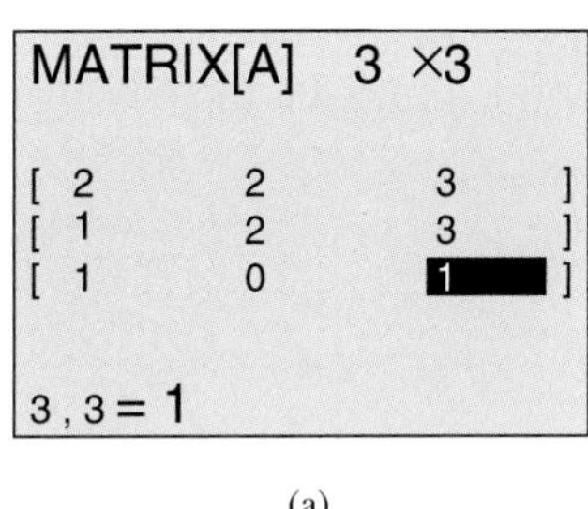

(a)

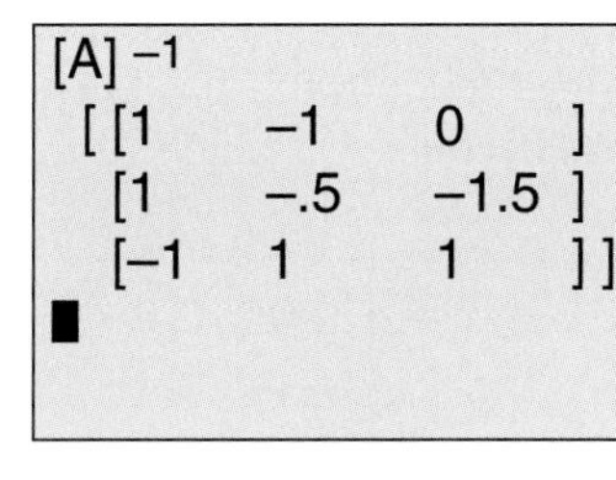

(b)

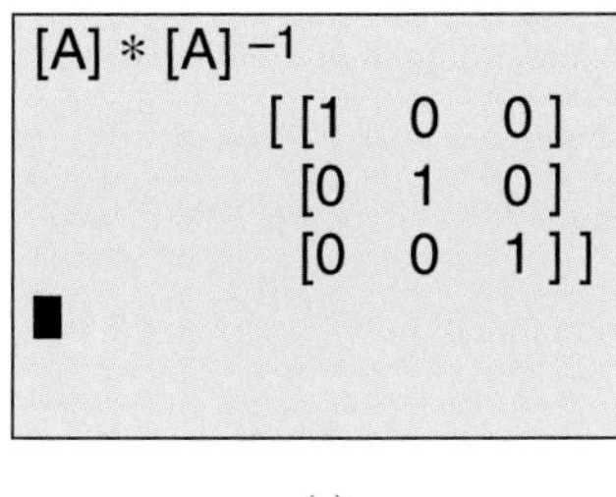

(c)

Figure 6-7

■ Solving a System of Equations

If we multiply the matrices on the left-hand side of the equation

$$\begin{bmatrix} 1 & 1 & 0 \\ 1 & 2 & 1 \\ 2 & 3 & 2 \end{bmatrix}\begin{bmatrix} x \\ y \\ z \end{bmatrix} = \begin{bmatrix} 20 \\ 30 \\ 55 \end{bmatrix}$$

and set the corresponding entries equal, we get the following system of equations:

$$\begin{cases} x + y = 20 \\ x + 2y + z = 30 \\ 2x + 3y + 2z = 55 \end{cases}$$

A system of equations can always be written as a matrix equation $AX = B$, where A is the coefficient matrix of the system, X is a column matrix of variables, and B is the column matrix of constants. If matrix A is invertible, the matrix equation $AX = B$ is easy to solve.

ACTIVE EXAMPLE 5 Solve: $\begin{bmatrix} 1 & 1 & 0 \\ 1 & 2 & 1 \\ 2 & 3 & 2 \end{bmatrix}\begin{bmatrix} x \\ y \\ z \end{bmatrix} = \begin{bmatrix} 20 \\ 30 \\ 55 \end{bmatrix}$.

Solution The 3×3 matrix on the left is the matrix whose inverse was found in Example 3. We multiply each side of the equation on the left by this inverse to obtain an equivalent system of equations.

COLLEGE
Algebra $f(x)$ **Now**™
Go to **http://1pass.thomson.com** or your CD to practice this example.

$$\begin{bmatrix} 1 & -2 & 1 \\ 0 & 2 & -1 \\ -1 & -1 & 1 \end{bmatrix}\begin{bmatrix} 1 & 1 & 0 \\ 1 & 2 & 1 \\ 2 & 3 & 2 \end{bmatrix}\begin{bmatrix} x \\ y \\ z \end{bmatrix} = \begin{bmatrix} 1 & -2 & 1 \\ 0 & 2 & -1 \\ -1 & -1 & 1 \end{bmatrix}\begin{bmatrix} 20 \\ 30 \\ 55 \end{bmatrix}$$

$$\begin{bmatrix} 1 & 0 & 0 \\ 0 & 1 & 0 \\ 0 & 0 & 1 \end{bmatrix}\begin{bmatrix} x \\ y \\ z \end{bmatrix} = \begin{bmatrix} 15 \\ 5 \\ 5 \end{bmatrix}$$ **Multiply the matrices. On the left, remember that $A^{-1}A = I$.**

$$\begin{bmatrix} x \\ y \\ z \end{bmatrix} = \begin{bmatrix} 15 \\ 5 \\ 5 \end{bmatrix}$$ $IX = X$.

The solution of this system can be read directly from the matrix on the right-hand side. Verify that the values $x = 15$, $y = 5$, $z = 5$ satisfy the original equations.

Self Check Solve: $\begin{bmatrix} 2 & -4 \\ 4 & -7 \end{bmatrix}\begin{bmatrix} x \\ y \end{bmatrix} = \begin{bmatrix} -4 \\ 2 \end{bmatrix}$. (The inverse was found in Example 2.) ■

Example 5 suggests the following result.

Solving Systems of Equations

If A is invertible, the solution of the matrix equation $AX = B$ is

$$X = A^{-1}B$$

Accent on Technology — SOLVING EQUATIONS

```
[A]-1 * [B]
          [[15 ]
           [5  ]
           [5  ]]
█
```

Figure 6-8

To use a graphing calculator to solve the system of Example 5, we enter the matrix $\begin{bmatrix} 1 & 1 & 0 \\ 1 & 2 & 1 \\ 2 & 3 & 2 \end{bmatrix}$ into the calculator as matrix A and the matrix $\begin{bmatrix} 20 \\ 30 \\ 55 \end{bmatrix}$ as matrix B and press 2nd QUIT. We then press MATRIX 1 x^{-1} × MATRIX 2 ENTER, to get Figure 6-8. The solution is $x = 15$, $y = 5$, and $z = 5$. Always check the results in the original equation.

This method is especially useful for finding solutions of several systems of equations that differ from each other only in the column matrix B. If the coefficient matrix A remains unchanged from one system of equations to the next, then A^{-1} needs to be found only once. The solution of each system is found by a single matrix multiplication, $A^{-1}B$.

An Application of Matrix Inversion

ACTIVE EXAMPLE 6 A company that manufactures medical equipment spends time on paperwork, manufacture, and testing for each of three versions of a circuit board. The times spent on each and the total time available are given in the following tables.

	Hours Required per Unit		
	Product A	**Product B**	**Product C**
Paperwork	1	1	0
Manufacture	1	2	1
Testing	2	3	2

Hours Available	
Paperwork	20
Manufacture	30
Testing	55

Solution We can let x, y, and z represent the number of units of products A, B, and C to be manufactured, respectively. We can then set up the following system of equations:

COLLEGE **Algebra** $f(x)$ **Now**™
Go to **http://1pass.thomson.com** or your CD to practice this example.

Paperwork: $x + y = 20$ — One hour is needed for every A, and one hour for every B.

Manufacture: $x + 2y + z = 30$ — One hour is needed for every A and C, and two hours for every B.

Testing: $2x + 3y + 2z = 55$ — Two hours are needed for every A and C, and three hours for every B.

In matrix form, the system becomes

$$\begin{bmatrix} 1 & 1 & 0 \\ 1 & 2 & 1 \\ 2 & 3 & 2 \end{bmatrix} \begin{bmatrix} x \\ y \\ z \end{bmatrix} = \begin{bmatrix} 20 \\ 30 \\ 55 \end{bmatrix}$$

We solved this equation in Example 5 to get $x = 15$, $y = 5$, and $z = 5$. To use all of the available time, the company should manufacture 15 units of product A and 5 units each of products B and C. ■

Self Check Answers

1. yes **2.** $\begin{bmatrix} 3 & -2 \\ -4 & 3 \end{bmatrix}$ **3.** $\begin{bmatrix} 5 & 1 & -3 \\ -2 & -1 & 2 \\ -2 & 0 & 1 \end{bmatrix}$ **4.** not possible **5.** $\begin{bmatrix} x \\ y \end{bmatrix} = \begin{bmatrix} 18 \\ 10 \end{bmatrix}$

6.4 Exercises

VOCABULARY AND CONCEPTS ***Fill in the blanks.***

1. Matrices A and B are multiplicative inverses if $\underline{AB = BA = I}$.
2. A nonsingular matrix $\underline{\text{is}}$ (is, is not) invertible.
3. If A is invertible, elementary row operations can change $[A \mid I]$ into $\underline{[I \mid A^{-1}]}$.
4. If A is invertible, the solution of $AX = B$ is $\underline{X = A^{-1}B}$.

PRACTICE *Find the inverse of each matrix, of possible.*

5. $\begin{bmatrix} 3 & -4 \\ -2 & 3 \end{bmatrix}$ $\begin{bmatrix} 3 & 4 \\ 2 & 3 \end{bmatrix}$

6. $\begin{bmatrix} 2 & 3 \\ 3 & 5 \end{bmatrix}$ $\begin{bmatrix} 5 & -3 \\ -3 & 2 \end{bmatrix}$

7. $\begin{bmatrix} 3 & 7 \\ 2 & 5 \end{bmatrix}$ $\begin{bmatrix} 5 & -7 \\ -2 & 3 \end{bmatrix}$

8. $\begin{bmatrix} 1 & -2 \\ 2 & -5 \end{bmatrix}$ $\begin{bmatrix} 5 & -2 \\ 2 & -1 \end{bmatrix}$

9. $\begin{bmatrix} 1 & 0 & 3 \\ -1 & 1 & 3 \\ -2 & 1 & 1 \end{bmatrix}$ $\begin{bmatrix} -2 & 3 & -3 \\ -5 & 7 & -6 \\ 1 & -1 & 1 \end{bmatrix}$

10. $\begin{bmatrix} 2 & 1 & -1 \\ 2 & 2 & -1 \\ -1 & -1 & 1 \end{bmatrix}$ $\begin{bmatrix} 1 & 0 & 1 \\ -1 & 1 & 0 \\ 0 & 1 & 2 \end{bmatrix}$

11. $\begin{bmatrix} 3 & 2 & 1 \\ 1 & 1 & -1 \\ 4 & 3 & 1 \end{bmatrix}$ $\begin{bmatrix} 4 & 1 & -3 \\ -5 & -1 & 4 \\ -1 & -1 & 1 \end{bmatrix}$

12. $\begin{bmatrix} -2 & 1 & -3 \\ 2 & 3 & 0 \\ 1 & 0 & 1 \end{bmatrix}$ $\begin{bmatrix} 3 & -1 & 9 \\ -2 & 1 & -6 \\ -3 & 1 & -8 \end{bmatrix}$

13. $\begin{bmatrix} 1 & 3 & 5 \\ 0 & 1 & 6 \\ 1 & 4 & 11 \end{bmatrix}$ no inverse

14. $\begin{bmatrix} 1 & 2 & 3 \\ 4 & 5 & 6 \\ 7 & 8 & 9 \end{bmatrix}$ no inverse

15. $\begin{bmatrix} 1 & 2 & 3 \\ 0 & 1 & 2 \\ 0 & 0 & 1 \end{bmatrix}$ $\begin{bmatrix} 1 & -2 & 1 \\ 0 & 1 & -2 \\ 0 & 0 & 1 \end{bmatrix}$

16. $\begin{bmatrix} 1 & 2 & 3 \\ 0 & 1 & 1 \\ 0 & -1 & 0 \end{bmatrix}$ $\begin{bmatrix} 1 & -3 & -1 \\ 0 & 0 & -1 \\ 0 & 1 & 1 \end{bmatrix}$

17. $\begin{bmatrix} 1 & 6 & 4 \\ 1 & -2 & -5 \\ 2 & 4 & -1 \end{bmatrix}$ no inverse

18. $\begin{bmatrix} 1 & 1 & 1 \\ 1 & 0 & -1 \\ 1 & 2 & 3 \end{bmatrix}$ no inverse

Use a graphing calculator to find the inverse of each matrix.

19. $\begin{bmatrix} 1 & 2 & 3 & 4 \\ 0 & 1 & 2 & 3 \\ 0 & 0 & 1 & 2 \\ 0 & 0 & 0 & 1 \end{bmatrix}$ $\begin{bmatrix} 1 & -2 & 1 & 0 \\ 0 & 1 & -2 & 1 \\ 0 & 0 & 1 & -2 \\ 0 & 0 & 0 & 1 \end{bmatrix}$

20. $\begin{bmatrix} 1 & 0 & 0 & 0 \\ 1 & 1 & 0 & 0 \\ 1 & 1 & 1 & 0 \\ 1 & 2 & 2 & 1 \end{bmatrix}$ $\begin{bmatrix} 1 & 0 & 0 & 0 \\ -1 & 1 & 0 & 0 \\ 0 & -1 & 1 & 0 \\ 1 & 0 & -2 & 1 \end{bmatrix}$

21. $\begin{bmatrix} 1 & 1 & -1 \\ 0.5 & 1 & 0.5 \\ 1 & 1 & -1.5 \end{bmatrix}$ $\begin{bmatrix} 8 & -2 & -6 \\ -5 & 2 & 4 \\ 2 & 0 & -2 \end{bmatrix}$

22. $\begin{bmatrix} -2 & -1 & 1 \\ 0.5 & -1.5 & -0.5 \\ 0 & 1 & 0.5 \end{bmatrix}$ $\begin{bmatrix} -.2 & 1.2 & 1.6 \\ -.2 & -.8 & -.4 \\ .4 & 1.6 & 2.8 \end{bmatrix}$

23. $\begin{bmatrix} 3 & 3 & -3 & 2 \\ 1 & -4 & 3 & -5 \\ 3 & 0 & -2 & -1 \\ -1 & 5 & -3 & 6 \end{bmatrix}$ $\begin{bmatrix} -2.5 & 5 & 3 & 5.5 \\ 5.5 & -8 & -6 & -9.5 \\ -1 & 3 & 1 & 3 \\ -5.5 & 9 & 6 & 10.5 \end{bmatrix}$

24. $\begin{bmatrix} 1 & 0 & 0 & 0 \\ 2 & 1 & 0 & 0 \\ 3 & 2 & 1 & 0 \\ 4 & 3 & 2 & 1 \end{bmatrix}$ $\begin{bmatrix} 1 & 0 & 0 & 0 \\ -2 & 1 & 0 & 0 \\ 1 & -2 & 1 & 0 \\ 0 & 1 & -2 & 1 \end{bmatrix}$

Use the method of Example 5 to solve each system of equations. Note that several systems have the same coefficient matrix.

25. $\begin{cases} 3x - 4y = 1 \\ -2x + 3y = 5 \end{cases}$ $x = 23, y = 17$

26. $\begin{cases} 3x - 4y = -1 \\ -2x + 3y = 3 \end{cases}$ $x = 9, y = 7$

27. $\begin{cases} 3x - 4y = 0 \\ -2x + 3y = 0 \end{cases}$ $x = 0, y = 0$

28. $\begin{cases} 3x - 4y = -3 \\ -2x + 3y = -2 \end{cases}$ $x = -17, y = -12$

29. $\begin{cases} 2x + y - z = 2 \\ 2x + 2y - z = 4 \\ -x - y + z = -1 \end{cases}$ $x = 1, y = 2, z = 2$

30. $\begin{cases} 2x + y - z = 3 \\ 2x + 2y - z = -1 \\ -x - y + z = 4 \end{cases}$ $x = 7, y = -4, z = 7$

31. $\begin{cases} -2x + y - 3z = 2 \\ 2x + 3y = -3 \\ x + z = 5 \end{cases}$ $x = 54, y = -37, z = -49$

32. $\begin{cases} -2x + y - 3z = 5 \\ 2x + 3y = 1 \\ x + z = -2 \end{cases}$ $x = -4, y = 3, z = 2$

Use a graphing calculator to solve each system of equations. Use the method of Example 5.

33. $\begin{cases} 5x + 3y = 13 \\ -7x + 5y = -9 \end{cases}$ $x = 2, y = 1$

34. $\begin{cases} 8x - 3y = 7 \\ -3x + 2y = 0 \end{cases}$ $x = 2, y = 3$

35. $\begin{cases} 5x + 2y + 3z = 12 \\ 2x + 5z = 7 \\ 3x + z = 4 \end{cases}$ $x = 1, y = 2, z = 1$

36. $\begin{cases} 3x + 2y - z = 0 \\ 5x - 2y = 5 \\ 3x + y + z = 6 \end{cases}$ $x = 1, y = 0, z = 3$

APPLICATIONS

37. Manufacturing and testing The numbers of hours required to manufacture and test each of two models of heart monitor are given in the first table, and the numbers of hours available each week for manufacturing and testing are given in the second table.

	Hours Required per Unit	
	Model A	**Model B**
Manufacturing	23	27
Testing	21	22

Hours Available	
Manufacturing	127
Testing	108

How many of each model can be manufactured each week? **2 of model A, 3 of model B**

38. Making clothes A clothing manufacturer makes coats, shirts, and slacks. The time required for cutting, sewing, and packaging each item is shown in the table. How many of each should be made to use all available labor hours?

	Coats	Shirts	Slacks
Cutting	20 min	15 min	10 min
Sewing	60 min	30 min	24 min
Packaging	5 min	12 min	6 min

Time available	
Cutting	115 hr
Sewing	280 hr
Packaging	65 hr

120 coats, 200 shirts, 150 slacks

39. Cryptography The letters of a message, called **plaintext,** are assigned values 1–26 (for a–z) and are written in groups of 2 as 2×1 matrices. To write the message in **cyphertext,** each 2×1 matrix is multiplied by a matrix A, where

$$A = \begin{bmatrix} 1 & 1 \\ 2 & 3 \end{bmatrix}$$

The cyphertext of one message is

$$AB = \begin{bmatrix} 17 \\ 43 \end{bmatrix}$$

Find the message. **Hi**

40. Cryptography The letters of a message, called **plaintext,** are assigned values 1–26 (for a–z) and are written in groups of 3 as 3×1 matrices. To write the message in **cyphertext,** each 3×1 matrix is multiplied by matrix A, where

$$A = \begin{bmatrix} 1 & 1 & 0 \\ 2 & 3 & 3 \\ 1 & 1 & 1 \end{bmatrix}$$

The cyphertext of one message is

$$AY = \begin{bmatrix} 30 \\ 122 \\ 49 \end{bmatrix}$$

Find the plaintext. **yes**

DISCOVERY AND WRITING *Use examples chosen from* 2×2 *matrices to support each answer.*

41. Does $(AB)^{-1} = A^{-1}B^{-1}$? **no**

42. Does $(AB)^{-1} = B^{-1}A^{-1}$? **yes**

Let $A = \begin{bmatrix} -1 & -1 \\ 1 & 1 \end{bmatrix}$.

43. Show that $A^2 = 0$.

44. Show that the inverse of $I - A$ is $I + A$.

Let $A = \begin{bmatrix} 3 & 0 & 0 \\ -2 & -1 & -2 \\ 3 & 6 & 3 \end{bmatrix}$ *and* $X = \begin{bmatrix} x \\ y \\ z \end{bmatrix}$. *Solve each equation. Each solution is called an* **eigenvector** *of the matrix A.*

45. $(A - 2I)X = \mathbf{0}$ $\quad X = \begin{bmatrix} 0 \\ 0 \\ 0 \end{bmatrix}$

46. $(A - 3I)X = \mathbf{0}$ $\quad X = \begin{bmatrix} 0 \\ 0 \\ 0 \end{bmatrix}$

47. Suppose that A, B, and C are $n \times n$ matrices and A is invertible. If $AB = AC$, prove that $B = C$.

48. Prove that $\begin{bmatrix} a & b \\ c & d \end{bmatrix}$ has an inverse if and only if $ad - bc \neq 0$. (*Hint:* Try to find the inverse and see what happens.)

49. Suppose that B is any matrix for which $B^2 = \mathbf{0}$. Show that $I - B$ is invertible by showing that the inverse of $I - B$ is $I + B$.

50. Suppose that C is any matrix for which $C^3 = \mathbf{0}$. Show that $I - C$ is invertible by showing that the inverse of $I - C$ is $I + C + C^2$.

REVIEW *Find the domain of each function.*

51. $y = \dfrac{3x - 5}{x^2 - 4}$ all reals except 2 and -2

52. $y = \dfrac{3x - 5}{x^2 + 4}$ all reals

53. $y = \dfrac{3x - 5}{\sqrt{x^2 + 4}}$ all reals

54. $y = \dfrac{3x - 5}{\sqrt{x^2 - 4}}$ $x > 2$ or $x < -2$

Find the range of each function.

55. $y = x^2$ $y \geq 0$

56. $y = x^3$ all reals

57. $y = \log x$ all reals

58. $y = 2^x$ $y > 0$

6.5 Determinants

In this section, you will learn about

- Evaluating Determinants of Higher-Order Matrices
- Properties of Determinants
- Using Determinants to Solve Systems of Equations
- Writing Equations of Lines
- Finding Areas of Triangles

The **determinant function** associates a number with any square matrix. The function is written as det(A) or as $|A|$.

Determinants

If a, b, c, and d are numbers, the determinant of $A = \begin{bmatrix} a & b \\ c & d \end{bmatrix}$ is

$$\det(A) = \begin{vmatrix} a & b \\ c & d \end{vmatrix} = ad - bc$$

Comment Do not confuse the notation $|A|$ with absolute value symbols.

EXAMPLE 1 **a.** $\begin{vmatrix} 1 & 2 \\ 3 & 4 \end{vmatrix} = 1 \cdot 4 - 2 \cdot 3$

$= 4 - 6$

$= -2$

b. $\begin{vmatrix} -2 & 3 \\ -\pi & \frac{1}{2} \end{vmatrix} = (-2)\left(\dfrac{1}{2}\right) - (3)(-\pi)$

$= -1 + 3\pi$

Self Check Evaluate: $\begin{vmatrix} 3 & -2 \\ 5 & -4 \end{vmatrix}$. ■

Evaluating Determinants of Higher-Order Matrices

To evaluate determinants of higher-order matrices, we must define the **minor** and the **cofactor** of an element in a matrix.

Minor and Cofactor of a Matrix

Let $A = [a_{ij}]$ be a square matrix of size (or order) $n \geq 2$.

1. The **minor** of a_{ij}, denoted as M_{ij}, is the determinant of the $n - 1 \times n - 1$ matrix formed by deleting the ith row and the jth column of A.
2. The **cofactor** of a_{ij}, denoted as C_{ij}, is $\begin{cases} M_{ij} \text{ when } i + j \text{ is even} \\ -M_{ij} \text{ when } i + j \text{ is odd} \end{cases}$.

EXAMPLE 2 In $A = \begin{bmatrix} 1 & 2 & 3 \\ 4 & 5 & 6 \\ 7 & 8 & 9 \end{bmatrix}$, find the minor and cofactor of **a.** a_{31} and **b.** a_{12}.

Solution **a.** The minor M_{31} is the minor of $a_{31} = 7$ appearing in row 3, column 1. It is found by deleting row 3 and column 1:

$$M_{31} = \begin{vmatrix} 1 & 2 & 3 \\ 4 & 5 & 6 \\ 7 & 8 & 9 \end{vmatrix} = \begin{vmatrix} 2 & 3 \\ 5 & 6 \end{vmatrix} = -3$$

Because $i + j$ is even $(3 + 1 = 4)$, the cofactor of the minor M_{31} is M_{31}:

$$C_{31} = M_{31} = \begin{vmatrix} 2 & 3 \\ 5 & 6 \end{vmatrix} = 2 \cdot 6 - 3 \cdot 5 = 12 - 15 = -3$$

b. The minor M_{12} is the minor of $a_{12} = 2$ appearing in row 1, column 2. It is found by deleting row 1 and column 2.

$$M_{12} = \begin{vmatrix} 1 & 2 & 3 \\ 4 & 5 & 6 \\ 7 & 8 & 9 \end{vmatrix} = \begin{vmatrix} 4 & 6 \\ 7 & 9 \end{vmatrix} = -6$$

Because $i + j$ is odd $(1 + 2 = 3)$, the cofactor of the minor M_{12} is $-M_{12}$:

$$C_{12} = -M_{12} = -\begin{vmatrix} 4 & 6 \\ 7 & 9 \end{vmatrix} = -(4 \cdot 9 - 6 \cdot 7) = -(36 - 42) = -(-6) = 6$$

Self Check Find the cofactor of a_{23}. ■

We are now ready to evaluate determinants of higher-order matrices.

Value of a Determinant

If A is a square matrix of order $n \geq 2$, $|A|$ is the sum of the products of the elements in any row (or column) and the cofactors of those elements.

In Example 3, we evaluate a 3×3 determinant by expanding the determinant in three ways: along two different rows and along a column. This method is called **expanding a determinant by minors.**

ACTIVE EXAMPLE 3 Evaluate $\begin{vmatrix} 1 & 2 & -3 \\ -1 & 0 & 1 \\ -2 & 2 & 1 \end{vmatrix}$ by expanding along the designated row or column.

COLLEGE **Algebra f(x) Now™**

Explore this Active Example and test yourself on the steps followed here at **http://1pass.thomson.com**. You can also find this example on the Interactive Video Skillbuilder CD-ROM.

Expanding on row 1:

$$\begin{vmatrix} \mathbf{1} & \mathbf{2} & \mathbf{-3} \\ -1 & 0 & 1 \\ -2 & 2 & 1 \end{vmatrix} = a_{11}C_{11} + a_{12}C_{12} + a_{13}C_{13}$$

$$= \mathbf{1}\begin{vmatrix} 0 & 1 \\ 2 & 1 \end{vmatrix} + \mathbf{2}\left[-\begin{vmatrix} -1 & 1 \\ -2 & 1 \end{vmatrix}\right] + (\mathbf{-3})\begin{vmatrix} -1 & 0 \\ -2 & 2 \end{vmatrix}$$

$$= 1(-2) + 2(-1) - 3(-2)$$

$$= -2 - 2 + 6$$

$$= 2$$

Expanding on row 3:

$$\begin{vmatrix} 1 & 2 & -3 \\ -1 & 0 & 1 \\ \mathbf{-2} & \mathbf{2} & \mathbf{1} \end{vmatrix} = a_{31}C_{31} + a_{32}C_{32} + a_{33}C_{33}$$

$$= \mathbf{-2}\begin{vmatrix} 2 & -3 \\ 0 & 1 \end{vmatrix} + \mathbf{2}\left[-\begin{vmatrix} 1 & -3 \\ -1 & 1 \end{vmatrix}\right] + \mathbf{1}\begin{vmatrix} 1 & 2 \\ -1 & 0 \end{vmatrix}$$

$$= -2(2) + 2(+2) + 1(2)$$

$$= -4 + 4 + 2$$

$$= 2$$

Gabriel Cramer
(1704–1752)

Cramer, a Swiss mathematician, made contributions to the fields of geometry, analysis, the study of algebraic curves, the history of mathematicians, and determinants. Although other mathematicians had previously worked with determinants, it was the work of Cramer that popularized them.

Expanding on column 2:

$$\begin{vmatrix} 1 & \mathbf{2} & -3 \\ -1 & \mathbf{0} & 1 \\ -2 & \mathbf{2} & 1 \end{vmatrix} = a_{12}C_{12} + a_{22}C_{22} + a_{32}C_{32}$$

$$= \mathbf{2}\left[-\begin{vmatrix} -1 & 1 \\ -2 & 1 \end{vmatrix}\right] + \mathbf{0}\begin{vmatrix} 1 & -3 \\ -2 & 1 \end{vmatrix} + \mathbf{2}\left[-\begin{vmatrix} 1 & -3 \\ -1 & 1 \end{vmatrix}\right]$$

$$= 2(-1) + 0(-5) + 2(+2)$$

$$= -2 + 4$$

$$= 2$$

In each case, the result is 2.

Self Check Evaluate $\begin{vmatrix} 1 & 0 & 1 \\ 2 & -1 & 0 \\ 3 & 1 & -1 \end{vmatrix}$ by expanding on its first row and second column. ■

EXAMPLE 4 Evaluate: $\begin{vmatrix} 0 & 0 & 2 & 0 \\ 1 & 2 & 17 & -3 \\ -1 & 0 & 28 & 1 \\ -2 & 2 & -37 & 1 \end{vmatrix}$.

Solution Because row 1 contains three 0's, we expand the determinant along row 1. Then only one cofactor needs to be evaluated.

$$\begin{vmatrix} 0 & 0 & 2 & 0 \\ 1 & 2 & 17 & -3 \\ -1 & 0 & 28 & 1 \\ -2 & 2 & -37 & 1 \end{vmatrix} = 0|?| - 0|?| + 2\begin{vmatrix} 1 & 2 & -3 \\ -1 & 0 & 1 \\ -2 & 2 & 1 \end{vmatrix} - 0|?|$$

$$= 2(2) \quad \text{See Example 3.}$$

$$= 4$$

Self Check Evaluate: $\begin{vmatrix} 0 & 1 & 0 \\ 2 & 11 & -2 \\ 1 & 13 & 1 \end{vmatrix}$. ■

Example 4 suggests the following theorem.

Zero Row or Column Theorem If every entry in a row or column of a square matrix A is 0, then $|A| = 0$.

Accent on Technology EVALUATING DETERMINANTS

Many graphing calculators are able to evaluate determinants. Simply enter a square matrix (say, A) and press det(and A. For example, to evaluate the determinant

$$\begin{vmatrix} 2 & 3 & 0.5 & 6 \\ 1 & 4 & -2 & -3 \\ 3 & 4 & -3 & -2 \\ -0.7 & 6 & 2 & 1 \end{vmatrix}$$

on a TI-83 Plus calculator, enter the matrix as shown in Figure 6-9(a). Note that you have to scroll left and right to see the entire screen. Then press 2nd QUIT MATRIX, select MATH, and press 1. Then press MATRIX, 1, and ENTER to obtain Figure 6-9(b). The value of the determinant is 99.3.

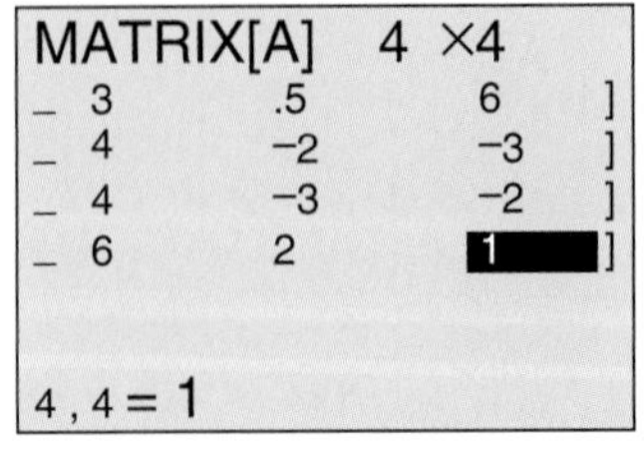

(a)

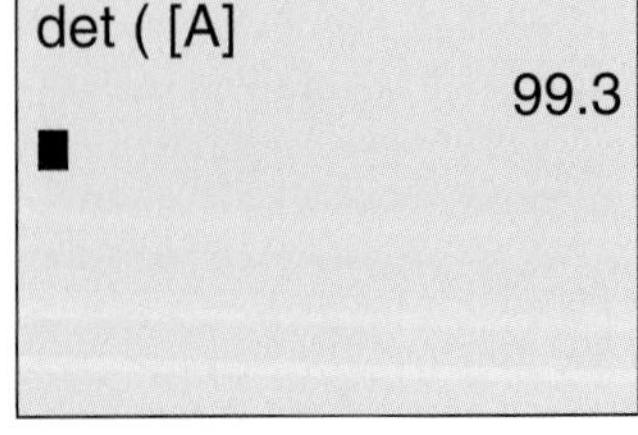

(b)

Figure 6-9

Properties of Determinants

We have seen that there are row operations for transforming matrices. There are similar row and column operations for transforming determinants.

Row and Column Operations for Determinants

Let A be a square matrix and k be a real number.

1. If a matrix B is obtained from matrix A by interchanging two rows (or columns), then $|B| = -|A|$.
2. If B is obtained from A by multiplying every element in a row (or column) of A by k, then $|B| = k|A|$.
3. If B is obtained from A by adding k times any row (or column) of A to another row (or column) of A, then $|B| = |A|$.

We illustrate each row operation by showing that it is true for the 2×2 determinant

$$|A| = \begin{vmatrix} a & b \\ c & d \end{vmatrix}$$

Interchanging two rows: Let B be obtained from A by interchanging its two rows. Then

$$|B| = \begin{vmatrix} c & d \\ a & b \end{vmatrix} = cb - da = -(ad - bc) = -|A|$$

Multiplying every element in a row by k: Let B be obtained from A by multiplying its second row by k. Then

$$|B| = \begin{vmatrix} a & b \\ kc & kd \end{vmatrix} = akd - bkc = k(ad - bc) = k|A|$$

Adding k times any row of A to another row of A: Let B be obtained from A by adding k times its first row to its second row. Then

$$|B| = \begin{vmatrix} a & b \\ ka + c & kb + d \end{vmatrix} = a(kb + d) - b(ka + c) = akb + ad - bka - bc = ad - bc = |A|$$

We will use several row and column operations in the next example.

ACTIVE EXAMPLE 5 Evaluate: $\begin{vmatrix} 10 & 20 & -10 & 20 \\ 2 & 1 & 1 & 1 \\ 1 & 2 & -3 & 2 \\ 2 & -1 & -1 & 1 \end{vmatrix}$.

Solution To get smaller numbers in the first row, we can use a type 2 row operation and multiply each entry in the first row by $\frac{1}{10}$. However, we must then multiply the resulting determinant by 10 to retain its original value.

COLLEGE **Algebra** $f(x)$ **Now**™
Go to **http://1pass.thomson.com** or your CD to practice this example.

(1) $$\begin{vmatrix} 10 & 20 & -10 & 20 \\ 2 & 1 & 1 & 1 \\ 1 & 2 & -3 & 2 \\ 2 & -1 & -1 & 1 \end{vmatrix} \overset{\left(\frac{1}{10}\right)R1 \to}{=} 10\begin{vmatrix} 1 & 2 & -1 & 2 \\ 2 & 1 & 1 & 1 \\ 1 & 2 & -3 & 2 \\ 2 & -1 & -1 & 1 \end{vmatrix}$$

To get three 0's in the first row of the second determinant in Equation 1, we perform a type 3 row operation and expand the new determinant along its first row.

$$10\begin{vmatrix} 1 & 2 & -1 & 2 \\ 2 & 1 & 1 & 1 \\ 1 & 2 & -3 & 2 \\ 2 & -1 & -1 & 1 \end{vmatrix} \begin{matrix} (-1)R3 + R1 \rightarrow \\ = 10 \\ \\ \end{matrix} \begin{vmatrix} 0 & 0 & 2 & 0 \\ 2 & 1 & 1 & 1 \\ 1 & 2 & -3 & 2 \\ 2 & -1 & -1 & 1 \end{vmatrix}$$

(2) $$= 10(2)\begin{vmatrix} 2 & 1 & 1 \\ 1 & 2 & 2 \\ 2 & -1 & 1 \end{vmatrix}$$

To introduce 0's into the 3×3 determinant in Equation 2, we perform a column operation on the 3×3 determinant and expand the result on its first column.

$$20\begin{vmatrix} 2 & 1 & 1 \\ 1 & 2 & 2 \\ 2 & -1 & 1 \end{vmatrix} \overset{(-2)C3 + C1}{=} 20\begin{vmatrix} 0 & 1 & 1 \\ -3 & 2 & 2 \\ 0 & -1 & 1 \end{vmatrix}$$

$$= 20\left[-(-3)\begin{vmatrix} 1 & 1 \\ -1 & 1 \end{vmatrix}\right]$$

$$= 20(3)[1 - (-1)]$$

$$= 60(2)$$

$$= 120$$

Self Check Evaluate: $\begin{vmatrix} 15 & 20 & 5 & 10 \\ 1 & 2 & 2 & 3 \\ 0 & 3 & -1 & 1 \\ 1 & 0 & 0 & -1 \end{vmatrix}$. ■

We will consider one final theorem.

Theorem If A is a square matrix with two identical rows (or columns), then $|A| = 0$.

Proof If the square matrix A has two identical rows (or columns), we can apply a type 3 row (or column) operation to zero out one of those rows (or columns). Since the matrix would then have an all-zero row (or column), its determinant would be 0, by the zero row or column theorem. ■

■ Using Determinants to Solve Systems of Equations

We can solve the system $\begin{cases} ax + by = e \\ cx + dy = f \end{cases}$ by multiplying the first equation by d, multiplying the second equation by $-b$, and adding to get

$$\begin{aligned} adx + bdy &= ed \\ -bcx - bdy &= -bf \\ \hline adx - bcx &= ed - bf \end{aligned}$$

If $ad \neq bc$, we can solve the resulting equation for x:

$$\begin{aligned} adx - bcx &= ed - bf \\ (ad - bc)x &= ed - bf && \text{Factor out } x. \\ x &= \frac{ed - bf}{ad - bc} && \text{Divide both sides by } ad - bc. \end{aligned} \tag{3}$$

If $ad \neq bc$, we can also solve the system for y to get

$$y = \frac{af - ec}{ad - bc} \tag{4}$$

We can write the values of x and y in Equations 3 and 4 using determinants.

$$x = \frac{\begin{vmatrix} e & b \\ f & d \end{vmatrix}}{\begin{vmatrix} a & b \\ c & d \end{vmatrix}} = \frac{ed - bf}{ad - bc} \qquad y = \frac{\begin{vmatrix} a & e \\ c & f \end{vmatrix}}{\begin{vmatrix} a & b \\ c & d \end{vmatrix}} = \frac{af - ec}{ad - bc}$$

If we compare these formulas with the original system,

$$\begin{cases} ax + by = e \\ cx + dy = f \end{cases}$$

we see that the denominators are the determinant of the coefficient matrix:

$$\text{Denominator determinant} = \begin{vmatrix} a & b \\ c & d \end{vmatrix}$$

To find the numerator determinant for x, we replace the a and c in the first column of the denominator determinant with the constants e and f.

To find the numerator determinant for y, we replace the b and d in the second column of the denominator determinant with the constants e and f.

$$x = \frac{\begin{vmatrix} e & b \\ f & d \end{vmatrix}}{\begin{vmatrix} a & b \\ c & d \end{vmatrix}} \qquad y = \frac{\begin{vmatrix} a & e \\ b & f \end{vmatrix}}{\begin{vmatrix} a & b \\ c & d \end{vmatrix}}$$

This method of using determinants to solve systems of equations is called **Cramer's rule.**

Cramer's Rule for Two Equations in Two Variables

If the system $\begin{cases} ax + by = e \\ cx + dy = f \end{cases}$ has a single solution, it is given by

$$x = \frac{D_x}{D} \quad \text{and} \quad y = \frac{D_y}{D}$$

where $D = \begin{vmatrix} a & b \\ c & d \end{vmatrix}$, $D_x = \begin{vmatrix} e & b \\ f & d \end{vmatrix}$, and $D_y = \begin{vmatrix} a & e \\ c & f \end{vmatrix}$.

If D, D_x, and D_y are all 0, the system is consistent, but the equations are dependent. If $D = 0$ and $D_x \neq 0$ or $D_y \neq 0$, the system is inconsistent.

EXAMPLE 6 Use Cramer's rule to solve $\begin{cases} 3x + 2y = 7 \\ -x + 5y = 9 \end{cases}$.

Solution

$$x = \frac{\begin{vmatrix} 7 & 2 \\ 9 & 5 \end{vmatrix}}{\begin{vmatrix} 3 & 2 \\ -1 & 5 \end{vmatrix}} = \frac{7 \cdot 5 - 2 \cdot 9}{3 \cdot 5 - 2(-1)} = \frac{35 - 18}{15 + 2} = \frac{17}{17} = 1$$

$$y = \frac{\begin{vmatrix} 3 & 7 \\ -1 & 9 \end{vmatrix}}{\begin{vmatrix} 3 & 2 \\ -1 & 5 \end{vmatrix}} = \frac{3 \cdot 9 - 7(-1)}{3 \cdot 5 - 2(-1)} = \frac{27 + 7}{15 + 2} = \frac{34}{17} = 2$$

Verify that the pair (1, 2) satisfies both of the equations in the system.

Self Check Solve: $\begin{cases} 2x + 5y = 9 \\ 3x + 7y = 13 \end{cases}$. ■

We can use Cramer's rule to solve systems of n equations in n variables where each equation has the form

$$a_1x_1 + a_2x_2 + \cdots + a_nx_n = c$$

To do so, we let D be the determinant of the coefficient matrix of the system and let D_{x_i} be the determinant formed by replacing the ith column of D by the column of constants from the right of the equal signs. If $D \neq 0$, Cramer's rule provides the following solution:

$$x_1 = \frac{D_{x_1}}{D}, \quad x_2 = \frac{D_{x_2}}{D}, \quad . \; . \; . \; , \quad x_n = \frac{D_{x_n}}{D}$$

ACTIVE EXAMPLE 7 Use Cramer's rule to solve the system $\begin{cases} 2x - y + 2z = 3 \\ x - y + z = 2 \\ x + y + 2z = 3 \end{cases}$.

Solution Each of the values x, y, and z is the quotient of two 3×3 determinants. The denominator of each quotient is the determinant consisting of the nine coefficients of the variables. The numerators for x, y, and z are modified copies of this denominator determinant. We substitute the column of constants for the coefficients of the variable for which we are solving.

COLLEGE **Algebra** *f(x)* **Now**™
Go to **http://1pass.thomson.com** or your CD to practice this example.

$$\begin{cases} 2x - y + 2z = 3 \\ x - y + z = 2 \\ x + y + 2z = 3 \end{cases}$$

$$x = \frac{\begin{vmatrix} 3 & -1 & 2 \\ 2 & -1 & 1 \\ 3 & 1 & 2 \end{vmatrix}}{\begin{vmatrix} 2 & -1 & 2 \\ 1 & -1 & 1 \\ 1 & 1 & 2 \end{vmatrix}} = \frac{3\begin{vmatrix} -1 & 1 \\ 1 & 2 \end{vmatrix} - (-1)\begin{vmatrix} 2 & 1 \\ 3 & 2 \end{vmatrix} + 2\begin{vmatrix} 2 & -1 \\ 3 & 1 \end{vmatrix}}{2\begin{vmatrix} -1 & 1 \\ 1 & 2 \end{vmatrix} - (-1)\begin{vmatrix} 1 & 1 \\ 1 & 2 \end{vmatrix} + 2\begin{vmatrix} 1 & -1 \\ 1 & 1 \end{vmatrix}} = \frac{2}{-1} = -2$$

$$y = \frac{\begin{vmatrix} 2 & 3 & 2 \\ 1 & 2 & 1 \\ 1 & 3 & 2 \end{vmatrix}}{\begin{vmatrix} 2 & -1 & 2 \\ 1 & -1 & 1 \\ 1 & 1 & 2 \end{vmatrix}} = \frac{2\begin{vmatrix} 2 & 1 \\ 3 & 2 \end{vmatrix} - 3\begin{vmatrix} 1 & 1 \\ 1 & 2 \end{vmatrix} + 2\begin{vmatrix} 1 & 2 \\ 1 & 3 \end{vmatrix}}{-1} = \frac{1}{-1} = -1$$

$$z = \frac{\begin{vmatrix} 2 & -1 & 3 \\ 1 & -1 & 2 \\ 1 & 1 & 3 \end{vmatrix}}{\begin{vmatrix} 2 & -1 & 2 \\ 1 & -1 & 1 \\ 1 & 1 & 2 \end{vmatrix}} = \frac{2\begin{vmatrix} -1 & 2 \\ 1 & 3 \end{vmatrix} - (-1)\begin{vmatrix} 1 & 2 \\ 1 & 3 \end{vmatrix} + 3\begin{vmatrix} 1 & -1 \\ 1 & 1 \end{vmatrix}}{-1} = \frac{-3}{-1} = 3$$

Verify that the triple $(-2, -1, 3)$ satisfies each equation in the system. ■

■ Writing Equations of Lines

If we are given the coordinates of two points in the xy-plane, we can use determinants to write the equation of the line passing through those points.

Two-Point Form of the Equation of a Line

The equation of the line passing through points $P(x_1, y_1)$ and $Q(x_2, y_2)$ is given by

$$\begin{vmatrix} x & y & 1 \\ x_1 & y_1 & 1 \\ x_2 & y_2 & 1 \end{vmatrix} = 0$$

EXAMPLE 8 Write the equation of the line passing through $P(-2, 3)$ and $Q(4, -5)$.

Solution We set up the equation $\begin{vmatrix} x & y & 1 \\ -2 & 3 & 1 \\ 4 & -5 & 1 \end{vmatrix} = 0$ and expand along the first row to get

$$[3(1) - 1(-5)]x - [-2(1) - 1(4)]y + [(-2)(-5) - 3(4)]1 = 0$$
$$8x + 6y - 2 = 0$$
$$4x + 3y = 1 \quad \text{Add 2 to both sides and divide both sides by 2.}$$

The equation of the line is $4x + 3y = 1$.

Self Check Find the equation of the line passing through (1, 3) and (3, 5). ■

Finding Areas of Triangles

Area of a Triangle

If points $P(x_1, y_1)$, $Q(x_2, y_2)$, and $R(x_3, y_3)$ are the vertices of a triangle, then the area of the triangle is given by

$$A = \pm\frac{1}{2}\begin{vmatrix} x_1 & y_1 & 1 \\ x_2 & y_2 & 1 \\ x_3 & y_3 & 1 \end{vmatrix} \quad \text{Pick either + or − to make the area positive.}$$

EXAMPLE 9 Find the area of the triangle shown in Figure 6-10.

Solution We set up the equation

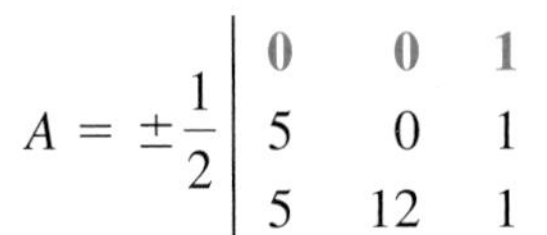

$$A = \pm\frac{1}{2}\begin{vmatrix} 0 & 0 & 1 \\ 5 & 0 & 1 \\ 5 & 12 & 1 \end{vmatrix}$$

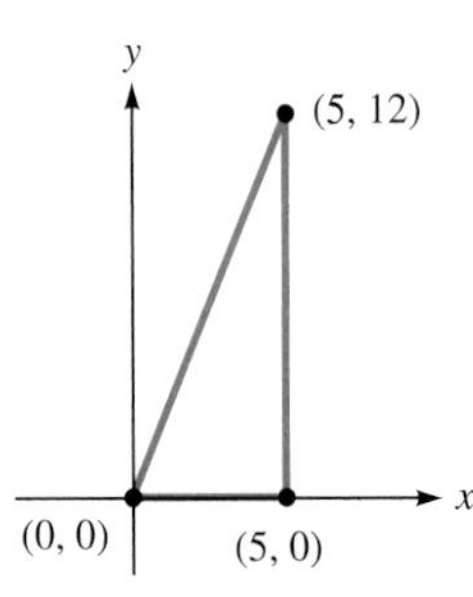

Figure 6-10

and expand the determinant along the first row to get

$$= \pm\frac{1}{2}\left[1\begin{vmatrix} 5 & 0 \\ 5 & 12 \end{vmatrix}\right]$$

$$= \pm\frac{1}{2}(60 - 0)$$

$$= 30$$

The area of the triangle is 30 square units.

Self Check Find the area of the triangle with vertices at (1, 2), (2, 3), and (−1, 4). ■

Self Check Answers

1. −2 **2.** 6 **3.** 6, 6 **4.** −4 **5.** 30 **6.** (2, 1) **8.** $x - y = -2$ **9.** 2

6.5 Exercises

VOCABULARY AND CONCEPTS *Fill in the blanks.*

1. The determinant of a square matrix A is written as $|A|$ or det A.
2. $\begin{vmatrix} a & b \\ c & d \end{vmatrix} = ad - bc$
3. If every entry in one row or one column of A is zero, then $|A| =$ 0.
4. If a matrix B is obtained from matrix A by adding one row to another, then $|B| = |A|$.
5. If two columns of A are identical, then $|A| =$ 0.
6. In Cramer's rule, the denominator is the determinant of the coefficient matrix.

PRACTICE *Evaluate each determinant.*

7. $\begin{vmatrix} 2 & 1 \\ -2 & 3 \end{vmatrix}$ 8
8. $\begin{vmatrix} -3 & -6 \\ 2 & -5 \end{vmatrix}$ 27
9. $\begin{vmatrix} 2 & -3 \\ -3 & 5 \end{vmatrix}$ 1
10. $\begin{vmatrix} 5 & 8 \\ -6 & -2 \end{vmatrix}$ 38

In Exercises 11–18, $A = \begin{vmatrix} 1 & -2 & 3 \\ 4 & 5 & -6 \\ -7 & 8 & 9 \end{vmatrix}$. *Find each minor or cofactor.*

11. M_{21} $\begin{vmatrix} -2 & 3 \\ 8 & 9 \end{vmatrix}$
12. M_{13} $\begin{vmatrix} 4 & 5 \\ -7 & 8 \end{vmatrix}$
13. M_{33} $\begin{vmatrix} 1 & -2 \\ 4 & 5 \end{vmatrix}$
14. M_{32} $\begin{vmatrix} 1 & 3 \\ 4 & -6 \end{vmatrix}$

15. C_{21} $-\begin{vmatrix} -2 & 3 \\ 8 & 9 \end{vmatrix}$

16. C_{13} $\begin{vmatrix} 4 & 5 \\ -7 & 8 \end{vmatrix}$

17. C_{33} $\begin{vmatrix} 1 & -2 \\ 4 & 5 \end{vmatrix}$

18. C_{32} $-\begin{vmatrix} 1 & 3 \\ 4 & -6 \end{vmatrix}$

Evaluate each determinant.

19. $\begin{vmatrix} 2 & -3 & 5 \\ -2 & 1 & 3 \\ 1 & 3 & -2 \end{vmatrix}$ −54

20. $\begin{vmatrix} 1 & 3 & 1 \\ -2 & 5 & 3 \\ 3 & -2 & -2 \end{vmatrix}$ 0

21. $\begin{vmatrix} 1 & -1 & 2 \\ 2 & 1 & 3 \\ 1 & 1 & -1 \end{vmatrix}$ −7

22. $\begin{vmatrix} 1 & 3 & 1 \\ 2 & 1 & -1 \\ 2 & -1 & 1 \end{vmatrix}$ −16

23. $\begin{vmatrix} 2 & 1 & -1 \\ 1 & 3 & 5 \\ 2 & -5 & 3 \end{vmatrix}$ 86

24. $\begin{vmatrix} 3 & 1 & -2 \\ -3 & 2 & 1 \\ 1 & 3 & 0 \end{vmatrix}$ 14

25. $\begin{vmatrix} 0 & 1 & -3 \\ -3 & 5 & 2 \\ 2 & -5 & 3 \end{vmatrix}$ −2

26. $\begin{vmatrix} 1 & -7 & -2 \\ -2 & 0 & 3 \\ -1 & 7 & 1 \end{vmatrix}$ 14

27. $\begin{vmatrix} 0 & 0 & 1 & 0 \\ -2 & 1 & 0 & 1 \\ 1 & 0 & 1 & 2 \\ 2 & 0 & 1 & 2 \end{vmatrix}$ 2

28. $\begin{vmatrix} 1 & 0 & -2 & 1 \\ 0 & 1 & 0 & 1 \\ 0 & 3 & -1 & 2 \\ 0 & -1 & 0 & 1 \end{vmatrix}$ −2

29. $\begin{vmatrix} 1 & 2 & 1 & 3 \\ -2 & 1 & -3 & 1 \\ -1 & 0 & 1 & -2 \\ 2 & -1 & -1 & 3 \end{vmatrix}$ 12

30. $\begin{vmatrix} -1 & 3 & -2 & 5 \\ 2 & 1 & 0 & 1 \\ 1 & 3 & -2 & 5 \\ 2 & -1 & 0 & -1 \end{vmatrix}$ 0

Determine whether each statement is true. Do not evaluate the determinants.

31. $\begin{vmatrix} 1 & 3 & -4 \\ -2 & 1 & 3 \\ 1 & 3 & 2 \end{vmatrix} = -\begin{vmatrix} -2 & 1 & 3 \\ 1 & 3 & -4 \\ 1 & 3 & 2 \end{vmatrix}$ true

32. $\begin{vmatrix} 4 & 6 & 8 \\ 10 & 5 & 15 \\ 20 & 5 & 10 \end{vmatrix} = \begin{vmatrix} 2 & 3 & 4 \\ 10 & 5 & 15 \\ 20 & 5 & 10 \end{vmatrix}$ false

33. $\begin{vmatrix} -2 & -3 & -4 \\ 5 & -1 & 2 \\ 1 & 2 & 3 \end{vmatrix} = -\begin{vmatrix} 2 & 3 & 4 \\ -5 & 1 & -2 \\ 1 & 2 & 3 \end{vmatrix}$ false

34. $\begin{vmatrix} 1 & 2 & 3 \\ 4 & 5 & 6 \\ 7 & 8 & 9 \end{vmatrix} = \begin{vmatrix} 5 & 7 & 9 \\ 4 & 5 & 6 \\ 7 & 8 & 9 \end{vmatrix}$ true

If $\begin{vmatrix} a & b & c \\ d & e & f \\ g & h & i \end{vmatrix}$ ***= 3, find the value of each determinant.***

35. $\begin{vmatrix} d & e & f \\ a & b & c \\ -g & -h & -i \end{vmatrix}$ 3

36. $\begin{vmatrix} 5a & 5b & 5c \\ -d & -e & -f \\ 3g & 3h & 3i \end{vmatrix}$ −45

37. $\begin{vmatrix} a+g & b+h & c+i \\ d & e & f \\ g & h & i \end{vmatrix}$ 3

38. $\begin{vmatrix} g & h & i \\ a & b & c \\ d & e & f \end{vmatrix}$ 3

Use Cramer's rule to find the solution of each system, if possible.

39. $\begin{cases} 3x + 2y = 7 \\ 2x - 3y = -4 \end{cases}$ (1, 2)

40. $\begin{cases} x - 5y = -6 \\ 3x + 2y = -1 \end{cases}$ (−1, 1)

41. $\begin{cases} x - y = 3 \\ 3x - 7y = 9 \end{cases}$ (3, 0)

42. $\begin{cases} 2x - y = -6 \\ x + y = 0 \end{cases}$ (−2, 2)

43. $\begin{cases} x + 2y + z = 2 \\ x - y + z = 2 \\ x + y + 3z = 4 \end{cases}$ (1, 0, 1)

44. $\begin{cases} x + 2y - z = -1 \\ 2x + y - z = 1 \\ x - 3y - 5z = 17 \end{cases}$ $\left(\frac{4}{17}, -\frac{30}{17}, -\frac{39}{17}\right)$

45. $\begin{cases} 2x - y + z = 5 \\ 3x - 3y + 2z = 10 \\ x + 3y + z = 0 \end{cases}$ (1, −1, 2)

46. $\begin{cases} x - y - z = 2 \\ x + y + z = 2 \\ -x - y + z = -4 \end{cases}$ (2, 1, −1)

47. $\begin{cases} \frac{x}{2} + \frac{y}{3} + \frac{z}{2} = 11 \\ \frac{x}{3} + y - \frac{z}{6} = 6 \\ \frac{x}{2} + \frac{y}{6} + z = 16 \end{cases}$ (6, 6, 12)

48. $\begin{cases} \frac{x}{2} + \frac{y}{5} + \frac{z}{3} = 17 \\ \frac{x}{5} + \frac{y}{2} + \frac{z}{5} = 32 \\ x + \frac{y}{3} + \frac{z}{2} = 30 \end{cases}$ (10, 60, 0)

49. $\begin{cases} 2p - q + 3r - s = 0 \\ p + q - s = -1 \\ 3p - r = 2 \\ p - 2q + 3s = 7 \end{cases}$ $\left(\frac{5}{6}, \frac{2}{3}, \frac{1}{2}, \frac{5}{2}\right)$

50. $\begin{cases} a + b + c + d = 8 \\ a + b + c + 2d = 7 \\ a + b + 2c + 3d = 3 \\ a + 2b + 3c + 4d = 4 \end{cases}$

(7, 5, −3, −1)

Use determinants to write the equation of the line that passes through the given points.

51. $P(0, 0)$, $Q(4, 6)$ **$3x - 2y = 0$**

52. $P(2, 3)$, $Q(6, 8)$ **$5x - 4y = -2$**

53. $P(-2, 3)$, $Q(5, -3)$ **$6x + 7y = 9$**

54. $P(1, -2)$, $Q(-4, 3)$ **$x + y = -1$**

Use determinants to find the area of each triangle with vertices at the given points.

55. $P(0, 0)$, $Q(12, 0)$, $R(12, 5)$ **30 sq. units**

56. $P(0, 0)$, $Q(0, 5)$, $R(12, 5)$ **30 sq. units**

57. $P(2, 3)$, $Q(10, 8)$, $R(0, 20)$ **73 sq. units**

58. $P(1, 1)$, $Q(6, 6)$, $R(2, 10)$ **20 sq. units**

In Exercises 59–61, illustrate each column operation by showing that it is true for the determinant $\begin{vmatrix} a & b \\ c & d \end{vmatrix}$.

59. Interchanging two columns

60. Multiplying each element in a column by k

61. Adding k times any column to another column

62. Use the method of addition to solve $\begin{cases} ax + by = e \\ cx + dy = f \end{cases}$ for y, and thereby show that $y = \dfrac{af - ec}{ad - bc}$.

Expand the determinants and solve for x.

63. $\begin{vmatrix} 3 & x \\ 1 & 2 \end{vmatrix} = \begin{vmatrix} 2 & -1 \\ x & -5 \end{vmatrix}$ **8**

64. $\begin{vmatrix} 4 & x^2 \\ 1 & -1 \end{vmatrix} = \begin{vmatrix} x & 4 \\ 2 & 3 \end{vmatrix}$ **−4, 1**

65. $\begin{vmatrix} 3 & x & 1 \\ x & 0 & -2 \\ 4 & 0 & 1 \end{vmatrix} = \begin{vmatrix} 2 & x \\ x & 4 \end{vmatrix}$ **−1**

66. $\begin{vmatrix} x & -1 & 2 \\ -2 & x & 3 \\ 4 & -3 & -1 \end{vmatrix} = \begin{vmatrix} 2 & 2 \\ 5 & x \end{vmatrix}$ **−4, 3**

APPLICATIONS

67. Investing A student wants to average a 6.6% return by investing \$20,000 in the three stocks listed in the table. Because HiTech is a high-risk investment, he wants to invest three times as much in SaveTel and OilCo combined as he invests in HiTech. How much should he invest in each stock?
\$5,000 in HiTech, \$8,000 in SaveTel, \$7,000 in OilCo

Stock	Rate of return
HiTech	10%
SaveTel	5%
OilCo	6%

68. Ice skating The illustration shows three circles traced out by a figure skater during her performance. If the centers of the circles are the given distances apart, find the radius of each circle.
3 yd, 7 yd, 11 yd

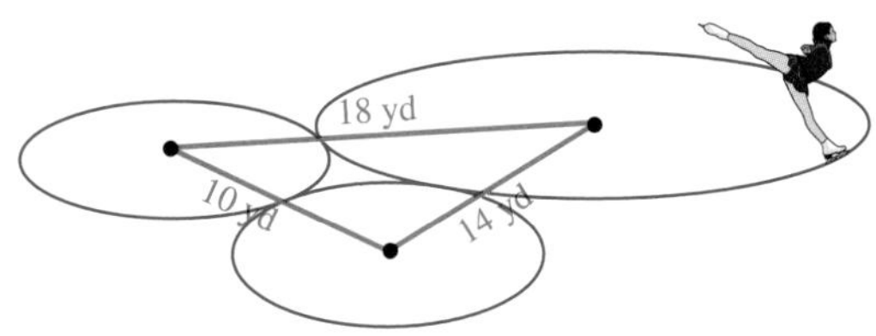

DISCOVERY AND WRITING ***Evaluate each determinant. What do you discover?***

69. $\begin{vmatrix} 1 & 3 & 4 \\ 0 & 5 & 2 \\ 0 & 0 & 2 \end{vmatrix}$ **10**

70. $\begin{vmatrix} 2 & 1 & -2 \\ 0 & 3 & 4 \\ 0 & 0 & -1 \end{vmatrix}$ **−6**

71. $\begin{vmatrix} 1 & 2 & 4 & 3 \\ 0 & 2 & 2 & 1 \\ 0 & 0 & 3 & 2 \\ 0 & 0 & 0 & 4 \end{vmatrix}$ **24**

72. $\begin{vmatrix} 2 & 1 & -2 & 1 \\ 0 & 2 & 2 & -1 \\ 0 & 0 & 3 & 1 \\ 0 & 0 & 0 & 2 \end{vmatrix}$ **24**

73. Use an example chosen from 2×2 matrices to show that the determinant of the product of two matrices is the product of the determinants of those two matrices.

74. Find an example among 2×2 matrices to show that the determinant of a sum of two matrices is not equal to the sum of the determinants of those matrices.

75. A determinant is a function that associates a number with every square matrix. Give the domain and the range of that function.
domain: $n \times n$ matrices; range: reals

76. Use an example chosen from 2×2 matrices to show that for $n \times n$ matrices A and B, $AB \neq BA$ but $|AB| = |BA|$.

77. If A and B are matrices and $|AB| = 0$, must $|A| = 0$ or $|B| = 0$? Explain. **yes**

78. If A and B are matrices and $|AB| = 0$, must $A = \mathbf{0}$ or $B = \mathbf{0}$? Explain. **no**

Use a graphing calculator to evaluate each determinant.

79. $\begin{vmatrix} 2.3 & 5.7 & 6.1 \\ 3.4 & 6.2 & 8.3 \\ 5.8 & 8.2 & 9.2 \end{vmatrix}$ **21.468**

80. $\begin{vmatrix} .32 & -7.4 & -6.7 \\ 3.3 & 5.5 & -0.27 \\ -8 & -0.13 & 5.47 \end{vmatrix}$ **−164.716332**

REVIEW ***Factor each polynomial, if possible.***

81. $x^2 + 3x - 4$
$(x - 1)(x + 4)$

82. $2x^2 - 5x - 12$
$(x - 4)(2x + 3)$

83. $9x^3 - x$
$x(3x + 1)(3x - 1)$

84. $x^2 + 3x + 5$
prime

Add the fractions.

85. $\dfrac{1}{x - 2} + \dfrac{2}{2x - 1}$
$\dfrac{4x - 5}{(x - 2)(2x - 1)}$

86. $\dfrac{2}{x} + \dfrac{3}{x^2} - \dfrac{1}{x - 1}$
$\dfrac{x^2 + x - 3}{x^2(x - 1)}$

87. $\dfrac{2}{x^2 + 1} + \dfrac{1}{x}$
$\dfrac{x^2 + 2x + 1}{x(x^2 + 1)}$

88. $\dfrac{2x}{x^2 + 1} + \dfrac{1}{x}$
$\dfrac{3x^2 + 1}{x(x^2 + 1)}$

6.6 Partial Fractions

In this section, you will learn about

- When the Denominator Has Distinct Linear Factors
- When the Denominator Has Distinct Quadratic Factors
- When the Denominator Has Repeated Linear Factors
- When the Denominator Has Repeated Quadratic Factors
- When the Degree of $P(x)$ Is Equal to or Greater Than the Degree of $Q(x)$

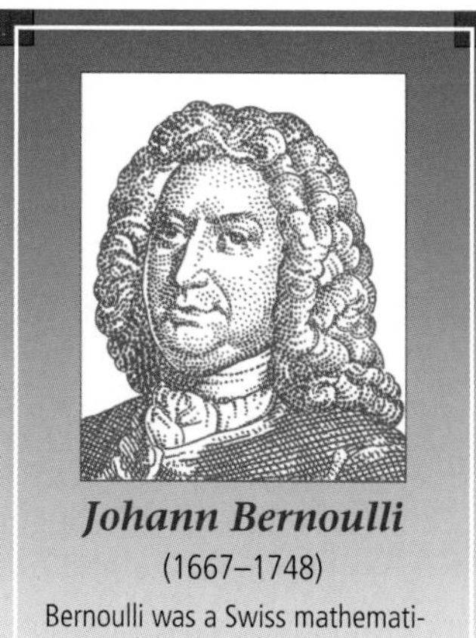

Johann Bernoulli
(1667–1748)
Bernoulli was a Swiss mathematician and teacher of Euler. His method of decomposition by partial fractions was a major contribution to the calculus.

In this section, we discuss how to write complicated fractions as sums of simpler fractions. We begin by reviewing how to add fractions. For example, to find the sum

$$\frac{2}{x} + \frac{6}{x + 1} + \frac{-1}{(x + 1)^2}$$

we write each fraction with an LCD of $x(x + 1)^2$, add the fractions by adding their numerators and keeping the common denominator, and simplify.

$$\begin{aligned} \frac{2}{x} + \frac{6}{x + 1} + \frac{-1}{(x + 1)^2} &= \frac{2(x + 1)^2}{x(x + 1)^2} + \frac{6x(x + 1)}{(x + 1)x(x + 1)} + \frac{-1x}{(x + 1)^2 x} \\ &= \frac{2x^2 + 4x + 2 + 6x^2 + 6x - x}{x(x + 1)^2} \\ &= \frac{8x^2 + 9x + 2}{x(x + 1)^2} \end{aligned}$$

To reverse the addition process and write a fraction as the sum of simpler fractions with denominators of smallest possible degree, we must **decompose a fraction into partial fractions.** For example, to decompose the fraction

$$\frac{8x^2 + 9x + 2}{x(x + 1)^2}$$

into partial fractions, we will assume that there are constants A, B, and C such that

$$\frac{8x^2 + 9x + 2}{x(x + 1)^2} = \frac{A}{x} + \frac{B}{x + 1} + \frac{C}{(x + 1)^2}$$

After writing the terms on the right-hand side as fractions with an LCD of $x(x + 1)^2$, we add the fractions to get

$$\frac{8x^2 + 9x + 2}{x(x + 1)^2} = \frac{A(x + 1)^2}{x(x + 1)^2} + \frac{Bx(x + 1)}{x(x + 1)(x + 1)} + \frac{Cx}{(x + 1)^2x}$$

$$= \frac{Ax^2 + 2Ax + A + Bx^2 + Bx + Cx}{x(x + 1)^2}$$

(1) $$\frac{8x^2 + 9x + 2}{x(x + 1)^2} = \frac{(A + B)x^2 + (2A + B + C)x + A}{x(x + 1)^2}$$ Factor out x^2 from $Ax^2 + Bx^2$. Factor x from $2Ax + Bx + Cx$.

Since the fractions on the left- and right-hand sides of Equation 1 are equal and their denominators are equal, the coefficients of their polynomial numerators are equal.

$$\begin{cases} A + B = 8 & \text{These are the coefficients of } x^2. \\ 2A + B + C = 9 & \text{These are the coefficients of } x. \\ A = 2 & \text{These are the constants.} \end{cases}$$

We can solve this system to find that $A = 2$, $B = 6$, and $C = -1$. Then we know that

$$\frac{8x^2 + 9x + 2}{x(x + 1)^2} = \frac{2}{x} + \frac{6}{x + 1} + \frac{-1}{(x + 1)^2}$$

The method of partial fractions uses the following theorem.

Polynomial Factorization Theorem

The factorization of any polynomial $Q(x)$ with real coefficients is the product of polynomials of the forms

$$(ax + b)^n \qquad \text{and} \qquad (ax^2 + bx + c)^n$$

where n is a positive integer and $ax^2 + bx + c$ is irreducible over the real numbers.

The example and this theorem illustrate the topic of this section. We begin with an algebraic fraction—like $\frac{P(x)}{Q(x)}$, the quotient of two polynomials—and write it as the sum of two or more fractions with simpler denominators. By the theorem, we know that they will be either first-degree or irreducible second-degree polynomials, or powers of those. We consider each case separately.

We suppose that $P(x)$ and $Q(x)$ have real coefficients, that the degree of $P(x)$ is less than the degree of $Q(x)$, and that the fraction $\frac{P(x)}{Q(x)}$ is in lowest terms.

When the Denominator Has Distinct Linear Factors

ACTIVE EXAMPLE 1 Decompose $\dfrac{9x + 2}{(x + 2)(3x - 2)}$ into partial fractions.

Solution Since each denominator is linear, there are constants A and B such that

$$\frac{9x + 2}{(x + 2)(3x - 2)} = \frac{A}{x + 2} + \frac{B}{3x - 2}$$

COLLEGE **Algebra** $f(x)$ **Now**™
Explore this Active Example and test yourself on the steps followed here at **http://1pass.thomson.com**. You can also find this example on the Interactive Video Skillbuilder CD-ROM.

After writing the terms on the right-hand side as fractions with an LCD of $(x + 2)(3x - 2)$, we add the fractions to get

$$\frac{9x + 2}{(x + 2)(3x - 2)} = \frac{A(3x - 2)}{(x + 2)(3x - 2)} + \frac{B(x + 2)}{(x + 2)(3x - 2)}$$

$$\frac{9x + 2}{(x + 2)(3x - 2)} = \frac{3Ax - 2A + Bx + 2B}{(x + 2)(3x - 2)}$$

(2) $$\frac{9x + 2}{(x + 2)(3x - 2)} = \frac{(3A + B)x - 2A + 2B}{(x + 2)(3x - 2)}$$ Factor x from $3Ax + Bx$.

Since the fractions in Equation 2 are equal, the coefficients of their polynomial numerators are equal, and we have

$$\begin{cases} 9 = 3A + B & \text{These are the coefficients of } x. \\ 2 = -2A + 2B & \text{These are the constants.} \end{cases}$$

After solving this system to find that $A = 2$ and $B = 3$, we have

$$\frac{9x + 2}{(x + 2)(3x - 2)} = \frac{2}{x + 2} + \frac{3}{3x - 2}$$

Self Check Decompose $\dfrac{2x + 2}{x(x + 2)}$ into partial fractions. ■

When the Denominator Has Distinct Quadratic Factors

If $Q(x)$ has a prime quadratic factor of the form $ax^2 + bx + c$, the partial fraction decomposition of $\frac{P(x)}{Q(x)}$ will have a corresponding term of the form

$$\frac{Bx + C}{ax^2 + bx + c}$$

Comment Be sure that the quadratic factor is not factorable over the real numbers—check it with the discriminant. If $b^2 - 4ac$ is negative, the factor is prime.

ACTIVE EXAMPLE 2 Decompose $\dfrac{2x^2 + x + 1}{x^3 + x}$ into partial fractions.

Solution Since the denominator can be written as a product of a linear factor and a prime quadratic factor, the partial fractions have the form

$$\begin{aligned}\frac{2x^2 + x + 1}{x(x^2 + 1)} &= \frac{A}{x} + \frac{Bx + C}{x^2 + 1} \\ &= \frac{A(x^2 + 1)}{x(x^2 + 1)} + \frac{(Bx + C)x}{x(x^2 + 1)} \\ &= \frac{Ax^2 + A + Bx^2 + Cx}{x(x^2 + 1)} \\ &= \frac{(A + B)x^2 + Cx + A}{x(x^2 + 1)} \qquad \text{Factor } x^2 \text{ from } Ax^2 + Bx^2.\end{aligned}$$

We can equate the corresponding coefficients of the numerators $2x^2 + x + 1$ and $(A + B)x^2 + Cx + A$ to get the system

$$\begin{cases} A + B = 2 & \text{These are the coefficients of } x^2. \\ C = 1 & \text{These are the coefficients of } x. \\ A = 1 & \text{These are the constants.} \end{cases}$$

with solutions $A = 1, B = 1, C = 1$. The partial fraction decomposition is

$$\begin{aligned}\frac{2x^2 + x + 1}{x(x^2 + 1)} &= \frac{A}{x} + \frac{Bx + C}{x^2 + 1} \\ &= \frac{1}{x} + \frac{1x + 1}{x^2 + 1} \\ &= \frac{1}{x} + \frac{x + 1}{x^2 + 1}\end{aligned}$$

Self Check Decompose $\dfrac{2x^2 + 1}{x^3 + x}$ into partial fractions. ■

■ When the Denominator Has Repeated Linear Factors

If $Q(x)$ has n linear factors of $ax + b$, then $(ax + b)^n$ is a factor of $Q(x)$. Each factor of the form $(ax + b)^n$ generates the following sum of n partial fractions:

$$\frac{A}{ax + b} + \frac{B}{(ax + b)^2} + \frac{C}{(ax + b)^3} + \cdots + \frac{D}{(ax + b)^n}$$

EXAMPLE 3 Decompose $\dfrac{3x^2 - x + 1}{x(x - 1)^2}$ into partial fractions.

Solution Here, each factor in the denominator is linear. The linear factor x appears once, and the linear factor $x - 1$ appears twice. Thus, there are constants A, B, and C such that

$$\frac{3x^2 - x + 1}{x(x - 1)^2} = \frac{A}{x} + \frac{B}{x - 1} + \frac{C}{(x - 1)^2}$$

After writing the terms on the right-hand side as fractions with an LCD of $x(x-1)^2$, we combine them to get

$$\begin{aligned}\frac{3x^2-x+1}{x(x-1)^2} &= \frac{A(x-1)^2}{x(x-1)^2}+\frac{Bx(x-1)}{x(x-1)(x-1)}+\frac{Cx}{(x-1)^2x}\\ &= \frac{Ax^2-2Ax+A+Bx^2-Bx+Cx}{x(x-1)^2}\\ &= \frac{(A+B)x^2+(-2A-B+C)x+A}{x(x-1)^2}\end{aligned}$$

Factor x^2 from $Ax^2 + Bx^2$.
Factor x from $-2Ax - Bx + Cx$.

Since the fractions are equal, the coefficients of the polynomial numerators are equal, and we have

$$\begin{cases} A+B=3 & \text{These are the coefficients of } x^2.\\ -2A-B+C=-1 & \text{These are the coefficients of } x.\\ A=1 & \text{These are the constants.}\end{cases}$$

We can solve this system to find that $A = 1$, $B = 2$, and $C = 3$. So, we have

$$\begin{aligned}\frac{3x^2-x+1}{x(x-1)^2} &= \frac{A}{x}+\frac{B}{x-1}+\frac{C}{(x-1)^2}\\ &= \frac{1}{x}+\frac{2}{x-1}+\frac{3}{(x-1)^2}\end{aligned}$$

Self Check Decompose $\dfrac{3x^2+7x+1}{x(x+1)^2}$ into partial fractions. ■

■ When the Denominator Has Repeated Quadratic Factors

If $Q(x)$ has n prime factors of $ax^2 + bx + c$, then $(ax^2 + bx + c)^n$ is a factor. Each factor of the form $(ax^2 + bx + c)^n$ generates a sum of n partial fractions of the form

$$\frac{Ax+B}{ax^2+bx+c}+\frac{Cx+D}{(ax^2+bx+c)^2}+\cdots+\frac{Ex+F}{(ax^2+bx+c)^n}$$

ACTIVE EXAMPLE 4 Decompose $\dfrac{3x^2+5x+5}{(x^2+1)^2}$ into partial fractions.

Solution Since the quadratic factor $x^2 + 1$ is used twice, we must find constants A, B, C, and D such that

COLLEGE **Algebra** *f(x)* **Now**™
Go to **http://1pass.thomson.com** or your CD to practice this example.

$$\frac{3x^2+5x+5}{(x^2+1)^2}=\frac{Ax+B}{x^2+1}+\frac{Cx+D}{(x^2+1)^2}$$

We add the fractions on the right-hand side to get

$$\frac{3x^2+5x+5}{(x^2+1)^2}=\frac{(Ax+B)(x^2+1)}{(x^2+1)(x^2+1)}+\frac{Cx+D}{(x^2+1)^2}$$

$$= \frac{Ax^3 + Ax + Bx^2 + B + Cx + D}{(x^2 + 1)^2}$$

$$= \frac{Ax^3 + Bx^2 + (A + C)x + B + D}{(x^2 + 1)^2} \quad \text{Factor } x \text{ from } Ax + Cx.$$

If we add the term $0x^3$ to the numerator on the left-hand side, we can equate the corresponding coefficients in the numerators to get

$$\begin{aligned} A &= 0 && \text{These are the coefficients of } x^3. \\ B &= 3 && \text{These are the coefficients of } x^2. \\ A + C &= 5 && \text{These are the coefficients of } x. \\ B + D &= 5 && \text{These are the constants.} \end{aligned}$$

with a solution of $A = 0$, $B = 3$, $C = 5$, and $D = 2$. So we have

$$\frac{3x^2 + 5x + 5}{(x^2 + 1)^2} = \frac{Ax + B}{x^2 + 1} + \frac{Cx + \mathbf{D}}{(x^2 + 1)^2}$$

$$= \frac{0x + 3}{x^2 + 1} + \frac{5x + \mathbf{2}}{(x^2 + 1)^2}$$

$$= \frac{3}{x^2 + 1} + \frac{5x + 2}{(x^2 + 1)^2}$$

Self Check Decompose $\dfrac{x^3 + 2x + 3}{(x^2 + 2)^2}$ into partial fractions. ■

■ When the Degree of $P(x)$ Is Equal to or Greater Than the Degree of $Q(x)$

When the degree of $P(x)$ is equal to or greater than the degree of $Q(x)$ in the fraction $\frac{P(x)}{Q(x)}$, we do a long division before decomposing the fraction into partial fractions.

EXAMPLE 5 Decompose $\dfrac{x^2 + 4x + 2}{x^2 + x}$ into partial fractions.

Solution Because the degree of the numerator and denominator are the same, we must do a long division and express the fraction in quotient + $\frac{\text{remainder}}{\text{divisor}}$ form:

$$\begin{array}{r} 1 \\ x^2 + x \overline{) x^2 + 4x + 2} \\ \underline{x^2 + x } \\ 3x + 2 \end{array}$$

So we can write

$$\frac{x^2 + 4x + 2}{x^2 + x} = 1 + \frac{3x + 2}{x^2 + x} \tag{3}$$

Because the degree of the numerator of the fraction on the right-hand side of Equation 3 is less than the degree of the denominator, we can find its partial fraction decomposition:

$$\begin{aligned}\frac{3x+2}{x^2+x} &= \frac{3x+2}{x(x+1)}\\ &= \frac{A}{x} + \frac{B}{x+1}\\ &= \frac{A(x+1)+Bx}{x(x+1)}\\ &= \frac{(A+B)x+A}{x(x+1)}\end{aligned}$$

We equate the corresponding coefficients in the numerator and solve the resulting system of equations to find that the solution is $A = 2$, $B = 1$. So we have

$$\frac{x^2+4x+2}{x^2+x} = 1 + \frac{2}{x} + \frac{1}{x+1}$$

Self Check Decompose $\frac{x^2+x-1}{x^2-x}$ into partial fractions. ■

Self Check Answers

1. $\frac{1}{x} + \frac{1}{x+2}$ 2. $\frac{1}{x} + \frac{x}{x^2+1}$ 3. $\frac{1}{x} + \frac{2}{x+1} + \frac{3}{(x+1)^2}$ 4. $\frac{x}{x^2+2} + \frac{3}{(x^2+2)^2}$ 5. $1 + \frac{1}{x} + \frac{1}{x-1}$

6.6 Exercises

VOCABULARY AND CONCEPTS ***Fill in the blanks.***

1. A polynomial with real coefficients factors as the product of first-degree and second-degree factors or powers of those.
2. The second-degree factors of a polynomial with real coefficients are prime, which means they don't factor further over the real numbers.

PRACTICE ***Decompose each fraction into partial fractions.***

3. $\frac{3x-1}{x(x-1)}$ $\quad \frac{1}{x} + \frac{2}{x-1}$
4. $\frac{4x+6}{x(x+2)}$ $\quad \frac{3}{x} + \frac{1}{x+2}$
5. $\frac{2x-15}{x(x-3)}$ $\quad \frac{5}{x} - \frac{3}{x-3}$
6. $\frac{5x+21}{x(x+7)}$ $\quad \frac{3}{x} + \frac{2}{x+7}$
7. $\frac{3x+1}{(x+1)(x-1)}$ $\quad \frac{1}{x+1} + \frac{2}{x-1}$
8. $\frac{9x-3}{(x+1)(x-2)}$ $\quad \frac{4}{x+1} + \frac{5}{x-2}$
9. $\frac{-2x+11}{x^2-x-6}$ $\quad \frac{1}{x-3} - \frac{3}{x+2}$
10. $\frac{7x+2}{x^2+x-2}$ $\quad \frac{4}{x+2} + \frac{3}{x-1}$
11. $\frac{3x-23}{x^2+2x-3}$ $\quad \frac{8}{x+3} - \frac{5}{x-1}$
12. $\frac{-x-17}{x^2-x-6}$ $\quad \frac{3}{x+2} - \frac{4}{x-3}$
13. $\frac{9x-31}{2x^2-13x+15}$ $\quad \frac{5}{2x-3} + \frac{2}{x-5}$
14. $\frac{-2x-6}{3x^2-7x+2}$ $\quad \frac{4}{3x-1} - \frac{2}{x-2}$
15. $\frac{4x^2+4x-2}{x(x^2-1)}$ $\quad \frac{2}{x} + \frac{3}{x-1} - \frac{1}{x+1}$
16. $\frac{x^2-6x-13}{(x+2)(x^2-1)}$ $\quad \frac{1}{x+2} - \frac{3}{x-1} + \frac{3}{x+1}$

17. $\dfrac{x^2 + x + 3}{x(x^2 + 3)}$ $\quad \dfrac{1}{x} + \dfrac{1}{x^2 + 3}$

18. $\dfrac{5x^2 + 2x + 2}{x^3 + x}$ $\quad \dfrac{2}{x} + \dfrac{3x + 2}{x^2 + 1}$

19. $\dfrac{3x^2 + 8x + 11}{(x + 1)(x^2 + 2x + 3)}$ $\quad \dfrac{3}{x + 1} + \dfrac{2}{x^2 + 2x + 3}$

20. $\dfrac{-3x^2 + x - 5}{(x + 1)(x^2 + 2)}$ $\quad \dfrac{1}{x^2 + 2} - \dfrac{3}{x + 1}$

21. $\dfrac{5x^2 + 9x + 3}{x(x + 1)^2}$ $\quad \dfrac{3}{x} + \dfrac{2}{x + 1} + \dfrac{1}{(x + 1)^2}$

22. $\dfrac{2x^2 - 7x + 2}{x(x - 1)^2}$ $\quad \dfrac{2}{x} - \dfrac{3}{(x - 1)^2}$

23. $\dfrac{-2x^2 + x - 2}{x^2(x - 1)}$ $\quad \dfrac{1}{x} + \dfrac{2}{x^2} - \dfrac{3}{x - 1}$

24. $\dfrac{x^2 + x + 1}{x^3}$ $\quad \dfrac{1}{x} + \dfrac{1}{x^2} + \dfrac{1}{x^3}$

25. $\dfrac{3x^2 - 13x + 18}{x^3 - 6x^2 + 9x}$ $\quad \dfrac{2}{x} + \dfrac{1}{x - 3} + \dfrac{2}{(x - 3)^2}$

26. $\dfrac{3x^2 + 13x + 20}{x^3 + 4x^2 + 4x}$ $\quad \dfrac{5}{x} - \dfrac{2}{x + 2} - \dfrac{3}{(x + 2)^2}$

27. $\dfrac{x^2 - 2x - 3}{(x - 1)^3}$ $\quad \dfrac{1}{x - 1} - \dfrac{4}{(x - 1)^3}$

28. $\dfrac{x^2 + 8x + 18}{(x + 3)^3}$ $\quad \dfrac{1}{x + 3} + \dfrac{2}{(x + 3)^2} + \dfrac{3}{(x + 3)^3}$

29. $\dfrac{x^3 + 4x^2 + 2x + 1}{x^4 + x^3 + x^2}$ $\quad \dfrac{1}{x} + \dfrac{1}{x^2} + \dfrac{2}{x^2 + x + 1}$

30. $\dfrac{3x^3 + 5x^2 + 3x + 1}{x^2(x^2 + x + 1)}$ $\quad \dfrac{2}{x} + \dfrac{1}{x^2} + \dfrac{x + 2}{x^2 + x + 1}$

31. $\dfrac{4x^3 + 5x^2 + 3x + 4}{x^2(x^2 + 1)}$ $\quad \dfrac{3}{x} + \dfrac{4}{x^2} + \dfrac{x + 1}{x^2 + 1}$

32. $\dfrac{2x^2 + 1}{x^4 + x^2}$ $\quad \dfrac{1}{x^2} + \dfrac{1}{x^2 + 1}$

33. $\dfrac{-x^2 - 3x - 5}{x^3 + x^2 + 2x + 2}$ $\quad -\dfrac{1}{x + 1} - \dfrac{3}{x^2 + 2}$

34. $\dfrac{-2x^3 + 7x^2 + 6}{x^2(x^2 + 2)}$ $\quad \dfrac{3}{x^2} + \dfrac{-2x + 4}{x^2 + 2}$

35. $\dfrac{x^3 + 4x^2 + 3x + 6}{(x^2 + 2)(x^2 + x + 2)}$ $\quad \dfrac{x + 1}{x^2 + 2} + \dfrac{2}{x^2 + x + 2}$

36. $\dfrac{x^3 + 3x^2 + 2x + 4}{(x^2 + 1)(x^2 + x + 2)}$ $\quad \dfrac{1}{x^2 + 1} + \dfrac{x + 2}{x^2 + x + 2}$

37. $\dfrac{2x^4 + 6x^3 + 20x^2 + 22x + 25}{x(x^2 + 2x + 5)^2}$ $\quad \dfrac{1}{x} + \dfrac{x}{x^2 + 2x + 5} + \dfrac{x + 2}{(x^2 + 2x + 5)^2}$

38. $\dfrac{x^3 + 3x^2 + 6x + 6}{(x^2 + x + 5)(x^2 + 1)}$ $\quad \dfrac{1}{x^2 + x + 5} + \dfrac{x + 1}{x^2 + 1}$

39. $\dfrac{x^3}{x^2 + 3x + 2}$ $\quad x - 3 - \dfrac{1}{x + 1} + \dfrac{8}{x + 2}$

40. $\dfrac{2x^3 + 6x^2 + 3x + 2}{x^3 + x^2}$ $\quad 2 + \dfrac{1}{x} + \dfrac{2}{x^2} + \dfrac{3}{x + 1}$

41. $\dfrac{3x^3 + 3x^2 + 6x + 4}{3x^3 + x^2 + 3x + 1}$ $\quad 1 + \dfrac{2}{3x + 1} + \dfrac{1}{x^2 + 1}$

42. $\dfrac{x^4 + x^3 + 3x^2 + x + 4}{(x^2 + 1)^2}$ $\quad 1 + \dfrac{x + 1}{x^2 + 1} + \dfrac{2}{(x^2 + 1)^2}$

43. $\dfrac{x^3 + 3x^2 + 2x + 1}{x^3 + x^2 + x}$ $\quad 1 + \dfrac{1}{x} + \dfrac{x}{x^2 + x + 1}$

44. $\dfrac{x^4 - x^3 + x^2 - x + 1}{(x^2 + 1)^2}$ $\quad 1 - \dfrac{x + 1}{x^2 + 1} + \dfrac{1}{(x^2 + 1)^2}$

45. $\dfrac{2x^4 + 2x^3 + 3x^2 - 1}{(x^2 - x)(x^2 + 1)}$ $\quad 2 + \dfrac{1}{x} + \dfrac{3}{x - 1} + \dfrac{2}{x^2 + 1}$

46. $\dfrac{x^4 - x^3 + 5x^2 + x + 6}{(x^2 + 3)(x^2 + 1)}$ $\quad 1 - \dfrac{2x}{x^2 + 3} + \dfrac{x + 1}{x^2 + 1}$

DISCOVERY AND WRITING

47. Is the polynomial $x^3 + 1$ prime?
No; it's the sum of two cubes.

48. Decompose $\dfrac{1}{x^3 + 1}$ into partial fractions.
$\dfrac{\frac{1}{3}}{x + 1} + \dfrac{-\frac{1}{3}x + \frac{2}{3}}{x^2 - x + 1}$

REVIEW *Simplify each radical expression. Use absolute value symbols if necessary.*

49. $\sqrt{8a^3b}$ $\quad 2|a|\sqrt{2ab}$

50. $\sqrt{x^2 + 6x + 9}$ $\quad |x + 3|$

51. $\sqrt{18x^5} + x^2\sqrt{50x} - 5x^2\sqrt{2x}$ $\quad 3x^2\sqrt{2x}$

52. $\dfrac{\sqrt{x^2y^3}}{\sqrt{xy^5}}$ $\quad \dfrac{\sqrt{x}}{|y|}$

Solve each equation.

53. $\sqrt{x - 5} = x - 7$ $\quad x = 9$

54. $x - 5 = (x - 7)^2$ $\quad x = 6, 9$

6.7 Graphs of Linear Inequalities

In this section, you will learn about

- Graphs of Inequalities
- Graphs of Systems of Inequalities

Graphs of Inequalities

The **graph of an inequality** in x and y is the graph of all ordered pairs (x, y) that satisfy the inequality. In this section, we will consider graphs of **linear inequalities**—inequalities that can be expressed in a form such as

$$Ax + By < C, \qquad Ax + By > C, \qquad Ax + By \le C, \qquad \text{or} \qquad Ax + By \ge C$$

To graph the inequality $y > 3x + 2$, we note that one of the following statements is true:

$$y = 3x + 2, \qquad y < 3x + 2, \qquad \text{or} \qquad y > 3x + 2$$

The graph of $y = 3x + 2$ is a line, as shown in Figure 6-11(a). The graphs of the inequalities are half-planes, one on each side of that line. We can think of the graph of $y = 3x + 2$ as a boundary separating the two half-planes. The graph of $y = 3x + 2$ is drawn with a broken line, to show that it is not part of the graph of $y > 3x + 2$.

To find which half-plane is the graph of $y > 3x + 2$, we can substitute the coordinates of any point on one side of the line—say, the origin (0, 0)—into the inequality and simplify:

$$y > 3x + 2$$
$$0 > 3(0) + 2$$
$$0 > 2$$

Since $0 > 2$ is false, the coordinates (0, 0) do not satisfy the inequality, and the origin is not in the half-plane that is the graph of $y > 3x + 2$. Thus, the graph is the half-plane located on the other side of the broken line. The graph of the inequality $y > 3x + 2$ is shown in Figure 6-11(b).

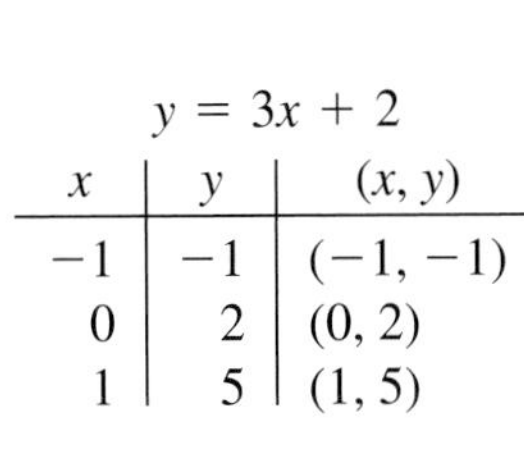
$y = 3x + 2$

x	y	(x, y)
−1	−1	(−1, −1)
0	2	(0, 2)
1	5	(1, 5)

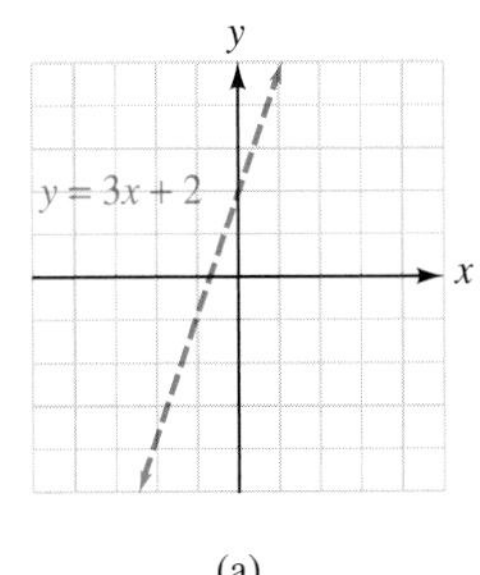

(a)

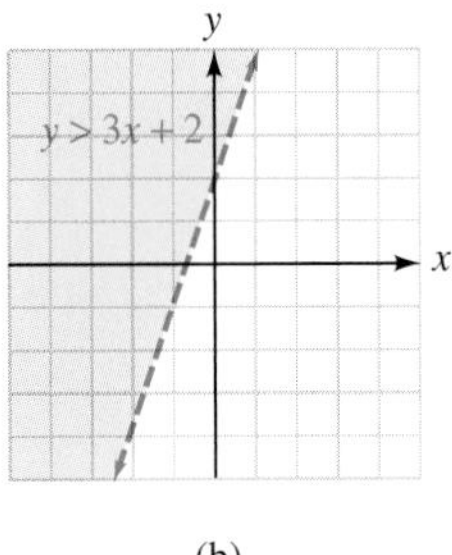

(b)

Figure 6-11

ACTIVE EXAMPLE 1 Graph the inequality: $2x - 3y \le 6$.

Solution This inequality is the combination of $2x - 3y < 6$ and $2x - 3y = 6$. We start by graphing $2x - 3y = 6$ to establish the boundary that separates two half-planes. This time we draw a solid line, because equality is permitted. See Figure 6-12(a).

COLLEGE **Algebra** $f(x)$ **Now**™

Explore this Active Example and test yourself on the steps followed here at **http://1pass.thomson.com**. You can also find this example on the Interactive Video Skillbuilder CD-ROM.

To decide which half-plane represents $2x - 3y \leq 6$, we check whether the coordinates of the origin satisfy the inequality:

$$2x - 3y \leq 6$$
$$2(\mathbf{0}) - 3(\mathbf{0}) \leq 6$$
$$0 \leq 6$$

Because $0 \leq 6$ is true, the origin lies in the graph of $2x - 3y \leq 6$. The graph is shown in Figure 6-12(b).

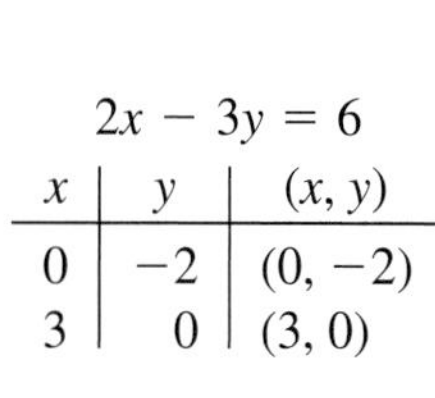

$2x - 3y = 6$

x	y	(x, y)
0	−2	(0, −2)
3	0	(3, 0)

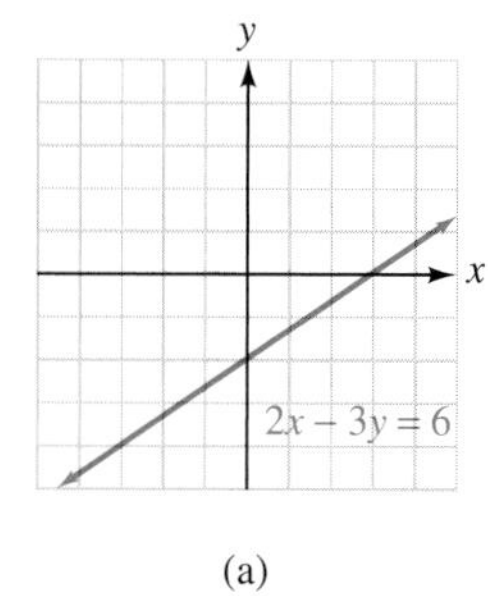

(a)

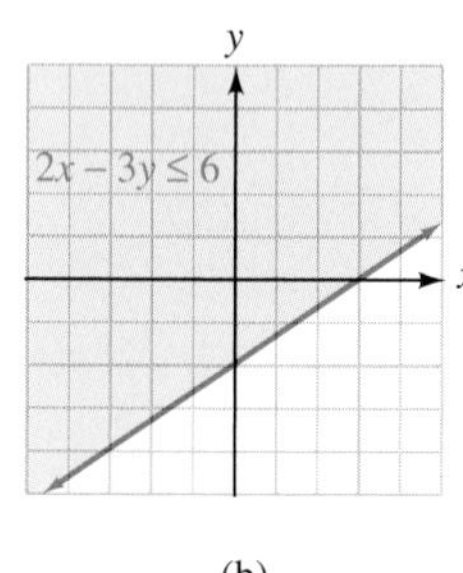

(b)

Figure 6-12

Self Check Graph: $3x + 2y \leq 6$. ■

EXAMPLE 2 Graph the inequality: $y < 2x$.

Solution Since the graph of $y = 2x$ is not part of the graph of the inequality, we graph the boundary with a broken line, as in Figure 6-13(a).

To decide which half-plane represents the graph $y < 2x$, we check whether the coordinates of some fixed point satisfy the inequality. This time, we cannot use the origin as a test point, because the boundary line passes through the origin. So we choose some other point—say, (3, 1)—for a test point:

$$y < 2x$$
$$\mathbf{1} < 2(\mathbf{3})$$
$$1 < 6$$

Since $1 < 6$, the point (3, 1) lies in the graph, and we have the graph shown in Figure 6-13(b).

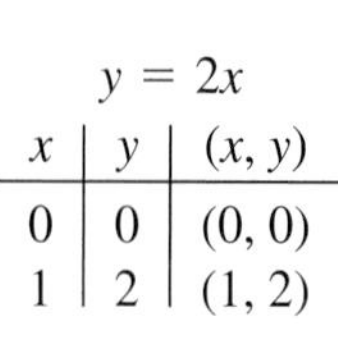

$y = 2x$

x	y	(x, y)
0	0	(0, 0)
1	2	(1, 2)

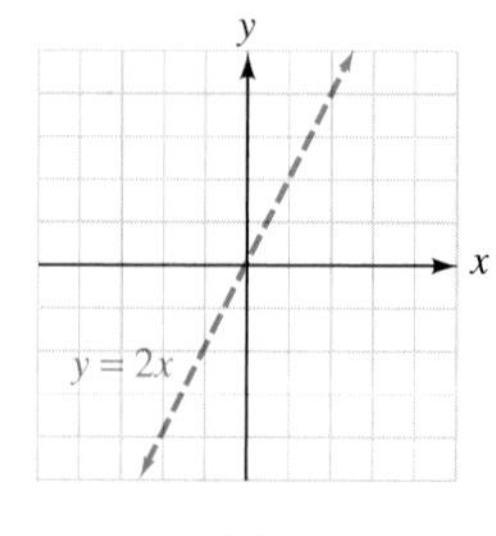

(a)

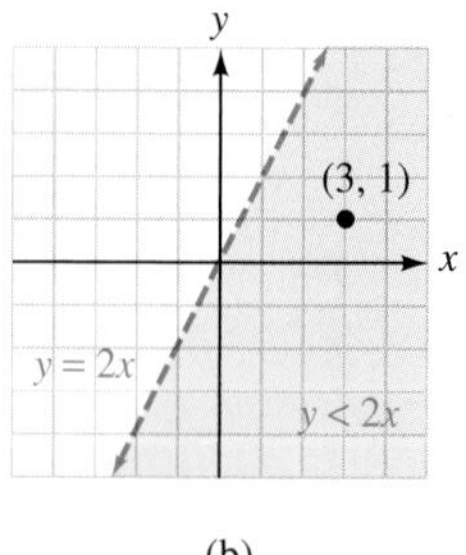

(b)

Figure 6-13

Self Check Graph: $y > 3x$. ■

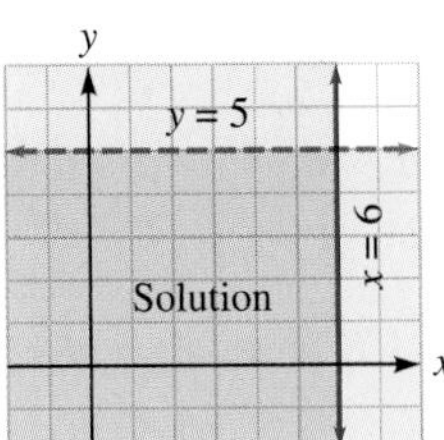

Figure 6-14

Graphs of Systems of Inequalities

We now consider systems of inequalities. To graph the solution set of the system

$$\begin{cases} y < 5 \\ x \leq 6 \end{cases}$$

we graph each inequality on the same set of coordinate axes, as in Figure 6-14. The graph of the inequality $y < 5$ is the half-plane that lies below the line $y = 5$. The graph of the inequality $x \leq 6$ includes the half-plane that lies to the left of the line $x = 6$ together with the line $x = 6$.

The portion of the xy-plane where the two graphs intersect is the graph of the system. Any point that lies in the doubly shaded region has coordinates that satisfy both inequalities in the system.

ACTIVE EXAMPLE 3 Graph the solution set of $\begin{cases} x + y \leq 1 \\ 2x - y > 2 \end{cases}$.

Solution On the same set of coordinate axes, we graph each inequality, as in Figure 6-15. The graph of $x + y \leq 1$ includes the graph of $x + y = 1$ and all points below it. Because the boundary line is included, we draw it as a solid line.

COLLEGE **Algebra f(x) Now**™
Go to **http://1pass.thomson.com** or your CD to practice this example.

$x + y = 1$

x	y	(x, y)
0	1	(0, 1)
1	0	(1, 0)

$2x - y = 2$

x	y	(x, y)
0	−2	(0, −2)
1	0	(1, 0)

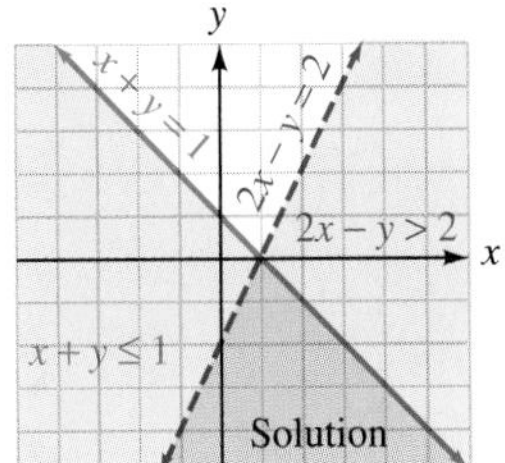

Figure 6-15

The graph of $2x - y > 2$ contains only those points below the line graph of $2x - y = 2$. Because the boundary line is not included, we draw it as a broken line.

The area that is shaded twice represents the solution of the system of inequalities. Any point in the doubly shaded region has coordinates that satisfy both inequalities.

Self Check Graph the solution set of $\begin{cases} x + y < 2 \\ x - 2y \geq 2 \end{cases}$. ■

ACTIVE EXAMPLE 4 Graph the solution set of $\begin{cases} y < x^2 \\ y > \dfrac{x^2}{4} - 2 \end{cases}$.

Solution The graph of $y = x^2$ is a parabola opening upward with vertex at the origin, as shown in Figure 6-16. The points with coordinates that satisfy the inequality are the points below the parabola.

COLLEGE **Algebra f(x) Now**™
Go to **http://1pass.thomson.com** or your CD to practice this example.

The graph of $y = \frac{x^2}{4} - 2$ is also a parabola opening upward. This time, the points that satisfy the inequality are the points above the parabola. The graph of the solution set is the shaded area between the two parabolas.

$y = x^2$

x	y	(x, y)
0	0	(0, 0)
1	1	(1, 1)
−1	1	(−1, 1)
2	4	(2, 4)
−2	4	(−2, 4)

$y = \frac{x^2}{4} - 2$

x	y	(x, y)
0	−2	(0, −2)
2	−1	(2, −1)
−2	−1	(−2, −1)
4	2	(4, 2)
−4	2	(−4, 2)

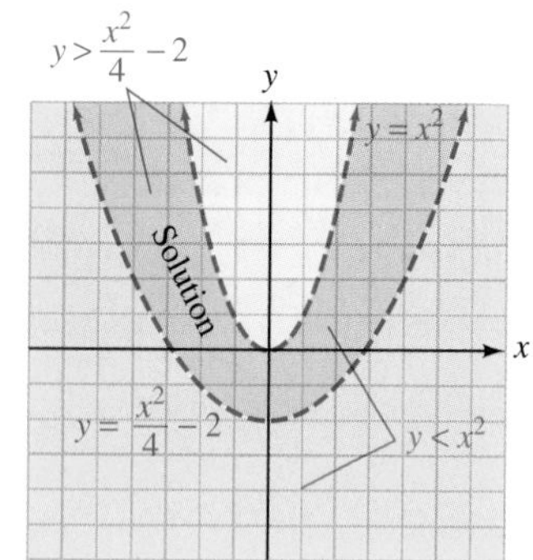

Figure 6-16

Self Check Graph the solution set of $\begin{cases} y \leq 4 - x^2 \\ y > \frac{x^2}{4} \end{cases}$.

EXAMPLE 5 Graph the solution set of $\begin{cases} x + y \leq 4 \\ x - y \leq 6. \\ x \geq 0 \end{cases}$

Solution We graph each inequality, as in Figure 6-17. The graph of $x + y \leq 4$ includes the line $x + y = 4$ and all points below it. Because the boundary line is included, we draw it as a solid line. The graph of $x - y \leq 6$ contains the line $x - y = 6$ and all points above it.

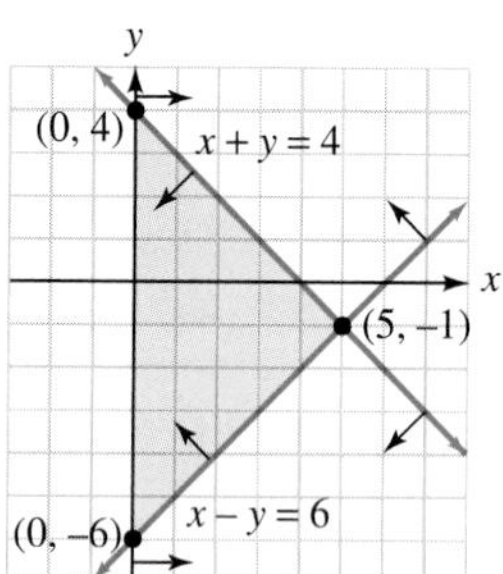

Figure 6-17

The graph of the inequality $x \geq 0$ contains the y-axis and all points to the right of the y-axis. The solution of the system of inequalities is the shaded area in the figure.

The coordinates of the corner points of the shaded area are (0, 4), (0, −6), and (5, −1).

Self Check Graph the solution set of $\begin{cases} x + y \leq 5 \\ x - 3y \leq -3. \\ x \geq 0 \end{cases}$

EXAMPLE 6 Graph the solution set of the system $\begin{cases} x \geq 1 \\ y \geq x \\ 4x + 5y < 20 \end{cases}$.

Solution The graph of the solution set of $x \geq 1$ includes those points on the graph of $x = 1$ and to the right. See Figure 6-18(a).

The graph of the solution set of $y \geq x$ includes those points on the graph of $y = x$ and above it. See Figure 6-18(b).

The graph of the solution set of $4x + 5y < 20$ includes those points below the graph of $4x + 5y = 20$. See Figure 6-18(c).

If these graphs are merged onto a single set of coordinate axes, as in Figure 6-18(d), the graph of the original system of inequalities includes those points within the shaded triangle together with the points on the sides of the triangle drawn as solid lines. The coordinates of the corner points are $\left(1, \frac{16}{5}\right)$, $(1, 1)$, and $\left(\frac{20}{9}, \frac{20}{9}\right)$.

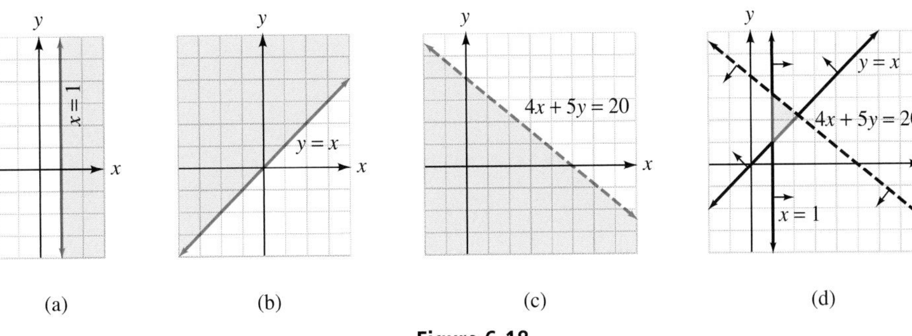

Figure 6-18

Self Check Answers

1. 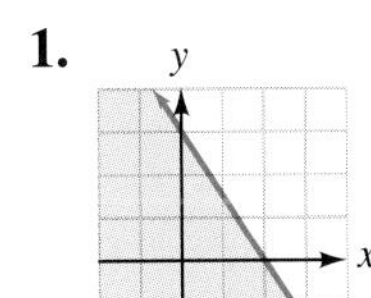

2. 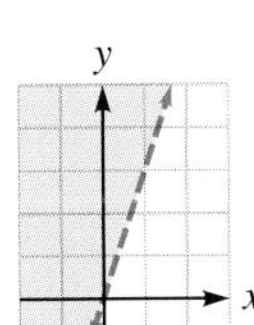

3. 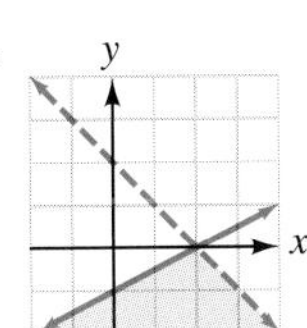

4. 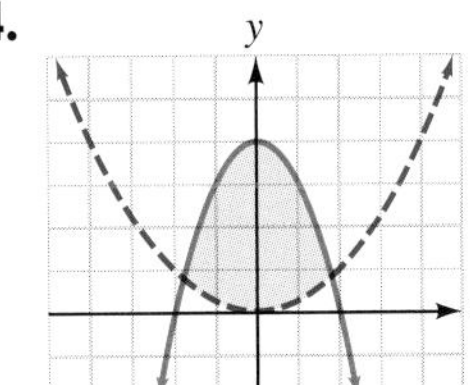

5. 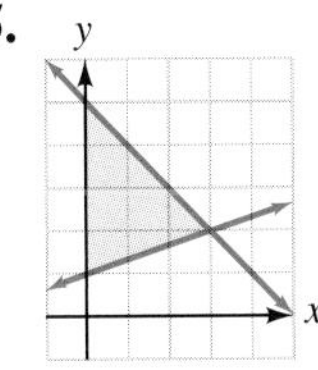

6.7 Exercises

VOCABULARY AND CONCEPTS ***Fill in the blanks.***

1. The graph of $Ax + By = C$ is a line. The graph of $Ax + By \leq C$ is a **half-plane**. The line is its **boundary**.
2. The boundary of the graph $Ax + By < C$ is **excluded** (included, excluded) from the graph.
3. The origin **is not** (is, is not) included in the graph of $3x - 4y > 4$.
4. The origin **is** (is, is not) included in the graph of $4x + 3y \leq 5$.

PRACTICE ***Graph each linear inequality.***

5. $2x + 3y < 12$

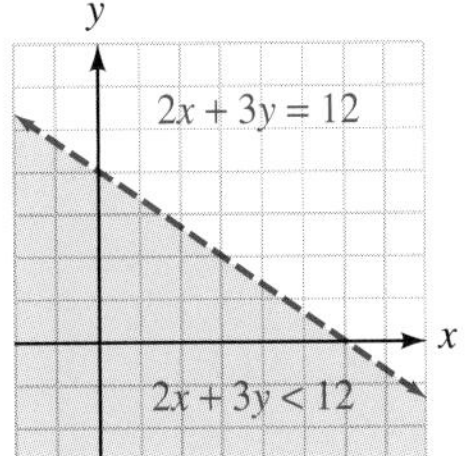

6. $4x - 3y > 6$

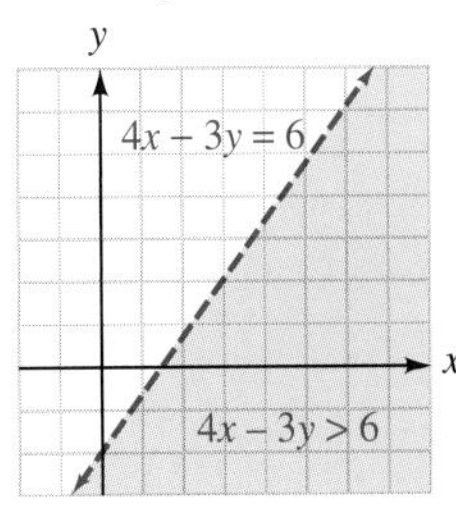

7. $x < 3$

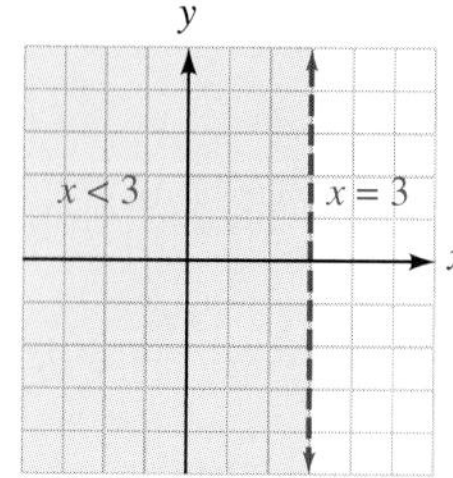

8. $y > -1$

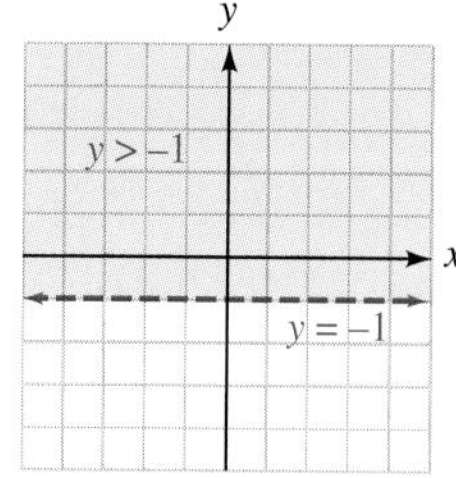

9. $4x - y > 4$

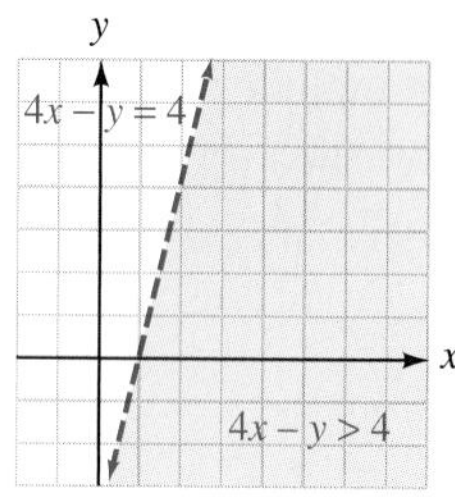

10. $x - 2y < 5$

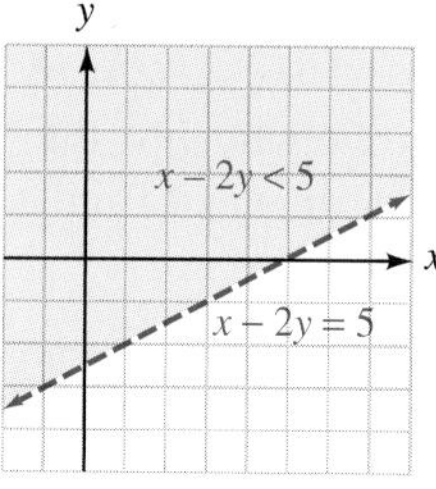

11. $y > 2x$

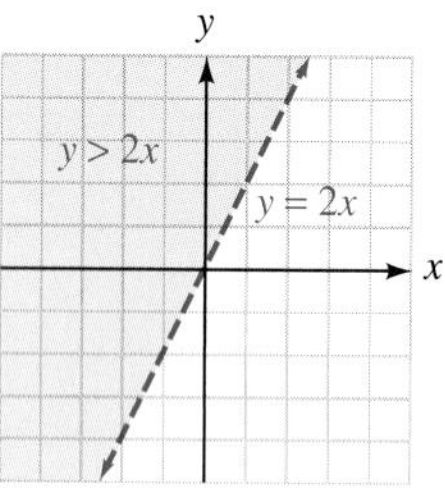

12. $y < 3x$

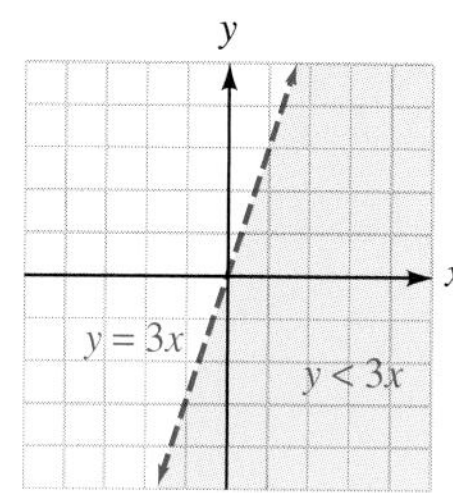

13. $y \leq \frac{1}{2}x + 1$

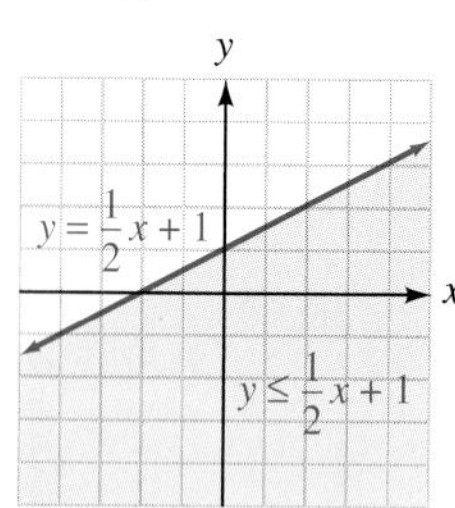

14. $y \geq \frac{1}{3}x - 1$

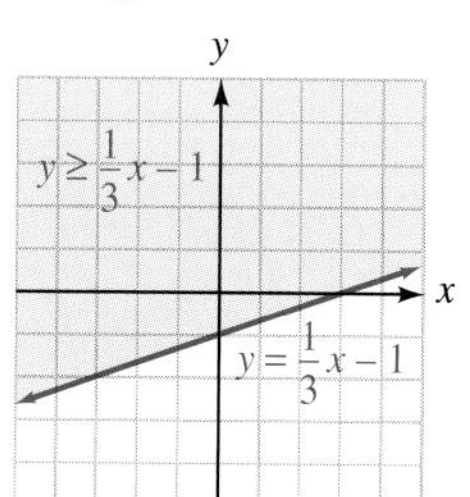

15. $2y \geq 3x - 2$

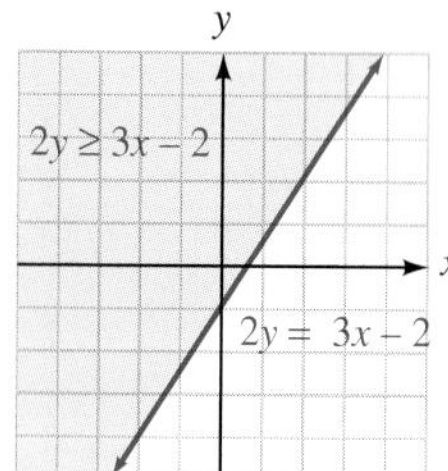

16. $3y \leq 2x + 3$

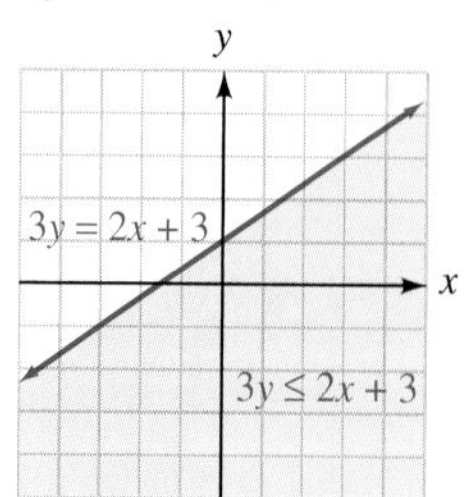

Graph the solution set of each system.

17. $\begin{cases} y < 3 \\ x \geq 2 \end{cases}$

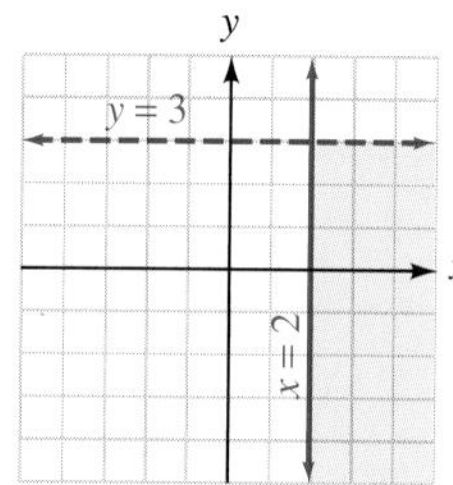

18. $\begin{cases} y \geq -2 \\ x < 0 \end{cases}$

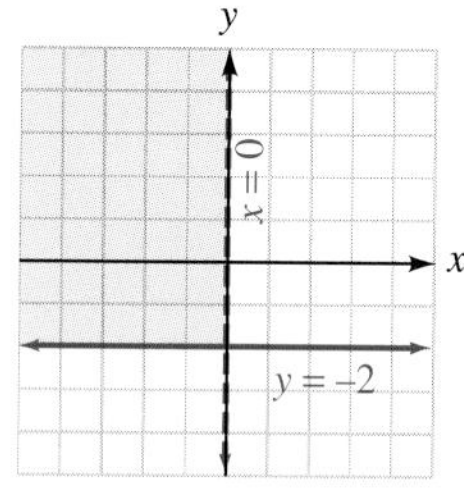

19. $\begin{cases} y \geq 1 \\ x < 2 \end{cases}$

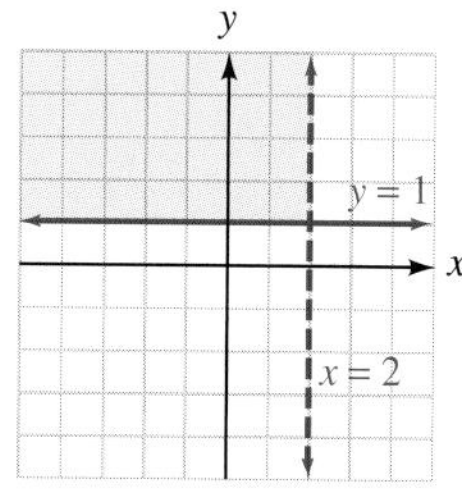

20. $\begin{cases} y \leq -1 \\ x > -1 \end{cases}$

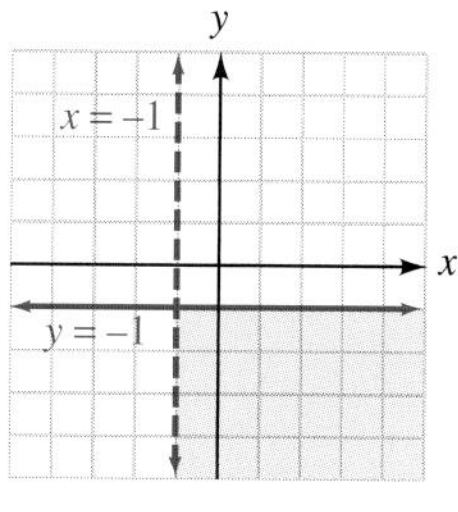

21. $\begin{cases} y \leq x - 2 \\ y \geq 2x + 1 \end{cases}$

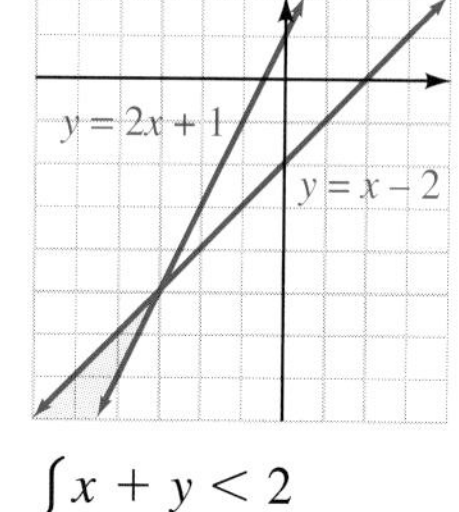

22. $\begin{cases} y < 3x + 2 \\ y < -2x + 3 \end{cases}$

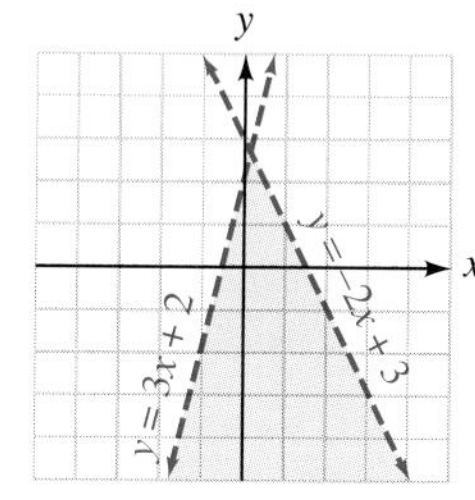

23. $\begin{cases} x + y < 2 \\ x + y \leq 1 \end{cases}$

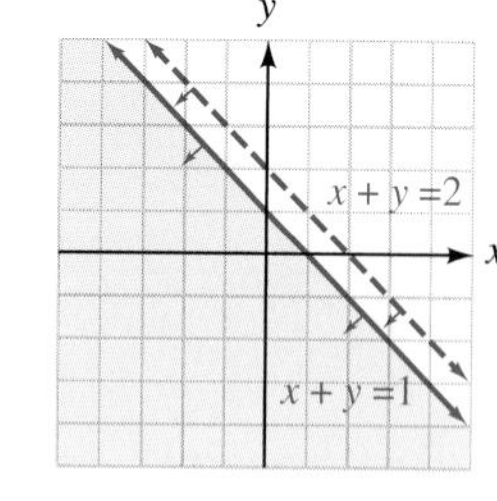

24. $\begin{cases} 3x + 2y \geq 6 \\ x + 3y \leq 2 \end{cases}$

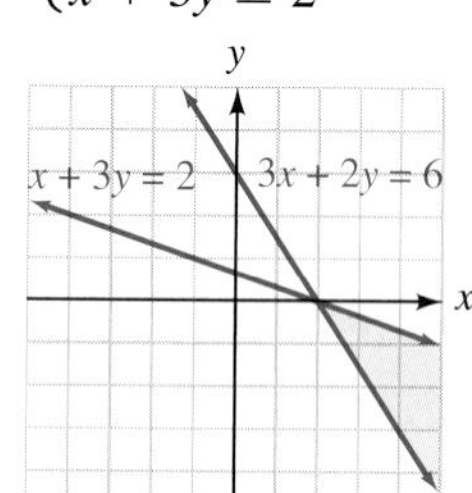

25. $\begin{cases} x + 2y < 3 \\ 2x - 4y < 8 \end{cases}$

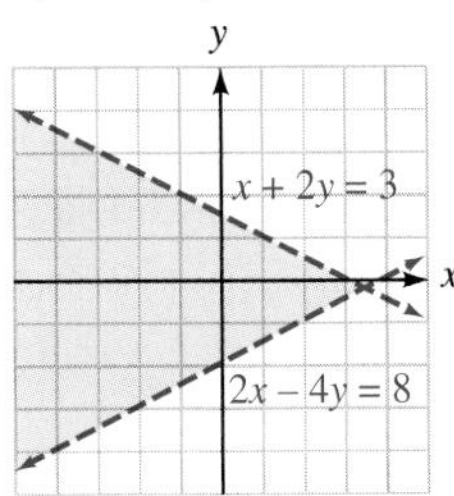

26. $\begin{cases} 3x + y \le 1 \\ -x + 2y \ge 9 \end{cases}$

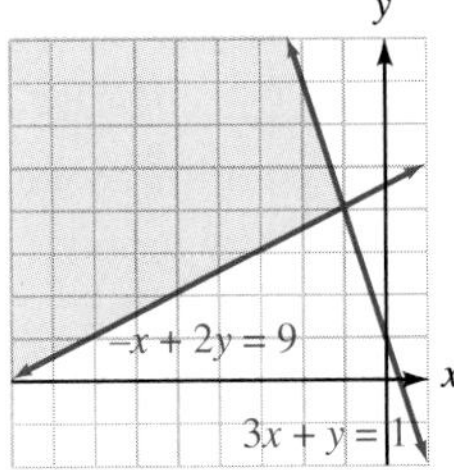

27. $\begin{cases} 2x - 3y \ge 6 \\ 3x + 2y < 6 \end{cases}$

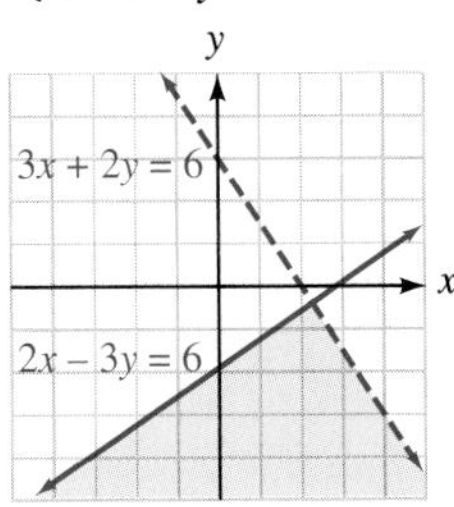

28. $\begin{cases} 4x + 2y \le 6 \\ 2x - 4y \ge 10 \end{cases}$

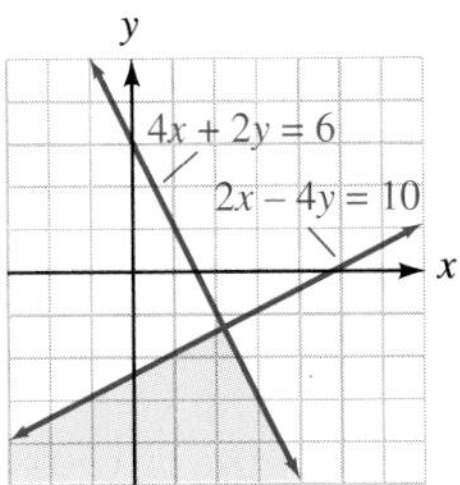

29. $\begin{cases} y \ge x^2 - 4 \\ y \le \dfrac{1}{2}x \end{cases}$

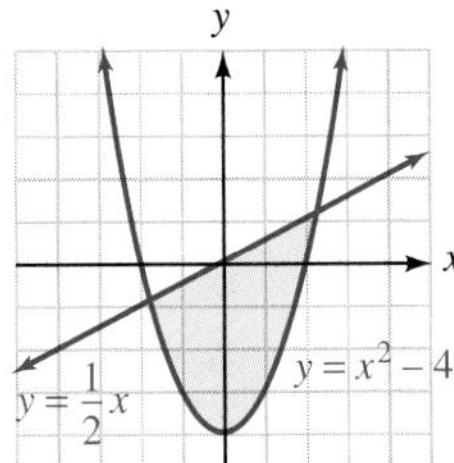

30. $\begin{cases} y \le -x^2 + 4 \\ y > -x - 1 \end{cases}$

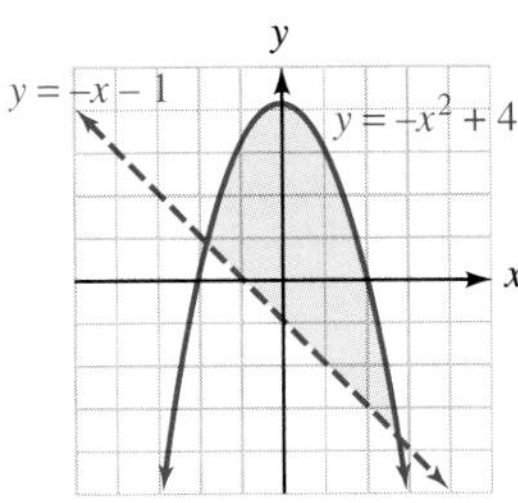

31. $\begin{cases} y \ge x^2 \\ y < 4 - x^2 \end{cases}$

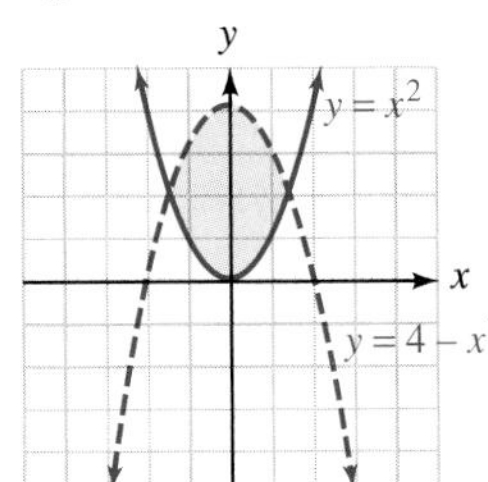

32. $\begin{cases} x^2 + y \le 1 \\ y - x^2 \ge -1 \end{cases}$

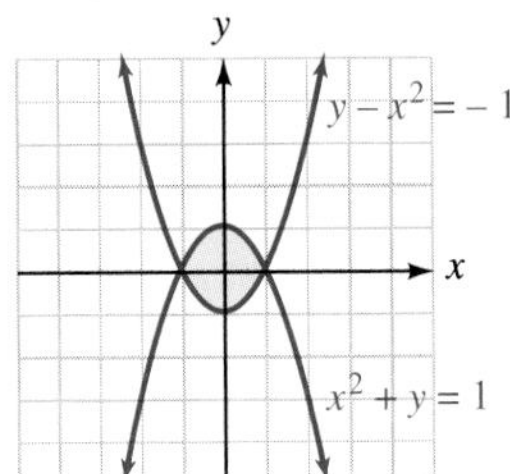

33. $\begin{cases} 2x - y \le 0 \\ x + 2y \le 10 \\ y \ge 0 \end{cases}$

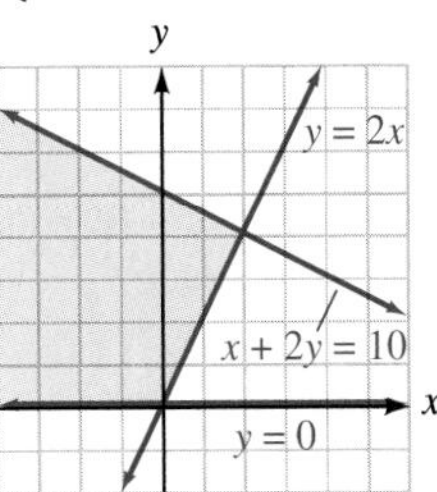

34. $\begin{cases} 3x - 2y \ge 5 \\ 2x + y \ge 8 \\ x \le 5 \end{cases}$

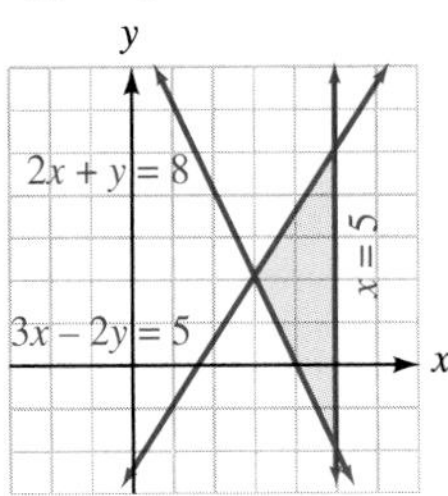

35. $\begin{cases} x - 2y \ge 0 \\ x - y \le 2 \\ x \ge 0 \end{cases}$

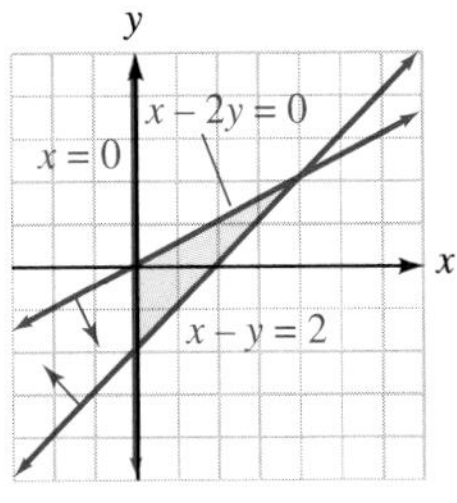

36. $\begin{cases} 2x + 3y \le 6 \\ x - y \ge 4 \\ y \ge -4 \end{cases}$

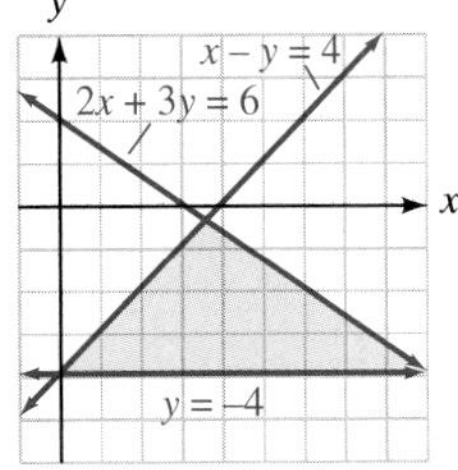

37. $\begin{cases} x + y \le 4 \\ x - y \le 4 \\ x \ge 0 \\ y \ge 0 \end{cases}$

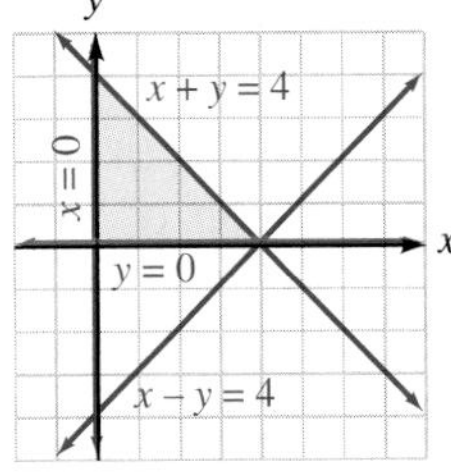

38. $\begin{cases} 2x + 3y \ge 12 \\ 2x - 3y \le 6 \\ x \ge 0 \\ y \le 4 \end{cases}$

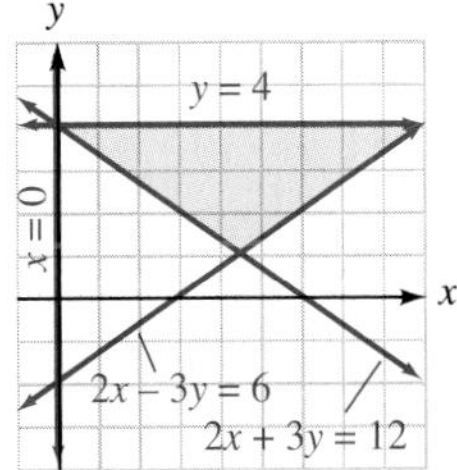

39. $\begin{cases} 3x - 2y \le 6 \\ x + 2y \le 10 \\ x \ge 0 \\ y \ge 0 \end{cases}$

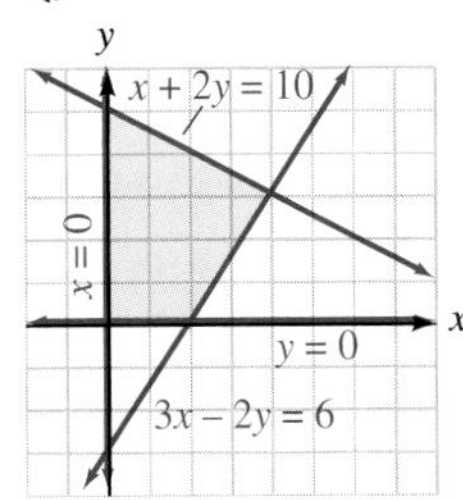

40. $\begin{cases} 3x + 2y \ge 12 \\ 5x - y \le 15 \\ x \ge 0 \\ y \le 4 \end{cases}$

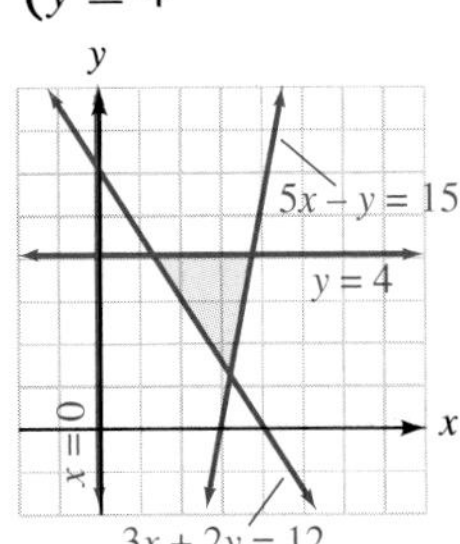

DISCOVERY AND WRITING

41. When graphing a linear inequality, explain how to determine the boundary.

42. When graphing a linear inequality, explain how to decide whether the boundary is included.

43. When graphing a linear inequality, explain how to decide which side of the boundary to shade.

44. Does the method you describe in Exercise 43 work for the inequality $3x - 2y \le 0$? Explain.

REVIEW

45. State the remainder theorem.

46. State the factor theorem.

Fill in the blanks.

47. According to Descartes' rule of signs, $53x^4 + 13x^2 - 12 = 0$ has one positive solution(s) and one negative solution(s).

48. A root of the polynomial equation $P(x) = 0$ is a zero of the polynomial $P(x)$.

49. If $x^{37} + 3x^{19} - 4$ is divided by $x - 1$, the remainder is 0.

50. Is $x - 2$ a factor of $x^4 - 7x - 2$? yes

6.8 Linear Programming

In this section, you will learn about

- Linear Programming
- Applications of Linear Programming

Linear Programming

Linear programming is a mathematical technique used to find the optimal allocation of resources in the military, business, telecommunications, and other fields. It got its start during World War II when it became necessary to move huge quantities of people, materials, and supplies as efficiently and economically as possible.

To solve linear programming problems, we maximize (or minimize) a function (called the **objective function**) subject to given conditions on its variables. These conditions (called **constraints**) are usually given as a system of linear inequalities. For example, suppose that the annual profit (in millions of dollars) earned by a business is given by the equation $P = y + 2x$ and that x and y are subject to the following constraints:

$$\begin{cases} 3x + y \le 120 \\ x + y \le 60 \\ x \ge 0 \\ y \ge 0 \end{cases}$$

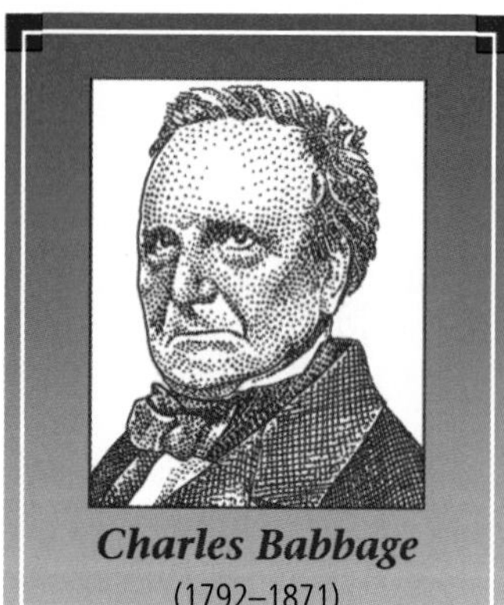

Charles Babbage
(1792–1871)
In 1823, Babbage built a steam-powered digital calculator, which he called a *difference engine.* Thought to be a crackpot by his London neighbors, Babbage was a visionary. His machine embodied principles still used in modern computers.

To find the maximum profit P that can be earned by the business, we solve the system of inequalities as shown in Figure 6-19(a) and find the coordinates of each corner point of the region R. This region is often called a **feasibility region.** We can then write the profit equation

$$P = y + 2x \qquad \text{in the form} \qquad y = -2x + P$$

The equation $y = -2x + P$ is the equation of a set of parallel lines, each with a slope of -2 and a y-intercept of P. The graph of $y = -2x + P$ for three values of P is shown as red lines in Figure 6-19b. To find the red line that passes through region R and provides the maximum value of P, we locate the red line with the greatest y-intercept. Since line l has the greatest y-intercept and intersects region R at the corner point (30, 30), the maximum value of P (subject to the given constraints) is

$$\begin{aligned} P &= y + 2x \\ &= 30 + 2(30) \\ &= 90 \end{aligned}$$

Thus, the maximum profit P that can be earned is \$90 million. This profit occurs when $x = 30$ and $y = 30$.

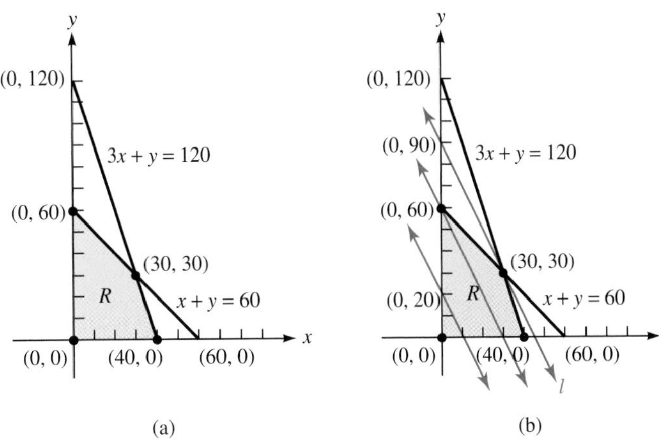

Figure 6-19

The preceding discussion illustrates the following important fact.

Maximum or Minimum of an Objective Function

> If a linear function, subject to the constraints of a system of linear inequalities in two variables, attains a maximum or a minimum value, that value will occur at a corner point or along an entire edge of the region R that represents the solution of the system.

ACTIVE EXAMPLE 1

If $P = 2x + 3y$, find the maximum value of P subject to the following constraints:

$$\begin{cases} x + y \leq 4 \\ 2x + y \leq 6 \\ x \geq 0 \\ y \geq 0 \end{cases}$$

COLLEGE **Algebra** $f(x)$ **Now™**
Explore this Active Example and test yourself on the steps followed here at **http://1pass.thomson.com**. You can also find this example on the Interactive Video Skillbuilder CD-ROM.

Solution We solve the system of inequalities to find the feasibility region R shown in Figure 6-20. The coordinates of its corner points are (0, 0), (3, 0), (0, 4), and (2, 2).

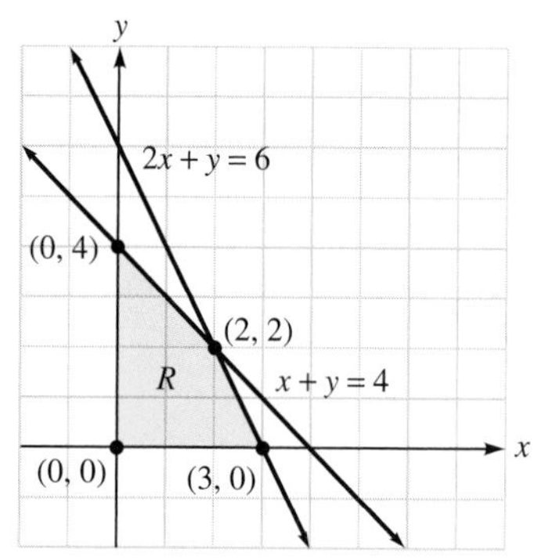

Figure 6-20

Since the maximum value of P will occur at a corner of R, we substitute the coordinates of each corner point into the objective function $P = 2x + 3y$ and find the one that gives the maximum value of P.

Point	$P = 2x + 3y$
(0, 0)	$P = 2(0) + 3(0) = 0$
(3, 0)	$P = 2(3) + 3(0) = 6$
(2, 2)	$P = 2(2) + 3(2) = 10$
(0, 4)	$P = 2(0) + 3(4) = 12$

The maximum value $P = 12$ occurs when $x = 0$ and $y = 4$.

Self Check Find the maximum value of $P = 4x + 3y$, subject to the constraints of Example 1. ■

ACTIVE EXAMPLE 2 If $P = 3x + 2y$, find the minimum value of P subject to the following constraints:

COLLEGE **Algebra f(x) Now™**
Go to **http://1pass.thomson.com** or your CD to practice this example.

$$\begin{cases} x + y \geq 1 \\ x - y \leq 1 \\ x - y \geq 0 \\ x \leq 2 \end{cases}$$

Solution We refer to the feasibility region shown in Figure 6-21 with corner points at $\left(\frac{1}{2}, \frac{1}{2}\right)$, (2, 2), (2, 1), and (1, 0).

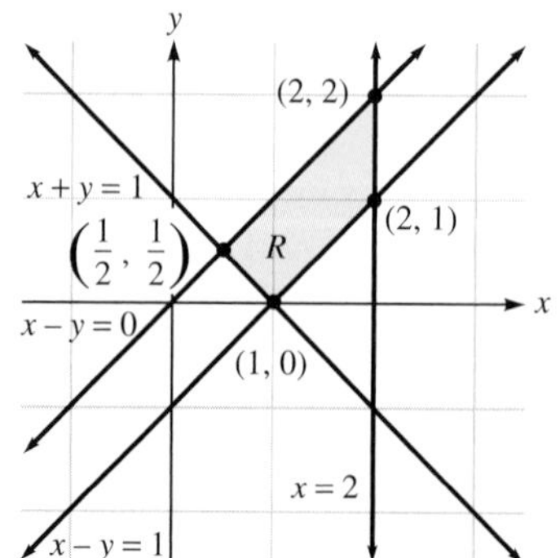

Figure 6-21

Since the minimum value of P occurs at a corner point of region R, we substitute the coordinates of each corner point into the objective function $P = 3x + 2y$ and find the one that gives the minimum value of P.

Point	$P = 3x + 2y$
$\left(\frac{1}{2}, \frac{1}{2}\right)$	$P = 3\left(\frac{1}{2}\right) + 2\left(\frac{1}{2}\right) = \frac{5}{2}$
$(2, 2)$	$P = 3(2) + 2(2) = 10$
$(2, 1)$	$P = 3(2) + 2(1) = 8$
$(1, 0)$	$P = 3(1) + 2(0) = 3$

The minimum value $P = \frac{5}{2}$ occurs when $x = \frac{1}{2}$ and $y = \frac{1}{2}$.

Self Check Find the minimum value of $P = 2x + y$, subject to the constraints of Example 2. ■

■ Applications of Linear Programming

Linear programming problems can be very complex and involve hundreds of variables. In this section, we will consider a few simple problems. Since they involve only two variables, we can solve them using graphical methods.

ACTIVE EXAMPLE 3 An accountant prepares tax returns for individuals and for small businesses. On average, each individual return requires 3 hours of her time and 1 hour of computer time. Each business return requires 4 hours of her time and 2 hours of computer time. Because of other business considerations, her time is limited to 240 hours, and the computer time is limited to 100 hours. If she earns a profit of $80 on each individual return and a profit of $150 on each business return, how many returns of each type should she prepare to maximize her profit?

COLLEGE **Algebra** *f(x)* **Now**™
Go to **http://1pass.thomson.com** or your CD to practice this example.

Solution First, we organize the given information into a table.

	Individual tax return	Business tax return	Time available
Accountant's time	3	4	240 hours
Computer time	1	2	100 hours
Profit	$80	$150	

Then we solve the problem using the following steps.

Find the objective function

Suppose that x represents the number of individual returns to be completed and y represents the number of business returns to be completed. Since each of the x individual returns will earn an $80 profit, and each of the y business returns will earn a $150 profit, the total profit is given by the equation

$$P = 80x + 150y$$

Find the feasibility region

Since the number of individual returns and business returns cannot be negative, we know that $x \geq 0$ and $y \geq 0$.

Since each of the x individual returns will take 3 hours of her time, and each of the y business returns will take 4 hours of her time, the total number of hours she will work will be $(3x + 4y)$ hours. This amount must be less than or equal to her available time, which is 240 hours. Thus, the inequality $3x + 4y \leq 240$ is a constraint on the accountant's time.

Since each of the x individual returns will take 1 hour of computer time, and each of the y business returns will take 2 hours of computer time, the total number of hours of computer time will be $(x + 2y)$ hours. This amount must be less than or equal to the available computer time, which is 100 hours. Thus, the inequality $x + 2y \leq 100$ is a constraint on the computer time.

We have the following constraints on the values of x and y.

$$\begin{cases} x \geq 0 & \text{The number of individual returns is nonnegative.} \\ y \geq 0 & \text{The number of business returns is nonnegative.} \\ 3x + 4y \leq 240 & \text{The accountant's time must be less than or equal to 240 hours.} \\ x + 2y \leq 100 & \text{The computer time must be less than or equal to 100 hours.} \end{cases}$$

To find the feasibility region, we graph each of the constraints to find region R, as in Figure 6-22. The four corner points of this region have coordinates of (0, 0), (80, 0), (40, 30), and (0, 50).

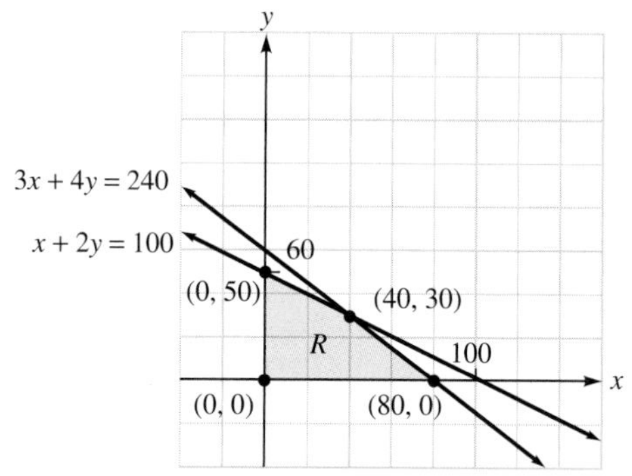

Figure 6-22

Find the maximum profit

To find the maximum profit, we substitute the coordinates of each corner point into the objective function $P = 80x + 150y$.

Point	$P = 80x + 150y$
(0, 0)	$P = 80(0) + 150(0) = 0$
(80, 0)	$P = 80(80) + 150(0) = 6{,}400$
(40, 30)	$P = 80(40) + 150(30) = 7{,}700$
(0, 50)	$P = 80(0) + 150(50) = 7{,}500$

From the table, we can see that the accountant will earn a maximum profit of \$7,700 if she prepares 40 individual returns and 30 business returns. ■

EXAMPLE 4 Vigortab and Robust are two diet supplements. Each Vigortab tablet costs 50¢ and contains 3 units of calcium, 20 units of vitamin C, and 40 units of iron. Each Robust tablet costs 60¢ and contains 4 units of calcium, 40 units of vitamin C, and 30 units of iron. At least 24 units of calcium, 200 units of vitamin C, and 120 units of iron are required for the daily needs of one patient. How many tablets of each supplement should be taken daily for a minimum cost? Find the daily minimum cost.

Solution First, we organize the given information into a table.

	Vigortab	Robust	Amount required
Calcium	3	4	24
Vitamin C	20	40	200
Iron	40	30	120
Cost	50¢	60¢	

Find the objective function

We can let x represent the number of Vigortab tablets to be taken daily and y the corresponding number of Robust tablets. Because each of the x Vigortab tablets will cost 50¢, and each of the y Robust tablets will cost 60¢, the total cost will be given by the equation

$$C = 0.50x + 0.60y \quad \text{50¢ = \$0.50 and 60¢ = \$0.60.}$$

Find the feasibility region

Since there are requirements for calcium, vitamin C, and iron, there is a constraint for each. Note that neither x nor y can be negative.

$$\begin{cases} 3x + 4y \geq 24 & \text{The amount of calcium must be greater than or equal to 24 units.} \\ 20x + 40y \geq 200 & \text{The amount of vitamin C must be greater than or equal to 200 units.} \\ 40x + 30y \geq 120 & \text{The amount of iron must be greater than or equal to 120 units.} \\ x \geq 0, y \geq 0 & \text{The number of tablets taken must be greater than or equal to 0.} \end{cases}$$

We graph the inequalities to find the feasibility region and the coordinates of its corner points, as in Figure 6-23.

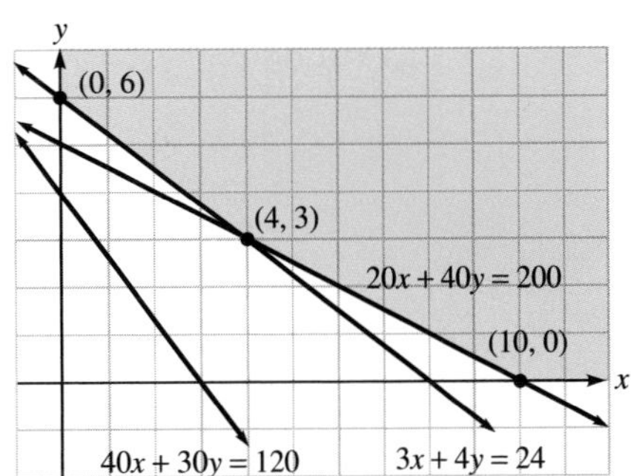

Figure 6-23

Find the minimum cost

In this case, the feasibility region is not bounded on all sides. The coordinates of the corner points are (0, 6), (4, 3), and (10, 0). To find the minimum cost, we substitute each pair of coordinates into the objective function.

Point	$C = 0.50x + 0.60y$
(0, 6)	$C = 0.50(0) + 0.60(6) = 3.60$
(4, 3)	$C = 0.50(4) + 0.60(3) = 3.80$
(10, 0)	$C = 0.50(10) + 0.60(0) = 5.00$

A minimum cost will occur if no Vigortab and 6 Robust tablets are taken daily. The minimum daily cost is \$3.60. ■

EXAMPLE 5 A television program director must schedule comedy skits and musical numbers for prime-time variety shows. Each comedy skit requires 2 hours of rehearsal time, costs \$3,000, and brings in \$20,000 from the show's sponsors. Each musical number requires 1 hour of rehearsal time, costs \$6,000, and generates \$12,000. If 250 hours are available for rehearsal, and \$600,000 is budgeted for comedy and music, how many segments of each type should be produced to maximize income? Find the maximum income.

Solution First, we organize the given information into a table.

	Comedy	Musical	Available
Rehearsal time (hours)	2	1	250
Cost (in \$1,000s)	3	6	600
Generated income (in \$1,000s)	20	12	

Find the objective function

We can let x represent the number of comedy skits and y the number of musical numbers to be scheduled. Since each of the x comedy skits generates \$20 thousand, the income generated by the comedy skits is \$$20x$ thousand. The musical numbers produce \$$12y$ thousand. The objective function to be maximized is

$$V = 20x + 12y$$

Find the feasibility region

Since there are limits on rehearsal time and budget, there is a constraint for each. Note that neither x nor y can be negative.

$$\begin{cases} 2x + y \leq 250 & \text{The total rehearsal time must be less than or equal to 250 hours.} \\ 3x + 6y \leq 600 & \text{The total cost must be less than or equal to \$600 thousand.} \\ x \geq 0,\ y \geq 0 & \text{The numbers of skits and musical numbers must be greater than or equal to 0.} \end{cases}$$

We graph the inequalities to find the feasibility region shown in Figure 6-24 and find the coordinates of each corner point.

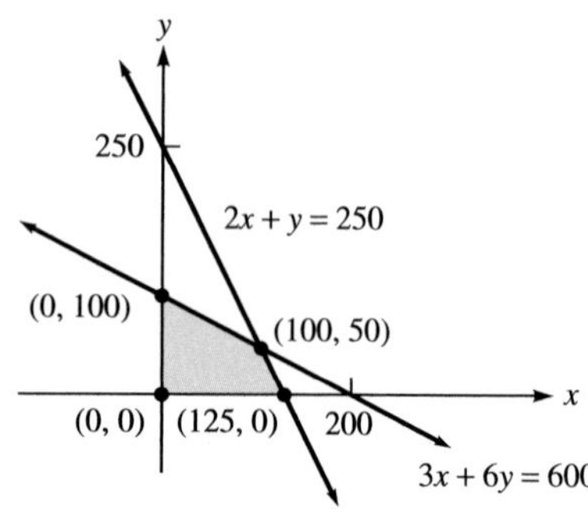

Figure 6-24

Find the maximum income

The coordinates of the corner points of the feasible region are (0, 0), (0, 100), (100, 50), and (125, 0). To find the maximum income, we substitute each pair of coordinates into the objective function.

Corner point	$V = 20x + 12y$
(0, 0)	$V = 20(0) + 12(0) = 0$
(0, 100)	$V = 20(0) + 12(100) = 1{,}200$
(100, 50)	$V = 20(100) + 12(50) = 2{,}600$
(125, 0)	$V = 20(125) + 12(0) = 2{,}500$

Maximum income will occur if 100 comedy skits and 50 musical numbers are scheduled. The maximum income will be 2,600 thousand dollars, or $2,600,000. ■

Self Check Answers

1. 14 **2.** $\frac{3}{2}$

6.8 Exercises

VOCABULARY AND CONCEPTS ***Fill in the blanks.***

1. In a linear program, the inequalities are called constraints.

2. Ordered pairs that satisfy the constraints of a linear program are called feasible solutions.

3. The function to be maximized (or minimized) in a linear program is called the objective function.

4. The objective function of a linear program attains a maximum (or minimum), subject to the constraints, at a corner or along an edge of the feasibility region.

PRACTICE ***Maximize P subject to the following constraints.***

5. $P = 2x + 3y$

$$\begin{cases} x \geq 0 \\ y \geq 0 \\ x + y \leq 4 \end{cases}$$

$P = 12$ at $(0, 4)$

6. $P = 3x + 2y$

$$\begin{cases} x \geq 0 \\ y \geq 0 \\ x + y \leq 4 \end{cases}$$

$P = 12$ at $(4, 0)$

7. $P = y + \frac{1}{2}x$

$$\begin{cases} x \geq 0 \\ y \geq 0 \\ 2y - x \leq 1 \\ y - 2x \geq -2 \end{cases}$$

$P = \frac{13}{6}$ at $\left(\frac{5}{3}, \frac{4}{3}\right)$

8. $P = 4y - x$

$$\begin{cases} x \leq 2 \\ y \geq 0 \\ x + y \geq 1 \\ 2y - x \leq 1 \end{cases}$$

$P = 4$ at $\left(2, \frac{3}{2}\right)$

9. $P = 2x + y$

$$\begin{cases} y \geq 0 \\ y - x \leq 2 \\ 2x + 3y \leq 6 \\ 3x + y \leq 3 \end{cases}$$

$P = \frac{18}{7}$ at $\left(\frac{3}{7}, \frac{12}{7}\right)$

10. $P = x - 2y$

$$\begin{cases} x + y \leq 5 \\ y \leq 3 \\ x \leq 2 \\ x \geq 0 \\ y \geq 0 \end{cases}$$

$P = 2$ at $(2, 0)$

11. $P = 3x - 2y$

$$\begin{cases} x \leq 1 \\ x \geq -1 \\ y - x \leq 1 \\ x - y \leq 1 \end{cases}$$

$P = 3$ at $(1, 0)$

12. $P = x - y$

$$\begin{cases} 5x + 4y \leq 20 \\ y \leq 5 \\ x \geq 0 \\ y \geq 0 \end{cases}$$

$P = 4$ at $(4, 0)$

Minimize P subject to the following constraints.

13. $P = 5x + 12y$

$$\begin{cases} x \geq 0 \\ y \geq 0 \\ x + y \leq 4 \end{cases}$$

$P = 0$ at $(0, 0)$

14. $P = 3x + 6y$

$$\begin{cases} x \geq 0 \\ y \geq 0 \\ x + y \leq 4 \end{cases}$$

$P = 0$ at $(0, 0)$

15. $P = 3y + x$

$$\begin{cases} x \geq 0 \\ y \geq 0 \\ 2y - x \leq 1 \\ y - 2x \geq -2 \end{cases}$$

$P = 0$ at $(0, 0)$

16. $P = 5y + x$

$$\begin{cases} x \leq 2 \\ y \geq 0 \\ x + y \geq 1 \\ 2y - x \leq 1 \end{cases}$$

$P = 1$ at $(1, 0)$

17. $P = 6x + 2y$

$$\begin{cases} y \geq 0 \\ y - x \leq 2 \\ 2x + 3y \leq 6 \\ 3x + y \leq 3 \end{cases}$$

$P = -12$ at $(-2, 0)$

18. $P = 2y - x$

$$\begin{cases} x \geq 0 \\ y \geq 0 \\ x + y \leq 5 \\ x + 2y \geq 2 \end{cases}$$

$P = -5$ at $(5, 0)$

19. $P = 2x - 2y$

$$\begin{cases} x \leq 1 \\ x \geq -1 \\ y - x \leq 1 \\ x - y \leq 1 \end{cases}$$

$P = -2$ at the edge joining $(1, 2)$ and $(-1, 0)$

20. $P = y - 2x$

$$\begin{cases} x + 2y \leq 4 \\ 2x + y \leq 4 \\ x + 2y \geq 2 \\ 2x + y \geq 2 \end{cases}$$

$P = -4$ at $(2, 0)$

APPLICATIONS ***Write the objective function and the inequalities that describe the constraints in each problem. Graph the feasibility region, showing the corner points. Then find the maximum or minimum value of the objective function.***

21. Making furniture Two woodworkers, Tom and Carlos, get \$100 for making a table and \$80 for making a chair. On average, Tom must work 3 hours and Carlos 2 hours to make a chair. Tom must work 2 hours and Carlos 6 hours to make a table. If neither wishes to work more than 42 hours per week, how many tables and how many chairs should they make each week to maximize their income? Find the maximum income.

	Table	Chair	Time available
Income (\$)	100	80	
Tom's time (hr)	2	3	42
Carlos's time (hr)	6	2	42

3 tables, 12 chairs, \$1,260

22. Making crafts Two artists, Nina and Rob, make yard ornaments. They get \$80 for each wooden snowman they make and \$64 for each wooden Santa Claus. On average, Nina must work 4 hours and Rob 2 hours to make a snowman. Nina must work 3 hours and Rob 4 hours to make a Santa Claus. If neither wishes to work more than 20 hours per week, how many of each ornament should they make each week to maximize their income? Find the maximum income.

	Snowman	Santa Claus	Time available
Income (\$)	80	64	
Nina's time (hr)	4	3	20
Rob's time (hr)	2	4	20

2 snowmen, 4 Santas, \$416

23. Inventories An electronics store manager stocks from 20 to 30 IBM-compatible computers and from 30 to 50 Apple computers. There is room in the store to stock up to 60 computers. The manager receives a commission of \$50 on the sale of each IBM-compatible computer and \$40 on the sale of each Apple computer. If the manager can sell all of the computers, how many should she stock to maximize her commissions? Find the maximum commission.

Inventory	IBM	Apple
Minimum	20	30
Maximum	30	50
Commission	\$50	\$40

30 IBMs, 30 Macs, \$2,700

24. Diet problems A diet requires at least 16 units of vitamin C and at least 34 units of vitamin B complex. Two food supplements are available that provide these nutrients in the amounts and costs shown in the table. How much of each should be used to minimize the cost?

Supplement	Vitamin C	Vitamin B	Cost
A	3 units/g	2 units/g	3¢/g
B	2 units/g	6 units/g	4¢/g

2 g of A, 5 g of B

25. Production Manufacturing VCRs and TVs requires the use of the electronics, assembly, and finishing departments of a factory, according to the following schedule:

	Hours for VCR	Hours for TV	Hours available per week
Electronics	3	4	180
Assembly	2	3	120
Finishing	2	1	60

Each VCR has a profit of \$40, and each TV has a profit of \$32. How many VCRs and TVs should be manufactured weekly to maximize profit? Find the maximum profit. **15 VCRs, 30 TVs, \$1,560**

26. **Production problems** A company manufactures one type of computer chip that runs at 2.0 GHz and another that runs at 2.8 GHz. The company can make a maximum of 50 fast chips per day and a maximum of 100 slow chips per day. It takes 6 hours to make a fast chip and 3 hours to make a slow chip, and the company's employees can provide up to 360 hours of labor per day. If the company makes a profit of \$20 on each 2.8-GHz chip and \$27 on each 2.0-GHz chip, how many of each type should be manufactured to earn the maximum profit?
10 fast chips and 100 slow chips

27. **Financial planning** A stockbroker has \$200,000 to invest in stocks and bonds. She wants to invest at least \$100,000 in stocks and at least \$50,000 in bonds. If stocks have an annual yield of 9% and bonds have an annual yield of 7%, how much should she invest in each to maximize her income? Find the maximum return.
\$150,000 in stocks, \$50,000 in bonds; \$17,000

28. **Production** A small country exports soybeans and flowers. Soybeans require 8 workers per acre, flowers require 12 workers per acre, and 100,000 workers are available. Government contracts require that there be at least 3 times as many acres of soybeans as flowers planted. It costs \$250 per acre to plant soybeans and \$300 per acre to plant flowers, and there is a budget of \$3 million. If the profit from soybeans is \$1,600 per acre and the profit from flowers is \$2,000 per acre, how many acres of each crop should be planted to maximize profit? Find the maximum profit.
10,000 acres of soybeans, 1,667 acres of flowers; $\$19\frac{1}{3}$ million.

29. **Band trips** A high school band trip will require renting buses and trucks to transport no fewer than 100 students and 18 or more large instruments. Each bus can accommodate 40 students plus three large instruments; it costs \$350 to rent. Each truck can accommodate 10 students plus 6 large instruments and costs \$200 to rent. How many of each type of vehicle should be rented for the cost to be minimum? Find the minimum cost. **2 buses, 2 trucks; \$1,100**

30. **Making ice cream** An ice cream store sells two new flavors: Fantasy and Excess. Each barrel of Fantasy requires 4 pounds of nuts and 3 pounds of chocolate and has a profit of \$500. Each barrel of Excess requires 4 pounds of nuts and 2 pounds of chocolate and has a profit of \$400. There are 16 pounds of nuts and 18 pounds of chocolate in stock, and the owner does not want to buy more for this batch. How many barrels of each should be made for a maximum profit? Find the maximum profit.
4 barrels of Fantasy, 0 barrels of Excess; \$2,000

DISCOVERY AND WRITING

31. Does the objective function attain a maximum at the corners of a region defined by nonlinear inequalities? Attempt to maximize $P(x) = x + y$ on the region.

$$\begin{cases} x \geq 0 \\ y \geq 0 \\ y \leq 4 - x^2 \end{cases}$$

and write a paragraph on your findings.

32. Attempt to minimize the objective function of Exercise 31.

REVIEW *Write each matrix in reduced row echelon form. Problem 33 cannot be done with a calculator.*

33. $\begin{bmatrix} 1 & 2 & 3 \\ 1 & -2 & 3 \\ 0 & 2 & -3 \\ 2 & 0 & 6 \end{bmatrix}$ $\begin{bmatrix} 1 & 0 & 0 \\ 0 & 1 & 0 \\ 0 & 0 & 1 \\ 0 & 0 & 0 \end{bmatrix}$

34. $\begin{bmatrix} 1 & 3 & -2 & 1 \\ 3 & 9 & -3 & 2 \end{bmatrix}$ $\begin{bmatrix} 1 & 3 & 0 & \frac{1}{3} \\ 0 & 0 & 1 & -\frac{1}{3} \end{bmatrix}$

35. The matrix in Exercise 34 is the system matrix of a system of equations. Find the general solution of the system. $\left(\frac{1}{3} - 3y, y, -\frac{1}{3}\right)$

36. Find the inverse of $\begin{bmatrix} 7 & 2 & 5 \\ 3 & 1 & 2 \\ 3 & 1 & 3 \end{bmatrix}$.

$\begin{bmatrix} 1 & -1 & -1 \\ -3 & 6 & 1 \\ 0 & -1 & 1 \end{bmatrix}$

PROBLEMS AND PROJECTS

1. Solve $\begin{cases} 16x + 21y = 14 \\ 19x + 25y = 14 \end{cases}$ both graphically and algebraically. Which method is easier, and why?

2. Solve $\begin{cases} \dfrac{2}{x} + \dfrac{5}{y} = -1 \\ \dfrac{3}{x} - \dfrac{1}{y} = 7 \end{cases}$.

3. Solve $\begin{cases} 3.7x + 2.9y = 6.61 \\ 4.2x + 3.3y = 7.49 \end{cases}$.

4. Solve $\begin{cases} 3.7x + 2.9y = 6.6 \\ 4.2x + 3.3y = 7.5 \end{cases}$, which is the system of Problem 3, with the constants on the right-hand side rounded to the nearest tenth. Does this solution approximate the solution to Problem 3?

5. Let $A = \begin{bmatrix} 1 & 1 & 0 \\ 0 & 1 & 1 \end{bmatrix}$ and $B = \begin{bmatrix} 1 & -1 \\ 0 & 1 \\ 0 & 0 \end{bmatrix}$.

   Verify that the product AB is the identity matrix. Is A invertible? Explain.

6. If $A = \begin{bmatrix} 0 & 0 & 1 \\ 0 & 1 & 0 \\ 1 & 0 & 0 \end{bmatrix}$, find $A^2, A^3, \ldots$. Describe the pattern.

Project 1

A 13-foot and a 15-foot ladder lean against opposite walls of a 12-foot-wide alley, as in the illustration. How far above the ground do they cross? (*Hint:* Set up a coordinate system.)

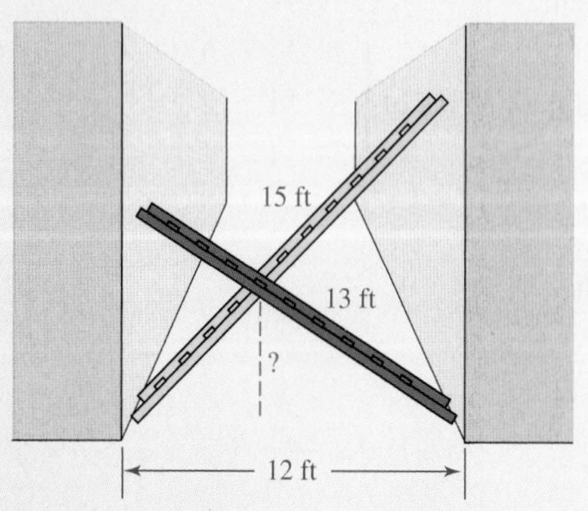

Project 2

The appearance of motion in computer graphics is accomplished, point-by-point, by matrix multiplication. For example, to reflect the point (x, y) in the y-axis, as in the illustration, the graphics software multiplies the point's **coordinate matrix,** $\begin{bmatrix} x \\ y \end{bmatrix}$, by $A = \begin{bmatrix} -1 & 0 \\ 0 & 1 \end{bmatrix}$, as follows:

$$A\begin{bmatrix} x \\ y \end{bmatrix} = \begin{bmatrix} -1 & 0 \\ 0 & 1 \end{bmatrix}\begin{bmatrix} x \\ y \end{bmatrix} = \begin{bmatrix} -x \\ y \end{bmatrix}$$

1. Describe geometrically the effect on a point (x, y) that results from multiplying its coordinate matrix by A.

   a. $A = \begin{bmatrix} 1 & 0 \\ 0 & -1 \end{bmatrix}$ b. $A = \begin{bmatrix} -1 & 0 \\ 0 & -1 \end{bmatrix}$

   c. $A = \begin{bmatrix} 3 & 0 \\ 0 & 3 \end{bmatrix}$ d. $A = \begin{bmatrix} 0 & 1 \\ 1 & 0 \end{bmatrix}$

2. To rotate the point (4, 3) clockwise until it reaches the x-axis at (5, 0), the coordinate matrix $\begin{bmatrix} 4 \\ 3 \end{bmatrix}$ is multiplied by a rotation matrix of the form $A = \begin{bmatrix} a & -b \\ b & a \end{bmatrix}$. See the illustration. Find the matrix A such that $A\begin{bmatrix} 4 \\ 3 \end{bmatrix} = \begin{bmatrix} 5 \\ 0 \end{bmatrix}$

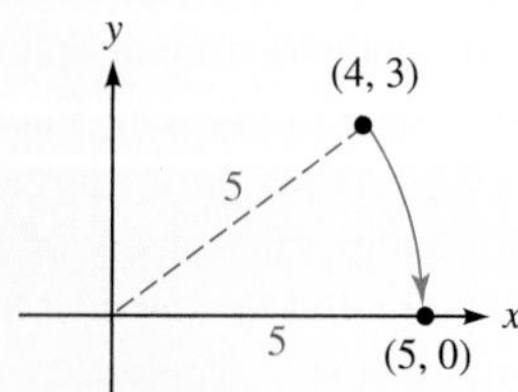

Project 3

The temperatures of the edges of a steel plate are shown in the illustration. The temperature at each other point is the average of the temperatures of four surrounding points. Find the temperatures of the six points on the plate in the illustration.

$\left(\textit{Hint}: \text{The temperature } T_1 \text{ of point 1 is given by } T_1 = \dfrac{0° + 10° + T_2 + T_3}{4}. \right)$

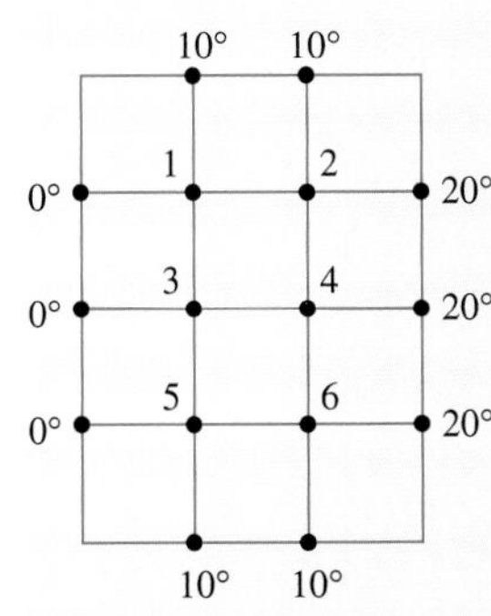

Project 4

Some graphing calculators can display the graphs of linear inequalities. Explore the capabilities of your calculator, and see if you can produce the display shown here. It is the graph of the system

$$\begin{cases} y \le x + 2 \\ y \ge 2 - x \\ y \le 4 - 2x \end{cases}$$

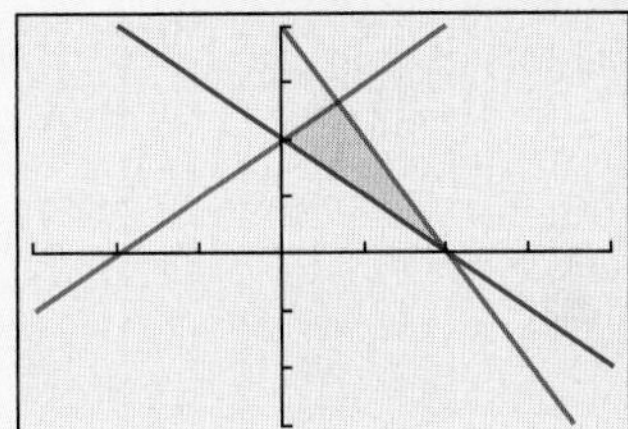

CHAPTER SUMMARY

CONCEPTS

REVIEW EXERCISES

6.1 Systems of Linear Equations

To solve a system of equations by graphing, find the coordinates of the point where all graphs intersect.

Solve each system by graphing.

1. $\begin{cases} 2x - y = -1 \\ x + y = 7 \end{cases}$

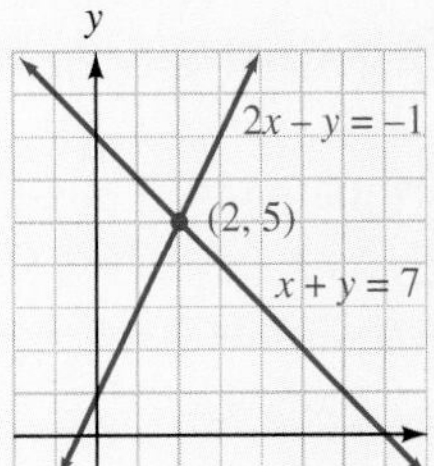

2. $\begin{cases} 5x + 2y = 1 \\ 2x - y = -5 \end{cases}$

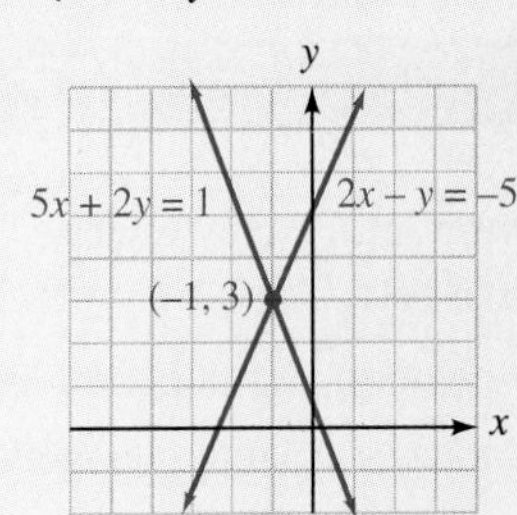

3. $\begin{cases} y = 5x + 7 \\ x = y - 7 \end{cases}$

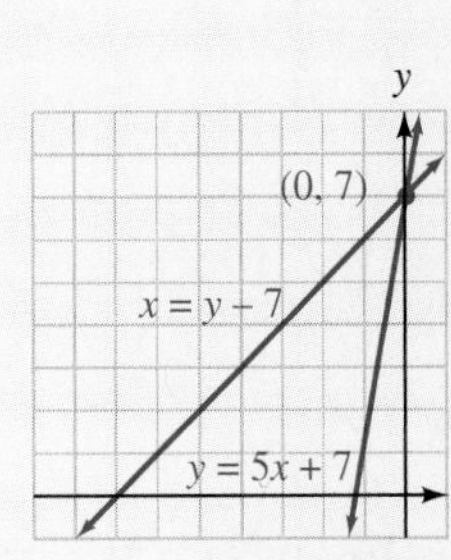

To solve a system by substitution, solve one equation for one variable, and substitute that result into the other equation.

Solve by substitution:

4. $\begin{cases} 2y + x = 0 \\ x = y + 3 \end{cases}$ **$x = 2, y = -1$**

5. $\begin{cases} 2x + y = -3 \\ x - y = 3 \end{cases}$ **$x = 0, y = -3$**

6. $\begin{cases} \dfrac{x + y}{2} + \dfrac{x - y}{3} = 1 \\ y = 3x - 2 \end{cases}$ **$x = 1, y = 1$**

To solve a system by addition, multiply one or both equations by suitable constants so that when the results are added, one variable will drop out.

Solve by addition:

7. $\begin{cases} x + 5y = 7 \\ 3x + y = -7 \end{cases}$
$x = -3, y = 2$

8. $\begin{cases} 2x + 3y = 11 \\ 3x - 7y = -41 \end{cases}$
$x = -2, y = 5$

9. $\begin{cases} 2(x + y) - x = 0 \\ 3(x + y) + 2y = 1 \end{cases}$
$x = 2, y = -1$

Solve by any method:

10. $\begin{cases} 3x + 2y - z = 2 \\ x + y - z = 0 \\ 2x + 3y - z = 1 \end{cases}$
$x = 1, y = 0, z = 1$

11. $\begin{cases} 5x - y + z = 3 \\ 3x + y + 2z = 2 \\ x + y = 2 \end{cases}$
$x = 1, y = 1, z = -1$

12. $\begin{cases} 2x - y + z = 1 \\ x - y + 2z = 3 \\ x - y + z = 1 \end{cases}$
$x = 0, y = 1, z = 2$

13. The buyer for a large department store must order 40 coats, some fake fur and some leather. He is unsure of the expected sales. He can buy 25 fur coats and the rest leather for $9,300, or 10 fur coats and the rest leather for $12,600. How much does he pay if he decides to split the order evenly?
$10,400

14. Adult tickets for the championship game are usually $5, but on Seniors' Day, seniors paid $4. Children's tickets were $2.50. Sales of 1,800 tickets totaled $7,425, and children and seniors accounted for one-half of the tickets sold. How many of each were sold?
900 adult tickets, 450 senior tickets, 450 children's tickets

6.2 Gaussian Elimination and Matrix Methods

The three elementary row operations for matrices:

1. In a type 1 row operation, any two rows of a matrix are interchanged.
2. In a type 2 row operation, the elements of any row of a matrix are multiplied by any nonzero constant.
3. In a type 3 row operation, any row of a matrix is altered by adding to it any multiple of another row.

Solve the matrix methods, if possible:

15. $\begin{cases} 2x + 5y = 7 \\ 3x - y = 2 \end{cases}$ **$x = 1, y = 1$**

16. $\begin{cases} x + 3y - z = 8 \\ 2x + y - 2z = 11 \\ x - y + 5z = -8 \end{cases}$ **$x = 3, y = 1, z = -2$**

17. $\begin{cases} x + 3y + z = 3 \\ 2x - y + z = -11 \\ 3x + 2y + 3z = 2 \end{cases}$ **$x = -10, y = 1, z = 10$**

18. $\begin{cases} x + y + z = 4 \\ 3x - 2y - 2z = -3 \\ 4x - y - z = 0 \end{cases}$ **no solution**

6.3 Matrix Algebra

Matrices are equal if and only if they are the same size and have the same corresponding entries.

Two $m \times n$ matrices are added by adding the corresponding elements of those matrices.

Two $m \times n$ matrices are subtracted by the rule $A - B = A + (-B)$

The product AB of the $m \times n$ matrix A and the $n \times p$ matrix B is the $m \times p$ matrix C. The ith-row, jth-column entry of C is found by keeping a running total of the products of the elements in the ith row of A with the corresponding elements in the jth column of B.

19. Solve for x and y. $\begin{bmatrix} 1 & -4 \\ x & 2 \\ 0 & x+7 \end{bmatrix} = \begin{bmatrix} 1 & x \\ -4 & 2 \\ x+4 & y \end{bmatrix}$ **$x = -4, y = 3$**

Perform the matrix operations, if possible:

20. $\begin{bmatrix} 3 & 2 & 1 \\ 3 & 2 & 1 \end{bmatrix} + \begin{bmatrix} -2 & 1 & 3 \\ 1 & -2 & 1 \end{bmatrix}$ $\quad \begin{bmatrix} \mathbf{1} & \mathbf{3} & \mathbf{4} \\ \mathbf{4} & \mathbf{0} & \mathbf{2} \end{bmatrix}$

21. $\begin{bmatrix} 2 & 3 & 5 \\ 1 & -2 & 4 \\ 2 & 1 & -2 \end{bmatrix} - \begin{bmatrix} 0 & -2 & 1 \\ 3 & 4 & -2 \\ 6 & -4 & 1 \end{bmatrix}$ $\quad \begin{bmatrix} \mathbf{2} & \mathbf{5} & \mathbf{4} \\ \mathbf{-2} & \mathbf{-6} & \mathbf{6} \\ \mathbf{-4} & \mathbf{5} & \mathbf{-3} \end{bmatrix}$

22. $\begin{bmatrix} 1 & -2 \\ -3 & 1 \end{bmatrix}\begin{bmatrix} 2 & 3 \\ -1 & 2 \end{bmatrix}$ $\quad \begin{bmatrix} \mathbf{4} & \mathbf{-1} \\ \mathbf{-7} & \mathbf{-7} \end{bmatrix}$

23. $\begin{bmatrix} -2 & 3 & 5 \\ 1 & -2 & -3 \end{bmatrix}\begin{bmatrix} 2 & 1 \\ -1 & 2 \\ -2 & 3 \end{bmatrix}$ $\quad \begin{bmatrix} \mathbf{-17} & \mathbf{19} \\ \mathbf{10} & \mathbf{-12} \end{bmatrix}$

24. $[1 \quad -3 \quad 2]\begin{bmatrix} 2 \\ 1 \\ 3 \end{bmatrix}$ $\quad \mathbf{[5]}$

25. $\begin{bmatrix} 1 \\ 2 \\ 1 \\ 5 \end{bmatrix}[2 \quad -1 \quad 1 \quad 3]$ $\quad \begin{bmatrix} \mathbf{2} & \mathbf{-1} & \mathbf{1} & \mathbf{3} \\ \mathbf{4} & \mathbf{-2} & \mathbf{2} & \mathbf{6} \\ \mathbf{2} & \mathbf{-1} & \mathbf{1} & \mathbf{3} \\ \mathbf{10} & \mathbf{-5} & \mathbf{5} & \mathbf{15} \end{bmatrix}$

26. $\begin{bmatrix} 1 & -5 & 3 \\ 2 & 1 & -1 \end{bmatrix}\begin{bmatrix} 2 \\ -2 \\ 3 \end{bmatrix}\begin{bmatrix} 1 & -1 \\ -1 & 3 \end{bmatrix}\begin{bmatrix} 1 \\ -2 \end{bmatrix}$ **not possible**

27. $[1 \quad -3 \quad 2]\begin{bmatrix} 2 \\ 1 \\ -5 \end{bmatrix} + [1 \quad -3]\begin{bmatrix} 2 \\ 5 \end{bmatrix}$ $\quad \mathbf{[-24]}$

28. $\left(\begin{bmatrix} 1 & -3 \\ 3 & 1 \end{bmatrix} + \begin{bmatrix} -1 & 3 \\ 1 & 1 \end{bmatrix}\right)\begin{bmatrix} 1 \\ -5 \end{bmatrix}$ $\quad \begin{bmatrix} \mathbf{0} \\ \mathbf{-6} \end{bmatrix}$

6.4 Matrix Inversion

The inverse of an $n \times n$ matrix A is A^{-1}, where

$$AA^{-1} = A^{-1}A = I$$

Use elementary row operations to transform $[A \mid I]$ into $[I \mid A^{-1}]$, where I is an identity matrix. If A cannot be transformed into I, then A is singular.

Find the inverse of each matrix, if possible.

29. $\begin{bmatrix} 2 & 3 \\ 3 & 5 \end{bmatrix}$ $\quad \begin{bmatrix} \mathbf{5} & \mathbf{-3} \\ \mathbf{-3} & \mathbf{2} \end{bmatrix}$

30. $\begin{bmatrix} 1 & 0 & 0 \\ 2 & 0 & -2 \\ 1 & 2 & 2 \end{bmatrix}$ $\quad \begin{bmatrix} \mathbf{1} & \mathbf{0} & \mathbf{0} \\ -\frac{3}{2} & \frac{1}{2} & \frac{1}{2} \\ \mathbf{1} & -\frac{1}{2} & \mathbf{0} \end{bmatrix}$

31. $\begin{bmatrix} 1 & 0 & 8 \\ 3 & 7 & 6 \\ 1 & 2 & 3 \end{bmatrix}$ $\begin{bmatrix} 9 & 16 & -56 \\ -3 & -5 & 18 \\ -1 & -2 & 7 \end{bmatrix}$

32. $\begin{bmatrix} 4 & 4 & 1 \\ 1 & 1 & 1 \\ -1 & -1 & 0 \end{bmatrix}$ **No inverse exists.**

If A is invertible, then the solution of $AX = B$ is $X = A^{-1}B$.

Use the inverse of the coefficient matrix to solve each system of equations.

33. $\begin{cases} 4x - y + 2z = 0 \\ x + y + 2z = 1 \\ x + z = 0 \end{cases}$ $x = 1, y = 2, z = -1$

34. $\begin{cases} w + 3x + y + 3z = 1 \\ w + 4x + y + 3z = 2 \\ x + y = 1 \\ w + 2x - y + 2z = 1 \end{cases}$ $w = 1, x = 1, y = 0, z = -1$

6.5 Determinants

The determinant:

$\det(A) = |A| =$

$\begin{vmatrix} a & b \\ c & d \end{vmatrix} = ad - bc$

The determinant of an $n \times n$ matrix A is the sum of the products of the elements of any row (or column) and the cofactors of those elements.

Evaluate each determinant.

35. $\begin{vmatrix} 3 & -2 \\ 1 & -3 \end{vmatrix}$ -7

36. $\begin{vmatrix} 1 & -2 & 3 \\ 2 & -1 & 3 \\ 1 & -1 & 0 \end{vmatrix}$ -6

37. $\begin{vmatrix} 1 & 3 & -1 \\ 1 & 2 & 1 \\ 1 & 0 & 2 \end{vmatrix}$ 3

38. $\begin{vmatrix} 1 & 2 & 3 & 4 \\ -1 & 3 & -3 & 2 \\ 0 & 0 & 0 & -1 \\ 3 & 3 & 4 & 3 \end{vmatrix}$ -25

Cramer's rule:
Form quotients of two determinants. The denominator is the determinant of the coefficient matrix, A. The numerator is the determinant of a modified coefficient matrix; when solving for the ith variable, replace the ith column of A with a column of constants, B.

Use Cramer's rule to solve each system.

39. $\begin{cases} x + 3y = -5 \\ -2x + y = -4 \end{cases}$ $x = 1, y = -2$

40. $\begin{cases} x - y + z = -1 \\ 2x - y + 3z = -4 \\ x - 3y + z = -1 \end{cases}$ $x = 1, y = 0, z = -2$

41. $\begin{cases} x - 3y + z = 7 \\ x + y - 3z = -9 \\ x + y + z = 3 \end{cases}$ $x = 1, y = -1, z = 3$

42. $\begin{cases} w + x - y + z = 4 \\ 2w + x + z = 4 \\ x + 2y + z = 0 \\ w + y + z = 2 \end{cases}$ $w = 1, x = 0, y = -1, z = 2$

Properties of Determinants:
If two rows (or columns) of a matrix are interchanged, the sign of its determinant is reversed.

If a row (or column) of a matrix is multiplied by a constant k, the value of its determinant is multiplied by k.

If a row (or column) of a matrix is altered by adding to it a multiple of another row (or column), the value of its determinant is unchanged.

If $\begin{vmatrix} a & b & c \\ d & e & f \\ g & h & i \end{vmatrix} = 7$, evaluate each determinant.

43. $\begin{vmatrix} 3a & 3b & 3c \\ d & e & f \\ g & h & i \end{vmatrix}$ 21

44. $\begin{vmatrix} a & b & c \\ d+g & e+h & f+i \\ g & h & i \end{vmatrix}$ 7

6.6 Partial Fractions

The fraction $\frac{P(x)}{Q(x)}$ can be written as the sum of simpler fractions with denominators determined by the prime factors of $Q(x)$.

Decompose into partial fractions.

45. $\dfrac{7x+3}{x^2+x}$

$\dfrac{3}{x} + \dfrac{4}{x+1}$

46. $\dfrac{4x^3+3x+x^2+2}{x^4+x^2}$

$\dfrac{3}{x} + \dfrac{2}{x^2} + \dfrac{x-1}{x^2+1}$

47. $\dfrac{x^2+5}{x^3+x^2+5x}$

$\dfrac{1}{x} - \dfrac{1}{x^2+x+5}$

48. $\dfrac{x^2+1}{(x+1)^3}$

$\dfrac{1}{x+1} - \dfrac{2}{(x+1)^2} + \dfrac{2}{(x+1)^3}$

6.7 Graphs of Linear Inequalities

Systems of inequalities in two variables can be solved by graphing.

The solution is represented by a plane region with boundaries determined by graphing the inequalities as if they were equations.

Solve each system by graphing.

49. $\begin{cases} 3x+2y \le 6 \\ x-y > 3 \end{cases}$

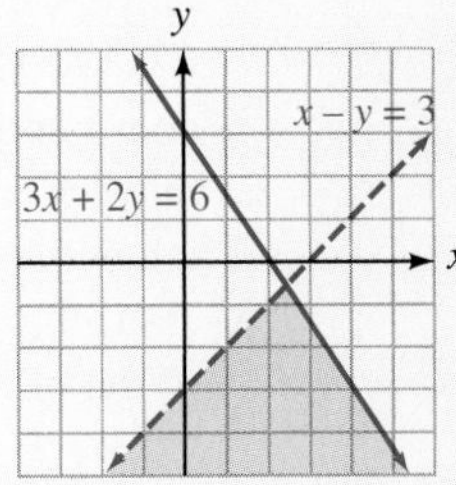

50. $\begin{cases} y \le x^2+1 \\ y \ge x^2-1 \end{cases}$

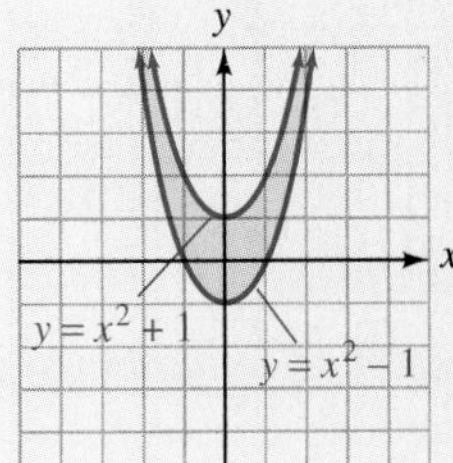

6.8 Linear Programming

The maximum and the minimum values of a linear function in two variables, subject to the constraints of a system of linear inequalities, are attained at a corner or along an entire edge of the region determined by the system of inequalities.

Maximize P subject to the given conditions.

51. $P = 2x + y$

$\begin{cases} x \ge 0 \\ y \ge 0 \\ x+y \le 3 \end{cases}$ $P = 6$ at $(3, 0)$

52. $P = 2x - 3y$

$\begin{cases} x \ge 0 \\ y \le 3 \\ x-y \le 4 \end{cases}$ $P = 12$ at $(0, -4)$

53. $P = 3x - y$

$$\begin{cases} y \geq 1 \\ y \leq 2 \\ y \leq 3x + 1 \\ x \leq 1 \end{cases}$$

$P = 2$ at (1, 1)

54. $P = y - 2x$

$$\begin{cases} x + y \geq 1 \\ x \leq 1 \\ y \leq \dfrac{x}{2} + 2 \\ x + y \leq 2 \end{cases}$$

$P = 3$ at $\left(-\frac{2}{3}, \frac{5}{3}\right)$

55. A company manufactures two fertilizers, x and y. Each 50-pound bag of fertilizer requires three ingredients, which are available in the limited quantities shown in the table. The profit on each bag of fertilizer x is \$6 and on each bag of y, \$5. How many bags of each product should be produced to maximize the profit? **1,000 bags of x, 1,400 bags of y**

Ingredient	Number of pounds in fertilizer x	Number of pounds in fertilizer y	Total number of pounds available
Nitrogen	6	10	20,000
Phosphorus	8	6	16,400
Potash	6	4	12,000

CHAPTER TEST

Solve each system of equations by the graphing method.

1. $\begin{cases} x - 3y = -5 \\ 2x - y = 0 \end{cases}$

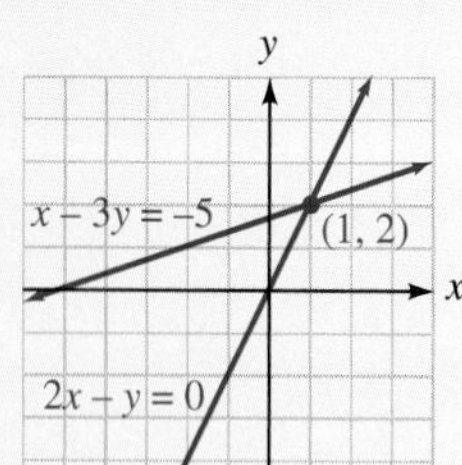

2. $\begin{cases} x = 2y + 5 \\ y = 2x - 4 \end{cases}$

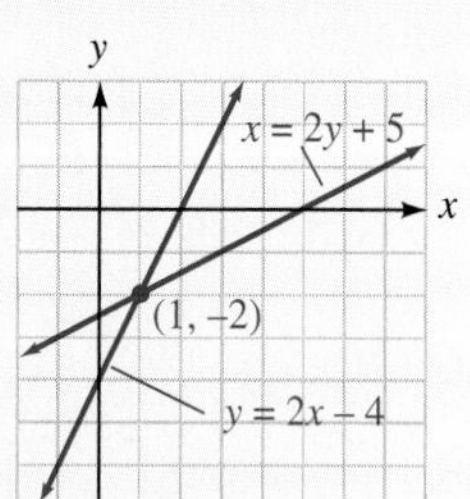

Solve each system of equations by the substitution or addition method.

3. $\begin{cases} 3x + y = 0 \\ 2x - 5y = 17 \end{cases}$ **$x = 1, y = -3$**

4. $\begin{cases} \dfrac{x + y}{2} + x = 7 \\ \dfrac{x - y}{2} - y = -6 \end{cases}$ **$x = 3, y = 5$**

5. Mixing solutions A chemist has two solutions, one has a 20% concentration and the other a 45% concentration. How many liters of each must she mix to obtain 10 liters of 30% concentration?
6 liters of 20% solution, 4 liters of 45% solution

6. Wholesale distribution Ace Electronics, Hi-Fi Stereo, and CD World buy a total of 175 VCR/DVD players from the same distributor each month. Because CD World buys 25 more units than the other two stores combined, CDW's cost is only \$160 per unit. The players cost Hi-Fi \$165 each and Ace \$170 each. How many players does each retailer buy each month if the distributor receives \$28,500 each month from the sale of the players to the three stores?
CD World 100 units, Ace 25 units, HiFi 50 units

Write each system of equations as a matrix and solve it by Gaussian elimination.

7. $\begin{cases} 3x - 2y = 4 \\ 2x + 3y = 7 \end{cases}$ **$x = 2, y = 1$**

8. $\begin{cases} x + 3y - z = 6 \\ 2x - y - 2z = -2 \\ x + 2y + z = 6 \end{cases}$ $\quad x = 1, y = 2, z = 1$

Write each system of equations as a matrix and solve it by Gauss–Jordan elimination. If the system has infinitely many solutions, show how each can be determined.

9. $\begin{cases} x + 2y + 3z = -5 \\ 3x + y - 2z = 7 \\ y - z = 2 \end{cases}$

$x = 1, y = 0, z = -2$

10. $\begin{cases} x + 2y + z = 0 \\ 3x - 2y - 2z = 7 \\ 4x - z = 7 \end{cases}$

$x = -\frac{2}{5}y + \frac{7}{5}$, $z = -\frac{8}{5}y - \frac{7}{5}$, $y =$ any number

Perform the operations.

11. $3\begin{bmatrix} 2 & -3 & 5 \\ 0 & 3 & -1 \end{bmatrix} - 5\begin{bmatrix} -2 & 1 & -1 \\ 0 & 3 & 2 \end{bmatrix}$

$\begin{bmatrix} 16 & -14 & 20 \\ 0 & -6 & -13 \end{bmatrix}$

12. $[1 \quad 2 \quad 3]\begin{bmatrix} 2 & -2 \\ -2 & 2 \\ 1 & 0 \end{bmatrix}\begin{bmatrix} 3 \\ -2 \end{bmatrix}$ $\quad [-1]$

Find the inverse of each matrix, if possible.

13. $\begin{bmatrix} 5 & 19 \\ 2 & 7 \end{bmatrix}$

$\begin{bmatrix} -\frac{7}{3} & \frac{19}{3} \\ \frac{2}{3} & -\frac{5}{3} \end{bmatrix}$

14. $\begin{bmatrix} -1 & 3 & -2 \\ 4 & 1 & 4 \\ 0 & 3 & -1 \end{bmatrix}$

$\begin{bmatrix} -13 & -3 & 14 \\ 4 & 1 & -4 \\ 12 & 3 & -13 \end{bmatrix}$

Use the inverses found in Questions 13 and 14 to solve each system.

15. $\begin{cases} 5x + 19y = 3 \\ 2x + 7y = 2 \end{cases}$

$x = \frac{17}{3}, y = -\frac{4}{3}$

16. $\begin{cases} -x + 3y - 2z = 1 \\ 4x + y + 4z = 3 \\ 3y - z = -1 \end{cases}$

$x = -36, y = 11, z = 34$

Evaluate each determinant.

17. $\begin{vmatrix} 3 & -5 \\ -3 & 1 \end{vmatrix}$

-12

18. $\begin{vmatrix} 3 & 5 & -1 \\ -2 & 3 & -2 \\ 1 & 5 & -3 \end{vmatrix}$

-24

Use Cramer's rule to solve each system for y.

19. $\begin{cases} 3x - 5y = 3 \\ -3x + y = 2 \end{cases}$

$-\frac{5}{4}$

20. $\begin{cases} 3x + 5y - z = 2 \\ -2x + 3y - 2z = 1 \\ x + 5y - 3z = 0 \end{cases}$

1

Decompose each fraction into partial fractions.

21. $\dfrac{5x}{2x^2 - x - 3}$

$\dfrac{3}{2x - 3} + \dfrac{1}{x + 1}$

22. $\dfrac{3x^2 + x + 2}{x^3 + 2x}$

$\dfrac{1}{x} + \dfrac{2x + 1}{x^2 + 2}$

Graph the solution set of each system.

23. $\begin{cases} x - 3y \geq 3 \\ x + 3y \leq 3 \end{cases}$

24. 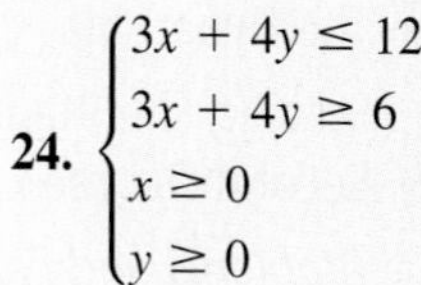
$\begin{cases} 3x + 4y \leq 12 \\ 3x + 4y \geq 6 \\ x \geq 0 \\ y \geq 0 \end{cases}$

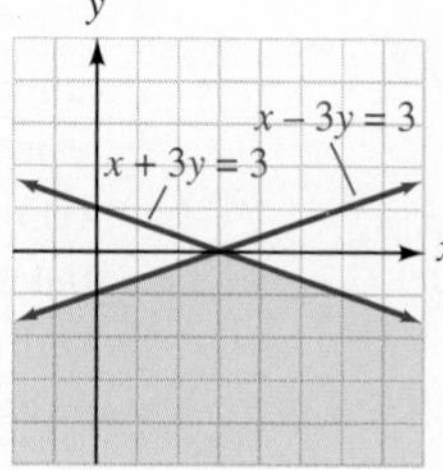

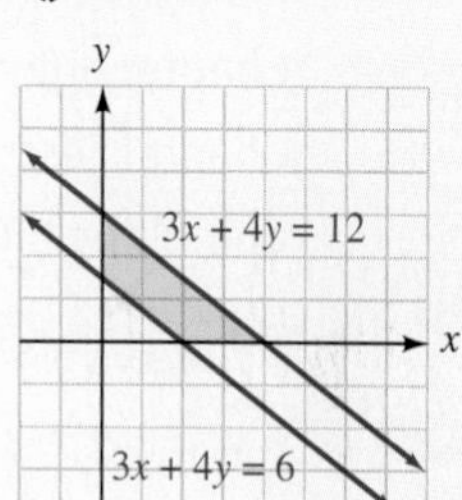

25. Maximize $P = 3x + 2y$ subject to

$\begin{cases} y \geq 0 \\ x \geq 0 \\ 2x + y \leq 4 \\ y \leq 2 \end{cases}$

$P = 7$ at $(1, 2)$

26. Minimize $P = y - x$ subject to

$\begin{cases} x \geq 0 \\ y \geq 0 \\ x + y \leq 8 \\ 2x + y \geq 2 \end{cases}$

$P = -8$ at $(8, 0)$

7 Conic Sections and Quadratic Systems

COLLEGE **Algebra $f(x)$ Now™**

Throughout the chapter, this icon introduces resources on the College AlgebraNow Web Site, accessed through **http://1pass.thomson.com**, that will:

- Help you test your knowledge of the material in this chapter prior to reading it,
- Allow you to take a pre-exam quiz, and
- Provide a Personalized Learning Plan targeting areas you should study.

Careers and Mathematics

ASTRONOMER

Mark Richards/PhotoEdit

Astronomers use the principles of physics and mathematics to learn about the fundamental nature of the universe, including the sun, moon, planets, stars, and galaxies. They also apply their knowledge to solve problems in navigation, space flight, and satellite communications, and to develop the instrumentation and techniques used to observe and collect astronomical data. Because most jobs are in basic research and development, a doctoral degree is the usual educational requirement for astronomers.

Some astronomers operate large space- or ground-based telescopes. Astronomers who make observations using ground-based telescopes may spend long periods in observatories; this work usually involves travel to remote locations. However, astronomers may spend only a few weeks each year making observations with optical telescopes, radio telescopes, and other instruments.

Astronomers held about 1,000 jobs in 2002.

JOB OUTLOOK

Employment of astronomers is projected to grow more slowly than the average for all occupations, through the year 2012. The federal government funds numerous noncommercial research facilities. If federal research and development funding continues to grow, job opportunities for astronomers dependent on federal research grants should be better than they have been in many years.

Median annual earnings of astronomers in 2002 were $85,020.

For a sample application, see Example 4 in Section 7.2.

In this chapter, we will study second-degree equations in x and y. They have many applications in such fields as navigation, astronomy, and satellite communications.

7.1 The Circle and the Parabola

In this section, you will learn about

■ The Circle ■ The Parabola ■ Graphing Equations of Parabolas

Second-degree equations in x and y have the form

$$Ax^2 + Bxy + Cy^2 + Dx + Ey + F = 0$$

where at least one of the coefficients A, B, and C is not zero. The graphs of these equations fall into one of several categories: a point, a pair of lines, a circle, a parabola, an ellipse, a hyperbola, or no graph at all. These graphs are called **conic sections,** because each one is the intersection of a plane and a right-circular cone, as shown in Figure 7-1.

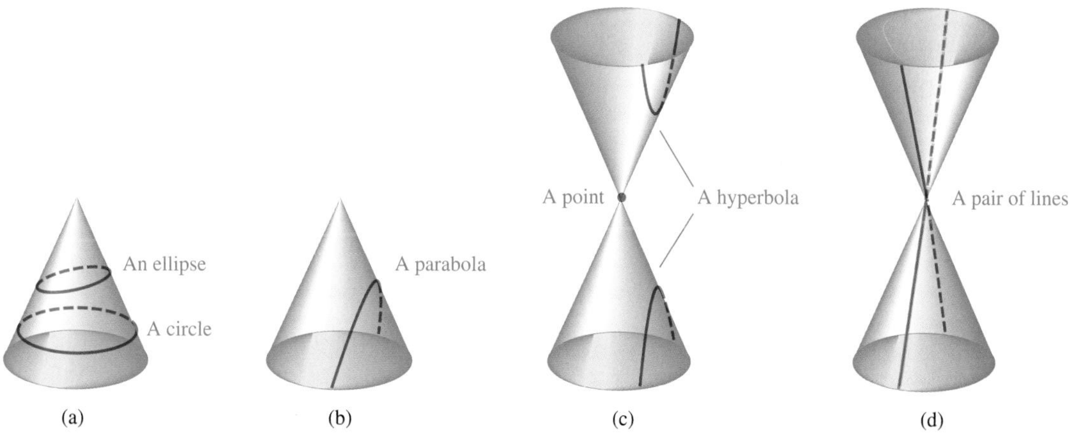

Figure 7-1

Although these shapes have been known since the time of the ancient Greeks, it wasn't until the 17th century that René Descartes (1596–1650) and Blaise Pascal (1623–1662) developed the mathematics needed to study them in detail.

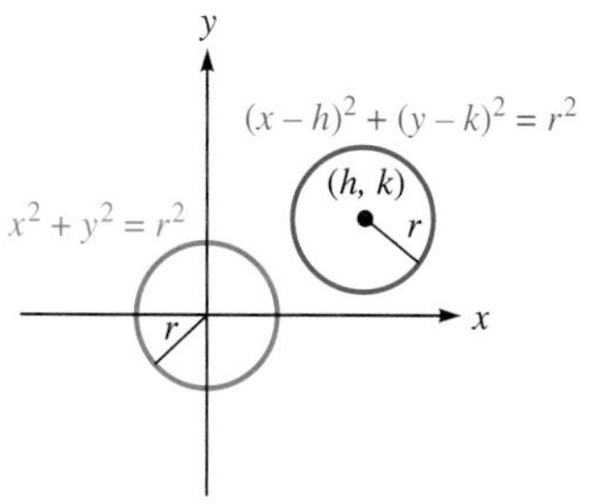

■ The Circle

In Section 2.4, we saw that the graph of any equation that can be written in the form

$$(x - h)^2 + (y - k)^2 = r^2$$

is a circle with radius r and center at point (h, k). This is called **the standard equation of the circle.**

We have also seen that the graph of $x^2 + y^2 = r^2$ is a circle with radius r and center at the origin. Both circles appear in Figure 7-2.

Figure 7-2

ACTIVE EXAMPLE 1 Find the equation of the circle shown in Figure 7-3.

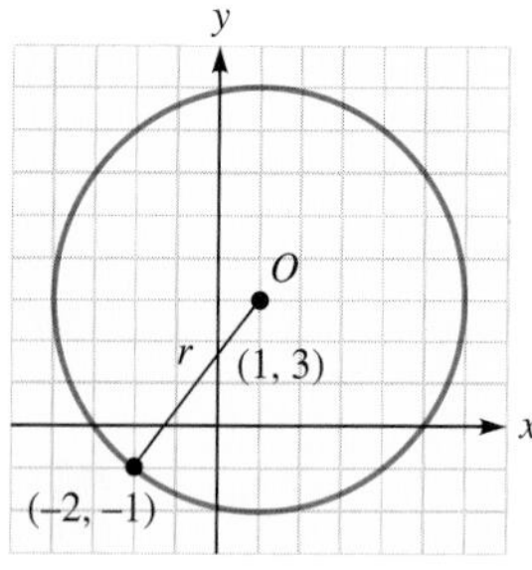

Figure 7-3

Solution To find the radius of the circle, we substitute the coordinates of the points in the distance formula and simplify:

COLLEGE **Algebra** f(x) **Now**™

Explore this Active Example and test yourself on the steps taken here at **http://1pass.thomson.com**. You can also find this example on the Interactive Video Skillbuilder CD-ROM.

$$r = \sqrt{(x_2 - x_1)^2 + (y_2 - y_1)^2}$$

$$r = \sqrt{(-2 - \mathbf{1})^2 + (-\mathbf{1} - \mathbf{3})^2}$$ **Substitute 1 for x_1, 3 for y_1, −2 for x_2, and −1 for y_2.**

$$= \sqrt{(-3)^2 + (-4)^2}$$
$$= \sqrt{9 + 16}$$
$$= \sqrt{25}$$
$$= 5$$

To find the equation of a circle with radius 5 and center at (1, 3), we substitute 1 for h, 3 for k, and 5 for r in the standard equation of the circle.

$$(x - h)^2 + (y - k)^2 = \boldsymbol{r}^2$$
$$(x - 1)^2 + (y - 3)^2 = \mathbf{5}^2$$

To write the equation in general form, we square the binomials and simplify.

$$x^2 - 2x + 1 + y^2 - 6y + 9 = 25$$
$$x^2 + y^2 - 2x - 6y - 15 = 0$$ **Subtract 25 from both sides and simplify.**

Self Check Find the equation of a circle with center at $O(-2, 1)$ and radius of 4. ■

The final equation in Example 1 can be written as

$$1x^2 + 0xy + 1y^2 - 2x - 6y - 15 = 0$$

which illustrates that the graph of

$$Ax^2 + Bxy + Cy^2 + Dx + Ey + F = 0$$

is a circle whenever $B = 0$ and $A = C$.

EXAMPLE 2 Graph the circle whose equation is $2x^2 + 2y^2 + 4x + 3y = 3$.

Solution To find the coordinates of the center and the radius, we complete the square on x and y. We begin by dividing both sides of the equation by 2 and rearranging terms to get

$$x^2 + 2x + y^2 + \frac{3}{2}y = \frac{3}{2}$$

We add 1 and $\frac{9}{16}$ to both sides to complete the square on x and y.

$$x^2 + 2x + \mathbf{1} + y^2 + \frac{3}{2}y + \frac{\mathbf{9}}{\mathbf{16}} = \frac{3}{2} + \mathbf{1} + \frac{\mathbf{9}}{\mathbf{16}}$$

Then we factor $x^2 + 2x + 1$ and $y^2 + \frac{3}{2}y + \frac{9}{16}$.

$$(x + 1)^2 + \left(y + \frac{3}{4}\right)^2 = \frac{49}{16}$$

$$[x - (-1)]^2 + \left[y - \left(-\frac{3}{4}\right)\right]^2 = \left(\frac{7}{4}\right)^2$$

From the equation, we see that the coordinates of the center of the circle are $h = -1$ and $k = -\frac{3}{4}$ and that the radius of the circle is $\frac{7}{4}$. The graph is shown in Figure 7-4.

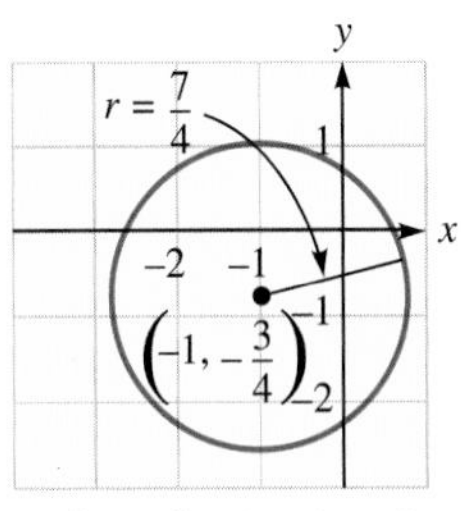

Figure 7-4

Self Check Graph the circle whose equation is $x^2 + y^2 - 6x - 2y = -6$. ■

Accent on Technology GRAPHING CIRCLES

Since the graphs of circles fail the vertical line test, their equations do not represent functions. It is somewhat more difficult to use a graphing calculator to graph equations that are not functions. For example, to graph the circle described by $(x - 1)^2 + (y - 2)^2 = 4$, we must split the equation into two functions and graph each one separately. We begin by solving the equation for y.

$$(x - 1)^2 + (y - 2)^2 = 4$$

$$(y - 2)^2 = 4 - (x - 1)^2 \quad \text{Subtract } (x - 1)^2 \text{ from both sides.}$$

$$y - 2 = \pm\sqrt{4 - (x - 1)^2} \quad \text{Take the square root of both sides.}$$

$$y = 2 \pm \sqrt{4 - (x - 1)^2} \quad \text{Add 2 to both sides.}$$

This equation defines two functions. If we use window settings of $[-3, 5]$ for x and $[-3, 5]$ for y and graph the functions

$$y = 2 + \sqrt{4 - (x - 1)^2} \quad \text{and} \quad y = 2 - \sqrt{4 - (x - 1)^2}$$

we get the distorted circle shown in Figure 7-5(a). To get a better circle, we can use the graphing calculator's squaring feature, which gives an equal unit distance on both the x- and y-axes. Using this feature, we get the circle shown in Figure 7-5(b). Sometimes the two arcs will not join because of approximations made by the calculator at each endpoint.

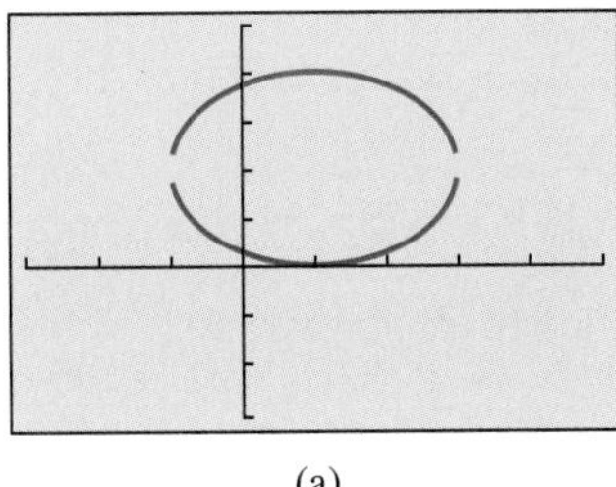

(a)

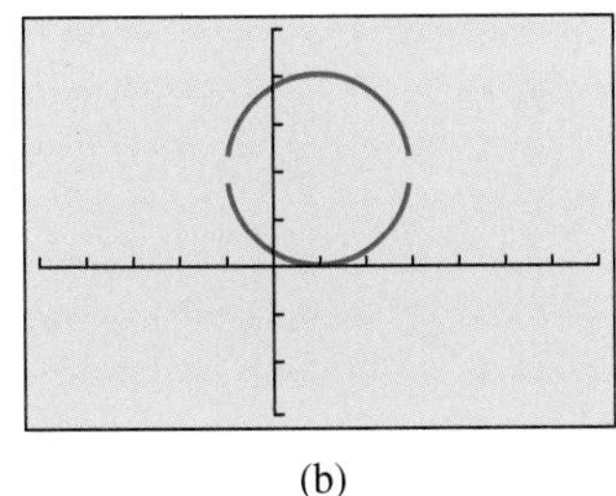

(b)

Figure 7-5

 **Comment** Note that in this case, it is easier to graph the circle by hand.

EXAMPLE 3 The effective broadcast area of a radio station is bounded by the circle

$$x^2 + y^2 = 2{,}500$$

where x and y are measured in miles. Another radio station's broadcast area is bounded by the circle

$$(x - 100)^2 + (y - 100)^2 = 900$$

Is there any location that can receive both stations?

Solution It is possible to receive both stations only if their circular broadcast areas overlap. (See Figure 7-6.) This happens when the sum of the radii, r and r', of the two circles is greater than the distance, d, between their centers. That is, $r + r' > d$.

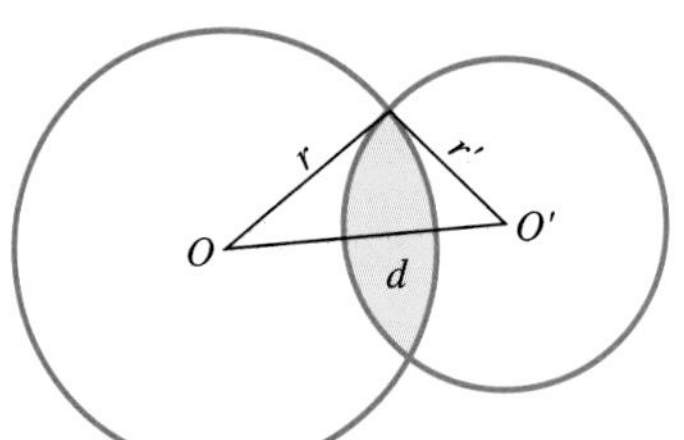

Figure 7-6

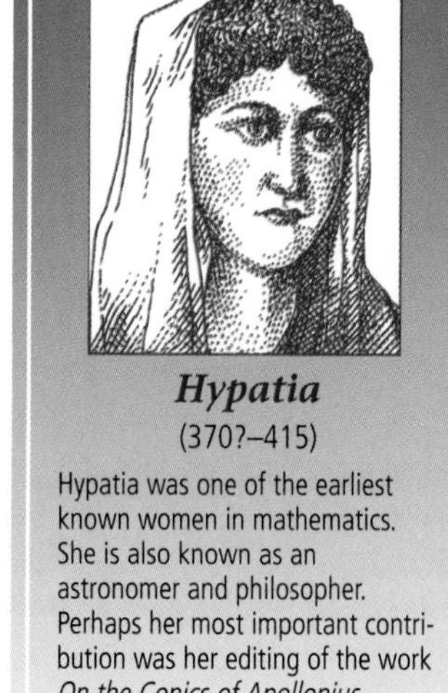

Hypatia

(370?–415)

Hypatia was one of the earliest known women in mathematics. She is also known as an astronomer and philosopher. Perhaps her most important contribution was her editing of the work *On the Conics of Apollonius,* which divides a cone into parts when cut by a plane. This concept led to the study of parabolas, ellipses, and hyperbolas.

The center of the circle $x^2 + y^2 = 2500$ is $(x_1, y_1) = (0, 0)$, and its radius r is 50 miles. The center of the circle $(x - 100)^2 + (y - 100)^2 = 900$ is $(x_2, y_2) = (100, 100)$, and its radius r' is 30 miles.

We can use the distance formula to find the distance d between the centers.

$$d = \sqrt{(x_2 - x_1)^2 + (y_2 - y_1)^2}$$

$$d = \sqrt{(\mathbf{100} - \mathbf{0})^2 + (\mathbf{100} - \mathbf{0})^2}$$

$$= \sqrt{100^2 + 100^2}$$

$$= \sqrt{100^2 \cdot 2}$$

$$= 100\sqrt{2}$$

$$\approx 141 \text{ miles}$$

The sum of the radii, $r + r'$, of the two circles is $(50 + 30)$ miles, or 80 miles. Since this is less than the distance d between their centers (141 miles), there is no location where both stations can be received.

Self Check In Example 3, if the station at the origin boosted its power to cover the area within

$$x^2 + y^2 = 9{,}000$$

would the coverage overlap? ■

■ The Parabola

In Chapter 3, we saw that the graphs of some functions are parabolas that open up or down. We now discuss parabolas that open to the left or to the right and examine the properties of all parabolas in greater detail.

The Parabola

A **parabola** is the set of all points in a plane equidistant from a line l (called the **directrix**) and a fixed point F (called the **focus**) that is not on the line. See Figure 7-7.

Pierre de Fermat
(1601–1665)

Pierre de Fermat shares the honor of discovering analytic geometry (with Descartes) and of developing the theory of probability (with Pascal). But to Fermat alone goes credit for founding number theory. He is probably most famous for a theorem called Fermat's last theorem. It states that if n represents a number greater than 2, there are no whole numbers a, b, and c that satisfy the equation $a^n + b^n = c^n$. This theorem was proved by Andrew Wiles in 1995.

The point on the parabola that is closest to the directrix is called the **vertex,** and the line passing through the vertex and the focus is called the **axis.** In this section, we consider parabolas that have a vertex at point (h, k) and open to the left, to the right, up, or down.

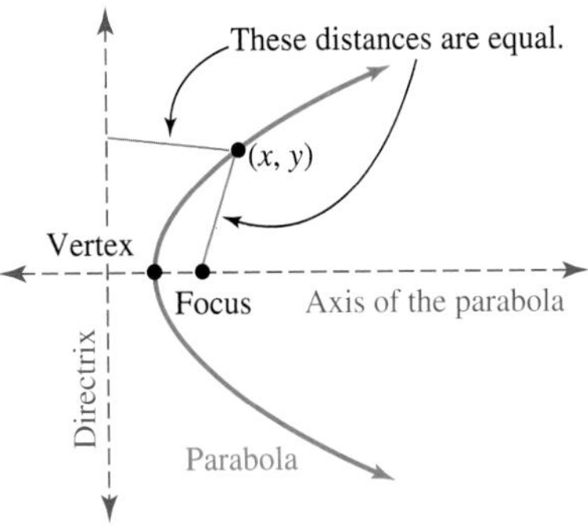

Figure 7-7

The parabola shown in Figure 7-8 opens to the right and passes through its vertex $V(h, k)$ and some point $P(x, y)$. Since each point is equidistant from the focus (point F) and the directrix, we can let $d(DV) = d(VF) = p$, where p is a directed distance.

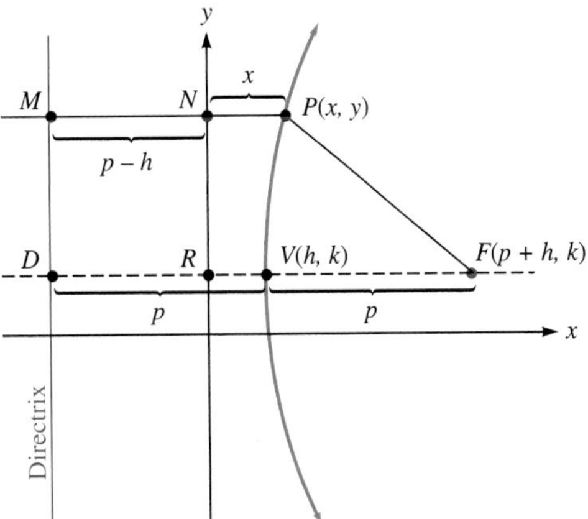

Figure 7-8

Because of the geometry of the figure,

$$d(MP) = p - h + x$$

and by the distance formula,

$$d(PF) = \sqrt{[x - (p + h)]^2 + (y - k)^2}$$

By the definition of a parabola, $d(MP) = d(PF)$. Thus,

$$p - h + x = \sqrt{[x - (p + h)]^2 + (y - k)^2}$$

$$(p - h + x)^2 = [x - (p + h)]^2 + (y - k)^2 \qquad \text{Square both sides.}$$

Finally, we expand the expression on each side of the equation and simplify:

$$p^2 - ph + px - ph + h^2 - hx + px - hx + x^2 = x^2 - 2px - 2hx + p^2 + 2ph + h^2 + (y - k)^2$$

$$-2ph + 2px = -2px + 2ph + (y - k)^2$$

$$4px - 4ph = (y - k)^2$$

$$4p(x - h) = (y - k)^2 \qquad (1)$$

Equation 1 is one of four **standard equations of a parabola.**

If $p < 0$ in Equation 1, the equation will be a parabola that opens to the left.

Standard Equation of a Parabola with Vertex at (h, k) That Opens to the Right or Left

The standard equation of a parabola with vertex at $V(h, k)$ and opening to the right or left is

$$(y - k)^2 = 4p(x - h)$$

where p is the distance from the vertex to the focus. If $p > 0$, the parabola opens to the right. If $p < 0$, the parabola opens to the left.

If the parabola has its vertex at the origin, both h and k are 0, and we have the following result.

Standard Equation of a Parabola with Vertex at the Origin That Opens to the Right or Left

The standard equation of a parabola with vertex at the origin and opening to the right or left is

$$y^2 = 4px$$

where p is the distance from the vertex to the focus. If $p > 0$, the parabola opens to the right. If $p < 0$, the parabola opens to the left.

Equations of parabolas that open up or down have the following standard equations.

Standard Equation of a Parabola That Opens Up or Down

Parabola opening	*Vertex at the origin*	*Vertex at (h, k)*
up or down	$x^2 = 4py$	$(x - h)^2 = 4p(y - k)$

If $p > 0$, the parabola opens up. If $p < 0$, the parabola opens down.

EXAMPLE 4 Find the equation of the parabola with vertex at the origin and focus at $(3, 0)$.

Solution A sketch of the parabola is shown in Figure 7-9. Because the focus is to the right of the vertex, the parabola opens to the right, and because the vertex is the origin, the standard equation is $y^2 = 4px$. The distance between the focus and the vertex is $p = 3$. We can substitute 3 for p in the standard equation to get

$$y^2 = 4px$$

$$y^2 = 4(3)x$$

$$y^2 = 12x$$

The equation of the parabola is $y^2 = 12x$.

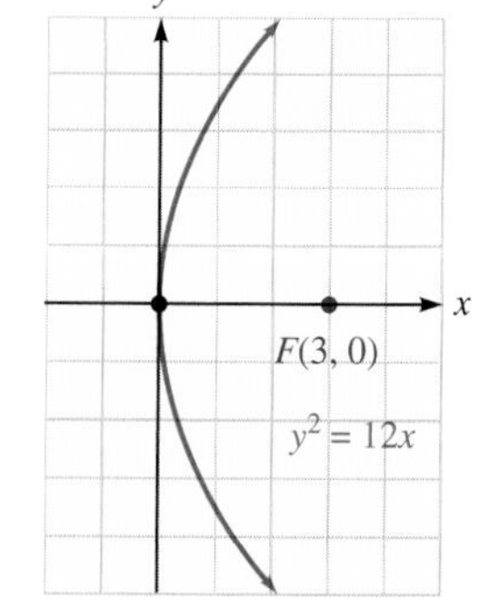

Figure 7-9

Self Check Find the equation of the parabola with vertex at the origin and focus at $(-3, 0)$.

ACTIVE EXAMPLE 5 Find the equation of the parabola that opens up, has vertex at the point (4, 5), and passes through the point (0, 7).

Solution Because the parabola opens up, we use the equation

$$(x - h)^2 = 4p(y - k)$$

Since $(h, k) = (4, 5)$ and the point (0, 7) is on the curve, we can substitute 4 for h, 5 for k, 0 for x, and 7 for y in the standard equation and solve for p.

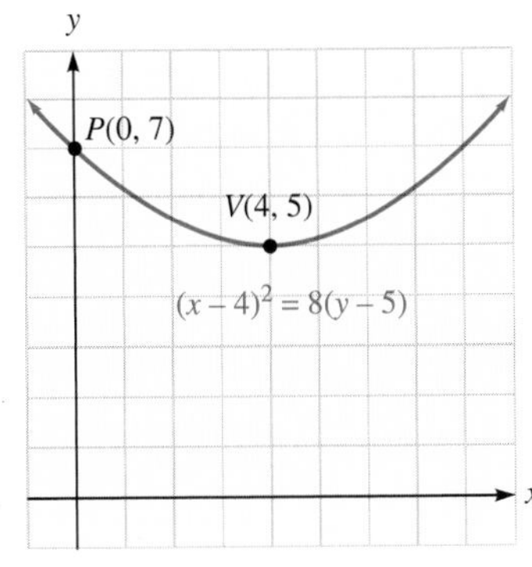

Figure 7-10

$$(x - \mathbf{h})^2 = 4p(y - \mathbf{k})$$
$$(\mathbf{0} - \mathbf{4})^2 = 4p(\mathbf{7} - \mathbf{5})$$
$$16 = 8p$$
$$2 = p$$

To find the equation of the parabola, we substitute 4 for h, 5 for k, and 2 for p in the standard equation and simplify:

$$(x - \mathbf{h})^2 = 4\mathbf{p}(y - \mathbf{k})$$
$$(x - \mathbf{4})^2 = 4 \cdot \mathbf{2}(y - \mathbf{5})$$
$$(x - 4)^2 = 8(y - 5)$$

The graph of the equation appears in Figure 7-10.

Self Check Find the equation of the parabola that opens up, has vertex at (4, 5), and passes through (0, 9). ■

EXAMPLE 6 Find the equations of two parabolas with a vertex at (2, 4) that pass through (0, 0).

Solution The two parabolas are shown in Figure 7-11.

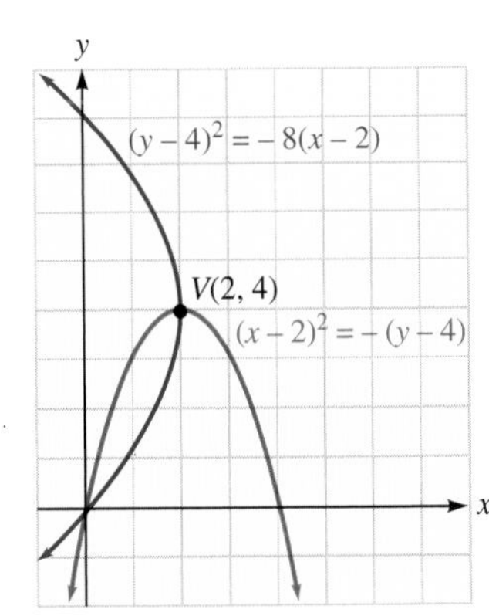

Figure 7-11

Part 1: To find the parabola that opens to the left, we use the equation $(y - k)^2 = 4p(x - h)$. Since the curve passes through the point $(x, y) = (0, 0)$ and the vertex $(h, k) = (2, 4)$, we substitute 0 for x, 0 for y, 2 for h, and 4 for k in the standard equation and solve for p:

$$(y - \mathbf{k})^2 = 4p(\mathbf{x} - \mathbf{h})$$
$$(\mathbf{0} - \mathbf{4})^2 = 4p(\mathbf{0} - \mathbf{2})$$
$$16 = -8p$$
$$-2 = p$$

Since $h = 2$, $k = 4$, $p = -2$, and the parabola opens to the left, its equation is

$$(y - \mathbf{k})^2 = 4\mathbf{p}(x - \mathbf{h})$$
$$(y - \mathbf{4})^2 = 4(\mathbf{-2})(x - \mathbf{2})$$
$$(y - 4)^2 = -8(x - 2)$$

Part 2: To find the equation of the parabola that opens down, we use the equation $(x - h)^2 = 4p(y - k)$ and substitute 2 for h, 4 for k, 0 for x, and 0 for y and solve for p:

$$(\mathbf{x} - \mathbf{h})^2 = 4p(\mathbf{y} - \mathbf{k})$$
$$(\mathbf{0} - \mathbf{2})^2 = 4p(\mathbf{0} - \mathbf{4})$$
$$4 = -16p$$
$$p = -\frac{1}{4}$$

Since $h = 2$, $k = 4$, $p = -\frac{1}{4}$, the equation is

$$(x - h)^2 = 4p(y - k)$$

$$(x - 2)^2 = 4\left(-\frac{1}{4}\right)(y - 4) \quad \text{Substitute 2 for } h, -\tfrac{1}{4} \text{ for } p, \text{ and 4 for } k.$$

$$(x - 2)^2 = -(y - 4)$$

■

■ Graphing Equations of Parabolas

ACTIVE EXAMPLE 7 Find the vertex and y-intercepts of the parabola $y^2 + 8x - 4y = 28$ and graph it.

Solution We can complete the square on y to write the equation in standard form:

COLLEGE **Algebra** $f(x)$ **Now**™
Go to **http://1pass.thomson.com** or your CD to practice this example.

$$y^2 + 8x - 4y = 28$$

$$y^2 - 4y = -8x + 28 \quad \text{Subtract } 8x \text{ from both sides.}$$

$$y^2 - 4y + 4 = -8x + 28 + 4 \quad \text{Add 4 to both sides.}$$

(2) $$(y - 2)^2 = -8(x - 4) \quad \text{Factor both sides.}$$

Equation 2 represents a parabola opening to the left with vertex at (4, 2). To find the y-intercepts, we substitute 0 for x in Equation 2 and solve for y.

$$(y - 2)^2 = -8(x - 4)$$

$$(y - 2)^2 = -8(0 - 4) \quad \text{Substitute 0 for } x.$$

$$y^2 - 4y + 4 = 32 \quad \text{Remove parentheses.}$$

(3) $$y^2 - 4y - 28 = 0$$

We can use the quadratic formula to find that the roots of Equation 3 are $y \approx 7.7$ and $y \approx -3.7$. So the points with coordinates of approximately $(0, 7.7)$ and $(0, -3.7)$ lie on the graph of the parabola. We can use this information and the knowledge that the graph opens to the left and has vertex at (4, 2) to draw the graph, as shown in Figure 7-12.

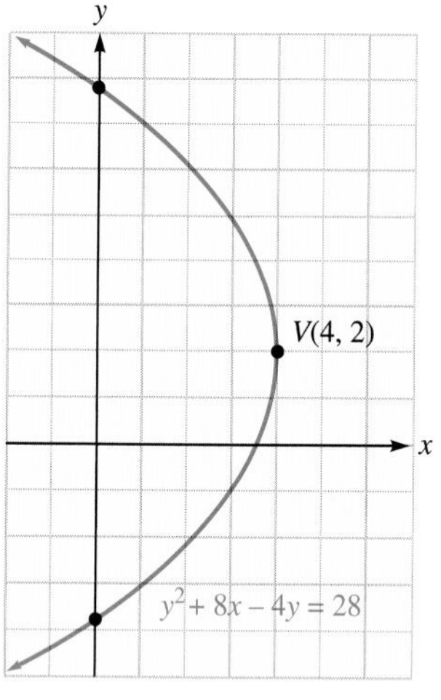

Figure 7-12

Self Check Find the vertex and y-intercepts of the parabola $y^2 - x + 2y = 3$. Then graph it.

■

To graph the equation $y^2 + 8x - 4y = 28$ (Example 7) with a graphing calculator, we first solve the equation for y.

$$y^2 + 8x - 4y = 28$$

$$y^2 - 4y = -8x + 28$$

$$y^2 - 4y + 4 = -8x + 28 + 4$$

$$(y - 2)^2 = -8x + 32$$

$$y - 2 = \pm\sqrt{-8x + 32}$$

$$y = 2 \pm \sqrt{-8x + 32}$$

If we use window settings of $[-2, 7]$ for x and $[-5, 9]$ for y and graph the functions

$$y = 2 + \sqrt{-8x + 32} \qquad \text{and} \qquad y = 2 - \sqrt{-8x + 32}$$

we will get a graph similar to the one shown in Figure 7-12.

EXAMPLE 8 If a stone is thrown straight up into the air, the equation $s = 128t - 16t^2$ expresses its height in feet t seconds after it is thrown. Find the maximum height reached by the stone.

Solution The graph of $s = 128t - 16t^2$, which expresses the height of the stone t seconds after it is thrown, is the parabola shown in Figure 7-13. To find the maximum height reached by the stone, we find the s-coordinate k of the vertex of the parabola. To find k, we write the equation of the parabola in standard form by completing the square on t.

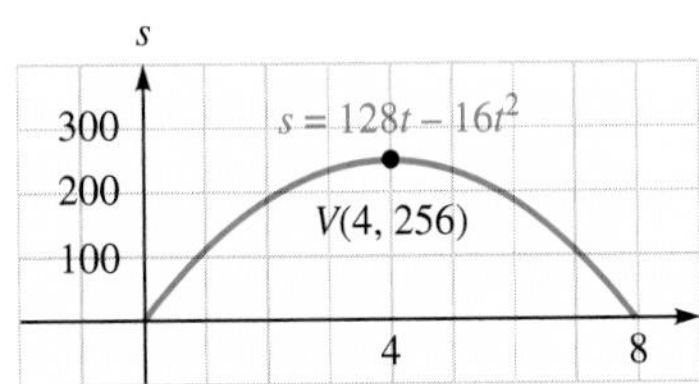

Figure 7-13

$$s = 128t - 16t^2$$

$$16t^2 - 128t = -s \qquad \text{Multiply both sides by } -1.$$

$$t^2 - 8t = \frac{-s}{16} \qquad \text{Divide both sides by 16.}$$

$$t^2 - 8t + 16 = \frac{-s}{16} + 16 \qquad \text{Add 16 to both sides to complete the square.}$$

$$(t - 4)^2 = \frac{-s + 256}{16} \qquad \text{Factor } t^2 - 8t + 16 \text{ and combine like terms.}$$

$$(t - 4)^2 = -\frac{1}{16}(s - 256) \qquad \text{Factor out } -\tfrac{1}{16}.$$

This equation indicates that the maximum height is 256 feet.

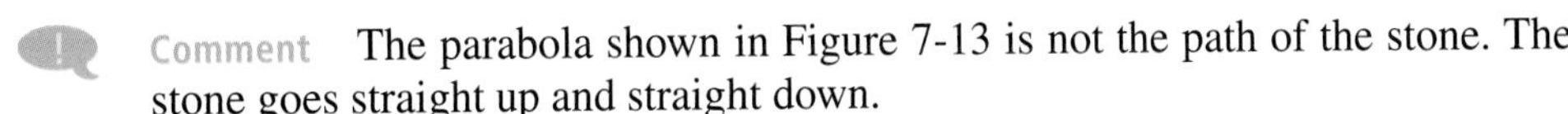

Comment The parabola shown in Figure 7-13 is not the path of the stone. The stone goes straight up and straight down.

Self Check At what time will the stone strike the ground? (*Hint:* Find the t-intercept.) ■

Self Check Answers

1. $x^2 + y^2 + 4x - 2y - 11 = 0$ **2.**

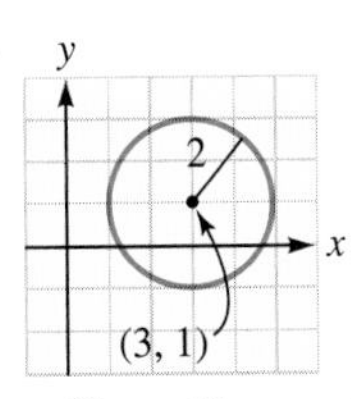

3. no **4.** $y^2 = -12x$ **5.** $(x - 4)^2 = 4(y - 5)$

7. vertex $(-4, -1)$; y-intercepts $(0, 1)$, $(0, -3)$ **8.** $t = 8$ seconds

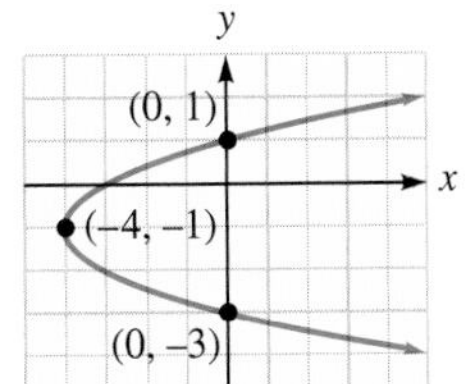

7.1 Exercises

VOCABULARY AND CONCEPTS *Give the coordinates of the circle's center and its radius.*

1. $(x - 2)^2 + (y + 5)^2 = 9$: center (2, −5); radius 3
2. $x^2 + y^2 - 36 = 0$: center (0, 0); radius 6
3. $x^2 + y^2 = 5$: center (0, 0); radius $\sqrt{5}$
4. $2(x - 9)^2 + 2y^2 = 7$: center (9, 0); radius $\sqrt{\frac{7}{2}}$

Determine whether the parabolic graph of the equation opens up, down, to the left, or to the right.

5. $y^2 = -4x$: opens to the left
6. $y^2 = 10x$: opens to the right
7. $x^2 = -8(y - 3)$: opens down
8. $(x - 2)^2 = (y + 3)$: opens up

Fill in the blanks.

9. A parabola is the set of all points in a plane equidistant from a line, called the directrix, and a fixed point not on the line, called the focus.
10. The general form of a second-degree equation in the variables x and y is $Ax^2 + Bxy + Cy^2 + Dx + Ey + F = 0$.

PRACTICE *Write the equation of each circle.*

11.

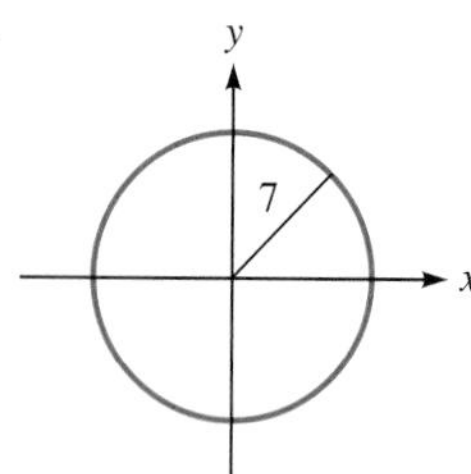

$x^2 + y^2 = 49$

12.

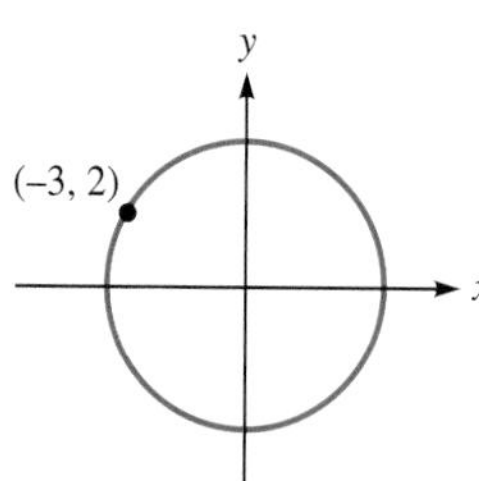

$x^2 + y^2 = 13$

13.

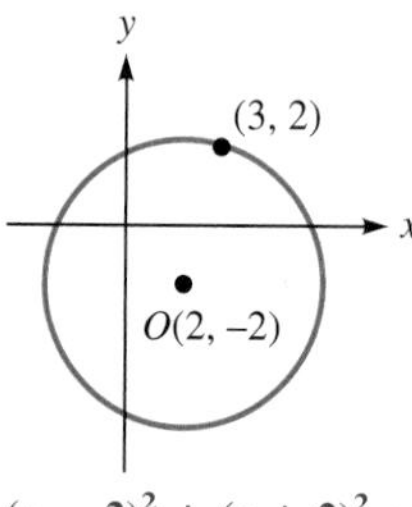

$(x - 2)^2 + (y + 2)^2 = 17$

14.

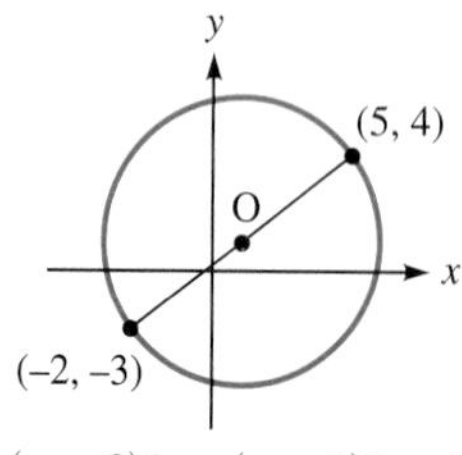

$\left(x - \frac{3}{2}\right)^2 + \left(y - \frac{1}{2}\right)^2 = \frac{49}{2}$

15. Radius of 6; center at the intersection of $3x + y = 1$ and $-2x - 3y = 4$ $(x - 1)^2 + (y + 2)^2 = 36$
16. Radius of 8; center at the intersection of $x + 2y = 8$ and $2x - 3y = -5$ $(x - 2)^2 + (y - 3)^2 = 64$

Graph each equation.

17. $x^2 + y^2 = 4$

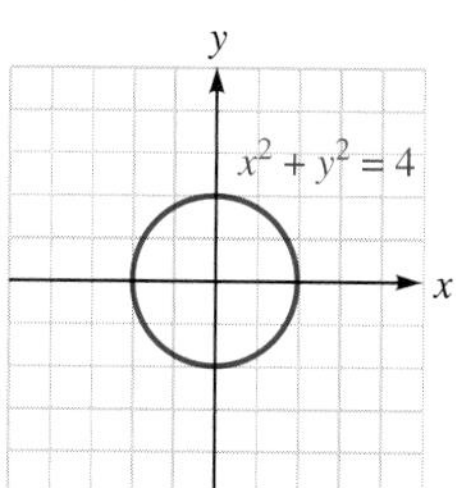

18. $x^2 - 2x + y^2 = 15$

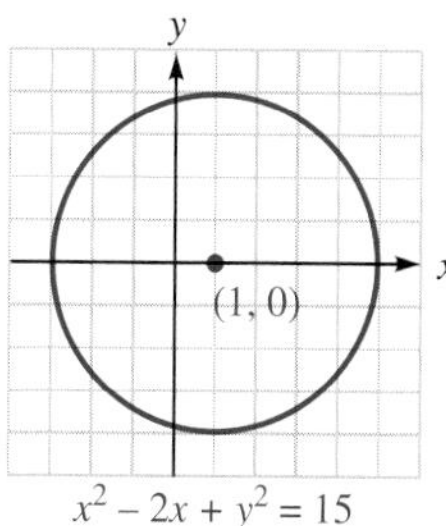

19. $3x^2 + 3y^2 - 12x - 6y = 12$

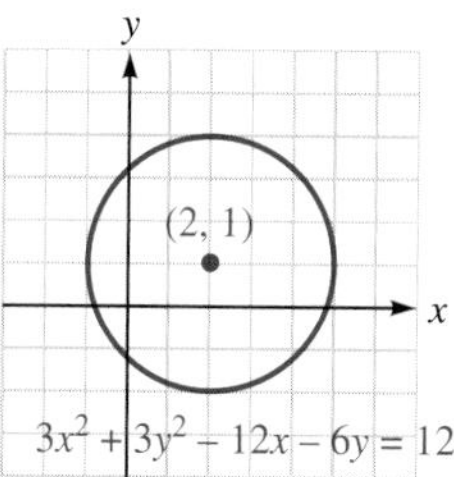

20. $2x^2 + 2y^2 + 4x - 8y + 2 = 0$

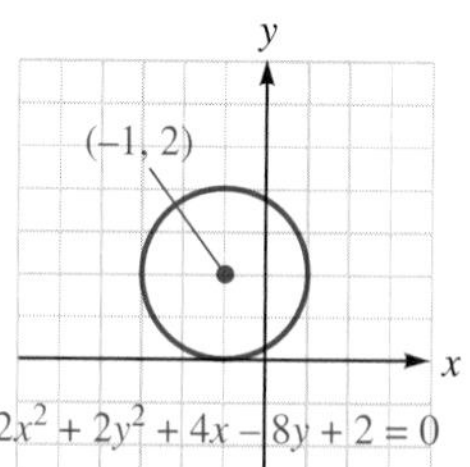

Find the equation of each parabola.

21. Vertex at (0, 0); focus at (0, 3) $x^2 = 12y$
22. Vertex at (0, 0); focus at (0, −3) $x^2 = -12y$
23. Vertex at (0, 0); focus at (−3, 0) $y^2 = -12x$
24. Vertex at (0, 0); focus at (3, 0) $y^2 = 12x$
25. Vertex at (3, 5); focus at (3, 2) $(x - 3)^2 = -12(y - 5)$

26. Vertex at (3, 5); focus at (−3, 5)
$(y - 5)^2 = -24(x - 3)$

27. Vertex at (3, 5); focus at (3, −2)
$(x - 3)^2 = -28(y - 5)$

28. Vertex at (3, 5); focus at (6, 5) $(y - 5)^2 = 12(x - 3)$

29. Vertex at (2, 2); passes through (0, 0)
$(y - 2)^2 = -2(x - 2)$ or $(x - 2)^2 = -2(y - 2)$

30. Vertex at (−2, −2); passes through (0, 0)
$(x + 2)^2 = 2(y + 2)$ or $(y + 2)^2 = 2(x + 2)$

31. Vertex at (−4, 6); passes through (0, 3)
$(x + 4)^2 = -\frac{16}{3}(y - 6)$ or $(y - 6)^2 = \frac{9}{4}(x + 4)$

32. Vertex at (−2, 3); passes through (0, −3)
$(x + 2)^2 = -\frac{2}{3}(y - 3)$ or $(y - 3)^2 = 18(x + 2)$

33. Vertex at (6, 8); passes through (5, 10) and (5, 6)
$(y - 8)^2 = -4(x - 6)$

34. Vertex at (2, 3); passes through $\left(1, \frac{13}{4}\right)$ and $\left(-1, \frac{21}{4}\right)$
$(x - 2)^2 = 4(y - 3)$

35. Vertex at (3, 1); passes through (4, 3) and (2, 3)
$(x - 3)^2 = \frac{1}{2}(y - 1)$

36. Vertex at (−4, −2); passes through (−3, 0) and $\left(\frac{9}{4}, 3\right)$
$(y + 2)^2 = 4(x + 4)$

Change each equation to standard form and graph it.

37. $y = x^2 + 4x + 5$

38. $2x^2 - 12x - 7y = 10$

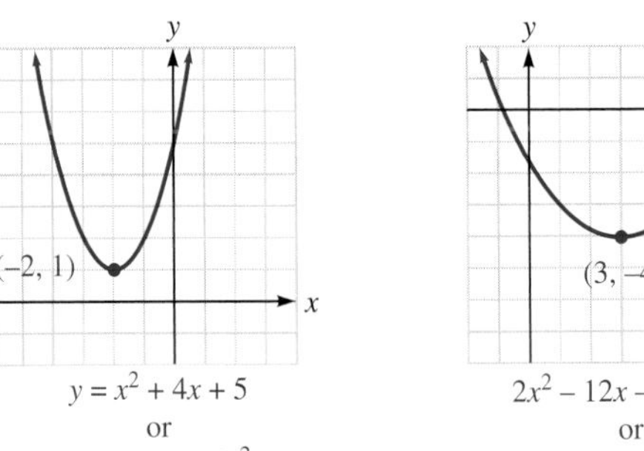

$y = x^2 + 4x + 5$
or
$y - 1 = (x + 2)^2$

$2x^2 - 12x - 7y = 10$
or
$(x - 3)^2 = \frac{7}{2}(y + 4)$

39. $y^2 + 4x - 6y = -1$

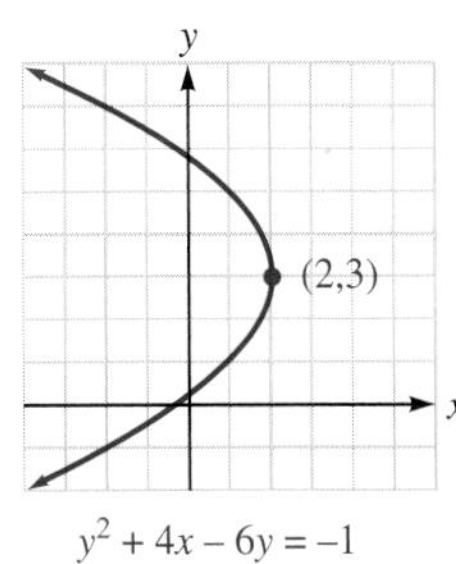

$y^2 + 4x - 6y = -1$
or
$(y - 3)^2 = -4(x - 2)$

40. $x^2 - 2y - 2x = -7$

(1, 3)

$x^2 - 2y - 2x = -7$
or
$(x - 1)^2 = 2(y - 3)$

41. $y^2 + 2x - 2y = 5$

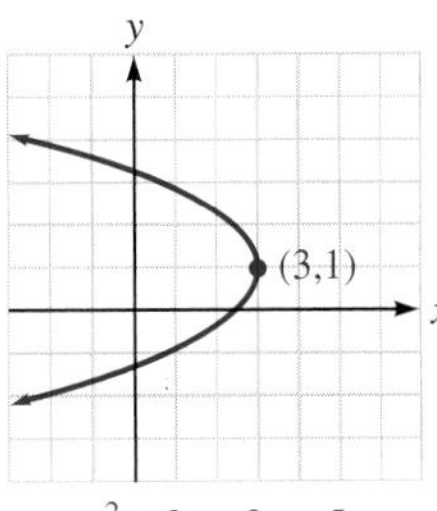

$y^2 + 2x - 2y = 5$
or
$(y - 1)^2 = -2(x - 3)$

42. $y^2 - 4y = 4x - 8$

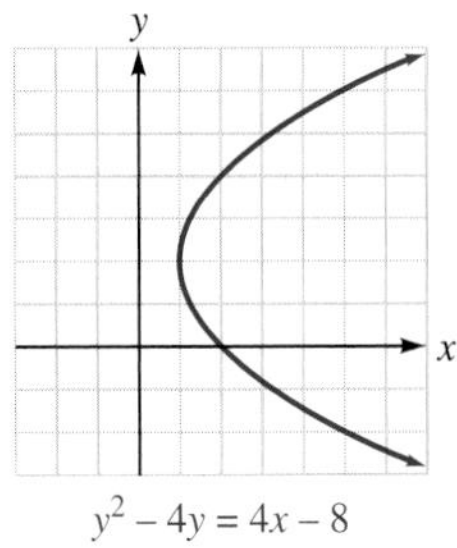

$y^2 - 4y = 4x - 8$
or
$(y - 2)^2 = 4(x - 1)$

43. $y^2 - 4y = -8x + 20$

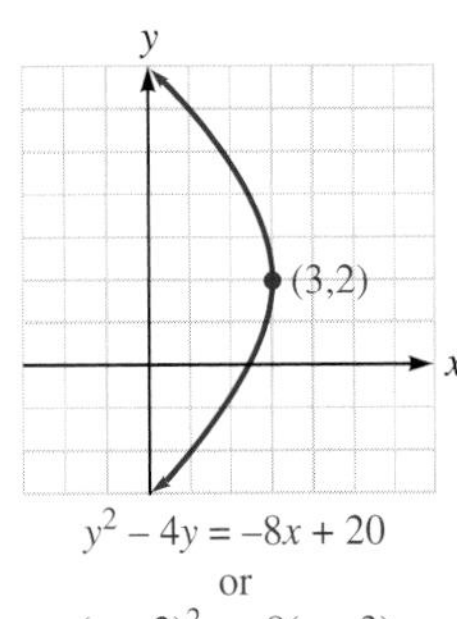

$y^2 - 4y = -8x + 20$
or
$(y - 2)^2 = -8(x - 3)$

44. $y^2 - 2y = 9x + 17$

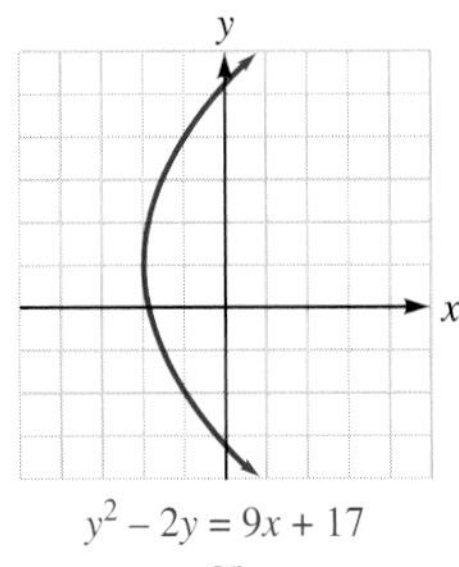

$y^2 - 2y = 9x + 17$
or
$(y - 1)^2 = 9(x + 2)$

45. $x^2 - 6y + 22 = -4x$

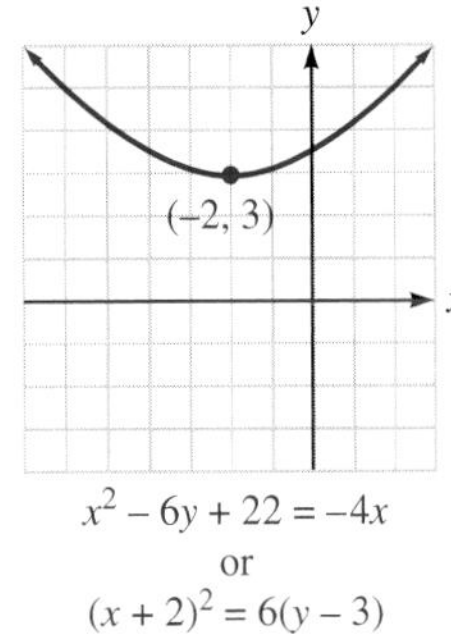

$x^2 - 6y + 22 = -4x$
or
$(x + 2)^2 = 6(y - 3)$

46. $4y^2 - 4y + 16x = 7$

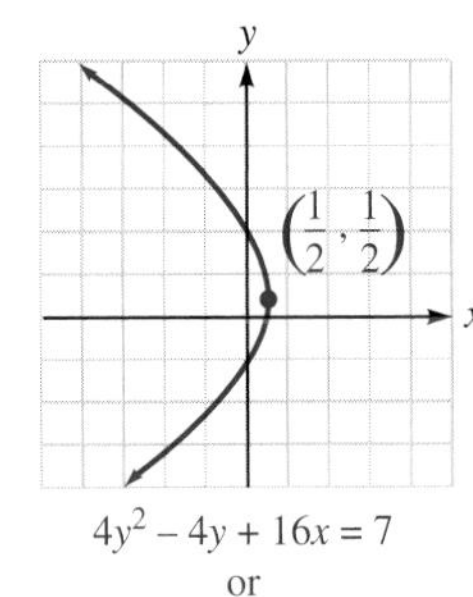

$4y^2 - 4y + 16x = 7$
or
$\left(y - \frac{1}{2}\right)^2 = -4\left(x - \frac{1}{2}\right)$

47. $4x^2 - 4x + 32y = 47$

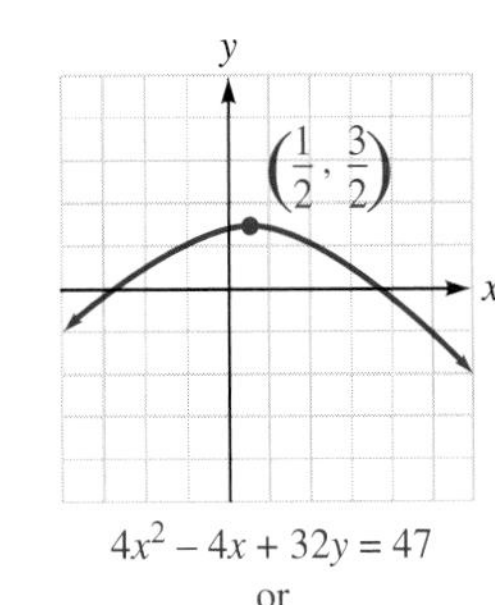

$4x^2 - 4x + 32y = 47$
or
$\left(x - \frac{1}{2}\right)^2 = -8\left(y - \frac{3}{2}\right)$

48. $4y^2 - 16x + 17 = 20y$

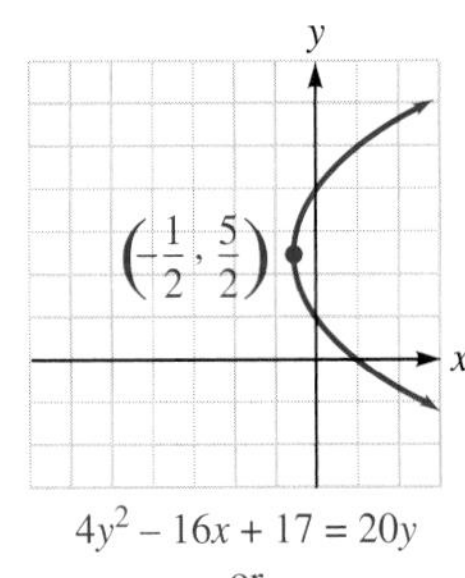

$4y^2 - 16x + 17 = 20y$
or
$\left(y - \frac{5}{2}\right)^2 = 4\left(x + \frac{1}{2}\right)$

APPLICATIONS

49. Broadcast range A television tower broadcasts a signal with a circular range, as shown in the illustration. Can a city 50 miles east and 70 miles north of the tower receive the signal? **yes**

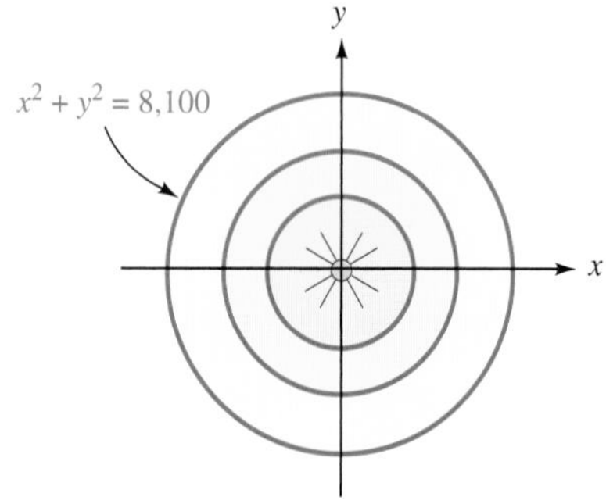

50. Warning sirens A tornado warning siren can be heard in the circular range shown in the illustration. Can a person 4 miles west and 5 miles south of the siren hear its sound? **no**

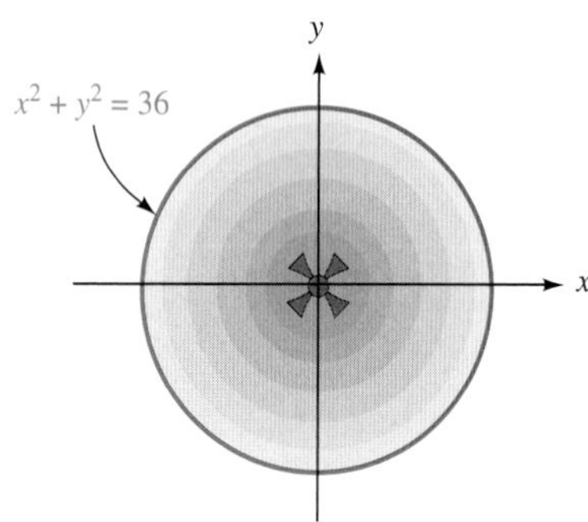

51. Radio translators Some radio stations extend their broadcast range by installing a translator—a remote device that receives the signal and retransmits it. A station with a broadcast range given by $x^2 + y^2 = 1,600$, where x and y are in miles, installs a translator with a broadcast area bounded by $x^2 + y^2 - 70y + 600 = 0$. Find the greatest distance from the main transmitter that the signal can be received. **60 mi**

52. Ripples in a pond When a stone is thrown into the center of a pond, the ripples spread out in a circular pattern, moving at a rate of 3 feet per second. If the stone is dropped at the point (0, 0) in the illustration, when will the ripple reach the seagull floating at the point (15, 36)? **in 13 sec**

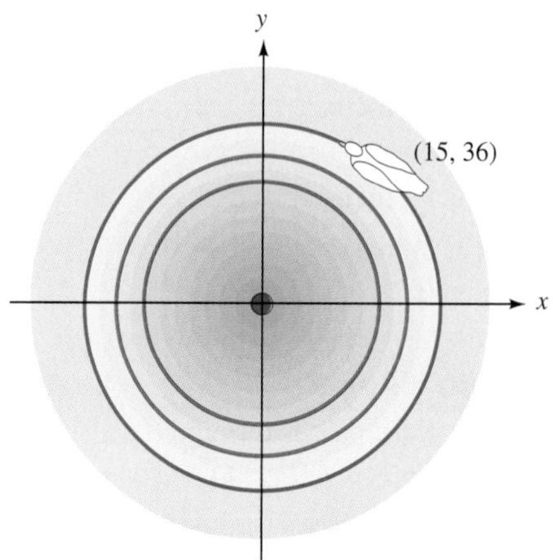

53. Writing the equation of a circle Find the equation of the outer rim of the circular arch shown in the illustration. $\mathbf{(x - 4)^2 + y^2 = 16}$

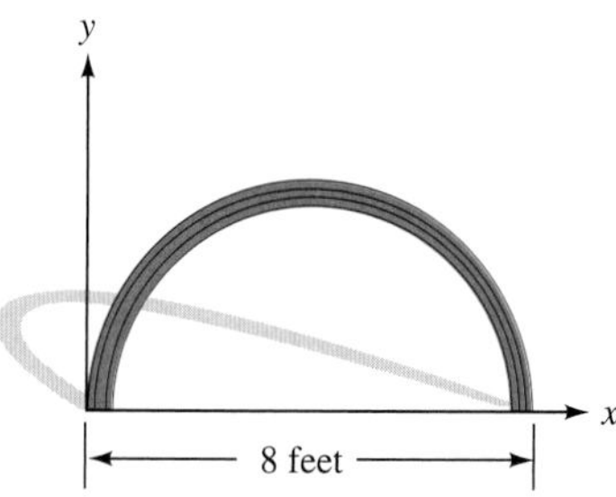

54. Writing the equation of a circle The shape of the window shown is a combination of a rectangle and a semicircle. Find the equation of the circle of which the semicircle is a part. $\mathbf{(x - 5)^2 + (y - 14)^2 = 25}$

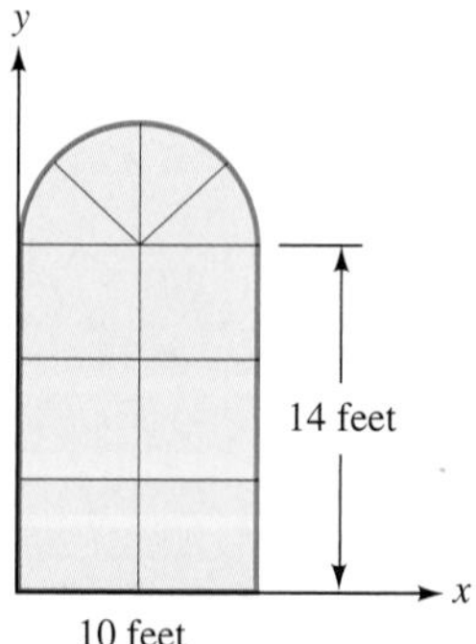

55. Meshing gears For design purposes, the large gear is described by the circle $x^2 + y^2 = 16$. The smaller gear is a circle centered at (7, 0) and tangent to the larger circle. Find the equation of the smaller gear. $(x - 7)^2 + y^2 = 9$

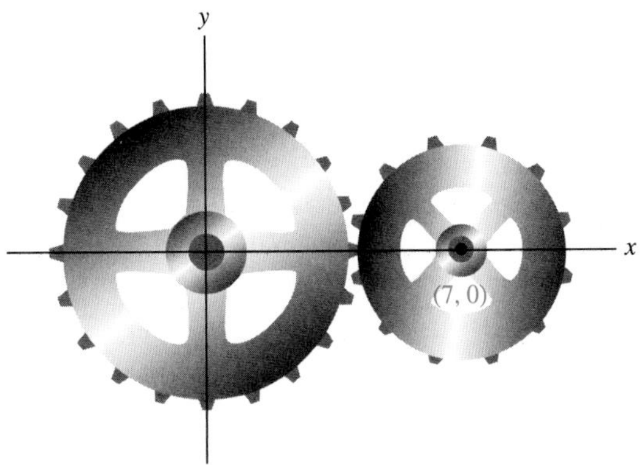

56. Walkways The walkway shown is bounded by the two circles $x^2 + y^2 = 2{,}500$ and $(x - 10)^2 + y^2 = 900$, measured in feet. Find the largest and the smallest width of the walkway. 30 ft and 10 ft

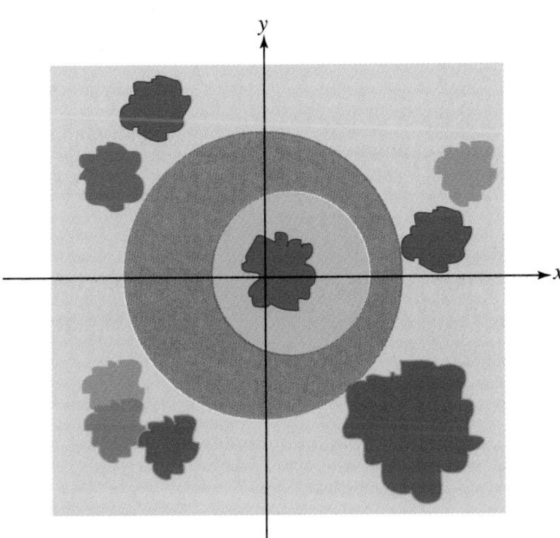

57. Solar furnaces A parabolic mirror collects rays of the sun and concentrates them at its focus. In the illustration, how far from the vertex of the parabolic mirror will it get the hottest? (All measurements are in feet.) 2 ft

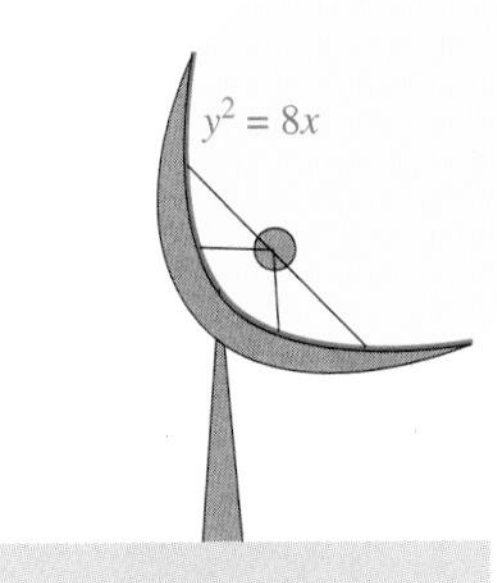

58. Searchlight reflectors A parabolic mirror reflects light in a beam when the light source is placed at its focus. In the illustration, how far from the vertex of the parabolic reflector should the light source be placed? (All measurements are in feet.) 3 ft

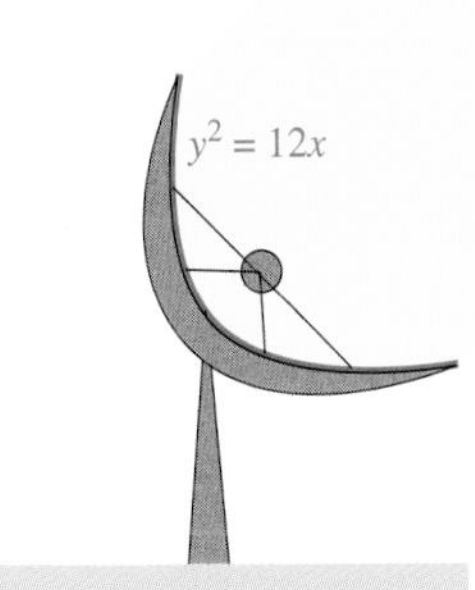

59. Writing the equation of a parabola Derive the equation of the parabolic arch shown. $x^2 = \frac{-45}{2}y$

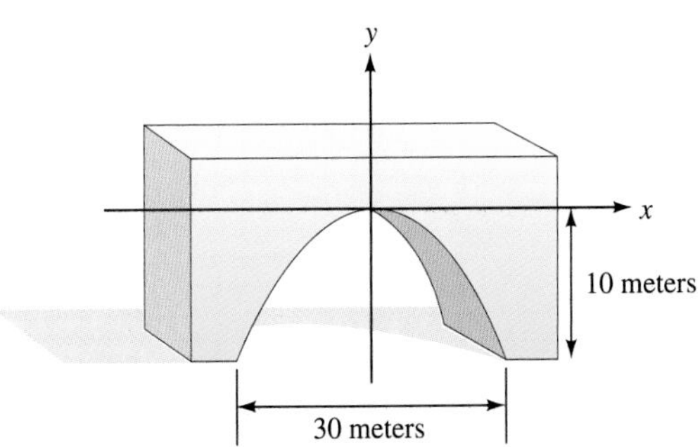

60. Projectiles The cannonball in the illustration follows the parabolic trajectory $y = 30x - x^2$. How far short of the castle does it land? **5 ft**

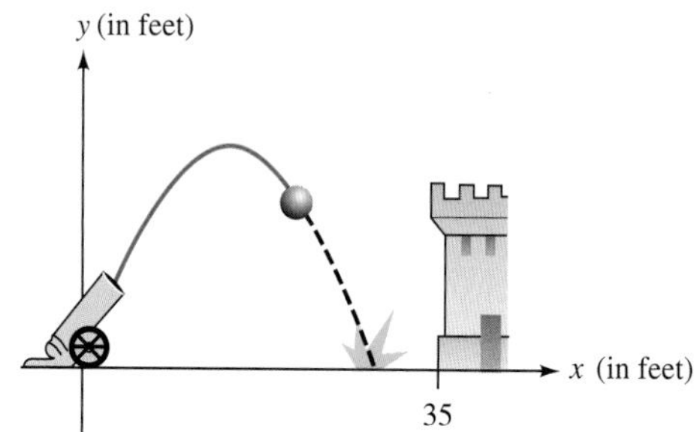

61. Satellite antennas The cross section of the satellite antenna in the illustration is a parabola given by the equation $y = \frac{1}{16}x^2$, with distances measured in feet. If the dish is 8 feet wide, how deep is it? **1 ft**

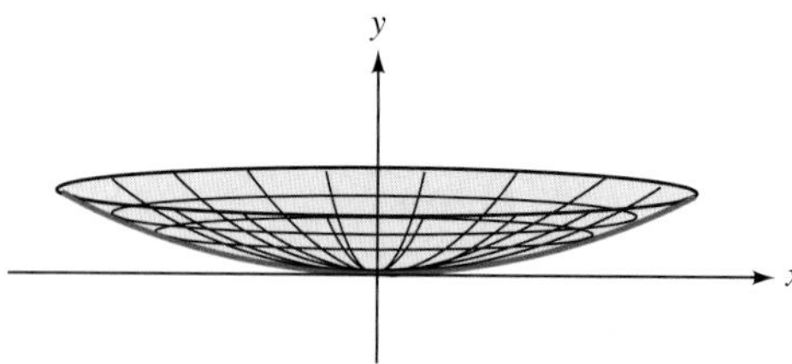

62. Design of a satellite antenna The cross section of the satellite antenna shown is a parabola with the pickup at its focus. Find the distance d from the pickup to the center of the dish. $\frac{9}{4}$ **ft**

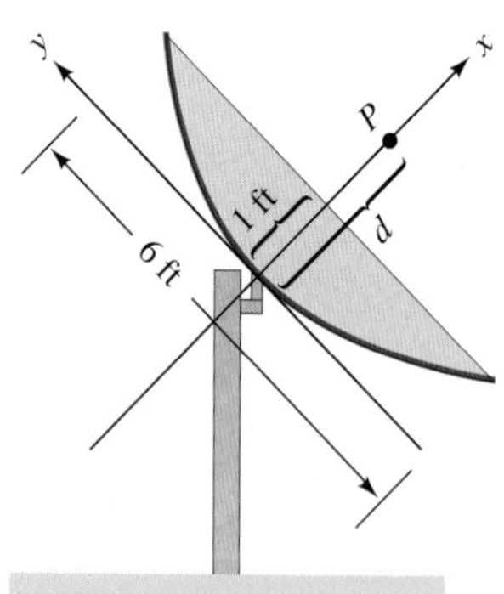

63. Operating a resort A resort owner plans to build and rent n cabins for d dollars per week. The price d that she can charge for each cabin depends on the number of cabins she builds, where $d = -45\left(\frac{n}{32} - \frac{1}{2}\right)$. Find the number of cabins she should build to maximize her weekly income. **8**

64. Toy rockets A toy rocket is s feet above the Earth at the end of t seconds, where $s = -16t^2 + 80\sqrt{3}t$. Find the maximum height of the rocket. **300 ft**

65. Design of a parabolic reflector Find the outer diameter (the length $\overline{AB}$) of the parabolic reflector shown. **about 12.6 cm**

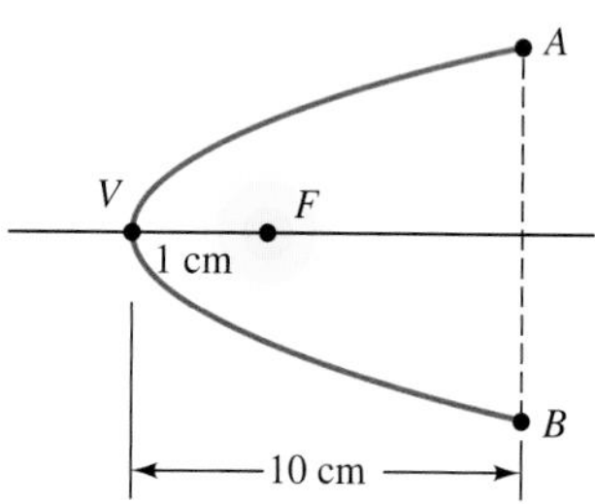

66. Design of a suspension bridge The cable between the towers of the suspension bridge shown in the illustration has the shape of a parabola with vertex 15 feet above the roadway. Find the equation of the parabola. $x^2 = \frac{13,500}{7}(y - 15)$

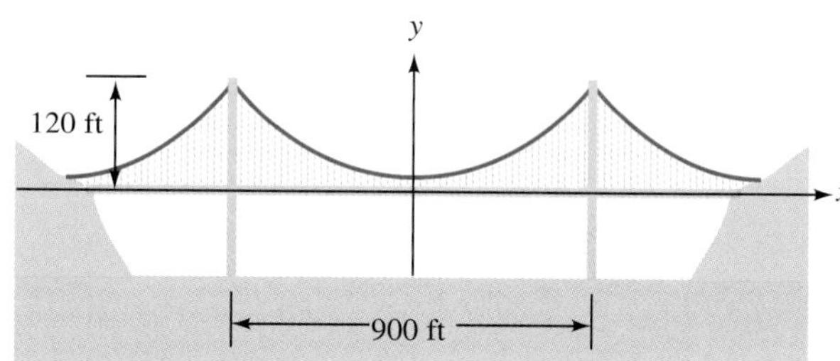

67. Gateway Arch The Gateway Arch in St. Louis has a shape that approximates a parabola. (See the illustration.) Find the width w of the arch 200 feet above the ground. **about 520 ft**

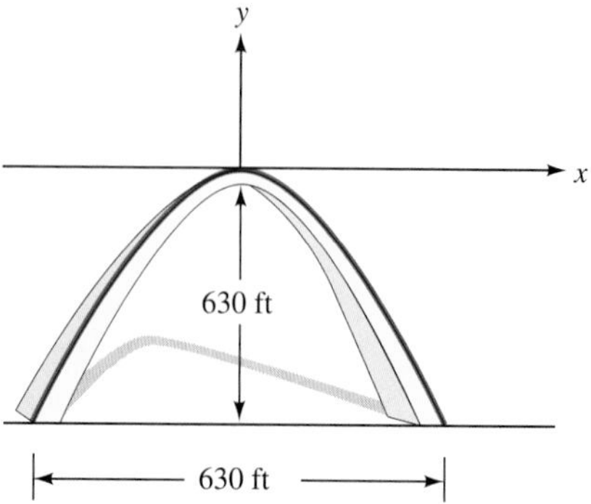

68. Building tunnels A construction firm plans to build a tunnel whose arch is in the shape of a parabola. (See the illustration.) The tunnel will span a two-lane highway 8 meters wide. To allow safe passage for vehicles, the tunnel must be 5 meters high at a distance of 1 meter from the tunnel's edge. Find the maximum height of the tunnel. $\frac{80}{7}$ m

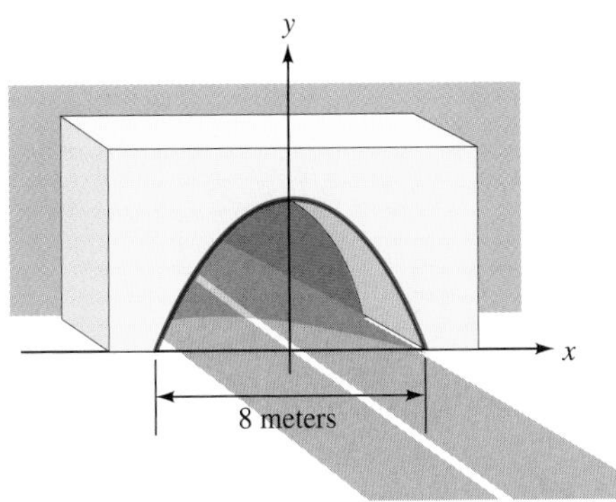

DISCOVERY AND WRITING

69. Show that the standard form of the equation of a parabola $(y - 2)^2 = 8(x - 1)$ is a special case of the general form of a second-degree equation in two variables. $0x^2 + 0xy + y^2 - 8x - 4y + 12 = 0$

70. Show that the standard form of the equation of a circle $(x + 2)^2 + (y - 5)^2 = 36$ is a special case of the general form of a second-degree equation in two variables. $x^2 + 0xy + y^2 + 4x - 10y - 7 = 0$

Find the equation, in the form $(x - h)^2 + (y - k)^2 = r^2$, of the circle passing through the given points.

71. (0, 8), (5, 3), and (4, 6) $x^2 + (y - 3)^2 = 25$

72. (−2, 0), (2, 8), and (5, −1) $(x - 2)^2 + (y - 3)^2 = 25$

Find the equation of the parabola passing through the given points. Give the equation in the form $y = ax^2 + bx + c$.

73. (1, 8), (−2, −1), and (2, 15) $y = x^2 + 4x + 3$

74. (1, −3), (−2, 12), and (−1, 3) $y = 2x^2 - 3x - 2$

75. Ballistics A stone tossed upward is s feet above the Earth after t seconds, where $s = -16t^2 + 128t$. Show that the stone's height x seconds after it is thrown is equal to its height x seconds before it hits the ground.

76. Ballistics Show that the stone in Exercise 75 reaches its greatest height in one-half of the time it takes until it strikes the ground.

REVIEW ***Find the number that must be added to make each binomial a perfect square trinomial.***

77. $x^2 + 4x +$ $\underline{4}$

78. $y^2 - 12y +$ $\underline{36}$

79. $x^2 - 7x +$ $\underline{\frac{49}{4}}$

80. $y^2 + 11y +$ $\underline{\frac{121}{4}}$

Solve each equation.

81. $x^2 + 4x = 5$ $x = 1, -5$

82. $y^2 - 12y = 13$ $y = -1, 13$

83. $x^2 - 7x - 18 = 0$ $x = -2, 9$

84. $y^2 + 11y = -18$ $y = -2, -9$

7.2 The Ellipse

In this section, you will learn about

■ The Ellipse ■ Graphing Equations of Ellipses

A third important conic is the ellipse.

■ The Ellipse

The Ellipse

An **ellipse** is the set of all points P in a plane such that the sum of the distances from P to two other fixed points F and F' is a positive constant.

We can illustrate this definition by showing how to construct an ellipse. To do so, we place two thumbtacks fairly close together, as in Figure 7-14(a). We then tie each end of a piece of string to a thumbtack, catch the loop with the point of a pencil, and (while keeping the string taut) draw the ellipse. Note that $d(F_1P) + d(PF_2)$ will be a positive constant.

An ellipse has an interesting property. Any light or sound that starts at one focus will be reflected through the other. This property is the basis of a medical procedure for treating kidney stones, called **lithotripsy.** The patient is placed in an elliptical tank of water with the kidney stone at one focus. Shock waves from a small controlled explosion at the other focus are concentrated on the stone, pulverizing it. See Figure 7-14(b).

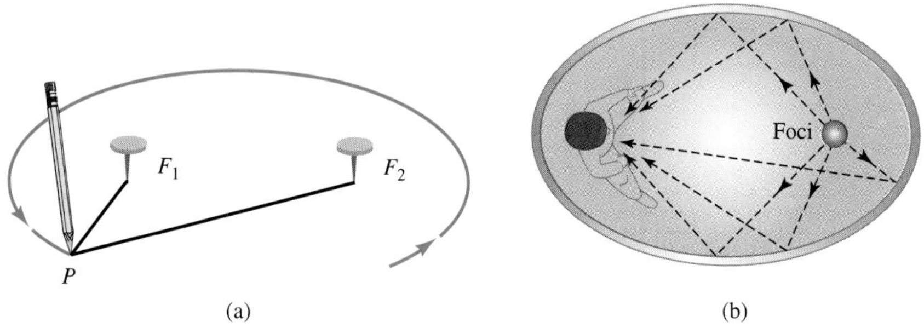

Figure 7-14

In the ellipse shown in Figure 7-15(a), the fixed points F and F' are called **foci** (each is a **focus**), the midpoint of the chord FF' is called the **center,** and the chord VV' is called the **major axis.** Each of the endpoints V and V' of the major axis is called a **vertex.** The chord BB', perpendicular to the major axis and passing through the center C of the ellipse, is called the **minor axis.**

To derive the equation of the ellipse shown in Figure 7-15(b), we note that point O is the midpoint of chord FF' and let $d(OF) = d(OF') = c$, where $c > 0$. Then the coordinates of point F are $(c, 0)$, and the coordinates of F' are $(-c, 0)$. We also let $P(x, y)$ be any point on the ellipse.

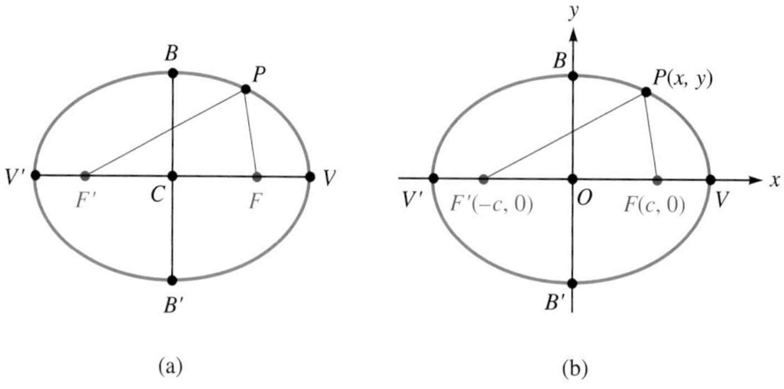

Figure 7-15

By the definition of an ellipse, $d(F'P) + d(PF)$ must be a positive constant, which we will call $2a$. Thus,

(1) $$d(F'P) + d(PF) = 2a$$

We can use the distance formula to compute the lengths of $F'P$ and PF,

$$d(F'P) = \sqrt{[x - (-c)]^2 + y^2} \quad \text{and} \quad d(PF) = \sqrt{(x - c)^2 + y^2}$$

and substitute these values into Equation 1 to obtain

$$\sqrt{[x - (-c)]^2 + y^2} + \sqrt{(x - c)^2 + y^2} = 2a$$

or

(2) $$\sqrt{[x + c]^2 + y^2} = 2a - \sqrt{(x - c)^2 + y^2}$$ **Subtract $\sqrt{(x - c)^2 + y^2}$ from both sides.**

We can square both sides of Equation 2 and simplify to get

$$(x + c)^2 + y^2 = 4a^2 - 4a\sqrt{(x - c)^2 + y^2} + [(x - c)^2 + y^2]$$
$$x^2 + 2cx + c^2 + y^2 = 4a^2 - 4a\sqrt{(x - c)^2 + y^2} + x^2 - 2cx + c^2 + y^2$$
$$4cx = 4a^2 - 4a\sqrt{(x - c)^2 + y^2}$$
$$cx = a^2 - a\sqrt{(x - c)^2 + y^2}$$
$$cx - a^2 = -a\sqrt{(x - c)^2 + y^2}$$

We square both sides again and simplify to get

$$c^2x^2 - 2a^2cx + a^4 = a^2[(x - c)^2 + y^2]$$
$$c^2x^2 - 2a^2cx + a^4 = a^2(x^2 - 2cx + c^2 + y^2)$$
$$c^2x^2 - 2a^2cx + a^4 = a^2x^2 - 2a^2cx + a^2c^2 + a^2y^2$$
$$c^2x^2 + a^4 = a^2x^2 + a^2c^2 + a^2y^2$$
$$a^4 - a^2c^2 = a^2x^2 - c^2x^2 + a^2y^2$$

(3) $$a^2(a^2 - c^2) = (a^2 - c^2)x^2 + a^2y^2$$

Because the shortest distance between two points is a line segment, $d(F'P) + d(PF) > d(F'F)$. Therefore, $2a > 2c$. Thus, $a > c$, and $a^2 - c^2$ is a positive number, which we will call b^2. Letting $b^2 = a^2 - c^2$ and substituting into Equation 3, we have

(4) $$a^2b^2 = b^2x^2 + a^2y^2$$

Dividing both sides of Equation 4 by a^2b^2 gives the equation

$$\frac{x^2}{a^2} + \frac{y^2}{b^2} = 1 \qquad \text{where } a > b > 0$$

To find the coordinates of the vertices V and V', we substitute 0 for y and solve for x:

$$\frac{x^2}{a^2} + \frac{y^2}{b^2} = 1$$
$$\frac{x^2}{a^2} + \frac{0^2}{b^2} = 1$$
$$\frac{x^2}{a^2} = 1$$
$$x^2 = a^2$$
$$x = a \quad \text{or} \quad x = -a$$

Since the coordinates of V are $(a, 0)$ and the coordinates of V' are $(-a, 0)$, a is the distance between the center of the ellipse and either of its vertices. Thus, the center of the ellipse is the midpoint of the major axis.

To find the coordinates of B and B', we substitute 0 for x and solve for y:

$$\frac{x^2}{a^2} + \frac{y^2}{b^2} = 1$$
$$\frac{0^2}{a^2} + \frac{y^2}{b^2} = 1$$
$$y^2 = b^2$$
$$y = b \quad \text{or} \quad y = -b$$

Since the coordinates of B are $(0, b)$ and the coordinates of B' are $(0, -b)$, the distance between the center of the ellipse and either endpoint of the minor axis is b. We have the following results.

The Ellipse: Major Axis on x-Axis, Center at (0, 0)

COLLEGE **Algebra** $f(x)$ **Now**™
Explore this Active Figure and test your understanding at **http://1pass.thomson.com**.

The standard equation of an ellipse with center at the origin and major axis on the x-axis is

$$\frac{x^2}{a^2} + \frac{y^2}{b^2} = 1 \qquad \text{where } a > b > 0$$

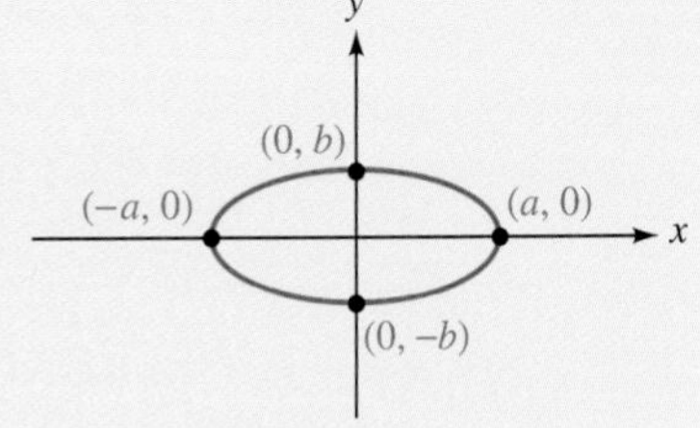

Vertices:

$V(a, 0)$ and $V'(-a, 0)$

Ends of the minor axis:

$B(0, b)$ and $B'(0, -b)$

The Ellipse: Major Axis on y-Axis, Center at (0, 0)

COLLEGE **Algebra** $f(x)$ **Now**™
Explore this Active Figure and test your understanding at **http://1pass.thomson.com**.

The standard equation of an ellipse with center at the origin and major axis on the y-axis is

$$\frac{y^2}{a^2} + \frac{x^2}{b^2} = 1 \qquad \text{where } a > b > 0$$

Vertices:

$V(0, a)$ and $V'(0, -a)$

Ends of the minor axis:

$B(b, 0)$ and $B'(-b, 0)$

To translate the ellipse to a new position centered at the point (h, k) instead of the origin, we replace x and y in the equations with $x - h$ and $y - k$, respectively, to get the following results.

The Ellipse: Major Axis Horizontal, Center at (h, k)

The standard equation of an ellipse with center at (h, k) and major axis horizontal is

$$\frac{(x-h)^2}{a^2} + \frac{(y-k)^2}{b^2} = 1 \qquad \text{where } a > b > 0$$

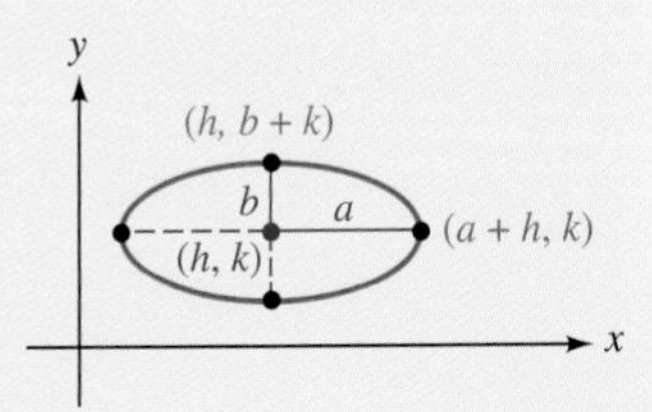

Vertices:

$V(a + h, k)$ and $V'(-a + h, k)$

Ends of the minor axis:

$B(h, b + k)$ and $B'(h, -b + k)$

The Ellipse: Major Axis Vertical, Center at (h, k)

The standard equation of an ellipse with center at (h, k) and major axis vertical is

$$\frac{(y-k)^2}{a^2} + \frac{(x-h)^2}{b^2} = 1 \quad \text{where } a > b > 0$$

Vertices:

$V(h, a + k)$ and $V'(h, -a + k)$

Ends of the minor axis:

$B(b + h, k)$ and $B'(-b + h, k)$

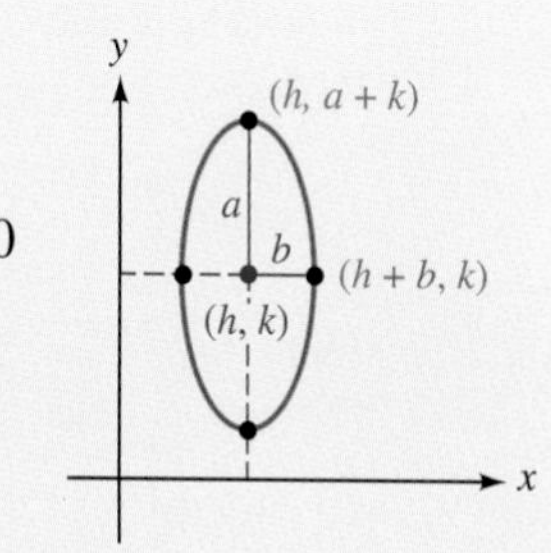

In all cases, the length of the major axis is $2a$, and the length of the minor axis is $2b$.

EXAMPLE 1 Use string and thumbtacks to construct an ellipse that is 10 inches wide and 6 inches high.

Solution To do the construction, we must find the length of string to use and the distance between the thumbtacks. To do so, we note that c represents the distance between the center of the ellipse and either focus. When the pencil point is at vertex V, the necessary length of string is $c + a + (a - c) = 2a$ inches. See Figure 7-16. Because $2a$ is the width of the ellipse (which is given to be 10 inches), the string should be 10 inches long.

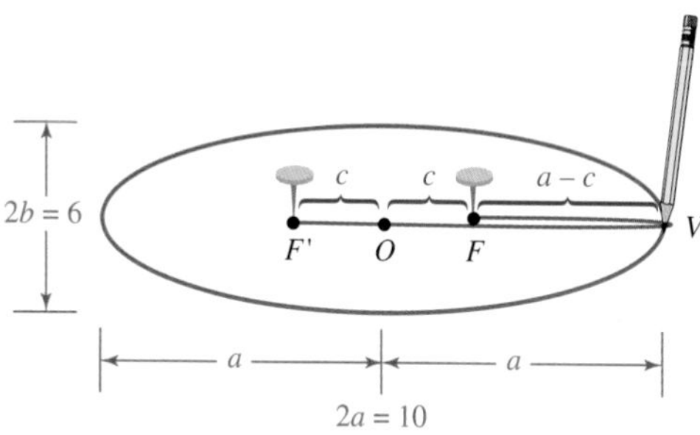

Figure 7-16

Since b is one-half the height of the ellipse, $b = 3$. Since $2a = 10$, $a = 5$. To find c, we can substitute 5 for a and 3 for b in the formula $b^2 = a^2 - c^2$.

$b^2 = a^2 - c^2$ In an ellipse, $b^2 = a^2 - c^2$.

$3^2 = 5^2 - c^2$ Substitute 3 for b and 5 for a.

$-16 = -c^2$ Subtract 5^2 from both sides and simplify.

$c = 4$ Divide by -1 and take the positive square root of both sides.

Since $c = 4$, the distance between the thumbtacks should be 8 inches. We can construct the ellipse by tying a 10-inch string to thumbtacks that are 8 inches apart. ■

ACTIVE EXAMPLE 2 Find the equation of the ellipse with center at the origin, major axis of length 6 units located on the x-axis, and minor axis of length 4 units.

Solution Since the center is the origin and the length of the major axis is 6, $a = 3$ and the coordinates of the vertices are $(3, 0)$ and $(-3, 0)$, as shown in Figure 7-17.

COLLEGE **Algebra** $f(x)$ **Now**™
Explore this Active Example and test yourself on the steps taken here at **http://1pass.thomson.com**. You can also find this example on the Interactive Video Skillbuilder CD-ROM.

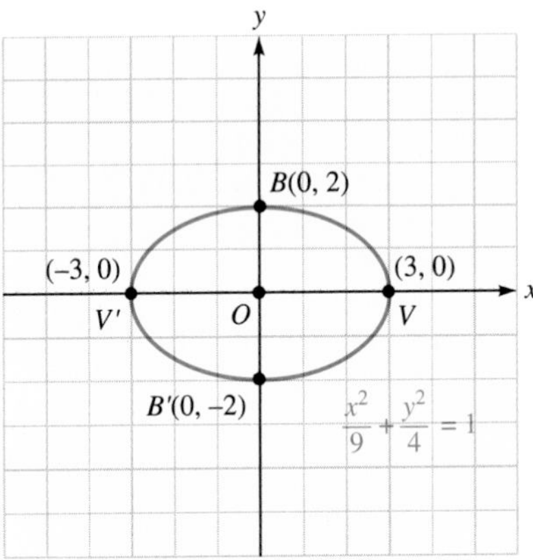

Active Figure 7-17

COLLEGE **Algebra** $f(x)$ **Now**™
Explore this Active Figure and test your understanding at **http://1pass.thomson.com**.

Since the length of the minor axis is 4, $b = 2$ and the coordinates of B and B' are $(0, 2)$ and $(0, -2)$. To find the equation, we substitute 3 for a and 2 for b in the standard equation of an ellipse with center at the origin and major axis on the x-axis.

$$\frac{x^2}{a^2} + \frac{y^2}{b^2} = 1$$

$$\frac{x^2}{3^2} + \frac{y^2}{2^2} = 1$$

$$\frac{x^2}{9} + \frac{y^2}{4} = 1$$

Self Check Find the equation of the ellipse with center at the origin, major axis of length 10 on the x-axis, and minor axis of length 8. ■

ACTIVE EXAMPLE 3 Find the equation of the ellipse with focus at $(0, 3)$ and vertices $V(3, 3)$ and $V'(-5, 3)$.

Solution Since the midpoint of the major axis is the center of the ellipse, the coordinates of the center are $(-1, 3)$, as in Figure 7-18. Because the major axis is parallel to the x-axis, the standard equation to use is

COLLEGE **Algebra** $f(x)$ **Now**™
Go to **http://1pass.thomson.com** or your CD to practice this example.

$$\frac{(x - h)^2}{a^2} + \frac{(y - k)^2}{b^2} = 1 \qquad \text{where } a > b > 0$$

From Figure 7-18, we see that the distance between the center of the ellipse and either vertex is $a = 4$. We also see that the distance between the focus and the center is $c = 1$.

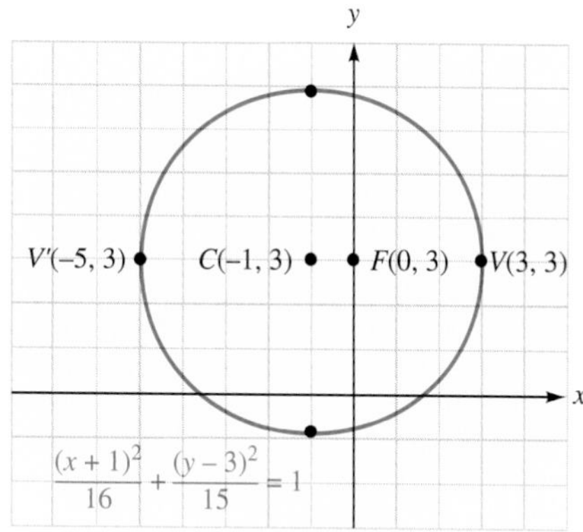

Figure 7-18

Since $b^2 = a^2 - c^2$ in an ellipse, we can substitute 4 for a and 1 for c and find b^2:

$$\begin{aligned} b^2 &= a^2 - c^2 \\ &= 4^2 - 1^2 \\ &= 15 \end{aligned}$$

To find the equation of the ellipse, we substitute -1 for h, 3 for k, 16 for a^2, and 15 for b^2 in the standard equation and simplify:

$$\frac{(x-h)^2}{\mathbf{a^2}} + \frac{(y-k)^2}{\mathbf{b^2}} = 1$$

$$\frac{[x-(-1)]^2}{\mathbf{16}} + \frac{(y-3)^2}{\mathbf{15}} = 1$$

$$\frac{(x+1)^2}{16} + \frac{(y-3)^2}{15} = 1$$

Self Check Find the equation of the ellipse with focus at (3, 1) and vertices at $V(5, 1)$ and $V'(-5, 1)$. ■

Everyday Connections — Eccentricity of an Ellipse

"Nothing in life is to be feared. It is only to be understood." Marie Curie, Scientist, awarded Nobel Prize for Physics (1903) and for Chemistry (1911)

Explaining Eccentricities

The peculiar orbits of three planets looping around a faraway star can be explained only if an unseen fourth planet blundered through and knocked them out of their circular orbits, according to a new study by researchers at the University of California, Berkeley, and Northwestern University.

The conclusion is based on computer extrapolations from 13 years of observations of planet motions around the star Upsilon Andromedae. It suggests that the non-circular and often highly elliptical orbits of many of the

Continued

extrasolar planets discovered to date may be the result of planets scattering off one another. In such a scenario, the perturbing planet could be shot out of the system entirely or could be kicked into a far-off orbit, leaving the inner planets with eccentric orbits.
Source: **http://www.astrobio.net, April 15, 2005**

Eccentricity vs. semimajor axis for extrasolar planets The 75 planets shown were found in a Doppler survey of 1,300 FGKM main sequence stars using the Lick, Keck, and AAT telescopes. The survey was carried out by the California-Carnegie planet search team.

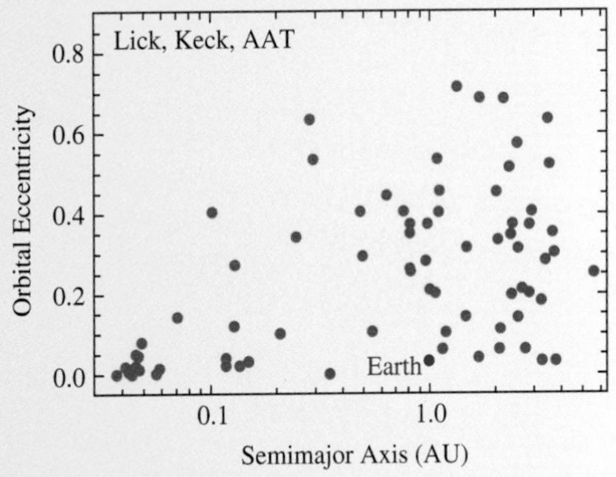

Source: **http://exoplanets.org/newsframe.html**

The eccentricity of an ellipse provides a measure of how much the curve resembles a true circle. Specifically, the eccentricity of a true circle equals 0. Use the data plot above to estimate how many of the 75 planets shown follow orbits that are true circles. **4 of the 75 planets shown have eccentricity 0**

EXAMPLE 4 The orbit of the Earth is approximately an ellipse, with the sun at one focus. The ratio of c to a (called the **eccentricity** of the ellipse) is about $\frac{1}{62}$, and the length of the major axis is approximately 186,000,000 miles. How close does the Earth get to the sun?

Solution We will assume that the ellipse has its center at the origin and vertices V' and V at $(-93{,}000{,}000, 0)$ and $(93{,}000{,}000, 0)$, as shown in Figure 7-19. Because the eccentricity $\frac{c}{a}$ is given to be $\frac{1}{62}$ and $a = 93{,}000{,}000$, we have

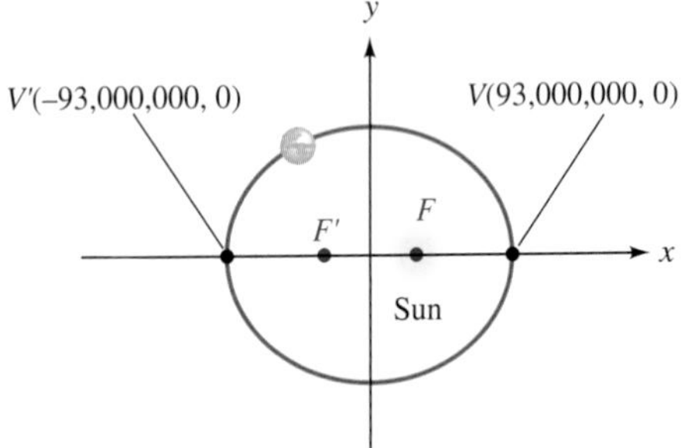

Figure 7-19

$$\frac{c}{a} = \frac{1}{62}$$

$$c = \frac{1}{62}a$$

$$c = \frac{1}{62}(\mathbf{93{,}000{,}000})$$

$$= 1{,}500{,}000$$

The distance $d(FV)$ is the shortest distance between the Earth and the sun. (You'll be asked to prove this in the exercises.) Thus,

$$d(FV) = a - c = 93{,}000{,}000 - 1{,}500{,}000 = 91{,}500{,}000 \text{ mi}$$

The Earth's point of closest approach to the sun (called the **perihelion**) is approximately 91.5 million miles. ■

We can use the eccentricity of an ellipse to judge its shape. If the eccentricity is close to 1, the ellipse is relatively flat, as in Figure 7-20(a). If the eccentricity is close to 0, the ellipse is more circular, as in Figure 7-20(b). Since the eccentricity of the Earth's orbit is $\frac{1}{62}$, the Earth's orbit is almost a circle.

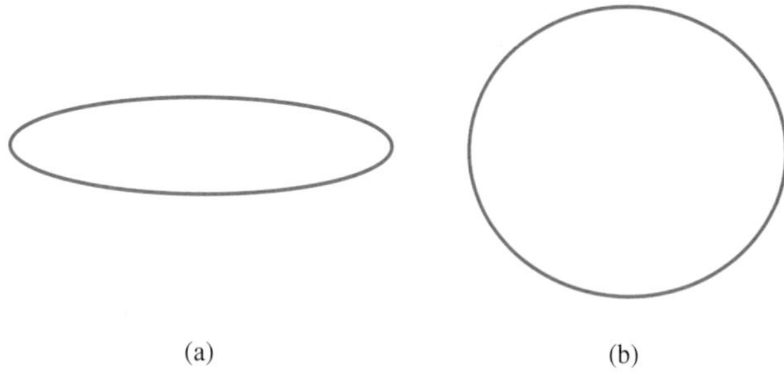

Figure 7-20

Graphing Equations of Ellipses

EXAMPLE 5 Graph the ellipse: $\dfrac{(x + 2)^2}{4} + \dfrac{(y - 2)^2}{9} = 1.$

Solution The center is at $(-2, 2)$, and the major axis is parallel to the y-axis. Because $a = 3$, the vertices are 3 units above and below the center at points $(-2, 5)$ and $(-2, -1)$. Because $b = 2$, the endpoints of the minor axis are 2 units to the right and left of the center at points $(0, 2)$ and $(-4, 2)$. Using these points, we can sketch the ellipse shown in Figure 7-21.

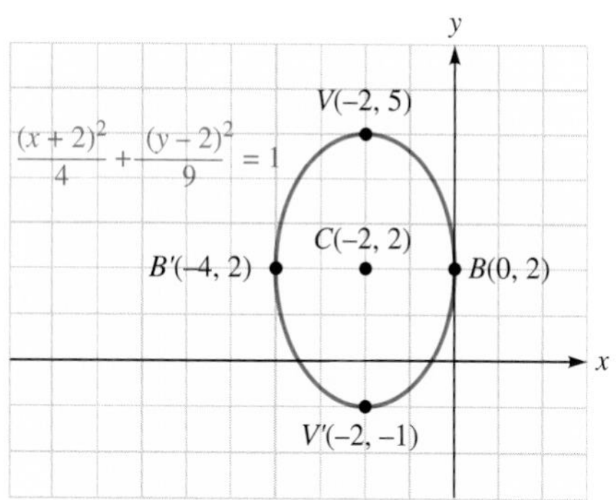

Figure 7-21

Self Check Graph the ellipse: $\dfrac{(x - 2)^2}{9} + \dfrac{(y + 1)^2}{25} = 1.$ ■

ACTIVE EXAMPLE 6 Graph: $4x^2 + 9y^2 - 16x - 18y = 11$.

Solution We write the equation in standard form by completing the square on x and y:

COLLEGE **Algebra** $f(x)$ **Now**™
Go to **http://1pass.thomson.com** or your CD to practice this example.

$$4x^2 + 9y^2 - 16x - 18y = 11$$
$$4x^2 - 16x + 9y^2 - 18y = 11$$
$$4(x^2 - 4x) + 9(y^2 - 2y) = 11$$
$$4(x^2 - 4x + 4) + 9(y^2 - 2y + 1) = 11 + 16 + 9$$
$$4(x - 2)^2 + 9(y - 1)^2 = 36$$
$$\frac{(x - 2)^2}{9} + \frac{(y - 1)^2}{4} = 1$$

We can now see that the graph is an ellipse with center at (2, 1) and major axis parallel to the x-axis. Because $a = 3$, the vertices are at $(-1, 1)$ and $(5, 1)$. Because $b = 2$, the endpoints of the minor axis are at $(2, -1)$ and $(2, 3)$. Using these points, we can sketch the ellipse shown in Figure 7-22.

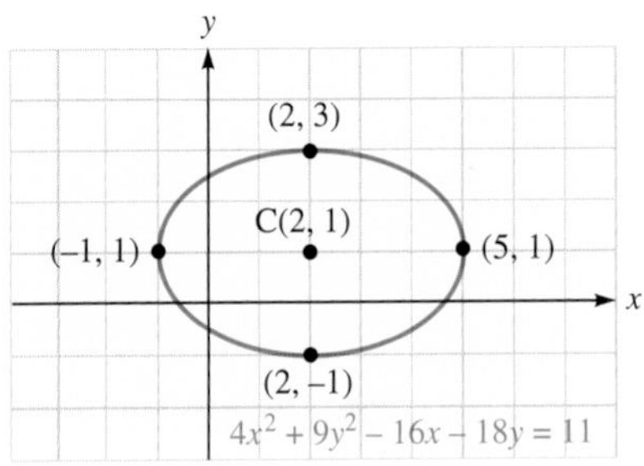

Figure 7-22

Self Check Graph: $4x^2 + 9y^2 - 8x + 36y = -4$. ■

Accent on Technology GRAPHING ELLIPSES

To use a graphing calculator to graph

$$\frac{(x + 2)^2}{4} + \frac{(y - 1)^2}{25} = 1$$

we first clear the equation of fractions by multiplying both sides by 100 and solving for y.

$$25(x + 2)^2 + 4(y - 1)^2 = 100 \quad \text{Multiply both sides by 100.}$$
$$4(y - 1)^2 = 100 - 25(x + 2)^2 \quad \text{Subtract } 25(x + 2)^2 \text{ from both sides.}$$
$$(y - 1)^2 = \frac{100 - 25(x + 2)^2}{4} \quad \text{Divide both sides by 4.}$$
$$y - 1 = \pm\frac{\sqrt{100 - 25(x + 2)^2}}{2} \quad \text{Take the square root of both sides.}$$
$$y = 1 \pm \frac{\sqrt{100 - 25(x + 2)^2}}{2} \quad \text{Add 1 to both sides.}$$

Continued

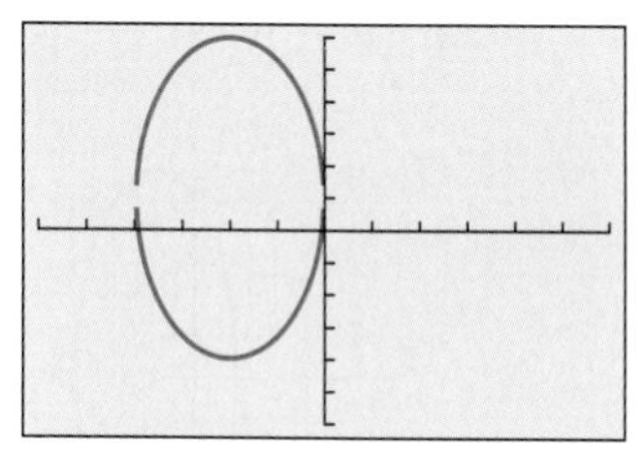

Figure 7-23

If we use window settings $[-6, 6]$ for x and $[-6, 6]$ for y and graph the functions

$$y = 1 + \frac{\sqrt{100 - 25(x + 2)^2}}{2} \quad \text{and} \quad y = 1 - \frac{\sqrt{100 - 25(x + 2)^2}}{2}$$

we will obtain the ellipse shown in Figure 7-23.

Self Check Answers

2. $\frac{x^2}{25} + \frac{y^2}{16} = 1$ **3.** $\frac{x^2}{25} + \frac{(y - 1)^2}{16} = 1$ **5.**

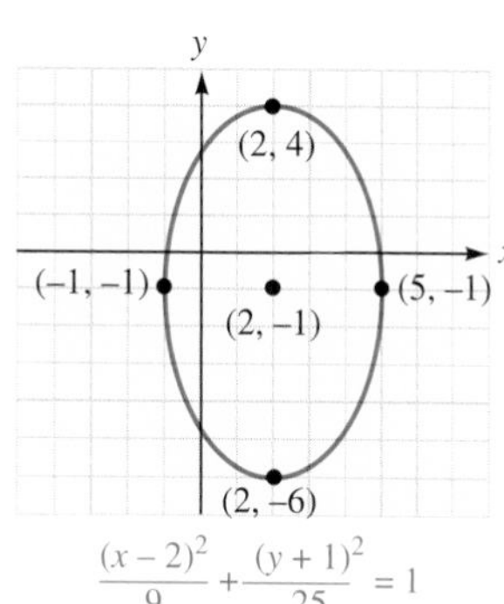

6.

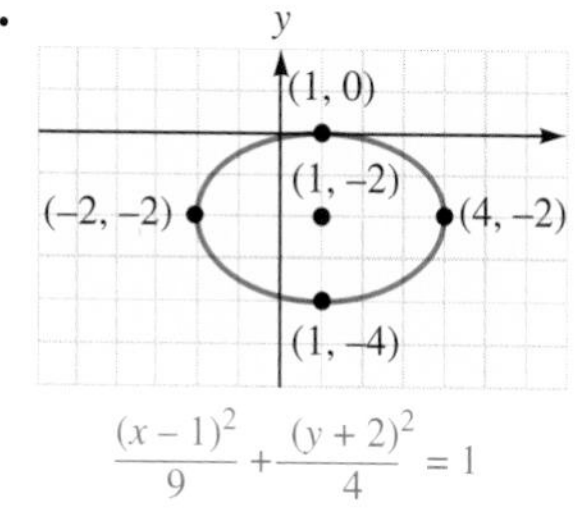

7.2 Exercises

VOCABULARY AND CONCEPTS ***Fill in the blanks.***

1. An ellipse is the set of all points in the plane such that the **sum** of the distances from two fixed points is a positive **constant**.

2. Each of the two fixed points in the definition of an ellipse is called a **focus** of the ellipse.

3. The chord that joins the **vertices** is called the major axis of the ellipse.

4. The chord through the center of an ellipse and perpendicular to the major axis is called the **minor** axis.

5. In the ellipse $\frac{x^2}{a^2} + \frac{y^2}{b^2} = 1$ $(a > b)$, the vertices are $V(\underline{a}, \underline{0})$ and $V'(\underline{-a}, \underline{0})$.

6. In an ellipse, the relationship between a, b, and c is $\underline{b^2 = a^2 - c^2}$.

7. To draw an ellipse that is 26 inches wide and 10 inches tall, how long should the piece of string be, and how far apart should the two thumbtacks be?
26-in. string; thumbtacks 24 in. apart

8. To draw an ellipse that is 20 centimeters wide and 12 centimeters tall, how long should the piece of string be, and how far apart should the two thumbtacks be? **20-cm string; thumbtacks 16 cm apart**

PRACTICE ***Write the equation of the ellipse that has its center at the origin.***

9. Focus at (3, 0); vertex at (5, 0) $\frac{x^2}{25} + \frac{y^2}{16} = 1$

10. Focus at (0, 4); vertex at (0, 7) $\frac{x^2}{33} + \frac{y^2}{49} = 1$

11. Focus at (0, 1); $\frac{4}{3}$ is one-half the length of the minor axis $\frac{9x^2}{16} + \frac{9y^2}{25} = 1$

12. Focus at (1, 0); $\frac{4}{3}$ is one-half the length of the minor axis $\frac{9x^2}{25} + \frac{9y^2}{16} = 1$

13. Focus at (0, 3); major axis equal to 8 $\frac{x^2}{7} + \frac{y^2}{16} = 1$

14. Focus at (5, 0); major axis equal to 12 $\frac{x^2}{36} + \frac{y^2}{11} = 1$

Write the equation of each ellipse.

15. Center at (3, 4); $a = 3$, $b = 2$; major axis parallel to the y-axis $\frac{(x-3)^2}{4} + \frac{(y-4)^2}{9} = 1$

16. Center at (3, 4); passes through (3, 10) and (3, −2); $b = 2$ $\frac{(x-3)^2}{4} + \frac{(y-4)^2}{36} = 1$

17. Center at (3, 4); $a = 3$, $b = 2$; major axis parallel to the x-axis $\frac{(x-3)^2}{9} + \frac{(y-4)^2}{4} = 1$

18. Center at (3, 4); passes through (8, 4) and (−2, 4); $b = 2$ $\frac{(x-3)^2}{25} + \frac{(y-4)^2}{4} = 1$

19. Foci at (−2, 4) and (8, 4); $b = 4$
$\frac{(x-3)^2}{41} + \frac{(y-4)^2}{16} = 1$

20. Foci at (8, 5) and (4, 5); $b = 3$
$\frac{(x-6)^2}{13} + \frac{(y-5)^2}{9} = 1$

21. Vertex at (6, 4); foci at (−4, 4) and (4, 4)
$\frac{x^2}{36} + \frac{(y-4)^2}{20} = 1$

22. Center at (−4, 5); $\frac{c}{a} = \frac{1}{3}$; vertex at (−4, −1)
$\frac{(x+4)^2}{32} + \frac{(y-5)^2}{36} = 1$

23. Foci at (6, 0) and (−6, 0); $\frac{c}{a} = \frac{3}{5}$ $\frac{x^2}{100} + \frac{y^2}{64} = 1$

24. Vertices at (2, 0) and (−2, 0); $\frac{2b^2}{a} = 2$
$\frac{x^2}{4} + \frac{y^2}{2} = 1$

Graph each ellipse.

25. $\frac{x^2}{25} + \frac{y^2}{49} = 1$

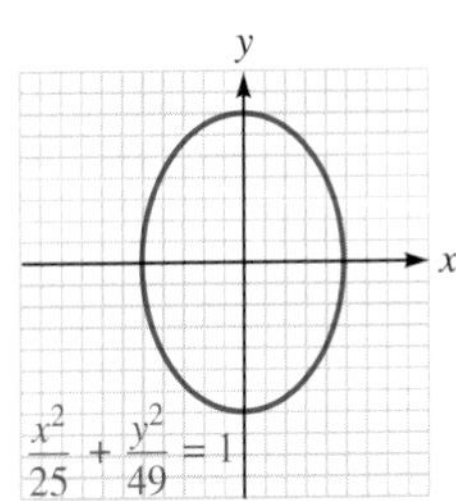

26. $4x^2 + y^2 = 4$

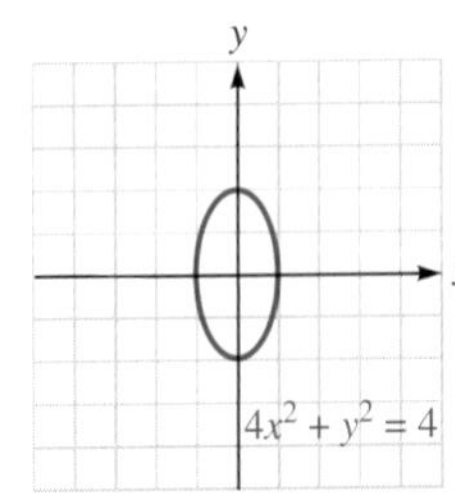

27. $\frac{x^2}{16} + \frac{(y+2)^2}{36} = 1$

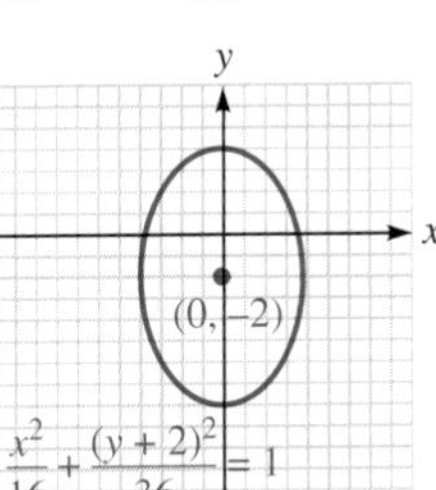

28. $(x-1)^2 + \frac{4y^2}{25} = 4$

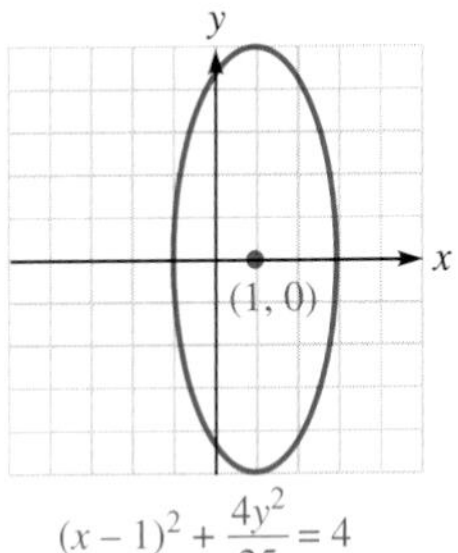

29. $x^2 + 4y^2 - 4x + 8y + 4 = 0$

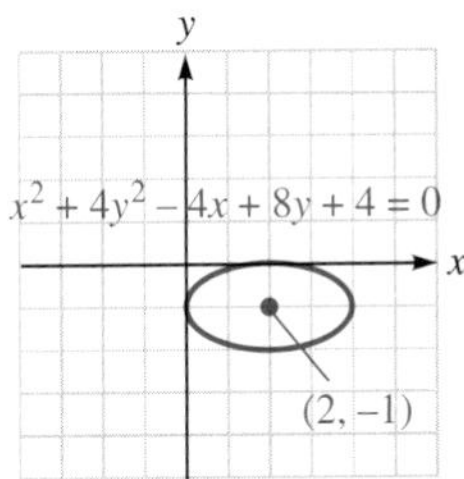

30. $x^2 + 4y^2 - 2x - 16y = -13$

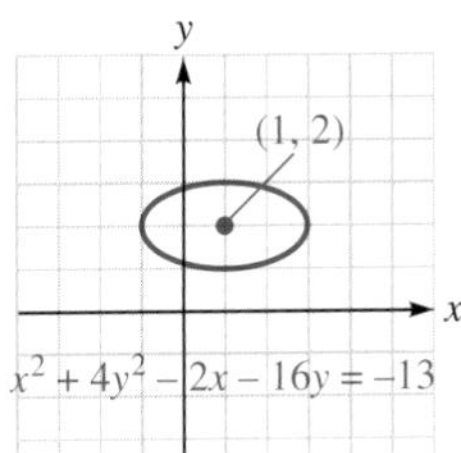

31. $16x^2 + 25y^2 - 160x - 200y + 400 = 0$

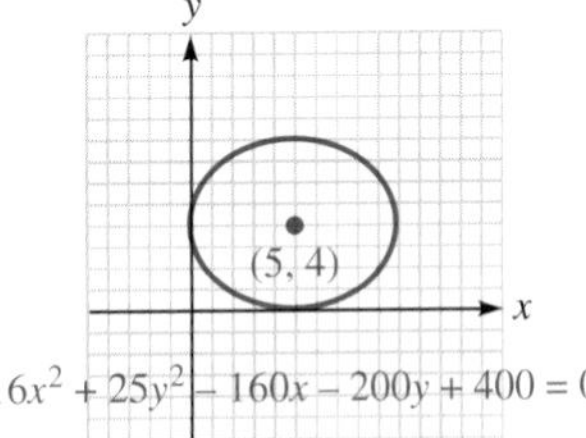

32. $3x^2 + 2y^2 + 7x - 6y = -1$

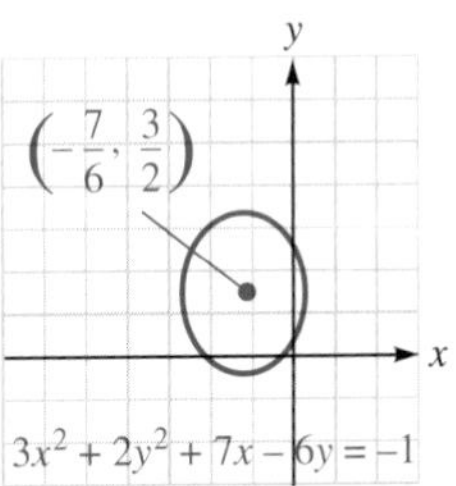

APPLICATIONS

33. Pool tables Find the equation of the outer edge of the elliptical pool table shown below.
$\frac{x^2}{900} + \frac{y^2}{400} = 1$

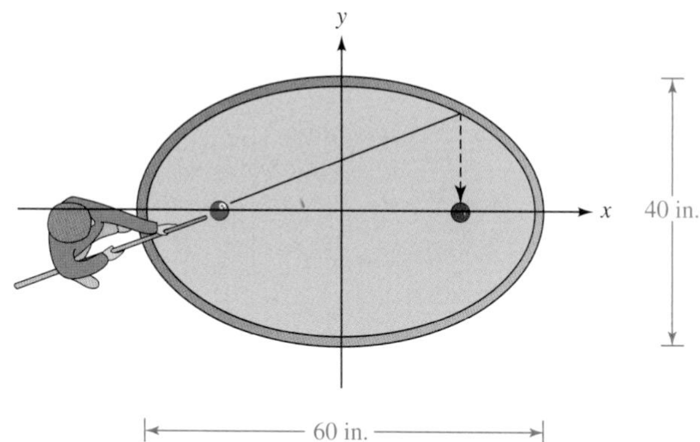

34. Astronomy The moon has an orbit that is an ellipse, with the Earth at one focus. If the major axis of the orbit is 378,000 miles and the ratio of c to a is approximately $\frac{11}{200}$, how far does the moon get from the Earth? (This farthest point in an orbit is called the **apogee.**) **199,395 mi**

35. Equation of an arch An arch is a semiellipse 12 meters wide and 5 meters high. Write the equation of the ellipse if the ellipse is centered at the origin.
$\frac{x^2}{36} + \frac{y^2}{25} = 1$

36. Design of a track A track is built in the shape of an ellipse with a maximum length of 100 meters and a maximum width of 60 meters. Write the equation of the ellipse and find its **focal width.** That is, find the length of a chord that is perpendicular to the major axis and passes through either focus of the ellipse.
$\frac{x^2}{2,500} + \frac{y^2}{900} = 1$; **36 m**

37. Whispering galleries Any sound from one focus of an ellipse reflects off the ellipse directly back to the other focus. This property explains whispering galleries such as Statuary Hall in Washington, D.C. The whispering gallery shown has the shape of a semiellipse. Find the distance sound travels as it leaves focus F and returns to focus F'. **26 m**

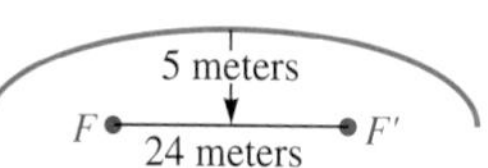

38. Finding the width of a mirror The oval mirror shown is in the shape of an ellipse. Find the width of the mirror 12 inches above its base. **about 20.8 in.**

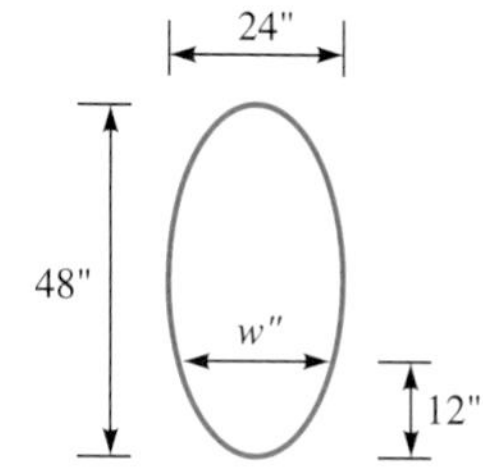

39. Finding the height of a window The window shown has the shape of an ellipse. Find the height of the window 20 inches from one end. **about 26.9 in.**

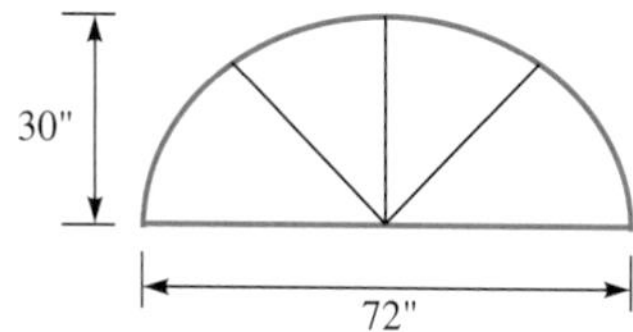

40. Area of an ellipse The area A of the ellipse

$$\frac{x^2}{a^2} + \frac{y^2}{b^2} = 1$$

is given by $A = \pi ab$. Find the area of the ellipse $9x^2 + 16y^2 = 144$. **12π sq. units ≈ 37.7 sq. units**

DISCOVERY AND WRITING

41. If F is a focus of the ellipse shown and B is an endpoint of the minor axis, use the distance formula to prove that the length of segment FB is a. (*Hint:* In an ellipse, $a^2 - c^2 = b^2$.)

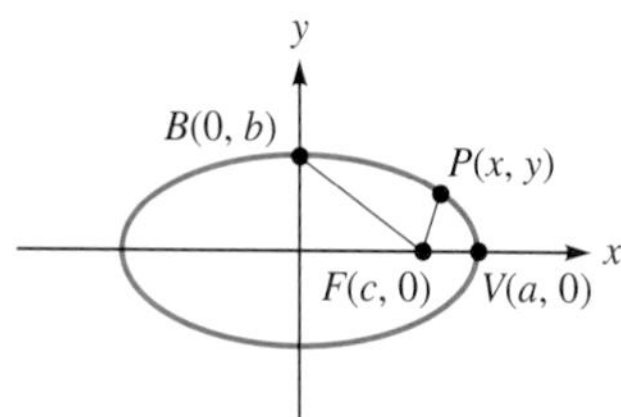

42. If F is a focus of the ellipse shown and P is any point on the ellipse, use the distance formula to show that the length of FP is $a - \frac{c}{a}x$. (*Hint:* In an ellipse, $a^2 - c^2 = b^2$.)

43. Finding the focal width In the ellipse shown, chord AA' passes through the focus F and is perpendicular to the major axis. Show that the length of AA' (called the **focal width**) is $\frac{2b^2}{a}$.

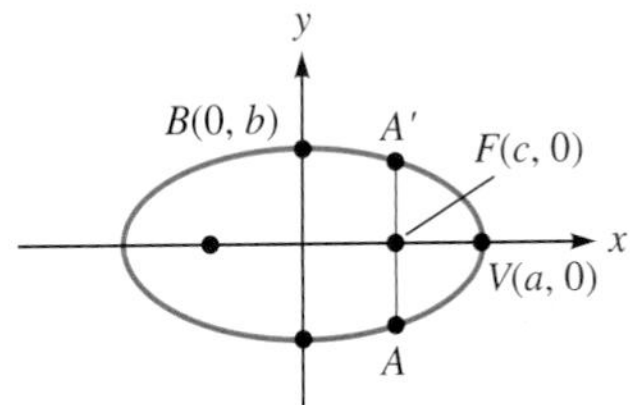

44. Prove that segment FV in Example 4 is the shortest distance between the Earth and the sun. (*Hint:* Refer to Exercise 42.)

45. Constructing an ellipse The ends of a piece of string 6 meters long are attached to two thumbtacks that are 2 meters apart. A pencil catches the loop and draws it tight. As the pencil is moved about the thumbtacks (always keeping the tension), an ellipse is produced, with the thumbtacks as foci. Write the equation of the ellipse. (*Hint:* You'll have to establish a coordinate system.) $\frac{x^2}{9} + \frac{y^2}{8} = 1$

46. The distance between point $P(x, y)$ and the point $(0, 2)$ is $\frac{1}{3}$ of the distance of point P from the line $y = 18$. Find the equation of the curve on which point P lies. $\frac{x^2}{32} + \frac{y^2}{36} = 1$

47. Prove that $a > b$ in the development of the standard equation of an ellipse.

48. Show that the expansion of the standard equation of an ellipse is a special case of the general second-degree equation in two variables.

REVIEW *Let* $A = \begin{bmatrix} 3 & -1 & 2 \\ 0 & 2 & -1 \\ 3 & 1 & 1 \end{bmatrix}$, $B = \begin{bmatrix} 1 & 2 \\ 2 & 0 \\ -1 & 1 \end{bmatrix}$, *and* $C = \begin{bmatrix} 0 & 1 \\ -1 & 1 \\ -2 & 0 \end{bmatrix}$. ***Find each of the following, if possible.***

49. AB $\begin{bmatrix} -1 & 8 \\ 5 & -1 \\ 4 & 7 \end{bmatrix}$

50. $B + C$ $\begin{bmatrix} 1 & 3 \\ 1 & 1 \\ -3 & 1 \end{bmatrix}$

51. $5B - 2C$ $\begin{bmatrix} 5 & 8 \\ 12 & -2 \\ -1 & 5 \end{bmatrix}$

52. $AC + B$ $\begin{bmatrix} -2 & 4 \\ 2 & 2 \\ -4 & 5 \end{bmatrix}$

53. The additive inverse of B $\begin{bmatrix} -1 & -2 \\ -2 & 0 \\ 1 & -1 \end{bmatrix}$

54. The multiplicative inverse of A **It doesn't exist.**

7.3 The Hyperbola

In this section, you will learn about

- Asymptotes of a Hyperbola
- Graphing Equations of Hyperbolas

Hyperbolas are the basis of a navigational system known as LORAN (LOng RAnge Navigation). They are also used to find the source of a distress signal, are the basis for the design of hypoid gears, and describe the orbits of some comets.

The definition of a hyperbola is similar to the definition of an ellipse, except that we require a constant difference of $2a$ instead of a constant sum.

The Hyperbola

A **hyperbola** is the set of all points P in a plane such that the absolute value of the difference of the distances from point P to two other points in the plane is a positive constant.

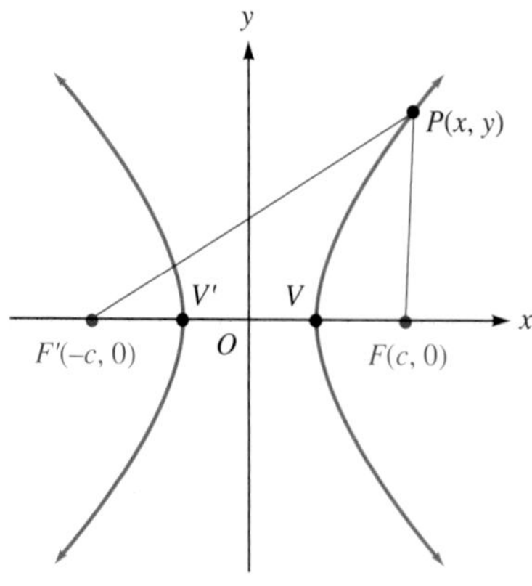

Figure 7-24

Points F and F' shown in Figure 7-24 are called the **foci** of the hyperbola, and the midpoint of chord FF' is called the **center.** The points V and V', where the hyperbola intersects FF', are called **vertices.** The segment VV' is called the **transverse axis.**

To develop the equation of the hyperbola centered at the origin, we note that the origin is the midpoint of chord FF', and we let $d(F'O) = d(OF) = c$, where $c > 0$. Then F is at $(c, 0)$, and F' is at $(-c, 0)$. The definition of a hyperbola requires that $d(F'P) - d(PF) = 2a$, where $2a$ is a positive constant. We use the distance formula to compute the lengths of $F'P$ and PF:

$$d(F'P) = \sqrt{[x - (-c)]^2 + y^2}$$
$$d(PF) = \sqrt{(x - c)^2 + y^2}$$

Substituting these values into the equation $|d(F'P) - d(PF)| = 2a$ gives

$$\sqrt{(x + c)^2 + y^2} - \sqrt{(x - c)^2 + y^2} = 2a$$

or

$$\sqrt{(x + c)^2 + y^2} = 2a + \sqrt{(x - c)^2 + y^2}$$

After squaring, we have

$$(x + c)^2 + y^2 = 4a^2 + 4a\sqrt{(x - c)^2 + y^2} + (x - c)^2 + y^2$$
$$x^2 + 2cx + c^2 + y^2 = 4a^2 + 4a\sqrt{(x - c)^2 + y^2} + x^2 - 2cx + c^2 + y^2$$
$$4cx = 4a^2 + 4a\sqrt{(x - c)^2 + y^2}$$
$$cx - a^2 = a\sqrt{(x - c)^2 + y^2}$$

Squaring both sides again and simplifying gives

$$c^2x^2 - 2a^2cx + a^4 = a^2(x^2 - 2cx + c^2 + y^2)$$
$$c^2x^2 - 2a^2cx + a^4 = a^2x^2 - 2a^2cx + a^2c^2 + a^2y^2$$
$$c^2x^2 + a^4 = a^2x^2 + a^2c^2 + a^2y^2$$

(1) $$(c^2 - a^2)x^2 - a^2y^2 = a^2(c^2 - a^2)$$

Because $c > a$ (you will be asked to prove this in the exercises), $c^2 - a^2$ is a positive number. So we can let $b^2 = c^2 - a^2$ and substitute b^2 for $c^2 - a^2$ in Equation 1 to get

$$b^2x^2 - a^2y^2 = a^2b^2 \qquad (b^2 = c^2 - a^2)$$

We divide both sides of the previous equation by a^2b^2 to get the standard equation for a hyperbola with center at the origin and foci on the x-axis:

$$\frac{x^2}{a^2} - \frac{y^2}{b^2} = 1$$

To find the x-intercepts of the graph, we let $y = 0$ and solve for x. We get

$$\frac{x^2}{a^2} = 1$$
$$x^2 = a^2$$
$$x = a \quad \text{or} \quad x = -a$$

We now know that the x-intercepts are the vertices $V(a, 0)$ and $V'(-a, 0)$. The distance between the center of the hyperbola and either vertex is a, and the center of the hyperbola is the midpoint of the segment $V'V$ as well as that of the segment FF'.

We attempt to find the y-intercepts by letting $x = 0$. Then the equation becomes

$$\frac{-y^2}{b^2} = 1 \qquad \text{or} \qquad y^2 = -b^2$$

Because $-b^2$ represents a negative number, and y^2 cannot be negative, the equation has no real solutions. Since there are no y-values corresponding to $x = 0$, the hyperbola does not intersect the y-axis.

This discussion suggests the following results.

Hyperbola: Foci on x-Axis, Center at (0, 0)

The standard equation of a hyperbola with center at the origin and foci on the x-axis is

$$\frac{x^2}{a^2} - \frac{y^2}{b^2} = 1$$

where $a^2 + b^2 = c^2$.

Vertices: $V(a, 0)$ and $V'(-a, 0)$
Foci: $F(c, 0)$ and $F'(-c, 0)$

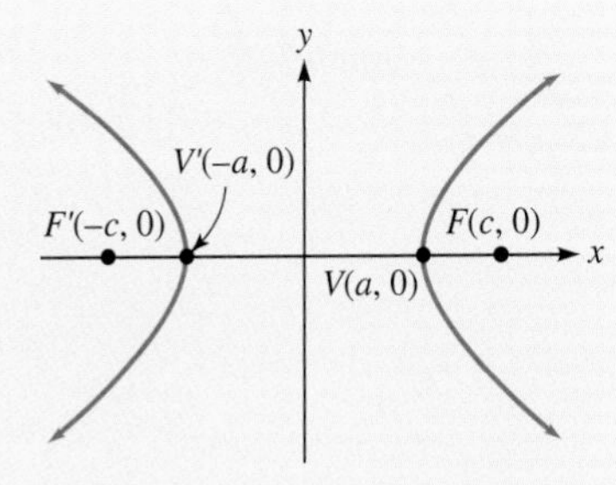

If the foci are on the y-axis, a similar equation results.

Hyperbola: Foci on y-Axis, Center at (0, 0)

The standard equation of a hyperbola with center at the origin and foci on the y-axis is

$$\frac{y^2}{a^2} - \frac{x^2}{b^2} = 1$$

where $a^2 + b^2 = c^2$.

Vertices: $V(0, a)$ and $V'(0, -a)$
Foci: $F(0, c)$ and $F'(0, -c)$

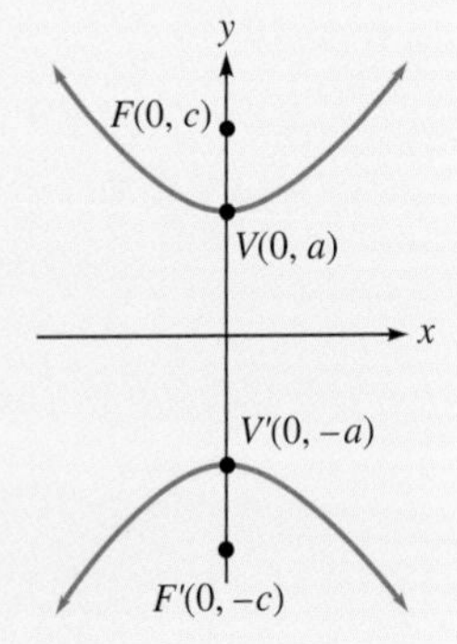

To translate the hyperbola to a new position centered at the point (h, k) instead of the origin, we replace x and y with $x - h$ and $y - k$, respectively. We get the following results.

Hyperbola: Transverse Axis Horizontal, Center at (h, k)

The standard equation of a hyperbola with center at the (h, k) and foci on a line parallel to the x-axis is

$$\frac{(x - h)^2}{a^2} - \frac{(y - k)^2}{b^2} = 1$$

where $a^2 + b^2 = c^2$.

Vertices: $V(a + h, k)$ and $V'(-a + h, k)$
Foci: $F(c + h, k)$ and $F'(-c + h, k)$

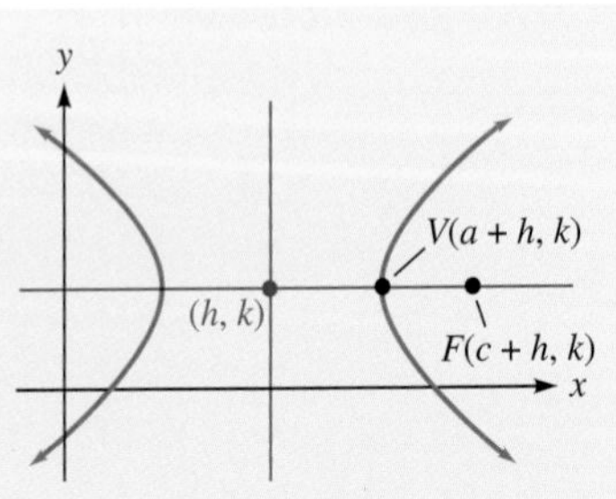

Hyperbola: Transverse Axis Vertical, Center at (h, k)

The standard equation of a hyperbola with center at (h, k) and foci on a line parallel to the y-axis is

$$\frac{(y-k)^2}{a^2} - \frac{(x-h)^2}{b^2} = 1$$

where $a^2 + b^2 = c^2$.

Vertices: $V(h, a + k)$ and $V'(h, -a + k)$
Foci: $F(h, c + k)$ and $F'(h, -c + k)$

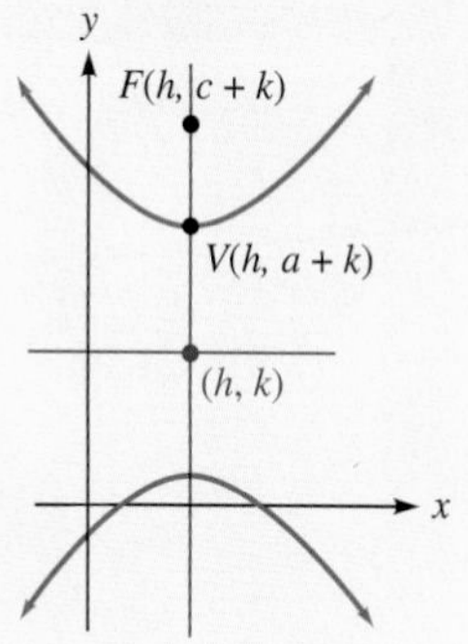

ACTIVE EXAMPLE 1 Write the equation of the hyperbola with vertices $(3, -3)$ and $(3, 3)$ and a focus at $(3, 5)$.

Solution First, we plot the vertices and focus, as shown in Figure 7-25. Because the foci lie on a vertical line, the equation to use is

$$\frac{(y-k)^2}{a^2} - \frac{(x-h)^2}{b^2} = 1$$

COLLEGE **Algebra Now™**
Explore this Active Example and test yourself on the steps taken here at **http://1pass.thomson.com**. You can also find this example on the Interactive Video Skillbuilder CD-ROM.

The center of the hyperbola is midway between the vertices V and V'. Thus, the center is point $(3, 0)$; $h = 3$; and $k = 0$. The distance between the vertex and the center is $a = 3$, and the distance between the focus and the center is $c = 5$. We can find b^2 by substituting 3 for a and 5 for c in the following equation to get

$$b^2 = c^2 - a^2 \quad \text{In a hyperbola, } b^2 = c^2 - a^2.$$
$$b^2 = 5^2 - 3^2$$
$$b^2 = 16$$

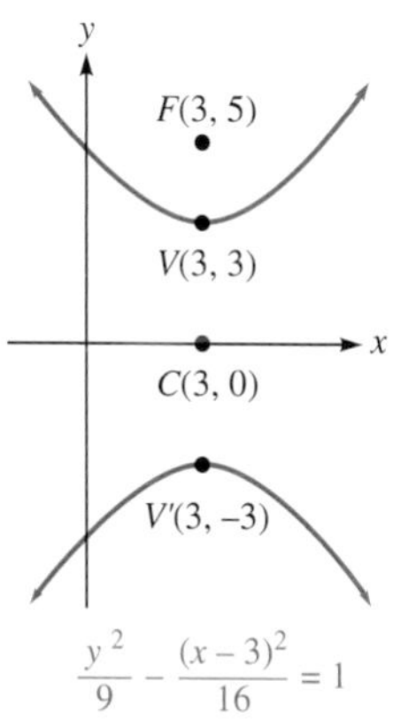

Figure 7-25

Substituting the values for h, k, a^2, and b^2 in the standard equation gives the equation of the hyperbola:

$$\frac{(y-k)^2}{\mathbf{a^2}} - \frac{(x-\mathbf{h})^2}{\mathbf{b^2}} = 1$$
$$\frac{(y-\mathbf{0})^2}{\mathbf{9}} - \frac{(x-\mathbf{3})^2}{\mathbf{16}} = 1$$
$$\frac{y^2}{9} - \frac{(x-3)^2}{16} = 1$$

Self Check Write the equation of the hyperbola with vertices $(3, 1)$ and $(-3, 1)$ and a focus at $(5, 1)$. ■

■ Asymptotes of a Hyperbola

The values of a and b are important aids in graphing hyperbolas. To see their value, we consider the hyperbola

$$\frac{x^2}{a^2} - \frac{y^2}{b^2} = 1$$

Andrew Wiles
(1953–)

Wiles first learned of Fermat's last theorem (see page 531) in 1963, when he was 10 years old. He was so intrigued by the problem that he decided to study mathematics. While on the faculty at Princeton University, he would isolate himself in the attic of his home and work on the problem. After seven years of hard work, he announced a solution in June 1993. After another year of work to fix some gaps in the solution, his 130-page proof was published in 1995. After 350 years, Fermat's last theorem has been proved.

with center at the origin and vertices at $V(a, 0)$ and $V'(-a, 0)$. We can plot points V, V', $B(0, b)$, and $B'(0, -b)$ and form rectangle $RSQP$, called the **fundamental rectangle,** as shown in Figure 7-26. We can show that the extended diagonals of this rectangle are asymptotes of the hyperbola. In the exercises, you will be asked to show that the equations of these two lines are

$$y = \frac{b}{a}x \qquad \text{and} \qquad y = -\frac{b}{a}x$$

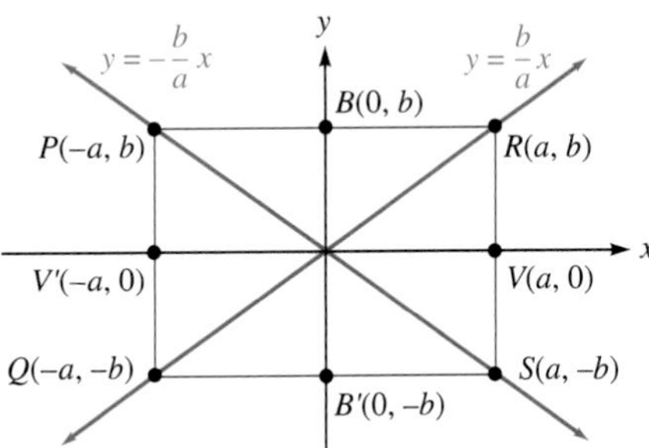

Figure 7-26

To show that the extended diagonals are asymptotes of the hyperbola, we solve the equation

$$\frac{x^2}{a^2} - \frac{y^2}{b^2} = 1$$

for y and modify its form as follows:

$$\frac{x^2}{a^2} - \frac{y^2}{b^2} = 1$$

$$b^2x^2 - a^2y^2 = a^2b^2 \qquad \text{Multiply both sides by } a^2b^2.$$

$$y^2 = \frac{b^2x^2 - a^2b^2}{a^2} \qquad \text{Subtract } b^2x^2 \text{ from both sides and divide both sides by } -a^2.$$

$$y^2 = \frac{b^2x^2}{a^2}\left(1 - \frac{a^2}{x^2}\right) \qquad \text{Factor out the fraction } \frac{b^2x^2}{a^2}.$$

(2) $$y = \pm\frac{bx}{a}\sqrt{1 - \frac{a^2}{x^2}} \qquad \text{Take the square root of both sides.}$$

In Equation 2, if a is constant and $|x|$ approaches ∞, then $\frac{a^2}{x^2}$ approaches 0, and $\sqrt{1 - \frac{a^2}{x^2}}$ approaches 1. Thus, the hyperbola approaches the lines

$$y = \frac{b}{a}x \qquad \text{and} \qquad y = -\frac{b}{a}x$$

Knowing the asymptotes makes it easy to sketch a hyperbola. We simply convert its equation into standard form, find the coordinates of its vertices, and construct the fundamental rectangle with its extended diagonals. Using the vertices and the asymptotes as guides, we can sketch the hyperbola shown in Figure 7-27. The segment BB' is called the **conjugate axis** of the hyperbola.

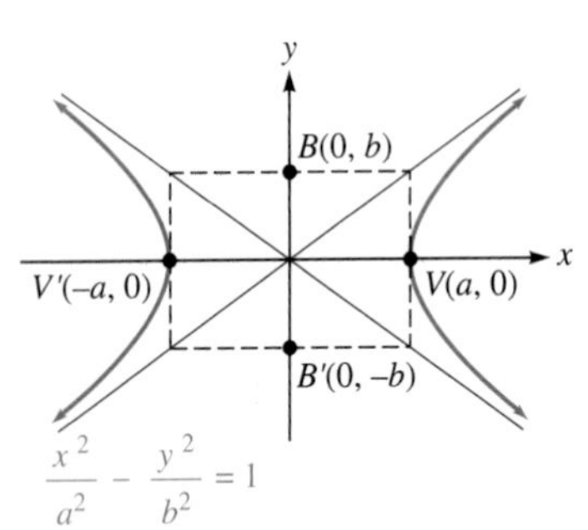

Figure 7-27

■ Graphing Equations of Hyperbolas

ACTIVE EXAMPLE 2 Graph the hyperbola: $x^2 - y^2 - 2x + 4y = 12$.

Solution We complete the square on x and y to write the equation into standard form:

COLLEGE **Algebra** *f(x)* **Now**™
Go to **http://1pass.thomson.com** or your CD to practice this example.

$$x^2 - 2x - y^2 + 4y = 12$$
$$x^2 - 2x - (y^2 - 4y) = 12$$
$$x^2 - 2x + 1 - (y^2 - 4y + 4) = 12 + 1 - 4$$
$$(x - 1)^2 - (y - 2)^2 = 9$$
$$\frac{(x-1)^2}{9} - \frac{(y-2)^2}{9} = 1$$

From the standard equation of a hyperbola, we see that the center is $(1, 2)$, that $a = 3$ and $b = 3$, and that the vertices are on a line segment parallel to the x-axis, as shown in Figure 7-28. The vertices V and V' are 3 units to the right and left of the center and have coordinates of $(4, 2)$ and $(-2, 2)$. Points B and B', 3 units above and below the center, have coordinates of $(1, 5)$ and $(1, -1)$. After using points V, V', B, and B' to construct the fundamental rectangle and its extended diagonals, we sketch the graph.

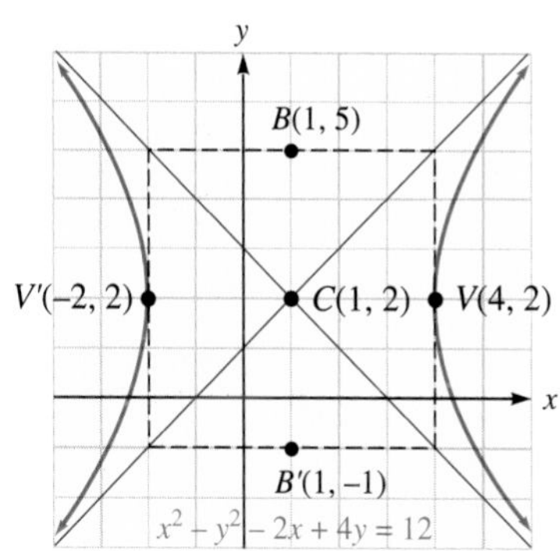

Figure 7-28

Self Check Graph the hyperbola: $x^2 - y^2 + 6x + 2y = 8$. ■

To graph the equation $x^2 - y^2 - 2x + 4y = 12$ with a graphing calculator, we first solve the equation for y.

$$x^2 - y^2 - 2x + 4y = 12$$
$$-x^2 + y^2 + 2x - 4y = -12$$
$$y^2 - 4y = x^2 - 2x - 12$$
$$y^2 - 4y + 4 = x^2 - 2x - 12 + 4$$
$$(y - 2)^2 = x^2 - 2x - 8$$
$$y - 2 = \pm\sqrt{x^2 - 2x - 8}$$
$$y = 2 \pm \sqrt{x^2 - 2x - 8}$$

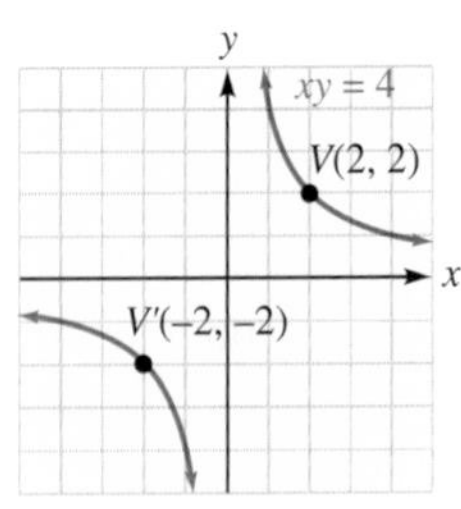

Figure 7-29

If we use window settings of $[-4, 6]$ for x and $[-3, 7]$ for y and graph the functions

$$y = 2 + \sqrt{x^2 - 2x - 8} \quad \text{and} \quad y = 2 - \sqrt{x^2 - 2x - 8}$$

we will get a graph similar to the one shown in Figure 7-28.

We have considered only those hyperbolas with a major axis that is horizontal or vertical. However, some hyperbolas have nonhorizontal or nonvertical major axes. For example, the graph of the equation $xy = 4$ is a hyperbola with vertices at (2, 2) and $(-2, -2)$, as shown in Figure 7-29.

Self Check Answers

1. $\dfrac{x^2}{9} - \dfrac{(y-1)^2}{16} = 1$ **2.** $x^2 - y^2 + 6x + 2y = 8$

B(−3, 5)
C(−3, 1)
V'(−7, 1)
V(1, 1)
B'(−3, −3)

7.3 Exercises

VOCABULARY AND CONCEPTS ***Fill in the blanks.***

1. A hyperbola is the set of all points in the plane such that the absolute value of the difference of the distances from two fixed points is a positive constant.

2. Each of the two fixed points in the definition of a hyperbola is called a focus of the hyperbola.

3. The vertices of the hyperbola $\dfrac{x^2}{a^2} - \dfrac{y^2}{b^2} = 1$ are $V(a, 0)$ and $V'(-a, 0)$.

4. The vertices of the hyperbola $\dfrac{y^2}{a^2} - \dfrac{x^2}{b^2} = 1$ are $V(0, a)$ and $V'(0, -a)$.

5. The chord that joins the vertices is called the transverse axis of the hyperbola.

6. In a hyperbola, the relationship between a, b, and c is $b^2 = c^2 - a^2$.

PRACTICE ***Write the equation of each hyperbola.***

7. Vertices (5, 0) and $(-5, 0)$; focus (7, 0)
$\dfrac{x^2}{25} - \dfrac{y^2}{24} = 1$

8. Focus (3, 0); vertex (2, 0); center (0, 0)
$\dfrac{x^2}{4} - \dfrac{y^2}{5} = 1$

9. Center (2, 4); $a = 2$, $b = 3$; transverse axis is horizontal $\dfrac{(x-2)^2}{4} - \dfrac{(y-4)^2}{9} = 1$

10. Center $(-1, 3)$; vertex (1, 3); focus (2, 3)
$\dfrac{(x+1)^2}{4} - \dfrac{(y-3)^2}{5} = 1$

11. Center (5, 3); vertex (5, 6); passes through (1, 8)
$\dfrac{(y-3)^2}{9} - \dfrac{(x-5)^2}{9} = 1$

12. Foci (0, 10) and $(0, -10)$; $\dfrac{c}{a} = \dfrac{5}{4}$ $\dfrac{y^2}{64} - \dfrac{x^2}{36} = 1$

13. Vertices (0, 3) and $(0, -3)$; $\dfrac{c}{a} = \dfrac{5}{3}$ $\dfrac{y^2}{9} - \dfrac{x^2}{16} = 1$

14. Focus (4, 0); vertex (2, 0); center (0, 0)
$\dfrac{x^2}{4} - \dfrac{y^2}{12} = 1$

15. Center $(1, -3)$; $a^2 = 4$; $b^2 = 16$
$\dfrac{(x-1)^2}{4} - \dfrac{(y+3)^2}{16} = 1$ or $\dfrac{(y+3)^2}{4} - \dfrac{(x-1)^2}{16} = 1$

16. Center (1, 4); focus (7, 4); vertex (3, 4)

$$\frac{(x-1)^2}{4} - \frac{(y-4)^2}{32} = 1$$

17. Center at the origin; passes through (4, 2) and (8, −6)

$$\frac{x^2}{10} - \frac{3y^2}{20} = 1$$

18. Center (3, −1); y-intercept −1; x-intercept $3 + \dfrac{3\sqrt{5}}{2}$

$$\frac{(x-3)^2}{9} - \frac{(y+1)^2}{4} = 1$$

Find the area of the fundamental rectangle of each hyperbola.

19. $4(x-1)^2 - 9(y+2)^2 = 36$ **24 sq. units**

20. $x^2 - y^2 - 4x - 6y = 6$ **4 sq. units**

21. $x^2 + 6x - y^2 + 2y = -11$ **12 sq. units**

22. $9x^2 - 4y^2 = 18x + 24y + 63$ **24 sq. units**

Write the equation of each hyperbola.

23. Center (−2, −4); $a = 2$; area of fundamental rectangle is 36 square units

$$\frac{(x+2)^2}{4} - \frac{4(y+4)^2}{81} = 1 \text{ or } \frac{(y+4)^2}{4} - \frac{4(x+2)^2}{81} = 1$$

24. Center (3, −5); $b = 6$; area of fundamental rectangle is 24 square units

$$\frac{(x-3)^2}{1} - \frac{(y+5)^2}{36} = 1 \text{ or } \frac{(y+5)^2}{1} - \frac{(x-3)^2}{36} = 1$$

25. Vertex (6, 0); one end of conjugate axis at $\left(0, \dfrac{5}{4}\right)$

$$\frac{x^2}{36} - \frac{16y^2}{25} = 1$$

26. Vertex (3, 0); focus (−5, 0); center (0, 0)

$$\frac{x^2}{9} - \frac{y^2}{16} = 1$$

Graph each hyperbola.

27. $\dfrac{x^2}{9} - \dfrac{y^2}{4} = 1$

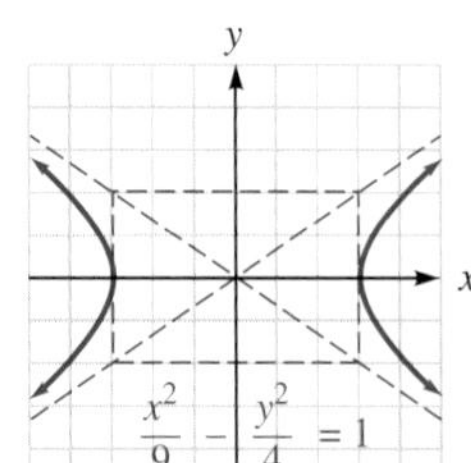

$\frac{x^2}{9} - \frac{y^2}{4} = 1$

28. $\dfrac{y^2}{4} - \dfrac{x^2}{9} = 1$

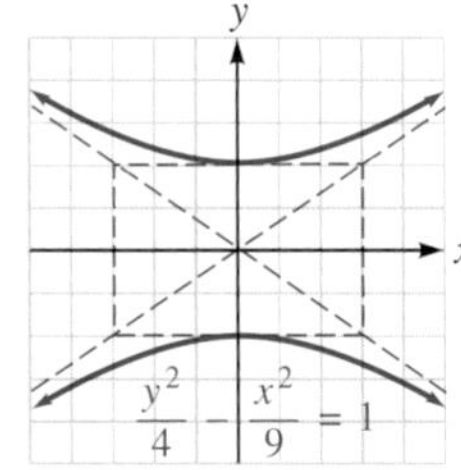

$\frac{y^2}{4} - \frac{x^2}{9} = 1$

29. $4x^2 - 3y^2 = 36$

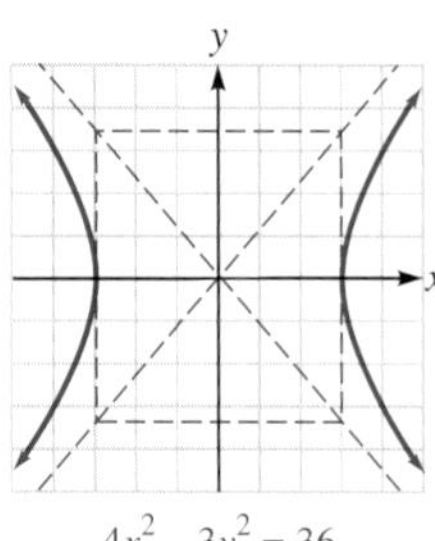

$4x^2 - 3y^2 = 36$

30. $3x^2 - 4y^2 = 36$

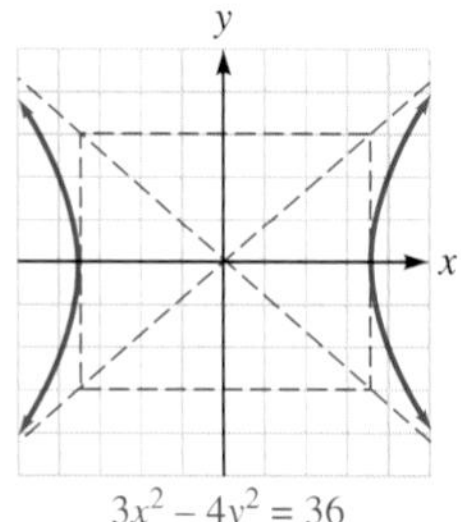

$3x^2 - 4y^2 = 36$

31. $y^2 - x^2 = 1$

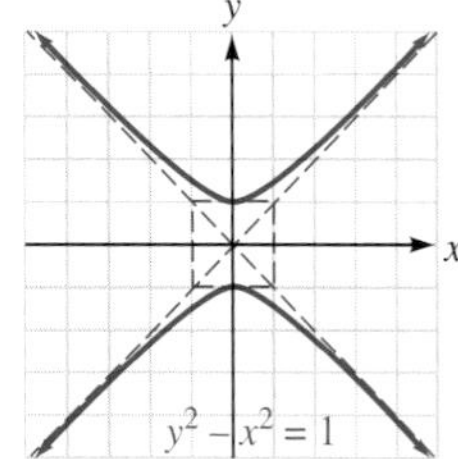

$y^2 - x^2 = 1$

32. $9(y+2)^2 - 4(x-1)^2 = 36$

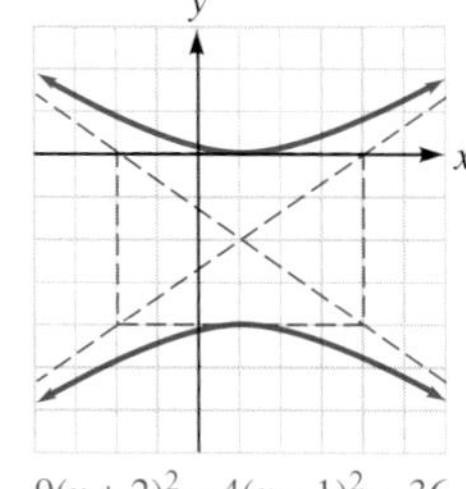

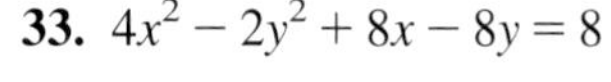

$9(y+2)^2 - 4(x-1)^2 = 36$

33. $4x^2 - 2y^2 + 8x - 8y = 8$

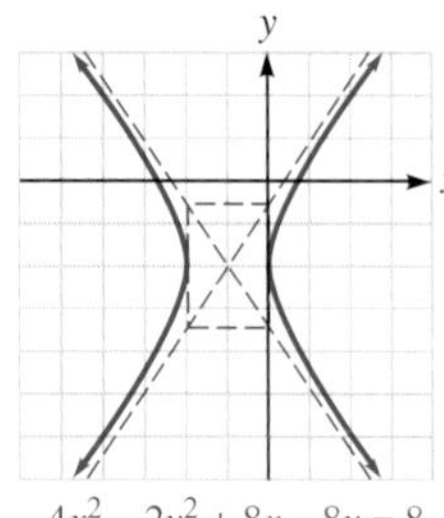

$4x^2 - 2y^2 + 8x - 8y = 8$

34. $x^2 - y^2 - 4x - 6y = 6$

$x^2 - y^2 - 4x - 6y = 6$

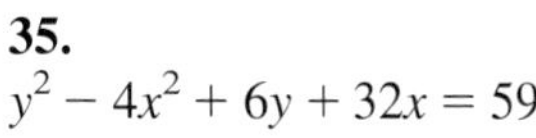

35. $y^2 - 4x^2 + 6y + 32x = 59$

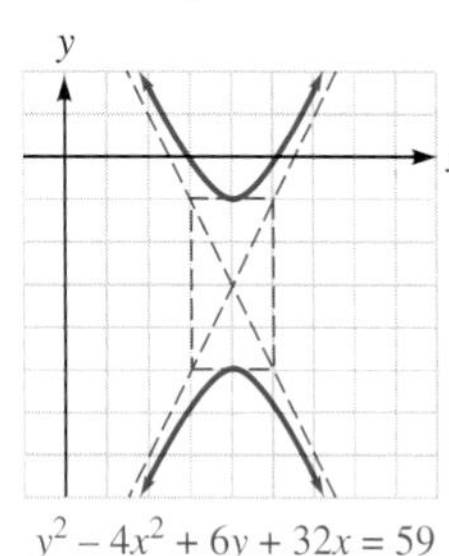

$y^2 - 4x^2 + 6y + 32x = 59$

36. $x^2 + 6x - y^2 + 2y = -11$

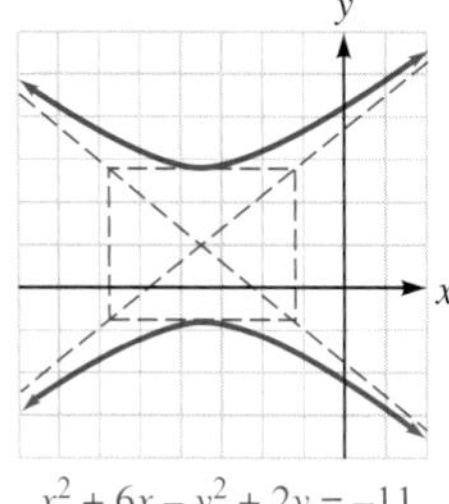

$x^2 + 6x - y^2 + 2y = -11$

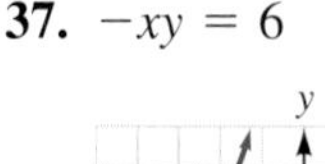

37. $-xy = 6$

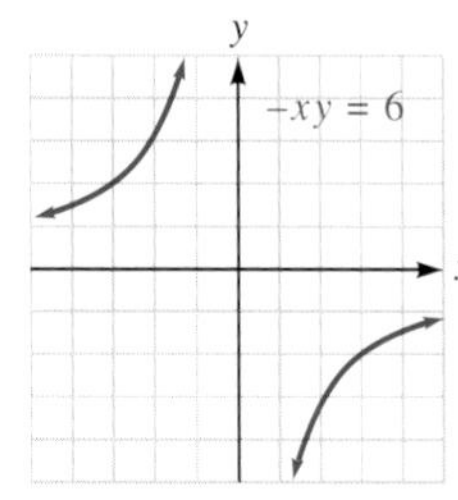

$-xy = 6$

38. $xy = 20$

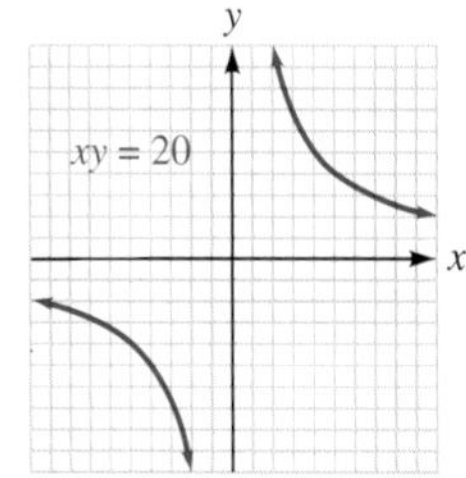

$xy = 20$

Find the equation of the curve on which point P lies.

39. The difference of the distances between $P(x, y)$ and the points $(-2, 1)$ and $(8, 1)$ is 6.
$\frac{(x-3)^2}{9} - \frac{(y-1)^2}{16} = 1$

40. The difference of the distances between $P(x, y)$ and the points $(3, -1)$ and $(3, 5)$ is 5.
$\frac{4(y-2)^2}{25} - \frac{4(x-3)^2}{11} = 1$

41. The distance between point $P(x, y)$ and the point $(0, 3)$ is $\frac{3}{2}$ of the distance between P and the line $y = -2$. $4x^2 - 5y^2 - 60y = 0$

42. The distance between point $P(x, y)$ and the point $(5, 4)$ is $\frac{5}{3}$ of the distance between P and the line $x = -3$. $\frac{4\left(x + \frac{15}{2}\right)^2}{225} - \frac{(y-4)^2}{100} = 1$

APPLICATIONS

43. Fluids See the illustration below. Two glass plates in contact at the left, and separated by about 5 millimeters on the right, are dipped in beet juice, which rises by capillary action to form a hyperbola. The hyperbola is modeled by an equation of the form $xy = k$. If the curve passes through the point (12, 2), what is k? **24**

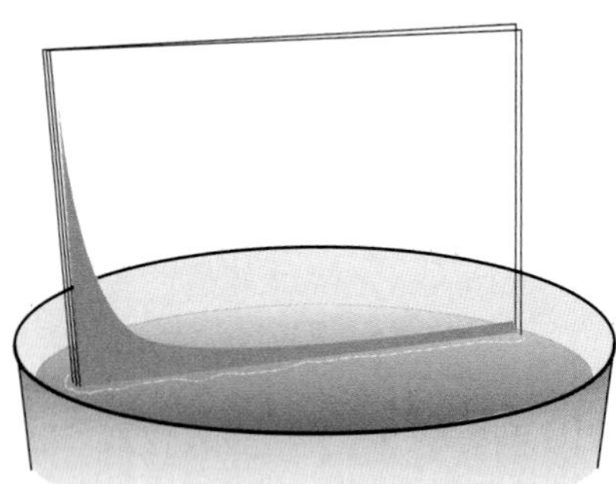

44. Astronomy Some comets have a hyperbolic orbit, with the sun as one focus. When the comet shown in the illustration is far away from Earth, it appears to be approaching Earth along the line $y = 2x$. Find the equation of its orbit if the comet comes within 100 million miles of the Earth.
$\frac{x^2}{100{,}000{,}000^2} - \frac{y^2}{200{,}000{,}000^2} = 1$

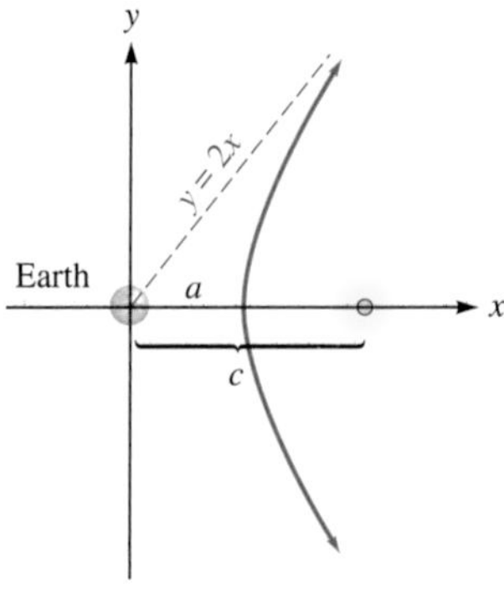

45. Alpha particles The particle in the illustration approaches the nucleus at the origin along the path $9y^2 - x^2 = 81$ in the coordinate system shown. How close does the particle come to the nucleus? **3 units**

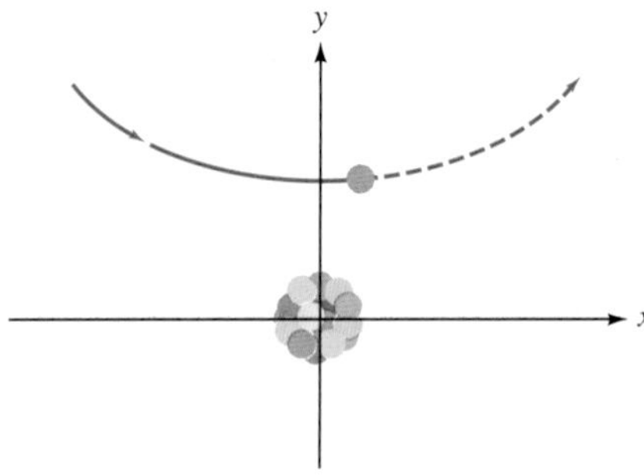

46. Physics Parallel beams of similarly charged particles are shot from two atomic accelerators 20 meters apart, as shown in the illustration. If the particles were not deflected, the beams would be 2.0×10^{-4} meter apart. However, because the charged particles repel each other, the beams follow the hyperbolic path $y = \frac{k}{x}$, for some k. Find k. $k = 1.0 \times 10^{-3}$

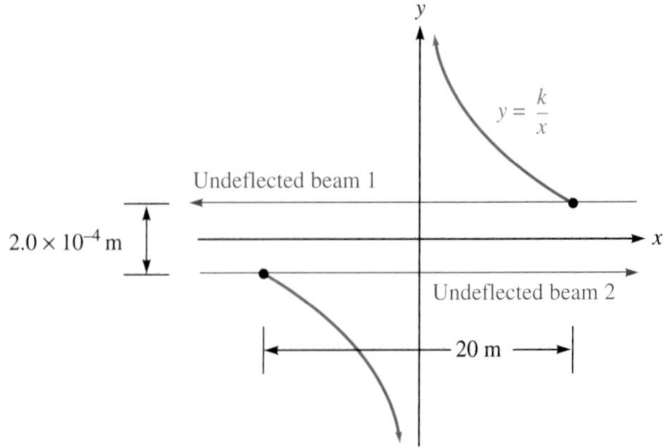

47. Navigation The LORAN system (LOng RAnge Navigation) in the illustration uses two radio transmitters 26 miles apart to send simultaneous signals. The navigator on a ship at $P(x, y)$ receives the closer signal first, and determines that the difference of the distances between the ship and each transmitter is 24 miles. That places the ship on a certain curve. Identify the curve and find its equation.
hyperbola; $\frac{x^2}{144} - \frac{y^2}{25} = 1$

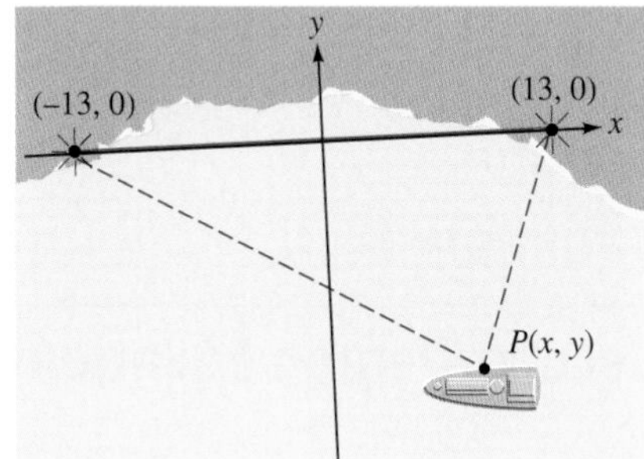

48. Navigation By determining the difference of the distances between the ship in the illustration and two radio transmitters, the LORAN navigation system places the ship on the hyperbola $x^2 - 4y^2 = 576$ in the coordinate system shown. If the ship is 5 miles out to sea, find its coordinates.
$x \approx 26$ **mi,** $y = 5$ **mi; (26, 5)**

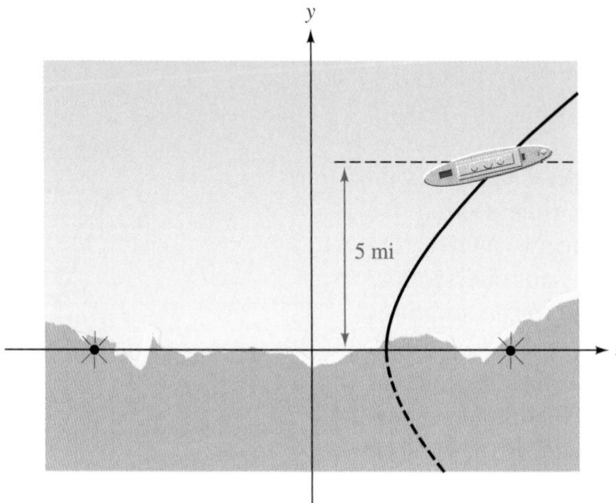

49. Wave propagation Stones dropped into a calm pond at points A and B create ripples that propagate in widening circles. In the illustration, points A and B are 20 feet apart, and the radii of the circles differ by 12 feet. The point $P(x, y)$ where the circles intersect moves along a curve. Identify the curve and find its equation. **hyperbola;** $\frac{x^2}{36} - \frac{y^2}{64} = 1$

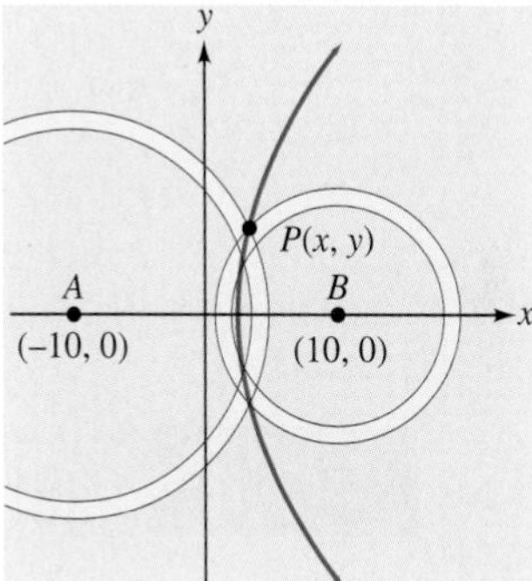

50. Sonic boom The position of a sonic boom caused by the faster-than-sound aircraft is one branch of the hyperbola $y^2 - x^2 = 25$ in the coordinate system shown. How wide is the hyperbola 5 miles from its vertex? $10\sqrt{3}$ **miles**

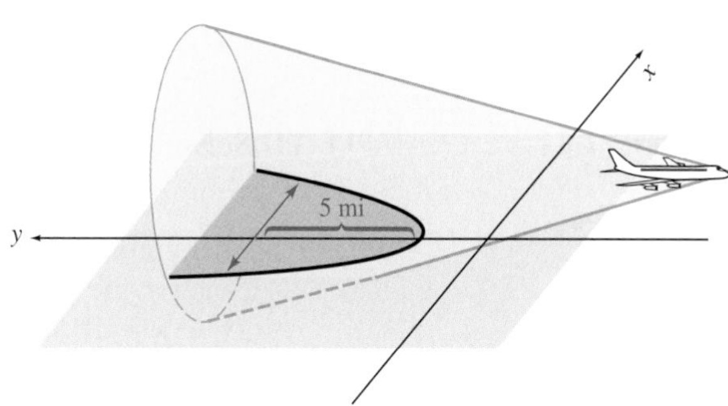

DISCOVERY AND WRITING

51. Prove that $c > a$ for a hyperbola with center at (0, 0) and line segment FF' on the x-axis.

52. Show that the extended diagonals of the fundamental rectangle of the hyperbola $\frac{x^2}{a^2} - \frac{y^2}{b^2} = 1$ are $y = \frac{b}{a}x$ and $y = -\frac{b}{a}x$.

53. Show that the expansion of the standard equation of a hyperbola is a special case of the general equation of second degree with $B = 0$.

54. Write a paragraph describing how you can tell from the equation of a hyperbola whether the transverse axis is vertical or horizontal.

REVIEW ***Find the inverse of each function. Write the answer in $y = f^{-1}(x)$ form.***

55. $f(x) = 3x - 2$
$f^{-1}(x) = \dfrac{x+2}{3}$

56. $f(x) = \dfrac{x+1}{x}$
$f^{-1}(x) = \dfrac{1}{x-1}$

57. $f(x) = \dfrac{5x}{x+2}$
$f^{-1}(x) = \dfrac{2x}{5-x}$

58. $f(x) = x$
$f^{-1}(x) = x$

Let $f(x) = x^2 + 1$ and $g(x) = (x + 1)^2$. Find each composite function.

59. $f(g(x))$
$f(g(x)) = (x + 1)^4 + 1$

60. $g(f(x))$
$g(f(x)) = (x^2 + 2)^2$

61. $f(f(x))$
$f(f(x)) = (x^2 + 1)^2 + 1$

62. $g(g(x))$
$g(g(x)) = [(x + 1)^2 + 1]^2$

7.4 Solving Simultaneous Second-Degree Equations

In this section, you will learn about

- Solving Systems by Graphing
- Solving Systems Algebraically

We now discuss techniques for solving systems of two equations in two variables, where at least one of the equations is of second degree.

Solving Systems by Graphing

ACTIVE EXAMPLE 1 Solve $\begin{cases} x^2 + y^2 = 25 \\ 2x + y = 10 \end{cases}$ by graphing.

Solution The graph of $x^2 + y^2 = 25$ is a circle with center at the origin and radius of 5. The graph of $2x + y = 10$ is a straight line. Depending on whether the line is a secant (intersecting the circle at two points) or a tangent (intersecting the circle at one point) or does not intersect the circle at all, there are two, one, or no solutions to the system, respectively.

COLLEGE **Algebra** f(x) **Now**™
Explore this Active Example and test yourself on the steps taken here at **http://1pass.thomson.com**. You can also find this example on the Interactive Video Skillbuilder CD-ROM.

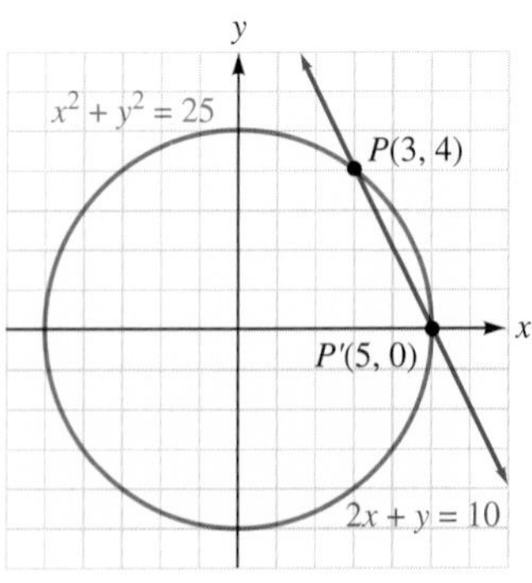

Figure 7-30

After graphing the circle and the line, as shown in Figure 7-30, we see that there are two intersection points, $P(3, 4)$ and $P'(5, 0)$. The solutions to the system are

$$\begin{cases} x = 3 \\ y = 4 \end{cases} \quad \text{and} \quad \begin{cases} x = 5 \\ y = 0 \end{cases}$$

Verify that these are exact solutions.

Self Check Solve: $\begin{cases} x^2 + y^2 = 25 \\ 2x - y = 5 \end{cases}$. ■

Accent on Technology — SOLVING SYSTEMS OF EQUATIONS

To solve the system of equations in Example 1 on a graphing calculator, we must first solve each of the equations for y. From the first equation, $x^2 + y^2 = 25$, we get two equations to graph:

$$y_1 = \sqrt{25 - x^2}$$
$$y_2 = -\sqrt{25 - x^2}$$

From the other equation, $2x + y = 10$, we get a third equation:

$$y_3 = 10 - 2x$$

The graphs of these equations will be similar to those shown in Figure 7-31(a). To find the top solution, we use ZOOM and TRACE to read the coordinates of the point of intersection, as shown in Figure 7-31(b): $x \approx 3$ and $y \approx 4$. Because the calculator graph is not complete at the other point of intersection shown in Figure 7-31(c), we must use our best judgment. We read the coordinates to be $x \approx 5$ and $y \approx 0$.

We can also find the coordinates of the intersection point by using the INTERSECT feature.

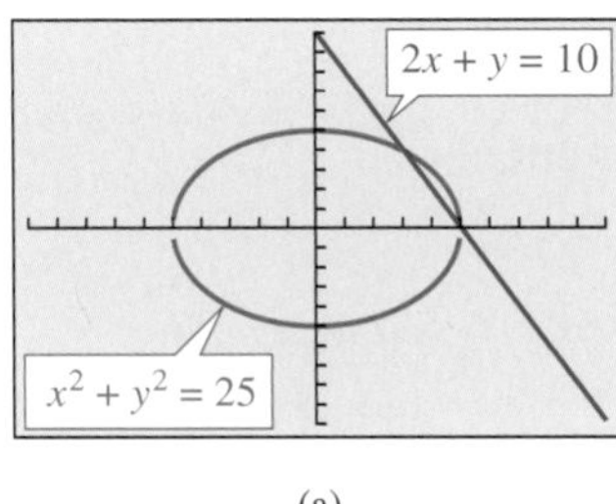

(a)

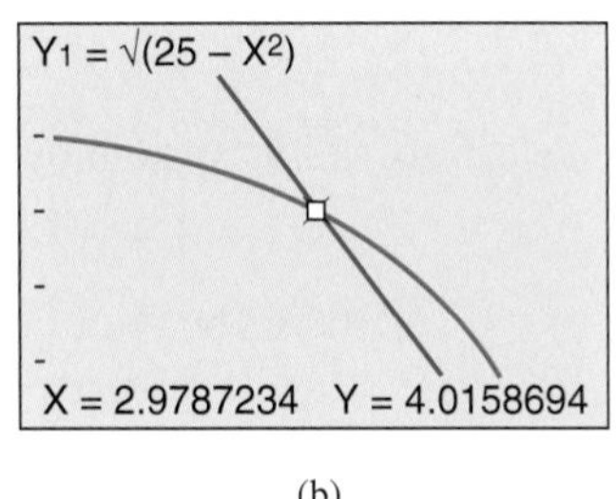

(b)

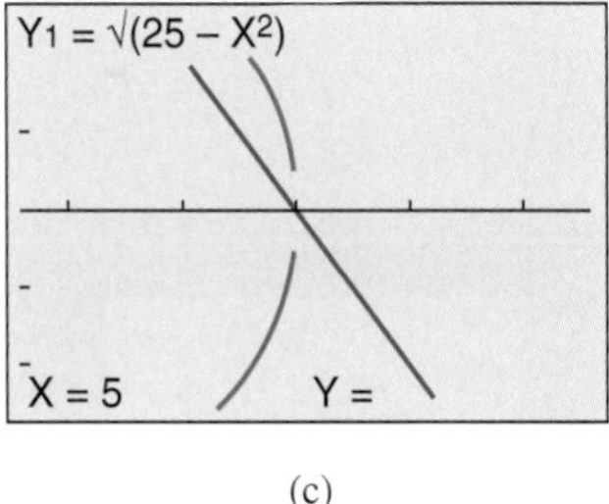

(c)

Figure 7-31

Solving Systems Algebraically

Algebraic methods can be used to find exact solutions.

ACTIVE EXAMPLE 2 Solve $\begin{cases} x^2 + y^2 = 25 \\ 2x + y = 10 \end{cases}$ algebraically.

Solution If a system contains one equation of second degree and another of first degree, we can solve it by substitution. Solving the linear equation for y gives

$$2x + y = 10$$
$$y = \mathbf{-2x + 10}$$

We can now substitute $-2x + 10$ for y in the second-degree equation and solve the resulting quadratic equation for x:

COLLEGE **Algebra** $f(x)$ **Now**™
Go to **http://1pass.thomson.com** or your CD to practice this example.

$$x^2 + y^2 = 25$$
$$x^2 + (-2x + 10)^2 = 25$$
$$x^2 + 4x^2 - 40x + 100 = 25 \qquad \text{Remove parentheses.}$$
$$5x^2 - 40x + 75 = 0 \qquad \text{Subtract 25 from both sides and combine like terms.}$$
$$x^2 - 8x + 15 = 0 \qquad \text{Divide both sides by 5.}$$
$$(x - 5)(x - 3) = 0 \qquad \text{Factor } x^2 - 8x + 15.$$
$$x - 5 = 0 \quad \text{or} \quad x - 3 = 0$$
$$x = 5 \quad \big| \quad x = 3$$

Because $y = -2x + 10$, if $x = 5$, then $y = 0$; and if $x = 3$, then $y = 4$. The two solutions are

$$\begin{cases} x = 5 \\ y = 0 \end{cases} \quad \text{and} \quad \begin{cases} x = 3 \\ y = 4 \end{cases}$$

Self Check Solve: $\begin{cases} x^2 + y^2 = 25 \\ 2x - y = 5 \end{cases}$. ■

EXAMPLE 3 Solve: $\begin{cases} 4x^2 + 9y^2 = 5 \\ y = x^2 \end{cases}$.

Solution We can solve this system by substitution.

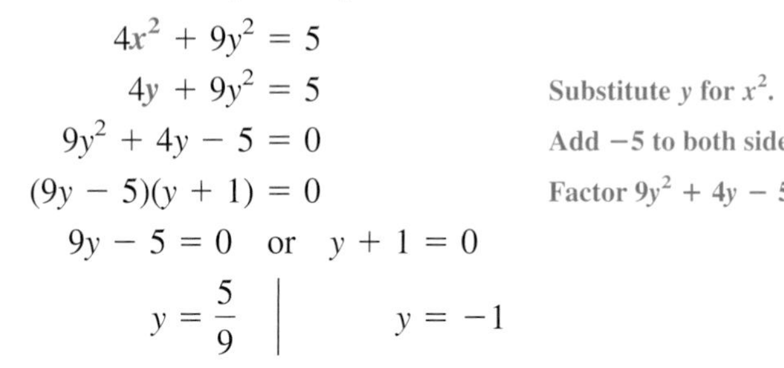

$$4x^2 + 9y^2 = 5$$
$$4y + 9y^2 = 5 \qquad \text{Substitute } y \text{ for } x^2.$$
$$9y^2 + 4y - 5 = 0 \qquad \text{Add } -5 \text{ to both sides.}$$
$$(9y - 5)(y + 1) = 0 \qquad \text{Factor } 9y^2 + 4y - 5.$$
$$9y - 5 = 0 \quad \text{or} \quad y + 1 = 0$$
$$y = \frac{5}{9} \quad \Big| \quad y = -1$$

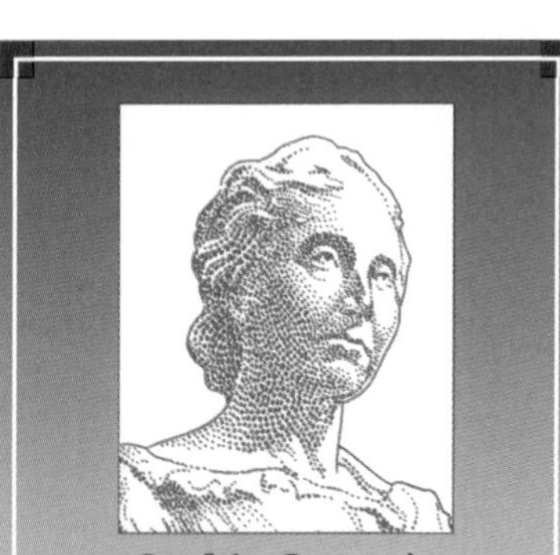

Sophie Germain
(1776–1831)
Sophie Germain was 13 years old during the French Revolution. Because of dangers caused by the insurrection in Paris, she was kept indoors and spent most of her time reading about mathematics in her father's library. Since interest in mathematics was considered inappropriate for a woman at that time, much of her work was written under the pen name of M. LeBlanc.

Because $y = x^2$, we can find x by solving the equations

$$x^2 = \frac{5}{9} \quad \text{and} \quad x^2 = -1$$

The solutions of $x^2 = \frac{5}{9}$ are

$$x = \frac{\sqrt{5}}{3} \quad \text{and} \quad x = -\frac{\sqrt{5}}{3}$$

The equation $x^2 = -1$ has no real solutions. Thus, the solutions of the system are

$$\left(\frac{\sqrt{5}}{3}, \frac{5}{9}\right) \quad \text{and} \quad \left(-\frac{\sqrt{5}}{3}, \frac{5}{9}\right)$$

Self Check Solve: $\begin{cases} x^2 + 3y^2 = 13 \\ x = y^2 - 1 \end{cases}$. ■

ACTIVE EXAMPLE 4 Solve: $\begin{cases} 3x^2 + 2y^2 = 36 \\ 4x^2 - y^2 = 4 \end{cases}$.

Solution When we have two second-degree equations of the form $ax^2 + by^2 = c$, we can solve the system by eliminating one of the variables by addition. To eliminate the terms involving y^2, we copy the first equation and multiply the second equation by 2 to obtain the following equivalent system.

COLLEGE **Algebra f(x) Now™**
Go to **http://1pass.thomson.com** or your CD to practice this example.

$$\begin{cases} 3x^2 + 2y^2 = 36 \\ 8x^2 - 2y^2 = 8 \end{cases}$$

We can then add the equations and solve the resulting equation for x:

$$11x^2 = 44$$
$$x^2 = 4$$
$$x = 2 \quad \text{or} \quad x = -2$$

To find y, we substitute 2 for x and then -2 for x in the first equation.

For $x = 2$

$$3x^2 + 2y^2 = 36$$
$$3(2)^2 + 2y^2 = 36$$
$$12 + 2y^2 = 36$$
$$2y^2 = 24$$
$$y^2 = 12$$
$$y = \sqrt{12} \quad \text{or} \quad y = -\sqrt{12}$$
$$y = 2\sqrt{3} \quad | \quad y = -2\sqrt{3}$$

For $x = -2$

$$3x^2 + 2y^2 = 36$$
$$3(-2)^2 + 2y^2 = 36$$
$$12 + 2y^2 = 36$$
$$2y^2 = 24$$
$$y^2 = 12$$
$$y = \sqrt{12} \quad \text{or} \quad y = -\sqrt{12}$$
$$y = 2\sqrt{3} \quad | \quad y = -2\sqrt{3}$$

The four solutions of this system are

$$(2, 2\sqrt{3}), \quad (2, -2\sqrt{3}), \quad (-2, 2\sqrt{3}), \quad \text{and} \quad (-2, -2\sqrt{3})$$

Self Check Solve: $\begin{cases} 2x^2 + y^2 = 23 \\ 3x^2 - 2y^2 = 17 \end{cases}$. ■

Self Check Answers

1. 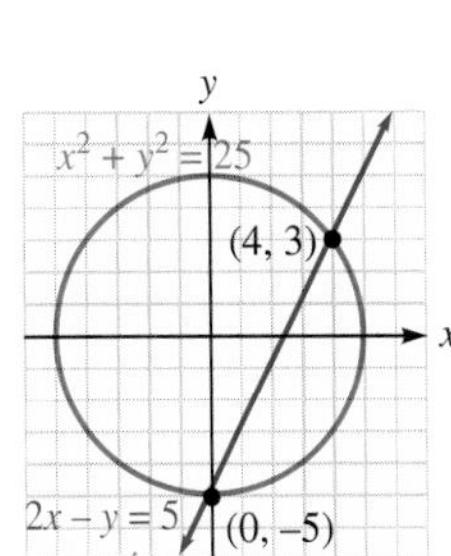

$\begin{cases} x = 4 \\ y = 3 \end{cases}$ and $\begin{cases} x = 0 \\ y = -5 \end{cases}$ **2.** $\begin{cases} x = 4 \\ y = 3 \end{cases}$ and $\begin{cases} x = 0 \\ y = -5 \end{cases}$ **3.** $(2, \sqrt{3})$ and $(2, -\sqrt{3})$

4. $(3, \sqrt{5}), (3, -\sqrt{5}), (-3, \sqrt{5}), (-3, -\sqrt{5})$

7.4 Exercises

VOCABULARY AND CONCEPTS *Fill in the blanks.*

1. Solutions of systems of second-degree equations are the points of intersection of the **graphs** of conic sections.
2. Approximate solutions of systems of second-degree equations can be found **graphically**, and exact solutions can be found **algebraically**.

PRACTICE *Solve each system of equations by graphing.*

3. $\begin{cases} 8x^2 + 32y^2 = 256 \\ x = 2y \end{cases}$

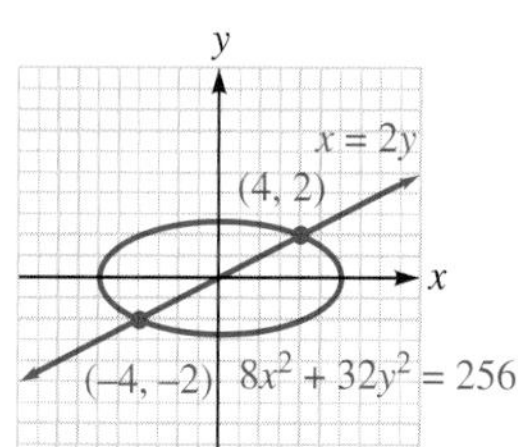

4. $\begin{cases} x^2 + y^2 = 2 \\ x + y = 2 \end{cases}$

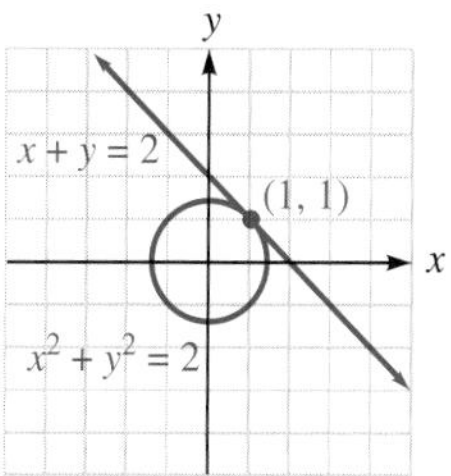

5. $\begin{cases} x^2 + y^2 = 90 \\ y = x^2 \end{cases}$

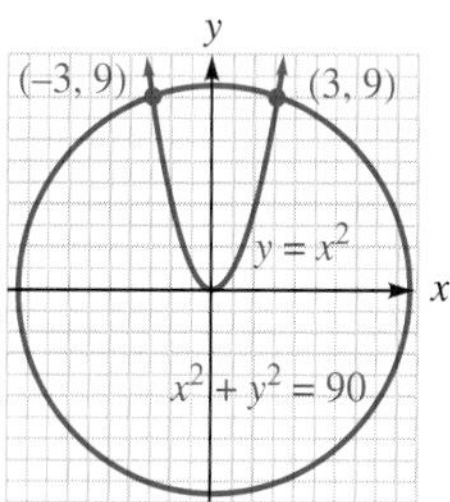

6. $\begin{cases} x^2 + y^2 = 5 \\ x + y = 3 \end{cases}$

y
x + y = 3
(1, 2)
(2, 1)
x
$x^2 + y^2 = 5$

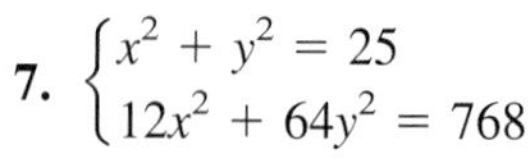

7. $\begin{cases} x^2 + y^2 = 25 \\ 12x^2 + 64y^2 = 768 \end{cases}$

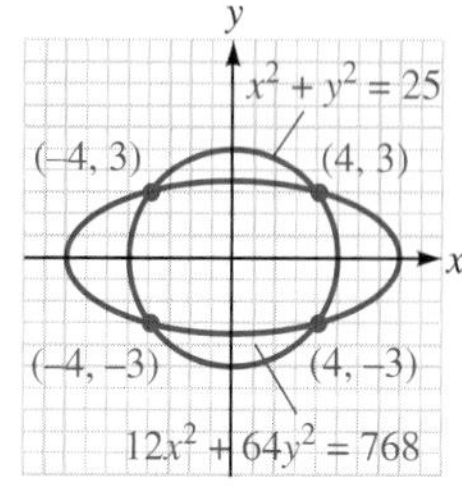

8. $\begin{cases} x^2 + y^2 = 13 \\ y = x^2 - 1 \end{cases}$

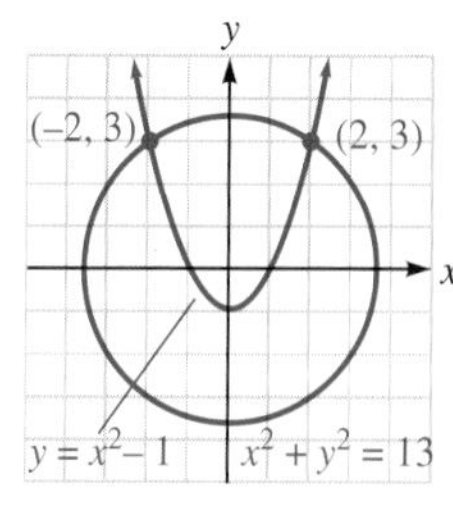

9. $\begin{cases} x^2 - 13 = -y^2 \\ y = 2x - 4 \end{cases}$

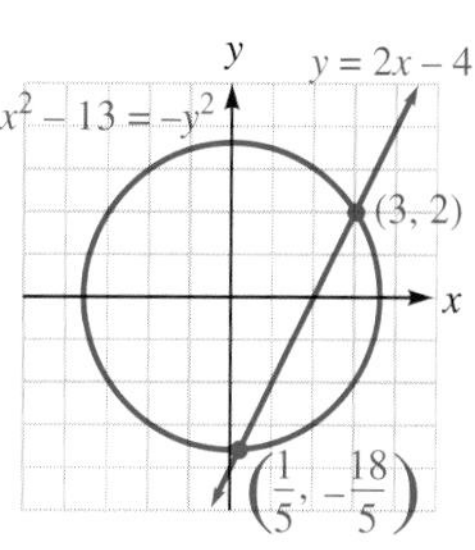

10. $\begin{cases} x^2 + y^2 = 20 \\ y = x^2 \end{cases}$

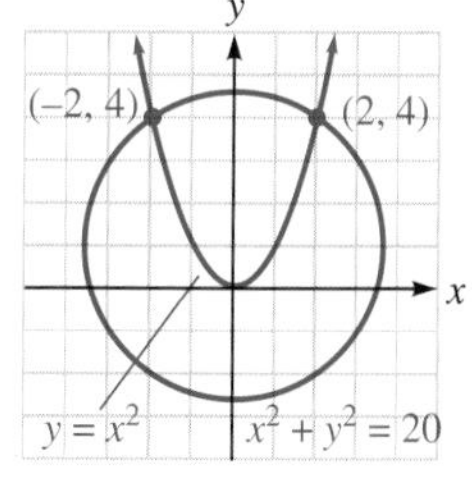

11. $\begin{cases} x^2 - 6x - y = -5 \\ x^2 - 6x + y = -5 \end{cases}$

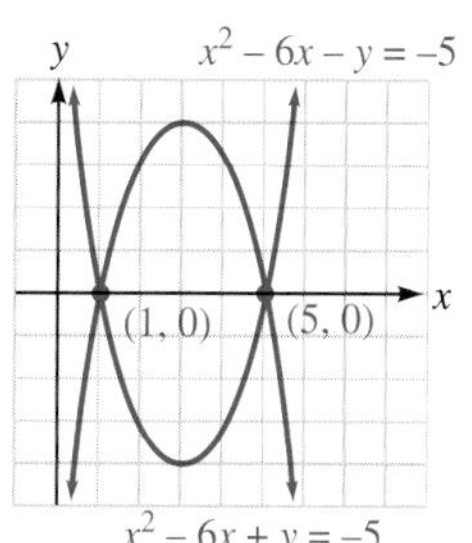

12. $\begin{cases} x^2 - y^2 = -5 \\ 3x^2 + 2y^2 = 30 \end{cases}$

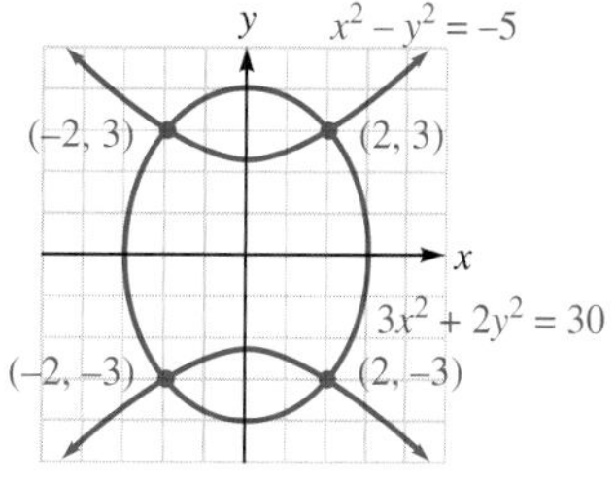

Use a graphing calculator to solve each system of equations.

13. $\begin{cases} y = x + 1 \\ y = x^2 + x \end{cases}$

(1, 2), (−1, 0)

14. $\begin{cases} y = 6 - x^2 \\ y = x^2 - x \end{cases}$

(2, 2), (−1.5, 3.75)

15. $\begin{cases} 6x^2 + 9y^2 = 10 \\ 3y - 2x = 0 \end{cases}$

(1, 0.67), (−1, −0.67)

16. $\begin{cases} x^2 + y^2 = 68 \\ y^2 - 3x^2 = 4 \end{cases}$

(4, 7.2), (4, −7.2), (−4, 7.2), (−4, −7.2)

Solve each system of equations algebraically for real values of x and y.

17. $\begin{cases} 25x^2 + 9y^2 = 225 \\ 5x + 3y = 15 \end{cases}$

(3, 0), (0, 5)

18. $\begin{cases} x^2 + y^2 = 20 \\ y = x^2 \end{cases}$

(2, 4), (−2, 4)

19. $\begin{cases} x^2 + y^2 = 2 \\ x + y = 2 \end{cases}$

(1, 1)

20. $\begin{cases} x^2 + y^2 = 36 \\ 49x^2 + 36y^2 = 1{,}764 \end{cases}$

(6, 0), (−6, 0)

21. $\begin{cases} x^2 + y^2 = 5 \\ x + y = 3 \end{cases}$

(1, 2), (2, 1)

22. $\begin{cases} x^2 - x - y = 2 \\ 4x - 3y = 0 \end{cases}$

$\left(-\frac{2}{3}, -\frac{8}{9}\right)$, (3, 4)

23. $\begin{cases} x^2 + y^2 = 13 \\ y = x^2 - 1 \end{cases}$
(−2, 3), (2, 3)

24. $\begin{cases} x^2 + y^2 = 25 \\ 2x^2 - 3y^2 = 5 \end{cases}$
(4, 3), (−4, 3), (4, −3), (−4, −3)

25. $\begin{cases} x^2 + y^2 = 30 \\ y = x^2 \end{cases}$
$(\sqrt{5}, 5), (-\sqrt{5}, 5)$

26. $\begin{cases} 9x^2 - 7y^2 = 81 \\ x^2 + y^2 = 9 \end{cases}$
(3, 0), (−3, 0)

27. $\begin{cases} x^2 + y^2 = 13 \\ x^2 - y^2 = 5 \end{cases}$
(3, 2), (3, −2), (−3, 2), (−3, −2)

28. $\begin{cases} 2x^2 + y^2 = 6 \\ x^2 - y^2 = 3 \end{cases}$
$(\sqrt{3}, 0), (-\sqrt{3}, 0)$

29. $\begin{cases} x^2 + y^2 = 20 \\ x^2 - y^2 = -12 \end{cases}$
(2, 4), (2, −4), (−2, 4), (−2, −4)

30. $\begin{cases} xy = -\dfrac{9}{2} \\ 3x + 2y = 6 \end{cases}$
$(-1, \frac{9}{2}), (3, -\frac{3}{2})$

31. $\begin{cases} y^2 = 40 - x^2 \\ y = x^2 - 10 \end{cases}$
$(-\sqrt{15}, 5), (\sqrt{15}, 5)$, **(−2, −6), (2, −6)**

32. $\begin{cases} x^2 - 6x - y = -5 \\ x^2 - 6x + y = -5 \end{cases}$
(5, 0), (1, 0)

33. $\begin{cases} y = x^2 - 4 \\ x^2 - y^2 = -16 \end{cases}$
(0, −4), (−3, 5), (3, 5)

34. $\begin{cases} 6x^2 + 8y^2 = 182 \\ 8x^2 - 3y^2 = 24 \end{cases}$
(3, 4), (3, −4), (−3, 4), (−3, −4)

35. $\begin{cases} x^2 - y^2 = -5 \\ 3x^2 + 2y^2 = 30 \end{cases}$
(−2, 3), (2, 3), (−2, −3), (2, −3)

36. $\begin{cases} \dfrac{1}{x} + \dfrac{1}{y} = 5 \\ \dfrac{1}{x} - \dfrac{1}{y} = -3 \end{cases}$
$(1, \frac{1}{4})$

37. $\begin{cases} \dfrac{1}{x} + \dfrac{2}{y} = 1 \\ \dfrac{2}{x} - \dfrac{1}{y} = \dfrac{1}{3} \end{cases}$ **(3, 3)**

38. $\begin{cases} \dfrac{1}{x} + \dfrac{3}{y} = 4 \\ \dfrac{2}{x} - \dfrac{1}{y} = 7 \end{cases}$ $(\frac{7}{25}, 7)$

39. $\begin{cases} 3y^2 = xy \\ 2x^2 + xy - 84 = 0 \end{cases}$
(6, 2), (−6, −2), $(\sqrt{42}, 0), (-\sqrt{42}, 0)$

40. $\begin{cases} x^2 + y^2 = 10 \\ 2x^2 - 3y^2 = 5 \end{cases}$
$(\sqrt{7}, \sqrt{3}), (\sqrt{7}, -\sqrt{3}), (-\sqrt{7}, \sqrt{3}), (-\sqrt{7}, -\sqrt{3})$

41. $\begin{cases} xy = \dfrac{1}{6} \\ y + x = 5xy \end{cases}$
$(\frac{1}{2}, \frac{1}{3}), (\frac{1}{3}, \frac{1}{2})$

42. $\begin{cases} xy = \dfrac{1}{12} \\ y + x = 7xy \end{cases}$
$(\frac{1}{4}, \frac{1}{3}), (\frac{1}{3}, \frac{1}{4})$

APPLICATIONS

43. **Geometry** The area of a rectangle is 63 square centimeters, and its perimeter is 32 centimeters. Find the dimensions of the rectangle. **7 cm by 9 cm**

44. **Fencing pastures** The rectangular pasture shown below is to be fenced in along a riverbank. If 260 feet of fencing is to enclose an area of 8,000 square feet, find the dimensions of the pasture.
80 ft by 100 ft or 50 ft by 160 ft

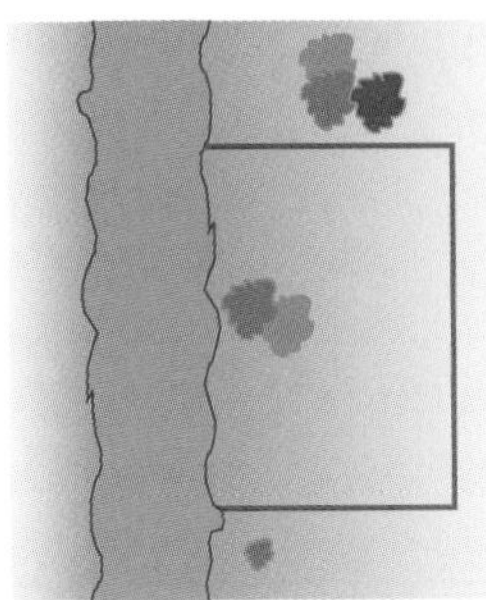

45. **Investments** Grant receives \$225 annual income from one investment. Jeff invested \$500 more than Grant, but at an annual rate of 1% less. Jeff's annual income is \$240. Find the amount and rate of Grant's investment. **\$2,500, 9%**

46. **Investments** Carol receives \$67.50 annual income from one investment. John invested \$150 more than Carol at an annual rate of $1\frac{1}{2}$% more. John's annual income is \$94.50. Find the amount and rate of Carol's investment. (*Hint:* There are two answers.)
either \$750 at 9% or \$900 at 7.5%

47. **Finding the rate and time** Jim drove 306 miles. Jim's brother made the same trip at a speed 17 miles per hour slower than Jim did and required an extra $1\frac{1}{2}$ hours. Find Jim's rate and time. **68 mph, 4.5 hr**

48. **Artillery** See the illustration. A shell fired from the base of a hill follows the parabolic path $y = -\frac{1}{6}x^2 + 2x$, with distances measured in miles. The hill has a slope of $\frac{1}{3}$. How far from the cannon is the point of impact? (*Hint:* Find the coordinates of the point and then the distance.) $\frac{10}{3}\sqrt{10}$ **mi**

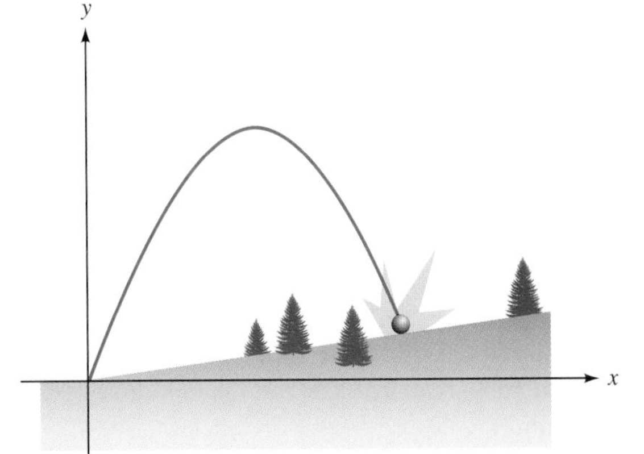

49. **Radio reception** A radio station located 120 miles due east of Collinsville has a listening radius of 100 miles. A straight road joins Collinsville with Harmony, a town 200 miles to the east and 100 miles north. See the illustration. If a driver leaves Collinsville and heads toward Harmony, how far from Collinsville will the driver pick up the station?
about 23 mi

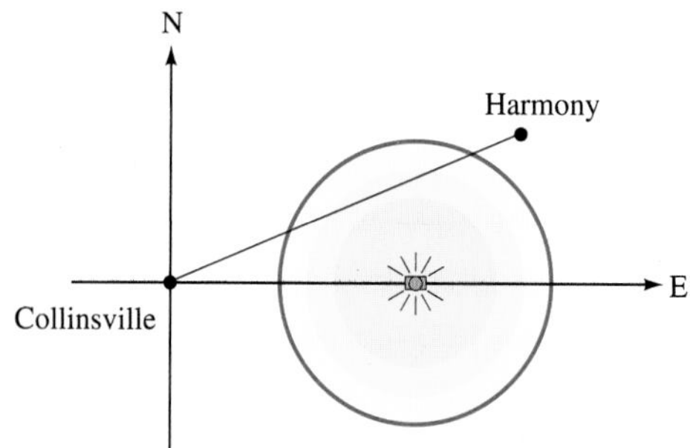

50. **Listening range** For how many miles will a driver in Exercise 49 continue to receive the signal? **about 169 mi**

DISCOVERY AND WRITING

51. Is it possible for a system of second-degree equations to have no common solution? If so, sketch the graphs of the equations of such a system.

52. Can a system have exactly one solution? If so, sketch the graphs of the equations of such a system.

53. Can a system have exactly two solutions? If so, sketch possible graphs.

54. Can a system have three solutions? If so, sketch the graphs.

55. Can it have exactly four solutions? If so, sketch the graphs.

56. Can a system have more than four solutions? If so, sketch the graphs.

REVIEW ***Find the vertical, horizontal, or slant asymptotes of each of the following rational functions.***

57. $y = \dfrac{3x + 1}{x - 1}$
vertical: $x = 1$; horizontal: $y = 3$

58. $y = \dfrac{x^2 + 3}{x - 1}$
vertical: $x = 1$; slant: $y = x + 1$

59. $y = \dfrac{3x + 1}{x^2 - 1}$
vertical: $x = 1, x = -1$; horizontal: $y = 0$

60. $y = \dfrac{x^2 + 3}{x^2 + 1}$
vertical: none; horizontal: $y = 1$

Determine whether the graph of each function has x-axis, y-axis, or origin symmetry, or none of these symmetries.

61. $f(x) = \dfrac{3x^2 + 5}{5x^2 + 3}$ **y-axis**

62. $f(x) = \dfrac{3x^3 + 5}{5x^2 + 3}$ **none**

63. $f(x) = \dfrac{3x^3}{5x^2 + 3}$
origin

64. $f(x) = \dfrac{3x^4 + 8}{2x^2 - 5}$
y-axis

PROBLEMS AND PROJECTS

1. In the equation of a hyperbola,

$$\frac{x^2}{a^2} - \frac{y^2}{b^2} = 1$$

replace the 1 on the right-hand side with 0. Solve the resulting equation for y by factoring the left-hand side (it is the difference of two squares) and setting each factor equal to 0. What are the graphs of these two equations, and what is their relationship to the hyperbola?

2. The 2-inch-diameter pipe in the illustration has been cut at a 45° angle. The cut edge is an ellipse. Find its equation.

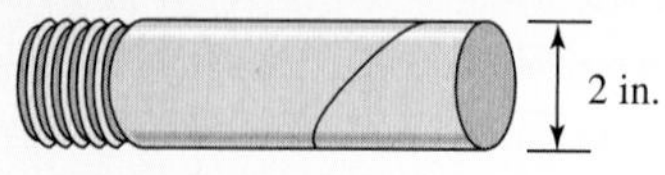

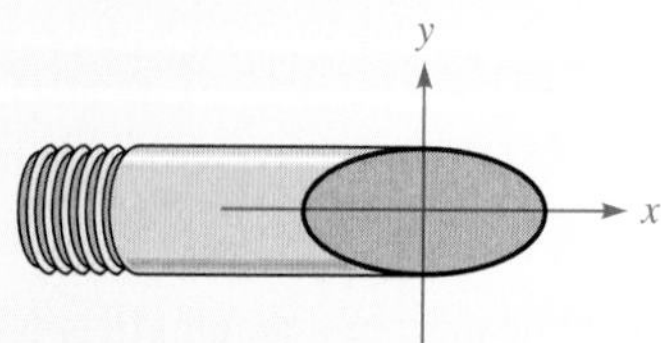

In the equations in Problems 3–6, each different value of k produces a different graph. The collection of all these graphs is called a **family of curves.** By choosing several values of k, graph a few members of each family. What common characteristic do members of the family share?

3. A family of circles $(x - k)^2 + (y - k)^2 = k^2$

4. A family of parabolas $y = x^2 + k$

5. A family of ellipses $x^2 + \dfrac{y^2}{k^2} = 1$

6. A family of hyperbolas $xy = k$

Project 1

To draw a hyperbola, tie a small loop in a long piece of string, large enough to hold the point of a pencil. As in the illustration, arrange the string around nails at A and B and hold a pencil in the loop at P. While keeping the string taut, carefully pull both ends together. Explain why the pencil traces a hyperbola. From the measured distance between A and B and the *difference* of lengths PA and PB, find the equation of the curve. How would you draw the other branch of the hyperbola?

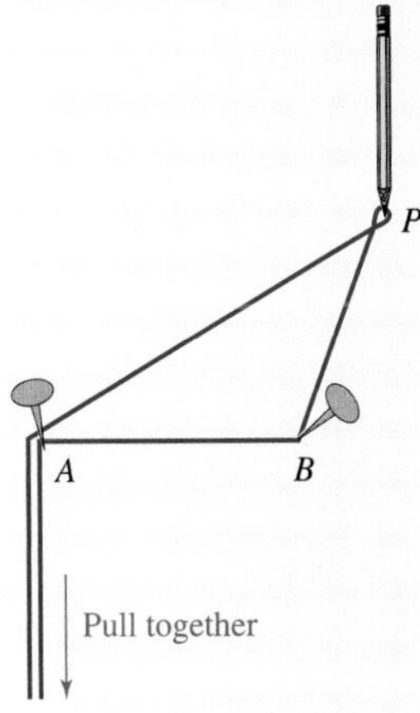

Project 2

Draw a line on a sheet of paper and mark a point P about 2 inches away, as in illustration (a). (Use wax paper, and draw the line and point with a fine-tip marker.) Fold the paper over to place the point on the line and make a sharp crease, as in (b). Fold the paper several times, placing P at different locations on the line. What curve do these creases seem to describe? What is the relationship of the point and the line to that curve? What happens if point P is closer to the line? Write a paragraph to describe your findings.

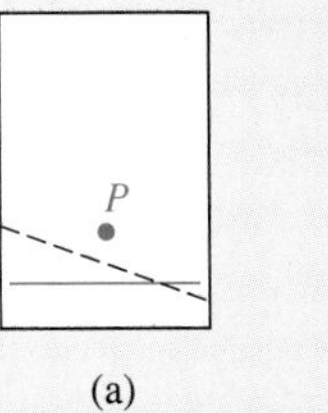

(a)

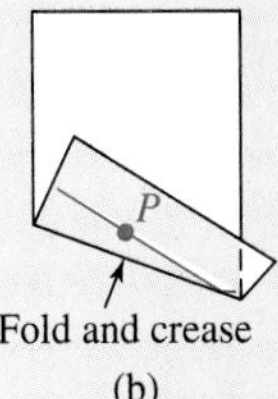

(b)

Project 3

Draw a circle on a sheet of paper and mark point P inside. Fold and crease several times, placing P at several locations on the circle, as in the illustration. What curve do the creases describe? What is the significance of point P?

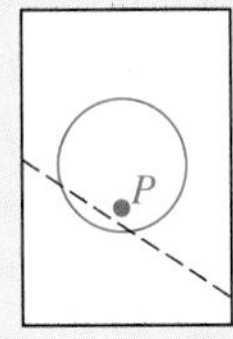

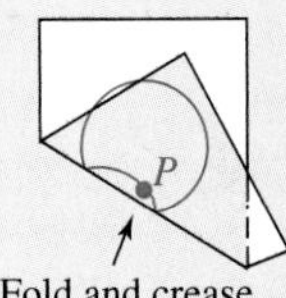

Project 4

This time, draw a circle and mark point P on the outside, as in the illustration. Fold and crease as before. What curve is described by these creases, and what is the significance of P? Do you get two branches of the curve?

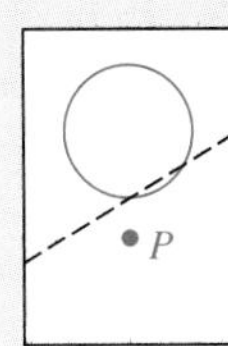

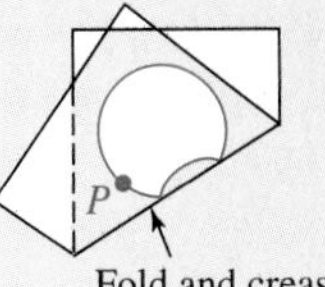

CHAPTER SUMMARY

CONCEPTS

7.1

Circle, center at (0, 0), radius r:

$x^2 + y^2 = r^2$

Circle, center at (h, k), radius r:

$(x - h)^2 + (y - k)^2 = r^2$

General form of a second-degree equation in x and y:

$Ax^2 + Bxy + Cy^2 + Dx + Ey + F = 0$

Parabola

Parabola opening	Vertex at origin
Right	$y^2 = 4px \quad (p > 0)$
Left	$y^2 = 4px \quad (p < 0)$
Up	$x^2 = 4py \quad (p > 0)$
Down	$x^2 = 4py \quad (p < 0)$

Parabola opening	Vertex at $V(h, k)$
Right	$(y - k)^2 = 4p(x - h) \ (p > 0)$
Left	$(y - k)^2 = 4p(x - h) \ (p < 0)$
Up	$(x - h)^2 = 4p(y - k) \ (p > 0)$
Down	$(x - h)^2 = 4p(y - k) \ (p < 0)$

REVIEW EXERCISES

The Circle and the Parabola

Write the equation of each circle.

1. Center (0, 0); radius 4 $x^2 + y^2 = 16$
2. Center (0, 0); passes through (6, 8) $x^2 + y^2 = 100$
3. Center (3, −2); radius 5 $(x - 3)^2 + (y + 2)^2 = 25$
4. Center (−2, 4); passes through (1, 0) $(x + 2)^2 + (y - 4)^2 = 25$
5. Endpoints of diameter (−2, 4) and (12, 16) $(x - 5)^2 + (y - 10)^2 = 85$
6. Endpoints of diameter (−3, −6) and (7, 10) $(x - 2)^2 + (y - 2)^2 = 89$

Write the equation of each circle in standard form.

7. $x^2 + y^2 - 6x + 4y = 3$ $(x - 3)^2 + (y + 2)^2 = 16$
8. $x^2 + 4x + y^2 - 10y = -13$ $(x + 2)^2 + (y - 5)^2 = 16$

Write the equation of each parabola.

9. Vertex (0, 0); passes through (−8, 4) and (−8, −4) $y^2 = -2x$
10. Vertex (0, 0); passes through (−8, 4) and (8, 4) $x^2 = 16y$
11. Find the equation of the parabola with vertex at (−2, 3), curve passing through point (−4, −8), and opening down. $(x + 2)^2 = -\frac{4}{11}(y - 3)$

Graph each equation.

12. $x^2 - 4y - 2x + 9 = 0$

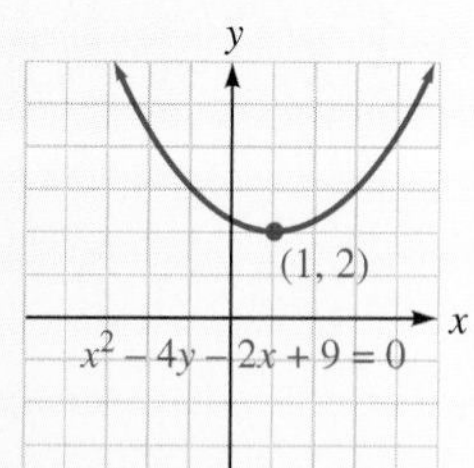

13. $y^2 - 6y = 4x - 13$

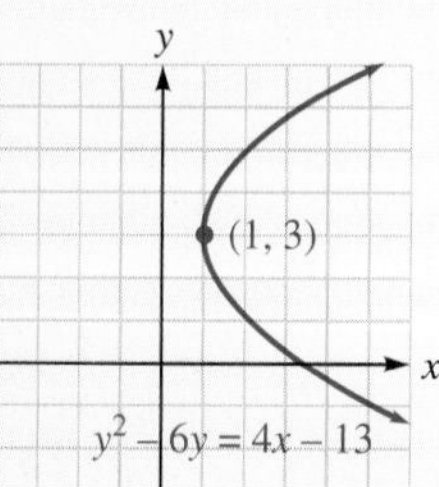

7.2 The Ellipse

Ellipse, center at (0, 0):

Major axis is $2a$, minor axis is $2b$ $(a > b > 0)$.

Center-to-focus distance is c, where $b^2 = a^2 - c^2$.

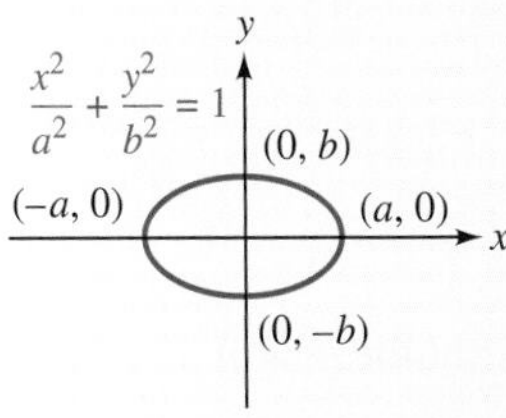

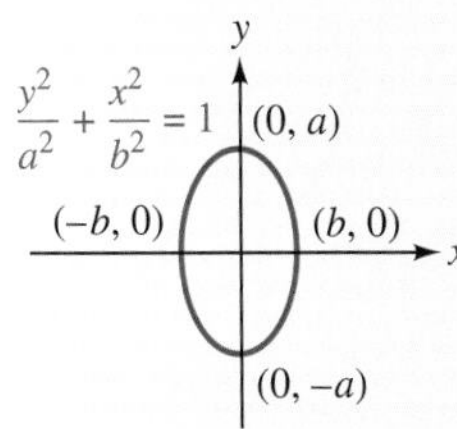

14. Write the equation of the ellipse with center at the origin, major axis that is horizontal and 12 units long, and minor axis 8 units long.

$$\frac{x^2}{36} + \frac{y^2}{16} = 1$$

15. Write the equation of the ellipse with center at the origin, major axis that is vertical and 10 units long, and minor axis 4 units long.

$$\frac{y^2}{25} + \frac{x^2}{4} = 1$$

Ellipse, center at (h, k):

Major axis is $2a$, minor axis is $2b$ $(a > b > 0)$.

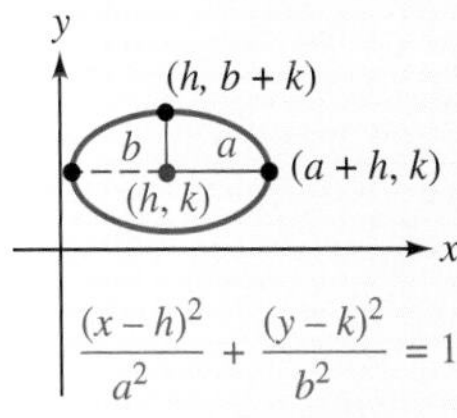

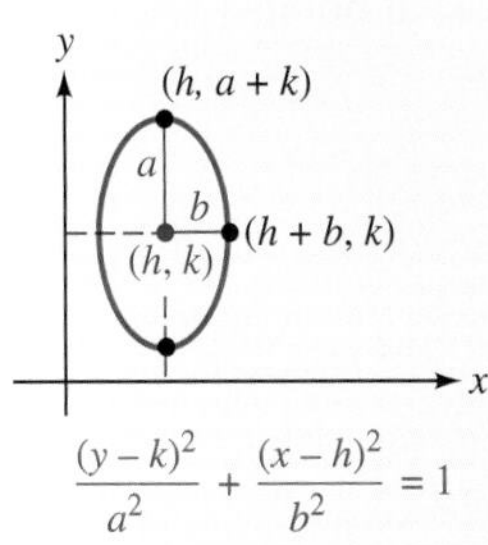

16. Write the equation of the ellipse with center at point $(-2, 3)$ and curve passing through points $(-2, 0)$ and $(2, 3)$. $\frac{(x+2)^2}{16} + \frac{(y-3)^2}{9} = 1$

17. Write the equation in standard form and graph it.

$$4x^2 + y^2 - 16x + 2y = -13$$

$$\frac{(x-2)^2}{1} + \frac{(y+1)^2}{4} = 1$$

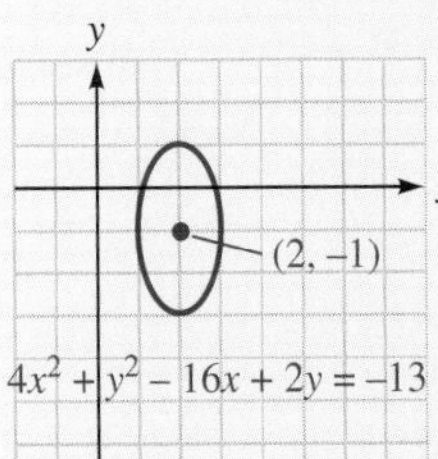

7.3 The Hyperbola

Hyperbola:

Center at (0, 0), center-to-focus distance is c, where $b^2 = c^2 - a^2$.

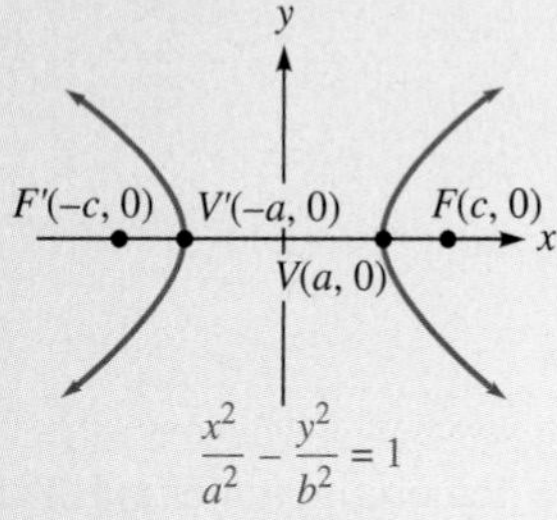

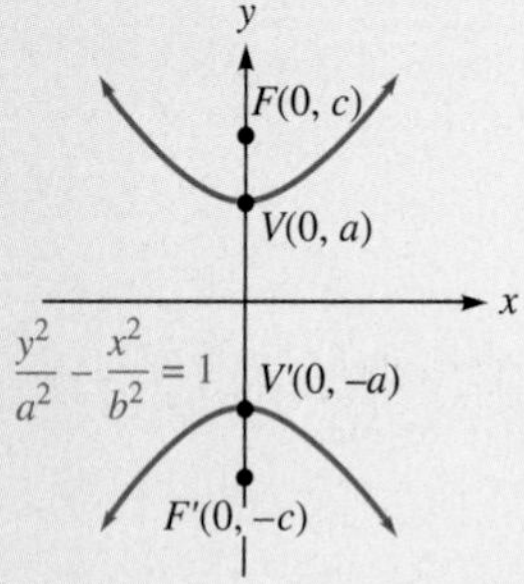

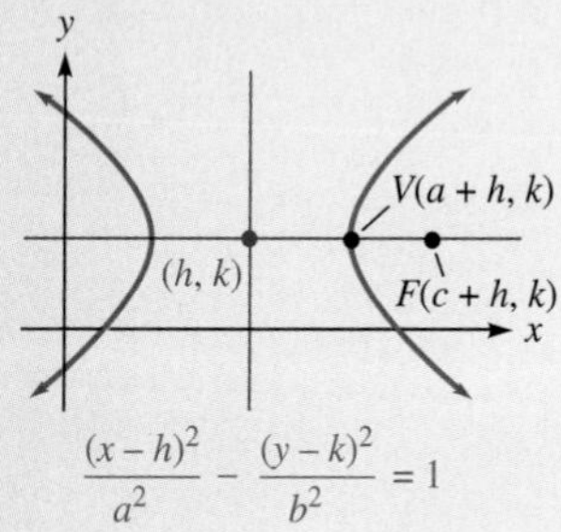

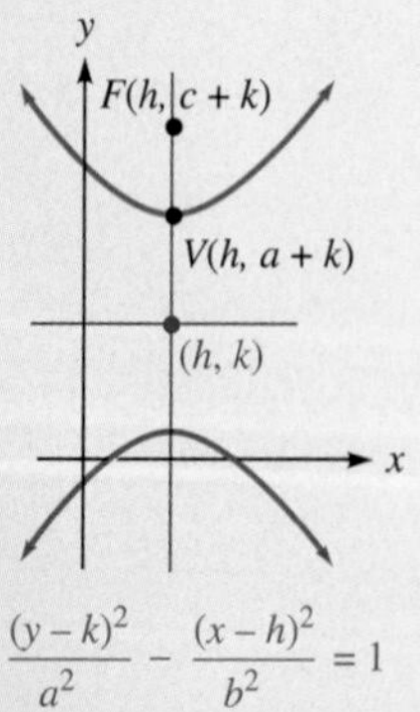

The extended diagonals of the fundamental rectangle are asymptotes of the graph of a hyperbola.

18. Write the equation of the hyperbola with center at the origin, passing through points $(-2, 0)$ and $(2, 0)$, and having a focus at $(4, 0)$.

$\frac{x^2}{4} - \frac{y^2}{12} = 1$

19. Write the equation of the hyperbola with center at the origin, one focus at $(0, 5)$, and one vertex at $(0, 3)$. $\frac{y^2}{9} - \frac{x^2}{16} = 1$

20. Write the equation of the hyperbola with vertices at points $(-3, 3)$ and $(3, 3)$ and a focus at point $(5, 3)$. $\frac{x^2}{9} - \frac{(y-3)^2}{16} = 1$

21. Write the equation of the hyperbola with vertices at points $(3, -3)$ and $(3, 3)$ and a focus at point $(3, 5)$. $\frac{y^2}{9} - \frac{(x-3)^2}{16} = 1$

22. Write the equation of the asymptotes of the hyperbola $\frac{x^2}{25} - \frac{y^2}{16} = 1$.

$y = \pm\frac{4}{5}x$

23. Write the equation in standard form and graph it.

$9x^2 - 4y^2 - 16y - 18x = 43$

$\frac{(x-1)^2}{4} - \frac{(y+2)^2}{9} = 1$

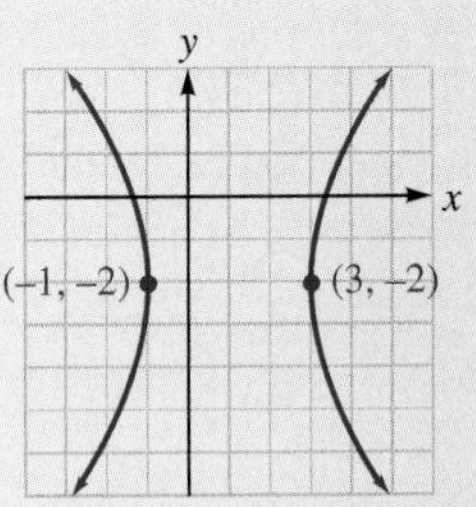

24. Graph: $4xy = 1$.

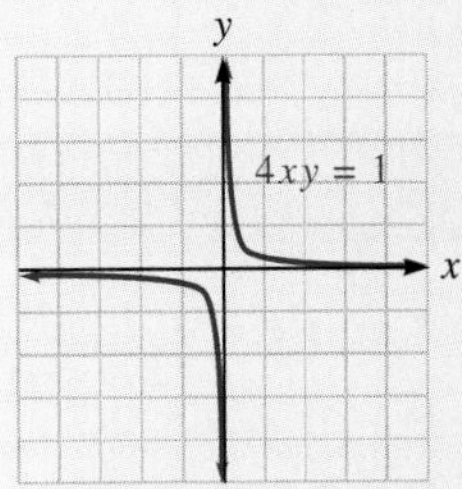

7.4 Solving Simultaneous Second-Degree Equations

Good estimates for solutions of systems of simultaneous second-degree equations can be found by graphing.

25. Solve by graphing: $\begin{cases} x^2 + y^2 = 16 \\ y = x + 4 \end{cases}$.

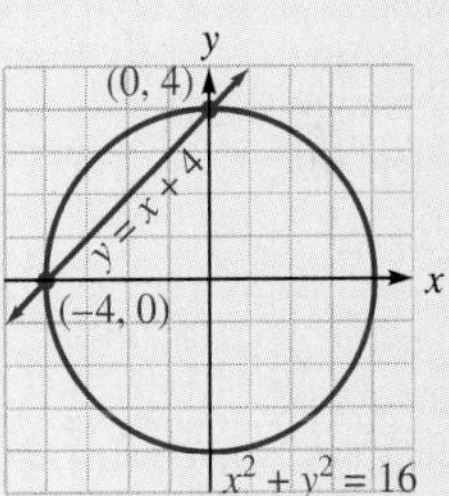

26. Solve by graphing: $\begin{cases} 3x^2 + y^2 = 52 \\ x^2 - y^2 = 12 \end{cases}$.

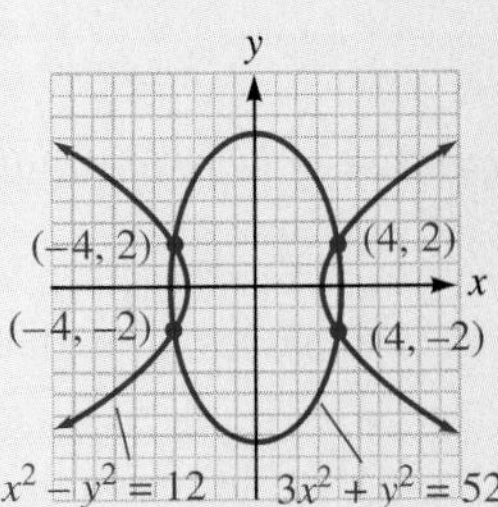

27. Solve by graphing: $\begin{cases} \frac{x^2}{16} + \frac{y^2}{12} = 1 \\ x^2 - \frac{y^2}{3} = 1 \end{cases}$.

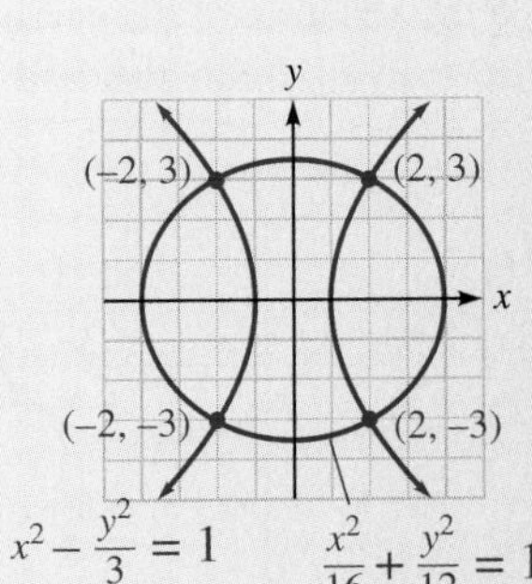

Exact solutions of systems of simultaneous second-degree equations can often be found by algebraic techniques.

28. Solve algebraically: $\begin{cases} 3x^2 + y^2 = 52 \\ x^2 - y^2 = 12 \end{cases}$. $(-4, 2), (-4, -2), (4, 2), (4, -2)$

29. Solve algebraically: $\begin{cases} x^2 + y^2 = 16 \\ -\sqrt{3}y + 4\sqrt{3} = 3x \end{cases}$. $(0, 4), \left(2\sqrt{3}, -2\right)$

30. Solve algebraically: $\begin{cases} \dfrac{x^2}{16} + \dfrac{y^2}{12} = 1 \\ x^2 - \dfrac{y^2}{3} = 1 \end{cases}$. $(-2, 3), (-2, -3), (2, 3), (2, -3)$

CHAPTER TEST

COLLEGE **Algebra** Now™ Preparing for an exam? Test yourself on key content at **http://1pass.thomson.com**

Write the equation of each circle.

1. Center $(2, 3)$; $r = 3$ $(x - 2)^2 + (y - 3)^2 = 9$

2. Ends of diameter at $(-2, -2)$ and $(6, 8)$
$(x - 2)^2 + (y - 3)^2 = 41$

3. Center $(2, -5)$, passes through $(7, 7)$
$(x - 2)^2 + (y + 5)^2 = 169$

4. Change the equation of the circle $x^2 + y^2 - 4x + 6y + 4 = 0$ to standard form and graph it.

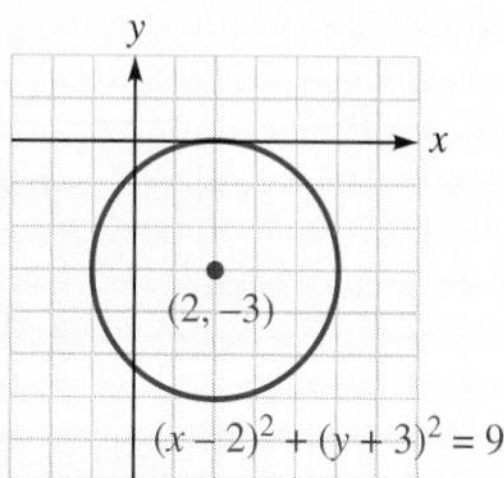

Find the equation of each parabola.

5. Vertex $(3, 2)$; focus at $(3, 6)$ $(x - 3)^2 = 16(y - 2)$

6. Vertex $(4, -6)$; passes through $(3, -8)$ and $(3, -4)$
$(y + 6)^2 = -4(x - 4)$

7. Vertex $(2, -3)$; passes through $(0, 0)$
$(x - 2)^2 = \frac{4}{3}(y + 3)$ or $(y + 3)^2 = -\frac{9}{2}(x - 2)$

8. Change the equation of the parabola $x^2 - 6x - 8y = 7$ into standard form and graph it.

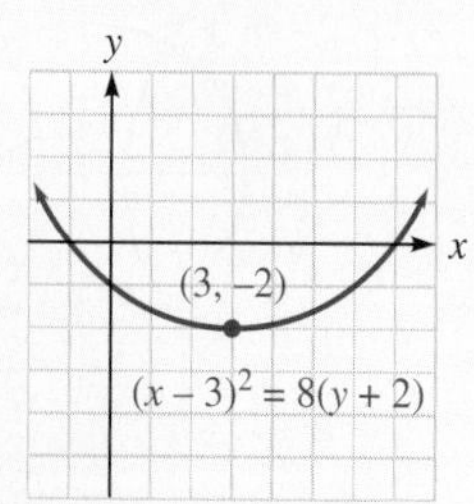

Find the equation of each ellipse.

9. Vertex $(10, 0)$, center at the origin, focus at $(6, 0)$
$\dfrac{x^2}{100} + \dfrac{y^2}{64} = 1$

10. Minor axis 24, center at the origin, focus at $(5, 0)$
$\dfrac{x^2}{169} + \dfrac{y^2}{144} = 1$

11. Center $(2, 3)$; passes through $(2, 9)$ and $(0, 3)$
$\dfrac{x^2}{4} + \dfrac{y^2}{36} = 1$

12. Change the equation of the ellipse $9x^2 + 4y^2 - 18x - 16y - 11 = 0$ into standard form and graph it.

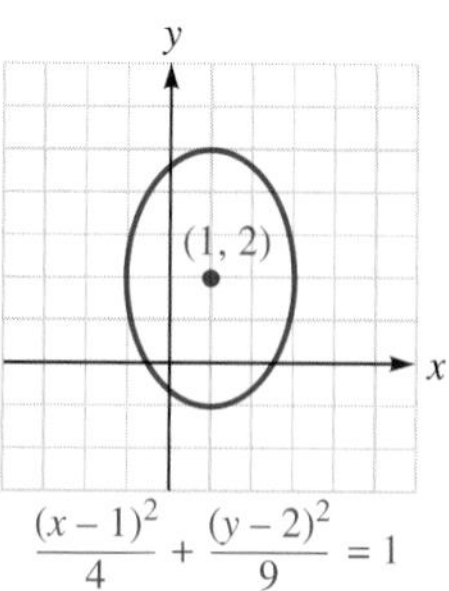

Find the equation of each hyperbola.

13. Center at the origin, focus at $(13, 0)$, vertex at $(5, 0)$
$\dfrac{x^2}{25} - \dfrac{y^2}{144} = 1$

14. Vertices $(6, 0)$ and $(-6, 0)$; $\dfrac{c}{a} = \dfrac{13}{12}$ $\dfrac{x^2}{36} - \dfrac{4y^2}{25} = 1$

15. Center $(2, -1)$, major axis horizontal and of length 16, distance of 20 between foci
$\dfrac{(x - 2)^2}{64} - \dfrac{(y + 1)^2}{36} = 1$

16. Change the equation of the hyperbola $x^2 - 4y^2 + 16y = 8$ into standard form and graph it.

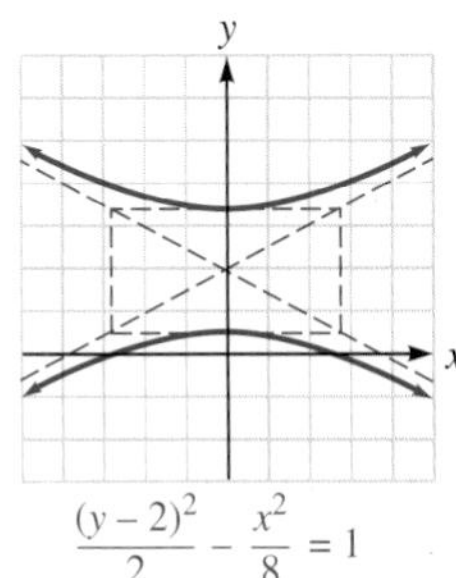

Solve each system algebraically.

17. $\begin{cases} x^2 + y^2 = 23 \\ y = x^2 - 3 \end{cases}$ $(\sqrt{7}, 4), (-\sqrt{7}, 4)$

18. $\begin{cases} 2x^2 - 3y^2 = 9 \\ x^2 + y^2 = 27 \end{cases}$
$(3\sqrt{2}, 3), (-3\sqrt{2}, 3), (3\sqrt{2}, -3), (-3\sqrt{2}, -3)$

Complete the square to write each equation in standard form, and identify the curve.

19. $y^2 - 4y - 6x - 14 = 0$
$(y - 2)^2 = 6(x + 3)$; parabola

20. $2x^2 + 3y^2 - 4x + 12y + 8 = 0$
$\frac{(x-1)^2}{3} + \frac{(y+2)^2}{2} = 1$; ellipse

CUMULATIVE REVIEW EXERCISES

Simplify each expression. Assume that all variables represent positive numbers, and write answers without using negative exponents.

1. $64^{2/3}$ 16

2. $8^{-1/3}$ $\frac{1}{2}$

3. $\frac{y^{2/3}y^{5/3}}{y^{1/3}}$ y^2

4. $\frac{x^{5/3}x^{1/2}}{x^{3/4}}$ $x^{17/12}$

5. $(x^{2/3} - x^{1/3})(x^{2/3} + x^{1/3})$ $x^{4/3} - x^{2/3}$

6. $(x^{-1/2} + x^{1/2})^2$ $\frac{1}{x} + 2 + x$

7. $\sqrt[3]{-27x^3}$ $-3x$

8. $\sqrt{48t^3}$ $4t\sqrt{3t}$

9. $\sqrt[3]{\frac{128x^4}{2x}}$
$4x$

10. $\sqrt{x^2 + 6x + 9}$
$x + 3$

11. $\sqrt{50} - \sqrt{8} + \sqrt{32}$ $7\sqrt{2}$

12. $-3\sqrt[4]{32} - 2\sqrt[4]{162} + 5\sqrt[4]{48}$ $-12\sqrt[4]{2} + 10\sqrt[4]{3}$

13. $3\sqrt{2}(2\sqrt{3} - 4\sqrt{12})$
$-18\sqrt{6}$

14. $\frac{5}{\sqrt[3]{x}}$ $\frac{5\sqrt[3]{x^2}}{x}$

15. $\frac{\sqrt{x}+2}{\sqrt{x}-1}$
$\frac{x + 3\sqrt{x} + 2}{x - 1}$

16. $\sqrt[6]{x^3y^3}$
$\sqrt{xy}$

Solve each equation.

17. $5\sqrt{x+2} = x + 8$ 2, 7

18. $\sqrt{x} + \sqrt{x+2} = 2$ $\frac{1}{4}$

19. Use the method of completing the square to solve the equation $2x^2 + x - 3 = 0$. $1, -\frac{3}{2}$

20. Use the quadratic formula to solve the equation $3x^2 + 4x - 1 = 0$. $\frac{-2 \pm \sqrt{7}}{3}$

Write each complex number in a + bi form.

21. $(3 + 5i) + (4 - 3i)$ $7 + 2i$

22. $(7 - 4i) - (12 + 3i)$ $-5 - 7i$

23. $(2 - 3i)(2 + 3i)$
$13 + 0i$

24. $(3 + i)(3 - 3i)$
$12 - 6i$

25. $(3 - 2i) - (4 + i)^2$
$-12 - 10i$

26. $\frac{5}{3 - i}$ $\frac{3}{2} + \frac{1}{2}i$

Find each value.

27. $|3 + 2i|$ $\sqrt{13}$

28. $|5 - 6i|$ $\sqrt{61}$

29. For what values of k will the solutions of $2x^2 + 4x = k$ be equal? -2

30. Find the coordinates of the vertex of the graph of the equation $y = \frac{1}{2}x^2 - x + 1$. $(1, \frac{1}{2})$

Give the result in interval notation.

31. Solve: $x^2 - x - 6 > 0$. $(-\infty, -2) \cup (3, \infty)$

32. Solve: $x^2 - x - 6 \le 0$. $[-2, 3]$

Let $f(x) = 3x^2 + 2$ and $g(x) = 2x - 1$. Find each value or function.

33. $f(-1)$ 5

34. $(g \circ f)(2)$ 27

35. $(f \circ g)(x)$
$12x^2 - 12x + 5$

36. $(g \circ f)(x)$
$6x^2 + 3$

37. Write $y = \log_2 x$ in exponential notation. $2^y = x$

38. Write $3^b = a$ in logarithmic notation. $\log_3 a = b$

Find x.

39. $\log_x 25 = 2$ 5

40. $\log_5 125 = x$ 3

41. $\log_3 x = -3$ $\frac{1}{27}$

42. $\log_5 x = 0$ 1

43. Find the inverse of $y = \log_2 x$. $y = 2^x$

44. If $\log_{10} 10^x = y$, then y equals what quantity? x

Log 7 = 0.8451 and log 14 = 1.1461. Evaluate each expression without using a calculator or tables.

45. log 98 1.9912

46. log 2 0.301

47. log 49 1.6902

48. $\log \frac{7}{5}$ (*Hint:* log 10 = 1.) 0.1461

49. Solve: $2^{x+2} = 3^x$. $\frac{2 \log 2}{\log 3 - \log 2}$

50. Solve: $2 \log 5 + \log x - \log 4 = 2$. 16

Use a calculator.

51. Boat depreciation How much will a $9,000 boat be worth after 9 years if it depreciates 12% per year?
$2,848.31

52. Find $\log_6 8$. 1.16056

53. Use graphing to solve $\begin{cases} 2x + y = 5 \\ x - 2y = 0 \end{cases}$. (2, 1)

54. Use substitution to solve $\begin{cases} 3x + y = 4 \\ 2x - 3y = -1 \end{cases}$. (1, 1)

55. Use addition to solve $\begin{cases} x + 2y = -2 \\ 2x - y = 6 \end{cases}$. (−2, −2)

56. Use any method to solve $\begin{cases} \frac{x}{10} + \frac{y}{5} = \frac{1}{2} \\ \frac{x}{2} - \frac{y}{5} = \frac{13}{10} \end{cases}$. (3, 1)

57. Evaluate: $\begin{vmatrix} 3 & -2 \\ 1 & -1 \end{vmatrix}$. −1

58. Use Cramer's rule and solve for y only:

$$\begin{cases} 4x - 3y = -1 \\ 3x + 4y = -7 \end{cases}$$ −1

59. Solve: $\begin{cases} x + y + z = 1 \\ 2x - y - z = -4 \\ x - 2y + z = 4 \end{cases}$. (−1, −1, 3)

60. Solve: $\begin{cases} x + 2y + 3z = 6 \\ 3x + 2y + z = 6 \\ 2x + 3y + z = 6 \end{cases}$ (1, 1, 1)

8 Natural-Number Functions and Probability

COLLEGE **Algebra** $f(x)$ **Now**™

Throughout the chapter, this icon introduces resources on the College AlgebraNow Web Site, accessed through **http://1pass.thomson.com**, that will:

- Help you test your knowledge of the material in this chapter prior to reading it,
- Allow you to take an exam-prep quiz, and
- Provide a Personalized Learning Plan targeting areas you should study.

Careers and Mathematics

STATISTICIAN

Statistics is the scientific application of mathematical principles to the collection, analysis, and presentation of numerical data. Statisticians contribute to scientific inquiry by applying their mathematical knowledge to the design of surveys and experiments; collection, processing, and analysis of data; and interpretation of the results. Statisticians work in a variety of subject areas, such as biology, economics, engineering, medicine, public health, and psychology.

Justin Sullivan/AP/Wide World Photos

Statisticians held about 20,000 jobs in 2002. Eighteen percent of these were in the federal government. Most of the remaining jobs were in private industry, especially in the research and testing services and management and public relations industries. In addition, many professionals with a background in statistics were among the 20,000 full-time mathematics faculty in colleges and universities in 2000.

JOB OUTLOOK

Graduates with a bachelor's or master's degree in statistics with a strong background in an allied field, such as finance, engineering, or computer science, should have good prospects of finding jobs related to their field of study. Federal agencies will hire statisticians in many fields, including demography, agriculture, consumer and producer surveys, Social Security, health care, and environmental quality.

Median annual earnings of statisticians were $57,080 in 2002.

Billy Beane (see above), General Manager of the Oakland As baseball team, uses statistics to improve the performance of his team.

For a sample application, see Example 6 in Section 8.4.

In this chapter, we introduce a way to expand powers of binomials of the form $(x + y)^n$. This method leads to ideas needed when working with probability and statistics.

8.1 The Binomial Theorem

In this section, you will learn about

- Pascal's Triangle
- Factorial Notation
- The Binomial Theorem
- Finding a Particular Term of a Binomial Expansion

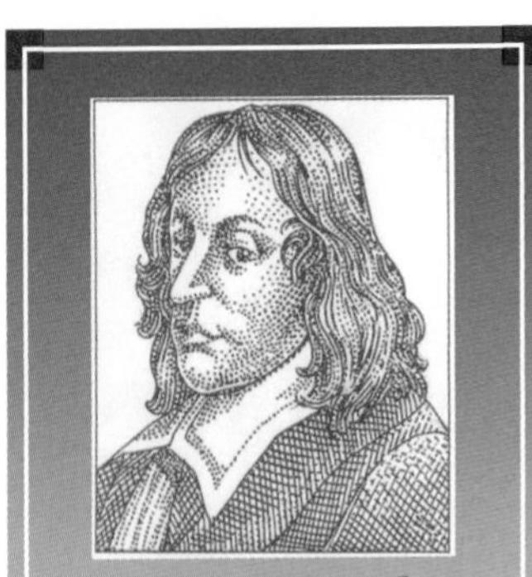

Blaise Pascal

(1623–1662)

Pascal was torn between religion and mathematics. Each surfaced at times in his life to dominate his interest. In mathematics, Pascal made contributions to the study of conic sections, probability, and differential calculus. At the age of 19, he invented a calculating machine. He is best known for a triangular array of numbers that bears his name.

We begin this chapter by introducing a way to expand binomials of the form $(x + y)^n$, where n is a natural number. This method, called the **binomial theorem,** is important when working with probability and statistics. Consider the following expansions.

$$(a + b)^0 = 1$$
$$(a + b)^1 = a + b$$
$$(a + b)^2 = a^2 + 2ab + b^2$$
$$(a + b)^3 = a^3 + 3a^2b + 3ab^2 + b^3$$
$$(a + b)^4 = a^4 + 4a^3b + 6a^2b^2 + 4ab^3 + b^4$$
$$(a + b)^5 = a^5 + 5a^4b + 10a^3b^2 + 10a^2b^3 + 5ab^4 + b^5$$
$$(a + b)^6 = a^6 + 6a^5b + 15a^4b^2 + 20a^3b^3 + 15a^2b^4 + 6ab^5 + b^6$$

Four patterns are apparent in the above expansions:

1. Each expansion has one more term than the power of the binomial.
2. The degree of each term in each expansion equals the exponent of the binomial.
3. The first term in each expansion is a raised to the power of the binomial.
4. The exponents of a decrease by 1 in each successive term, and the exponents on b, beginning with b^0 in the first term, increase by 1 in each successive term.

Pascal's Triangle

To see another pattern, we write the coefficients of the above expansions in a triangular array:

$(a + b)^0 =$ 1

$(a + b)^1 =$ 1 1

$(a + b)^2 =$ 1 2 1

$(a + b)^3 =$ 1 3 3 1

$(a + b)^4 =$ 1 4 6 4 1

$(a + b)^5 =$ 1 5 10 10 5 1

$(a + b)^6 =$ 1 6 15 20 15 6 1

In this array, each entry other than the 1's is the sum of the closest pair of numbers in the line above it. For example, the 20 in the bottom row is the sum of the 10's above it. The first 3 in the fourth row is the sum of the 1 and 2 above it.

This array, called **Pascal's triangle** after Blaise Pascal (1623–1662), continues with the same pattern forever. The next two lines are

$$(a + b)^7 = \quad 1 \quad 7 \quad 21 \quad 35 \quad 35 \quad 21 \quad 7 \quad 1$$
$$(a + b)^8 = \quad 1 \quad 8 \quad 28 \quad 56 \quad 70 \quad 56 \quad 28 \quad 8 \quad 1$$

ACTIVE EXAMPLE 1 Expand: $(x + y)^6$.

Solution The first term is x^6, and the exponents on x will decrease by 1 in each successive term. A y will appear in the second term, and the exponents on y will increase by 1 in each successive term, concluding when the term y^6 is reached. The variables in the expansion are

COLLEGE **Algebra** $f(x)$ **Now**™

Explore this Active Example and test yourself on the steps taken here at **http://1pass.thomson.com**. You can also find this example on the Interactive Video Skillbuilder CD-ROM.

$$x^6 \quad x^5y \quad x^4y^2 \quad x^3y^3 \quad x^2y^4 \quad xy^5 \quad y^6$$

We can use Pascal's triangle to find the coefficients of the variables. Because the binomial is raised to the sixth power, we choose the row in Pascal's triangle where the second entry is 6. The coefficients of the variables are the numbers in that row.

$$1 \quad 6 \quad 15 \quad 20 \quad 15 \quad 6 \quad 1$$

Putting this information together gives the expansion:

$$(x + y)^6 = x^6 + 6x^5y + 15x^4y^2 + 20x^3y^3 + 15x^2y^4 + 6xy^5 + y^6$$

Self Check Expand: $(p + q)^3$. ■

EXAMPLE 2 Expand: $(x - y)^6$.

Solution We write the binomial as $[x + (-y)]^6$ and substitute $-y$ for y in the result of Example 1.

$$\begin{aligned}[x + (-y)]^6 &\\ &= x^6 + 6x^5(-y) + 15x^4(-y)^2 + 20x^3(-y)^3 + 15x^2(-y)^4 + 6x(-y)^5 + (-y)^6\\ &= x^6 - 6x^5y + 15x^4y^2 - 20x^3y^3 + 15x^2y^4 - 6xy^5 + y^6\end{aligned}$$

In general, the signs in the expansion of $(x - y)^n$ alternate. The sign of the first term is $+$, the sign of the second term is $-$, and so on.

Self Check Expand: $(p - q)^3$. ■

■ Factorial Notation

To use the binomial theorem to expand a binomial, we will need **factorial notation.**

Factorial Notation

If n is a natural number, the symbol $n!$ (read either as **"*n* factorial"** or as **"factorial *n*"**) is defined as

$$n! = n(n - 1)(n - 2)(n - 3) \cdots (3)(2)(1)$$

EXAMPLE 3 Evaluate: **a.** $3!$, **b.** $6!$, and **c.** $10!$.

Solution **a.** $3! = 3 \cdot 2 \cdot 1 = 6$

b. $6! = 6 \cdot 5 \cdot 4 \cdot 3 \cdot 2 \cdot 1 = 720$

c. $10! = 10 \cdot 9 \cdot 8 \cdot 7 \cdot 6 \cdot 5 \cdot 4 \cdot 3 \cdot 2 \cdot 1 = 3{,}628{,}800$

Self Check Evaluate: **a.** $4!$ and **b.** $7!$ ■

There are two fundamental properties of factorials.

Properties of Factorials

1. By definition, $0! = 1$.
2. If n is a natural number, $n(n - 1)! = n!$.

EXAMPLE 4 Show that **a.** $6 \cdot 5! = 6!$ and **b.** $8 \cdot 7! = 8!$.

Solution **a.**
$$\begin{aligned} 6 \cdot 5! &= 6(5 \cdot 4 \cdot 3 \cdot 2 \cdot 1) \\ &= 6 \cdot 5 \cdot 4 \cdot 3 \cdot 2 \cdot 1 \\ &= 6! \end{aligned}$$

b.
$$\begin{aligned} 8 \cdot 7! &= 8(7 \cdot 6 \cdot 5 \cdot 4 \cdot 3 \cdot 2 \cdot 1) \\ &= 8 \cdot 7 \cdot 6 \cdot 5 \cdot 4 \cdot 3 \cdot 2 \cdot 1 \\ &= 8! \end{aligned}$$

Self Check Show that $4 \cdot 3! = 4!$. ■

■ The Binomial Theorem

We can now state the binomial theorem.

The Binomial Theorem

If n is any positive number, then

$$(a + b)^n = a^n + \frac{n!}{1!(n-1)!}a^{n-1}b + \frac{n!}{2!(n-2)!}a^{n-2}b^2 + \frac{n!}{3!(n-3)!}a^{n-3}b^3 + \cdots + \frac{n!}{r!(n-r)!}a^{n-r}b^r + \cdots + b^n$$

Comment In the expansion of $(a + b)^n$, the term containing b^r is given by

$$\frac{n!}{r!(n-r)!}a^{n-r}b^r$$

A proof of the binomial theorem appears in Appendix I.

In the binomial theorem, the exponents on the variables in each term on the right-hand side follow the familiar patterns:

1. The sum of the exponents on a and b in each term is n.
2. The exponents on a decrease by 1 in each successive term.
3. The exponents on b increase by 1 in each successive term.

However, the method of finding the coefficients is different. Except for the first and last terms, $n!$ is the numerator of each fractional coefficient. When the exponent on b is 2, the factors in the denominator of the fractional coefficient are $2!$ and $(n - 2)!$. When the exponent on b is 3, the factors in the denominator are $3!$ and $(n - 3)!$. When the exponent on b is r, the factors in the denominator are $r!$ and $(n - r)!$.

EXAMPLE 5 Expand: $(a + b)^5$.

Solution We substitute directly into the binomial theorem.

$$\begin{aligned}(a + b)^5 &= a^5 + \frac{5!}{1!(5-1)!}a^4b + \frac{5!}{2!(5-2)!}a^3b^2 + \frac{5!}{3!(5-3)!}a^2b^3 + \frac{5!}{4!(5-4)!}ab^4 + b^5 \\ &= a^5 + \frac{5 \cdot 4!}{1 \cdot 4!}a^4b + \frac{5 \cdot 4 \cdot 3!}{2 \cdot 1 \cdot 3!}a^3b^2 + \frac{5 \cdot 4 \cdot 3!}{3! \cdot 2 \cdot 1}a^2b^3 + \frac{5 \cdot 4!}{4! \cdot 1}ab^4 + b^5 \\ &= a^5 + 5a^4b + 10a^3b^2 + 10a^2b^3 + 5ab^4 + b^5\end{aligned}$$

We note that the coefficients are the same numbers as in the sixth row of Pascal's triangle.

Self Check Expand: $(p + q)^4$. ■

ACTIVE EXAMPLE 6 Expand: $(2x - 3y)^4$.

Solution We first find the expansion of $(a + b)^4$.

COLLEGE **Algebra** f(x) **Now**™
Go to **http://1pass.thomson.com** or your CD to practice this example.

$$\begin{aligned}(a + b)^4 &= a^4 + \frac{4!}{1!(4-1)!}a^3b + \frac{4!}{2!(4-2)!}a^2b^2 + \frac{4!}{3!(4-3)!}ab^3 + b^4 \\ &= a^4 + \frac{4 \cdot 3!}{1 \cdot 3!}a^3b + \frac{4 \cdot 3 \cdot 2!}{2 \cdot 1 \cdot 2!}a^2b^2 + \frac{4 \cdot 3!}{3! \cdot 1}ab^3 + b^4 \\ &= a^4 + 4a^3b + 6a^2b^2 + 4ab^3 + b^4\end{aligned}$$

We then substitute $2x$ for a and $-3y$ for b in the result.

$$\begin{aligned}(a + b)^4 &= a^4 + 4a^3b + 6a^2b^2 + 4ab^3 + b^4 \\ [2x + (-3y)]^4 &= (2x)^4 + 4(2x)^3(-3y) + 6(2x)^2(-3y)^2 + 4(2x)(-3y)^3 + (-3y)^4 \\ (2x - 3y)^4 &= 16x^4 - 96x^3y + 216x^2y^2 - 216xy^3 + 81y^4\end{aligned}$$

Self Check Expand: $(3x - 2y)^4$. ■

■ Finding a Particular Term of a Binomial Expansion

Suppose that we wish to find the fifth term of the expansion of $(a + b)^{11}$. It would be tedious to raise the binomial to the 11th power and then look at the fifth term. The binomial theorem provides an easier way.

EXAMPLE 7 Find the fifth term of the expansion of $(a + b)^{11}$.

Solution In the fifth term, the exponent on b is 4 (the exponent on b is always 1 less than the number of the term). Since the exponent on b added to the exponent on a equals 11, the exponent on a is 7. The variables of the fifth term are a^7b^4.

The number in the numerator of the fractional coefficient is $n!$, which in this case is $11!$. The factors in the denominator are $4!$ And $(11 - 4)!$. The complete fifth term is

$$\frac{11!}{4!(11-4)!}a^7b^4 = \frac{11!}{4!7!}a^7b^4$$
$$= \frac{11 \cdot 10 \cdot 9 \cdot 8 \cdot 7!}{4 \cdot 3 \cdot 2 \cdot 1 \cdot 7!}a^7b^4$$
$$= 330a^7b^4$$

Comment We can also use the formula for the term containing b^r, where $r = 4$ and $n = 11$.

$$\frac{n!}{r!(n-r)!}a^{n-r}b^r = \frac{11!}{4!(11-4)!}a^{11-4}b^4$$
$$= \frac{11!}{4!7!}a^7b^4$$
$$= 330a^7b^4$$

Self Check Find the sixth term of the expansion in Example 7. ■

EXAMPLE 8 Find the sixth term of the expansion of $(a + b)^9$.

Solution In the sixth term, the exponent on b is 5, and the exponent on a is $9 - 5$, or 4. The numerator of the fractional coefficient is 9!, and the factors in the denominator are 5! and $(9 - 5)!$. The sixth term of the expansion is

$$\frac{9!}{5!(9-5)!}a^4b^5 = \frac{9 \cdot 8 \cdot 7 \cdot 6 \cdot 5!}{5! \cdot 4!}a^4b^5$$
$$= \frac{9 \cdot 8 \cdot 7 \cdot 6}{4 \cdot 3 \cdot 2 \cdot 1}a^4b^5$$
$$= 126a^4b^5$$

Comment We can also use the formula for the term containing b^r, where $r = 5$ and $n = 9$.

$$\frac{n!}{r!(n-r)!}a^{n-r}b^r = \frac{9!}{5!(9-5)!}a^{9-5}b^5$$
$$= \frac{9!}{5!4!}a^4b^5$$
$$= 126a^4b^5$$

Self Check Find the fifth term of the expansion in Example 8. ■

ACTIVE EXAMPLE 9 Find the third term of the expansion of $(3x - 2y)^6$.

Solution We begin by finding the third term of the expansion of $(a + b)^6$.

(1)
$$\frac{6!}{2!(6-2)!}a^4b^2 = \frac{6 \cdot 5 \cdot 4!}{2 \cdot 1 \cdot 4!}a^4b^2 = 15a^4b^2$$

We can then substitute $3x$ for a and $-2y$ for b in Equation 1 to obtain the third term of the expansion of $(3x - 2y)^6$.

COLLEGE **Algebra** $f(x)$ **Now**™
Go to **http://1pass.thomson.com** or your CD to practice this example.

$$15a^4b^2 = 15(3x)^4(-2y)^2$$
$$= 15(3)^4(-2)^2x^4y^2$$
$$= 4{,}860x^4y^2$$

Self Check Find the fourth term of the expansion in Example 9. ■

Self Check Answers

1. $p^3 + 3p^2q + 3pq^2 + q^3$ **2.** $p^3 - 3p^2q + 3pq^2 - q^3$ **3. a.** 24 **b.** 5,040 **4.** $4 \cdot (3 \cdot 2 \cdot 1) = 4!$ **5.** $p^4 + 4p^3q + 6p^2q^2 + 4pq^3 + q^4$ **6.** $81x^4 - 216x^3y + 216x^2y^2 - 96xy^3 + 16y^4$ **7.** $462a^6b^5$ **8.** $126a^5b^4$ **9.** $-4{,}320x^3y^3$

8.1 Exercises

VOCABULARY AND CONCEPTS ***Fill in the blanks.***

1. In the expansion of a binomial, there will be one more term than the __power__ of the binomial.

2. The __degree__ of each term in a binomial expansion is the same as the exponent of the binomial.

3. The __first__ term in a binomial expansion is the first term raised to the power of the binomial.

4. In the expansion of $(p + q)^n$, the __exponents__ on p decrease by 1 in each successive term.

5. Expand 7!: __$7 \cdot 6 \cdot 5 \cdot 4 \cdot 3 \cdot 2 \cdot 1$__

6. $0! =$ __1__

7. $n \cdot$ __$(n-1)!$__ $= n!$

8. In the seventh term of $(a + b)^{11}$, the exponent on a is __5__.

PRACTICE ***Evaluate each expression.***

9. $4!$ 24

10. $-5!$ -120

11. $3! \cdot 6!$ 4,320

12. $0! \cdot 7!$ 5,040

13. $6! + 6!$ 1,440

14. $5! - 2!$ 118

15. $\frac{9!}{12!}$ $\frac{1}{1{,}320}$

16. $\frac{8!}{5!}$ 336

17. $\frac{5! \cdot 7!}{9!}$ $\frac{5}{3}$

18. $\frac{3! \cdot 5! \cdot 7!}{1!8!}$ 90

19. $\frac{18!}{6!(18-6)!}$ 18,564

20. $\frac{15!}{9!(15-9)!}$ 5,005

Use the binomial theorem to expand each binomial.

21. $(a + b)^3$ $a^3 + 3a^2b + 3ab^2 + b^3$

22. $(a + b)^4$ $a^4 + 4a^3b + 6a^2b^2 + 4ab^3 + b^4$

23. $(a - b)^5$ $a^5 - 5a^4b + 10a^3b^2 - 10a^2b^3 + 5ab^4 - b^5$

24. $(x - y)^4$ $x^4 - 4x^3y + 6x^2y^2 - 4xy^3 + y^4$

25. $(2x + y)^3$ $8x^3 + 12x^2y + 6xy^2 + y^3$

26. $(x + 2y)^3$ $x^3 + 6x^2y + 12xy^2 + 8y^3$

27. $(x - 2y)^3$ $x^3 - 6x^2y + 12xy^2 - 8y^3$

28. $(2x - y)^3$ $8x^3 - 12x^2y + 6xy^2 - y^3$

29. $(2x + 3y)^4$ $16x^4 + 96x^3y + 216x^2y^2 + 216xy^3 + 81y^4$

30. $(2x - 3y)^4$ $16x^4 - 96x^3y + 216x^2y^2 - 216xy^3 + 81y^4$

31. $(x - 2y)^4$ $x^4 - 8x^3y + 24x^2y^2 - 32xy^3 + 16y^4$

32. $(x + 2y)^4$ $x^4 + 8x^3y + 24x^2y^2 + 32xy^3 + 16y^4$

33. $(x - 3y)^5$
$x^5 - 15x^4y + 90x^3y^2 - 270x^2y^3 + 405xy^4 - 243y^5$

34. $(3x - y)^5$
$243x^5 - 405x^4y + 270x^3y^2 - 90x^2y^3 + 15xy^4 - y^5$

35. $\left(\frac{x}{2} + y\right)^4$ $\frac{x^4}{16} + \frac{x^3y}{2} + \frac{3x^2y^2}{2} + 2xy^3 + y^4$

36. $\left(x + \frac{y}{2}\right)^4$ $x^4 + 2x^3y + \frac{3x^2y^2}{2} + \frac{xy^3}{2} + \frac{y^4}{16}$

Find the required term in each binomial expansion.

37. $(a + b)^4$; 3rd term
$6a^2b^2$

38. $(a - b)^4$; 2nd term
$-4a^3b$

39. $(a + b)^7$; 5th term
$35a^3b^4$

40. $(a + b)^5$; 4th term
$10a^2b^3$

41. $(a - b)^5$; 6th term
$-b^5$

42. $(a - b)^8$; 7th term
$28a^2b^6$

43. $(a + b)^{17}$; 5th term
$2{,}380a^{13}b^4$

44. $(a - b)^{12}$; 3rd term
$66a^{10}b^2$

45. $(a - \sqrt{2})^4$; 2nd term
$-4\sqrt{2}a^3$

46. $(a - \sqrt{3})^8$; 3rd term
$84a^6$

47. $(a + \sqrt{3}b)^9$; 5th term
$1{,}134a^5b^4$

48. $(\sqrt{2}a - b)^7$; 4th term
$-140a^4b^3$

49. $\left(\frac{x}{2} + y\right)^4$; 3rd term
$\frac{3x^2y^2}{2}$

50. $\left(m + \frac{n}{2}\right)^8$; 3rd term
$7m^6n^2$

51. $\left(\frac{r}{2} - \frac{s}{2}\right)^{11}$; 10th term
$\frac{-55r^2s^9}{2{,}048}$

52. $\left(\frac{p}{2} - \frac{q}{2}\right)^9$; 6th term
$-\frac{63p^4q^5}{256}$

53. $(a + b)^n$; 4th term
$\frac{n!}{3!(n-3)!}a^{n-3}b^3$

54. $(a - b)^n$; 5th term
$\frac{n!}{4!(n-4)!}a^{n-4}b^4$

55. $(a + b)^n$; rth term $\frac{n!}{(r-1)!(n-r+1)!}a^{n-r+1}b^{r-1}$

56. $(a + b)^n$; $(r + 1)$th term $\frac{n!}{r!(n-r)!}a^{n-r}b^r$

DISCOVERY AND WRITING

57. Find the sum of the numbers in each row of the first ten rows of Pascal's triangle. Do you see a pattern?

58. Show that the sum of the coefficients in the binomial expansion of $(x + y)^n$ is 2^n. (*Hint:* Let $x = y = 1$.)

59. Find the constant term in the expansion of

$$\left(a - \frac{1}{a}\right)^{10}$$ -252

60. Find the coefficient of x^5 in the expansion of

$$\left(x + \frac{1}{x}\right)^9$$ 36

61. If we applied the pattern of coefficients to the coefficient of the first term in the binomial theorem, it would be $\frac{n!}{0!(n-0)!}$. Show that this expression equals 1.

62. If we applied the pattern of coefficients to the coefficient of the last term in the binomial theorem, it would be $\frac{n!}{n!(n-n)!}$. Show that this expression equals 1.

63. Define factorial notation and explain how to evaluate 10!.

64. With a calculator, evaluate 69!. Explain why you cannot find 70! with a calculator.

65. Explain how the rth term of a binomial expansion is constructed.

66. Explain the four patterns apparent in a binomial expansion.

REVIEW *Factor each expression.*

67. $3x^3y^2z^4 - 6xyz^5 + 15x^2yz^2$ $3xyz^2(x^2yz^2 - 2z^3 + 5x)$

68. $3z^2 - 15tz + 12t^2$ $3(z - t)(z - 4t)$

69. $a^4 - b^4$ $(a^2 + b^2)(a + b)(a - b)$

70. $3r^4 - 36r^3 - 135r^2$ $3r^2(r + 3)(r - 15)$

Simplify each complex fraction.

71. $\dfrac{\frac{1}{x} + \frac{1}{3}}{\frac{1}{x} - \frac{1}{3}}$ $\frac{3+x}{3-x}$

72. $\dfrac{x - \frac{1}{y}}{y - \frac{1}{x}}$ $\frac{x}{y}$

8.2 Sequences, Series, and Summation Notation

In this section, you will learn about

- Sequences
- Recursive Definition of a Sequence
- Series
- Summation Notation

In this section, we will introduce a function whose domain is the set of natural numbers. This function, called a **sequence,** is a list of numbers in a specific order.

Sequences

Sequences

A **sequence** is a function whose domain is the set of natural numbers.

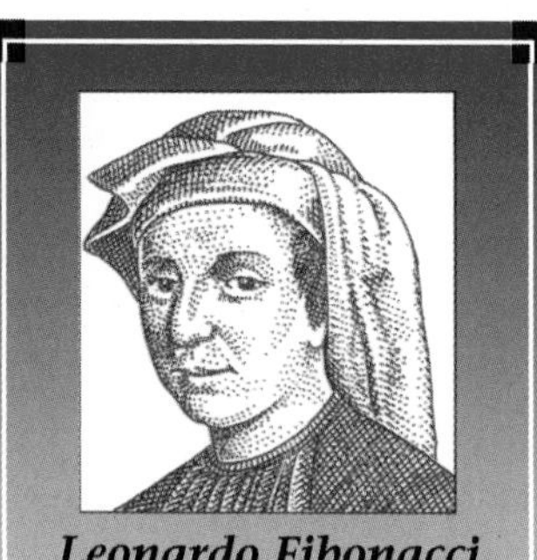

Leonardo Fibonacci
(late 12th and early 13th centuries)

Fibonacci, an Italian mathematician, is also known as Leonardo da Pisa. In his work *Liber abaci,* he advocated the adoption of Arabic numerals, the numerals that we use today.

Since a sequence is a function whose domain is the set of natural numbers, we can write its terms as a list of numbers. For example, if n is a natural number, the function defined by $f(n) = 2n - 1$ generates the sequence

$$1, 3, 5, \ldots, 2n - 1, \ldots$$

The number 1 is the first term, 3 is the second term, 5 is the third term, and $2n - 1$ is the **general,** or ***n*th term.** If n is a natural number, the function $f(n) = 3n^2 + 1$ generates the sequence

$$4, 13, 28, \ldots, 3n^2 + 1, \ldots$$

The number 4 is the first term, 13 is the second term, 28 is the third term, and $3n^2 + 1$ is the general term.

A constant function such as $g(n) = 1$ is a sequence, because it generates the sequence

$$1, 1, 1, \ldots$$

We seldom use function notation to denote a sequence, because it is often difficult or even impossible to write the general term. In such cases, if there is a pattern that is assumed to be continued, we simply list several terms of the sequence. Some examples of sequences are:

$$1^2, 2^2, 3^2, \ldots, n^2, \ldots$$

$$3, 9, 19, 33, \ldots, 2n^2 + 1, \ldots$$

$$1, 3, 6, 10, 15, 21, \ldots, \frac{n(n + 1)}{2}, \ldots$$

$1, 1, 2, 3, 5, 8, 13, 21, \ldots$ **(Fibonacci sequence)**

$2, 3, 5, 7, 11, 13, 17, 19, 23, \ldots$ **(prime numbers)**

The **Fibonacci sequence** is named after the 12th-century mathematician Leonardo of Pisa, also known as Fibonacci. After the two 1's in the Fibonacci sequence, each term is the sum of the two terms that immediately precede it. The Fibonacci sequence occurs in many fields, such as the growth patterns of plants, the reproductive habits of bees, and music.

Recursive Definition of a Sequence

A sequence can be defined **recursively** by giving its first term and a rule showing how to obtain the $(n + 1)$th term from the nth term. For example, the information

$a_1 = 5$ (the first term) and $a_{n+1} = 3a_n - 2$ (the rule showing how to get the $(n + 1)$th term from the nth term)

defines a sequence recursively. To find the first five terms of this sequence, we proceed as follows:

$a_1 = 5$

$a_2 = 3(a_1) - 2 = 3(5) - 2 = 13$ Substitute 5 for a_1 and simplify to get a_2.

$a_3 = 3(a_2) - 2 = 3(13) - 2 = 37$ Substitute 13 for a_2 and simplify to get a_3.

$a_4 = 3(a_3) - 2 = 3(37) - 2 = 109$ Substitute 37 for a_3 and simplify to get a_4.

$a_5 = 3(a_4) - 2 = 3(109) - 2 = 325$ Substitute 109 for a_4 and simplify to get a_5.

Series

To add the terms of a sequence, we replace each comma between its terms with a + sign to form a **series.** Because each sequence is infinite, the number of terms in the series associated with it is infinite also. Two examples of infinite series are

$$1^2 + 2^2 + 3^2 + \cdots + n^2 + \cdots$$

and

$$1 + 2 + 3 + 5 + 8 + 13 + 21 + \cdots$$

If the signs between successive terms of an infinite series alternate, the series is called an **alternating infinite series.** Two examples of alternating infinite series are

$$-3 + 6 - 9 + 12 - \cdots + (-1)^n 3n + \cdots$$

and

$$2 - 4 + 8 - 16 + \cdots + (-1)^{n+1} 2^n + \cdots$$

Summation Notation

Summation notation is a shorthand way to indicate the sum of the first n terms, or the ***n*th partial sum,** of a sequence. For example, the expression

$$\sum_{n=1}^{4} n$$ **The symbol Σ is the capital letter sigma in the Greek alphabet.**

indicates the sum of the four terms obtained when we successively substitute 1, 2, 3, and 4 for n.

$$\sum_{n=1}^{4} n = 1 + 2 + 3 + 4 = 10$$

The expression

$$\sum_{n=2}^{4} n^2$$

indicates the sum of the three terms obtained when we successively substitute 2, 3, and 4 for n^2.

$$\begin{aligned}\sum_{n=2}^{4} n^2 &= (2)^2 + (3)^2 + (4)^2 \\ &= 4 + 9 + 16 \\ &= 29\end{aligned}$$

The expression

$$\sum_{n=1}^{3} (2n^2 + 1)$$

indicates the sum of the three terms obtained if we successively substitute 1, 2, and 3 for n in the expression $2n^2 + 1$.

$$\begin{aligned}\sum_{n=1}^{3} (2n^2 + 1) &= [2(1)^2 + 1] + [2(2)^2 + 1] + [2(3)^2 + 1] \\ &= 3 + 9 + 19 \\ &= 31\end{aligned}$$

EXAMPLE 1 Evaluate: $\sum_{n=1}^{4}(n^2 - 1)$.

Solution Since n runs from 1 to 4, we substitute 1, 2, 3, and 4 for n in the expression $n^2 - 1$ and find the sum of the resulting terms:

$$\begin{aligned}\sum_{n=1}^{4}(n^2 - 1) &= (1^2 - 1) + (2^2 - 1) + (3^2 - 1) + (4^2 - 1)\\ &= 0 + 3 + 8 + 15\\ &= 26\end{aligned}$$

Self Check Evaluate: $\sum_{n=1}^{5}(n^2 - 1)$. ■

ACTIVE EXAMPLE 2 Evaluate: $\sum_{n=3}^{5}(3n + 2)$.

Solution Since n runs from 3 to 5, we substitute 3, 4, and 5 for n in the expression $3n + 2$ and find the sum of the resulting terms:

COLLEGE **Algebra** *f(x)* **Now**™
Explore this Active Example and test yourself on the steps taken here at **http://1pass.thomson.com**. You can also find this example on the Interactive Video Skillbuilder CD-ROM.

$$\begin{aligned}\sum_{n=3}^{5}(3n + 2) &= [3(\mathbf{3}) + 2] + [3(\mathbf{4}) + 2] + [3(\mathbf{5}) + 2]\\ &= 11 + 14 + 17\\ &= 42\end{aligned}$$

Self Check Evaluate: $\sum_{n=2}^{5}(3n + 2)$. ■

We can use summation notation to state the binomial theorem concisely. In Exercise 54, you will be asked to explain why the binomial theorem can be stated as

$$\sum_{r=0}^{n}\frac{n!}{r!(n - r)!}a^{n-r}b^r$$

There are three basic properties of summations. The first states that *the summation of a constant as k runs from 1 to n is n times the constant.*

Summation of a Constant

If c is a constant, then $\sum_{k=1}^{n} c = nc$.

Proof Because c is a constant, each term is c for each value of k as k runs from 1 to n.

$$\sum_{k=1}^{n} c = \overbrace{c + c + c + c + \cdots + c}^{n \text{ number of } c\text{'s}} = nc$$ ■

EXAMPLE 3 Evaluate: $\sum_{n=1}^{5} 13$.

Solution

$$\sum_{n=1}^{5} 13 = 13 + 13 + 13 + 13 + 13$$
$$= 5(13)$$
$$= 65$$

Self Check Evaluate: $\sum_{n=1}^{6} 12$. ■

A second property states that *a constant factor can be brought outside a summation sign.*

Summation of a Product

If c is a constant, then $\sum_{k=1}^{n} cf(k) = c\sum_{k=1}^{n} f(k)$.

Proof

$$\sum_{k=1}^{n} cf(k) = cf(1) + cf(2) + cf(3) + \cdots + cf(n)$$
$$= c[f(1) + f(2) + f(3) + \cdots + f(n)] \quad \text{Factor out } c.$$
$$= c\sum_{k=1}^{n} f(k)$$
■

ACTIVE EXAMPLE 4 Show that $\sum_{k=1}^{3} 5k^2 = 5\sum_{k=1}^{3} k^2$.

Solution

$$\sum_{k=1}^{3} 5k^2 = 5(1)^2 + 5(2)^2 + 5(3)^2$$
$$= 5 + 20 + 45$$
$$= 70$$

$$5\sum_{k=1}^{3} k^2 = 5[(1)^2 + (2)^2 + (3)^2]$$
$$= 5[1 + 4 + 9]$$
$$= 5(14)$$
$$= 70$$

The quantities are equal.

Self Check Evaluate: $\sum_{k=1}^{4} 3k$. ■

The third property states that *the summation of a sum is equal to the sum of the summations.*

Summation of a Sum

$$\sum_{k=1}^{n} [f(k) + g(k)] = \sum_{k=1}^{n} f(k) + \sum_{k=1}^{n} g(k)$$

Proof

$$\sum_{k=1}^{n} [f(k) + g(k)] = [f(1) + g(1)] + [f(2) + g(2)] + [f(3) + g(3)] + \cdots + [f(n) + g(n)]$$
$$= [f(1) + f(2) + f(3) + \cdots + f(n)] + [g(1) + g(2) + g(3) + \cdots + g(n)]$$
$$= \sum_{k=1}^{n} f(k) + \sum_{k=1}^{n} g(k)$$ ■

EXAMPLE 5 Show that $\sum_{k=1}^{3} (k + k^2) = \sum_{k=1}^{3} k + \sum_{k=1}^{3} k^2$.

Solution

$$\sum_{k=1}^{3} (k + k^2) = (1 + 1^2) + (2 + 2^2) + (3 + 3^2)$$
$$= 2 + 6 + 12$$
$$= 20$$

$$\sum_{k=1}^{3} k + \sum_{k=1}^{3} k^2 = (1 + 2 + 3) + (1^2 + 2^2 + 3^2)$$
$$= 6 + 14$$
$$= 20$$

Self Check Evaluate: $\sum_{k=1}^{3} (k^2 + 2k)$. ■

ACTIVE EXAMPLE 6 Evaluate $\sum_{k=1}^{5} (2k - 1)^2$ directly. Then expand the binomial, apply the previous properties, and evaluate the expression again.

Solution

Part 1: $\sum_{k=1}^{5} (2k - 1)^2 = 1 + 9 + 25 + 49 + 81 = 165$

COLLEGE **Algebra** $f(x)$ **Now**™
Go to **http://1pass.thomson.com** or your CD to practice this example.

Part 2: $\sum_{k=1}^{5} (2k - 1)^2 = \sum_{k=1}^{5} (4k^2 - 4k + 1)$

$$= \sum_{k=1}^{5} 4k^2 + \sum_{k=1}^{5} (-4k) + \sum_{k=1}^{5} 1$$ The summation of a sum is the sum of the summations.

$$= 4\sum_{k=1}^{5} k^2 - 4\sum_{k=1}^{5} k + \sum_{k=1}^{5} 1$$ Bring the constant factors outside the summation signs.

$$= 4\sum_{k=1}^{5} k^2 - 4\sum_{k=1}^{5} k + 5$$ The summation of a constant as k runs from 1 to 5 is 5 times that constant.

$$= 4(1 + 4 + 9 + 16 + 25) - 4(1 + 2 + 3 + 4 + 5) + 5$$
$$= 4(55) - 4(15) + 5$$
$$= 220 - 60 + 5$$
$$= 165$$

Either way, the sum is 165.

Self Check Evaluate: $\sum_{k=1}^{4} (2k - 1)^2$. ■

Self Check Answers

1. 50 **2.** 50 **3.** 72 **4.** 30 **5.** 26 **6.** 84

8.2 Exercises

VOCABULARY AND CONCEPTS ***Fill in the blanks.***

1. A sequence is a function whose domain is the set of natural numbers.

2. A series is formed when we add the terms of a sequence.

3. Summation notation is a shorthand way to indicate the sum of the first n terms of a sequence.

4. The symbol $\sum_{k=1}^{5} (k^2 - 3)$ indicates the sum of the five terms obtained when we successively substitute 1, 2, 3, 4, and 5 for k.

5. $\sum_{k=1}^{5} 6k^2 = 6 \sum_{k=1}^{5} k^2$

6. $\sum_{k=1}^{5} (k^2 + 3k) = \sum_{k=1}^{5} k^2 + 3\sum_{k=1}^{5} k$

7. $\sum_{k=1}^{5} c$, where c is a constant, equals $5c$.

8. The summation of a sum is equal to the sum of the summations.

PRACTICE ***Write the first six terms of the sequence defined by each function.***

9. $f(n) = 5n(n - 1)$ **0, 10, 30, 60, 100, 150**

10. $f(n) = n\left(\frac{n-1}{2}\right)\left(\frac{n-2}{3}\right)$ **0, 0, 1, 4, 10, 20**

Find the next term of each sequence.

11. 1, 6, 11, 16, . . . **21**

12. 1, 8, 27, 64, . . . **125**

13. $a, a + d, a + 2d, a + 3d, \ldots$ $a + 4d$

14. $a, ar, ar^2, ar^3, \ldots$ ar^4

15. 1, 3, 6, 10, . . . **15**

16. 20, 17, 13, 8, . . . **2**

Find the sum of the first five terms of the sequence with the given general term.

17. n **15**

18. $2k$ **30**

19. 3 **15**

20. $4k^0$ **20**

21. $2\left(\frac{1}{3}\right)^n$ $\frac{242}{243}$

22. $(-1)^n$ **−1**

23. $3n - 2$ **35**

24. $2k + 1$ **35**

Assume that each sequence is defined recursively. Find the first four terms of each sequence.

25. $a_1 = 3$ and $a_{n+1} = 2a_n + 1$ **3, 7, 15, 31**

26. $a_1 = -5$ and $a_{n+1} = -a_n - 3$ **−5, 2, −5, 2**

27. $a_1 = -4$ and $a_{n+1} = \frac{a_n}{2}$ $-4, -2, -1, -\frac{1}{2}$

28. $a_1 = 0$ and $a_{n+1} = 2a_n^2$ **0, 0, 0, 0**

29. $a_1 = k$ and $a_{n+1} = a_n^2$ k, k^2, k^4, k^8

30. $a_1 = 3$ and $a_{n+1} = ka_n$ $3, 3k, 3k^2, 3k^3$

31. $a_1 = 8$ and $a_{n+1} = \frac{2a_n}{k}$ $8, \frac{16}{k}, \frac{32}{k^2}, \frac{64}{k^3}$

32. $a_1 = m$ and $a_{n+1} = \frac{a_n^2}{m}$ m, m, m, m

Determine whether each series is an alternating infinite series.

33. $-1 + 2 - 3 + \cdots + (-1)^n n + \cdots$
an alternating series

34. $a + \frac{a}{b} + \frac{a}{b^2} + \cdots + a\left(\frac{1}{b}\right)^{n-1} + \cdots; b = 4$
not an alternating series

35. $a + a^2 + a^3 + \cdots + a^n + \cdots ; a = 3$
not an alternating series

36. $a + a^2 + a^3 + \cdots + a^n + \cdots ; a = -2$
an alternating series

Evaluate each sum.

37. $\sum_{k=1}^{5} 2k$ **30**

38. $\sum_{k=3}^{6} 3k$ **54**

39. $\sum_{k=3}^{4} (-2k^2)$ **−50**

40. $\sum_{k=1}^{100} 5$ **500**

41. $\sum_{k=1}^{5} (3k - 1)$ **40**

42. $\sum_{n=2}^{5} (n^2 + 3n)$ **96**

43. $\sum_{k=1}^{1,000} \frac{1}{2}$ **500**

44. $\sum_{x=4}^{5} \frac{2}{x}$ $\mathbf{\frac{9}{10}}$

45. $\sum_{x=3}^{4} \frac{1}{x}$ $\mathbf{\frac{7}{12}}$

46. $\sum_{x=2}^{6} (3x^2 + 2x) - 3\sum_{x=2}^{6} x^2$ **40**

47. $\sum_{x=1}^{4} (4x + 1)^2 - \sum_{x=1}^{4} (4x - 1)^2$ **160**

48. $\sum_{x=0}^{10} (2x - 1)^2 + 4\sum_{x=0}^{10} x(1 - x)$ **11**

49. $\sum_{x=6}^{8} (5x - 1)^2 + \sum_{x=6}^{8} (10x - 1)$ **3,725**

50. $\sum_{x=2}^{7} (3x + 1)^2 - 3\sum_{x=2}^{7} x(3x + 2)$ **6**

DISCOVERY AND WRITING

51. Find a counterexample to disprove the proposition that the summation of a product is the product of the summations. In other words, prove that

$$\sum_{k=1}^{n} f(k)g(k) \neq \sum_{k=1}^{n} f(k) \sum_{k=1}^{n} g(k)$$

52. Find a counterexample to disprove the proposition that the summation of a quotient is the quotient of the summations. In other words, prove that

$$\sum_{k=1}^{n} \frac{f(k)}{g(k)} \neq \frac{\sum_{k=1}^{n} f(k)}{\sum_{k=1}^{n} g(k)}$$

53. Explain what it means to define something recursively.

54. Explain why the binomial theorem can be stated as

$$\sum_{r=0}^{n} \frac{n!}{r!(n-r)!} a^{n-r} b^r$$

REVIEW *The triangles are similar. Find the value of x.*

55.

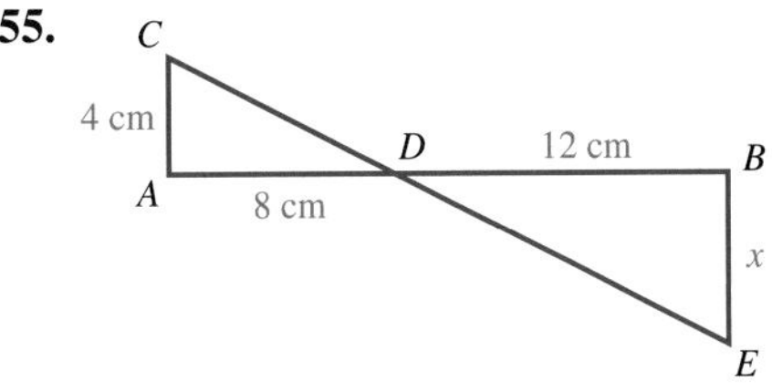

6 cm

56.

A, 12 m, D, 9 m, B, x, C, 21 m, E

28 m

Find the third side of each right triangle.

57. $AB = 24$ ft, $BC = 10$ ft

26 ft

58. 3 in., 7 in., x

$\mathbf{2\sqrt{10}}$ **in.**

8.3 Arithmetic Sequences

In this section, you will learn about

- Arithmetic Sequences
- Arithmetic Means
- Sum of the First *n* Terms of an Arithmetic Sequence
- Applications

The German mathematician Carl Friedrich Gauss (1777–1855) was once a student in the class of a strict teacher. One day, the teacher asked the students to add together all of the natural numbers from 1 through 100. Gauss immediately recognized that in the sum

$$1 + 2 + 3 + \cdots + 98 + 99 + 100$$

the first number (1) added to the last number (100) is 101, the second number (2) added to the second from the last number (99) is 101, and the third number (3) added to the third from the last number (98) is 101. He reasoned that there would be fifty pairs of such numbers, and that there would be fifty sums of 101. He multiplied 101 by 50 to get the correct answer of 5,050.

This story illustrates a problem involving the sum of the terms of a sequence, called an **arithmetic sequence,** in which each term except the first is found by adding a constant to the preceding term.

Arithmetic Sequences

Arithmetic Sequences

An **arithmetic sequence** is a sequence of the form

$$a, \quad a + d, \quad a + 2d, \quad a + 3d, \ldots, a + (n - 1)d, \ldots$$

where a is the **first term,** $a + (n - 1)d$ is the ***n*th term,** and d is the **common difference.**

In this definition, the second term has an addend of d, the third term has an addend of $2d$, the fourth term has an addend of $3d$, and so on. This is why the nth term has an addend of $(n - 1)d$.

EXAMPLE 1 Write the first six terms and the 21st term of an arithmetic sequence with a first term of 7 and a common difference of 5.

Solution Since the first term a is 7 and the common difference d is 5, the first six terms are

$$7, \quad 7 + 5, \quad 7 + 2(5), \quad 7 + 3(5), \quad 7 + 4(5), \quad 7 + 5(5)$$

or

$$7, \quad 12, \quad 17, \quad 22, \quad 27, \quad 32$$

To find the 21st term, we substitute 21 for n in the formula for the nth term:

$$\begin{aligned} n\text{th term} &= a + (n - 1)d \\ 21\text{st term} &= 7 + (21 - 1)5 \\ &= 7 + (20)5 \\ &= 107 \end{aligned}$$

The 21st term is 107.

Self Check Write the first five terms and the 18th term of an arithmetic sequence with a first term of 3 and a common difference of 6. ■

ACTIVE EXAMPLE 2 Find the 98th term of an arithmetic sequence whose first three terms are 2, 6, and 10.

Solution Here $a = 2$, $n = 98$, and $d = 6 - 2 = 10 - 6 = 4$. Because we want to find the 98th term, we substitute these numbers into the formula for the nth term:

COLLEGE **Algebra** $f(x)$ **Now**™
Explore this Active Example and test yourself on the steps taken here at **http://1pass.thomson.com**. You can also find this example on the Interactive Video Skillbuilder CD-ROM.

$$\begin{aligned} n\text{th term} &= a + (n - 1)d \\ 98\text{th term} &= 2 + (98 - 1)4 \\ &= 2 + (97)4 \\ &= 390 \end{aligned}$$

Self Check Write the 50th term of the arithmetic sequence whose first three terms are 3, 8, and 13. ■

■ Arithmetic Means

Numbers inserted between a first and last term to form a segment of an arithmetic sequence are called **arithmetic means.** When finding arithmetic means, we consider the last term, a_n, to be the nth term:

$$a_n = a + (n - 1)d$$

ACTIVE EXAMPLE 3 Insert three arithmetic means between -3 and 12.

Solution Since we are inserting three arithmetic means between -3 and 12, the total number of terms is five. Thus, $a = -3$, $a_n = 12$, and $n = 5$. To find the common difference, we substitute -3 for a, 12 for a_n, and 5 for n in the formula for the last term and solve for d:

COLLEGE **Algebra** $f(x)$ **Now**™
Go to **http://1pass.thomson.com** or your CD to practice this example.

$$\begin{aligned} a_n &= a + (n - 1)d \\ 12 &= -3 + (5 - 1)d \\ 15 &= 4d && \text{Add 3 to both sides and simplify.} \\ \frac{15}{4} &= d && \text{Divide both sides by 4.} \end{aligned}$$

Once we know d, we can find the other terms of the sequence:

$$\begin{aligned} a + d &= -3 + \frac{15}{4} = \frac{3}{4} \\ a + 2d &= -3 + 2\left(\frac{15}{4}\right) = -3 + \frac{30}{4} = 4\frac{1}{2} \\ a + 3d &= -3 + 3\left(\frac{15}{4}\right) = -3 + \frac{45}{4} = 8\frac{1}{4} \end{aligned}$$

The three arithmetic means are $\frac{3}{4}$, $4\frac{1}{2}$, and $8\frac{1}{4}$.

Self Check Find three arithmetic means between -5 and 23. ■

Sum of the First n Terms of an Arithmetic Sequence

To find the sum of the first n terms of an arithmetic sequence, we use the following formula.

Sum of the First n Terms of an Arithmetic Sequence

The formula

$$S_n = \frac{n(a + a_n)}{2}$$

gives the sum of the first n terms of an arithmetic sequence. In this formula, a is the first term, a_n is the last (or nth) term, and n is the number of terms.

Proof We write the first n terms of an arithmetic sequence (letting S_n represent their sum). Then we write the same sum in reverse order and add the equations term by term:

$$\begin{array}{rcl} S_n &=& a + (a + d) + \cdots + [a + (n - 2)d] + [a + (n - 1)d] \\ S_n &=& [a + (n - 1)d] + [a + (n - 2)d] + \cdots + (a + d) + a \\ \hline 2S_n &=& [2a + (n - 1)d] + [2a + (n - 1)d] + \cdots + [2a + (n - 1)d] + [2a + (n - 1)d] \end{array}$$

Because there are n equal terms on the right-hand side of the previous equation,

$$2S_n = n[2a + (n - 1)d]$$

or

(1) $$2S_n = n\{a + [a + (n - 1)d]\} \quad \text{Write } 2a \text{ as } a + a.$$

We can substitute a_n for $a + (n - 1)d$ on the right-hand side of Equation 1 and divide both sides by 2 to get

$$S_n = \frac{n(a + a_n)}{2}$$ ■

ACTIVE EXAMPLE 4 Find the sum of the first 30 terms of the arithmetic sequence 5, 8, 11, ...

Solution Here $a = 5$, $n = 30$, $d = 3$, and $a_n = 5 + 29(3) = 92$. Substituting these values into the formula for the sum of an arithmetic sequence gives

$$S_n = \frac{n(a + a_n)}{2}$$

$$S_{30} = \frac{30(5 + 92)}{2}$$

$$= 15(97)$$

$$= 1{,}455$$

The sum of the first 30 terms is 1,455.

Self Check Find the sum of the first 50 terms of the arithmetic sequence −2, 5, 12, ... ■

Applications

EXAMPLE 5 A student deposits \$50 in a non-interest-bearing account and plans to add \$7 a week. How much will she have in the account one year after her first deposit?

Solution Her weekly balances form an arithmetic sequence:

50, 57, 64, 71, 78, . . .

with a first term of 50 and a common difference of 7. To find her balance in one year (52 weeks), we substitute 50 for a, 7 for d, and 52 for n in the formula for the last term.

$$\begin{aligned} a_n &= a + (n - 1)d \\ &= \mathbf{50} + (\mathbf{52} - 1)\mathbf{7} \\ &= 50 + (51)7 \\ &= 407 \end{aligned}$$

After one year, the balance will be \$407.

Self Check How much will she have in the account after 60 weeks? ■

EXAMPLE 6 The equation $s = 16t^2$ represents the distance s (in feet) that an object will fall in t seconds.

- In 1 second, the object will fall 16 feet.
- In 2 seconds, the object will fall 64 feet.
- In 3 seconds, the object will fall 144 feet.

The object fell 16 feet during the first second, 48 feet during the next second, and 80 feet during the third second. Find the distance the object will fall during the 12th second.

Solution The sequence 16, 48, 80, . . . is an arithmetic sequence with $a = 16$ and $d = 32$. To find the 12th term, we substitute these values into the formula for the last term.

$$\begin{aligned} a_n &= a + (n - 1)d \\ &= \mathbf{16} + (\mathbf{12} - 1)\mathbf{32} \\ &= 16 + 11(32) \\ &= 368 \end{aligned}$$

During the 12th second, the object falls 368 feet.

Self Check How far will the object fall during the 20th second? ■

Self Check Answers

1. 3, 9, 15, 21, 27; 105 **2.** 248 **3.** 2, 9, 16 **4.** 8,475 **5.** \$463 **6.** 624 ft

8.3 Exercises

VOCABULARY AND CONCEPTS ***Fill in the blanks.***

1. An arithmetic sequence is a sequence of the form

$a, a + d, a + 2d, a + 3d, \ldots a + \underline{(n-1)}d$

2. In an arithmetic sequence, a is the <u>first</u> term, d is the common <u>difference</u>, and n is the <u>number</u> of terms.

3. The last term of an arithmetic sequence is given by the formula $\underline{a_n = a + (n-1)d}$.

4. The formula for the sum of the first n terms of an arithmetic sequence is given by the formula $\underline{S = \frac{n(a + a_n)}{2}}$

5. <u>Arithmetic means</u> are numbers inserted between a first and last term of a sequence to form an arithmetic sequence.

6. The formula $\underline{s = 16t^2}$ gives the distance (in feet) that an object will fall in t seconds.

PRACTICE ***Write the first six terms of the arithmetic sequences with the given properties.***

7. $a = 1; d = 2$ 1, 3, 5, 7, 9, 11

8. $a = -12; d = -5$ $-12, -17, -22, -27, -32, -37$

9. $a = 5$; 3rd term is 2 $5, \frac{7}{2}, 2, \frac{1}{2}, -1, -\frac{5}{2}$

10. $a = 4$; 5th term is 12 4, 6, 8, 10, 12, 14

11. 7th term is 24; common difference is $\frac{5}{2}$
$9, \frac{23}{2}, 14, \frac{33}{2}, 19, \frac{43}{2}$

12. 20th term is -49; common difference is -3
$8, 5, 2, -1, -4, -7$

Find the missing term in each arithmetic sequence.

13. Find the 40th term of an arithmetic sequence with a first term of 6 and a common difference of 8. 318

14. Find the 35th term of an arithmetic sequence with a first term of 50 and a common difference of -6.
-154

15. The 6th term of an arithmetic sequence is 28, and the first term is -2. Find the common difference. 6

16. The 7th term of an arithmetic sequence is -42, and the common difference is -6. Find the first term.
-6

17. Find the 55th term of an arithmetic sequence whose first three terms are -8, -1, and 6. 370

18. Find the 37th term of an arithmetic sequence whose second and third terms are -4 and 6. 346

19. If the fifth term of an arithmetic sequence is 14 and the second term is 5, find the 15th term. 44

20. If the fourth term of an arithmetic sequence is 13 and the second term is 3, find the 24th term. 113

Find the required means.

21. Insert three arithmetic means between 10 and 20.
$\frac{25}{2}, 15, \frac{35}{2}$

22. Insert five arithmetic means between 5 and 15.
$\frac{20}{3}, \frac{25}{3}, 10, \frac{35}{3}, \frac{40}{3}$

23. Insert four arithmetic means between -7 and $\frac{2}{3}$.
$-\frac{82}{15}, -\frac{59}{15}, -\frac{12}{5}, -\frac{13}{15}$

24. Insert three arithmetic means between -11 and -2.
$-\frac{35}{4}, -\frac{13}{2}, -\frac{17}{4}$

Find the sum of the first n terms of each arithmetic sequence.

25. $5 + 7 + 9 + \cdots$ (to 15 terms) 285

26. $-3 + (-4) + (-5) + \cdots$ (to 10 terms) -75

27. $\sum_{n=1}^{20}\left(\frac{3}{2}n + 12\right)$ 555

28. $\sum_{n=1}^{10}\left(\frac{2}{3}n + \frac{1}{3}\right)$ 40

Solve each problem.

29. Find the sum of the first 30 terms of an arithmetic sequence with 25th term of 10 and a common difference of $\frac{1}{2}$. $157\frac{1}{2}$

30. Find the sum of the first 100 terms of an arithmetic sequence with 15th term of 86 and first term of 2.
29,900

31. Find the sum of the first 200 natural numbers.
20,100

32. Find the sum of the first 1,000 natural numbers.
500,500

APPLICATIONS

33. Interior angles The sums of the angles of several polygons are given in the table. Assuming that the pattern continues, complete the table.

Figure	Number of sides	Sum of angles
Triangle	3	180°
Quadrilateral	4	360°
Pentagon	5	540°
Hexagon	6	720°
Octagon	8	**1,080°**
Dodecagon	12	**1,800°**

34. Borrowing money To pay for college, a student borrows \$5,000 interest-free from his father. If he pays his father back at the rate of \$200 per month, how much will he still owe after 12 months? **\$2,600**

35. Borrowing money If Juanita borrows \$5,500 interest-free from her mother to buy a new car and agrees to pay her mother back at the rate of \$105 per month, how much will she still owe after 4 years? **\$460**

36. Jogging One day, some students jogged $\frac{1}{2}$ mile. Because it was fun, they decided to increase the jogging distance each day by a certain amount. If they jogged $6\frac{3}{4}$ miles on the 51st day, how much was the distance increased each day? **$\frac{1}{8}$ mi**

37. Sales The year it incorporated, a company had sales of \$237,500. Its sales were expected to increase by \$150,000 annually for the next several years. If the forecast was correct, what will sales be in 10 years? **\$1,587,500**

38. Falling objects Find how many feet a brick will travel during the 10th second of its fall. **304 ft**

39. Falling objects If a rock is dropped from the Golden Gate Bridge, how far will it fall in the third second? **80 ft**

40. Designing patios Each row of bricks in the following triangular patio is to have one more brick than the previous row, ending with the longest row of 150 bricks. How many bricks will be needed? **11,325**

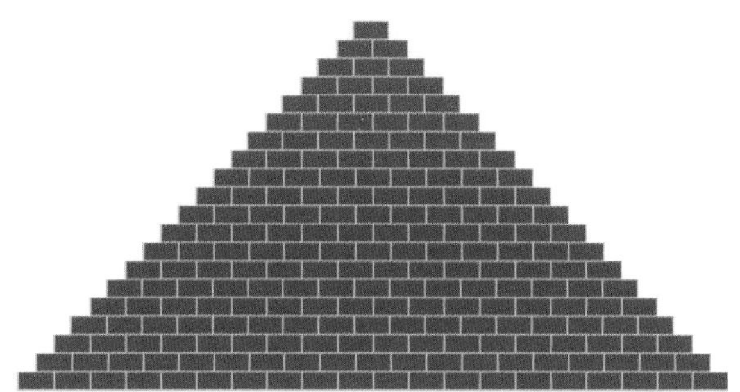

41. Pile of logs Several logs are stored in a pile with 20 logs on the bottom layer, 19 on the second layer, 18 on the third layer, and so on. If the top layer has one log, how many logs are in the pile? **210**

42. Theater seating The first row in a movie theater contains 24 seats. As you move toward the back, each row has 1 additional seat. If there are 30 rows, what is the capacity of the theater? **1,155**

DISCOVERY AND WRITING

43. Can an arithmetic sequence have a first term of 4, a 25th term of 126, and a common difference of $4\frac{1}{4}$? Explain.

44. In an arithmetic sequence, can a and d be negative, but a_n positive?

45. Can an arithmetic sequence be an alternating sequence? Explain.

46. Between 5 and $10\frac{1}{3}$ are three arithmetic means. One of them is 9. Find the other two. **$6\frac{1}{3}, 7\frac{2}{3}$**

REVIEW *Solve each equation.*

47. $x + \sqrt{x + 3} = 9$
6

48. $\dfrac{x + 3}{x} + \dfrac{x - 3}{5} = 2$
3, 5

49. $\dfrac{1}{x} + \dfrac{2}{x^2} + \dfrac{1}{x^3} = 0$
−1

50. $x + 3\sqrt{x} - 10 = 0$
4

51. $x^4 - 1 = 0$
$1, -1, i, -i$

52. $x^4 - 29x^2 + 100 = 0$
2, −2, 5, −5

8.4 Geometric Sequences

In this section, you will learn about

- Geometric Sequences
- Geometric Means
- Sum of the First *n* Terms of a Geometric Sequence
- Infinite Geometric Sequences
- Applications

Another common sequence is the *geometric sequence.*

Geometric Sequences

A **geometric sequence** is a sequence in which each term, except the first, is found by multiplying the preceding term by a constant.

Geometric Sequences

A **geometric sequence** is a sequence of the form

$$a, \quad ar, \quad ar^2, \quad ar^3, \ldots, ar^{n-1}, \ldots$$

where a is the **first term,** ar^{n-1} is the ***n*th term,** and r is the **common ratio.**

Comment The nth term of a geometric sequence is given by

$$a_n = ar^{n-1}$$

In this definition, the second term of the sequence has a factor of r^1, the third term has a factor of r^2, the fourth term has a factor of r^3, and so on. This explains why the nth term has a factor of r^{n-1}.

EXAMPLE 1 Write the first six terms and the 15th term of the geometric sequence whose first term is 3 and whose common ratio is 2.

Solution We first write the first six terms of the geometric sequence:

$$3, \quad 3(2), \quad 3(2)^2, \quad 3(2)^3, \quad 3(2)^4, \quad 3(2)^5$$

or

$$3, \quad 6, \quad 12, \quad 24, \quad 48, \quad 96$$

To find the 15th term, we substitute 15 for n, 3 for a, and 2 for r in the formula for the nth term:

$$\begin{aligned} n\text{th term} &= ar^{n-1} \\ 15\text{th term} &= 3(2)^{15-1} \\ &= 3(2)^{14} \\ &= 3(16{,}384) \\ &= 49{,}152 \end{aligned}$$

Self Check Write the first five terms of the geometric sequence whose first term is 2 and whose common ratio is 3. Find the 10th term. ■

EXAMPLE 2 Find the eighth term of a geometric sequence whose first three terms are 9, 3, and 1.

Solution Here $a = 9$, $r = \frac{1}{3}$, and $n = 8$. To find the eighth term, we substitute these values into the formula for the nth term.

$$n\text{th term} = ar^{n-1}$$
$$8\text{th term} = 9\left(\frac{1}{3}\right)^{8-1}$$
$$= 9\left(\frac{1}{3}\right)^{7}$$
$$= \frac{1}{243}$$

Self Check Find the eighth term of a geometric sequence whose first three terms are $\frac{1}{3}$, 1, and 3. ■

Geometric Means

Numbers inserted between a first and last term to form a segment of a geometric sequence are called **geometric means.** When finding geometric means, we consider the last term, a_n, to be the nth term.

ACTIVE EXAMPLE 3 Insert two geometric means between 4 and 256.

Solution The first term is $a = 4$, and because 256 is the fourth term, $n = 4$ and $a_n = 256$. To find the common ratio, we substitute these values into the formula for the nth term and solve for r:

COLLEGE **Algebra f(x) Now™**
Explore this Active Example and test yourself on the steps taken here at **http://1pass.thomson.com**. You can also find this example on the Interactive Video Skillbuilder CD-ROM.

$$ar^{n-1} = a_n$$
$$4r^{4-1} = 256$$
$$r^3 = 64$$
$$r = 4$$

The common ratio is 4. The two geometric means are the second and third terms of the geometric sequence:

$$ar = 4 \cdot 4 = 16$$
$$ar^2 = 4 \cdot 4^2 = 4 \cdot 16 = 64$$

The first four terms of the geometric sequence are 4, 16, 64, and 256. The two geometric means between 4 and 256 are 16 and 64.

Self Check Insert two geometric means between -3 and 192. ■

Sum of the First *n* Terms of a Geometric Sequence

To find the sum of the first n terms of a geometric sequence, we use the following formula.

Sum of the First *n* Terms of a Geometric Sequence

The formula

$$S_n = \frac{a - ar^n}{1 - r} \qquad (r \neq 1)$$

gives the sum of the first n terms of a geometric sequence. In the formula, S_n is the sum, a is the first term, r is the common ratio, and n is the number of terms.

Proof We write the sum of the first n terms of the geometric sequence:

(1) $$S_n = a + ar + ar^2 + \cdots + ar^{n-3} + ar^{n-2} + ar^{n-1}$$

Multiplying both sides of this equation by r gives

(2) $$S_n r = ar + ar^2 + \cdots + ar^{n-2} + ar^{n-1} + ar^n$$

We now subtract Equation 2 from Equation 1 and solve for S_n:

$$S_n - S_n r = a - ar^n$$

$$S_n(1 - r) = a - ar^n \quad \text{Factor out } S_n.$$

$$S_n = \frac{a - ar^n}{1 - r} \quad \text{Divide both sides by } 1 - r.$$

■

ACTIVE EXAMPLE 4 Find the sum of the first six terms of the geometric sequence 8, 4, 2, ...

Solution Here $a = 8$, $n = 6$, and $r = \frac{1}{2}$. Substituting these values in the formula for the sum of the first n terms of a geometric sequence gives

COLLEGE **Algebra** $f(x)$ **Now**™
Go to **http://1pass.thomson.com** or your CD to practice this example.

$$S_n = \frac{a - ar^n}{1 - r}$$

$$S_6 = \frac{8 - 8\left(\frac{1}{2}\right)^6}{1 - \frac{1}{2}}$$

$$= 2\left(\frac{63}{8}\right)$$

$$= \frac{63}{4}$$

The sum of the first six terms is $\frac{63}{4}$.

Self Check Find the sum of the first eight terms of the geometric sequence 81, 27, 9, ...

■

■ Infinite Geometric Sequences

Under certain conditions, we can find the sum of all of the terms in an **infinite geometric sequence.** To define this sum, we consider the geometric sequence

$$a, ar, ar^2, \ldots$$

- The first partial sum, S_1, of the sequence is $S_1 = a$.
- The second partial sum, S_2, of the sequence is $S_2 = a + ar$.
- The nth partial sum, S_n, of the sequence is $S_n = a + ar + ar^2 + \cdots + ar^{n-1}$.

If the nth partial sum, S_n, of an infinite geometric sequence approaches some number S as n approaches ∞, then S is called the **sum of the infinite geometric sequence.** The following symbol denotes the sum, S, of an infinite geometric sequence, provided the sum exists.

$$S = \sum_{n=1}^{\infty} ar^{n-1}$$

To develop a formula for finding the sum of all the terms in an infinite geometric sequence, we consider the formula

$$(3) \qquad S_n = \frac{a - ar^n}{1 - r} \qquad (r \neq 1)$$

If $|r| < 1$ and a is a constant, then as n approaches ∞, ar^n approaches 0, and the term ar^n in Equation 3 can be dropped. This argument gives the following formula.

Sum of the Terms of an Infinite Geometric Sequence

If $|r| < 1$, the sum of the terms of an infinite geometric sequence is given by

$$S_\infty = \frac{a}{1 - r}$$

where a is the first term and r is the common ratio.

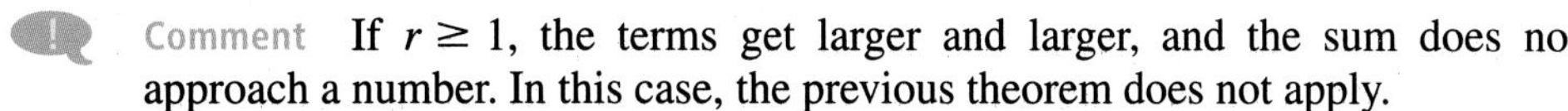

Comment If $r \geq 1$, the terms get larger and larger, and the sum does not approach a number. In this case, the previous theorem does not apply.

ACTIVE EXAMPLE 5 Change $0.\overline{4}$ to a common fraction.

Solution We write the decimal as an infinite geometric series and find its sum:

COLLEGE **Algebra** *f(x)* **Now™**
Go to http://1pass.thomson.com or your CD to practice this example.

$$\begin{aligned} S_\infty &= \frac{4}{10} + \frac{4}{100} + \frac{4}{1{,}000} + \frac{4}{10{,}000} + \cdots \\ &= \frac{4}{10} + \frac{4}{10}\left(\frac{1}{10}\right) + \frac{4}{10}\left(\frac{1}{10}\right)^2 + \frac{4}{10}\left(\frac{1}{10}\right)^3 + \cdots \end{aligned}$$

Since the common ratio r equals $\frac{1}{10}$ and $\left|\frac{1}{10}\right| < 1$, we can use the formula for the sum of an infinite geometric series:

$$S_\infty = \frac{a}{1 - r} = \frac{\frac{4}{10}}{1 - \frac{1}{10}} = \frac{\frac{4}{10}}{\frac{9}{10}} = \frac{4}{9}$$

Comment Using a generalization of this method, we can write any repeating decimal as a fraction.

Long division will verify that $\frac{4}{9} = 0.\overline{4}$.

Self Check Change $0.\overline{7}$ to a common fraction. ■

■ Applications

Many of the exponential growth problems discussed in Chapter 4 can be solved using the concepts of geometric sequences.

EXAMPLE 6 A statistician knows that a town with a population of 3,500 people has a predicted growth rate of 6% per year for the next 20 years. What should she predict the population to be 20 years from now?

Solution Let p_0 be the initial population of the town. After 1 year, the population p_1 will be the initial population (p_0) plus the growth (the product of p_0 and the rate of growth, r).

$$\begin{aligned} p_1 &= p_0 + p_0 r \\ &= p_0(1 + r) \qquad \textbf{Factor out } p_0. \end{aligned}$$

The population p_2 at the end of 2 years will be

$$\begin{aligned} p_2 &= p_1 + p_1 r \\ &= p_1(1 + r) && \textbf{Factor out } p_1. \\ &= p_0(1 + r)(1 + r) && \textbf{Substitute } p_0(1 + r) \textbf{ for } p_1. \\ &= p_0(1 + r)^2 \end{aligned}$$

The population at the end of the third year will be $p_3 = p_0(1 + r)^3$. Writing the terms in a sequence gives

$$p_0, \quad p_0(1 + r), \quad p_0(1 + r)^2, \quad p_0(1 + r)^3, \quad p_0(1 + r)^4, \ldots$$

This is a geometric sequence with p_0 as the first term and $1 + r$ as the common ratio. In this example, $p_0 = 3{,}500$, $1 + r = 1.06$, and (since the population after 20 years will be the value of the 21st term of the geometric sequence) $n = 21$. We can substitute these values into the formula for the last term of a geometric sequence to get

$$\begin{aligned} a_n &= ar^{n-1} \\ &= 3{,}500(1.06)^{21-1} \\ &= 3{,}500(1.06)^{20} \\ &\approx 11{,}224.97415 && \textbf{Use a calculator.} \end{aligned}$$

The population after 20 years will be approximately 11,225. ■

EXAMPLE 7 A student deposits $2,500 in a bank at 7% annual interest, compounded daily. If the investment is left untouched for 60 years, how much money will be in the account?

Solution We let the initial amount in the account be a_0 and r be the rate. At the end of the first day, the account is worth

$$a_1 = a_0 + a_0\left(\frac{r}{365}\right) = a_0\left(1 + \frac{r}{365}\right)$$

After the second day, the account is worth

$$a_2 = a_1 + a_1\left(\frac{r}{365}\right) = a_1\left(1 + \frac{r}{365}\right) = a_0\left(1 + \frac{r}{365}\right)^2$$

The daily amounts form the following geometric sequence

$$a_0, \quad a_0\left(1 + \frac{r}{365}\right), \quad a_0\left(1 + \frac{r}{365}\right)^2, \quad a_0\left(1 + \frac{r}{365}\right)^3, \ldots$$

where a_0 is the initial deposit and r is the annual rate of interest.

Because interest is compounded daily for 60 years (21,900 days), the amount at the end of 60 years will be the 21,901th term of the sequence.

$$a_{21,901} = 2,500\left(1 + \frac{0.07}{365}\right)^{21,900}$$

We can use a calculator to find that $a_{21,901} \approx \$166,648.71$. ■

EXAMPLE 8 A pump can remove 20% of the gas in a container with each stroke. Find the percentage of gas that remains in the container after six strokes.

Solution We let V represent the volume of the container. Because each stroke of the pump removes 20% of the gas, 80% remains after each stroke, and we have the geometric sequence

$$V, \quad 0.80V, \quad 0.80(0.80V), \quad 0.80[0.80(0.80V)], \ldots$$

or

$$V, \quad 0.80V, \quad (0.8)^2V, \quad (0.8)^3V, \quad (0.8)^4V, \ldots$$

The amount of gas remaining after six strokes is the seventh term, a_n, of the sequence:

$$\begin{aligned} a_n &= ar^{n-1} \\ &= V(\mathbf{0.8})^{\mathbf{7}-1} \\ &= V(0.8)^6 \end{aligned}$$

We can use a calculator to find that approximately 26% of the gas remains after six strokes. ■

Self Check Answers

1. 2, 6, 18, 54, 162; 39,366 **2.** 729 **3.** 12, −48 **4.** $\frac{3,280}{27}$ **5.** $\frac{7}{9}$

8.4 Exercises

VOCABULARY AND CONCEPTS ***Fill in the blanks.***

1. A geometric sequence is a sequence of the form $a, ar, ar^2, ar^3, \ldots$. The nth term is $a(\underline{r^{n-1}})$.

2. In a geometric sequence, a is the first term, r is the common ratio, and n is the number of terms.

3. The last term of a geometric sequence is given by the formula $a_n = \underline{ar^{n-1}}$

4. The formula for the sum of the first n terms of a geometric sequence is given by

$$S_n = \frac{a - ar^n}{1 - r} \quad (r \neq 1)$$

5. Geometric means are numbers inserted between a first and a last term to form a geometric sequence.

6. If $|r| < 1$, the formula $S_\infty = \frac{a}{1-r}$ gives the sum of the terms of an infinite geometric sequence.

PRACTICE ***Write the first four terms of each geometric sequence with the given properties.***

7. $a = 10; r = 2$
10, 20, 40, 80

8. $a = -3; r = 2$
−3, −6, −12, −24

9. $a = -2$ and $r = 3$
−2, −6, −18, −54

10. $a = 64; r = \frac{1}{2}$
64, 32, 16, 8

11. $a = 3; r = \sqrt{2}$ $3, 3\sqrt{2}, 6, 6\sqrt{2}$

12. $a = 2; r = \sqrt{3}$ $2, 2\sqrt{3}, 6, 6\sqrt{3}$

13. $a = 2$; 4th term is 54 2, 6, 18, 54

14. 3rd term is 4; $r = \frac{1}{2}$ 16, 8, 4, 2

Find the requested term of each geometric sequence.

15. Find the sixth term of the geometric sequence whose first three terms are $\frac{1}{4}$, 1, and 4. 256

16. Find the eighth term of the geometric sequence whose second and fourth terms are 0.2 and 5. 3,125

17. Find the fifth term of a geometric sequence whose second term is 6 and whose third term is -18. -162

18. Find the sixth term of a geometric sequence whose second term is 3 and whose fourth term is $\frac{1}{3}$. $\frac{1}{27}$

Solve each problem.

19. Insert three positive geometric means between 10 and 20. $10\sqrt[4]{2}, 10\sqrt{2}, 10\sqrt[4]{8}$

20. Insert five geometric means between -5 and 5, if possible. not possible

21. Insert four geometric means between 2 and 2,048. 8, 32, 128, 512

22. Insert three geometric means between 162 and 2. (There are two possibilities.) 54, 18, 6, or -54, 18, -6

Find the sum of the indicated terms of each geometric sequence.

23. 4, 8, 16, . . . (to 5 terms) 124

24. 9, 27, 81, . . . (to 6 terms) 3,276

25. 2, -6, 18, . . . (to 10 terms) $-29{,}524$

26. $\frac{1}{8}, \frac{1}{4}, \frac{1}{2}, \ldots$ (to 12 terms) $\frac{4{,}095}{8}$

27. $\sum_{n=1}^{6} 3\left(\frac{3}{2}\right)^{n-1}$ $\frac{1{,}995}{32}$

28. $\sum_{n=1}^{6} 12\left(-\frac{1}{2}\right)^{n-1}$ $\frac{63}{8}$

Find the sum of each infinite geometric sequence.

29. $6 + 4 + \frac{8}{3} + \cdots$ 18

30. $8 + 4 + 2 + 1 + \cdots$ 16

31. $\sum_{n=1}^{\infty} 12\left(-\frac{1}{2}\right)^{n-1}$ 8

32. $\sum_{n=1}^{\infty} \left(\frac{1}{3}\right)^{n-1}$ $\frac{3}{2}$

Change each decimal to a common fraction.

33. $0.\overline{5}$ $\frac{5}{9}$

34. $0.\overline{6}$ $\frac{2}{3}$

35. $0.\overline{25}$ $\frac{25}{99}$

36. $0.\overline{37}$ $\frac{37}{99}$

APPLICATIONS ***Use a calculator to help solve each problem.***

37. Staffing a department The number of students studying algebra at State College is 623. The department chair expects enrollment to increase 10% each year. How many professors will be needed in 8 years to teach algebra if one professor can handle 60 students? 23

38. Bouncing balls On each bounce, the rubber ball in the illustration rebounds to a height one-half of that from which it fell. Find the total vertical distance the ball travels. 30 m

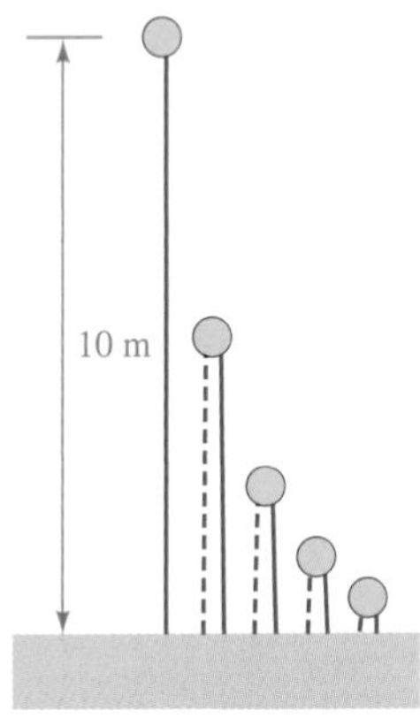

39. Bouncing balls A Super Ball rebounds to approximately 95% of the height from which it is dropped. If the ball is dropped from a height of 10 meters, how high will it rebound after the 13th bounce? 5.13 m

40. Genealogy The following family tree spans 3 generations and lists 7 people. How many names would be listed in a family tree that spans 10 generations? $2^{10} - 1 = 1{,}023$

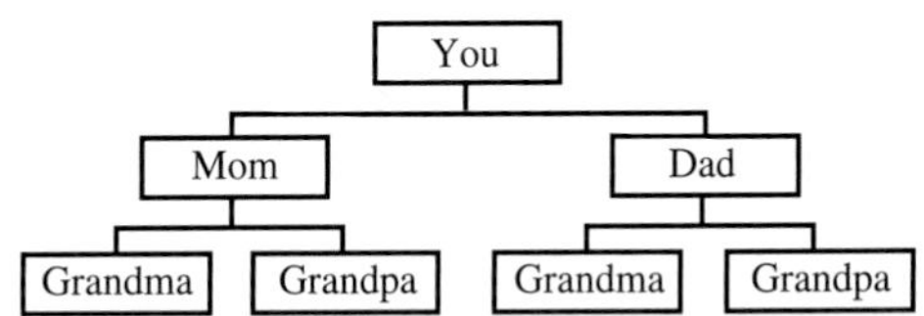

41. Investing money If a married couple invests \$1,000 in a 1-year certificate of deposit at $6\frac{3}{4}\%$ annual interest, compounded daily, how much interest will be earned during the year? \$69.82

42. Biology If a single cell divides into two cells every 30 minutes, how many cells will there be at the end of 10 hours? 1,048,576

43. Depreciation A lawn tractor, costing c dollars when new, depreciates 20% of the previous year's value each year. How much is the lawn tractor worth after 5 years? about $\frac{1}{3}c$

44. Financial planning Maria can invest $1,000 at $7\frac{1}{2}$%, compounded annually, or at $7\frac{1}{4}$%, compounded daily. If she invests the money for a year, which is the better investment? **the $7\frac{1}{4}$% investment**

45. Population study If the population of the Earth were to double every 30 years, approximately how many people would there be in the year 3010? (Consider the population in 1990 to be 5 billion and use 1990 as the base year.) **about 8.6×10^{19}**

46. Investing money If Linda deposits $1,300 in a bank at 7% interest, compounded annually, how much will be in the bank 17 years later? (Assume that there are no other transactions on the account.) **$4,106.46**

47. Real estate appreciation If a house purchased for $50,000 in 1988 appreciates in value by 6% each year, how much will the house be worth in the year 2010? **$180,176.87**

48. Compound interest Find the value of $1,000 left on deposit for 10 years at an annual rate of 7%, compounded annually. **$1,967.15**

49. Compound interest Find the value of $1,000 left on deposit for 10 years at an annual rate of 7%, compounded quarterly. **$2,001.60**

50. Compound interest Find the value of $1,000 left on deposit for 10 years at an annual rate of 7%, compounded monthly. **$2,009.66**

51. Compound interest Find the value of $1,000 left on deposit for 10 years at an annual rate of 7%, compounded daily. **$2,013.62**

52. Compound interest Find the value of $1,000 left on deposit for 10 years at an annual rate of 7%, compounded hourly. **$2,013.75**

53. Saving for retirement When John was 20 years old, he opened an individual retirement account by investing $2,000 at 11% interest, compounded quarterly. How much will his investment be worth when he is 65 years old? **$264,094.58**

54. Biology One bacterium divides into two bacteria every 5 minutes. If two bacteria multiply enough to completely fill a petri dish in 2 hours, how long will it take one bacterium to fill the dish? **2 hr and 5 min**

55. Pest control To reduce the population of a destructive moth, biologists release 1,000 sterilized male moths each day into the environment. If 80% of these moths alive one day survive until the next, then after a long time the population of sterile males is the sum of the infinite geometric series

$$1{,}000 + 1{,}000(0.8) + 1{,}000(0.8)^2 + 1{,}000(0.8)^3 + \cdots$$

Find the long-term population. **5,000**

56. Pest control If mild weather increases the day-to-day survival rate of the sterile male moths in Exercise 55 to 90%, find the long-term population. **10,000**

57. Mathematical myth A legend tells of a king who offered to grant the inventor of the game of chess any request. The inventor said, "Simply place one grain of wheat on the first square of a chessboard, two grains on the second, four on the third, and so on, until the board is full. Then give me the wheat." The king agreed. How many grains did the king need to fill the chessboard? **1.8447×10^{19} grains**

58. Mathematical myth Estimate the size of the wheat pile in Exercise 57. (*Hint:* There are about one-half million grains of wheat in a bushel.) **3.689×10^{13} bushels**

DISCOVERY AND WRITING

59. Does 0.999999 = 1? Explain. **no**

60. Does 0.999 . . . = 1? Explain. **yes**

REVIEW *Perform each operation.* $(i = \sqrt{-1}.)$

61. $(3 + 2i) + (2 - 5i)$ $5 - 3i$

62. $(7 + 8i) - (2 - 5i)$ $5 + 13i$

63. $(3 + i)(3 - i)$ $10 + 0i$

64. $(3 + \sqrt{3}i)(3 - \sqrt{3}i)$ $12 + 0i$

65. $\dfrac{2 + 3i}{2 - i}$ $\frac{1}{5} + \frac{8}{5}i$

66. $(7 + 3i)^2$ $40 + 42i$

67. i^{127} $-i$

68. i^{-127} i

8.5 Mathematical Induction

In this section, you will learn about

- **The Method of Mathematical Induction**

In Section 8.4, we developed the following formula for the sum of the first n terms of an arithmetic sequence:

$$S_n = \frac{n(a + a_n)}{2}$$

If we apply this formula to the arithmetic series $1 + 2 + 3 + \cdots + n$, we have

$$S_n = 1 + 2 + 3 + \cdots + n = \frac{n(1 + n)}{2}$$

To see that this formula is true, we can check it for some positive numbers n:

For $n = 1$: $\mathbf{1} = \frac{\mathbf{1(1 + 1)}}{2}$ is a true statement, because $1 = 1$.

For $n = 2$: $1 + \mathbf{2} = \frac{\mathbf{2(1 + 2)}}{2}$ is a true statement, because $3 = 3$.

For $n = 3$: $1 + 2 + \mathbf{3} = \frac{\mathbf{3(1 + 3)}}{2}$ is a true statement, because $6 = 6$.

For $n = 6$: $1 + 2 + 3 + 4 + 5 + \mathbf{6} = \frac{\mathbf{6(1 + 6)}}{2}$ is a true statement, because $21 = 21$.

However, because the set of positive numbers is infinite, it is impossible to prove the formula by verifying it for all positive numbers. To verify this sequence formula for all positive numbers n, we must have a method of proof called **mathematical induction,** a method first used extensively by Giuseppe Peano (1858–1932).

The Method of Mathematical Induction

Suppose that we are standing in line for a movie and are worrying about whether we will be admitted. Even if the first person gets in, our worries would still be justified. Perhaps there is only room in the theater for a few people.

It would be good news to hear the theater manager say, "If anyone gets in, the next person in line will get in also." However, this promise does not guarantee that anyone will be admitted. Perhaps the theater is already full, and no one will get in.

However, when we see the first person in line walk in, we can conclude that everyone will be admitted, because we know two things:

- The first person was admitted.
- Because of the promise, if the first person is admitted, then so is the second, and when the second person is admitted, then so is the third, and so on until everyone gets in.

Figure 8-1

This situation is similar to a game played with dominoes. Suppose that some dominoes are placed on end, as in Figure 8-1. When the first domino is knocked over, it knocks over the second. The second domino, in turn, knocks over the third, which knocks over the fourth, and so on until all of the dominoes fall. Two things must happen to guarantee that all of the dominoes fall:

- The first domino must be knocked over.
- Every domino that falls must knock over the next one.

When both conditions are met, it is certain that all of the dominoes will fall.

The preceding examples illustrate the basic principle of mathematical induction.

Mathematical Induction

If a statement involving the natural number n has the following two properties

1. The statement is true for $n = 1$, and
2. If the statement is true for $n = k$, then it is true for $n = k + 1$, then the statement is true for all natural numbers.

Mathematical induction provides a way to prove many formulas. Any proof by induction involves two parts. First, we must show that the formula is true for the number 1. Second, we must show that, if the formula is true for any natural number k, then it also is true for the natural number $k + 1$. A proof by induction is complete only when both of these properties are established.

ACTIVE EXAMPLE 1 Use induction to prove that the following formula is true for every natural number n.

$$1 + 2 + 3 + \cdots + n = \frac{n(n + 1)}{2}$$

Solution **Part 1:** Verify that the formula is true for $n = 1$. When $n = 1$, there is a single term, the number 1, on the left-hand side of the equation. Substituting 1 for n on the right-hand side, we have

$$1 = \frac{n(n + 1)}{2}$$

$$1 = \frac{(1)(1 + 1)}{2}$$

$$1 = 1$$

The formula is true when $n = 1$. Part 1 of the proof is complete.

COLLEGE **Algebra f(x) Now™**
Explore this Active Example and test yourself on the steps taken here at **http://1pass.thomson.com**. You can also find this example on the Interactive Video Skillbuilder CD-ROM.

Part 2: We assume that the given formula is true when $n = k$. By this assumption, called the **induction hypothesis,** we accept that

$$(1) \qquad 1 + 2 + 3 + \cdots + k = \frac{k(k + 1)}{2}$$

is a true statement. We must show that the induction hypothesis forces the given formula to be true when $n = k + 1$. We can show this by verifying the statement

$$(2) \qquad 1 + 2 + 3 + \cdots + k + (k + 1) = \frac{(k + 1)[(k + 1) + 1]}{2}$$

which is obtained from the given formula by replacing n with $k + 1$.

Comparing the left-hand sides of Equations 1 and 2 shows that the left-hand side of Equation 2 contains an extra term of $k + 1$. Thus, we add $k + 1$ to both sides of Equation 1 (which was assumed to be true) to obtain the equation

$$1 + 2 + 3 + \cdots + k + (k + 1) = \frac{k(k + 1)}{2} + (k + 1)$$

Because both terms on the right-hand side of this equation have a common factor of $k + 1$, the right-hand side factors, and the equation can be written as follows:

$$\begin{aligned} 1 + 2 + 3 + \cdots + k + (k + 1) &= (k + 1)\left(\frac{k}{2} + 1\right) \\ &= (k + 1)\left(\frac{k + 2}{2}\right) \\ &= \frac{(k + 1)(k + 2)}{2} \\ &= \frac{(k + 1)[(k + 1) + 1]}{2} \end{aligned}$$

This final result is Equation 2. Because the truth of Equation 1 implies the truth of Equation 2, part 2 of the proof is complete. Parts 1 and 2 together establish that the formula is true for any natural number n. ■

ACTIVE EXAMPLE 2 Use induction to prove the following formula for all natural numbers n.

$$1 + 5 + 9 + \cdots + (4n - 3) = n(2n - 1)$$

Solution **Part 1:** First we verify the formula for $n = 1$. When $n = 1$, there is a single term, the number 1, on the left-hand side of the equation. Substituting 1 for n on the right-hand side, we have

COLLEGE **Algebra** *f(x)* **Now**™
Go to http://1pass.thomson.com or your CD to practice this example.

$$\begin{aligned} 1 &= 1[2(1) - 1] \\ 1 &= 1 \end{aligned}$$

The formula is true for $n = 1$. Part 1 of the proof is complete.

Part 2: We assume that the formula is true for $n = k$. Hence,

$$1 + 5 + 9 + \cdots + (4k - 3) = k(2k - 1) \quad (3)$$

is a true statement. To show that the induction hypothesis guarantees the truth of the formula for $k + 1$ terms, we add the $(k + 1)$th term to both sides of Equation 3. Because the terms on the left-hand side increase by 4, the $(k + 1)$th term is $(4k - 3) + 4$, or $4k + 1$. Adding $4k + 1$ to both sides of Equation 3 gives

$$1 + 5 + 9 + \cdots + (4k - 3) + (4k + 1) = k(2k - 1) + (4k + 1)$$

We can simplify the right-hand side and write the previous equation as follows:

$$\begin{aligned} 1 + 5 + 9 + \cdots + (4k - 3) + [4(k + 1) - 3] &= 2k^2 + 3k + 1 \\ &= (k + 1)(2k + 1) \\ &= (k + 1)[2(k + 1) - 1] \end{aligned}$$

Since this result has the same form as the given formula, except that $k + 1$ replaces n, the truth of the formula for $n = k$ implies the truth of the formula for $n = k + 1$. Part 2 of the proof is complete.

Because both of the induction requirements are true, the formula is true for all natural numbers n. ■

ACTIVE EXAMPLE 3 Prove that $\frac{1}{2} + \frac{1}{4} + \frac{1}{8} + \cdots + \frac{1}{2^n} < 1$.

Solution **Part 1:** We verify the formula for $n = 1$. When $n = 1$, there is a single term, the fraction $\frac{1}{2}$, on the left-hand side of the equation. Substituting 1 for n on the right-hand side, we have the following true statement:

COLLEGE Algebra f(x) Now™
Go to http://1pass.thomson.com or your CD to practice this example.

$$\frac{1}{2} < 1$$

The formula is true for $n = 1$. Part 1 of the proof is complete.

Part 2: We assume that the inequality is true for $n = k$. Thus,

$$\frac{1}{2} + \frac{1}{4} + \frac{1}{8} + \cdots + \frac{1}{2^k} < 1$$

We can multiply both sides of the above inequality by $\frac{1}{2}$ to get

$$\frac{1}{2}\left(\frac{1}{2} + \frac{1}{4} + \frac{1}{8} + \cdots + \frac{1}{2^k}\right) < 1\left(\frac{1}{2}\right)$$

or

$$\frac{1}{4} + \frac{1}{8} + \frac{1}{16} + \cdots + \frac{1}{2^{k+1}} < \frac{1}{2}$$

We now add $\frac{1}{2}$ to both sides of this inequality to get

$$\frac{1}{2} + \frac{1}{4} + \frac{1}{8} + \frac{1}{16} + \cdots + \frac{1}{2^{k+1}} < \frac{1}{2} + \frac{1}{2}$$

or

$$\frac{1}{2} + \frac{1}{4} + \frac{1}{8} + \frac{1}{16} + \cdots + \frac{1}{2^{k+1}} < 1$$

The resulting inequality is the same as the original inequality, except that $k + 1$ appears in place of n. Thus, the truth of the inequality for $n = k$ implies the truth of the inequality for $n = k + 1$. Part 2 of the proof is complete.

Because both of the induction requirements have been verified, this inequality is true for all natural numbers. ■

Some statements are not true when $n = 1$ but are true for all natural numbers equal to or greater than some given natural number (say, q). In these cases, we verify the given statements for $n = q$ in part 1 of the induction proof. After establishing part 2 of the induction proof, the given statement is proved for all natural numbers that are greater than q.

8.5 Exercises

VOCABULARY AND CONCEPTS *Fill in the blanks.*

1. Any proof by induction requires two parts.
2. Part 1 is to show that the statement is true for $n = 1$.
3. Part 2 is to show that the statement is true for $n = k + 1$ whenever it is true for $n = k$.
4. When we assume that a formula is true for $n = k$, we call the assumption the induction hypothesis.

PRACTICE *Verify each formula for n = 1, 2, 3, and 4.*

5. $5 + 10 + 15 + \cdots + 5n = \frac{5n(n + 1)}{2}$
6. $1^2 + 2^2 + 3^2 + \cdots + n^2 = \frac{n(n + 1)(2n + 1)}{6}$
7. $7 + 10 + 13 + \cdots + (3n + 4) = \frac{n(3n + 11)}{2}$
8. $1(3) + 2(4) + 3(5) + \cdots + n(n + 2) = \frac{n}{6}(n + 1)(2n + 7)$

Prove each formula by induction, if possible.

9. $2 + 4 + 6 + \cdots + 2n = n(n + 1)$
10. $1 + 3 + 5 + \cdots + (2n - 1) = n^2$
11. $3 + 7 + 11 + \cdots + (4n - 1) = n(2n + 1)$
12. $4 + 8 + 12 + \cdots + 4n = 2n(n + 1)$
13. $10 + 6 + 2 + \cdots + (14 - 4n) = 12n - 2n^2$
14. $8 + 6 + 4 + \cdots + (10 - 2n) = 9n - n^2$
15. $2 + 5 + 8 + \cdots + (3n - 1) = \frac{n(3n + 1)}{2}$
16. $3 + 6 + 9 + \cdots + 3n = \frac{3n(n + 1)}{2}$
17. $1^2 + 2^2 + 3^2 + \cdots + n^2 = \frac{n(n + 1)(2n + 1)}{6}$
18. $1 + 2 + 3 + \cdots + (n - 1) + n + (n - 1) + \cdots + 3 + 2 + 1 = n^2$
19. $\frac{1}{3} + 2 + \frac{11}{3} + \cdots + \left(\frac{5}{3}n - \frac{4}{3}\right) = n\left(\frac{5}{6}n - \frac{1}{2}\right)$
20. $\frac{1}{1 \cdot 2} + \frac{1}{2 \cdot 3} + \frac{1}{3 \cdot 4} + \cdots + \frac{1}{n(n + 1)} = \frac{n}{n + 1}$
21. $\frac{1}{2} + \frac{1}{4} + \frac{1}{8} + \cdots + \left(\frac{1}{2}\right)^n = 1 - \left(\frac{1}{2}\right)^n$
22. $\frac{1}{3} + \frac{2}{9} + \frac{4}{27} + \cdots + \frac{1}{3}\left(\frac{2}{3}\right)^{n-1} = 1 - \left(\frac{2}{3}\right)^n$
23. $2^0 + 2^1 + 2^2 + 2^3 + \cdots + 2^{n-1} = 2^n - 1$
24. $1^3 + 2^3 + 3^3 + \cdots + n^3 = \left[\frac{n(n + 1)}{2}\right]^2$
25. Prove that $x - y$ is a factor of $x^n - y^n$. (*Hint:* Consider subtracting and adding xy^k to the binomial $x^{k+1} - y^{k+1}$.)
26. Prove that $n < 2^n$.
27. There are 180° in the sum of the angles of any triangle. Prove that $(n - 2)180°$ is the sum of the angles of any simple polygon when n is its number of sides. (*Hint:* If a polygon has $k + 1$ sides, it has $k - 2$ sides plus three more sides.)
28. Consider the equation
$$1 + 3 + 5 + \cdots + 2n - 1 = 3n - 2$$
   a. Is the equation true for $n = 1$? yes
   b. Is the equation true for $n = 2$? yes
   c. Is the equation true for all natural numbers n? no
29. If $1 + 2 + 3 + \cdots + n = \frac{n}{2}(n + 1) + 1$ were true for $n = k$, show that it would be true for $n = k + 1$. Is it true for $n = 1$? no
30. Prove that $n + 1 = 1 + n$ for each natural number n.
31. If n is any natural number, prove that $7^n - 1$ is divisible by 6.
32. Prove that $1 + 2n < 3^n$ for $n > 1$.
33. Prove that, if r is a real number where $r \neq 1$, then
$$1 + r + r^2 + \cdots + r^n = \frac{1 - r^{n+1}}{1 - r}$$
34. Prove the formula for the sum of the first n terms of an arithmetic sequence:
$$a + [a + d] + [a + 2d] + [a + (n - 1)d] = \frac{n(a + a_n)}{2}$$
where $a_n = a + (n - 1)d$.

DISCOVERY AND WRITING

35. The expression a^m, where m is a natural number, was defined in Section 0.2. An alternative definition of a^m is (part 1) $a^1 = a$ and (part 2) $a^{m+1} = a^m \cdot a$. Use induction on n to prove the product rule for exponents, $a^m a^n = a^{m+n}$.

36. Use induction on n to prove the power rule for exponents, $(a^m)^n = a^{mn}$. (See Exercise 35.)

37. **Tower of Hanoi** A well-known problem in mathematics is "The Tower of Hanoi," first attributed to Edouard Lucas in 1883. In this problem, several disks, each of a different size and with a hole in the center, are placed on a board, with progressively smaller disks going up the stack. The object is to transfer the stack of disks to another peg by moving only one disk at a time and never placing a disk over a smaller one.
   a. Find the minimum number of moves required if there is only one disk. **1**
   b. Find the minimum number of moves required if there are two disks. **3**
   c. Find the minimum number of moves required if there are three disks. **7**
   d. Find the minimum number of moves required if there are four disks. **15**

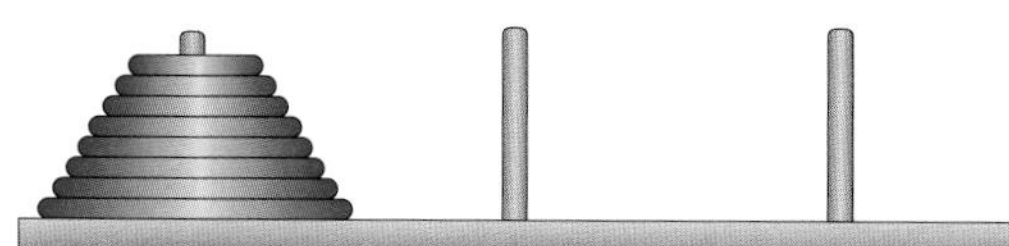

38. **Tower of Hanoi** The results in Exercise 37 suggest that the minimum number of moves required to transfer n disks from one peg to another is given by the formula $2^n - 1$. Use the following outline to prove that this result is correct using mathematical induction.
   a. Verify the formula for $n = 1$.
   b. Write the induction hypothesis.
   It takes $2^k - 1$ moves to transfer k disks.
   c. How many moves are needed to transfer all but the largest of $k + 1$ disks to another peg? $\mathbf{2^k - 1}$
   d. How many moves are needed to transfer the largest disk to an empty peg? **1**
   e. How many moves are needed to transfer the first k disks back onto the largest one? $\mathbf{2^k - 1}$
   f. How many moves are needed to accomplish steps **c, d,** and **e?** $\mathbf{2^k - 1 + 1 + 2^k - 1}$
   g. Show that part **f** can be written in the form $2^{(k+1)} - 1$.
   h. Write the conclusion of the proof.

REVIEW *Graph each inequality.*

39. $3x + 4y \le 12$

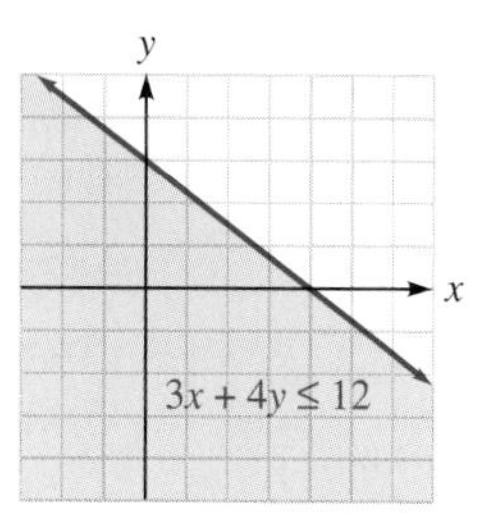

40. $4x - 3y < 12$

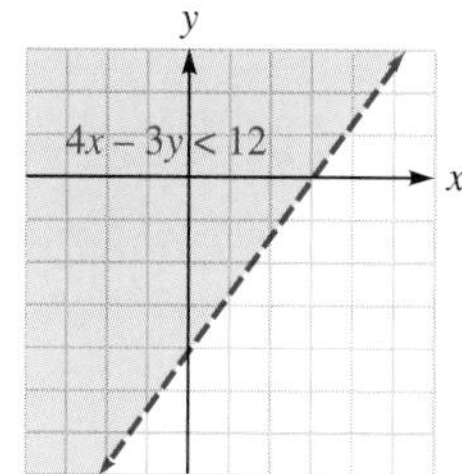

41. $2x + y > 5$

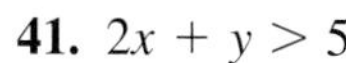

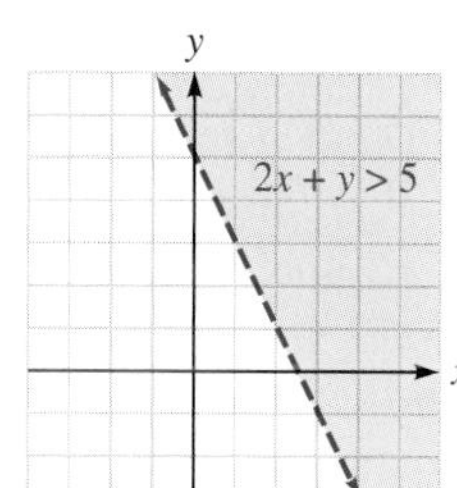

42. $x + 3y \ge 7$

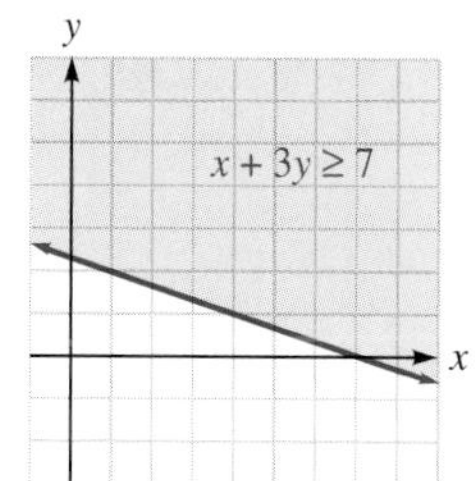

8.6 Permutations and Combinations

In this section, you will learn about

- The Multiplication Principle for Events
- Permutations
- Formulas for Permutations
- Combinations
- Formulas for Combinations
- The Binomial Theorem
- Distinguishable Words

The Multiplication Principle for Events

Lydia plans to go to dinner and attend a movie. If she has a choice of four restaurants and three movies, in how many ways can she spend her evening? There are four choices of restaurants and, for any one of these choices, there are three choices of movies, as shown in the tree diagram in Figure 8-2.

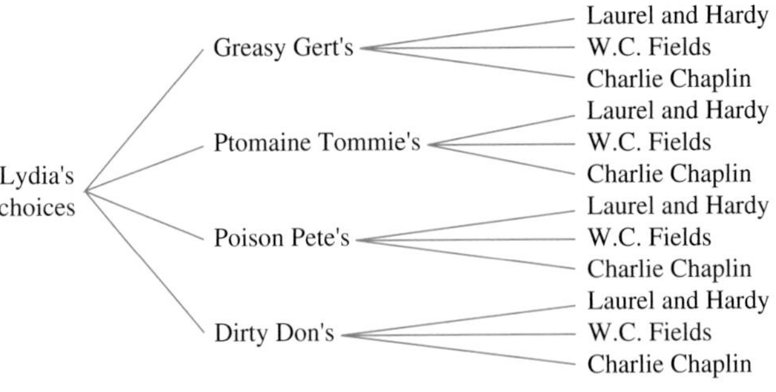

Figure 8-2

The diagram shows that Lydia has 12 ways to spend her evening. One possibility is to eat at Ptomaine Tommie's and watch W. C. Fields. Another is to eat at Dirty Don's and watch Laurel and Hardy.

Any situation that has several outcomes is called an **event.** Lydia's first event (choosing a restaurant) can occur in 4 ways. Her second event (choosing a movie) can occur in 3 ways. Thus, she has $4 \cdot 3$, or 12, ways to spend her evening. This example illustrates the **multiplication principle for events.**

Multiplication Principle for Events

Let E_1 and E_2 be two events. If E_1 can be done in a_1 ways, and if—after E_1 has occurred—E_2 can be done in a_2 ways, then the event "E_1 followed by E_2" can be done in $a_1 \cdot a_2$ ways.

The multiplication principle can be extended to n events.

EXAMPLE 1 If a traveler has 4 ways to go from New York to Chicago, 3 ways to go from Chicago to Denver, and 6 ways to go from Denver to San Francisco, in how many ways can she go from New York to San Francisco?

Solution We can let E_1 be the event "going from New York to Chicago," E_2 be the event "going from Chicago to Denver," and E_3 be the event "going from Denver to San Francisco." Since there are 4 ways to accomplish E_1, 3 ways to accomplish E_2, and 6 ways to accomplish E_3, the number of routes available is

$$4 \cdot 3 \cdot 6 = 72$$

Self Check If a man has 4 sweaters and 5 pairs of slacks, how many outfits can he wear? ■

■ Permutations

Suppose we want to arrange 7 books in order on a shelf. We can fill the first space with any one of the 7 books, the second space with any of the remaining 6 books, the third space with any of the remaining 5 books, and so on, until there is only one space left to fill with the last book. According to the multiplication principle, the number of ordered arrangements of the books is

$$7 \cdot 6 \cdot 5 \cdot 4 \cdot 3 \cdot 2 \cdot 1 = 5{,}040$$

When finding the number of possible ordered arrangements of books on a shelf, we are finding the number of **permutations.** The number of permutations of

7 books, using all the books, is 5,040. The symbol $P(n, r)$ is read as "the number of permutations of n things r at a time." Thus, $P(7, 7) = 5{,}040$.

EXAMPLE 2 Assume that there are 7 signal flags of 7 different colors to hang on a mast. How many different signals can be sent when 3 flags are used?

Solution We are asked to find $P(7, 3)$, the number of permutations (ordered arrangements) of 7 things using 3 of them. Any one of the 7 flags can hang in the top position on the mast. Any one of the 6 remaining flags can hang in the middle position, and any one of the remaining 5 flags can hang in the bottom position. By the multiplication principle, we have

$$P(7, 3) = 7 \cdot 6 \cdot 5 = 210$$

It is possible to send 210 different signals.

Self Check How many signals can be sent if two of the seven flags are missing? ■

■ Formulas for Permutations

Although it is correct to write $P(7, 3) = 7 \cdot 6 \cdot 5$, we will change the form of the answer to obtain a convenient formula. To derive this formula, we proceed as follows:

$$\begin{aligned} P(7, 3) &= 7 \cdot 6 \cdot 5 \\ &= \frac{7 \cdot 6 \cdot 5 \cdot 4 \cdot 3 \cdot 2 \cdot 1}{4 \cdot 3 \cdot 2 \cdot 1} && \text{Multiply numerator and denominator by } 4 \cdot 3 \cdot 2 \cdot 1. \\ &= \frac{7!}{4!} \\ &= \frac{7!}{(7-3)!} \end{aligned}$$

The generalization of this idea gives the following formula.

Formula for $P(n, r)$

The number of permutations of n things r at a time is given by

$$P(n, r) = \frac{n!}{(n-r)!}$$

EXAMPLE 3 Find: **a.** $P(8, 4)$, **b.** $P(n, n)$, and **c.** $P(n, 0)$.

Solution **a.** $P(8, 4) = \dfrac{8!}{(8-4)!} = \dfrac{8 \cdot 7 \cdot 6 \cdot 5 \cdot 4!}{4!} = 1{,}680$

b. $P(n, n) = \dfrac{n!}{(n-n)!} = \dfrac{n!}{0!} = n!$

c. $P(n, 0) = \dfrac{n!}{(n-0)!} = \dfrac{n!}{n!} = 1$

Self Check Find: **a.** $P(7, 5)$ and **b.** $P(6, 0)$. ■

Parts b and c of Example 3 establish the following formulas.

Formulas for $P(n, n)$ and $P(n, 0)$

The number of permutations of n things n at a time and n things 0 at a time are given by the formulas

$$P(n, n) = n! \qquad \text{and} \qquad P(n, 0) = 1$$

ACTIVE EXAMPLE 4 In how many ways can a baseball manager arrange a batting order of 9 players if there are 25 players on the team?

Solution To find the number of permutations of 25 things 9 at a time, we substitute 25 for n and 9 for r in the formula for finding $P(n, r)$.

COLLEGE **Algebra** $f(x)$ **Now**™

Explore this Active Example and test yourself on the steps taken here at **http://1pass.thomson.com**. You can also find this example on the Interactive Video Skillbuilder CD-ROM.

$$P(n, r) = \frac{n!}{(n - r)!}$$

$$P(25, 9) = \frac{25!}{(25 - 9)!}$$

$$= \frac{25!}{16!}$$

$$= \frac{25 \cdot 24 \cdot 23 \cdot 22 \cdot 21 \cdot 20 \cdot 19 \cdot 18 \cdot 17 \cdot 16!}{16!}$$

$$\approx 741{,}354{,}768{,}000$$

The number of permutations is approximately 741,354,768,000.

Self Check In how many ways can the manager arrange a batting order if 2 players can't play? ■

EXAMPLE 5 In how many ways can 5 people stand in a line if 2 people refuse to stand next to each other?

Solution The total number of ways that 5 people can stand in line is

$$P(5, 5) = 5! = 5 \cdot 4 \cdot 3 \cdot 2 \cdot 1 = 120$$

To find the number of ways that 5 people can stand in line if 2 people insist on standing together, we consider the two people as one person. Then there are 4 people to stand in line, and this can be done in $P(4, 4) = 4! = 24$ ways. However, because either of the two who are paired could be first, there are two arrangements for the pair who insist on standing together. Thus, there are $2 \cdot 4!$, or 48 ways that 5 people can stand in line if 2 people insist on standing together.

The number of ways that 5 people can stand in line if 2 people refuse to stand together is $5! = 120$ (the total number of ways to line up 5 people) minus $2 \cdot 4! = 48$ (the number of ways to line up the 5 people if 2 do stand together):

$$120 - 48 = 72$$

There are 72 ways to line up the people.

Self Check In how many ways can 5 people stand in a line if one person demands to be first? ■

EXAMPLE 6 In how many ways can 5 people be seated at a round table?

Solution If we were to seat 5 people in a row, there would be 5! possible arrangements. However, at a round table, each person has a neighbor to the left and to the right. If each person moves one, two, three, four, or five places to the left, everyone has the same neighbors and the arrangement has not changed. Thus, we must divide 5! by 5 to get rid of these duplications. The number of ways that 5 people can be seated at a round table is

$$\frac{5!}{5} = 4! = 4 \cdot 3 \cdot 2 \cdot 1 = 24$$

Self Check In how many ways can 6 people be seated at a round table? ■

The results of Example 6 suggest the following fact.

Circular Arrangements There are $(n - 1)!$ ways to arrange n things in a circle.

Combinations

Suppose that a class of 12 students selects a committee of 3 persons to plan a party. With committees, order is not important. A committee of John, Maria, and Raul is the same as a committee of Maria, Raul, and John. However, if we assume for the moment that order is important, we can find the number of permutations of 12 things 3 at a time.

$$P(12, 3) = \frac{12!}{(12 - 3)!} = \frac{12 \cdot 11 \cdot 10 \cdot 9!}{9!} = 1{,}320$$

Everyday Connections **World Series of Poker**

"Money was never a big motivation for me, except as a way to keep score. The real excitement is playing the game." Donald Trump

Richest Sporting Event on the Planet Begins Today in Las Vegas (June 2, 2005) Thousands of poker enthusiasts from around the globe will begin descending on the Nevada desert today to compete for some $75 million in total prize money at the 36th annual World Series of Poker. The largest and most prestigious event of its kind, the World Series of Poker is expected to generate more than 15,000 player registrations by the time its main event begins July 7.
Source: www.worldseriesofpoker.com

In poker, a full house is a hand of 5 cards containing three cards of one kind and two cards of a second kind (for example, a hand containing three 6's and two queens). A standard deck of 52 cards contains 4 different suits, and each suit contains 13 kinds of cards (2, 3, 4, 5, 6, 7, 8, 9, 10, J, Q, K, A). How many possible full-house hands are there? **3,744**

However, since we do not care about order, this result of 1,320 ways is too large. Because there are 6 ways ($3! = 6$) of ordering every committee of 3 students, the result of $P(12, 3) = 1{,}320$ is exactly 6 times too big. To get the correct number of committees, we must divide $P(12, 3)$ by 6:

$$\frac{P(12, 3)}{6} = \frac{1{,}320}{6} = 220$$

In cases of selection where order is not important, we are interested in **combinations,** not permutations. The symbols $C(n, r)$ and $\binom{n}{r}$ both mean the number of combinations of n things r at a time.

Formulas for Combinations

If a committee of r people is chosen from a total of n people, the number of possible committees is $C(n, r)$, and there will be $r!$ arrangements of each committee. If we consider the committee as an ordered grouping, the number of orderings is $P(n, r)$. Thus, we have

(1) $$r!C(n, r) = P(n, r)$$

We can divide both sides of Equation 1 by $r!$ to obtain the formula for finding $C(n, r)$.

$$C(n, r) = \binom{n}{r} = \frac{P(n, r)}{r!} = \frac{n!}{r!(n-r)!}$$

Comment When discussing permutations, order counts. When discussing combinations, order doesn't count.

Formula for $C(n, r)$

The number of combinations of n things r at a time is given by

$$C(n, r) = \binom{n}{r} = \frac{n!}{r!(n-r)!}$$

In the exercises, you will be asked to prove the following formulas.

Formulas for $C(n, n)$ and $C(n, 0)$

If n is a whole number, then

$$C(n, n) = 1 \quad \text{and} \quad C(n, 0) = 1$$

ACTIVE EXAMPLE 7 If Carla must read 4 books from a reading list of 10 books, how many choices does she have?

Solution Because the order in which the books are read is not important, we find the number of combinations of 10 things 4 at a time:

$$C(10, 4) = \frac{10!}{4!(10-4)!} = \frac{10 \cdot 9 \cdot 8 \cdot 7 \cdot 6!}{4 \cdot 3 \cdot 2 \cdot 1 \cdot 6!}$$

$$= \frac{10 \cdot 9 \cdot 8 \cdot 7}{4 \cdot 3 \cdot 2}$$

$$= 210$$

COLLEGE **Algebra** $f(x)$ **Now**™
Go to **http://1pass.thomson.com** or your CD to practice this example.

Carla has 210 choices.

Self Check How many choices would Carla have if she had to read 5 books? ■

ACTIVE EXAMPLE 8 A class consists of 15 men and 8 women. In how many ways can a debate team be chosen with 3 men and 3 women?

Solution There are $C(15, 3)$ ways of choosing 3 men and $C(8, 3)$ ways of choosing 3 women. By the multiplication principle, there are $C(15, 3) \cdot C(8, 3)$ ways of choosing members of the debate team:

COLLEGE **Algebra** $f(x)$ **Now**™
Go to **http://1pass.thomson.com** or your CD to practice this example.

$$\begin{aligned} C(15, 3) \cdot C(8, 3) &= \frac{15!}{3!(15-3)!} \cdot \frac{8!}{3!(8-3)!} \\ &= \frac{15 \cdot 14 \cdot 13}{6} \cdot \frac{8 \cdot 7 \cdot 6}{6} \\ &= 25{,}480 \end{aligned}$$

There are 25,480 ways to choose the debate team.

Self Check In how many ways can the debate team be chosen if it is to have 4 men and 2 women? ■

■ The Binomial Theorem

The formula

$$C(n, r) = \frac{n!}{r!(n-r)!}$$

gives the coefficient of the $(r + 1)$th term of the binomial expansion of $(a + b)^n$. This implies that the coefficients of a binomial expansion can be used to solve problems involving combinations. The binomial theorem is restated below—this time listing the $(r + 1)$th term and using combination notation.

Binomial Theorem

If n is any positive integer, then

$$(a+b)^n = \binom{n}{0}a^n + \binom{n}{1}a^{n-1}b + \binom{n}{2}a^{n-2}b^2 + \cdots + \binom{n}{r}a^{n-r}b^r + \cdots + \binom{n}{n}b^n$$

Comment In the expansion of $(a + b)^n$, the term containing b^r is given by

$$\binom{n}{r}a^{n-r}b^r$$

EXAMPLE 9 Use Pascal's triangle to compute $C(7, 5)$.

Solution Consider the eighth row of Pascal's triangle and the corresponding combinations:

$$\begin{array}{cccccccc} 1 & 7 & 21 & 35 & 35 & 21 & 7 & 1 \\ \binom{7}{0} & \binom{7}{1} & \binom{7}{2} & \binom{7}{3} & \binom{7}{4} & \binom{7}{5} & \binom{7}{6} & \binom{7}{7} \end{array}$$

$C(7, 5) = \binom{7}{5} = 21.$

Self Check Use Pascal's triangle to compute $C(6, 5)$. ■

Distinguishable Words

A "word" is a distinguishable arrangement of letters. For example, six words can be formed with the letters a, b, and c if each letter is used exactly once. The six words are

abc, acb, bac, bca, cab, and cba

If there are n distinct letters and each letter is used once, the number of distinct words that can be formed is $n! = P(n, n)$. It is more complicated to compute the number of distinguishable words that can be formed with n letters when some of the letters are duplicates.

EXAMPLE 10 Find the number of "words" that can be formed if each of the 6 letters of the word *little* is used once.

Solution For the moment, we assume that the letters of the word *little* are distinguishable: "LitTle." The number of words that can be formed using each letter once is $6! = P(6, 6)$. However, in reality we cannot tell the l's or the t's apart. Therefore, we must divide by a number to get rid of these duplications. Because there are 2! orderings of the two l's and 2! orderings of the two t's, we divide by $2! \cdot 2!$. The number of words that can be formed using each letter of the word *little* is

$$\frac{P(6, 6)}{2! \cdot 2!} = \frac{6!}{2! \cdot 2!} = \frac{6 \cdot 5 \cdot 4 \cdot 3 \cdot 2 \cdot 1}{2 \cdot 1 \cdot 2 \cdot 1} = 180$$

Self Check How many words can be formed if each letter of the word *balloon* is used once? ■

Example 10 illustrates the following general principle.

Distinguishable Words

If a word with n letters has a of one letter, b of another letter, and so on, the number of distinguishable words that can be formed using each letter of the n-letter word exactly once is

$$\frac{n!}{a!b! \cdots}$$

Self Check Answers

1. 20 **2.** 60 **3. a.** 2,520 **b.** 1 **4.** approximately 296,541,907,200 **5.** 24 **6.** 120 **7.** 252 **8.** 38,220 **9.** 6 **10.** 1,260

8.6 Exercises

VOCABULARY AND CONCEPTS ***Fill in the blanks.***

1. If E_1 and E_2 are two events and E_1 can be done in 4 ways and E_2 can be done in 6 ways, then the event E_1 followed by E_2 can be done in __24__ ways.
2. An arrangement of n objects is called a __permutation__.
3. $P(n, r) = \underline{\frac{n!}{(n-r)!}}$
4. $P(n, n) = \underline{n!}$
5. $P(n, 0) = \underline{1}$

6. There are $(n-1)!$ ways to arrange n things in a circle.

7. $C(n, r) = \dfrac{n!}{r!(n-r)!}$

8. Using combination notation, $C(n, r) = \binom{n}{r}$.

9. $C(n, n) = 1$

10. $C(n, 0) = 1$

11. If a word with n letters has a of one letter, b of another letter, and so on, the number of different words that can be formed is
$$\frac{n!}{a! \cdot b! \cdot \cdot \cdot}$$

12. Where the order of selection is not important, we are interested in combinations, not permutations.

PRACTICE *Evaluate each expression.*

13. $P(7, 4)$ **840**

14. $P(8, 3)$ **336**

15. $C(7, 4)$ **35**

16. $C(8, 3)$ **56**

17. $P(5, 5)$ **120**

18. $P(5, 0)$ **1**

19. $\binom{5}{4}$ **5**

20. $\binom{8}{4}$ **70**

21. $\binom{5}{0}$ **1**

22. $\binom{5}{5}$ **1**

23. $P(5, 4) \cdot C(5, 3)$ **1,200**

24. $P(3, 2) \cdot C(4, 3)$ **24**

25. $\binom{5}{3}\binom{4}{3}\binom{3}{3}$ **40**

26. $\binom{5}{5}\binom{6}{6}\binom{7}{7}\binom{8}{8}$ **1**

27. $\binom{68}{66}$ **2,278**

28. $\binom{100}{99}$ **100**

APPLICATIONS

29. **Choosing lunch** A lunchroom has a machine with eight kinds of sandwiches, a machine with four kinds of soda, a machine with both white and chocolate milk, and a machine with three kinds of ice cream. How many different lunches can be chosen? (Consider a lunch to be one sandwich, one drink, and one ice cream.) **144**

30. **Manufacturing license plates** How many six-digit license plates can be manufactured if no license plate number begins with 0? **900,000**

31. **Available phone numbers** How many different seven-digit phone numbers can be used in one area code if no phone number begins with 0 or 1? **8,000,000**

32. **Arranging letters** In how many ways can the letters of the word *number* be arranged? **720**

33. **Arranging letters with restrictions** In how many ways can the letters of the word *number* be arranged if the e and r must remain next to each other? **240**

34. **Arranging letters with restrictions** In how many ways can the letters of the word *number* be arranged if the e and r cannot be side by side? **480**

35. **Arranging letters with repetitions** How many ways can five Scrabble tiles bearing the letters, F, F, F, L, and U be arranged to spell the word *fluff*? **6**

36. **Arranging letters with repetitions** How many ways can six Scrabble tiles bearing the letters B, E, E, E, F, and L be arranged to spell the word *feeble*? **6**

37. **Placing people in line** In how many arrangements can 8 women be placed in a line? **40,320**

38. **Placing people in line** In how many arrangements can 5 women and 5 men be placed in a line if the women and men alternate? **28,800**

39. **Placing people in line** In how many arrangements can 5 women and 5 men be placed in a line if all the men line up first? **14,400**

40. **Placing people in line** In how many arrangements can 5 women and 5 men be placed in a line if all the women line up first? **14,400**

41. **Combination locks** How many permutations does a combination lock have if each combination has 3 numbers, no two numbers of the combination are the same, and the lock dial has 30 notches? **24,360**

42. **Combination locks** How many permutations does a combination lock have if each combination has 3 numbers, no two numbers of the combination are the same, and the lock dial has 100 notches? **970,200**

43. **Seating at a table** In how many ways can 8 people be seated at a round table? **5,040**

44. **Seating at a table** In how many ways can 7 people be seated at a round table? **720**

45. **Seating at a table** In how many ways can 6 people be seated at a round table if 2 of the people insist on sitting together? **48**

46. **Seating arrangements with conditions** In how many ways can 6 people be seated at a round table if 2 of the people refuse to sit together? **72**

47. **Arrangements in a circle** In how many ways can 7 children be arranged in a circle if Sally and John want to sit together and Martha and Peter want to sit together? **96**

48. Arrangements in a circle In how many ways can 8 children be arranged in a circle if Laura, Scott, and Paula want to sit together? 720

49. Selecting candy bars In how many ways can 4 candy bars be selected from 10 different candy bars? 210

50. Selecting birthday cards In how many ways can 6 birthday cards be selected from 24 different cards? 134,596

51. Circuit wiring A wiring harness containing a red, a green, a white, and a black wire must be attached to a control panel. In how many different orders can the wires be attached? 24

52. Grading homework A professor grades homework by randomly checking 7 of the 20 problems assigned. In how many different ways can this be done? 77,520

53. Forming words with distinct letters How many words can be formed from the letters of the word *plastic* if each letter is to be used once? 5,040

54. Forming words with repeated letters How many words can be formed from the letters of the word *banana* if each letter is to be used once? 60

55. Manufacturing license plates How many license plates can be made using two different letters followed by four different digits if the first digit cannot be 0 and the letter O is not used? 2,721,600

56. Planning class schedules If there are 7 class periods in a school day, and a typical student takes 5 classes, how many different time patterns are possible for the student? 21

57. Selecting golf balls From a bucket containing 6 red and 8 white golf balls, in how many ways can we draw 6 golf balls of which 3 are red and 3 are white? 1,120

58. Selecting a committee In how many ways can you select a committee of 3 Republicans and 3 Democrats from a group containing 18 Democrats and 11 Republicans? 134,640

59. Selecting a committee In how many ways can you select a committee of 4 Democrats and 3 Republicans from a group containing 12 Democrats and 10 Republicans? 59,400

60. Drawing cards In how many ways can you select a group of 5 red cards and 2 black cards from a deck containing 10 red cards and 8 black cards? 7,056

61. Planning dinner In how many ways can a husband and wife choose 2 different dinners from a menu of 17 dinners? 272

62. Placing people in line In how many ways can 7 people stand in a row if 2 of the people refuse to stand together? 3,600

63. Geometry How many lines are determined by 8 points if no 3 points lie on a straight line? 28

64. Geometry How many lines are determined by 10 points if no 3 points lie on a straight line? 45

65. Coaching basketball How many different teams can a basketball coach start if the entire squad consists of 10 players? (Assume that a starting team has 5 players and each player can play all positions.) 252

66. Managing baseball How many different teams can a manager start if the entire squad consists of 25 players? (Assume that a starting team has 9 players and each player can play all positions.) 2,042,975

67. Selecting job applicants There are 30 qualified applicants for 5 openings in the sales department. In how many different ways can the group of 5 be selected? 142,506

68. Sales promotions If a customer purchases a new stereo system during the spring sale, he may choose any 6 CDs from 20 classical and 30 jazz selections. In how many ways can the customer choose 3 of each? 4,628,400

69. Guessing on matching questions Ten words are to be paired with the correct 10 out of 12 possible definitions. How many ways are there of guessing? 66

70. Guessing on true-false exams How many possible ways are there of guessing on a 10-question true-false exam, if it is known that the instructor will have 5 true and 5 false responses? 252

DISCOVERY AND WRITING

71. Prove that $C(n, n) = 1$.

72. Prove that $C(n, 0) = 1$.

73. Prove that $\binom{n}{r} = \binom{n}{n-r}$.

74. Show that the binomial theorem can be expressed in the form

$$(a + b)^n = \sum_{k=0}^{n} \binom{n}{k} a^{n-k} b^k$$

75. Explain how to use Pascal's triangle to find $C(8, 5)$.

76. Explain how to use Pascal's triangle to find $C(10, 8)$.

REVIEW *Find the value of x.*

77. $\log_x 16 = 4$ 2

78. $\log_\pi x = \frac{1}{2}$ $\sqrt{\pi}$

79. $\log_{\sqrt{7}} 49 = x$ 4

80. $\log_x \frac{1}{2} = -\frac{1}{3}$ 8

Determine whether the statement is true or false.

81. $\log_{17} 1 = 0$ true

82. $\log_5 0 = 1$ false

83. $\log_b b^b = b$ true

84. $\frac{\log_7 A}{\log_7 B} = \log_7 \frac{A}{B}$ false

8.7 Probability

In this section, you will learn about

- Probability ■ Multiplication Property of Probabilities

Probability

The probability that an event will occur is a measure of the likelihood of that event. A tossed coin, for example, can land in two ways, either heads or tails. Because one of these two equally likely outcomes is heads, we expect that out of several tosses, about half will be heads. We say that the probability of obtaining heads in a single toss of the coin is $\frac{1}{2}$.

If records show that out of 100 days with weather conditions like today's, 30 have received rain, the weather service will report, "There is a $\frac{30}{100}$ or 30% probability of rain today."

An **experiment** is a process for which the outcome is uncertain. Tossing a coin, rolling a die, drawing a card, and predicting rain are examples of experiments. For any experiment, the set of all possible outcomes is called a **sample space.**

The sample space, S, for the experiment of tossing two coins is the set

$$S = \{(\text{H}, \text{H}), (\text{H}, \text{T}), (\text{T}, \text{H}), (\text{T}, \text{T})\}$$

where the ordered pair (H, T) represents the outcome "heads on the first coin and tails on the second coin." Because there are two possible outcomes for the first coin and two for the second coin, we know (by the multiplication principle for events) that there are $2 \cdot 2 = 4$ possible outcomes. Since there are 4 elements in the sample space S, we write

$n(S) = 4$ Read as "The number of elements in set S is 4."

An **event** associated with an experiment is any subset of the sample space of that experiment. For example, if E is the event "getting at least one heads" in the experiment of tossing two coins, then

$$E = \{(\text{H}, \text{H}), (\text{H}, \text{T}), (\text{T}, \text{H})\}$$

and $n(E) = 3$. Because the outcome of getting at least one heads can occur in 3 out of 4 possible ways, we say that the **probability** of a favorable outcome is $\frac{3}{4}$.

$$P(E) = P(\text{at least one heads}) = \frac{3}{4}$$

We define the probability of an event as follows.

Probability of an Event

If S is the sample space of an experiment with n distinct and equally likely outcomes, and E is an event that occurs in s of those ways, then the **probability of E** is

$$P(E) = \frac{n(E)}{n(S)} = \frac{s}{n}$$

Because $0 \le s \le n$, it follows that $0 \le \frac{s}{n} \le 1$. This implies that all probabilities have values from 0 to 1. An event that cannot happen has probability 0. An event that is certain to happen has probability 1.

To say that the probability of tossing heads on one toss of a coin is $\frac{1}{2}$ means that if a fair coin is tossed a large number of times, the ratio of the number of heads to the total number of tosses is nearly $\frac{1}{2}$.

To say that the probability of rolling 5 on one roll of a die is $\frac{1}{6}$ means that as the number of rolls approaches infinity, the ratio of the number of favorable outcomes (rolling a 5) to the total number of outcomes (rolling a 1, 2, 3, 4, 5, or 6) approaches $\frac{1}{6}$.

EXAMPLE 1 Show the sample space of the experiment "rolling two dice a single time."

Solution We can list ordered pairs and let the first number be the result on the first die and the second number the result on the second die. The sample space, S, is the set with the following elements:

$$\begin{array}{cccccc} (1, 1) & (1, 2) & (1, 3) & (1, 4) & (1, 5) & (1, 6) \\ (2, 1) & (2, 2) & (2, 3) & (2, 4) & (2, 5) & (2, 6) \\ (3, 1) & (3, 2) & (3, 3) & (3, 4) & (3, 5) & (3, 6) \\ (4, 1) & (4, 2) & (4, 3) & (4, 4) & (4, 5) & (4, 6) \\ (5, 1) & (5, 2) & (5, 3) & (5, 4) & (5, 5) & (5, 6) \\ (6, 1) & (6, 2) & (6, 3) & (6, 4) & (6, 5) & (6, 6) \end{array}$$

Since there are 6 possible outcomes with the first die and 6 possible outcomes with the second die, we expect $6 \cdot 6 = 36$ equally likely possible outcomes, and we have $n(S) = 36$.

Self Check How many pairs in the above sample space have a sum of 7? ■

EXAMPLE 2 Find the probability of the event "rolling a sum of 7 on one roll of two dice."

Solution The sample space is listed in Example 1. We let E be the set of favorable outcomes, those that give a sum of 7:

$$E = \{(1, 6), (2, 5), (3, 4), (4, 3), (5, 2), (6, 1)\}$$

Since there are 6 favorable outcomes among the 36 equally likely outcomes, $n(E) = 6$, and

$$P(E) = P(\text{rolling a 7}) = \frac{n(E)}{n(S)} = \frac{6}{36} = \frac{1}{6}$$

Self Check Find the probability of rolling a sum of 4. ■

A standard playing deck of 52 cards has two red suits, hearts and diamonds, and two black suits, clubs and spades. Each suit has 13 cards, including an ace, a king, a queen, a jack, and cards numbered from 2 to 10. We will refer to a standard deck of cards in many examples and exercises.

ACTIVE EXAMPLE 3 Find the probability of drawing 5 cards, all hearts, from a standard deck of cards.

Solution Since the number of ways to draw 5 hearts from the 13 hearts is $C(13, 5)$, we have $n(E) = C(13, 5)$. Since the number of ways to draw 5 cards from the deck is $C(52, 5)$, we have $n(S) = C(52, 5)$. The probability of drawing 5 hearts is the ratio of the number of favorable outcomes to the number of possible outcomes.

COLLEGE **Algebra** $f(x)$ **Now**™
Explore this Active Example and test yourself on the steps taken here at **http://1pass.thomson.com**. You can also find this example on the Interactive Video Skillbuilder CD-ROM.

$$P(5\text{ hearts}) = \frac{C(13, 5)}{C(52, 5)}$$

$$P(5\text{ hearts}) = \frac{\frac{13!}{5!8!}}{\frac{52!}{5!47!}}$$

$$= \frac{13!}{5!8!} \cdot \frac{5!47!}{52!}$$

$$= \frac{13 \cdot 12 \cdot 11 \cdot 10 \cdot 9 \cdot 8!}{8!} \cdot \frac{47!}{52 \cdot 51 \cdot 50 \cdot 49 \cdot 48 \cdot 47!}$$

$$= \frac{13 \cdot 12 \cdot 11 \cdot 10 \cdot 9}{52 \cdot 51 \cdot 50 \cdot 49 \cdot 48}$$

$$= \frac{33}{66{,}640}$$

The probability of drawing 5 hearts is $\frac{33}{66{,}640}$.

Self Check Find the probability of drawing 6 cards, all diamonds, from the deck. ■

■ Multiplication Property of Probabilities

There is a property of probabilities that is similar to the multiplication principle for events. In the following theorem, we read $P(A \cap B)$ as "the probability of A and B" and $P(B \mid A)$ as "the probability of B given A." If A and B are events, the set $A \cap B$ contains the outcomes that are in both A and B.

Multiplication Property of Probabilities

If $P(A)$ represents the probability of event A, and $P(B \mid A)$ represents the probability that event B will occur after event A, then

$$P(A \cap B) = P(A) \cdot P(B \mid A)$$

ACTIVE EXAMPLE 4 A box contains 40 cubes of the same size. Of these cubes, 17 are red, 13 are blue, and the rest are yellow. If 2 cubes are drawn at random, without replacement, find the probability that 2 yellow cubes will be drawn.

Solution Of the 40 cubes in the box, 10 are yellow. The probability of getting a yellow cube on the first draw is

COLLEGE **Algebra** *f(x)* **Now**™
Go to **http://1pass.thomson.com** or your CD to practice this example.

$$P(\text{yellow cube on the first draw}) = \frac{10}{40} = \frac{1}{4}$$

Because there is no replacement after the first draw, 39 cubes remain in the box, and 9 of these are yellow. The probability of drawing a yellow cube on the second draw is

$$P(\text{yellow cube on the second draw}) = \frac{9}{39} = \frac{3}{13}$$

The probability of drawing 2 yellow cubes in succession is the product of the probability of drawing a yellow cube on the first draw and the probability of drawing a yellow cube on the second draw.

$$P(\text{drawing two yellow cubes}) = \frac{1}{4} \cdot \frac{3}{13} = \frac{3}{52}$$

Self Check Find the probability that 2 blue cubes will be drawn. ■

EXAMPLE 5 Repeat Example 3 using the multiplication property of probabilities.

Solution The probability of drawing a heart on the first draw is $\frac{13}{52}$. The probability of drawing a heart on the second draw *given that we got a heart on the first draw* is $\frac{12}{51}$. The probability is $\frac{11}{50}$ on the third draw, $\frac{10}{49}$ on the fourth draw, and $\frac{9}{48}$ on the fifth draw. By the multiplication property of probabilities,

$$P(\text{5 hearts in a row}) = \frac{13}{52} \cdot \frac{12}{51} \cdot \frac{11}{50} \cdot \frac{10}{49} \cdot \frac{9}{48}$$

$$= \frac{33}{66{,}640}$$

Self Check Find the probability of drawing 6 cards, all diamonds, from the deck. ■

ACTIVE EXAMPLE 6 In a school, 30% of the students are gifted in mathematics and 10% are gifted in art and mathematics. If a student is gifted in mathematics, find the probability that the student is also gifted in art.

Solution Let $P(M)$ be the probability that a randomly chosen student is gifted in mathematics, and let $P(M \cap A)$ be the probability that the student is gifted in both art and mathematics. We must find $P(A \mid M)$, the probability that the student is gifted in art, given that he or she is gifted in mathematics. To do so, we substitute the given values

COLLEGE **Algebra** *f(x)* **Now**™
Go to **http://1pass.thomson.com** or your CD to practice this example.

$$P(M) = 0.3 \qquad \text{and} \qquad P(M \cap A) = 0.1$$

in the formula for multiplication of probabilities and solve for $P(A \mid M)$:

$$
\begin{aligned}
P(M \cap A) &= P(M) \cdot P(A \mid M) \\
0.1 &= (0.3)P(A \mid M) \\
P(A \mid M) &= \frac{0.1}{0.3} \\
&= \frac{1}{3}
\end{aligned}
$$

If a student is gifted in mathematics, there is a probability of $\frac{1}{3}$ that he or she is also gifted in art.

Self Check If 40% of the students are gifted in art, find the probability that a student gifted in art is also gifted in mathematics. ■

Self Check Answers

1. 6 **2.** $\frac{1}{12}$ **3.** $\frac{33}{391{,}510}$ **4.** $\frac{1}{10}$ **5.** $\frac{33}{391{,}510}$ **6.** $\frac{1}{4}$

8.7 Exercises

VOCABULARY AND CONCEPTS *Fill in the blanks.*

1. An experiment is any process for which the outcome is uncertain.

2. A list of all possible outcomes for an experiment is called a sample space.

3. The probability of an event E is defined as $P(E) = \frac{n(E)}{n(S)} = \frac{s}{n}$.

4. $P(A \cap B) = P(A) \cdot P(B \mid A)$

PRACTICE *List the sample space of each experiment.*

5. Rolling one die and tossing one coin
{(1, H), (2, H), (3, H), (4, H), (5, H), (6, H), (1, T), (2, T), (3, T), (4, T), (5, T), (6, T)}

6. Tossing three coins
{(H, H, H), (H, H, T), (H, T, H), (H, T, T), (T, H, H), (T, H, T), (T, T, H), (T, T, T)}

7. Selecting a letter of the alphabet
{a, b, c, d, e, f, g, h, i, j, k, l, m, n, o, p, q, r, s, t, u, v, w, x, y, z}

8. Picking a one-digit number
{0, 1, 2, 3, 4, 5, 6, 7, 8, 9}

An ordinary die is rolled. Find the probability of each event.

9. Rolling a 2 $\frac{1}{6}$

10. Rolling a number greater than 4 $\frac{1}{3}$

11. Rolling a number larger than 1 but less than 6 $\frac{2}{3}$

12. Rolling an odd number $\frac{1}{2}$

Balls numbered from 1 to 42 are placed in a container and stirred. If one is drawn at random, find the probability of each result.

13. The number is less than 20. $\frac{19}{42}$

14. The number is less than 50. 1

15. The number is a prime number. $\frac{13}{42}$

16. The number is less than 10 or greater than 40. $\frac{11}{42}$

If the spinner shown below is spun, find the probability of each event. Assume that the spinner never stops on a line.

17. The spinner stops on red. $\frac{3}{8}$

18. The spinner stops on green. $\frac{1}{4}$

19. The spinner stops on orange. 0

20. The spinner stops on yellow. $\frac{1}{8}$

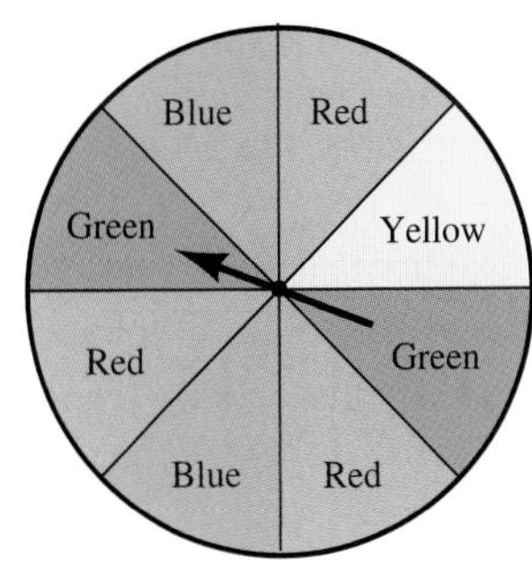

Find the probability of each event.

21. Rolling a sum of 4 on one roll of two dice $\frac{1}{12}$

22. Drawing a diamond on one draw from a card deck $\frac{1}{4}$

23. Drawing two aces in succession from a card deck if the card is replaced and the deck is shuffled after the first draw $\frac{1}{169}$

24. Drawing two aces from a card deck without replacing the card after the first draw $\frac{1}{221}$

25. Drawing a red egg from a basket containing 5 red eggs and 7 blue eggs $\frac{5}{12}$

26. Getting 2 red eggs in a single scoop from a bucket containing 5 red eggs and 7 yellow eggs $\frac{5}{33}$

27. Drawing a bridge hand of 13 cards, all of one suit about 6.3×10^{-12}

28. Drawing 6 diamonds from a card deck without replacing the cards after each draw $\frac{33}{391{,}510}$

29. Drawing 5 aces from a card deck without replacing the cards after each draw 0

30. Drawing 5 clubs from the black cards in a card deck $\frac{9}{460}$

31. Drawing a face card (king, queen, or jack) from a card deck $\frac{3}{13}$

32. Drawing 6 face cards in a row from a card deck without replacing the cards after each draw $\frac{33}{727{,}090}$

33. Drawing 5 orange cubes from a bowl containing 5 orange cubes and 1 beige cube $\frac{1}{6}$

34. Rolling a sum of 4 with one roll of three dice $\frac{1}{72}$

35. Rolling a sum of 11 with one roll of three dice $\frac{1}{8}$

36. Picking, at random, 5 Republicans from a group containing 8 Republicans and 10 Democrats $\frac{1}{153}$

37. Tossing 3 heads in 5 tosses of a fair coin $\frac{5}{16}$

38. Tossing 5 heads in 5 tosses of a fair coin $\frac{1}{32}$

Assume that the probability that an airplane engine will fail during a torture test is $\frac{1}{2}$ and that the aircraft in question has 4 engines.

39. Construct a sample space for the torture test.
(S = survive, F = fail) {SSSS, SSSF, SSFS, SFSS, FSSS, SSFF, SFSF, FSSF, SFFS, FSFS, FFSS, SFFF, FSFF, FFSF, FFFS, FFFF}

40. Find the probability that all engines will survive the test. $\frac{1}{16}$

41. Find the probability that exactly 1 engine will survive. $\frac{1}{4}$

42. Find the probability that exactly 2 engines will survive. $\frac{3}{8}$

43. Find the probability that exactly 3 engines will survive. $\frac{1}{4}$

44. Find the probability that no engines will survive. $\frac{1}{16}$

45. Find the sum of the probabilities in Exercises 40 through 44. 1

Assume that a survey of 282 people is taken to determine the opinions of doctors, teachers, and lawyers on a proposed piece of legislation, with the results as shown in the table. A person is chosen at random from those surveyed. Refer to the table to find each probability.

	Number that favor	Number that oppose	Number with no opinion	Total
Doctors	70	32	17	119
Teachers	83	24	10	117
Lawyers	23	15	8	46
Total	176	71	35	282

46. The person favors the legislation. $\frac{88}{141}$

47. A doctor opposes the legislation. $\frac{32}{119}$

48. A person who opposes the legislation is a lawyer. $\frac{15}{71}$

49. Quality control In a batch of 10 tires, 2 are known to be defective. If 4 tires are chosen at random, find the probability that all 4 tires are good. $\frac{1}{3}$

50. Medicine Out of a group of 9 patients treated with a new drug, 4 suffered a relapse. Find the probability that 3 patients of this group, chosen at random, will remain disease-free. $\frac{5}{42}$

Use the multiplication property of probabilities.

51. If $P(A) = 0.3$ and $P(B \mid A) = 0.6$, find $P(A \cap B)$. 0.18

52. If $P(A \cap B) = 0.3$ and $P(B \mid A) = 0.6$, find $P(A)$. 0.5

53. Conditional probability The probability that a person owns a luxury car is 0.2, and the probability that the owner of such a car also owns a personal computer is 0.7. Find the probability that a person, chosen at random, owns both a luxury car and a computer. 0.14

54. **Conditional probability** If 40% of the population have completed college, and 85% of college graduates are registered to vote, what percent of the population are both college graduates and registered voters? **34%**

55. **Conditional probability** About 25% of the population watches the evening television news coverage as well as the morning soap operas. If 75% of the population watches the news, what percent of those who watch the news also watch the soaps? **about 33%**

56. **Conditional probability** The probability of rain today is 0.40. If it rains, the probability that Bill will forget his raincoat is 0.70. Find the probability that Bill will get wet. **0.28**

DISCOVERY AND WRITING

57. If $P(A \cap B) = 0.7$, is it possible that $P(B \mid A) = 0.6$? Explain. **no**

58. Is it possible that $P(A \cap B) = P(A)$? Explain. **yes**

REVIEW *Solve each equation.*

59. $|x + 3| = 7$ **−10, 4**

60. $|3x - 2| = |2x - 3|$ **−1, 1**

Graph each inequality.

61. $|x - 3| < 7$

-4 10

62. $|3x - 2| \geq |2x - 3|$

-1 1

8.8 Computation of Compound Probabilities

In this section, you will learn about

- Compound Events
- Independent Events

Compound Events

Sometimes we must find the probability of one event *or* another, or the probability of one event *and* another. Such events are called **compound events.** As we have seen, if A and B are two events, the probability that A and B will both occur is $P(A \cap B)$. In this section, we also discuss $P(A \cup B)$, the probability that either A or B will occur.

Suppose we want to find the probability of drawing a king or a heart from a standard card deck. If K is the event "drawing a king" and H is the event "drawing a heart," then $P(K) = \frac{4}{52}$, and $P(H) = \frac{13}{52}$. However, the probability of drawing a king *or* a heart is not the sum of these two probabilities. Because the king of hearts was counted twice, once as a king and once as a heart, and because the probability of drawing the king of hearts is $\frac{1}{52}$, we must subtract $\frac{1}{52}$ from the sum of $\frac{4}{52}$ and $\frac{13}{52}$ to get the correct probability.

$$\begin{aligned} P(\text{king } or \text{ heart}) &= P(\text{king}) + P(\text{heart}) - P(\text{king of hearts}) \\ P(K \cup H) &= P(K) + P(H) - P(K \cap H) \\ &= \frac{4}{52} + \frac{13}{52} - \frac{1}{52} \\ &= \frac{16}{52} \\ &= \frac{4}{13} \end{aligned}$$

In general, we have the following rule.

$P(A \cup B)$

If A and B are two events, then

$$P(A \cup B) = P(A) + P(B) - P(A \cap B)$$

If events A and B have no outcomes in common, then $A \cap B = \emptyset$ and $P(A \cap B) = P(\emptyset) = 0$. Such events are **mutually exclusive** (if one event occurs, the other cannot).

$P(A \cup B)$

If A and B cannot occur simultaneously, then

$$P(A \cup B) = P(A) + P(B)$$

The event $\bar{A}$ (read as "not A") contains all outcomes of the sample space that are not elements of event A. Because the events A and $\bar{A}$ are mutually exclusive,

$$P(A \cup \bar{A}) = P(A) + P(\bar{A})$$

Because either event A or event $\bar{A}$ must happen, $P(A \cup \bar{A}) = 1$. Thus,

$$\begin{aligned} P(A \cup \bar{A}) &= 1 \\ P(A) + P(\bar{A}) &= 1 \\ P(\bar{A}) &= 1 - P(A) \quad \text{Add } -P(A) \text{ to both sides.} \end{aligned}$$

This result gives another property of compound probabilities.

$P(\bar{A})$

If A is any event, then

$$P(\bar{A}) = 1 - P(A)$$

ACTIVE EXAMPLE 1 A counselor tells a student that his probability of earning a grade of D in algebra is $\frac{1}{5}$ and his probability of earning an F is $\frac{1}{25}$. Find the probability that the student earns a C or better.

Solution Because "earning a D" and "earning an F" are mutually exclusive, the probability of earning a D or F is given by

$$\begin{aligned} P(D \cup F) &= P(D) + P(F) \quad \text{Note that } P(D \text{ and } F) = 0. \\ &= \frac{1}{5} + \frac{1}{25} \\ &= \frac{6}{25} \end{aligned}$$

COLLEGE **Algebra** $f(x)$ **Now**™

Explore this Active Example and test yourself on the steps taken here at **http://1pass.thomson.com**. You can also find this example on the Interactive Video Skillbuilder CD-ROM.

The probability that the student will receive a C or better is

$$\begin{aligned} P(C \text{ or better}) &= 1 - P(D \cup F) \\ &= 1 - \frac{6}{25} \\ &= \frac{19}{25} \end{aligned}$$

The probability of earning a C or better is $\frac{19}{25}$.

Self Check Find the probability that the student passes (earns a D or better). ■

Independent Events

If two events do not influence each other, they are called **independent events.**

Independent Events

The events A and B are said to be **independent events** if and only if $P(B) = P(B \mid A)$.

In the previous section, we discussed the multiplication property for probabilities:

$$P(A \cap B) = P(A) \cdot P(B \mid A)$$

Substituting $P(B)$ for $P(B \mid A)$ in this property gives a formula for computing probabilities of compound independent events.

Formula for $P(A \cap B)$

If A and B are independent events, then

$$P(A \cap B) = P(A) \cdot P(B)$$

The event A of "drawing an ace from a standard deck of cards" and the event B of "tossing heads" on one toss of a coin are independent events, because neither event influences the other. Consequently,

$$\begin{aligned} P(A \cap B) &= P(A) \cdot P(B) \\ P(\text{drawing an ace and tossing heads}) &= P(\text{drawing an ace}) \cdot P(\text{tossing heads}) \\ &= \frac{4}{52} \cdot \frac{1}{2} \\ &= \frac{1}{26} \end{aligned}$$

EXAMPLE 2 The probability that a baseball player can get a hit is $\frac{1}{3}$. Find the probability that she will get three hits in a row.

Solution Assume that the three times at bat are independent events: One time at bat does not influence her chances of getting a hit on another turn at bat. Because $P(E_1) = \frac{1}{3}$, $P(E_2) = \frac{1}{3}$, and $P(E_3) = \frac{1}{3}$,

$$P(E_1 \cap E_2 \cap E_3) = \frac{1}{3} \cdot \frac{1}{3} \cdot \frac{1}{3} = \frac{1}{27}$$

The probability that she will get three hits in a row is $\frac{1}{27}$.

Self Check The probability that another player can get a hit is $\frac{1}{4}$. Find the probability that she will get four hits in a row. ■

ACTIVE EXAMPLE 3 A die is rolled three times. Find the probability that the outcome is 6 on the first roll, an even number on the second roll, and an odd prime number on the third roll.

Solution The probability of a 6 on any roll is $P(6) = \frac{1}{6}$. Because there are three even integers represented on a die, the probability of rolling an even number is $P(\text{even number}) = P(E) = \frac{3}{6} = \frac{1}{2}$. Since 3 and 5 are the only odd prime numbers on a die, the probability of rolling an odd prime is $P(\text{odd prime}) = P(O) = \frac{2}{6} = \frac{1}{3}$.

Because these three events are independent, the probability of the events happening in succession is the product of the probabilities:

$$\begin{aligned} P(\text{six and even number and odd prime}) &= P(6 \cap E \cap O) \\ &= P(6) \cdot P(E) \cdot P(O) \\ &= \frac{1}{6} \cdot \frac{1}{2} \cdot \frac{1}{3} \\ &= \frac{1}{36} \end{aligned}$$

Self Check Find the probability that the outcome is five on the first roll, an odd number on the second roll, and two on the third roll. ■

ACTIVE EXAMPLE 4 The probability that a drug will cure dandruff is $\frac{1}{8}$. However, if the drug is used, the probability that it will cause side effects is $\frac{1}{6}$. Find the probability that a patient who uses the drug will be cured and will suffer no side effects.

Solution The probability that the drug will cure dandruff is $P(C) = \frac{1}{8}$. The probability of having side effects is $P(E) = \frac{1}{6}$. The probability that the patient will have no side effects is

COLLEGE **Algebra** *f(x)* **Now**™
Go to **http://1pass.thomson.com** or your CD to practice this example.

$$P(\overline{E}) = 1 - P(E) = 1 - \frac{1}{6} = \frac{5}{6}$$

Since these events are independent,

$$\begin{aligned} P(\text{cure and no side effects}) &= P(C \cap \overline{E}) \\ &= P(C) \cdot P(\overline{E}) \\ &= \frac{1}{8} \cdot \frac{5}{6} \\ &= \frac{5}{48} \end{aligned}$$

Self Check Find the probability that the patient will be cured and suffer side effects. ■

Self Check Answers

1. $\frac{24}{25}$ **2.** $\frac{1}{256}$ **3.** $\frac{1}{72}$ **4.** $\frac{1}{48}$

8.8 Exercises

VOCABULARY AND CONCEPTS ***Fill in the blanks.***

1. A <u>**compound**</u> event is one event *or* another or one event *followed by* another.

2. $P(A \cup B) =$ <u>$P(A) + P(B) - P(A \cap B)$</u>

3. If A and B are <u>**mutually exclusive**</u>, then $P(A \cup B) = P(A) + P(B)$.

4. The event $\overline{A}$ is read as "<u>**not** ***A***</u>."

5. $P(\overline{A}) =$ <u>$1 - P(A)$</u>

6. Two events, A and B, are called independent events when <u>$P(B) = P(B|A)$</u>.

7. If A and B are independent events, then $P(A \cap B) =$ <u>$P(A) \cdot P(B)$</u>.

8. If two events do not influence each other, they are called <u>**independent**</u> events.

PRACTICE *Assume that you draw one card from a standard card deck. Find the probability of each event.*

9. Drawing a black card $\frac{1}{2}$
10. Drawing a jack $\frac{1}{13}$
11. Drawing a black card or an ace $\frac{7}{13}$
12. Drawing a red card or a face card $\frac{8}{13}$

Assume that you draw two cards from a card deck, without replacement. Find the probability of each event.

13. Drawing two aces $\frac{1}{221}$
14. Drawing three aces **0**
15. Drawing a club and then another black card $\frac{25}{204}$
16. Drawing a heart and then a spade $\frac{13}{204}$

Assume that you roll two dice once. Find the probability of each result.

17. Rolling a sum of 7 or 6 $\frac{11}{36}$
18. Rolling a sum of 5 or an even sum $\frac{11}{18}$
19. Rolling a sum of 10 or an odd sum $\frac{7}{12}$
20. Rolling a sum of 12 or 1 $\frac{1}{36}$

Assume that you have a bucket containing 7 beige capsules, 3 blue capsules, and 6 green capsules. You make a single draw from the bucket, taking one capsule. Find the probability of each result.

21. Drawing a beige or a blue capsule $\frac{5}{8}$
22. Drawing a green capsule $\frac{3}{8}$
23. Not drawing a blue capsule $\frac{13}{16}$
24. Not drawing either a beige or a blue capsule $\frac{3}{8}$

Assume that you are using the same bucket of capsules as in Exercises 21–24.

25. On two draws from the bucket, find the probability of drawing a beige capsule followed by a green capsule. (Assume that the capsule is returned to the bucket after the first draw.) $\frac{21}{128}$
26. On two draws from the bucket, find the probability of drawing one blue capsule and one green capsule. (Assume that the capsule is not returned to the bucket after the first draw.) $\frac{3}{20}$
27. On three successive draws from the bucket (without replacement), find the probability of failing to draw a beige capsule. $\frac{3}{20}$
28. Jeff rolls a die and draws one card from a card deck. Find the probability of his rolling a 4 and drawing a four. $\frac{1}{78}$
29. **Birthday problem** Three people are in an elevator together. Find the probability that all three were born on the same day of the week. $\frac{1}{49}$
30. **Birthday problem** Three people are on a bus together. Find the probability that at least one was born on a different day of the week from the others. $\frac{48}{49}$
31. **Birthday problem** Five people are in a room together. Find the probability that all five were born on a different day of the year. **0.973**
32. **Birthday problem** Five people are on a bus together. Find the probability that at least two of them were born on the same day of the year. **0.027**
33. **Sharing homework** If the probability that Rick will solve a problem is $\frac{1}{4}$ and the probability that Dinah will solve it is $\frac{2}{5}$, find the probability that at least one of them will solve it. $\frac{11}{20}$
34. **Signaling** A bugle is used for communication at camp. The call for dinner is based on four pitches and is five notes long. If a child can play these four pitches on a bugle, find the probability that the first five notes that the child plays will call the camp to dinner. (Assume that the child is equally likely to play any of the four pitches each time a note is blown.) $\frac{1}{1{,}024}$

A woman visits her cabin in Canada. The probability that her lawnmower will start is $\frac{1}{2}$, the probability that her gas-powered saw will start is $\frac{1}{3}$, and the probability that her outboard motor will start is $\frac{3}{4}$. Find each probability.

35. That all three will start $\frac{1}{8}$
36. That none will start $\frac{1}{12}$
37. That exactly one will start $\frac{3}{8}$
38. That exactly two will start $\frac{5}{12}$

APPLICATIONS

39. **Immigration** Three children leave Thailand to start a new life in either the United States or France. The probability that May Xao will go to France is $\frac{1}{3}$, that Tou Lia will go to France is $\frac{1}{2}$, and that May Moua will go to France is $\frac{1}{6}$. Find the probability that exactly two of them will end up in the United States. $\frac{17}{36}$

40. Preparing for the GED The administrators of a program to prepare people for the high school equivalency exam have found that 80% of the students require tutoring in math, 60% need help in English, and 45% need work in both math and English. Find the probability that a student selected at random needs help with either math or English. **0.95**

41. Insurance losses The insurance underwriters have determined that in any one year, the probability that George will have a car accident is 0.05, and that if he has an accident, the probability that he will be hospitalized is 0.40. Find the probability that George will have an accident but not be hospitalized. **0.03**

42. Grading homework One instructor grades homework by randomly choosing 3 out of the 15 problems assigned. Bill did only 8 problems. What is the probability that he won't get caught? $\frac{8}{65}$

DISCOVERY AND WRITING

43. Explain the difference between dependent and independent events, and give examples of each.

44. Explain why **a.** $P(A \mid A) = 1$ and **b.** $P(A \mid \overline{A}) = 0$.

REVIEW ***Maximize P subject to the given constraints.***

45. $P = 2x + y$

$$\begin{cases} x \geq 0 \\ y \geq 0 \\ x \leq 4 \\ x + 2y \leq 8 \end{cases}$$

$P = 10$ at (4, 2)

46. $P = 3x - y$

$$\begin{cases} x \geq 0 \\ y \geq 0 \\ y \leq 2 \\ x \leq 2 \\ x + y \leq 3 \end{cases}$$

$P = 6$ at (2, 0)

8.9 Odds and Mathematical Expectation

In this section, you will learn about

■ Odds ■ Mathematical Expectation

A concept closely related to probability is **mathematical odds.**

■ Odds

Odds

The **odds for an event** is the probability of a favorable outcome divided by the probability of an unfavorable outcome.

The **odds against an event** is the probability of an unfavorable outcome divided by the probability of a favorable outcome.

ACTIVE EXAMPLE 1 The probability that a horse will win a race is $\frac{1}{4}$. Find the odds for and the odds against the horse.

Solution Because the probability that the horse will win is $\frac{1}{4}$, the probability that the horse will not win is $\frac{3}{4}$. The odds for the horse are

$$\frac{\text{probability of a win}}{\text{probability of a loss}} = \frac{\frac{1}{4}}{\frac{3}{4}} = \frac{1}{3}$$

or 1 to 3. The odds against the horse are

COLLEGE **Algebra** f(x) **Now**™

Explore this Active Example and test yourself on the steps taken here at **http://1pass.thomson.com.** You can also find this example on the Interactive Video Skillbuilder CD-ROM.

$$\frac{\text{probability of a loss}}{\text{probability of a win}} = \frac{\frac{3}{4}}{\frac{1}{4}} = 3$$

or 3 to 1. The odds for an event is the reciprocal of the odds against the event.

Self Check The probability that a horse will lose a race is $\frac{1}{5}$. Find the odds for and the odds against the horse. ■

Mathematical Expectation

Suppose we have a chance to play a simple game with the following rules:

1. Roll a single die once.
2. If a 6 appears, win $3.
3. If a 5 appears, win $1.
4. If any other number appears, win 50¢.
5. The cost to play (one roll of the die) is $1.

In this game, the probability of any one of the six outcomes—rolling a 6, 5, 4, 3, 2, or 1—is $\frac{1}{6}$, and the winnings are $3, $1, and 50¢. The expected winnings can be found by using the following equation and simplifying the right-hand side:

$$\begin{aligned} E &= \frac{1}{6}(3) + \frac{1}{6}(1) + \frac{1}{6}(0.50) + \frac{1}{6}(0.50) + \frac{1}{6}(0.50) + \frac{1}{6}(0.50) \\ &= \frac{1}{6}(3 + 1 + 0.50 + 0.50 + 0.50 + 0.50) \\ &= \frac{1}{6}(6) \\ &= 1 \end{aligned}$$

Over the long run, we could expect to win $1 with every play of the game. Since it costs $1 to play the game, the expected gain or loss is 0. Because the expected winnings are equal to the admission price, the game is fair.

Mathematical Expectation

If a certain event has n different outcomes with probabilities $P_1, P_2, \ldots, P_n$ and the winnings assigned to each outcome are $x_1, x_2, \ldots x_n$, the expected winnings, or **mathematical expectation,** E is given by

$$E = P_1x_1 + P_2x_2 + \cdots + P_nx_n$$

ACTIVE EXAMPLE 2 It costs $1 to play the following game: Roll two dice; collect $5 if we roll a sum of 7, and collect $2 if we roll a sum of 11. All other numbers pay nothing. Is it wise to play the game?

COLLEGE Algebra f(x) Now™
Go to **http://1pass.thomson.com** or your CD to practice this example.

Solution The probability of rolling a 7 on a single roll of two dice is $\frac{6}{36}$, the probability of rolling an 11 is $\frac{2}{36}$, and the probability of rolling something else is $\frac{28}{36}$. The mathematical expectation is

$$E = \frac{6}{36}(5) + \frac{2}{36}(2) + \frac{28}{36}(0) = \frac{17}{18} \approx \$0.944$$

By playing the game for a long period of time, we can expect to get back about 95¢ for every dollar spent. For the fun of playing, the cost is about 5¢ a game. If the game is enjoyable, it might be worth the expected loss. However, the game is slightly unfair.

Self Check Is it wise to play the game if you win \$4 when you roll a sum of 7 and \$6 when you roll a sum of 11? ■

Self Check Answers

1. 4 to 1; 1 to 4 **2.** The game is even.

8.9 Exercises

VOCABULARY AND CONCEPTS ***Fill in the blanks.***

1. The odds for an event is the probability of a favorable outcome divided by the probability of an unfavorable outcome.

2. The odds against an event is the probability of an unfavorable outcome divided by the probability of a favorable outcome.

3. If the odds for an event are 1 to 4, the probability of winning is $\frac{1}{5}$.

4. Mathematical expectation E is given by $E = P_1x_1 + P_2x_2 + P_3x_3 + \cdots + P_nx_n$.

PRACTICE ***Assume a single roll of a die.***

5. Find the probability of rolling a 6. $\frac{1}{6}$

6. Find the odds in favor of rolling a 6. 1 to 5

7. Find the odds against rolling a 6. 5 to 1

8. Find the probability of rolling an even number. $\frac{1}{2}$

9. Find the odds in favor of rolling an even number. 1 to 1

10. Find the odds against rolling an even number. 1 to 1

Assume a single roll of two dice.

11. Find the probability of rolling a sum of 6. $\frac{5}{36}$

12. Find the odds in favor of rolling a sum of 6. 5 to 31

13. Find the odds against rolling a sum of 6. 31 to 5

14. Find the probability of rolling an even sum. $\frac{1}{2}$

15. Find the odds in favor of rolling an even sum. 1 to 1

16. Find the odds against rolling an even sum. 1 to 1

Assume that you are drawing one card from a standard card deck.

17. Find the odds in favor of drawing a queen. 1 to 12

18. Find the odds against drawing a black card. 1 to 1

19. Find the odds in favor of drawing a face card. 3 to 10

20. Find the odds against drawing a diamond. 3 to 1

21. If the odds in favor of victory are 5 to 2, find the probability of victory. $\frac{5}{7}$

22. If the odds in favor of victory are 5 to 2, find the odds against victory. 2 to 5

23. If the odds against winning are 90 to 1, find the odds in favor of winning. 1 to 90

24. Find the odds in favor of rolling a 7 on a single roll of two dice. 1 to 5

25. Find the odds against tossing four heads in a row with a fair nickel. 15 to 1

26. Find the odds in favor of a couple having four girl babies in succession. (Assume $P(\text{girl}) = \frac{1}{2}$.) 1 to 15

27. The odds against a horse are 8 to 1. Find the probability that the horse will win. $\frac{1}{9}$

28. The odds against a horse are 1 to 1. Find the probability that the horse will lose. $\frac{1}{2}$

29. It costs \$2 to play the following game:
   a. Draw one card from a card deck.
   b. Collect \$5 if an ace is drawn.
   c. Collect \$4 if a king is drawn.
   d. Collect nothing for all other cards drawn.

   Is it wise to play this game? Explain.
   No; the expected winnings are \$$\frac{9}{13}$.

30. **Lottery tickets** One thousand tickets are sold for a lottery with two grand prizes of \$800. Find a fair price for the tickets. **\$1.60**

31. Find the odds against a couple having three baby boys in a row. (Assume $P(\text{boy}) = \frac{1}{2}$.) **7 to 1**

32. Find the odds in favor of tossing at least three heads in five tosses of a fair coin. **1 to 1**

33. Suppose you toss a coin five times and collect \$5 if you toss five heads, \$4 if you toss four heads, \$3 if you toss three heads, and no money for any other combination. How much should you pay to play the game if the game is to be fair? **\$1.72**

34. If you roll two dice one time and collect \$10 for double 6's and \$1 for double 1's, what is a fair price for playing the game? **\$0.31**

35. Counting an ace as 1, a face card as 10, and all others at their numerical value, find the expected value if you draw one card from a card deck. **6.54**

36. Find the expected sum of one roll of two dice. **7**

37. A multiple-choice test of eight questions gives five possible answers for each question. Only one of the answers for each question is right. Find the probability of getting seven right answers by simple guessing.
   $\frac{32}{390{,}625}$

38. In the situation described in Exercise 37, find the odds in favor of getting seven answers right.
   32 to 390,593

DISCOVERY AND WRITING

39. If "the odds are against an event," what can be said about its probability? **less than $\frac{1}{2}$**

40. To disguise the unlikely chance of winning, a contest promoter publishes the odds in favor of winning as 0.0000000372 to 1. What are the odds against winning? **26,881,720 to 1**

REVIEW

41. Find the equation of the line perpendicular to $3x - 2y = 9$ and passing through the point $(1, 1)$.
   $2x + 3y = 5$

42. Find the equations of the parabolas with vertex at the point $(3, -2)$ and passing through the origin.
   $(x - 3)^2 = \frac{9}{2}(y + 2)$ **or** $(y + 2)^2 = -\frac{4}{3}(x - 3)$

43. Find the equation of the circle with center at $(3, 5)$ and tangent to the x-axis. $(x - 3)^2 + (y - 5)^2 = 25$

44. Write the equation
   $$2x^2 + 4y^2 + 12x - 24y + 46 = 0$$
   in standard form, and identify the curve.
   $\frac{(x + 3)^2}{4} + \frac{(y - 3)^2}{2} = 1$; **ellipse**

PROBLEMS AND PROJECTS

1. Solve for n: $\dfrac{n!}{(n - 2)!} = 9{,}900$.

2. Subtract: $\dfrac{1}{(n - 1)!} - \dfrac{1}{n!}$.

3. Use mathematical induction to prove that
   $$3^0 + 3^1 + 3^2 + \cdots + 3^{n-1} = \frac{3^n - 1}{2}$$

4. What is the sum of all possible three-digit numbers that can be formed from the digits 2, 5, and 7?

5. If $P(n, r) = 5{,}040$ and $C(n, r) = 210$, find r. (*Hint:* $P(n, r) = r!C(n, r)$.)

6. In a bus accident, 3 out of the 20 passengers were injured. Four of the 20 were politicians. What is the probability that all the injured were politicians?

Project 1

By the binomial theorem,

$$(1 + x)^m = 1 + mx + \frac{m(m - 1)}{2!}x^2 + \frac{m(m - 1)(m - 2)}{3!}x^3 + \cdots + \frac{m(m - 1)\cdots(m - n + 1)}{n!}x^n + \cdots$$

- If m is not a positive integer or zero, explain why the coefficients of the terms of the binomial theorem never become zero. In this case, the formula is called the **binomial series.**
- In the binomial series, let $x = -\frac{1}{2}$ and $m = \frac{1}{2}$, so that the left-hand side becomes

$$\left(1 - \frac{1}{2}\right)^{1/2} = \left(\frac{1}{2}\right)^{1/2} = \sqrt{\frac{1}{2}} = \frac{\sqrt{2}}{2}$$

Use a calculator to evaluate this expression.

- On the right-hand side of the equation, also let $x = -\frac{1}{2}$ and $m = \frac{1}{2}$. Calculate the first 4 terms of the series. How close is their sum to the value in the previous question? Is the sum of the first 5 terms more accurate?

Project 2

If n, a, b, and c are nonnegative integers and $a + b + c = n$, then the **multinomial coefficient** $\binom{n}{a,\, b,\, c}$ is defined as $\dfrac{n!}{a!b!c!}$.

- Expand $(x + y + z)^3$.
- In this expansion, verify that the coefficient of x^2y is $\binom{3}{2,\, 1,\, 0}$ and that the coefficient of xyz is $\binom{3}{1,\, 1,\, 1}$. What is the multinomial coefficient of x^3? (*Hint:* $x^3 = x^3y^0z^0$.) What is the multinomial coefficient of xz^2?
- In the expansion of $(w + x + y + z)^7$, what is the coefficient of w^2x^3yz?
- In the expansion of $(x + y)^n$, explain why the multinomial coefficients are just the binomial coefficients.

Project 3

- Note the hockey-stick pattern in the numbers of Pascal's triangle in Illustration (a). What is the missing number in the rightmost hockey stick? Does the pattern work with larger hockey sticks? Experiment and report your conclusions.
- In Illustration (b), find the pattern in the sums of increasingly larger portions of Pascal's triangle. Find the sum of all of the numbers up to and including the row that begins 1 10 45 . . . .
- There are many other patterns hidden in Pascal's triangle. Find some more and share them with your class. Illustration (c) is an idea to get you started.

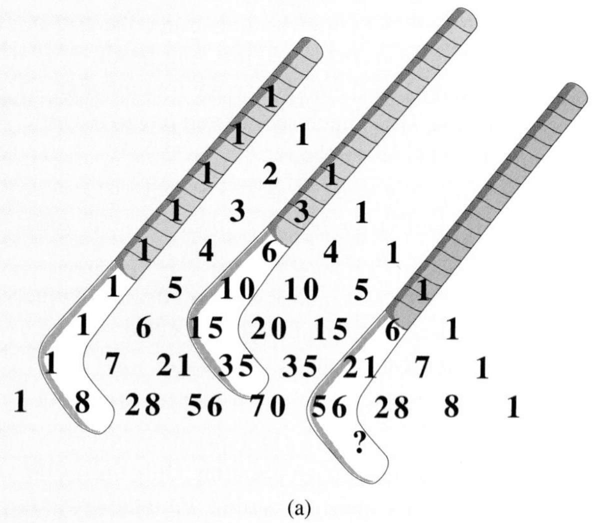

(a)

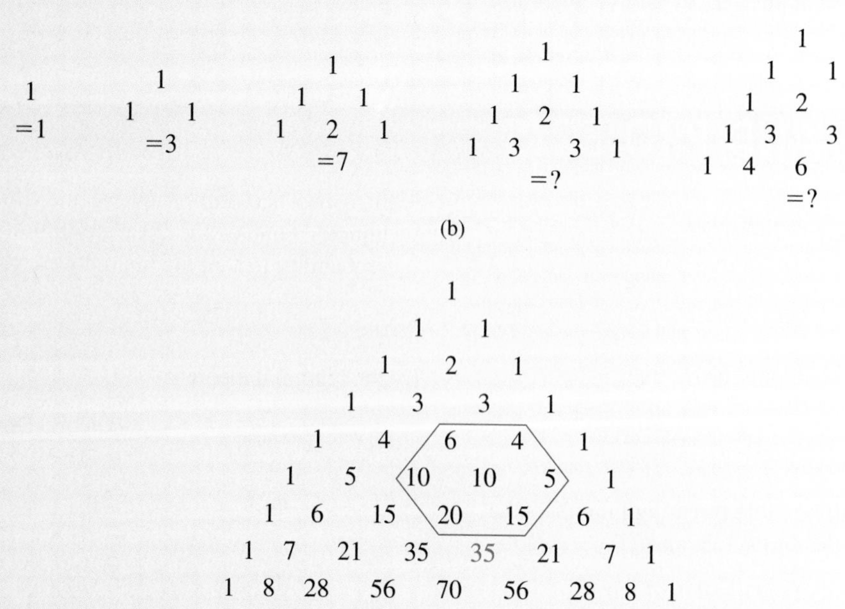

(b)

(c)

CHAPTER SUMMARY

CONCEPTS / REVIEW EXERCISES

8.1 The Binomial Theorem

$n! = n(n-1)(n-2)\cdots 3\cdot 2\cdot 1$

$0! = 1 \quad n(n-1)! = n!$

The binomial theorem:
If n is any positive integer, then

$$(a+b)^n = a^n + \frac{n!}{1!(n-1)!}a^{n-1}b + \frac{n!}{2!(n-2)!}a^{n-2}b^2 + \frac{n!}{3!(n-3)!}a^{n-3}b^3 + \frac{n!}{r!(n-r)!}a^{n-r}b^r + \cdots + b^n$$

Find each value.

1. $6!$ 720
2. $7!\cdot 0!\cdot 1!\cdot 3!$ 30,240
3. $\frac{8!}{7!}$ 8
4. $\frac{5!\cdot 7!\cdot 8!}{6!\cdot 9!}$ $\frac{280}{3}$

Expand each expression.

5. $(x+y)^3$
$x^3 + 3x^2y + 3xy^2 + y^3$
6. $(p+q)^4$
$p^4 + 4p^3q + 6p^2q^2 + 4pq^3 + q^4$
7. $(a-b)^5$
$a^5 - 5a^4b + 10a^3b^2 - 10a^2b^3 + 5ab^4 - b^5$
8. $(2a-b)^3$
$8a^3 - 12a^2b + 6ab^2 - b^3$

Find the required term of each expansion.

9. $(a+b)^8$; 4th term $56a^5b^3$
10. $(2x-y)^5$; 3rd term $80x^3y^2$
11. $(x-y)^9$; 7th term $84x^3y^6$
12. $(4x+7)^6$; 4th term $439{,}040x^3$

8.2 Sequences, Series, and Summation Notation

A **sequence** is a function whose domain is the set of natural numbers.

If c is a constant, then

$$\sum_{k=1}^{n} c = nc$$

If c is a constant, then

$$\sum_{k=1}^{n} cf(k) = c\sum_{k=1}^{n} f(k)$$

$$\sum_{k=1}^{n}[f(k)+g(k)] = \sum_{k=1}^{n} f(k) + \sum_{k=1}^{n} g(k)$$

Write the fourth term in each sequence.

13. $0, 7, 26, \ldots, n^3 - 1, \ldots$
63
14. $\frac{3}{2}, 3, \frac{11}{2}, \ldots, \frac{n^2+2}{2}, \ldots$
9

Find the first four terms of each sequence.

15. $a_1 = 5$ and $a_{n+1} = 3a_n + 2$ 5, 17, 53, 161
16. $a_1 = -2$ and $a_{n+1} = 2a_n{}^2$ −2, 8, 128, 32,768

Evaluate each expression.

17. $\sum_{k=1}^{4} 3k^2$ 90
18. $\sum_{k=1}^{10} 6$ 60
19. $\sum_{k=5}^{8}(k^3 + 3k^2)$ 1,718
20. $\sum_{k=1}^{30}\left(\frac{3}{2}k - 12\right) - \frac{3}{2}\sum_{k=1}^{30} k$ −360

8.3 Arithmetic Sequences

If a is the first term, n is the number of terms, and d is the common difference, then the formula $a_n = a + (n - 1)d$ gives the last (or nth) term of an arithmetic sequence.

Find the required term of each arithmetic sequence.

21. 5, 9, 13, . . .; 29th term
117

22. 8, 15, 22, . . .; 40th term
281

23. 6, −1, −8, . . .; 15th term
−92

24. $\frac{1}{2}, -\frac{3}{2}, -\frac{7}{2}, \ldots$; 35th term
$-\frac{135}{2}$

25. Find three arithmetic means between 2 and 8. $\frac{7}{2}, 5, \frac{13}{2}$

26. Find five arithmetic means between 10 and 100. **25, 40, 55, 70, 85**

The formula

$$S_n = \frac{n(a + a_n)}{2}$$

gives the sum of the first n terms of an arithmetic sequence, where S_n is the sum, a is the first term, a_n is the last (or nth) term, and n is the number of terms.

Find the sum of the first 40 terms in each sequence.

27. 5, 9, 13, . . . **3,320**

28. 8, 15, 22, . . . **5,780**

29. 6, −1, −8, . . . **−5,220**

30. $\frac{1}{2}, -\frac{3}{2}, -\frac{7}{2}, \ldots$ **−1,540**

8.4 Geometric Sequences

If a is the first term and r is the common ratio, then the formula $a_n = ar^{n-1}$ gives the nth (or last) term of a geometric sequence.

Find the required term of each geometric sequence.

31. 81, 27, 9, . . .; 11th term
$\frac{1}{729}$

32. 2, 6, 18, . . .; 9th term
13,122

33. $9, \frac{9}{2}, \frac{9}{4}, \ldots$ 15th term
$\frac{9}{16{,}384}$

34. $8, -\frac{8}{5}, \frac{8}{25}, \ldots$; 7th term
$\frac{8}{15{,}625}$

35. Find three positive geometric means between 2 and 8. $2\sqrt{2}, 4, 4\sqrt{2}$

36. Find four geometric means between −2 and 64. **4, −8, 16, −32**

37. Find the positive geometric mean between 4 and 64. **16**

The formula

$$S_n = \frac{a - ar^n}{1 - r} \quad (r \neq 1)$$

gives the sum of the first n terms of a geometric sequence, where S_n is the sum, a is the first term, r is the common ratio, and n is the number of terms.

Find the sum of the first 8 terms in each sequence.

38. 81, 27, 9, . . . $\frac{3{,}280}{27}$

39. 2, 6, 18, . . . **6,560**

40. $9, \frac{9}{2}, \frac{9}{4}, \ldots$ $\frac{2{,}295}{128}$

41. $8, -\frac{8}{5}, \frac{8}{25}, \ldots$ $\frac{520{,}832}{78{,}125}$

42. Find the sum of the first eight terms of the sequence $\frac{1}{3}$, 1, 3, ...
$\frac{3{,}280}{3}$

43. Find the seventh term of the sequence $2\sqrt{2}, 4, 4\sqrt{2}, \ldots$.
$16\sqrt{2}$

If $|r| < 1$, the formula

$$S_\infty = \frac{a}{1 - r}$$

gives the sum of the terms of an infinite geometric sequence, where S_∞ is the sum, a is the first term, and r is the common ratio.

Find the sum of each infinite sequence, if possible.

44. $\frac{1}{3}, \frac{1}{6}, \frac{1}{12}, \ldots$ $\frac{2}{3}$

45. $\frac{1}{5}, -\frac{2}{15}, \frac{4}{45}, \ldots$ $\frac{3}{25}$

46. $1, \frac{3}{2}, \frac{9}{4}, \ldots$ no sum

47. $0.5, 0.25, 0.125, \ldots$ 1

Change each decimal into a common fraction.

48. $0.\overline{3}$ $\frac{1}{3}$

49. $0.\overline{9}$ 1

50. $0.\overline{17}$ $\frac{17}{99}$

51. $0.\overline{45}$ $\frac{5}{11}$

52. Investment problem If Leonard invests $3,000 in a 6-year certificate of deposit at the annual rate of 7.75%, compounded daily, how much money will be in the account when it matures? **$4,775.81**

53. College enrollments The enrollment at Hometown College is growing at the rate of 5% over each previous year's enrollment. If the enrollment is currently 4,000 students, what will it be 10 years from now? What was it 5 years ago? **6,516; 3,134**

54. House trailer depreciation A house trailer that originally cost $10,000 depreciates in value at the rate of 10% per year. How much will the trailer be worth after 10 years? **$3,486.78**

8.5 Mathematical Induction

Mathematical induction: If a statement involving the natural number n has the two properties that

1. the statement is true for $n = 1$, and
2. if the statement is true for $n = k$, then it is true for $n = k + 1$,

the statement is true for all natural numbers.

55. Verify the following formula for $n = 1, n = 2, n = 3$, and $n = 4$:

$$1^3 + 2^3 + 3^3 + \cdots + n^3 = \frac{n^2(n + 1)^2}{4}$$

56. Prove the formula given in Exercise 55 by mathematical induction.

8.6 Permutations and Combinations

$$P(n, r) = \frac{n!}{(n - r)!}$$

$$P(n, n) = n!$$

$$P(n, 0) = 1$$

$$C(n, r) = \binom{n}{r} = \frac{n!}{r!(n - r)!}$$

$$C(n, n) = 1$$

$$C(n, 0) = 1$$

Evaluate each expression.

57. $P(8, 5)$ 6,720

58. $C(7, 4)$ 35

59. $0! \cdot 1!$ 1

60. $P(10, 2) \cdot C(10, 2)$ 4,050

61. $P(8, 6) \cdot C(8, 6)$ 564,480

62. $C(8, 5) \cdot C(6, 2)$ 840

63. $C(7, 5) \cdot P(4, 0)$ 21

64. $C(12, 10) \cdot C(11, 0)$ 66

65. $\frac{P(8, 5)}{C(8, 5)}$ 120

66. $\frac{C(8, 5)}{C(13, 5)}$ $\frac{56}{1,287}$

67. $\frac{C(6, 3)}{C(10, 3)}$ $\frac{1}{6}$

68. $\frac{C(13, 5)}{C(52, 5)}$ $\frac{33}{66,640}$

There are $(n-1)!$ ways to place n things in a circle.

If an n-letter word has a of one letter, b of another letter, and so on, the number of distinguishable words that can be formed using each letter exactly once is

$$\frac{n!}{a!b!\cdots}$$

69. In how many ways can 10 teenagers be seated at a round table if 2 girls wish to sit with their boyfriends? **20,160**

70. How many distinguishable words can be formed from the letters of the word *casserole* if each letter is used exactly once? **90,720**

8.7 Probability

An event that cannot happen has a probability of 0. An event that is certain to happen has a probability of 1. All other events have probabilities between 0 and 1.

If S is the sample space of an experiment with n distinct and equally likely outcomes, and if E is an event that occurs in s of those ways, then the probability of E is

$$P(E) = \frac{n(E)}{n(S)} = \frac{s}{n}$$

$$P(A \cap B) = P(A) \cdot P(B \mid A)$$

71. Make a tree diagram (like Figure 8-2 on page 614) to illustrate the possible results of tossing a coin four times.

72. In how many ways can you draw a 5-card poker hand of 3 aces and 2 kings? **24**

73. Find the probability of drawing the hand described in Exercise 72. $\frac{1}{108{,}290}$

74. Find the probability of not drawing the hand described in Exercise 72. $\frac{108{,}289}{108{,}290}$

75. Find the probability of having a 13-card bridge hand consisting of 4 aces, 4 kings, 4 queens, and 1 jack. **about 6.3×10^{-12}**

76. Find the probability of choosing a committee of 3 men and 2 women from a group of 8 men and 6 women. $\frac{60}{143}$

77. Find the probability of drawing a club or a spade on one draw from a card deck. $\frac{1}{2}$

78. Find the probability of drawing a black card or a king on one draw from a card deck. $\frac{7}{13}$

79. Find the probability of getting an ace-high royal flush in hearts (ace, king, queen, jack, and ten of hearts) in poker. $\frac{1}{2{,}598{,}960}$

80. Find the probability of being dealt 5 cards of one suit in a poker hand. $\frac{33}{16{,}660}$

81. Find the probability of getting 3 heads or fewer on 4 tosses of a fair coin. $\frac{15}{16}$

8.8 Computation of Compound Probabilities

If A and B are two events, then

$P(A \cup B) = P(A) + P(B) - P(A \cap B)$

If A and B cannot occur simultaneously, then

$$P(A \cup B) = P(A) + P(B)$$

If A is any event, then

$$P(\overline{A}) = 1 - P(A)$$

If A and B are independent events, then

$$P(A \cap B) = P(A) \cdot P(B)$$

82. On one draw from a standard card deck, find the probability of drawing a two or a spade. $\frac{4}{13}$

83. If the probability that a drug cures a disease is 0.83, and we give the drug to 800 people with the disease, find the expected number of people who will not be cured. **136**

8.9 Odds and Mathematical Expectation

The concepts of probabilities and odds are related.

84. Find the odds against a horse if the probability that the horse will win is $\frac{7}{8}$.
1 to 7

85. Find the odds in favor of a couple having 4 baby girls in a row. **1 to 15**

86. If the probability that Joe will marry is $\frac{5}{6}$ and the probability that John will marry is $\frac{3}{4}$, find the odds against either one becoming a husband.
1 to 23

87. If the odds against Priscilla's graduation from college are $\frac{10}{11}$, find the probability that she will graduate. $\frac{11}{21}$

A game is fair if its cost to play equals the expected winnings.

88. Find the expected earnings if you collect \$1 for every heads you get when you toss a fair coin 4 times. **\$2**

89. If the total number of subsets that a set with n elements can have is 2^n, explain why

$$\binom{n}{0} + \binom{n}{1} + \binom{n}{2} + \cdots + \binom{n}{n} = 2^n$$

CHAPTER TEST

Find each value.

1. $3! \cdot 0! \cdot 4! \cdot 1!$ **144**

2. $\dfrac{2! \cdot 4! \cdot 6! \cdot 8!}{3! \cdot 5! \cdot 7!}$ **384**

Find the required term in each expansion.

3. $(x + 2y)^5$; 2nd term
$\mathbf{10x^4y}$

4. $(2a - b)^8$; 7th term
$\mathbf{112a^2b^6}$

Find each sum.

5. $\sum_{k=1}^{3} (4k + 1)$ **27**

6. $\sum_{k=2}^{4} (3k - 21)$ **−36**

Find the sum of the first ten terms of each sequence.

7. 2, 5, 8, . . . **155**

8. 5, 1, −3, . . . **−130**

9. Find three arithmetic means between 4 and 24.
9, 14, 19

10. Find two geometric means between -2 and -54.
$-6, -18$

Find the sum of the first ten terms of each sequence.

11. $\frac{1}{4}, \frac{1}{2}, 1, \ldots$ **255.75**

12. $6, 2, \frac{2}{3}, \ldots$ **about 9**

13. A car costing $\$c$ when new depreciates 25% of the previous year's value each year. How much is the car worth after 3 years? **about $\$0.42c$**

14. A house costing $\$c$ when new appreciates 10% of the previous year's value each year. How much will the house be worth after 4 years? **about $\$1.46c$**

15. Prove by induction:

$$3 + 4 + 5 + \cdots + (n + 2) = \tfrac{1}{2}n(n + 5)$$

16. How many six-digit license plates can be made if no plate begins with 0 or 1? **800,000**

Find each value.

17. $P(7, 2)$ **42**

18. $P(4, 4)$ **24**

19. $C(8, 2)$ **28**

20. $C(12, 0)$ **1**

21. How many ways can 4 men and 4 women stand in line if all the women are first? **576**

22. How many different ways can 6 people be seated at a round table? **120**

23. How many different words can be formed from the letters of the word *bluff* if each letter is used once?
60

24. Show the sample space of the experiment: toss a fair coin three times.
{(H, H, H), (H, H, T), (H, T, H), (H, T, T), (T, H, H), (T, H, T), (T, T, H), (T, T, T)}

Find each probability.

25. Rolling a 5 on one roll of a die $\frac{1}{6}$

26. Drawing a jack or a queen from a standard card deck
$\frac{2}{13}$

27. Receiving 5 hearts for a 5-card poker hand $\frac{33}{66{,}640}$

28. Tossing 2 heads in 5 tosses of a fair coin $\frac{5}{16}$

29. If 30% of the population enjoy jazz, 80% enjoy popular music, and 30% don't like either, what percent of the people like both types of music? **40%**

30. If the probability that a person is sick is 0.1, find the probability that the person is well. **0.9**

Find each probability.

31. Drawing a black card or a face card on one draw from a standard card deck $\frac{8}{13}$

32. Rolling a sum of 3 or an even sum with one roll of two dice $\frac{5}{9}$

33. If the odds for a horse are 3 to 1, find the probability that the horse will win. $\frac{3}{4}$

34. Assume a game with the following rules:
Roll a single die once.
If 6 appears, win \$4.
If 5 appears, win \$2.
Find the fair price to play the game. **\$1**

9 The Mathematics of Finance

COLLEGE **Algebra** $f(x)$ **Now**™

Throughout the chapter, this icon introduces resources on the College AlgebraNow Web Site, accessed through **http://1pass.thomson.com**, that will:

- Help you test your knowledge of the material in this chapter prior to reading it,
- Allow you to take an exam-prep quiz, and
- Provide a Personalized Learning Plan targeting areas you should study.

Careers and Mathematics

ACTUARY

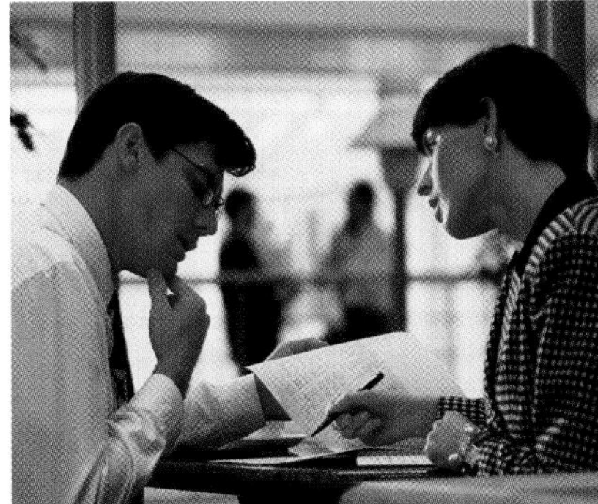

Tim Brown/Getty Images

Actuaries determine future risk, make price decisions, and formulate investment strategies. Some actuaries also design insurance, financial, and pension plans and ensure that these plans are maintained on a sound financial basis. Most actuaries specialize in life and health or property and casualty insurance; others work primarily in finance or employee benefits.

Most actuaries are employed in the insurance industry, in which they estimate the amount a company will pay in claims. For example, property/casualty actuaries calculate the expected amount of claims resulting from automobile accidents, which varies depending on the insured person's age, sex, driving history, type of car, and other factors. Actuaries ensure that the price, or premium, charged for such insurance will enable the company to cover claims and other expenses.

Actuaries held about 15,000 jobs in 2002. Of these, more than 50% were wage and salary workers employed in the insurance industry.

JOB OUTLOOK

Employment of actuaries is expected to grow as fast as the average for all occupations through 2012.

Median annual earnings of actuaries were $69,970 in 2002.

For a sample application, see Example 3 in Section 9.3.

In this chapter, we discuss the mathematics of finance—the rules that govern investing and borrowing money.

9.1 Interest

In this section, you will learn about

- Simple Interest ■ Compound Interest ■ Bank Notes
- Effective Rate of Interest ■ Present Value

When money is borrowed, the lender expects to be paid back the amount of the loan plus an additional charge for the use of the money. This additional charge is called **interest.** When money is deposited in a bank, the bank pays the depositor for the use of the money. The money the deposit earns is also called *interest.*

Interest can be computed in two ways: either as *simple interest* or as *compound interest.*

■ Simple Interest

Simple interest is computed by finding the product of the **principal** (the amount of money on deposit), the **rate of interest** (usually written as a decimal), and the **time** (usually expressed in years).

$$\text{Interest} = \text{principal} \cdot \text{rate} \cdot \text{time}$$

This word equation suggests the following formula.

Simple Interest

The simple interest I earned on a principal P in an account paying an annual interest rate r for a length of time t is given by the formula

$$I = Prt$$

ACTIVE EXAMPLE 1 Find the simple interest earned on a deposit of \$5,750 that is left on deposit for $3\frac{1}{2}$ years and earns an annual interest rate of $4\frac{1}{2}\%$.

Solution We write $3\frac{1}{2}$ and $4\frac{1}{2}\%$ as decimals and substitute the given values in the formula for simple interest.

COLLEGE **Algebra** $f(x)$ **Now**™

Explore this Active Example and test yourself on the steps taken here at **http://1pass.thomson.com.** You can also find this example on the Interactive Video Skillbuilder CD-ROM.

$I = Prt$ The formula for simple interest.

$I = 5{,}750 \cdot 0.045 \cdot 3.5$ Substitute 5,750 for P, 0.045 for r, and 3.5 for t.

$I = 905.625$ Perform the multiplications.

In $3\frac{1}{2}$ years, the account will earn \$905.63 in simple interest.

Self Check Find the simple interest earned on a deposit of \$12,275 that is left on deposit for $5\frac{1}{4}$ years and earns an annual interest rate of $3\frac{3}{4}\%$. ■

EXAMPLE 2 Three years after investing \$15,000 a retired couple received a check for \$3,375 in simple interest. Find the annual interest rate their money earned during that time.

Solution The couple invested \$15,000 (the principal) for 3 years (the time) and made \$3,375 (the simple interest). We must find the annual interest rate r.

$$
\begin{aligned}
I &= Prt \\
3{,}375 &= \mathbf{15{,}000} \cdot r \cdot \mathbf{3} && \text{Substitute 3,375 for } I\text{, 15,000 for } P\text{, and 3 for } t. \\
3{,}375 &= 45{,}000r && \text{Simplify.} \\
\frac{3{,}375}{\mathbf{45{,}000}} &= \frac{45{,}000}{\mathbf{45{,}000}}r && \text{Divide both sides by 45,000.} \\
0.075 &= r && \text{Perform the divisions.} \\
r &= 7.5\% && \text{Write 0.075 as a percent.}
\end{aligned}
$$

The couple received an annual rate of 7.5% for the 3-year period.

Self Check Find the length of time it will take for the interest to grow to \$9,000. ■

■ Compound Interest

When interest is left in an account and also earns interest, we say that the account earns **compound interest.**

EXAMPLE 3 Yolanda deposits \$10,000 in a savings account paying 6% interest, compounded annually. Find the balance in her account after each of the first three years.

Solution At the end of the first year, the interest earned is 6% of the principal, or

$$0.06(\$10{,}000) = \$600$$

This interest is added to the principal; after one year, the balance is \$10,600.

The second year's earned interest is 6% of the current balance, or

$$0.06(\$10{,}600) = \$636$$

This interest is also added to the principal, giving a second-year balance of \$11,236.

The interest earned during the third year is

$$0.06(\$11{,}236) = \$674.16$$

giving Yolanda a balance of \$11,236 + \$674.16, or \$11,910.16, after three years.

Self Check What will be the balance in Yolanda's account after two more years? ■

We can generalize this example to find a formula for compound interest calculations. Suppose that the original deposit, the **principal,** is P dollars, that interest is paid at an **annual rate** r, and that the **accumulated amount** or the **future value** in the account at the end of the first year is A_1. Then the interest earned that year is Pr, and

The amount after one year	equals	the original deposit	plus	the interest earned.

$$A_1 = P + Pr$$
$$= P(1 + r) \quad \text{Factor out the common factor, } P.$$

The amount, A_1, at the end of the first year is also the deposit at the beginning of the second year. The amount at the end of the second year, A_2, is

The amount after two years	equals	the amount after one year	plus	the interest earned on the amount after one year.

$$A_2 = A_1 + A_1 r$$
$$= A_1(1 + r) \quad \text{Factor out the common factor, } A_1.$$
$$= P(1 + r)(1 + r) \quad \text{Substitute } P(1 + r) \text{ for } A_1.$$
$$= P(1 + r)^2 \quad \text{Simplify.}$$

By the end of the third year, the amount will be

$$A_3 = P(1 + r)^3$$

The pattern continues, and we have the following result.

Compound Interest (Annual Compounding)

A single deposit P, called the **principal,** earning compound interest for n years at an annual rate r, will grow to a **future value** FV according to the formula

$$FV = P(1 + r)^n$$

EXAMPLE 4 For their newborn child, parents deposit \$10,000 in a college account that pays 8% interest, compounded annually. How much will the account be worth on the child's 17th birthday?

Solution We substitute $P = 10{,}000$, $r = 0.08$, and $n = 17$ in the compound interest formula and use a calculator to find the future value.

$$FV = P(1 + r)^n$$
$$FV = \mathbf{10{,}000}(1 + \mathbf{0.08})^{\mathbf{17}}$$
$$= 10{,}000(1.08)^{17}$$
$$\approx 37{,}000.18054801$$

To the nearest cent, \$37,000.18 will be available on the child's 17th birthday.

Self Check If the parents leave the money on deposit for two more years, what amount will be available? ■

Interest compounded once each year is **compounded annually.** Many financial institutions compound interest more often. For example, instead of paying an annual rate of 8% once a year, they might pay 4% twice each year, or 2% four times each year. The annual rate, 8%, is also called the **nominal rate,** and the time between interest calculations is the **conversion period.** If there are k periods each year, interest is paid at the **periodic rate** given by the following formula.

Periodic Rate

$$\text{Periodic rate} = \frac{\text{annual rate}}{\text{number of periods per year}}$$

This formula is often written as

$$i = \frac{r}{k}$$

where i is the periodic interest rate, r is the annual rate, and k is the number of times interest is paid each year.

If interest is calculated k times during each year, in n years there will be kn conversions. Each conversion is at the periodic rate i. This leads to another form of the compound interest formula.

Compound Interest Formula

A principal P, earning interest compounded k times a year for n years at an annual rate r, will grow to the future value FV according to the formula

$$FV = P(1 + i)^{kn}$$

where $i = \frac{r}{k}$ is the periodic interest rate.

Interest paid twice each year is called **semiannual** compounding, four times each year **quarterly** compounding, twelve times each year **monthly** compounding, and 360 or 365 times each year **daily** compounding.

ACTIVE EXAMPLE 5 If the parents of Example 4 invested that \$10,000 in an account paying 8%, but compounded quarterly, how much more money would be available after 17 years?

Solution We first calculate the periodic rate, i.

COLLEGE **Algebra** f(x) **Now**™
Go to **http://1pass.thomson.com** or your CD to practice this example.

$$i = \frac{r}{k}$$

$$i = \frac{\mathbf{0.08}}{\mathbf{4}} \qquad \textbf{Substitute } r = \mathbf{0.08} \textbf{ and } k = \mathbf{4}.$$

$$i = 0.02$$

We then substitute $P = 10{,}000$, $i = 0.02$, $k = 4$, and $n = 17$ in the compound interest formula, and use a calculator.

$$FV = P(1 + i)^{kn}$$
$$FV = \mathbf{10{,}000}(1 + \mathbf{0.02})^{\mathbf{4 \cdot 17}}$$
$$= 10{,}000(1.02)^{68}$$
$$\approx 38{,}442.50502546$$

To the nearest cent, \$38,442.51 will be available, an increase of \$1,442.33 over annual compounding.

Self Check **a.** What would \$10,000 become in 17 years if compounded monthly at a nominal rate of 8%? **b.** How does this compare with quarterly compounding? ■

Accent on Technology GROWTH OF MONEY

We can use a graphing calculator to find the time it would take a \$10,000 investment to triple, assuming an 8% annual rate, compounded quarterly.

In n years, \$10,000 earning 8% interest, compounded quarterly, will become the future value

$$10{,}000(1.02)^{4n}$$

To watch this value grow, we enter the function

$$Y_1 = 10000 * 1.02\text{^}(4 * X)$$

in a graphing calculator, and set the window to $0 \leq X \leq 10$ (for 10 years) and $0 \leq Y \leq 40000$ (for the dollar amount).

The graph appears in Figure 9-1(a). To find the time it would take for the investment to triple, we use TRACE to move to the point with a Y-value close to 30,000. The X-value in Figure 9-1(b) shows that the investment would triple in about 13.9 years.

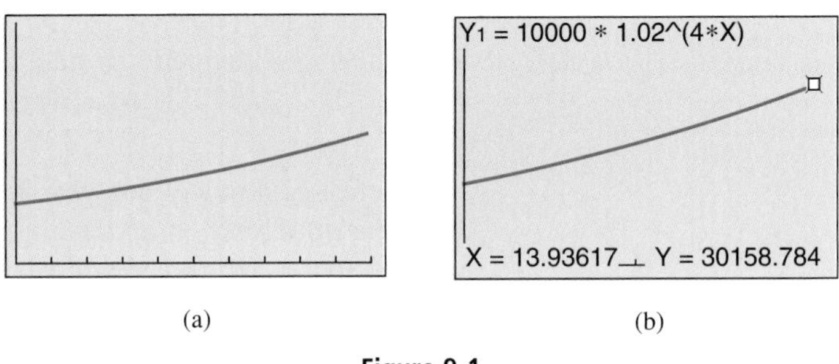

(a) (b)

Figure 9-1

Bank Notes

When a customer borrows money from a bank, the bank is making an investment in that person. The amount of the loan is the bank's deposit, and the bank expects to be repaid with interest in a single **balloon payment** at a later date. These loans, called **notes,** are based on a 360-day year, and they are usually written for 30 days, 90 days, or 180 days. We use the formula for compound interest to calculate the terms of the loan.

EXAMPLE 6 A student needs \$4,000 for tuition. His bank writes a 9%, 180-day note, with interest compounded daily. What will he owe at the end of 180 days?

Solution In making the loan, the bank invests the principal, \$4,000. The amount to be repaid is the expected future value,

$$FV = P(1 + i)^{kn}$$

where the principal P is \$4,000; the frequency of compounding k is 360; the periodic rate i is $\frac{0.09}{360} = 0.00025$; and the term n is 0.5 (180 days is one-half of 360 days.)

$$FV = P(1 + i)^{kn}$$
$$FV = 4{,}000(1 + 0.00025)^{360 \cdot 0.5}$$
$$= 4{,}000(1.00025)^{180}$$
$$\approx 4{,}184.087907996 \quad \text{Use a calculator.}$$
$$\approx 4{,}184.09 \quad \text{Round to the nearest cent.}$$

The student must repay \$4,184.09.

Self Check A woman borrows \$7,500 for 90 days at 12%. If interest is compounded daily, how much will she owe at the end of 90 days? ■

Effective Rate of Interest

The true performance of an investment depends on the compounding frequency as well as the annual rate. To help investors compare different savings plans, financial institutions are required by law to provide the **effective rate**—the rate that, if compounded annually, would provide the same yield as the plan that is more frequently compounded.

To derive a formula for effective rate, we assume that P dollars are invested for n years at an annual rate r, compounded k times per year. That same principal P, compounded annually at the effective rate R, would produce the same accumulated value. We have the equation

Accumulated amount at effective rate R, compounded annually	equals	accumulated amount at annual rate r, compounded k times per year.

$$P(1 + R)^n = P(1 + i)^{kn} \quad i \text{ is the periodic rate, } i = \tfrac{r}{k}.$$

We can solve this equation for R.

$$P(1 + R)^n = P(1 + i)^{kn}$$
$$(1 + R)^n = (1 + i)^{kn} \quad \text{Divide both sides by } P.$$
$$[(1 + R)^n]^{1/n} = [(1 + i)^{kn}]^{1/n} \quad \text{Raise both sides to the } 1/n \text{ power.}$$
$$1 + R = (1 + i)^k \quad \text{Multiply the exponents.}$$
$$R = (1 + i)^k - 1 \quad \text{Subtract 1 from both sides.}$$

Effective Rate of Interest

The **effective rate of interest** R for an account paying a nominal rate r, compounded k times per year, is

$$R = (1 + i)^k - 1$$

where i is the periodic rate, $i = \frac{r}{k}$.

ACTIVE EXAMPLE 7 A bank offers the savings plans shown in the table. Calculate the effective interest rates.

COLLEGE **Algebra** $f(x)$ **Now**™
Go to **http://1pass.thomson.com** or your CD to practice this example.

	a. Money market fund	b. Certificate of deposit
Annual rate	6.5%	7%
Compounding	quarterly	monthly
Effective rate		

Solution **a.** For the money market fund, $r = 0.065$ and $k = 4$, so $i = \frac{r}{k} = \frac{0.065}{4} = 0.01625$. To find the effective rate, we substitute $k = 4$ and $i = 0.01625$ in the formula for effective rate.

$$R = (1 + i)^k - 1$$
$$R = (1 + 0.01625)^4 - 1$$
$$\approx 0.0666016088 \quad \text{Use a calculator.}$$
$$\approx 0.0666 \quad \text{Round to the nearest ten thousandth.}$$

As a percent, the effective rate is 6.66%, or approximately $6\frac{2}{3}\%$.

b. For the certificate of deposit, $r = 0.07$ and $k = 12$, so

$$i = \frac{0.07}{12} \approx 0.00583333$$
$$i = (1 + 0.00583333)^{12} - 1$$
$$\approx 0.0722900809 \quad \text{Use a calculator.}$$
$$\approx 0.0723 \quad \text{Round to the nearest ten thousandth.}$$

As a percent, the effective rate is 7.23%.

Self Check A passbook savings account offers daily compounding (365 days per year) at an annual rate of 6%. Find the effective rate to the nearest hundredth. ■

■ Present Value

From the initial deposit (the principal P), the compound interest formula predicts the future value n years later. This is the situation suggested by Figure 9-2, where we know the beginning amount and need to find the future value.

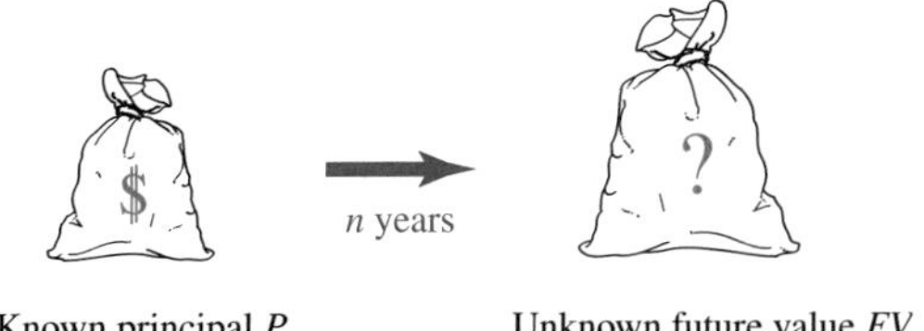

Figure 9-2

Often, the situation is reversed: We need to make a deposit now that will become a specific amount several years from now—perhaps enough to buy a car or pay tuition. As Figure 9-3 suggests, we need to know what single deposit *now* will accomplish that goal: What **present value** PV will yield a *specific* future value?

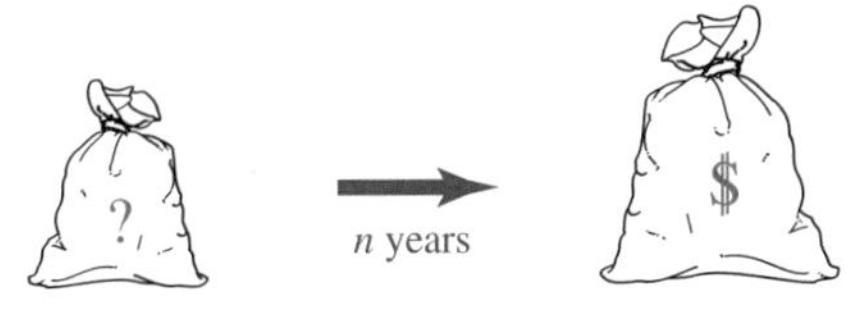

Figure 9-3

To derive the formula for present value, we replace P in the compound interest formula with PV and solve for PV.

$$FV = PV(1 + i)^{kn} \quad \text{The principal is now the present value, } PV.$$

$$\frac{FV}{(1 + i)^{kn}} = \frac{PV(1 + i)^{kn}}{(1 + i)^{kn}} \quad \text{Divide both sides by } (1 + i)^{kn}.$$

$$\frac{FV}{(1 + i)^{kn}} = PV \quad \text{Simplify.}$$

$$FV(1 + i)^{-kn} = PV \quad \text{Use the definition of negative exponent: } \frac{1}{a^x} = a^{-x}.$$

Present Value

The **present value** PV that must be deposited now to provide a **future value,** FV, n years from now is given by the formula

$$PV = FV(1 + i)^{-kn}$$

where interest is compounded k times per year at an annual rate r (i is the periodic rate, $\frac{r}{k}$).

ACTIVE EXAMPLE 8 When a medical student graduates in 8 years, she will need \$25,700 to buy medical equipment. What amount must she deposit now (at 8%, compounded twice per year) to meet this future obligation?

Solution Use the annual rate ($r = 0.08$) and the frequency of compounding ($k = 2$) to find the periodic rate:

COLLEGE **Algebra** *f(x)* **Now**™

Go to **http://1pass.thomson.com** or your CD to practice this example.

$$i = \frac{r}{k} = \frac{0.08}{2} = 0.04$$

In the present value formula, we substitute

the number of years, $n = 8$,
the periodic rate, $i = 0.04$,
the frequency of compounding, $k = 2$, and
the future value, $FV = 25{,}700$.

$$PV = FV(1 + i)^{-kn}$$

$$PV = 25{,}700(1 + 0.04)^{-2 \cdot 8}$$

$$= 25{,}700(1.04)^{-16} \quad \text{Use a calculator.}$$

$$\approx 13{,}721.44 \quad \text{Round to the nearest cent.}$$

She must deposit \$13,721.44 now to have \$25,700 in 8 years.

Self Check If the student decides to take two extra years to complete medical school, her obligation will be \$27,000. What present value will meet her goal? ■

Self Check Answers

1. \$2,416.64 **2.** 8 yr **3.** \$13,382.26 **4.** \$43,157.01 **5. a.** \$38,786.48 **b.** \$343.97 more than with quarterly compounding **6.** \$7,728.37 **7.** 6.18% **8.** \$12,322.45

9.1 Exercises

VOCABULARY AND CONCEPTS ***Fill in the blanks.***

1. A bank pays <u>interest</u> for the privilege of using your money.
2. If interest is left on deposit to earn more interest, the account earns <u>compound</u> interest.
3. Interest compounded once each year is called <u>annual</u> compounding.
4. The initial deposit is called the <u>principal</u> or the <u>present</u> value.
5. After a specific time, the principal grows to a <u>future</u> value.
6. Interest is calculated as a <u>percent</u> of the amount on deposit.
7. Future value = principal + <u>interest</u> earned
8. The annual rate is also called the <u>nominal</u> rate.
9. <u>**Periodic rate**</u> $= \dfrac{\text{annual rate}}{\text{number of periods per year}}$
10. The time between interest calculations is the <u>conversion</u> period.
11. In the future value formula $FV = P(1 + i)^{kn}$,
    P is the <u>principal</u>,
    i is the <u>periodic rate</u>,
    k is the <u>frequency of compounding</u>,
    and n is the <u>number of years</u>.
12. In the present value formula $PV = FV(1 + i)^{-kn}$,
    FV is the <u>future value</u>,
    i is the <u>periodic rate</u>,
    k is the <u>frequency of compounding</u>,
    and n is the <u>number of years</u>.
13. To help consumers compare savings plans, banks advertise the <u>effective</u> rate of interest.
14. If after one year, \$100 grows to \$110, the effective rate is <u>10</u> %.

PRACTICE

15. Find the simple interest earned in an account where \$4,500 is on deposit for 4 years at $3\frac{1}{4}\%$ annual interest. **\$585**
16. Find the simple interest earned in an account where \$12,400 is on deposit for $8\frac{1}{4}$ years at $4\frac{1}{2}\%$ annual interest. **\$4,603.50**
17. Find the principal necessary to earn \$814 in simple interest if the money is to be left on deposit for 4 years and earns $5\frac{1}{2}\%$ annual interest. **\$3,700**
18. Find the time necessary for a deposit of \$11,500 to earn \$3,450 in simple interest if the money is to earn $3\frac{3}{4}\%$ annual interest. **8 yr**
19. Find the annual rate necessary for a deposit of \$50,000 to earn \$7,500 in simple interest if the money is to be left on deposit for $2\frac{1}{2}$ years. **6%**
20. Find the time necessary for a deposit of \$5,000 to double in an account paying $6\frac{1}{4}\%$ simple interest. **16 yr**

Assume that \$1,200 is deposited in an account in which interest is compounded annually at a rate of 8%. Find the accumulated amount after the given number of years.

21. 1 year **\$1,296**
22. 3 years **\$1,511.65**
23. 5 years **\$1,763.19**
24. 20 years **\$5,593.15**

Assume that \$1,200 is deposited in an account in which interest is compounded annually at the given rate. Find the accumulated amount after 10 years.

25. 3% **\$1,612.70**
26. 5% **\$1,954.67**
27. 9% **\$2,840.84**
28. 12% **\$3,727.02**

Assume that \$1,200 is deposited in an account in which interest is compounded at the given frequency, at an annual rate of 6%. Find the accumulated amount after 15 years.

29. $k = 2$ **\$2,912.71**
30. $k = 4$ **\$2,931.86**
31. $k = 12$ **\$2,944.91**
32. $k = 365$ **\$2,951.31**

Find the effective interest rate with the given annual rate r and compounding frequency k.

33. $r = 6\%, k = 4$ **6.14%**
34. $r = 8\%, k = 12$ **8.30%**
35. $r = 9\frac{1}{2}\%, k = 2$ **9.73%**
36. $r = 10\%, k = 360$ **10.52%**

Find the present value of \$20,000 due in 6 years, at the given annual rate and compounding frequency.

37. 6%, semiannually **\$14,027.60**
38. 8%, quarterly **\$12,434.43**
39. 9%, monthly **\$11,678.47**
40. 7%, daily (360 days/year) **\$13,141.47**

APPLICATIONS

41. Small business To start a mobile dog-grooming service, a woman borrowed \$2,500. If the loan was for 2 years and the amount of interest was \$175, what simple interest rate was she charged? $3\frac{1}{2}\%$

42. Banking Three years after opening an account that paid 6.45% simple interest, a depositor withdrew the \$3,483 in interest earned. How much money was left in the account? **\$18,000**

43. Saving for college At the birth of their child, the Fieldsons deposited \$7,000 in an account paying 6% interest, compounded quarterly. How much will be available when the child turns 18? **\$20,448.10**

44. Planning a celebration When the Fernandez family made reservations at the end of 2003 for the December 2009 New Year's celebration in Paris, they placed \$5,700 into an account paying 8% interest, compounded monthly. What amount will be available at the time of the celebration? **\$9,196.96**

45. Planning for retirement When Jim retires in 12 years, he expects to live lavishly on the money in a retirement account that is earning $7\frac{1}{2}\%$ interest, compounded semiannually. If the account now contains \$147,500, how much will be available at retirement? **\$356,867.13**

46. Pension fund management The managers of a pension fund invested \$3 million in government bonds paying 8.73% annual interest, compounded semiannually. After 8 years, what will the investment be worth? **\$5.943 million**

47. Real estate investing Property values in the suburbs have been appreciating about 11% annually. If this trend continues, what will a \$137,000 home be worth in four years? Give the result to the nearest dollar. **\$207,976**

48. Real estate investing Property in suburbs closer to the city is appreciating about 8.5% annually. If this trend continues, what will a \$47,000 one-acre lot be worth in five years? Give the result to the nearest dollar. **\$70,672**

49. Gas consumption The gas utilities expect natural gas consumption to increase at 7.2% per year for the next decade. Monthly consumption for one county is currently 4.3 million cubic feet. What monthly demand for gas is expected in ten years? **8.62 million ft^3**

50. Comparing banks Bank One offers a passbook account with 4.35% annual rate, compounded quarterly. Bank Two offers a money market account at 4.3%, compounded monthly. Which account provides the better growth? (*Hint:* Find the effective rates.) **Bank One**

51. Comparing accounts A savings and loan offers the two accounts shown in the table. Find the effective rates.

	Annual rate	Compounding	Effective rate
NOW account	7.2%	quarterly	**7.4%**
Money market	6.9%	monthly	**7.12%**

52. Comparing accounts A credit union offers the two accounts shown in the table. Find the effective rates.

	Annual rate	Compounding	Effective rate
Certificate of deposit	6.2%	semiannually	**6.3%**
Passbook	5.25%	quarterly	**5.35%**

53. Car repair Craig borrows \$1,230 for unexpected car repair costs. His bank writes a 90-day note at 12%, with interest compounded daily. What will Craig owe? **\$1,267.45**

54. Fly now, pay later For a 7-day Hawaii vacation, Beth borrowed \$2,570 for 9 months at an annual rate of 11.4%, compounded monthly. What did she owe? **\$2,798.27**

55. Buying a computer A man estimates that the computer he plans to buy in 18 months will cost \$4,200. To meet this goal, how much should he deposit in an account paying 5.75%, compounded monthly? **\$3,853.73**

56. Buying a copier An accounting firm plans to deposit enough money now in an account paying 7.6% interest, compounded quarterly, to finance the purchase of a \$2,780 copier in 18 months. What should be the amount of that deposit? **\$2,483.13**

DISCOVERY AND WRITING

57. Adding to an investment To prepare for his retirement in 14 years, Jay deposited \$12,000 in an account paying 7.5% annual interest, compounded monthly. Ten years later, he deposited another \$12,000. How much will be available at retirement? **\$50,363.13**

58. Changing rates Ten years ago, a man invested \$1,100 in a 5-year certificate of deposit paying 10%, compounded monthly. When the CD matured, he invested the proceeds in another 5-year CD paying 8%, compounded semiannually. How much is available now? **\$2,679**

59. The power of time "A young person's most powerful money-making scheme," said an investment advisor, "is *time.*" Write a paragraph explaining what the advisor meant.

60. **Watching money grow** \$10,000 is invested at 10%, compounded annually. Use a graphing calculator to find how long it will take for the accumulated value to exceed \$1 million. **about $48\frac{1}{2}$ years**

61. Explain why the compound interest formula on page 649 is equivalent to the one on page 337.

62. Explain why the present value formula on page 653 is equivalent to the compound interest formula on page 649.

REVIEW ***Simplify each expression. Assume that all variables represent positive numbers.***

63. $\dfrac{x^2 - 2x - 15}{2x^2 - 9x - 5}$ $\quad \dfrac{x+3}{2x+1}$

64. $3x(x^2 - 5) - (x^3 - 2x)$ $\quad 2x^3 - 13x$

65. $\dfrac{(3 - x)(x + 3)}{-x^2 + 9}$ $\quad 1$

66. $-\sqrt{x^2 - 6x + 9}$ $(x \geq 3)$ $\quad -(x - 3)$

9.2 Annuities and Future Value

In this section, you will learn about

- Future Value of an Annuity
- Sinking Funds

■ Future Value of an Annuity

Financial plans that involve a series of payments are called **annuities.** Monthly mortgage payments, for example, are part of an annuity, as are regular contributions to a retirement plan.

Annuity A plan involving payments made at regular intervals is called an **annuity.**

Future Value The **future value** of an annuity is the sum of all the payments and the interest those payments earn.

Term The time over which the payments are made is called the **term** of the annuity.

Ordinary Annuity In an **ordinary annuity,** the payments are made at the *end* of each time interval.

In this book, we will only consider ordinary annuities with equal periodic payments made for a fixed term.

To understand how an annuity works, imagine a savings account that pays 12% annual interest, compounded monthly. Its periodic rate is $\frac{12\%}{12}$, or 1%. Imagine also that each month for the next year, we deposit \$100 in that account.

To determine the future value of this annuity, we think of each monthly payment as a one-time initial contribution to a compound-interest savings account. The future value of the annuity is the sum of the accumulated values of 12 individual accounts. As Figure 9-4 suggests, the first deposit earns interest for 11 months, the second for 10 months, and so on. Because the last deposit is made at the end of the year, it earns no interest. A few minute's work with a calculator shows the total to be \$1,268.25.

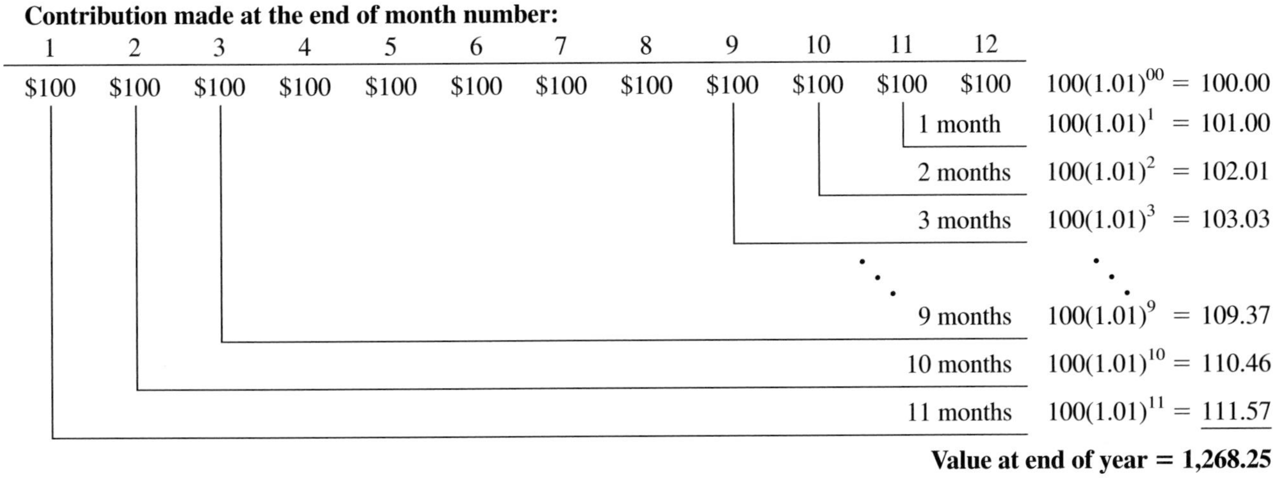

Contribution made at the end of month number:

1	2	3	4	5	6	7	8	9	10	11	12		
\$100	\$100	\$100	\$100	\$100	\$100	\$100	\$100	\$100	\$100	\$100	\$100	$100(1.01)^{00}$	= 100.00
											1 month	$100(1.01)^{1}$	= 101.00
											2 months	$100(1.01)^{2}$	= 102.01
											3 months	$100(1.01)^{3}$	= 103.03
											⋱		⋱
											9 months	$100(1.01)^{9}$	= 109.37
											10 months	$100(1.01)^{10}$	= 110.46
											11 months	$100(1.01)^{11}$	= 111.57

Value at end of year = 1,268.25

Figure 9-4

We will now generalize this example to derive a formula for the future value of an annuity. Consider an annuity with regular deposits of P dollars and with interest compounded k times per year for n years, at an annual rate r $\left(\text{periodic rate } i = \frac{r}{k}\right)$. During n years, there will be kn periods and kn deposits.

The last deposit, made at the end of the last period, earns no interest. The next-to-last deposit earns interest for one period, and so on. The first deposit, made at the end of the first period, earns compound interest for $kn - 1$ periods.

Future value of the annuity = future value of the last deposit + future value of the next-to-last deposit + ⋯ + future value of the first deposit.

$$FV = P + P(1 + i)^1 + P(1 + i)^2 + P(1 + i)^3 + \cdots + P(1 + i)^{kn-1}$$

This is a geometric sequence with first term P and common ratio $(1 + i)$. The sum is given by

$$FV = \frac{P[(1 + i)^{kn} - 1]}{i}$$

Recall that the sum of the terms of the geometric sequence $S = a + ar + ar^2 + ar^3 + \cdots + ar^{n-1}$ is $S = \frac{a(r^n - 1)}{r - 1}$.

We summarize this result.

Future Value of an Annuity

The future value FV of an ordinary annuity with deposits of P dollars made regularly k times each year for n years, with interest compounded k times per year at an annual rate r, is

$$FV = \frac{P[(1+i)^{kn} - 1]}{i}$$

where i is the periodic rate, $i = \frac{r}{k}$.

ACTIVE EXAMPLE 1 Verify the future value of the annuity outlined in Figure 9-4 on page 657.

Solution From Figure 9-4, we find

the term, in years:	$n = 1$
the frequency of compounding:	$k = 12$
the annual rate:	$r = 0.12$
the regular deposit:	$P = 100$

COLLEGE **Algebra** f(x) **Now**™
Explore this Active Example and test yourself on the steps taken here at **http://1pass.thomson.com**. You can also find this example on the Interactive Video Skillbuilder CD-ROM.

We then calculate the periodic interest rate: $i = \frac{r}{k} = \frac{0.12}{12} = 0.01$, and substitute these numbers in the formula for the future value of an annuity.

$$FV = \frac{P[(1+i)^{kn} - 1]}{i}$$

$$FV = \frac{100[(1+0.01)^{12\cdot 1} - 1]}{0.01}$$

$$= \frac{100[(1.01)^{12} - 1]}{0.01}$$

$$\approx 1{,}268.25030132 \quad \text{Use a calculator.}$$

$$\approx 1{,}268.25 \quad \text{Round to the nearest cent.}$$

The annuity of Figure 9-4 will provide \$1,268.25 by the end of the year.

Self Check Under the payroll savings plan, Hogan contributes \$50 a month to an ordinary annuity paying $7\frac{1}{2}\%$ annual interest, compounded monthly. How much will he have in 5 years? ■

Accent on Technology GROWTH OF MONEY

The value of an annuity accumulates rapidly. To watch the accumulated amount in the annuity of Example 1 grow, we graph the function

$$FV = \frac{100[(1.01)^{12n} - 1]}{0.01}$$

$$= 10{,}000(1.01^{12n} - 1) \quad \frac{100}{0.01} = 10{,}000.$$

Continued

We enter the function

$$Y_1 = 10000*(1.01\wedge(12*X) - 1)$$

on a graphing calculator, set the window values to be $0 \leq X \leq 50$ and $0 \leq Y \leq 1{,}000{,}000$, and graph it, as in Figure 9-5(a). Note how the amount increases more rapidly as the years go by. Using TRACE, as in Figure 9-5(b), we see that over \$1 million accumulates in just 39 years.

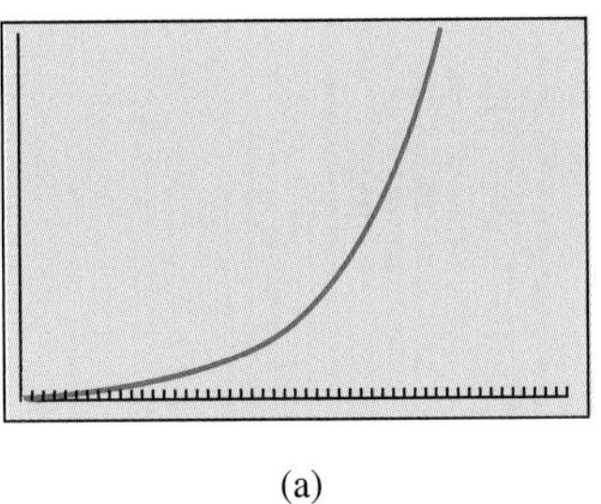

(a)

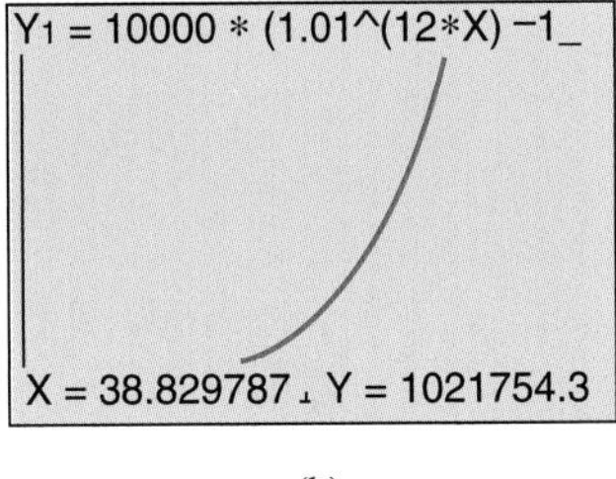

(b)

Figure 9-5

Sinking Funds

If we know the amount of each deposit, we can calculate the future value of an annuity using the formula on page 658. The situation is often reversed: What regular deposits, made now, will provide a specific future amount? An annuity created to produce a fixed future value is called a **sinking fund.** To determine the required periodic payment P, we solve the future value formula for P.

$$FV = \frac{P[(1 + i)^{kn} - 1]}{i}$$

$$FV\frac{i}{(1 + i)^{kn} - 1} = \frac{P[(1 + i)^{kn} - 1]}{i} \cdot \frac{i}{(1 + i)^{kn} - 1}$$ To isolate P, multiply both sides by $\frac{i}{(1 + i)^{kn} - 1}$.

$$\frac{FVi}{(1 + i)^{kn} - 1} = P$$ Simplify.

Sinking Fund Payment

For an annuity to provide a future value FV, regular deposits P are made k times per year for n years, with interest compounded k times per year at an annual rate r. The payment P is given by

$$P = \frac{FVi}{(1 + i)^{kn} - 1}$$

where i is the periodic rate, $i = \frac{r}{k}$.

ACTIVE EXAMPLE 2 An accounting firm will need \$17,000 in 5 years to replace its computer system. What periodic deposits to an annuity paying quarterly interest at a 9% annual rate will achieve that goal?

Solution We substitute the given values in the formula to find the sinking fund payment:

future value:	$FV = \$17{,}000$
annual rate:	$r = 0.09$
term:	$n = 5$
number of periods per year:	$k = 4$
periodic rate:	$i = \frac{r}{k} = \frac{0.09}{4} = 0.0225$

$$P = \frac{FVi}{(1+i)^{kn} - 1}$$

$$= \frac{(17{,}000)(0.0225)}{(1 + 0.0225)^{4 \cdot 5} - 1}$$

$$\approx 682.415203056 \quad \text{Use a calculator.}$$

$$\approx 682.42 \quad \text{Round to the nearest cent.}$$

Quarterly payments of \$682.42 will accumulate to \$17,000 in 5 years.

Self Check What quarterly deposits to the above account are required to raise the \$50,000 startup cost of a branch office in 7 years? ■

Self Check Answers

1. \$3,626.36 **2.** \$1,301.26

9.2 Exercises

VOCABULARY AND CONCEPTS ***Fill in the blanks.***

1. Plans involving payments made at regular intervals are called <u>annuities</u>.

2. In an ordinary annuity, payments are made at the <u>end</u> of each period.

3. The future value of an annuity is the sum of all the <u>payments</u> and <u>interest</u>.

4. The time over which the payments are made is the <u>term</u> of the annuity.

5. In the future value formula, $FV = \frac{P[(1+i)^{kn} - 1]}{i}$

P is the <u>regular deposit</u>,
i is the <u>periodic rate</u>,
k is the <u>frequency of compounding</u>,
and n is the <u>number of years</u>.

6. An annuity created to fund a specific future obligation is a <u>sinking</u> fund.

PRACTICE ***Assume that \$100 is deposited at the end of each year in an account in which interest is compounded annually at a rate of 6%. Find the accumulated amount after the given number of years.***

7. 10 years **\$1,318.08** **8.** 5 years **\$563.71**

9. 3 years **\$318.36** **10.** 20 years **\$3,678.56**

Assume that \$100 is deposited at the end of each year into an account in which interest is compounded annually at the given rate. Find the accumulated amount after 10 years.

11. 4% **\$1,200.61** **12.** 7% **\$1,381.64**

13. 9.5% **\$1,556.03** **14.** 8.5% **\$1,483.51**

Assume that* $100 *is deposited at the end of each period in an account in which interest is compounded at the given frequency, at an annual rate of* 8%. *Find the accumulated amount after* 15 *years.

15. $k = 2$ $5,608.49

16. $k = 4$ $11,405.15

17. $k = 12$ $34,603.82

18. $k = 1$ $2,715.21

Find the amount of each regular payment to provide* $20,000 *in* 10 *years, at the given annual rate and compounding frequency.

19. 4%, annually $1,665.81

20. 6%, quarterly $368.54

21. 9%, semiannually $637.52

22. 8%, monthly $109.32

APPLICATIONS

23. Saving for a vacation For next year's vacation, the Phelps family is saving $200 each month in an account paying 6% annual interest, compounded monthly. How much will be available a year from now? $2,467.11

24. Planning for retirement Hank's regular $1,300 quarterly contributions to his retirement account have earned 6.5% annual interest, compounded quarterly, since he started 21 years ago. How much is in his account now? $229,839.59

25. Pension fund management The managers of a company's pension fund invest the monthly employee contributions of $135,000 into a government fund paying 8.7%, compounded monthly. To what value will the fund grow in 20 years? $86,803,923.58

26. Saving for college A mother has been saving regularly for her daughter's college—$25 each month for 11 years. The money has been earning $7\frac{1}{2}$% annual interest, compounded monthly. How much is now in the account? $5,104.12

27. Buying office machines A company's new corporate headquarters will be completed in $2\frac{1}{2}$ years. At that time, $750,000 will be needed for office equipment. How much should be invested monthly to fund that expense? Assume 9.75% interest, compounded monthly. $22,177.71

28. Retirement lifestyle A woman would like to receive a $500,000 lump-sum distribution from her retirement account when she retires in 25 years. She begins making monthly contributions now to an annuity paying 8.5%, compounded monthly. Find the amount of that monthly contribution. $484.47

29. Comparing accounts Which account will require the lower annual contributions to fund a $10,000 obligation in 20 years? (*Hint:* Compare the yearly total contributions.)

Bank A	5.5%; annually	
Bank B	5.35%; monthly	Bank B

30. Avoiding a balloon payment The last payment of a home mortgage is a balloon payment of $47,000, which the owner is scheduled to pay in 12 years. How much *extra* should he start including in each monthly payment to eliminate the balloon payment? His mortgage is at 10.2%, compounded monthly. $167.63

DISCOVERY AND WRITING

31. Retirement strategy Jim will retire in 30 years. He will invest $100 each month for 15 years and then let the accumulated value continue to grow for the next 15 years. How much will be available at retirement? Assume 8%, compounded monthly. $114,432.12

32. Retirement strategy (See Exercise 31.) Jim's brother Jack will also retire in 30 years. He plans on doing nothing during the first 15 years, then contributing twice as much—$200 monthly—to "catch up." How much will be available at retirement? Assume 8%, compounded monthly. $69,207.64

33. Changing plans A woman needs $13,500 in 10 years. She would like to make regular annual contributions for the first 5 years and then let the amount grow at compound interest for the next 5 years. What should her contributions be? Assume 9%, compounded annually. $1,466.08

34. Talking financial sense How would you explain to a friend who has just been hired for her first job that now is the time to start thinking about retirement?

REVIEW *Solve each equation.*

35. $\dfrac{2(5x - 12)}{x} = 8$ 12

36. $\dfrac{2(5x - 12)}{x} = x$ 4, 6

37. $\sqrt{2x + 3} = 3$ 3

38. $\sqrt{2x + 3} = x$ 3

9.3 Present Value of an Annuity; Amortization

In this section, you will learn about

- Present Value of an Annuity
- Amortization

Present Value of an Annuity

Instead of using an annuity to create a future value A, we might ask, "What *single* deposit made *now* would create that same future value?" The one deposit that gives the same final result as an annuity is called the **present value** of that annuity.

To find a formula for the present value of an annuity, we combine two previous formulas. A series of regular payments of P dollars for n years will grow to a future value FV given by

$$FV = \frac{P[(1+i)^{kn} - 1]}{i} \tag{1}$$

From a formula in Section 9.1, the present value of a future asset is given by

$$PV = FV(1+i)^{-kn} \tag{2}$$

We find the present value of a series of future payments by substituting the right-hand side of Equation 1 into Equation 2.

$$PV = FV(1+i)^{-kn} \qquad \text{This is Equation 2.}$$

$$PV = \frac{P[(1+i)^{kn} - 1]}{i}(1+i)^{-kn} \qquad \text{Substitute } \frac{P[(1+i)^{kn} - 1]}{i} \text{ for } FV \text{ in Equation 2.}$$

$$= \frac{P[(1+i)^{kn}(1+i)^{-kn} - 1(1+i)^{-kn}]}{i} \qquad \text{Use the distributive property.}$$

$$= \frac{P[1 - (1+i)^{-kn}]}{i} \qquad \text{Simplify: } x^m x^{-m} = x^0 = 1.$$

Present Value of an Annuity

The present value PV of an annuity with payments of P dollars made k times per year for n years, with interest compounded k times per year at an annual rate r, is

$$PV = \frac{P[1 - (1+i)^{-kn}]}{i}$$

where i is the periodic rate, $i = \frac{r}{k}$.

EXAMPLE 1 To buy a boat in 2 years, the Higgins family plans to save \$200 a month in an account that pays 12% interest, compounded monthly. **a.** What will be the total of the payments? **b.** What will the value of the account be in 2 years? **c.** What single deposit in that account now would do as well?

Solution **a.** At \$200 per month for 24 months, the total amount contributed is

$$\$200(24) = \$4,800$$

b. To find the value after 2 years, we use the formula for future value of an annuity:

$$FV = \frac{P[(1+i)^{kn} - 1]}{i}$$

$$FV = \frac{\mathbf{200}[\mathbf{1.01}^{\mathbf{12\cdot 2}} - 1]}{\mathbf{0.01}}$$

$$\approx 5,394.692971 \qquad \text{Use a calculator.}$$

$$\approx 5,394.69 \qquad \text{Round to the nearest cent.}$$

c. To find the present value of the annuity, we substitute

the term, in years:	$n = 2$
the frequency of compounding:	$k = 12$
the annual rate:	$r = 0.12$
the payment:	$P = 200$
the periodic interest rate:	$i = \frac{r}{k} = \frac{0.12}{12} = 0.01$

in the present value formula.

$$PV = \frac{P[1 - (1+i)^{-kn}]}{i}$$

$$PV = \frac{\mathbf{200}[1 - (1 + \mathbf{0.01})^{-\mathbf{12\cdot 2}}]}{\mathbf{0.01}}$$

$$= \frac{200[1 - (1.01)^{-24}]}{0.01} \qquad \text{Simplify.}$$

$$\approx 4,248.677451 \qquad \text{Use a calculator.}$$

$$\approx 4,248.68 \qquad \text{Round to the nearest cent.}$$

\$4,248.68 is the present value of the annuity; that one deposit now will provide the same final amount, \$5,394.69, as the annuity.

Self Check For his retirement in 30 years, a man plans to make monthly contributions of \$25 to an ordinary annuity paying $8\frac{1}{2}\%$ annually, compounded monthly. **a.** What will be the total of his contributions? **b.** What single deposit now will provide the same retirement benefit? ■

State lottery winnings are usually paid as a 20-year annuity. That is to the state's advantage, because it can fund the annuity with a single amount that is much smaller than the total prize.

ACTIVE EXAMPLE 2 Britta won the lottery. She will receive \$75,000 per month for the next 20 years—a total of \$18 million. What single deposit should the lottery commission make now to fund Britta's annuity? Assume 8.4% annual interest, compounded monthly.

Solution The lottery commission finds the present value of the annuity, with

COLLEGE **Algebra** $f(x)$ **Now**™
Explore this Active Example and test yourself on the steps taken here at **http://1pass.thomson.com**. You can also find this example on the Interactive Video Skillbuilder CD-ROM.

the payment:	$P = 75{,}000$
the annual rate:	$r = 0.084$
the frequency of compounding:	$k = 12$
the periodic rate:	$i = \frac{r}{k} = \frac{0.084}{12} = 0.007$
the term, in years:	$n = 20$

These values are used in the formula for the present value of an annuity.

$$PV = \frac{P[1 - (1 + i)^{-kn}]}{i}$$

$$PV = \frac{\mathbf{75{,}000}[1 - (\mathbf{1.007})^{-\mathbf{12 \cdot 20}}]}{\mathbf{0.007}}$$

$$\approx 8{,}705{,}700.365 \qquad \text{Use a calculator.}$$

$$\approx 8{,}705{,}700.37 \qquad \text{Round to the nearest cent.}$$

To fund the $18 million prize, the commission must deposit $8,705,700.37.

Self Check The lottery pays a total prize of $120,000 in monthly installments, as a 10-year annuity. Assuming 8.4% interest, compounded monthly, what current deposit is needed to fund the annuity? ■

EXAMPLE 3 As a settlement in an automobile injury lawsuit, Robyn will receive $30,000 each year for the next 25 years, for a total of $750,000. The insurance company is offering a one-payment settlement of $300,000, now. Should she accept? Assume that the money can be invested at 9% annual interest.

Solution Robyn should calculate the present value of an annuity with

the payment:	$P = 30{,}000$
the annual rate:	$r = 0.09$
the frequency of compounding:	$k = 1$ (annual)
the periodic rate:	$i = \frac{r}{k} = \frac{0.09}{1} = 0.09$
the term, in years:	$n = 25$

She should use these values in the formula for the present value of an annuity.

$$PV = \frac{P[1 - (1 + i)^{-kn}]}{i}$$

$$PV = \frac{\mathbf{30{,}000}[1 - (\mathbf{1.09})^{-\mathbf{1 \cdot 25}}]}{\mathbf{0.09}}$$

$$\approx 294{,}677.3881 \qquad \text{Use a calculator.}$$

$$\approx 294{,}677.39 \qquad \text{Round to the nearest cent.}$$

The annuity is worth only $294,677.39, and the company is offering $300,000. Robyn should accept the $300,000.

Self Check If Robyn could invest the settlement at 8% interest, should she still accept the lump-sum offer? ■

When a worker is employed, regular contributions are usually made to a retirement fund. After retirement, those funds are given back, either as an annuity or as a lump-sum distribution.

ACTIVE EXAMPLE 4 Carlos wants to fund an annuity to supplement his retirement income. How much should he deposit now to generate retirement income of \$1,000 a month for the next 20 years? Assume that he can get $9\frac{3}{4}\%$ interest, compounded monthly.

COLLEGE **Algebra** *f(x)* **Now**™
Go to **http://1pass.thomson.com** or your CD to practice this example.

Solution Carlos must calculate the present value of a future stream of income, with

the payment:	$P = 1{,}000$
the annual rate:	$r = 0.0975$
the frequency of compounding:	$k = 12$
the periodic rate:	$i = \frac{r}{k} = \frac{0.0975}{12} = 0.008125$
the term, in years:	$n = 20$

He should use these values in the formula for the present value of an annuity.

$$PV = \frac{P[1 - (1 + i)^{-kn}]}{i}$$

$$PV = \frac{\mathbf{1{,}000}[1 - (\mathbf{1.008125})^{-\mathbf{12 \cdot 20}}]}{\mathbf{0.008125}}$$

$$\approx 105{,}428$$

\$105,428 put on deposit now will provide Carlos with \$1,000 per month in retirement income for 20 years.

Self Check If Carlos can invest at $8\frac{3}{4}\%$, what deposit is needed now? ■

Everyday Connections Mortgage Rates

"It has long been felt that the benefits of becoming a homeowner are not limited to the new owner, but also spill over to other members of society. Spillovers that are commonly cited include the fruits of greater participation in civic affairs, reductions in crime, and improved scholastic performance of children." William Shew and Irwin M. Selzer of Hudson Institute

Anemic Job Growth Keeps Mortgage Rates Down

By Laura Bruce, June 9, 2005

Mortgage rates have reacted predictably to Friday's way-softer-than-expected employment report: They fell, according to Bankrate.com's weekly survey of large lenders.

The news keeps the mortgage market in good shape for borrowers, at least for now. It shouldn't be presumed that rates will continue to stay so low. The economy may not be a runaway locomotive, but job growth on a

Continued

month-to-month basis remains fairly strong, according to economists, and the unemployment rate continues to decline. That said, let's look at the highlights of this week's survey.

The benchmark 30-year fixed-rate mortgage fell 4 basis points to 5.61 percent. A basis point is one-hundredth of 1 percentage point. The mortgages in this week's survey had an average total of 0.32 discount and origination points. One year ago, the 30-year fixed rate stood at 6.36 percent.

The 15-year fixed-rate mortgage fell 2 basis points to 5.24 percent. The five-year adjustable-rate mortgage fell 1 basis point to 5.14 percent.
Source: http://www.bankrate.com

Suppose a prospective homeowner obtains a mortgage loan with the following terms:

- Mortgage amount $215,000.00
- Mortgage term 30 years
- Annual interest rate (fixed) 5.25%

Calculate the monthly mortgage payment (principle and interest). **$1,187.24**

Amortization

Before a bank will allow you to borrow money, you must sign a **promissory note** indicating that you promise to pay the money back. We discussed one-payment notes in Section 9.1. Most loans, however, are repaid in installments instead of all at once. Spreading the repayment over several equal payments is called **amortization.**

When a loan is made, the lending institution is buying an annuity from the borrower—the lender pays the borrower an agreed-upon amount and expects regular payments in return. To calculate the amount of these regular installment payments, we solve the present value formula for P to get

$$P = \frac{PVi}{1 - (1 + i)^{-kn}}$$

In this context, the present value is the amount of the loan. Using the letter A (for *amount*) instead of PV, we have the following formula.

Installment Payments

The periodic payment P required to repay an amount A is given by

$$P = \frac{Ai}{1 - (1 + i)^{-kn}}$$

where

r is the annual rate,

k is the frequency of compounding (usually monthly),

i is the periodic rate, $i = \frac{r}{k}$, and

n is the term of the loan.

ACTIVE EXAMPLE 5 The Almondi family takes a 15-year mortgage of \$200,000 for their new home, at 10.8%, compounded monthly. **a.** What will be the amount of their monthly payment? **b.** What is the total of their payments over the full term?

Solution **a.** In the formula for calculating the installment payment, we substitute

COLLEGE **Algebra** $f(x)$ **Now™**
Go to **http://1pass.thomson.com** or your CD to practice this example.

the amount:	$A = 200{,}000$
the annual rate:	$r = 0.108$
the frequency of compounding:	$k = 12$
the periodic rate:	$i = \frac{r}{k} = \frac{0.108}{12} = 0.009$
the term, in years:	$n = 15$

$$P = \frac{Ai}{1 - (1 + i)^{-kn}}$$

$$P = \frac{(\mathbf{200{,}000})(\mathbf{0.009})}{1 - (\mathbf{1.009})^{-\mathbf{12 \cdot 15}}}$$

$$\approx 2{,}248.14$$

Each monthly mortgage payment will be \$2,248.14.

b. There are $12 \cdot 15 = 180$ payments of \$2,248.14 each, for a total of \$404,665—more than twice the amount borrowed!

Self Check Instead of a 15-year mortgage, the Almondis considered a 30-year mortgage. Answer the two questions again. ■

Self Check Answers

1. a. \$9,000 **b.** \$3,251.34 **2.** about \$81,000 **3.** No; the annuity is now worth more than \$320,243. **4.** \$113,159 **5. a.** \$1,874.48 **b.** \$674,814

9.3 Exercises

VOCABULARY AND CONCEPTS ***Fill in the blanks.***

1. The current worth of a future stream of income is the **present value** of an annuity.
2. The amount required now to produce a future stream of income is the **present value** of an annuity.
3. A loan is called a **promissory note** because you promise to repay it.
4. Often, loan repayment is spread out over several **installments**.
5. Spreading repayment of a loan over several *equal* payments is called **amortizing** the loan.
6. An amortized loan is also called a **mortgage**.

PRACTICE ***Find the present value of an annuity with the given terms.***

7. Annual payments of \$3,500 at 5.25%, compounded annually for 25 years **\$48,116.14**
8. Semiannual payments of \$375 at a 4.92% annual rate, compounded semiannually for 10 years **\$5,868.09**

Find the periodic payment required to repay a loan with the given terms.

9. \$25,000 repaid over 15 years, with monthly payments at a 12% annual rate **\$300.04**
10. \$1,750 repaid in 18 monthly installments, at an annual rate of 19% **\$112.50**

11. Funding retirement Instead of making quarterly contributions of \$700 to a retirement fund for the next 15 years, a man would rather make only one contribution, now. How much should that be? Assume $6\frac{1}{4}\%$ annual interest, compounded quarterly. **\$27,128.43**

12. Funding a lottery To fund Jamie's lottery winnings of \$15,000 per month for the next 20 years, the lottery commission needs to make a single deposit now. Assuming 9.2% compounded monthly, what should the deposit be? **\$1,643,603.78**

13. Money up front Instead of receiving an annuity of \$12,000 each year for the next 15 years, a young woman would like a one-time payment, now. Assuming she could invest the proceeds at $8\frac{1}{2}\%$, what would be a fair amount? **\$99,650.84**

14. Funding retirement What single amount deposited now into an account paying $7\frac{2}{3}\%$ annual interest, compounded quarterly, would fund an annuity paying \$5,000 quarterly for the next 25 years? **\$221,794.27**

15. Buying a car The Jepsens are buying a \$21,700 car and financing it over the next 4 years. They secure an 8.4% loan. What will their monthly payments be? **\$533.84**

16. Total cost of buying a car What will be the total amount the Jepsens will pay over the life of the loan? (See Exercise 15.) **\$25,624.32**

17. Choosing a mortgage One lender offers two mortgages—a 15-year mortgage at 12%, and a 20-year mortgage at 11%. For each, find the monthly payment to repay \$130,000. **15-yr: \$1,560.22; 20-yr: \$1,341.84**

18. Total cost of a mortgage For each of the mortgages in Exercise 17, find the total of the monthly payments. **15-yr: \$280,839.60; 20-yr: \$322,041.60**

DISCOVERY AND WRITING

19. Getting an early start As Jorge starts working now at the age of 20, he decides to make regular contributions to a savings account. He wants to accumulate enough by age 55 to fund an annuity of \$5,000 per month until age 80. What should his monthly contributions be? Assume that both accounts pay 8.75%, compounded monthly. **\$220.13**

20. Comparing annuities Which of these 20-year plans is best, and why? All are at 8% annually.
- **a.** \$1,000 each year for 10 years, and then let the accumulated amount grow for 10 years.
- **b.** \$500 each year for 20 years.
- **c.** Do nothing for 10 years, and then contribute \$2,000 each year for 10 years.
- **d.** One payment of \$8,000 now, and let it grow.

21. Changing the payment A woman contributed \$500 per quarter for the first 10 years of an annuity, but changed to quarterly payments of \$1,500 for the last 10 years. Assuming $7\frac{1}{4}\%$ annual interest compounded quarterly, what is her accumulated value? **\$146,506.50**

22. Changing the rate A woman contributed \$150 per month for 10 years to an account that paid 5% for the first 5 years, but 6.5% for the last 5 years. How much has she saved? **\$24,707.09**

REVIEW ***Simplify each expression. Assume that all variables represent positive numbers.***

23. $\dfrac{6\sqrt{30}}{3\sqrt{5}}$ $2\sqrt{6}$

24. $\dfrac{6}{\sqrt{7}-2}$ $2(\sqrt{7}+2)$

25. $3\sqrt{5x}+5\sqrt{20x}$ $13\sqrt{5x}$

26. $\sqrt{\dfrac{x^3y^5}{x^5y^6}}$ $\dfrac{\sqrt{y}}{xy}$

PROBLEMS AND PROJECTS

Write a brief paragraph explaining the given terms.

1. *interest* and *interest rate*
2. *present value* and *future value*
3. *compound interest* and *simple interest*
4. *annual rate* and *effective rate*
5. *annual rate* and *periodic rate*
6. *annuity* and *sinking fund*
7. The United States gross federal debt in 1997 was approximately \$5.5 trillion. If the interest on that debt was 5%, how much did each of the country's 260 million people owe in interest each year? **\$1,057.70**
8. With an annual inflation rate of 5%, \$100 today will have the purchasing power (in today's dollars) of $\$100(1.05)^{-1}$, or \$95.24, a year from now. At a 5% inflation rate, what monthly income would be required in 20 years to provide the purchasing power of \$2,000 per month today? **\$5,306.60**

Project 1

In an installment loan, the borrower agrees to pay back principal and interest in a series of installments for a specified period of time. For a mortgage, the interest is calculated as compound interest. Another type of installment loan uses simple interest, which is added on to the amount of the loan. This method, called **add-on interest,** is often used for car loans.

For example, you need to borrow \$1,000 for 2 years at an add-on interest rate of 12%. The simple interest is \$240, because

$$I = Prt$$

$$I = (1{,}000)(0.12)(2) \quad P = 1{,}000,\ r = 0.12,\ t = 2.$$

$$= 240$$

The amount to be repaid is \$1,240, the sum of the interest and the principal. The required monthly payment A is

$$A = \frac{1{,}240}{24} \quad \text{In 2 years, there are 24 monthly payments.}$$

$$= 51.67$$

Each month's payment is \$51.67.

- If this had been a mortgage loan, what would the payment have been? **\$47.07**
- Notice that you have paid 2 years' interest on the full amount, even though you will not owe the full amount for 2 years. Explain the effect this has on the real interest rate.
- To help consumers compare interest rates, the Truth-in-Lending Act requires that all lenders state the true annual interest, called the **Annual Percentage Rate (APR).** It is calculated by the formula

$$\text{APR} = \frac{2Nr}{N+1}$$

where N is the total number of payments and r is the add-on interest rate. Find the APR for this loan, if satisfied in 2 years. **APR = 23.04%**

- By doubling each monthly payment, this loan would be satisfied in 1 year. Is it wise to pay off this loan early? Calculate the APR and explain. (*Hint:* \$240 interest on \$1,000 for 1 year is no longer 12%.)
  APR = 44.3%, if there is no refund of prepaid interest.

Project 2

Another method for calculating the terms of an installment loan is called the **rule of 78.** It is based on the fact that the sum of the fractions

$$\frac{12}{78} + \frac{11}{78} + \frac{10}{78} + \frac{9}{78} + \cdots + \frac{1}{78}$$

is equal to 1. For a one-year loan, the interest charged is simple interest, but $\frac{12}{78}$ of that interest is repaid the first month, $\frac{11}{78}$ of the interest the second month, and so on.

- Show that $\frac{12}{78} + \frac{11}{78} + \frac{10}{78} + \frac{9}{78} + \cdots + \frac{1}{78} = 1$.
- For a loan of \$1,000 for one year at 12%, simple interest of \$120 would be added to the principal, and each month's payment would be $\frac{1}{12}$ of \$1,120, or \$99.33. Of this first month's payment, $\frac{12}{78}$ of \$120, or \$18.46, would be interest. Of the *last* month's payment, $\frac{1}{78}$ of \$120, or only \$1.54, would be interest.

  Make a table showing how the monthly payment is divided between interest and principal for each of the 12 months.
- If a "rule of 78" loan is paid back early, the borrower is entitled to a return of the unpaid interest. Suppose the previous loan (\$1,000 at 12% for one year) is repaid in 6 months. What interest will be refunded? Why is this not one-half of \$120?
  \$32.31

CHAPTER SUMMARY

CONCEPTS	REVIEW EXERCISES
9.1	**Interest**
Funds in a savings account earn **simple interest** at an **annual rate** r. The amount deposited is the **principal,** P. $I = Prt$	**1.** \$2,000 is deposited in an account that earns 9% simple interest. Find the value of the account in 5 years. **\$2,900** **2.** \$2,000 is deposited in an account in which interest is compounded annually at 9%. Find the value in 5 years. **\$3,077.25**

The **accumulated amount** or the **future value** FV after the **term** of n years is found by the formula

$FV = P(1 + r)^n$
(annual compounding)

3. Brian borrows \$2,350 for medical bills. The bank writes a 60-day note at 14%, with interest compounded daily. What will Brian owe? **\$2,405.47**

If interest is calculated k times each year, then the **periodic rate** is $i = \frac{r}{k}$, and the future value is

$FV = P(1 + i)^{kn}$
(compounding k times per year)

4. \$2,000 earns interest, compounded quarterly, at an annual rate of 7.6% for 16 years. Find the future value. **\$6,670.80**

The **effective rate** R is used to compare different savings plans.

$R = (1 + i)^k - 1$

5. BigBank advertises a savings account at a 6.3% rate, compounded quarterly. BestBank offers 6.21%, compounded daily. Calculate each effective rate and choose the better account.
BigBank, 6.45%; BestBank, 6.4%; BigBank

The **present value** PV is the single deposit *now* that will yield a specific future value, FV.

$PV = FV(1 + i)^{-kn}$

6. What amount deposited now in an account paying 5.75% interest, compounded semiannually, will yield \$7,900 in 6 years? **\$5,622.23**

9.2 Annuities and Future Value

An **annuity** is a series of periodic payments made at regular intervals. The **future value** FV of an annuity is the sum of all the payments and the interest those payments earn. The time over which the payments are made is called the **term** of the annuity. In an **ordinary annuity,** the payments are made at the *end* of each time interval.

The future value FV of an ordinary annuity with deposits of P dollars made regularly k times each year for n years, with interest compounded k times per year at an annual rate r is

$$FV = \frac{P[(1 + i)^{kn} - 1]}{i}$$

where i is the periodic rate, $i = \frac{r}{k}$.

An annuity with the purpose of funding a future obligation is a **sinking fund.**

7. \$500 is deposited at the end of each year into an annuity in which interest is compounded annually at 5%. Find the accumulated amount after 13 years.
\$8,856.49

8. \$150 is deposited monthly into an account that pays 8% annual interest, compounded monthly. Find the future value after 20 years.
\$88,353.06

To yield a specific future value FV, regular deposits P are made k times per year for n years, with interest compounded k times per year at an annual rate r. The payment P is

$$P = \frac{FVi}{(1 + i)^{kn} - 1}$$

where i is the periodic rate, $i = \frac{r}{k}$.

9. The owners of a small dry cleaning shop will need $40,700 to open a second shop in 7 years. What monthly payments to a sinking fund earning 7.5% interest, compounded monthly, will meet that obligation?
$369.89

9.3 Present Value of an Annuity; Amortization

The **present value** of an annuity is the current worth of a future stream of income.

The present value PV of an annuity with payments of P dollars made k times per year for n years, with interest compounded k times per year at an annual rate r, is

$$PV = \frac{P[1 - (1 + i)^{-kn}]}{i}$$

where i is the periodic rate, $i = \frac{r}{k}$.

10. An annuity pays $250 semiannually for 20 years. At a semiannually compounded rate of 6.5%, what is the present value? **$5,552.11**

11. The lottery must fund a 20-year annuity of $50,000 per year. At 9.6%, compounded annually, what must be invested now? **$437,563.50**

Loans are often paid off in **installments.** If equal installments are paid over a fixed time, the payments are **amortized.**

The periodic payment P required to repay an amount A is given by

$$P = \frac{Ai}{1 - (1 + i)^{-kn}}$$

where

r is the annual rate,

k is the frequency of compounding,

i is the periodic rate, $i = \frac{r}{k}$, and

n is the term of the loan.

12. What are the monthly payments for a $150,500, 15-year, 10.75% mortgage? What is the total amount paid? **$1,687.03; $303,665.40**

13. Answer the previous question, but for a 30-year mortgage.
$1,404.89; $505,760.40

CHAPTER TEST

Fill in the blanks.

1. When interest is left on deposit to earn more interest, the account earns **compound** interest.
2. The annual rate of interest divided by the number of periods is called the **periodic** interest rate.
3. To compare different savings plans, compare the **effective** rates of interest.
4. The nominal rate of interest is also called the **annual** rate.
5. Plans involving regular periodic payments are called **annuities**.
6. An annuity to fund a specific future obligation is a **sinking fund**.
7. The current value of a series of future payments is the **present value** of an annuity.
8. Repaying a loan over several regular, equal installments is called **amortizing** the loan.
9. $1,300 is deposited in a new account that earns 5% simple interest. What will the account be worth in 10 years? **$1,950**
10. $1,300 is deposited in a new account that earns 5% interest, compounded annually. What will the account be worth in 10 years? **$2,117.56**
11. $1,300 is deposited in an account that earns 5% annual interest, compounded monthly. What will it be worth in 10 years? **$2,141.11**
12. What is the effective rate of the savings plan in Problem 11? **5.116%**
13. What single deposit now will yield $5,000 in 10 years? Assume 7% annual interest, compounded quarterly. **$2,498.00**
14. Each month for 5 years, a student made $700 payments to an account paying 7.3% annual interest, compounded monthly. Find the accumulated amount. **$50,506.10**
15. What monthly payment to a sinking fund will raise $8,000 in 5 years? Assume 6.5% annual interest, compounded monthly. **$113.20**
16. Find the present value of an annuity that pays $1,000 each month for 15 years. Assume 6.8% annual interest, compounded monthly. **$112,652.71**
17. What are the monthly payments for a 15-year, $90,000 mortgage at 8.95%? **$910.16**

CUMULATIVE REVIEW EXERCISES

Solve each system by graphing.

1. $\begin{cases} 2x + y = 8 \\ x - 2y = -1 \end{cases}$

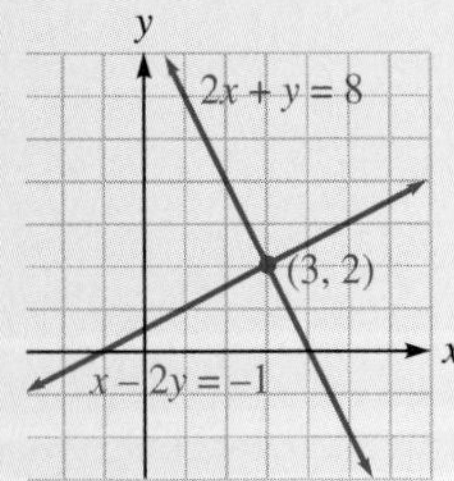

2. $\begin{cases} 3x = -y + 2 \\ y + x - 4 = -2x \end{cases}$

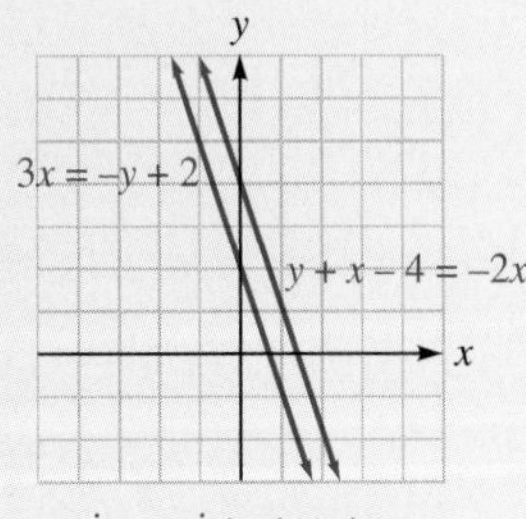

inconsistent system

Solve each system.

3. $\begin{cases} 5x = 3y + 12 \\ 2x - 3y = 3 \end{cases}$ **$x = 3, y = 1$**

4. $\begin{cases} 2x + y - z = 7 \\ x - y + z = 2 \\ x + y - 3z = 2 \end{cases}$ **$x = 3, y = 2, z = 1$**

Solve each system using matrices.

5. $\begin{cases} 2x + y - z = 0 \\ x - y + z = 3 \\ x + y - 3z = -5 \end{cases}$ **$x = 1, y = 0, z = 2$**

6. $\begin{cases} 2x - 2y + 3z + t = 2 \\ x + y + z + t = 5 \\ -x + 2y - 3z + 2t = 2 \\ x + y + 2z - t = 4 \end{cases}$

$x = 1, y = 2, z = 1, t = 1$

Let $A = \begin{bmatrix} 2 & 1 \\ 1 & 4 \end{bmatrix}$, $B = \begin{bmatrix} -1 & 2 \\ 2 & 3 \end{bmatrix}$, *and* $C = \begin{bmatrix} 2 & 0 & -1 \\ -1 & 2 & 2 \end{bmatrix}$. *Find each matrix.*

7. $A + B$

$\begin{bmatrix} 1 & 3 \\ 3 & 7 \end{bmatrix}$

8. $B - A$

$\begin{bmatrix} -3 & 1 \\ 1 & -1 \end{bmatrix}$

9. AC

$\begin{bmatrix} 3 & 2 & 0 \\ -2 & 8 & 7 \end{bmatrix}$

10. $B^2 + 2A$

$\begin{bmatrix} 9 & 6 \\ 6 & 21 \end{bmatrix}$

Find the inverse of each matrix, if possible.

11. $\begin{bmatrix} 2 & 6 \\ 2 & 4 \end{bmatrix}$

$\begin{bmatrix} -1 & \frac{3}{2} \\ \frac{1}{2} & -\frac{1}{2} \end{bmatrix}$

12. $\begin{bmatrix} 1 & -1 & 1 \\ 1 & 4 & 0 \\ 2 & 4 & 1 \end{bmatrix}$

$\begin{bmatrix} 4 & 5 & -4 \\ -1 & -1 & 1 \\ -4 & -6 & 5 \end{bmatrix}$

Evaluate each determinant.

13. $\begin{vmatrix} -3 & 5 \\ 4 & 7 \end{vmatrix}$

-41

14. $\begin{vmatrix} 2 & -3 & 2 \\ 0 & 1 & -1 \\ 1 & -2 & 1 \end{vmatrix}$

-1

Set up the determinants to find x and y in the system $\begin{cases} \mathbf{4x + 3y = 11} \\ \mathbf{-2x + 5y = 24} \end{cases}$. *Do not evaluate the determinants.*

15. $x = \dfrac{\begin{vmatrix} 11 & 3 \\ 24 & 5 \end{vmatrix}}{\begin{vmatrix} 4 & 3 \\ -2 & 5 \end{vmatrix}}$

16. $y = \dfrac{\begin{vmatrix} 4 & 11 \\ -2 & 24 \end{vmatrix}}{\begin{vmatrix} 4 & 3 \\ -2 & 5 \end{vmatrix}}$

Decompose each fraction into partial fractions.

17. $\dfrac{-x + 1}{(x + 1)(x + 2)}$

$\dfrac{2}{x + 1} - \dfrac{3}{x + 2}$

18. $\dfrac{x - 4}{(2x - 5)^2}$

$\dfrac{\frac{1}{2}}{2x - 5} - \dfrac{\frac{3}{2}}{(2x - 5)^2}$

Find each solution by graphing.

19. $y \leq 2x + 6$

20. 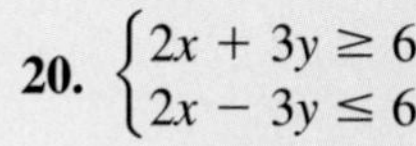
$\begin{cases} 2x + 3y \geq 6 \\ 2x - 3y \leq 6 \end{cases}$

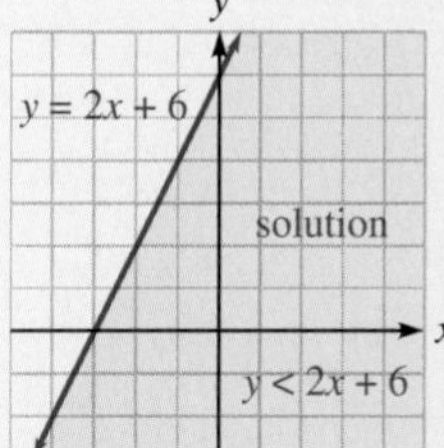

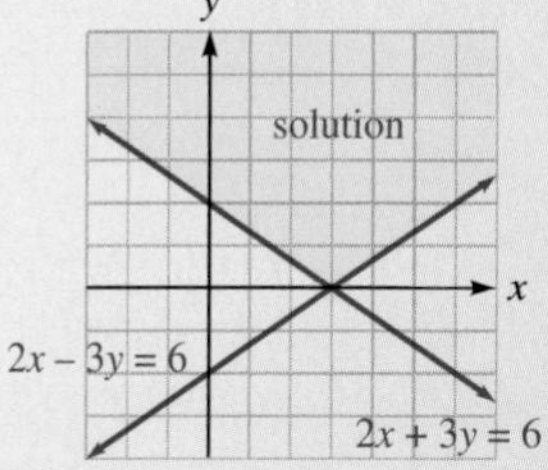

Write the equation of each circle with the given center O and radius r.

21. $O(0, 0)$; $r = 4$ $\quad x^2 + y^2 = 16$

22. $O(2, -3)$; $r = 11$ $\quad (x - 2)^2 + (y + 3)^2 = 121$

Complete the square on x and/or y and graph each equation.

23. $x^2 + y^2 - 4y = 12$

24. $x^2 - 2y - 2x = -7$

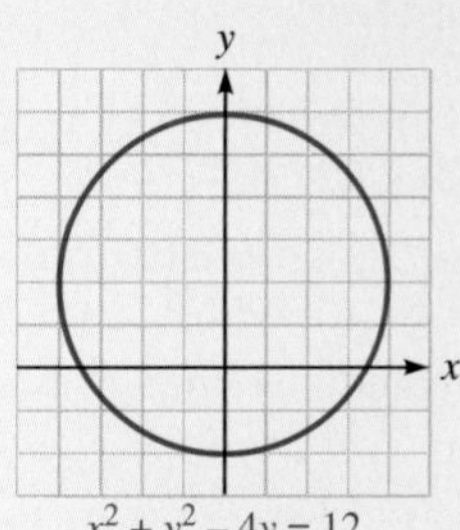

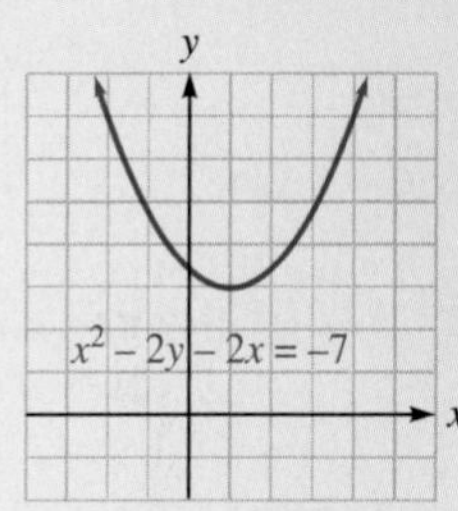

25. $x^2 + 4y^2 + 2x = 3$

26. 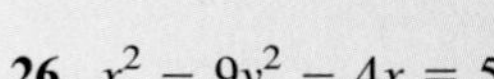
$x^2 - 9y^2 - 4x = 5$

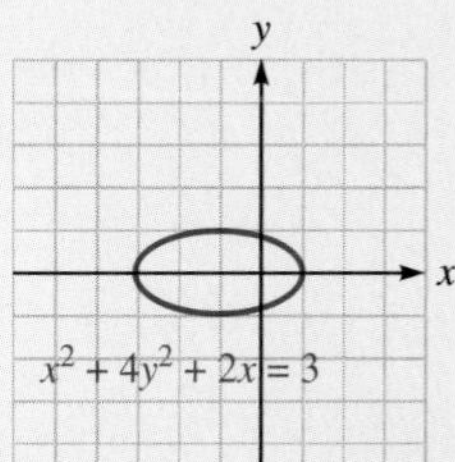

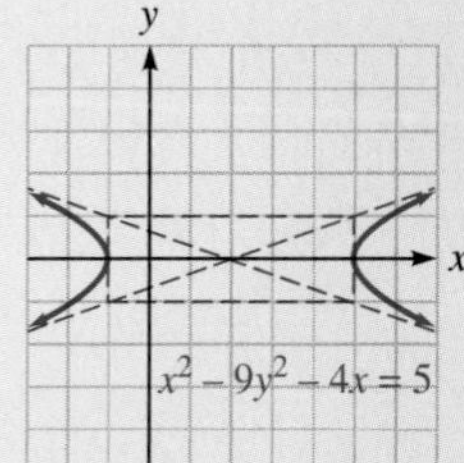

Write the equation of each ellipse.

27. Center (0, 0); horizontal major axis of 12; minor axis of 8 $\quad \dfrac{x^2}{36} + \dfrac{y^2}{16} = 1$

28. Center (2, 3); $a = 5$; $c = 2$, major axis vertical

$\dfrac{(x - 2)^2}{21} + \dfrac{(y - 3)^2}{25} = 1$

Write the equation of each hyperbola.

29. Center (0, 0); focus (3, 0); vertex (2, 0) $\frac{x^2}{4} - \frac{y^2}{5} = 1$

30. Center (2, 4); area of fundamental rectangle is 36 square units; $a = b$; transverse axis parallel to y-axis $\frac{(y-4)^2}{9} - \frac{(x-2)^2}{9} = 1$

Find the required term of the expansion of $(x + 2y)^8$.

31. 2nd term $16x^7y$

32. 6th term $1{,}792x^3y^5$

Find each sum.

33. $\sum_{k=1}^{5} 2$ 10

34. $\sum_{k=2}^{6} (3x + 1)$ 65

Find the sum of the first six terms of each sequence.

35. $-2, 1, 4, \ldots$ 33

36. $\frac{1}{9}, \frac{1}{3}, 1, \ldots$ $\frac{364}{9}$

Find each value.

37. $P(8, 4)$ 1,680

38. $P(24, 0)$ 1

39. $C(12, 10)$ 66

40. $P(4, 4) \cdot C(6, 6)$ 24

41. In how many ways can 6 men and 4 women be placed in a line if the women line up first? 17,280

42. In how many ways can a committee of 4 people be selected from a group of 12 people? 495

Find each probability.

43. Rolling 11 on one roll of two dice $\frac{1}{18}$

44. Being dealt an all-red 5-card poker hand from a standard deck $\frac{506}{19{,}992} = \frac{253}{9{,}996}$

45. If the probability that a person is married is 0.6 and the probability that a married person has children is 0.8, find the probability that a randomly chosen person is married with children. 0.48

46. The probability that a student earns an A in algebra is 0.3, and the probability that a student earns a B is 0.4. Find the probability that a student earns a C or less. 0.3

47. Prove the formula by induction:

$$4 + 7 + 10 + \cdots + (3n + 1) = \frac{n(3n + 5)}{2}$$

48. Compute the fair cost to play the following game.

1. Draw a card from a standard card deck.
2. If you draw an ace, you receive \$100.
3. If you draw a king, you receive \$10.
4. All other cards pay nothing. about \$8.46

49. What single deposit made now in an account that pays $8\frac{1}{2}\%$ interest, compounded annually, will grow to \$10,000 in 12 years? \$3,757.02

50. A bank offers a \$110,000, 20-year mortgage at 8.75%. Find the monthly payment. \$972.08

APPENDIX I

A Proof of the Binomial Theorem

The binomial theorem can be proved for positive-integer exponents using mathematical induction.

The Binomial Theorem

If n is a positive integer, then

$$(a+b)^n = a^n + \frac{n!}{1!(n-1)!}a^{n-1}b^1 + \frac{n!}{2!(n-2)!}a^{n-2}b^2 + \cdots + \frac{n!}{r!(n-r)!}a^{n-r}b^r + \cdots + b^n$$

Proof As in all induction proofs, there are two parts.

Part 1: Substituting the number 1 for n on both sides of the equation, we have

$$\begin{aligned}(a+b)^1 &= a^1 + \frac{1!}{1!(1-1)!}a^{1-1}b^1\\ a+b &= a + a^0 b\\ a+b &= a+b\end{aligned}$$

and the theorem is true when $n = 1$. Part 1 is complete.

Part 2: We write expressions for two general terms in the statement of the induction hypothesis. We assume that the theorem is true for $n = k$:

$$(a+b)^k = a^k + \frac{k!}{1!(k-1)!}a^{k-1}b + \frac{k!}{2!(k-2)!}a^{k-2}b^2 + \cdots + \frac{k!}{(r-1)!(k-r+1)!}a^{k-r+1}b^{r-1} + \frac{k!}{r!(k-r)!}a^{k-r}b^r + \cdots + b^k$$

We multiply both sides of this equation by $a + b$ and hope to obtain a similar equation in which the quantity $k + 1$ replaces all of the n values in the binomial theorem:

$$
\begin{aligned}
&(a + b)^k(a + b) \\
&\quad = (a + b)\left[a^k + \frac{k!}{1!(k-1)!}a^{k-1}b + \frac{k!}{2!(k-2)!}a^{k-2}b^2 + \cdots \right. \\
&\quad\quad \left. + \frac{k!}{(r-1)!(k-r+1)!}a^{k-r+1}b^{r-1} + \frac{k!}{r!(k-r)!}a^{k-r}b^r + \cdots + b^k\right]
\end{aligned}
$$

We distribute the multiplication first by a and then by b:

$$
\begin{aligned}
(a + b)^{k+1} = &\left[a^{k+1} + \frac{k!}{1!(k-1)!}a^k b + \frac{k!}{2!(k-2)!}a^{k-1}b^2 + \cdots \right. \\
&\left. + \frac{k!}{(r-1)!(k-r+1)!}a^{k-r+2}b^{r-1} + \frac{k!}{r!(k-r)!}a^{k-r+1}b^r + \cdots + ab^k\right] \\
&+ \left[a^k b + \frac{k!}{1!(k-1)!}a^{k-1}b^2 + \frac{k!}{2!(k-2)!}a^{k-2}b^3 + \cdots \right. \\
&\left. + \frac{k!}{(r-1)!(k-r+1)!}a^{k-r+1}b^r + \frac{k!}{r!(k-r)!}a^{k-r}b^{r+1} + \cdots + b^{k+1}\right]
\end{aligned}
$$

Combining like terms, we have

$$
\begin{aligned}
(a + b)^{k+1} = &\, a^{k+1} + \left[\frac{k!}{1!(k-1)!} + 1\right]a^k b \\
&+ \left[\frac{k!}{2!(k-2)!} + \frac{k!}{1!(k-1)!}\right]a^{k-1}b^2 + \cdots \\
&+ \left[\frac{k!}{r!(k-r)!} + \frac{k!}{(r-1)!(k-r+1)!}\right]a^{k-r+1}b^r + \cdots + b^{k+1}
\end{aligned}
$$

These results may be written as

$$
\begin{aligned}
(a + b)^{k+1} = &\, a^{k+1} + \frac{(k+1)!}{1!(k+1-1)!}a^{(k+1)-1}b + \frac{(k+1)!}{2!(k+1-2)!}a^{(k+1)-2}b^2 \\
&+ \cdots + \frac{(k+1)!}{r!(k+1-r)!}a^{(k+1)-r}b^r + \cdots + b^{k+1}
\end{aligned}
$$

This formula has precisely the same form as the binomial theorem, with the quantity $k + 1$ replacing all of the original n values. Therefore, the truth of the theorem for $n = k$ implies the truth of the theorem for $n = k + 1$. Because both parts of the axiom of mathematical induction are verified, the theorem is proved. ■

An Alternate Approach to Circles and Parabolas

We have graphed first-degree equations in two variables, such as $y = 3x + 8$ and $4x - 3y = 12$. Their graphs are lines. In this appendix, we will graph two types of second-degree equations in two variables, such as $x^2 + y^2 = 25$ and $y + 13 = -3x^2 - 12x$. The graphs of these equations are called *conic sections.*

■ Conic Sections

The curves formed by the intersection of a plane and a right-circular cone are called **conic sections.** These curves have four basic shapes, called **circles, parabolas, ellipses,** and **hyperbolas,** as shown in Figure II-1. In this appendix, we will discuss circles and parabolas.

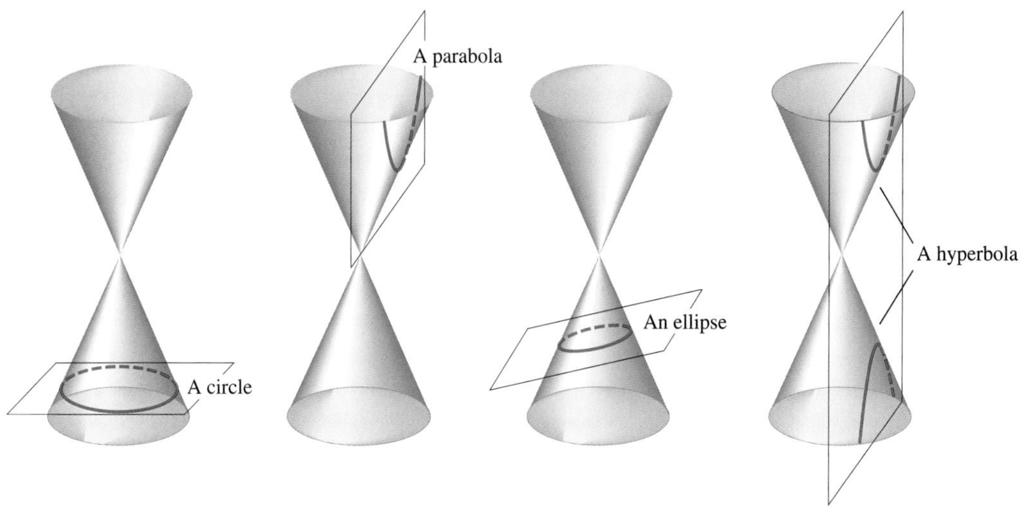

Figure II-1

In Section 1 of this appendix, we will discuss the most familiar of these conic sections, the circle.

II.1 Circles

In this section, you will learn about

- Circles
- Problem Solving Using Circles

Circles have many applications. For example, everyone is familiar with circular wheels and gears, pizza cutters, hula hoops, and Ferris Wheels.

Circles

Circles

A **circle** is the set of all points in a plane that are a fixed distance from a point called its **center.** The fixed distance is called the **radius of the circle.**

To find the equation of a circle with radius r and center at $C(h, k)$, we must find all points $P(x, y)$ in the xy-plane such that the length of segment PC is r. (See Figure II-2.) We can use the distance formula to find the length of PC, which is r.

$$r = \sqrt{(x - h)^2 + (y - k)^2}$$

After squaring both sides of the equation, we get

$$r^2 = (x - h)^2 + (y - k)^2$$

This result is called the ***standard form of the equation of a circle with radius r and center at (h, k).***

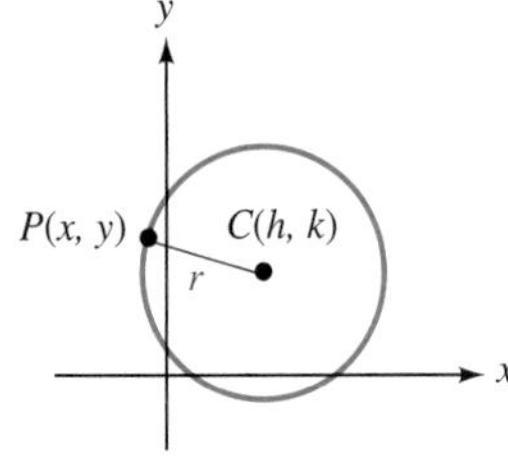

Figure II-2

The Standard Equation of a Circle with Center at (h, k)

The **standard form of the equation of a circle** with radius r and center at (h, k) is

$$(x - h)^2 + (y - k)^2 = r^2$$

If $r = 0$, the circle is a single point called a **point circle.** If the center of a circle is the origin, then $(h, k) = (0, 0)$, and we have the following result.

The Standard Equation of a Circle with Center at (0, 0)

The **standard form of the equation of a circle** with radius r and center at $(0, 0)$ is

$$x^2 + y^2 = r^2$$

EXAMPLE 1 Find the center and the radius of each circle and graph it:
a. $(x - 4)^2 + (y - 1)^2 = 9$, **b.** $x^2 + y^2 = 25$, and
c. $(x + 3)^2 + y^2 = 12$.

Solution **a.** It is easy to determine the center and the radius of a circle when its equation is written in standard form.

$$\begin{array}{ccc} (x - 4)^2 + (y - 1)^2 = \mathbf{9} \\ \uparrow \quad\quad \uparrow \quad \uparrow \\ (x - h)^2 + (y - k)^2 = r^2 \end{array}$$

In this equation, $h = 4$, $k = 1$, and $r^2 = 9$. Since the radius of a circle must be positive, $r = 3$.

Thus, the center of the circle is $(h, k) = (4, 1)$ and the radius is 3.

To plot four points on the circle, we move up, down, left, and right 3 units from the center, as shown in Figure II-3(a). We then draw a circle through the points to get the graph of $(x - 4)^2 + (y - 1)^2 = 9$, as shown in Figure II-3(b).

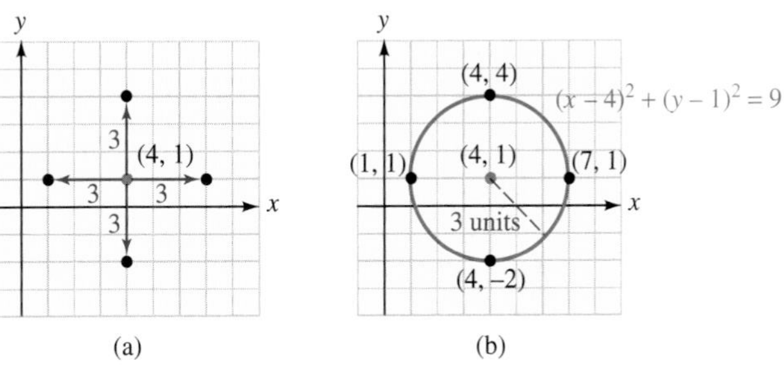

Figure II-3

b. To determine h and k, we can write $x^2 + y^2 = 25$ in the following way:

$$\begin{array}{ccc} (x - 0)^2 + (y - 0)^2 = \mathbf{25} \\ \uparrow \quad\quad \uparrow \quad \uparrow \\ h \quad\quad k \quad r^2 \end{array}$$

In this equation, $h = 0$, $k = 0$, and $r^2 = 25$. Since the radius must be positive, $r = 5$.

Thus, the center of the circle is at (0, 0), the origin, and the radius is 5.

To plot four points on the circle, we move up, down, left, and right 5 units from the center. Then we draw a circle through the points to get the graph of $x^2 + y^2 = 25$, as shown in Figure II-4.

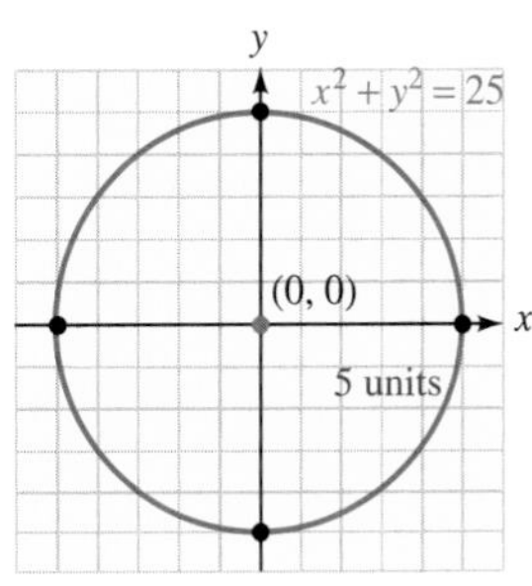

Figure II-4

c. To determine h, we can write $x + 3$ as $x - (-3)$.

Standard form requires a minus here.
↓

$$\begin{array}{ccc} [x - (-3)]^2 + (y - 0)^2 = \mathbf{12} \\ \uparrow \quad\quad \uparrow \quad \uparrow \\ h \quad\quad k \quad r^2 \end{array}$$

In this equation, $h = -3$, $k = 0$, and $r^2 = 12$.

Since $r^2 = 12$, we have

$$r = \pm\sqrt{12} = \pm 2\sqrt{3} \qquad \pm\sqrt{12} = \pm\sqrt{4 \cdot 3} = \pm\sqrt{4}\sqrt{3} = \pm 2\sqrt{3}.$$

Since the radius cannot be negative, we have $r = 2\sqrt{3}$. The center of the circle is at $(-3, 0)$ and its radius is $2\sqrt{3}$.

To plot four points on the circle, we move up, down, left, and right $2\sqrt{3} \approx 3.5$ units from the center. Then we draw a circle through the points to get the graph of $(x + 3)^2 + y^2 = 12$, as shown in Figure II-5.

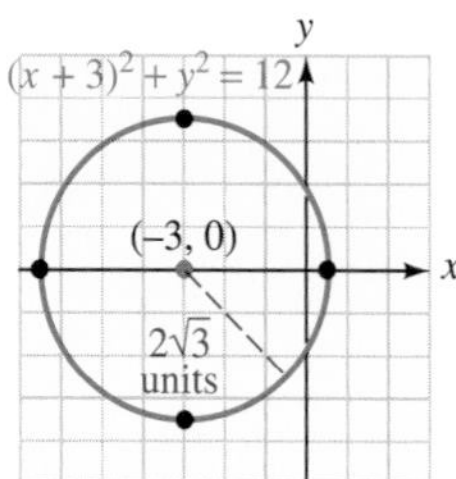

Figure II-5

EXAMPLE 2 Find the equation of a circle with a radius of 9 and a center at $(6, -5)$.

Solution We substitute 6 for h, -5 for k, and 9 for r in the standard equation of a circle and simplify.

$$(x - h)^2 + (y - k)^2 = r^2$$
$$(x - 6)^2 + [y - (-5)]^2 = 9^2$$
$$(x - 6)^2 + (y + 5)^2 = 81$$

The equation of the circle is $(x - 6)^2 + (y + 5)^2 = 81$.

In Example 2, the result was written in standard form as $(x - 6)^2 + (y + 5)^2 = 81$. If we square $x - 6$ and $y + 5$, we will obtain a different form for the circle's equation.

$$(x - 6)^2 + (y + 5)^2 = 81$$
$$x^2 - 12x + 36 + y^2 + 10y + 25 = 81$$
$$x^2 - 12x + y^2 + 10y - 20 = 0 \qquad \text{Subtract 81 from both sides and simplify.}$$
$$x^2 + y^2 - 12x + 10y - 20 = 0$$

This result is called the *general form of the equation of a circle.*

The General Form of the Equation of a Circle

The **general form of the equation of a circle** is

$$x^2 + y^2 + cx + dy + e = 0$$

where c, d, and e are constants.

We can convert from the general form to the standard form of the equation of a circle by completing the square.

EXAMPLE 3 Write in standard form: $x^2 + y^2 - 4x + 2y - 11 = 0$.

Solution To write the equation in standard form, we complete the square twice.

$$x^2 + y^2 - 4x + 2y - 11 = 0$$

$$x^2 - 4x + y^2 + 2y - 11 = 0 \quad \text{Write the } x\text{-terms together and the } y\text{-terms together.}$$

$$x^2 - 4x + y^2 + 2y = 11 \quad \text{Add 11 to both sides.}$$

To complete the square on $x^2 - 4x$, we note that $\frac{1}{2}(-4) = -2$ and $(-2)^2 = 4$. To complete the square on $y^2 + 2y$, we note that $\frac{1}{2}(2) = 1$ and $1^2 = 1$. We add **4** and **1** to both sides of the equation.

$$x^2 - 4x + \mathbf{4} + y^2 + 2y + \mathbf{1} = 11 + \mathbf{4} + \mathbf{1}$$

$$(x - 2)^2 + (y + 1)^2 = 16 \quad \text{Factor } x^2 - 4x + 4 \text{ and } y^2 + 2y + 1 \text{ and simplify.}$$

This equation can also be written as $(x - 2)^2 + (y + 1)^2 = 4^2$. ■

EXAMPLE 4 Find the general form of the equation of a circle with radius of 5 and center at (3, 2).

Solution We substitute 5 for r, 3 for h, and 2 for k in the standard equation of a circle and simplify.

$$(x - h)^2 + (y - k)^2 = r^2$$

$$(x - 3)^2 + (y - 2)^2 = 5^2$$

$$x^2 - 6x + 9 + y^2 - 4y + 4 = 25 \quad \text{Remove parentheses.}$$

$$x^2 + y^2 - 6x - 4y - 12 = 0 \quad \text{Subtract 25 from both sides and simplify.}$$

The general form of the equation is $x^2 + y^2 - 6x - 4y - 12 = 0$. ■

EXAMPLE 5 Find the general form of the equation of the circle with endpoints of its diameter at $(8, -3)$ and $(-4, 13)$.

Solution We can find the center $O(h, k)$ of the circle by finding the midpoint of its diameter. By the midpoint formula and $(x_1, y_1) = (8, -3)$ and $(x_2, y_2) = (-4, 13)$, we have

$$h = \frac{x_1 + x_2}{2} \qquad k = \frac{y_1 + y_2}{2}$$

$$h = \frac{8 + (-4)}{2} \qquad k = \frac{-3 + 13}{2}$$

$$= \frac{4}{2} \qquad = \frac{10}{2}$$

$$= 2 \qquad = 5$$

The center is $O(h, k) = O(2, 5)$.

To find the radius, we find the distance between the center (2, 5) and one endpoint, say $(8, -3)$, of the diameter.

$$r = \sqrt{(x_2 - x_1)^2 + (y_2 - y_1)^2} \quad \text{This is the distance formula.}$$

$$r = \sqrt{(2 - 8)^2 + [5 - (-3)]^2} \quad \text{Substitute 8 for } x_1, -3 \text{ for } y_1, 2 \text{ for } x_2, \text{ and 5 for } y_2.$$

$$= \sqrt{(-6)^2 + (8)^2}$$

$$= \sqrt{36 + 64}$$

$$= 10$$

To find the equation of the circle with center at (2, 5) and radius of 10, we substitute 2 for h, 5 for k, and 10 for r in the standard equation of a circle and simplify:

$$(x - h)^2 + (y - k)^2 = r^2$$

$$(x - 2)^2 + (y - 5)^2 = 10^2$$

$$x^2 - 4x + 4 + y^2 - 10y + 25 = 100 \quad \text{Square each binomial.}$$

$$x^2 + y^2 - 4x - 10y - 71 = 0 \quad \text{Subtract 100 from both sides and simplify.}$$

■

EXAMPLE 6 Graph the circle whose equation is $2x^2 + 2y^2 - 8x + 4y = 40$.

Solution We first divide both sides of the equation by 2 to make the coefficients of x^2 and y^2 equal to 1.

$$2x^2 + 2y^2 - 8x + 4y = 40$$

$$x^2 + y^2 - 4x + 2y = 20 \quad \text{Divide both sides by 2.}$$

To find the coordinates of the center and the radius, we write the previous equation in standard form by completing the square on x and y.

$$x^2 + y^2 - 4x + 2y = 20$$

$$x^2 - 4x + y^2 + 2y = 20$$

$$x^2 - 4x + 4 + y^2 + 2y + 1 = 20 + 4 + 1 \quad \text{Add 4 and 1 to both sides to complete the squares.}$$

$$(x - 2)^2 + (y + 1)^2 = 25 \quad \text{Factor } x^2 - 4x + 4 \text{ and } y^2 + 2y + 1 \text{ and simplify.}$$

$$(x - 2)^2 + [(y - (-1)]^2 = 5^2 \quad y + 1 = y - (-1).$$

From the equation of the circle, we see that its radius is 5 and that the coordinates of the center are $h = 2$ and $k = -1$. The graph is shown in Figure II-6.

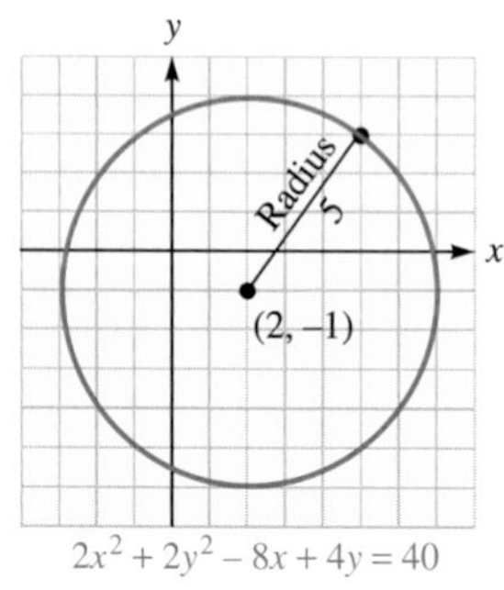

Figure II-6

■

Accent on Technology GRAPHING CIRCLES

To use a graphing calculator to graph $(x - 2)^2 + (y + 1)^2 = 25$, we must solve the equation for y.

$$(x - 2)^2 + (y + 1)^2 = 25$$
$$(y + 1)^2 = 25 - (x - 2)^2$$
$$y + 1 = \pm\sqrt{25 - (x - 2)^2}$$
$$y = -1 \pm \sqrt{25 - (x - 2)^2}$$

The last expression represents two equations: $y = -1 + \sqrt{25 - (x - 2)^2}$ and $y = -1 - \sqrt{25 - (x - 2)^2}$. We graph both of these equations separately on the same coordinate axes by entering the first equation as Y_1 and the second as Y_2.

$$Y_1 = -1 + \sqrt{}(25 - (X - 2)\text{^}2)$$
$$Y_2 = -1 - \sqrt{}(25 - (X - 2)\text{^}2)$$

Depending on the setting of the maximum and minimum values of x and y, the graph may not appear to be a circle and might look like Figure II-7(a). Most graphing calculators can be set to display equal-sized divisions on the x- and y-axes, producing a better graph such as the one shown in Figure II-7(b).

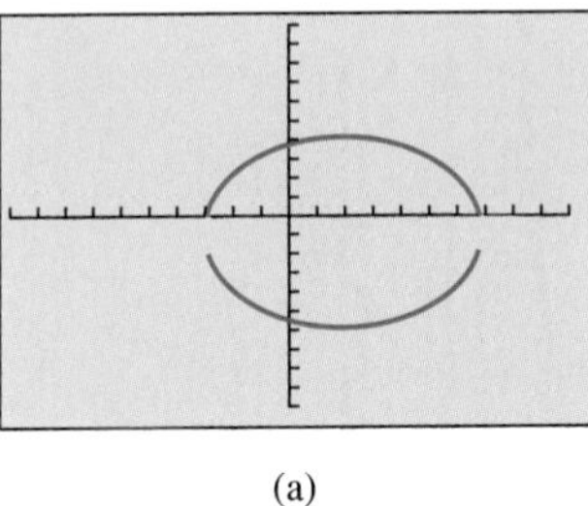

(a)

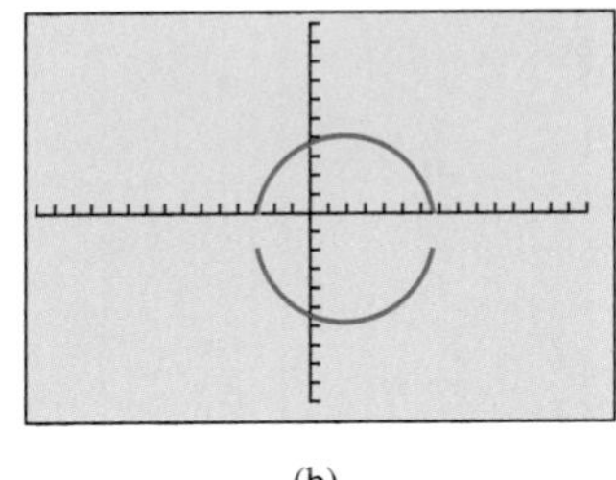

(b)

Figure II-7

Note that in this example, it is easier to graph the circle by hand than with a calculator.

Problem Solving Using Circles

EXAMPLE 7 The broadcast area of a television station is bounded by the circle $x^2 + y^2 = 3{,}600$, where x and y are measured in miles. A translator station picks up the signal and retransmits it from the center of a circular area bounded by

$$(x + 30)^2 + (y - 40)^2 = 1{,}600$$

Find the location of the translator and the greatest distance from the main transmitter that the signal can be received.

Solution The coverage of the TV station is bounded by $x^2 + y^2 = 3{,}600$, a circle at the origin with a radius of 60 miles shown in yellow in Figure II-8. Since the translator is at the center of the circle $(x + 30)^2 + (y - 40)^2 = 1{,}600$, it is located at $(-30, 40)$, a point 30 miles west and 40 miles north of the station. The radius of the translator's coverage is $\sqrt{1{,}600}$ miles, or 40 miles.

As shown in the figure, the greatest distance of reception is the sum of d, the distance from the translator to the television station, and 40 miles, the radius of the translator's coverage.

To find d, we use the distance formula to find the distance between $(x_1, y_1) = (-30, 40)$ and the origin $(x_2, y_2) = (0, 0)$.

$$d = \sqrt{(x_1 - x_2)^2 + (y_1 - y_2)^2}$$
$$d = \sqrt{(-30 - 0)^2 + (40 - 0)^2}$$
$$= \sqrt{(-30)^2 + 40^2}$$
$$= \sqrt{900 + 1{,}600}$$
$$= \sqrt{2{,}500}$$
$$= 50$$

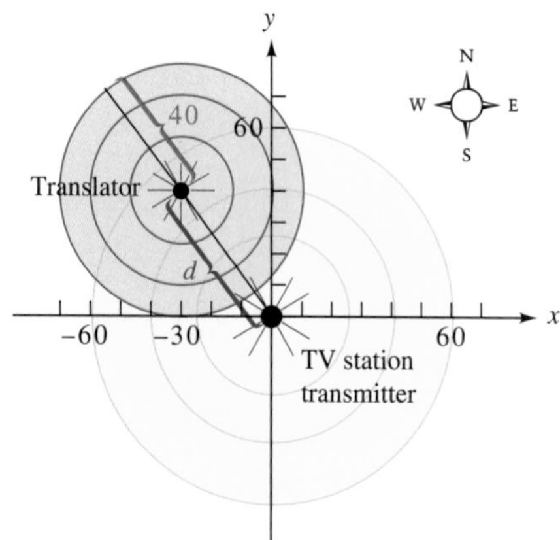

Figure II-8

The translator is located 50 miles from the television station, and it can broadcast the signal 40 miles. The greatest reception distance from the main TV station is $50 + 40$ miles, or 90 miles. ■

II-1 Exercises

VOCABULARY AND CONCEPTS ***Fill in the blanks.***

1. A circle is the set of all points in a plane that are a fixed distance from a point called its center.
2. A radius is the distance from the center of a circle to a point on the circle.
3. The standard equation of a circle with center at the origin and radius r is $x^2 + y^2 = r^2$.
4. The standard equation of a circle with center at (h, k) and radius r is $(x - h)^2 + (y - k)^2 = r^2$.

Determine the coordinates of each circle's center and radius.

5. $(x - 1)^2 + (y - 2)^2 = 16$ **(1, 2), 4**
6. $(x - 5)^2 + (y - 4)^2 = 100$ **(5, 4), 10**
7. $(x - 2)^2 + (y + 5)^2 = 9$ **(2, −5), 3**
8. $x^2 + y^2 - 36 = 0$ **(0, 0), 6**
9. $x^2 + y^2 = 5$ **(0, 0), $\sqrt{5}$**
10. $2(x - 9)^2 + 2y^2 = 7$ **(9, 0), $\sqrt{\frac{7}{2}}$ or $\frac{\sqrt{14}}{2}$**

PRACTICE *Find the equation of each circle.*

11.

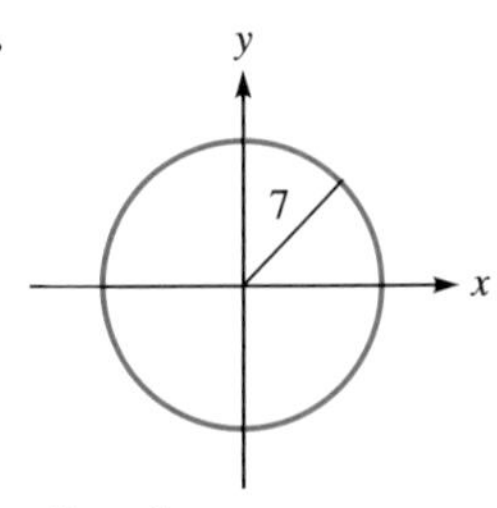

$x^2 + y^2 = 49$

12.

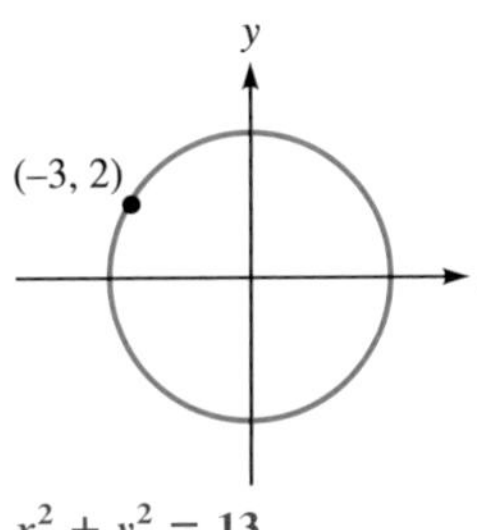

$x^2 + y^2 = 13$

13.

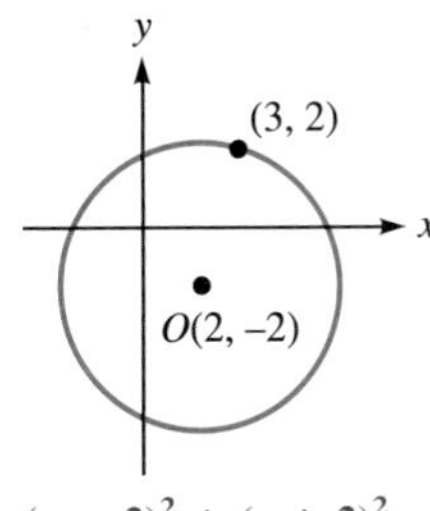

$(x - 2)^2 + (y + 2)^2 = 17$

14.

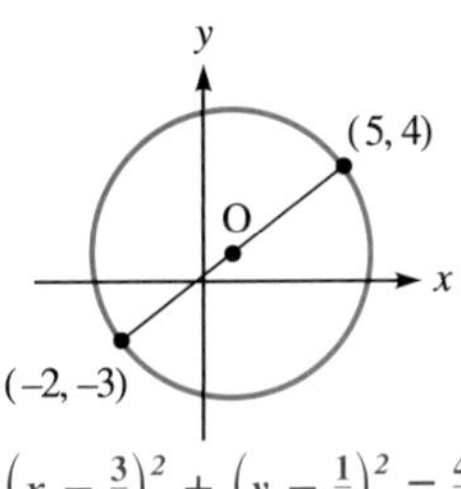

$\left(x - \frac{3}{2}\right)^2 + \left(y - \frac{1}{2}\right)^2 = \frac{49}{2}$

Find the equation of each circle with the given properties. Give the answer in general form.

15. Center at the origin; $r = 1$ $x^2 + y^2 - 1 = 0$

16. Center at the origin; $r = 4$ $x^2 + y^2 - 16 = 0$

17. Center at (6, 8); $r = 4$ $x^2 + y^2 - 12x - 16y + 84 = 0$

18. Center at (5, 3); $r = 2$ $x^2 + y^2 - 10x - 6y + 30 = 0$

19. Center at $(3, -4)$; $r = \sqrt{2}$
$x^2 + y^2 - 6x + 8y + 23 = 0$

20. Center at $(-9, 8)$; $r = 2\sqrt{3}$
$x^2 + y^2 + 18x - 16y + 133 = 0$

21. Ends of diameter at $(3, -2)$ and $(3, 8)$
$x^2 + y^2 - 6x - 6y - 7 = 0$

22. Ends of diameter at (5, 9) and $(-5, -9)$
$x^2 + y^2 - 106 = 0$

23. Center at $(-3, 4)$ and passing through the origin
$x^2 + y^2 + 6x - 8y = 0$

24. Center at $(-2, 6)$ and passing through the origin
$x^2 + y^2 + 4x - 12y = 0$

25. Radius of 6; center at the intersection of $3x + y = 1$ and $-2x - 3y = 4$ $(x - 1)^2 + (y + 2)^2 = 36$

26. Radius of 8; center at the intersection of $x + 2y = 8$ and $2x - 3y = -5$ $(x - 2)^2 + (y - 3)^2 = 64$

Graph each equation.

27. $x^2 + y^2 - 25 = 0$ **28.** $x^2 + y^2 - 8 = 0$

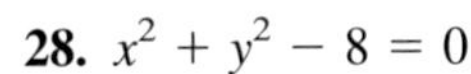

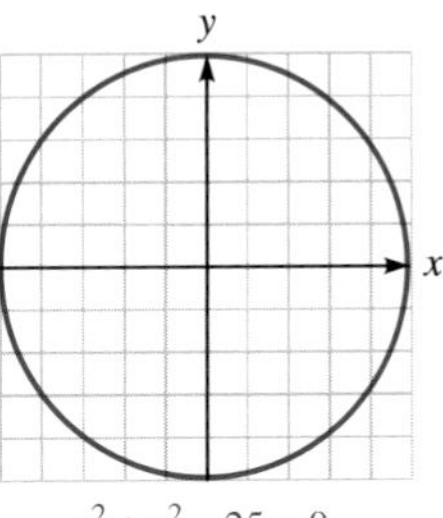

$x^2 + y^2 - 25 = 0$

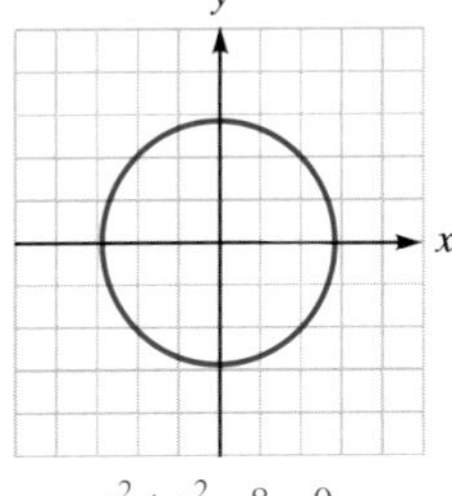

$x^2 + y^2 - 8 = 0$

29. $(x - 1)^2 + (y + 2)^2 = 4$ **30.** $(x + 1)^2 + (y - 2)^2 = 9$

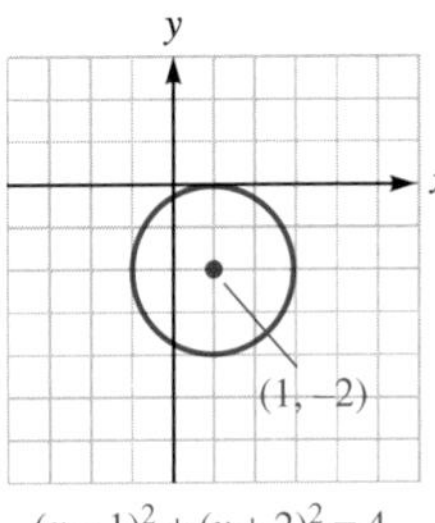

$(x - 1)^2 + (y + 2)^2 = 4$

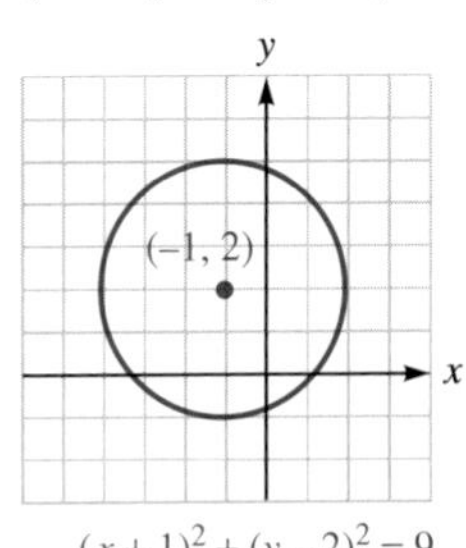

$(x + 1)^2 + (y - 2)^2 = 9$

31. $x^2 + y^2 + 2x - 24 = 0$ **32.** $x^2 + y^2 - 4y = 12$

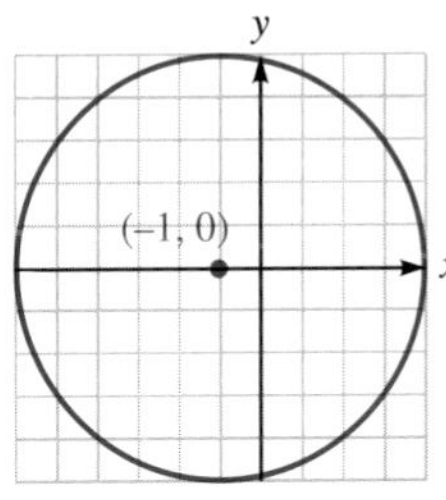

$x^2 + y^2 + 2x - 24 = 0$

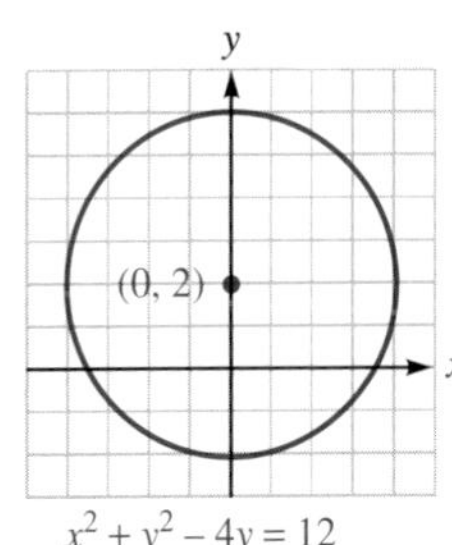

$x^2 + y^2 - 4y = 12$

33. $9x^2 + 9y^2 - 12y = 5$ **34.** $4x^2 + 4y^2 + 4y = 15$

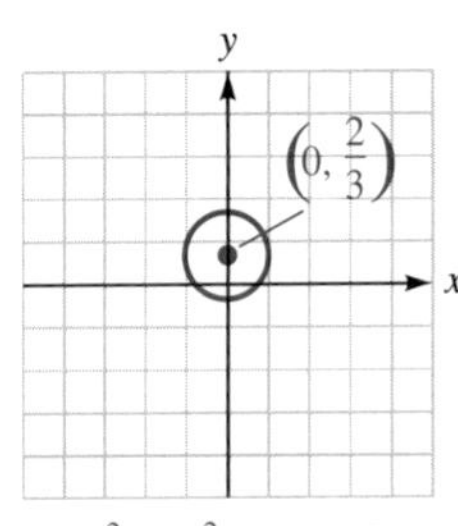

$9x^2 + 9y^2 - 12y = 5$

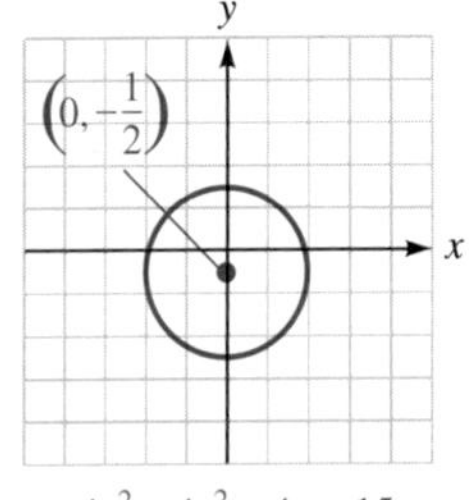

$4x^2 + 4y^2 + 4y = 15$

35. $4x^2 + 4y^2 - 4x + 8y + 1 = 0$

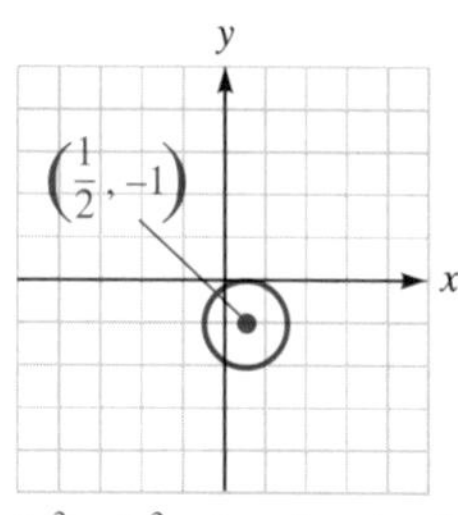

36. $9x^2 + 9y^2 - 6x + 18y + 1 = 0$

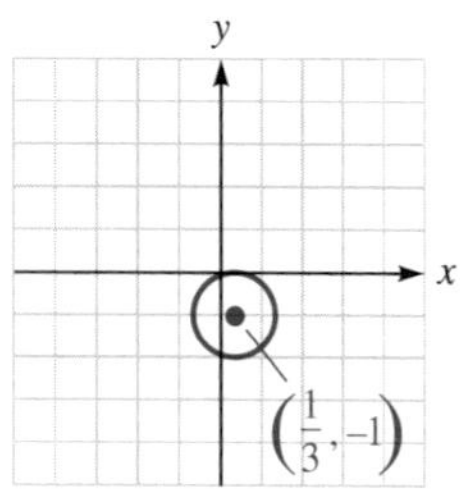

APPLICATIONS

37. **Broadcast ranges** A TV station broadcasts a signal with a circular range. Can a city 50 miles east and 70 miles north of the tower receive the signal? **yes**

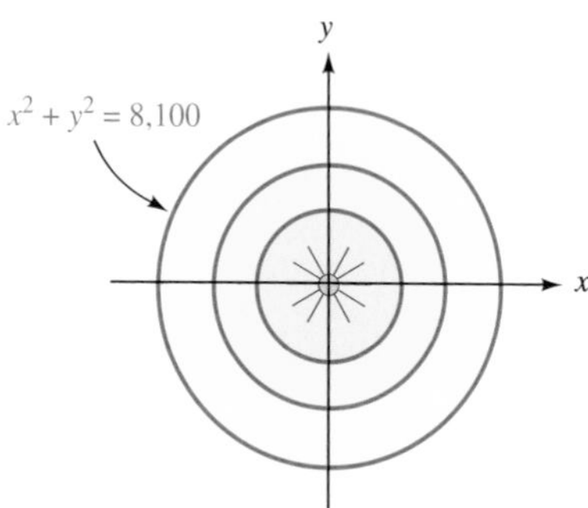

38. **Warning sirens** A tornado siren can be heard in the circular range shown below. Can a person 4 miles west and 5 miles south of the siren hear its sound? **no**

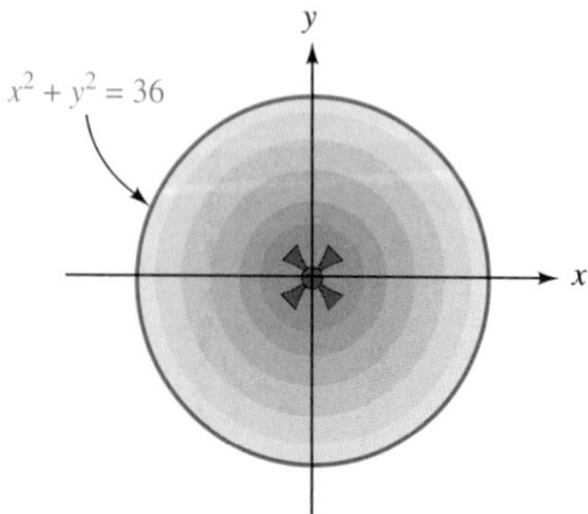

39. **Meshing gears** For design purposes, the large gear shown below is described by the circle $x^2 + y^2 = 16$. The smaller gear is a circle centered at (7, 0) and tangent to the larger circle. Find the equation of the outer edge of the smaller gear. $(x - 7)^2 + y^2 = 9$

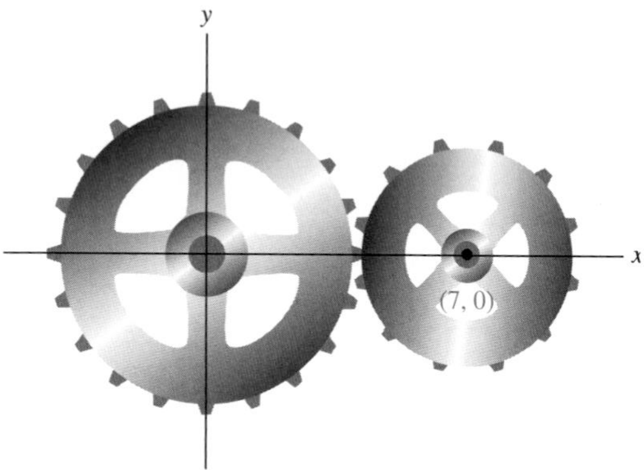

40. **Walkways** A walkway is bounded by the two circles $x^2 + y^2 = 2{,}500$ and $(x - 10)^2 + y^2 = 900$, where the radius is measured in feet. Find the larger and the smaller width of the walkway. **30 ft and 10 ft**

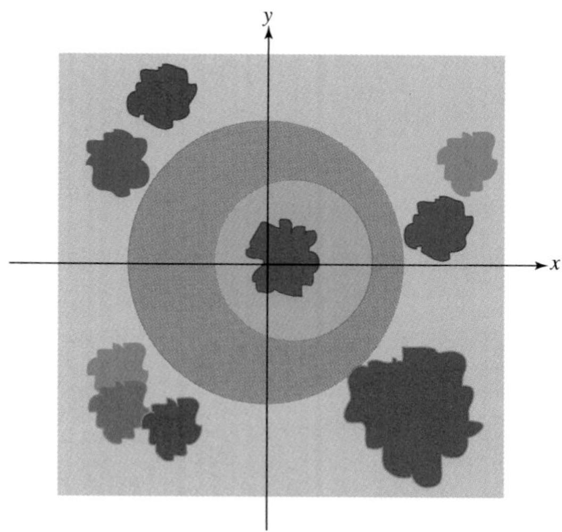

41. **Broadcast ranges** Radio stations applying for licensing cannot use the same frequency if their broadcast areas overlap. One station's coverage is bounded by $x^2 + y^2 - 8x - 20y + 16 = 0$ and the other's by $x^2 + y^2 + 2x + 4y - 11 = 0$. Can they apply for the same frequency? **no**

42. Highway designs Engineers want to join two sections of highway with a curve that is one-fourth of a circle. The equation of the circle is $x^2 + y^2 - 16x - 20y + 155 = 0$, where the distances are measured in kilometers. Find the location (relative to the center of the town) of the intersections of the highway with State and Main. **11 km, 13 km**

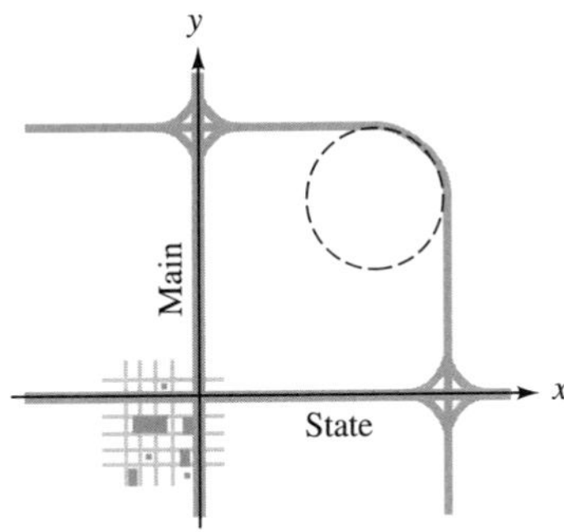

43. CB radios The CB radio of a trucker covers the circular area shown below. Find the equation of the circle in general form. $x^2 + y^2 - 14x - 8y + 40 = 0$

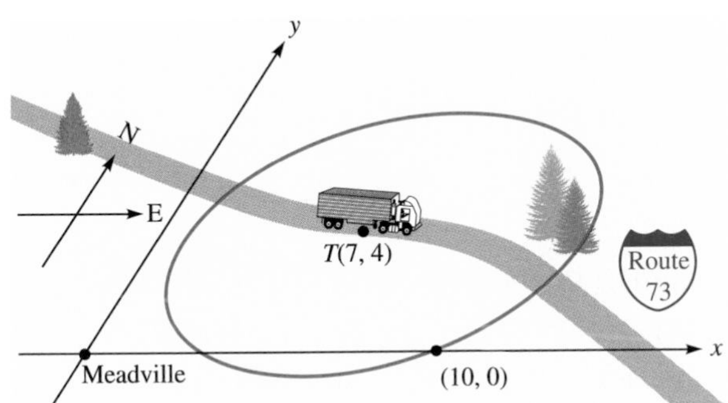

44. Tires Two 24-inch-diameter tires stand against a wall. Find the equations of the circular boundaries of the tires.
$x^2 + y^2 - 24x - 24y + 144 = 0$;
$x^2 + y^2 - 72x - 24y + 1{,}296 = 0$

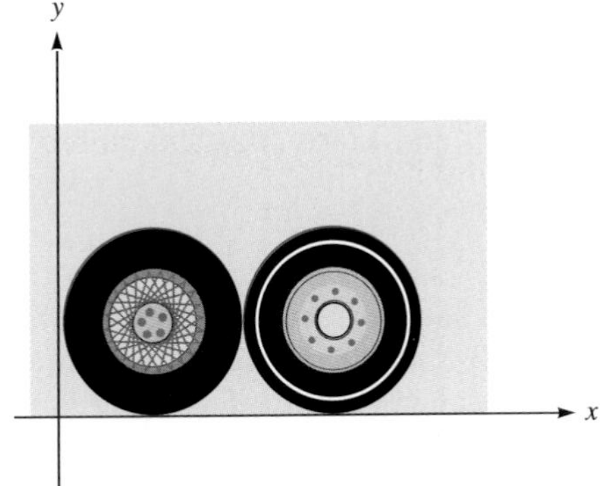

45. Radio translators Some radio stations extend their broadcast ranges by installing a translator. A station with a broadcast range given by $x^2 + y^2 = 1{,}600$, with x and y given in miles, installs a translator with a broadcast area bounded by $x^2 + y^2 - 70y + 600 = 0$. Find the greatest distance from the main transmitter that the signal can be received. **60 mi**

46. Ripples in a pond When a stone is thrown into a pond, the ripples spread out in a circular pattern, moving at the rate of 3 feet per second. If the stone is dropped at the point (0, 0) in the illustration, when will the ripple reach the seagull floating at point (15, 36)? **in 13 sec**

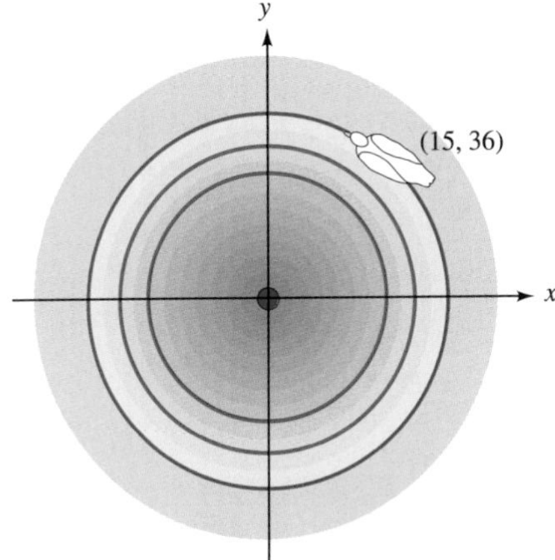

47. Writing the equation of a circle Find the equation of the outer rim of the circular arch shown in the illustration. $(x - 4)^2 + y^2 = 16$

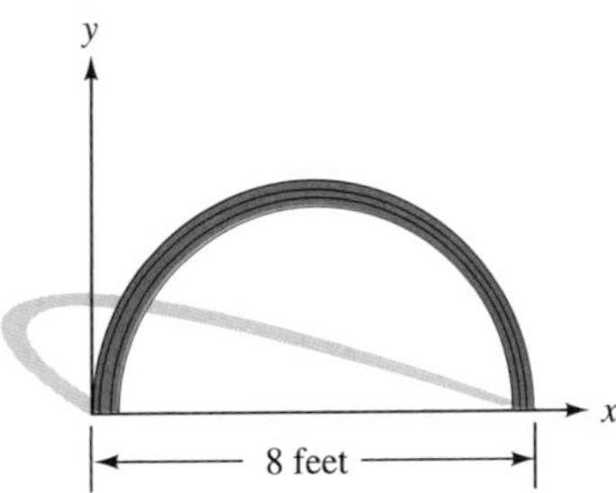

48. Writing the equation of a circle The shape of the window shown is a combination of a rectangle and a semicircle. Find the equation of the circle of which the semicircle is a part. $(x - 5)^2 + (y - 14)^2 = 25$

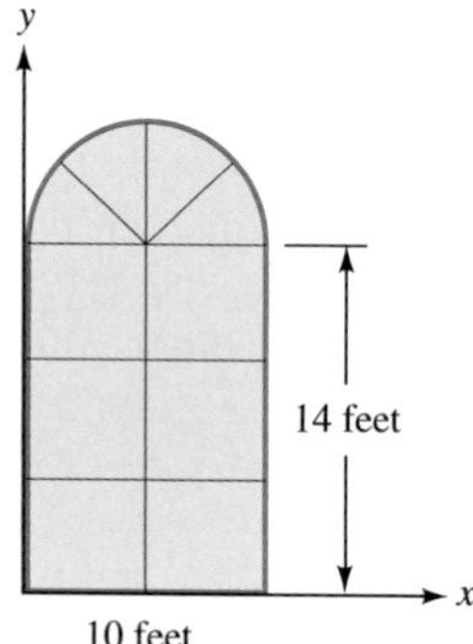

DISCOVERY AND WRITING *Find the equation of the circle passing through the given points. Give the answer in standard form.*

49. (0, 8), (5, 3), and (4, 6) $x^2 + (y - 3)^2 = 25$

50. (−2, 0), (2, 8), and (5, −1) $(x - 2)^2 + (y - 3)^2 = 25$

51. Explain why the standard form of the equation of the circle

$$(x + 2)^2 + (y - 5)^2 = 36$$

is a special case of the general form of a second-degree equation in two variables.

52. The following sign was posted in front of a polling place on election day. Explain what it means.

No electioneering within 1,500-foot radius of this polling place.

REVIEW *Solve each equation.*

53. $|3x - 4| = 11$ $5, -\frac{7}{3}$

54. $\left|\frac{4 - 3x}{5}\right| = 12$ $\frac{64}{3}, -\frac{56}{3}$

55. $|3x + 4| = |5x - 2|$ $3, -\frac{1}{4}$

56. $|6 - 4x| = |x + 2|$ $\frac{4}{5}, \frac{8}{3}$

II.2 Parabolas

In this section, you will learn about

- Parabolas that Open Upward or Downward
- Graphing Parabolas
- Solving Problems Involving Parabolas
- Parabolas that Open to the Left or Right
- Graphing Parabolas

Another type of conic section is the parabola. In Section 3.2, we saw that graphs of quadratic functions are parabolas that open upward or downward. In that section, we defined quadratic functions, graphed them, found the vertex of their graphs, and solved maximum and minimum problems.

In this appendix, we will take another look at parabolas that open upward or downward and consider parabolas that open to the left or right.

Parabolas that Open Upward or Downward

In Section 3.2, we showed that if a, h, and k are constants and $a \neq 0$, the graph of

$$y = a(x - h)^2 + k \quad (1)$$

is a parabola with vertex at (h, k). If $a > 0$, the parabola will open upward, as in Figure II-9(a). If $a < 0$, the parabola will open downward, as in Figure II-9(b).

In Figure II-9(a), the line $x = 1$ is called an **axis of symmetry.** In Figure II-9(b), the line $x = -2$ is an axis of symmetry.

$y = (x - 1)^2 - 4$

x	$(x, f(x))$
-2	$(-2, 5)$
-1	$(-1, 0)$
0	$(0, -3)$
1	$(1, -4)$
2	$(2, -3)$
3	$(3, 0)$
4	$(4, 5)$

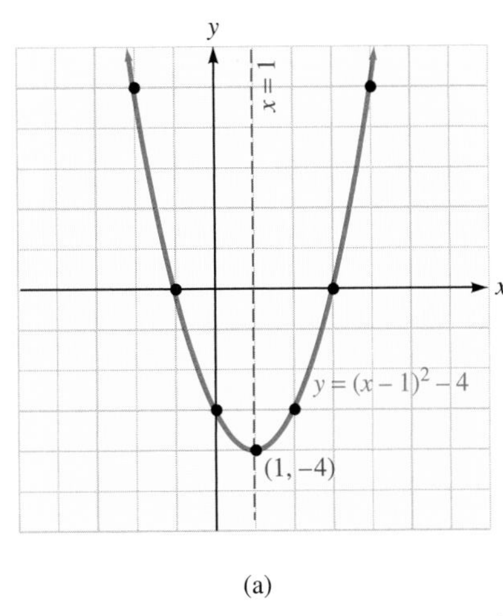

(a)

$y = -2(x + 2)^2 + 5$

x	$(x, f(x))$
-4	$(-4, -3)$
-3	$(-3, 3)$
-2	$(-2, 5)$
-1	$(-1, 3)$
0	$(0, -3)$

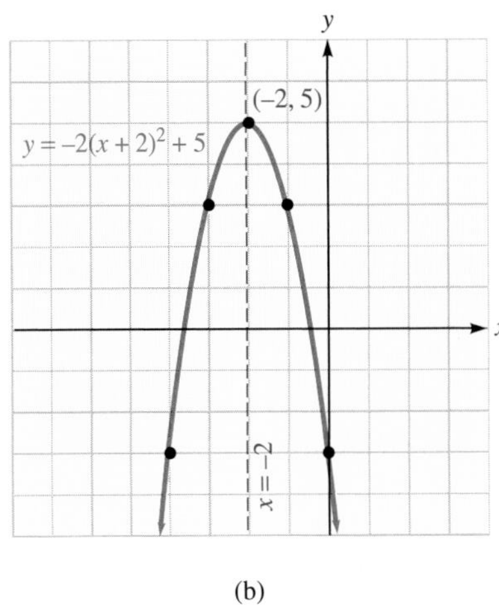

(b)

Figure II-9

To write the equation of a parabola in a form similar to the standard equation of a circle, we subtract k from both sides of Equation 1.

The Standard Equation of a Parabola that Opens Upward or Downward

The **standard equation of a parabola** that opens upward or downward is

$$y - k = a(x - h)^2$$

If $a > 0$, the parabola opens upward. If $a < 0$, the parabola opens downward. The vertex of the parabola is at (h, k).

If the vertex is at the origin, the standard equation is

$$y = ax^2$$

If parentheses are removed in the equation $y - k = a(x - h)^2$ and the equation is solved for y, the equation will have the form $y = ax^2 + bx + c$, where $a \neq 0$. This is called the **general form** of a parabola. It will open upward when $a > 0$ and open downward when $a < 0$.

■ Graphing Parabolas

EXAMPLE 1 Find the standard equation of the parabola that opens upward, has a vertex at the point (4, 5), and passes through the point (0, 7). Then graph it.

Solution Because the parabola opens upward, we can use the equation $y - k = a(x - h)^2$. In this equation, we can substitute 4 for h, 5 for k, 0 for x, 7 for y, and solve for a.

(2)
$$y - k = a(x - \boldsymbol{h})^2$$
$$7 - 5 = a(\mathbf{0} - \mathbf{4})^2$$
$$2 = 16a$$
$$\frac{1}{8} = a$$

To obtain the standard equation, we can substitute 4 for h, 5 for k, and $\frac{1}{8}$ for a in Equation 2 to get

$$y - 5 = \frac{1}{8}(x - 4)^2$$

The graph appears in Figure II-10.

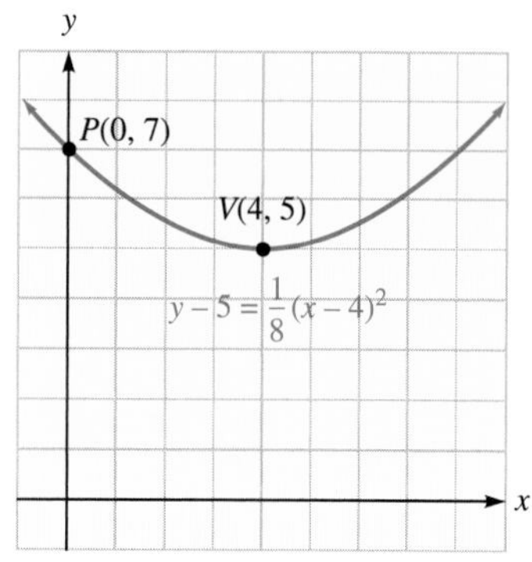

Figure II-10

EXAMPLE 2 Find the equation of the parabola opening downward with vertex at (2, 4) that passes through (0, 0).

Solution The parabola is shown in Figure II-11. To find its equation, we use the equation $y - k = a(x - h)^2$ and substitute 2 for h, 4 for k, 0 for x, 0 for y, and solve for a.

$$\begin{aligned} y - k &= a(x - h)^2 \\ 0 - 4 &= a(0 - 2)^2 \\ -4 &= 4a \\ -1 &= a \end{aligned}$$

Since $h = 2$, $k = 4$, and $a = -1$, the equation is $y - 4 = -1(x - 2)^2$.

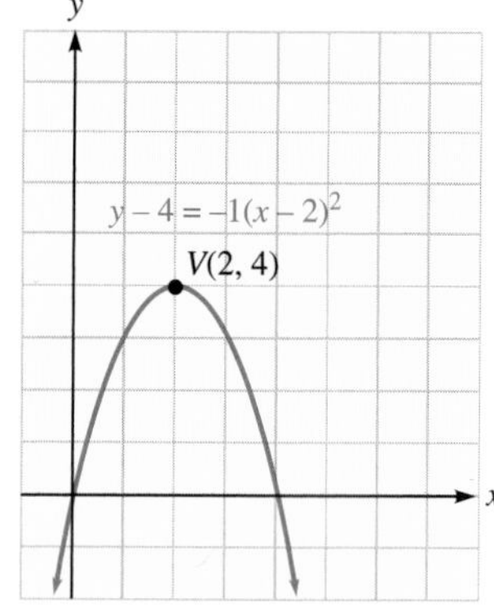

Figure II-11

Solving Problems Involving Parabolas

EXAMPLE 3 If a stone is thrown straight up into the air, the equation $s = 128t - 16t^2$ expresses its height in feet t seconds after it is thrown. Find the maximum height reached by the stone.

Solution The graph of $s = 128t - 16t^2$, which expresses the height in feet of the stone t seconds after it is thrown, is the parabola shown in Figure II-12. To find the maximum height reached by the stone, we find the s-coordinate k of the vertex of the parabola. To find k, we write the equation of the parabola in standard form by completing the square on t.

$$s = 128t - 16t^2$$

$$s = -16(t^2 - 8t) \qquad \text{Factor out } -16.$$

$$s = -16(t^2 - 8t + 16 - 16) \qquad \text{Add 16 to complete the square on } t. \text{ To counteract this addition, subtract 16.}$$

$$s = -16(t - 4)^2 + 256 \qquad \text{Distribute the multiplication by } -16 \text{ and factor } t^2 - 8t + 16.$$

(3) $$s - 256 = -16(t - 4)^2 \qquad \text{Subtract 256 from both sides.}$$

From Equation 3, we can see that the vertex of the parabola is at (4, 256). This indicates that the stone will reach its maximum height of 256 feet in 4 seconds.

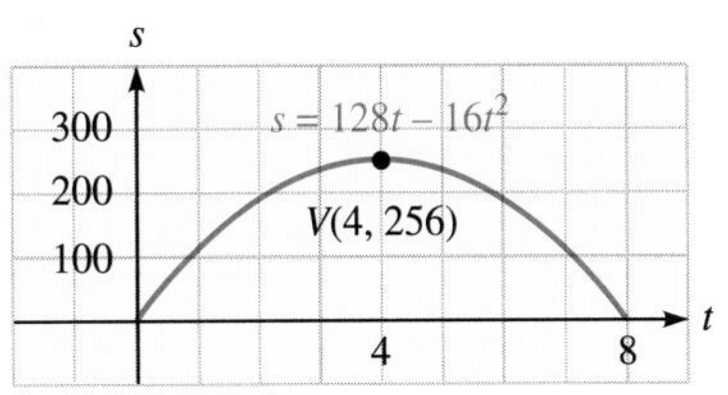

Figure II-12

Parabolas that Open to the Left or Right

Parabolas can also open to the left or right, but their equations are not functions because their graphs fail the vertical line test. The standard equation of a parabola that opens to the left or right is as follows.

The Standard Equation of a Parabola that Opens to the Left or Right

The **standard equation of a parabola** that opens to the left or right is

$$x - h = a(y - k)^2$$

If $a > 0$, the parabola opens to the right. If $a < 0$, the parabola opens to the left. The vertex of the parabola is at (h, k).

If the vertex is at the origin, the standard equation is

$$x = ay^2$$

Graphing Parabolas

EXAMPLE 4 Graph: $x = -3y^2 - 12y - 13$.

Solution To write the equation in standard form, we complete the square on y.

$$x = -3y^2 - 12y - 13$$

$$x = -3(y^2 + 4y \qquad) - 13 \qquad \text{Factor } -3 \text{ from } -3y^2 - 12y.$$

$$x = -3(y^2 + 4y + 4) - 13 + 12 \qquad \text{Complete the square on } y. \text{ Since this step adds } -3 \cdot 4 = -12, \text{ we must counteract this addition of } -12 \text{ by adding 12.}$$

$$x + 1 = -3(y + 2)^2 \qquad \text{Factor } y^2 + 4y + 4, \text{ combine terms, and add 1 to both sides.}$$

This final equation can be written in standard form as $x - (-1) = -3[(y - (-2)]^2$. When we compare this equation to the standard equation of a parabola $x - h = a(y - k)^2$, we see that $h = -1$, $a = -3$, and $k = -2$. The graph of this equation is a parabola that opens to the left with vertex at $(-1, -2)$ and an axis of symmetry of $y = -2$.

We can construct a table of values and use symmetry to plot several points on the parabola. Then we draw a smooth curve through the points to get the graph shown in Figure II-13.

$$x = -3y^2 - 12y - 13$$
or
$$x + 1 = -3(y + 2)^2$$

x	y
-4	-1
-13	0

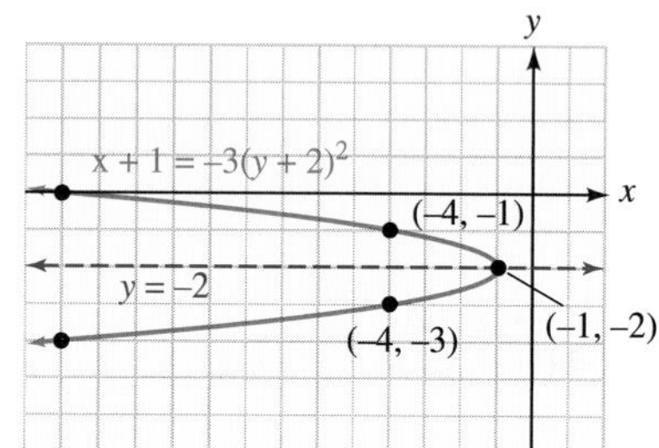

Figure II-13

■

EXAMPLE 5 Find the equation of the parabola, opening to the left with vertex at (2, 4), that passes through (0, 0).

Solution The parabola is shown in Figure II-14. To find the equation of the parabola that opens to the left, we use the equation $x - h = a(y - k)^2$ and substitute 2 for h, 4 for k, 0 for x, and 0 for y, and solve for a.

$$x - h = a(y - k)^2$$
$$0 - 2 = a(0 - 4)^2$$
$$-2 = 16a$$
$$-\frac{1}{8} = a$$

Since $h = 2$, $k = 4$, and $a = -\frac{1}{8}$, the equation is $x - 2 = -\frac{1}{8}(y - 4)^2$.

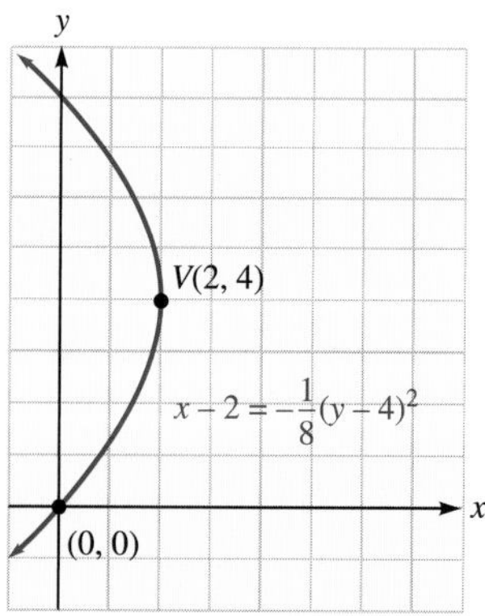

Figure II-14

■

II-2 Exercises

VOCABULARY AND CONCEPTS *Fill in the blanks.*

1. The graph of the equation $y - k = a(x - h)^2$ is a **parabola** with vertex at (h, k).
2. The parabola in Exercise 1 will open upward when $a > 0$ and open downward when $a < 0$.
3. The vertex of the parabolic graph of $y - 5 = 2(x - 3)^2$ will be at $(3, 5)$.
4. The parabolic graph of the equation $y - k = a(x - h)^2$ will open **downward** when $a < 0$.
5. The parabolic graph of the equation $x - h = a(y - k)^2$ will open **to the right** when $a > 0$.
6. The parabolic graph of the equation $x - h = a(y - k)^2$ will open **to the left** when $a < 0$.

PRACTICE *Find the equation of each parabola.*

7. Vertex at (0, 0), passing through (6, 3), opening upward $y = \frac{1}{12}x^2$
8. Vertex at (0, 0), passing through (6, −3), opening downward $y = -\frac{1}{12}x^2$
9. Vertex at (2, 2), passing through (0, 0), opening downward $y - 2 = -\frac{1}{2}(x - 2)^2$
10. Vertex at (−2, −2), passing through (0, 0), opening upward $y + 2 = \frac{1}{2}(x + 2)^2$
11. Vertex at (3, 5), passing through (9, 2), opening downward $y - 5 = -\frac{1}{12}(x - 3)^2$
12. Vertex at (−2, 2), passing through (4, −7), opening downward $y - 2 = -\frac{1}{4}(x + 2)^2$
13. Vertex at (−4, 6), passing through (0, 3), opening downward $y - 6 = -\frac{3}{16}(x + 4)^2$
14. Vertex at (−2, 3), passing through (0, 9), opening upward $y - 3 = \frac{3}{2}(x + 2)^2$
15. Vertex at (0, 0), passing through (−3, 6), opening to the left $x = -\frac{1}{12}y^2$
16. Vertex at (0, 0), passing through (3, −6), opening to the right $x = \frac{1}{12}y^2$
17. Vertext at (3, 5), passing through (2, 3), opening to the left $x - 3 = -\frac{1}{4}(y - 5)^2$
18. Vertex at (3, −5), passing through (6, 1), opening to the right $x - 3 = \frac{1}{12}(y + 5)^2$
19. Vertex at (−2, 2), passing through (0, 0), opening to the right $x + 2 = \frac{1}{2}(y - 2)^2$
20. Vertex at (2, 2), passing through (0, 0), opening to the left $x - 2 = -\frac{1}{2}(y - 2)^2$
21. Vertex at (−4, 6), passing through (0, 3), opening to the right $x + 4 = \frac{4}{9}(y - 6)^2$
22. Vertex at (−2, 3), passing through (0, −3), opening to the right $x + 2 = \frac{1}{18}(y - 3)^2$

Change each equation to standard form and graph it.

23. $y = x^2 + 4x + 5$

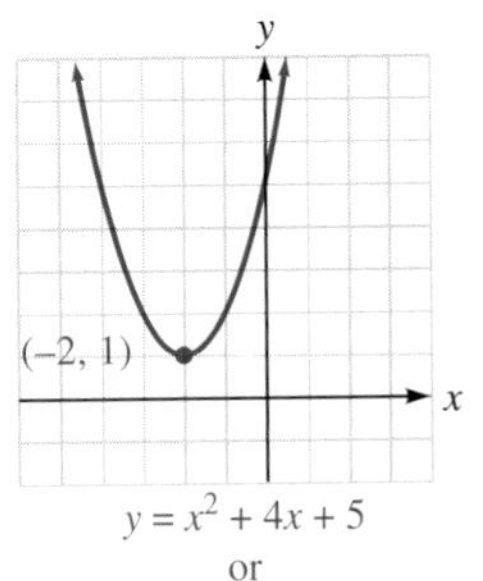

$y = x^2 + 4x + 5$
or
$y - 1 = (x + 2)^2$

24. 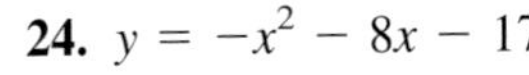
$y = -x^2 - 8x - 17$

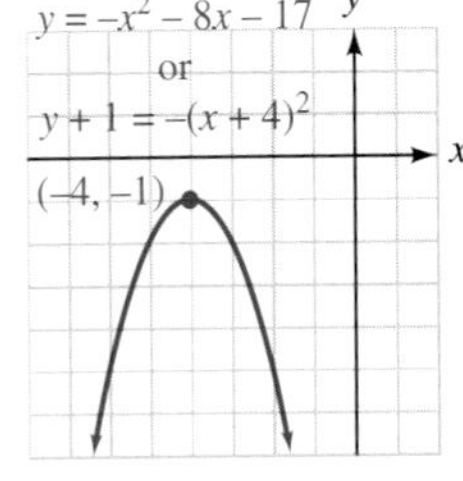

25. $y = -x^2 + 6x - 8$

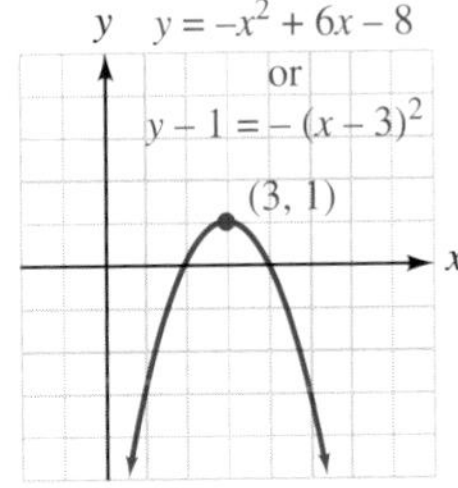

26. 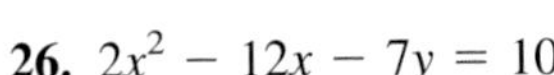
$2x^2 - 12x - 7y = 10$

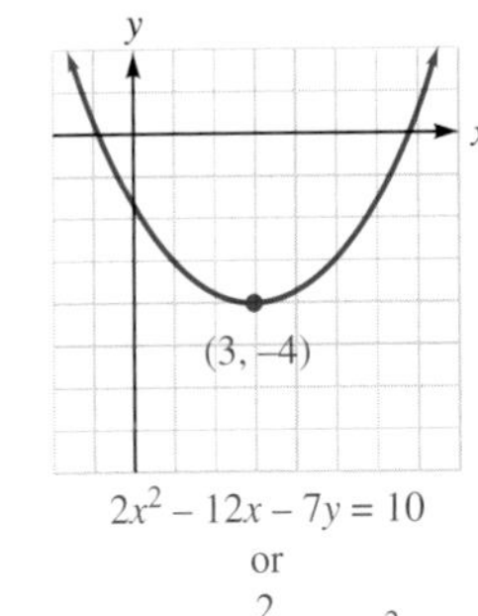

$2x^2 - 12x - 7y = 10$
or
$y + 4 = \frac{2}{7}(x - 3)^2$

27. $x^2 - 2y - 2x = -7$

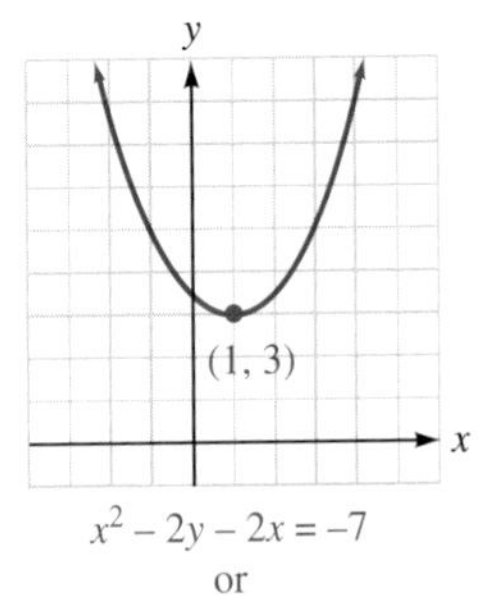

$x^2 - 2y - 2x = -7$
or
$y - 3 = \frac{1}{2}(x - 1)^2$

28. 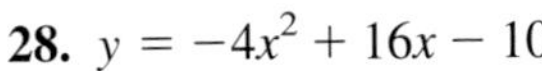
$y = -4x^2 + 16x - 10$

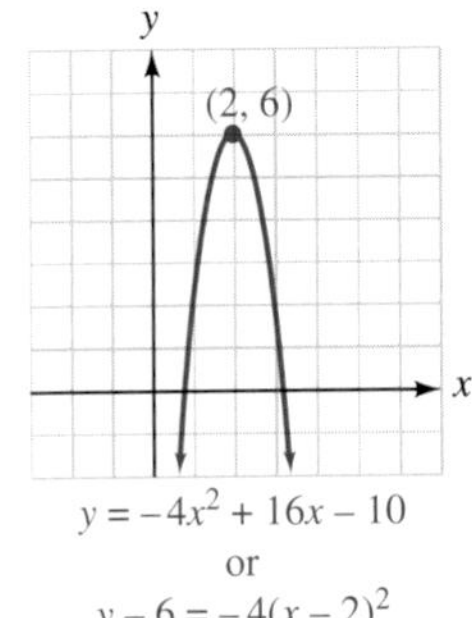

$y = -4x^2 + 16x - 10$
or
$y - 6 = -4(x - 2)^2$

29. $y = -2x^2 + 4x + 3$

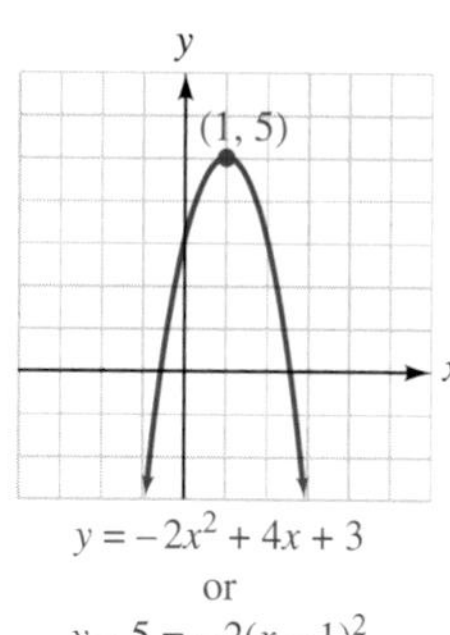

30. $x^2 - 6y + 22 = -4x$

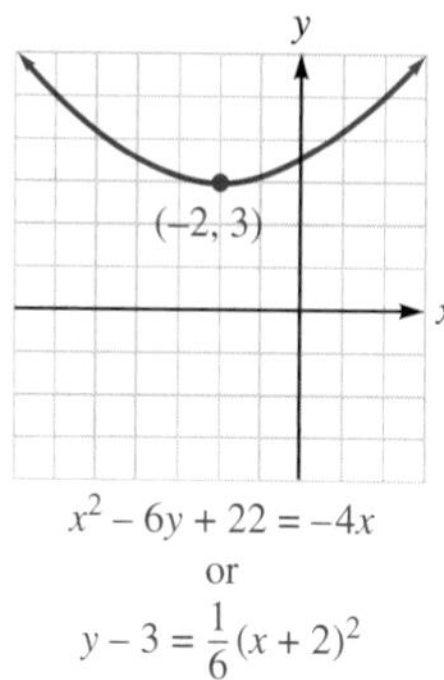

31. $y^2 + 4x - 6y = -1$

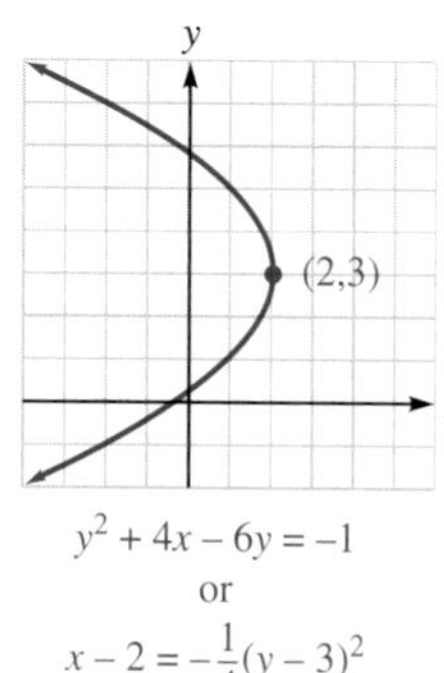

32. $x = 2y^2 + 4y + 5$

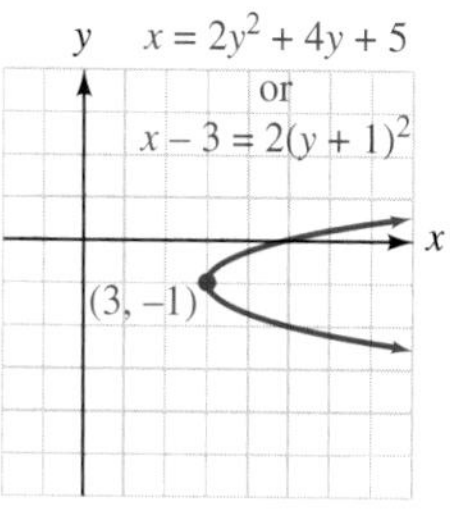

33. $x = 3y^2 - 12y + 11$

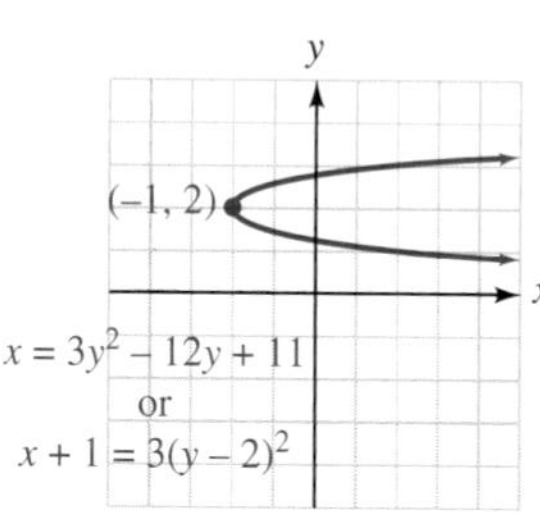

34. $y^2 + 2x - 2y = 5$

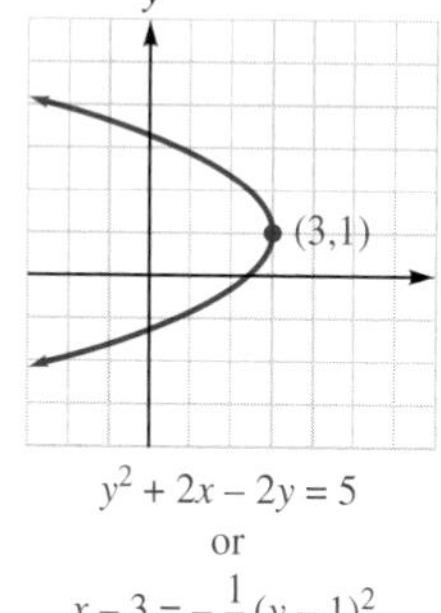

35. $y^2 - 4y = -8x + 20$

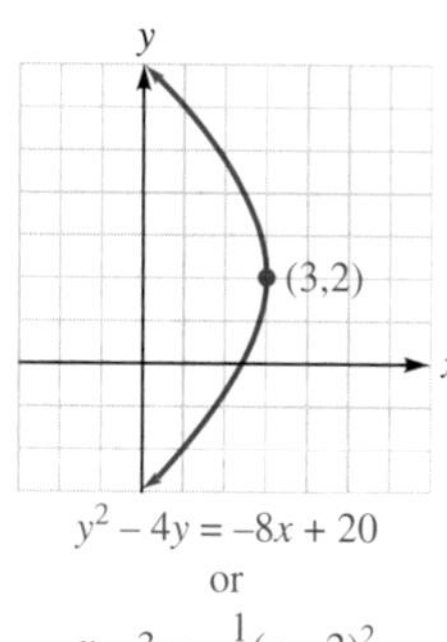

36. $x = \frac{1}{2}y^2 + 2y$

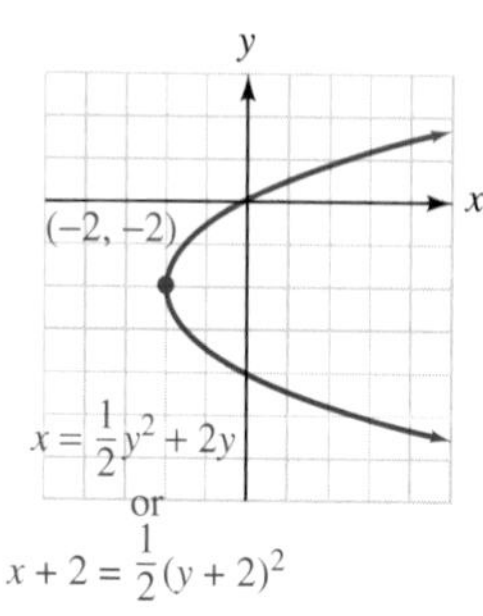

37. $4y^2 - 4y + 16x = 7$

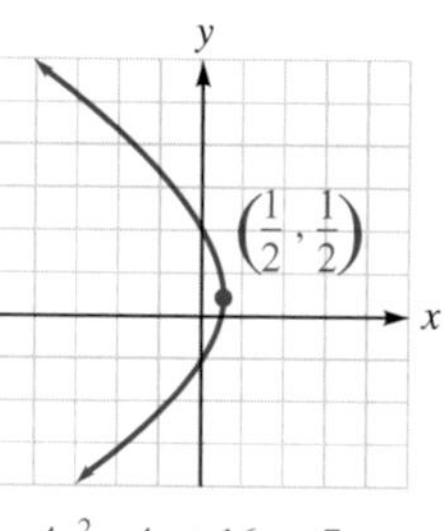

38. $4y^2 - 16x + 17 = 20y$

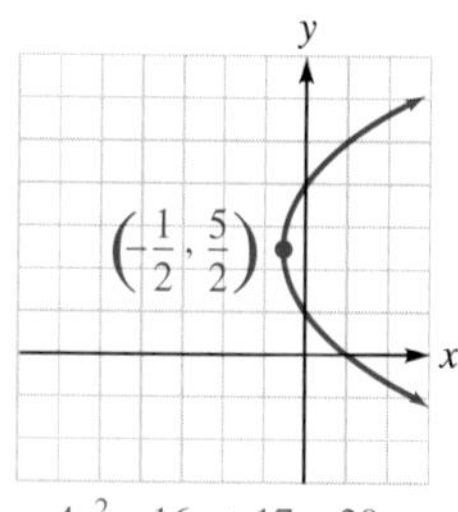

APPLICATIONS

39. Fireworks A fireworks shell is shot straight up with an initial velocity of 120 feet per second. Its height y after x seconds is given by the equation $y = 120x - 16x^2$. If the shell is designed to explode when it reaches its maximum height, how long after being fired, and at what height, will the fireworks appear in the sky? **3.75 sec, 225 ft**

40. Writing the equation of a parabola Derive the equation of the parabolic arch shown. $x^2 = -\frac{45}{2}y$

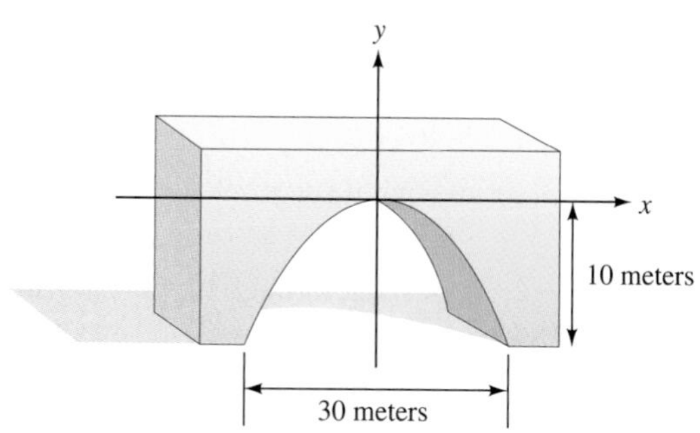

41. Projectiles The cannonball in the illustration follows the parabolic trajectory $y = 30x - x^2$. How far short of the castle does it land? **5 ft**

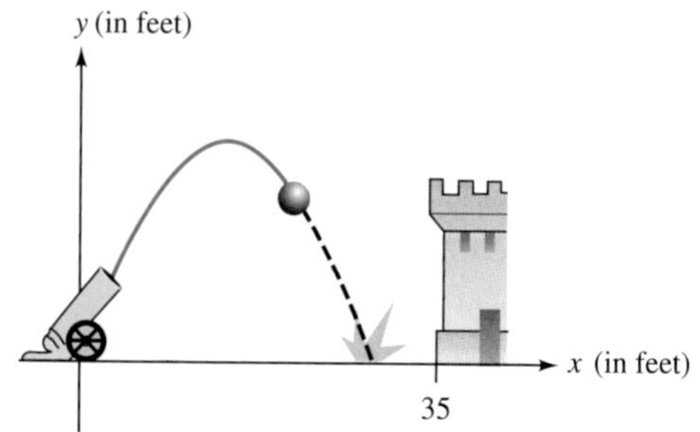

42. In Exercise 41, how high does the cannonball get? **225 ft**

43. Comets If the orbit of a comet is approximated by the equation $2y^2 - 9x = 18$, how far is it from the Sun at the vertex of its orbit? Distances are measured in astronomical units (AU). **2 AU**

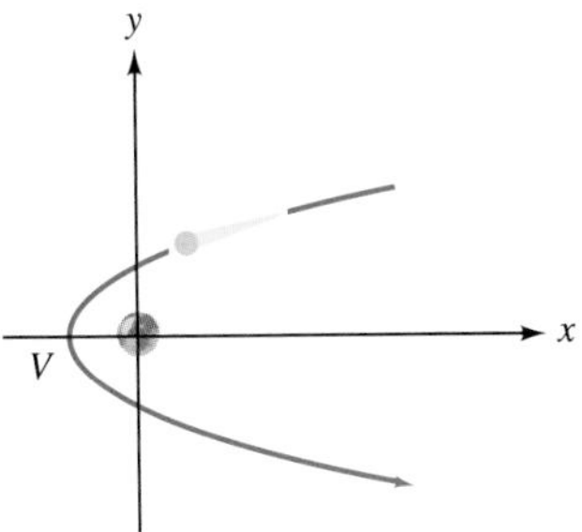

44. Architecture A parabolic arch has an equation of $x^2 + 20y - 400 = 0$, where x is measured in feet. Find the maximum height of the arch. **20 ft**

45. Ballistics An object is thrown from the origin of a coordinate system with the x-axis along the ground and the y-axis vertical. Its path, or **trajectory,** is given by the equation $y = 400x - 16x^2$. Find the object's maximum height. **2,500 units**

46. Operating resorts A resort owner plans to build and rent n cabins for d dollars per week. The price d that she can charge for each cabin depends on the number of cabins she builds, where $d = -45(\frac{n}{32} - \frac{1}{2})$. Find the number of cabins she should build to maximize her weekly income. **8**

47. Design of a suspension bridge The cable between the towers of the suspension bridge shown in the illustration has the shape of a parabola with vertex 15 feet above the roadway. Find the equation of the parabola. $x^2 = \frac{13{,}500}{7}(y - 15)$

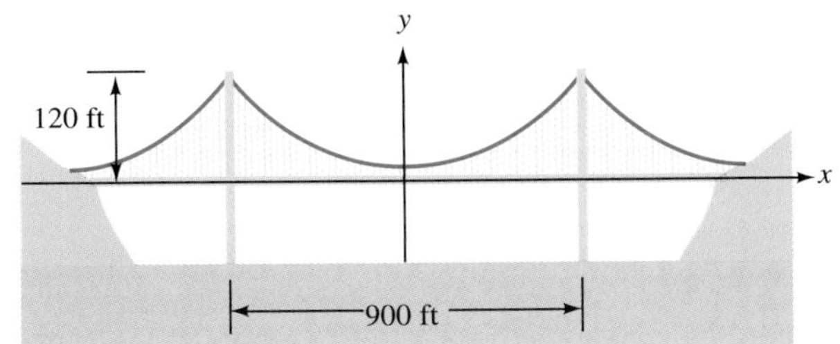

48. Gateway Arch The Gateway Arch in St. Louis has a shape that approximates a parabola. See the illustration. Find the width w of the arch 200 feet above the ground. **about 520 ft**

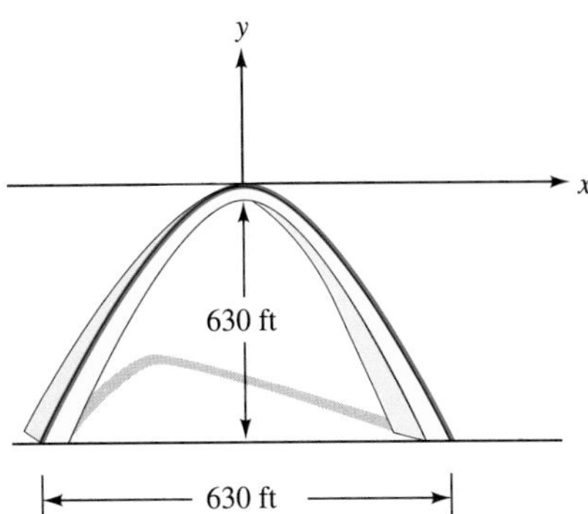

49. Building tunnels A construction firm plans to build a tunnel whose arch is in the shape of a parabola. See the illustration. The tunnel will span a two-lane highway 8 meters wide. To allow safe passage for vehicles, the tunnel must be 5 meters high at a distance of 1 meter from the tunnel's edge. Find the maximum height of the tunnel. $\frac{80}{7}$ **m**

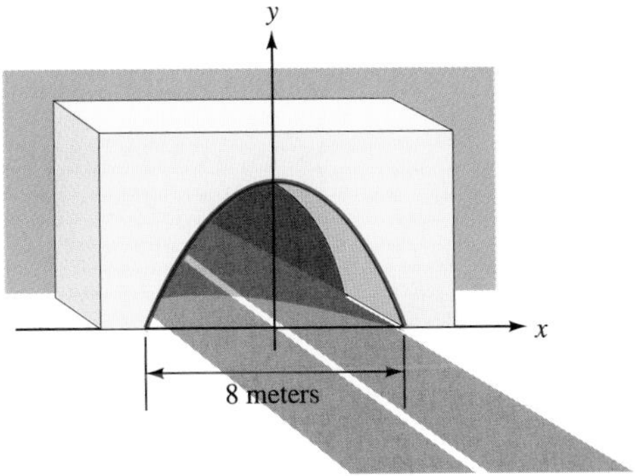

50. Ballistics A stone tossed upward is s feet above the ground in t seconds, where $s = -16t^2 + 128t$. Show that the stone's height x seconds after it is thrown is equal to its height x seconds before it hits the ground. Then show that the stone reaches its maximum height in one-half of the time it takes before it hits the ground.

DISCOVERY AND WRITING

51. Under what conditions will the graph of $x = a(y - k)^2 + h$ have no y-intercepts?
$a < 0$ and $h < 0$; $a > 0$ and $h > 0$

52. Explain how to determine from its equation whether the graph of a parabola opens up, down, right, or left.

REVIEW *Find each product.*

53. $3x^{-2}y^2(4x^2 + 3y^{-2})$ $12y^2 + \frac{9}{x^2}$

54. $(2a^{-2} - b^{-2})(2a^{-2} + b^{-2})$ $\frac{4}{a^4} - \frac{1}{b^4}$

Simplify each complex fraction.

55. $\dfrac{x^{-2} + y^{-2}}{x^{-2} - y^{-2}}$ $\dfrac{y^2 + x^2}{y^2 - x^2}$

56. $\dfrac{2x^{-3} - 2y^{-3}}{4x^{-3} + 4y^{-3}}$ $\dfrac{y^3 - x^3}{2(y^3 + x^3)}$

Tables

Table A Powers and Roots

n	n^2	$\sqrt{n}$	n^3	$\sqrt[3]{n}$	n	n^2	$\sqrt{n}$	n^3	$\sqrt[3]{n}$
1	1	1.000	1	1.000	51	2,601	7.141	132,651	3.708
2	4	1.414	8	1.260	52	2,704	7.211	140,608	3.733
3	9	1.732	27	1.442	53	2,809	7.280	148,877	3.756
4	16	2.000	64	1.587	54	2,916	7.348	157,464	3.780
5	25	2.236	125	1.710	55	3,025	7.416	166,375	3.803
6	36	2.449	216	1.817	56	3,136	7.483	175,616	3.826
7	49	2.646	343	1.913	57	3,249	7.550	185,193	3.849
8	64	2.828	512	2.000	58	3,364	7.616	195,112	3.871
9	81	3.000	729	2.080	59	3,481	7.681	205,379	3.893
10	100	3.162	1,000	2.154	60	3,600	7.746	216,000	3.915
11	121	3.317	1,331	2.224	61	3,721	7.810	226,981	3.936
12	144	3.464	1,728	2.289	62	3,844	7.874	238,328	3.958
13	169	3.606	2,197	2.351	63	3,969	7.937	250,047	3.979
14	196	3.742	2,744	2.410	64	4,096	8.000	262,144	4.000
15	225	3.873	3,375	2.466	65	4,225	8.062	274,625	4.021
16	256	4.000	4,096	2.520	66	4,356	8.124	287,496	4.041
17	289	4.123	4,913	2.571	67	4,489	8.185	300,763	4.062
18	324	4.243	5,832	2.621	68	4,624	8.246	314,432	4.082
19	361	4.359	6,859	2.668	69	4,761	8.307	328,509	4.102
20	400	4.472	8,000	2.714	70	4,900	8.367	343,000	4.121
21	441	4.583	9,261	2.759	71	5,041	8.426	357,911	4.141
22	484	4.690	10,648	2.802	72	5,184	8.485	373,248	4.160
23	529	4.796	12,167	2.844	73	5,329	8.544	389,017	4.179
24	576	4.899	13,824	2.884	74	5,476	8.602	405,224	4.198
25	625	5.000	15,625	2.924	75	5,625	8.660	421,875	4.217
26	676	5.099	17,576	2.962	76	5,776	8.718	438,976	4.236
27	729	5.196	19,683	3.000	77	5,929	8.775	456,533	4.254
28	784	5.292	21,952	3.037	78	6,084	8.832	474,552	4.273
29	841	5.385	24,389	3.072	79	6,241	8.888	493,039	4.291
30	900	5.477	27,000	3.107	80	6,400	8.944	512,000	4.309
31	961	5.568	29,791	3.141	81	6,561	9.000	531,441	4.327
32	1,024	5.657	32,768	3.175	82	6,724	9.055	551,368	4.344
33	1,089	5.745	35,937	3.208	83	6,889	9.110	571,787	4.362
34	1,156	5.831	39,304	3.240	84	7,056	9.165	592,704	4.380
35	1,225	5.916	42,875	3.271	85	7,225	9.220	614,125	4.397
36	1,296	6.000	46,656	3.302	86	7,396	9.274	636,056	4.414
37	1,369	6.083	50,653	3.332	87	7,569	9.327	658,503	4.431
38	1,444	6.164	54,872	3.362	88	7,744	9.381	681,472	4.448
39	1,521	6.245	59,319	3.391	89	7,921	9.434	704,969	4.465
40	1,600	6.325	64,000	3.420	90	8,100	9.487	729,000	4.481
41	1,681	6.403	68,921	3.448	91	8,281	9.539	753,571	4.498
42	1,764	6.481	74,088	3.476	92	8,464	9.592	778,688	4.514
43	1,849	6.557	79,507	3.503	93	8,649	9.644	804,357	4.531
44	1,936	6.633	85,184	3.530	94	8,836	9.695	830,584	4.547
45	2,025	6.708	91,125	3.557	95	9,025	9.747	857,375	4.563
46	2,116	6.782	97,336	3.583	96	9,216	9.798	884,736	4.579
47	2,209	6.856	103,823	3.609	97	9,409	9.849	912,673	4.595
48	2,304	6.928	110,592	3.634	98	9,604	9.899	941,192	4.610
49	2,401	7.000	117,649	3.659	99	9,801	9.950	970,299	4.626
50	2,500	7.071	125,000	3.684	100	10,000	10.000	1,000,000	4.642

Table B Base-10 Logarithms

N	0	1	2	3	4	5	6	7	8	9
1.0	.0000	.0043	.0086	.0128	.0170	.0212	.0253	.0294	.0334	.0374
1.1	.0414	.0453	.0492	.0531	.0569	.0607	.0645	.0682	.0719	.0755
1.2	.0792	.0828	.0864	.0899	.0934	.0969	.1004	.1038	.1072	.1106
1.3	.1139	.1173	.1206	.1239	.1271	.1303	.1335	.1367	.1399	.1430
1.4	.1461	.1492	.1523	.1553	.1584	.1614	.1644	.1673	.1703	.1732
1.5	.1761	.1790	.1818	.1847	.1875	.1903	.1931	.1959	.1987	.2014
1.6	.2041	.2068	.2095	.2122	.2148	.2175	.2201	.2227	.2253	.2279
1.7	.2304	.2330	.2355	.2380	.2405	.2430	.2455	.2480	.2504	.2529
1.8	.2553	.2577	.2601	.2625	.2648	.2672	.2695	.2718	.2742	.2765
1.9	.2788	.2810	.2833	.2856	.2878	.2900	.2923	.2945	.2967	.2989
2.0	.3010	.3032	.3054	.3075	.3096	.3118	.3139	.3160	.3181	.3201
2.1	.3222	.3243	.3263	.3284	.3304	.3324	.3345	.3365	.3385	.3404
2.2	.3424	.3444	.3464	.3483	.3502	.3522	.3541	.3560	.3579	.3598
2.3	.3617	.3636	.3655	.3674	.3692	.3711	.3729	.3747	.3766	.3784
2.4	.3802	.3820	.3838	.3856	.3874	.3892	.3909	.3927	.3945	.3962
2.5	.3979	.3997	.4014	.4031	.4048	.4065	.4082	.4099	.4116	.4133
2.6	.4150	.4166	.4183	.4200	.4216	.4232	.4249	.4265	.4281	.4298
2.7	.4314	.4330	.4346	.4362	.4378	.4393	.4409	.4425	.4440	.4456
2.8	.4472	.4487	.4502	.4518	.4533	.4548	.4564	.4579	.4594	.4609
2.9	.4624	.4639	.4654	.4669	.4683	.4698	.4713	.4728	.4742	.4757
3.0	.4771	.4786	.4800	.4814	.4829	.4843	.4857	.4871	.4886	.4900
3.1	.4914	.4928	.4942	.4955	.4969	.4983	.4997	.5011	.5024	.5038
3.2	.5051	.5065	.5079	.5092	.5105	.5119	.5132	.5145	.5159	.5172
3.3	.5185	.5198	.5211	.5224	.5237	.5250	.5263	.5276	.5289	.5302
3.4	.5315	.5328	.5340	.5353	.5366	.5378	.5391	.5403	.5416	.5428
3.5	.5441	.5453	.5465	.5478	.5490	.5502	.5514	.5527	.5539	.5551
3.6	.5563	.5575	.5587	.5599	.5611	.5623	.5635	.5647	.5658	.5670
3.7	.5682	.5694	.5705	.5717	.5729	.5740	.5752	.5763	.5775	.5786
3.8	.5798	.5809	.5821	.5832	.5843	.5855	.5866	.5877	.5888	.5899
3.9	.5911	.5922	.5933	.5944	.5955	.5966	.5977	.5988	.5999	.6010
4.0	.6021	.6031	.6042	.6053	.6064	.6075	.6085	.6096	.6107	.6117
4.1	.6128	.6138	.6149	.6160	.6170	.6180	.6191	.6201	.6212	.6222
4.2	.6232	.6243	.6253	.6263	.6274	.6284	.6294	.6304	.6314	.6325
4.3	.6335	.6345	.6355	.6365	.6375	.6385	.6395	.6405	.6415	.6425
4.4	.6435	.6444	.6454	.6464	.6474	.6484	.6493	.6503	.6513	.6522
4.5	.6532	.6542	.6551	.6561	.6571	.6580	.6590	.6599	.6609	.6618
4.6	.6628	.6637	.6646	.6656	.6665	.6675	.6684	.6693	.6702	.6712
4.7	.6721	.6730	.6739	.6749	.6758	.6767	.6776	.6785	.6794	.6803
4.8	.6812	.6821	.6830	.6839	.6848	.6857	.6866	.6875	.6884	.6893
4.9	.6902	.6911	.6920	.6928	.6937	.6946	.6955	.6964	.6972	.6981
5.0	.6990	.6998	.7007	.7016	.7024	.7033	.7042	.7050	.7059	.7067
5.1	.7076	.7084	.7093	.7101	.7110	.7118	.7126	.7135	.7143	.7152
5.2	.7160	.7168	.7177	.7185	.7193	.7202	.7210	.7218	.7226	.7235
5.3	.7243	.7251	.7259	.7267	.7275	.7284	.7292	.7300	.7308	.7316
5.4	.7324	.7332	.7340	.7348	.7356	.7364	.7372	.7380	.7388	.7396

Table B *(continued)*

N	0	1	2	3	4	5	6	7	8	9
5.5	.7404	.7412	.7419	.7427	.7435	.7443	.7451	.7459	.7466	.7474
5.6	.7482	.7490	.7497	.7505	.7513	.7520	.7528	.7536	.7543	.7551
5.7	.7559	.7566	.7574	.7582	.7589	.7597	.7604	.7612	.7619	.7627
5.8	.7634	.7642	.7649	.7657	.7664	.7672	.7679	.7686	.7694	.7701
5.9	.7709	.7716	.7723	.7731	.7738	.7745	.7752	.7760	.7767	.7774
6.0	.7782	.7789	.7796	.7803	.7810	.7818	.7825	.7832	.7839	.7846
6.1	.7853	.7860	.7868	.7875	.7882	.7889	.7896	.7903	.7910	.7917
6.2	.7924	.7931	.7938	.7945	.7952	.7959	.7966	.7973	.7980	.7987
6.3	.7993	.8000	.8007	.8014	.8021	.8028	.8035	.8041	.8048	.8055
6.4	.8062	.8069	.8075	.8082	.8089	.8096	.8102	.8109	.8116	.8122
6.5	.8129	.8136	.8142	.8149	.8156	.8162	.8169	.8176	.8182	.8189
6.6	.8195	.8202	.8209	.8215	.8222	.8228	.8235	.8241	.8248	.8254
6.7	.8261	.8267	.8274	.8280	.8287	.8293	.8299	.8306	.8312	.8319
6.8	.8325	.8331	.8338	.8344	.8351	.8357	.8363	.8370	.8376	.8382
6.9	.8388	.8395	.8401	.8407	.8414	.8420	.8426	.8432	.8439	.8445
7.0	.8451	.8457	.8463	.8470	.8476	.8482	.8488	.8494	.8500	.8506
7.1	.8513	.8519	.8525	.8531	.8537	.8543	.8549	.8555	.8561	.8567
7.2	.8573	.8579	.8585	.8591	.8597	.8603	.8609	.8615	.8621	.8627
7.3	.8633	.8639	.8645	.8651	.8657	.8663	.8669	.8675	.8681	.8686
7.4	.8692	.8698	.8704	.8710	.8716	.8722	.8727	.8733	.8739	.8745
7.5	.8751	.8756	.8762	.8768	.8774	.8779	.8785	.8791	.8797	.8802
7.6	.8808	.8814	.8820	.8825	.8831	.8837	.8842	.8848	.8854	.8859
7.7	.8865	.8871	.8876	.8882	.8887	.8893	.8899	.8904	.8910	.8915
7.8	.8921	.8927	.8932	.8938	.8943	.8949	.8954	.8960	.8965	.8971
7.9	.8976	.8982	.8987	.8993	.8998	.9004	.9009	.9015	.9020	.9025
8.0	.9031	.9036	.9042	.9047	.9053	.9058	.9063	.9069	.9074	.9079
8.1	.9085	.9090	.9096	.9101	.9106	.9112	.9117	.9122	.9128	.9133
8.2	.9138	.9143	.9149	.9154	.9159	.9165	.9170	.9175	.9180	.9186
8.3	.9191	.9196	.9201	.9206	.9212	.9217	.9222	.9227	.9232	.9238
8.4	.9243	.9248	.9253	.9258	.9263	.9269	.9274	.9279	.9284	.9289
8.5	.9294	.9299	.9304	.9309	.9315	.9320	.9325	.9330	.9335	.9340
8.6	.9345	.9350	.9355	.9360	.9365	.9370	.9375	.9380	.9385	.9390
8.7	.9395	.9400	.9405	.9410	.9415	.9420	.9425	.9430	.9435	.9440
8.8	.9445	.9450	.9455	.9460	.9465	.9469	.9474	.9479	.9484	.9489
8.9	.9494	.9499	.9504	.9509	.9513	.9518	.9523	.9528	.9533	.9538
9.0	.9542	.9547	.9552	.9557	.9562	.9566	.9571	.9576	.9581	.9586
9.1	.9590	.9595	.9600	.9605	.9609	.9614	.9619	.9624	.9628	.9633
9.2	.9638	.9643	.9647	.9652	.9657	.9661	.9666	.9671	.9675	.9680
9.3	.9685	.9689	.9694	.9699	.9703	.9708	.9713	.9717	.9722	.9727
9.4	.9731	.9736	.9741	.9745	.9750	.9754	.9759	.9763	.9768	.9773
9.5	.9777	.9782	.9786	.9791	.9795	.9800	.9805	.9809	.9814	.9818
9.6	.9823	.9827	.9832	.9836	.9841	.9845	.9850	.9854	.9859	.9863
9.7	.9868	.9872	.9877	.9881	.9886	.9890	.9894	.9899	.9903	.9908
9.8	.9912	.9917	.9921	.9926	.9930	.9934	.9939	.9943	.9948	.9952
9.9	.9956	.9961	.9965	.9969	.9974	.9978	.9983	.9987	.9991	.9996

Table C Base-*e* Logarithms

N	0	1	2	3	4	5	6	7	8	9
1.0	.0000	.0100	.0198	.0296	.0392	.0488	.0583	.0677	.0770	.0862
1.1	.0953	.1044	.1133	.1222	.1310	.1398	.1484	.1570	.1655	.1740
1.2	.1823	.1906	.1989	.2070	.2151	.2231	.2311	.2390	.2469	.2546
1.3	.2624	.2700	.2776	.2852	.2927	.3001	.3075	.3148	.3221	.3293
1.4	.3365	.3436	.3507	.3577	.3646	.3716	.3784	.3853	.3920	.3988
1.5	.4055	.4121	.4187	.4253	.4318	.4383	.4447	.4511	.4574	.4637
1.6	.4700	.4762	.4824	.4886	.4947	.5008	.5068	.5128	.5188	.5247
1.7	.5306	.5365	.5423	.5481	.5539	.5596	.5653	.5710	.5766	.5822
1.8	.5878	.5933	.5988	.6043	.6098	.6152	.6206	.6259	.6313	.6366
1.9	.6419	.6471	.6523	.6575	.6627	.6678	.6729	.6780	.6831	.6881
2.0	.6931	.6981	.7031	.7080	.7129	.7178	.7227	.7275	.7324	.7372
2.1	.7419	.7467	.7514	.7561	.7608	.7655	.7701	.7747	.7793	.7839
2.2	.7885	.7930	.7975	.8020	.8065	.8109	.8154	.8198	.8242	.8286
2.3	.8329	.8372	.8416	.8459	.8502	.8544	.8587	.8629	.8671	.8713
2.4	.8755	.8796	.8838	.8879	.8920	.8961	.9002	.9042	.9083	.9123
2.5	.9163	.9203	.9243	.9282	.9322	.9361	.9400	.9439	.9478	.9517
2.6	.9555	.9594	.9632	.9670	.9708	.9746	.9783	.9821	.9858	.9895
2.7	.9933	.9969	1.0006	.0043	.0080	.0116	.0152	.0188	.0225	.0260
2.8	1.0296	.0332	.0367	.0403	.0438	.0473	.0508	.0543	.0578	.0613
2.9	.0647	.0682	.0716	.0750	.0784	.0818	.0852	.0886	.0919	.0953
3.0	1.0986	.1019	.1053	.1086	.1119	.1151	.1184	.1217	.1249	.1282
3.1	.1314	.1346	.1378	.1410	.1442	.1474	.1506	.1537	.1569	.1600
3.2	.1632	.1663	.1694	.1725	.1756	.1787	.1817	.1848	.1878	.1909
3.3	.1939	.1969	.2000	.2030	.2060	.2090	.2119	.2149	.2179	.2208
3.4	.2238	.2267	.2296	.2326	.2355	.2384	.2413	.2442	.2470	.2499
3.5	1.2528	.2556	.2585	.2613	.2641	.2669	.2698	.2726	.2754	.2782
3.6	.2809	.2837	.2865	.2892	.2920	.2947	.2975	.3002	.3029	.3056
3.7	.3083	.3110	.3137	.3164	.3191	.3218	.3244	.3271	.3297	.3324
3.8	.3350	.3376	.3403	.3429	.3455	.3481	.3507	.3533	.3558	.3584
3.9	.3610	.3635	.3661	.3686	.3712	.3737	.3762	.3788	.3813	.3838
4.0	1.3863	.3888	.3913	.3938	.3962	.3987	.4012	.4036	.4061	.4085
4.1	.4110	.4134	.4159	.4183	.4207	.4231	.4255	.4279	.4303	.4327
4.2	.4351	.4375	.4398	.4422	.4446	.4469	.4493	.4516	.4540	.4563
4.3	.4586	.4609	.4633	.4656	.4679	.4702	.4725	.4748	.4770	.4793
4.4	.4816	.4839	.4861	.4884	.4907	.4929	.4951	.4974	.4996	.5019
4.5	1.5041	.5063	.5085	.5107	.5129	.5151	.5173	.5195	.5217	.5239
4.6	.5261	.5282	.5304	.5326	.5347	.5369	.5390	.5412	.5433	.5454
4.7	.5476	.5497	.5518	.5539	.5560	.5581	.5602	.5623	.5644	.5665
4.8	.5686	.5707	.5728	.5748	.5769	.5790	.5810	.5831	.5851	.5872
4.9	.5892	.5913	.5933	.5953	.5974	.5994	.6014	.6034	.6054	.6074
5.0	1.6094	.6114	.6134	.6154	.6174	.6194	.6214	.6233	.6253	.6273
5.1	.6292	.6312	.6332	.6351	.6371	.6390	.6409	.6429	.6448	.6467
5.2	.6487	.6506	.6525	.6544	.6563	.6582	.6601	.6620	.6639	.6658
5.3	.6677	.6696	.6715	.6734	.6752	.6771	.6790	.6808	.6827	.6845
5.4	.6864	.6882	.6901	.6919	.6938	.6956	.6974	.6993	.7011	.7029

Use the properties of logarithms and ln 10 $\approx$ 2.3026 to find logarithms of numbers less than 1 or greater than 10.

Table C (*continued*)

N	0	1	2	3	4	5	6	7	8	9
5.5	1.7047	.7066	.7084	.7102	.7120	.7138	.7156	.7174	.7192	.7210
5.6	.7228	.7246	.7263	.7281	.7299	.7317	.7334	.7352	.7370	.7387
5.7	.7405	.7422	.7440	.7457	.7475	.7492	.7509	.7527	.7544	.7561
5.8	.7579	.7596	.7613	.7630	.7647	.7664	.7681	.7699	.7716	.7733
5.9	.7750	.7766	.7783	.7800	.7817	.7834	.7851	.7867	.7884	.7901
6.0	1.7918	.7934	.7951	.7967	.7984	.8001	.8017	.8034	.8050	.8066
6.1	.8083	.8099	.8116	.8132	.8148	.8165	.8181	.8197	.8213	.8229
6.2	.8245	.8262	.8278	.8294	.8310	.8326	.8342	.8358	.8374	.8390
6.3	.8405	.8421	.8437	.8453	.8469	.8485	.8500	.8516	.8532	.8547
6.4	.8563	.8579	.8594	.8610	.8625	.8641	.8656	.8672	.8687	.8703
6.5	1.8718	.8733	.8749	.8764	.8779	.8795	.8810	.8825	.8840	.8856
6.6	.8871	.8886	.8901	.8916	.8931	.8946	.8961	.8976	.8991	.9006
6.7	.9021	.9036	.9051	.9066	.9081	.9095	.9110	.9125	.9140	.9155
6.8	.9169	.9184	.9199	.9213	.9228	.9242	.9257	.9272	.9286	.9301
6.9	.9315	.9330	.9344	.9359	.9373	.9387	.9402	.9416	.9430	.9445
7.0	1.9459	.9473	.9488	.9502	.9516	.9530	.9544	.9559	.9573	.9587
7.1	.9601	.9615	.9629	.9643	.9657	.9671	.9685	.9699	.9713	.9727
7.2	.9741	.9755	.9769	.9782	.9796	.9810	.9824	.9838	.9851	.9865
7.3	.9879	.9892	.9906	.9920	.9933	.9947	.9961	.9974	.9988	2.0001
7.4	2.0015	.0028	.0042	.0055	.0069	.0082	.0096	.0109	.0122	.0136
7.5	2.0149	.0162	.0176	.0189	.0202	.0215	.0229	.0242	.0255	.0268
7.6	.0281	.0295	.0308	.0321	.0334	.0347	.0360	.0373	.0386	.0399
7.7	.0412	.0425	.0438	.0451	.0464	.0477	.0490	.0503	.0516	.0528
7.8	.0541	.0554	.0567	.0580	.0592	.0605	.0618	.0631	.0643	.0656
7.9	.0669	.0681	.0694	.0707	.0719	.0732	.0744	.0757	.0769	.0782
8.0	2.0794	.0807	.0819	.0832	.0844	.0857	.0869	.0882	.0894	.0906
8.1	.0919	.0931	.0943	.0956	.0968	.0980	.0992	.1005	.1017	.1029
8.2	.1041	.1054	.1066	.1078	.1090	.1102	.1114	.1126	.1138	.1150
8.3	.1163	.1175	.1187	.1199	.1211	.1223	.1235	.1247	.1258	.1270
8.4	.1282	.1294	.1306	.1318	.1330	.1342	.1353	.1365	.1377	.1389
8.5	2.1401	.1412	.1424	.1436	.1448	.1459	.1471	.1483	.1494	.1506
8.6	.1518	.1529	.1541	.1552	.1564	.1576	.1587	.1599	.1610	.1622
8.7	.1633	.1645	.1656	.1668	.1679	.1691	.1702	.1713	.1725	.1736
8.8	.1748	.1759	.1770	.1782	.1793	.1804	.1815	.1827	.1838	.1849
8.9	.1861	.1872	.1883	.1894	.1905	.1917	.1928	.1939	.1950	.1961
9.0	2.1972	.1983	.1994	.2006	.2017	.2028	.2039	.2050	.2061	.2072
9.1	.2083	.2094	.2105	.2116	.2127	.2138	.2148	.2159	.2170	.2181
9.2	.2192	.2203	.2214	.2225	.2235	.2246	.2257	.2268	.2279	.2289
9.3	.2300	.2311	.2322	.2332	.2343	.2354	.2364	.2375	.2386	.2396
9.4	.2407	.2418	.2428	.2439	.2450	.2460	.2471	.2481	.2492	.2502
9.5	2.2513	.2523	.2534	.2544	.2555	.2565	.2576	.2586	.2597	.2607
9.6	.2618	.2628	.2638	.2649	.2659	.2670	.2680	.2690	.2701	.2711
9.7	.2721	.2732	.2742	.2752	.2762	.2773	.2783	.2793	.2803	.2814
9.8	.2824	.2834	.2844	.2854	.2865	.2875	.2885	.2895	.2905	.2915
9.9	.2925	.2935	.2946	.2956	.2966	.2976	.2986	.2996	.3006	.3016

Index

4.1 Exponential Functions and Their Graphs

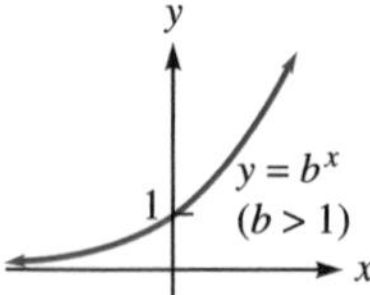

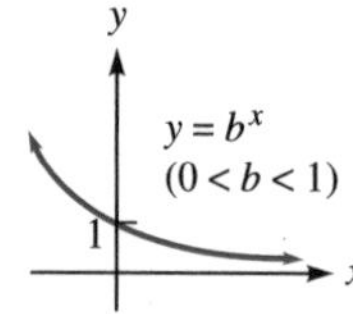

Formula for compound interest: $A = P\left(1 + \frac{r}{k}\right)^{kt}$

$e = 2.718281828 \ldots$

4.2 Applications of Exponential Functions

Formula for radioactive decay: $A = A_0 2^{-t/h}$

Formula for continuous compound interest: $A = Pe^{rt}$

Formula for population growth: $P = P_0 e^{kt}$

4.3 Logarithmic Functions and Their Graphs

$y = \log_b x$ is equivalent to $x = b^y$.

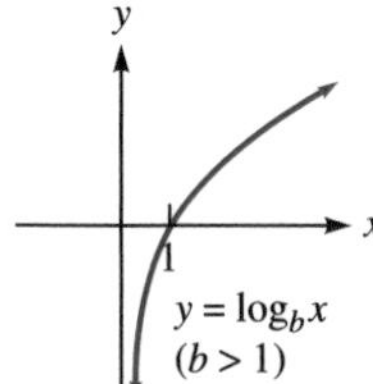

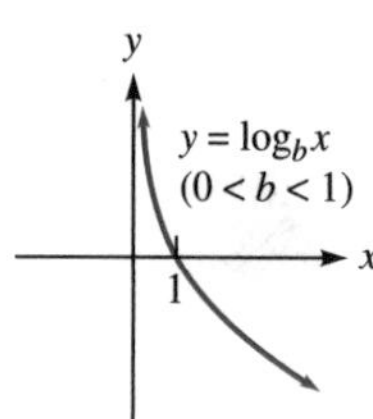

4.5 Properties of Logarithms

$\log_b 1 = 0$ $\qquad \log_b b = 1$

$\log_b b^x = x$ $\qquad b^{\log_b x} = x$

$\log_b MN = \log_b M + \log_b N$

$\log_b \frac{M}{N} = \log_b M - \log_b N$

$\log_b M^p = p \log_b M$

If $\log_b x = \log_b y$, then $x = y$.

The change-of-base formula: $\log_b y = \frac{\log_a y}{\log_a b}$

6.5 Determinants

$$\begin{vmatrix} a & b \\ c & d \end{vmatrix} = ad - bc$$

7.1 The Circle and the Parabola

$(x - h)^2 + (y - k)^2 = r^2$ Circle with center at (h, k) and radius r

$x^2 + y^2 = r^2$ Circle with center at the origin and radius r

$$\left.\begin{array}{l}(y - k)^2 = 4p(x - h) \\ (x - h)^2 = 4p(y - k)\end{array}\right\} \text{ Parabola with vertex at } (h, k)$$

7.2 The Ellipse

$$\left.\begin{array}{l}\dfrac{(x - h)^2}{a^2} + \dfrac{(y - k)^2}{b^2} = 1 \\[2ex] \dfrac{(y - k)^2}{a^2} + \dfrac{(x - h)^2}{b^2} = 1\end{array}\right\} \begin{array}{l}\text{Ellipse with center at } (h, k) \\ (0 < b < a)\end{array}$$

7.3 The Hyperbola

$$\left.\begin{array}{l}\dfrac{(x - h)^2}{a^2} - \dfrac{(y - k)^2}{b^2} = 1 \\[2ex] \dfrac{(y - k)^2}{a^2} - \dfrac{(x - h)^2}{b^2} = 1\end{array}\right\} \text{ Hyperbola with center at } (h, k)$$

8.1 The Binomial Theorem

$n! = n(n - 1)(n - 2) \cdot \cdots \cdot 3 \cdot 2 \cdot 1$

$0! = 1$

$n(n - 1)! = n!$

$$(a + b)^n = a^n + \frac{n!}{1!(n - 1)!}a^{n-1}b + \frac{n!}{2!(n - 2)!}a^{n-2}b^2 + \cdots + b^n$$

8.2 Sequences, Series, and Summation Notation

If c is a constant, then

$$\sum_{k=1}^{n} c = nc$$

$$\sum_{k=1}^{n} cf(k) = c\sum_{k=1}^{n} f(k)$$

$$\sum_{k=1}^{n} [f(k) + g(k)] = \sum_{k=1}^{n} f(k) + \sum_{k=1}^{n} g(k)$$